# ADVANCED INORGANIC CHEMISTRY

[ Volume I ]

AF824940

S. P. Banerjee

CELEBRATING 60 YEARS 1960-2020

Since BOOKS AND ALLIED 1960

*[ Meant for Chemistry (Hons./Major) & Advanced students.*
*Also useful for NET, SET, GATE & other competitive examinations. ]*

---

# ADVANCED
# INORGANIC CHEMISTRY

## [ Volume I ]

**Sankar Prosad Banerjee**
*Formerly Associate Professor and Head of the Department of Chemistry,*
*Vivekananda Mahavidyalaya, Burdwan.*
*Ex-Part Time Teacher, Institute of Science Education (1980-2011),*
*The University of Burdwan, Golapbag, Burdwan.*

**BOOKS & ALLIED (P) LTD.**
NO. 1E(1), "SHUBHAM PLAZA"
83/1, DR. S.C. BANERJEE ROAD, BELIAGHATA, KOLKATA 700010
***Tel*** : (033)2374-0062 • ***e-mail*** : booksandallied1960@gmail.com
***website*** : www.booksandallied.com

# *ADVANCED*
# INORGANIC CHEMISTRY
[ Volume I ]

**First Published : August 2013**
Reprinted : 2016
**Second Edition** (thoroughly revised) **: 2017**
Reprinted : 2018
**Revised Reprinted : 2022**



***Publisher*** : Ruba Sen
BOOKS & ALLIED (P) LTD.
No. 1E(1), "SHUBHAM PLAZA"
83/1, Dr. S.C. Banerjee Road, Beliaghata, Kolkata 700010

***Typesetter*** : BOOKS & ALLIED (P) LTD.
No. 1E(1), "SHUBHAM PLAZA"
83/1, Dr. S.C. Banerjee Road, Beliaghata, Kolkata 700010

***Printer*** : EUREKA
91/A, Baithakkhana Road, Kolkata 700009

***Project Supervisor*** : Abhijit Ghosal

ISBN **978-93-84294-34-2**

*Price* : **₹ 675.00**

*An honest man's the noblest work of God.*

***—Alexander Pope***

*To the memory of my late father Radhika Prosad Banerjee*
*whose soul left his mortal body for eternal bliss in October, 1975*

and

*to my beloved father in-law Kartick Chandra Ghosal*
*who left us for heavenly abode on 23rd July, 2014*

and

*to my mentor (Guru) Prof. R. L. Dutta*

# About the Book

The book is divided into seven chapters. First one comprises concept of atom followed by long form of the periodic table in the second chapter as per recommendations of 2005 and the permanent naming and their symbols are discussed from atomic number 100 to 111 as also element number 112. Inert pair effect is explianed with reference to relativistic effect.

For the interest of inquisitive students the orientation of hybrid orbitals are briefly discussed in the third chapter. Some interesting application of molecular orbital theory is included in this chapter.

Elaborate discussion of the concept of acids and bases have been made in the fourth chapter with brief resume of the terrestrial distribution of elements (Sec. 4.5.5). Symbiosis, some important application of SHAB principle and correlation of acidity of metal cations has also been discussed in this chapter. In order to cover the syllabus of some parts of analytical chemistry, the non-aqueous titrations included in the fifth chapter. The important theoretical discussion regarding redox processes in analytical chemistry as also a brief resume of masking and demasking is the content of sixth chapter.

Largest number of solved problems and exercises have been incorporated in the last chapter (coordination chemistry) which also has a discussion on coordination polymer.

# About the Author

Dr. Sankar Prosad Banerjee did his M.Sc. with specialisation in Inorganic Chemistry in 1969 from the University of Burdwan. He joined the same university as full-time Research Scholar in 1971 and received his Ph.D degree in 1977 under the supervision of Prof. R. L. Dutta.

He has published several research papers and a review article in the area of Coordination Chemistry.

He joined the Vivekananda Mahavidyalaya, Burdwan in July 1973 as lecturer and served the college till his retirement in January, 2008 from the position as Associate Professor and Head of the Department of Chemistry (U.G. and P.G.). Presently the author is associated with teaching of Postgraduate students of Vivekananda Mahavidyalaya, Burdwan.

# Preface to the Second Edition

The text **Advanced Inorganic Chemistry, Vol. I** provides a sound foundation for undergraduate and graduate students of inorganic chemistry. After the first publication of the book in 2013, the revision work is necessitated both from the modification of curriculum of syllabus of Indian Universities as well as to fulfill the demands of my large number of students to answer the questions set in different undergraduate examinations of 2013 onwards.

In this revision work of second edition the pivotal role taken by Mr. Arunabha Sen, MS(USA), Director, Books & Allied (P) Ltd., whose enthusiastic attitude gear up me always for speedy revision work. I thank him heartily.

All the seven chapters are thoroughly revised with the inclusion of many solved questions and hints in the exercise of different undergraduate examinations of Calcutta and Burdwan University for the years 2013, 2014, 2015 and partly for 2016. Chapterwise worked-out NET/GATE questions are also included for the said years. The splitting of f-orbitals in crystal field and some modifications of the section of colour of coordination compounds have been incorporated in chapter 7. This will benefit the students as well as the teachers of inorganic chemistry.

On going from the first to second edition I would like to acknowledge the help rendered by all the teachers of the Department of Chemistry, Vivekananda Mahavidyalaya, Burdwan. Finally this writing project would not have been complete without input from my wife, Sm. Gita Banerjee, MA(Double) whose boundless energy acclerate me immensely.

Students are the last word to criticize a book and so I dedicate this new edition to my innumerable number of beloved students in India and abroad.

**S. P. BANERJEE**

*Burdwan*
*Vivekananda Birth Day, 12th January, 2017*
*E-mail :* sankarprasadbanerjee@rediff.com

# Preface to the First Edition

This textbook is intended for B.Sc. (Honours) advanced inorganic chemistry course for different Indian Universities as well as NET, SLET, SSC, etc. In all the seven chapters the subject matter is represented in a logical way with special emphasis on coordination chemistry. Nomenclature of coordination compounds have been made as per recommendations of 2005.

Present undergraduate students are facing problems to answer the questions set in B.Sc.(Hons.) examinations of different Universities. In order to stimulate the interest of the students, I have tried to solve as many questions as possible for different B.U. and C.U. examinations. At the end of each chapter adequate number of questions are given for self evaluation of students.

The author is thankful to Prof. B. K. Ghosh of Burdwan University for inspiring the author to write a book of this type. I express my indebtness to Prof. A. K. Das, Visva Bharati University for helping me in various ways to answer some difficult questions with his scholastic knowledge. The author affectionately acknowledged the help rendered by all the teachers of the Department of Chemistry, Vivekananda Mahavidyalaya, Burdwan, specially Sri P. Mondal, Dr. R. Pathak, Dr. B. Halder, Dr. P.S. Sengupta, Dr. S. Mondal and Smt. N. Singh. Warm thanks are due to Mr. Arunabha Sen, MS(USA), Managing Director, Books & Allied (P) Ltd. for taking keen interest of this book. I extend my thanks to the DTP operator, Mr. Abhijit Ghosal who untiringly worked through my manuscript to give the final shape of the present work. The sincere effort of Sri P. P. Singha Roy, Executive-Business Development is specially acknowledged. Finally the author is grateful to his wife Smt. Gita Banerjee M.A.(double) for various types of help in preparing the manuscript.

Lastly the present endeavour will be fruitful to the author if the students and readers are benefited. Inspite of sincere efforts there might be some errors crept in for which the author apologised. Constructive critisim from all the readers for the improvement of the book is solicited.

*Burdwan* **S. P. BANERJEE**
*Subha Mahalaya, 15th October, 2012*
*E-mail :* sankarprasadbanerjee@rediff.com

# Contents

# 1

# CONCEPT OF ATOM

What is the ultimate constitution of matter? The answer to this was provided by Greek philosopher *Demokritos* and Indian philosopher *Kanad,* but there was no scientific record of their statements. First scientific statement of the ultimate consitution of matter was given by John Dalton. Daltonian concept of matter are :

1. Every matter composed of tiny minute particles called atom which is not divisible.
2. All the atoms of the similar matter are identical in all respect of chemical and physical properties.
3. Atoms taking part in chemical reactions and they unite together in simple ratios to form chemical compounds.

Statements (1) and (2) are no longer tenable in the present situation. Atoms are now divisible to subatomic fundamental particles i.e. electron, proton, neutron etc. as also discovery of isotope has made the statement (2) wrong. Statement (3) is the valid one. Regarding statement (3) a question arises how the Daltonian concept can be applied in case of compounds like $TiH_{1.73}$, $PdH_{0.9}$, $VH_{0.6}$ etc. This is not the limitation of the Daltonian concept because these are not true stoichiometric chemical compounds.

## 1.1 SUBATOMIC FUNDAMENTAL PARTICLES

**The electron :** We known that gases are poor conductors of electricity at ordinary presssure. When pressure is lowered then electricity is passed through gasses and several complicated phenomena arises in the discharge tube. But if the pressure is still lowered then the former all complicated phenomena vanishes and at this stage a streak of rays emanate from the cathode and impinges on the anode. These rays are called **cathode rays.** Experimentally it has been verified that cathode rays are nothing but the streams of negatively charged particles i.e. flow of electrons which can be justified from the properties of cathode rays :

1. The rays travel in straight line.
2. They are able to heat bodies by impact.
3. They are able to ionise a gas.
4. They possess a considerable amount of momentum.
5. They can be deflected by the application of electric and magnetic field. From the direction of deflection it was concluded that the charge accompanying the cathode rays is negative.

Properties (4) and (5) suggest that electron has considerable amount of mass and charge. The next question arises what is the magnitude of charge and mass of electron?

In fact there is no direct method for the determination of mass of the electon. J.J. Thomson determined the e/m ratio of electron. The accepted value for e/m ratio is $1.76 \times 10^8$ $Cg^{-1}$ or in SI unit $1.76 \times 10^{11}$ C $kg^{-1}$.

### 1.1.1 The Charge of the electron

Two methods are available :

1. Coulometric determination of electronic charge.
2. Millikan's oil drop experiment.

The accurate method of determination of electronic charge is done by Millikan's oil drop experiment. The discussion of this method is beyond the scope of this book. Coulometric determination of electronic charge based on Faraday's laws of electrolysis. If one Faraday of electricity is passed through an electrolyte solution, 1 gm-equivalent of substantance will be dissolved or deposited at the respective electrodes. Say one Faraday of electricity is passed through 1 gm equivalent of $AgNO_3$ solution. Then at cathode one gram equivalent of silver will be deposited. One gram equivalent of silver is equal to one gram atom of silver. One gram atom of silver corresponds to Avogadro's number (N) of silver atoms. For the deposition of one atom of silver only one electron is associated. So for deposition of N number of silver atoms, total number of electrons is N. If the charge of one electron is e, then total charge = Ne. Therefore we can write

$$F = Ne, \text{ so that, } e = \frac{F}{N} = \frac{96500 \text{ coulomb}}{6.023 \times 10^{23}}$$

$$= 1.60 \times 10^{-19} \text{ coulomb}$$

$$= 4.8 \times 10^{-10} \text{ esu.}$$

### 1.1.2 The mass of the electron

$$m = \frac{1.60 \times 10^{-19} \text{ C}}{1.76 \times 10^{8} \text{ C g}^{-1}} = 9.108 \times 10^{-28} \text{ g} = 9.108 \times 10^{-31} \text{ kg}$$

### 1.1.3 Other subatomic nuclear particles

In order to maintain electroneutrality of the atom there must be positive charged species within the atom. Using perforated cathode in the discharge tube a rays opposite to cathode rays emerged which is known as *positive rays*. The charge and mass of the positive rays particles depend on the gases present inside the discharge tube. The lightest positive particle was obtained when hydrogen gas was taken in the discharge tube and in that case the positive particle was known as *proton* ($H^+$). It has a mass $1.6727 \times 10^{-27}$ kg (1.00728 amu).

Neutron is another fundamental neutral particle discovered by Chadwick (1932) by the nuclear reaction :

$${}^{9}_{4}Be + {}^{4}_{2}He \rightarrow {}^{12}_{6}C + {}^{1}_{0}n$$

which has a mass of $1.6748 \times 10^{-27}$ kg (1.00866 amu)

Subsequently other extremely short lived positive particle **positron** was discovered by Anderson (1932). It has the same mass and charge as that of the electron. *Meson* was discovered by Yukawa (1935). It may be positive, negative or neutral and has mass in between that of proton and electron. A neutral particle of negligible mass *neutrino* was predicted by Pauli (1931).

## 1.2 THE ATOMIC MODEL

The presence of free electrons was observed during discharge in gases and it was also established at that time that metal on being heated give off electrons. These proved that electrons were present in atoms. But an atom as a whole is electroneutral. Hence there must exist within the atom positive electrical charge. The emission of α-particles from the atoms of radioactive substances pointed to the possibility of the existence of positively charge units within. Therefore there was the necessity of a satisfactory atomic model.

J.J. Thomson (1904) proposed an atomic model but this was discarded due to the failure of the explanation of the atomic spectra.

### 1.2.1 The Rutherford's atom model (Solar or Planetary model of Atom) Scattering of α-particles through matter

An atomic model was proposed by Rutherford with the analogy of the solar system. In the central portion of the atom there was a central positive charge and surrounding the nucleus there was a planetary revolving electron. An enormous amount of empty space exist in between the nucleus and the revolving electron.

In 1911 Rutherford proposed a new theory of the scattering of α-particles through matter that greatly influenced our conception of atom. The α particles (designated as $He^{2+}$) were obtained from spontaneously disintegrating polonium and was allowed to strike on a thin metal foil of gold or platinum. The number of α-particles passing through the gold or platinum foil was traced by fluorescent zinc sulphide screen placed at a certain distance from the gold or platinum foil. The number of α-particles approaching towards the foil was noted by G.M. counter. It was found that most of the α–particles passed staraight through the foil or suffer very little deflection. But among 20,000 α-particles only one or two particles was rebounded back

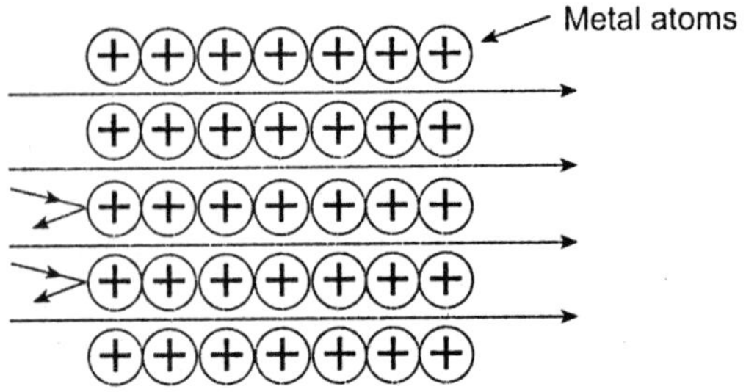


*Figure 1.1 : Scattering of α-particles through matter.*

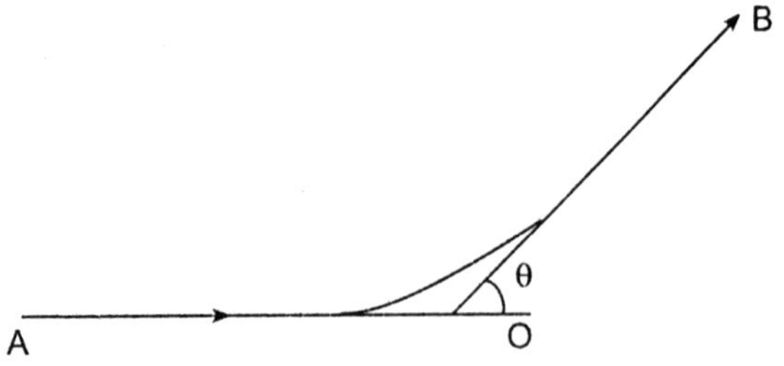


*Figure 1.2 : Direction of α-particle*

(Fig.1.1) which proves the empty space within the atom. The small central part of the atom which is responsible for the very large scattering of very few α–particles (1 or 2 in 20,000) is very dense and positively charged. This small central part was called the *atomic nucleus* by Rutherford.

If we take a fairly energetic α-particle, this will not directly hit the nucleus. As the α-particle is approaching towards the nucleus at O (Fig. 1.2) from A, there will be a strong repulsive force operated between the nucleus and the α-particle. At the distance of closest approach of the nucleus, the α-particle will change its track from AO to OB with the angle θ which is known as angle of scattering. Rutherford shown that the number of α-particles scattered through a given angle is directly proportional to the thickness of the foil, is proportional to the square of the nuclear charge and inversely proportional to the fourth power of the velocity of α-particles:

$$N = N_0\, n\, t\, \frac{(Ze)^2 e^2}{R^2 (mv^2)^2 \sin^4 \frac{\theta}{2}}$$

where N = number of α-particles striking per unit area of ZnS screen placed at a distance of R from the Au or Pt foil.

$N_0$ = Number of α-particles striking per unit area of Au or Pt foil

Z = Atomic number of the metal

e = electronic charge

θ = angle of scattering

t = thickness of the foil

m = mass of α-particles

$v$ = velocity of α-particles

n = number of atoms per unit volume.

Rutherford was able to determine the dimension of the nucleus. The charge on the nucleus and the α-particle is Ze and 2e respectively. In a head on collison α-particle will not directly hit the nucleus so that the distance of closest approach speaks about the nuclear dimension (Fig. 1.3) if we take a fairly energetic α-particle. The potential energy at a distance R from the

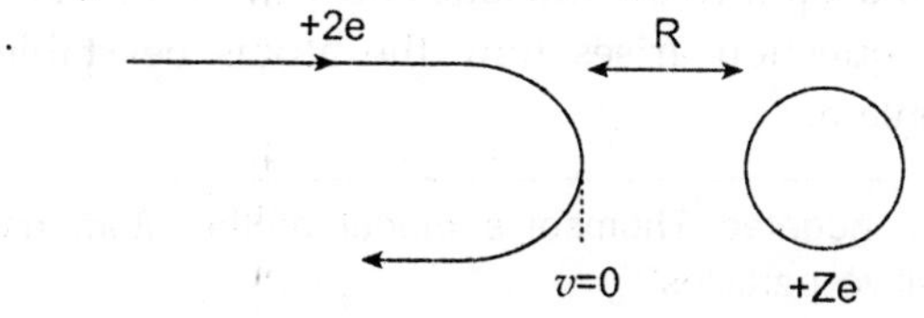


***Figure 1.3 :*** *Calculation of nuclear dimension.*

nucleus is $\frac{2Ze^2}{R}$. The kinetic energy of the α-particle is $\frac{1}{2}mv^2$. It is reasonable to assume that when the kinetic energy of the α-particle will be equal to the potential energy of the system, the particle will reverse its path. Therefore,

$$\frac{1}{2}mv^2 = \frac{2Ze^2}{R}$$

i.e., $R = \frac{4Ze^2}{mv^2}$

$$= \frac{4 \times 79\ (4.8 \times 10^{-10}\ \text{esu})^2}{6.62 \times 10^{-24}\ \text{g}\ (1.6 \times 10^9\ \text{cm})^2} \approx 4.0 \times 10^{-13}\ \text{cm}$$

i.e., nuclear dimension of the order of ~$10^{-13}$ cm. Atomic dimension was found to be ~ $10^{-8}$ cm from various measurements which indicates that atoms possess an empty structure.

The following conclusions can be drawn from Rutherford's atom model :

1. All the positive charge and almost the entire mass of the atom was concentrated in a small central part called atomic nucleus.
2. The radius of an atomic nucleus is ~$10^{-13}$ cm which is the same as that of the electron. Since the radius of an atom is ~$10^{-8}$ cm it is obvious that an atom must have a very empty space within.
3. An atomic model was proposed by Rutherford. A very small positive nucleus was considered surrounded by planetary electrons in analogy with planetes. Such a system cannot be stable if the electrons were at rest. Therefore it was proposed that the electrons were moving in circular orbit around the nucleus so that coulombic attraction between the nucleus and the electron was equal to the centrifugal force on the electron.

### 1.2.2 Objection against the model

1. How could a packed combination of protons and electrons give a stable nucleus.
2. It is a general phenomena that whenever a body moves in a circular orbit, it is being acclerated towards the centre of the orbit. Hence the revolution of a negatively charged electron in the field of positively charged nucleus with an acclerated motion must result in a loss of energy of motion resulting the radius of revolving orbit smaller and smaller

and ultimately the electron will drop into the nucleus and will be a breakdown of the postulated atomic structure. Therefore a question arises how the atoms be stable? Bohr suggested the necessary solution to this problem.

**Problem 1.1** Which experiment negated Thomson's model of the atom as an intimate mixture of negative and positive particles? **[B.U. 2005]**

**Ans. :** Rutherford's experiment with scattering of $\alpha$-particles through metal foil.

**Problem 1.2** Why in the Rutherford's experiment gold or plalimum foil was taken?

**Ans. :** Gold has atomic number 79 i.e. the nucleus of gold has 79 protons. If we take fairly energetic $\alpha$-particles then it will not directly hit the nucleus because in that case there will be a strong repulsive force operated. If low atomic number metal foil say aluminium is taken then there might be the possibility of the following type nuclear transmutation reaction

$$^{27}_{13}Al + ^{4}_{2}He \rightarrow ^{30}_{15}P + ^{1}_{0}n\,;\quad ^{30}_{15}P \rightarrow ^{30}_{14}Si + ^{0}_{+1}e$$

and in that case the whole scattering experiment will be null and void.

Instead of gold or platinum, if more high atomic number metals were taken, then the experiment will be more easier. But as we take more high atomic number elements, then the nucleus of those elements are itself unstable and may naturally radioactive.

Considering all these aspects gold or platinum metal was taken by Rutherford in his famous scattering experiment.

## 1.3 MOSLEY'S WORK ON ATOMIC NUMBER

Rutherford's experiment suggested that within the atom positive charge exist in the nucleus. The next question arises what is the magnitude of the total positive charge within the nucleus? X-ray spectrograph which is used for this purpose is shown below (Fig. 1.4) :

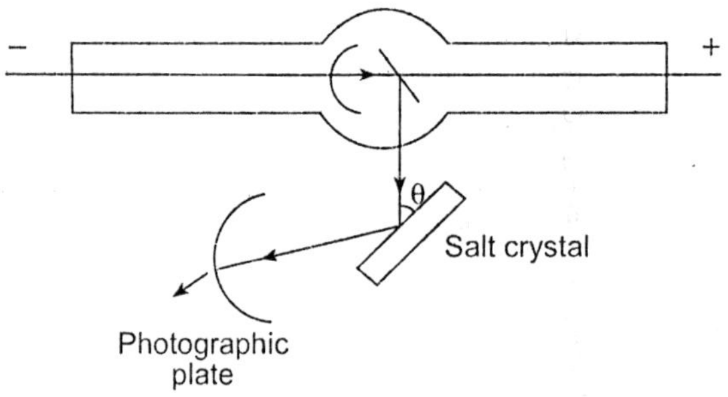


*Figure 1.4 : Schematic diagram of X-ray spectrograph.*

Cathode rays on impinging a metal target produces X-rays which has the characteristic of the target material and thereby it is called characteristic X-rays. This ray falls on the surface of a salt crystal. Since the spacing of atoms and ions in crystals is of the same magnitude as the wave length of the characteristic X-rays, therefore, before Mosley's work Bragg suggested that crystals can act as diffraction gratings for X-rays. According to Bragg, when a beam of X-rays falls on a crystal, each atom in that crystal begins to act as a secondary source of waves and the resultant wave front is obtained according to Hygen's construction in optics.

The X-rays produced in the cathode tube are sorted into a beam by passing them through two slits. The resulting beam strikes the surface of the salt crystal. As the atoms in a given crystal are arranged in a regular pattern, some X-rays are diffracted in the successive layers of the crystal whereby the intensity of the diffracted light will be much less than that of the orignal intensity of X-rays. For maximum intensity of the diffracted light we get,

$$2d \sin\theta = n\lambda$$

where, $\theta$ = angle between the incident ray with that of the salt crystal

$d$ = distance between successive layers

$\lambda$ = wavelength of the diffracted light

$n$ = order of the spectrum.

If we know the value for d and $\theta$, $\lambda$ can be calculated from the above formula.

The characteristic radiation i.e. the radiation which is diffracted from the salt crystal is of wide importance because this radiation appears as monochromatic X-rays as K, L, M, N etc. series of lines in the spectrogram.

In 1913–1914 Mosley was able to determine the atomic number of thirty eight different elements beginning with that of aluminium and ending with gold. Where the element is metal this was used directly as anticathode but where the element is non metal then in that case a suitable compound of the element is coated on a metal and is used as anticathode. He used potassium ferrocyanide as the salt crystal.

Mosley's experiment suggested that the spectral lines exhibited by those elements belonged to two different series of lines namely K and L lines and further it was noticed that for lighter elements only K lines are obtained and these K lines are also splitted into two components i.e. $K_\alpha$ and $K_\beta$ and these lines are directed towards the shorter wavelength as the atomic number of the element increases. For the heavier elements L lines are obtained which are also splitted into two components (e.g. $L_\alpha$ and $L_\beta$). Mosley was able to cinchronised his data in terms of a mathematical expression connecting the observed frequency or wavelength of a particular line with a quantity which is known as atomic number (Z) i.e.,

$$\nu = a(Z - \sigma)^2$$

$$\text{i.e., } \nu^{\frac{1}{2}} = Za^{\frac{1}{2}} - \sigma a^{\frac{1}{2}}$$

where, a is a proportionality constant and $\sigma$ is a screening constant determined from Slater's rule. When Mosley plotted $\sqrt{\nu}$ of the characteristic frequency with that of Z, approximately straight line was obtained (Fig. 1.5).

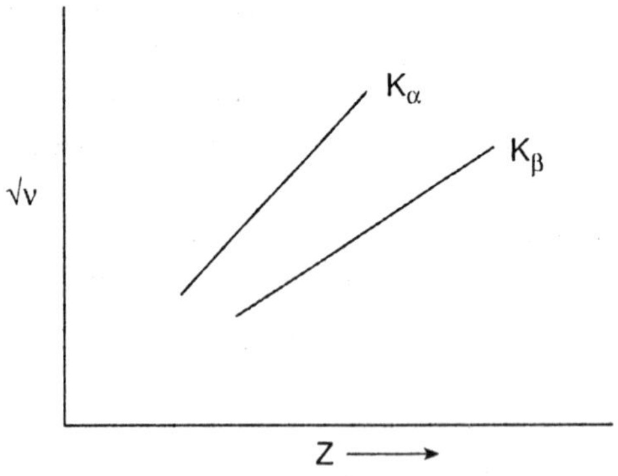


*Figure 1.5 : X-ray spectra of K lines.*

Mosley's mathematical expression suggested that the quantity v increases linearly with atomic number i.e., as we pass from one element to the next, the successive differences between the values of $\sqrt{v}$ of $K_\alpha$ lines clustering closely around the mean 0.029 (Table 1.1).

**Table 1.1 : Some values of wavelength measurements of $K_\alpha$ lines**

| Element | $\lambda_{K_\alpha}$ (A°) | $\frac{1}{\sqrt{\lambda}} K_\alpha (A^{0^{-\frac{1}{2}}})$ | Successive difference X $10^{-3}$ | Atomic weight | Atomic number |
|---|---|---|---|---|---|
| Ca | 3.35 | 0.546 | — | 40.08 | 20 |
| Sc | 3.03 | 0.575 | 29 | 44.96 | 21 |
| Ti | 2.74 | 0.604 | 29 | 47.96 | 22 |
| V | 2.50 | 0.633 | 29 | 50.94 | 23 |
| Cr | 2.29 | 0.662 | 29 | 52.00 | 24 |

Mosley established firmly that the really important quantity in explaining the periodicity of properties of an element is its atomic number and not the atomic weight and he suggested the perioidic law as *the properties of the elements are the periodic functions of their atomic number.*

### 1.3.1 Achievements of Mosley's work

1. Atomic number was taken as more fundamental than atomic weight in periodic classification of elemetns.
2. For the determination of atomic number Mosley provided an experimental method.
3. An explanation was provided for the anomolous placement of the four pairs of misfit elements (e.g. Ar, K; Co, Ni etc.) in the periodic table were regularised.
4. Undiscovered elements were pin pointed and these have since been discovered.

**Problem 1.3** Mendeleef's statement of the periodic law had to be modified after Mosley's discovery — explain. **[B.U. 1995]**

**Hint. :** The characteristic X-ray frequency is related to the atomic number of the element

$$\sqrt{\nu} = Z\sqrt{a} - \sigma\sqrt{a}$$

which leads to the conclusion that atomic number is more fundamental property than atomic weight in periodic classification of elements.

**Problem 1.4** How are X-ray frequencies related to atomic number of elements? **[B.U. 1997, 2005]**

**Ans. :** Problem 1.3.

## 1.4 ELECTROMAGNETIC SPECTRUM, ELECTROMAGNETIC RADIATIONS OR ELECTROMAGNETIC WAVES

We know that light is an electromagnetic radiation. In the nineteenth century a wave theory of light was accepted. According to which light has the *wave parameters* and its profile is given in figure 1.6.

(a) **Wavelength (λ)** : It is the distance between the two adjacent crests or troughs in a particular wave expressed in centimetre, cm (in c.g.s. unit) and metre, m (in S.I. units).

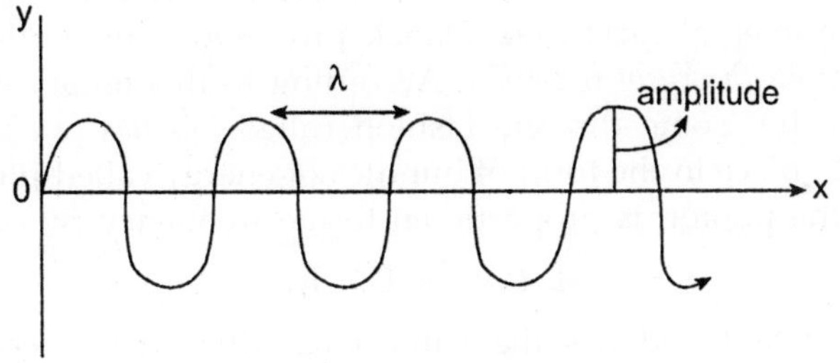


***Figure 1.6 :*** *Wave profile of light.*

(b) **Amplitude :** It is the height of the wave measured from the mean position (along the y axis) and is expressed in same unit of wave length.

(c) **Frequency (ν) :** It is defined as the number of waves or cycles which pass through a point per second. This is expressed in cycles per second (cps) thereby its unit is second$^{-1}$($s^{-1}$) or Hertz (Hz) in both the system of units. It is evident that frequency and wavelength are inversely related i.e. $\nu \alpha \frac{1}{\lambda}$ which follows that $\nu\lambda$ = constant. This constant is denoted by the velocity of light (c). This means that velocity of any electromagnetic radiation should not exceed the velocity of light.

Frequency is also expressed in wave number ($\overline{\nu}$) which states the number of waves per unit length i.e.,

$$\overline{\nu} = \frac{1}{\lambda} \quad \text{as} \quad \nu\lambda = c$$

$$\therefore \overline{\nu} = \frac{\nu}{c}$$

(d) **Velocity** : It states the distance travelled by the wave per second and is expressed as cm $s^{-1}$ (c.g.s. unit) or $ms^{-1}$ (S.I. unit).

Newton had shown that when sunlight is passed through a prism there will be a continuous separation of seven colours called VIBGYOR. A light pattern consisting of this type is called *continuous spectra* since the colours are diffused i.e. there is no sharp line of demarcation between the colours. On the other hand there are light sources which consist of some lines separated by dark spaces. This type of spectra is called *discontinuous spectra* or line spectra or sometimes called *atomic spectra* since this type of spectra is produced from atoms.

If we accept the wave theory of light there will be two difficulties in the explanation of the following facts:

1. How line spectra produced from atoms.
2. The explanation of the photoelectric effect i.e., only light of certain minimum energy could knock electrons from a metal surface.

Classical wave theory of light is inadequate to explain these experimental facts. In order to explain the above experimental facts Max Planck proposed a revolutionary theory which is known as *Planck's quantum theory of radiation.* According to this theory energy in matter is not in the continuous form but some discrete fashion called *bundles of energy* i.e. absorption or emission of energy takes place in the form of bundles of energy called **photon**. Planck proposed that the energy (E) of the photon is proportional to the frequency ($\nu$) of the photon i.e.

$$E \propto \nu \quad \text{or} \quad E = h\nu$$

where h is Planck's constant and has the value $6.627 \times 10^{-27}$ erg– second (in c.g.s. unit) or $6.627 \times 10^{-34}$ Joule-second (in SI unit). This means that the amount of energy emitted or absorbed by an oscillator is an integral multiple of $h\nu$ (quantum of energy) i.e. $E = nh\nu$ where n = 1,2,3 ....... . This is known as *quantisation of energy.*

This model of light affords an explanation of line spectra given by various elements. It should be realised that the line spectra of atom proves that an atom does not take or emit any hapazzard amount of energy but only some *quantised amount.* Quantum of energy or *quanta* are the basic *currency* of light energy and can be absorbed or emitted from atoms in definite whole number units. There must be some mechanism hidden in the electronic structure of atoms that must prevent it to absorb or emit any amount of energy.

### 1.4.1 Photoelectric effect

*When a beam of light of some suitable wavelength or frequency is falling on the surface of a metal, the electrons are ejected from the surface of the metal. This phenomena is known as photoelectric effect* and the emitted electrons are known as *photoelectrons.*

Einstein explained photoelectric effect with the help of Planck's *quantum theory of radiation.* In this theory radiation is regarded as a bundles of photon each of energy hν moving in space with a velocity of light. Electron emission could occur only when energy of the photon of the exciting light is greater than some minimum energy which is called *threshold frequency* or *cut-off frequency* represented by $\nu_0$. Electron emission can not be possible if the radiation has a frequency smaller than the threshold frequency. Now assuming that collison between a photon and an atom results in absorption of the whole photon by the atom and the consequent emission of photoelectron. Before the electron can escape, it must perform work (w) to liberate itself from the attraction of the metal, the remainder of the energy appears in the form of kinetic energy. According to the law of conservation of energy we can write

$$h\nu = w + \tfrac{1}{2}mv^2$$

where $v$ = velocity of the ejected electron

With light of frequency ($\nu_0$) just sufficient to emit electron ($v$ = 0), then,

$$h\nu_0 = w$$

hence $$h\nu = h\nu_0 + \tfrac{1}{2}mv^2 \qquad \text{...(1.1)}$$

Equation (1.1) is known as Einstein's photoelectric law. The value of Planck's constant was determined by Millikan with the help of equation (1.1). He determined kinetic energy corresponding to the light of frequency ν and $\nu_0$. With the help of these kinetic energy values he calculated the value of h from equation (1.1). The value was found to be $6.570 \times 10^{-34}$ Js which is in good agreement with that of the experimental value $6.627 \times 10^{-34}$ Js.

---

**Problem 1.5** The critical wavelength of light producing photoelectric effect in tungsten is 2600Å. What is the energy of a quantum of this wavelength? How many quanta correspond to 1 kcal of energy?

**Ans. :** Quantum energy $E = h\nu = \frac{hc}{\lambda}$ $\quad [\because \nu\lambda = c \therefore \nu = \frac{c}{\lambda}]$

$$= \frac{6.627 \times 10^{-27}\ \text{erg - sec} \times 3 \times 10^{10}\ \text{cm/sec.}}{2{,}600 \times 10^{-8}\ \text{cm}}$$

$= 0.007646 \times 10^{-9}$ erg

$= 0.007646 \times 10^{-16}$ Joule $\quad [\because 1\ \text{Joule} = 10^7\ \text{erg}]$

$= 7.646 \times 10^{-19}$ Joule $\quad [1\ \text{kcal} = 4.184 \times 10^3\ \text{Joule}]$

$$\therefore \text{Number of quanta} = \frac{4.184 \times 10^3\ \text{Joule}}{7.646 \times 10^{-19}\ \text{Joule}}$$

$= 5.472 \times 10^{21}$

---

**Problem 1.6** The frequency of violet light is $7.31 \times 10^{14}$ s$^{-1}$. What will be the wavelength and energy of this light?

**Ans. :** We know that $\nu\lambda = c$

i.e. $\lambda = \frac{c}{v} = \frac{3 \times 10^{10}\ cms^{-1}}{7.31 \times 10^{14}\ s^{-1}}$

$= 0.41 \times 10^{-4}$ cm

$= 4.1 \times 10^{-7}$ m

$= 410$ nm [$\because$ 1 nm $= 10^{-9}$ m]

Since E = hv

$= (6.627 \times 10^{-34}$ Js$)(7.31 \times 10^{14}$ s$^{-1})$

$= 4.84 \times 10^{-19}$ J

### 1.4.2 Wave nature of light

The wave nature of light experimentally verified by the production of continuous spectra (VIBGYOR) which indicates that light propagates in space in the form of wave. Moreover light shows the properties of reflection, refraction and diffraction which indicates the wave character of light.

### 1.4.3 Particle nature of light

The most definite proof of the particle nature of light comes from *photo electric effect*. Another convincing evidence comes from *Compton effect*. There is a change in wavelength of X-rays when these are scattered upon impact with elements of low atomic weight. X-rays are an electromagnetic radiation which is composed of *photons* as proposed by Planck. Theoretically particle nature of light forms the basis of *Planck's quantum theory of radiation*. This concept laid the foundation of a new concept known as *wave mechanics* or *quantum mechanics* discarding the old concept of *classical mechanics*. Quantum mechanics states that both light and electron has dual character.

## 1.5 EXPERIMENTAL FINDINGS OF HYDROGEN SPECTRA

Possibly the most important tool in determining the electronic structure of atom is to study the radiations emitted or absorbed by atoms. These radiations when passed through a suitable prism, are found to consist of a large number of lines which are characteristic of that atom. Hydrogen is the simplest of all atoms and therefore we are studying the spectra of hydrogen first before going to the complex atomic system. The experimental arrangement is shown in the following figure 1.7.

The electric discharge produces a stream of fast moving electrons which bombards the hydrogen molecules so that they dissociate into atoms. The discharge further excites some of the atoms so that they are capable of emitting light. This emitted light is passed through the spectroscope which analyse the light and the spectra recorded in the spectrogram. Hydrogen gives four lines ($H_\alpha$, $H_\beta$, $H_\gamma$ and $H_\delta$) in the visible region (i.e. in the wavelength 4000Å to 8000Å) which gradually close up (Table 1.2).

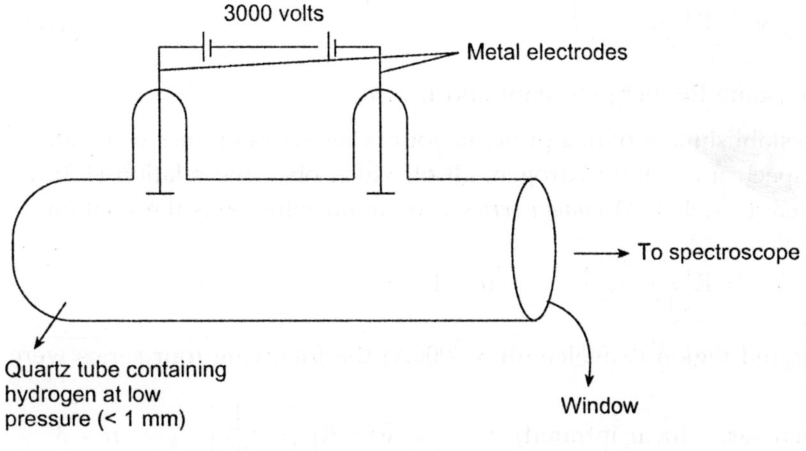


***Figure 1.7** : Apparatus to study hydrogen spectra.*

**Table 1.2 : Spectral lines of hydrogen in the visible region**

| Name | Colour | Position λ(Å) | $\bar{\nu}$ (cm$^{-1}$) |
|---|---|---|---|
| $H_\alpha$ | Red | 6565 | 15234 |
| $H_\beta$ | Green-blue | 4862 | 20565 |
| $H_\gamma$ | Deep blue | 4342 | 23033 |
| $H_\delta$ | Violet | 4103 | 24374 |

In search for other lines in the hydrogen spectra, a considerable amount of spectral data was accumulated and in order to establish some sequence among these lines Balmer in 1885 proposed an emperical relationship between the wave number ($\bar{\nu}$) and the position of the lines in the spectrogram, since then it was called *Balmer series* obtained in the visible region. The relation is :

$$\bar{\nu} = R\left[\frac{1}{2^2} - \frac{1}{n^2}\right] \qquad \text{......(1.2)}$$

where R is a constant called Rydberg constant experimentally found to be 109678 cm$^{-1}$ and n is an integer which has values greater than 2.

The nature of this relation prompted Balmer and Ritz to realise that there might exist other series of spectral lines in which the first term of equation (1.2) would be different from $\frac{1}{2^2}$. In 1908 Balmer and Ritz discovered a principle of combination which is known as *Ritz combination principle* in which the wave number of any line in the spectrogram can be given by the following general equation

$$\bar{\nu} = R\left[\frac{1}{m^2} - \frac{1}{n^2}\right] \quad ......(1.3)$$

where R is the same Rydberg constant and $n > m$.

After the establishment of this principle four other series of lines were later been discovered in the entire spectrum of the hydrogen, all of which obey the relation (1.3). In the ultraviolet region (wavelength < 4000Å) *Lyman series* was found which has the relation

$$\bar{\nu} = R\left[\frac{1}{1^2} - \frac{1}{n^2}\right] \qquad n > 1$$

In the infrared region (wavelength > 8000Å) the following four series were found.

Paschen series (near infrared) : $\bar{\nu} = R\left[\frac{1}{3^2} - \frac{1}{n^2}\right]$ $n > 3$

Brackett series (far infrared) : $\bar{\nu} = R\left[\frac{1}{4^2} - \frac{1}{n^2}\right]$ $n > 4$

Pfund series (far infrared) : $\bar{\nu} = R\left[\frac{1}{5^2} - \frac{1}{n^2}\right]$ $n > 5$

Humphreys series (far infrared) : $\bar{\nu} = R\left[\frac{1}{6^2} - \frac{1}{n^2}\right]$ $n > 6$

In fact there arises several questions :

1) What is the theoretical basis behind the Ritz combination principle?
2) Why the position of the lines is given by the difference of two square terms?

The answer to these come from the Bohr's theory of the hydrogen atom.

## 1.6 BOHR'S THEORY OF ATOMIC SPECTRA

In order to explain the stability of the atom as also to explain the failure of Rutherford's atom model Bohr in 1913 developed his theory which accounted many of the properties of atomic spectra of hydrogen. He has taken the Rutherford's atom model in which a single electron revolve round the nucleus under the action of coulomb field of force. Discarding the laws of classical mechanics he applied the quantum theory of radiation as propsed by Planck and Einstein to the Rutherford's atom model. His theory based on the following postulates :

1) The electron cannot rotate in any orbit. The permissible orbits are those for which the angular momentum of the electron is an integral multiple of $\frac{h}{2\pi}$ i.e.

$$p_\phi = n\left(\frac{h}{2\pi}\right)$$

where $p_\phi$ = angular momentum = $mvr$, m = mass of the electron, $v$ = velocity of the electron and r = radius of curvature, h = Planck's constant, n = principal quantum number = 1,2,3.......

These are termed stationary orbit and no energy is radiated nor absorbed by that electron when rotating in these stationary orbits.*

2) During emission or absorption of radiant energy Planck-Einstein equation $\varepsilon = h\nu$ is obeyed where $\varepsilon$ = energy difference between the two states = $E_2 - E_1$ ($E_2 > E_1$) and $\nu$ is the frequency of the light.

3) When an electron is rotating in any energy states other than the least energy state, the electron and the atom as a whole is said to be in the excited state. An excited electron has always the tendency to revert back to the non excited ground state.

4) The dynamic equilibrium of the system in the stationary state is governed by the laws of classical mechanics.

### 1.6.1 Calculation of Bohr radius of hydrogen and hydrogen like systems

Let us consider an electron of charge (–e) is revolving round the nucleus of charge (+Ze) in a circular orbit of radius r. The mass of the electron is m and velocity is $v$ (Fig. 1.8).

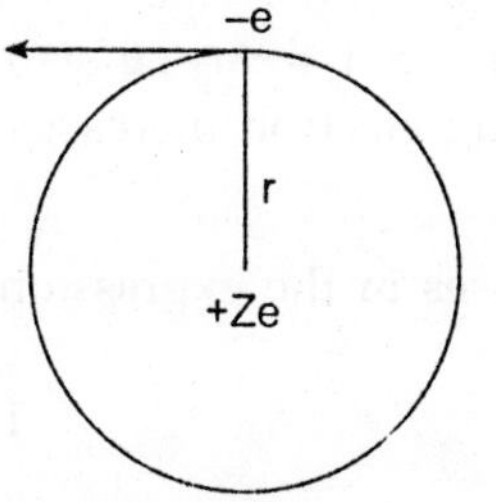


*Figure 1.8* : Bohr model of atom.

The centrifugal force (F) = $\dfrac{mv^2}{r}$ and

the coulomb's force (F′) = $\dfrac{Ze^2}{r^2}$

In a stationary orbit these two forces balance each other, so that,

$$\frac{mv^2}{r} = \frac{Ze^2}{r^2} \qquad \therefore v^2 = \frac{Ze^2}{mr} \qquad ...(1.4)$$

According to Bohr's first postulate

$$mvr = \frac{nh}{2\pi} \qquad \therefore v = \frac{nh}{2\pi mr} \qquad ...(1.5)$$

* The defect of the Rutherford's atom model is removed in this postulate because when an electron rotating in the stationary orbit, no energy is radiated by that electron. So there is no question to reduce the radius of the revolving electron and thereby there will be no possibility to drop the electron into the nucleus.

From equation (1.4) and (1.5) we get,

$$\frac{Ze^2}{mr} = \frac{n^2h^2}{4\pi^2m^2r^2}$$

$$\text{i.e.} \quad r = \frac{n^2h^2}{4\pi^2Ze^2m} \qquad ...(1.6)$$

For hydrogen atom Z = 1, Therefore,

$$r = \frac{n^2h^2}{4\pi^2e^2m} \qquad ...(1.7)$$

On putting the values $n = 1$, $h = 6.627 \times 10^{-27}$ erg-second, $\pi = 3.14$, $e = 4.8 \times 10^{-10}$ esu and $m = 9.108 \times 10^{-28}$ g we get,

$$r = \frac{1^2(6.627\times10^{-27}\ \text{erg-second})^2}{4(3.14)^2(4.8\times10^{-10}\ \text{esu})^2\, 9.108\times10^{-28}\ \text{g}}$$

$$= 0.529\times10^{-8}\ \text{cm}^* = 0.529\text{Å} = a_0 = \text{1st Bohr radius.}$$

When $n = 2$, $r = 4a_0$ and when $n = 3$, $r = 9a_0$ i.e. as the principal quantum number value increases, the distance of the revolving electron increases.

---

* How the dimension of length arises in the expression of r?

$$r = \frac{n^2h^2}{4\pi^2e^2m}$$

Let the dimension of length = L

dimension of mass = M

and dimension of time = T

$$= \frac{[h^2]}{[e^2][m]}$$

We know tht $\varepsilon = h\nu$ $\quad \therefore h = \dfrac{\varepsilon}{\nu} = \dfrac{\text{Energy}}{\text{Frequency}} = \dfrac{\text{Work done}}{\text{Frequency}}$

$$= \frac{\text{Force} \times \text{Distance}}{\text{Frequency}} = \frac{\text{Mass} \times \text{Accleration} \times \text{Distance}}{\text{Frequency}}$$

$$= \frac{M \times LT^{-2} \times L}{\frac{1}{T}} = ML^2T^{-1} \quad \text{i.e.} \quad \boxed{h = ML^2T^{-1}}$$

Let the two charges (e and e) separated by a distance d

$$\therefore \text{Coulomb force (F)} = \frac{e^2}{d^2}$$

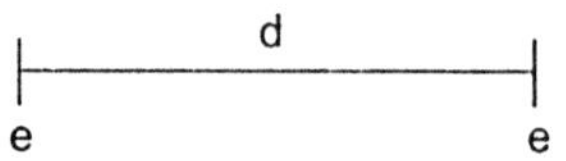


i.e. $e^2 = Fd^2 = \text{Mass} \times \text{accleration} \times (\text{distance})^2$

$$= M \times LT^{-2}L^2 = ML^3T^{-2}$$

$$\text{i.e.} \quad \boxed{e^2 = ML^3T^{-2}}$$

$$\therefore r = \frac{(ML^2T^{-1})^2}{ML^3T^{-2}M} = \frac{M^2L^4T^{-2}}{M^2L^3T^{-2}} = L$$

**Problem 1.7** In case of hydrogen atom an electron jumps from first orbit to third orbit. What will be the percentage increase in radius?

**Ans.** : For $n = 1$ $r = a_0$

$n = 3$ $r = 9a_0$

$\therefore$ Increase in radius = $8a_0$

So that percentage increase in radius = $\frac{8a_0}{a_0} \times 100 = 800\%$

## Calculation of radius in S.I. unit

As before the centrifugal force (F) = $\frac{mv^2}{r}$. But the Coulomb's force (F) i.e. the electrostatic force of attraction which exist between the nucleus and the revolving electron. In c.g.s. system this force = $\frac{Ze^2}{r^2}$ but in S.I. system this force = $\frac{Ze^2}{r^2} \times \frac{1}{4\pi\varepsilon_0}$ .Where $\varepsilon_0$ is the permittivity of the medium which has a value $8.854 \times 10^{-12}$ $J^{-1}C^2m^{-1}$.

$$\therefore \frac{mv^2}{r} = \frac{Ze^2}{r^2 4\pi\varepsilon_0}, \text{ so that}$$

$$v^2 = \frac{Ze^2}{mr} \times \frac{1}{4\pi\varepsilon_0} \qquad \text{...(1.8)}$$

From equation (1.5) and (1.8) we can write $r = \frac{n^2h^2\varepsilon_0}{\pi me^2 Z}$ ...(1.9)

On putting all the values for h, $\varepsilon_0$ etc. we get,

$$r = \frac{n^2(6.627\times10^{-34}\text{ Js})^2 \times (8.854\times10^{-12}\text{ J}^{-1}\text{C}^2\text{m}^{-1})}{3.14(9.108\times10^{-31}\text{ kg})(1.602\times10^{-19}\text{C})^2 Z}$$

$$= \frac{n^2 \times 5.29\times10^{-11} \times \text{J}^2\text{s}^2\text{J}^{-1}\text{C}^2\text{m}^{-1}}{\text{kgC}^2 \times Z}$$

$$= \frac{n^2 \times 5.29\times10^{-11}\text{ Js}^2\text{m}^{-1}}{\text{kg}\times Z}$$

$$= \frac{n^2 \times 5.29\times10^{-11}\text{ kgm}^2\text{s}^{-2}\text{s}^2\text{m}^{-1}}{\text{kg}\times Z} \qquad [\because \text{J} = \text{kgm}^2\text{s}^{-2}]$$

$$= \frac{0.529\times10^{-10}\times n^2}{Z}\text{ m}$$

For hydrogen atom Z = 1 and n = 1.

$\therefore$ $r = 0.529 \times 10^{-10}$ m $= 0.529 \times 10^{-8}$ cm.

### 1.6.2 Calculation of the velocity of the electron in the nth orbit of the hydrogen and hydrogen like systems

(a) **In c.g.s. unit**

According to the following relations : $mvr = n\left(\frac{h}{2\pi}\right)$, $r = \frac{n^2h^2}{4\pi^2Ze^2m}$

$$\therefore v = \frac{nh}{2\pi mr} = \frac{nh}{2\pi m} \times \frac{4\pi^2Ze^2m}{n^2h^2} = \frac{2\pi Ze^2}{nh}$$

For H atom Z = 1, so that $v = \frac{2\pi e^2}{nh}$

This expression indicates that velocity decreases as n increases. The velocity of the electron in the first Bohr orbit is $2.183 \times 10^8$ cm $s^{-1}$ and that of the second orbit is $\frac{2.183 \times 10^8}{2}$ cm $s^{-1}$ and so on.

(b) **In S.I. unit**

We have the relation $v = \frac{nh}{2\pi mr}$ In S.I. unit $r = \frac{n^2h^2\varepsilon_0}{\pi me^2Z}$

$$\therefore v = \frac{nh}{2\pi m} \times \frac{\pi me^2Z}{n^2h^2\varepsilon_0} = \frac{Ze^2}{2\varepsilon_0 nh}$$

For hydrogen atom Z = 1

$$\therefore v = \frac{e^2}{2\varepsilon_0 nh}$$

If we put the value e = $1.602 \times 10^{-19}$ C, $\varepsilon_0 = 8.85 \times 10^{-12}$ $J^{-1}C^2m^{-1}$ and h = $6.627 \times 10^{-34}$ Js and for J = $kgm^2s^{-2}$, we get

$$v = \frac{(1.602 \times 10^{-19} C)^2}{2(8.85 \times 10^{-12} J^{-1}C^2m^{-1}) \times n \times (6.627 \times 10^{-34} Js)}$$

$$= \frac{0.2187 \times 10^7}{n} ms^{-1}$$

---

**Problem 1.8** What will be the velocity of an electron rotating in the third orbit of the hydrogen atom? How much this velocity is greater or less than the velocity of light?

**Ans.** : For third orbit $v = \frac{2.183 \times 10^8}{3}$ cm $s^{-1}$

Velocity of light (c) = $3 \times 10^{10}$ cm $s^{-1}$

$$\frac{\text{Velocity of light}}{\text{Velocity of electron}} = \frac{3\times 10^{10}\ \text{cm s}^{-1}}{\dfrac{2.183\times 10^{8}}{3}\ \text{cm s}^{-1}} = \frac{9\times 10^{2}}{2.183} = 4.12\times 10^{2}$$

∴ The velocity of light is 412 times greater than the velocity of the electron in the 3rd orbit.

### 1.6.3 Revolutions per second in nth orbit of hydrogen and hydrogen like systems

**In c.g.s. unit**

$$\text{Number of revolutions} = \frac{v}{2\pi r}$$

$$= \frac{\dfrac{2.183\times 10^{8}\ \text{cm s}^{-1}\times Z}{n}}{\dfrac{2\times 3.14\times 0.529\times 10^{-8}\times n^{2}\ \text{cm}}{Z}}$$

$$= \frac{2.183\times 10^{16}\times Z^{2}}{2\times 3.14\times 0.529\times n^{3}}\ \text{s}^{-1}$$

$$= \frac{0.65711\times 10^{16}\times Z^{2}}{n^{3}}\ \text{s}^{-1}$$

In case of hydrogen atom Z = 1 and for first orbit n = 1.

∴ Revolutions per second = $0.65711\times 10^{16} = 65.711\times 10^{14}$

In S.I. unit the expression will be the same as in c.g.s. unit.

### 1.6.4 Calculation of the total energy of the system

According to the Bohr model of the atom (Fig. 1.8) the total energy (E) is the sum of the kinetic energy (T) and potential energy (V)

$$\therefore\ E = T + V,\ T = \frac{1}{2}mv^{2} = \frac{Ze^{2}}{2r}\ \text{(from equation 1.4)}$$

$$V = \int_{\infty}^{r}\frac{Ze^{2}}{r^{2}}\,dr = -\frac{Ze^{2}}{r}$$

$$\text{Thus the total energy } E = \frac{Ze^{2}}{2r} - \frac{Ze^{2}}{r} = -\frac{Ze^{2}}{2r}.$$

Now putting the value of r from equation (1.6) we get,

$$E = -\frac{Ze^{2}}{2}\times\frac{4\pi^{2}Ze^{2}m}{n^{2}h^{2}} = -\frac{2\pi^{2}e^{4}mZ^{2}}{n^{2}h^{2}} \qquad \text{...(1.10)}$$

For hydrogen atom Z = 1, so that

$$E = -\frac{2\pi^2 e^4 m}{n^2 h^2} \quad \text{...(1.11)}$$

The energy of the electron in the first shell is obtained by putting n = 1 and the values of π, e, m etc.

$E = -2.18 \times 10^{-11}$erg = –13.595ev = –313.6kcal.

It is evident from equation (1.10) and (1.11) that as the distance of the revolving electron increases, the electron will go to the higher energy state. These equations (1.10) and (1.11) correspond to the total energy of hydrogen like systems and hydrogen atom respectively in c.g.s. unit.

**In S.I. unit :**

$E = T + V = \frac{1}{2}mv^2 + V$. In this unit force of attractioin is equal to $\frac{Ze^2}{4\pi\varepsilon_0 r^2}$, but the centrifugal force is the same as in c.g.s. system i.e. $\frac{mv^2}{r}$. Since these forces are equal

$$\therefore \frac{mv^2}{r} = \frac{Ze^2}{4\pi\varepsilon_0 r^2} \qquad \therefore v^2 = \frac{Ze^2}{4\pi\varepsilon_0 mr} \quad \text{...(1.12)}$$

$$\text{Here } V = \int_{\infty}^{r} \frac{Ze^2}{4\pi\varepsilon_0 r^2} dr = -\frac{Ze^2}{4\pi\varepsilon_0 r} \quad \text{...(1.13)}$$

$$\therefore \text{Total energy, } E = \frac{1}{2}m\frac{Ze^2}{4\pi\varepsilon_0 mr} - \frac{Ze^2}{4\pi\varepsilon_0 r} \quad \text{(from equation 1.12 and 1.13)}$$

$$= \frac{Ze^2}{8\pi\varepsilon_0 r} - \frac{Ze^2}{4\pi\varepsilon_0 r}$$

$$= -\frac{Ze^2}{8\pi\varepsilon_0 r} = -\frac{Ze^2}{8\pi\varepsilon_0} \times \frac{\pi me^2 Z}{n^2 h^2 \varepsilon_0} \quad \text{(Putting the value of r from equation 1.9)}$$

$$= -\frac{me^4 Z^2}{8\varepsilon_0^2 n^2 h^2} \quad \text{...(1.14)}$$

For hydrogen atom Z = 1, so that,

$$E = -\frac{me^4}{8\varepsilon_0^2 n^2 h^2} \quad \text{...(1.15)}$$

$$= -\frac{(9.108\times10^{-31}\text{ kg})\times(1.602\times10^{-19}\text{C})^4}{8\times(8.85\times10^{-12}\text{J}^{-1}\text{C}^2\text{m}^{-1})^2\times(6.627\times10^{-34}\text{ Js})^2\times1^2}$$

$= -2.18 \times 10^{-18}$ J

### 1.6.5 Explanation of the spectra in terms of Bohr's theory

Let us consider the two energy states $E_1$ and $E_2$ with corresponding quantum numbers $n_1$ and $n_2$ respectively. Then according to equation (1.10) i.e. in c.g.s. unit

$$E_1 = -\frac{2\pi^2 e^4 m}{n_1^2 h^2} \qquad E_2 = -\frac{2\pi^2 e^4 m}{n_2^2 h^2}$$

Now when the electron jumps from the higher orbit $n_2$ to the lower orbit $n_1$, there is an emission of energy according to Planck-Einstein equation,

$$h\nu = E_2 - E_1 = \frac{2\pi^2 e^4 m}{h^2}\left[\frac{1}{n_1^2} - \frac{1}{n_2^2}\right] \quad n_2 > n_1$$

i.e. $$\nu = \frac{2\pi^2 e^4 m}{h^3}\left[\frac{1}{n_1^2} - \frac{1}{n_2^2}\right]$$

We know that $\bar{\nu} = \frac{\nu}{c}$, then

$$\bar{\nu} = \frac{2\pi^2 e^4 m}{ch^3}\left[\frac{1}{n_1^2} - \frac{1}{n_2^2}\right]$$

All the terms in the fraction $\frac{2\pi^2 e^4 m}{ch^3}$ are constant and if we put the values for $\pi$, e, m etc. the value obtained 109,700 $cm^{-1}$. This constant is known as Rydberg constant. Thus we have

$$\bar{\nu} = R\left[\frac{1}{n_1^2} - \frac{1}{n_2^2}\right] \quad ...(1.16)$$

where $R = \frac{2\pi^2 e^4 m}{ch^3}$ for hydrogen atom and for hydrogen like systems $R = \frac{2\pi^2 e^4 m Z^2}{ch^3}$.

The quation (1.16) is identical with that of the general equation of Ritz combination principle (equation 1.3) and this equation (1.16) explains why the position of a line in the spectrogram is given by a difference of two square terms.

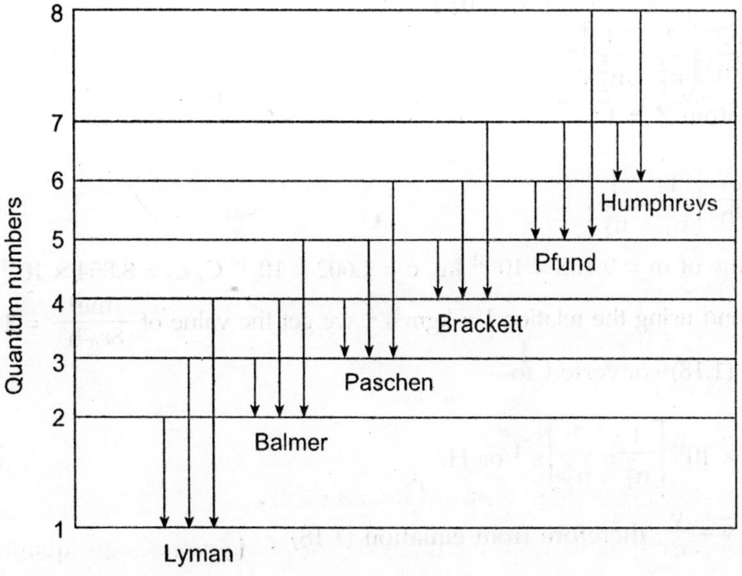


*Figure 1.9 : Emission spectra of hydrogen.*

Spectral lines are produced by the radiation of photons and the position of the lines is determined by the frequencies of the photons emitted. Experimental spectra of hydrogen exhibited several series of lines (Fig. 1.9). Transitions of an electron to the innermost orbit, represented by n = 1 from n = 2, 3, 4 etc. excited states produces Lyman series in the ultraviolet region. Likewise Balmer, Paschen, Brackett and Pfund series is produced (Fig. 1.9).

**Problem 1.9** Why the Rydberg constant value of hydrogen has the unit $cm^{-1}$ or the dimension is length$^{-1}$ ?

**Ans.** : $R = \frac{2\pi^2 e^4 m}{ch^3} = \frac{[e^4][m]}{[c][h^3]}$

From page 16, $R = \frac{M^2L^6T^{-4} \times M}{LT^{-1}M^3L^6T^{-3}} = \frac{M^3L^6T^{-4}}{M^3L^7T^{-4}}$

$= \frac{1}{L} = L^{-1}$

### 1.6.6 Calculation of frequency and wave number in S.I. unit

According to equation (1.14)

$$E_1 = -\frac{me^4Z^2}{8\varepsilon_0^2 n_1^2 h^2} \qquad E_2 = -\frac{me^4Z^2}{8\varepsilon_0^2 n_2^2 h^2}$$

$$\varepsilon = E_2 - E_1 = h\nu = \frac{me^4Z^2}{8\varepsilon_0^2 h^2}\left[\frac{1}{n_1^2} - \frac{1}{n_2^2}\right]$$

$$\therefore \nu = \frac{me^4Z^2}{8\varepsilon_0^2 h^3}\left[\frac{1}{n_1^2} - \frac{1}{n_2^2}\right] \qquad ...(1.17)$$

For hydrogen atom Z = 1.

$$\therefore \nu = \frac{me^4}{8\varepsilon_0^2 h^3}\left[\frac{1}{n_1^2} - \frac{1}{n_2^2}\right] \qquad ...(1.18)$$

Putting the value of $m = 9.108 \times 10^{-31}$ kg, $e = 1.602 \times 10^{-19}$ C, $\varepsilon_0 = 8.854 \times 10^{-12}\ J^{-1}C^2m^{-1}$ and $h = 6.627 \times 10^{-34}$ Js and using the relation $J = kgm^2s^{-2}$, we get the value of $\frac{me^4}{8\varepsilon_0^2 h^3} = 3.2906 \times 10^{15}\ s^{-1}$.

Thus equation (1.18) converted to—

$$\nu = 3.2906 \times 10^{15}\left[\frac{1}{n_1^2} - \frac{1}{n_2^2}\right] s^{-1} \text{ or Hz} \qquad ...(1.19)$$

Wave number $\bar{\nu} = \frac{\nu}{c}$, therefore from equation (1.18)

$$\bar{\nu} = \frac{me^4}{8\varepsilon_0^2 h^3 c}\left[\frac{1}{n_1^2} - \frac{1}{n_2^2}\right] \qquad ...(1.20)$$

Therefore in S.I. unit the expression of Rydberg constant (R) for hydrogen atom :

$$R = \frac{me^4}{8\varepsilon_0^2 h^3 c} \qquad ...(1.21)$$

and for hydrogen like system

$$R = \frac{me^4 Z^2}{8\varepsilon_0^2 h^3 c} \qquad ...(1.22)$$

According to equation (1.21), if we put all the values of m, e etc. in S.I. unit, we get the value equal to $1.0976 \times 10^7$ $m^{-1}$. Therefore equation (1.20) is transformed to :

$$\bar{v} = 1.0976 \times 10^7 \left[\frac{1}{n_1^2} - \frac{1}{n_2^2}\right] m^{-1} \qquad ...(1.23)$$

This equation (1.23) also explain the experimental findings of hydrogen spectra (Fig. 1.9).

### 1.6.7 Merit of Bohr model

It can explain the spectra for one electron hydrogen and hydrogen like systems e.g. $He^+$, $Li^{2+}$, $Be^{3+}$ etc.

### 1.6.8 Defects of Bohr's theory

1. This theory fails to explain the spectra for polyelectron systems e.g. sodium, calcium etc.

2. It has been observed that when a single spectral line is placed in a strong magnetic field (*Zeeman effect*) or placed in an electric field (*Stark effect*), the single lines further split into the closely spaced lines which is known as fine structure of spectral lines. This fine structure cannot be explained by Bohr's theory because for a particular principal quantum number (n) there is a single energy state so that transition between the two energy states, the line will be single but the observed fine structure suggested that for a particular energy state there may be closely spaced energy levels so that transition between these energy levels the line will not be a single, but closely spaced lines.

3. The fundamental objection against Bohr's theory is that it involves the use of two theories which are essentially opposed each other. Quantum theory governed the existance of stationary orbit, whereas motion of electron in Bohr orbit is given by classical mechanics.

### 1.6.9 Experimental evidences in favour of Bohr's theory

1. The experimental value of the frequencies of the spectral lines of the hydrogen are in close agreement with that calculated from Bohr's theory.

2. The experimental value of Rydberg constant for hydrogen are in close agreement with that obtained from Bohr's theory.

3. The probability of finding the electron in 1s orbital of hydrogen is in agreement with that of Bohr radius (0.53Å) calculated from Bohr model of atom.

## 1.7 SOMMERFELD THEORY

In order to explain fine structure of spectral lines Sommerfeld modified Bohr's theory which suggests that electronic transition from one definite energy state ($E_1$) to other energy state ($E_2$) will produce a single line. But the observed fine structure of spectral lines suggest that for a particular energy state there might be closely spaced associated energy levels, so that transition between these energy levels, the line will not be a single one but consisted of several very closely spaced lines.

Sommerfeld guided by this suggestion modified Bohr's theory in two respects :

(1) Concerning the geometry of the electronic path.

(2) Relativistic variation of the mass of the electron.

As regards to the first modification he suggested—

(a) The electronic paths around the nucleus are elliptic with different eccentricities of which Bohr's circular orbit is a special case, and

(b) In elliptical orbits the electron has two degrees of freedom—radius vector r and azimuthal angle $\phi$. Corresponding to these two degrees of freedom two quantum conditions can be applied for permitted elliptical paths.

(1) $\oint p_\phi d\phi = kh$ (2) $\oint p_r dr = n_r h$

$$p_\phi = k\left(\frac{h}{2\pi}\right) \qquad p_r = n_r\left(\frac{h}{2\pi}\right)$$

where $p_\phi$ and $p_r$ are the angular and radial momenta of the electron with coordinates $\phi$ and r, h is Planck constant, k and $n_r$ are two positive integers called the azimuthal and the radial quantum numbers respectively.

Keplar's second law of planetary motion states that $p_\phi$ is unvariant so

$$\oint p_\phi d\phi = p_\phi \int_0^{2\pi} d\phi = kh$$

$$\therefore\ p_\phi = k\left(\frac{h}{2\pi}\right)$$

This result is similar to Bohr's quantum conditions for circular orbits.

Let an electron of mass m and charge (–e) moving in an elliptical path with velocity $v$ with the nucleus of charge (+e) at its focus S (Fig. 1.10).

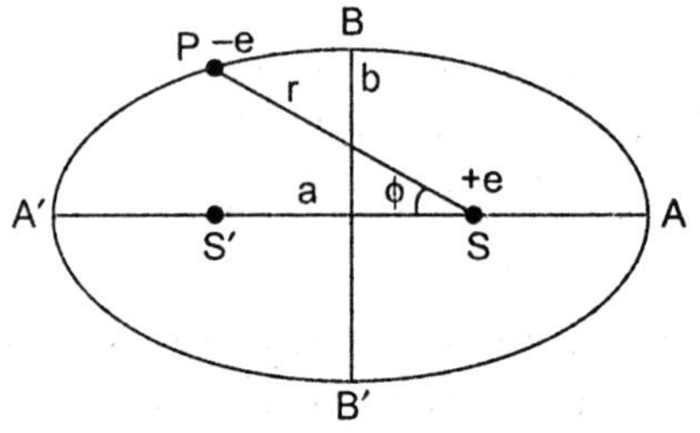


***Figure 1.10** : Elliptical orbit of Sommerfeld.*

The eccentricity ε of the ellipse is related to the quantum number as :

$$1 - \varepsilon^2 = \frac{k^2}{(k+n_r)^2}$$

From the geometry of the ellipse we can write

$$1 - \varepsilon^2 = \frac{b^2}{a^2}$$

where a and b are the length of the semimajor and semiminor axes. Therefore,

$$\sqrt{1-\varepsilon^2} = \frac{b}{a} = \frac{k}{k+n_r} = \frac{k}{n} \qquad ...(1.24)$$

where n is the total quantum number equal to $k+n_r$ is an integer. So according to Sommerfeld theory the shape of the ellipse is also quantised. Accordingly out of all the classically permitted ellipses only those for which ratio of minor and major axes is that of two integers, are allowed by quantum conditions.

(1) The highest permissible value of k is n so that $1 - \varepsilon^2 = 1$, $\varepsilon = 0$ so that $\frac{b}{a} = 1$ i.e. the ellipse turns into a circle.

(2) The lowest permissible value of k = 0, so that $1 - \varepsilon^2 = 0$, $\frac{b}{a} = 0$ i.e. the ellipse turns into a straight line. Hence this result means that the electron will oscillate on a straight line through the nucleus. This have no physical meaning so that $k \neq 0$. Hence k can have the values 1,2,3...... and $n_r$ may have the values 0,1,2,3,...... . The sum of $k+n_r$ = n can have the values 1,2,3,...... .

By introducing the above two quantum conditions, Sommerfeld calculated the total energy of the system as :

$$E = \frac{p_\phi^2(\varepsilon^2 + 2\varepsilon\cos\phi + 1)}{2ma^2(1-\varepsilon^2)^2} - \frac{e^2(1+\varepsilon\cos\phi)}{a(1-\varepsilon^2)} \qquad ...(1.25)$$

From the equation (1.25) it is evident that as the value of ϕ changes during rotation of the electron in the stationary orbit, the value of energy will change which is quite opposite to Bohr's theory because during rotation of electron in stationary orbit the total energy will remain constant. Thus for a conservative system the total energy is constant and is independent of time and the angle ϕ. Since the total energy is constant and cosϕ varies, the coefficient of cosϕ must vanish and the sum of the coefficient of cosϕ will be zero.

Therefore, $$\frac{p_\phi^2 2\varepsilon}{2ma^2(1-\varepsilon^2)^2} = \frac{e^2\varepsilon}{a(1-\varepsilon^2)}$$

so that $$p_\phi^2 = e^2am(1-\varepsilon^2)$$

$$\therefore\ a = \frac{p_\phi^2}{e^2m(1-\varepsilon^2)} = \frac{k^2h^2}{4\pi^2e^2m(1-\varepsilon^2)} \quad \left[\because p_\phi = k\left(\frac{h}{2\pi}\right)\right]$$

Again $\frac{b}{a} = \frac{k}{n} = \sqrt{1-\varepsilon^2}$

$$\therefore\ \frac{k^2}{n^2} = 1 - \varepsilon^2$$

So that $a = \frac{k^2h^2}{4\pi^2e^2m}\cdot\frac{n^2}{k^2} = \frac{n^2h^2}{4\pi^2e^2m} = a_0n^2 \quad \left[\because a_0 = \frac{h^2}{4\pi^2e^2m}\right]$

where $a_0$ is the first Bohr radius. Making the coefficient of $\cos\phi = 0$, the equation (1.25) transformed to—

$$E = \frac{e^2am(1-\varepsilon^2)(1+\varepsilon^2)}{2ma^2(1-\varepsilon^2)^2} - \frac{e^2}{a(1-\varepsilon^2)} \qquad ...(1.26)$$

$[\therefore p_\phi^2 = e^2am(1-\varepsilon^2)]$

$$= \frac{e^2(1+\varepsilon^2)}{2a(1-\varepsilon^2)} - \frac{e^2}{a(1-\varepsilon^2)}$$

$$= -\frac{e^2}{2a}\left[\frac{2-1-\varepsilon^2}{1-\varepsilon^2}\right] = -\frac{e^2}{2a} = -\frac{e^2}{2}\cdot\frac{4\pi^2e^2m}{n^2h^2}$$

$$= -\frac{2\pi^2e^4m}{n^2h^2} \qquad ...(1.27)$$

This equation (1.27) is of particular interest in that the energy is exactly the same as that obtained by Bohr's theory.

Now from the relation (1.24)

$$\frac{b}{a} = \frac{k}{n} = \sqrt{1-\varepsilon^2}$$

it is evident that for a particular value of n there can be only a limited number of elliptical orbits that are permitted, having the same value of the major axis but of different eccentricities.

Let n = 3, then k = 1,2,3

We know that $a = a_0n^2$

For n = 3, $a = 9a_0$

$$b = a\frac{k}{n} = a_0n^2\frac{k}{n} = a_0nk = 3a_0 \text{ when } k = 1 \text{ and } n = 3$$

$b = 6a_0$ when k = 2, n = 3

$b = 9a_0$ when k = 3, n = 3 i.e. in this case the ellipse turns into a circle.

Thus for n = 3, there will be three electronic paths of which two are elliptic and one is circular (Fig. 1.11) all of which corresponds to the same energy value corresponding to equation (1.27) but have different eccentricities and hence are indistinguishable. Again all electronic paths having the same value of semimajor axis 'a' possess the same energy, since the length of the semimajor axis is determined solely by the total quantum number n. This means that the theory of elliptical orbit, inspite of the two new quantum conditions proposed by Sommerfeld, fails to explain fine structure of spectral lines.

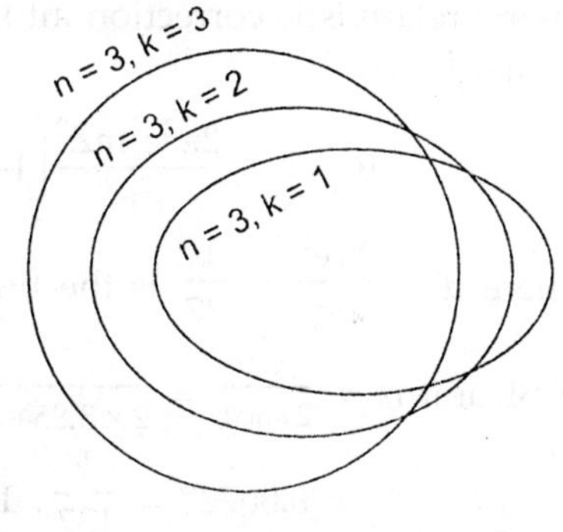

*Figure 1.11 : The possible orbits for n = 3.*

### 1.7.1 Sommerfeld's relativistic corrections

In order to explain fine structure Sommerfeld introduced relativistic correction to the electron mass. The velocity of an electron moving in an elliptical orbit varies considerably at different paths of the orbit being maximum when the electron is nearest to the nucleus (i.e. *perihelion position*) and velocity is minimum when it is farthesest from the nucleus (i.e. *aphelion position*). According to the special theory of relativity this variation of velocity means, the variation of mass of the electron. Due to different velocity of the electron in orbits of the same n and different k the mass of the electron changes and the energy levels are slightly different. If the rest mass of the electron is $m_0$ and moving mass is m then according to special theory of relatively

$$m = m_0\left(1-\frac{v^2}{c^2}\right)^{-\frac{1}{2}}$$

When $v << c$ (velocity of light) then $\frac{v^2}{c^2}$ term can be neglected and then $m = m_0$ and therefore, this relativistic effect will be appreciable only at very high velocities.

When relativistic variation of the mass of the electron is applied the electronic path is no longer a simple ellipse, but transformed to a complicated curve known as **rosette—a precessing ellipse** (Fig. 1.12). As the electron traverses the elliptical path, the major axis revolves slowly in the plane of the ellipse with a constant angular velocity around the nucleus situated at the focus of the ellipse. Such an ellipse is doubly periodic, besides the original period of revolution, there is also the period of precessional motion.

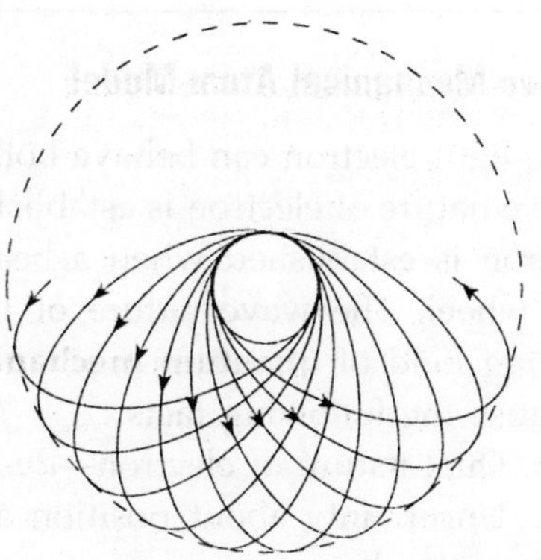
*Figure 1.12 : A precessional ellipse.*

Taking relativistic correction into account the energy or the term value has been deduced by Sommerfeld as :

$$E_{n,k} = -\frac{2\pi^2 e^4 m Z^2}{n^2 h^2}\left[1+\frac{\alpha^2 Z^2}{n^2}\left(\frac{n}{k}-\frac{3}{4}\right)\right] \quad ...(1.28)$$

where $\alpha = \frac{2\pi e^2}{hc} = \frac{1}{137}$ is the fine structure constant.

In SI unit $\alpha = \frac{e^2}{2\varepsilon_0 ch} = \frac{(1.6\times10^{-19}\,C)^2}{2\times 8.854\times10^{-12}\,J^{-1}C^2m^{-1}\times 3\times10^8\,ms^{-1}\times 6.627\times10^{-34}\,Js}$

$= 0.00727 = \frac{1}{137}$. Therefore $\alpha$ is a unitless quantity.

For hydrogen atom Z = 1, then

$$E_{n,k} = -\frac{2\pi^2 e^4 m}{n^2 h^2}\left[1+\frac{\alpha^2}{n^2}\left(\frac{n}{k}-\frac{3}{4}\right)\right] \quad ...(1.29)$$

So small alternations of energy values will occur for orbits having same n but different k values.

Fine structure of spectral lines now can be explained by transitions between orbits of different n and k. According to selection rule only those transitions are permitted which has $\Delta k = \pm 1$.

### 1.7.2 Merits and demerits

Though Sommerfeld theory can explain the defect of Bohr's theory i.e. fine structure of spectral lines is explained yet this theory cannot explain the correct number of finer spectral lines which are obtained experimentally. Sommerfeld's azimuthal quantum number (k) is different from the one (*l*) obtained from wave mechanics. It mainly deals with the hydrogen like systems, but for other systems it does not work well.

**Problem 1.10** (a) What led Sommerfeld to modify Bohr's theory?
(b) What was his modifications?
(c) How fine structure of spectral lines is explained? **[B.U. 2007, 2010, 2012]**

**Hint :** (a) First paragraph of Sec. 1.7.
(b) Second paragraph of Sec. 1.7.
(c) Equation (1.28) or (1.29).

## 1.8 Wave Mechanical Atom Model

Like light electron can behave both as a particle as well as wave (Sec. 1.4.2 and Sec. 1.4.3). The wave nature of electron is established by **Davisson and Germer** experiment. Particle nature of electron is established when a beam of electrons in cathode ray discharge tube rotates a paddle wheel. The wave nature of the electron marked the advent of a new horizon—the fascinating field of **quantum mechanics** or **wave mechanics**. This new mechanics (Sec. 1.4.3) systematise the following facts :

(1) Dual nature of electron—de-Broglie concept.
(2) Uncertainty about position and momentum of electron—the Heisenberg uncertainty principle.
(3) The wave motion of the electron—Schrodinger wave equation.
(4) Electronic arrangements of atoms—quantum numbers.

### 1.8.1 de-Broglie concept—duality of matter

This concept connects the particle character of the electron viz. mass, momentum etc. with that of the wave character i.e. wave length.

The famous Planck-Einstein equation is

$$E = h\nu$$

and according to the Einstein theory of mass-energy relationship

$$E = mc^2$$

So that $m = \frac{h\nu}{c^2}$

The momentum (p) = $mv$ = mc (where c is the velocity of light)

$$= \frac{h\nu}{c^2} \cdot c = \frac{h\nu}{c}$$

According to equation $\nu\lambda = c$ (page 10)

$$\therefore p = \frac{h\nu}{\nu\lambda} = \frac{h}{\lambda}$$

$$\text{i.e. } \lambda = \frac{h}{p} \qquad \text{...(1.30)}$$

This is the famous de-Broglie relationship. This relation (1.30) states that there must be a wave associated with the particle nature of the electron. This relationship can explain the Bohr's concept of stationary orbit where angular momentum of the electron is integral multiple of $\frac{h}{2\pi}$ i.e. $mvr = n\left(\frac{h}{2\pi}\right)$. Now according to de-Broglie the electron while rotating in stationary orbit is associated with *stationary wave* or standing wave which extends round the nucleus in a circular orbit (Fig. 1.13) i.e. the wave will have a stationary profile which means that the wave will not extend beyond the allowed space. If the two ends of an electron wave meet in the same phase to produce a regular series of crests and throughs i.e. there is a constructive interference of electron wave, then in that case the electron wave is said to be **non radiating standing wave.** According to the condition of stationary wave we can write for circular orbit of Bohr i.e.

$$2\pi r = n\lambda$$

From equation (1.30) we say that

$$2\pi r = n\frac{h}{p} = n\frac{h}{mv}$$

$$\text{i.e. } mvr = n\left(\frac{h}{2\pi}\right)$$

Therefore the wave-particle dualism of electron suggested by de-Broglie are in agreement of Bohr concept. de-Broglie wave concept was supported by electron diffraction experiments.

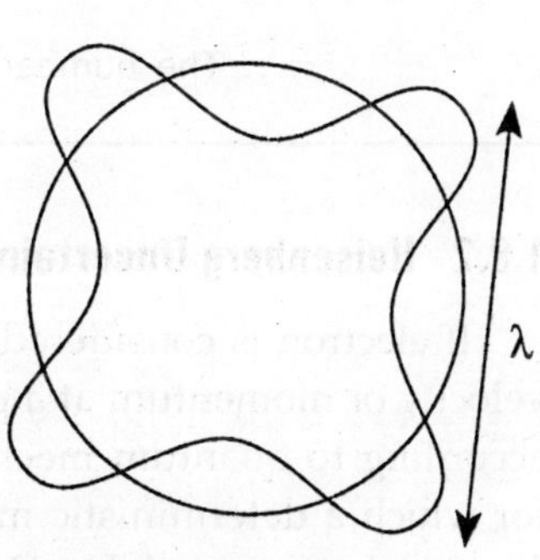


***Figure 1.13** : Stationary wave of electron n = 4.*

**Problem 1.11** Calculate the de-Broglie wavelength of an electron travelling at 10% of the speed of light. Mass of the electron (m) = $9.1 \times 10^{-31}$ kg.

**Ans. :** Velocity of the electron which is 10% of the speed of light

$$= 3 \times 10^{8}\ ms^{-1} \times \frac{10}{100} = 3 \times 10^{7}\ ms^{-1}$$

Momentum of the electron ($mv$) = $9.1 \times 10^{-31}\ kg \times 3 \times 10^{7}\ ms^{-1}$

$= 2.73 \times 10^{-23}\ kg\ ms^{-1}$

We know $\lambda = \frac{h}{mv} = \frac{6.627 \times 10^{-34}\ Js\ \text{or}\ kg\ m^2 s^{-1}}{2.73 \times 10^{-23} kg\ ms^{-1}}$

$= 2.427 \times 10^{-11}$ m

$= 0.02427$ nm [$\because$ 1 nm = $10^{-9}$ m]

**Problem 1.12** The velocity of an electron is $2 \times 10^{8}$ cm.sec.$^{-1}$. Calculate its wave length in nm.

**Ans. :** $$\lambda = \frac{h}{mv} = \frac{6.627 \times 10^{-34}\ kg\ m^2\ s^{-1}}{9.1 \times 10^{-31}\ kg \times \frac{2 \times 10^{8}}{10^{2}} ms^{-1}}$$

$$= \frac{6.627 \times 10^{-34}}{9.1 \times 10^{-31} \times 2 \times 10^{6}}\ m$$

$= 0.364 \times 10^{-9}$ m

$= 0.364$ nm.

**Problem 1.13** Find out the number of waves made by a Bohr electron in one complete revolution in its third orbit.

**Ans. :** Velocity of the electron in third orbit, $v = \frac{2\pi e^2}{nh} = \frac{2\pi e^2}{3h}$.

In one complete revolution the distance traversed = $2\pi r$.

$\therefore$ The number of waves = $\dfrac{\frac{2\pi e^2}{3h}}{2\pi r} = \frac{e^2}{3hr}$ cycles/second.

## 1.8.2 Heisenberg Uncertainty Principle

If electron is considered as particle or large *macroscopic* particle then the exact position and velocity or momentum at a particular time can be specified according to classical mechanics. But according to quantum mechanics electron has a dual character and is a *submicroscopic* particle for which a deterministic model of simultaneous two properties is impossible. An explanation for this is provided by Heisenberg (1927) in his uncertainty principle. It states that, **it is impossible to determine exactly both the position and momentum of an electron simultaneously or of moving particle like that of electron at the same time.**

It means that if an electron is acting as a particle, its position can be determined accurately but there will be an uncertainty in velocity or momentum as the electron has associated wave property with that of particle nature. On the other hand if the velocity or momentum of the electron is fixed, there should be an uncertainty in position. Mathematically this can be expressed as :

$$\left.\begin{aligned} \Delta x.\Delta p \geq \frac{h}{4\pi} \\ \Delta E.\Delta T \geq \frac{h}{4\pi} \end{aligned}\right\}^{*} \qquad \ldots(1.31)$$

where $\Delta x$ = uncertainty in position

$\Delta p$ = uncertainty in momentum

$\Delta E$ = uncertainty in energy

$\Delta T$ = uncertainty in the time of passage at a particular point.

The verification of the mathematical forms can be made by Compton effect.

---

**Problem 1.14** According to Bohr's theory of hydrogen atom, the velocity of an electron in the first orbit is $2.183 \times 10^{6}$ $ms^{-1}$. What will be the uncertainty in velocity of the electron if the uncertainty in position is 5 pm?

**Ans. :** $\Delta x.\Delta p = \dfrac{h}{4\pi}$

$$\therefore \Delta p = \frac{h}{4\pi\,\Delta x} = \frac{6.627 \times 10^{-34}\,\text{Js}}{4 \times 3.14 \times 5 \times 10^{-12}\,\text{m}} \qquad [\because 1\text{ pm} = 10^{-12}\text{ m}]$$

We know that $\Delta p$ = mass × uncertainty in velocity ($\Delta v$)

$= 9.108 \times 10^{-31}$ kg $\times \Delta v$

$$\therefore \Delta v = \frac{6.627 \times 10^{-34}\,\text{Js}}{4 \times 3.14 \times 5 \times 10^{-12}\,\text{m} \times 9.108 \times 10^{-31}\,\text{kg}}$$

$= 0.1159 \times 10^{8}$ $ms^{-1}$

---

### 1.8.3 Schrodinger Wave Equation

Since thre is an analogy between electron and light (Sec. 1.4.3) and since an electron behaves like a wave, it is necessary to construct an wave equation for electron wave which is known as Schrodinger wave equation.

---

* Different authors give different expression e.g. $\Delta x.\Delta p \geq h$, $\Delta x.\Delta p \geq \frac{h}{2\pi}$ but the most plausable is $\Delta x.\Delta p \geq \frac{h}{4\pi}$.

Light waves confirm to an equation of the type—

$$\Psi_{(x,t)} = Ae^{2\pi i \frac{x}{\lambda}} e^{-2\pi i v t} \quad ...(1.32)$$

where $\Psi$ = wave function or amplitude function, A is a constant, x = distance traversed in the direction of wave motion, $\lambda$ = wave length, $v$ = frequency and t = time. This equation (1.32) has the space and time dependent part. We are concerning the **stationary profile** of electron wave so that the equation will contain time as variable. The time dependent wave equation finds application in the field of radiation whereby the problems related to the energy of the electron utilise the time independent equation. The time independent wave equation is :

$$\psi = Ae^{-2\pi i \frac{x}{\lambda}}$$

According to Euler's theorem

$$e^{ix} = \underset{\text{real part}}{\cos x} + \underset{\text{imaginary part}}{i\sin x}$$

$$\therefore \Psi = A\cos 2\pi\frac{x}{\lambda} + Ai\sin 2\pi\frac{x}{\lambda}$$

$$\text{i.e. } \Psi = A\cos 2\pi\frac{x}{\lambda} \text{ (taking the real part)} \quad ...(1.33)$$

Here $\Psi$ is called the wave function or amplitude function which is varying along x direction, the constant A which gives the maximum amplitude. As motion of light or electron wave is described by second order differential equation therefore on differentiating equation (1.33) w.r.t x

$$\frac{d\Psi}{dx} = -\left(A\sin 2\pi\frac{x}{\lambda}\right)\frac{2\pi}{\lambda} \quad ...(1.34)$$

Again on differentiating equation(1.34)

$$\frac{d^2\Psi}{dx^2} = -\left(A\cos 2\pi\frac{x}{\lambda}\right)\frac{2\pi}{\lambda}\cdot\frac{2\pi}{\lambda} = -\frac{4\pi^2\Psi}{\lambda^2}$$

$$\therefore \quad \frac{d^2\Psi}{dx^2} + \frac{4\pi^2\Psi}{\lambda^2} = 0 \quad ...(1.35)$$

The equation (1.35) is the **classical wave equation** for the stationary wave.

The total energy for hydrogen like system (E) = K.E.(T) = P.E.(V).

Now from de-Broglie concept $\lambda = \frac{h}{p} = \frac{h}{mv}$. $\therefore v = \frac{h}{\lambda m}$.

$$\text{K.E.(T)} = \frac{1}{2}mv^2 = \frac{1}{2}m\frac{h^2}{\lambda^2 m^2} = \frac{h^2}{2m\lambda^2}$$

$$\text{So that } \lambda^2 = \frac{h^2}{2mT} \quad ...(1.36)$$

Putting this value of $\lambda^2$ in equation (1.35) we get

$$\frac{d^2\Psi}{dx^2} = -\frac{8\pi^2 m}{h^2}T\Psi$$

$$\therefore\ T\Psi = -\frac{h^2}{8\pi^2 m}\cdot\frac{d^2\Psi}{dx^2}$$

Now the K.E.(T) = Total energy (E) – Potential energy (V)

So that $$\frac{d^2\Psi}{dx^2} + \frac{8\pi^2 m}{h^2}(E - V)\Psi = 0 \qquad ...(1.37)$$

This is the time independent Schrodinger wave equation in one dimension. If we generalise this equation in three dimensions (x,y,z cordinates), we have,

$$\left.\begin{aligned} &\frac{\partial^2\Psi}{\partial x^2} + \frac{\partial^2\Psi}{\partial y^2} + \frac{\partial^2\Psi}{\partial z^2} + \frac{8\pi^2 m}{h^2}(E - V)\Psi = 0 \\ &\text{or, } \nabla^2\Psi + \frac{8\pi^2 m}{h^2}(E - V)\Psi = 0 \end{aligned}\right\} \qquad ...(1.38)$$

The $\nabla^2$ (said as del square) is called **Laplacian operator**

which is $$\nabla^2 = \frac{\partial^2}{\partial x^2} + \frac{\partial^2}{\partial y^2} + \frac{\partial^2}{\partial z^2}$$

Now from equation (1.38)

$$\nabla^2\Psi + \frac{8\pi^2 m}{h^2}E\Psi - \frac{8\pi^2 m}{h^2}V\Psi = 0$$

or, $$\frac{8\pi^2 m}{h^2}E\Psi = -\nabla^2\Psi + \frac{8\pi^2 m}{h^2}V\Psi$$

So that, $$E\Psi = \left[-\frac{h^2}{8\pi^2 m}\nabla^2 + V\right]\Psi$$

i.e. $$E\Psi = \hat{H}\,\Psi \qquad ...(1.39)$$

where $\hat{H}$ is called **Hamiltonian operator** = $-\frac{h^2}{8\pi^2 m}\nabla^2 + V$ , which is a function of potential energy as also coordinates. Equations (1.38) and (1.39) are different forms of Schrodinger wave equation.

### 1.8.4 What is wave function?

Since the motion of an electron has a wave character, quantum mechanics describes its motion in an atom by what is called a **wave function Ψ**. In other words a quantum mechanical system is denoted by specifying a function Ψ. This is a function of all coordinates and time. It is also called **state function**. Mathematically it is written as

$$\Psi = \Psi(x,y,z,t)$$

where x,y,z are the coordinates of a point and t is time. This function is different for various points of atomic space.

A wave function which include time is called **time dependent wave function**. A system whose properties are independent of time is said to be a **stationary state** and the corresponding wave function is said to be **stationary state wave function**. Such a state function is separable into the product of a space dependent part and the time dependent part i.e.

$$\Psi(x,y,z,t) = \Psi(x,y,z)f(t)$$

where $\Psi$ is the space dependent function and f is the time dependent function.

### 1.8.5 Charge cloud and Probability density

According to uncertainty principle it is expected that the wave function ($\Psi$) has some probabilistic approach. This probabilistic approach is necessary even for a single particle due to our inability in the simultaneous precise measurement of position and momentum of the particle i.e. the wave function describing a particle of a system is in some sense is a measure of the particle being present there. So, if $\Psi = 0$, there is no particle and if $\Psi \neq 0$, it may be stated that there particle may be found in that region. Therefore *a question arises whether the wave function $\Psi$ represents the probability of finding a particle at a point (x,y,z) at time t*. This is not true since the wave function can assume positive, negative or imaginary value but the probability of *finding the particle should always be positive*. Thus strictly speaking the wave function has no physical significance.

*Significance of $\Psi$*

We therefore need a quantity which would represent the probability of finding the particle. According to Heisenberg, Born and Bohr the probability of finding the particle at a given point at any given time is given by $\Psi^2$ if $\Psi$ is real and positive. But as stated before if $\Psi$ is imaginary, then $\Psi^2$ will also be imaginary. Then in that case $\Psi\Psi^*$ denotes the **probability density function** or probability of finding the electron where $\Psi^*$ is the complex conjugate of $\Psi$($\Psi$ = a+ib, $\Psi^*$ = a–ib $\therefore$ $\Psi\Psi^*$ = (a+ib)(a–ib) = $a^2 + b^2$ which is real and positive).

*Eigen values and Eigen functions*

The Schrodinger wave equation (1.38) is a differential equation of second order which has many solutions of $\Psi$. Some of the solutions are imaginary and are not valid. In this equation if the potential energy V of the system is known, then the energy E can be determined. Acceptable values of wave function must satisfy the following conditions which are called **eigen functions** or **wave function** and the corresponding energies are called **eigen values**.

(a) The wave function should be single valued i.e. there can never be more than one value for the amplitude function $\Psi$ at a given point.

(b) The wave function must be finite and contineous i.e. $\frac{\partial \Psi}{\partial x}, \frac{\partial \Psi}{\partial y}, \frac{\partial \Psi}{\partial z}$ should be continuous functions of x, y and z respectively.

(c) $\Psi$ should be normalised i.e. $\int_{-\infty}^{+\infty} \Psi^2 d\tau = 1$

where $d\tau$ is the small volume element (= dx dy dz) at a particular point defined by (x, y, z) in space.

### Why the probability density function (ρ) is related to $\Psi^2$ or $\Psi\Psi^*$?

We know that light or any electromagnetic radiation has dual nature and so also for electron (Sec. 1.4.3). According to classical mechanics, the intensity of radiation is proportional to the square of the amplitude i.e. $\Psi^2$. This is known as **intensity effect** of electromagnetic radiation. This intensity effect is related to photon density, since photon density is related to the probability of finding the same in the given volume which is called the probability density i.e. the probability density function

$$\rho \propto \Psi^2 \quad \text{i.e.} \quad \rho = k\Psi^2$$

where k is a proportionality constant.

If $\Psi$ is a solution of the Schrodinger wave equation then $k\Psi$ will also represent the same i.e.

$$\rho = \Psi^2 \quad \text{when } \Psi \text{ is real and } \rho = \Psi\Psi^*$$

where $\Psi$ is imaginary. The probability of finding the electron is given by $\int_{-\infty}^{+\infty} \Psi^2 d\tau$ or $\int_{-\infty}^{+\infty} \Psi\Psi^* d\tau$, where $\int_{-\infty}^{+\infty}$ represents the whole space and $d\tau$ is the small volume element.

### 1.8.6 Applications of Schrodinger wave equation in the following problems : (a) Particle in one dimensional box, (b) Particle in two dimensional box, (c) Hydrogen atom.

**Particle in one dimensional box**

Let us consider a rectangular box with dimensions abc lying along the x, y and z axes respectively is chosen and the particle is restricted so as to move only inside the box i.e. it has no existance out side the box. This restriction may be applied by allowing the potential energy to go to infinity at the sides of the box. Anywhere inside of the box the potential energy of the particle is zero so that it can move without any restriction within the box of infinite height.

For the sake of simplicity we shall consider a one dimensional box only (Fig. 1.14). In the three dimensional box, the wave function is represented by $\Psi(x,y,z)$ but in one dimensional box it should be $\Psi(x)$ and for this case the wave equation reduces to—

$$\frac{d^2\psi_{(x)}}{dx^2} + \frac{8\pi^2 m}{h^2}(E - V)\psi_{(x)} = 0$$

As long as the particle remains in the box its potential energy is zero and the equation transformed to—

$$\frac{d^2\psi_{(x)}}{dx^2} + \frac{8\pi^2 m}{h^2} E\psi_{(x)} = 0$$

$$\text{or,} \quad \frac{d^2\psi_{(x)}}{dx^2} + \alpha^2\psi_{(x)} = 0 \qquad \ldots(1.40)$$

$$\text{where} \quad \alpha^2 = \frac{8\pi^2 mE}{h^2}$$

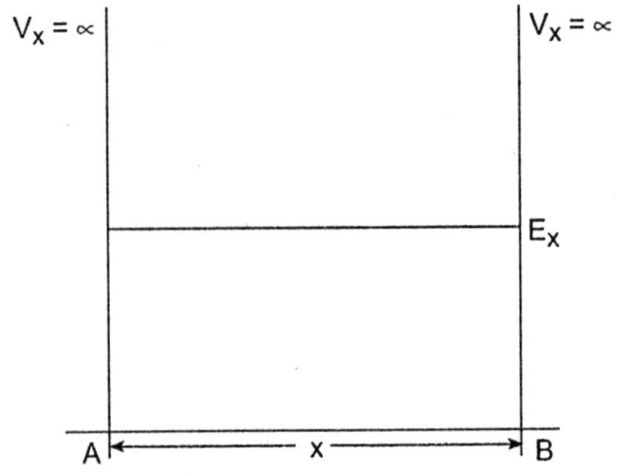


*Figure 1.14 : Particle in one dimensional box.*

It is evident the probability of finding the particle at the extremities is zero i.e. $\Psi^2 = 0$ at the two extremities A and B. Therefore $\Psi = 0$ at A and B.

*A and B are nodes hence half of the waves of more than one waves is possible but quarter of the wave is not possible.*

Let AB = L so L will be given by $L = n\frac{\lambda}{2}$ where n is a positive integer. Again from de-Broglie concept $\lambda = \frac{h}{mv}$,

$$\therefore L = \frac{nh}{2mv} \quad \text{or,} \quad mv = n\frac{h}{2L}$$

$$\text{i.e. } m^2v^2 = \frac{n^2h^2}{4L^2}$$

Now the total energy (E) = K.E.(T) + P.E.(V)

But V = 0, the minimum energy which is purely a kinetic energy of the particle within the box, so that,

$$E = T = \frac{1}{2}mv^2 = \frac{m^2v^2}{2m} = \frac{n^2h^2}{8L^2m} = n^2\frac{h^2}{8L^2m} \quad \text{...(1.41)}$$

Therefore E has some discrete values depending upon n

When $n = 1$, $L = \frac{\lambda}{2}$ and $E = \frac{h^2}{8L^2m}$

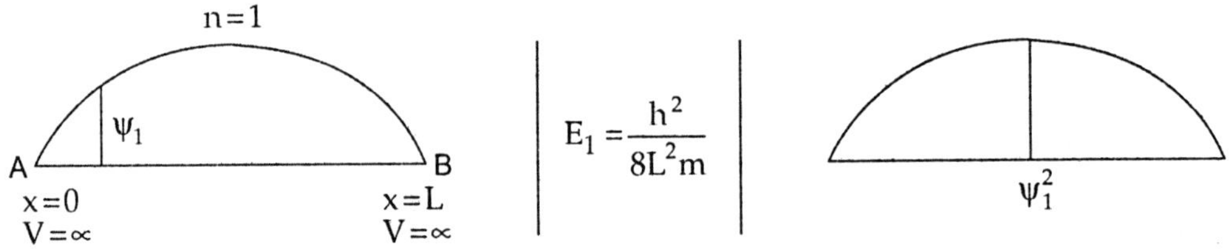

When n = 2, $L = 2\frac{\lambda}{2}$ and $E_2 = \frac{4h^2}{8L^2m} = 4E_1$

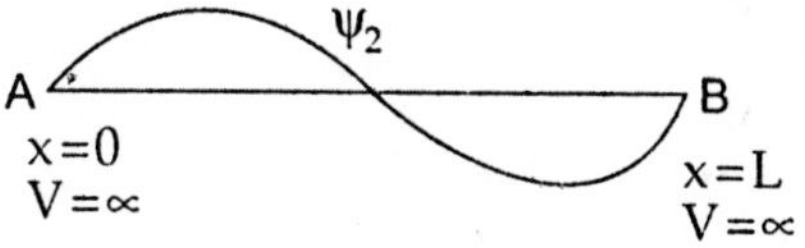


$$E_2 = \frac{4h^2}{8L^2m}$$

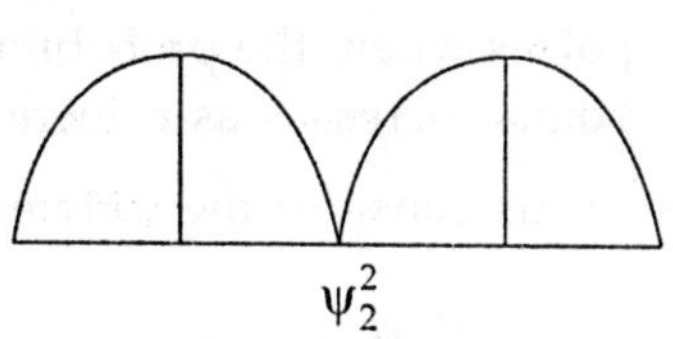


When n = 3, $L = 3\frac{\lambda}{2}$ and $E_3 = \frac{9h^2}{8L^2m} = 9E_1$

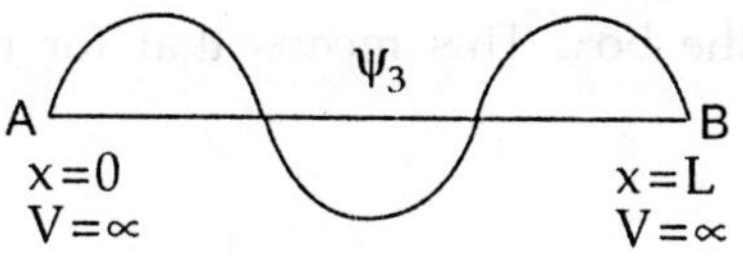


$$E_3 = \frac{9h^2}{8L^2m}$$

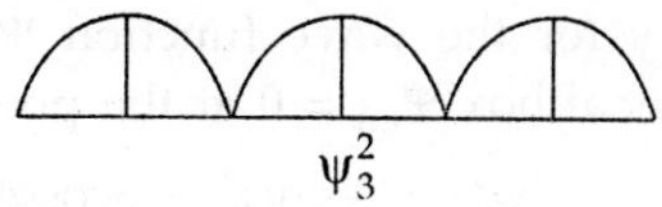


When n = 4, $L = 4\frac{\lambda}{2}$ and $E_4 = \frac{16h^2}{8L^2m} = 16E_1$

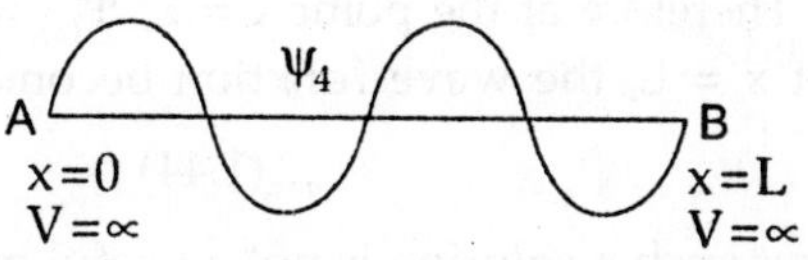


$$E_4 = \frac{16h^2}{8L^2m}$$

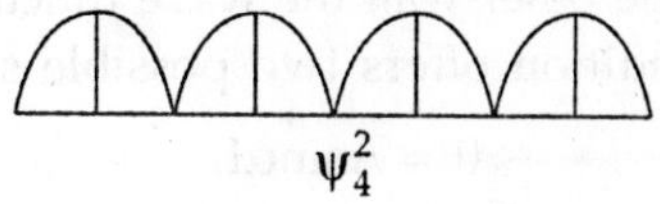


Thus it is seen that, energy cannot contineously vary, but vary with some particular multiple of $E_1$ so that the particle only have discrete eneigy values i.e. to say **energy is quantised.**

It is to be noted that n cannot be zero because in that case, the wave function will be zero everywhere and the probability of finding the particle anywhere in the box will also be zero. Since this is against the actual fact, it is not acceptable. The follwing points are to be noted :

(a) The probability of finding the electron depends on x i.e. L.

(b) The energy of the particle is quantised.

(c) E is inversely proportional to L, the length of the box. Thus as the box is smaller, the energy of the particle becomes larger. The energy difference (ΔE) between the successive energy levels may be written as :

$$\Delta E = E_{n+1} - E_n = \frac{(n+1)^2h^2}{8mL^2} - \frac{n^2h^2}{8mL^2} = \frac{(2n+1)h^2}{8mL^2}$$

This means that the energy difference decreases with increasing L.

(d) For lowest state i.e. for n = 1, there is no point except the boundaries for which the wave function is zero. For n = 2,3,4,......etc., there are one, two, three, ...... points respectively for which the corresponding wave function is zero. These are called **nodal points** where the probability of finding the particle is zero. It is to be noted that nodal points increases as n increases and wavelength decreases.

Now let us consider the differential equation (1.40)

$$\frac{d^2\psi_{(x)}}{dx^2} + \alpha^2 \psi_{(x)} = 0$$

and this equation has the solution

$$\Psi_{(x)} = A\sin\alpha x + B\cos\alpha x \quad ...(1.42)$$

where A and B are arbitrary constant. To this solution however certain restrictions can be applied which is called **boundary conditions** i.e. the particle cannot exist outside the box, it is necessary for the wave function $\Psi_{(x)} = 0$ at the walls of the box. This means that for one dimensional box $\Psi_{(x)} = 0$ at the point x = 0.

$$\therefore\ 0 = A\sin\alpha 0 + B\cos\alpha 0$$

$$\text{i.e. } 0 = A(0) + B(1)$$

$$\text{i.e. } B = 0$$

As a result of this boundary condition, the solution (1.42) reduces to—

$$\Psi_{(x)} = A\sin\alpha x \quad ...(1.43)$$

At the other wall the wave function again goes to zero. Therefore at the point x = L, $\Psi_{(x)} = 0$. This condition offers two possible solutions. For the point x = L, the wave function becomes

$$0 = A\sin\alpha L \quad ...(1.44)$$

If A = 0, the condition of equality of equation (1.44) holds but such a solution is not a useful one.

However, there is another way in which the equality of equation (1.44) is maintained. The sine of an angle is zero at any integral multiple of π, thus if $\alpha = \frac{n\pi}{L}$ where n is an integer, the equality can still be satisfied. As a result of these boundary conditions, the wave function for the particle now becomes :

$$\Psi_{(x)} = A\sin\frac{n\pi}{L}x \quad ...(1.45)$$

and hence $\Psi_{(x)}$ can be constructed. Now we have to evaluate the constant A. This can be done by normalisation condition of the wave function. Since the particle is within the box, therefore the probability of finding the electron in the box is unity i.e.

$$\int_0^L \Psi\Psi^* dx = 1$$

which means that $\int_0^L A^2 \sin^2 \alpha x dx = 1$*

$$\text{or,} \quad \frac{1}{A^2} = \int_0^L \sin^2 \alpha x dx \qquad ...(1.46)$$

If this expression is solved for A and the result substituted into the wave equation, the complete normalised wave function for the particle in a one dimensional box is found to be—

$$\Psi_{(x)} = \sqrt{\frac{2}{L}} \sin \frac{n\pi}{L} x \qquad ...(1.47)$$

---

**Problem 1.15** Calculate the ground state energy of the electron if it is confined in a one dimensional box of length 1Å.

**Ans.** : We know $E = \frac{n^2h^2}{8mL^2}$ Equation (1.41)

$$= \frac{1^2 (6.627 \times 10^{-34} Js)^2}{8(10^{-10} m)^2 (9.108 \times 10^{-31} kg)(1.602 \times 10^{-19} J/ev)}$$

= 37.62 ev.

---

### 1.8.7 Solution of the wave equation for hydrogen atom. Shapes of the atomic orbitals

In case of hydrogen atom or in differnt hydrogen like systems e.g. $He^+$, $Li^{2+}$, $Be^{3+}$ etc., since a single electron of charge –e is revolving round the nucleus of charge +Ze and the electron at a distance r from the nucleus, the potential energy (V) is given by $-\frac{Ze^2}{r}$ (for hydrogen like systems) or $-\frac{e^2}{r}$ (for hydrogen atom). In Schrodinger wave equation (1.38) the wave function Ψ depends on the Cartesian coordinates which represent the x, y and z axis in a three dimensional system and indicate the orientation of the nucleus in the atom in space. But we know that the electron around the nucleus has wave mechanical nature which is smeared out around the nucleus, therefore it is important to know the distribution of the electron around the nucleus rather than the orientation of the nucleus in space. In view of spherically symmetric field of electron that occurs around the nucleus, the Schrodinger wave equation is transformed to spherical polar coordinates. This is represented as r, θ and ϕ in Fig. 1.15.

---

* $\int_0^L A^2 \sin^2 \alpha x dx = 1$ We know $\sin^2\theta = \frac{1}{2}(1 - \cos 2\theta)$

$\therefore \frac{A^2}{2}\left[\int_0^L dx - \int_0^L \cos 2\alpha x dx\right] = 1$ or, $\frac{A^2}{2}[L - 0] = 1$ or, $A^2L = 2$ i.e. $A = \sqrt{\frac{2}{L}}$

In $\Delta$OPP′

$\angle$OPP′ = $\theta$

OP′ = perpendicular

PP′ = base

OP = hypotenuse

$$\therefore \frac{OP'}{OP} = \sin\theta$$

$\therefore$ OP′ = OPsin$\theta$ = rsin$\theta$

$$\frac{PP'}{OP} = \cos\theta$$

PP′ = OPcos$\theta$ = rcos$\theta$

Here PP′ = ON = z

$\therefore$ $\boxed{z = r\cos\theta}$

In $\Delta$OBP′, $\angle$OBP′ = 90°

OP′ = hypotenuse

BP′ = perpendicular

OB = base

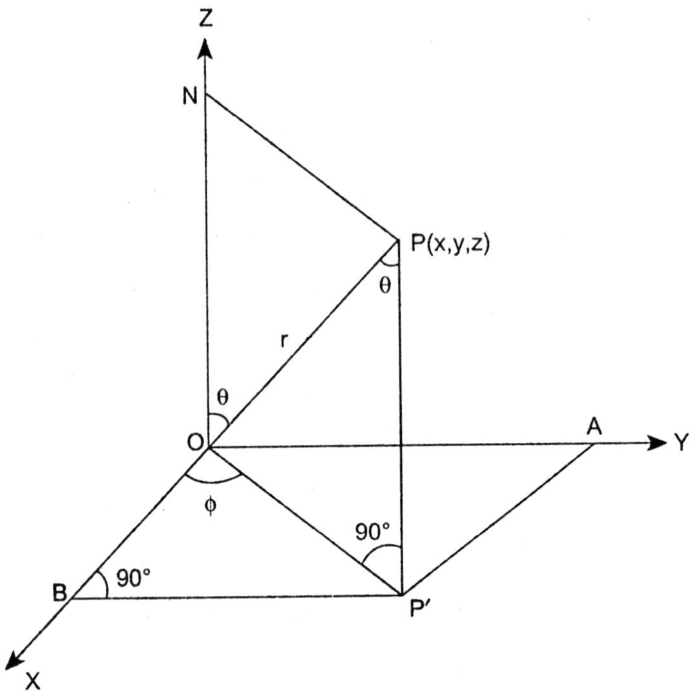


***Figure 1.15** : Polar system of coordinates with reference to Cartesian system.*

$$\therefore \frac{BP'}{OP'} = \sin\phi$$

$\therefore$ BP′ = OP′sin$\phi$ = rsin$\theta$sin$\phi$

BP′ = OA = y

$\therefore$ $\boxed{y = r\sin\theta\sin\phi}$

Again in $\Delta$OBP′, $\frac{OB}{OP'} = \cos\phi$

$\therefore$ OB = OP′cos$\phi$ = rsin$\theta$cos$\phi$

OB = x

$\therefore$ $\boxed{x = r\sin\theta\cos\phi}$

Substituting the values of x, y and z in Schrodinger equation (1.38) which transformed to spherical-polar coordinates as :

$$\frac{1}{r^2}\left[\frac{\partial}{\partial r}\left(r^2\frac{\partial\Psi}{\partial r}\right) + \frac{1}{\sin\theta}\cdot\frac{\partial}{\partial\theta}\left(\sin\theta\frac{\partial\Psi}{\partial\theta}\right) + \frac{1}{\sin^2\theta}\cdot\frac{\partial^2\Psi}{\partial\phi^2}\right] + \frac{8\pi^2 m}{h^2}(E-V)\Psi = 0 \qquad ...(1.48)$$

Complete solution of equation (1.48) is beyond the scope of this book. But the solution of this equation (1.48) is $\Psi$ which will be a functions of r, $\theta$ and $\phi$ instead of x, y, z i.e.

$$\left.\begin{aligned} \Psi(r,\theta,\phi) &= R(r)\,\Theta(\theta)\,\phi(\phi) \\ \text{or, } \Psi(r,\theta,\phi) &= \Psi_{radial}\,\Psi_{angular} \end{aligned}\right\} \quad ...(1.49)$$

The radial part of the wave function is determined by the distance whereas the angular part is determined by the angle $\theta$ and $\phi$. *The radial part of the wave function depends on principal quantum number*(n) and the orbital or azimuthal quantum number (l) while the angular part depends on the value of l and the magnetic quantum number (m).* The various combination of the solutions are also solutions to the differential equation(1.48). The wave function of an electron in an atom is termed an atomic orbital which is determined by three quantum numbers n, *l* and m.

### 1.8.7.1 Radial wave function and radial probability distribution

It is stated that $\Psi_{radial} = \Psi_{(r)}$. Accordingly $\Psi^2_{radial}$ represents the probability of finding the electron around the nucleus in space. The probability of finding the electron in a small volume element $d\tau$ is represented as $\Psi^2_{radial}\, d\tau$. *Now what will be the value of $d\tau$ with reference to the distance r.* Let us consider a thin concentric cell of thickness $\Delta r$ over the sphere of radius r (Fig. 1.16) where the centre (O) represents the nucleus of the atom. The volume of this concentric cell = $4\pi r^2 \Delta r$ and the probability of finding the electron in this concentric cell is $4\pi r^2 \Delta r \Psi^2_{radial}$. Since $d\tau$ is a small volume and $\Psi^2_{radial}$ denotes the probability per unit volume, therefore a plot of $4\pi r^2 \Psi^2_{radial}$ vs r indicates the variation of probability of finding the electron in the concentric cell of thickness $\Delta r$ at a distance r from the nucleus and these plots are called **radial probability distribution curves**. The function $4\pi r^2 \Psi^2_{radial}$ is called **radial probability density function** or **radial probability distribution function**. The radial probability distribution curves for 1s, 2s, 3s, 2p and 3d are shown in Fig. 1.17.

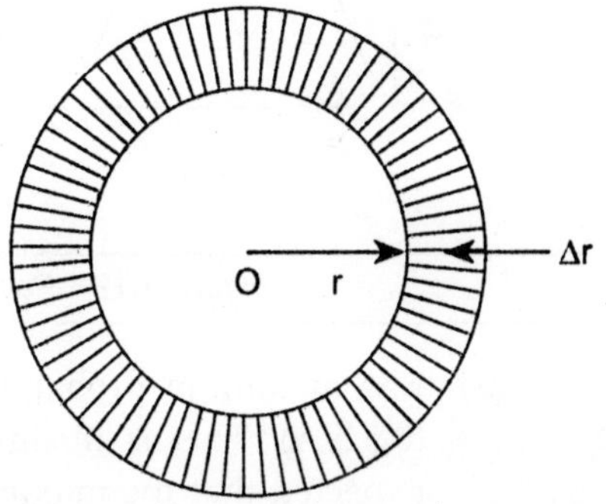


***Figure 1.16** : Depiction of radial function.*

The following points are to be noted for radial probability distribution curves :

(1) The probability is zero at the nucleus i.e. when r = 0.

(2) The number of minima appearing in a particular curve gives the number of *radical nodes* or sometimes called *nodal points. Radial node is that point where the probability of finding the electron is zero. Generally there are (n – l – 1) radial nodes and [(n – l – 1) + 1] maxima in the radial distribution curves.* For examples for 1s, 2s, 2p, 3s, 3p and 3d orbitals the number of radial nodes equal to 0, 1, 0, 2, 1, 0 respectively. *Greater the number of radial nodes, the larger will be the size of the orbital.*

* will be discussed later $R(r) = f_1(n,l)$, $\Theta(\theta) = f_2(l,m)$ and $\phi(\phi) = f_3(m)$.

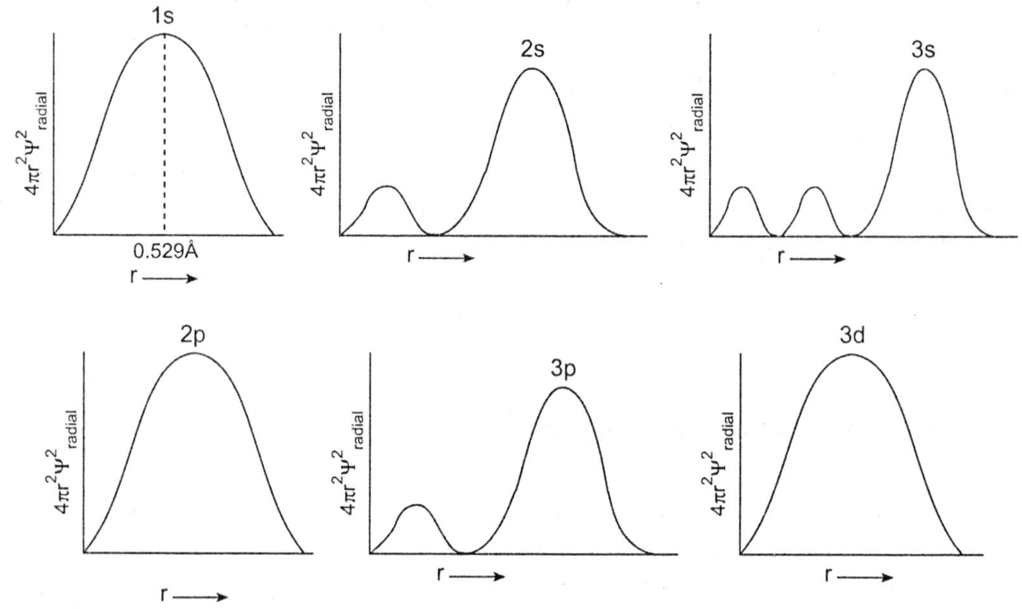


***Figure 1.17** : Radial probability distribution curves for different orbitals.*

(3) The maximum probability of finding the electron for 1s orbital is at a distance of $a_0$ (0.529Å). This is in agreement with that of Bohr's theory of hydrogen atom. It is expected that the maximum probability of finding the electron for 2s orbital will be at a distance of $4a_0$. But actually this value is at $4.91a_0$. Similarly for 3s orbital the expected value is $9a_0$ but the actual value is at $13.04a_0$. This is due to the repulsive force operated between the residual small probability of finding the electron at $a_0$ for 2s orbital (as indicated in the Fig. 1.17). This repulsive force is more pronounced in case of 3s orbital as there are two residual probability at $a_0$ and at $4.91a_0$ so that the maximum probability of finding the electron is more shifted from the expected value $9a_0$ to $13.04a_0$.

(4) It is evident from the Fig. 1.17 that the maximum probability of finding the electron for 3s, 3p and 3d orbital is in order 3s > 3p > 3d. The curve for 3s orbital shows that the small peaks are more closer to the nucleus than the small peak which exist in case of 3p orbital. Due to this additional peaks at lower distances for 3s orbital electrons exist larger fraction of time closer to the nucleus than that of 3p electron which is also larger than that of 3d electron. This means that the probability of occurance of a 3s electron nearest to the nucleus is greater that that for the 3p electrons. This is why the 3s electrons are more tightly bound to the nucleus than that of the 3p electron which in turn more tightly bound than that of 3d electron i.e. the **penetrating power** of 3s orbital is greater than 3p orbital and the sequence is ns > np > nd. It should be mentioned here that a correlation have to be made with Sommerfeld's ellipticity of the orbital (Sec. 1.7) with that of the radial distribution function.

### 1.8.7.2 Angular wave function, angular probability distribution function and shapes of atomic orbitals

Angular part of the wave function is more important than radial part because angular part is solely responsible for the orientation of the orbitals in space. As stated previously (Sec. 1.8.7) the total wave function $\Psi_{n,l,m}(r,\theta,\phi) = R_{n,l}(r)\,\Theta_{l,m}(\theta)\,\phi_m(\phi)$ (equation 1.49). The angular part of the wave function which is a product of $\Theta_{l,m}(\theta)\,\phi_m(\phi)$ is a particular class of composite wave function which is commonly known as **spherical harmonics** which determine the shapes of orbitals. Complete solution of spherical harmonics requires rigorous mathematical operations and when these are plotted against $\theta$ and $\phi$ the resulting figures describe the shape of the orbitals. The expressions for R(r), $\Theta(\theta)$ and $\phi(\phi)$ obtained by solving equation (1.48) are shown in Table 1.3 for n = 1, 2 and 3. When we put Z = 1 gives the orbitals for the hydrogen atom. Since the radial wave function gives the size of the orbital and the composite or angular function gives the shape, therefore the **complete wave function** i.e. the product of the radial and angular part of the wave function gives the total shape of the orbital.

**Shape of s-orbital :** For s orbital $l$ = 0 and so m = 0 and therefore the complete wave function = $R(r)\frac{1}{\sqrt{4\pi}}$. Since the wave function is independent of $\theta$ and $\phi$ therefore s orbital is spherically symmetrical (Fig. 1.18). In 1s orbital there will be no radial node but in 2s, 3s, 4s orbitals there will be one, two, three radial nodes respectively.

**Shape of p-orbitals :** For p-orbital $l$ = 1. When $l \neq 0$, the hydrogen like atomic wave function are not spherically symmetrical but depends on the angle $\theta$ and $\phi$. For $l$ = 1, m = 0, +1 and –1, when $l$ = 1, m = 0 i.e. $\Psi_{2p_z} = R(r)\left(\frac{3}{4\pi}\right)^{\frac{1}{2}}\cos\theta$. The angular functions with m $\neq$0 are more difficult to represent pictorially because they not only dependent on $\theta$ but also on $\phi$.

For $l$ = 1, m = +1 i.e. $\Psi_{2p_x} = R(r)\left(\frac{3}{8\pi}\right)\sin\theta\cos\phi$

and for $l$ = 1, m = –1 i.e. $\Psi_{2p_y} = R(r)\left(\frac{3}{8\pi}\right)\sin\theta\sin\phi$

The shapes of $p_x$, $p_y$ and $p_z$ orbitals are depicted in Fig. 1.18. It is to be noted that $2p_x$ orbital has no radial node but $3p_x$ and $4p_x$ orbital has one and two radial nodes respectively. The radial node = $n - l - 1$ and the angular node = $l$, so that total node = $n - 1$.

**Shapes of d-orbitals :** For d orbital $l$ = 2 and therefore m has values +2, +1, 0, –1, –2 and so there are five d orbitals

For $l$ = 2, m = 0 $\Psi_{3d_{z^2}} = R(r)\left(\frac{5}{16\pi}\right)^{\frac{1}{2}}(3\cos^2\theta - 1)$

For $l$ = 2, m = +1 $\Psi_{3d_{xz}} = R(r)\left(\frac{15}{4\pi}\right)^{\frac{1}{2}}\sin\theta\cos\theta\cos\phi$

For $l$ = 2, m = –1 $\Psi_{3d_{yz}} = R(r)\left(\frac{15}{4\pi}\right)^{\frac{1}{2}}\sin\theta\cos\theta\sin\phi$

For $l$ = 2, m = +2 $\quad \Psi_{3d_{x^2-y^2}} = R(r)\left(\frac{15}{16\pi}\right)^{\frac{1}{2}} \sin^2\theta \cos 2\phi$

For $l$ = 2, m = –2 $\quad \Psi_{3d_{xy}} = R(r)\left(\frac{15}{16\pi}\right)^{\frac{1}{2}} \sin^2\theta \sin 2\phi$

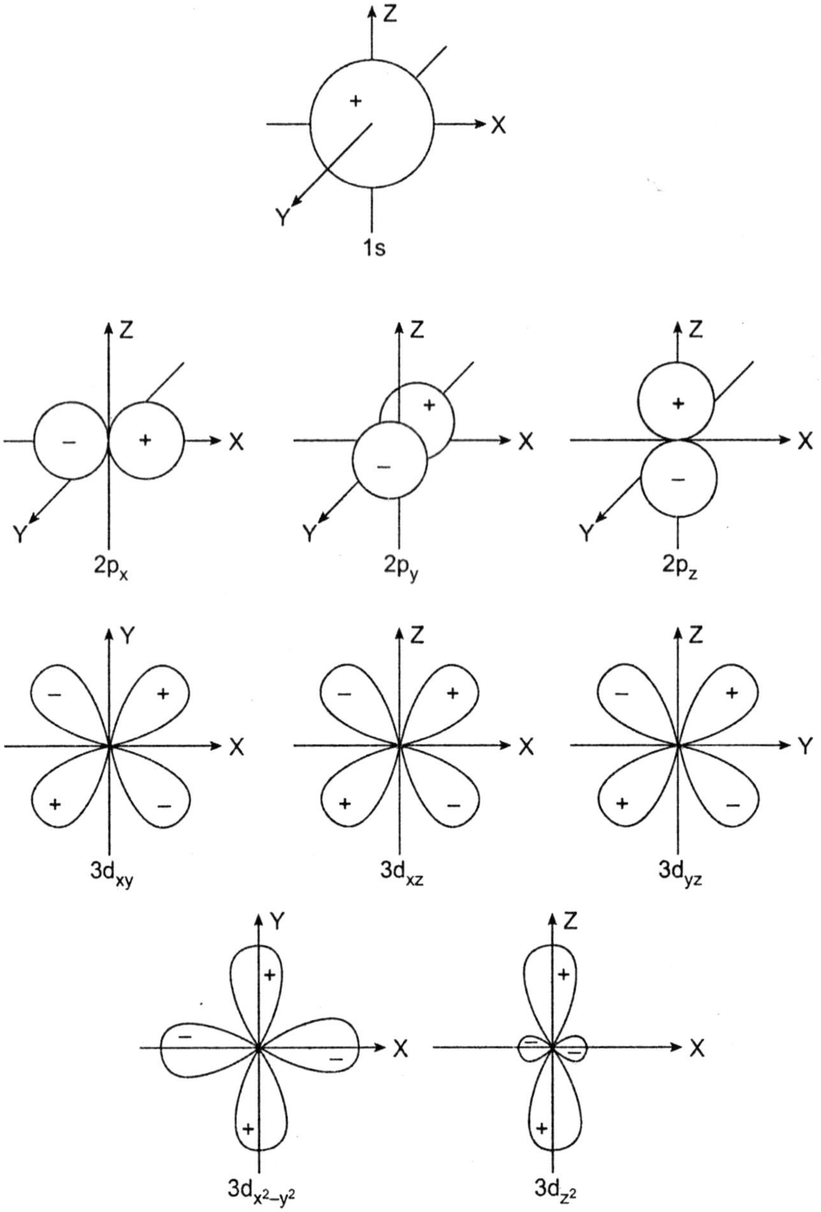


***Figure 1.18** : Boundary surfaces for the angular part of the wavefunction $\Psi(\theta,\phi)$ for the 1s, 2p, and 3d orbitals of the hydrogen atom shown in polar diagrams.*

The angular dependence of five 3d wave functions are given in Fig. 1.18. Notations of the d orbitals i.e. $d_{z^2}$ lies along z axis, $d_{x^2-y^2}$ lies aong x and y axes, $d_{xz}$ lies along x–z plane, $d_{yz}$ lies along y–z plane whereas $d_{xy}$ lies along x–y plane. It is evident that 3d orbitals has no radial node but 4d and 5d orbitals have one and two radial nodes respectively. It is worthwhile to mention that the probability contour maps for the hydrogen atomic orbitals i.e. 1s, 2s, $2p_x$, $3d_{xy}$, $3d_{yz}$ etc. enclose maximum 90 to 95% probability, there is a small but finite probability of finding the electron outside the orbital drawings of Fig. 1.18. For special interest the shapes of the seven 4f orbitals are depicted in Fig. 1.19.

### 1.8.7.3 Critique on radial part of the wave function, R(r) and radial probability distribution function, $4\pi r^2\Psi^2$ radial

Solution of the wave equation for H atom (Sec. 1.8.7, Sec. 1.8.7.1 and Table 1.3) indicates that radial part of the wave function depends only on the two quantum numbers i.e. principal quantum number (n) and azimuthal quantum number (*l*) so that it is very often expressed as $\Psi_{radial} = R_{n,l}^{(r)}$. Fig. 1.20 shows radial parts of the wave function, R(r) against distance, r, from the nucleus for 1s, 2s, 2p, 3p, 4p and 3d orbitals. The characteristic features of radial wave function may be summarised as :

(a) Table 1.3 shows that radial parts of the wave functions contains an **exponential decay term** like that of exponential decay of radioactivity which indicates that wave function

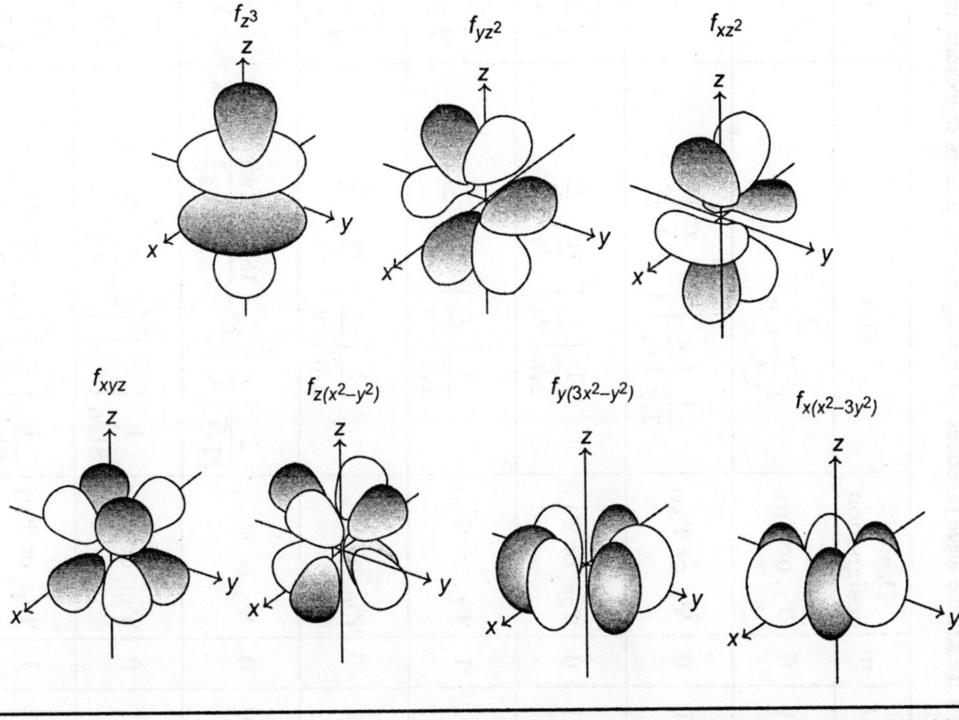


*Figure 1.19 : Shapes of general set of seven 4f orbitals. In each case shaded regions denote positive sign while the white spaces denote negative sign. (Wave function for the orbitals given in appendix-E).*

**Table 1.3 : Atomic wave functions for hydrogen like systems expressed as real functions for n = 1, 2 and 3, z = atmic number, $a_0$ = Bohr radius**

| n | *l* | m | Orbital designation | R(r) | Θ(θ) | Φ(φ) | Complete wave function |
|---|---|---|---|---|---|---|---|
| 1 | 0 | 0 | $\Psi_{1s}$ or $\Psi_{100}$ | $2\left(\frac{z}{a_0}\right)^{\frac{3}{2}} e^{-\frac{zr}{a_0}}$ | $\frac{1}{\sqrt{2}}$ | $\frac{1}{\sqrt{2\pi}}$ | $\Psi_{1s} = \frac{1}{\sqrt{\pi}}\left(\frac{z}{a_0}\right)^{\frac{3}{2}} e^{-\frac{zr}{a_0}}$ |
| 2 | 0 | 0 | $\Psi_{2s}$ or $\Psi_{200}$ | $\frac{1}{2\sqrt{2}}\left(\frac{z}{a_0}\right)^{\frac{3}{2}}\left(2-\frac{zr}{a_0}\right)e^{-\frac{zr}{2a_0}}$ | $\frac{1}{\sqrt{2}}$ | $\frac{1}{\sqrt{2\pi}}$ | $\Psi_{2s} = \frac{1}{4\sqrt{2\pi}}\left(\frac{z}{a_0}\right)^{\frac{3}{2}}\left(2-\frac{zr}{a_0}\right)e^{-\frac{zr}{2a_0}}$ |
| 2 | 1 | 0 | $\Psi_{2P_z}$ or $\Psi_{210}$ | $\frac{1}{2\sqrt{6}}\left(\frac{z}{a_0}\right)^{\frac{3}{2}}\frac{zr}{a_0}e^{-\frac{zr}{2a_0}}$ | $\sqrt{\frac{3}{2}}\cos\theta$ | $\frac{1}{\sqrt{2\pi}}$ | $\Psi_{2p_z} = \frac{1}{4\sqrt{2\pi}}\left(\frac{z}{a_0}\right)^{\frac{3}{2}}\frac{zr}{a_0}e^{-\frac{zr}{2a_0}}\cos\theta$ |
| 2 | 1 | 1 | $\Psi_{2p_x}$ or $\Psi_{211}$ | $\frac{1}{2\sqrt{6}}\left(\frac{z}{a_0}\right)^{\frac{3}{2}}\frac{zr}{a_0}e^{-\frac{zr}{2a_0}}$ | $\sqrt{\frac{3}{2}}\sin\theta$ | $\frac{1}{\sqrt{2\pi}}\cos\phi$ | $\Psi_{2p_x} = \frac{1}{4\sqrt{2\pi}}\left(\frac{z}{a_0}\right)^{\frac{3}{2}}\frac{zr}{a_0}e^{-\frac{zr}{2a_0}}\sin\theta\cos\phi$ |
| 2 | 1 | –1 | $\Psi_{2p_y}$ or $\Psi_{21-1}$ | $\frac{1}{2\sqrt{6}}\left(\frac{z}{a_0}\right)^{\frac{3}{2}}\frac{zr}{a_0}e^{-\frac{zr}{2a_0}}$ | $\sqrt{\frac{3}{2}}\sin\theta$ | $\frac{1}{\sqrt{2\pi}}\sin\phi$ | $\Psi_{2p_y} = \frac{1}{4\sqrt{2\pi}}\left(\frac{z}{a_0}\right)^{\frac{3}{2}}\frac{zr}{a_0}e^{-\frac{zr}{2a_0}}\sin\theta\sin\phi$ |
| 3 | 0 | 0 | $\Psi_{3s}$ or $\Psi_{300}$ | $\frac{2}{81\sqrt{3}}\left(\frac{z}{a_0}\right)^{\frac{3}{2}}\left(27-\frac{18zr}{a_0}+\frac{2z^2r^2}{a_0^2}\right)e^{-\frac{zr}{3a_0}}$ | $\sqrt{\frac{1}{2}}$ | $\frac{1}{\sqrt{2\pi}}$ | $\Psi_{3s} = \frac{1}{81\sqrt{3\pi}}\left(\frac{z}{a_0}\right)^{\frac{3}{2}}\left(27-\frac{18zr}{a_0}+\frac{2z^2r^2}{a_0^2}\right)e^{-\frac{zr}{3a_0}}$ |
| 3 | 1 | 0 | $\Psi_{3p_z}$ or $\Psi_{310}$ | $\frac{4}{81\sqrt{6}}\left(\frac{z}{a_0}\right)^{\frac{3}{2}}\left(6-\frac{zr}{a_0}\right)\frac{zr}{a_0}e^{-\frac{zr}{3a_0}}$ | $\sqrt{\frac{3}{2}}\cos\theta$ | $\frac{1}{\sqrt{2\pi}}$ | $\Psi_{3p_z} = \frac{\sqrt{2}}{81\sqrt{\pi}}\left(\frac{z}{a_0}\right)^{\frac{3}{2}}\frac{zr}{a_0}\left(6-\frac{zr}{a_0}\right)e^{-\frac{zr}{3a_0}}\cos\theta$ |
| 3 | 1 | 1 | $\Psi_{3p_x}$ or $\Psi_{311}$ | $\frac{4}{81\sqrt{6}}\left(\frac{z}{a_0}\right)^{\frac{3}{2}}\left(6-\frac{zr}{a_0}\right)\frac{zr}{a_0}e^{-\frac{zr}{3a_0}}$ | $\sqrt{\frac{3}{2}}\sin\theta$ | $\frac{1}{\sqrt{2\pi}}\cos\phi$ | $\Psi_{3p_x} = \frac{\sqrt{2}}{81\sqrt{\pi}}\left(\frac{z}{a_0}\right)^{\frac{3}{2}}\frac{zr}{a_0}\left(6-\frac{zr}{a_0}\right)e^{-\frac{zr}{3a_0}}\sin\theta\cos\phi$ |

*Table contd.*

| n | $l$ | m | Orbital designation | R(r) | θ(θ) | φ(φ) | Complete wave function |
|---|---|---|---|---|---|---|---|
| 3 | 1 | –1 | $\Psi_{3p_y}$ or $\Psi_{31-1}$ | $\frac{4}{81\sqrt{6}}\left(\frac{z}{a_0}\right)^{\frac{3}{2}}\left(6-\frac{zr}{a_0}\right)\frac{zr}{a_0}e^{-\frac{zr}{3a_0}}$ | $\sqrt{\frac{3}{2}}\sin\theta$ | $\frac{1}{\sqrt{2\pi}}\sin\phi$ | $\Psi_{3p_y}=\frac{\sqrt{2}}{81\sqrt{\pi}}\left(\frac{z}{a_0}\right)^{\frac{3}{2}}\frac{zr}{a_0}\left(6-\frac{zr}{a_0}\right)e^{-\frac{zr}{3a_0}}\sin\theta\sin\phi$ |
| 3 | 2 | 0 | $\Psi_{3d_{z^2}}$ or $\Psi_{320}$ | $\frac{4}{81\sqrt{30}}\left(\frac{z}{a_0}\right)^{\frac{3}{2}}\frac{z^2r^2}{a_0^2}e^{-\frac{zr}{3a_0}}$ | $\sqrt{\frac{5}{8}}(3\cos^2\theta-1)$ | $\frac{1}{\sqrt{2\pi}}$ | $\Psi_{3d_{z^2}}=\frac{1}{81\sqrt{6\pi}}\left(\frac{z}{a_0}\right)^{\frac{3}{2}}\frac{z^2r^2}{a_0^2}e^{-\frac{zr}{3a_0}}(3\cos^2\theta-1)$ |
| 3 | 2 | 1 | $\Psi_{3d_{xz}}$ or $\Psi_{321}$ | $\frac{4}{81\sqrt{30}}\left(\frac{z}{a_0}\right)^{\frac{3}{2}}\frac{z^2r^2}{a_0^2}e^{-\frac{zr}{3a_0}}$ | $\sqrt{\frac{15}{4}}\sin\theta\cos\theta$ | $\frac{1}{\sqrt{\pi}}\cos\phi$ | $\Psi_{3d_{xz}}=\frac{\sqrt{2}}{81\sqrt{\pi}}\left(\frac{z}{a_0}\right)^{\frac{3}{2}}\frac{z^2r^2}{a_0^2}e^{-\frac{zr}{3a_0}}\sin\theta\cos\theta\cos\phi$ |
| 3 | 2 | –1 | $\Psi_{3d_{yz}}$ or $\Psi_{32-1}$ | $\frac{4}{81\sqrt{30}}\left(\frac{z}{a_0}\right)^{\frac{3}{2}}\frac{z^2r^2}{a_0^2}e^{-\frac{zr}{3a_0}}$ | $\sqrt{\frac{15}{4}}\sin\theta\cos\phi$ | $\frac{1}{\sqrt{\pi}}\sin\phi$ | $\Psi_{3d_{yz}}=\frac{\sqrt{2}}{81\sqrt{\pi}}\left(\frac{z}{a_0}\right)^{\frac{3}{2}}\frac{z^2r^2}{a_0^2}e^{-\frac{zr}{3a_0}}\sin\theta\cos\theta\sin\phi$ |
| 3 | 2 | 2 | $\Psi_{3d_{x^2-y^2}}$ or $\Psi_{322}$ | $\frac{4}{81\sqrt{30}}\left(\frac{z}{a_0}\right)^{\frac{3}{2}}\frac{z^2r^2}{a_0^2}e^{-\frac{zr}{3a_0}}$ | $\sqrt{\frac{15}{16}}\sin^2\theta$ | $\frac{1}{\sqrt{\pi}}\cos2\phi$ | $\Psi_{3d_{x^2-y^2}}=\frac{1}{81\sqrt{2\pi}}\left(\frac{z}{a_0}\right)^{\frac{3}{2}}\frac{z^2r^2}{a_0^2}e^{-\frac{zr}{3a_0}}\sin^2\theta\cos2\phi$ |
| 3 | 2 | –2 | $\Psi_{3d_{xy}}$ or $\Psi_{32-2}$ | $\frac{4}{81\sqrt{30}}\left(\frac{z}{a_0}\right)^{\frac{3}{2}}\frac{z^2r^2}{a_0^2}e^{-\frac{zr}{3a_0}}$ | $\sqrt{\frac{15}{16}}\sin^2\theta$ | $\frac{1}{\sqrt{\pi}}\sin2\phi$ | $\Psi_{3d_{xy}}=\frac{1}{81\sqrt{2\pi}}\left(\frac{z}{a_0}\right)^{\frac{3}{2}}\frac{z^2r^2}{a_0^2}e^{-\frac{zr}{3a_0}}\sin^2\theta\sin2\phi$ |

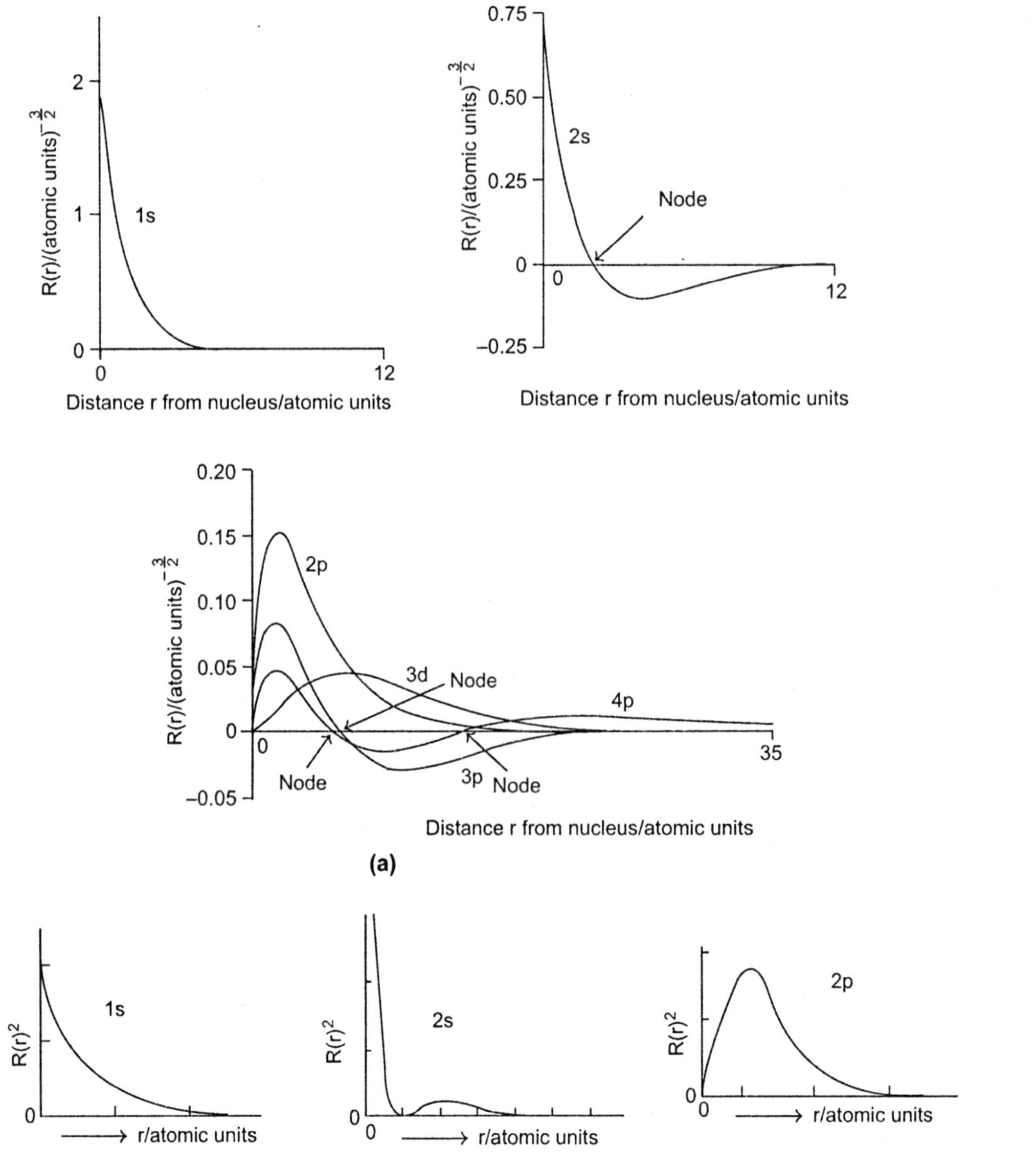


***Figure 1.20 : (a)** Plots of radial parts of the wave function R(r) against r for the 1s, 2s, 2p, 3p, 4p and 3d atomic orbitals, the nucleus is at r = 0. **(b)** Plots of radial probability density against r for 1s, 2s and 2p atomic orbitals.*

decay exponentially as r increases but the decay slower for n=2 than for n=1 i.e. the decay is slower for higher values of n. This means that plausably the electron being further from the nucleus increases as n increases. This pattern of decay curve continues for higher values of n which is justified from the curve (Fig. 1.20(a)).

(b) s atomic orbitals have a finite value of R(r) at the nucleus and for other orbitals R(r)=0 at the nucleus.

(c) For 2p, 3d, 4f orbitals R(r) is positive everywhere except at the origin.

(d) For orbitals (2s, 3p, 4d, 5f) the R(r) may be positive or negative but the wave function has only one sign changed at a particular point where R(r) = 0. These particular points are called radial nodal points. At the nodal point the value of the radial wave function changes its sign from positive to negative. For s orbital there are (n–1) radial nodes while for the p and d orbitals there are (n–2) and (n–3) radial nodes.

(e) In the case of 3s, 4p, 5d and 6f orbitals, the function R(r) has two sign changes i.e. it possesses two radial nodes.

Fig. 1.20 also denotes radial probability density as a function r i.e. $R(r)^2$ vs r.

Another aspect of radial part of wave function is to be considered as **radial distribution function (RDF)** or **radial probability distribution function** or **radial probability density.** According to Sec. 1.8.7.1 the square of the radial wave function gives the measure of probability of finding the electron at various points in space. Thus the probability of finding the electron in the small volume element $d\tau$ enclosed between the concentric sphere of radius r and r+dr is given by $\Psi^2_{radial}d\tau = R(r)^2 d\tau$ where $d\tau = 4\pi r^2 dr$. Therefore the radial probability of finding the electron in a shell of thickness dr at a distance r from the nucleus is given by, $4\pi r^2 \Psi^2_{radial} dr = 4\pi r^2 R(r)^2 dr$. Thus a plot of $4\pi r^2 R(r)^2$ vs r indicates the radial probability distribution curve as depicted in Fig. 1.17.

## 1.9 Quantum Numbers

The numbers used to identify the various quantum states that are available to electrons are termed quantum numbers. Quantum numbers describe values of conserved numbers in the dynamics of the quantum system. They often describe specifically the **energy of electrons** in atoms, but other possibilities include **angular momentum, spin** etc. Since any quantum mechanical system can have one or more quantum numbers, it is a futile job to list all possible quantum numbers.

**How many quantum numbers are needed to describe any quantum mechanical system** has no universal answer. The dynamics of any quantum mechanical system are described by a quantum Hamiltonian, H. There is one quantum number of the system corresponding to the energy i.e. eigen value of the Hamiltonian. There is also one quantum number for each operator O that commutes with the Hamiltonian operator i.e. satisfies the relation OH = HO. These are all the quantum numbers that the system can have. It is to be noted that the operator O defining the quantum numbers should be independent of each other. Often there is more than one way to choose a set of independent operators. Naturally in different situations different sets of quantum numbers may be used for the description of the same quantum mechanical system.

Conventionally the following four quantum numbers are described by the hydrogen like atom or Bohr atom systems :

**Principal quantum number (n) :** It denotes the eigenvalue of hydrogen like atom with the $J^2$ part removed (J = angular momentum of the electron). The electron orbits are located in shells, a shell conceptually being a spherical surface with the atomic nucleus located at the centre of the sphere. The locations available for electron orbits are in a series of concentric shells corresponding to the orbit number n which is referred to as the principal quantum number (n). It has the integral values 1, 2, 3, ...... This quantum number denotes the distance between the electron and the nucleus (i.e. the radial coordinate r). The distance increases with increasing value of n and hence quantum states with different n values are said to belong different orbitals. Therefore this quantum number designate the **size** of the orbital.

---

**Problem 1.16** Why principal quantum number has integral values?

**Ans. :** From the particle in one dimensional box (Sec. 1.8.6) we have

$$E = \frac{n^2h^2}{8L^2m}$$ where n is a positive integer.

Thus the energy values will be some discrete values and no other values. The energy of the particle thus **quantised** and the integer is called the principal quantum number. It is to be noted that quantisation arises naturally from the boundary conditions. From the quantum rules in a vibrating string for which the number of nodes is integral.

---

**Orbital or azimuthal quantum number (*l*) :** This quantum number states the rotation of electron orbit plane in space. This also states the energy due to the **orbital angular momentum** of the electron and therefore it is sometimes called **angular quantum number** or **subsidiary quantum number**. Orbital angular momentum of the electron is spatially quantised. This refers to the permitted relative tilts among the electron orbits of a shell. The spatial quantisation is that the projection of the angular momentum vector on an axis of measurement can only have integral values in the range 0 to (n–1) i.e. *l* has values 0,1,2,3,...... . This quantum number is very important in chemistry, since it specifies the **shape of the atomic** orbitals and strongly influence the formation of **chemical bonds** and also **bond angles**.

$$l = 0, 1, 2, 3, \ldots\ldots\ldots (n-1)$$

s p d f .........

The notations s, p, d and f indicates the spectral terms i.e. sharp, principal, diffuse and fundamental. In this context a question arises what is the relation between *l* and Sommerfeld azimuthal quantum number k (Sec. 1.7). As stated (Sec. 1.7) k has values 1, 2, 3, ......... . Therefore *l* and k are related as $l = k - 1$.

**Magnetic quantum number (m) :** This quantum number arises for the explanation of certain experimental facts such as Zeeman effect (application of strong magnetic field on the spectral lines) and Sterk effect (application of electric field on spectral lines). In both the cases the single spectral line is splitted into closely spaced lines. This quantum number denotes some orientation or tilts of electron orbit plane with respect to the applied magnetic or electric field.

This quantum number actually denotes the projection of the orbital angular momentum along the specified direction. The allowed projections of the angular momentum on a selected axis of measurement are each of the allowed values of $m\left(\frac{h}{2\pi}\right)$, where the various projections differ because of different tilting of the various orbits. The number of possible tilting of the total angular momentum is given by the total number of m values. Along the field direction m may take integral values from $+l$ to $-l$ including zero i.e. altogether $(2l+1)$ values are permitted for a particular value of $l$ e.g. when $l = 0$, m = 0; when $l = 1$, m = +1, 0, –1; when $l = 2$, m = +2, +1, 0, –1 and –2 etc. Possible orientations for $l = 2$ with respect to specified field H are shown in Fig. 1.21.

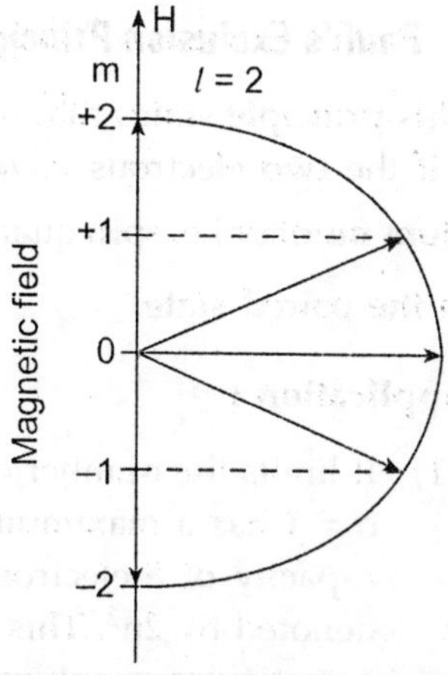

*Figure 1.21 : Possible projection of or orbital angular momentum for d orbital.*

**Spin quantum number (s)** : It is indicated from spectroscopy that upto two electrons can be accommodated in a single orbital. However two electrons can never have the same quantum state or have the same set of four quantum numbers. Therefore a fourth quantum number arises which is known as spin quantum number – the detail of which may be explained by **relativistic quantum mechanics.**

Simply it denotes the rotation of the electron in its own axis. The rotation may be denoted clockwise or anticlockwise. Like orbital angular momentum, spin angular momentum interact with the external magnetic field and precess around the field direction as in the case of orbital vector. This has also 2s+1 orientation and the values may be +s to –s but not including zero **due to the anomolous behaviour of spin** which assume the non-integral values $\pm\frac{1}{2}$. If we take the $+\frac{1}{2}$ value for clockwise rotation indicated by (↑) then the $-\frac{1}{2}$ value arbitrarily given by anticlockwise rotation denoted by (↓). When two electrons are in the same orbital, their spins are always opposite–one electron has given the $+\frac{1}{2}$ value and the other by $-\frac{1}{2}$ and denoted by (↑↓).

The following table denotes the description of the quantum numbers of the electrons (Table 1.4).

**Table 1.4 : Description of quantum numbers**

| Name | Symbol | Function | Range values | Example of values |
|---|---|---|---|---|
| Principal quantum number | n | size | $1 \le n$ | n = 1,2,3,...... |
| Orbital or azimuthal quantum number (angular momentum) | $l$ | shape | $0 \le l \le n-1$ | For n = 3 $l = 0,1,2$ |
| Magnetic quantum number (projection of angular momentum) | m | orientation | $-l \le m \le +l$ | For $l = 2$ m = +2,+1,0,–1,–2 |
| Spin quantum number | s | rotation of electron in its own axis | $\pm\frac{1}{2}$ | always $+\frac{1}{2}$ or $-\frac{1}{2}$ |

### 1.9.1 Pauli's Exclusion Principle (Wolfgang Pauli, 1925)

This principle states that *no two electrons in an atom has the same set of four quantum numbers.* Thus if the two electrons have same orbital wave functions (i.e. same n, $l$ and m), the fourth quantum number i.e. spin quantum number must differ by $+\frac{1}{2}$ and $-\frac{1}{2}$ so that the two electrons are in the paired state.

**Application :**

(1) It limits the number of electrons in a particular quantum shell. For example K shell i.e. n = 1 has a maximum of 2 electrons, L shell i.e. n = 2 has maximum accommodation capacity of 8 electrons i.e. maximum limit of accommodation in a particular shell is denoted by $2n^2$. This is also justified from the number of set of quantum numbers for a particular quantum shell e.g. n = 3 i.e. M shell has the following 18 set of quantum numbers so that M shell has a maximum accommodation capacity of 18 electrons.

| | | | |
|---|---|---|---|
| n = 3 | $l$ = 0 | m = 0 | $s = \pm\frac{1}{2}$ |
| n = 3 | $l$ = 1 | m = +1 | $s = \pm\frac{1}{2}$ |
| n = 3 | $l$ = 1 | m = 0 | $s = \pm\frac{1}{2}$ |
| n = 3 | $l$ = 1 | m = –1 | $s = \pm\frac{1}{2}$ |
| n = 3 | $l$ = 2 | m = +2 | $s = \pm\frac{1}{2}$ |
| n = 3 | $l$ = 2 | m = +1 | $s = \pm\frac{1}{2}$ |
| n = 3 | $l$ = 2 | m = 0 | $s = \pm\frac{1}{2}$ |
| n = 3 | $l$ = 2 | m = –1 | $s = \pm\frac{1}{2}$ |
| n = 3 | $l$ = 2 | m = –2 | $s = \pm\frac{1}{2}$ |

Similarly for K shell and L shell there will be 2 and 8 set of quantum numbers respectively.

(2) This principle limits the maximum number of electrons in a particular orbital i.e. s, p and d orbital has a maximum number of 2, 6 and 10 electrons respectively. Why p orbital has maximum of 6 electrons? For p orbital $l$ = 1 so that there will be following 6 set of quantum numbers (For n > 1).

| | | |
|---|---|---|
| $l$ = 1 | m = +1 | $s = +\frac{1}{2}$ |
| $l$ = 1 | m = +1 | $s = -\frac{1}{2}$ |
| $l$ = 1 | m = 0 | $s = +\frac{1}{2}$ |

$l = 1 \qquad m = 0 \qquad s = -\frac{1}{2}$

$l = 1 \qquad m = -1 \qquad s = +\frac{1}{2}$

$l = 1 \qquad m = -1 \qquad s = -\frac{1}{2}$

(3) This has an extreme importance in chemistry in determining the microstates for a particular electronic configuration. For example in the ground state electronic configuration of carbon i.e. $1s^2\ 2s^2\ 2p^2$ the microstates that are available without violating the Pauli's exclusion principle are $^1D_2$, $^3P_2$, $^3D_1$, $^3P_0$ and $^1S_0$ where S, P, D etc. denotes electronic states of the atom or ion.

(4) The idea of an **antisymmetric wave function** comes from this principle which is an another postulate of quantum mechanics which states that *all electronic wave functions must be antisymmetric under the interchange of any two electrons.*

(5) Description of quantum number of orbital electrons of elements may be stated by this principle e.g.

| Element | Atomic number | Electron number | Quantum numbers | | | |
|---|---|---|---|---|---|---|
| | | | n | $l$ | m | s |
| H | 1 | 1 | 1 | 0 | 0 | $-\frac{1}{2}$ |
| He | 2 | 1 | 1 | 0 | 0 | $-\frac{1}{2}$ |
| | | 2 | 1 | 0 | 0 | $+\frac{1}{2}$ |
| Li | 3 | 1 | 1 | 0 | 0 | $-\frac{1}{2}$ |
| | | 2 | 1 | 0 | 0 | $+\frac{1}{2}$ |
| | | 3 | 2 | 0 | 0 | $-\frac{1}{2}$ |
| Be | 4 | 1 | 1 | 0 | 0 | $-\frac{1}{2}$ |
| | | 2 | 1 | 0 | 0 | $+\frac{1}{2}$ |
| | | 3 | 2 | 0 | 0 | $-\frac{1}{2}$ |
| | | 4 | 2 | 0 | 0 | $+\frac{1}{2}$ |
| B | 5 | 1 | 1 | 0 | 0 | $-\frac{1}{2}$ |
| | | 2 | 1 | 0 | 0 | $+\frac{1}{2}$ |
| | | 3 | 2 | 0 | 0 | $-\frac{1}{2}$ |
| | | 4 | 2 | 0 | 0 | $+\frac{1}{2}$ |
| | | 5 | 2 | 1 | –1 | $-\frac{1}{2}$ |

| Element | Atomic number | Electron number | Quantum numbers | | | |
|---|---|---|---|---|---|---|
| | | | n | $l$ | m | s |
| C | 6 | 1 | 1 | 0 | 0 | $-\frac{1}{2}$ |
| | | 2 | 1 | 0 | 0 | $+\frac{1}{2}$ |
| | | 3 | 2 | 0 | 0 | $-\frac{1}{2}$ |
| | | 4 | 2 | 0 | 0 | $+\frac{1}{2}$ |
| | | 5 | 2 | 1 | –1 | $-\frac{1}{2}$ |
| | | 6 | 2 | 1 | –1 | $+\frac{1}{2}$ |
| Ne | 10 | 1 | 1 | 0 | 0 | $-\frac{1}{2}$ |
| | | 2 | 1 | 0 | 0 | $+\frac{1}{2}$ |
| | | 3 | 2 | 0 | 0 | $-\frac{1}{2}$ |
| | | 4 | 2 | 0 | 0 | $+\frac{1}{2}$ |
| | | 5 | 2 | 1 | –1 | $-\frac{1}{2}$ |
| | | 6 | 2 | 1 | –1 | $+\frac{1}{2}$ |
| | | 7 | 2 | 1 | 0 | $-\frac{1}{2}$ |
| | | 8 | 2 | 1 | 0 | $+\frac{1}{2}$ |
| | | 9 | 2 | 1 | 1 | $-\frac{1}{2}$ |
| | | 10 | 2 | 1 | 1 | $+\frac{1}{2}$ |
| Na | 11 | 1 | 1 | 0 | 0 | $-\frac{1}{2}$ |
| | | 2 | 1 | 0 | 0 | $+\frac{1}{2}$ |
| | | 3 | 2 | 0 | 0 | $-\frac{1}{2}$ |
| | | 4 | 2 | 0 | 0 | $+\frac{1}{2}$ |
| | | 5 | 2 | 1 | –1 | $-\frac{1}{2}$ |
| | | 6 | 2 | 1 | –1 | $+\frac{1}{2}$ |
| | | 7 | 2 | 1 | 0 | $-\frac{1}{2}$ |
| | | 8 | 2 | 1 | 0 | $+\frac{1}{2}$ |
| | | 9 | 2 | 1 | 1 | $-\frac{1}{2}$ |
| | | 10 | 2 | 1 | 1 | $+\frac{1}{2}$ |
| | | 11 | 3 | 0 | 0 | $-\frac{1}{2}$ |

| Element | Atomic number | Electron number | Quantum numbers | | | |
|---|---|---|---|---|---|---|
| | | | n | $l$ | m | s |
| Mg | 12 | 1 | 1 | 0 | 0 | $-\frac{1}{2}$ |
| | | 2 | 1 | 0 | 0 | $+\frac{1}{2}$ |
| | | 3 | 2 | 0 | 0 | $-\frac{1}{2}$ |
| | | 4 | 2 | 0 | 0 | $+\frac{1}{2}$ |
| | | 5 | 2 | 1 | –1 | $-\frac{1}{2}$ |
| | | 6 | 2 | 1 | –1 | $+\frac{1}{2}$ |
| | | 7 | 2 | 1 | 0 | $-\frac{1}{2}$ |
| | | 8 | 2 | 1 | 0 | $+\frac{1}{2}$ |
| | | 9 | 2 | 1 | 1 | $-\frac{1}{2}$ |
| | | 10 | 2 | 1 | 1 | $+\frac{1}{2}$ |
| | | 11 | 3 | 0 | 0 | $-\frac{1}{2}$ |
| | | 12 | 3 | 0 | 0 | $+\frac{1}{2}$ |

and so on for other elements.

### 1.9.2 Quantum mechanical approach of Pauli's exclusion principle, antisymmetric wave function

Let us consider a two electrons molecular system e.g. $H_2$ molecule where the electron 1 is associated with nucleus A and the electron 2 is associated with nucleus B. Though the two hydrogen atoms and the two electrons are indistinguishable yet for simplicity's shake we can write the wave function as :

$$\Psi_1 = \Psi_A(1)\,\Psi_B(2) \qquad \text{...(1.50)}$$

Since electrons are absolutely indistinguishable, it is more logical that electron 1 moves towards the nucleus B while electron 2 moves towards the nucleus A. Therefore, the wave function will be—

$$\Psi_2 = \Psi_A(2)\,\Psi_B(1) \qquad \text{...(1.51)}$$

Thus the wave function for the two electrons $H_2$ molecular system can be represented by **linear combination of atomic orbitals (LCAO)** as :

$$\left.\begin{aligned} \Psi &= C_1\Psi_1 + C_2\Psi_2 \\ \text{or,}\quad \Psi &= C_1\Psi_A(1)\Psi_B(2) + C_2\Psi_A(2)\Psi_B(1) \end{aligned}\right\} \qquad \text{...(1.52)}$$

where $C_1$ and $C_2$ are called mixing coefficients and represent how much extent $\Psi_1$ mixes with $\Psi_2$.

From the detailed calculation of energy (variation method) it has been suggested that $C_1 = C_2$ and $C_2 = -C_1$ so that the wave function (1.52) represented as :

$$\Psi_S = C_S[\Psi_A(1)\Psi_B(2) + \Psi_A(2)\Psi_B(1)] \quad ...(1.53)$$

and $$\Psi_A = C_A[\Psi_A(1)\Psi_B(2) - \Psi_A(2)\Psi_B(1)] \quad ...(1.54)$$

Wave function $\Psi_S$ (equation 1.53) shows that if the nucleus or electrons are interchanged, $\Psi_S$ remains the same. It is therefore said to be **symmetric wave function** with respect to the coordinates of both nuclei and electrons. On the other hand, if on changing the electrons and nucleus, the sign will be reversed, then the wave function is called **antisymmetric** $\Psi_A$ (equation 1.54). In these equations (1.53 and 1.54) $C_S$ and $C_A$ are **normalisation factor*** so select hat the probability of finding the electrons in space is unity.

Now the question arises how the **symmetric** and **antisymmetric** wave functions are related to the Pauli's exclusion principle? According to this principle there can not be two electrons in an **atomic** or **a molecular** system in which all four quantum numbers will be the same. *The quantum numbers states the form of wave function characterising the state of the electron.* Since the symmetric function remains the same when the electrons are interchanged, it indicates that these electrons are in the same state, which is contrary to this principle. To characterise the wave function completely, we have to consider the **spin wave function** of the electron so that the total wave function is given by

$$\Psi_{total} = \Psi_{orbital} \times \Psi_{spin} \quad ...(1.55)$$

Now the main theme of the **Pauli's antisymmetric** principle is based on the fact that the wave functions for many electron system can be obtained with respect to the exchange of permutation of the spin and orbital coordinates of electrons. The total wave function ($\Psi_{total}$) can be obtained by combination of two spatial wave functions $\Psi_S$ and $\Psi_A$ and four spin wave functions because the two eletrons have two types of spin namely $\pm\frac{1}{2}$. There will be eight such combinations :

$$\alpha_1 = \Psi_S \times \Psi_{spin_1} \qquad \alpha_5 = \Psi_A \times \Psi_{spin_1}$$

$$\alpha_2 = \Psi_S \times \Psi_{spin_2} \qquad \alpha_6 = \Psi_A \times \Psi_{spin_2}$$

$$\alpha_3 = \Psi_S \times \Psi_{spin_3} \qquad \alpha_7 = \Psi_A \times \Psi_{spin_3}$$

$$\alpha_4 = \Psi_S \times \Psi_{spin_4} \qquad \alpha_8 = \Psi_A \times \Psi_{spin_4}$$

* Normalisation condition $\int_{-\infty}^{+\infty} \Psi_S^2 d\tau = 1$ and $\int_{-\infty}^{+\infty} \Psi_A^2 d\tau = 1$ where $d\tau$ is a small volume element = dxdydz. If $C_S$ is the normalisation factor, then $C_S^2 \int_{-\infty}^{+\infty} \Psi_S^2 d\tau = 1$. Here the normalising factor $C_S$ and $C_A$ comes out to be $\frac{1}{\sqrt{2}}$.

where $\alpha_1$ to $\alpha_8$ denotes the total wave function with spin orbit interaction. Here $\Psi_{spin_1}$, $\Psi_{spin_2}$, $\Psi_{spin_3}$ are symmetric wave function and $\Psi_{spin_4}$ is antisymmetric wave function constructed like that of $\Psi_S$ and $\Psi_A$ for orbital wave function. **It is experimentally established that for many electron wave function must be antisymmetric with respect to the interchange of any pair of electrons between the spatial and spin coordinates.** Hence according to this experimental fact among the above eight wave functions $\alpha_4$, $\alpha_5$, $\alpha_6$ and $\alpha_7$ are acceptable. Thus the Pauli's exclusion principle may be stated in the following forms :

(1) Wave functions must be antisymmetric in the interchange of any two electrons.

(2) No two spin and orbital wave functions used to construct a many electron wave function can be identical.

(3) Total wave function for a system having two electrons must be antisymmetric.

(4) Spin-orbital wave function for two electrons must not be the same.

The antisymmetric principle can be applied to particles like electrons, protons etc. i.e. for fermions which have half integral spin.

### 1.9.3 Hund's Rule (German spectroscopist Friederich Hund)

In a multielectron system the electrons are filled in different orbitals such as s, p, d and f orbitals for a particular value of n. The maximum number of electrons in a particular orbital = $2(2l + 1)$. The spin state of electrons can be stated by this rule. It may have the following forms :

(1) So long orbitals are available electrons will be in free spin.

(2) Pairing of electrons is unfavourable process, energy should be needed in order to pair the electrons.

#### Application :

(1) This rule is applied to determine the minimum energy configuration for degenerate orbitals i.e. for orbitals having same energy. For example p orbital is three fold degenerate and if there is three electrons then the possible modes of filling electrons are :

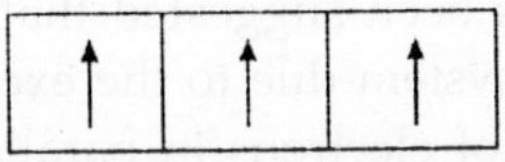

$P_x$ $P_y$ $P_z$

(A) (Correct)

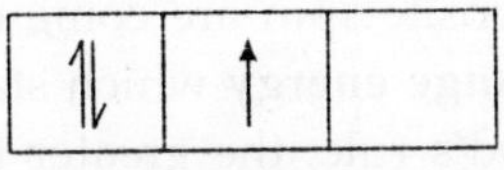

$P_x$ $P_y$ $P_z$

(B) (Wrong)

In (A) the three electrons are in free spin but in (B), remaining one orbital vacant the electron is paired—this is against this rule so that this rule is known as Hund's rule of **maximum spin multiplicity** i.e. the correct electronic configuration for N is $1s^2\ 2s^2\ 2p_x^1\ 2p_y^1\ 2p_z^1$ and not as $1s^2\ 2s^2\ 2p_x^2 2p_y^1$. The higher stability of maximum spin multiplicity configuration can be attributed by the term **exchange integral** in quantum mechanics which suggest that $1s^2\ 2s^2\ 2p_x^1\ 2p_y^1\ 2p_z^1$ is the minimum energy configuration for N.

(2) This rule is used to determine the term symbol of the ground electronic state for multielectron system.

## 1.10 THE AUFBAU PRINCIPLE

(aufbau = a German word meaning building up)

This principle states that *obeying Pauli's exclusion principle, orbitals with lowest energy will be filled up first then the orbitals of higher energy will be filled by electrons.*

The lowest energy orbital is determined by an emperical (n+$l$) rule. *The higher the (n+l) value, the higher is the energy. When two or more orbitals have the same (n+l) values the energy of the orbital increases with increasing value of n.* According to the (n+$l$) rule the relative ordering of orbitals emerges as :

$$1s < 2s < 2p < 3s < 3p < 4s < 3d < 4p < 5s < 4d < 5p < 6s$$
$$< 4f < 5d < 6p < 7s < 5f < 6d < 7p < 8s$$

This (n+$l$) rule will hold good for elements having atomic number up to 90. For the heavier elements when the effective nuclear charge increases the above sequence of orbitals will be modified as follows :

$$1s < 2s < 2p < 3s < 3p < 3d < 4s < 4p < 4d < 4f < 5s < 5p < 5d$$
$$< 5f < 6s < 6p < 6d < 7s < 7p < 8s$$

## 1.11 THE STABILITY OF HALF-FILLED OR FULL FILLED ORBITALS. THE EXCHANGE ENERGY

The stability of half filled or full filled orbitals are more compared to the partially filled orbitals can be explained by **exchange energy** and **spin correlation term**. According to the prevaling belief, the spin of the electron or of some other particle has a mysterious internal angular momentum for which no definite physical picture is available and for which there is no classical analog. The physical concept of **exchange energy** and spin correlation can be explained by a quantum mechanical phenomena.

The spin correlation states that electrons of the same spin repell each other and keep away while electrons of opposite spin are come closer. It has been suggested that electrons of same spin develop an **exchange energy** which stabilises the system due to the **exchange interaction.** According to the Hund's rule, the greater the number of electrons in parallel spin the **greater is the exchange interaction** and **greater will be the stability of the system.**

Now the electrons (in absence of electric and magnetic field) in an atom with the same n and $l$ values are **indistinguishable.** Each electron is associated with a wave function and they can exchange their positions when the wave function of the individual electrons are exchangable and generated an **exchange energy.** The exchange energy ($E_{ex}$) for a set of two electrons is represented as $^{n}C_2$ where n is the number of electrons with parallel spin i.e.

$$E_{ex} = {}^{n}C_2 = \frac{n!}{2!(n-2)!} = \frac{n(n-1)}{2}$$

Actually exchange energy is proportional to $\frac{n(n-1)}{2}$ i.e.

$$E_{ex} \propto \frac{n(n-1)}{2}$$

$$\therefore\ E_{ex} = K\frac{n(n-1)}{2}$$ where K is a constant.

In case of $d^4$, $d^5$ and $d^6$ configurations the exchange energies are 6K, 10K and 10K+P respectively where

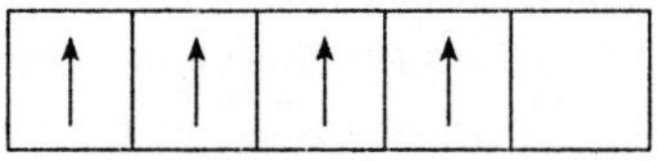

$d^4$

$E_{ex} \propto {}^4C_2$

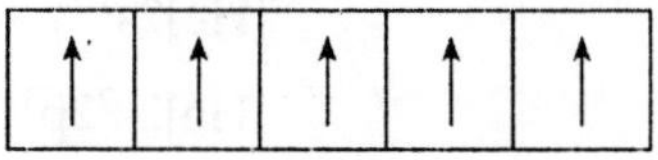

$d^5$

$E_{ex} \propto {}^5C_2$

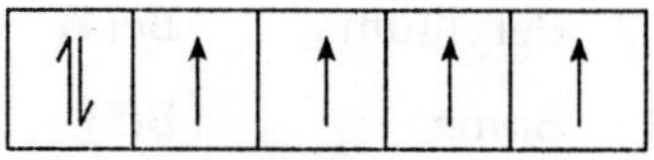

$d^6$ (Here n = 5)

$E_{ex} \propto ({}^5C_2 + {}^1C_2)$

P is the pairing energy. Though the exchange energy for $d^5$ and $d^6$ configurations are same but due to the pairing of one electron in $d^6$ configuration it is destabilised i.e. exchange energy is maximum in case of $d^5$ configuration so that $d^5$ configuration is more stable than the nearest half filled configurations.

In a similar way we can consider the $d^8$, $d^9$ and $d^{10}$ configurations.

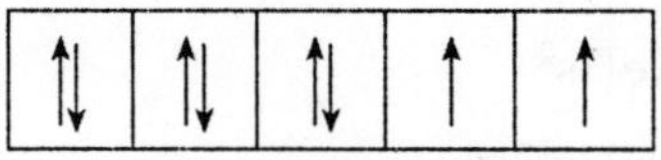

$d^8$

$E_{ex} \propto ({}^5C_2 + {}^3C_2)$ $\qquad E_{ex} = 13K + 3P$

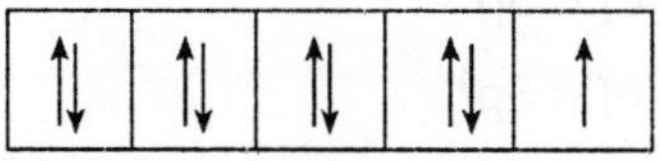

$d^9$

$E_{ex} \propto ({}^5C_2 + {}^4C_2)$ $\qquad E_{ex} = 16K + 4P$

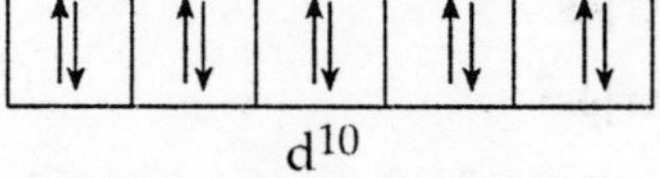

$d^{10}$

$E_{ex} \propto ({}^5C_2 + {}^5C_2)$ $\qquad E_{ex} = 20K + 5P$

As the exchange energy for $d^{10}$ configuration is greater than $d^9$ configuration, therefore $d^{10}$ configuration is more stable than $d^9$ configuration. This explain why the configuration of Cr and Cu are $4s^13d^5$ and $4s^13d^{10}$ instead of $4s^23d^4$ and $4s^23d^9$ respectively.

## 1.12 GROUND STATE ELECTRONIC CONFIGURATION OF ELEMENTS

By applying Pauli's exclusion principle, Hund's rule and aufbau principle the ground state electronic configuration of the elements are given in Table 1.5.

**Table 1.5 : Electronic configuration of the elements**

| Element, symbol (atomic number) | | Electronic configuration |
|---|---|---|
| Hydrogen | H(1) | $1s^1$ |
| Helium | He(2) | $1s^2$ |
| Lithium | Li(3) | $[He]2s^1$ |
| Beryllium | Be(4) | $[He]2s^2$ |
| Boron | B(5) | $[He]2s^22p^1$ |
| Carbon | C(6) | $[He]2s^22p^2$ |
| Nitrogen | N(7) | $[He]2s^22p^3$ |
| Oxygen | O(8) | $[He]2s^22p^4$ |
| Fluorine | F(9) | $[He]2s^22p^5$ |
| Neon | Ne(10) | $[He]2s^22p^6$ |
| Sodium | Na(11) | $[Ne]3s^1$ |
| Magnesium | Mg(12) | $[Ne]3s^2$ |
| Aluminium | Al(13) | $[Ne]3s^23p^1$ |
| Silicon | Si(14) | $[Ne]3s^23p^2$ |
| Phosphorus | P(15) | $[Ne]3s^23p^3$ |
| Sulphur | S(16) | $[Ne]3s^23p^4$ |
| Chlorine | Cl(17) | $[Ne]3s^23p^5$ |
| Argon | Ar(18) | $[Ne]3s^23p^6$ |
| Potassium | K(19) | $[Ar]4s^1$ |
| Calcium | Ca(20) | $[Ar]4s^2$ |
| Scandium | Sc(21) | $[Ar]3d^14s^2$ |
| Titanium | Ti(22) | $[Ar]3d^24s^2$ |
| Vanadium | V(23) | $[Ar]3d^34s^2$ |
| Chromium | Cr(24) | $[Ar]3d^54s^1$ |
| Manganese | Mn(25) | $[Ar]3d^54s^2$ |
| Iron | Fe(26) | $[Ar]3d^64s^2$ |
| Cobalt | Co(27) | $[Ar]3d^74s^2$ |

*Table contd. next page*

| Element, symbol (atomic number) | | Electronic configuration |
|---|---|---|
| Nickel | Ni(28) | $[Ar]3d^84s^2$ |
| Copper | Cu(29) | $[Ar]3d^{10}4s^1$ |
| Zinc | Zn(30) | $[Ar]3d^{10}4s^2$ |
| Gallium | Ga(31) | $[Ar]3d^{10}4s^24p^1$ |
| Germanium | Ge(32) | $[Ar]3d^{10}4s^24p^2$ |
| Arsenic | As(33) | $[Ar]3d^{10}4s^24p^3$ |
| Selenium | Se(34) | $[Ar]3d^{10}4s^24p^4$ |
| Bromine | Br(35) | $[Ar]3d^{10}4s^24p^5$ |
| Krypton | Kr(36) | $[Ar]3d^{10}4s^24p^6$ |
| Rubidium | Rb(37) | $[Kr]5s^1$ |
| Strontium | Sr(38) | $[Kr]5s^2$ |
| Yttrium | Y(39) | $[Kr]4d^15s^2$ |
| Zirconium | Zr(40) | $[Kr]4d^25s^2$ |
| Niobium | Nb(41) | $[Kr]4d^45s^1$ |
| Molybdenum | Mo(42) | $[Kr]4d^55s^1$ |
| Technetium | Tc(43) | $[Kr]4d^55s^2$ |
| Ruthenium | Ru(44) | $[Kr]4d^75s^1$ |
| Rhodium | Rh(45) | $[Kr]4d^85s^1$ |
| Palladium | Pd(46) | $[Kr]4d^{10}5s^0$ |
| Silver | Ag(47) | $[Kr]4d^{10}5s^1$ |
| Cadmium | Cd(48) | $[Kr]4d^{10}5s^2$ |
| Indium | In(49) | $[Kr]4d^{10}5s^25p^1$ |
| Tin | Sn(50) | $[Kr]4d^{10}5s^25p^2$ |
| Antimony | Sb(51) | $[Kr]4d^{10}5s^25p^3$ |
| Tellurium | Te(52) | $[Kr]4d^{10}5s^25p^4$ |
| Iodine | I(53) | $[Kr]4d^{10}5s^25p^5$ |
| Xenon | Xe(54) | $[Kr]4d^{10}5s^25p^6$ |
| Caesium | Cs(55) | $[Xe]6s^1$ |
| Barium | Ba(56) | $[Xe]6s^2$ |
| Lanthanum | La(57) | $[Xe]5d^16s^2$ |
| Cerium | Ce(58) | $[Xe]4f^26s^2$ |

*Table contd. next page*

| Element, symbol (atomic number) | | Electronic configuration |
|---|---|---|
| Praseodymium | Pr(59) | [Xe]$4f^3 6s^2$ |
| Neodymium | Nd(60) | [Xe]$4f^4 6s^2$ |
| Promethium | Pm(61) | [Xe]$4f^5 6s^2$ |
| Samarium | Sm(62) | [Xe]$4f^6 6s^2$ |
| Europium | Eu(63) | [Xe]$4f^7 6s^2$ |
| Gadolinium | Gd(64) | [Xe]$4f^7 5d^1 6s^2$ |
| Terbium | Tb(65) | [Xe]$4f^9 6s^2$ |
| Dysprosium | Dy(66) | [Xe]$4f^{10} 6s^2$ |
| Holmium | Ho(67) | [Xe]$4f^{11} 6s^2$ |
| Erbium | Er(68) | [Xe]$4f^{12} 6s^2$ |
| Thullium | Tm(69) | [Xe]$4f^{13} 6s^2$ |
| Yetterbium | Yb(70) | [Xe]$4f^{14} 6s^2$ |
| Lutetium | Lu(71) | [Xe]$4f^{14} 5d^1 6s^2$ |
| Hafnium | Hf(72) | [Xe]$4f^{14} 5d^2 6s^2$ |
| Tantalum | Ta(73) | [Xe]$4f^{14} 5d^3 6s^2$ |
| Tungsten | W(74) | [Xe]$4f^{14} 5d^4 6s^2$ |
| Rhenium | Re(75) | [Xe]$4f^{14} 5d^5 6s^2$ |
| Osmium | Os(76) | [Xe]$4f^{14} 5d^6 6s^2$ |
| Iridium | Ir(77) | [Xe]$4f^{14} 5d^7 6s^2$ |
| Platinum | Pt(78) | [Xe]$4f^{14} 5d^9 6s^1$ |
| Gold | Au(79) | [Xe]$4f^{14} 5d^{10} 6s^1$ |
| Mercury | Hg(80) | [Xe]$4f^{14} 5d^{10} 6s^2$ |
| Thallium | Tl(81) | [Xe]$4f^{14} 5d^{10} 6s^2 6p^1$ |
| Lead | Pb(82) | [Xe]$4f^{14} 5d^{10} 6s^2 6p^2$ |
| Bismuth | Bi(83) | [Xe]$4f^{14} 5d^{10} 6s^2 6p^3$ |
| Polonium | Po(84) | [Xe]$4f^{14} 5d^{10} 6s^2 6p^4$ |
| Astatine | At(85) | [Xe]$4f^{14} 5d^{10} 6s^2 6p^5$ |
| Radon | Rn(86) | [Xe]$4f^{14} 5d^{10} 6s^2 6p^6$ |
| Francium | Fr(87) | [Rn]$7s^1$ |
| Radium | Ra(88) | [Rn]$7s^2$ |
| Actinium | Ac(89) | [Rn]$6d^1 7s^2$ |
| Thorium | Th(90) | [Rn]$6d^2 7s^2$ |

*Table contd. next page*

| | | |
|---|---|---|
| Protactinium | Pa(91) | [Rn]$5f^26d^17s^2$ |
| Uranium | U(92) | [Rn]$5f^36d^17s^2$ |
| Neptunium | Np(93) | [Rn]$5f^57s^2$ |
| Plutonium | Pu(94) | [Rn]$5f^67s^2$ |
| Americium | Am(95) | [Rn]$5f^77s^2$ |
| Curium | Cm(96) | [Rn]$5f^76d^17s^2$ |
| Berkelium | Bk(97) | [Rn]$5f^86d^17s^2$ |
| Californium | Cf(98) | [Rn]$5f^{10}7s^2$ |
| Einsteinium | Es(99) | [Rn]$5f^{11}7s^2$ |
| Fermium | Fm(100) | [Rn]$5f^{12}7s^2$ |
| Mendelevium | Md(101) | [Rn]$5f^{13}7s^2$ |
| Nobelium | No(102) | [Rn]$5f^{14}7s^2$ |
| Lawrencium | Lr(103) | [Rn]$5f^{14}6d^17s^2$ |
| Ruthefordium | Rf(104) | [Rn]$5f^{14}6d^27s^2$ |
| Dubnium | Db(105) | [Rn]$5f^{14}6d^37s^2$ |
| Seaborgium | Sg(106) | [Rn]$5f^{14}6d^47s^2$ |
| Bohrium | Bh(107) | [Rn]$5f^{14}6d^57s^2$ |
| Hassium | Hs(108) | [Rn]$5f^{14}6d^67s^2$ |
| Meitnerium | Mt(109) | [Rn]$5f^{14}6d^77s^2$ |
| Dasmastadtium | Ds(110) | [Rn]$5f^{14}6d^87s^2$ |
| Roentgenium | Rg(111) | [Rn]$5f^{14}6d^97s^2$ |
| Copernicium | Cn(112) | [Rn]$5f^{14}6d^{10}7s^2$ |
| Trium | Uut(113) | [Rn]$5f^{14}6d^{10}7s^27p^1$ |
| Flerovium | Fl(114) | [Rn]$5f^{14}6d^{10}7s^27p^2$ |
| Pentium | Uup(115) | [Rn]$5f^{14}6d^{10}7s^27p^3$ |
| Livermorium | Lv(116) | [Rn]$5f^{14}6d^{10}7s^27p^4$ |
| Septium | Uus(117) | [Rn]$5f^{14}6d^{10}7s^27p^5$ |
| Octium | Uuo(118) | [Rn]$5f^{14}6d^{10}7s^27p^6$ |

**Problem 1.17** What would be the wavelength of emission spectrum when the electron jumps from the level with n=2 to ground state of $He^+$? **[B.U. 2002, 2008]**

($h = 6.6\times10^{-27}$ erg-sec., $m_e = 9.1\times10^{-28}$ g, 1 erg = $6.24\times10^{11}$ ev)

**Ans. :** $\frac{1}{\lambda} = Z^2R_H\left[\frac{1}{1^2} - \frac{1}{2^2}\right] = 4\times109700\ \text{cm}^{-1}\times\frac{3}{4}$

$\therefore \lambda = \frac{4}{4\times109700\times3}\text{cm} = 300\text{Å}$

**Problem 1.18** In an atom, the angular momentum of an electron is $\sqrt{6}\frac{h}{2\pi}$. What will be the minimum value of the principal quantum number of the electron? **[C.U. 2006]**

**Ans. :** Angular momentum is $\sqrt{l(l+1)}\ \frac{h}{2\pi}$, therefore $\sqrt{l(l+1)}\ \frac{h}{2\pi} = \sqrt{6}\frac{h}{2\pi}$ i.e. $l(l+1) = 6$, $l^2+l-6 = 0$, $l^2+3l-2l-6 = 0$.

So that, $(l+3)(l-2) = 0$ and so $l = 2$. Thus the minimum value of n = 3 for $l = 2$.

---

**Problem 1.19** An electron in a hydrogen atom in its ground state absorbs 1.5 times as much energy as the minimum required for its escape from the atom. What is the wavelength of the emitted electron? Given : 13.6 ev energy is needed for ionisation and m = $9.109\times10^{-28}$ g (m = mass of electron) **[C.U. 2006]**

**Ans. :** Energy absorbed ($\varepsilon$) = $13.6\times1.5 = 20.4$ ev

We know that $\varepsilon = h\nu = h\frac{c}{\lambda}$

$$\therefore \lambda = \frac{hc}{\varepsilon} = \frac{6.62\times10^{-34}\ \text{Js}\times3\times10^{8}\ \text{ms}^{-1}}{20.4\times1.6\times10^{-19}\ \text{J}} = 6.0845\times10^{-6}\ \text{m}\ [\because\ 1\ \text{ev} = 1.6\times10^{-19}\ \text{J}]$$

---

**Problem 1.20** "The intrinsic spin angular momentum of an electron is defined by two quantum numbers"—Explain the statement. **[C.U. 2006]**

**Ans. :** Spin angular momentum $\vec{S} = p_S = s\left(\frac{h}{2\pi}\right)$ in classical mechanics or $p_s = \sqrt{s(s+1)}\ \frac{h}{2\pi}$ in quantum mechanics. It actually measures the angular momentum of the spinning motion of electron. In a magnetic field it has got two quantisied orientations giving rise to spin quantum number half-integral values i.e. $+\frac{1}{2}$ or $-\frac{1}{2}$.

---

**Problem 1.21** The wave number of the first line in the Balmer series of hydrogen is x $cm^{-1}$, what is the corresponding wave number of $Be^{3+}$? **[B.U. 2001]**

**Ans. :** The equation for Balmer series of hydrogen atom is $\bar{\nu} = R\left[\frac{1}{2^2} - \frac{1}{n^2}\right]$ n > 2.

The value of $\bar{\nu}$ = x $cm^{-1}$. For $Be^{3+}$, $\bar{\nu} = R_{Be^{3+}}\left[\frac{1}{2^2} - \frac{1}{n^2}\right]$ n > 2

We know that $R_{Be^{3+}} = Z^2R_H$ (Sec. 1.6.5) where $R_H$ = Rydberg constant of hydrogen.

Therefore $\bar{\nu}$ for $Be^{3+}$ = 16x $cm^{-1}$.

---

**Problem 1.22** How many radial nodes each of the following orbitals possess? (a) 2s, (b) 4s, (c) 3p, (d) 5d, (e) 1s, (f) 4f.

**Ans. :** Radial nodes for ns orbitals = n – 1

Radial nodes for np orbitals = n – 2

Radial nodes for nd orbitals = n – 3

Radial nodes for nf orbitals = n – 4

Therefore (a) for 2s orbitals, radial nodes = 1, (b) = 3, (c) = 1, (d) = 2, (e) = 0, (f) = 2.

---

**Problem 1.23** Write the ground state electronic configuration of boron and give a set of quantum numbers that uniquely defines each electron.

**Ans. :** $1s^2 2s^2 2p^1$. Set of quantum numbers :

$n = 1 \; l = 0 \; m = 0 \; s = +\frac{1}{2}$

$n = 1 \; l = 0 \; m = 0 \; s = -\frac{1}{2}$

$n = 2 \; l = 0 \; m = 0 \; s = +\frac{1}{2}$

$n = 2 \; l = 0 \; m = 0 \; s = -\frac{1}{2}$

$n = 2 \; l = 1 \; m = 0, +1 \text{ or } -1 \; s = +\frac{1}{2} \text{ or } -\frac{1}{2}$

**Problem 1.24** Explain which one of the following $p^3$ electronic configuration is most stable. Show calculations. **[C.U. 2015]**

[↑↓ | ↑ |   ]   [↑ | ↑ | ↑]   [↑ | ↑ | ↓]

**Ans. :** Section 1.11. According to the principle of exchange energy

(a) the configuration [↑↓ | ↑ |   ] has $E_{ex} \propto ({}^2C_2 + {}^1C_2)$ i.e. $E_{ex} = K\frac{n(n-1)}{2}$

$= K\frac{2(2-1)}{2} + K\frac{1(1-1)}{2} = K,$

(b) the configuration [↑ | ↑ | ↑] has $E_{ex} \propto {}^3C_2$ i.e. $E_{ex} = K\frac{3(3-1)}{2} = 3K,$

(c) the configuration [↑ | ↑ | ↓] has $E_{ex} \propto ({}^2C_2 + {}^1C_2)$ i.e. $E_{ex} = K.$

Therefore from the above exchange energy calculations the distribution [↑ | ↑ | ↑] is most stable.

**Problem 1.25** What is lowest value of 'n that allows 'g' orbitals to exist. **[B.U. 2015]**

**Ans. :** n = 5.

**Problem 1.26** The energy required to carry out the following reaction $H_2 \rightarrow 2H$ is 104 $kcalmol^{-1}$. What will be the corresponding value of the wavelength of the photon required for this dissociation?

**Ans. :** For a mole of energy, E = Nhν, where N = Avogadro's number.

$E = Nh\nu = Nh\frac{c}{\lambda}$ so that,

$$\lambda = \frac{Nhc}{E} = \frac{(6.023\times10^{23}\,mol^{-1})(6.63\times10^{-34}\,Js)(3.0\times10^{8}\,ms^{-1})}{(104\;kcalmol^{-1})(4.184\times10^{3}\,J\;kcal^{-1})} = 275.3 \text{ nm.}$$

**Problem 1.27** An electron of mass $9.1\times10^{-31}$ kg moves in the first Bohr orbit with a speed of $2\times10^{6}$ $ms^{-1}$. If its momentum is measured with an accuracy of 1% what is the uncertainty of position?

**Ans. :** Momentum $p = mv = 9.1\times10^{-31}$ kg $\times$ 2 $\times$ $10^6$ $ms^{-1}$ = $18.2\times10^{-25}$ $kgms^{-1}$.

$\therefore$ Uncertainty in momentum $\Delta p = 18.2\times10^{-25}\times0.01 = 18\times2\times10^{-27}$ $kgms^{-1}$

We know that $\Delta x.\Delta p = h$, so that $\Delta x = \frac{h}{\Delta p} = \frac{6.63\times10^{-34}\,Js}{18.2\times10^{-27}\,kgms^{-1}} = 3.7\times10^{-8}$ m.

---

**Problem 1.28** A complete wave function must be 'normalised', which means that in the space available to the system, $\int\Psi^2 d\tau = 1$ (if $\Psi$ is real) or $\int\Psi^*\Psi d\tau = 1$ (if $\Psi$ is imiginary) where $d\tau$ is a small volume element. Given the function $\Psi(\theta) = \sin\theta$, where $\theta$ can change contineously from 0 to $2\pi$. Find the normalisation factor.

**Ans. :** $$\int_0^{2\pi}\Psi^2 d\theta = \int_0^{2\pi}\sin^2\theta d\theta = \int_0^{2\pi}\left(\frac{1-\cos 2\theta}{2}\right)d\theta$$

$$= \int_0^{2\pi}\frac{d\theta}{2} - \int_0^{2\pi}\frac{\cos 2\theta}{2}d\theta = \frac{2\pi}{2} - \left[\frac{\sin 2\theta}{4}\right]_0^{2\pi} = \pi$$

Since the integral is not equal to 1, the $\Psi$ is not normalised. Let N be the normalisation factor.

$$\therefore \int_0^{2\pi} N^2\sin^2\theta = 1 \text{ or } N^2\pi = 1, \text{ so that, } N = \frac{1}{\sqrt{\pi}}$$

---

**Problem 1.29** Electronic configuration for gadolinium (Gd) is $[Xe]4f^7 5d^1 6s^2$, whereas that of $Gd^{2+}$ is—
1. $[Xe]4f^6 5d6s^2$, 2. $[Xe]4f^6 6s^2$, 3. $[Xe]4f^6 5d^1 6s^1$, 4. $[Xe]4f^7 5d^1$. **[NET 2011]**

**Ans. :** 4 i.e. $[Xe]4f^7 5d^1$.

---

**Problem 1.30** The orbital angular momentum (in units of $\frac{h}{2\pi}$) of the electron in the 3d orbital is—
1. 2, 2. 3, 3. $2^{1/2}$, 4. $6^{1/2}$.

**Ans. :** 4. Angular momentum = $\sqrt{l(l+1)}\frac{h}{2\pi}$. For 3d orbital $l$ = 2, so $\sqrt{2(2+1)}\frac{h}{2\pi} = \sqrt{6}\frac{h}{2\pi}$.

---

**Problem 1.31** 5f orbital has— (a) one node, (b) two nodes, (c) three nodes, (d) four nodes.

**Ans. :** (a) i.e. one node.

---

**Problem 1.32** What is the wavelength of the electron in the nth Bohr orbit?
(a) $\lambda$ = (n $\times$ 0.529)Å, (b) $\lambda$ = (n $\times$ 2 $\times$ 3.146 $\times$ 0.529)Å,
(c) $\lambda$ = (n $\times$ 2 $\times$ 3.146 $\times$ 0.529)$^2$Å, (d) None of these.

**Ans. :** (b). We know $\lambda = 2\pi r$ where r = radius of the 1st Bohr orbit.

$\therefore$ $\lambda$= (n $\times$ 2 $\times$ 3.146 $\times$ 0.529)Å.

---

**Problem 1.33** The principal and azimuthal quantum numbers of electrons in 4f orbitals are—
(a) n = 4, $l$ = 2, (b) n = 4, $l$ = 4, (c) n = 4, $l$ = 3, (d) n = 3, $l$ = 4.

**Ans. :** (c) i.e. n = 4, $l$ = 3.

---

**Problem 1.34** The wave character of electron was experimentally verified by—
(a) de-Broglie, (b) Einstein, (c) Germer, (d) Schrodinger.

**Ans. :** (c) i.e. Germer.

**Problem 1.35** If the energy of an electron in the second energy level of hydrogen atom is –E, its value in the third energy level will be—

(a) $-\frac{2E}{4}$, (b) $-\frac{9E}{4}$, (c) $-\frac{4E}{9}$, (d) $-4E \times 9$.

**Ans. :** (c) i.e. $-\frac{4E}{9}$.

**Problem 1.36** If an electron travels with a velocity of $\frac{1}{100}$ th the speed of light(c) in the first Bohr orbit, its velocity in the 5th Bohr orbit is—

(a) 0.002c, (b) 0.1c, (c) 0.5c, (d) 0.7c.

**Ans. :** (a) i.e. 0.002c. The velocity ($v$) of the electron in the nth orbit = $\frac{2\pi e^2}{nh}$

In case of first Bohr orbit $v = \frac{2\pi e^2}{1h}$

$$\therefore \frac{c}{100} = \frac{2\pi e^2}{1h}$$

The velocity of the electron is the 5th orbit = $\frac{2\pi e^2}{5h} = \frac{1}{5} \times \frac{2\pi e^2}{h} = \frac{c}{100} \times \frac{1}{5} = 0.002c.$

**Problem 1.37** The radial part of an atomic orbital wave function depends on—

(a) n, (b) n and $l$, (c) n, $l$ and m, (d) $l$ and m.

**Ans. :** (b) i.e. depends on n and $l$.

**Problem 1.38** The orbital with two radial and two angular nodes is—

1. 3p, 2. 5d, 3. 5f, 4. 8d. **[NET 2013]**

**Ans. :** 2 i.e. 5d.

## EXERCISE

1. Using Bohr's theory, calculate the ionisation energy of the single election in the $Be^{3+}$ ion in electron volt and the radius of this ion in its ground state. **[B.U. 1996]**

   **Hint :** Use equation I.E. = $\frac{Z^2me^4}{8\varepsilon_0^2h^2}$ (Section 2.5.1) and $r = \frac{n^2h^2}{4\pi^2Ze^2m}$

   or in S.I. unit $r = \frac{n^2h^2\varepsilon_0}{\pi e^2mZ}$ (Section 1.6.1)

2. The potential energy of an electron in a hydrogen like atom is given by $V = \frac{1}{2}m\omega^2r^2$, where $\omega$ = constant and r = distance of electron from the origin. Assuming Bohr model of quantisation of angular momentum and circular orbit is valid, show that radius of nth orbit is proportional to $\sqrt{n}$. **[B.U. 2002]**

   **Hint :** $mvr = n\left(\frac{h}{2\pi}\right)$, $v = \omega r$, so that $m\omega r^2 = n\left(\frac{h}{2\pi}\right)$

   i.e. $r^2 = \frac{nh}{2\pi} \cdot \frac{1}{m\omega}$, so that $r \propto \sqrt{n}$ when others are constant.

3. Assign the principal quantum number that would be associated with the first 'g' subshell. **[B.U. 2005, 2013]**

   **Ans. :** n = 5, 'O'-shell.

4. "The magnitude of the charge on the electron is $4.8 \times 10^{-10}$ esu". What is the charge on the nucleus of a He atom? **[B.U. 2006]**

   **Ans. :** $2 \times 4.8 \times 10^{-10}$ esu positive charge.

5. The Aufbau configuration and the actual configuration of elemental Cu differ. Write each one and mention the reason for such difference in the actual configuration. **[B.U. 2006, 2016]**

   **Hint :** Sec. 1.10 and Sec. 1.11. Aufbau configuration [Ar]$3d^9 4s^2$, actual configuration is [Ar]$3d^{10}4s^1$.$3d^{10}$ configuration is more stable than $3d^9$ configuration. Fulfilled d orbital is more stable than partially filled one.

6. Which element is the first to have a completely filled 4f, 5d and 6s orbitals. Indicate the position of the element in the periodic table. **[B.U. 2006, 2008]**

   **Hint :** Hg(80) → [Xe]$4f^{14}5d^{10}6s^2$ period-6, group-12.

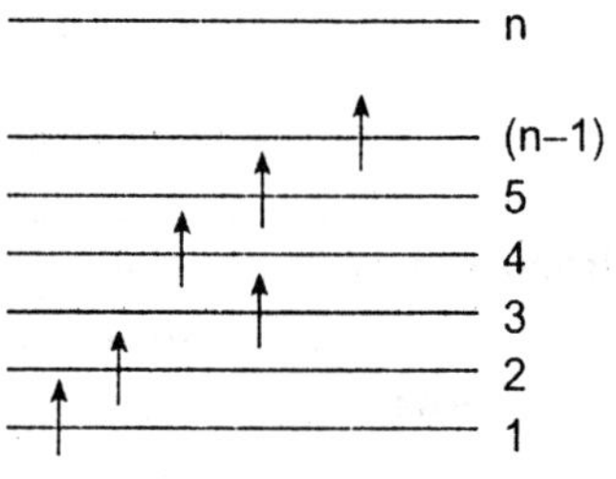


7. Find the total number of observable lines when the electron of hydrogen atom is excited gradually to a level of principal quantum number n. **[B.U. 2006]**

   **Hint :** (n–1) i.e. one less than n.

8. Chemical reactivities of elements depend on extranuclear electrons—explain.

9. Elemental Cr has four unpaired electrons—comment. **[B.U. 2014]**

   **Hint :** Cr(24) : $1s^2 2s^2 2p^6 3s^2 3p^6 4s^1 3d^5$ which indicates 6 unpaired electrons. For $1s^2 2s^2 2p^6 3s^2 3p^6 4s^2 3d^4$ configuration we can expect 4 unpaired electrons but this is no longer tenable as the stability of half-filled d orbital ($d^5$) is more compared to partially filled orbital ($d^4$).

10. How many nodal planes do each of the following orbitals have? How many nodal spheres? Does the orbital have any inner lobes? (a) 6s, (b) 9s, (c) 11d, (d) 5g, (e) 1s.

    **Hint :** (a) No nodal planes, 5 spheres, yes. (b) No nodal planes, 8 spheres, yes. (c) 2 nodal planes, 8 spheres, yes. (d) 4 nodal planes, no sphere, no. (e) No nodal plane, no sphere, no inner lobes.

11. How many maximum number of electrons may be accommodated in 'g' subshell?

    **Ans. :** 9 × 2 = 18.

12. Which of the following electronic transition will exhibit highest energy? $n_3 \rightarrow n_2$, $n_4 \rightarrow n_3$, $n_5 \rightarrow n_4$ and $n_6 \rightarrow n_5$.

    **Hint :** According to Bohr's theory as the principal quantum number (n) value increases the electrons will go to the higher energy state and the distance of the revolving electron increases as also the higher energy states are narrowed down. So that $E_2 - E_1 > E_3 - E_2 > E_4 - E_3 > E_5 - E_4$ ...... etc. Therefore $n_3 \rightarrow n_2$ will record highest energy.

13. In hydrogen spectra the shortest wavelength of the Lymann series is x, then what will be the longest wavelength in Balmer series for $He^+$?

    **Hint :** $\overline{v} = \frac{1}{\lambda} = R^2 Z^2 \left[\frac{1}{n_1^2} - \frac{1}{n_2^2}\right]$ $n_2 > n_1$. Shortest wave length means highest energy.

For Lymann series in case of hydrogen atom, $Z = 1$, $n_1 = 1$, $n_2 = \infty$ and $\lambda = x$ (given) so that $\frac{1}{x} = R$.

Longest wavelength denotes minimum energy. In case of $He^+$ for Balmer series, $Z = 2$, $n_1 = 2$ and $n_2 = 3$.

$$\therefore \frac{1}{\lambda} = R \times 2^2 \left[\frac{1}{2^2} - \frac{1}{3^2}\right] = \frac{2^2}{x} \times \frac{5}{36} = \frac{5}{9x}.$$

14. What will be the total number of orbitals in n = 5?

**Hint :** For n = 5. We get 5s, 5p, 5d, 5f and 5g, so total orbitals is 25.

15. Calculate the wave number ($\overline{\nu}$) and energy of the yellow light having wavelength 580 nm ($c = 3 \times 10^8\ ms^{-1}$ and $h = 6.627 \times 10^{-34}$ Js)

**Hint :** $\overline{\nu} = \frac{1}{\lambda} = \frac{1}{580 \times 10^{-9}}\ m^{-1} = 1.724 \times 10^6\ m^{-1}$ [1 nm = $10^{-9}$ m]

$$E = h\nu = h\frac{c}{\lambda} = \frac{6.627 \times 10^{-34} \times 3 \times 10^8}{580 \times 10^{-9}} = 3.428 \times 10^{-19}\ J.$$

16. Offer an explanation for the fact that the spectrum of atomic hydrogen consists of a series of lines even though hydrogen has only one electron. **[B.U. 1991]**

**Hint :** During the study of the emission spectra of hydrogen there are several gaseous atoms of hydrogen exist. Many of these hydrogen atoms will absorb different amounts of energy from the external source and the ground state electron of the hydrogen atoms are excited to different energy states. Now the excited electron of different hydrogen atoms will be revert back to the different nonexcited lower energy states so that the emitted energy will produce different lines. For example when electron revert from n = 2, 3, 4, 5 etc. to n = 1 will produce Lyman series. In this way other series of lines are obtained as depicted in Fig. 1.9.

17. Explain what are meant by azimuthal and principal quantum numbers in the Sommerfeld theory of the hydrogen atom. What is the relation between the principal and azimuthal quantum members?

18. How many orbitals are there in a shell with quantum number 'n'?

19. What is precessional motion of electron according to Sommerfeld theory?

**Hint :** According to Sommerfeld theory the mass and velocity of the electron is always changing during revolution in the elliptical orbit i.e. there is always a relative increase or decrease of mass of the electron. As a result of this change in mass which is greatest at perihelion and greatest for the most, elliptical orbits, there is an advance of the perihelion or a precession of the electron orbit plane and generated a shape given in Fig. 1.12. This displacement of electron orbit plane is known as precessional motion of electron.

20. What is the exchange energy? From the concept of exchange pair of electrons how ground state electronic configuration of chromium can be determined? **[B.U. 2004]**

**Ans. :** Let us consider the arrangements of two electrons in $p^2$ configuration. The p orbital has three orbitals and the two electrons may be arranged in fifteen ways of which three are depicted below :

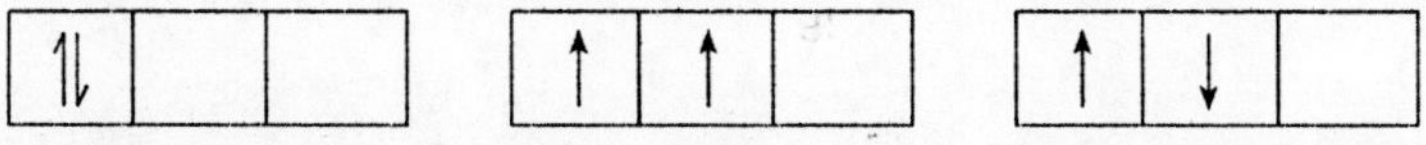

$p_x$ $p_y$ $p_z$ (a) $p_x$ $p_y$ $p_z$ (b) $p_x$ $p_y$ $p_z$ (c)

Arrangement (a) is discarded from the standpoint of Hund's rule of maximum multiplicity. Among (b) and (c) which one is more stable can be decided from **exchange energy.** Quantum mechanical calculations and coulombic energy denotes that when two electrons are in parallel spin (either ↑ or ↓) exchange their positions in space and thereby there will be no change in the electronic arrangement of the system and it would lead to decrease in energy. Such a pair of electrons is called **exchange pair** and as the number of exchange pair of electrons increases, there will be greater decrease in energy. **The energy decrease per exchange pair of electrons is called exchange energy.** The exchange energy has a negative sign since it stabilises the system and so sometimes called **stabilisation energy.** Arrangement (b) has one exchange pair but arrangement (c) has no exchange pair and so arrangement (b) is more stable than that of (c).

The valence-shell of chromium has six electrons which may be distributed in 4s and 3d orbitals in several ways :

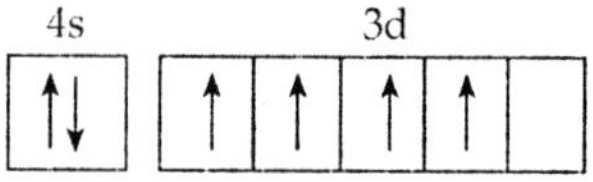

(I)

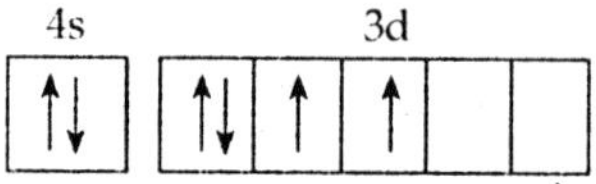

(II)

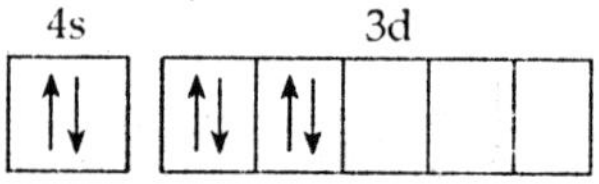

(III)

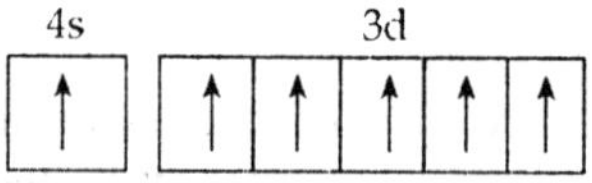

(IV)

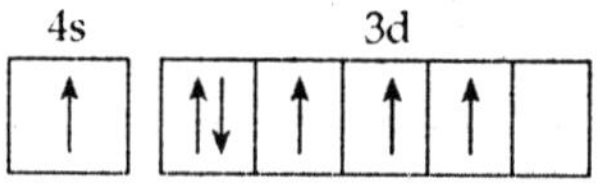

(V)

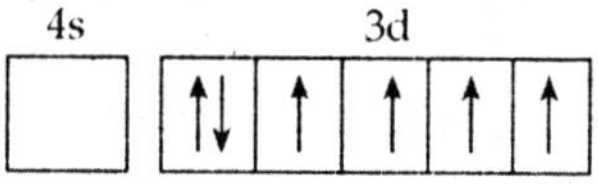

(VI)

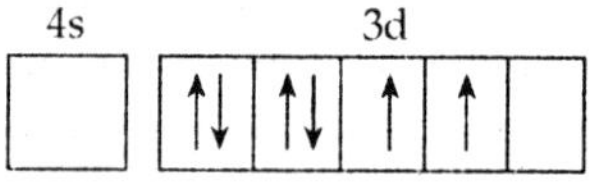

(VII)

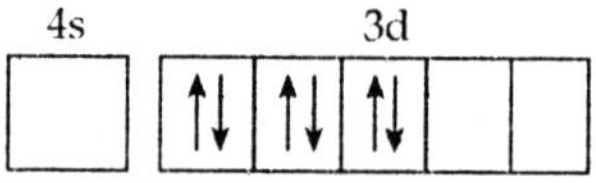

(VIII)

Among these arrangements (I) and (IV) are probable and others are improbable due to Hund's rule and aufbau principle.

The number of exchange pairs of electrons in arrangement (I) = $^5C_2 + {}^1C_2 = 10 + 0$ (Since there are 5 electrons in same and one electron in opposite spin). Hence the exchange energy associated with this arrangement = –10E where E is the energy per exchange pair and as one electron is paired in this arrangement so the total energy $E_I = -10E + P$ where P = pairing energy.

The number of exchange pairs for arrangement (IV) = $^6C_2 = 15$ and hence the exchange energy = –15E. Since the arrangement (IV) can be achieved by promotion of one electron from 3d to 4s obital and so the promotional energy ΔE will be associated in this configuration and the energy associated with this arrangement will be $E_{IV} = -15E + \Delta E$.

Now since the difference in energy for 4s and 3d orbital is small, so we can take $\Delta E \approx P$ and hence arrangement (IV) having greater amount of exchange energy (–15E) is more stable than that of arrangement (I) having less amount of exchange energy (–10E). Hence the ground state electron configuration of chromium is $4s^1 3d^5$ and not $4s^2 3d^4$.

21. The potential energy of interaction between two point charges separated by distance r is proportional to 1. $\frac{1}{r^2}$ 2. $\frac{1}{r}$ 3. $\frac{1}{r^3}$ 4. $\frac{1}{r^4}$. **[NET 2000]**

**Ans. :** 2.

22. A radial wave function R(r) of the hydrogen atom is plotted against r below : [NET 2001]

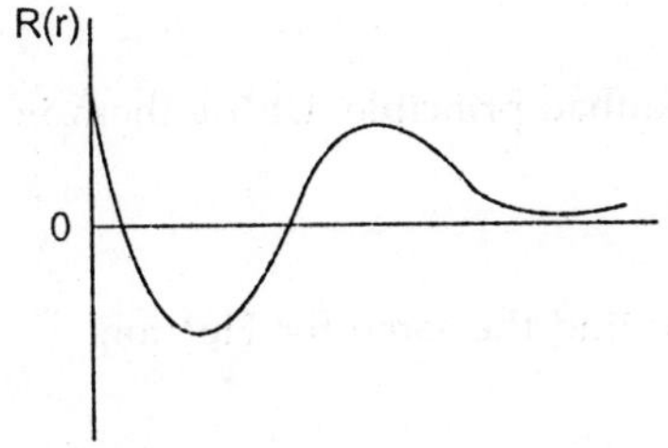


This corresponds to the orbital : 1. 3d 2. 3p 3. 3s 4. 2s.

**Ans. :** 3.

23. An atomic orbital having the quantum numbers n = 4 and $l$ = 1 will have the shape of—
1. sphere 2. dumb-bell 3. cone 4. ellipsoid. [NET 2001]

**Ans. :** 2.

24. Show that Bohr's postulate of quantised angular momentum for an electron moving in a circular orbit can be derived by the application of de-Broglie's hypothesis. [C.U. 1997, 2007]

**Hint :** Sec. 1.8.1 $m\upsilon r = n\left(\frac{h}{2\pi}\right)$.

25. What optical transition in $He^+$ spectrum would have the same wavelength as the first Lyman transition of hydrogen. (Neglect the effect of reduced mass.) [C.U. 1997, 2013, B.U. 2014]

**Hint :** For $He^+$, the transition $\frac{1}{m^2} - \frac{1}{n^2} = \frac{1}{2^2} - \frac{1}{4^2}$.

26. Show that the de Broglie wave lengths λ, for the electron wave in the nth Bohr orbit of hydrogen atom can be represented as $\lambda = 2\pi n a_0$ where $a_0$ is the radius of first Bohr orbit in hydrogen atom. [C.U. 2007, 2005]

**Hint :** Sec. 1.8.1. $2\pi r = n\lambda$ $r = n^2 a_0$

$\therefore 2\pi n^2 a_0 = n\lambda$.

27. The wave number of a line in the Lyman series of hydrogen is equal to the wave number of a line in the Balmer series of $He^+$. Assign the transition involved in each case. The small difference between $R_H$ and $R_{He^+}$ may be ignored. [C.U. 2007]

28. Explain 'Radial Function' and 'Radial Distribution Function' of an orbital. Compare the radial distribution plots for 2s and 2p orbitals and hence comment on their relative penetrating power. [C.U. 2007]

**Hint :** Sec. 1.8.7.1.

29. Assuming Bohr's model is applicable, deduce an expression for radius of nth orbit for the movement of a particle having mass 208 times that of electron and charge equal to that of electron (e), around a nucleus of infinite mass and charge +3e (particle moves in circular orbit). [B.U. 2003, 2013]

**Ans. :** According to Bohr's theory, the radius expression for nth orbit is :

$$r = \frac{n^2h^2}{4\pi^2 Ze^2 m'}$$

According to problem the mass (m') = 208 m, where m is the mass of electron and Z = +3e.

$$\therefore\ r = \frac{n^2h^2}{4\pi^2 3ee^2 208m} = \frac{n^2h^2}{2496\pi^2 e^3 m}.$$

30. State Pauli exclusion principle, Hund's rule and Aufbau principle. Utilise them in predicting ground state electron distribution of Si, P and S. **[B.U. 2009, 2012]**

**Hint :** Sec. 1.9.1, 1.9.3, 1.10 and 1.12.

31. The first Bohr radius of hydrogen atom is 0.529Å; find the same for $He^+$ ion. **[B.U. 2009]**

**Hint :** $r_{He^+} = \frac{0.529}{2}$ Å

32. Identify the possible Bohr-Sommerfeld orbits for n = 4. **[B.U. 2009]**

**Hint :** n = 4, then k = 1, 2, 3, 4

$a = 16a_0$, $b = 4a_0$; $a = 16a_0$, $b = 8a_0$; $a = 16a_0$, $b = 12a_0$ and $a = 16a_0$, $b = 16a_0$

33. Point out the difference in the shape and size of 2s and 2p orbitals. **[B.U. 2010, 2015]**

**Hint :** Sec. 1.8.7.2. Size of 2s and 2p are same but shapes are different. 2s orbital is spherical having eccentricity while 2p orbitals have lobes along the three different axes i.e. x, y and z (Fig. 1.18).

34. State the hypothesis of de-Broglie. Calculate the de-Broglie wavelength of the following :

(i) A rifle bullet ($m = 2\times10^{-3}$ kg) moving with a speed of 300 $ms^{-1}$

(ii) An electron moving with a speed nearly $^1/_{10}$th of that of light. **[C.U. 2009]**

**Hint :** (i) $m = 2\times10^{-3}$ kg, $v = 300\ ms^{-1}$

$mv = 2\times10^{-3}\ kg \times 300\ ms^{-1} = 600\times10^{-3}\ kg\ ms^{-1}$

$h = 6.62\times10^{-34}$ Js or kg $m^2s^{-1}$ [$\because J = kgm^2s^{-2}$]

$$\lambda = \frac{h}{mv} = \frac{6.62\times10^{-34}\,kgm^2s^{-1}}{600\times10^{-3}\,kgms^{-1}} = 0.0110333\times10^{-31}\ m.$$

(ii) See problem 1.11.

35. State Heisenberg uncertainty principle. Calculate theoretical uncertainty in its position within 1 m/s for an electron moving at 100 m/s. **[C.U. 2009]**

**Hint :** $\Delta x.\Delta p = \frac{h}{4\pi}$ or $\Delta x.m\Delta v = \frac{h}{4\pi}$ ($\because p = mv$)

Here the velocity has an uncertainty $2ms^{-1}$ i.e. from 99 to 101 $ms^{-1}$.

$$\Delta x = \frac{h}{4\pi}\cdot\frac{1}{\Delta p} = \frac{h}{4\pi}\cdot\frac{1}{m\Delta v} = \frac{6.62\times10^{-34}\,kgm^2s^{-1}}{4\times3.14\times9.11\times10^{-31}\,kg\times2ms^{-1}} = 3\times10^{-5}\ m.$$

36. Calculate the uncertainty in position of an electron whose velocity is $3.0\times10^4$ cm $s^{-1}$ accurate upto 0.001%. Mass of an electron = $9.1\times10^{-28}$ g.

**Hint :** $\Delta v = \frac{0.001}{100}\times3.0\times10^4$ cm $s^{-1}$ = 0.3 cm $s^{-1}$

$$\Delta x = \frac{h}{4\pi}\cdot\frac{1}{\Delta p} = \frac{h}{4\pi m\Delta v} = \frac{6.62\times10^{-27}\,erg}{4\times3.14\times9.10\times10^{-28}\,g\times0.3cms^{-1}} = 1.92\ cm.$$

37. Compare the nature of radial probability function of 2s, 2p and 3s orbitals of hydrogen atom. Why the probability function is always positive? **[C.U. 2009]**

**Hint :** Sec. 1.8.7.1 and Fig. 1.17.

38. Justify that ground state configuration of chromium is $3d^5 4s^1$, not $3d^4 4s^2$. [C.U. 2009]

**Hint :** Sec. 1.11 and exercise problem 20.

39. 'The principal quantum number, the azimuthal quantum number and the magnetic quantum number respectively signify size, shape and orientation of orbitals in an atom'. Comment. [B.U. 2013, 2006, 2002]

**Hint :** Sec. 1.9.

40. State the Heisenberg uncertainty principle. Suppose an electron is confined within the nucleus of diameter $10^{-14}$ metre. Find the uncertainty in determination of its velocity. Hence, show that an electron can never reside inside the nucleus. [C.U. 2008]

**Hint :** Sec. 1.8.2. $\Delta x.\Delta p = \frac{h}{4\pi}$

The electron is confined within the nucleus of diameter $10^{-14}$ m, so that its uncertainty in position cannot exceed the limit $10^{-14}$ m i.e. $\Delta x = 10^{-14}$ m.

Now, $\Delta x.\Delta p = \frac{h}{4\pi}$ i.e. $\Delta x.m\Delta v = \frac{h}{4\pi}$

$$\text{So that } \Delta v = \frac{h}{4\pi \times \Delta x \times m} = \frac{6.62\times10^{-34}\,\text{kgm}^2\text{s}^{-1}}{4\times3.14\times10^{-14}\,\text{m}\times9.11\times10^{-31}\,\text{kg}}$$

$$= 0.0578 \times 10^{11} \text{ m s}^{-1}$$

$$= 5.78 \times 10^{9} \text{ m s}^{-1}$$

This huge uncertainty in the velocity of the electron suggests that it cannot exist in the nucleus.

$$\text{Moreover the uncertainty in momentum } \Delta p = \frac{h}{4\pi\Delta x} = \frac{6.62\times10^{-34}\,\text{kgm}^2\text{s}^{-1}}{4\times3.14\times10^{-14}\,\text{m}}$$

$$= 0.52 \times 10^{-20} \text{ kg m s}^{-1}$$

Then the relativistic kinetic energy of the electron

$$= 0.52 \times 10^{-20} \text{ kg m s}^{-1} \times 3 \times 10^{8} \text{ m s}^{-1}$$

$$= 1.56 \times 10^{-12} \text{ J} = 9.75 \text{ Mev } (1 \text{ ev} = 1.6\times10^{-19} \text{ J})$$

Thus if electron will be a nuclear particle then its energy will be more than 9.75 Mev but experimentally this huge amount of energy can never be associated with the electron. So it can be concluded that an electron can never exist in the nucleus.

41. If a $1\times10^{-3}$ kg body is travelling along the X-axis at 1 metre/sec. within 0.01 metre/sec. Calculate the theoretical uncertainty in position.

**Hint :** The velocity has an uncertainty of 0.02 metre/sec. i.e. from 0.99 to 1.01 metre/sec.

$$\therefore \Delta x = \frac{h}{4\pi\Delta p} = \frac{h}{4\pi\Delta(mv)} = \frac{6.62\times10^{-34}\,\text{kgm}^2\text{s}^{-1}}{4\times3.14\times1\times10^{-3}\,\text{kg}\times0.02\text{ ms}^{-1}} = 2.6 \times 10^{-30} \text{ metre}$$

42. What will be the largest and the smallest wavelengths in the Paschen series of hydrogen atom?

**Hint :** For smallest wavelength : transition $n = \infty$ to $n = 3$

For largest wavelength : transition $n = 4$ to $n = 3$.

43. The energy of an electron in an excited hydrogen atom is –3.4 ev. Calculate the angular momentum of electron according to Bohr's theory.

Given : Mass of an electron = $9.1 \times 10^{-31}$ kg
Rydberg's constant = $1.09737 \times 10^{7}$ m$^{-1}$
Planck's constant = $6.626176 \times 10^{-34}$ Js
Speed of light = $3 \times 10^{8}$ m s$^{-1}$ **[C.U. 2008]**

**Hint :** According to quation (1.14)

$$E = -\frac{me^4Z^2}{8\varepsilon_0^2 n^2 h^2}$$

$m = 9.1 \times 10^{-31}$ kg,

$e = 1.602 \times 10^{-19}$ C,

$\varepsilon_0 = 8.85 \times 10^{-12}$ J$^{-1}$ C$^2$ m$^{-1}$,

$h = 6.626176 \times 10^{-34}$ Js

$$E = -\frac{9.10\times10^{-31}\,\text{kg}\times\left(1.602\times10^{-19}\,\text{C}\right)^4}{8\times\left(8.85\times10^{-12}\,\text{J}^{-1}\text{C}^2\text{m}^{-1}\right)^2\times\left(6.626176\times10^{-34}\,\text{Js}\right)^2}\cdot\frac{Z^2}{n^2}$$

$$= -\,2.18\times10^{-18}\,\frac{Z^2}{n^2}\cdot\frac{\text{kgC}^4}{\left(\text{J}^{-1}\text{C}^2\text{m}^{-1}\right)^2(\text{Js})^2}$$

$$= -\,2.18\times10^{-18}\,\frac{Z^2}{n^2}\cdot\frac{\text{kgC}^4}{\text{J}^{-2}\text{C}^4\text{m}^{-2}\text{J}^2\text{s}^2}$$

$$= -\,2.18\times10^{-18}\,\frac{Z^2}{n^2}\,\text{kgm}^2\text{s}^{-2}$$

Now according to condition, $E = -3.4$ ev $= -3.4 \times 1.6 \times 10^{-19}$ J [$\because$ 1 ev $= 1.6 \times 10^{-19}$ J]

$= -3.4 \times 1.6 \times 10^{-19}$ kgm$^2$s$^{-2}$ [$\because$ J = kg m$^2$ s$^{-2}$]

$$\therefore -\,2.18\times10^{-18}\,\frac{Z^2}{n^2}\,\text{kgm}^2\text{s}^{-2} = -3.4 \times 1.6 \times 10^{-19}\ \text{kg m}^2\,\text{s}^{-2}$$

For hydrogen atom $Z = 1$, $n^2 = 4$ so that $n = 2$

$$\text{Therefore angular momentum} = n\left(\frac{h}{2\pi}\right) = \frac{2\times6.626176\times10^{-34}\,\text{Js}}{2\times3.14}$$

$= 2.11 \times 10^{-34}$ kg m$^2$ s$^{-1}$

44. Which of the following arrangement would be more stable and why?

| ↑ | ↑ | ↑ |
|---|---|---|
| $p_x$ | $p_y$ | $p_z$ |

| ↑ | ↑ | ↓ |
|---|---|---|
| $p_x$ | $p_y$ | $p_z$ |

| ↑↓ | ↑ | |
|---|---|---|
| $p_x$ | $p_y$ | $p_z$ |

**[C.U. 2007, B.U. 2012]**

**Hint :**

| ↑ | ↑ | ↑ |
|---|---|---|

45. For the hydrogen spectrum show that Lyman series occurs between 912Å and 1216Å and Balmer series occurs between 3647Å and 6564Å ($R = 1.0968 \times 10^{7}$ m$^{-1}$). **[C.U. 2007]**

**Hint :** For Lyman series : Longest wavelength : $\bar{\nu} = \frac{1}{\lambda} = R\left[\frac{1}{1^2} - \frac{1}{2^2}\right]$

i.e. $\lambda = \frac{4}{3R} = \frac{4}{3 \times 1.0968 \times 10^7 \text{ m}^{-1}} = 1216\text{Å}$

and shortest wavelength : $\bar{\nu} = R\left[\frac{1}{1^2} - \frac{1}{\infty^2}\right]$

so that $\lambda = \frac{1}{R} = \frac{1}{1.0968 \times 10^7 \text{ m}^{-1}} = 912\text{Å}$

For Balmer Series : Longest wavelength : $\bar{\nu} = \frac{1}{\lambda} = R\left[\frac{1}{2^2} - \frac{1}{3^2}\right] = \frac{5}{36}R$

so that $\lambda = \frac{36}{5R} = \frac{36}{5 \times 1.0968 \times 10^7 \text{ m}^{-1}} = 6564.5\text{Å}$

shortest wavelength : $\frac{1}{\lambda} = R\left[\frac{1}{2^2} - \frac{1}{\infty^2}\right]$

i.e. $\lambda = \frac{4}{R} = \frac{4}{1.0968 \times 10^7 \text{m}^{-1}} = 3647\text{Å}$

46. For a particular transition in Bohr's model say from n = 3 to n = 2 only one possibility is permitted while in Sommerfeld's model there might be different possible ways for a particular transiton— Comment.

**Hint :** Sec. 1.7 and 1.7.1. In Bohr model for a particular quantum number there is a single energy state but in Sommerfeld model there is a closely spaced levels for a particular quantum number.

47. Starting from Bohr's equation for the energy of an electron in the hydrogen atom derive Rydberg's equation for the frequency of lines in the hydrogen spectrum. Hence calculate the frequency of radiation emitted when an electron drops from the third to the first Bohr orbit. [Rydberg constant = 109677 $\text{cm}^{-1}$] **[B.U. 1992]**

**Hint :** $\bar{\nu} = \frac{\nu}{c} = R\left[\frac{1}{1^2} - \frac{1}{3^2}\right]$ so that $\nu = cR\left[\frac{1}{1^2} - \frac{1}{3^2}\right]$

$= 3 \times 10^{10} \text{ cm/sec.} \times 109677 \text{ cm}^{-1} \times \frac{8}{9}$

$= 292472 \times 10^{10}$ cycles/sec.

48. What are the limitations of Bohr's theory? **[B.U. 2003, 2008, 2013, 2015]**

**Hint :** Sec. 1.6.8.

49. State the Pauli's antisymmetric principle. **[B.U. 2006, 2015]**

**Hint :** Sec. 1.9.2.

50. What is the maximum number of electrons that can be accommodated in atom in which highest principal quantum number value is 4? **[B.U. 2001]**

**Hint :** 60 electrons.

51. What is the significance of the term "orbital magnetic quantum number"? Draw the vector orientation of the $m_l$ values corresponding to $l$ = 2 in magnetic field. Terms are all of their usual significance. **[B.U. 2007]**

**Hint :** Sec. 1.9 and Fig. 1.21.

52. Write down the relation of Rydberg constants of hydrogen atom and monocationic helium ion. **[B.U. 2008, 2012, 2014, 2016]**

**Hint :** $R_{He^+} = 4R_H$.

53. Give the qualitative pictures of the shapes of the d-orbitals on x, y, z axes, showing proper signs of lobes. **[B.U. 2008, 2012, 2014]**

**Hint :** Sec. 1.8.7.2 and Fig. 1.18.

54. 'Though the (n+*l*) rule to determine the order of energy of different subshells is useful in most cases, there are some exception'—Justify the statement with an example. **[C.U. 2012]**

**Hint :** Sec. 1.10.

55. 'The orbital angular momentum of an electron does not give its total angular momentum'—Explain. Which quantum number is used to supplement the orbital angular momentum? **[C.U. 2012]**

**Hint :** Sec. 1.9. Orbital angular momentum = $\sqrt{l(l+1)}\frac{h}{2\pi}$ and spin angular momentum = $\sqrt{s(s+1)}\frac{h}{2\pi}$.

56. An electron circles a nucleus of charge Ze. Of the two orbits 1 and 2 of radii $r_1$ and $r_2$ respectively, its total energy is greater while in orbit 1. Prove that $r_1 > r_2$. Also show that the velocity and acceleration in orbit 2 are greater than those in orbit 1. **[C.U. 2012]**

**Hint :**

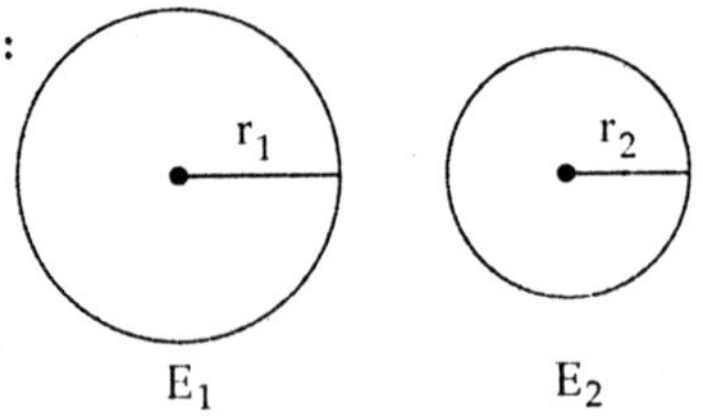


From equations (1.6) and (1.10) we can prove $\frac{E_1}{E_2} = \frac{r_1}{r_2}$.

As $E_1 > E_2$ so $r_1 > r_2$.

From Sec. 1.6.2, $v = \frac{2\pi ze^2}{nh}$, $v_2 > v_1$ can be proved and acceleration (f) = $\frac{v^2}{r}$ which suggests $f_2 > f_1$.

57. An electron in a hydrogen atom in its ground state absorbs 1.6 times as much energy as the minimum required for it to escape from the atom. Calculate the the wavelength of the emitted electron. [Ionisation energy = 13.6 ev and m = $9.11 \times 10^{-28}$ gm] **[C.U. 2013]**

**Hint :** Problem 1.19.

58. How do the shapes of s and p orbitals can be obtained from angular functions? Give reasons. **[C.U. 2013]**

**Hint :** Sec. 1.8.7.2.

59. State Pauli's exclusion principle. Apply this to predict the maximum capacity of 3rd quantum shell for accommodating electrons. **[C.U. 2011]**

**Hint :** Sec. 1.9.1. Maximum 18 electrons will be accommodated.

60. Calculate the shortest wavelength in the absorption spectrum of deuterium. (Rydberg constant = 109,737 $cm^{-1}$) **[C.U. 2011]**

**Hint :** According to equation (1.16)

$$\bar{v} = \frac{1}{\lambda} = R\left[\frac{1}{n_1^2} - \frac{1}{n_2^2}\right]$$

Shortest wavelength means $n_1 = 1$ and $n_2 = \infty$

$$\therefore \quad \bar{\nu} = \frac{1}{\lambda} = R\left[\frac{1}{1^2} - \frac{1}{\infty^2}\right] = R[1 - 0] = R$$

$$\therefore \lambda = \frac{1}{R} = \frac{1}{109737} \text{ cm} = 9.11 \times 10^{-6} \text{ cm.}$$

61. Discuss the origin and physical significance of magnetic quantum number. **[C.U. 2013, 2015]**

**Hint :** Sec. 1.9.

62. Out of the following configurations, which will be more stable one? (i) $(n-1)d^4ns^2$, (ii) $(n-1)d^5ns^1$— Justify from viewpoint of energy. **[C.U. 2011]**

**Hint :** Sec. 1.11 and question number 20 and 38.

63. Show the radial probability distribution function diagram of the orbitals of 3s, 3p, 3d in a hydrogen atom. Interpret the above diagrams. **[C.U. 2011]**

**Hint :** Sec. 1.8.7.1 and Fig. 1.17 and also question number 37.

64. How do the shapes of s and p orbitals can be obtained from angular functions? Give reasons. **[C.U. 2013]**

**Hint :** Sec. 1.8.7.2 and question number 33.

65. Draw the curves indicated below for the 2s orbital of hydrogen like atoms : (i) R vs r, (ii) $R^2$ vs r and (iii) $4\pi r^2R^2$ vs r. **[C.U. 2012]**

**Hint :** Sec. 1.8.7.3.

66. Show diagramatically the variation of radial distribution function with 'r' for the orbitals 3s and 3p in hydrogenic systems. Give interpretations. **[C.U. 2014]**

**Hint :** Sec. 1.8.7.1 and Sec. 1.8.7.3.

67. How do the four quantum numbers signify the state of an electron in H atom? **[C.U. 2014]**

**Hint :** Sec. 1.9 and Sec. 1.9.1.

68. Explain the differences between the functions R(r), $R(r)^2$ and $4\pi r^2R(r)^2$ for 1s electron of H atom. Represent graphically by plotting against r. **[C.U. 2015]**

**Hint :** Sec. 1.8.7.1 and Sec. 1.8.7.3.

69. Comment on differences between plots of R(r) against r and $4\pi r^2R(r)^2$ against R for each of the following atomic orbitals of an H atom : (a) 1s, (b) 4s, (c) 3p.

**Hint :** Sec. 1.8.7.1 and Sec. 1.8.7.3.

70. Explain how Aufbau principle is violated in the process of electron filling of lanthanum. **[C.U. 2014]**

**Hint :** The ground state electronic configuration of lanthanides are determined only with difficulty because of the complexity of their atomic spectra, but it is generally agreed that they are generally of the type $[Xe]4f^n5d^06s^2$. Violation of Aufbau principle or exception of the general pattern of the electronic configuration is observed in the following cases :

(a) For Ce the increase in effective nuclear charge after La is not sufficient to stabilise the $4f^25d^0$ electron configuration compared to $4f^15d^1$. This is due to the sudden contraction and reduction in energy of the 4f orbitals immediately after La is not yet sufficient to avoid occupancy of the 5d orbital.

(b) Gd has the configuration $4f^75d^1$ which is due to the stability of the half-filled 4f orbital.

(c) Lu has $4f^{14}5d^1$ configuration where the last electron is added beyond the capacity of the 4f orbital i.e. 4f orbital is filled in Yb(70).

71. State Sommerfeld's modifications on Bohr's concept of hydrogen atom and comment on the new energy states. State the demerits of this theory. **[B.U. 2014, 2016]**

**Hint :** Sec. 1.7, Sec. 1.7.1, Sec. 1.7.2.

72. Write down the Schrodinger wave equation indicating the meaning of each term involved. **[B.U. 2015]**

**Hint :** Sec. 1.8.3.

73. State Pauli exclusion principle, Hund's rule and Aufbau principle. Utilise these in predicting ground state electron distribution of C, N and O. **[B.U. 2015]**

**Hint :** Question number 30 (Exercise).

74. State the origin and physical significance of the magnetic quantum number with pictorial diagram. What will be the values of l and $m_l$ for $p_z$ and $d_{z^2}$ orbitals? **[B.U. 2015]**

**Hint :** Sec. 1.9

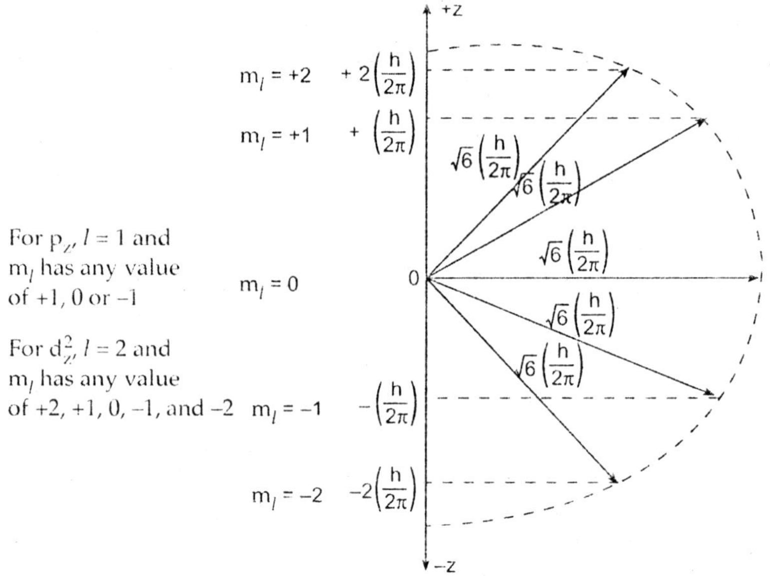


75. State Heisenberg's uncertainty principle in words and in mathematical form. Point out its significance. **[B.U. 2015]**

**Hint :** Sec. 1.8.2

76. Show that the de-Broglie hypothesis applied to an electron moving in a circular orbit leads to Bhor's postulate of quantisation of angular momentum. **[B.U. 2015]**

**Hint :** Sec. 1.8.1. According to de-Broglie concept $\lambda = \frac{h}{p}$. Now according to dual nature of electron when the electron is rotating in circular orbit according to the particle picture, it is associated with stationary wave according to Fig. 1.13.

Therefore we can write,

$$2\pi r = n\lambda = n\frac{h}{p} = n\frac{h}{mv}$$

$$\therefore\ mvr = n\left(\frac{h}{2\pi}\right),$$

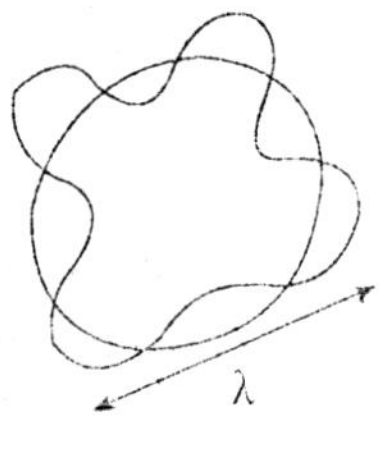


which is the quantisation of angular momentum according to Bohr's theory.

77. 'Bohr's orbits are known as energy levels and stationary states as well'.—Explain. **[B.U. 2015]**

**Hint :** According to Bohr's theory $E = -\frac{2\pi e^4 mZ^2}{n^2h^2}$

n has values 1, 2, 3, ........... so that the values obtained for $E_1$, $E_2$, $E_3$ ...... etc. are called energy levels.

According to the postulates of Bohr, $mvr = n\left(\frac{h}{2\pi}\right)$. This is the quantisation of angular momentum in case of Bohr's stationary orbit. Thus Bohr's energy levels are also called stationary states.

78. Point out the significance of $\Psi$ and $\Psi^2$. **[B.U. 2015]**

**Hint :** Sec. 1.8.4 and Sec. 1.8.5.

79. Enumerate the principles that control the distribution of ten 3d-electrons from scandium to zinc. **[B.U. 2015]**

**Hint :** Pauli's exclusion principle (Sec. 1.9.1) and Hund's rule (Sec. 1.9.3).

80. The energy of a photon associated with light of wavelength 200 nm is— (a) $\sim 10^{-16}$ J, (b) $\sim 10^{-17}$ J, (c) $\sim 10^{-18}$ J, (d) $\sim 10^{-19}$ J.

**Ans. :** (d).

$\lambda = 200$ nm $\nu = \frac{c}{\lambda} = \frac{3.0 \times 10^8 \text{ ms}^{-1}}{200 \times 10^{-9} \text{ m}} = 1.5 \times 10^{15}$ $s^{-1}$.

$E = h\nu = (6.63 \times 10^{-34}$ Js$)(1.5 \times 10^{15}$ $s^{-1}) = 9.94 \times 10^{-19}$ J.

81. An electron travels with a velocity of v $ms^{-1}$. For a photon to have the same de-Broglie wavelength, the velocity will be approximately (a) v/1840, (b) 1840/v (c) 1840v, (d) v.

**Ans. :** (a) i.e. v/1840.

82. The number of nodal plane in a $p_x$ orbital is—(a) one, (b) two, (c) three, (d) four.

**Ans. :** (a) i.e. one.

83. Calculate the exchange energy of $p^4$ configuration with maximum pairing and maximum unpairing.

**Hint :** Maximum unpairing configuration is | ↑↓ | ↑ | ↑ |

Here exchange energy $= K\frac{3(3-1)}{2} + K\frac{1(1-1)}{2} = 3K$

Maximum pairing configuration is | ↑↓ | ↑↓ | |

Here exchange energy $= K\frac{2(2-1)}{2} + K\frac{2(2-1)}{2} = 2K$

84. Name a man made halogen element. What is its atomic number? **[B.U. 2013]**

**Hint. :** Astatine, atomic number 85.

85. What are the ground state electronic configurations of $V^{3+}$ and $Eu^{3+}$? **[B.U. 2013]**

**Hint :** $V^{3+}$ : $1s^2 2s^2 2p^6 3s^2 3p^6 3d^2$

$Eu^{3+}$ : $1s^2 2s^2 2p^6 3s^2 3p^6 4s^2 3d^{10} 4p^6 4d^{10} 4f^6 5s^2 5p^6$ or $[Xe]4f^6$.

86. The most probable value of r for an electron in 1s orbital of hydrogen atom is—1. $a_0\sqrt{2}$, 2. $a_0$, 3. $\sqrt{2}a_0$, 4. $3a_0\sqrt{2}$. **[NET 2013]**

**Ans. :** 2 i.e. $a_0$.

87. Wavelength (λ in nm) of the Lyman series for an one-electron ion is in the range $24 \le \lambda \le 30$. The ionisation energy of the ion will be closest to $\left(1J = \frac{10^{19}}{1.6} ev\right)$

1. 32 ev, 2. 42 ev. 3. 52 ev, 4. 62 ev. **[NET 2014]**

**Ans. :** 2. i.e. 42 ev.

$$\text{Ionisation energy} = \frac{hc}{\lambda} = \frac{6.626 \times 10^{-34} \times 3 \times 10^{8} \times 10^{19}}{24 \times 10^{-9} \times 1.6} = 52 \text{ ev.}$$

$$\text{For other side wavelength range} = \frac{6.626 \times 10^{-34} \times 3 \times 10^{8} \times 10^{19}}{30 \times 10^{-9} \times 1.6} = 42 \text{ ev.}$$

88. Name the six series in the hydrogen spectrum indicating which ones were discovered first and last. **[B.U. 2016]**

**Hint :** The six series of hydrogen spectrum with their year of discovery are :

(a) Lyman (1915) in ultraviolet region, (b) Blamer (1835) in the visible region, (c) Paschen (1896) in the near infrared region, (d) Brackett (1922) in the far infrared region, (e) Pfund (1925) in the far infrared region, (f) Humphreys (after 1925) in the far infrared region.

Therefore Balmer series was discovered first and Humphreys series was discovered last.

89. Calculate the wavelength of the first line and the last line in the Balmer series of hydrogen spectrum ($R = 1.1 \times 10^7 \ m^{-1}$). **[B.U. 2016]**

**Hint :** In case of Balmer series $\bar{\nu} = R\left[\frac{1}{2^2} - \frac{1}{n^2}\right]$ where n > 2

First first line $\bar{\nu} = R\left[\frac{1}{2^2} - \frac{1}{3^2}\right] = \frac{5}{36} R \quad \therefore \lambda = \frac{36}{5} R = \frac{36}{5 \times 1.1 \times 10^7 \ m^{-1}} = 6.54 \times 10^{-7}$ m.

For last line $\bar{\nu} = R\left[\frac{1}{2^2} - \frac{1}{\infty^2}\right] = R\left[\frac{1}{4} - 0\right] = \frac{R}{4}$

$\therefore \lambda = \frac{4}{R} = \frac{4}{1.1 \times 10^7 \ m^{-1}} = 3.64 \times 10^{-7}$ m.

90. Define radial wave function and angular wave function. Express total wave function in terms of these two. **[B.U. 2016]**

**Hint :** Sec. 1.8.7, Sec. 1.8.7.1, Sec. 1.8.7.2 and Sec. 1.8.7.3.

In order to understand the distribution of electrons in a given volume of space about the nucleus, one needs to know about the wave functions (Ψ). An orbital wave function (Ψ) can be written as the product of two wave functions, the radial wave funciton ($\Psi_{radial}$) and the angular wave function ($\Psi_{angular}$). The former is a function of distance (r) only and the latter is a funciton of two angle variables (θ, ϕ). Thus $\Psi_{radial} = R(r)$ and $\Psi_{angular} = \theta(\theta)\phi(\phi)$ is a particular class of composite wave function which is commonly known as spherical harmonics. So complete wave function is $\Psi = R(r)\theta(\theta)\phi(\phi)$ in which the radial part R(r) depends on quantum numbers n and *l* and it tells about the size of the orbital, whereas the angular part $\theta(\theta)\phi(\phi)$ depends on quantum numbers *l* and m and tells about the shape of the atomic orbitals.

91. What do you mean by Rydberg constant? **[B.U. 2016]**

**Hint :** Sec. 1.6.5. The nineteenth-century spectroscopist Johann Rydberg found from the spectroscopic observation on hydrogen atoms suggested that all the waelengths can be described by the expression,

$$\frac{1}{\lambda} = R\left[\frac{1}{\eta_1^2} - \frac{1}{\eta_2^2}\right]$$

where R is the Rydberg constant, an emperical constant which have the value $1.097 \times 10^7 \text{ m}^{-1}$. Latter Niels Bohr in 1913 from his theory obtained the expression for Rydberg constant for hydrogen as $R = \frac{2\pi^2 e^4 m}{ch^3}$ and for hydrogen like systems as $R = \frac{2\pi^2 e^4 mZ^2}{ch^3}$ where the symbols have their usual meaning.

92. Which quantum number does solely determine the allowed energies for a hydrogenic system? Write down the expression for energy and explain the terms involved therein. Note the significance of the sign involved in the energy expression and the condition that corresponds to zero energy. **[B.U. 2016]**

**Hint :** Sec. 1.6.4. The principal quantum number n determine the allowed energies for a hydrogenic system. The energy expression is $E = -\frac{2\pi^2 e^4 mZ^2}{n^2 h^2}$ where negative sign indicates that as the value of n increases the revolving electron will go to the higher energy state which signify that the energy of the electron in a bound state is lower than a widely separated stationary electron and nucleus. Positive values of energy correspond to unbound states of the electron in which it may travel with any velocity and hence possess any energy. Moreover, because the energy is proportional to $1/n^2$, the energy levels converge as the energy increases i.e. becomes less negative.

The zero energy when $n = \infty$ corresponds to the electron and nucleus being widely separated and stationary which corresponds to ground state.

93. The transition that belongs to the Lyman series in the hydrogen atom spectrum is—
1. 1s ← 4s; 2. 1s ← 4p; 3. 2s ← 4s; 4. 2s ← 4p **[NET 2016]**

**Ans. :** 2 i.e. 1s ← 4s.

# 2

# PERIODIC TABLE AND PERIODIC PROPERTIES

Modern periodic law states that **properties of elements and their compounds are periodic function of the atomic number of the elements.** All the existing elements are arranged in an extended form of periodic table in accordance with the modern periodic law. **Periodicity** is the regular manner of variation of physical and chemical properties of the elements with atomic number. **Aperiodicity** is the reverse of periodicity which is also called secondary periodicity and is related with anomalous periodic properties.

## 2.1 THE MODERN PERIODIC TABLE

The long form of periodic table according to the 1990 and 2005 recommendations of the nomenclature of Inorganic Chemistry nominated by IUPAC is given in Table 2.1. In this table the notations for subgroup A and B are omitted and there is a running number from 1 to 18 which indicates the group and there are seven periods.

IUPAC (1990) recommended a thirtytwo column periodic table and recommended to propose any format of the modern periodic table by the chemists, which can explain the electronic structure and chemical and physical properties of elements, may be accepted.

## 2.2 ELECTRONIC CONFIGURATION OF ELEMENTS AND PERIODIC TABLE

There is an intimate connection between the electronic configuration of the elements and placement of the elements in the periodic table. Generally the outer quantum shell number denotes the period number and the total number of electrons in that quantum shell denotes the group number e.g.

**Table 2.1 : Modern periodic table according to 2005 recommendation of IUPAC (Eighteen-Column)**

| | 1<br>IA | 2<br>IIA | 3<br>IIIA | 4<br>IVA | 5<br>VA | 6<br>VIA | 7<br>VIIA | 8 | 9<br>VIIIA | 10 | 11<br>IB | 12<br>IIB | 13<br>IIIB | 14<br>IVB | 15<br>VB | 16<br>VIB | 17<br>VIIB | 18<br>VIIIB | IUPAC 2005<br>IUPAC 1970 |
|---|---|---|---|---|---|---|---|---|---|---|---|---|---|---|---|---|---|---|---|
| Period I | H<br>1 | | | | | | | | | | | | | | | | | He<br>2 | |
| Period II | Li<br>3 | Be<br>4 | | | | | | | | | | | B<br>5 | C<br>6 | N<br>7 | O<br>8 | F<br>9 | Ne<br>10 | |
| Period III | Na<br>11 | Mg<br>12 | | | | | | | | | | | Al<br>13 | Si<br>14 | P<br>15 | S<br>16 | Cl<br>17 | Ar<br>18 | |
| Period IV | K<br>19 | Ca<br>20 | Sc<br>21 | Ti<br>22 | V<br>23 | Cr<br>24 | Mn<br>25 | Fe<br>26 | Co<br>27 | Ni<br>28 | Cu<br>29 | Zn<br>30 | Ga<br>31 | Ge<br>32 | As<br>33 | Se<br>34 | Br<br>35 | Kr<br>36 | |
| Period V | Rb<br>37 | Sr<br>38 | Y<br>39 | Zr<br>40 | Nb<br>41 | Mo<br>42 | Tc<br>43 | Ru<br>44 | Rh<br>45 | Pd<br>46 | Ag<br>47 | Cd<br>48 | In<br>49 | Sn<br>50 | Sb<br>51 | Te<br>52 | I<br>53 | Xe<br>54 | |
| Period VI | Cs<br>55 | Ba<br>56 | La-Lu*<br>57-71 | Hf<br>72 | Ta<br>73 | W<br>74 | Re<br>75 | Os<br>76 | Ir<br>77 | Pt<br>78 | Au<br>79 | Hg<br>80 | Tl<br>81 | Pb<br>82 | Bi<br>83 | Po<br>84 | At<br>85 | Rn<br>86 | |
| Period VII | Fr<br>87 | Ra<br>88 | Ac-Lr**<br>89-103 | Rf<br>104 | Db<br>105 | Sg<br>106 | Bh<br>107 | Hs<br>108 | Mt<br>109 | Ds<br>110 | Rg<br>111 | Uub<br>112 | Uut<br>113 | Uuq<br>114 | Uup<br>115 | Uuh<br>116 | Uus<br>117 | Uuo<br>118 | |

| | | | | | | | | | | | | | | | |
|---|---|---|---|---|---|---|---|---|---|---|---|---|---|---|---|
| * | La<br>57 | Ce<br>58 | Pr<br>59 | Nd<br>60 | Pm<br>61 | Sm<br>62 | Eu<br>63 | Gd<br>64 | Tb<br>65 | Dy<br>66 | Ho<br>67 | Er<br>68 | Tm<br>69 | Yb<br>70 | Lu<br>71 |
| ** | Ac<br>89 | Th<br>90 | Pa<br>91 | U<br>92 | Np<br>93 | Pu<br>94 | Am<br>95 | Cm<br>96 | Bk<br>97 | Cf<br>98 | Es<br>99 | Fm<br>100 | Md<br>101 | No<br>102 | Lr<br>103 |

| | Electronic configuration | Period number | Group number |
|---|---|---|---|
| s-block elements | $1s^22s^1$ | 2 | 1 |
| | $1s^22s^22p^63s^1$ | 3 | 1 |
| | $1s^22s^22p^63s^23p^64s^23d^{10}4p^65s^1$ | 5 | 1 |
| | $1s^22s^2$ | 2 | 2 |
| | $1s^22s^22p^63s^23p^64s^23d^{10}4p^65s^2$ | 5 | 2 |
| p-block elements | $1s^22s^22p^63s^23p^3$ | 3 | 5/15 |
| | $1s^22s^22p^63s^23p^4$ | 3 | 6/16 |
| | $1s^22s^22p^63s^23p^64s^23d^{10}4p^1$ | 4 | 3/13 |
| | $1s^22s^22p^63s^23p^64s^23d^{10}4p^3$ | 4 | 5/15 |
| | $1s^22s^22p^63s^23p^5$ | 3 | 7/17 |
| | $1s^22s^22p^6$ | 2 | 8/18 |
| | $1s^22s^22p^63s^23p^64s^23d^{10}4p^6$ | 4 | 8/18 |
| | $1s^22s^22p^63s^23p^64s^23d^{10}4p^2$ | 4 | 4/14 |
| | $1s^22s^22p^63s^23p^64s^23d^{10}4p^64d^{10}5s^25p^5$ | 5 | 7/17 |

The above rules hold good in case of s and p block elements. But in case of d-block transition elements the placement is followed by following :

(a) When the d-orbital is incompletely filled then the total number of s and d electrons is its group number followed by the outer quantum shell number is its period number e.g.

Cr(24) → $1s^22s^22p^63s^23p^64s^13d^5$ Period-4 Group-6

Ru(44) → $1s^22s^22p^63s^23p^64s^23d^{10}4p^65s^14d^7$ Period-5 Group-8

Ni(28) → $1s^22s^22p^63s^23p^64s^23d^8$ Period-4 Group-10

(b) When the d-orbital is complete then in that case the total number of electrons in s orbital is its group number e.g.

Cu(29) → $1s^22s^22p^63s^23p^64s^13d^{10}$ Period-4 Group-1/11

Zn(30) → $1s^22s^22p^63s^23p^64s^23d^{10}$ Period-4 Group-2/12

---

**Problem 2.1** What are transition elements? How many transition series present in the periodic table?

**Ans. :** Transition elements are those in which there is incompletely filled d or f orbital in the elemental state or in the naturally occuring oxidation state.

In the periodic table there are three transition series and two inner transition series :

First transition series : Sc(21) — Cu(29), here electrons are added to 3d orbital.

Second " " : Y(39) — Ag(47), here electrons are added to 4d orbital.

Third " " : Hf(72) — Au(79), here electrons are added to 5d orbital.

First inner transition series : La(57) — Lu(71), here electrons are added to 4f orbital.

Second " " " : Ac(89) — Lr(103), here electrons are added to 5f orbital.

**Problem 2.2** Why La(57)-Lu(71) or Ac(89)-Lr(103) elements are called inner transition elements?

**Ans. :** Here electrons are added to 4f or 5f orbital. These orbitals are called inner orbitals because radial distribution function shows that 4f orbital is more nearer to the nucleus than 3d orbital so that when electrons are added to the 4f orbital, with increasing nuclear charge, this will attract the 4f electrons more strongly and thereby a special type of attraction is operated which is known as **lanthanide contraction**. As electrons are added to inner 4f or 5f orbitals, so that these elements are called inner transition elements.

## 2.3 SYSTEMATIC NAMING AND SYMBOLS OF ELEMENTS OF ATOMIC NUMBER GREATER THAN 100

1990 and 2005 recommendations of IUPAC used the following roots for systematic naming of elements :

| 0 | 1 | 2 | 3 | 4 | 5 | 6 | 7 | 8 | 9 |
|---|---|---|---|---|---|---|---|---|---|
| nil | un | bi | tri | quad | pent | hex | sept | oct | enn |

With the help of the above numerical roots the naming of the elements are terminated by 'ium'. The last 'n' of 'enn' is dropped when it occurs before 'nil' and the 'i' of 'bi' and 'tri' would also be dropped when it occurs before 'ium'. The symbol of an element is composed of the **initial letters of numerical roots** as follows (as per 1990 recommendations) :

| Atomic number | Numerical roots | Naming | Symbol |
|---|---|---|---|
| 101 | un, nil, un | unnilunium | Unu |
| 102 | un, nil, bi | unnilbium | Unb |
| 103 | un, nil, tri | unniltrium | Unt |
| 104 | un, nil, quad | unnilquadium | Unq |
| 105 | un, nil, pent | unnilpentium | Unp |
| 106 | un, nil, hex | unnilhexium | Unh |
| 107 | un, nil, sept | unnilseptium | Uns |
| 108 | un, nil, oct | unniloctium | Uno |
| 109 | un, nil, enn | unnilennium | Une |

In 2005 recommendations, element number 100 to 111 are assigned permanent naming with their symbol as follows :

| Element number | Name | Symbol |
|---|---|---|
| 100 | fermium | Fm |
| 101 | mendelevium | Md |
| 102 | nobelium | No |
| 103 | lawrencium | Lr |
| 104 | rutherfordium | Rf |
| 105 | dubnium | Db |
| 106 | seaborgium | Sg |
| 107 | bohrium | Bh |
| 108 | hasium | Hs |
| 109 | meitnerium | Mt |
| 110 | darmstadtium | Ds |
| 111 | roentgenium | Rg |
| 112 | copernicium | Cn* |

In this latest recommendations the newly discovered elements** referred to the scientific literature are given temporary names and symbols from element number 113 onwards as :

| Element number | Name | Symbol |
|---|---|---|
| 113 | ununtrium | Uut |
| 114 | ununquadium | Uuq |
| 115 | ununpentium | Uup |
| 116 | ununhexium | Uuh |
| 117 | ununseptium | Uus |
| 118 | ununoctium | Uuo |
| 119 | ununennium | Uue |
| 120 | unbinilium | Ubn |
| 121 | unbiunium | Ubu |
| 122 | unbibium | Ubb |
| 200 | binilnilium | Bnn |
| 201 | binilunium | Bnu |
| 202 | binilbium | Bnb |
| 900 | ennilnilium | Enn |

* *The Statesman 7.11.2011* ** *Appendix–I*

## 2.4 CAPACITIES OF PERIODS WITH REFERENCE TO PAULI'S EXCLUSION PRINCIPLE AND AUFBAU PRINCIPLE

According to Pauli's exclusion principle the maximum accommodation capacity of a period is dictated by $2n^2$ where n denotes the principal quantum number. For period I and II the maximum accommodation capacity is 2 and 8 elements which is in accord with $2n^2$ formula. But for n=3 i.e. for period III, the maximum accommodation capacity should be 18 but actually there exist 8 elements. This is due to the fact that when n = 3, $l$ = 0, 1, 2 and the orbital designations will be 3s, 3p and 3d. But according to aufbau principle the orbitals 3s and 3p will be filled making a maximum accommodation capacity of 2+6 = 8 electrons i.e. maximum 8 elements will be accommodated. Similarly for periods IV, V, VI and VII are :

| Periods | Orbital designation | Actual orbitals filled | Accommodation |
|---|---|---|---|
| IV (n=4) | 4s, 4p, 4d, 4f | 4s, 4p and 3d | 2+6+10 = 18 |
| V (n=5) | 5s, 5p, 5d, 5f | 5s, 5p and 4d | 2+6+10 = 18 |
| VI (n=6) | 6s, 6p, 6d, 6f | 6s, 6p, 5d and 4f | 2+6+10+14 = 32 |
| VII (n=7) | 7s, 7p, 7d, 7f | 7s, 7p, 6d and 5f | 2+6+10+14 = 32 |

Therefore in the modern form of the periodic table threre will be a maximum accommodation capacity of 2+8+8+18+18+32+32 = 118 elements i.e. the atomic number of last element is 118. If there might be the discovery of an element having atomic number 119, then there will be a new period i.e. period VIII will start and will be accommodated below the last element of alkali metals.

## 2.5 PERIODIC PROPERTIES

There are some properties of atoms and molecules which vary periodically are called periodic properties e.g. ionisation potential, electronegativity etc.

### 2.5.1 Ionisation potential/Ionisation energy

It is defined as the **minimum amount of energy required to release the most loosely bound electron from a neutral gaseous atom.**

Atom + Ionisation energy or

Ionisation Potential (I.P.) ⟶ Unipositive ion + electron

i.e. $A + I.P. \longrightarrow A^+ + e$

This is called the first ionisation poteitial ($I_1$) of the atom A. If energy is supplied to $A^+$ then there will be the formation of $A^{2+}$ and the energy is called second ionisation potential ($I_2$). In a similar way we can consider third ($I_3$), fourth ($I_4$) ionisation potentials leading to nth ($I_n$) ionisation potential. Naturally the energy sequence will be :

$$I_1 < I_2 < I_3 < I_4 \ldots\ldots\ldots < I_n$$

When no mention is made for the value of ionisation energy, this implied that this is the value for first ionisation energy.

Generally ionisation potential is measured in electron volt (ev) unit. Electron volt is defined as the amount of energy gained by an electron falling through a potential difference of one volt.

1 eV atom$^{-1}$ (electron volt per atom) = electronic charge × 1 volt atom$^{-1}$

$= 1.6 \times 10^{-19}$ coulomb × 1 volt atom$^{-1}$

$= 1.6 \times 10^{-19}$ coulomb-volt atom$^{-1}$

$= 1.6 \times 10^{-19}$ Joule atom$^{-1}$ (as 1 coulomb-volt = 1 Joule)

$= 1.6 \times 10^{-19} \times 6.023 \times 10^{23}$ Joule mole$^{-1}$

$= 1.6 \times 10^{-19} \times 6.023 \times 10^{23} \times 10^{7}$ erg mole$^{-1}$ (as 1 erg = $10^7$ Joule)

$= \dfrac{1.6 \times 6.023 \times 10^{11}}{4.184 \times 10^{7}}$ cal mole$^{-1}$ (as 1 cal = 4.184 Joule)

$= \dfrac{1.6 \times 6.023 \times 10^{11}}{4.184 \times 10^{7} \times 10^{3}}$ kcal mole$^{-1}$

$= 23.06$ kcal mole$^{-1}$

∴ 1 eV per atom = 23.06 kcal mole$^{-1}$

In SI or MKS unit, 1 eV atom$^{-1}$ = $\dfrac{1.6 \times 10^{-19} \times 6.023 \times 10^{23}}{10^{3}}$ kJ mol$^{-1}$

$= 96.48$ kJ mol$^{-1}$

The term ionisation potential is very often mistakenly used because the process actually involves the excitation energy of the electron from n = 1 to n = ∝, so it is more plausable to express the process as **ionisation energy** instead of **ionisation potential**. The term potential are very often mistakenly used because it is experimentally determined by applying a potential gradient. Ionisation energy is an **extensive property** whereas potential is an intensive property expressed in volt. The properties of matter which depends on the amount of matter is called **extensive property** e.g. internal energy, volume, kinetic energy, entropy etc. On the other hand those properties of matter which does not depend on the amount of matter is called **intensive property** e.g. density, temperature, viscosity, refractive index etc.

**Factors influencing the magnitude of ionisation energy :**

(a) **The distance of the removing electron :** Remaining the nuclear charge constant as the distance of the removing electron from the nucleus increases, the attraction between the nucleus and the removing electron decreases thereby the ionisation energy decreaes.

(b) **Nuclear charge :** As nuclear charge increases (keeping the distance of the removing electrons constant) the attraction between the nucleus and the outer most electron increases and ionisation energy increases.

These two factors influence the variation of ionisation energy in a period and in a group in the periodic table. In a period from left to right electrons are added to the same quantum shell and as the nuclear charge increases one by one, therefore the value of the ionisation energy increases in a period from left to right. On the other hand in a group from top to bottom

**Table 2.2 : Variation of first ionisation energy of elements (eV)**

| | | | | | | | | | | | | | | | | | |
|---|---|---|---|---|---|---|---|---|---|---|---|---|---|---|---|---|---|
| H | | | | | | | | | | | | | | | | | He |
| 13.59 | | | | | | | | | | | | | | | | | 24.58 |
| Li | Be | | | | | | | | | | | B | C | N | O | F | Ne |
| 5.39 | 9.32 | | | | | | | | | | | 8.30 | 11.26 | 14.53 | 13.61 | 17.42 | 21.56 |
| Na | Mg | | | | | | | | | | | Al | Si | P | S | Cl | Ar |
| 5.14 | 7.64 | | | | | | | | | | | 5.98 | 8.15 | 10.48 | 10.36 | 13.01 | 15.75 |
| K | Ca | Sc | Ti | V | Cr | Mn | Fe | Co | Ni | Cu | Zn | Ga | Ge | As | Se | Br | Kr |
| 4.34 | 6.11 | 6.54 | 6.82 | 6.74 | 6.76 | 7.43 | 7.90 | 7.86 | 7.63 | 7.72 | 9.39 | 6.00 | 7.88 | 9.81 | 9.75 | 11.84 | 14.00 |
| Rb | Sr | Y | Zr | Nb | Mo | Tc | Ru | Rh | Pd | Ag | Cd | In | Sn | Sb | Te | I | Xe |
| 4.18 | 5.69 | 6.38 | 6.84 | 6.88 | 7.10 | 7.28 | 7.36 | 7.46 | 8.33 | 7.57 | 8.99 | 5.78 | 7.34 | 8.64 | 9.01 | 10.45 | 12.13 |
| Cs | Ba | La | Hf | Ta | W | Re | Os | Ir | Pt | Au | Hg | Tl | Pb | Bi | Po | At | Rn |
| 3.89 | 5.21 | 5.61 | 6.65 | 7.88 | 7.98 | 7.87 | 8.7 | 9.0 | 9.0 | 9.22 | 10.43 | 6.10 | 7.41 | 7.29 | 8.43 | 9.2 | 10.75 |

1 ev atom$^{-1}$ = 96.48 kJ mole$^{-1}$

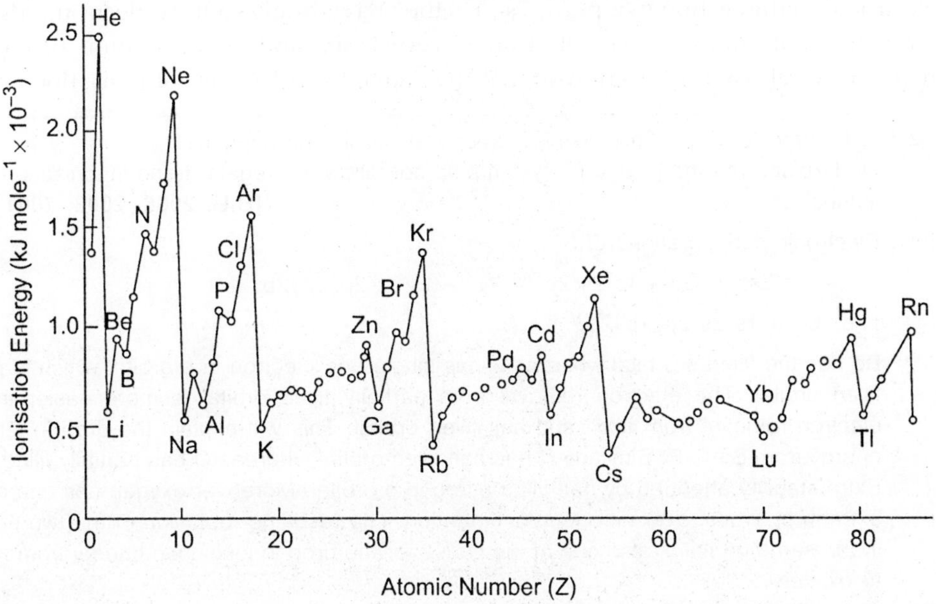


***Figure 2.1** : Periodic trends of ionisation energy of the elements.*

electrons are added to different quantum shell, here both nuclear charge and distance increases, the predominating factor is distance, therefore as distance increases the ionisation energy decreases. Table 2.2 and Fig. 2.1 depict the ionisation energy variation of the elements.

The following points should have to be noted for the variation of ionisation energies :

(1) The inert gases register highest values i.e. they occupy highest position in the curve (Fig. 2.1).

(2) The alkali metals have the lowest values.

(3) For transition series, there is a slight increasing tendency but the end elements (Zn, Cd, Hg) have a sharp increasing trend.

(4) In case of lanthanides there is also a slight increasing trend.

It is worthwhile to explain why the inert gases have the highest values of the ionisation energy whereas the alkali metals have the lowest values. Inert gases have the general electronic configuration $ns^2np^6$ i.e. all the electrons are paired. Therefore if we want to get the value of ionisation energy the expenditure of energy will be in two phases :

(1) Energy needed to unpair the paired electrons.

(2) Secondly the energy required to knock off an unpaired electron.

Therefore as the expenditure of energy in two phases thereby inert gases register highest values of ionisation energies. On the other hand alkali metals have the electronic configuration $ns^2np^6 (n+1)s^1$ i.e. they have single solitary electron outside the inert gas configuration. This electron is very labile and small amount of energy is required to release this electron so that they have lowest values of ionisation energies.

---

**Problem 2.3** Why there is a drop in ionisation energy at boron and oxygen?
Or, Explain why the $p^1$ and $p^4$ systems do not follow the regular trend in ionisation energy values? **[B.U. 2006, 2003, 2001, 1996]**

**Hint** : Electronic configurations of

$Be \rightarrow 1s^22s^2$ $\quad B \rightarrow 1s^22s^22p^1$ $\quad N \rightarrow 1s^22s^22p_x^12p_y^12p_z^1$

and $O \rightarrow 1s^22s^22p_x^22p_y^12p_z^1$

Be has the filled s orbital whereas B has single one electron in p orbital which is partially filled orbital. The electron removal from partially filled orbital is more easier than the electron removal from filled and half filled orbital. This will explain the drop in ionisation energy from Be to B. Similarly N has half filled orbital whereas O has partially filled orbital. Extra stability attended by half filled nitrogen p orbital. Moreover oxygen has one electron extra in $p_x$ orbital over nitrogen configuration. The resulting repulsion of the two electrons in the same orbital is also one of the cause for the drop in ionisation energy from nitrogen to oxygen.

---

(c) **The screening effect of the lower lying inner shell :** Lower lying inner shell will screen the nucleus and thereby the attractive force between the nucleus and the outermost electron will vary leading to variation of ionisation energy. For a particular quantum shell the radial distribution function shows that (Sec. 1.8.7.1) the penetrating power of the orbital has the sequence ns > np > nd > nf. Screening efficiency also falls off in the order ns > np > nd > nf. In the case of first ionisation process (IE) for Be, B and Mg, Al, there will be a greater variation. For B and Al electron is removed from 2p and 3p orbitals respectively whereas for Be and Mg this is removed from 2s and 3s orbitals respectively. For knocking off s electron requires more energy compared to the p electron therefore, I.E. of Be > I.E. of B and I.E. of Mg > I.E. of Al.

From Table 2.2 it is evident that I.E. of $Cu(3d^{10}4s^1)$ < I.E. of $Zn(3d^{10}4s^2)$ > I.E. of $Ga(3d^{10}4s^24p^1)$

For Zn, electron removal will require greater energy because electron will be knocked off from filled 4s orbital (paired electrons). On the other hand for Cu the electron will be removed from half filled 4s orbital (unpaired single electron) whereas for Ga the electron will be removed from 4p orbital (single unpaired electron). From screening efficiency electron removal from s orbital will require more energy than the p orbital and hence is the above sequence i.e. I.E of Cu > I.E. of Ga.

(d) **Overall charge of the ionising species :** For isoelectronic systems as the overall charge of the ionising species increases the I.E. increases, e.g.

| $Li$ | $Be^+$ | $B^{2+}$ | $C^{3+}$ | $N^{4+}$ | $O^{5+}$ | $F^{6+}$ |
|---|---|---|---|---|---|---|
| 5.39 | 18.21 | 37.92 | 64.48 | 97.86 | 138.10 | 185.14 eV per atom |

This is because from Li to F the nuclear charge increases from 3 to 9 and all the species have the same electronic configuration $1s^22s^1$ thereby the overall charge increases, so that increasing trend in I.E. is registered. In this context it is reasonable to connect a relationship between nuclear charge (Z) and I.E. From Bohr's theory (Sec. 1.6.4) the expression for I.E. (SI unit)

$$\text{I.E.} = E_2 - E_1 = \frac{Z^2me^4}{8\varepsilon_0^2h^2}\left[\frac{1}{n_1^2} - \frac{1}{n_2^2}\right]$$

$$= \frac{Z^2me^4}{8\varepsilon_0^2h^2}\left[\frac{1}{1^2} - \frac{1}{\infty^2}\right] \qquad [\because n_1 = 1 \text{ and } n_2 = \infty \text{ for I.E.}]$$

$$= \frac{Z^2me^4}{8\varepsilon_0^2h^2}$$

Therefore keeping the other terms constant, the I.E. will be proportional to the square of the nuclear charge i.e.

$$\text{I.E.} \propto Z^2$$

so that $\sqrt{\text{I.E.}} \propto Z$

If we plot the $\sqrt{\text{I.E.}}$ vs Z for the above isoelectronic systems, we expect a straight line curve passing through the origin. But the actual plot is in Fig. 2.2. The discrepancy is due to the application of Bohr's theory for polyelectron systems.

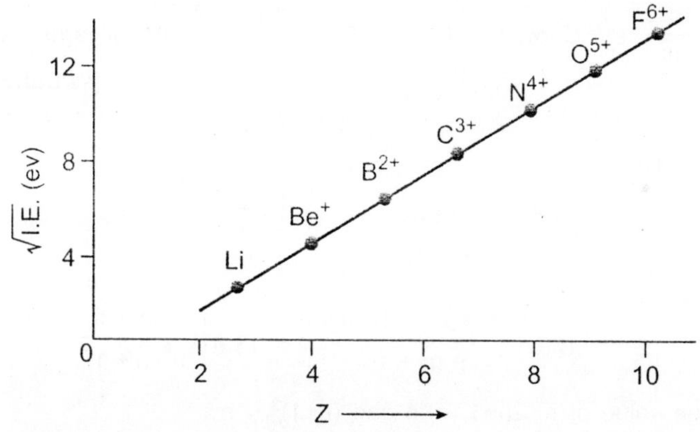


***Figure 2.2** : Variation of ionisation energies for isoelectronic atoms or ions.*

The application of I.E. data are useful in deciding the metallic character of the elements. Generally the I.E. values decrease down the group which is a pointer that metallic character increases down the group. Moreover low values of I.E. indicate greater electron loosing tendency whereas high values indicate less tendency to loose the electrons e.g. alkali metals have low values of I.E. whereas inert gases have high I.E. values.

| Li | Na | K | Rb | Cs | |
|---|---|---|---|---|---|
| 5.39 | 5.14 | 4.34 | 4.18 | 3.89 | eV atom$^{-1}$ |

| He | Ne | Ar | Kr | Xe | |
|---|---|---|---|---|---|
| 24.58 | 21.56 | 15.75 | 14.00 | 12.13 | eV atom$^{-1}$ |

**Problem 2.4** Why the first ionisation energy of oxygen is less than that of nitrogen though the nuclear charge of the former is greater?

**Ans. :** Similar to problem 2.3.

**Problem 2.5** The second ionisation energy of sodium should be larger than the second ionisation energy of magnesium. Comment. **[B.U. 2000]**

**Hint :** Na $\rightarrow 1s^2 2s^2 2p^6 3s^1$ Mg $\rightarrow 1s^2 2s^2 2p^6 3s^2$

The second ionisation energy of sodium means the electron will be removed from inert gas configuration whereas the same value for Mg means electron will be removed from $1s^2 2s^2 2p^6 3s^1$ i.e. from Na configuration and hence the difference.

**Problem 2.6** The fourth ionisation potential of boron is abnormally high compared to the other three. Explain. **[B.U. 2015, 2009, 2007, 1995, 1994]**

**Hint :** B $\rightarrow 1s^2 2s^2 2p^1$

The fourth ionisation potential means the electron will be removed from $1s^2$ configuration i.e. from inert gas configuration so that the value is abnormally high.

**Problem 2.7** In hydrogen atom energy of the electron in the nth orbit is given as $E_n = -\frac{13.6}{n^2}$ eV. Show that $E_{(n+1)} - E_n = \frac{13.6 \times 2}{n^3}$ eV for large values of n.

**[W.B.J.E. 2001, S.S.C. 2013]**

**Ans. :** $E_{(n+1)} = -\frac{13.6}{(n+1)^2}$

$$\therefore E_{(n+1)} - E_n = \frac{13.6}{n^2} - \frac{13.6}{(n+1)^2} = 13.6\left[\frac{(n+1)^2 - n^2}{n^2(n+1)^2}\right]$$

$$= 13.6\left[\frac{n^2 + 2n + 1 - n^2}{n^2(n+1)^2}\right] = 13.6\left[\frac{2n+1}{n^2(n+1)^2}\right]$$

For large value of n, $2n+1 \approx 2n$ and $(n+1)^2 \approx n^2$

$$\therefore E_{(n+1)} - E_n = \frac{13.6 \times 2n}{n^2 \times n^2} = \frac{13.6 \times 2}{n^3} \text{ (Proved)}$$

**Problem 2.8** Calculate the I.E. of hydrogen atom.

**Ans. :** According to Bohr's theory (Sec. 1.6.4) the energy of the hydrogen atom is given by the expression $E_n = -\frac{2\pi^2 e^4 m}{n^2 h^2}$

So the expression for I.E. for hydrogen atom

$$\text{I.E.} = E_1 - E_\infty = \frac{2\pi^2 e^4 m}{h^2}\left[\frac{1}{1^2} - \frac{1}{\infty^2}\right] = \frac{2\pi^2 e^4 m}{h^2}[1 - 0] = \frac{2\pi^2 e^4 m}{h^2}$$

$$= \frac{2 \times (3.14)^2 (4.8 \times 10^{-10} \text{ esu})^4 \times 9.108 \times 10^{-28} \text{ gm}}{(6.63 \times 10^{-27} \text{ erg-sec})^2}$$

$$= \frac{2(3.14)^2(4.8)^4 \times 9.108 \times 10^{-14}}{(6.63)^2} \text{ erg}$$

$$= 2168.953 \times 10^{-14} \text{ erg}$$

$$= \frac{2168.953 \times 10^{-14}}{1.6 \times 10^{-12}} \text{ev eV} \quad [\because 1\text{eV} = 1.6 \times 10^{-12} \text{ erg}]$$

$$= 1355.59 \times 10^{-2} \text{ eV}$$

$$= 13.5559 \text{ eV}$$

## 2.5.2 Electron affinity

*Electron affinity is defined as the amount of energy released when an isolated gaseous species (may be atom, ion or mole) captures an electron.*

Atom + Electron ⟶ Uninegative ion + Electron affinity (E.A.)

I.E. denotes some expenditure of energy i.e. endoergic process whereas electron affinity denotes the exoergic one. All I.E. values are given by positive sign so that all the electron affinity data should be a negative sign. But all electron affinity values are given by positive sign. *Negative electron affinity means the system does not like to have any more electron.* Therefore in order to distinguish a system having affinity for electron and a system having less affinity for electron, one is attended by positive sign and the other have negative sign respectively. But one should note that absolute magnitude of all electron affinity data is negative.

It is quite plausable from the definition of I.E. and E.A. processes, one is the reverse of the other i.e.

$$-\text{E.A. of X (g)} = +\text{I.E. of } X^- \text{ (g) ion}$$

In fact I.E. denotes some accumulation of positive charge(s) whereas E.A. indicates some accumulation of negative charge(s).

Like ionisation energies, we can also consider the successive electron affinities i.e. $EA_1$, $EA_2$ ...... etc. Second electron affinity ($EA_2$) means the amount of energy required to add an electron to the uninegative gaseous ion of an element so that :

$$M(g) + e \longrightarrow M^-(g) \quad \Delta H = \text{energy released} = +EA_1$$

$$M^-(g) + e \longrightarrow M^{2-}(g) \quad \Delta H = \text{energy required} = -EA_2$$

*Therefore second electron affinity is endoergic process like that of I.E. The second electron affinity is a difficult process because more work has to be done to place an electron to an already negative species. This is why higher electron affinities are always thermodynamically unfavourable* (e.g. $O^- \rightarrow O^{2-}$; $S^- \rightarrow S^{2-}$ Table 2.3).

Electron affinity value is difficult to determine. This is obtained in the indirect way in a thermodynamic cycle known as Born-Haber cycle where lattice energy (Sec. 3.1.3) of an ionic compound is determined in which electron affinity is one of the process in the cycle.

Electron affinity is sometimes referred to as **electron attachment enthalpy** and some of the values are given in Table 2.3. Variation of electron affinity as a function of atomic number is depicted in Fig. 2.3.

**Table 2.3 : Electron affinity values in eV per atom**

| 1 | 2 | 13 | 14 | 15 | 16 | 17 | 18 |
|---|---|---|---|---|---|---|---|
| H<br>0.747 | | | | | | | He<br>0 |
| Li<br>0.54 | Be<br>–0.60 | B<br>0.20 | C<br>1.25 | N<br>–0.10 | O<br>1.47<br>$O \rightarrow O^{2-}$, –7.41 | F<br>3.45 | Ne<br>0 |
| Na<br>0.74 | Mg<br>–0.40 | Al<br>0.60 | Si<br>1.63 | P<br>0.70 | S<br>2.07<br>$S \rightarrow S^{2-}$, –3.91 | Cl<br>3.61 | Ar<br>0 |
| K<br>0.70 | Ca<br>–0.30 | Ga<br>0.30 | Ge<br>1.20 | As<br>0.78 | Se<br>1.70<br>$Se \rightarrow Se^{2-}$, –4.21 | Br<br>3.36 | Kr<br>0 |
| Rb<br>0.49 | Sr<br>–0.30 | In<br>0.30 | Sn<br>1.25 | Sb<br>1.05 | Te<br>1.97 | I<br>3.06 | Xe<br>0 |
| Cs<br>0.47 | Ba<br>–0.30 | Tl<br>0.30 | Pb<br>0.36 | Bi<br>0.95 | Po<br>1.89 | At<br>2.81 | Rn<br>0 |

Like I.E. the E.A. values are expressed in eV atom$^{-1}$, kJ mole$^{-1}$ or kcal mole$^{-1}$. E.A. is also an **extensive property** like that of I.E. In general electron affinity will decrease in a group from

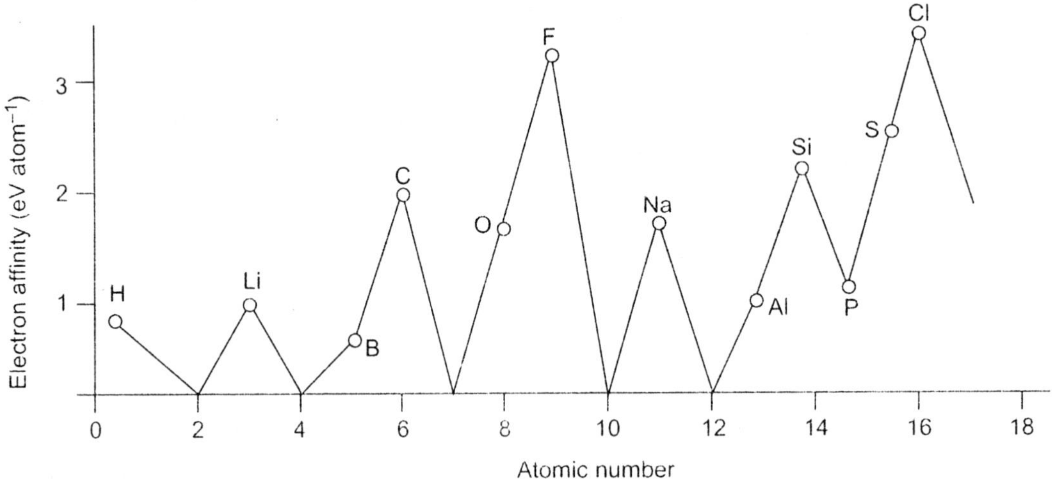


*Figure 2.3 : Variation of electron affinity with atomic number.*

top to bottom and increases in a period from left to right but there might be some exception. There is a complete lack of evidence of electron affinity data for inert gases. This is due to the fact that for inert gases the additional electron will be added to the next quantum shell remaining the nuclear charge intact and thereby this additional one electron will not have any attraction with the nucleus, so that electron affinity value will not be available for inert gases.

---

**Problem 2.9** Why the electron affinity value of Cl is greater than that of F? **[C.U. 2000]**

**Hint :** $F \longrightarrow 1s^2 2s^2 2p^5$ $\quad$ $Cl \longrightarrow 1s^2 2s^2 2p^6 3s^2 3p^5$

$F + e \longrightarrow F^-$ $\quad$ $Cl + e \longrightarrow Cl^-$

F and Cl has outer quantum shell n = 2 and n = 3 respectively. Cl has larger size than that of F, therefore the repulsion caused due to the entry of the electron in the Cl atomic orbital is less than the repulsion caused by the electron due to adding in small F atomic orbital and hence the larger electron affinity of Cl than that of F.

---

**Problem 2.10** The second electron affinity of oxygen and sulphur is negative, which indicates that the systems are not energetically favourable, yet a question arises how the stable CuO and CuS are formed? **[B.U. 2010]**

**Ans. :** Because I.E. and E.A. are not the sole criteria for the formation of ionic compounds like CuO and CuS. In case of ionic compounds an energy is associated which is called lattice energy, which is exoergic and which brings about the energetically unfavourable processes (like $O \rightarrow O^{2-}$ and $S \rightarrow S^{2-}$) energetically favourable and the compounds are formed.

---

The formation of $O \rightarrow O^{2-}$ and $S \rightarrow S^{2-}$ involves the following steps involving the second electron affinity.

$O(g) + e \rightarrow O^-(g)$ $\quad$ $EA_1 = +141.1 \text{ kJ mol}^{-1}$

$O^-(g) + e \rightarrow O^{2-}(g)$ $\quad$ $EA_2 = -844.0 \text{ kJ mol}^{-1}$

---

$O(g) + 2e \rightarrow O^{2-}(g)$ $\quad$ $\Delta H = EA_1 + EA_2 = 141.1 - 844.0 = -702.9 \text{ kJ mol}^{-1}$

---

$S(g) + e \rightarrow S^-(g)$ $\quad$ $EA_1 = +200.43 \text{ kJ mol}^{-1}$

$S^-(g) + e \rightarrow S^{2-}(g)$ $\quad$ $EA_2 = -333.0 \text{ kJ mol}^{-1}$

---

$S(g) + 2e \rightarrow S^{2-}(g)$ $\quad$ $\Delta H = EA_1 + EA_2 = 200.43 - 333.0 = -132.57 \text{ kJ mol}^{-1}$

---

So the conversions of $O(g) \rightarrow O^{2-}(g)$ and $S(g) \rightarrow S^{2-}(g)$ are endothermic processes.

Following are the factors influencing electron affinity :

(a) size of the atom

(b) nuclear charge

(c) penetrating power of the orbital to which the electron is accommodated

(d) electronic configuration of the atom.

As the size of the atom increases, E.A decreases whereas with increasing nuclear charge E.A. increases. When the incoming electron is added to the orbital which has greater penetrating power so that the added electron come closer to the nucleus and thereby the EA increases. In the same quantum shell the penetrating power is ns > np > nd > nf thereby when electron is added to the s orbital, E.A will be highest, when electron is added to f orbital the E.A will be lowest.

The atom or ion having filled or half filled electronic configuration is more stable. In these cases E.A decreases because in those cases the new electron will destroy the stable electronic configuration e.g. the E.A of alkaline earth metals (Be, Mg) is endoergic (Table 2.3) because electron is to be added to the stable electronic configuration ($ns^2$).

In case of transition and inner transition elements the variation of E.A in most of the cases is opposite to that of representative elements. Due to the lanthanide contraction, the post-lanthanide elements will have higher E.A, e.g. gold has the highest E.A (223 kJ $mole^{-1}$) among the metals and the existance of $Au^-$ is proved.

### 2.5.3 Electronegativity

It is defined by different approaches. Most important one is Pauling's approach. Pauling defined *electronegativity is the electron attracting power of an atom when it is covalently bonded in a molecule*. More electronegative element will drag the bonded electron pair more efficiently thereby there will be accumulation of partial negative charge on it. From the definition of electronegativity and electron affinity, both are the index of electron attracting power, but the later indicates the electron attracting power of an isolated atom whereas the former indicates the same when the atom is associated with a chemical bond. It is to be noted that **electronegativity of an atom is not the inherent property, but electron affinity is the inherent property**. Because electronegativity (E.N) of an atom depends on the bonding pattern as also the nature and number of atoms attached to that atom e.g. the electronegativity of carbon and sulphur in $CH_4$, $C_2H_4$, $C_2H_2$ and $H_2S$, $SF_4$, $SF_6$ respectively are different. Both I.E. and E.A. has same units (e.g. eV $atom^{-1}$ or kJ $mole^{-1}$) but E.N. has no such unit because it represents the relative attracting power of electron of an atom in a chemical bond.

There are several approaches developed to measure the electronegativity :

(1) **Pauling's method :** This is by far the best one of all the methods. If an atom A combine with atom B to form a chemical bont A–B, then the electron pair comprising to form the chemical bond may assume the following dispositions :

| A : B | $A^{-}: B^{+}$ | $A^{+} :B^{-}$ |
|---|---|---|
| (I) Here E.N. of A and B are same and the electron pair residing midway beween A-B bond. | (II) Here E.N. of A is more than that of B so that A assumes negative charge and B assumes positive charge. | (III) Here E.N. of B is more than that of A so that B assumes negative charge and A assumes positive charge. |

(I) structure denotes covalent contribution whereas structures (II) and (III) denote ionic contributions. Because of ionic and covalent contribution to the chemical bond A–B, the bond energy or the bond dissociation energy $D_{A-B}$ is always greater than the corresponding hypothetical 100% covalent bond energy of the molecule. The bond dissociation energy expressed in kcal mole$^{-1}$. The 100% covalent bond energy of the molecule is defined by Pauling as the geometric mean bond energy of the molecule $A_2(D_{A-A})$ and $B_2(D_{B-B})$ i.e.

$$D_{A-B(cov)} = \sqrt{D_{A-A}D_{B-B}}$$

As the difference in E.N. between the atoms A and B increases, $D_{A-B}$ increases from $D_{A-B(cov)}$ i.e. the difference in energy between $D_{A-B}$ and $D_{A-B(cov)}$ increases. The difference in energy $\Delta$ is called the **resonance energy** and is defined as a measure of the electronegativity difference beween the atom A and B. If $X_A$ and $X_B$ denotes the E.N. of the atom A and B, then Pauling proposed an emperical relationship :

$$X_A - X_B \propto \sqrt{\Delta} \qquad \text{where } \Delta = D_{A-B} - D_{A-B(cov)}$$

As bond dissociation energy is expressed in kcal mol$^{-1}$ and since 1 eV = 23.06 kcal mol$^{-1}$, the above expression denotes the square of the electronegativity difference in electron volt. If the bond energy is expressed in kJ mol$^{-1}$ then the above relationship takes the form as :

$$X_A - X_B = 0.102\sqrt{\Delta} \qquad \ldots(2.1)$$

where 0.102 is taken as the proportionality constant in SI unit. The equation (2.1) denotes that if the E.N. of one element is known then the E.N. of other element is to be calculated and then by repeating the process we may calculate the E.N. of other elements. In this process Pauling arbitrarily had to fix fluorine as the highest electronegative atom and has given the value 4.0. Assigning fluorine as the highest electronegative atom, the values of E.N. which are calculated for other elements is known as Pauling scale of E.N. and depicted in Table 2.4.

(2) **Mulliken's approach** : R.S. Mulliken (1934) developed the idea of the electronegativity of the atom A has some connection between its ionisation energy and electron affinity. He suggested that ionisation energy and electron affinity denotes the electron releasing and electron accepting tendency respectively of an isolated atom so that when the same atom is joined in a chemical bond, the electron attracting tendency of the same atom will be the average value of the ionisation energy and electron affinity i.e.

$$X_M = \frac{\text{I.E.} + \text{E.A.}}{2} \qquad \ldots(2.2)$$

where $X_M$ denotes the electronegativity in Mulliken method. He explained the formation of a pure covalent bond between A and B, when the hypothetical structures $A^+B^-$ and $A^-B^+$ should have equal energies. Let us consider the energy involved in the formation of the species $A^+B^-$ and $A^-B^+$.

$$A + \text{I.E.}_A \longrightarrow A^+ + e$$
$$B + e \longrightarrow B^- + \text{E.A.}_B$$
$$\overline{A + B + (\text{I.E.}_A - \text{E.A.}_B) \longrightarrow A^+B^-}$$

Similarly

$$B + \text{I.E.}_B \longrightarrow B^+ + e$$
$$A + e \longrightarrow A^- + \text{E.A.}_A$$
$$\overline{A + B + (\text{I.E.}_B - \text{E.A.}_A) \longrightarrow A^-B^+}$$

**Table 2.4 : Pauling electronegativities (H=2.1)**

| Group 1 | 2 | 3 | 4 | 5 | 6 | 7 | 8 | 9 | 10 | 11 | 12 | 13 | 14 | 15 | 16 | 17 |
|---|---|---|---|---|---|---|---|---|---|---|---|---|---|---|---|---|
| Li 1.0 | Be 1.5 | | | | | | | | | | | B 2.0 | C 2.5 | N 3.0 | O 3.5 | F 4.0 |
| Na 0.9 | Mg 1.2 | | | | | | | | | | | Al 1.5 | Si 1.8 | P 2.1 | S 2.5 | Cl 3.0 |
| K 0.8 | Ca 1.0 | Sc 1.3 | Ti 1.5 | V 1.6 | Cr 1.6 | Mn 1.5 | Fe 1.8 | Co 1.8 | Ni 1.8 | Cu 1.9 | Zn 1.6 | Ga 1.6 | Ge 2.0 | As 2.0 | Se 2.4 | Br 2.8 |
| Rb 0.8 | Sr 1.0 | Y 1.2 | Zr 1.4 | Nb 1.6 | Mo 1.8 | Tc 1.9 | Ru 2.2 | Rh 2.9 | Pd 2.2 | Ag 1.9 | Cd 1.7 | In 1.7 | Sn 2.0 | Sb 1.9 | Te 2.1 | I 2.5 |
| Cs 0.7 | Ba 0.9 | La-Lu 1.1-1.2 | Hf 1.3 | Ta 1.5 | W 1.7 | Re 1.9 | Os 2.2 | Ir 2.2 | Pt 2.2 | Au 2.4 | Hg 1.9 | Tl 1.5 | Pb 1.8 | Bi 1.9 | Po 2.0 | At 2.2 |
| Fr 0.7 | Ra 0.9 | Ac 1.1 | Th 1.3 | Pa 1.5 | U 1.7 | Np 1.3 | | | | | | | | | | |

Therefore in the formation of the species $A^+B^-$ the energy involved is ($I.E._A - E.A._B$) whereas in the formation of the specie $A^-B^+$ the corresponding energy is ($I.E._B - E.A._A$). Let the species $A^+B^-$ is more stable than $A^-B^+$ then (1) the ionisation energy of A is less than that of B i.e. $I.E._A < I.E._B$ and (2) the electron affinity of B is more than that of A i.e. $E.A._B > E.A._A$ and hence the energy involved in the formation of $A^+B^-$ is less than the energy involved in the formation of $A^-B^+$ i.e.

$$(I.E._A - E.A._B) < (I.E._B - E.A._A)$$

$$\text{or} \quad (I.E._A + E.A._A) < (I.E._B + E.A._B)$$

Similarly $A^-B^+$ formation is favoured when $(I.E._A + E.A._A) > (I.E._B + E.A._B)$ *Mulliken therefore suggested that (I.E.+E.A.) value of an atom is a measure of the tendency of an atom to pull electrons towards itself or according to Mulliken (I.E.+E.A.) is a measure of absolute electronegativity of an atom ($X_M$) given by equation (2.2).* It is to be noted that atoms having high ionisation energy, also have high electron affinity and such system is highly electronegative

As both I.E. and E.A. are given in eV atom$^{-1}$, Mulliken's electronegativity ($X_M$) is related to Pauling's electronegativity (X) as :

$$X = 0.168\,(I.E. + E.A. - 1.23)$$
$$= 0.336\,(X_M - 0.615)$$
$$= 0.336\,X_M - 0.2 \qquad \ldots(2.3)$$

Thus Mulliken's and Pauling electronegativity is related linearly. One important feature of Mulliken's approach is that it denotes the **orbital electronegativity** because ionisation energy and electron affinity depends on the nature of the orbital involved. The electronegativity increases in the sequence ns > np > nd and for hybrid orbitals $sp > sp^2 > sp^3$.

For example :

| | $sp^3$ | $sp^2$ | sp | p | | s | p |
|---|---|---|---|---|---|---|---|
| C | 2.52 | 2.80 | 3.34 | 1.74 | F | 10.4 | 3.95 |
| N | 3.74 | 4.18 | 5.14 | 2.52 | Cl | 6.3 | 3.00 |

Practical disadvantages of this method is that : (a) the electron affinity values of the elements is not always available (b) the electron affinity and ionisation energy values depend on the nature of the orbital involved i.e. the valence state of the species has to be considered. Hence to define electronegativity, **valence state ionisation energy** and **valence state electron affinity** is to be considered, both are dependent on the oxidation state and state of hybridisation of the element. For example in case of transition elements having $sp^3$ and $d^3s$ hybridisation assume same tetrahedral geometry, but the electronegativity of two hybrid orbitals are widely different. In such situation measurement of bond angle is of no use to denote the nature of hybridisation.

(3) **Allred-Rochow method :** Allred and Rochow (1958) stated electronegativity of an atom A is the electrostatic force of attraction experienced by the bonding electron towards the nucleus. The electrostatic force of attraction depends on the effective nuclear charge ($Z^*$) and the covalent radius ($r_{cov}$). Therefore

$$(X_A)_{AR} = F = \frac{e^2Z^*}{r_{cov}^2} \text{ (CGS system)} \quad ...(2.4)$$

where $(X_A)_{AR}$ = electronegativity of the atom A in Allred-Rochow scale.

e = electronic charge = $4.8 \times 10^{-10}$ esu.

$Z^*$ = effective nuclear charge = $Z - \sigma$

Z = total nuclear charge

$\sigma$ = screening constant, the value of which is obtained from Slater's rule

Allred and Rochow used the single bond covalent radius ($r_{cov}$) in angstrom units and they used certain correction parameter thereby the equation comes as :

$$(X_A)_{AR} = \frac{0.36Z^*}{r_{cov}^2} \quad ...(2.5)$$

In order to correlate Allred-Rochow electronegativity with that of the Pauling scale, the following emperical equation is obtained :

$$(X_A)_{AR} = \frac{0.36Z^*}{r_{cov}^2} + 0.74 \quad ...(2.6)$$

The computed Allred-Rochow and experimental Pauling electronegativities of the elements are in good agreement in the second, third and fourth period elements. However this method has some limitations due to nonavailability of reliable data for covalent radius.

**Problem 2.11** Compare the Mulliken and Pauling electronegativity according to the following data :

| Atom | Ionisation energy I.E., kcal mol$^{-1}$ | Electron affinity E.A, kcal mol$^{-1}$ | Mulliken's electronegativity $X_M$ | Pauling's electronegativity X |
|---|---|---|---|---|
| F | 401.7 | 78.4 | 4.0 | 4.0 |
| Cl | 299.0 | 83.4 | 3.2 | 3.0 |
| Br | 272.3 | 77.7 | 2.9 | 2.8 |
| I | 241.0 | 70.6 | 2.6 | 2.5 |

**Ans. :** $X_M = \frac{I.E.+E.A.}{2}$, since 1 eV = 23.06 kcal mol$^{-1}$.

$\therefore X_M = \frac{I.E.+E.A.}{23.06 \times 2}$, this value of $X_M$ is divided by 2.6 in order to adjust the value with that of Pauling's scale i.e. 2.6 is termed as scale adjustment factor. For example in case of F,

$$X = \frac{401.7 + 78.4}{23.06 \times 2} = \frac{10.4098}{2.6} = 4.0$$

In this way other values are calculated and inserted above.

(4) **Jaffe-Hinze concept (1962)** : Jaffe et. al. extended the idea of Mullikan that electronegativity not only depends on the nature of the orbital but also on the charge (q) of the species concerned. They defined electronegativity as :

$$X_{J\text{-}H} = \frac{dE}{dq} \quad ...(2.7)$$

where E is the total energy of the species (ionisation energy, electron affinity etc.) which is related by q in the following way :

$$E = \alpha q + \beta q^2 \quad ...(2.8)$$

And on differentiation we get,

$$X_{J\text{-}H} = \frac{dE}{dq} = \alpha + 2\beta q = a + bq \quad ...(2.9)$$

where $a = \alpha$ and $b = 2\beta$

Thus electronegativity is linearly related to q i.e. the charge of the species concerned. For neutral atom the electronegativity is given by a, b is called **charge coefficient** and represents the rate of change of electronegativity with charge. Generally larger atoms have smaller values of b while small atoms have larger values. The electronegativity of sulphur in $SF_4$ is more than that of sulphur in $SCl_4$. Similarly the electronegativity of oxygen in ClOH is greater than that of oxygen in $C_2H_5OH$.

It is to be noted that Mulliken and Jaffe et. al. concept is more theoretically sound compared to that of Pauling and Allred-Rochow concept which are based on some emperical considerations.

(5) **Group electronegativity** : *It is defined as the electronegativity of the atom adjusted for the presence of substituents are called group electronegativities.* This is dependent on the inductive ability of a substituent group. Group electronegativities can be calculated directly from the Mulliken-Jaffe approach considering kinetic data, dipole moments and atomic electronegativities. The central atom of a group may attain a large amount of positive or negative charge. This means that groups can donate or accept charge better than that would be indicated by their electronegativity values e.g. methyl group is slightly more electronegative than hydrogen yet methyl group is considered to be better donor than hydrogen atom. Some of the group electronegativity values are : $CH_3$(2.3), $CF_3$(3.47), $CCl_3$(2.95), OH(2.82), CN(3.32), $OCH_3$(2.52).

(6) **Sanderson's concept (1955)** : In this concept electronegativity is defined as **stability ratio (SR)** which gives the measure of compactness of the electron cloud in the valence shell of the atom i.e.

$$SR = \frac{E.D}{E.D_i} \quad ...(2.10)$$

where E.D = electron density of the atom concerned which is called average electron density

$E.D_i$ = ideal electron density of the atom which is calculated by the linear interpolation between the E.D values of the next higher and next lower inert gas atom.

E.D is expressed as :

$$E.D = \frac{\text{Number of electrons i.e. atomic number}}{\frac{4}{3}\pi r^3} \quad ...(2.11)$$

where r = covalent radius of the atom. For higher electronegative atom, the electron density will be higher i.e. r will be small and therefore stability ratio give the measure of electronegativity ($X_s$) i.e.

$$X_s = SR = \frac{E.D}{E.D_i} = \frac{r_i^3}{r^3} \quad ...(2.12)$$

where $r_i$ = ideal covalent radius obtained as defined by $E.D_i$.

### 2.5.4 Application of Electronegativity values

(a) **Periodic trends :** The overall trends in any scale of electronegativity is that it generally **increases from left to right** in a period and **decreases top to bottom in a group** of the periodic table like that of ionisation energy. For s-block elements group trends are retained but going down groups of f- and d- block elements there is a significant **increase in electronegativity.** As we go down the groups 13 and 14, the values first **decrease,** then **increase** and then **decrease** again. This is due to the participitation of d and f orbitals which have poor shielding capability. It is to be noted that electronegativity is an important predictor of chemical properties, we should also expect unusual trends in chemical reactivity. This non linear variation of electronegativity as also other periodic properties in a group is termed as **secondary periodicity.**

(b) **Electronegativity and nature of the bond :**

(1) If the two atoms have similar electronegativities then the bond type will be covalent i.e. homonuclear diatomic molecules e.g. $H_2$, $Cl_2$ etc.

(2) When the electronegativity of the atom $A(X_A)$ is low and that of $B(X_B)$ is high or vice-versa then $X_A - X_B$ or $X_B - X_A$ is high, then there will be considerable ionic character of the chemical bond.

In this case Pauling introduced the idea of ionic character of a chemical bond (say A–B) and suggested the following emperical equation connecting the electronegativity difference of the combining atoms as :

$$\text{Percentage ionic character} = 100\left[1-\exp\left\{-\frac{1}{4}(X_A \sim X_B)^2\right\}\right] \quad ...(2.13)$$

Pauling's equation (2.13) has some drawbacks and this was further modified by Hanny and Smyth as :

$$\text{Percentage ionic character} = 16(X_A \sim X_B) + 3.5(X_A \sim X_B)^2 \quad ...(2.14)$$

Percentage ionic character as calculated by equation (2.14) as :

| | | | | | |
|---|---|---|---|---|---|
| $X_A \sim X_B$ | 1 | 1.5 | 2.0 | 2.5 | 3.0 |
| % ionic character | 19.5 | 31.9 | 46.0 | 61.9 | 79.5 |

It is evident that greater the electronegativity difference between the combining atoms, the greater is the percentage ionic character. Electronegativity difference between F and K is 3.2(4.0 – 0.8 = 3.2) which indicates 87.04% ionic character, the value is very high. On the other hand C and H has a poor electronegativity difference (2.5 – 2.1 = 0.4) which induces only 7% ionic character. The electronegativity difference between H and F is 1.9, therefore HF has 43% ionic character. Similarly HCl, HBr and HI has ~17%, ~13% and ~7% respectively corresponding electronegativity differences are 0.9, 0.7 and 0.4 respectively. A plot of percentage ionic character with that of electronegativity difference is shown in Fig. 2.4.

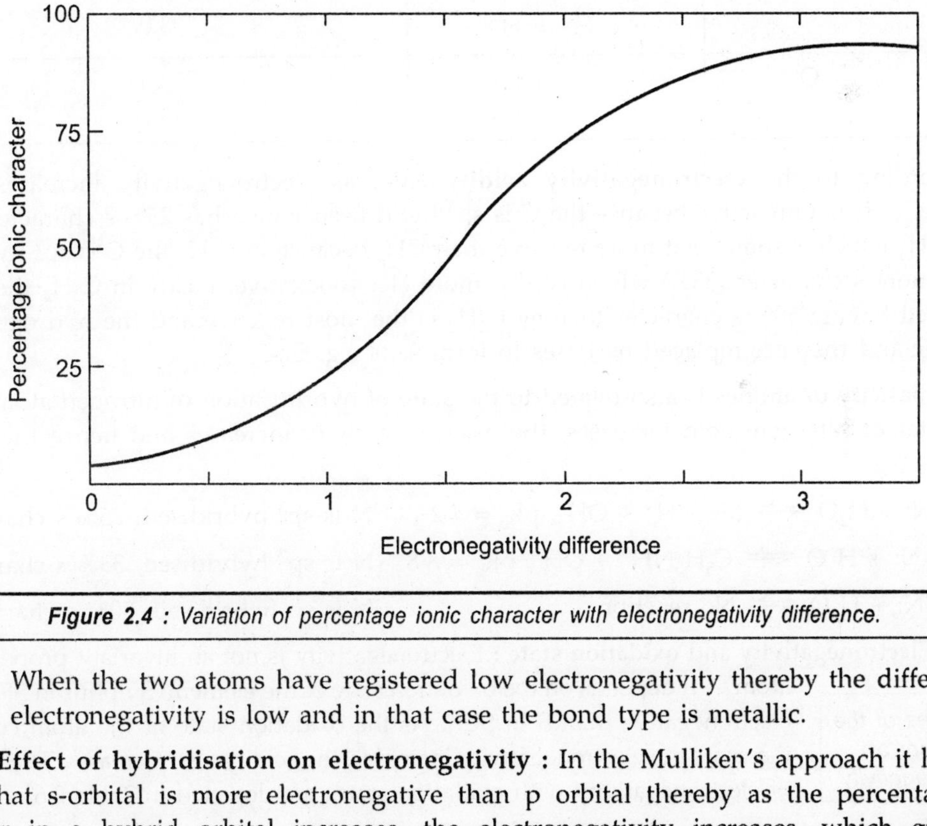


***Figure 2.4** : Variation of percentage ionic character with electronegativity difference.*

(3) When the two atoms have registered low electronegativity thereby the difference in electronegativity is low and in that case the bond type is metallic.

(c) **Effect of hybridisation on electronegativity** : In the Mulliken's approach it has been stated that s-orbital is more electronegative than p orbital thereby as the percentage of s character in a hybrid orbital increases, the electronegativity increases, which gives the electronegativity sequence as sp > $sp^2$ > $sp^3$. Variation of electronegativity values of some elements with different hybrid orbitals are given in Table 2.5.

**Table 2.5 : Variation of electronegativity values with variation of hybridisation**

| Element | Hybridisation | Electronegativity (Pauling) |
|---|---|---|
| Be | sp | 1.5 |
| B | $sp^2$<br>$sp^3$ | 2.0<br>1.9 |
| C | sp<br>$sp^2$<br>$sp^3$ | 3.0<br>2.7<br>2.5 |
| N | sp<br>$sp^2$<br>$sp^3$ | 3.7<br>3.3<br>3.0 |
| O | $sp^2$<br>$sp^3$ | 3.9<br>3.7 |

According to the **electronegativity acidity rule**, as electronegativity increases acidity increases. $CH_4$ is unreactive because the C is $sp^3$ hybridised which has 25% s character. As we go to $C_2H_4$ which is somewhat more reactive than $CH_4$ because in $C_2H_4$ the C is $sp^2$ hybridised having more s character (33%) which is also more electronegative. Lastly in $C_2H_2$ the C is sp hybridised having 50% s character thereby $C_2H_2$ is the most reactive and the hydrogen atoms are acidic and they are replaced by bases to form salts e.g. $CaC_2$.

The basicity of amines is also related to the state of hybridisation of nitrogen atom. As the s character of nitrogen atom increases, the electronegativity increases and hence the basicity decreases e.g.

$Me_3N: + H_2O \rightleftharpoons Me_3NH^+ + OH^-$, $pk_b = 4.2$ N is $sp^3$ hybridised, 25% s character

$C_5H_5N: + H_2O \rightleftharpoons C_5H_5NH^+ + OH^-$, $pk_b = 8.8$ N is $sp^2$ hybridised, 33% s character

$MeCN: + H_2O \rightleftharpoons$ No reaction N is sp hybridised, 50% s character

(d) **Electronegativity and oxidation state :** Electronegativity is not an invariant property of an atom — it depends on the oxidation state of the element. Sulphur in $SF_6$ is more electronegative than S in $SCl_2$. As the oxidation state of the atom concerned increases the electronegativity increases, the acidity also increases. The variation of electronegativity with oxidation states are depicted in Table 2.6.

*Limitations of the concept of electronegativity*

(e) **Effect of substituent on electronegativity :** The electronegativity of an atom can vary with respect to the partial charge accumulated by substituent atoms or groups. As for example $CH_3I$ and $CF_3I$ hydrolyses in different way

$$CH_3I + OH^- \longrightarrow CH_3OH + I^-$$

$$CF_3I + OH^- \longrightarrow CF_3H + IO^-$$

**Table 2.6 : Variation of Pauling electronegativity with oxidation states**

| Atom (oxidation state) | Electronegativity |
|---|---|
| Fe(II) | 1.83 |
| Fe(III) | 1.96 |
| Sn(II) | 1.80 |
| Sn(IV) | 1.96 |
| Cu(I) | 1.90 |
| Cu(II) | 2.00 |
| Pb(II) | 1.87 |
| Pb(IV) | 2.33 |
| Mo(II) | 2.16 |
| Mo(III) | 2.19 |
| Mo(IV) | 2.24 |
| Mo(V) | 2.27 |
| Mo(VI) | 2.35 |

In $CH_3I$ the negative $OH^-$ attacks the most positive carbon atom and thereby the $I^-$ is displaced. On the other hand in $CF_3I$ the F atoms induce some positive charge on the carbon atom and the electronegativity increases above that of iodine, which is attacked by $OH^-$ to form HOI. HOI then loses an $H^+$ ion in alkaline medium to form $IO^-$.

The following reactions also illustrate the variation of polarity of C–I bonds which depends on the substituent of the carbon atom :

$$CH_3I + Na[Mn(CO)_5] \longrightarrow NaI + CH_3Mn(CO)_5$$

$$CF_3I + Na[Mn(CO)_5] \longrightarrow Mn(CO)_5I + CF_3Na$$

$$CF_3Na + CF_3I \longrightarrow NaI + C_2F_6$$

$$2CF_3I + Na[Mn(CO)_5] \longrightarrow NaI + C_2F_6 + Mn(CO)_5I$$

In case of boron halides it is expected that electronegativity of B in $BF_3$ is highest and least in case of $BI_3$ i.e. according to electronegativity acidity rule the acidity order should be $BF_3 > BCl_3 > BBr_3 > BI_3$ but the actual order of Lewis acidity is the reverse of what is expected i.e. $BI_3 > BBr_3 > BCl_3 > BF_3$. This is due to the structure of boron halides. In case of $BF_3$, the B-F bond has multiple bonding character and has the following resonating structures :

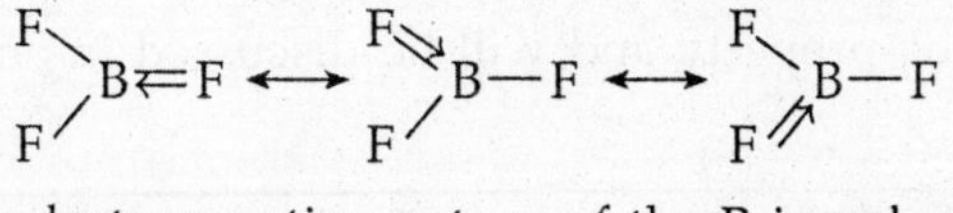

Due to this, electronegative nature of the B is reduced and thereby Lewis acidity property is somewhat quenched. On the other hand in case of $BI_3$ there is least scope of forming multiple bonding character between B and I due to the larger size of iodine thereby B in $BI_3$ is electron greedy, so that $BI_3$ will have highest acidity among the boron halides.

(f) **Effective nuclear charge and electronegativity :** In case of Allred-Rochow concept, the electronegativity is proportional to the effective nuclear charge ($Z_{eff}$) at the periphery of the atom. This factor is responsible for the variation of electronegativity in a period and in a group.

(g) **Charge transfer transitions and electronegativity :** Sometimes the colour of chemical compounds can be explained by **charge transfer transitions/redox transitions** particularly ligand to metal charge transfer transition (LMCT). This is favoured when the metal is highly polarising i.e. in the high oxidation state having high electronegativity and the ligand is highly polarisable i.e. low electronegativity. With increasing polarisability i.e. decreasing electronegativity of the ligand, the charge transfer band undergoing a **red shift** i.e. absorption occurs at a lower frequency e.g.

AgCl (white), AgBr (pale yellow), AgI (yellow) and $Ag_2S$ (black). For transition metal ions (oxospecies) as the oxidation state increases (higher electronegativity) there is a gradual decrease of frequency e.g.

$VO_4^{3-}$ (white), $CrO_4^{2-}$ (yellow), $MnO_4^{2-}$ (purple)

The energy of the LMCT transition is related to a certain property of the filled ligand π-orbitals and the metal $t_{2g}$ orbitals. This property is known as **optical electronegativity** of ligand ($EN_L$) and metal ($EN_M$) which is related to the wave number ( $\bar{\nu}$ ) of the LMCT band as :

$$\bar{\nu} = C(EN_L - EN_M)$$

when C = 30,000 $cm^{-1}$ then $EN_L$ and $EN_M$ coresponds to the Pauling scale of electronegativity.

**Problem 2.12** Define spectroscopic electronegativity.

**Ans. :** This concept proposed by L.C. Allen (1989) and is based on the average energy of the s and p electrons in an atom.

$$X_s = \frac{N_pE_p + N_sE_s}{N_p + N_s}$$

Where $X_s$ is spectroscopic electronegativity, $E_p$ and $E_s$, $N_p$ and $N_s$ are the energies and number of electrons in the p and s orbitals of the valency shell, respectively. Practical application of $X_s$ is that electronegativities of noble gases can be calculated for which no thermochemical data is available for calculation of Pauling electronegativity.

## 2.5.5 Electronegativity related to Hardness and Softness

R.G. Pearson (1988) defined **absolute hardness** (η) as :

$$\eta = \frac{I.E. - E.A.}{2}$$

This is another fundamental periodic property and will be discussed in acid base chapter (Sec. 4.5.4).

**Problem 2.13** Calculate the electronegativity of fluorine from the following data : $D_{H-H}$ = 104.2 kcal $mol^{-1}$, $D_{F-F}$ = 36.6 kcal $mol^{-1}$ and $D_{H-F}$ = 135.6 kcal $mol^{-1}$.

**Ans. :** According to Pauling's method,

$$\Delta = D_{H-F} - \sqrt{D_{H-H}D_{F-F}} = 135.6 - \sqrt{104.2 \times 36.6}$$

= 135.6 – 61.755 = 73.845 kcal = 73.845 × 4.184 = 308.967 kJ $mol^{-1}$ **

Now from equation (2.1) $X_F - X_H = 0.102\sqrt{308.967} = 1.79$.

$\therefore X_F = 1.79 + X_H = 1.79 + 2.1 = 3.89$

** Since in equation (2.1), the bond energy is expressed in kJ $mol^{-1}$ and the conversion factor is 1 eV = 96.48 kJ $mol^{-1}$ = 23.06 kcal $mol^{-1}$.

$\therefore$ 1 kcal = 4.184 kJ.

## 2.6 WORK FUNCTION AND IONISATION ENERGY / POTENTIAL

In the photoelectric effect (Sec. 1.4.1) a photon of light of frequency $\nu_0$ called **threshold frequency** can be able to knock off an electron from the metal surface. The most loosely bound electrons at the surface of the metal atoms require minimum energy ($h\nu_0$) to be released from the binding forces of the nucleus. This minimum energy is called **work function** or **threshold energy**. Ionisation potential/energy also require minimum amount of energy to release the most loosely bound electron from the metal atom in an **isolated gaseous condition**. *Thus both work function and ionisation energy/potential are idex of the minimum amount of energy to knock off the most loosely bound electron, the work function term denotes the minimum energy of the metal in the solid state whereas the ionisation energy/potential denotes the same in the gaseous state.* It is evident that the amount of energy in case of work function is always less compared to that of ionisation energy/potential. Some values are given in Table 2.7.

**Table 2.7 : Values of work function and ionisation potential**

| Metal | Work function (ev) | Ionisation energy/potential (eV) |
|---|---|---|
| Na | 2.50 | 5.14 |
| K | 2.26 | 4.34 |
| Cs | 1.87 | 3.89 |
| Cu | 4.30 | 7.72 |
| Zn | 3.44 | 9.39 |
| Ag | 4.73 | 7.57 |
| Fe | 4.71 | 7.90 |

**Problem 2.14** When sodium metal surface is exposed to radiation of 300 nm, the electrons with kinetic energy (K.E.) 1.68 × $10^5$ $Jmole^{-1}$ are emitted. What is the minimum energy needed to remove an electron from the metal surface? What is the maximum wavelength that will cause a photo-electron to be emitted?

**Ans.** : K.E. per electron = $\dfrac{\text{Total K.E.}}{\text{Avogadro's number}} = \dfrac{1.68 \times 10^5 \text{ J}}{6.02 \times 10^{23}} = 2.79 \times 10^{-19}$ J

From equation (1.1)

$$h\nu = h\nu_0 + \frac{1}{2}mv^2$$

i.e. $h\nu_0 = h\nu - K.E.$, we know that $E_{min} = h\nu_0$

$\therefore E_{min} = h\nu - K.E.,$ $\qquad h\nu = h \times \frac{c}{\lambda}$ $[\nu\lambda = c; \therefore \nu = \frac{c}{\lambda}]$

$= 6.627 \times 10^{-19} - 2.79 \times 10^{-19}$ J $\qquad = \frac{(6.627 \times 10^{-34} Js)(3 \times 10^8 ms^{-1})}{300 \times 10^{-9} m}$

$= 3.837 \times 10^{-19}$ J $\qquad = 6.627 \times 10^{-19}$ J

$\therefore h\nu_0 = 3.837 \times 10^{-19}$ J.

We know that $h\nu_0 = h \times \frac{c}{\lambda_{max}}$

$$\therefore \frac{6.627 \times 10^{-34} Js \times 3 \times 10^8 ms^{-1}}{\lambda_{max}} = 3.837 \times 10^{-19} \text{ J.}$$

$$\therefore \lambda_{max} = \frac{6.627 \times 10^{-34} Js \times 3 \times 10^8 ms^{-1}}{3.837 \times 10^{-19} J} = 5.181 \times 10^{-7} \text{ m.}$$

## 2.7 ATOMIC OR COVALENT RADII

These terms are similar. *In a homonuclear diatomic molecule e.g. $H_2$, $Cl_2$ etc., one half of the distance between the nuclei of two covalently bonded atoms is taken as the covalent radius of that atom.* The distance between the nuclei of the two atoms is taken as the bond length. In this concept atoms are considered as **hard sphere** model and they are touching each other. But in the practical case, the atoms may not touch each other and also due to the multiple bonding character, there will be some shrinkage of the bond length, thereby covalent radius will be somewhat shorter than can be expected for single bond. The distance between the centres of any two neighbouring atoms can be measured by X-ray diffraction or spectroscopic studies. The bond length of $H_2$ molecule is measured as 0.74Å so that atomic radius of hydrogen is 0.37Å. Similar measurements give radius of chlorine and bromine as 0.99Å and 1.14Å respectively.

In case of heteroatom systems e.g. A–B, the situation may be complicated. In general we can expect additive property of the atomic radius in order to have the bond length value. For example single bond covalent radius of C is measured as 0.77Å so that C–H bond length will be 0.77+0.37 = 1.14Å which is in agreement with that obtained spectroscopically. However in heteroatom systems the measured bond length may sometimes not tally with the calculated values because of the electronegativity difference as also other factors. We shall consider some factors governing variation of atomic radii :

### (1) State of hybridisation and bond multiplicity

*Limitations of the concept of atomic radii*

As bond multiplicity increases bond length decreases as also atomic radius decreases. Bond multiplicity also related to the state of hybridisaton. As the s character of the hybrid orbital increases, the bond multiplicity increases as also covalent radius decreases e.g.

| | | | | | |
|---|---|---|---|---|---|
| N–N > | N = N > | N ≡ N | C ≡ C < | C = C < | C – C |
| $sp^3$ | $sp^2$ | sp | sp | $sp^2$ | $sp^3$ |

Variation of bond multiplicity and bond length/atomic radius are depicted in Table 2.8. In the case of multiple bonding we can calculate the atomic/covalent radius in the same way as that of single bond radius i.e. bond lengths are related to multiple bond lengths e.g.

$$r_{C\equiv N} = \frac{r_{C\equiv C} + r_{N\equiv N}}{2} \quad ...(2.15)$$

**Table 2.8 : Variation of covalent/atomic radii with bond multiplicity**

| Covalent bond | Bond length (pm)* | Bond energy kJ mol$^{-1}$ | Covalent/atomic radius (pm) |
|---|---|---|---|
| C–C | 154 | 356 | 77 |
| C=C | 134 | 682 | 67 |
| C≡C | 120 | 962 | 60 |
| N–N | 140 | 240 | 70 |
| N=N | 125 | 450 | 62 |
| N≡N | 110 | 942 | 55 |
| O–O | 148 | 138 | 74 |
| O=O | 121 | 493 | 60 |
| P=P | 200 | — | 100 |
| P≡P | 186 | — | 93 |
| Si–Si | 234 | — | 117 |
| C–O | 143 | 351 | — |
| C=O | 127 | 724 | — |
| C≡O | 110 | — | — |
| S=S | 188 | — | 94 |
| F–F | 142** | 159 | 64 |

** 1Å = 100 pm (pm = picometre)*

*** The increased bond length in F–F is probably due to the repulsive force operated between the non bonding electrons of each F atoms. kcal mole$^{-1}$ = 4.18 kJ mol$^{-1}$.*

The calculated value of C≡N according to equation (2.15) is $\frac{120+110}{2}$ = 115 pm which is quite agreement with that obtained from measurement of C≡N in $CH_3CN$ (116 pm). Similar is true for C–Si bond (calculated value is 194 pm, experimental value of C–Si bond length in $(CH_3)_4Si$ is 193 pm). In some cases there may be shrinkage or lengthening of bond lengths may occur. Bent's rule (Sec. 3.3.3) can explain the relative shortage of bond length. The $PCl_5$ is an example of non-equivalent hybrid orbitals. Here the axial P-Cl bonds are longer than the equatorial P-Cl bonds thereby the atomic radius of phosphorous will be different in two cases.

*Limitations of the concept of atomic radii*

### (2) Effect of π-bonding or Synergic effect

When the metals are in zero oxidation states the radius of the metal is sometimes called **metallic radius** which is similar to covalent radius. Metals like Ni, Fe, Cr and Mo etc. form carbonyl complexes like $Ni(CO)_4$, $Fe(CO)_5$, $Cr(CO)_6$ and $Mo(CO)_6$ respectively. In these complexes CO acting as a σ donor and π-acceptor ligand e.g.

$$M \underset{\pi}{\overset{\sigma}{\rightleftarrows}} :CO$$

This type of interaction is known as **synergic effect.** Due to this type of effect the M–C bond distance have considerable shrinkage so that atomic radius of metal have some reduced value than can be expected. This aspect will be discussed in the chapter of complexes of π-acid ligands.

### (3) Effect of geometry

For a particular atom, the state of hybridisation, the stereochemistry and the number of covalent bonds are interrelated. Whether the arrangement is linear (sp) or tetrahedral ($sp^3$ or $d^3s$) or octahedral ($sp^3d^2$ or $d^2sp^3$) or square planar ($dsp^2$) depends entirely on the element concerned as also on the attached groups or ligands. The atomic radius of an atom is different for different stereochemistry. For example,

Octahedral radius ≈ square planar radius > tetrahedral radius > linear radius

### (4) Effect of electronegativity difference

The additivity of atomic radius to produce the bond length is applicable so long as the two atoms have same electronegativity or very little difference in electronegativity. If A is more electronegative than B, then A–B bond has considerable ionic character, then in that case A will assume some negative charge and B will have some positive charge. *In this way there results an electrostatic attraction beween A and B and thereby the two atoms will be much closer.* In other words the bond between A and B have considerable shrinkage than can be expected from the sum of the atomic radii of A and B. For example in HCl the expected bond length = 0.37 + 0.99 = 1.36Å but the experimental value is 1.27Å. In this case chlorine having a bulkier atom will induce some shrinkage of atomic radius of hydrogen and here the atomic radius of hydrogen = 1.27 – 0.99 = 0.28Å. Some of the variation of bond lengths are depicted in Table 2.9.

**Table 2.9 : Values of calculated and experimental bond lengths**

| Compound | Chemical bond | Calculated value (pm) | Experimental value (pm) |
|---|---|---|---|
| HF | H–F | 108 | 92 |
| HCl | H–Cl | 136 | 128 |
| HBr | H–Br | 151 | 142 |
| HI | H–I | 170 | 161 |
| $CH_4$ | C–H | 114 | 109 |
| $CF_4$ | C–F | 148 | 136 |

This shrinkage of the chemical bond (A–B) is explained by the Schomaker and Stevenson relation :

$$r_{A-B} = r_A + r_B - C|X_A - X_B| \quad ...(2.16)$$

where $r_{A-B}$, $r_A$ and $r_B$ expressed in picometre, C is the Schomaker-Stevenson coefficient usually has the value 9 but may vary from system to system, $X_A$ and $X_B$ are the electronegativities of the atoms A and B respectively.

**Problem 2.15** Calculate the N–O and C–O bond length. The atomic radius of N, O and C are 70, 74 and 77 pm respectively while electronegativities are 3.0, 3.5 and 2.5 respectively.

**Ans.** : According to equation (2.16)

$$r_{N-O} = r_N + r_O - 9|X_O - X_N|$$
$$= 70 + 74 - 9(3.5 - 3) = 139.5 \text{ pm.}$$

$$r_{C-O} = r_C + r_O - 9|X_O - X_C|$$
$$= 77 + 74 - 9(3.5 - 2.5) = 142 \text{ pm.}$$

[Experimental C–O bond length 143 pm. Table 2.8]

## 2.8 IONIC RADII

**Ionic radii corresponds to the radii of the cation and anion in ionic crystals.** Isolated ions are somewhat different from the ions in ionic crystals. Ionic crystals are considered as **hard sphere** model where the cations and anions are considered as spherical and are touching each other in the equilibrium positon. The internuclear distance ($r_{C-A}$) between the cation (C) and the anion (A) in an ionic crystal can be measured which is logically the sum of the radius of the cation ($r_C$) and anion ($r_A$).

$$r_{C-A} = r_C + r_A$$

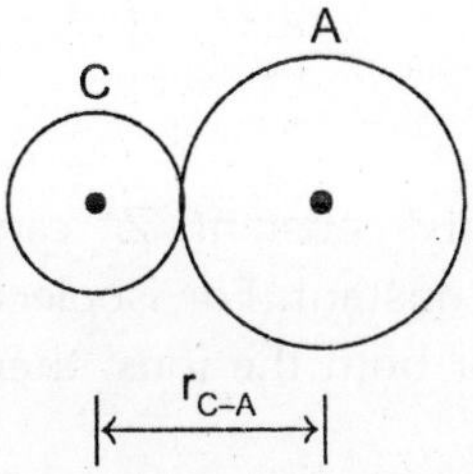


It is evident that cations have smaller size than the neutral atoms whereas the anions have larger size compared to that of neutral atoms. Therefore a problem arises how we will compute the individual ionic radius of cation and anion from the equilibrium internuclear distance ($r_{C-A}$). There are several approaches :

### (1) X-ray studies

X-ray studies can give us data precisely for the equilibrium internuclear distance for oppositely charged ions but cannot be able to provide information about the individual ionic radii of cations and anions. For example in KI the $r_{K^+-I^-} = 353$ pm and in NaCl the $r_{Na^+-Cl^-} = 271$ pm.

From these data individual ionic radius cannot be calculated unless one is known. Lande (1920) assumed that since $Li^+$ is the smallest monopositive metal cation and $I^-$ has a very large size, therefore in LiI crystal X-ray data can give us idea about the $r_{Li^+-I^-}$ distance (274 pm) as also $r_{I^--I^-}$ distance (426 pm) as depicted in the Fig. 2.5.

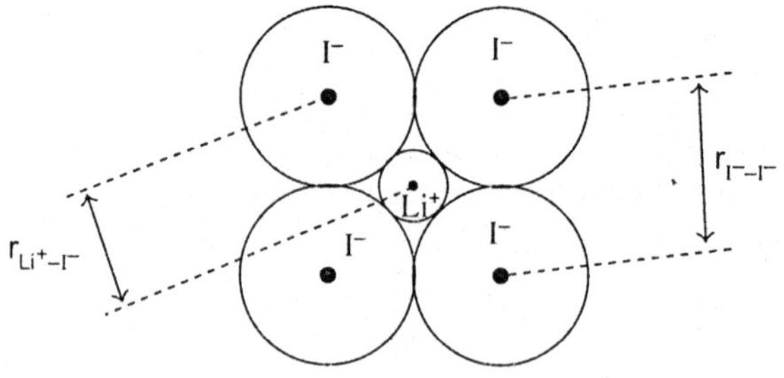


***Figure 2.5** : Determination of ionic radius of $I^-$ in LiI crystal.*

Thus the ionic radius of $I^-$ will be equal to the half of the $I^-$–$I^-$ distance and is equal to 213 pm. Thus using the value of $I^-$ the ionic radius of $Li^+$ = 274–213 = 61 pm and the ionic radius of $K^+$ in KI = $r_{K^+-I^-} - r_{I^-}$ = 353 − 213 = 140 pm.

In this way by physical measurements the ionic radius of other cations and anions can be calculated.

## (2) Pauling's approach

Pauling proposed that ionic radius of an ion should be inversely proportional to the effective nuclear charge. As effective nuclear charge (Z*) increases, the radius decreases i.e.

$$r = \frac{C_n}{Z^*} \qquad \ldots(2.17)$$

where $C_n$ is a proportionality constant, Z* can be calculated from Slater's rule where $Z^* = Z - \sigma$ and $\sigma$ is a screening constant. For isoelectronic systems e.g. NaF, KCl, RbBr and CsI the constant term $C_n$ is same for both the ions, therefore we can write,

$$\frac{r_+}{r_-} = \frac{Z^*_{anion}}{Z^*_{cation}} \qquad \ldots(2.18)$$

For NaF both $Na^+$ and $F^-$ has the electronic configuration $1s^22s^22p^6$. The screening constant $(\sigma) = 8 \times 0.35 + 2 \times 0.85 = 4.50$ so that—

$$\frac{Z^*_{Na^+}}{Z^*_{F^-}} = \frac{9-4.5}{11-4.5} = \frac{4.5}{6.5}$$

The internuclear distance $r_{Na^+-F^-}$ is determined from X-ray crystallography which is equal to 231 pm. Therefore,

$$r_{Na^+} = 231 \times \frac{4.5}{4.5+6.5} = 94.5 \text{ pm}$$

$$r_{F^-} = 231 \times \frac{6.5}{4.5+6.5} = 136.5 \text{ pm}$$

Similarly in MgO the $r_{O^{2-}}$ and $r_{Mg^{2+}}$ calculated as 140 pm and 65 pm respectively. Some values of ionic radius are given in Table 2.10.

**Table 2.10 : Pauling ionic radius of some representative ions in picometre**

| | | | | | | |
|---|---|---|---|---|---|---|
| $Li^+$(60) | $Be^{2+}$(31) | | | $N^{3-}$(132)* | $O^{2-}$(140) | $H^-$(208) |
| $Na^+$(95) | $Mg^{2+}$(65) | $Al^{3+}$(50) | $Si^{4+}$(41) | $P^{5+}$(52) | $S^{2-}$(184) | $F^-$(136) |
| $K^+$(133) | $Ca^{2+}$(99) | $Ga^{3+}$(62) | $Ge^{4+}$(53) | $As^{3+}$(69) | $Se^{2-}$(198) | $Cl^-$(181) |
| $Rb^+$(148) | $Sr^{2+}$(113) | $In^{3+}$(81) | $Sn^{4+}$(71) | $Sb^{3+}$(90) | $Te^{2-}$(221) | $Br^-$(195) |
| | $Ba^{2+}$(135) | $Tl^{3+}$(95) | $Pb^{4+}$(84) | $Bi^{3+}$(120) | | $I^-$(216) |
| | | | | | | $At^-$(220) |

* *Shannon-Prewitt crystal ionic radii*

### (3) Yatsimirski's thermochemical radii

For non-spherical ions X-ray studies and Pauling's approach give us erronous result. The radius of the non-spherical ions are determined by Yatsimirski's approach. In this concept radius of the ions are determined from thermochemical lattice energy data. Some typical values are $CN^-$(177 pm), $OH^-$(119 pm), $CO_3^{2-}$(164 pm), $NO_3^-$(165 pm), $SO_4^{2-}$(244 pm), $N_3^-$(181 pm), $PO_4^{3-}$(238 pm), $CH_3COO^-$(148 pm), $NO_2^-$(155 pm).

## 2.9 VAN DER WAALS RADII

This radii arises due to the presence of weak intermolecular forces operated between the real gas molecules. These forces are collectively known as van der Waals forces and will be discussed in the chapter of chemical bonding. These forces are not truely a chemical bond and are therefore sometimes called non-bonded forces.

*Difference between atomic and van der Waals radius*

Due to the formation of electron pair bond covalent radii arises but due to the weak non-bonded forces van der Waals radii arises. *The internucler distance between the atoms of two different molecules which come closer together due to the weak dispersion forces exist, is the sum of van der Waals radii of the atoms under consideration (Fig. 2.6) and half of this distance is known as van der Waals radius or non-bonded radius.*

*Limitation of van der Waals radius*

Pictorially Van der Waals radius and covalent radius are apparently the same but actually they always differ. Van der Waals radius is not an invariant property of an element, it depends on the compression of environmental forces. For example the van der Waals radius of H = 120–145 pm, C = 165–170 pm,

N = 150 pm, O = 150 pm, F = 155 pm, Cl = 180 pm, Br = 190 pm and I = 204 pm. Since there are weak intermolecular forces operated beween the molecules, therefore van der Waals radius are always greater than covalent radius.

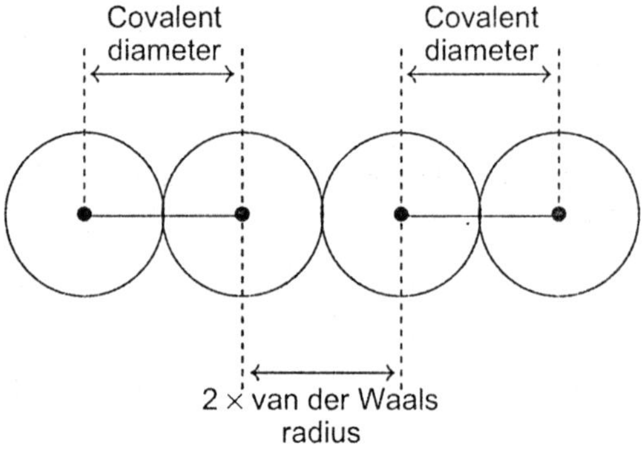


*Figure 2.6 : van der Waals and covalent radius.*

## 2.10 NOTES ON DIFFERENT RADII VALUES

1. Generally the radius values (atomic covalent, ionic and van der Waals radii) in a period of periodic table from left to right decreases and in a group from top to bottom increases for regular elements. Table 2.11 shows a long list of atomic/covalent/metallic radius, ionic radius and van der Waals radius of elements. It is evident that both Al and Ga has the same atomic radius though they have wide difference in atomic number 13 and 31 respectively belonging to the same group. This is also true in case of Si and Ge belonging to group–14 (Si 117pm; Ge 122 pm). Al/Ga and Si/Ge are p-block elements having electronic configurations Al($1s^22s^22p^63s^23p^1$) and Si($1s^22s^22p^63s^23p^2$). It is expected that as we go from Al/Si to Ca($1s^22s^22p^63s^23p^64s^2$) the radius must increase. The atomic radius of Ca is 197 pm. After Ca the electrons are added to the 3d orbital. The radial distribution function of 3d and 4s orbitals (Sec. 1.8.7.1) suggest that 3d orbital is more nearer to the nucleus. Therefore whatever number of electrons are added to the 3d orbital, this will not bring in increase in radius, instead increasing nuclear charge from Sc(21) to Zn(30) brings about a slow contracton in size. After Zn the electron is added to 4p orbital of Ga($1s^22s^22p^63s^23p^64s^23d^{10}4p^1$) thereby some compensation of the radius occurs and so Al/Ga and Si/Ge has almost equal atomic radius. This effect is referred to as **scandide contraction** of the elements Ga through Kr. This scandide contraction can also be explained by effective nuclear charge (Slater's rule) (Sec. 2.11).

Table 2.11 shows that for first inner transition elements [La(57) to Lu(71)], the ionic radius of +3 cation has a slow decrease with increasing atomic number. This is known as **lanthanide contraction** (Problem 2.2). This is due to the filling of electrons in inner 4f orbital.

2. The radius of the positive ions always less than the neutral atoms. This is due to the fact that when electron(s) are removed from the neutral atom, the nucleus will attract the outermost electron more strongly thereby positive ions will have lower radius than neutral atoms. For example the radius of Na and $Na^+$ is 186 pm and 95 pm respectively.

**Table 2.11 : Values of atomic, ionic and van der Waals radius in picometre**

| Atomic/covalent/metallic radius | | Ionic radius | | van der Waals radius | |
|---|---|---|---|---|---|
| H(37/28) | He(93) | $H^-$(208) | $Li^+$(60) | H(120) | He(120) |
| Li(152) | Be(111) | $N^{3-}$(150) | $O^{2-}$(140) | N(150) | O(140) |
| B(90) | C(77) | $F^-$(136) | $Na^+$(95) | F(135) | Ne(160) |
| N(70) | O(60) | $Mg^{2+}$(65) | $Al^{3+}$(50) | Na(230) | Mg(170) |
| F(64) | Ne(112) | $Si^{4+}$(41) | $P^{3-}$(190) | P(185) | S(180) |
| Na(186) | Mg(160) | $S^{2-}$(184) | $Cl^-$(181) | F(155) | Cl(180) |
| Al(143) | Si(117) | $K^+$(133) | $Ca^{2+}$(99) | Br(190) | I(204) |
| P(110) | S(102) | $Ti^{2+}$(100) | $Ti^{3+}$(81) | Ga(190) | Se(190) |
| Cl(99) | Ar(154) | $V^{2+}$(93) | $V^{3+}$(78) | Te(210) | Ar(190) |
| K(235) | Ca(197) | $Cr^{2+}$(89) | $Cr^{3+}$(76) | | |
| Sc(164) | Ti(147) | $Mn^{2+}$(80) | $Fe^{2+}$(76) | | |
| V(135) | Cr(129) | $Fe^{3+}$(70) | $Co^{2+}$(73) | | |
| Mn(127) | Fe(126) | $Co^{3+}$(68) | $Ni^{2+}$(72) | | |
| Co(125) | Ni(125) | $Cu^{2+}$(87) | $Zn^{2+}$(62) | | |
| Cu(128) | Zn(137) | $La^{3+}$(117) | $Ce^{3+}$(115) | | |
| Ga(140) | Br(114) | $Pr^{3+}$(113) | $Nd^{3+}$(112) | | |
| As(122) | Sb(143) | $Pm^{3+}$(111) | $Sm^{3+}$(109) | | |
| I(133) | Cs(262) | $Eu^{3+}$(108) | $Gd^{3+}$(108) | | |
| Y(182) | Zr(160) | $Tb^{3+}$(106) | $Dy^{3+}$(105) | | |
| Nb(147) | Mo(140) | $Au^{3+}$(99) | | | |
| Tc(135) | Ru(134) | | | | |
| Rh(134) | Pd(137) | | | | |
| Ag(144) | Cd(152) | | | | |
| La(188) | Ce(182) | | | | |
| Pr(183) | Nd(182) | | | | |
| Sm(180) | Eu(204) | | | | |
| Gd(180) | Tb(178) | | | | |
| Dy(177) | Au(144) | | | | |

3. Conversely the radius of the anions are greater than the neutral atoms. For example the radius of $Cl^-$ (181 pm) is greater than the atomic radius of Cl (99 pm). This is due to the fact that in neutral atom there is a balance between the number of electrons and number of protons in the nucleus but in negative anion there is a repulsive force operated in the electron shell thereby the radius of anions are larger than that of neutral atoms.

4. With increasing positive charge on metal cation, the radius decreases. For example $Sn^{4+}$ (71 pm) is less than that of $Sn^{2+}$ (93 pm). This is due to the fact that with increasing charge on

the metal cation, extra shringkage operated between the nucleus with that of the outermost electron shell.

5. Generally the anions have larger size than cation so that radius ratio ($r_{cation}/r_{anion}$) is less than 1 for ionic crystals and in determining radius ratio the radius of the cation is more important than anion. This radius ratio rule dictate the coordination number of the cation. It is evident that radius of a cation increases as coordination number increases. This is due to the fact that as more anions crowd around the cation, they repel each other more and thereby the radius of a cation increases e.g.

| | | | | | | |
|---|---|---|---|---|---|---|
| Cordination number of $Na^+$ | 4 | 5 | 6 | 7 | 8 | 12 |
| Radius of $Na^+$ (pm) | 113 | 114 | 116 | 126 | 132 | 138 |

6. Transition metals normally a good complex former and they form high spin (h.s.) and law spin (l.s.) complexes. The radius of the cation in high spin state is greater than that of low spin state. This will be discussed in coordination chemistry chapter.

7. Lastly we may select a series of isoelectronic ions of fifth period e.g. $Rb^+$, $Sr^{2+}$, $Y^{3+}$, $Zr^{4+}$, $Nb^{5+}$, $Mo^{6+}$ and $Tc^{7+}$. As the positive charge of the nucleus increases from Rb(37) to Tc(43), the attraction for the 36 electrons become stronger, resulting a contraction in cationic radii with increasing positive charge from +1 to +7. The Shannon-Prewitt crystal ionic radii values in picometre are : $Rb^+$(166), $Sr^{2+}$(132), $Y^{3+}$(104), $Zr^{4+}$(86), $Nb^{5+}$(78), $Mo^{6+}$(73) and $Tc^{7+}$(70).

---

**Problem 2.16** Show an example of a complex where metal-carbon bond multiplicity increases from single bond to triple bond.

**Ans. :** Following complex shows that as bond multiplicity increases bond length decreases.

W ≡ C bond length = 181 pm

W = C bond length = 198 pm

W – C bond length = 226 pm

Bond multiplicity determined by X-ray structure determination of the crystalline substance.

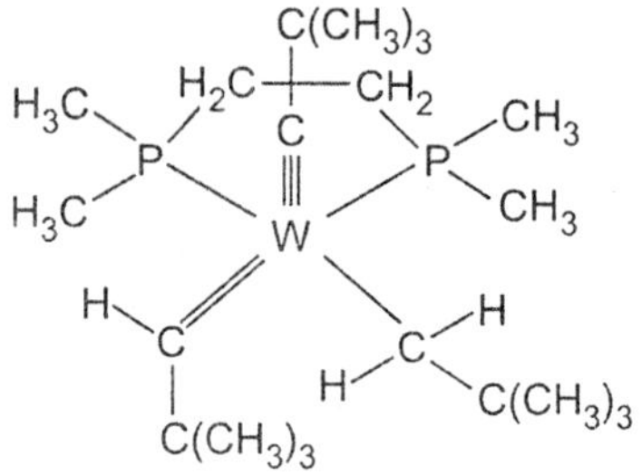


---

**Problem 2.17** Arrange each of the following sets in order of increasing size (a) Li, C, F, Ne; (b) Be, Ca, Ba, Ra; (c) B, Al, Ga, In, Tl; (d) V, Nb, Ta, Db (element number 105).

**Ans. :** (a) Ne < F < C < Li, (b) Be < Ca < Ba < Ra, (c) B < Al ≈ Ga < In ≈ Tl, (d) V < Nb = Ta ≈ Db.

---

**Problem 2.18** The first ionisation energy of Au is larger than that of Ag, but the reverse is true for the second ionisation energy.

**Hint :** Ag(47) ⟶ $[Kr]4d^{10}5s^1$, Au(79) ⟶ $[Xe]4f^{14}5d^{10}6s^1$

The increase of first I.E. value of Au is due to the participitation of inner 4f orbital but the increase of second I.E. in case of Ag is due to the greater amount of energy required to unpair 4d electron compared to that of 5d electron.

**Problem 2.19** The single bonded metallic radius of Na is 157 pm. Give a reasonable estimate of the following radius assuming that increment/decrement between the radii of different magnitude is 60 pm. (a) covalent radius (b) anionic radius (c) cationic radius.

**Ans.** : (a) metallic and covalent radius of Na is same i.e. 157 pm.

(b) anionic radius of Na i.e. $Na^-$ = 157 + 60 = 217 pm.

(c) cation radius of Na i.e. $Na^+$ = 157 – 60 = 97 pm.

## 2.11 EFFECTIVE NUCLEAR CHARGE* (SLATER'S RULE)

According to the radial distribution function (Sec. 1.8.7.1), for a particular quantum shell, the attractive force towards the nucleus is in the order : s electron > p electron > d-electron > f electron i.e. as *the ellipticity of the orbital increases, the attractive force of the electron with the nucleus increases and also as the ellipticity increases the penetrating power also increases.* Now the energy of an electron in Bohr's model is given by equation (1.14) as :

$$E = -\frac{me^4Z^2}{8\varepsilon_0^2n^2h^2}$$

i.e. the energy of the electron is a function of $\frac{Z^2}{n^2}$. It is evident from this equation (1.14) that as the principal quantum number n increases, the energy of the electron rapidly increases. Thus let us consider that $I_H$ as the ionisation energy of the electron (13.6 eV), then ionisation energy of polyelectron systems e.g. $He(1s^2)$, $Li(1s^22s^1)$ and $Be(1s^22s^2)$ will be :

$$I_{He} = 4I_H = 4 \times 13.6 = 54.4 \text{ eV} \qquad I_{Obs} = 24.4 \text{ eV}$$

$$I_{Li} = \frac{9}{4}I_H = \frac{9 \times 13.6}{4} = 30.6 \text{ eV} \qquad I_{Obs} = 5.40 \text{ eV}$$

$$I_{Be} = \frac{16}{4}I_H = \frac{16 \times 13.6}{4} = 54.4 \text{ eV} \qquad I_{Obs} = 9.32 \text{ eV}$$

It is evident from equation (1.14) if we calculate the nuclear charge from the $I_{Obs}$ value for He, the value is 1.35 which is less than the actual nuclear charge (i.e. 2). Similarly the values for Li and Be are 1.59 and 1.7 respectively i.e. the actual nuclear charge is somewhat reduced that can be expected. In case of Li there is one electron in the 2s orbital which is the outermost orbital and there is inner two electrons in 1s orbital which screen the outermost 2s electron thereby reducing the nuclear charge. In other words the outermost electron will not experiencing the full nuclear charge for Z = 3 for Li. It is rather experiencing so lower value namely $Z^* = Z - \sigma$ where $Z^*$ is the effective nuclear charge and $\sigma$ is called the shielding or screening constant (in units of e). *Screening constant for an electron indicates what fraction of one unit of nuclear charge it reduces.*

*Concept of penetration and screening*

J.S. Slater (1930) proposed some emperical rules for calculating screening constant ($\sigma$) for different polyelectronic systems :

(1) First we have to write the electronic configuration of the atom according to aufbau principle, Hund's rule etc. Then the electronic configuration should be made in the following groupings and order :

(1s)(2s,2p)(3s,3p)(3d)(4s,4p)(4d)(4f)(5s,5p)......

* Effective nuclear charge can also be calculated by Self Consistent Field (SCF) method.

(2) Electrons to the right of the (ns,np) or (nd) or (nf) group are not effective in screening the ns or np electron and contribute nothing to σ i.e. *higher energy electron is ineffective of screening the lower energy electron.*

(3) All other electrons in the same group (ns,np) as the electron in question shield that electron to a value of 0.35 unit of nuclear charge. For 1s orbital electron, a value of 0.30 will be taken instead of 0.35 in order to have a better result.

(4) (a) All electrons in the (n–1)th shell contribute 0.85 unit of nuclear charge for each electron. (b) All electrons in the (n–2)th shell or lower than the electron in question contribute 1.00 nuclear charge each i.e. the electrons in the (n–2)th and lower shells completely screen electrons in the nth shell.

These rules are applicable for ns and np electrons but for d or f electrons rules (1) to (3) remain as it is but rule (4) is replaced by rule (5).

(5) All the electrons below nd and nf level contribute 1.00 unit of nuclear charge each to the screening constant. This implies the poor penetrating power of the d or f electron.

(6) The value of screening constant (σ) is calculated from rule (1-5) and the value is subtracted from the true nuclear charge (Z) and we get the value of effective nuclear charge (Z*).

(7) Slater calculated Z* by the values of true principal quantum number n and n* where n* = n–quantum defect.** The values of n and n* are related as :

| | | | | | | |
|---|---|---|---|---|---|---|
| n | 1 | 2 | 3 | 4 | 5 | 6 |
| n* | 1 | 2 | 3 | 3.7 | 4.0 | 4.2 |

It is evident that for higher values of n, the quantum defect is maximum. n* is called effective principal quantum number.

** We have seen according to equation (1.16) that the position of the line in spectrogram is given by the difference between two square terms i.e.

$$\bar{v} = \frac{R}{n_1^2} - \frac{R}{n_2^2}$$ where $n_2 > n_1$ are the principal quantum numbers.

In case of hydrogen atom the optical behaviour of the single electron is somewhat different compared to that of alkali metals or any other polyelectronic systems. In case of hydrogen atom the effective nuclear charge is one thereby the optical electron is only experiencing the nuclear charge. But for other systems the optical electrons are in the influence of lower lying inner shell and therefore there will be different values of effective nuclear charge for different systems and the attractive force on the optical electron by the nucleus will be somewhat different than can be expected. In those cases the above equation (1.16) is modified in the following way :

$$\bar{v} = \frac{R}{(n_1 - \mu_1)^2} - \frac{R}{(n_2 - \mu_2)^2}$$ where $\mu_1$ and $\mu_2$ are called small constants.

This equation is also known as the general formula of Ritz combination principle. In this equation principal quantum numbers of equation (1.16) are replaced by n–μ term where μ is called **quantum defect** so that effective quantum number n* = n–μ. μ is maximum for s electron. This can also be explained by the penetrating power of orbitals.

Slater's rules can be systematised in the following way :

| Electron group | For higher groups | For same groups | For groups (n–1) | For groups ≤ (n–2) |
|---|---|---|---|---|
| 1s | 0 | 0.30 | — | — |
| ns, np | 0 | 0.35 | 0.85 | 1.00 |
| nd, nf | 0 | 0.35 | 1.00 | 1.00 |

## 2.11.1 Application of Slater's rule

(I) **Calculation of effective nuclear charge.**

(a) Effective nuclear charge for 1s electron of hydrogen

Here $\sigma = 0$ so $Z^* = 1.0 - 0 = 1.0$

(b) $Z^*$ for He

$\sigma = 1 \times 0.30 = 0.30$

$\therefore Z^* = 2 - 0.30 = 1.70$

(c) $Z^*$ for Li, electronic configuration $1s^2 2s^1$

$\sigma = 2 \times 0.85 = 1.70$

$Z^* = 3 - 1.70 = 1.30$

(d) (i) $Z^*$ for N : $1s^2 2s^2 2p^3$

i.e. $(1s)^2(2s,2p)^5$

$\sigma = 4 \times 0.35 + 2 \times 0.85 = 3.10$

$\therefore Z^* = 7 - 3.10 = 3.90$

(ii) $Z^*$ for 1s electron of N

$\sigma = 1 \times 0.30 = 0.30$

$\therefore Z^* = 7 - 0.30 = 6.70$

(e) (i) $Z^*$ for Mg : $1s^2 2s^2 2p^6 3s^2$ i.e. $(1s)^2(2s,2p)^8(3s)^2$

$\sigma = 1 \times 0.35 + 8 \times 0.85 + 2 \times 1.00 = 9.15$

$Z^* = 12 - 9.15 = 2.85$

(ii) $Z^*$ for $Mg^{2+}$ : $(1s)^2(2s,2p)^8$

$\sigma = 7 \times 0.35 + 2 \times 0.85 = 4.15$

$Z^* = 12 - 4.15 = 7.85$

(f) (i) $Z^*$ for Cl : $1s^2 2s^2 2p^6 3s^2 3p^5$ i.e. $(1s)^2(2s,2p)^8(3s,3p)^7$

$\sigma = 6 \times 0.35 + 8 \times 0.85 + 2 \times 1.00 = 10.90$

$Z^* = 17 - 10.90 = 6.10$

(ii) $Z^*$ for $Cl^-$ : $1s^2 2s^2 2p^6 3s^2 3p^6$ i.e. $(1s)^2(2s,2p)^8(3s,3p)^8$

$\sigma = 7 \times 0.35 + 8 \times 0.85 + 2 \times 1.00 = 11.25$

$Z^* = 17 - 11.25 = 5.75$

(g) $Z^*$ for $F^-$ : $1s^2 2s^2 2p^6$ i.e. $(1s)^2(2s,2p)^8$

$\sigma = 7 \times 0.35 + 2 \times 0.85 = 4.15$

$Z^* = 9 - 4.15 = 4.85$

(h) (i) $Z^*$ for Sc : $(1s)^2(2s,2p)^8(3s,3p)^8(3d)^1(4s)^2$

$\sigma = 1 \times 0.35 + 1 \times 0.85 + 8 \times 0.85 + 8 \times 1.00 + 2 \times 1.00 = 18.00$

$Z^* = 21 - 18.00 = 3.00$

(ii) $Z^*$ for 3d electron of Sc

$\sigma = 8 \times 1.00 + 8 \times 1.00 + 2 \times 1.00 = 18.00$

$Z^* = 21 - 18.00 = 3.00$

(i) (i) $Z^*$ for Cu (for 4s electron) : $(1s)^2(2s,2p)^8(3s,3p)^8(3d)^{10}(4s)^1$

$\sigma = 10 \times 0.85 + 8 \times 0.85 + 8 \times 1.00 + 2 \times 1.00 = 25.3$

$Z^* = 29 - 25.3 = 3.70$

(ii) $Z^*$ for 3d electron of Cu

$\sigma = 9 \times 0.35 + 8 \times 1.00 + 8 \times 1.00 + 2 \times 1.00 = 21.15$

$Z^* = 29 - 21.15 = 7.85$

(iii) $Z^*$ for 3d electron of $Cu^{2+}$ : $(1s)^2(2s,2p)^8(3s,3p)^8(3d)^9$

$\sigma = 8 \times 0.35 + 8 \times 1.00 + 8 \times 1.00 + 2 \times 1.00 = 20.80$

$Z^* = 29 - 20.8 = 8.2$

(j) (i) $Z^*$ at the periphery of Cr atom.

In this case we have to consider all the electrons to calculate the screening constant value ($\sigma$)

Cr : $(1s)^2(2s,2p)^8(3s,3p)^8(3d)^5(4s)^1$

$\sigma = 1 \times 0.35 + 5 \times 0.85 + 8 \times 0.85 + 8 \times 1.00 + 2 \times 1.00 = 21.40$

$Z^* = 24 - 21.40 = 2.60$

(ii) $Z^*$ at the periphery of Cu atom.

$\sigma = 1 \times 0.35 + 10 \times 0.85 + 8 \times 0.85 + 8 \times 1.00 + 2 \times 1.00 = 25.65$

$Z^* = 29 - 25.65 = 3.35$

(k) $Z^*$ for 4p and 3d electron of Se : $(1s)^2(2s,2p)^8(3s,3p)^8(3d)^{10}(4s,4p)^6$

For 4p electron : $\sigma = 5 \times 0.35 + 10 \times 0.85 + 8 \times 0.85 + 8 \times 1.00 + 2 \times 1.00 = 27.05$

$Z^* = 34 - 27.05 = 6.95$

For 3d electron : $\sigma = 9 \times 0.35 + 18 \times 1.00 = 21.15$

$Z^* = 34 - 21.15 = 12.85$

(I) $Z^*$ for $O^{2-}$ ion : $(1s)^2(2s,2p)^8$

$\sigma = 7 \times 0.35 + 2 \times 0.85 = 4.15$

$Z^* = 8 - 4.15 = 3.85$

(II) **For isoelectronic systems (e.g. Li, $Be^+$, $B^{2+}$, $C^{3+}$, $N^{4+}$ $O^{5+}$, $F^{6+}$), the value of σ is same for all the species, hence Z* will be different for the different systems.**

(III) **The cation is smaller than the parent neutral atom whereas an anion is larger than the neutral atom.**

$Z^*$ for F $= 9 - (6 \times 0.35 + 2 \times 0.85) = 5.20$

$Z^*$ for $F^- = 4.85$

As Z* for $F^-$ is less than that of F atom thereby $F^-$ will have a larger size compared to that of F atom, since the electrostatic attraction between the nucleus and the electrons is more in fluorine atom than that of $F^-$ ion.

$Z^*$ for Na $= 11 - (8 \times 0.85 + 2 \times 1.00) = 11 - 8.8 = 2.2$

$Z^*$ for $Na^+ = 11 - (7 \times 0.35 + 2 \times 0.85) = 11 - 4.15 = 6.85$

Since Z* for $Na^+$ ion is greater than that of Na atom, therefore the nucleus of $Na^+$ will attract the outermost electron more strongly than that of Na atom and hence there will be a decrease in radius as we go from Na to $Na^+$.

(IV) **According to aufbau principle the energy order of orbital is 4s < 3d but why during ionisation process the 4s electron(s) are lost first then the 3d electron.**

This can be explained with reference to Z* values for 4s and 3d electron. Let us consider for Cu.

$Z^*$ for 4s electron $= 3.70$

$Z^*$ for 3d electron $= 7.85$

As the Z* value for 3d electron is higher compared to that of 4s electron therefore the nucleus will attract the 3d electron more strongly compared to that of 4s electron and hence during ionisation 4s electron will be lost first then 3d electron. Similarly this is applied for other transition elements also.

In this context we can divide the electrons and orbitals in two classes : (a) The **core electrons** or orbitals in which the electrons are tightly bound having high Z* value (b) the **valence electrons** or orbitals which are higher energy and loosely bound electron having low Z* value.

(V) **General expression for effective nuclear charge for second period and transition elements.**

Let us consider the electronic configuration of a second period element in the general form i.e. $(1s)^2(2s,2p)^n$

$\sigma = (n-1) \times 0.35 + 2 \times 0.85 = 1.35 + 0.35n$

Atomic number of the element $= 2+n$

$\therefore Z^* = (2+n) - (1.35+0.35n)$

$= \boxed{0.65 + 0.65n}$

This indicates that *as we go from left to right across a second period element the Z* increases by an amount of 0.65 unit of nuclear charge* and thereby this is also a reason why atomic radius decreases in a period from left to right.

For d-block transition element let us consider the general electronic configuration $(1s)^2(2s,2p)^8(3s,3p)^8(3d)^n(4s)^2$

$$\sigma = 1 \times 0.35 + n \times 0.85 + 8 \times 0.85 + 8 \times 1.0 + 2 \times 1.0 = 17.15 + 0.85n$$

Atomic number of the element = 20 + n

So that $Z^* = (20+n) - (17.15 + 0.85n) = \boxed{2.85 + 0.15n}$

Thus the Z* value increases more slowly in case of d block elements (i.e. 0.15 units of nuclear charge per element) compared to p-block elements. This indicates the poor shielding capability of d orbital compared to that of s and p orbital.

---

**Problem 2.20** What will be the expected trend in Z* on going from C to O?

**Ans.** : Z* for 2p electron for C, N and O are 3.25, 3.90 and 4.55 respectively i.e. effective nuclear charge increases from C to O but the increase in Z* between N and O (4.55 – 3.90 = 0.65) is the same as the increase in Z* between C and N (3.90 – 3.25 = 0.65).

C : $1s^22s^22p^2$

N : $1s^22s^22p^3$

O : $1s^22s^22p^4$

On going from C to N, the additional one electron is occupied to the empty 2p orbital, but on going from N to O the additional one electron is occupied in the 2p orbital which is already occupied by one electron and thereby the incoming electron will experience a repulsive force. Therefore, Z* for N = Z* for C + one nuclear charge – screening due to one 2p electron

= Z* for C + 1 – 0.35 = Z* for C + 0.65

Similarly Z* for O = Z* for N + one nuclear charge – screening due to 2p electron

= Z* for N + 0.65

Thus the increase in Z* is the same in two cases.

---

## 2.11.2 Drawbacks of Slater's rule

1. Radial probability distribution curves (Sec. 1.8.7.1) of ns, np and nd orbitals show that the orbitals have different penetrating power as also different shielding capability. Then a question arises how all the electrons in the lower shell s, p, d orbitals have equal screening efficiency to higher shell electrons or even in the same quantum shell.

2. Penetrating power of the higher energy orbital into the core orbital particularly for s orbital has not been taken into account in calculating the screening constant value.

---

**Problem 2.21** The ionisation energy of Li is approximately 2/5ths that of hydrogen. Assuming Bohr's theory holds good for Li atom calculate the effective nuclear charge of Li.

**[B.U. 1994, C.U. 1999, 2014]**

**Ans.** : According to Slater the energy of an electron is given by Bohr's model (equation 1.14).

$$E = -\frac{me^4}{8\varepsilon_0^2 h^2}\left(\frac{Z^*}{n^*}\right)^2 = -13.6\left(\frac{Z^*}{n^*}\right)^2 \text{eV}$$

The ionisation energy of Li means

$$Li(1s^2 2s^1) \xrightarrow{-e} Li^+(1s^2)$$

The energy of all the electrons in an atom or ion is obtained by summation of $\left(\frac{Z^*}{n^*}\right)^2$ terms for each electron i.e. $\Sigma - 13.6\left(\frac{Z^*}{n^*}\right)^2$ eV. The ionisation energy of an atom or ion is equal to the difference in energy between the energies before and after ionisation. Therefore ionisation energy (I.E.) of Li is

$$I.E. = E_{Li^+} - E_{Li}$$

$$E_{Li} = -13.6\left(\frac{Z^*}{n^*}\right)^2_{2s^1} - 13.6 \times 2\left(\frac{Z^*}{n^*}\right)^2_{1s^2}$$

$$E_{Li^+} = -13.6 \times 2\left(\frac{Z^*}{n^*}\right)^2_{1s^2}$$

$$\text{So that I.E.} = -13.6 \times 2\left(\frac{Z^*}{n^*}\right)^2_{1s^2} + 13.6\left(\frac{Z^*}{n^*}\right)^2_{2s^1} + 13.6 \times 2\left(\frac{Z^*}{n^*}\right)^2_{1s^2}$$

$$= 13.6\left(\frac{Z^*}{n^*}\right)^2_{2s^1} \text{eV}$$

The I.E. of hydrogen = 13.6 eV, so that according to condition I.E. of Li = $\frac{2}{5} \times 13.6$ eV.

$$\therefore \frac{2}{5} \times 13.6 \text{ ev.} = 13.6\left(\frac{Z^*}{n^*}\right)^2_{2s^1} \text{eV}$$

For 2s electron n = 2 and $n^*$ = 2

$$\text{i.e. } \frac{2}{5} \times 13.6 = 13.6\frac{Z^{*2}}{4}$$

$\therefore Z^* = 1.264$ unit of nuclear charge.

---

**Problem 2.22** Calculate the ionisation energy of oxygen using Slater's rule.

**Ans. :** O : $(1s)^2(2s,2p)^6$ $\qquad$ $I.E. = E_{O^+} - E_O$

$O^+$ : $(1s)^2(2s,2p)^5$ $\qquad$ $E_O = -13.6 \times 2\left(\frac{Z^*}{n^*}\right)^2_{1s^2} - 13.6 \times 2\left(\frac{Z^*}{n^*}\right)^2_{2s^2} - 13.6 \times 4\left(\frac{Z^*}{n^*}\right)^2_{2p^4}$

$$E_{O^+} = -13.6 \times 2\left(\frac{Z^*}{n^*}\right)^2_{1s^2} - 13.6 \times 2\left(\frac{Z^*}{n^*}\right)^2_{2s^2} - 13.6 \times 3\left(\frac{Z^*}{n^*}\right)^2_{2p^3}$$

$$\therefore I.E. = -13.6 \times 2\left(\frac{Z^*}{n^*}\right)^2_{1s^2} - 13.6 \times 2\left(\frac{Z^*}{n^*}\right)^2_{2s^2} - 13.6 \times 3\left(\frac{Z^*}{n^*}\right)^2_{2p^3} + 13.6 \times 2\left(\frac{Z^*}{n^*}\right)^2_{1s^2}$$

$$+ 13.6 \times 2\left(\frac{Z^*}{n^*}\right)^2_{2s^2} + 13.6 \times 4\left(\frac{Z^*}{n^*}\right)^2_{2p^4}$$

The effective nuclear charge for 1s electron remains the same in case of O and $O^+$ so the contribution is equal and thereby cancel. We have to consider 2s and 2p electrons to calculate the energy of the two systems.

$Z^*$ for $O^+$ = 8 – (4 × 0.35 + 2 × 0.85) = 4.9 unit of nuclear charge.

$Z^*$ for O = 8 – (5 × 0.35 + 2 × 0.85) = 4.55 unit of nuclear charge.

$$\text{I.E.} = -13.6 \times 5\left(\frac{Z^*}{n^*}\right)^2_{O^+} + 13.6 \times 6\left(\frac{Z^*}{n^*}\right)^2_{O}$$

$$= -13.6 \times 5\left(\frac{4.9}{2}\right)^2_{O^+} + 13.6 \times 6\left(\frac{4.55}{2}\right)^2_{O}$$

$$= -408.17 + 422.331 = 14.161 \text{ eV.}$$

This value is close to the experimental value (13.61 eV).

---

**Problem 2.23** Calculate the electron affinity of F by using Slater's rule. **[B.U. 2016]**

**Ans. :** $F(g) + e \rightarrow F^-(g)$

$(EA)_F = E_{F^-} - E_F$

$F : 1s^2 2s^2 2p^5$

$F^- : 1s^2 2s^2 2p^6$

$$E_{F^-} = -13.6 \times 2\left(\frac{Z^*}{n^*}\right)^2_{1s^2} - 13.6 \times 8\left(\frac{Z^*}{n^*}\right)^2_{2s^2 2p^6}$$

$$E_F = -13.6 \times 2\left(\frac{Z^*}{n^*}\right)^2_{1s^2} - 13.6 \times 7\left(\frac{Z^*}{n^*}\right)^2_{2s^2 2p^5}$$

As before the effective nuclear charge for 1s orbital is same for both the systems $F^-$ and F and thereby neglected.

$$(EA)_F = -13.6 \times 8\left(\frac{Z^*}{n^*}\right)^2_{F^-} + 13.6 \times 7\left(\frac{Z^*}{n^*}\right)^2_{F}$$

$Z^*$ for $F^-$ = 4.85 and F = 5.20

$$(EA)_F = -13.6 \times 8\left(\frac{4.85}{2}\right)^2 + 13.6 \times 7\left(\frac{5.20}{2}\right)^2$$

$$= -639.812 + 643.552 = 3.74 \text{ eV (Experimental value = 3.45 eV)}$$

Similarly the electron affinity value of Cl can also be calculated.

---

**Problem 2.24** Bond energies of $H_2$, $I_2$ and HI molecules are 104.2, 36.1 and 71.4 kcal/mol respectively, the Pauling's electronegativity of hydrogen = 2.1. Calculate Pauling's electronegativity of iodine using these data. **[B.U. 1997]**

**Ans. :** $\Delta = D_{H-I} - \sqrt{D_{H-H} D_{I-I}}$

$= 71.4 - \sqrt{104.2 \times 36.1}$

$= 71.4 - 61.3$

$= 10.1 \text{ kcal mol}^{-1}$

$= 42.23 \text{ kJ mol}^{-1}$ [∵ $1 \text{ kcal mol}^{-1} = 4.184 \text{ kJ mol}^{-1}$]

Now $X_I - X_H = 0.102\sqrt{42.23} = 0.6628$

$\therefore X_I = 0.6628 + X_H = 0.6628 + 2.1 = 2.7628.$

---

## 2.12 INERT PAIR EFFECT AND RELATIVISTIC EFFECTS ON ORBITALS

The wave function of an electron has two parts—radial and angular part (Sec. 1.8.7) and the corresponding velocity is known as radial and angular velocity of an electron.

According to Bohr's theory (Sec. 1.6.1 and 1.6.2) the radius (equation 1.9) of the nth orbit and the velocity of the electron in the nth orbit is given by—

$$r = \frac{n^2h^2\varepsilon_0}{\pi me^2Z} \quad \text{and} \quad v = \frac{Ze^2}{2\varepsilon_0 nh}$$

It is evident from these expressions that as n increases the radius of the revolving electron increases and the velocity decreases but for a particular n value the velocity of the electron increases with the nuclear charge (Z). It is also plausable that for high nuclear charge the n value will also be high. Electron is treated as particle and thereby we can expect the relativistic variation of the mass of the electron for high atomic number elements. The radial velocity of the electron approaches to the speed of light for heavy elements. According to the theory of relativity the rest mass ($m_0$) and moving mass (m) of the electron is related by Einstein's equation (Sec. 1.7.1) as :

$$m = m_0\left(1 - \frac{v^2}{c^2}\right)^{-\frac{1}{2}}$$

It has been suggested that the radial velocity approaches to the velocity of light (c = 137 a.u., a.u = atomic unit) when the nuclear charge Z = 80 i.e. mercury. Thus for heavy elements e.g. mercury the radial velocity of the s electron approaches to the velocity of light. For mercury the velocity of the 1s electron is—

$$v = \frac{Ze^2}{2\varepsilon_0 nh} = \frac{Z}{n} \text{ a.u} = 80 \text{ a.u}$$

The average radial velocity of 1s electron of an atom is the ratio of the nuclear charge to that of 137. Therefore for mercury the 1s electron moves at a speed of $\frac{80 \text{ a.u}}{137 \text{ a.u}} = 0.58$ times that of light. Hence the mass will be—

$$m = m_0\left(1 - (0.58)^2\right)^{-\frac{1}{2}} = 1.2\ m_0$$

So that the mass of 1s electron of mercury is 1.2 times that of rest mass i.e. the mass of the 1s electron is increased by about 20%. On the other hand according to Bohr's theory the radius is inversely proportional to the mass of the electron (equation 1.9). Therefore the radius of 1s electron will be about 20% reduced than can be expected. This **radial effect** is known as **relativistic contraction.** This relativistic contraction can also be applied for 2s and higher s orbitals depending upon the radial velocity of the electron and roughly the rate of contraction decreases. This effect is small in case of p orbitals and is very poor in case of d or f orbitals. In fact in case of d or f orbitals there will be **relativistic expansion** by the contracted s and p orbitals. Indeed the sequence of contraction runs roughly in the order s > p > d > f. This relativistic contraction is the cause of inert pair effect.

Since this type of contraction is maximum in case of sixth period elements having 6s orbital of gold and mercury (Au : [Xe]$4f^{14}5d^{10}6s^1$; Hg : [Xe]$4f^{14}5d^{10}6s^2$) and some anomaly arises in case of these elements. One of the anomolous behaviour is the enhance electronegativity of gold (2.4) and mercuy (1.9) as compared to silver (1.9) and cadmium (1.7) respectively. As 6s orbital is tightly bound due to relativistic contraction, its energy separation from 6p orbital is abnormally large so that 6p orbital are not the valence orbital but are considered as post valence orbital. Mercury having $5d^{10}6s^2$ configuration with d and s orbitals filled, essentially being treated as noble liquid metal. Consequently mercury ($6s^2$) is less reactive than cadmium ($5s^2$) or zinc ($4s^2$).

*Cause of inert pair effect*

Beyond gold and mercury the 6s electrons are difficult to ionise and consequently show instability of group oxidation state and thereby these elements prefer to exhibit oxidation state which is two units less than the group oxidation state. This is known as **inert pair effect.** *In fact in case of heavy post transition elements having $ns^2np^2$ configuration, since the two np electrons are further away from the nucleus than the two ns electrons and thereby the two np electrons are easily lost and the two ns electrons remain as inert pair and the effect caused by this inert pair electrons are known as inert pair effect.* Due to the inert pair effect the element will show redundant value of oxidaiton state and this **inert pair effect will be more prominant to the lower member of the group.** If the group valence of the elements is V then V–2 covalency is most stable for the lower member of the group e.g.

(1) Group 13 elements ($ns^2np^1$) : B, Al, Ga and In will show +3 oxidation state whereas for Tl, the +1 oxidation state is most stable, $Tl^{3+}$ is unstable.

(2) Group 14 elements ($ns^2np^2$) : C, Si, Ge and Sn exhibit the +4 oxidation state whereas for Pb the +2 oxidation state is stable. The tetravalent lead compounds ($PbCl_4$) is easily reduced to divalent ($PbCl_2$) state.

(3) Group 15 elements ($ns^2np^3$) : Except nitrogen other congeners show +5 group oxidation state but $Bi^{5+}$ compounds are unstable and easily reduced to +3 oxidation state.

(4) Group 12 elements ($ns^2(n-1)d^{10}$) : Zn and Cd will show consistent bipositive state in their compounds. But in case of Hg the +1 oxidation state is exhibitated in a very unique way by forming a metal-metal bond i.e. $Hg_2^{2+}$. This is the unique feature of Hg.

Now a question arises what will be the stereochemical feature of $s^2$ pair of electrons responsible for inert pair effect. It has been suggested that the inert pair of electron is residing in equatorial position of a distorted pentagonal bipyramidal arrangement of some crystalline compounds of $Pb^{2+}$ and $Bi^{3+}$. Finally inert pair effect is not the **intrinsic property**, which does not imply intrinsic inertness as is evident from the ionisation energies of the s electrons. The ionisation energies of $s^2$ pair of electrons is less in case of Pb compared to Si.

*Limitation*

**Problem 2.25**

| Element : | Zn | Cd | Hg |
|---|---|---|---|
| Ionisation energy in eV : (first and second) | 9.39, 17.96 | 8.99, 16.90 | 10.43, 18.75 |

Explain why the first and second ionisation energies decreases from Zn to Cd but increases from Cd to Hg?

**Hint :** This is due to the radial effect of s electrons whereby more relativistic contraction of the s orbitals compared to the d and f orbitals.

**Problem 2.26** Using Slater's rules calculate Z* for the valence electron of gold. Assuming Allred-Rochow electronegativity of gold is 2.54, calculate Z* using Allred-Rochow electronegativity equation (covalent radius of gold is 140 pm).

**Ans. :** Au : $(1s)^2(2s,2p)^8(3s,3p)^8(3d)^{10}(4s,4p)^8(4d)^{10}(4f)^{14}(5s,5p)^8(5d)^{10}(6s)^1$

$$Z^* = 79 - (10 \times 0.85 + 8 \times 0.85 + 60 \times 1.0)$$

$$= 79 - 75.3 = 3.7 \text{ unit of nuclear charge.}$$

According to equation (2.6)

$$(X_A)_{AR} = \frac{0.36Z^*}{r_{cov}^2} + 0.74 \quad \text{Here } r_{cov} = 140 \text{ pm} = 1.4\text{Å}$$

$$\text{So that } Z^* = \frac{[(X_A)_{AR} - 0.74]r_{cov}^2}{0.36} = 9.8$$

The value of Z* is considerably larger than the value calculated from Slater's rules.

**Problem 2.27** Explain the concept of electronegativity. Is it an intrinsic property? State its usefulness in predicting the bond types, polarity of molecules and molecular association. **[B.U. 2005]**

**Ans. :** For electronegativity and intrinsic property see Sec. 2.5.3 and question number 2 (exercise) of this chapter.

For bond type see Sec. 2.5.4(b).

**Polarity of molecules :** Electronegativity provides useful guidance of the electron distribution in a chemical bond. The relative electronegativity values show which will be the positive and negative ends in a bond e.g. $Na^+H^-$ and $H^+Cl^-$. A more convenient way to represent bond polarity in polyatomic molecules is represented as :

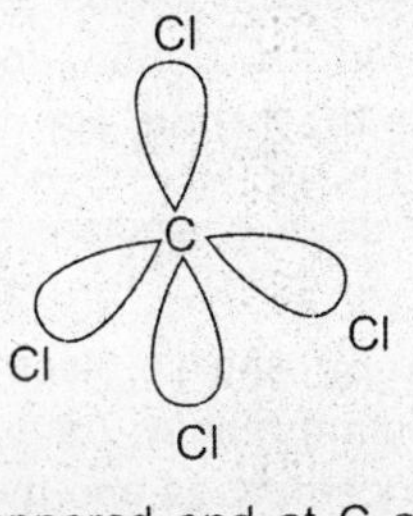


Tappered end at C and swelled end at Cl

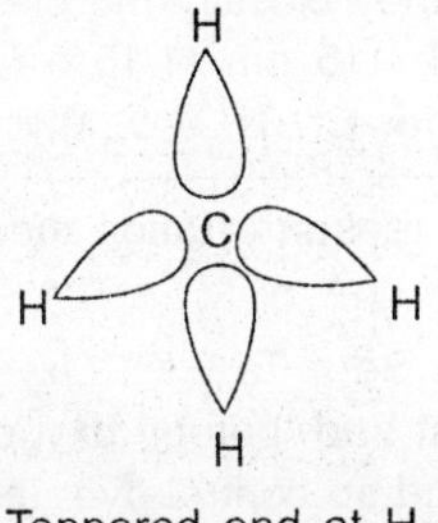


Tappered end at H and swelled end at C

This difference in pictorial bond polarity direction is due to electronegativity order i.e. H < C < Cl.

**Molecular association :** The very common way to show this in the hydrogen bond. If hydrogen is connected with a electronegative element, it becomes positive end of a polar system and tries to be associated with electron rich centre e.g. in $H_2O$ or HF

```
 H      H      H      H
 |      |      |      |
····O—H····O—H····O—H····O—H····
```

H—F····H—F····H—F····H—F

resulting abnormal value of melting and boiling points.

**Problem 2.28** The F–F bond distance in $F_2$ is 141.3 pm. Calculate Allred-Rochow electronegativity of fluorine using Slater's rule. **[C.U. 1997]**

**Ans. :** According to equation (2.6) electronegativity of fluorine will be—

$$X_F = \frac{0.36Z^*}{r_{cov}^2} + 0.74 \qquad \text{Here } r_{cov} = \frac{141.3}{2} = 70.65 \text{ pm} = 0.7065\text{Å}.$$

$F \rightarrow 1s^2 2s^2 2p^5$

The screening constant $\sigma$ for F = 6 × 0.35 + 2 × 0.85 = 3.8.

$\therefore Z^* = 9 - 3.8 = 5.2$

$$\therefore X_F = \frac{0.36 \times 5.2}{(0.7065)^2} + 0.74 = 4.49.$$

**Problem 2.29** Define Bragg–Slater radii. **[C.U. 2009]**

**Ans. :** This is one of the classification of radii for a series of elements. Bragg (1927) assumed that the radius of $O^{2-}$ ion is half of the internuclear distance of the adjacent oxide ions in silicates, so that $r_{O^{2-}} = {}^1/_2$(O–O) distance. In this way he proposed a set of radii of elements whose sum gave the observed internuclear distance for several hundred ionic or covalent crystals and the value lie within 6 pm (0.06Å). We know that cations have smaller size compared to that of neutral atom while anions have larger size than that of neutral atom. Slater (1964) therefore proposed an emperical rule in which he stated that increase of ionic radius for anions are about 85 pm (0.85Å) and the decrease of ionic radius for cations are about 85 pm (0.85Å) compared to that of the respective atomic or covalent radii. He thus suggested a unique radii for the element in multiples of 5 pm (0.05Å) which gave the observed internuclear distances in the solid crystals which lie within 6 pm (0.06Å). Thus the combined effort of Bragg and Slater is known as Bragg-Slater radii and they classify the elements with some typical values which are : 100 pm (1Å) for P, S, Cl; 115 pm (1.15Å) for As, Se, Br; 135 pm (1.35Å) for Co, Ni, Zn, Cu; 140 pm (1.40Å) for Cr, Ti, V, I, Mn, Fe; 150 pm (1.50Å) for Mg, Hg etc.

**Problem 2.30** Adducing proper reason arrange the following in the order of decreasing electron affinity : O, Se, Te, S. **[B.U. 2010, 2006]**

**Ans. :** The order is S > Se > Te > O. Oxygen has no d-orbital and so electron holding due to its uptake is least and thereby oxygen has least electron affinity. As size increases from S to Te via Se and so control over incoming electron decreases and thus electron affinity decreases.

**Problem 2.31** During ionisation of vanadium, the 4s electron comes out first–Establish it using Slater's rules. **[C.U. 2008, B.U. 2010, 2016]**

**Ans. :** For Slater's rules see Sec. 2.11.

V : $(1s)^2(2s, 2p)^8(3s, 3p)^8(3d)^3(4s)^2$

For 4s electron $\sigma = 1\times0.35 + 3\times0.85 + 8\times0.85 + 8\times1.0 + 2\times1.0 = 19.70$

$Z^* = Z - \sigma = 23 - 19.70 = 3.30$

For 3d electron $\sigma = 2\times0.35 + 8\times1.0 + 8\times1.0 + 2\times1.0 = 18.70$

$Z^* = 23 - 18.70 = 4.30$

As the effective nuclear charge for 3d electron is more compared to that of 4s electron, so 3d electron is more tightly held than the 4s electron in case of vanadium and thus during ionisation 4s electron will be removed first than that of 3d electron.

**Problem 2.32** Calculate the screening constant of the first valence electron in Mn(Z=25) with the help of Slater's rule. **[B.U. 2014]**

**Ans. :** Sec. 2.11.1.

Mn : $(1s)^2(2s,2p)^8(3s,3p)^8(3d)^5(4s)^2$

$\sigma = 1\times0.35 + 5\times0.85 + 8\times0.85 + 8\times1.00 + 2\times1.00 = 21.4.$

**Problem 2.33** The F–F bond distance in $F_2$ molecule is 141.3 pm. Calculate the Allred-Rochow electronegativity of fluorine using Slater's rule. **[B.U. 2013]**

**Ans. :** As per equation (2.6)

$$(\chi_A)_{AR} = \frac{0.36Z^*}{r_{cov}^2} + 0.74$$

Z* for fluorine = 9 – 3.80 = 5.2

$$r_{cov} = \frac{141.3}{2} = 70.65 \text{ pm} = 0.7065\text{Å}$$

$$\therefore\ (\chi_F)_{AR} = \frac{0.36\times 5.2}{(0.7065)^2} + 0.74 = 4.49$$

**Problem 2.34** Why atomic radii of zirconium and hafnium are almost same? **[B.U. 2013]**

**Ans. :** The electronic configuration of Zr and Hf are $[Kr]4d^25s^2$ and $[Xe]4f^{14}5d^26s^2$ respectively. Poor shielding of 4f orbital and filling of 14 electrons in 4f orbitals hardly cope with increase in nuclear charge in Hf and strong attraction of nuclear charge effects rapid contraction in atomic size like that of lanthanide contraction. Thus Zr and Hf have similar atomic radii. This is justified from their similar chemistry of Zr and Hf.

**Problem 2.35** What optical transition in $He^+$ spectrum would have the same wavelength as the first Lyman transition of hydrogen (Ignore the effect of reduced mass). **[B.U. 2014]** Also calculate second ionisation potential of He and first Bohr orbit for $He^+$. ($e = 1.6\times10^{-19}$ coulomb, $m = 9.1\times10^{-31}$ kg, $h = 6.626\times10^{-34}$ J.sec., $c = 2.997\times10^8$ m/sec. and $\varepsilon_0 = 8.854\times10^{-12}$ coulomb²/newton.metre²).

**Ans. :** The upper part of the problem is question number 25 (exercise) of first chapter.

As $He^+$ has only one electron, Bohr equation may be applicable. The Rydberg constant (R) for H and $He^+$ will be same as the reduced mass effect is neglected.

For hydrogen the first Lyman transition i.e. when electron jumps from n = 2 to n = 1, we have,

$$\bar{\nu} = \frac{1}{\lambda} = RZ^2\left(\frac{1}{n_1^2} - \frac{1}{n_2^2}\right) = R\left[\frac{1}{1^2} - \frac{1}{2^2}\right] = \frac{3R}{4}$$

where Z = atomic number of the respective element. Since wavelength of the radiation emitted in optical transition of $He^+$ is the same for hydrogen, therefore for $He^+$(Z=2), so that,

$$\frac{1}{\lambda} = RZ^2\left(\frac{1}{n_1^2} - \frac{1}{n_2^2}\right) = 4R\left[\frac{1}{n_1^2} - \frac{1}{n_2^2}\right]$$

According to the condition, $\frac{1}{\lambda} = 4R\left[\frac{1}{n_1^2} - \frac{1}{n_2^2}\right] = \frac{3R}{4}$

$\therefore$ $n_1$ and $n_2$ have to be 2 and 4 respectively as $\frac{1}{\lambda} = 4R\left[\frac{1}{2^2} - \frac{1}{4^2}\right] = \frac{3R}{4}$

As the second ionisation potential of He is the energy required to remove the second electron from first orbit to $\infty$, so that second ionisation potential of He = $\frac{Z^2e^4m}{8\varepsilon_0^2h^2}\left[\frac{1}{1^2} - \frac{1}{\infty^2}\right]$

$$= \frac{Z^2e^4m}{8\varepsilon_0^2h^2} = \frac{2^2\left(1.6\times10^{-10}\right)^4\left(9.1\times10^{-31}\right)}{8\left(8.854\times10^{-12}\right)^2\left(6.626\times10^{-34}\right)^2} = 8.66 \times 10^{-18}\ J$$

The first Bohr orbit of $He^+(Z=2)$ may be calculated as,

$$r = \frac{\varepsilon_0 n^2h^2}{\pi mZe^2} = \frac{\left(8.854\times10^{-12}\right)(1)^2\left(6.626\times10^{-34}\right)^2}{3.14\left(9.1\times10^{-31}\right)(2)\left(1.6\times10^{-19}\right)^2} = 2.65 \times 10^{-11}\ \text{metre}$$

---

**Problem 2.36** Define aperiodicity and illustrate. **[B.U. 2015]**

**Ans. :** Aperiodicity is the reverse of periodicity is also called secondary periodicity and is related with abnormal periodic properties. As for example in a period from left to right generally the ionisation energy will increase but there is a drop in ionisation energy from Be to B as also N to O. Similarly in a group from top to bottom generally atomic radius will increase. But Al and Ga has the same atomic radius values though there is a wide difference in atomic numbers. These are the examples of aperiodicity i.e. anomalous periodic properties.*

---

**Problem 2.37** Justify the following data : **[C.U. 2015]**

| Element | 1st ionisation energy ($kJ\ mol^{-1}$) | 2nd ionisation energy ($kJ\ mol^{-1}$) |
|---|---|---|
| N | 1403 | 2856 |
| O | 1314 | 3388 |

**Ans. :** $N \rightarrow 1s^22s^22p^3$ $N^+ \rightarrow 1s^22s^22p^2$

$O \rightarrow 1s^22s^22p^4$ $O^+ \rightarrow 1s^22s^22p^3$

In case of N the p orbital is half-filled while in case O the p orbital is partially filled. The stability of half-filled orbital is more compared to that of partially filled orbital so that electron removal is more easier from O orbital than that from N orbital thereby there will be drop in 1st ionisation energy from N to O.

But in case of 2nd ionisation energy the $O^+$ has half-filled orbital while $N^+$ has partially filled orbital. Hence reverse order is observed i.e. second ionisation energy of oxygen is greater than that of oxygen.

---

**Problem 2.38** Calculate the Allred – Rochow electronegativity of As and Br from the given data : $r_{As}$ = 1.22Å and $r_{Br}$ = 1.14Å. **[C.U. 2015]**

**Ans. :** As(33) $\rightarrow 1s^22s^22p^63s^23p^63d^{10}4s^24p^3$

i.e. $(1s)^2(2s,2p)^8\ (3s,3p)^8(3d)^{10}(4s,4p)^5$

$\sigma$ = 4×0.35 + 10×0.85 + 8×0.85 + 8×1.00 + 2×1.00 = 26.7.

$\therefore$ $Z^*$ = 33 – 26.7 = 6.3 unit of nuclear charge.

---

* Secondary periodicity has been proposed for the observed fluctuation of different properties of the elements in the fourth and sixth period in a group due to interposition of d and f subshells e.g. Ga–Ge–As and Tl–Pb–Bi.

Electronegativity according to Allred-Rochow

$$(\chi_A)_{AR} = \frac{0.36Z^*}{r^2_{cov}} + 0.74 = \frac{0.36 \times 6.3}{(1.22)^2} + 0.74 = 2.26.$$

Therefore electronegativity of As according to Allred-Rochow = 2.26.

Br(35) : $(1s)^2(2s,2p)^8(3s,3p)^8(3d)^{10})(4s,4p)^7$

$\sigma$ = 6×0.35 + 10×0.85 + 8×0.85 + 8×1.00 + 2×1.00 = 27.4

Z* = 35 – 27.4 = 7.6 unit of nuclear charge.

$$\therefore \text{electronegativity of Br} = \frac{0.36 \times 7.3}{(1.14)^2} + 0.74 = 2.84.$$

---

**Problem 2.39** How does the effective nuclear charge change for a 2p electron in passing from B to C and that of 2s electron in going from Li to Be? Give reasons. **[C.U. 2015]**

**Ans.** : B : $(1s)^2(2s,2p)^3$ $\sigma$ =2×0.35 + 2×0.85 = 2.4

Z* = 5 – 2.4 = 2.6

C : $(1s)^2(2s,2p)^4$ $\sigma$ = 3×0.35 + 2.×0.85 = 2.75

Z* = 6 – 2.75 = 3.25

Li : $(1s)^2(2s)^1$ $\sigma$ = 2×0.85 = 1.70

Z* = 3 – 1.70 = 1.30

Be : $(1s)^2(2s)^2$ $\sigma$ = 1×0.35 + 2×0.85 = 2.05

Z* = 4 – 2.05 = 1.95

In all cases the valence orbitals n = 2. As more and more n = 2 quantum shell populated the effective nuclear charge increases from Li to C.

---

**Problem 2.40** Among $F^-$, $Na^+$, $O^{2-}$ and $Mg^{2+}$ ions, those having the highest and the lowest radii respectively are—

(a) $O^{2-}$ and $Na^+$, (b) $F^-$ and $Mg^{2+}$, (c) $O^{2-}$ and $Mg^{2+}$, (d) $Mg^{2+}$ and $O^{2-}$ **[NET 2014]**

**Ans.** : (c) i.e. $O^{2-}$ and $Mg^{2+}$. All are isoelectronic having 10 electrons. Highest negative charged $O^{2-}$ will have highest ionic radius and the highest positive charged $Mg^{2+}$ species will have lowest radius.

## EXERCISE

1. Explain the large decrease in electron affinity between Li(0.61 eV) and Be(–0.5 eV), despite of increase in nuclear charge. **[B.U. 2013]**

**Hint :** The electron affinity of an element is defined as the amount of energy released when a neutral gaseous atom captures an electron. The electronic configuration of Li and Be are $1s^22s^1$ and $1s^22s^2$ respectively i.e. Li has unfilled 2s orbital while Be has filled 2s orbital. Hence in case of Be the added electron will go to the less tightly bound 2p orbital making sharp fall of electron affinity from Li to Be.

2. Electronegativity is an intrinsic property or not? **[B.U. 2005]**

**Hint :** It is not an intrinsic property of an element as it is not possible to find absolute magnitude of electronegativity of an isolated atom. In heteronuclear bond like A–B, A–C etc., the relative strength of electron attracting power is changing.

3. The ionic radius of $S^{2-}$ and $Te^{2-}$ are 1.84 and 2.21Å respectively. Predict the ionic radius of $Se^{2-}$.

**Ans. :** $\frac{1.84 + 2.21}{2} = 2.025$Å.

4. What is the meaning of a negative electron affinity? Does any element have a negative ionisation potential? [B.U. 2006]

**Hint :** Negative electron affinity means the system does not like to have any more electron which means that energy is required rather than liberated during addition of an electron to a gaseous chemical species to form a gaseous anion. Negative ionisation potential is not possible.

5. Why the first ionisation potential of Cu is higher than that for K whereas the second ionisation potentials are in the reverse order? [B.U. 2010, 2006]

**Hint :** Cu(29) : $[Ar]3d^{10}4s^1$; K(19) : $[Ar]4s^1$. Due to poor screening of d-electrons, effective nucler charge in Cu is greater than that of K and thereby the first ionisation potential of Cu is greater than that of K, but the second ionisation potential for K involves the loss of an electron from an octet ($3s^23p^6$ configuration) whereas that for Cu involves the more easily ionisable penultimate configuration ($3s^23p^63d^{10}$ configuration) and so there is a reverse order in second ionisation potential.

6. Calculate Z* for the valence electron and the 3d electron of Zn. [C.U. 2005]

**Ans. :** 4.35 and 8.85 unit of nuclear charge.

7. Compare the 5s, 5p, 5f, 5d and 5g orbitals in order of increasing (a) penetrating ability (b) shielding ability and (c) degree to which they are shielded by inner electrons.

**Hint :** (a) 5s > 5p > 5d > 5f > 5g
(b) 5s > 5p > 5d > 5f > 5g
(c) 5s < 5p < 5d < 5f < 5g

8. Calculate effective nuclear charge for the 6s and 4f electrons of Ce.

**Ans. :** 2.85 and 11.65 unit of nuclear charge.

9. Account for the smaller increase in effective nuclear charge for a 2p electron in going from N to O than in going from C to N. [B.U. 1999, 1995]

**Hint :** On going from C to N the additional one electron is added to the empty 2p orbital whereas on going from N to O the additional one electron is also added to the 2p orbital but this is already occupied by one electron. In this case therefore the added electron will experience greater repulsive force thereby the increase in effective nuclear charge is more cancelled than that between C and N.

10. Why Al (Z = 13) and Ga (Z = 31) have the same atomic radius? [B.U. 1994]

**Hint :** Sec. 2.10(1).

11. Account for the decrease of first ionisation energy between P and S.

**Hint :** P : $[Ne]3s^23p_x^1 3p_y^1 3p_z^1$ S : $[Ne]3s^23p_x^2 3p_y^1 3p_z^1$

The resulting repulsion of the two electrons in the same $3p_x$ orbital of S is the cause of decrease of first ionisation energy from P to S.

12. The ionisation energy of hydrogen is 13.6 eV. What will be the difference in energy between n = 1 and n = 6 levels?

**Ans. :** According to Bohr's theory the expression for ionisation energy of hydrogen :

$$\text{I.E.} = E_1 - E_\infty = \frac{2\pi^2e^4m}{h^2}\left[\frac{1}{1^2} - \frac{1}{\infty^2}\right] = \frac{2\pi^2e^4m}{h^2}$$

$$\therefore \frac{2\pi^2e^4m}{h^2} = 13.6 \text{ eV.}$$

Now the difference in energy between n = 1 and n = 6 level can be given by—

$$E = \frac{2\pi^2e^4m}{h^2}\left[\frac{1}{1^2}-\frac{1}{6^2}\right] = 13.6\left[\frac{1}{1^2}-\frac{1}{6^2}\right] = \frac{13.6\times35}{36} = 13.2 \text{ eV.}$$

13. Explain why ionisation energy of transition metals has no appreciable variation?

14. Which of the following isoelectronic anionic species has the largest radii? Give reasons. $N^{3-}$, $O^{2-}$, $F^-$ and $C^{4-}$.

15. Arrange C, N, O and F in the decreasing order of their second ionisation potential values.

16. Explain why the electron affinity of inert gases are zero?

    **Hint :** This is due to closed electron shell.

17. Calculate the amount of energy released in kcal when 2g of chlorine is converted to $Cl^-$(g) ion. The electron affinity of chlorine is 3.7 eV.

**Ans. :** No. of atoms of chlorine in 2g of chlorine = $\frac{6.023\times10^{23}\times2}{35.5}$.

Electron affinity means $Cl(g) + e \longrightarrow Cl^-(g)$

$\therefore$ The energy released = $\frac{1}{6.023\times10^{23}}\times\frac{6.023\times10^{23}\times2\times3.7}{35.5}$

= 0.208 eV = 0.208 × 23.06 = 4.80 kcal [1 eV = 23.06 kcal]

18. The electron affinity of chlorine is 350 kJ $mol^{-1}$. What will be the amount of energy released when 1.775 g of chlorine is converted to chloride ion in the gaseous state?

    **Ans. :** 17.5 kJ.

19. How much amount of energy (kJ) is required to convert 3.45 mg sodium vapour to sodium ion? (Ionisation energy of sodium is 490 kJ $mol^{-1}$)

**Ans. :** No. of moles of sodium = $\frac{3.45\times10^{-3}}{23} = 1.5\times10^{-4}$.

Ionisation energy means $Na(g) + I.E. \longrightarrow Na^+(g)$.

$\therefore$ The amount of energy required = $490 \times 1.5 \times 10^{-4}$ = 0.0735 kJ.

20. Account for the large difference in first (141 kJ $mol^{-1}$) and second (–770 kJ $mol^{-1}$) electron affinity of oxygen.

21. Atomic radius of an element is not an unified property of an element—it depends on the environment. Justify the statement.

    **Hint :** Sec. 2.7(1).

22. The ionic radii of $Li^+$, $Be^{2+}$ and $B^{3+}$ follow the order—
    1. $Be^{2+} > B^{3+} > Li^+$ 2. $Li^+ > B^{3+} > Be^{2+}$ 3. $B^{3+} > Be^{2+} > Li^+$ 4. $Li^+ > Be^{2+} > B^{3+}$

    **Ans. :** 4. **[NET 2001]**

23. The ionisation energy of boron is less than that of beryllium because—
    1. beryllium has a higher nuclear charge than boron
    2. beryllium has a lower nuclear charge than boron
    3. the outermost electron in boron occupies a 2p orbital

4. the 2s and 2p orbitals of boron are degenerate. [NET 2000]

**Ans. :** 3.

24. Electron affinity of $Mn^{3+}$ is greater than that of $Fe^{3+}$—explain. [C.U. 2006, B.U. 2016]

**Hint :** $Fe^{3+} \longrightarrow 3d^5$, $Mn^{3+} \longrightarrow 3d^4$

$3d^5$ system is more stable than $3d^4$ and so electron affinity of $Mn^{3+} > Fe^{3+}$. The electron addition in half-filled $3d^5$ configuration results some electron-electron repulsion.

25. State Slater's rules and apply these to find the effective nuclear charge on a valence electron of $Cl^-$ ion. [C.U. 2007]

26. Calculate the Pauling's electronegativity of fluorine from the following data :

Bond energy of $H_2$ = 104 kcal mol$^{-1}$
Bond energy of $F_2$ = 38 kcal mol$^{-1}$
Bond energy of HF = 134 kcal mol$^{-1}$
Electronegativity of H = 2.1

**Hint :** Problem 2.13.

27. $SnCl_4$ is relatively more stable than $SnCl_2$ but the reverse is true for the corresponding lead compounds—Comment. [B.U. 1998]

**Hint :** Sec. 2.12. This is due to inert pair effect.

28. Of nitrogen and oxygen atoms which one has higher ionisation potential and why?

**Hint :** See Problem 2.4. [SSC 2009]

29. Comment on the relative ionic radii of $O^{2-}$, $F^-$ and $Na^+$. [C.U. 2009]

**Hint :** $O^{2-} > F^- > Na^+$.

30. Write in words the meaning of a negative value of electron affinity. Does any element have a negative ionisation energy? "Second electron affinity of oxygen is negative, yet we notice a number of oxides."—Explain. [B.U. 2010, 2015]

**Hint :** Sec. 2.5.2.

31. Explain why ionisation energy of mercury is comparable to an inert gas. [B.U. 2010, 2016]

**Hint :** Hg : [Xe] $4f^{14}5d^{10}6s^2$ : I.P. 10.43 eV.

Rn : [Xe] $4f^{14}5d^{10}6s^26p^6$ : I.P. 10.75 eV.

The inert gases have stable duplet (for He) or octet (for others) configuration which makes electron withdrawal very difficult and thereby the ionisation energy is high for an inert gas. From the electronic configuration of Hg a good degree of stability or inertness is associated to $6s^2$ electrons due to inert pair effect as the core electron shields this electron pair. So ionisation energy of mercury is comparable to that of an inert gas.

32. In a hydrogenic ion the energy required to excite the electron from its ground state to first excited state is 40.8 eV. Find the energy required to release the electron from the ion.

[B.U. 2009, 2012 C.U. 2005]

**Hint :** Energy $\varepsilon = E_2 - E_1 = \dfrac{2\pi^2 e^4 m Z^2}{h^2}\left[\dfrac{1}{n_1^2} - \dfrac{1}{n_2^2}\right]$ For ground state $n_1 = 1$ and for first excited state $n_2 = 2$

$$\therefore \varepsilon = \frac{2\pi^2 e^4 m Z^2}{h^2}\left[\frac{1}{1^2} - \frac{1}{2^2}\right] \text{ i.e. } \varepsilon = \frac{2\pi^2 e^4 m Z^2}{h^2} \times \frac{3}{4}$$

Accordingly, $\frac{2\pi^2 e^4 m Z^2}{h^2} \times \frac{3}{4} = 40.8$ eV, i.e. $\frac{2\pi^2 e^4 m Z^2}{h^2} = 54.4$

Now during release of electron from the ion $n_2 = \infty$ and in that case the energy corresponds to ionisation energy I, so that,

$$I = \frac{2\pi^2 e^4 m Z^2}{h^2}\left[\frac{1}{1^2} - \frac{1}{\infty^2}\right] = \frac{2\pi^2 e^4 m Z^2}{h^2} = 54.4 \text{ eV.}$$

33. Locate the positions of Rh and Ce in the modern form of the Mendeleev-Seaborg periodic table. **[B.U. 2010]**

**Hint :** Rh : Period 5, Group 9; Ce : Period 6, Group 3.

34. The wave number/wave length of the first line of the Lymann series of hydrogen is identical to that of the second line of the Balmer series for some hydrogen like ion 'X'. Identify the ion and also find its ionisation energy. (Given ionisation energy of hydrogen = 13.6 eV.] **[C.U. 2009, 2015, B.U. 2016]**

**Hint :** $\frac{1}{\lambda_H} = R_H\left[\frac{1}{1^2} - \frac{1}{2^2}\right]$ so that $\lambda_H = \frac{4}{3R_H}$

$\frac{1}{\lambda_X} = R_X\left[\frac{1}{2^2} - \frac{1}{4^2}\right]$ so that $\lambda_X = \frac{16}{3R_X}$

Therefore, $\frac{4}{3R_H} = \frac{16}{3R_X} = \frac{16}{3Z^2R_H}$, i.e. Z = 2

So that the ion is $He^+$. Its ionisation energy = 4×13.6 eV.

35. Write down the IUPAC name of the elements having atomic number 109 and 111. What would be their symbol? **[C.U. 2009]**

**Hint :** Sec. 2.3.

36. State Slater's rules and using this calculate the effective nuclear charge for one 3d and one 4s electron of Fe. **[C.U. 2009, B.U. 2012]**

**Hint :** Sec. 2.11 and 2.11.1.

37. Discuss the factors governing the electronegativity of an element. **[C.U. 2009]**

**Hint :** Sec. 2.5.4.

38. Calculate the electronegativity of chlorine in Mulliken's scale and hence find out the electronegativity of the same element in Pauling's scale. (Given : Electron affinity of chlorine = 4.0 eV per atom, Ionisation energy of chlorine = 13.0 eV per atom). **[C.U. 2009]**

**Hint :** $\chi_M = \frac{4.0 + 13.0}{2} = 8.5$.

Pauling electronegativity, $\chi = \frac{8.5}{2.6} = 3.26$ as per problem 2.11.

39. Establish a relation between electron affinity of X(g) atom and ionisation potential of $X^-$ (g) ion. **[C.U. 2009]**

**Hint :** Sec. 2.5.2.

40. Portray the modern form of the periodic table and state the number of groups and periods. **[B.U. 2010]**

**Hint :** Sec. 2.1 and Table 2.1. A sketch of periods 1 to 7 and groups 1 to 18 with A and B.

41. Using Born-Haber cycle and the following–calculate the electron affinity of chlorine :

$\Delta H_f(RbCl) = -102.9$ kcal mol$^{-1}$

IP(Rb) = 95 kcal mol$^{-1}$

$\Delta H_{sub}(Rb) = 20.5$ kcal mol$^{-1}$
$D_{Cl_2} = 54$ kcal mol$^{-1}$
$U_{RbCl} = -166$ kcal mol$^{-1}$ [C.U. 2009]

**Hint :** From Sec. 3.1.3.

$$\Delta H_f = S + I + 0.5D + E + U$$

$$-102.9 = 20.5 + 95 + 27 + E - 166$$

so that $E = -79.4$ kcal mol$^{-1}$ = $-3.44$ eV.

42. Calculate the screening constant for a 4s electron in Zn. [B.U. 1996, 2008]

**Hint :** Sec. 2.11.1 and Problem 6 (exercise).

43. Compare feasibility of the following decomposition reactions. Give probable reasons.
$SnCl_4 \rightarrow SnCl_2 + Cl_2$, $PbCl_4 \rightarrow PbCl_2 + Cl_2$ [C.U. 2004]

**Hint :** Sec. 2.12. Due to the inert pair effect the second decomposition reaction is feasible but the first is not.

44. State and explain the factors influencing ionisation energies of elements. The valence shell electronic configuration of both Ca and Zn is $4s^2$ but first ionisation energy of Ca is lower than Zn—Explain. [C.U. 2007]

**Hint :** Sec. 2.5.1 for first part.

Ca : $[Ar]4s^2$ and Zn : $[Ar]3d^{10}4s^2$. The radial distribution function of 4s and 3d orbitals suggest that 3d orbital is more nearer to the nucleus. Therefore whatever number of electrons are added to the 3d orbital after Ca, this will not bring an increase in size, instead increasing nuclear charge brings about a slow contraction in size thereby atomic/metallic radius of Ca and Zn are 197 pm and 137 pm respectively i.e. the size of Ca is larger than that of Zn atom so that first I.E. of Ca will be lower than Zn.

45. Rationalise the trends in specified atomic properties in the following cases :

| Properties | Li | Be | B |
|---|---|---|---|
| First ionisation energy (ev) | 5.39 | 9.32 | 8.30 |
| | C | N | O |
| First electron affinity (ev) | 1.27 | −0.07 | 1.46 |

[B.U. 2012]

**Hint :** Ionisation energy increases from left to right in a period as electrons are added in the same quantum shell. Electron affinity also increases from left to right in a period but N has half filled electron configuration and so it registered negative electron affinity which indicates it has least tendency to take additional electron.

46. Why $Au^-$ is stable but $Cu^-$ does not exist? [C.U. 2012]

**Hint :** Sec. 2.5.2. Au has the highest electron afinity value (223 kJ mole$^{-1}$).

47. Calculate the electronegativity of bromine (At. No. 35) in the Allred-Rochow scale. $r_{covalent} = 114$ pm. [C.U. 2012]

**Hint :** Problem 2.28.

48. Arrange N, O, $N^+$, $O^+$ in order of increaing ionisation potential. Explain the order.[B.U. 2014, 2016]

**Hint :** The electronic configurations of respective species are :

N : $1s^22s^22p^3$ which has half-filled p-orbital

O : $1s^22s^22p^4$ which has partially filled p-orbital

$N^+$ : $1s^22s^22p^2$ which has partially filled p-orbital having one positive charge.

$O^+$ : $1s^22s^22p^3$ which has half-filled p-orbital having one positive charge. Therefore gradation of ionisation potential will be $O < N < N^+ < O^+$.

49. What is the position of the recently named Lv (livermorium) having atomic number 116? **[B.U. 2014]**

**Hint :** Lv have the electronic configuration $[Rn]5f^{14}6d^{10}7s^27p^4$ so that its position in the periodic table will be period 7 and group 6/16.

50. Second ionisation potential of some period 4 elements are (in eV units) Ca(11.87), Sc(12.80), Ti(13.58), V(14.15), Cr(16.5) and Mn(15.64). Identify the orbitals from which ionisation occurs and account for the trends in values. **[B.U.. 2013]**

| **Hint :** | Ca | Sc | Ti | V | Cr | Mn |
|---|---|---|---|---|---|---|
| Second I.P. | 11.87 | 12.80 | 13.58 | 14.15 | 16.5 | 15.64 eV |
| Electronic configuration | $[Ar]4s^2$ | $[Ar]4s^23d^1$ | $[Ar]4s^23d^2$ | $[Ar]4s^23d^3$ | $[Ar]4s^13d^5$ | $[Ar]4s^23d^5$ |

Although in all cases the first electron is released from the outermost 4s electron but the release of second electron is somewhat different. As the nuclear charge increases from Ca to V, the second I.P. increases slowly, but there is a sudden jump from V to Cr and a little fall is observed at Mn. This is due to release of second electron from the half-filled 3d orbital in case of $Cr(4s^13d^5)$ resulting high second I.P. value.

51. Comment on the positions of the s, p, d and f block elements in the modern form of the Mendeleev-Seaborg periodic table. Write in one or two lines what might be its possible extension. **[B.U. 2015]**

**Hint :** The arrangement of the elements in the modern periodic table is governed by aufbau principle. Each period starts with an element having its last electron in an ns orbital; n = 1 for H, n = 2 for Li and so on. The period ends with an element whose last electron completes $ns^2np^6$ or $ns^2(n-1)d^{1-10}np^6$ configuration. With this consideration the long form of the periodic table may be designed into four regions. Each region corersponds to the filling of the s, p, d and f orbitals and thus the regions are designated by the term s, p, d and f block elements as shown in the adjacent figure. The eight group elements belonging to the s and p blocks are often called **representative elements,** those in the d block are called the **transition elements** and the members of the f block are termed **inner transition elements**. However Zn, Cd, Hg and some f block elements are not treated as transition elements.

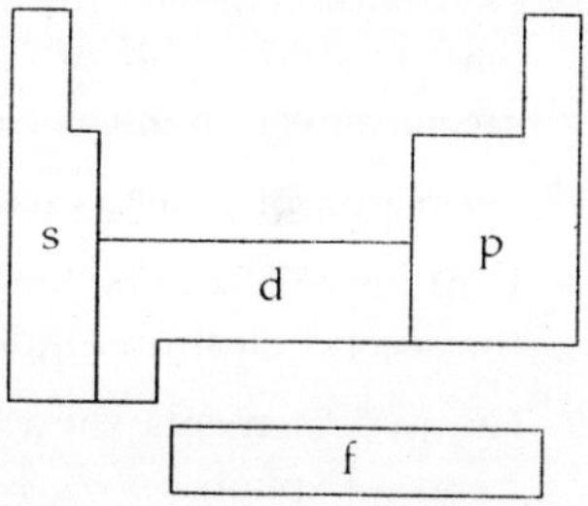


As regards the possible extension of the modern periodic table as per the IUPAC (1990) recommendation may be converted to thirty two column instead of eighteen column. Moreover any element beyond atomic number 118 will be placed by starting a new period 8 and will be accommodated below the last element of alkali metals.

52. Locate the positions of irridium and thorium in the modern form of the Mendeleev–Seaborg periodic table. **[B.U. 2015]**

**Hint :** Ir(77) : $[Xe]4f^{14}5d^76s^2$ Period 6 ; Group 9

Th(90) : $[Rn]6d^27s^2$ Period 7; Group 4

53. State the basis of Pauling's electronegativity scale. **[B.U. 2015]**

**Hint :** Sec. 2.5.3. The basis of Pauling's electronegativity is the bond dissociation energy of a chemical bond A–B. In this electronegativity scale fluorine is taken as highest electronegative and is assigned the value 4.0.

54. Define work function and mention its difference with ionisation energy. **[B.U. 2015]**

**Hint :** Sec. 2.6.

55. Explain the basis of electronegativity as described by Allred-Rochow and correlate it with the Pauling's scale. **[C.U. 2013]**

**Hint :** Sec. 2.5.8. The basis of Allred-Rochow electronegativity is the electrostatic force of attraction between the bonded electron with that of the nucleus. This electrostatic force of attraction depends on the effective nuclear charge with that of the covalent radius.

56. Explain the stabilities of common oxidation states of Bi and Pb in the light of a suitable theory. **[C.U. 2013]**

**Hint :** Sec. 2.12. The stabilities of common oxidation states of last member of a periodic group is reduced from group oxidation states and can be explained from inert pair effect.

57. The first ionisation potentials of the coinage metals follow the order :
Cu > Ag < Au – Explain. **[C.U. 2013]**

**Hint :** Cu(29) : $[Ar]3d^{10}4s^1$

Ag(47) : $[Kr]4d^{10}5s^1$

Au(79) : $[Xe]4f^{14}5d^{10}6s^1$

The lesser I.P. value in case of Ag is due to the removal of electron from larger size 5s orbital in case of Ag compared to that of the removal of electron from 4s orbital in case of Cu. In case of Au the electron is removed from 6s orbital so that the I.P. value expected to be less than Ag but due to the participitation of 4f orbital in case of Au which has poor screening efficiency and thus experience greater effective nuclear charge thereby registers greater value of I.P.

58. Formation of gaseous oxide and sulfide ions are endothermic processes.—Explain. **[C.U. 2013]**

**Hint :** Sec. 2.5.2. The formation of $O^- \rightarrow O^{2-}$ and $S^- \rightarrow S^{2-}$ is not energetically favourable as because the second electron affinity is negative so that the processes are endothermic.

59. Group 3 elements show regular trend in their atomic radii but Group 4 elements show irregular trend.—Explain. **[C.U. 2013]**

**Hint :** The outermost electronic configuration of Group 3 and Group 4 elements are $ns^2np^1$ and $ns^2np^2$ respectively but underneath due to the involvement of d and f orbitals anomaly arises in case of atomic radii.

60. Using Slater's rules, calculate the effective nuclear charge for 4s and 3d electron of Zn. **[C.U. 2013]**

**Hint :** Sec. 2.11 and Sec. 2.11.1.

61. Explain inert pair effect with reference to Grop 14 elements. **[C.U. 2013]**

**Hint :** Sec. 2.12.

62. During $Mn^{2+}$ formation from Mn, electron loss takes place from 4s orbital rather than 3d orbital.—Explain. **[C.U. 2014]**

**Hint :** Sec. 2.11.1(IV).

63. Explain the electron affinity values (in eV) of the following elements : Ca(–0.3), Mn(–0.5), O(1.46), S(2.08). **[C.U. 2014]**

**Hint :** Sec. 2.5.2.

Ca : $1s^2 2s^2 2p^6 3s^2 3p^6 4s^2$

Mn : $1s^2 2s^2 2p^6 3s^2 3p^6 4s^2 3d^5$

O : $1s^2 2s^2 2p^4$

S : $1s^2 2s^2 2p^6 3s^2 3p^4$

Ca has filled 4s orbital and Mn has stable half-filled 3d orbital and due to the stability of filled and half-filled orbital electron attachment enthalpy is negative. O and S has partially filled p orbital so that they have a tendency to attach with additional electron to form uninegative ion so that they have positive electron affinity value.

64. Calculate Mulliken electronegativity of F and Cl from the given data; I.E. (kcal $mol^{-1}$) : F(401.7), Cl(299.1); E.A.(kcal $mol^{-1}$) : F(78.4), Cl(83.5). Convert the values to those of Pauling values using a suitable factor. **[C.U. 2014]**

**Hint :** Sec. 2.5.3. Problem 2.11 and question number 38(Exercise).

65. Explain the sequence of ionisation energies (kJ $mol^{-1}$) : Cu(746), Zn(906), Ga(579). **[C.U. 2014]**

**Hint :** The electronic configuration of the elements are :

Cu : $1s^2 2s^2 2p^6 3s^2 3p^6 4s^1 3d^{10}$

Zn : $1s^2 2s^2 2p^6 3s^2 3p^6 4s^2 3d^{10}$

Ga : $1s^2 2s^2 2p^6 3s^2 3p^6 4s^2 3d^{10} 4p^1$

In case of Cu the unpaired electron is removed from the 4s orbital but in case of Zn the 4s electrons are paired and so Zn will register higher ionisation energy compared to Cu. There is also increase of nuclear charge from Cu to Zn. In case of Ga the single unpaired electron will be removed during ionisation from 4p orbital and thus a drop in ionisation energy from Zn to Ga.

66. Atomic radii of Nb and Ta are almost identical but that of Cs is larger than Rb.—Explain. **[C.U. 2014]**

**Hint :** Sec. 2.7.

Nb : $[Kr]4d^3 5s^2$

Ta : $[Xe]4f^{14} 5d^3 6s^2$

Nb is a member of second transition series while Ta is a member of third transition series. There is a close similarity between second and third transition series elements are observed compared to that of first transition series elements because 4d and 5d orbitals are more spread out in space than the 3d orbitals of the first transition series. This leads to smaller interelectronic repulsion since 5d orbital has larger size compared to that of 4d orbital. Moreover in case of Ta we have 4f electrons and hence a lanthanide type of contraction is observed. Cs : $[Xe]6s^1$ and Rb : $[Kr]5s^1$. In case of Cs and Rb the outer quantum shell number are 6 and 5 respectively which suggests Cs will have a larger size compared to that of Rb.

67. Account for the stability of oxidation states of Tl and Pb. **[C.U. 2015]**

**Hint :** Sec. 2.12. According to inert pair effect the oxidation states of Tl and Pb are +1 and +2 respectively.

68. The electron affinity of Au is abnormally high and it may exist as auride.—Justify. **[C.U. 2015]**

**Hint :** Sec. 2.5.2. Au has the electronic configuration $[Xe]4f^{14}5d^{10}6s^1$. Because of the contraction and stabilisation of the 6s orbital, the outermost, or the valence shell of Au is formed by both the 5d and 6s orbitals. Indeed, electronically, Au is halogen like with one electron missing from the pseudo noble gas configuration. Hence similar to the existance of halogen molecule $X_2$, gold also forms covalent $Au_2$ molecule. Moreover, Au also forms ionic compounds such as RbAu and CsAu in which the $Au^-$ has the pseudo-noble gas configuration. Due to the lanthanide type contraction, the post lanthanide elements will have higher electron affinity and among the elements Au registers highest electron affinity (223 kJ $mol^{-1}$) and the existance of auride ion ($Au^-$) like that of halide ion ($X^-$) has been proved in the compounds e.g. RbAu, CsAu etc.

69. Account for the electron affinity (eV) change in the following pair : C(1.26), N(–0.07). **[C.U. 2015]**

**Hint :** Electronic configurations of C and N are $1s^22s^22p^2$ and $1s^22s^22p^3$ respectively. N has half-filled p orbital while the p orbital for C is partially filled. Greater stability of half-filled p orbital is registered by the negative electron affinity while less stable p orbital of C has greater tendency to be half-filled which is shown by positive value of electron affinity.

70. If the ionisation energy of H atom is x, the ionisation energy of $Li^{2+}$ is :
(a) 2x, (b) 3x, (c) 9x, (d) 27x. **[NET 2014]**

**Hint :** (c). According to Bohr's theory, ionisation energy of H and hydrogen like species is given by—

$$I.E. = \frac{2\pi^2 e^4 m Z^2}{n^2 h^2} \quad \text{(Problem 2.8)}$$

Thus I.E. $\propto Z^2$. For $Li^{2+}$, Z = 3, therefore

I.E. of $Li^{2+} = (3)^2\ x = 9x$

71. The correct order of the size of S, $S^{2-}$, $S^{2+}$ and $S^{4+}$ species is—
(a) $S > S^{2+} > S^{4+} > S^{2-}$, (b) $S^{2+} > S^{4+} > S^{2-} > S$
(c) $S^{2-} > S > S^{2+} > S^{4+}$, (d) $S^{4+} > S^{2-} > S > S^{2+}$ **[NET 2014]**

**Hint :** (c) i.e. $S^{2-} > S > S^{2+} > S^{4+}$. The order of size is anion > neutral atom > cation having low charge > cation having high charge.

72. The correct order of radii is—
(a) N < Be < B, (b) $F^- < O^{2-} < N^{3-}$, (c) Na < Li < K, (d) $Fe^{3+} < Fe^{2+} < Fe^{4+}$.

**Hint :** (b) i.e. $F^- < O^{2-} < N^{3-}$.

73. The smallest ionic radius among the following is—
(a) $K^+$, (b) $Ca^{2+}$, (c) $Sc^{3+}$, (d) $Ti^{3+}$, (e) $Ti^{4+}$

**Hint :** (e) i.e. $Ti^{4+}$.

74. Choose the correct order of bond energies in the following series :
(a) C = O > O = O > C – O, (b) C = O > C – O > O = O
(c) C – O > O = O > C = O, (d) O = O > C – O > C = O

**Hint :** (a) i.e. C = O > O = O > C – O.

75. The screening effect of inner electrons of the nucleus causes—
(a) Decreases the ionisation energy,
(b) Increase in the ionisation energy,
(c) No effect on the ionisation energy,
(d) Increases in the attraction of the nucleus to the electrons.

**Hint :** (a) i.e. decreases the ionisation energy.

76. The decreasing order of second ionisation energy of K, Ca and Ba is—
(a) K > Ca > Ba, (b) Ca > Ba > K, (c) Ba > K > Ca, (d) K > Ba > Ca

**Ans. :** (a) i.e. K > Ca > Ba.

77. The electron affinity of the following elements can be arranged—
(a) Cl > O > N > C, (b) Cl > O > C > N, (c) Cl > N > C > O, (d) Cl > C > O > N

**Hint :** (b) i.e. Cl > O > C > N. The electron affinity of Cl is highest. In a period from left to right electron affinity increases. Nitrogen has low value because of stable half-filled p orbital.

78. Why the electronegativity of Ga is greater than that of Al? **[B.U. 2013]**

**Hint :** Al has atomic number (13)(n=3) while Ga has atomic number 31(n=4). Therefore it is expected that Ga will have larger size compared to that of Al. In fact both Al and Ga has almost equal atomic radius values. It is expected that Ca will have greater atomic radius than that of Al but after Ca the electrons are added to 3d orbital which have poor shielding capability. Therefore whatever number of electrons are added to the 3d orbital this will not bring any increase in size instead increasing nuclear charge brings about a contraction in size. Due to this size effect the electronegativity of Ga is greater than that of Al.

79. Calculate the electronegativity of carbon atom in C–H bond, if $E_{C-H}$, $E_{H-H}$ and $E_{C-C}$ are 98.8 and 104 and 83 kcal/mole respectively. (Here E = bond energy). **[SSC 2010]**

**Hint :** Problem 2.13.

$\Delta = E_{C-H} - \sqrt{E_{H-H}E_{C-C}} = 98.8 - \sqrt{104 \times 83} = 98.8 - 93 = 5.8 \text{ kcal} = 5.8 \times 4.184 \text{ kJmol}^{-1}$

$= 24.26 \text{ kJmol}^{-1}$ [$\because$ 1 kcal = 4.184 kJ]

Now $\chi_C - \chi_H = 0.102\sqrt{24.26} = 0.502.$

$\therefore \chi_C = 0.502 + \chi_H = 0.502 + 2.1 = 2.602.$

80. Establish Pauling's electronegativity scale giving its theoretical basis. **[C.U. 2014]**

**Hint :** Sec. 2.5.3.

81. Ge(II) compounds are reducing agents, while Pb(IV) compounds are oxidising agents. Explain. **[C.U. 2014]**

**Hint :** Sec. 2.12. Due to the inert pair effect which is prominant to the lower member of the group the valence or oxidation state will be redundant. Thus in Group 14 Ge will exhibit +4 oxidation state while for Pb the +2 oxidation state is stable. Therefore Ge(II) compounds will act as a reducing agent while Pb(IV) compounds will be reduced to the Pb(II) compounds and thus acting as oxidising agent.

82. Calculate the electron affinity of iodine with the help of following data (given in kcalmol$^{-1}$)
$(\Delta H_{for})_{NaI} = -68.8$, $(\Delta H_{sub})_{Na} = 25.9$, $\frac{1}{2}(\Delta H_{diss})_{I_2} = 25.5$,
$(IP)_{Na} = 118.4$, $(U)_{NaI} = -165.4$. **[B.U. 2015]**

**Hint :** Sec. 3.1.3 and Question number 41.

$\Delta H_f = S + I + 0.5D + E + U$

$-68.8 = 25.9 + 118.4 + 25.5 + E - 165.4$

so that $E = -73.2 \text{ kcalmol}^{-1} = -3.17 \text{ eV}$.

83. The element with atomic number 87 belongs to—
(a) s-block, (b) p-block, (c) d-block, (d) f-block.

**Ans. :** (a) i.e. s-block. This is francium (87) i.e. $[Rn]7s^1$.

84. $PtF_6$ is known to form complexes with $O_2$; from the values of ionisation potential (eV) of oxygen and a set of inert gases, justify the gases expected to form such complexes : $O_2$(12.20), He(24.60), Ne(21.60), Ar(15.76). **[B.U. 2014]**

**Hint :** $PtF_6$ is an extremely strong fluorinating agent and will oxidise $O_2$ to $O_2^+$ thereby will be the formation of crystalline product $O_2^+[PtF_6^-]$. From the low ionisation potential vlaue of Ar it is expected that $O_2$ and Ar may form some complexes.

85. Locate the position of ruthenium, gadolinium and uranium in the modern periodic table. **[B.U. 2016]**

**Hint :** Ru(44) : $[Kr]4d^7 5s^1$ Period 5; Group 8

Gd(64) : $[Xe]4f^7 5d^1 6s^2$ Period 6; Group 3

U(92) : $[Rn]4f^3 6d^1 7s^2$ Period 7; Group 3.

86. 'Atomic radius of neon is greater than that of fluorine.'—Comment. **[B.U. 2016]**

**Hint :** In case of fluorine the atomic radius means the half of the F-F internuclear distance (Sec. 2.7). But as neon is an inert gas so in this case atomic radius means the van der Waals radius (Sec. 2.9). As van der Waals radius arises from the weak dispersion forces between the two molecules so van der Waals radius is always greater than that of atomic radius. Between fluorine and neon it is expected that neon will have lower value of radius than that of flurone as atomic radius will decrease from left to right in a period. But fluorine has the atomic radius value 64 pm while neon registers the radius value as 112 pm. Similarly atomic radius of Cl(99 pm) less than the radius of Ar(154 pm).

87. 'Electronegativity of nitrogen in Pauling's scale is 3.0 in all nitrogen compounds.'—Comment. **[B.U. 2016]**

**Hint :** Sec. 2.5.4(c). The electronegativity of nitrogen (3.0) will not be same for all nitrogen compounds. Because electronegativity of an element depends on the oxidation state and also nature of hybridisation. For example the following table will justify the variation of electronegativity of nitrogen in different compounds.

| Compound | State of hybridisation of nitrogen | Percentage of s character of hybrid orbital | Electronegativity |
|---|---|---|---|
| MeCN | sp | 50% | 3.7 |
| $C_5H_5N$ | $sp^2$ | 33% | 3.3 |
| $Me_3N$ | $sp^3$ | 25% | 3.0 |

Therefore we may conclude that as the s character of the hybrid orbital of nitrogen increases, electronegativity increases.

88. 'Second electron affinity of oxygen is negative, yet we observe a good number of oxides'.—Explain. **[B.U. 2016]**

**Hint :** Sec. 2.5.2 and Problem 2.10.

89. What do you mean by quantum defect? State its origin. **[B.U. 2016]**

**Hint :** Sec. 2.1(footnote). The origin of quantum defect arises due to the difference in the optical behaviour of single electron in case of hydrogen and alkali metals as also due to the different penetrating power of orbitals.

The optical behaviour of single electron in case of alkali metals is comparable to that of hydrogen and hydrogen-like systems. The alkali metals contain only one valency electron in the outermost shell, and the electron is involved in the optical transition according to the Balmer-Ritz equation :

$\bar{\nu} = R\left[\frac{1}{n_1^2} - \frac{1}{n_2^2}\right]$ where $n_2 > n_1$, are quantum numbers, R is Rydberg constant and $\bar{\nu}$ is wave number.

Here it can be assumed that all the inner electrons have nothing to do with the outermost optical electron and the outermost electron is experiencing only unit residual positive charge (i.e. Z=1) at the nucleus. Hence it expected that the energy of this optical electron should be expressed as in the case of hydrogen-like systems as given above. But in case of hydrogen atom the optical behaviour of the single electron is somewhat different compared to that of alkali metals or any other polyelectronic systems. In case of hydrogen atom the effective nuclear charge is zero thereby the optical electron is only experiencing the nuclear charge. But for other systems the optical electrons are in the influence of lower lying inner shell and therefore there will be different values of effective nuclear charge for different systems and the attractive force on the optical electron by the nucleus will be somewhat different than can be expected. Thus the above equation should have to be modified and from the spectral studies it is found as :

$\bar{\nu} = R\left[\frac{1}{(n_1 - \mu_1)^2} - \frac{1}{(n_2 - \mu_2)^2}\right]$ where $\mu_1$ and $\mu_2$ are small constants.

This equation is also known as the general formula of the Ritz combination principle. In this equation the principal quantum numbers i.e. $n_1$ and $n_2$ are replaced by $n-\mu$ where $\mu$ is called **quantum defect** so that effective quantum number is $n^* = n-\mu$, thus $n^*$ becomes nonintegral. This fact i.e. nonintegral principal quantum number, definitely strikes at the very fundamental root of the Bohr-Sommerfeld theory. This fact is also observed in many other polyelectronic systems besides the alkali metals. The extent of departure of $n^*$ from the integral values depends on the nature of the electron undergoing optical transition. The extent of deviation of $n^*$ follow the sequence : s-electron > p-electron > d-electron > f-electron i.e. $\mu$ is maximum for s-electron. This can be explained by the penetrating power of orbitals. Due to this penetration, the electron experiences more positive dharge at the nucleus. This enhanced attractive force towards the nucleus can be rationalised mathematically either by considering the higher value of Z (i.e. Z > 1) or by the lower value of the principal quantum number n, as it is evident that the lower the value of n, the higher is the attractive force towards the nucleus. Thus it can be concluded that the more the penetration, the more is the attractive force and hence the more the quantum defect.

90. Write down the general electronic configurations of s-, p-, d- and f-block elements adducing proper reason. **[B.U. 2016]**

**Hint : s-block elements :** $ns^2np^6(n+1)s^{1-2}$

**p-block elements :** $ns^2np^6(n+1)s^2(n+1)p^{1-6}$

**d-block elements :** $ns^2np^6(n+1)s^2nd^{1-10}$ (Having some exception) or have the configuration [Noble gas configuration] $(n-1)d^{1-10}ns^{1-2}$.

**f-block elements :** $ns^2np^6(n-1)f^{1-14}nd^n(n+1)s^2$ having several exception.

The f-block elements may have the general electronic configuration as $(n-2)f^{1-14}(n-1)d^{1-2}ns^2$ as the differentiating electron houses deep-seated f-orbital or penultimate d-orbital or the general electronic configuration of lanthanides (Ln) are either $[Xe]4f^n5d^06s^2$ or $[Xe]4f^n5d^16s^2$ and the actinides have the general electronic configuration as $[Rn]5f^{0-14}6d^{0-2}7s^2$.

# 3

# CHEMICAL BONDING and CHEMICAL FORCES

**A chemical bond is defined as a chemical force that holds together between two or more atoms to form a chemical compound or aggregate of molecular species.** There are three extreme types of chemical forces operative between the chemical compounds :

(1) Ionic or electrovalent bond

(2) Covalent or electron pair bond

(3) Metallic bond.

## 3.1 IONIC OR ELECTROVALENT BOND

It is evident that solid sodium and chlorine gas reacts to form solid sodium chloride. Crystallographic evidence of solid sodium chloride suggests that each $Na^+$ ion is surrounded by six $Cl^-$ ions or vice-versa in an non-ending pattern which is known as hexagonal close packed type lattice. Then a question arises what is the motive behind the formation of sodium chloride? Sodium has one electron outside the $2s^22p^6$ configuration whereas chlorine has one electron short of the next noble gas (Ar) configuration. The sodium loses one electron to form $Na^+$ ion and chlorine will gain one electron to form $Cl^-$ ion and thereby $Na^+$ and $Cl^-$ ions will assume the stable inert gas configuration. **This is one of the mechanism or the necessary and sufficient condition for the formation of ionic or electrovalent bond that each of the cations and anions will assume the $ns^2np^6$ configuration.** Then by electrostatic attraction between the cations and anions ionic or electrovalent compounds are formed. Therefore in the formation of ionic or electrovalent bond there is a complete transfer of electron(s) occurs from one atomic orbital to other (i.e. oxidation-reduction). Thus ionic bond is defined as **electrostatic force of attraction operated between the cations and anions having stable inert gas configurations.**

### 3.1.1 Factors influencing the formation of ionic bonds

1. The elements which have **low ionisation energies** are the best members to form the cations. The alkali and alkaline earth metals which are placed in the extreme left of the periodic table very easily form stable cationic species. These elements have one or two electrons excess outside the stable inert gas configuration. Low ionisation energy is responsible to knock off these electrons to form the stable cations.

2. The non-metallic partner which forms the anion should have **high electron affinity**. The halogens which are placed to the right of the periodic table are the best member to form the anion easily because they have just one electron short of the next noble gas configuration thereby they gain one electron easily to form the mononegative species e.g. $F^-$, $Cl^-$ etc. High electron afinity of these elements means high exoergic process greatly influence the formation of ionic compounds. It is to be noted that the anions having high charge e.g. $N^{3-}$, $P^{3-}$ are not suitable for the formation of ionic compounds as such ions are highly polarisable and favour covalency.

3. During the formation of ionic compounds various exoergic and endoergic processes involved. From the consideration of the total energy change it has been suggested that **low sublimation energy of the metal and low bond dissociation energy of the non-metallic partner influence the formation of an ionic compound.**

### 3.1.2 Lattice energy or Lattice enthalpy

Mere formation of oppositely charged ions will not led us the formation of an ionic compound. We know ionic compounds has specific geometry and to have the specific geometry each cation should have to be surrounded by specific number of anions or vice-versa and in order to do this an amount of energy will be released from the system and the system will be stabilised. The amount of energy which is released is sometimes called lattice energy or lattice enthalpy. Lattice energy is defined **as the amount of energy released for the formation of one mole of an ionic compound from the assembly of the constituent gaseous cations and anions when brought from infinity to their respective equlibrium position in the crystal lattice.**

The theoretical calculation of lattice energy can be made in the following way. Ionic crystals e.g. NaCl is composed of oppositely charged ions and electrostatic force of attraction predominantly operate between them. Apart from the attractive force there might be some repulsive force operative between the ions due to closer approach of the electron shells of the ions as also for the closer approach of the nucleus of the two ions. The attractive and repulsive forces balance at equilibrium and give rise to a stable position with a minimum potential energy of the system when the cation and the anion will be in the equilibrium interionic distance $r_0$(say). At minimum interionic distance the potential energy of the system will be minimum. Now during the formation of an ionic compound the cations and anions will come from infinity distance to their position in minimum interionic distance ($r_0$) at equilibrium. The decrease in energy in bringing the two ions in minimum interionic distance in the crystal lattice measures the potential energy of the system. **The lattice energy of an ionic solid is therefore is opposite in sign but numerically equal to the minimum potential energy of the system.** In this calculation of lattice energy, van der Waals or other type of intermolecular forces are ignored.

Let the charge of the cation is $Z_i$ and that of anion is $Z_j$ and let the distance between them is denoted by r.

The attraction energy i.e. the potential energy per ion pair is given by coulomb's law,

$$E_a = -\frac{Z_iZ_je^2}{r} \qquad \text{...(3.1)}$$

Due to the closer approach of oppositely charged ions, there will be a repulsive energy,

$$E_r = \frac{B}{r^n} \qquad \text{...(3.2)}$$

where B is a constant and n is called Born exponent the value of which depends on the principal quantum numbers of the electrons involved in the repulsion and can be determined from compressibility factor of an ionic compound. It is to be noted that repulsive term is due to a short range type force and is operative only when the ions extremely close to each other.

$$\therefore \text{ Potential energy } U = -\frac{Z_iZ_je^2}{r} + \frac{B}{r^n} \qquad \text{...(3.3)}$$

We have to evaluate the term B by considering that at equilibrium potential energy will be minimum so that $\frac{dU}{dr} = 0$.

For uni-univalent compound e.g. NaCl, $Z_i = Z_j = 1$ and in a crystal lattice we have to consider mutual potential energies of all the charges in the crystal structure. Due to the different geometry of different ionic crystals, a structural correction term is included in the electrostatic potential energy of the system and thereby equation (3.1) is modified as—

$$E_a = -\frac{AZ_iZ_je^2}{r} = -\frac{Ae^2}{r} \text{ (for } Z_i = Z_j = 1) \qquad \text{...(3.4)}$$

where A is called Madelung constant the value of which depends on the geometry of ionic solids but independent of ionic radius and charge of the ions. The values of the Madelung constant for NaCl type lattice (6 : 6 coordination) is 1.7475, for CsCl type lattice (8 : 8 coordination) is 1.7627, for ZnS type lattice (4 : 4 coordination) is 1.6381 and for $CaF_2$ type lattice (8 : 4 coordination) is 2.5194. Therefore,

$$U = -\frac{Ae^2}{r} + \frac{B}{r^n}$$

$$\text{Hence } \left(\frac{dU}{dr}\right)_{r=r_0} = \frac{Ae^2}{r_0^2} - \frac{Bn}{r_0^{n+1}} = 0$$

$$\therefore \quad \frac{Bn}{r_0^{n+1}} = \frac{Ae^2}{r_0^2} \qquad \text{so that } B = \frac{Ae^2 r_0^{n-1}}{n}$$

Hence at minimum interionic distance i.e. at minimum potential energy,

$$U = -\frac{Ae^2}{r_0} + \frac{Ae^2 r_0^{n-1}}{nr_0^n} = -\frac{Ae^2}{r_0}\left(1 - \frac{1}{n}\cdot\frac{r_0^{n-1}}{r_0^{n-1}}\right) = -\frac{Ae^2}{r_0}\left(1 - \frac{1}{n}\right)$$

Now the total energy for one mole of the compound is given by—

$$U = -\frac{N_0 Ae^2}{r_0}\left(1 - \frac{1}{n}\right) \quad \text{where } N_0 = \text{Avogadro's number}$$

So, the general expression of lattice energy can be written as—

$$U = -\frac{N_0 AZ_iZ_je^2}{r_0}\left(1 - \frac{1}{n}\right) \qquad ...(3.5)$$

According to Systeme International (SI)* unit attraction term (equaiton 3.1)

$$E_a = -\frac{Z_iZ_je^2}{4\pi\varepsilon_0 r} \qquad ...(3.6)$$

where $\varepsilon_0$ is the permittivity of vacuum. Similarly for repulsive term (equation 3.2) can be written as—

$$E_r = \frac{B}{4\pi\varepsilon_0 r^n} \qquad ...(3.7)$$

So that on similar treatment we get the expression for lattice energy as—

$$U = -\frac{N_0 AZ_iZ_je^2}{4\pi\varepsilon_0 r_0}\left(1 - \frac{1}{n}\right) \qquad ...(3.8)$$

Equation (3.5) or (3.8) is called **Born-Lande equation**. The equation (3.5) further simplified by incorporation of values of constant terms i.e. when the charges (e) is expressed in **esu** and distance in **centimetre**, then lattice energy is obtained in **ergs**. Therefore $N_0e^2 = 6.023 \times 10^{23} \times (4.8 \times 10^{-10})^2 = 138760$, so that equation (3.5) may be written as—

$$U = -\frac{138760 AZ_iZ_j}{r_0(\text{in cm})}\left(1 - \frac{1}{n}\right)\text{erg} \qquad ...(3.9)$$

Similarly for SI system if we put the value of $r_0$ in metre (m), permittivity $\varepsilon_0 = 8.854 \times 10^{-12}$ $C^2m^{-1}J^{-1}$, $e = 1.602 \times 10^{-19}$C, so that lattice energy is obtained in **Joule** unit. Therefore,

$$\frac{N_0e^2}{4\pi\varepsilon_0} = \frac{6.023 \times 10^{23}(1.602 \times 10^{-19}C)^2}{4 \times 3.142 \times 8.854 \times 10^{-12} C^2m^{-1}J^{-1}}$$

$$= 0.1389095 \times 10^{-3}$$

* *Abbreviated form of "Systeme International d'Unites".*

Therefore equation (3.8) can be modified as—

$$U = -\frac{0.1389095\times10^{-3}AZ_iZ_j}{r_0(\text{in metre})}\left(1-\frac{1}{n}\right)\text{ Joule} \qquad ...(3.10)$$

Calculated values of lattice energies of some ionic crystals are given in Table 3.1.

**Problem 3.1** Calculate the value of lattice energy of NaCl. A = 1.74, $r_0$ = 2.79Å, n = 8. Compare this value which can be calculated from SI unit.

**Ans.** : According to equation (3.9)

$$U = -\frac{138760\times1.74}{2.79\times10^{-8}\text{cm}}\left(1-\frac{1}{8}\right) = -75721.182 \times 10^8 \text{ erg}$$

$= -75721.182 \times 10^8 \times 10^{-7}$* Joule

$= -757211.82$ Joule

$= -757.22$ kJoule/mol (Experimental value = $-774.00$ kJoule/mol)

Now according to SI unit (equation 3.10)

$$U = -\frac{0.1389095\times10^{-3}\times1.74}{0.279\times10^{-9}\text{metre}}\left(1-\frac{1}{8}\right)$$

$= -758494.2$ Joule $= -758.50$ kJoule/mol

*1 erg = $2.39 \times 10^{-8}$ cal = $1 \times 10^{-7}$ Joule

### 3.1.3 The Born-Haber cycle—Experimental determination of lattice energy

**It is a thermodynamic cycle by which we may evaluate the lattice energy of an ionic compound in an indirect way by dissecting the heat of formation of an ionic crystal into a number of energy functions e.g. ionisation energy, electron affinity, heats of dissociation and sublimation etc. and applying Hess's Law of constant heat summation.** By algebric summation of energy terms the lattice energy can be calculated. This cyclic procedure is called Born-Haber cycle which is depicted below with reference to sodium chloride (Fig. 3.1). The sequence of steps are :

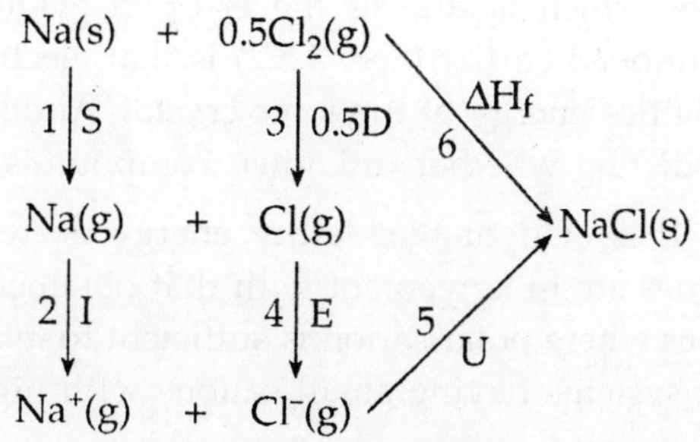


*Figure 3.1 : The Born-Haber cycle for the formation of NaCl.*

(1) Conversion of solid sodium to gaseous sodium atoms i.e. sublimation energy(S)

$Na(s) \longrightarrow Na(g) \qquad S = +107 \text{ kJ mol}^{-1}$

(2) By expending an amount of energy, the gaseous sodium atoms are converted to gaseous sodium ions. The energy involved is called the ionisation energy (I)

$Na(g) + I \longrightarrow Na^+(g) + e \quad I = +495.8 \text{ kJ mol}^{-1}$

(3) By the application of dissociation energy (D), 0.5 mole $Cl_2(g)$ is converted to gaseous chlorine atoms

$0.5Cl_2(g) + 0.5D \longrightarrow Cl(g) \quad 0.5D = +122 \text{ kJ mol}^{-1}$

(4) Conversion of gaseous chlorine atoms to gaseous chloride ions by the addition of one electron to gaseous chlorine atoms. In this case an amount of energy released which is known as electron affinity (E)

$Cl(g) + e \longrightarrow Cl^-(g) \qquad E = -348 \text{ kJ mol}^{-1}$

(5) In this step gaseous $Na^+$ and $Cl^-$ ions combine to form solid sodium chloride and an amount of energy released from the system which is called lattice energy (U)

$Na^+(g) + Cl^-(g) \longrightarrow NaCl(s)$ Lattice energy = –U which has to be determined

(6) Formation of solid sodium chloride from solid sodium and 0.5 mole gaseous chlorine. In this case an amount of energy is released from the system which is called the heat of formation ($\Delta H_f$)

$Na(s) + 0.5Cl_2(g) \longrightarrow NaCl(s) \qquad \Delta H_f = -411 \text{ kJ mol}^{-1}$

Then according to Hess's law

$$\Delta H_f = S + I + 0.5D + E + U$$

$$\therefore -411 \text{ kJ} = +107 \text{ kJ} + 495.8 \text{ kJ} + 122 \text{ kJ} - 348 \text{ kJ} + U$$

$$U = -787.8 \text{ kJ mol}^{-1}$$

Some experimental values of lattice energies are given in Table 3.1. The agreement between the calculated and experimental values are quite satisfactory but the values may differ in some cases which is due to the error in evaluating different thermodynamic terms associated as mentioned earlier (Sec. 2.5.2) is that electron affinity can be determined if we know the value of lattice energy of the ionic crystal. Another use of Born-Haber calculations give us an idea in predicting whether unknown compounds will be thermodynamically stable or not.

It is evident that lattice energy decreases as interionic distance increases. The calculated values are in agreement with that obtained experimentally but deviations are significant in the cases where polarisation is sufficient to induce covalency. Signifcant deviations are observed for the systems having small cations with high positive charge and with larger anions.

### 3.1.4 The importance of lattice energy

(a) **Solubility and Solvation energy (Hydration energy) :** Solubility properties of ionic compounds in water have some complexities. There is no hard and fast rule governed the solubility of ionic solids but a general idea can be made by the term 'like dissolves like' which indicates non polar covalent substances dissolve in non polar solvents and polar substances dissolve in polar solvents having high dielectric constant values. But this rule has several

**Table 3.1 : Lattice energies of some of the ionic solids**

| Ionic solid | Interionic distance (Å) | $-U_{exp}$ kJ $mol^{-1}$ | $-U_{cal}$ kJ $mol^{-1}$ (equation 3.10) |
|---|---|---|---|
| NaF | | 926.0 | 915.0 |
| NaCl | increases ↓ | 787.8 | 758.5 |
| NaBr | | 752.0 | 739.0 |
| NaI | | 705.0 | 629.0 |
| LiF | | 1034.0 | 1062.0 |
| LiCl | increases ↓ | 840.1 | 820.0 |
| LiBr | | 781.2 | 770.5 |
| LiI | | 753.0 | 707.1 |
| LiCl | | 845.0 | 820.0 |
| KCl | increases ↓ | 702.9 | 694.5 |
| RbCl | | 673.6 | 669.4 |
| CsCl | | 644.3 | 627.6 |
| BeO | | 4541.0 | — |
| MgO | | 3895.0 | — |
| CaO | increases ↓ | 3520.0 | — |
| SrO | | 3325.0 | — |
| BaO | | 3108.0 | — |
| KF | | 803.3 | 803.3 |
| KCl | increases ↓ | 702.9 | 694.5 |
| KBr | | 673.6 | 665.2 |
| KI | | 636.0 | 627.6 |
| $MgF_2$ | | 2907.8 | 2861.8 |
| $MgBr_2$ | increases ↓ | 2405.8 | 2146.4 |
| $MgI_2$ | | 2297.0 | 1981.6 |

1 kcal $mol^{-1}$ = 4.184 kJ $mol^{-1}$, 1Å = 100 picometre.

exceptions and no general solubility rules can give us the idea of solubility of ionic solids. However some generalisation can be made in this way.

(i) All carbonates are insoluble except those of Group I elements and $NH_4^+$ ion.

(ii) All hydroxides are insoluble except those of the Group I elements, $Sr^{2+}$ and $Ba^{2+}$.

(iii) All nitrates are soluble except $BiONO_3$.

(iv) Large non acidic cations and large non basic anions give insoluble salts. Generally acidic cations and basic anions produce insoluble salts.

The process of dissolution of ionic compounds in polar solvents involve breakdown of the crystal lattice and forms individual solvated ions. The polar solvent molecules interact with the cations and anions of the ionic solid and rupture the interionic forces between them and thereby dissolution occurs. As the interionic distance for a system of ionic solid increases (Table 3.1) the solubility generally increases because there occurs a diminished attractive force. During this dissolution process the lattice energy of the ionic solid is converted to **solvation energy. Solvation energy is the energy released when one mole of an ionic solid is solvated to a solvent**—for water this solvation energy corresponds to the sum of the hydration energy* of cations and anions. The hydration energy is defined as **the amount of energy liberated when a cation or an anion in the gaseous state interacts with a solvent to form a solvated cation and solvated anion.** When the lattice energy of an ionic solid is greater than that of solvation energy, the dissolution process is hindered and the compound will be insoluble in that solvent e.g. $BaSO_4$. On the other hand when the solvation energy is greater than lattice energy, the compound will easily dissolve in that solvent e.g. LiCl. In this case though the lattice energy of LiCl is high yet due to the small size of $Li^+$ ion there will be excessive hydration so that hydration energy will be high and thereby solvation energy overcomes the lattice energy easily, resulting easy dissolution of lithium salts in water (e.g. LiCl is more soluble in water than NaCl though the lattice energy of LiCl is greater than NaCl). In the third case when the lattice energy is equal to solvation energy the compound will slowly dissolve in that solvent e.g. $MgSO_4$.

The dissolution process of an ionic halide MX may be depicted in the following way according to a Born-Haber type cycle (Fig. 3.2) :

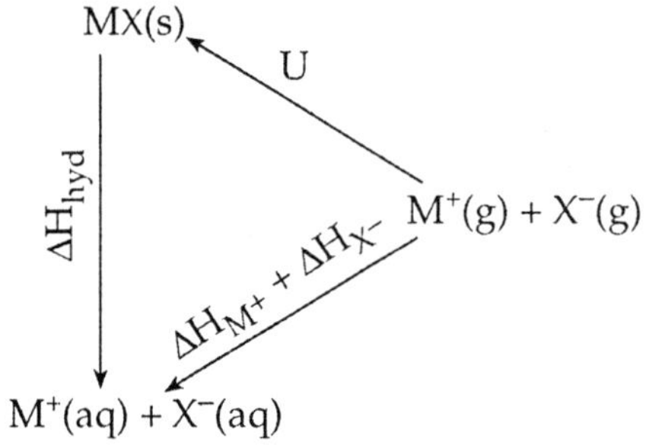


***Figure 3.2** : Schematic representation of dissolution of an ionic halide in water.*

* Hydration energy ($\Delta H_{hyd}$) of a cation is given by Latimer's equation

$$\Delta H_{hyd} = -\frac{60900\,Z^2}{r+50} \text{ kJ mol}^{-1}$$ where Z is the charge on the cation and r is the radius.

**Table 3.2 : Hydration enthalpy (energy) of some cations and anions**

| Ion | Hydration energy (kJ $mol^{-1}$) | Ion | Hydration energy (kJ $mol^{-1}$) |
|---|---|---|---|
| | +1 ions | | +3 ions |
| H | –1091 | Al | –4660 |
| Li | –515 | Ga | –4685 |
| Na | –405 | In | –4109 |
| K | –321 | Tl | –4184 |
| Rb | –296 | | –1 ions |
| Cs | –263 | F | –513 |
| | +2 ions | Cl | –370 |
| Be | | Br | –339 |
| Mg | –1922 | I | –206 |
| Ca | –1592 | OH | –460 |
| Sr | –1445 | $NO_3$ | –314 |
| Ba | –1304 | CN | –342 |
| Ra | –1259 | $N_3$ | –298 |
| | | $ClO_4$ | –235 |
| | | | –2 ions |
| | | S | –1372 |
| | | $CO_3$ | –1314 |
| | | $SO_4$ | –1059 |

The enthalpy of hydration or solvation ($\Delta H_{hyd}$) is related to the lattice energy (U) and the hydration enthalpy of cation ($\Delta H_{M^+}$) and anion ($\Delta H_{X^-}$) by the equation,

$$U - \Delta H_{M^+} - \Delta H_{X^-} = -\Delta H_{hyd}$$

$\Delta H_{hyd}$ may be positive or negative depending on the hydration process of cation and anion. Small cations will polarise the solvent molecules (water) more strongly thereby hydration enthalpy of cations will be large. For larger anions hydration energy (enthalpy) will have greater value (Table 3.2). Hydration of cations and anions occur through the formation of **primary and secondary hydration sphere.** Since oxygen is more electronegative than hydrogen, each O–H bonds in water is polar in which the bonded electrons are closer to oxygen, resulting a partial negative charge on oxygen and a partial positive charge on hydrogen (Fig. 3.3). As $H_2O$ is an angular molecule, the positive charge cation ($M^+$) in water surrounds itself with water moleculs through the oxygen end of water molecules and the negative charge anion ($X^-$) surrounds itself with the hydrogen ends of water molecules and thus **hydrated ions** are produced in water (Fig. 3.3). This formation of hydrated ions is referred to as **primary hydration sphere** or **inner sphere** (Fig. 3.3). The hydration of a cation depends on charge and radius of the ion as also on charge/radius ratio of the cation. Naturally small size cation will have greater **hydration number** i.e.

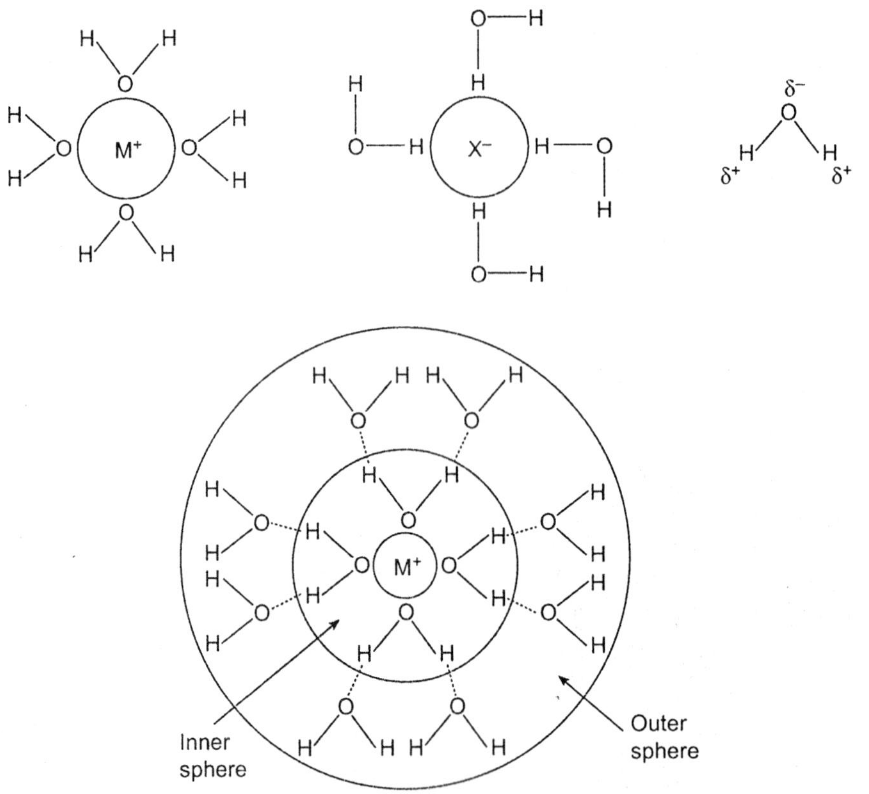


***Figure 3.3** : Hydration of cation, hydration of anion, polar $H_2O$ molecule and primary and secondary hydration sphere of a cation.*

greater number of water molecules are attached. The water molecules in the inner sphere exert an enhanced hydrogen-bonding attraction for other molecules resulting a **secondary hydration sphere** or **outer sphere**. Thus a multiple hydration layers occur for overall hydration of a cation.

**Problem 3.2** What is Madélung constant? What is its significance?

**Ans. :** It is a characteristic property of an ionic lattice, the value of which depends on the number and arrangement of the neighbouring ions around a particular counter ion i.e. its value dependent on the type of crystal lattice. In equation $U = -\frac{Ae^2}{r} + \frac{B}{r^n}$ (Section 3.1.2) where A is called Madelung constant.

The significance of A is obtained by the hard-sphere model of the NaCl type where $M^+$ and $X^-$ ions form the crystal lattice. The geometry of the sturcture is attended by assuming that the incompressible ions (i.e. $M^+$ and $X^-$) will be packed in such a way occupying minimum volume and least electrostatic energy. Sodium chloride forms an hexagonal close packed lattice where each $Na^+$ ion is surrounded by $6Cl^-$ ions or vice-versa. The positive ion $Na^+$ ($Z_i$) is surrounded by six nearest neighbour ions $Cl^-$ ($Z_j$) of opposite charge each at a distance of $r_0$ from $Z_i$. Twelve positive ions ($Na^+$) at a distance of $\sqrt{2}r_0$, eight negative ions

($Cl^-$) at a further distance of $\sqrt{3}r_0$, six positve ions at a further distance of $\sqrt{4}r_0$, twenty four negative ions are at $\sqrt{5}r_0$ and so on.

The energy of interaction between ions is equal to the product of their respective charges divided by their distance apart. Thus the electrostatic energy ($E_{st}$) for positive ion is given by—

$$E_{st} = \frac{6e^2}{r_0} Z_i Z_j + \frac{12e^2}{\sqrt{2}r_0} Z_i^2 + \frac{8e^2}{\sqrt{3}r_0} Z_i Z_j + \frac{6e^2}{\sqrt{4}r_0} Z_i^2 + ......$$

$$= -\frac{e^2}{r_0}|Z|^2\left[6 - \frac{12}{\sqrt{2}} + \frac{8}{\sqrt{3}} - \frac{6}{\sqrt{4}} + ......\right]$$

where the sum of the infinite series converges to a limiting value of 1.7475. This is the Madelung constant value for sodium chloride type lattice. Values of A for some of ionic lattice are given below :

| **Crystal structure** | **Madelung constant** |
|---|---|
| Sodium chloride (NaCl) | 1.7475 |
| Cesium chloride (CsCl) | 1.7627 |
| Zinc blende (ZnS) | 1.6381 |
| Wurtzite (ZnS) | 1.6413 |
| Rutile ($TiO_2$) | 2.4080 |
| Fluorite ($CaF_2$) | 2.5194 |

(b) **Computation of proton affinity** : The proton affinity (P) is defined as the energy liberated when a gaseous proton is added to a geseous species i.e.

$$X(g) + H^+(g) \longrightarrow XH^+(g) - P$$

Let us consider the process of $NH_3$, for which a thermodynamic cycle can be written for $NH_4Cl(s)$.

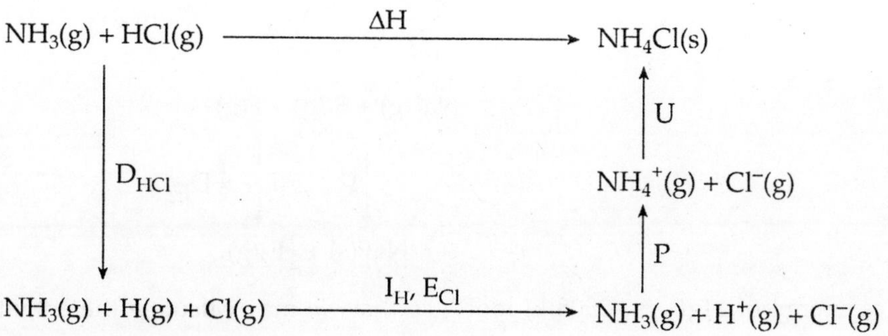


Then according to Hess's law

$$\Delta H = D_{HCl} + I_H + E_{Cl} + P + U$$

Where $D_{HCl}$ = bond dissociation energy of HCl = 431.0 kJ mol$^{-1}$

$I_H$ = ionisation energy of hydrogen = 1311.7 kJ mol$^{-1}$

$E_{Cl}$ = electron affinity of chlorine = –348.0 kJ mol$^{-1}$

U = lattice energy of $NH_4Cl(s)$ = –676.1 kJ mol$^{-1}$

$\Delta H$ = heat of reaction of $NH_3(g) + HCl(g)$ to form $NH_4Cl(s) = -176.15$ kJ $mol^{-1}$

$P$ = proton affinity of $NH_3$

$\therefore -176.15 = 431.0 + 1311.7 - 348.0 + P - 676.1$

i.e. $-176.15 = P + 718.6$, so that $P = -894.75$ kJ $mol^{-1}$

The proton affinity for $NH_4Br$ and $NH_4I$ has been calculated which has the values –894.53 kJ $mol^{-1}$ and –886.17 kJ $mol^{-1}$ respectively. The value of proton affinity for $NH_4F$ is not available because the data give misleading result due to hydrogen bonding in the crystal. The proton affinity of water is greater than that of phosphine so that phosphonium salts are readily hydrolysed by water.

---

**Problem 3.3** The radius of $F^-$ and $K^+$ are same, yet the hydration energy of $F^-$ is greater than that of $K^+$—explain. **[B.U. 1993]**

**Ans. :** Water is a polar molecule—the hydrogen atoms assume partial positive charge and oxygen atom assumes partial negative charge. The difference in hydration energy occurs due to the fact that more number of small hydrogen atoms with partial positive charge are attracted towards the $F^-$ ions whereas in case of $K^+$, the larger oxygen atom with partial negative charge is attracted. Thus there is a scope of strong hydrogen bonding in case of $F^-$ than that of $K^+$ thereby greater number of water molecules may be arranged around $F^-$. The hydration energies of $K^+$ and $F^-$ are –321 and –513 kJ $mol^{-1}$ respectively.

---

(c) **Possibility of the formation of hypothetical compounds and disproportionation to stable compounds :** Let us consider the possibility of formation of $NaF_2$, $NaCl_2$ and CaCl or other halides. It is assumed that lattice energy of hypothetical $NaF_2$ is approximately equal to that of the stable compound $MgF_2$ (similarly CaCl may be considered to the stable compound KCl). The formation of $NaF_2(s)$ may be dipicted as follows :

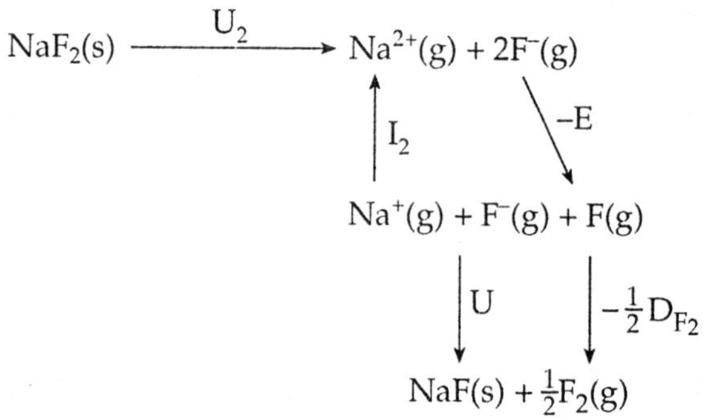


The lattice energy of $MgF_2$ is –2908 kJ $mol^{-1}$, so that we may calculate the enthalpy change ($\Delta H_f$) for the above process :

$$\Delta H_f = 2908.0 - 4560.5 + 334.7 - 908.0 - 77.4 = -2303.2 \text{ kJ mol}^{-1}$$

An enormous amount of energy is required for the second ionisation energy of sodium atom (4560.5 kJ $mol^{-1}$) which should have to be released for the formation of $NaF_2$ and thereby $NaF_2$ disproportionates to the stable NaF and elementary fluorine ($NaF_2 \longrightarrow NaF + \frac{1}{2}F_2$), so that $NaF_2$ does not exist.

In a similar way the stability of CaF may be considered by assuming that KF has the same structure with lattice energy 795 kJ mol$^{-1}$. Thus the enthalpy of formation ($\Delta H_f$) can be calculated as :

$$\Delta H_f = S_{Ca} + I_{Ca} + \frac{1}{2}D_{F_2} - E_{F_2} - U$$

$$= 201.0 + 591.0 + 79.6 - 336.0 - 795.0$$

$$= -259.4 \text{ kJ mol}^{-1}$$

Thus CaF disproportionates in the following way :

$$2CaF \longrightarrow CaF_2 + Ca$$

$$2\Delta H_f = -519 \quad \Delta H_f = -1245 \quad \Delta H_f = 0 \text{ kJ mol}^{-1}$$

Thus the net heat change for the process = –1245 – (–519) = –726 kJ

So that CaF will not exist and must be converted to stable $CaF_2$.

---

**Problem 3.4** $NH_4F$ crystallises as Wurtzite structure (Madelung constant 1.641 and Born exponent 8) in which the closest distance between $NH_4^+$ and $F^-$ ions is 2.63Å. Calculate the lattice energy of $NH_4F$. **[C.U. 1989]**

**Ans. :** According to equation (3.9)

$$U = -\frac{138760 AZ_iZ_j}{r_0(\text{in cm})}\left(1-\frac{1}{n}\right)\text{erg}$$

$$= -\frac{138760 \times 1.641}{2.63 \times 10^{-8}\text{ cm}}\left(1-\frac{1}{8}\right)$$

$$= -75757.418 \times 10^8 \text{ erg}$$

$$= -75757.418 \times 10^8 \times 10^{-7} \text{ Joule}$$

$$= -757.574 \text{ kJ mol}^{-1}$$

---

**Problem 3.5** (a) Calculate the lattice energy of NaCl(s) using the following data :

Madelung constant = 1.748

Born exponent = 8

Electronic charge = 4.8 × 10$^{-10}$ esu

Equilibrium interionic distance = 2.79Å

(b) Using the above value of lattice energy of NaCl(s), calculate the elecron affinity of Cl applying Born-Haber cycle. Given—

Ionisation potential of Na = 495 kJ/mol

Heat of sublimation of Na = 108 kJ/mol

Bond dissociation energy of $Cl_2$ = 240 kJ/mol

Heat of formation of NaCl = –380 kJ/mol **[B.U.1996]**

**Ans. :** (a) According to equation (3.5)

$$U = -\frac{N_0 AZ_iZ_je^2}{r_0}\left(1-\frac{1}{n}\right)$$

$$U = -\frac{6.023\times 10^{23}\times 1.748\times \left(4.8\times 10^{-10}\ \text{esu}\right)^2}{2.97\times 10^{-8}\ \text{cm}}\left(1-\frac{1}{8}\right)$$

$$= -\frac{6.023\times 10^{23}\times 1.748\times \left(4.8\times 10^{-10}\right)^2\times 7}{2.79\times 10^{-8}\times 8}\ \text{erg}$$

$= -76.07474 \times 10^{11}$ erg

$= -760.7475$ kJ mol$^{-1}$

(b) From Sec. 3.1.3

$\Delta H_f = S + I + 0.5D + E + U$

$\therefore -380 = 108 + 495 + 120 + E - 760.7475$

i.e. $E = -380 - 108 - 495 - 120 + 760.7475$

$= -342.2525$ kJ mol$^{-1}$

$= -3.547$ eV

[1 eV/atom = 96.48 kJ/mol
1 kcal mol$^{-1}$ = 4.184 kJ/mol$^{-1}$
1 eV = 23.06 kcal mol$^{-1}$]

---

**Problem 3.6** Calculate the electron affinity of iodine from the following data :

Lattice energy of KI(s) = 153.0 kcal mol$^{-1}$
Heat of formation of KI(s) = 78.3 kcal mol$^{-1}$
Sublimation energy of K(s) = 21.4 kcal mol$^{-1}$
Ionisation potential of K(g) = 100.0 kcal mol$^{-1}$
Dissociation energy of $I_2$(g) = 51.0 kcal mol$^{-1}$ **[B.U. 1992]**

**Ans. :** According to Sec. 3.1.3

$\Delta H_f = S + I + 0.5D + E + U$

$\therefore -78.3 = 21.4 + 100.0 + 25.5 + E - 153.0$

i.e. $E = -78.3 - 21.4 - 100.0 - 25.5 + 153.0$

$= -72.2$ kcal

$= -3.13$ eV

---

(d) **High oxidation states of metals :** Among the halogens fluorine exhibit the highest oxidation states of metals. So for halides of $MX_n$ type, n will be highest when X = F and for this case the metal fluoride will show highest lattice energy.

---

**Problem 3.7** The lattice energies of KF , KCl, KBr and KI follow the order—

1. KF > KCl > KBr > KI
2. KI > KBr > KCl > KF
3. KF > KCl > KI > KBr
4. KI > KBr > KF > KCl **[NET 2001]**

**Ans. :** 1.

---

**Problem 3.8** In the Born-Haber cycle for the formation of solid common salt (NaCl) the largest contribution comes from—

1. the low ionisaiton potential of Na
2. the high electron affinity of Cl
3. the low $\Delta H_{vap}$ of Na(s)
4. the lattice energy **[NET 2001]**

**Ans. :** 4.

**Problem 3.9** The lowest lattice energy is exhibited by—1. $MnF_2$, 2. $FeF_2$, 3. $NiF_2$, 4. $CuF_2$ **[NET 2000]**

**Ans.** : 1.

**Problem 3.10** The salt has the least solubility in water is—1. $BaCl_2$, 2. $BaNO_3$, 3. $MgSO_4$, 4. $BaSO_4$

**Ans.** : 4. **[NET 2000]**

## 3.2 COVALENT BOND

In the case of ionic bond there is a complete transfer of electron(s) from one atomic orbital to other. But covalent bond is formed by mutual sharing of electrons e.g. $H_2$, $Cl_2$, $NH_3$ etc. Concept to covalent bond developed by the celebrated scientist G. N. Lewis of the University of California. According to him **a chemical bond formed between two atoms (similar or different) through mutual sharing of one or more electron pairs (one electron is given by each of the two atoms) is called covalent bond.** Lewis stated that each pair of electrons involved in bonding lies between the atoms and is shared by two neighbouring atoms e.g. A–B or A–A, A=B or A=A or B=B and A≡B or A≡A or B≡B consists of one, two and three shared pairs of electrons respectively.

### 3.2.1 Octet rule and Lewis structures

In the case of ionic bond each of the cations and anions must assume the stable inert gas configuration i.e. **octet rule** is followed. Lewis assumed the existance of large number of molecules to follow the octet (e.g. $Cl_2$, $NH_3$ etc.) but this is not the basic criteria to form covalent molecules. There are several molecules where there will be short of octet (e.g. $H_2$, $BeCl_2$), some molecules have expansion of octet (e.g. $SF_6$, $PCl_5$). This type of molecules where there is an expansion of octet occurs are called **hypervalent compounds** and the phenomena is known as **hypervalence.** Generally hypervalence is exhibited by the elements of periods 3 to 6 where the element is capable of expanding its valence shell due to the availability of low-lying unfilled d-orbitals. The ionic species viz. $SO_3^{2-}$, $PO_4^{3-}$, $SO_4^{2-}$ and $ClO_4^{2-}$ exhibit hypervalence.

*Hypervalence and Hypervalent compounds*

Lewis suggested that the structure of a covalent compound may be expressed by writing a pair of dots between two atoms for each covalent bond, where a dot represents an electron and a pair of electron (:) is represented by (–). This fashion of writing the structure of a molecule is known as **Lewis structure** e.g. $Cl_2$ and HF molecules are represented as :

$$:\ddot{\underset{..}{Cl}}:\ddot{\underset{..}{Cl}}: \qquad H:\ddot{\underset{..}{F}}:$$

The construction of Lewis structures have the following steps :

(a) The valence electrons of the constituent atoms of a molecule or an ionic species are added up to get the total number of electrons. For ionic species having one negative charge corresponds to one electron is to be added and for having one positive charge, one electron is subtracted from the total number of electrons.

(b) From structural information the chemical symbol of the atoms have to be written showing which are bonded together. In this process generally the less electronegative atom constitute the central atom and the more electronegative atom(s) surrounded the central atom e.g. $CO_2$, $SO_4^{2-}$, $ClO_4^-$ and $PCl_5$ etc. But there may be some exceptions e.g. $CH_4$, $NH_3$, $H_2O$ etc.

(c) The total number of electrons are distributed among the atoms with the consideration that each pair of electron constitute the neighbouring bonded pair of atoms and the excess pair of electrons may form lone pairs or multiple bonds so that each atom may achieve octet. All bonding pair of electrons may be represented by a single line (–).

Let us consider some examples—

H : H or H – H, : N⋮⋮N : or :N≡N:, O :: C :: O or O=C=O

Now in case of $BF_4^-$ there are 3+4×7+1 = 32 electrons i.e. 16 pairs of electron will have to be accommodated around the five atoms as—

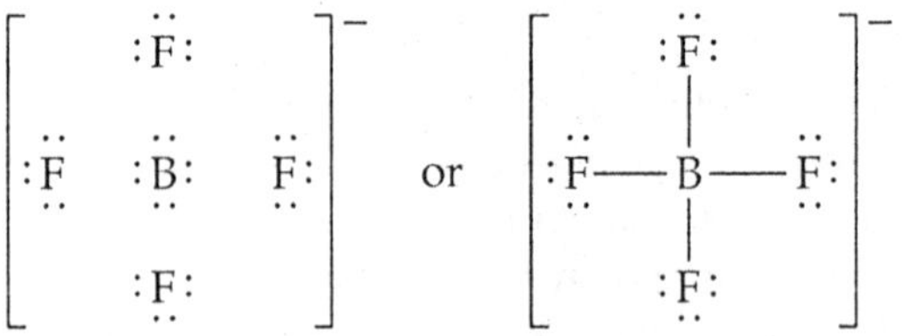

(d) The procedure for drawing Lewis structures become complicated when a given species may be represented by more than one Lewis structures. In that case the preferred structure may be drawn by consideration of **formal charge** (F.C.) of the constituent atoms which may be defined as :

F.C. = (Number of valence electrons on the central atom – number of lone pair electrons on the atom in the species) – $\frac{1}{2}$(the number of electrons shared by the central atom with its adjacent atoms in the species).

*The formal charge denotes the number of electrons that an atom gains or loses when it enters into ideal covalent bonding with other atoms.* This formal charge dictates the stability of alternative Lewis structures for a species. It has been stated *that lowest energy structure is one which have lowest formal charge.* The lowest energy structure is considered as one having negative formal charge for more electronegative atom and positive formal charge for less electronegative atom. The formal charge is determined on the basis of pure covalent view of a molecule so that the emperical charge (positive or negative) assigned to that element. Let us consider the following structures for $NO_3^-$ (5+3×6+1 = 24 electrons) and $BF_3$ (3+3×7 = 24 electrons)

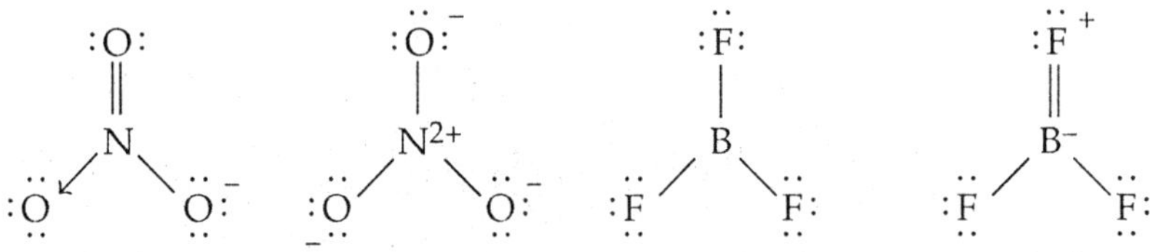

(I) (II) (III) (IV)

Among the structures (I) and (III) are the plausable structures for $NO_3^-$ and $BF_3$ respectively and we may reject the structures (II) and (IV).

(e) If there too many electrons to locate, the extra electrons are to be on the central atom and so there may be the violation of octet rule. Violation of octet rule may generate reasonance structures for some of the species.

(f) Finally we have to check the total number of electrons shown in the structures is in agreement with that calculated for valence electrons in step (a).

Lewis structures of some molecules/ionic species are depicted in Table 3.3.

**Table 3.3 : Lewis structures of some of the species**

| Molecule/ionic species | Valence electrons | Structure |
|---|---|---|
| NO | 5 + 6 = 11 |  |
| $NO_2$ | 5 + 2 × 6 = 17 | 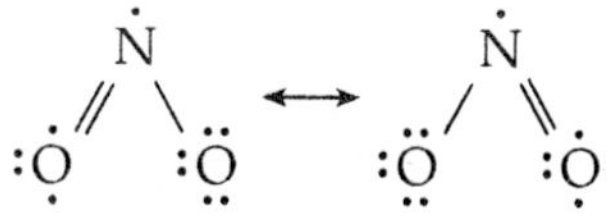 |
| $ClO_2$ | 7 + 2 × 6 = 19 | 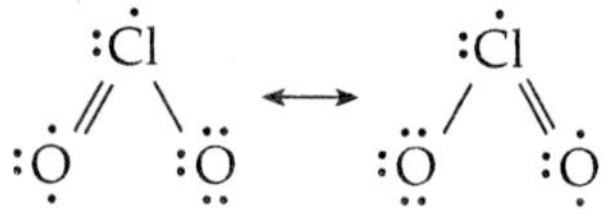 |
| $ClO_4^-$ | 7 + 4 × 6 + 1 = 32 | 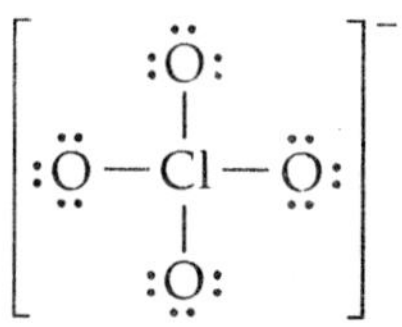 |
| $SO_4^{2-}$ | 6 + 4 × 6 + 2 = 32 | 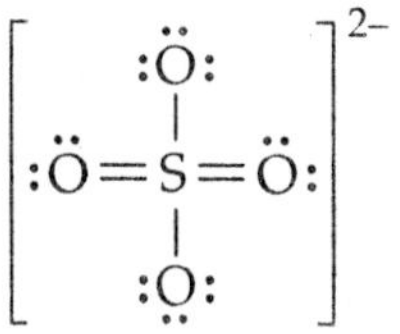 |
| $PO_4^{3-}$ | 5 + 4 × 6 + 3 = 32 | 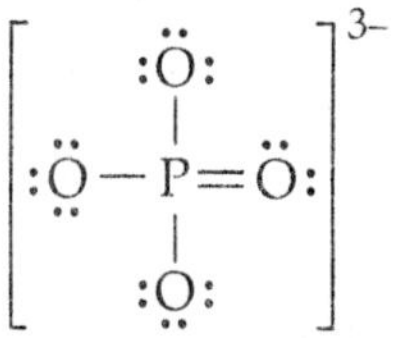 |
| $SO_2$ | 6 + 2 × 6 = 18 | 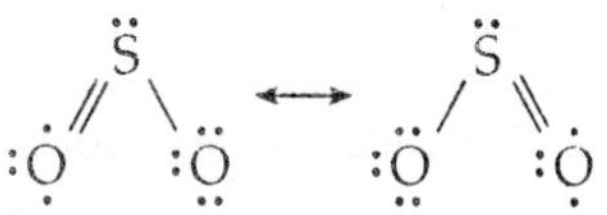 |
| $SO_3$ | 6 + 3 × 6 = 24 | 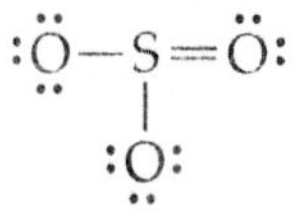 |
| $SO_3^{2-}$ | 6 + 3 × 6 + 2 = 26 | 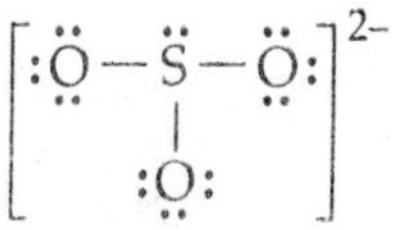 |
| $SbCl_5$ | 5 + 5 × 7 = 40 | 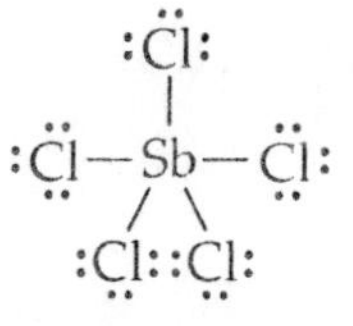 |

| Molecule/ionic species | Valence electrons | Structure | |
|---|---|---|---|
| $ClF_5$ | $7 + 5 \times 7 = 42$ | 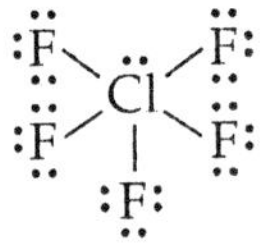 | |
| $ClF_4^-$ | $7 + 4 \times 7 + 1 = 36$ | 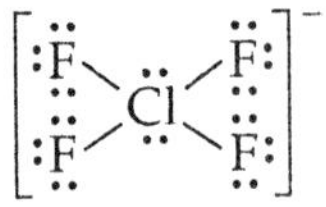 | |
| $IF_6^-$ | $7 + 6 \times 7 + 1 = 50$ | 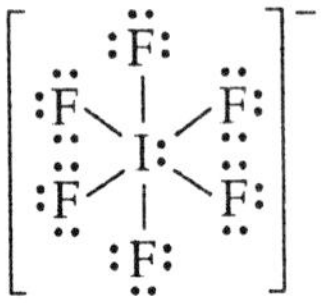 | |
| $NSF_3$ | $5 + 6 + 3 \times 7 = 32$ | 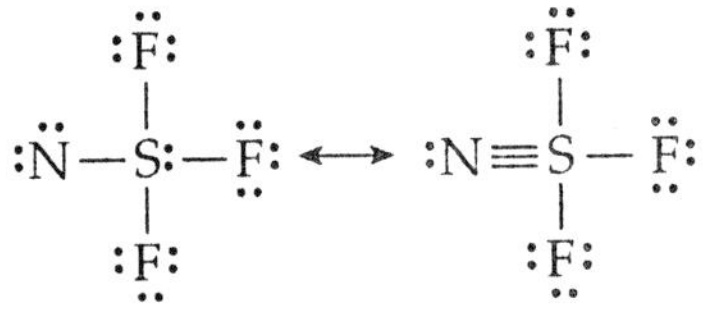 | |
| $ICl_2^+$ | $3 \times 7 - 1 = 20$ | 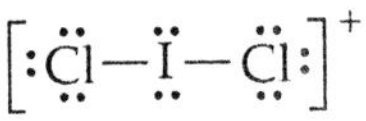 | |
| $SCN^-$ | $6 + 4 + 5 + 1 = 16$ | 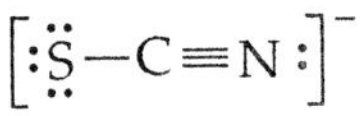 | May be possible other Lewis structures. |
| $CNO^-$ | $4 + 5 + 6 + 1 = 16$ | 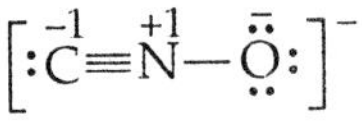 | Other structures are possible. |
| $OCN^-$ | $6 + 4 + 5 + 1 = 16$ | 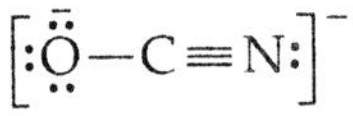 | 〃 |

**Problem 3.11** Write down the electron dot diagram for $SeO_3^{2-}$. **[B.U. 2005]**

**Ans. :** Valence electrons = $6 + 3 \times 6 + 2 = 26$

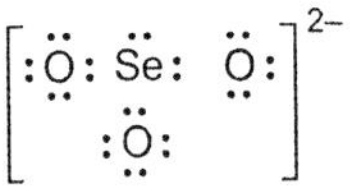

**Problem 3.12** Calculate the lattice energy of KCl [Given interionic separation = 3.14Å, n = 9, A = 1.746, $\varepsilon_0 = 8.85 \times 10^{-12}$ $C^2J^{-1}m^{-1}$, e = $1.6 \times 10^{-19}$ coulomb] **[B.U. 2004]**

**Ans. :** According to equation (3.8)

$$U = -\frac{N_0 A Z_i Z_j e^2}{4\pi\varepsilon_0 r_0}\left(1 - \frac{1}{n}\right)$$

$$= -\frac{6.023\times10^{23}\times1.746\times\left(1.6\times10^{-19}\text{C}\right)^2}{4\times3.142\times8.85\times10^{-12}\text{C}^2\text{m}^{-1}\text{J}^{-1}\times3.14\times10^{-8}\times10^{-2}\text{m}}$$

$= -770.828$ kJ mol$^{-1}$

**Problem 3.13** Establish Born-Haber cycle for the formation of MgS(s) starting from Mg(s) and $S_8$(s) and hence calculate the electron affinity of S(g) for the reaction S(g) + 2e $\longrightarrow$ $S^{2-}$(g) using the thermochemical data given below :

Enthalpy of formation = 345 kJ mol$^{-1}$

Enthalpy of sublimation of Mg(s) = 153 kJ mol$^{-1}$

Sum of 1st and 2nd ionisation potentials of Mg(g) = 2187 kJ mol$^{-1}$

Enthalpy of atomisation of $S_8$(s) = 559 kJ mol$^{-1}$

Lattice energy of MgS(s) = 2948 kJ mol$^{-1}$

[here (s) stands for solid and (g) stands for gas] **[C.U. 1996]**

**Ans. :**

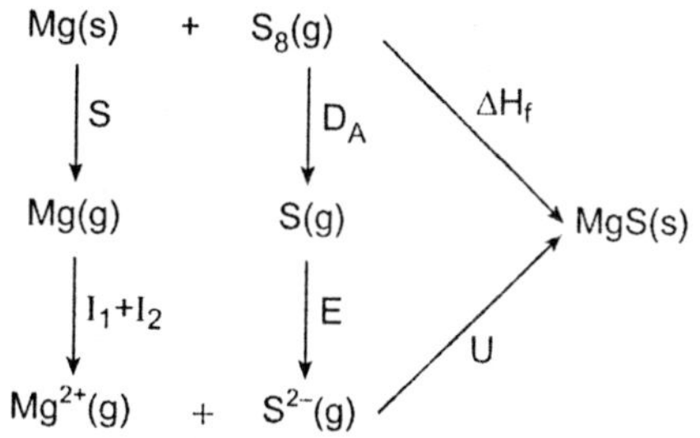


Where enthalpy of formation of MgS(s) = $\Delta H_f$ = 345 kJ mol$^{-1}$

Enthalpy of sublimation from Mg(s) to Mg(g) = S = 153 kJ mol$^{-1}$

Sum of the first and second ionisation potentials ($I_1$ + $I_2$) of Mg(g) to $Mg^{2+}$(g) = 2187 kJ mol$^{-1}$

Enthalpy of atomisation of $S_8$(s) = $D_A = \frac{1}{8} \times 559$ kJ mol$^{-1}$

Lattice energy of MgS(s) = U = 2948 kJ mol$^{-1}$

Electron affinity of S(g) = E?

$$\Delta H_f = S + (I_1 + I_2) + \frac{1}{8}D_A + E + U$$

$$\therefore -345 = 153 + 2187 + \frac{1}{8} \times 559 + E - 2948$$

$E = 193.1$ kJ mol$^{-1}$

### 3.2.1.1 Prediction of structures and shapes of molecules : Molecular topologies : Helferich's rule

Long before the development of bonding theories such as hybridisation concept, VSEPR theory, Bent's rule etc. Helferich proposed some emperical rules to establish the molecular shapes of different molecular species. These rules are called **Helferich's rules** which are stated as :

(a) Determine the possible representation of Lewis electron dot structure of a particular molecular species with the consideration that the less electronegative element is generally treated as the central element.

(b) Select the structural arrangement of the groups around the central atom in such a way that the valency shell electrons will be placed in such a geometry that the electron-electron repulsion in the valence shell will be minimised.

(c) The effects of the $\pi$-electrons are ignored in determining the geometry as $\pi$-bonds follow the $\sigma$-bonds.

The manifestation of Helferich's rules are the following :

(i) **Molecular species of type $AX_2$** : If the central atom A has no lone pair then the species will be **linear** and the $\angle XAX = 180°$. If lone pair residing on A then the molecule will assume **angular geometry** i.e. **V-shaped**. Thus **linear molecules** are : $O{=}C{=}O$, $N{\equiv}N^+{-}O^-$, $N^-{=}N^+{=}N^-$, Cl–Be–Cl and **angular or V-shaped molecules** are :

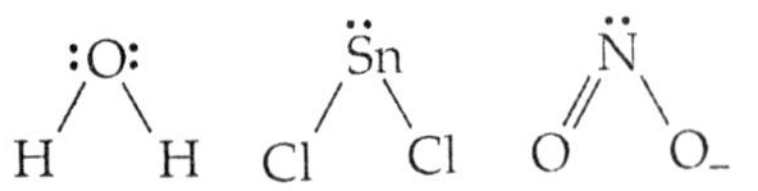

(ii) **Molecular species of type $AX_3$** : If the central atom has no lone pair then the species will assume **planar geometry** otherwise the species will exist as **trigonal pyramidal** geometry if the central atom has lone pair i.e.

**Trigonal planar :** 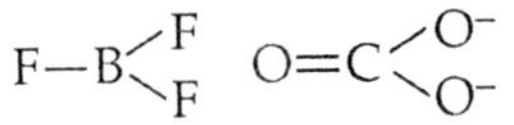 i.e. $CO_3^{2-}$

**Trigonal pyramidal :** 

(iii) **Molecular species of type $AX_4$** : If the central atom A has no lone pair then the species will be **tetrahedral** otherwise it will have the planar or other geometrical structure e.g.

**Tetrahedral :** $CH_4$, $NH_4^+$, $BF_4^-$, $BH_4^-$, $BeF_4^{2-}$

**Planar :** 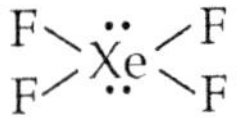

A large number of molecular species do not follow this rule e.g. $Ni(CO)_4$, $NiCl_4^{2-}$ are tetrahedral while $Ni(CN)_4^{2-}$ is planar. Hence these Helferich's rules are not of much importance in predicting the geometry of molecular species which can be interpreted by the development of VSEPR theory with the aid of hybridisation concept.

### 3.2.2 The Nature of Covalent Bond

Two theories have been considered to explain the formation of a covalent bond—valence bond theory developed by Heitler, London, Pauling etc. and the more plausable one is molecular orbital theory developed by Hund and Mullikan. Valence bond concept developed from the idea of pairing of electron spins and the maximum overlapping of atomic orbitals in the region of covalent bond whereby maximum electron density is observed along the bond direction of the two combing atoms. Molecular orbital theory considers the molecule as a whole and electrons are allocated into the molecular orbitals in the same very way as electrons are arranged in atomic orbitals. Neither of the two methods can be applied to all systems so we discuss both theories in brief.

### 3.2.3 Valence Bond Theory (VBT)

Let us consider the formation of hydrogen molecule by the combination of two hydrogen atoms. Though the two hydrogen atoms are indistinguishable yet for convenience we are considering two hydrogen atoms are distinguishable and designated as $H_A$ and $H_B$ (Fig. 3.4). Let electron 1 is associated with $H_A$ atom and electron 2 is associated with $H_B$ atom. The various distances of the electron 1 and 2 from the nucleus of $H_A$ and $H_B$ as also the distance between the two nucleus and between the two electrons are designated in Fig. 3.4. During the formation of hydrogen molecule from $H_A$ and $H_B$ there will be exchange of six forces—four will be attractive and two will be repulsive :

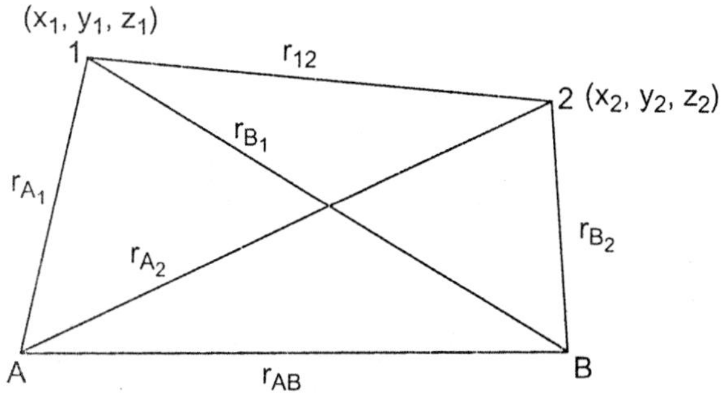


*Figure 3.4 : Combination of two hydrogen atoms to form hydrogen molecule.*

(a) Intrinsic attractive force beween the electron 1 and that of the nucleus of $H_A$.

(b) The attractive force between the electron 1 and that of the nucleus of $H_B$.

(c) The intrinsic attractive force between the electron 2 and that of the nucleus of $H_B$.

(d) The attractive force between the electron 2 and that of the nucleus of $H_A$.

(e) The repulsive force between the electron 1 and 2 and

(f) The repulsive force between the nucleus of $H_A$ and $H_B$.

When the two hydrogen atoms are wide apart there will be no exchange of forces and when they begin to come closer the potential energy of the $H_2$ molecule is lowered compared to the sum of the potential energy of the two hydrogen atoms. This means that $H_2$ molecule is more stable than the individual two hydrogen atoms. At the minimum internuclear distance (0.74Å) of the two hydrogen atoms, the repulsive energy is minimum (Fig. 3.5). The spins of the two electrons (1 and 2) can be represented in two ways by the potential energy curve (Fig. 3.5). There are two possible equivalent structures which are indistinguishable once the chemical bond is formed. Curve I represents that the two electrons are in parallel spin whreas curve II represents the electrons are in opposed spin. If the two electrons have parallel spins then the energy continues to rise as the atoms come closer together (curve I) and there is no possibility of bond formation. On the other hand when the spins of the two electrons are in the opposite direction (curve II) the potential energy curve has a minimum energy (435 kJ) at an optimum distance of 0.74Å. *This is also one of the condition for formation of covalent bond that the combing atoms to form*

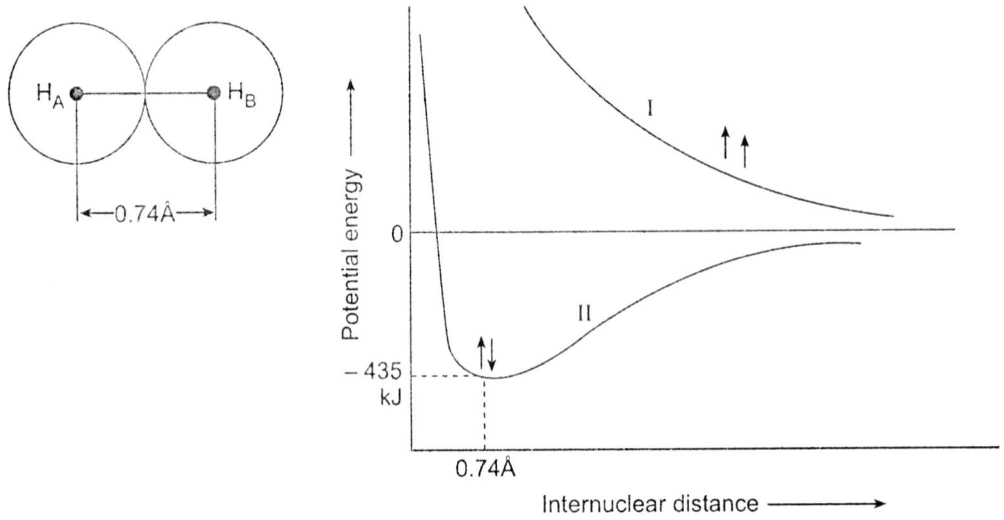


*Figure 3.5 : Minimum internuclear distance and potential energy curve for $H_2$ molecule.*

Condition for the formation of a covalent bond

*a covalent bond must possess an unpaired electron and that the two electrons must be paired in order to permit the Pauli's exclusion principle. That's why covalent bond sometimes called* **spin paired bond** or **two centre two electrons** (2c–2e) **bond.**

### 3.2.4 Wave mechanical treatment of valence bond theory

The formation of covalent bond based on wave mechanical approach was proposed originally by W. Heitler and F. London (1927) and was developed by L. Pauling and J. C. Slater (1931). Let us consider two non-interacting atoms A and B say two hydrogen atoms which are widely apart and let the wave function of these two systems are $\Psi_A$ and $\Psi_B$ respectively. In the case of $H_2$ molecule we can say that the two hydrogen atoms having one electron in 1s orbital overlap each other to form the $H_2$ molecule. The extent of overlap is represented by overlap integral $S_{AB} = \int \Psi_A \Psi_B d\tau$ where $d\tau$ represents the overlap region (Fig. 3.6). As the extent of overlap increases, the value of the overlap integral increases, so also the strength of the covalent bond increases. Initially the two atoms are wide apart so that there will be no interaction between the two wave functions $\Psi_A$ and $\Psi_B$ and the wave function for the $H_2$ system can be represented as—

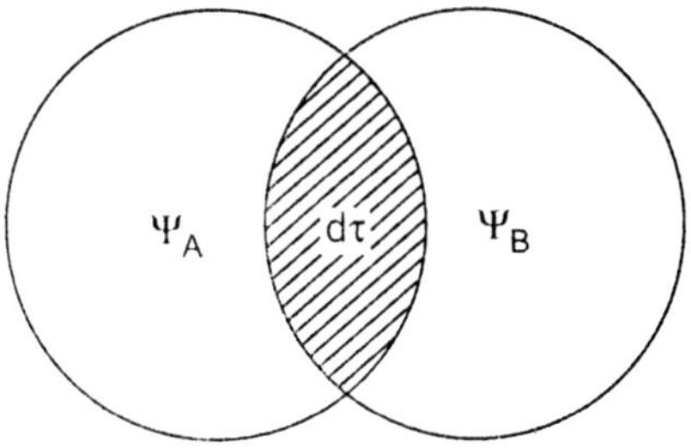


*Figure 3.6 : Extent of overlap.*

$$\Psi = \Psi_A \Psi_B \quad \text{...(3.11)}$$

and the total energy of the system is given by—

$$E = E_A + E_B = 2E_H \quad ...(3.12)$$

where $E_A$ and $E_B$ are the energies of the two individual atomic systems. Actually the energy of the $H_2$ molecule is somewhat less than that of $2E_H(E_H$ = energy of $H_A$ or $H_B)$

Now for every molecular system we have to solve the Schrodinger wave equation (Sec. 1.8.3) in order to get the value for the wave function $\Psi$ of the system. If we take the wave function $\Psi$ as in equation (3.11) the value of the energy and internuclear distance far deviated from the experimental value as depicted by Fig. 3.5. Therefore modification have to be made for the wave function $\Psi$ as represented by equation (3.11). We have to use several trial wave function for $H_2$ system and by using those trial wave functions we have to solve the Schrodinger wave equation by indirect way which is knwon as **variation method**. The variation method is nothing but the solving of wave equation in an indirect manner.

According to Sec. 3.2.3 electron 1 is associated with $H_A$ and electron 2 is associated with $H_B$ but there may be a finite probability so that electron 2 is associated with $H_A$ and electron 1 is associated with $H_B$, so that the wave functions can be represented as—

*Exchange of electrons*

$$\Psi_1 = \Psi_A(1)\Psi_B(2) \quad \text{and} \quad \Psi_2 = \Psi_A(2)\Psi_B(1)$$

$$\therefore \Psi_{cov} = C_1\Psi_1 + C_2\Psi_2 = C_1\Psi_A(1)\Psi_B(2) + C_2\Psi_A(2)\Psi_B(1) \quad ...(3.13)$$

where $C_1$ and $C_2$ are the two mixing coefficients. There is also some finite probability for both the electrons are attached either $H_A$ or $H_B$ and in that case the wave function can represent the ionic structure as—

$$\Psi_{ionic} = \Psi_A(1)\Psi_A(2) + \Psi_B(1)\Psi_B(2) \quad ...(3.14)$$

Hence the wave function for the $H_2$ molecule can be represented as :

$$\Psi = \Psi_{cov} + \lambda\Psi_{ionic} \quad ...(3.15)$$

where $\lambda$ is a mixing coefficient which represents how much ionic wave function contributes to the actual state of the $H_2$ molecule. Equations (3.15) and (3.13) are sometimes called **linear combination of atomic orbitals** (LCAO).

Now applying the trial wave function $\Psi$ according to equation (3.15) and using variation principle with reference to Schrodinger wave equation ($E\Psi = H\Psi$) (equation 1.39) we get the energy of the $H_2$ system as :

$$E = \frac{\int \Psi H \Psi d\tau}{\int \Psi^2 d\tau} \quad ...(3.16)$$

where $d\tau$ is the small overlap region i.e. the volume element dxdydz. Solving the equation (3.16) we get the energy values as :

$$E_+ = 2E_H + \frac{K+J}{1+S_{AB}^2} \quad ...(3.17)$$

and

$$E_- = 2E_H + \frac{K-J}{1-S_{AB}^2} \quad ...(3.18)$$

where K is called **coulomb integral** $= \int \Psi_{A(1)}\Psi_{B(2)}\left(-\frac{1}{r_{A_2}}-\frac{1}{r_{B_1}}+\frac{1}{r_{AB}}+\frac{1}{r_{12}}\right)\Psi_{A(1)}\Psi_{B(2)}d\tau$

$$= \int \Psi_{A(1)}\Psi_{B(2)}H\Psi_{A(1)}\Psi_{B(2)}d\tau$$

and J is called **exchange integral** or **resonance integral**

$$= \int \Psi_{A(1)}\Psi_{B(2)}\left(-\frac{1}{r_{A_1}}-\frac{1}{r_{B_2}}+\frac{1}{r_{AB}}+\frac{1}{r_{12}}\right)\Psi_{A(2)}\Psi_{B(1)}d\tau$$

$$= \int \Psi_{A(1)}\Psi_{B(2)}H\Psi_{A(2)}\Psi_{B(1)}d\tau$$

Both K and J are negative and further J > K ($K = \frac{1}{3}J$, found by calculation) so that,

$$E_+ < 2E_H \quad \text{and} \quad E_- > 2E_H$$

Thus when two hydrogen atoms each with energy $E_H$ is combined to form $H_2$ molecule, the total energy state of $H_2$ molecule splits up into two, one $E_-$ being higher by $\frac{K-J}{1-S_{AB}^2}$ amount and the other $E_+$ being lower by $\frac{K+J}{1+S_{AB}^2}$ amount. The variation of $E_+$ and $E_-$ with respect to internuclear distance $r_{AB}$ is represented in Fig. 3.7. The $E_+$ and $E_-$ are called symmetric and antisymmetric energy states respectively.

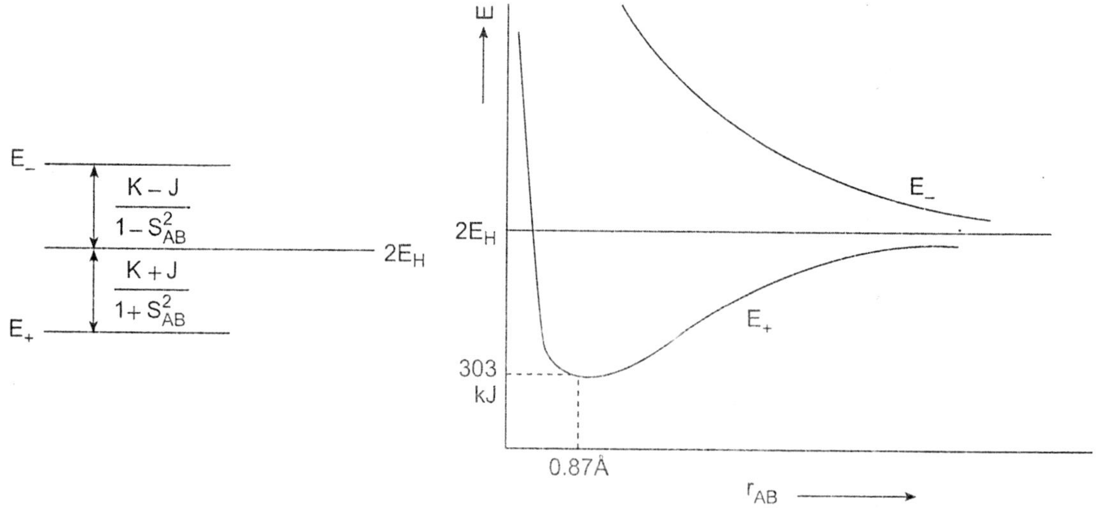


*Figure 3.7 : Formation of $H_2$ molecule according to valence bond theory.*

The values for energy and internuclear distance are in agreement with that obtained experimentally (Fig. 3.5) but the result is not excellent. Further refinement of the result can be made by introducing corrections due to **screening** and **polarisation effects**. Thus our trial wave function is correct qualitatively but not quantitatively and it still requires further modification to match the experimental value.

The wave functions for symmetric ($E_+$) and antisymmetric ($E_-$) energy states are called symmetric ($\Psi_s$) and antisymmetric ($\Psi_a$) wave function which may be obtained by taking the wave function according to equation (3.13) i.e.

$$\Psi_{cov} \text{ or } \Psi = C_1\Psi_1 + C_2\Psi_2$$

The mixing coefficients $C_1$ and $C_2$ are calculated as $\frac{1}{\sqrt{2}}$ so that the symmetric and antisymmetric wave functions may be represented as :

$$\Psi_s = \frac{1}{\sqrt{2}}(\Psi_1 + \Psi_2) = \frac{1}{\sqrt{2}}[\Psi_A(1)\Psi_B(2) + \Psi_A(2)\Psi_B(1)] \quad ...(3.19)$$

$$\Psi_a = \frac{1}{\sqrt{2}}(\Psi_1 - \Psi_2) = \frac{1}{\sqrt{2}}[\Psi_A(1)\Psi_B(2) - \Psi_A(2)\Psi_B(1)] \quad ...(3.20)$$

Comparison of the curves (Fig. 3.5 and 3.7) show that according to Pauli's exclusion principle the symmetric wave function ($\Psi_s$) essentially have paired spin. According to Pauli's exclusion principle the *total wave function for any system (atomic or molecular) will be antisymmetric with respect to the exchange of the electrons (Sec. 1.9.2). The antisymmetric wave function ($\Psi_a$) with symmetric spin function* (↑↑) (curve I of Fig. 3.5) *does not give minimum in the curve.* On the other hand **symmetric wave function** ($\Psi_s$) with antisymmetric spin function (↑↓) will produce minima at 0.74Å having minimum energy 435 kJ mol$^{-1}$. Hence the valence bond method sometimes called **the method of electron pairs**. Thus the spin pairing phenomena autometically arises to make the total wave function antisymmetric and this theory offers a quantum mechanical interpretation of Lewis theory of the electron pair bond.

---

**Problem 3.14** Two hydrogenic atoms A and B are described by the wave function $\Psi_A$ and $\Psi_B$ respectively. Electron (1) is associated with A and electron (2) is associated with B. Describe the formation of a bond between A and B in the light of **valence bond theory** and establish the wave functions for its covalent and ionic forms. **[C.U. 1996]**

**Ans. :** Sec. 3.2.4

$\Psi_{cov} = C_1\Psi_1 + C_2\Psi_2$ and equations (3.19) and (3.20)

$\Psi_{ionic} = \Psi_A(1)\Psi_A(2) + \Psi_B(1)\Psi_B(2)$

---

### 3.2.5 Overlap of atomic orbitals and concept of hybridisation

The mechanism of the formation of a covalent bond is obtained by the concept of overlap between the atomic orbitals of combining atoms. Atomic orbitals have definite shape in space (except s orbital) and so overlap of atomic orbitals must occur in definite direction so as to make maximum probability of finding the electron along the bond direction i.e. **the covalent bond must have a directional nature.** The directional nature of the covalent bond is a direct consequence of orbital overlap. The resulting covalent bonds will produce a **definite geometry** of each covalent molecule. For example $NH_3$, $H_2O$, $O_3$ etc. molecules have definite geometry. The ultimate geometry of covalent molecules may be established by different factors such as :

(a) Idea of **hybridisation** according to valence bond theory which is required to explain the number of bonds formed, the equivalence of the bonds (there may be exception), the geometry of the molecules and the better overlapping of atomic orbitals.

(b) Mutual repulsion of bonding electron pairs so as to make the covalent bonds as far apart as possible.

(c) Repulsion between non bonding or unshared pair of electrons if any which greatly influence the geometry of a molecule.

Let us consider the basic approach of hybridisation. The ground state electronic configuration of Be, B and C are $1s^22s^2$, $1s^22s^22p^1$ and $1s^22s^22p^2$ having zero, one and two unpaired electrons respectively. Therefore it is expected that Be will not form any compound, B will form a single covalent bond and C will form only two covalent bonds. But actually the common compounds of Be, B and C are $BeCl_2$, $BCl_3$ and $CH_4$. If we write the excited state electronic configuratons of Be, B and C as $1s^22s^12p_x^{\ 1}$, $1s^22s^12p_x^1p_y^1$ and $1s^22s^12p_x^12p_y^12p_z^1$ then there may be the formation of 2, 3 and 4 bonds around Be, B and C but the geometry of molecules $BeCl_2$, $BCl_3$ and $CH_4$ are not explained because they have linear, triangular planar and tetrahedral geometry respectively. Therefore in order to explain the geometry of a molecule a concept is developed which is known as hybridisation. *Hybridisation is a process by which pure atomic orbitals will redistribute their energies among themselves so as to make equivalent bonds (may be exception) before combination with other atoms.* According to this model the 2s and 2p orbitals of Be overlap each other and then split into two equivalent sp hybrid orbitals as :

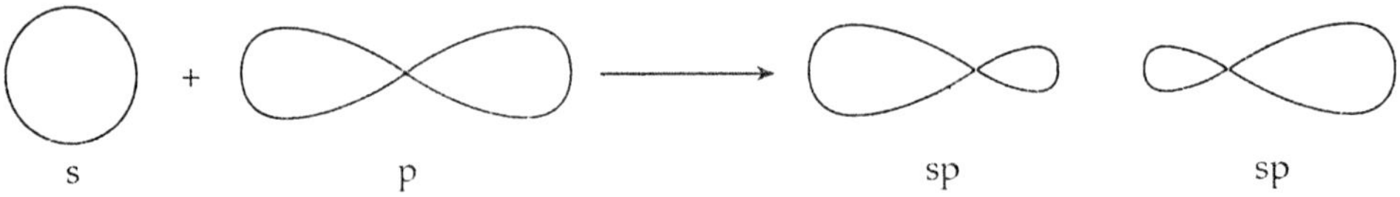


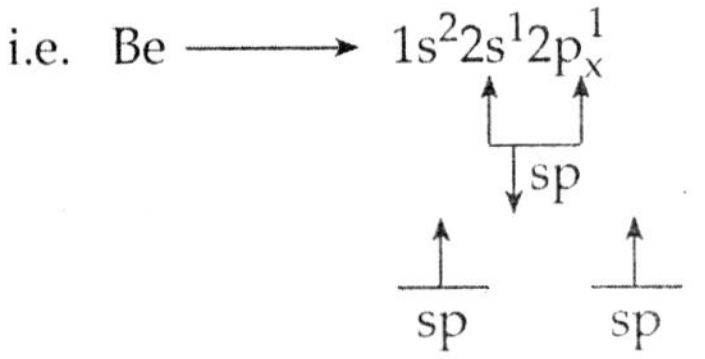


The two newly formed sp hybrid orbitals have identical energies but they have lower energies than the individual 2s and 2p orbitals. The two sp hybrid orbitals have a single unpaired electron in each orbital according to the Hund's rule and then the two chlorine atoms having single unpaired electron in 3p orbital ($Cl \rightarrow 1s^22s^22p^63s^23p_x^13p_y^23p_z^2$) overlap each other to form the $BeCl_2$ molecule as :

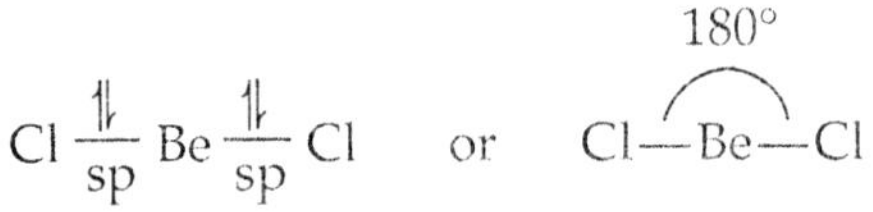


In this way there will be the formation of linear $BeCl_2$ molecule having Cl—Be—Cl bond angle 180°. Similarly the formation of $BCl_3$ and $CH_4$ can be depicted as :

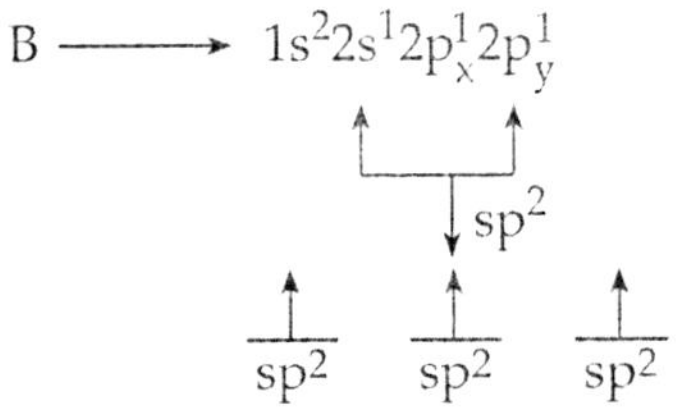


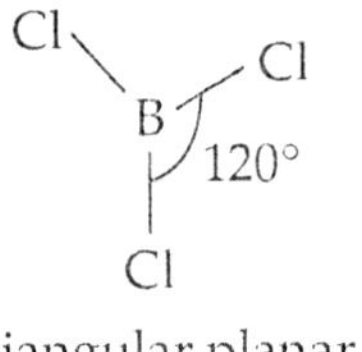


Triangular planar

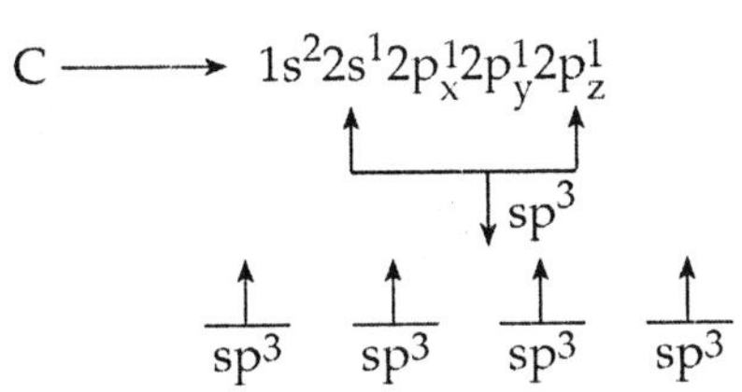


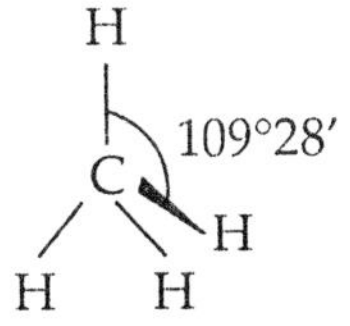


Tetrahedral

Similarly other hybrid orbitals viz. $dsp^2$ (square planar), $dsp^3$ or $sp^3d$ (trigonal bipyramid), $d^2sp^2$ or $sp^3d$ (square pyramid), $d^2sp^3$ (inner orbital octahedral) and $sp^3d^2$ (outer orbital octahedral) can be constructed.

For the formation of seven bonds there may be three geometries i.e. **pentagonal bipyramid, capped octahedra** and **tetragonally capped trigonal prismatic** (Fig. 3.8). Except pentagonal bipyramid geometry, the bond angles for other geometries will vary. The preferred geometry for the formation of seven bonds is pentagonal bipyramid where the repulsion is minimised. For **coordination number 8** or for the formation of eight bonds the two geometries (**square antiprism** and **dodecahedron**) are preferred (Fig. 3.8). For the formation of 9 bonds (coordination number 9) two geometries are available of which preferred one is **tricapped trigonal prism** (Fig. 3.8). Higher coordination numbers upto 14 are available for some of the large central atoms.

### 3.2.6 Equivalent and non-equivalent hybrid orbitals

Generally hybrid orbitals are equivalent e.g. sp, $sp^2$ and $sp^3$ etc. but there are certain hybrid orbitals which are non-equivalent. Equivalent hybrid orbitals results when the percentage contribution of individual atomic orbitals are the same in all the hybrid orbitals. In sp, $sp^2$ and $sp^3$ hybrid orbitals the percentage of s orbital contribution is 50, 33 and 25 percent respectively in all the 2, 3 and 4 hybrid orbitals of sp, $sp^2$ and $sp^3$ respectively. But in case of $sp^3d$ or $dsp^3$ the five bonds are not equivalent—the three equivalent hybrid orbitals formed between

*Equatorial and axial bonds*

$(s+p_x+p_y)$ lie in the xy **equatorial plane** which are at 120° apart and the other two equivalent hybrid orbitals formed between the overlap of $(p_z+d_{z^2})$ directed mutually 180° in the **axial directions** i.e. perpendicular to the xy plane. The equatorial bonds has s orbital conbribution but axial bonds has no s character and so the axial and equatorial bonds are called non-equivalent hybrids. In $PCl_5$ this is well accord with the experimental facts i.e. three equatorial chlorines i.e. three equatorial P-Cl bonds are shorter (2.04Å) than the two axial P-Cl bonds (2.19Å) having **trigonal bipyramidal** (TBP) geometry of $PCl_5$. Therefore this hybridisation can be split into **$sp^2$ for the three equatorial bonds** and dp for two axial bonds.

Five coordination can also be achieved through $d^2sp^2$ and $sp^3d$ hybridisation having square pyramid (SP) geometry e.g. $[VO(acac)_2]$ where acacH = acetylacetone. Here in $sp^3d$ process the combination can be represented as $s+p_x+p_y+d_{x^2-y^2}$ which forms four $sp^2d$ equivalent equatorial bonds in the xy plane and the fifth one is the pure $p_z$ orbital lying perpendicular to the xy plane (axial bond). In $AX_5$ molecule the energy difference between TBP and SP geometries are small so that there may be an equilibrium between the two geometries i.e. TBP $\rightleftharpoons$ SP. For this reason there are some molecular species e.g. $CdCl_5^{3-}$, $Ni(CN)_5^{3-}$ etc. may exist in both the configurations.

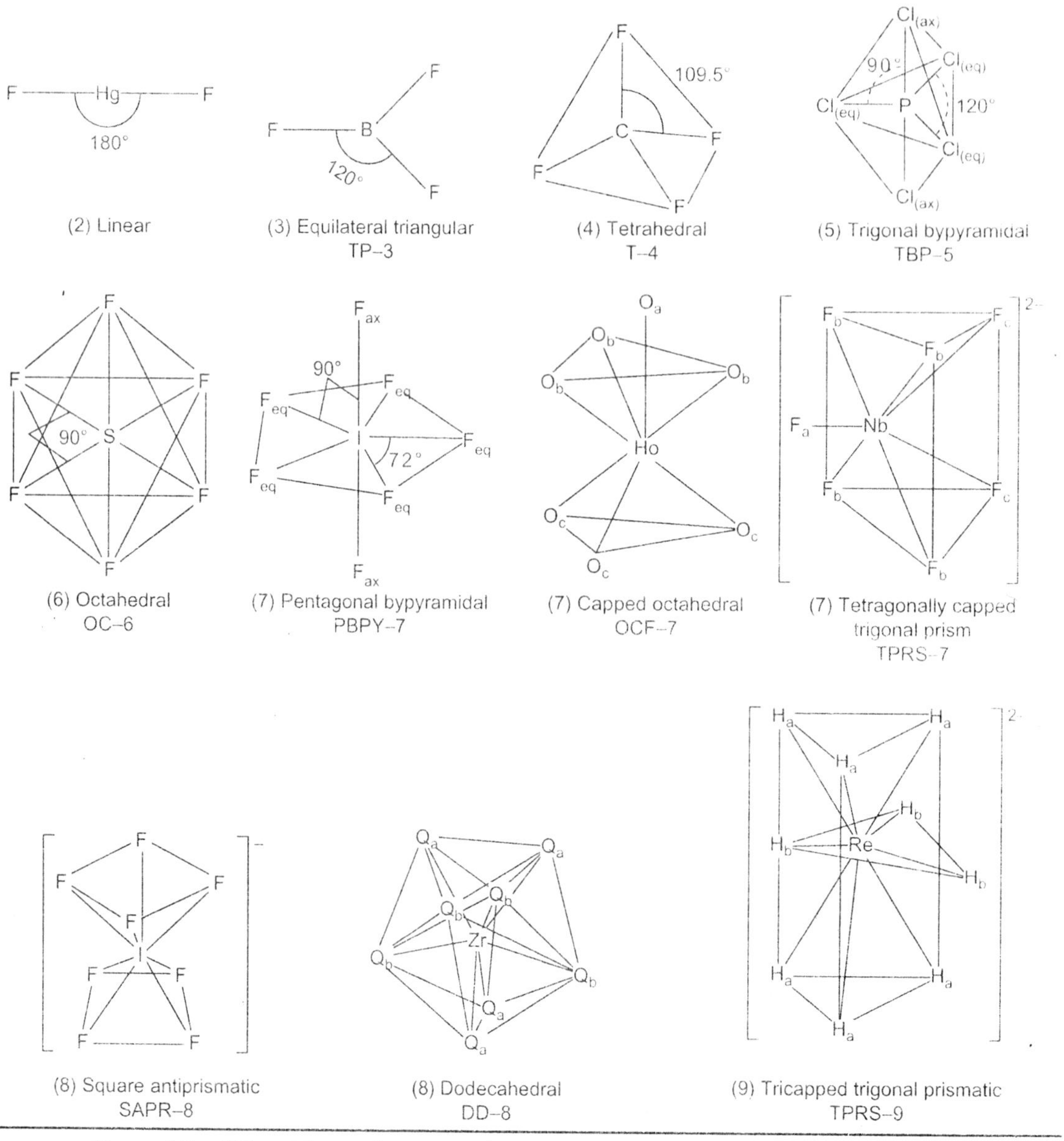


*Figure 3.8 : Different types of geometry from coordination number/number of bonds from 2 to 9. 'ax' for axial and 'eq' for equatorial or 'a' for axial 'b or c' for equatorial. The OCF-7 is a part of the structure of $[Ho(H_2O)(C_{15}H_{11}O_2)_3]$. DD-8 is the part of the complex $[Zr(C_5H_7O_2)_4]$.*

Whether the hybrid orbitals may be equivalent (tetrahedral $sp^3$) or non-equivalent (trigonal bipyramid or square pyramid $sp^3d$) will be dictated by the electronegativity difference of the central atom with that of the combining atoms. The equivalence of the hybrid orbitals are lost in some cases where some of the hybrid orbitals are occupied by lone pairs. In $CH_4$ and $CH_3Cl$ the central atom C in both the cases is $sp^3$ so that in case of $CH_4$ the four hybrid orbitals are equivalent having ideal tetrahedral angle 109°28′. But in case of $CH_3Cl$ the bond angle

H—C—H is 110° which is somewhat different from bond angle H—C—Cl. This variation in bond angle may be explained by Bent's rule (Sec. 3.3.3). This difference in bond angles attributed by the non uniform distribution of s and p character of different atomic orbitals in the bonding orbitals. The three hybrid orbitals of carbon that overlap with hydrogen has slight higher s character making higher bond angle (110°). The fourth hybrid orbital of carbon that overlaps with chlorine p orbital have slightly less s character making lower bond angle. In general with combination of s and p, with the increase of p character, the bond angle decreases. In $NH_3$ and $H_2O$, the central element nitrogen and oxygen respectively is $sp^3$ hybridised and the bond angles H—N—H and H—O—H are 107° and 104° respectively. This variation of bond angle will be discussed in VSEPR theory (Sec. 3.3.1).

### 3.2.7 Aspects of hybridisation in terms of wave mechanics

The orbital wave functions undergo a linear combination to form hybrid orbitals like that in case of molecular orbitals. The difference between hybrid orbitals and molecular orbitals is that in case of hybrid orbitals it involves between orbitals of same atom whereas in case of molecular orbitals, LCAO involves between different atoms.

*LCAO in hybridisation*

The basic approach of covalent bonding is the overlap of atomic orbitals—the shapes and sign of the wave functions must be such that the two atomic orbitals have positive overlap. In order to form hybrid orbitals overlap, the involving atomic orbitals **should not differ significantly in the radial parts of their wave functions**. On the other hand for effective hybridisation, the combining atomic orbitals should have comparable magnitude of energy. Hybridised orbitals can be constructed from the angular parts of the individual wave functions. We shall consider some salient feature of hybridisation :

(a) Hybrid orbital concept represents better overlap properties than do the pure s, p and d orbitals. For example the s and p orbital of the same atom can be mixed or hybridised to produce new **sp hybrid orbital** (Fig. 3.9) combination of crest (positive lobe) of s orbital with p orbital with its crest to the right produces sp hybrid orbital with most of the probability of finding the electron at the right [Fig. 3.9(a)]. On the other hand other possible combination of the same orbitals produces maximum overlap represented to the left [Fig. 3.9(b)]. In a similar way other hybrid orbitals can also be represented (Fig. 3.10).

(b) One or more electrons from fully filled orbitals of the atom in the ground state, will be promoted to the next available orbitals of higher energy resulting in an excited state electronic configuration. This requires energy. The orbitals in the excited state then undergo hybridisation to form equivalent hybrid orbitals. The energy difference between the unhybridised atomic orbitals and the hybridised atomic orbitals is known as energy of hybridisation ($E_{hy}$). The $E_{hy}$ gives a measure to carry out the process of hybridisation. Hence for an effective hybridisation to occur, the entire input of energy is more than recovered than the $E_{hy}$ when the hybrid orbitals combine with the orbitals of other atoms to form covalent bonds. Thus the energy released during covalent bond

*Energetics of hybridisation*

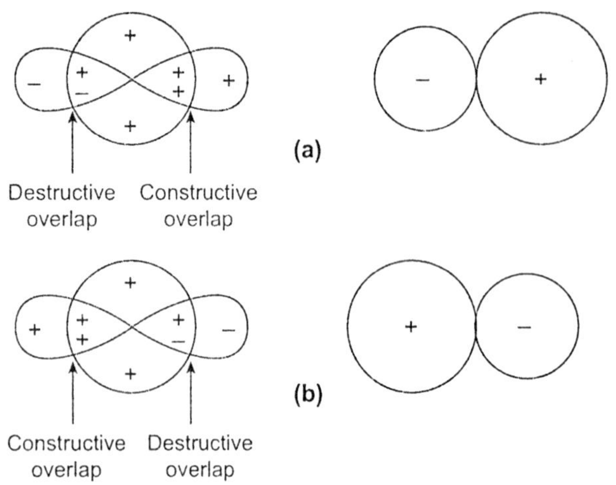


***Figure 3.9** : Pictorial representation of overlap of s and p orbital to form sp hybrid orbitals.*

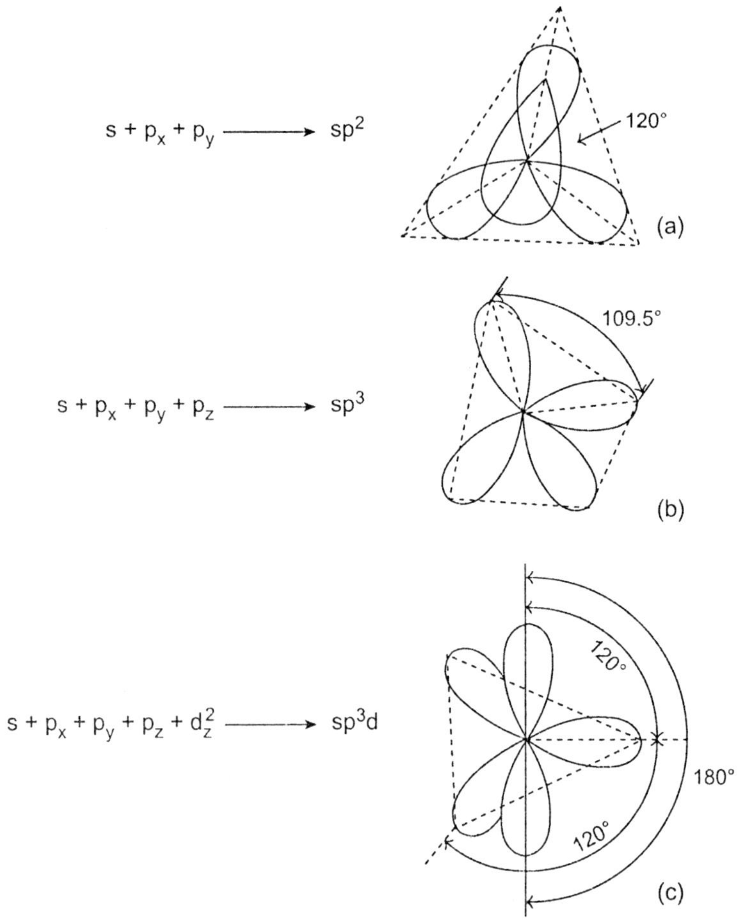


***Figure 3.10** : The shapes of different types of hybrid orbitals (a) $sp^2$, (b) $sp^3$, (c) $sp^3d$.*

formation involving hybrid orbitals is much more than the energy relased if the covalent bond is formed from pure atomic orbitals. Therefore covalent bonds formed by hybrid orbitals are more stable than those formed by pure atomic orbitals so that **atoms prefer to combine with hybrid orbitals rather than that of pure atomic orbitals.**

(c) Bonding potentiality of the hybrid orbitals are dictated by the angular strength of the orbitals which is related to the angular part of the wave function. s orbital is spherically symmetrical whereas p orbital is dumbell shaped having greater angular orientation is most suitable for overlap in projected direction. Angular strength of orbitals is related to the overlap integrals ($S_{AB}$) which are directed in proper direction of the hybrid orbitals.

(d) The number of hybrid orbitals are equal to the number of participating atomic orbitals.

(e) Hybridisation occurs between the orbitals of the same atom.

(f) Hybrid orbitals generally form sigma bonds or contain lone pair of electrons.

(g) As s character of the hybrid orbitals increases, electronegativity increases ($sp > sp^2 > sp^3$). Bond angle increases with increasing s character ($sp^3 < sp^2 < sp$) but decreases with increasing p character.

### 3.2.8 Wave functions for hybrid orbitals

The wave functions for hybrid orbitals e.g. sp, $sp^2$, $sp^3$ etc. can be constructed by the following rules :

(a) The wave function of the hybrid orbitals can be obtained by the linear combination of combining atomic orbitals viz.

$$\Psi_i = a_i\Psi_s + b_i\Psi_{p_x} + c_i\Psi_{p_y} + d_i\Psi_{p_z} \qquad ...(3.21)$$

where $\Psi_s$, $\Psi_{p_x}$, $\Psi_{p_y}$ and $\Psi_{p_z}$ are the individual atomic orbital wave functions for s, $p_x$, $p_y$ and $p_z$ orbitals respectively. The wave functions should be **orthonormal** i.e. they simultaneously **normalised** and mutally **orthogonal** i.e.

$$\int_{-\infty}^{+\infty}\Psi_m\Psi_m d\tau = 1 \text{ (normalisation condition Sec. 1.8.5)} \qquad ...(3.22)$$

and $$\int_{-\infty}^{+\infty}\Psi_n\Psi_n d\tau = 0 \text{ (orthogonal condition)} \qquad ...(3.23)$$

where $\Psi_m$ and $\Psi_n$ are two wave functions and $\Psi_i$ constitute the ith hybrid orbital wave function. The coefficients $a_i$, $b_i$, $c_i$ and $d_i$ can be calculated from the following conditions :

(i) Each wave function of a hybrid orbital is separately normalised so that

$$\int_{-\infty}^{+\infty} \Psi_i^2 d\tau = 1$$

For such an normalised wave function we can write,

$$a_i^2 + b_i^2 + c_i^2 + d_i^2 = 1 \quad ...(3.24)$$

(ii) Each hybrid orbital in the set of hybrid orbitals is orthogonal to the other hybrid orbitals i.e.

$$\int_{-\infty}^{+\infty} \Psi_i \Psi_j d\tau = 0$$

so that we can write $a_i a_j + b_i b_j + c_i c_j + d_i d_j = 0$ ...(3.25)

where $a_j$, $b_j$, $c_j$ and $d_j$ are the coefficients for the construction of jth hybrid orbital.

(iii) The squares of the coefficients of component wave functions which give contributions of the component wave functions summed over all the hybrid orbitals wherein they participate equal to unity i.e.

$$\Sigma a_i^2 = 1 \quad ...(3.26)$$

This is due to the fact that pure atomic orbitals must be completely utilised in the formation of hybrid orbitals.

(b) As s orbital is spherically symmetrical so each equivalent hybrid orbital has a contribution of $\frac{1}{\sqrt{n}}$ of the s orbital distributed in n hybrid orbitals which means that the coefficient of $\Psi_s$ in each hybrid orbital is $\frac{1}{\sqrt{n}}$.

(c) In forming the equivalent hybrid orbitals we have to specify certain coordinate axis (say x axis) for the formation of the ith hybrid orbital and place as many hybrid orbitals as possible in the plane containing this axis (say xy plane). This restriction is required to simplify the construction of hybrid orbitals, for calculating the orientation of hybrid orbitals in space and for determining the angle between the hybrid orbitals.

Let us consider the wave function for sp hybrid orbital e.g. $BeCl_2$. As the sp hybrid orbitals are to be placed along the x axis we shall consider the hybridisation of s and $p_x$ orbitals. The Be atom should have two bonds of the form

$$\Psi_i = a_i \Psi_s + b_i \Psi_{p_x}$$

where a and b are two coefficients which would produce $\Psi_i$ to have a maximum value in a certain direction. Now applying normalisation requirements we can say that for each function as well as a linear combination of them,

$$\int_{-\infty}^{+\infty} \Psi_i^2 d\tau = 1$$

i.e. $\int_{-\infty}^{+\infty} \Psi_s^2 d\tau = 1$ and $\int_{-\infty}^{+\infty} \Psi p_x^2 d\tau = 1$ ...(3.27)

From orthogonality condition of atomic orbitals, we have

$$\int_{-\infty}^{+\infty} \Psi_s \Psi_{p_x} d\tau = 0$$

Therefore from equation (3.27) we have

$$1 = \int_{-\infty}^{+\infty} \Psi_i^2 d\tau = \int_{-\infty}^{+\infty} \left(a_i \Psi_s + b_i \Psi_{p_x}\right)^2 d\tau$$

$$= a_i^2 \int \Psi_s^2 d\tau + b_i^2 \int \Psi_{p_x}^2 d\tau + 2a_i b_i \int \Psi_s \Psi_{p_x} d\tau$$

$$= a_i^2 + b_i^2 \qquad \left[\because \int \Psi_s^2 d\tau = 1,\ \int \Psi_{p_x}^2 d\tau = 1 \text{ and } \int \Psi_s \Psi_{p_x} = 0 \text{ as the wave functions } \Psi_s \text{ and } \Psi_{p_x} \text{ are mutually orthonormal}\right]$$

This means that $a_i$ and $b_i$ should be chosen in such a way that the sum of their squares is always unity.

$$\therefore\ b_i = \pm\sqrt{1 - a_i^2}$$

Now contribution of the orbital $\Psi_s$ towards the formation of the two hybrid orbitals should be equal to the sum of the contribution of same orbital to each of these bonds. In other words since the s orbital is equally distributed between the two hybrid orbitals and since the square of the coefficient of a wave function gives its contribution to the hybrid orbital [according to rule a(iii)], then

$$a_i^2 \int \Psi_s^2 d\tau + a_i^2 \int \Psi_s^2 d\tau = \int \Psi_s^2 d\tau = 1$$

i.e. $2a_i^2 = 1$ so that $a_i = \pm\frac{1}{\sqrt{2}}$

Since $\Psi_s$ cannot have negative value, we must take only the positive value of $a_i$, so that,

$$b_i = \pm\sqrt{1 - \left(\frac{1}{\sqrt{2}}\right)^2} = \pm\frac{1}{\sqrt{2}}$$

Hence the two sp hybrid orbitals can be represented as,

$$\Psi_1 = \frac{1}{\sqrt{2}}\Psi_s + \frac{1}{\sqrt{2}}\Psi_{p_x} \quad \text{and} \quad \Psi_2 = \frac{1}{\sqrt{2}}\Psi_s - \frac{1}{\sqrt{2}}\Psi_{p_x}$$

In a similar way the wave function for other hybrid orbitals e.g. $sp^2$, $sp^3$ etc. can be constructed.

### 3.2.9 Orientation of the hybrid orbitals

According to Sec. 1.8.7.2, s and p orbitals wave functons are—

$$\Psi_s = R(r)_{n,0}\frac{1}{2\sqrt{\pi}},\quad \Psi_{p_x} = R(r)_{n,1}\frac{3}{2\sqrt{\pi}}\sin\theta\cos\phi$$

$$\Psi_{p_y} = R(r)_{n,1}\frac{3}{2\sqrt{\pi}}\sin\theta\sin\phi,\quad \Psi_{p_z} = R(r)_{n,1}\frac{3}{2\sqrt{\pi}}\cos\theta$$

Since the radius of the ns and np orbitals are approximately the same, the radial wave function $R(r)_{n,0}$ and $R(r)_{n,1}$ may be taken as same and will be written simply as R(r).

Let us consider the sp hybrid orbitals where the two orbitals lie at an angle of 180°. We can say that the two bonds are formed in such a way that there is a formation of strongest possible bond and the overlap will be maximum. Let $\Psi_1$ is the wave function of one of the hybrid orbitals so that $\Psi_1$ will be as large as possible in the direction of the bond (i.e. z axis) and the overlap will be maximum i.e. $\frac{d\Psi}{d\theta} = 0$. Now for sp hybridisation

$$\Psi_1 = \frac{1}{\sqrt{2}}\Psi_s + \frac{1}{\sqrt{2}}\Psi_{p_z}$$

In terms of spherical polar coordinates we can write

$$\Psi_1 = \frac{1}{\sqrt{2}}\frac{1}{2\sqrt{\pi}}R(r) + \frac{1}{\sqrt{2}}\frac{3}{2\sqrt{\pi}}\cos\theta$$

Now making $\frac{d\Psi_1}{d\theta} = 0$,

$$0 + \frac{1}{\sqrt{2}}\frac{3}{2\sqrt{\pi}}(-\sin\theta)R(r) = 0$$

so that $\sin\theta = 0$ i.e. $\theta = 0°$ or $180°$

We get exactly the same result when we consider the wave function $\Psi_2$ of the second sp hybrid orbital.

Under the maximum overlap condition, θ is the the angle which the directions of the hybrid orbitals $\Psi_1$ and $\Psi_2$ make with z axis. This means that one of the sp hybrid orbitals is aligned along the +z axis (for θ = 0°) and the other orbital is aligned along the –z axis (for θ = 180°) so that the angle between the two sp hybrid orbitals is 180°.

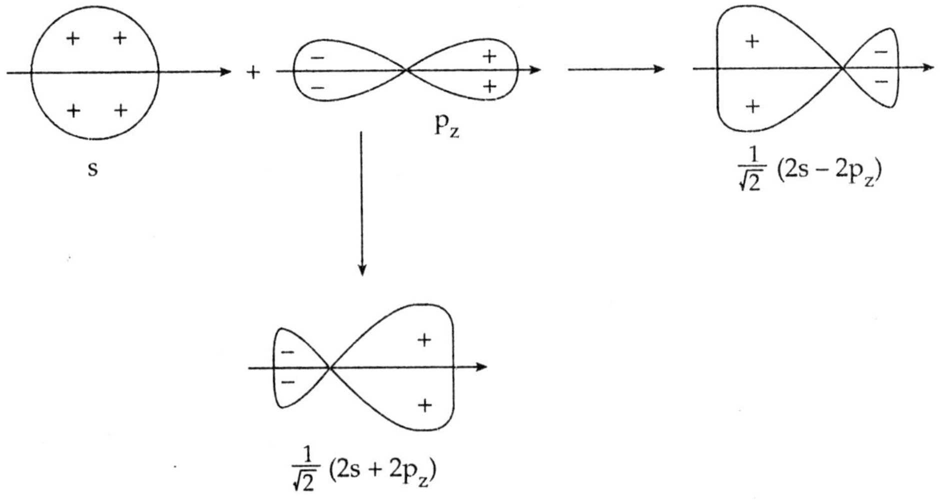


Similarly the orientation of other hybrid orbitals e.g. $sp^2$, $sp^3$ can also be calculated. Table 3.4 represnts various aspects of hybridisation with their specific geometries.

### 3.2.10 Hybridisation and Multiple bonding

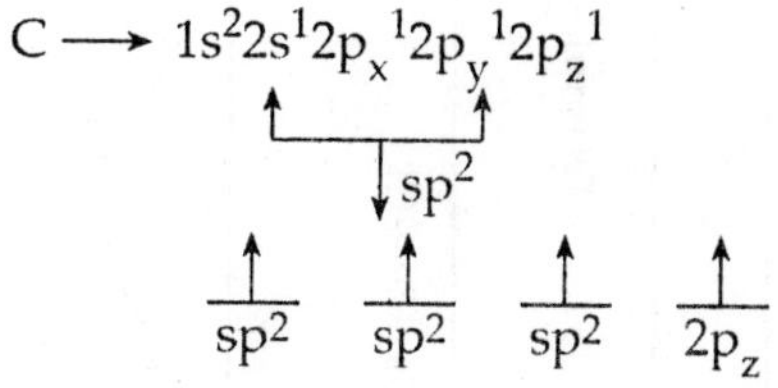


Carbon forms C–C, C=C and C≡C in different compounds. In $C_2H_4$ the two carbon atoms are $sp^2$ hybridised. The excited state electronic configuration of carbon is $sp^2$ hybridised—three $sp^2$ hybrid orbitals each having one unpaired electron and one unpaired electron in $2p_z$ orbital. Two $sp^2$ hybrid orbitals overlap with two hydrogen atoms to form two sigma (σ) bonds and also other $sp^2$ hybrid orbital having one unpaired electron overlap with other $sp^2$ hybrid orbital of other carbon to form another σ bond. **The σ bond is formed by the head-on overlap of two hybrid orbitals or one hybrid orbital and other atomic orbital.** In $C_2H_4$ there are five σ bonds. On the other hand the two $2p_z$ orbitals of carbon overlap sidewise to form one pi (π) bond i.e. in $C_2H_4$ the C=C has one σ and one π bond. **The π bond is formed by the sidewise or end-on overlap of two atomic orbitals.** The formation of $C_2H_4$ is depicted in Fig. 3.11.

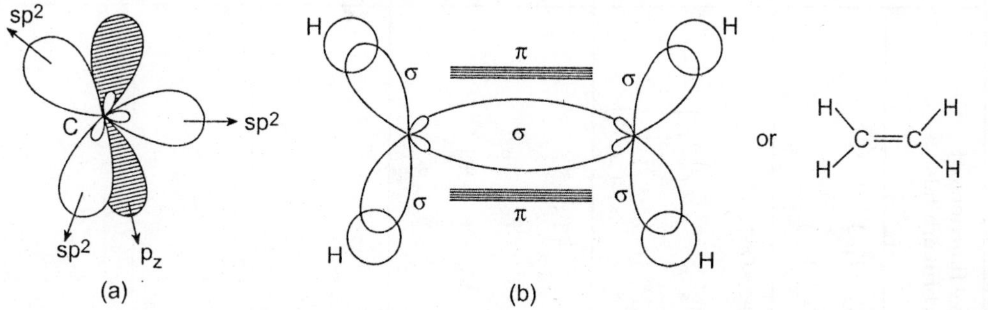


***Figure 3.11 :** Formation of $C_2H_4$ molecule(a) formation of $sp^2$ hybrid orbitals and $p_z$ orbital of carbon (b) formation of σ and π bonds.*

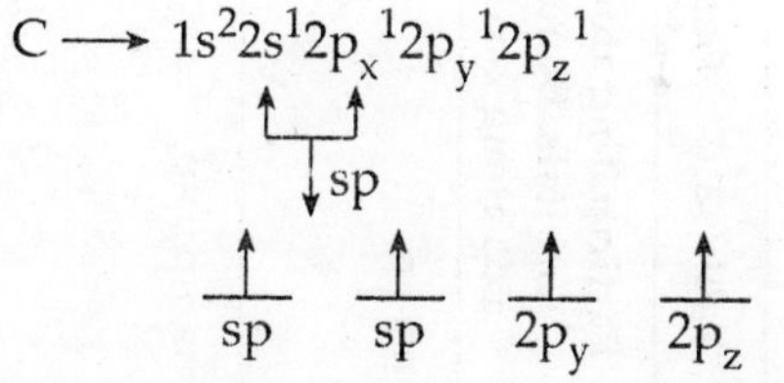


In $C_2H_2$ molecule each carbon is sp hybridised. Two sp hybrid orbitals of two carbon atoms each having single unpaired electron overlap with 1s orbital of each hydrogen having single unpaired electron to form two C–H σ bonds. The other sp hybrid orbital of each carbon overlap in a head-on position to form another C–C σ bond i.e. altogether there are three σ bonds in $C_2H_2$. The two $p_y$ and $p_z$ orbitals of each carbon overlap in a side-wise position to form two π bonds as depicted in Fig. 3.12. The superposition of these π-orbitals lead to cylindrical symmetry about the C–C bond axis.

**Table 3.4 : Geometry, wave function etc. of some of the hybrid orbitals**

| Type of hybridisation | No. of bonds/ coordination no. | Participating atomic orbitals. Bond axis along x-axis | Wave functions of hybrid orbitals | Geometry | Bond angle | Example |
|---|---|---|---|---|---|---|
| sp | 2 | ns, $np_x$ | $\Psi_1 = \frac{1}{\sqrt{2}}(\Psi_s + \Psi_{p_x})$<br>$\Psi_2 = \frac{1}{\sqrt{2}}(\Psi_s - \Psi_{p_x})$ | Linear | 180° | Cl—Be—Cl |
| $sp^2$ | 3 | ns, $np_x$, $np_y$ | $\Psi_1 = \frac{1}{\sqrt{3}}(\Psi_s + \sqrt{2}\Psi_{p_x})$<br>$\Psi_2 = \frac{1}{\sqrt{3}}\left(\Psi_s - \frac{1}{\sqrt{2}}\Psi_{p_x} + \sqrt{\frac{3}{2}}\Psi_{p_y}\right)$<br>$\Psi_3 = \frac{1}{\sqrt{3}}\left(\Psi_s - \frac{1}{\sqrt{2}}\Psi_{p_x} - \sqrt{\frac{3}{2}}\Psi_{p_y}\right)$ | Trigonal planar | 120° | 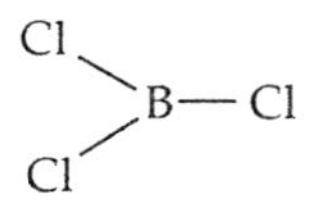 |
| $sp^3$ | 4 | ns, $np_x$, $np_y$, $np_z$ | $\Psi_1 = \frac{1}{2}(\Psi_s + \Psi_{p_x} + \Psi_{p_y} + \Psi_{p_z})$<br>$\Psi_2 = \frac{1}{2}(\Psi_s + \Psi_{p_x} - \Psi_{p_y} - \Psi_{p_z})$<br>$\Psi_3 = \frac{1}{2}(\Psi_s - \Psi_{p_x} + \Psi_{p_y} - \Psi_{p_z})$<br>$\Psi_4 = \frac{1}{2}(\Psi_s - \Psi_{p_x} - \Psi_{p_y} + \Psi_{p_z})$ | Tetrahedral ($T_d$) | 109°28′ | 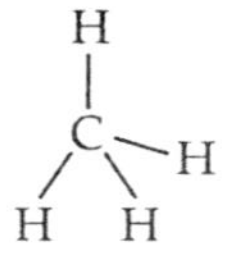 |
| $d^3s$ | 4 | $(n-1)d_{xy}$, $(n-1)d_{xz}$, $(n-1)d_{yz}$, ns | —— | Tetrahedral ($T_d$) | 109°28′ | 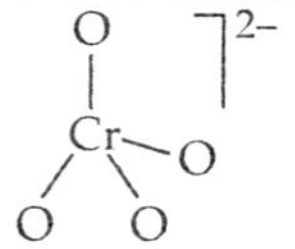 |
| $dsp^2$ | 4 | $(n-1)d_{x^2-y^2}$, ns, $np_x$, $np_y$ | —— | Square planar | 90° | 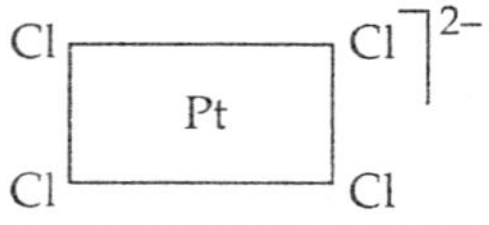 |
| $sp^2d$ | 4 | ns, $np_x$, $np_y$, $nd_{x^2-y^2}$ | —— | Square planar | 90° | 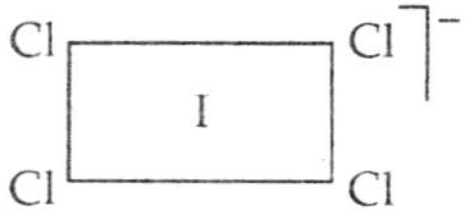 |

*Table contd. next page*

| Type of hybridisation | No. of bonds/ coordination no. | Participating atomic orbitals. Bond axis along x-axis | Wave functions of hybrid orbitals | Geometry | Bond angle | Example |
|---|---|---|---|---|---|---|
| $sp^3d$ | 5 | ns, $np_x$, $np_y$, $np_z$, $nd_{z^2}$ | $\Psi_1 = \frac{1}{\sqrt{3}}(\Psi_s + \sqrt{2}\Psi_{p_x})$ <br> $\Psi_2 = \frac{1}{\sqrt{3}}\left(\Psi_s - \frac{1}{\sqrt{2}}\Psi_{p_x} + \sqrt{\frac{3}{2}}\Psi_{p_y}\right)$ <br> $\Psi_3 = \frac{1}{\sqrt{3}}\left(\Psi_s - \frac{1}{\sqrt{2}}\Psi_{p_x} - \sqrt{\frac{3}{2}}\Psi_{p_y}\right)$ (equatorial) <br> $\Psi_4 = \frac{1}{\sqrt{2}}\left(\Psi_{p_z} + \Psi_{d_{z^2}}\right)$ <br> $\Psi_5 = \frac{1}{\sqrt{2}}\left(\Psi_{p_z} + \Psi_{d_{z^2}}\right)$ (axial) | Trigonal bipyramidal (TBP) | 180/90° | 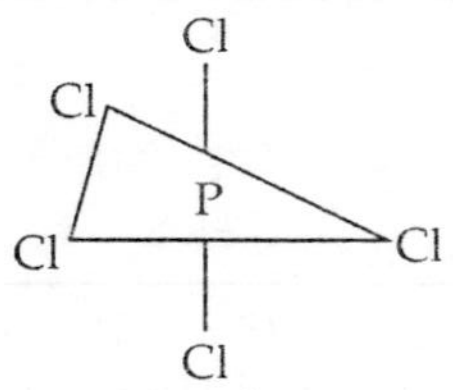  |
| $dsp^3$ | 5 | $(n-1)d_{z^2}$, ns, $np_x$, $np_y$, $np_z$ | —— | Trigonal bipyramidal (TBP) | 180/90° | 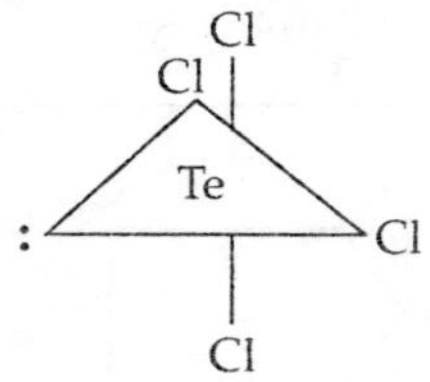  |
| $dsp^3$ | 5 | $(n-1)d_{x^2-y^2}$, ns, $np_x$, $np_y$, $np_z$ | —— | Square pyramid | 90° | |
| $sp^3d$ | 5 | ns, $np_x$, $np_y$, $np_z$, $nd_{x^2-y^2}$, | —— | Square pyramid | 90° | 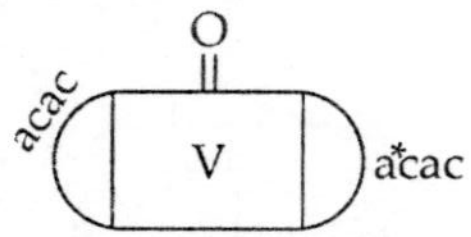  |
| $d^2sp^3$ | 6 | $(n-1)d_{x^2-y^2}$, $(n-1)d_{z^2}$, ns, $np_x$, $np_y$, $np_z$ | —— | Inner orbital octahedral ($O_h$) | 180/90° | 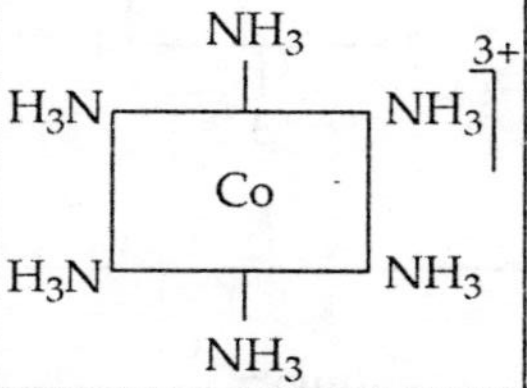  |

* acacH = acetylacetone

*Table contd. next page*

| Type of hybridisation | No. of bonds/ coordination no. | Participating atomic orbitals. Bond axis along x-axis | Wave functions of hybrid orbitals | Geometry | Bond angle | Example |
|---|---|---|---|---|---|---|
| $sp^3d^2$ | 6 | ns, $np_x$, $np_y$, $np_z$, $nd_{x^2-y^2}$, $nd_{z^2}$ | $\Psi_1 = \frac{1}{\sqrt{6}}(\Psi_s + \sqrt{2}\Psi d_{z^2} + \sqrt{3}\Psi_{p_z})$<br>$\Psi_2 = \frac{1}{\sqrt{6}}\left(\Psi_s - \frac{1}{\sqrt{2}}\Psi d_{z^2} + \sqrt{\frac{3}{2}}\Psi d_{x^2-y^2} + \sqrt{3}\Psi_{p_x}\right)$<br>$\Psi_3 = \frac{1}{\sqrt{6}}\left(\Psi_s - \frac{1}{\sqrt{2}}\Psi d_{z^2} - \sqrt{\frac{3}{2}}\Psi d_{x^2-y^2} + \sqrt{3}\Psi_{p_y}\right)$<br>$\Psi_4 = \frac{1}{\sqrt{6}}\left(\Psi_s - \frac{1}{\sqrt{2}}\Psi d_{z^2} + \sqrt{\frac{3}{2}}\Psi d_{x^2-y^2} - \sqrt{3}\Psi_{p_x}\right)$<br>$\Psi_5 = \frac{1}{\sqrt{6}}\left(\Psi_s - \frac{1}{\sqrt{2}}\Psi d_{z^2} - \sqrt{\frac{3}{2}}\Psi d_{x^2-y^2} - \sqrt{3}\Psi_{p_y}\right)$<br>$\Psi_6 = \frac{1}{\sqrt{6}}(\Psi_s + \sqrt{2}\Psi d_{z^2} - \sqrt{3}\Psi_{p_z})$ | Outer orbital octatedral ($O_h$) | 180/90° | 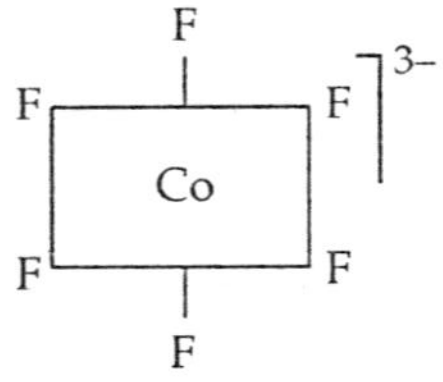 |
| $d^3sp^3$ | 7 | $(n-1)d_{x^2-y^2}$, $(n-1)d_{xy}$, $(n-1)d_{z^2}$, ns, $np_x$, $np_y$, $np_z$ | —— | Pentagonal bipyramid | 180/90° | 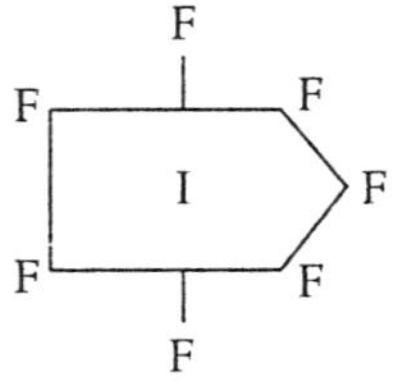 |
| $d^4sp^3$ | 8** | $(n-1)d_{z^2}$, $(n-1)d_{xy}$, $(n-1)d_{yz}$, $(n-1)d_{zx}$, ns, $np_x$, $np_y$, $np_z$ | —— | Dodeca-hedral | — | $[Cr(O_2)_4]^{3-}$ (Fig. 3.8) |
| $d^4sp^3$ | 8 | $(n-1)d_{x^2-y^2}$, $(n-1)d_{xy}$, $(n-1)d_{yz}$, $(n-1)d_{zx}$, ns, $np_x$, $np_y$, $np_z$ | —— | Square antiprism | — | $[M(acac)_4]$ M = Zr(IV), Ce(IV) (Fig. 3.8) |
| $d^5sp^3$ | 9 | $(n-1)d_{x^2-y^2}$, $(n-1)d_{xy}$, $(n-1)d_{yz}$, $(n-1)d_{xz}$, $(n-1)d_{z^2}$, ns, $np_x$, $np_y$, $np_z$ | —— | Tricapped trigonal prismatic | — | $ReH_9^{2-}$ (Fig. 3.8) |

** For coordination number 8 cubic geometry also available.

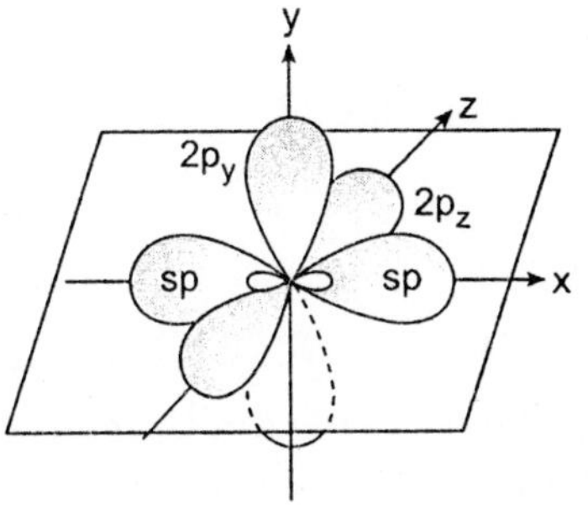


(a)

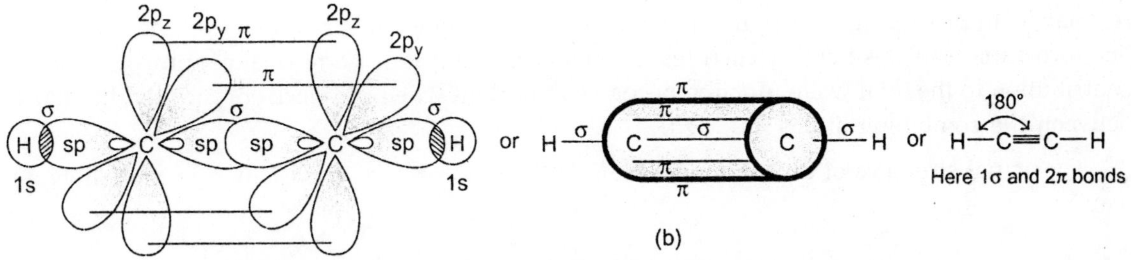


(b)

*Figure 3.12 : Bonding in acetylene molecule—(a) formation of sp hybrid orbitals, $p_y$ and $p_z$ orbitals of carbon, (b) formation of one σ and two π bonds.*

It is evident that as bond multiplicity increases, bond length decreases so that bond energy increases (Table 2.8).

### 3.2.11 Comparison between σ and π bonds

(a) Bonding molecular orbital that results from the positive overlap of two atomic orbitals is called σ bond which has **no nodal plane** through both the nuclei. On the other hand if the bonding molecular orbital has **one nodal plane** passing through both the nuclei is called π bond. σ bond is more stronger than π bond.

(b) σ bond can alone be formed, but π bond always forms in association with σ bond. For example carbon-carbon single bond is σ bond, carbon-carbon double bond consists of one σ plus one π, the carbon-carbon triple bond consists of one σ plus two π bonds.

(c) The π bond is cylindrical whereas σ bond is not.

(d) There will be maximum electron density along the bond axis in case of σ bond whereas in case of π bond there will be small electron density along the surface of the cylinder.

(e) σ bonds determine the geometry of a molecule whereas π bond is not.

(f) In all chemical reactions π bond will take part first then σ bond.

(g) Free rotation about C–C σ bond is permitted but rotation about C=C or C≡C is restricted.

(h) A σ bond has cylindrical symmetry along the bond axis whereas π bond possess a plane of symmetry often called as the nodal plane.

### 3.2.12 Resonance and delocalisation in valence bond theory

According to valence bond theory the $H_2$ molecule can be represented in ionic form (i.e. $H^+H^-$) with that of the covalent structure in the following form :

$$\underset{(I)}{H-H} \longleftrightarrow \underset{(II)}{H^-: H^+} \longleftrightarrow \underset{(III)}{H^+ : H^-}$$

The actual wave function of the molecule can be represented by equation (3.15) where $\lambda < 1$. Thus the true structure of $H_2$ molecule can never be represented either (I) or (II) but a mixture of (I), (II) and (III) i.e. a **resonance hybrid** of all the structures. Resonating forms of a molecule is represented by a double headed arrow ($\leftrightarrow$). This phenomena is known as **ionic-covalent resonance.** In describing a molecule in terms of covalent bonds it is possible to write more than one forms that satisfies Lewis structures. The molecule said to have resonance forms. The ionic contribution to the total wave function becomes important when the electronegativity difference between the combining atoms are high.

The actual structure of carbon monoxide and $CO_3^{2-}$ ion can be represented in the following way :

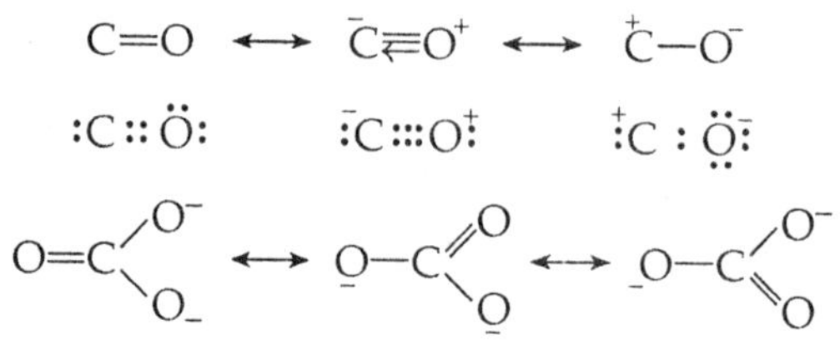

In these resonating structures only the position of $\pi$ bonds are shifted while the $\sigma$ bonding pairs are fixed within the nuclei. Each structure can be represented by a suitable wave function and the actual structure of the molecular species can be represented by a resonance hybrid which is known as **resonance or mesomerism.** Thus the actual wave function of a resonance hybrid can be represented by the linear combination of individual wave function of each structure i.e.

$$\Psi = C_1\Psi_1 + C_2\Psi_2 + C_3\Psi_3 + \ldots\ldots$$

where $\Psi_1$, $\Psi_2$, $\Psi_3$ ... etc. represent the individual wave function of each canonical forms. In the above structure of $CO_3^{2-}$ ion each structure has equal contribution to the total wave function ($\Psi$) so that we can say $C_1 = C_2 = C_3$. In the above canonical structure of $CO_3^{2-}$ only the electron cloud of $\pi$ bond is shifted from one structure to the other so that one-third $\pi$ bond order of each carbon-oxygen bond order is expected. Some of the resonating molecules are :

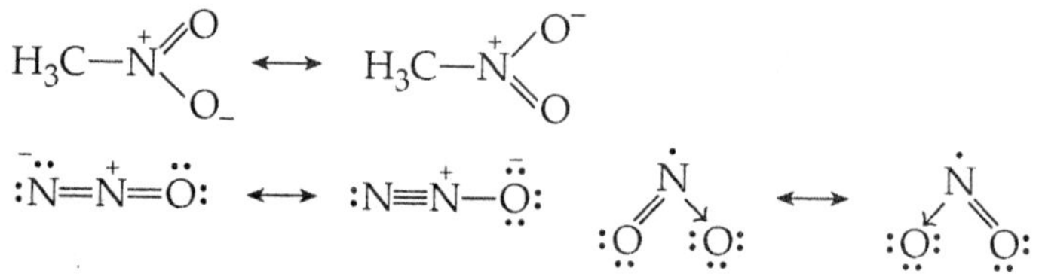

The actual energy of a resonance hybrid is always less than any of the contributing structures of the hybrid though all structures may not contribute equally to the resonance hybrid. The actual structure is thus more stable than any of the canonical forms. This extra stability is expressed as **resonance energy**.

Let $E_1(\Psi_1)$, $E_2(\Psi_2)$ and $E_3(\Psi_3)$ are the energies of the three hypothetical canonical structure for a molecule and among these $E_3$ is the lowest energy corresponding to the most stable contributing structure. If $E_0(\Psi)$ is the actual energy of the molecule which is determined experimentally then $E_0$ is always less than $E_1$, $E_2$ and $E_3$ and then the resonance energy = $E_3 - E_0$ (Fig. 3.13).

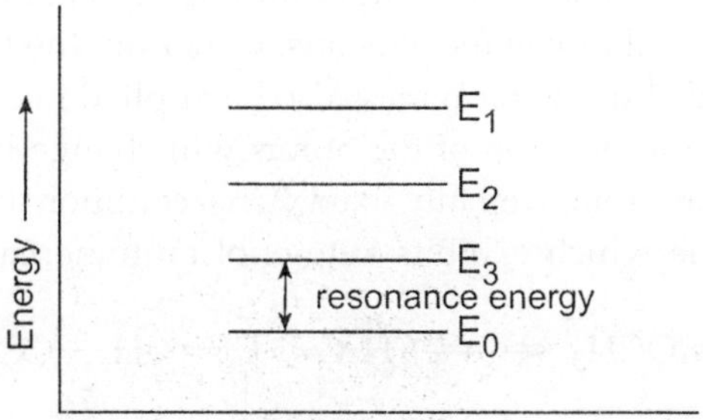


*Figure 3.13 : Representation of resonance energy of a molecule.*

Thus, resonance energy = energy of the most stable contributing structure – energy of the actual molecule = theoretically calculated energy – experimentally determined energy = observed heat of formation ~ calculated heat of formation.

*Resonance energy*

The measured heat of formation of $CO_2$ is 383 kcal mol$^{-1}$ whereas the energy for most stable structure of O=C=O is calculated theoretically as 346 kcal mol$^{-1}$, so that the resonance energy of the molecule is 37 kcal mol$^{-1}$. Thus resonance energy primarily **exchange energy** of different contributing structures, more the contributing structures more will be the value of resonance energy.

The concept of resonace may be followed by certain rules :

(a) This is not applied to different molecular species i.e. they must have the same atomic skeleton having the same position of the nuclei e.g. N—O—N and N—N—O cannot form a resonance hybrid.

(b) Concept of resonance can be applied when more than one Lewis structures can exist for a molecular species.

(c) All the resonating forms have the same number of unpaired electrons.

(d) Only the $\pi$ electron cloud is shifted from one structure to other.

(e) The canonical forms of a resonance hybrid may not contribute equally to the actual structure.

(f) Quantum mechanically the wave function ($\Psi$) for a canonical system having three resonating structures can be represented as $\Psi = C_1\Psi_1 + C_2\Psi_2 + C_3\Psi_3$ where $\Psi_1$, $\Psi_2$ and $\Psi_3$ represent the wave function for individual canonical system and $C_1$, $C_2$ and $C_3$ are called the mixing coefficients. By proper choice of $C_1$, $C_2$ and $C_3$ the energy of the system is minimum. The actual energy of a system is lower than this minimum energy and this difference in energy is known as **resonance energy**.

### 3.2.13 Resonance and Tautomerism

These terms are closely related and can be applied to a chemical compound. In case of tautomerism the different structural species of the same chemical compound is in dynamic equilibrium and the different structural species has real existance, may be separated by some means, the resonating forms of a chemical compound has no real existance and in no way they can be separated by any physical or chemical means. Generally the term tautomerism popularly known for organic compounds but the term may also be applied in case of inorganic compounds also. In case of tautomerism the position of the atoms will change in different structures but in resonance the position of the atoms remain fixed. One common ligand used in coordination chemistry is the acetylacetone which exhibits keto-enol tautomerism of the following type :

$$H_3CCOCH_2COCH_3 \rightleftharpoons H_3C-\underset{\displaystyle OH}{\underset{|}{C}}=CH-CO-CH_3$$

*Keto form* *Enol form*

Tautomerism is represented by a double headed arrow ($\rightleftharpoons$) between different structures. Resonance is a concept while tautomerism is a phenomena.

---

**Problem 3.15** 'Both $sp^3$ and $d^3s$ hybridisations may lead to tetrahedral geometry'—Give example. **[B.U. 2007]**

**Ans. :** Table 3.4.

---

**Problem 3.16** Depict the electron dot diagram of $BrO_3^-$. **[B.U. 2006]**

**Ans. :** 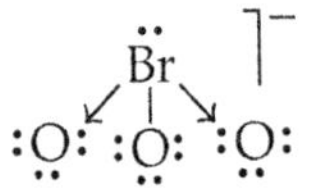


Total electrons = 26.

---

**Problem 3.17** Define resonance. Enumerate the condition necessary for this phenomena. State giving reasons whether the following pairs will show resonance or not?

(i) 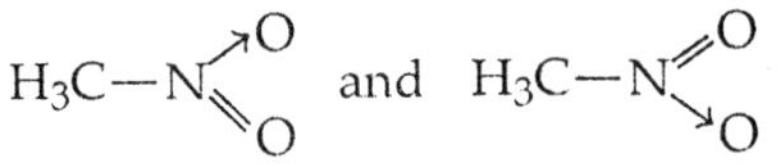


(ii) 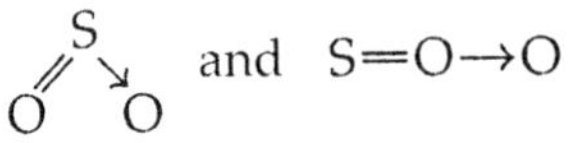


**[B.U. 2005, 2011, 2012]**

**Ans. :** If a chemical species can be represented in more than one canonical forms that are close in energy, the structures are called resonating structures and the concept is known as resonance.

**Conditions :**

(a) Molecular species must be represented by two or more Lewis structures.

(b) The position of the atoms should remain the same.

(c) The different structures (electronic) do not imply different kinds of chemcial species.

(d) The structures should have close energy.

(e) The different structures must have same number of bond pairs and lone pairs.

(i) In this case they are resonating forms as they obey all the above points.

(ii) Here the position of atoms change, so there is much difference in energy and cannot represent resonance.

## 3.3 GEOMETRY OF COVALENT MOLECULES

The concept of hybridisation is extremely useful in predicting the shapes of molecular species, the basic approach is the overlap of atomic orbitals of suitable symmetry. Let us consider the formation of $NH_3$ and $H_2O$ molecule. In both the molecules the central nitrogen and oxygen is $sp^3$ hybridised.

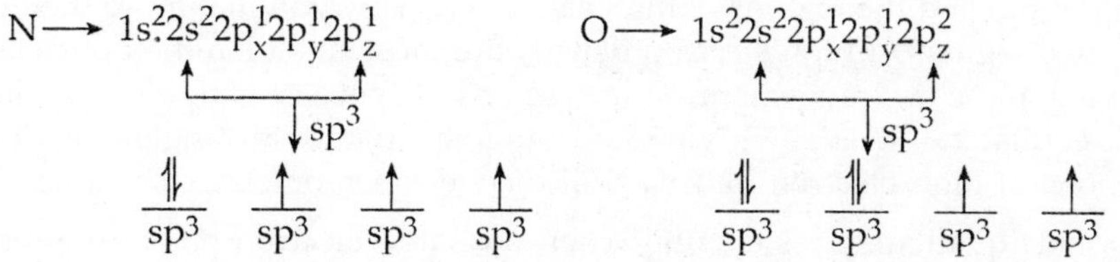


In $NH_3$, the three hydrogen atoms each having single unpaired electron in 1s orbital overlap with three $sp^3$ hybrid orbitals of nitrogen to form three N—H bonds around nitrogen and one $sp^3$ hybrid orbital houses a lone pair. Due to the presence of one lone pair the ideal bond angle for $sp^3$ hybridisation is deviated and the ∠HNH in $NH_3$ is 107° (Fig. 3.14). Similarly in $H_2O$, the two $sp^3$ hybrid orbitals of oxygen each having single unpaired electron overlap with two 1s electrons of two hydrogen atoms to form two O—H bonds and the two $sp^3$ hybrid orbitals houses two lone pairs. Presence of two lone pairs deviate the bond angle of HOH from 109° to 104° (Fig. 3.14). Since lone pair of electrons are expected to assume more space around the central atom, there will be a decrease in bond angle from the ideal geometry of the molecules as shown in Table 3.4. Bond angles for p block elements with unshared electron pairs are given in Table 3.5. In all cases bond angles are reduced for the formation of four bonds (coordination number 4).

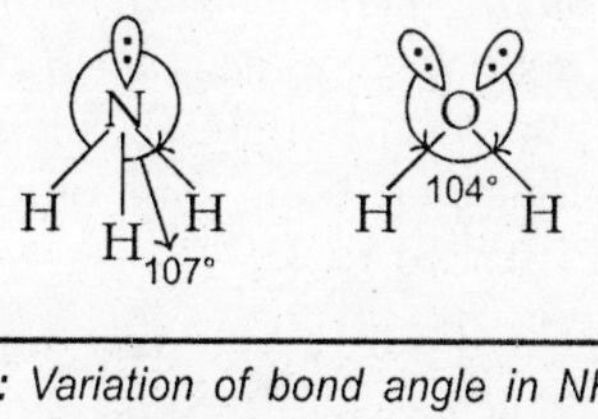


***Figure 3.14** : Variation of bond angle in $NH_3$ and $H_2O$.*

**Table 3.5 : Bond angles of molecules having lone pair of electrons for p block elements**

| Molecule : | $NH_3$ | $PH_3$ | $AsH_3$ | $SbH_3$ | $PCl_3$ | $AsCl_3$ | $SbCl_3$ | $NF_3$ | $PF_3$ | $AsF_3$ | $SbF_3$ | $H_2O$ | $H_2S$ | $H_2Se$ | $H_2Te$ |
|---|---|---|---|---|---|---|---|---|---|---|---|---|---|---|---|
| Bond angle : | 107° | 93° | 92° | 91° | 100° | 98° | 99° | 102° | 98° | 96° | 88° | 104° | 92° | 91° | 89.5° |

Therefore hybrid orbital model needs further refinement to explain the geometry of a molecular species.

### 3.3.1 Valence Shell Electron Pair Repulsion (VSEPR) Model

This model is a supplement to the valence bond theory of covalent bonding. This theory developed by R. J. Gillespie and R.S. Nyholm in 1957. This model based on the fact that the shapes of molecules or ions are determined by the repulsion of valence shell pair of electrons i.e. by the electrostatic repulsions of the negatively charged covalent bond electron pairs and lone pairs of hybridised valence electrons about the central atom. **Valence shell pair of electrons means the total pair of electrons in the valence shell of central atom i.e. the total number of electrons in the valence shell plus the number of electrons gained by the central atom.**

Valence shell electron pairs are arranged around the central atom in such a way that the repulsion will be minimum and the regions of high electron density will maximise their distance apart. For example, two regions of high electron density are most stable on the opposite side of the central atom giving a linear arrangement. Three regions of space of high electron density at the corners of an equilateral triangle giving a triangular planar arrangement. Thus the arrangement of regions of high electron density will dictate the geometry of a molecule.

VSEPR theory yields qualitative results and when unshared electron pairs are present this model predict slight deviation of bond angle from ideal geometry of Fig. 3.8. The deviation of bond angle with different coordination number/number of bonds with different geometries having different number of unshared pair of electrons are depicted in Fig. 3.15. It is evident that unshared electron pairs occupy more space about the central atom that an electron pair shared with an outer atom. It is to be noted from Fig. 3.15 that when the lone pair repulsions are balanced on all sides of a given bond pairs as in the case of $XeF_2$, an undistorted bond angle is retained.

In VSEPR theory it is assumed that inner electrons of the central element will not take part in valence shell repulsion and this theory provide the idea of geometries of molecules of non-transition elements as also symmetrical transition metals ($d^0$, $d^5$ and $d^{10}$ systems) having no crystal field stabilisation energy [Sec. 7.14.2(B)]. The refinement of bond angles due to lone pair (unshared pair)—bond pair repulsion is governed by certain rules :

**Rule I :** *Molecular species having no lone pair (l.p.) of electrons on the central atom will not exert repulsive force and the number of σ bonding pairs will determine the ideal geometry of the concerned species (Table 3.4). π bonding pair will not influence the geometry of the system.* In $CO_2$ there is no *l*.p. so ideal linear geometry having regular bond angle 180° is retained but in case of $SO_2$ there is one *l*.p. on central S and so due to lone pair-bond pair repulsion there will be slight deviation of bond angle from 120° ($sp^2$ hybridised S) to 119°.

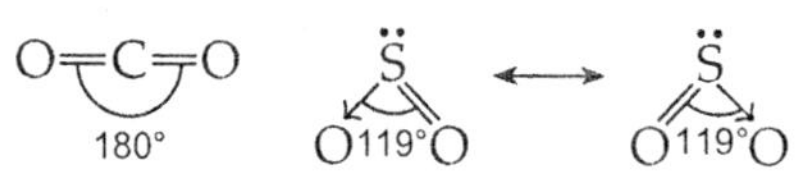

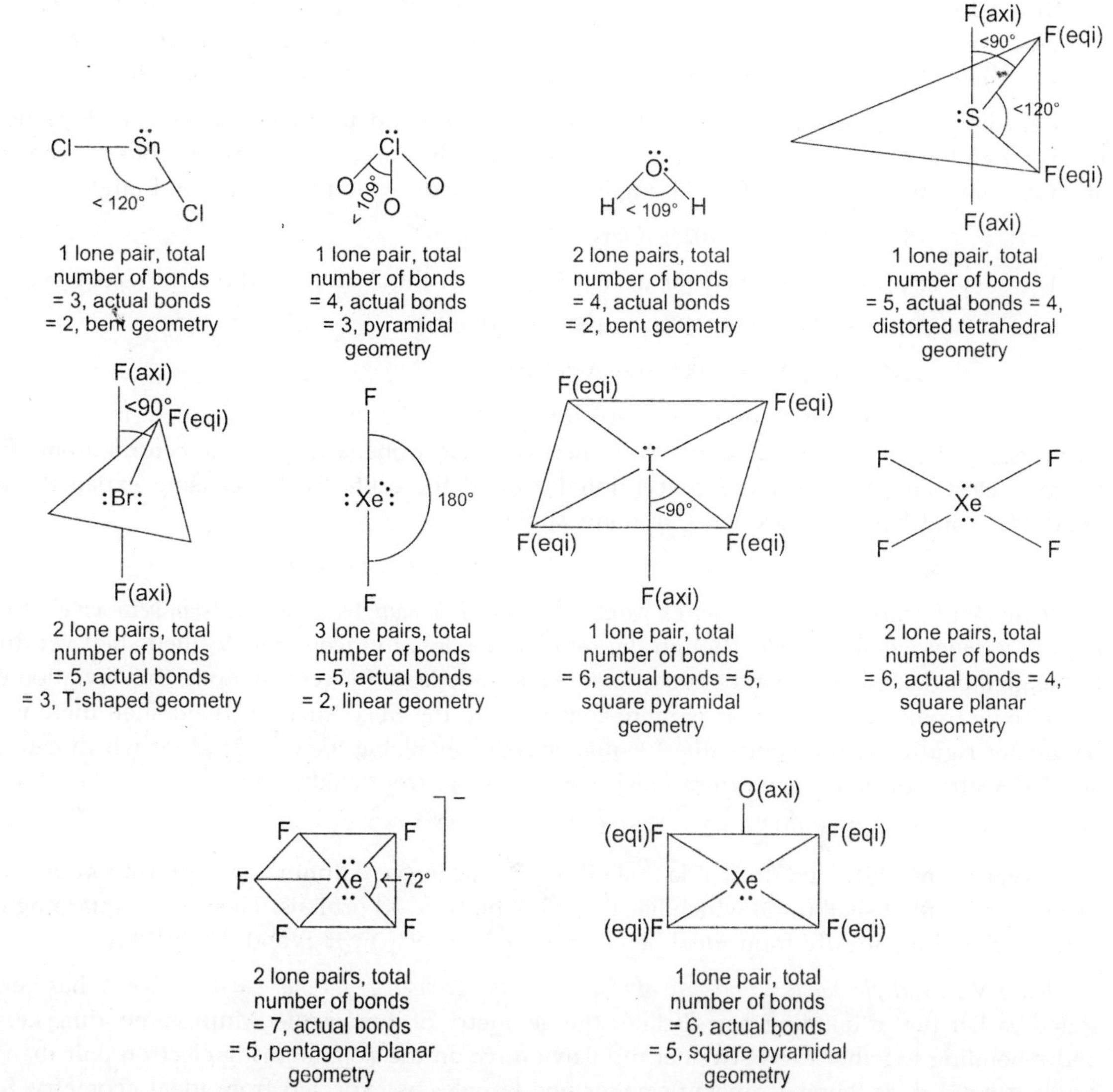


***Figure 3.15** : Distortions of bond angle due to lone pair of electrons in different geometries.*
*axi = axial, eqi = equatorial*

**Rule II** : *Lone pair of electrons repel the neighbouring electron pairs more strongly than bond pair of electrons* and the sequence of repulsion is :

lone pair (*l*.p.) – lone pair (*l*.p.) > lone pair (*l*.p.) – bond pair (b.p.) > bond pair (b.p.) – bond pair (b.p.)

This can be illustrated by the variation of bond angle $CH_4$ (109°), $NH_3$ (107°) and $H_2O$ (104°). In case of $CH_4$ there is no *l*.p., only b.p. – b.p. repulsion which is minimum. In $NH_3$ there is one *l*.p. and the *l*.p. – b.p. repulsion reduced the bond angle from 109° to 107°. More reduction of bond angle is observed in $H_2O$ where *l*.p. – *l*.p. repulsion exist which is stronger than *l*.p. – b.p. repulsion.

**Rule III :** *Repulsion exerted by the bond pair (b.p.) decreases as the electronegativity of the B atom increases in a $AB_n$ type of species where A is the central atom and n is the number of atoms attached to A. This can be stated as BAB bond angle decreases with increasing electronegativity of B.*

With increasing electronegativity of the B atom the bond pair electron density is pulled further from the central atom A and the charge cloud is displaced further away from the central atom causing more decrease of bond angles as illustrated in the following bond angles :

$H_3N$ 107°, $NF_3$ 102°; $OH_2$ 104°, $OF_2$ 103°

The same effect also observed for decreasing size of the B atom of the same group e.g.

$PF_3$ 98°, $PCl_3$ 100°, $PBr_3$ 101.5° and $PI_3$ 102°

$AsCl_3$ 98°, $AsBr_3$ 100° and $AsI_3$ 101°

The sequence of bond angle as depicted in Table 3.5 can be explained in terms of electronegativity of the central atom. With increasing electronegativity of the central atom, the bonding electron pairs are more concentrated around the central atom causing expansion of bond angle and hence the sequence of bond angle :

$H_2O > H_2S > H_2Se > H_2Te$; $NH_3 > PH_3 > AsH_3 > SbH_3$

**Rule IV :** *Repulsion between electron pairs in filled shell is more than the repulsion between electron pairs in incompletely filled shell.* Though the repulsion is less in incompletely filled shell yet due to the greater flexibility of bonds due to more space available for central atom, the deviation of bond angle from ideal behaviour is more compared to the filled shell. In filled shell there will be greater rigidity of the bonds due to smaller space available for central atom which causes small deviation of bond angle from ideal behaviour e.g. from Table 3.5 :

$H_2O$ 104°, $H_2S$ 92° $NH_3$ 107° $PH_3$ 93°

Oxygen and nitrogen has filled shell (n=2) whereas sulphur and phosphorous has incompletely filled shell (n=3) which has the availability of 3d orbitals. This causes quashing of bond angles dramatically from ideal behaviour (109°) to 92° in $H_2S$ and 93° in $PH_3$.

**Rule V :** *Multiple bonds do not grossly influence the geometry of a molecular species.* It has been stated earlier that σ bonding pair dictates the geometry of a molecule. Multiple bonding i.e. σ and π bonding together exert stronger repulsive force on the neighbouring electron pair than a single σ bond does thereby causing smaller bond angles as expected from ideal geometry, for example,

| | | | |
|---|---|---|---|
| $O = PF_3$ | $\angle FPF = 102.5°$ | $O = CF_2$ | $\angle FCF = 108°$ |
| $O = PCl_3$ | $\angle ClPCl = 103.5°$ | $O = CCl_2$ | $\angle ClCCl = 111°$ |
| $O = PBr_3$ | $\angle BrPBr = 108°$ | $O = C(CH_3)_2$ | $\angle CH_3CCH_3 = 119.6°$ |

In the case of $OPX_3$ molecules having tetrahedral geometry the ideal bond angle is 109° whereas in case of $OCX_2$ molecules the expected bond angle is 120° having triangular planar geometry. In both the cases large deviation occurs in case of fluorine. Due to the highest electronegativity of fluorine which pulls the bond pair of electrons farthest from phosphorous and carbon and thereby bond pair - bond pair repulsion minimises whereby repulsion of the double bond makes the ∠FPF and ∠FCF least.

The shapes of some typical molecules according to VSEPR model are given in Table 3.6.

**Table 3.6 : Geometry of some molecules having lone pair of electrons**

| Type of molecules | Hybridisation and distribution of orbitals | Types of electron pairs | Actual shape | Example |
|---|---|---|---|---|
| $AB_2$ | $sp^2$, triangular planar | 2 b.p. + 1 *l*.p. | V-shaped (bent) | $SnCl_2$, $PbCl_2$, $SO_2$ |
| | $sp^3$, tetrahedral | 2 b.p. + 2 *l*.p. | V-shaped (bent) | $H_2O$, $F_2O$, $Cl_2O$, $NO_2^-$, $XO_2^-$, (X=Cl, Br, I) |
| | $sp^3d$, trigonal bipyramid | 2 b.p. + 3 *l*.p. | Linear | $XeF_2$, $ICl_2^-$, $I_3^-$ |
| $AB_3$ | $sp^3$, tetrahedral | 3 b.p. + 1 *l*.p. | Trigonal pyramid | $NH_3$, $NF_3$, $SO_3^{2-}$, $H_3O^+$ $PX_3$, $XeO_3$, $XO_3^-$, (X=Cl, Br, I) |
| | $sp^3d$, trigonal bipyramid | 3 b.p. + 2 *l*.p. | T-shaped | $ClF_3$, $BrF_3$ |
| $AB_4$ | $sp^3d$, trigonal bipyramid | 4 b.p. + 1 *l*.p. | Distorted tetrahedral | $TeCl_4$, $TeBr_4$, $SF_4$, $XeO_2F_2$ |
| | $sp^3d^2$, octahedral | 4 b.p. + 2 *l*.p. | Square planar | $XeF_4$, $ICl_4^-$, $BrF_4^-$ |
| $AB_5$ | $sp^3d^2$, octahedral | 5 b.p. + 1 *l*.p. | Square pyramid | $IF_5$, $BrF_5$, $XeOF_4$ |
| | $sp^3d^3$, pentagonal bipyramid | 5 b.p. + 2 *l*.p. | Pentagonal planar | $XeF_5^-$ |

### 3.3.2 Limitations of VSEPR model

(a) This model cannot provide quantitative result of bond angles.

(b) This theory is not applicable in case of molecules having **fluxional behaviour** i.e. for **stereochemically non rigidity character** of a molecule like $XeF_6$. *Those molecules that have more than one equivalent minimum-energy conformation and that are readily able to change from one conformation to other,* are **called fluxional molecules.**

(c) It fails to explain the geometry of transition metal complexes where the crystal field stabilisation energy (CFSE) plays a pivotal role in structure determination.

---

**Problem 3.18** What is the evidence that oxygen in water is $sp^3$ hybridised?

**Ans. :** Structure determination of the $H_2O$ molecule shows that the molecule is angular having HOH bond angle 104°. The ground state electronic configuration of oxygen has two unpaired electrons in 2p orbitals. If these 2p orbitals overlap with two hydrogen atoms to form two O–H bonds the bond angle would be expected to be 90° which is far deviated from the measured bond angle. The actual bond angle is more closer to the tetrahedral bond angle (109°28') which support that the oxygen in water is $sp^3$ hybridised. Moreover this hybridisation explain the angular nature of $H_2O$ molecule.

---

**Problem 3.19** Explain the bond angle of $OH_2$ and $OF_2$. **[B.U. 2007]**

**Ans. :** According to VSEPR rule (Rule III), fluorine is more electronegative than hydrogen, so bonding electron pair will be more farther from the central atom causing greater deviation of bond angle from ideal behaviour.

**Problem 3.20** Write down the rules of VSEPR. '$NH_3$, $BCl_3$ and $BrF_3$ have comparable molecular formula but their shapes are different'—Explain. **[B.U. 2007]**

**Ans. :** For rules—Sec. 3.3.1. In $NH_3$, central nitrogen is $sp^3$ hybridised having one lone pair and three bond pairs on nitrogen. Due to *l*.p.-b.p. repulsion ∠HNH is 107° and the geometry is trigonal pyramid.

In $BCl_3$, the central B is $sp^2$ hybridised having 3 b.p. of electrons so that regular geometry is retained (triangular planar) having ∠ClBCl = 120°.

The central Br in $BrF_3$ is $sp^3d$ hybridised having 3 b.p. + 2 *l*.p. The two lone pairs and one fluorine occupy equatorial positions while two other fluorines occupy axial positions. Thus $BrF_3$ is T-shaped.

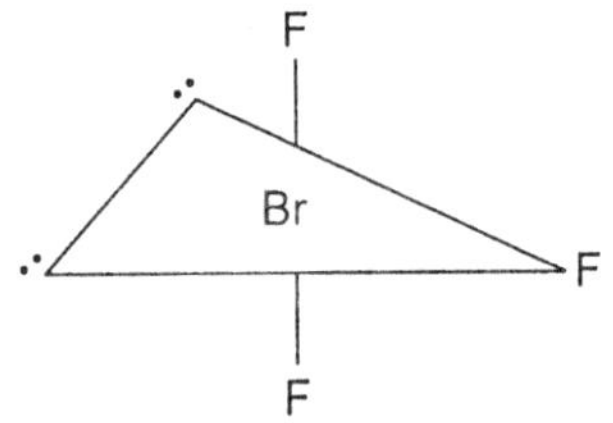


VSEPR theory indicates 5 pairs of electron (7+1+1+1 = 10). As there are two lone pairs the molecule is also called as Ψ-trigonal bipyramidal geometry.

**Problem 3.21** Arrange the following in the order of decreasing bond angles : $OF_2$, $OCl_2$ and $ON_2$. **[B.U. 2005]**

**Ans. :** Bond angles :

| $ON_2$ | > $OCl_2$ | > $OF_2$ |
|---|---|---|
| 180° | 111° | 103° |
| linear, sp hybridised nitrogen | V-shaped (bent) | |
| | $sp^3$ hybridised oxygen | |

In $ON_2$ there is no lone pair so regular geometry is retained for sp hybridisation. In $OCl_2$ and $OF_2$ the central oxygen has 2 *l*.p. + 2 b.p. Since fluorine is more electronegative than chlorine, the bond pairs of $F_2O$ have lesser electron density in the vicinity of oxygen than the bond pairs of $Cl_2O$ and hence ∠ClOCl > ∠FOF. This can also be due to the bulkier size of chlorine than that of fluorine.

**Problem 3.22** Note the type of hybridisation of the core element in $POCl_3$. **[B.U. 2006]**

**Ans. :** $P \longrightarrow 1s^2 2s^2 2p^6 3s^2 3p_x^1 3p_y^1 3p_z^1$

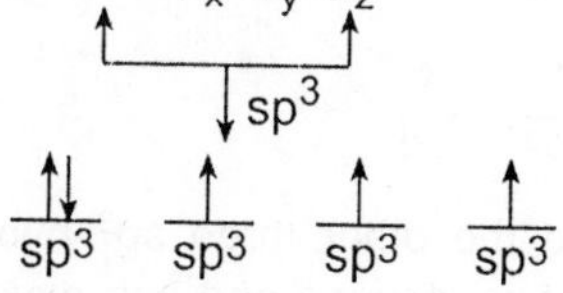


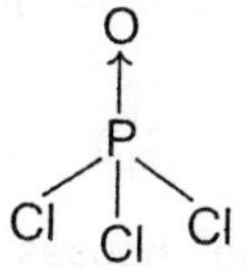


*or*

$P \longrightarrow 1s^2 2s^2 2p^6 3s^1 3p_x^1 3p_y^1 3p_z^1 3d^1$

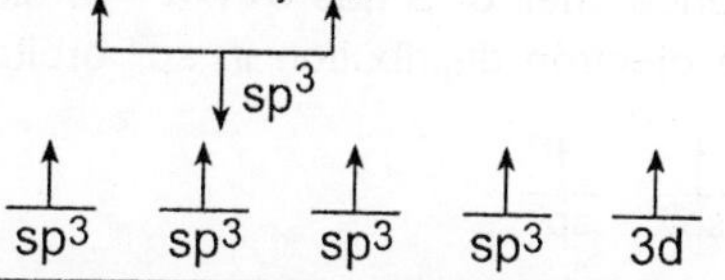


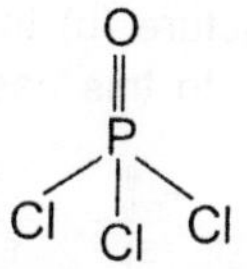


**Problem 3.23** Draw the most stable structure of $BrF_3$. **[B.U. 2006]**

**Ans. :** Central Br has 7+1+1+1 = 10 electrons having 5 electron pairs of which 3 b.p. + 2 *l*.p. A trigonal bipyramidal geometry having $sp^3d$ hybridisation is expected for central bromine. The two lone pairs and one fluorine occupy the equatorial sites since lone pairs are fatter than bond pairs and two fluorine atoms occupy axial positions.

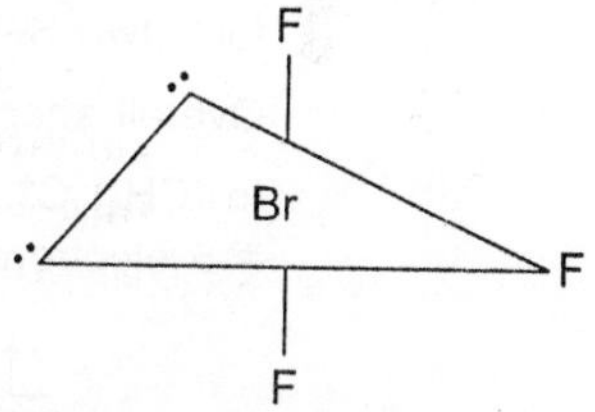


**Problem 3.24** How are $XeF_2$, $XeF_4$ and $XeF_6$ related structurally? **[B.U. 2006]**

**Ans. :**

| $XeF_2$ | $XeF_4$ | $XeF_6$ |
|---|---|---|
| Xe → 8 electrons | Xe → 8 electrons | Xe → 8 electrons |
| 2F → 2 " | 4F → 4 " | 6F → 6 " |
| 10 electrons | 12 electrons | 14 electrons |
| = 3 *l*.p. + 2 b.p. | = 4 b.p. + 2 *l*.p. | = 6 b.p. + 1 *l*.p. |
| Linear geometry, central Xe is $sp^3d$ hybridised having trigonal bipyramidal stucture | Square planar geometry derived from octahedral geometry having $sp^3d^2$ hybridisation of central Xe | Pentagonal bipyramidal geometry, one position is ocupied by lone pair |

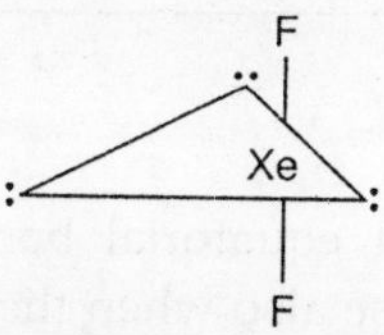


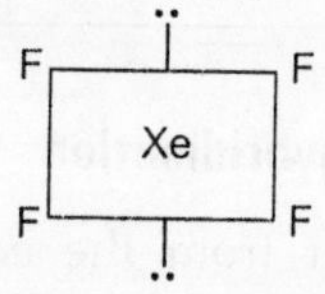


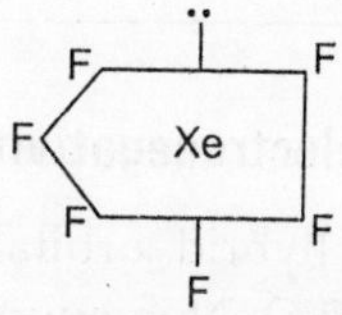


**Problem 3.25** Applying VSEPR theory predict the structures of $SOCl_2$ and $(CH_3)_3C^+$. **[B.U. 2008]**

**Ans. :** In $SOCl_2$ sulphur serves as the central element because which is less electronegative than oxygen and chlorine.

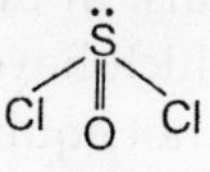


(a)

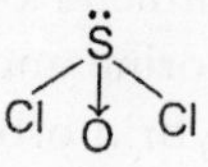


(b)

In structure(a) central S has 6+1+1+2 = 10 electrons i.e. five pairs of which one pair makes the $\pi$ bond which does not influence the shape of the molecule. In this case central S is roughly $sp^3$ hybridised.

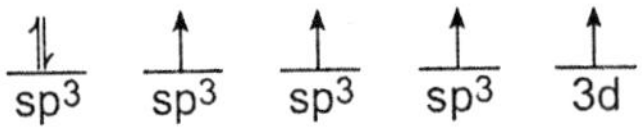


One $sp^3$ houses lone pair and the other three $sp^3$ hybrid orbitals each having single unpaired electron overlap with two chlorine and one oxygen to form three $\sigma$ bonds. One unpaired electron in 3d orbital overlap with one unpaired electron of oxygen in 2p orbital of oxygen to form a $\pi$ bond $[3d_{\pi(S)} - 2p_{\pi(O)}]$.

In structure (b) the valence shell of S has 6+1+1 = 8 electrons having 4 pairs (3 b.p. + 1 *l*.p.). In this case the electron distribution in $sp^3$ orbital of central S is—

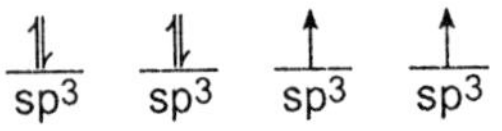


One $sp^3$ houses lone pair, one $sp^3$ forms a coordinate link with oxygen and the other two $sp^3$ hybrid orbitals having single unpaired electron overlap with two chlorine atoms to form two S–Cl $\sigma$ bonds.

Overall shape of the molecule is pyramidal. $\angle OSCl = 106°$ and $\angle ClSCl = 114°$.

In $(CH_3)_3C^+$, the central C has valence shell electrons = 4+3–1 (due to positive charge) = 6 electrons = 3 electron pairs (b.p.). The central carbon ($C^+$) is $sp^2$ hybridised.

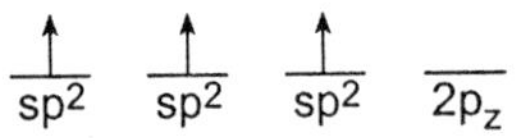


The $sp^2$ hybrid orbitals each having single unpaired electron overlap with suitable orbital of each methyl radical to form three bonds. Three bond pairs will assume maximum stability having triangular planar geometry with bond angle 120°.

---

**Problem 3.26** Using VSEPR theory predict the shapes of $XeO_2F_2$ and $BrF_4^-$. Indicate the state of hybridisation of the central atom in each case. **[C.U. 2005]**

**Ans.** : See Table 3.6. 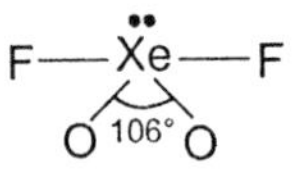

4 b.p. + 1 *l*.p.

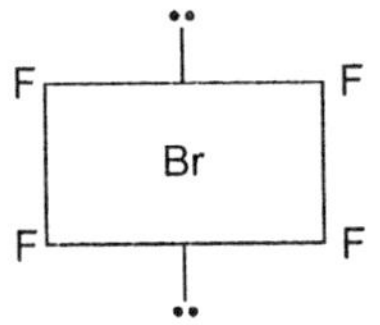

4 b.p. + 2 *l*.p.

---

### 3.3.3 Bent's rule–electronegativity and hybridisation

Non-equivalent hybrid orbitals result from the axial and equatorial bonds around the central atom (Sec. 3.2.6). Non equivalence of the bonds may arise also when the central atom is joined to different elements having different electronegativity e.g. $CHCl_3$ and $CH_3Cl$. In $CH_3Cl$ the $\angle H—C—H = 110°$ and is different from $\angle Cl—C—H$. In case of $CHCl_3$ the $\angle Cl—C—Cl = 110°$ which has a difference of bond angle of $\angle Cl—C—H$. This small variation of bond angles can be explained on the basis of the s and p character of hybrid orbitals. In the molecules $CHCl_3$ and $CH_3Cl$, the central carbon is joined to chlorine and hydrogen which have appreciable difference in electronegativity and thereby all the hybrid orbitals ($sp^3$) are not equivalent. Bent's rule gives us an idea why in the case of **isovalent hybridisation** all the hybrid orbitals are not equivalent. Bent's concept can be stated as :

(a) When an electronegative atom bound to the central atom, this will pull more electron density from the central atom.

(b) The greater the p character of the hybrid orbital less will be the electronegativity. Conversely the hybrid orbital having more s character is more electronegative.

(c) *The more electronegative substituents prefer those hybrid orbitals which have less s character while less electronegative or more electropositive substituents prefer hybrid orbitals with more s character.*

(d) The central atom will direct less p character and greater s character, into the hybrids directed towards less electronegative substituents.

The Bent's rule gives us some idea about the refinements of bond angle over the VSEPR theory. The illustration of Bent's rule can be made in the case of non-equivalent hybrid e.g. $PCl_5$ where the central P is $sp^3d$ hybridised. This hybridisation is a combination of $sp^2$ and pd hybridisation. The $sp^2$ forms trigonal planar equatorial bonds whereas pd forms perpendicular axial bonds. In case of mixed halides such as $PCl_3F_2$, the fluorine is more electronegative than chlorine. Therefore fluorine will occupy those hybrid orbital which has less s character i.e. pd hybrid orbital so that two fluorine atoms will occupy in the axial positions. As chlorine is less electronegative than fluorine so three chlorine atoms will occupy the hybrid orbitals which has more s character i.e. $sp^2$ hybrid orbitals so that three chlorine atoms will be in equatorial directions :

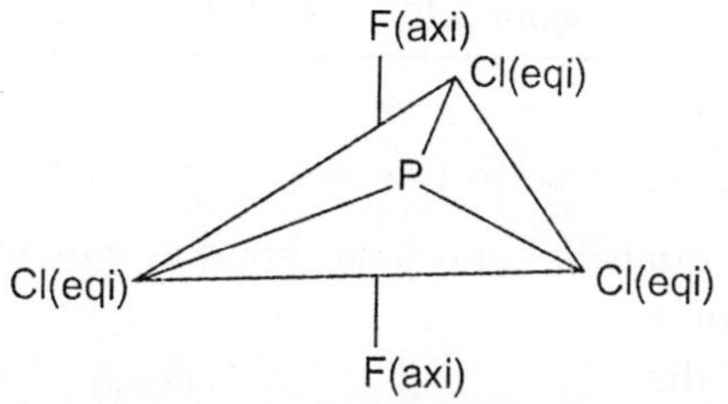


∠FPF = 180° whereas ∠ClPCl = 120°. The variation of s and p character of different s-p type hybrid orbitals are given below :

| Hybridisation | Percentage character | | Bond angle |
|---|---|---|---|
| | s | p | |
| sp | 50 | 50 | 180° |
| $sp^2$ | $33^1/_3$ | $66^2/_3$ | 120° |
| $sp^3$ | 25 | 75 | 109°28′ |
| $p^2$ | 0 | 100 | 90° |

The greater the s character greater is the bond angle while greater p character indicated smaller bond angle.

The variation of bond angle in $NH_3$ and $H_2O$ can be explained by this rule assuming that oxygen and nitrogen atom make use of s-p type hybrid orbitals. In $NH_3$ and $H_2O$ the ∠H—N—H and ∠H—O—H are 107° and 104° respectively. This difference in bond angle is due to the non-uniform distribution of s and p character of the bonding and non-bonding hybrid orbitals. In both the cases the central nitrogen and oxygen is $sp^3$ hybridised so that the ideal bond angle should be 109°28′. In $H_2O$ there is a reduction of about 5° angle whereas in $NH_3$ this angle is reduced to about 2°. The percentage of s character calculated in two adjacent O—H bonds is ~20% while in $NH_3$, the adjacent three N—H bonds has ~22% s character. Thus in $NH_3$ the bond angle will be higher than that of $H_2O$. Oxygen is more electronegative than nitrogen and so oxygen will have to combine with hybrid orbital having less s character while nitrogen will have a tendency to combine with orbital having more s character. s orbital tend to attract less electronegative substituents or more electronegative substituents tend to get towards those orbitals which have lesser tendency to withdraw electrons. This tendency of more electronegative substituents to seek out orbitals with less s character is called **apicophilicity**.

In case of multiple bonding e.g. $OSF_4$, Bent's rule may be applied with reference to $SF_4$. In $SF_4$ according to Table 3.6, there is 4 b.p. + 1 *l*.p. having distorted tetrahedral geometry. Here the *l*.p. will occupy the equatorial site (Fig. 3.16) where central S is $sp^3d$ hybridised. The axial ∠FSF = 173° which is 7° reduced from the ideal angle 180°. The equatorial ∠FSF = 101° which is about 19° reduced from the ideal angle 120° which is expected by the *l*.p. and b.p. repulsion according to VSEPR theory.

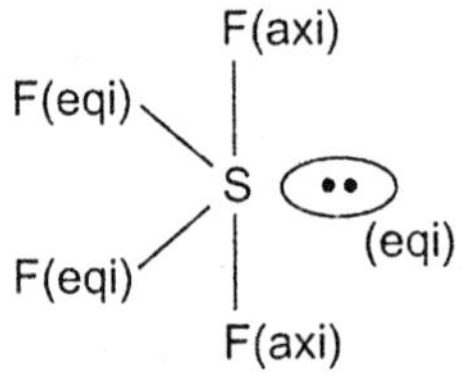


***Figure 3.16** : Distorted tetrahedral structure of $SF_4$.*

In $OSF_4$, the axial ∠FSF = 164° (Fig. 3.17) which is just 16° less than the ideal 180° angle while the equatorial ∠FSF = 115° which is quite close to the ideal angle 120°.

Bent's rule can explain the variation of equatorial and axial bond angles in $SF_4$ and $OSF_4$. The lone pair in $SF_4$ is under the control of a single nucleus which is considered to have the effective electronegativity zero. So it is more likely that this lone pair will attract the more s character than the equatorial bond pairs. Therefore in equatorial bond pairs, the p character will be more than that predicted by $sp^2$ hybridisation, thereby we expect greater decrease of equatorial bond angle (120° to 101°). In case of $OSF_4$ due to the formation of S=O, the axial bond angle decreases about 16° from the ideal bond angle which is quite plausable that the double bond will exert greater repulsive force. But the equatorial bond angle is quite close to 120° which is expected from Bent's rule. This is considered as the substitution of the lone pair in $SF_4$ by double bonded oxygen in $OSF_4$. As oxygen is more electronegative than lone pair, so p character will predominate than s character so that equatorial bonds will be more closer to $sp^2$ in $OSF_4$ than $SF_4$. Hence the equatorial bond angles are quite close to the ideal bond angle for $sp^2$ hybridisation.

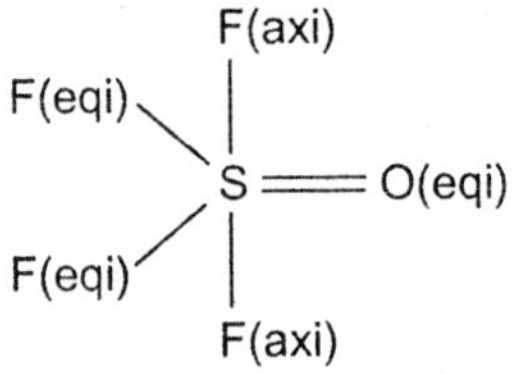


***Figure 3.17** : Structure of $OSF_4$.*

The variation of equatorial and axial bond angles in several cases can be explained by the application of Bent's rule e.g. in $ClF_3$ two lone pairs and one F occupy equatorial positions while two other fluorines occupy axial positions. The ∠FClF is slightly lower than ideal 90°. In $OClF_3$ the equatorial ∠OClF = 109° which is shorter than expected 120°.

In conclusion a statement of the hybridisation of the central element of some of the molecular species and their structures given in Table 3.7.

**Table 3.7 : Hybridisation of the central element and the structure of some molecular species**

| Molecular species | Hybridisation of the central element | Structure |
|---|---|---|
| $BX_3$(X=F, Cl) | B, $sp^2$ | 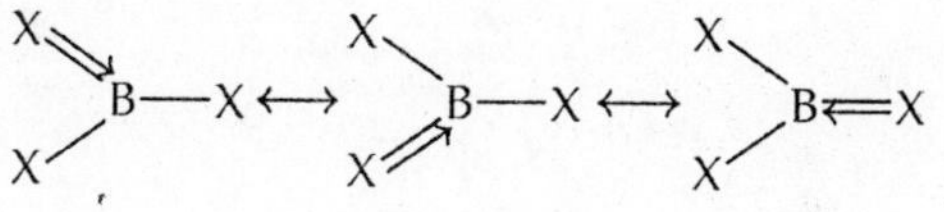 |
| $BeCl_2$ | Be, sp | Cl—Be—Cl |
| $B_2H_6$ | B, $sp^3$ | 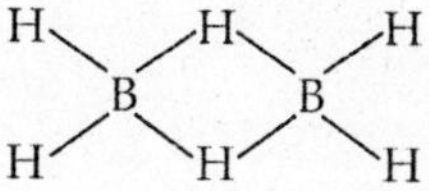 |
| $AlH_4^-$ | Al, $sp^3$ | 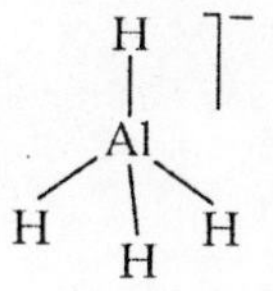 |
| $BF_4^-$ | B, $sp^3$ | 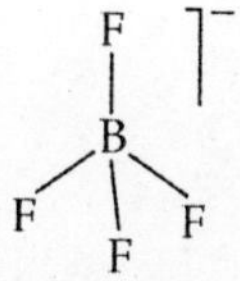 |
| $B_3N_3H_6$ (Inorganic benzene) | B and N, $sp^2$ | 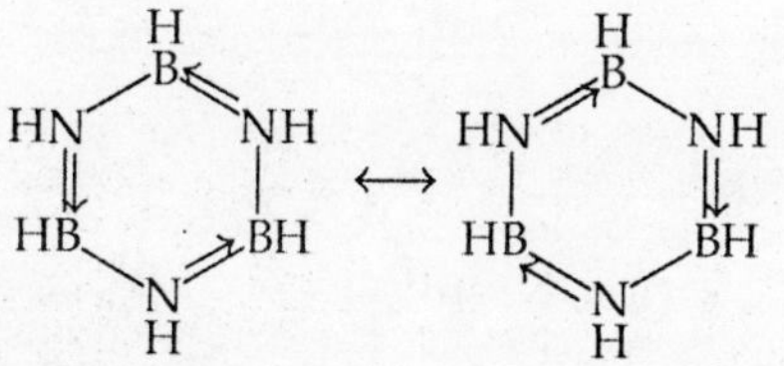 |
| BN (Inorganic graphite) | B and N, $sp^2$ | Graphite like structure |
| CO | C, sp | $:C \equiv O:$ |
| $CO_2$ | C, sp | O=C=O |

| Molecular species | Hybridisation of the central element | Structure |
|---|---|---|
| $(CN)_2$ | C and N, sp | $N\equiv C—C\equiv N$ |
| $[SbCl_6]^{3-}$ | Sb, $sp^3d^2$ | 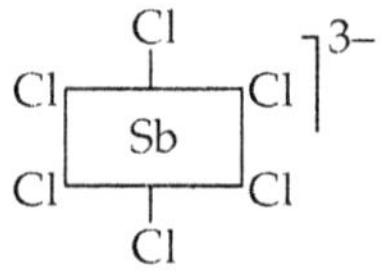 |
| $[PCl_6]^-$ | P, $sp^3d^2$ | 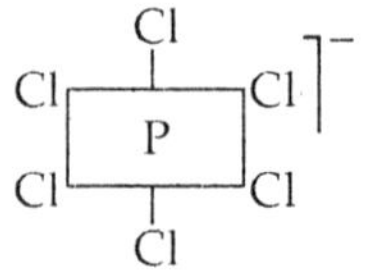 |
| $NCl_3$ | N, $sp^3$ | 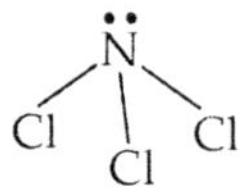 |
| $[AsF_6]^-$ | As, $sp^3d^2$ | 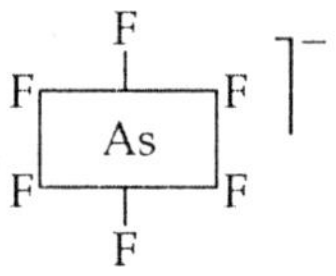 |
| $N_2$ | N, sp | $:N\equiv N:$ |
| $P_4$ | P, roughly $sp^3$ | 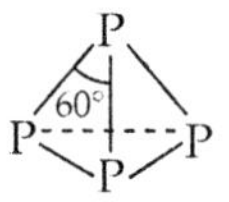 |
| $O_2$ | O, sp | $:\ddot{O}=\ddot{O}:$ |
| $N_2H_4$ | N, $sp^3$ | 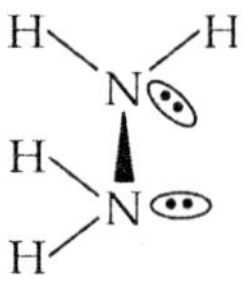 |
| $NH_2OH$ | N, $sp^3$ | 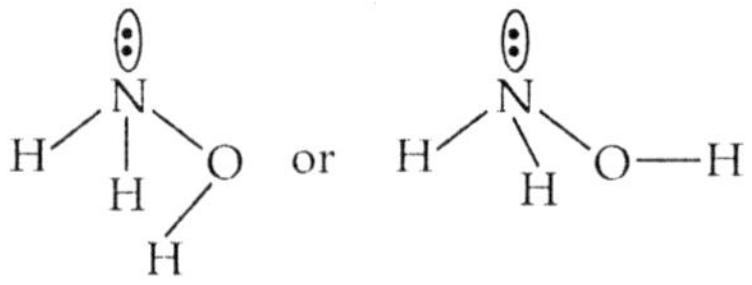 |
| $N_2O$ | N, sp | $:\overset{-}{\ddot{N}}=\overset{+}{\underset{\underset{sp}{\uparrow}}{N}}=\ddot{O}: \longleftrightarrow :N\equiv\overset{+}{\underset{\underset{sp}{\uparrow}}{N}}—\overset{-}{\underset{\cdot\cdot}{\ddot{O}}}:$ |
| NO | N, sp | $:\dot{N}=\ddot{O}: \longleftrightarrow :\overset{-}{\ddot{N}}=\overset{+}{\dot{O}}:$ |

| Molecular species | Hybridisation of the central element | Structure |
|---|---|---|
| $N_2O_3$ | N, $sp^2$ | 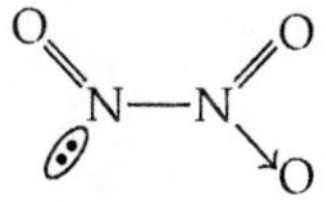 |
| $NO_2$ | N, $sp^2$ | 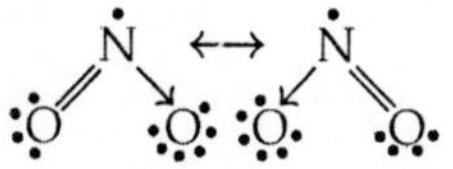 |
| $N_2O_5$ | Both N is $sp^2$, O is sp | 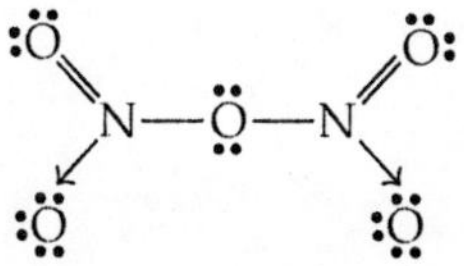 |
| $HNO_3$ | N, $sp^2$ | 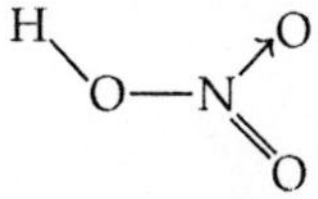 |
| $HNO_2$ | N, $sp^2$ | 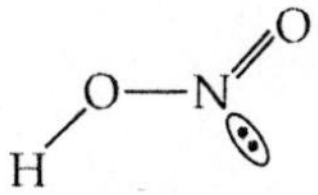 |
| $NO_3^-$ | N, $sp^2$ | 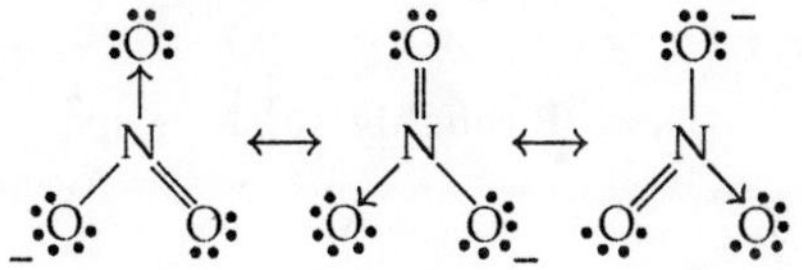 |
| $NO_2^-$ | N, $sp^2$ | 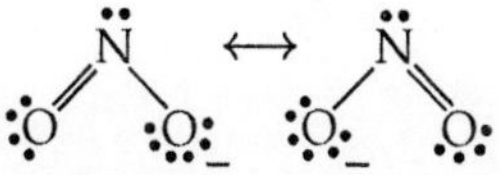 |
| $P_2O_3$ or $P_4O_6$ | P, roughly $sp^3$ | 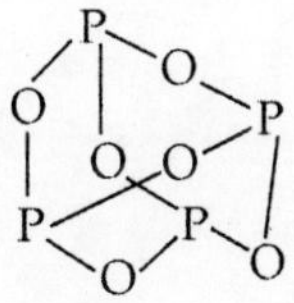 |
| $P_2O_4$ or $P_4O_8$ | P, roughly $sp^3$ | 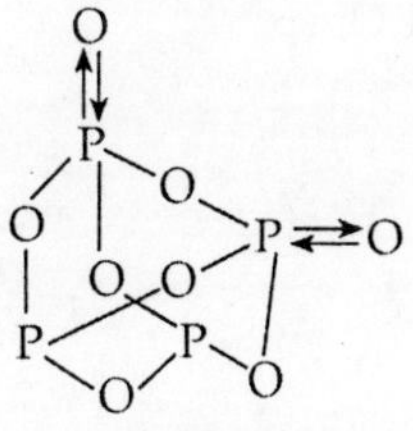 |

| Molecular species | Hybridisation of the central element | Structure |
|---|---|---|
| $P_2O_5$ or $P_4O_{10}$ | P, roughly $sp^3$ | 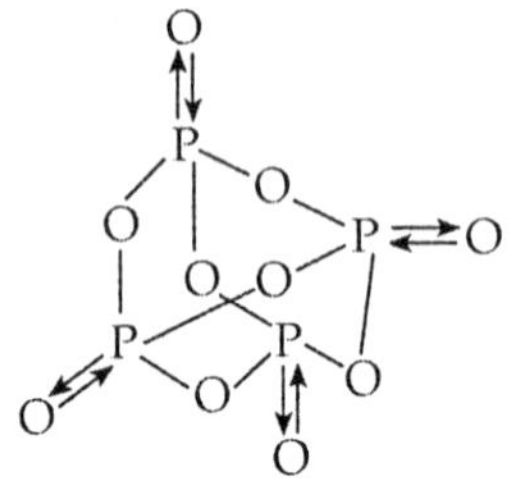 |
| $H_3PO_2$ | P, $sp^3$ | 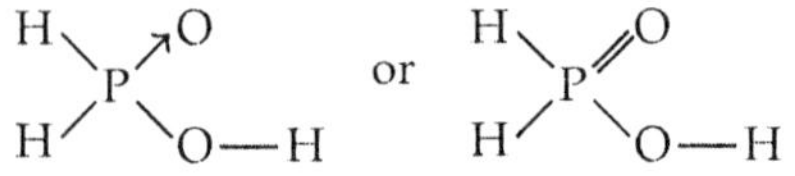 |
| $H_3PO_3$ | P, $sp^3$ | 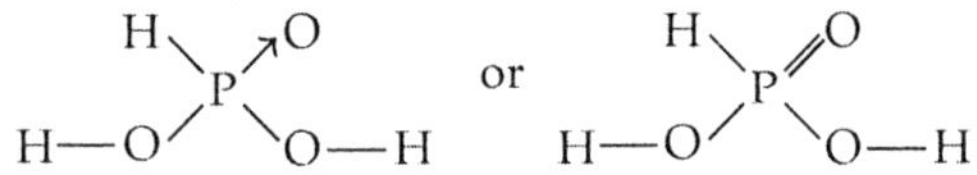 |
| $H_3PO_4$ | P, $sp^3$ | 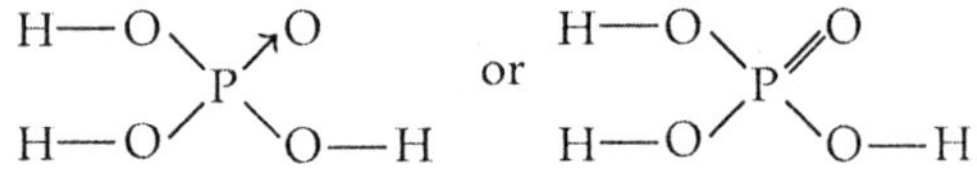 |
| $(PNCl_2)_3$ | P roughly $sp^3$, N is $sp^2$ | 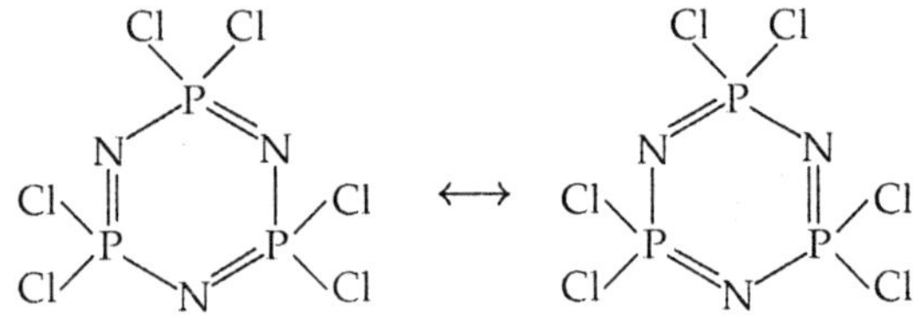 |
| $PCl_3$ | P, $sp^3$ | 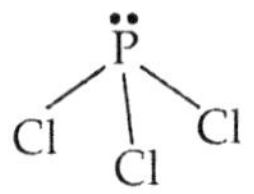 |
| $TeF_6$ | Te, $sp^3d^2$ | 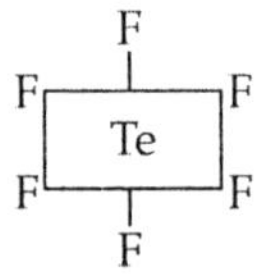 |
| $O_3$ | Central O, $sp^2$ | 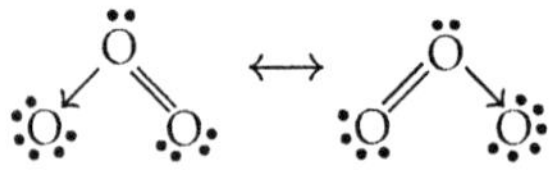 |
| $SO_2$ | S, $sp^2$ | 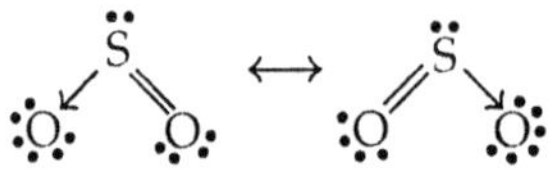 |
| $SO_3$ | S, $sp^2$ | 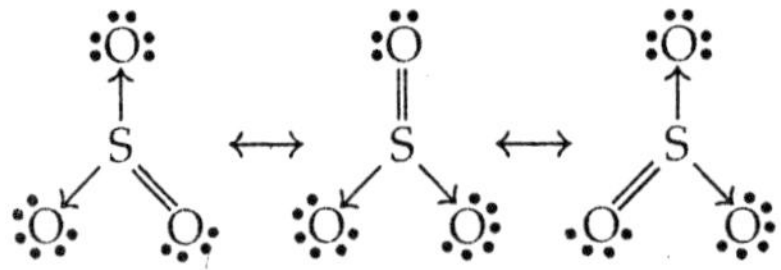 |

| Molecular species | Hybridisation of the central element | Structure |
|---|---|---|
| $H_2SO_3$ or $SO_3^{2-}$ | S, $sp^3$ | 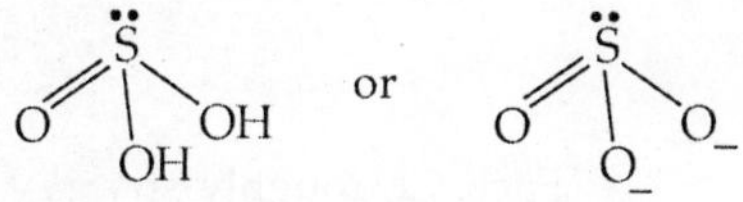 |
| $H_2SO_4$ or $SO_4^{2-}$ | S, $sp^3$ | 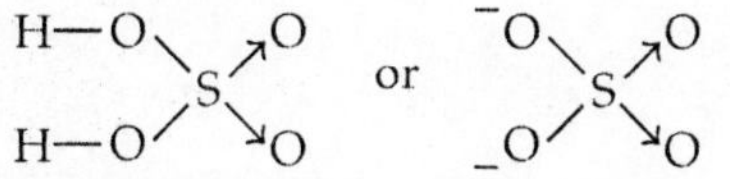 |
| $S_2O_3^{2-}$ | Central S, $sp^3$ | 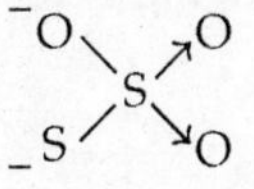 |
| $S_2O_6^{2-}$ | Two S, $sp^3$ | 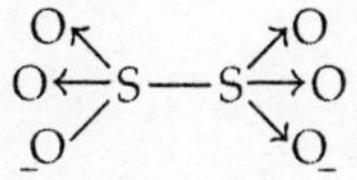 |
| $H_2O_2$ | — | Open book structure 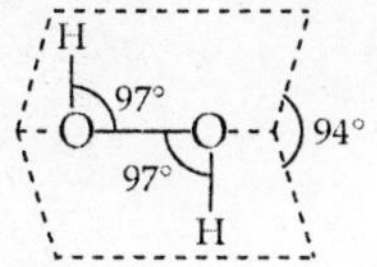 |
| $SO_2Cl_2$ | S, $sp^3$ | 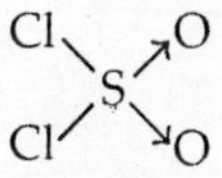 |
| $SOCl_2$ | S, $sp^3$ | 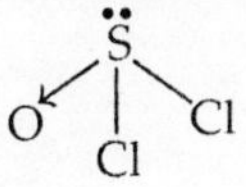 |
| $IF_7$ | I, $sp^3d^3$ | 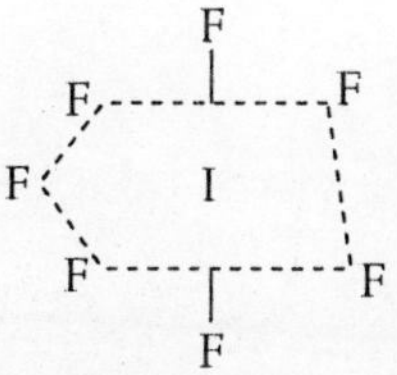 (Pentagonal bipyramid) |
| $Cl_2O$, $F_2O$ | O, $sp^3$ | 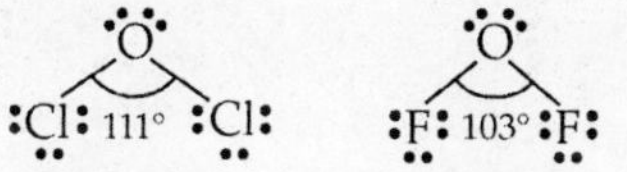 |
| $ClO_2$ | Cl, $sp^3$ | 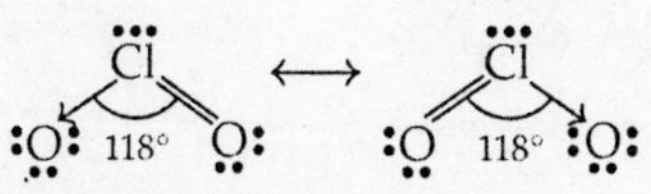 |
| $Cl_2O_6$ | Each Cl, $sp^3$ | 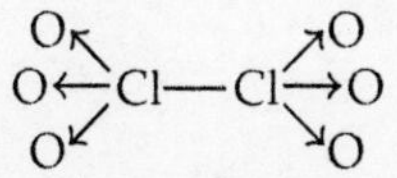 |

| Molecular species | Hybridisation of the central element | Structure |
|---|---|---|
| $Cl_2O_7$ | Each Cl, roughly $sp^3$ | 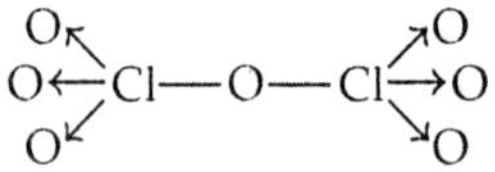 terminal ∠OClO = 115° bridging ∠ClOCl = 119° |
| $I_2O_5$ | Each I $sp^2$, bridging O sp | 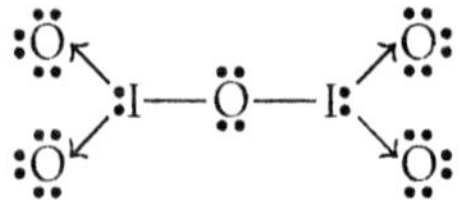 |
| $PF_4^-$ | P, $sp^3d$ | 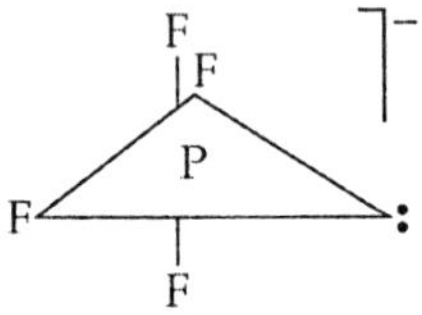 |
| $ClO_2^-$ | Cl, $sp^3$ | 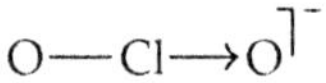 |
| $ClO_3^-$ | Cl, $sp^3$ | 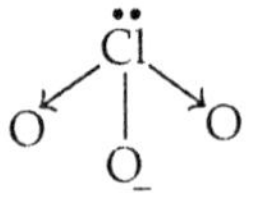 |
| $ClO_4^-$ | Cl, $sp^3$ | 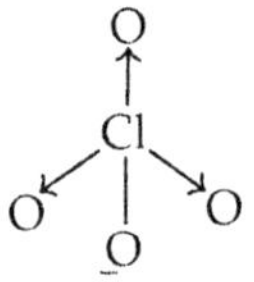 |
| $H_5IO_6$ | I, $sp^3d^2$ | 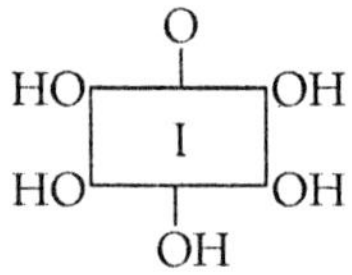 |
| NOCl | N, $sp^2$ | 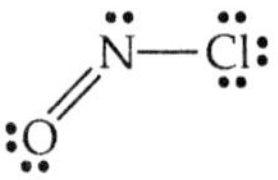 |
| $\dot{C}H_3$ | C, $sp^2$ | 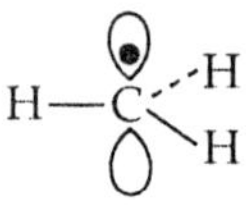 |
| $\dot{C}F_3$ | C, $sp^3$ | 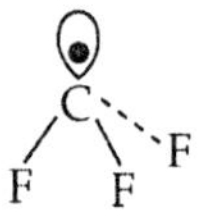 |

**Problem 3.27** Distinguish between $d^2sp^3$ and $sp^3d^2$ hybridisation. **[B.U. 2000]**

**Ans. :** (a) In both the cases the geometry is octahedral but in $d^2sp^3$ the inner d orbital is participated but in $sp^3d^2$ the outer d orbital taking part in hybridisation.

(b) In $d^2sp^3$ hybridisation the number of unpaired electron is reduced or sometimes zero having diamagnetic character e.g. $[Co(NH_3)_6]Cl_3$, whereas in case of $sp^3d^2$, the number of unpaired electrons are increased having paramagnetic character e.g. $[CoF_6]^{3-}$.

(c) In $d^2sp^3$ inner two $(n-1)d_{z^2}$ and $(n-1)d_{x^2-y^2}$ orbitals, outer one ns orbital and three np orbitals are used whereas in $sp^3d^2$, the inner one (n–1)s, three (n–1)p orbitals and outer two $nd_{z^2}$ and $nd_{x^2-y^2}$ are participated.

(d) The $d^2sp^3$ hybridisation generally given by strong field ligand and are sometimes called **inner orbital low spin complex** whereby $sp^3d^2$ hybridisation is called **outer orbital high spin complex**.

---

**Problem 3.28** The H—C—H bond angle in $CH_4$ is 109.5°. Due to lone pair repulsion, the H—O—H angle in $H_2O$ will—

1. remain the same, 2. increase, 3. decrease, 4. become 180°. **[NET 2001]**

**Ans. :** 3.

---

**Problem 3.29** The structure of the noble gas compound $XeF_4$ is—

1. square planar, 2. distorted tetrahedral, 3. tetrahedral, 4. octahedral. **[NET 2001]**

**Ans. :** 1.

---

**Problem 3.30** The preferred geometry of a complex in which the central atom is $dsp^2$ hybridised is—

1. Square planar, 2. Tetrahedron, 3. Square pyramid, 4. Trigonal bipyramid. **[NET 2001]**

**Ans. :** 1.

---

### 3.3.3.1 Stereochemical Nonrigidity : Berry Pseudorotation

So far the discussion have been made in VSEPR model (Sec.3.3.1) and Bent's rule (Sec. 3.3.3) applied to **structurally rigid** or **static** molecular species. In studying the particular molecular structure there are several instantaneous methods are available with the time scale limit $10^2$ to $10^{-20}$ second. However many molecules are **structurally nonrigid** or **stereochemically nonrigid**. If the rearrangement of structures of a molecule leads to configurations which are chemically equivalent having minimum energy and that they are readily transformable from one form to other, we may say that the molecule is **fluxional**. On the other hand if the rearrangement produces chemically distinguishable molecules, the phenomena is known to be **isomerism**. Fluxional molecules differ from other stereochemically nonrigid molecules in possessing more than a single configuration representing minimum energy. The possibility of stereochemical nonrigidity (fluxionality) is ascertained on the NMR spectroscopic time scale.

The ∠HCH in $CH_4$ is 109° and there are eight electrons around the central carbon (four electrons from carbon and one from each of H atom). This has a regular tetrahedral geometry. In $NH_3$ molecule also there exist eight electrons around the central N (five electrons from nitrogen and one from each of H atom) but there is a lone pair of electron residing on nitrogen whereby the ∠HNH recorded as 107°. In both the cases the central C and N are $sp^3$ hybridised but the variation of bond angle in case of $NH_3$ from regular tetrahedral geometry is due to the presence of lone pair/unshared pair of electron. The unshared pair of electron requires more space and forces the bonding pairs slightly closer together.

*Atomic inversion*

However this structural model is justified so long as we take a **static model** of $NH_3$ molecule. In fact $NH_3$ molecule undergoes an **atomic inversion** in vibrational motion like that of **Walden inversion** in the chiral centre of $sp^3$ carbon in $S_N^2$ process. In this inversion process the H atoms around N atom is inverted like the inversion of an umbrella in a high wind without breaking of N–H bonds :

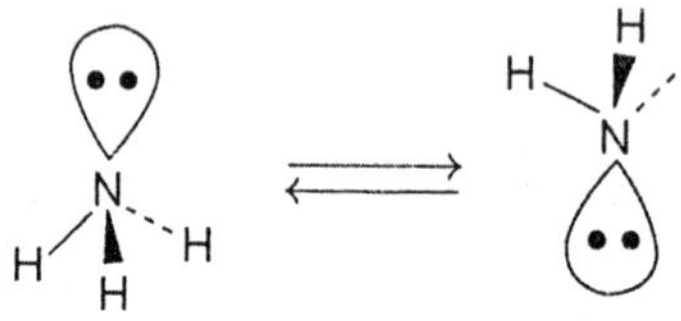

Actually this inversion is accompanied by vibrational motion and the molecule passes through a **trigonal planar** transition state as depicted below :

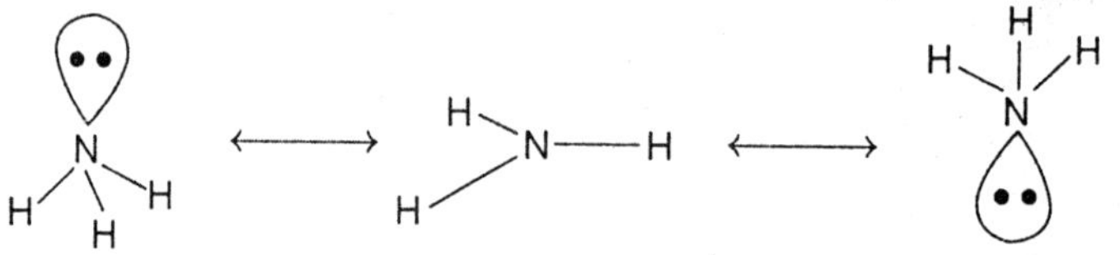

N($sp^3$) $C_{3v}$      N($sp^2$) $D_{3h}$      N($sp^3$) $C_{3v}$

This vibrational motion has a frequency ~$1010 s^{-1}$ having **inversion energy barrier** ~2076 $cm^{-1}$ but the difference between the first and second vibrational states is only 950 $cm^{-1}$ (~1.14kJ$mol^{-1}$). It is possible to calculate the population of the second vibrational state by using Boltzmann distribution law only to the extent of 0.0105 so clearly there is not a sufficient thermal energy available to cause **rapid inversion** if the molecule must pass over the energy barrier height of ~2076 $cm^{-1}$. In this case the inversion occurs through **quantum mechanical tunneling** which suggests that the molecule passes from one structure to the other without having to pass over the energy barrier. This inversion process of $NH_3$ molecule is utilised for the construction of **ammonia clock** because this inversion occurs at a fixed rate at a particular temperature. The vibrational frequency of its inversion is taken as the basis for time scale unit. Like ammonia several chiral amines and phophines undergo atomic inversion and in this inversion process the magnitude of inversion energy barrier depends on the energy required in the transition, $sp^3(C_{3v}) \longrightarrow sp^2(D_{3h})$. In case of phosphines and its derivatives the bond angle ~90° and hence in these compounds, in reaching the transition state, the bond angle distortion is much higher than that involved in the analogous nitrogen compounds. Thus the phosphine derivatives always show higher activation barrier energy and thus register higher stability.

On the other hand the five coordinated molecules e.g. $Fe(CO)_5$, $PF_5$, $BrF_5$ etc. undergo facile inversion because they register lower value for activation barrier energy in solution and these species are considered as stereochemically nonrigid :

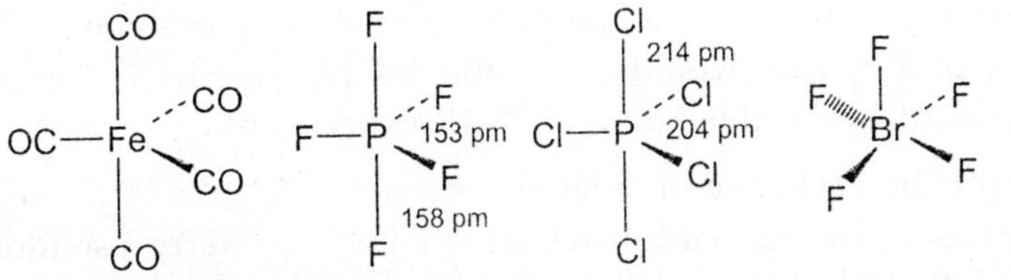


In IR spectroscopic timescale $Fe(CO)_5$ appears to be static while in 13C NMR spectroscopic experiment it is a dynamic molecule and exchange the axial and equatorial CO groups. The fluxional character of $BrF_5$ is established in NMR timescale. Depending on the observed nucleus the timescale of NMR spectroscopic technique is about $10^{-1}$ to $10^{-5}$ second.

In order to understand the mechanism of atomic inversion we have to invoke the VSEPR theory (Sec. 3.3.1) and Bent's rule (Sec. 3.3.3) with axial and equatorial bonds. Let us consider the $SF_4$ molecule which has the predicted structure of trigonal bipyramid but there the following two possible structures :

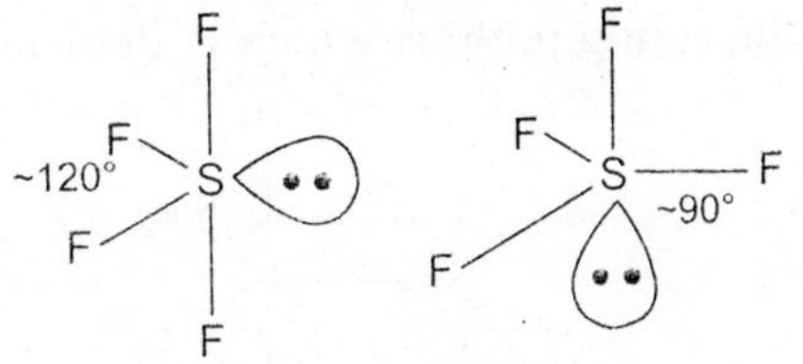


In the left structure the unshared pair of electrons is residing at ~90° from two bonding pairs and ~120° from the other two bonding pairs. In the right hand structure the unshared pair is ~90° from three bonding pairs and 180° from the other bonding pair. These two variations of location of unshared pair may not look very different, but the repulsion between electron pairs is inversely related to the distance of separation raised to an exponent that may be as large as 6. A small difference in distance leads to a substantial difference in repulsion. Thus the structure on the left is correct for $SF_4$ because the two bonding electron pairs at ~90° from the unshared pair is lower in energy. In **trigonal bipyramidal geometry (TBP) the unshared pair of electrons are found to exist in the equatorial positions.** One important feature of TBP geometry is the unequal bond lengths of axial and equatorial bonds as shown above for the molecules $PF_5$ and $PCl_5$. The two axial bonds are somewhat longer than three equatorial bonds whereby the central phosphorus is $sp^3d$ hybridised. This $sp^3d$ hybridisation is considered as $sp^2$+dp whereby $sp^2$ comprises three equatorial trigonal planar bonds and dp comprises two axial bonds directed 180° from each other. One important aspect of these molecules like $PF_5$, $PCl_5$ etc. having five pair of electrons is that the central atom having TBP geometry will form unequal hybrid having longer two axial bonds and three shorter equatorial bonds.

As regards the location of axial and equatorial bonds in TBP geometry it is plausable to assume axial bonds having high electronegative atoms or unshared pairs of electrons because the axial bonds are dp having less s character i.e. less electronegative while the equatorial bonds will be occupied by less electronegative element or bonding electron pairs as the equatorial bonds make use of $sp^2$ hybrids having more s-character i.e. more electronegativity. In the mixed halide $PCl_3F_2$ it was found that the two F atoms are found in axial positions while the three Cl atoms have a preference to exist in the equatorial positions in TBP geometry. According to this principle in the molecule $PCl_2F_3$ it is expected that the two F atoms will be in axial positions and two Cl atoms and one F atom will be occupied in equatorial positions. But NMR studies in different variation of temperatures established that all the F atoms are equivalent and undergo structural change thereby make the F atoms in equatorial positions. This is consistent with the time scale of NMR experiment with our earlier observation of $NH_3$ molecule having vibrational frequency of the order of ~1010s$^{-1}$. Now a question arises as to what type of structural change could occur in $PCl_2F_3$ or $PF_5$ that would make the fluorine atoms in equatorial and axial positions appear equivalent in the time scale of NMR experiment.

In order to highlight the exchange of axial and equatorial bonds in a five coordinate TBP geometry R.S. Berry proposed a mechanism which is known as **Berry pseudorotation.** In this process it is suggested that the trigonal bipyramid geometry ($sp^3d$ i.e. the combination of $s+p_x+p_y+p_z+d_{z^2}$) transformed through the square pyramid geometry ($sp^3d$ i.e. the combination of $s+p_x+p_y+p_z+d_{x^2-y^2}$) to another TBP geometry of different configuration. In this axial/equatorial exchange process the two axial groups are interchanging with two equatorial groups while the third equatorial group (taken as pivot group) remains as an equatorial group in both the configurations which is depicted below :

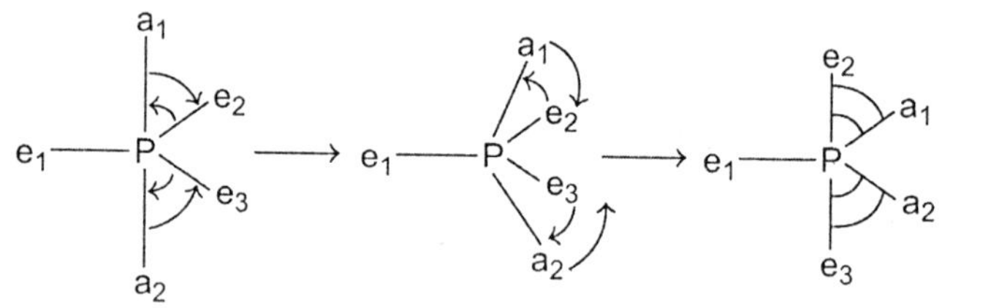


TBP ($D_{3h}$) (configuration I) Square pyramid ($C_{4v}$) TBP ($D_{3h}$) (configuration II)

In this exchange process the groups initially in the equatorial positions are designated as $e_1$, $e_2$, $e_3$ and those in the axial positions are designated as $a_1$, $a_2$. Here the transition state is square pyramid geometry which has $\angle a_1Pa_2 = \angle e_2Pe_3 = 150°$; $\angle e_1Pa_1 = \angle e_1Pe_2 = \angle e_1Pe_3 = \angle e_1Pa_2 = 105°$; $\angle a_1Pe_2 = \angle e_3Pa_2 = \angle e_2Pa_2 = \angle e_2Pa_1 = 87°$. In configuration I the original axial angle i.e. $\angle a_1Pa_2$ = 180 closes to 150° and at the same time, the equatorial angle i.e. $\angle e_2Pe_3$ opens up to 150° from 120°. This situation transformed the geometry from TBP to square pyramid. Then the angle $\angle e_2Pe_3$ opens up to 180° and the $\angle a_1Pa_2$ reduces to 120°. Thus in this process of change of axial and equatorial bonds of $PF_5$ the two configurations are identical and indistinguishable (configuration I and II). Thus the total energy change for the process is zero i.e. $\Delta G° = 0$. This mechanism is similar to the **atomic inversion** of the $NH_3$ molecule except for the fact that movement of atoms is by rotational motion. Lowering the temperature reduces molecular vibration enough that the fluorine resonance indicates fluorine atoms in two different environments i.e. axial and equatorial. At higher temperatures structural changes are rapid and only one fluorine environment is indicated as we saw for $NH_3$, not all molecules have static environment.

Like pseudorotation another type of rotation is to be considered which is called **turnstile** rotation. In this axial/equatorial exchange process a pair of groups rotates in a **turnstile manner** with respect to the remaining three groups as shown below :

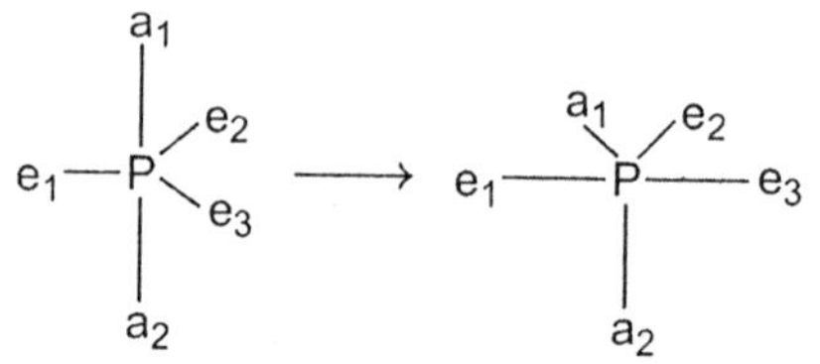

The NMR studies of $AB_5$ molecules cannot distinguish the two types of rotation i.e. Berry pseudorotation and turnstile rotation but the more plausable energetically is the Berry pseudorotation.

## 3.4 MOLECULAR ORBITAL THEORY (MOT)

In Sec. 3.2.2 the mention of MOT have been made for covalent bonding. The hybrid orbital concept according to VBT denotes the shape of a molecular species having certain approximation because **hybrid orbitals are still atomic orbitals.** In case of molecules **the orbitals must be molecular orbitals and not atomic orbitals**. The inadequacy of VBT in case of simple molecule like $O_2$ can be demonstrated. According to VBT $O_2$ molecule is diamagnetic (:Ö=Ö:) having each oxygen is sp hybridised. But experiment shows that $O_2$ molecule is paramagnetic to the extent of two unpaired electrons. This defect of VBT can be explained on the basis that $O_2$ is a molecule and in case of molecules the atomic orbital concept (sp hybridisation) is no longer tenable, in case of $O_2$ molecule the orbitals must be molecular orbital (MO). Same is true for $B_2$ molecule in which VBT denotes it will be diamagnetic but actually this is paramagnetic. This necessitated the modification of VBT and the concept of MOT developed.

*Defect of VBT*

Each electron in a molecule is described by a certain wave function $\Psi$ which because it represents the orbit of the electron in a moleucle, is properly called **molecular orbital**. The atomic orbitals are monocentric whereas molecular orbitals are polycentric. The interpretation of wavefunction $\Psi$ is that $\Psi^2 d\tau$ or $\Psi\Psi^* d\tau$ if $\Psi$ is complex ($\Psi^*$ is the complex conjugate of $\Psi$) denotes the relative probability that this electron is in the small volume $d\tau$. If $\Psi$ is normalised then $\int \Psi^2 d\tau = 1$ and then $\Psi^2 d\tau$ measures the exact probability of finding the electron in this small volume element $d\tau (d\tau = dxdydz)$.

*Molecular orbital*

Now some natural question arises :

(a) Why do atoms combine to form molecule?

(b) How do they combine?

Simply combination of atoms held together by the formation of a chemical bond. The total energy of a molecule is less than the total energy of the atoms i.e. a molecule is more stable than the combined atoms.

Let us consider the formation of a diatomic molecule AB from the atoms A and B whose wave functions are $\Psi_A$ and $\Psi_B$ respectively. The molecular wave function $\Psi$ of the AB molecule can be represented by the sum of the characteristic properties of the wave functions $\Psi_A$ and $\Psi_B$. Suppose $\Psi_A$ represents an s-type atomic orbital and $\Psi_B$ represents an $p_x$ type atomic orbital. Let an electric field is applied in the x-direction of an s-type atomic orbital, it will pull the electron density along the x-direction and ultimately the s-type atomic orbital is converted to $p_x$ type orbital. In this case we can describe the complete wave function as a sum of the combinations from each orbital wave functions of s and $p_x$ orbital i.e.

$$\Psi = C_1\Psi_A + C_2\Psi_B \quad ...(3.28)$$

where $C_1$ and $C_2$ are constants to be chosen so that energy of the system is minimised. This is known **as linear combination of atomic orbitals (LCAO).** This is most favourably written as,

$$\Psi = N(\Psi_A + \lambda\Psi_B) \qquad ...(3.29)$$

where N is a normalisation factor which is normally unity so that—

$$\Psi = \Psi_A + \lambda\Psi_B \qquad ...(3.30)$$

The coefficient $\lambda$ is dependent on the extent of overlap of two wave functions $\Psi_A$ and $\Psi_B$ and so $\lambda$ depends on the radial part of the wave function.

### 3.4.1 The sign of the wave function

In Sec. 1.8.7.2 we have seen that the orbital wave functions (s,p,d orbitals) have positive or negative sign. These signs do not have a physical significance which means that positive or negative sign does not mean that there is a high probability or less probability respectively of finding the electron because probability of finding the electron is denoted by $\Psi^2$ and not by $\Psi$.

*Constructive and destructive interference*

However, the sign of the wave function has special importance when the wave functions combine to form molecular orbital wave function. When two wave functions combine into the same region of space with the same sign, there will be an enhance wave function and this enhancement of amplitude is sometimes called **constructive interference (bonding)** (Fig. 3.18(a)). This aspect of wave function is of great importance in the explanation of chemical bonding. On the other hand positive + negative or negative + positive combination of wave cancelled each other causing reduced probability of finding the electron in that region and is called **destructive interference (antibonding)** [Fig. 3.18(b)].

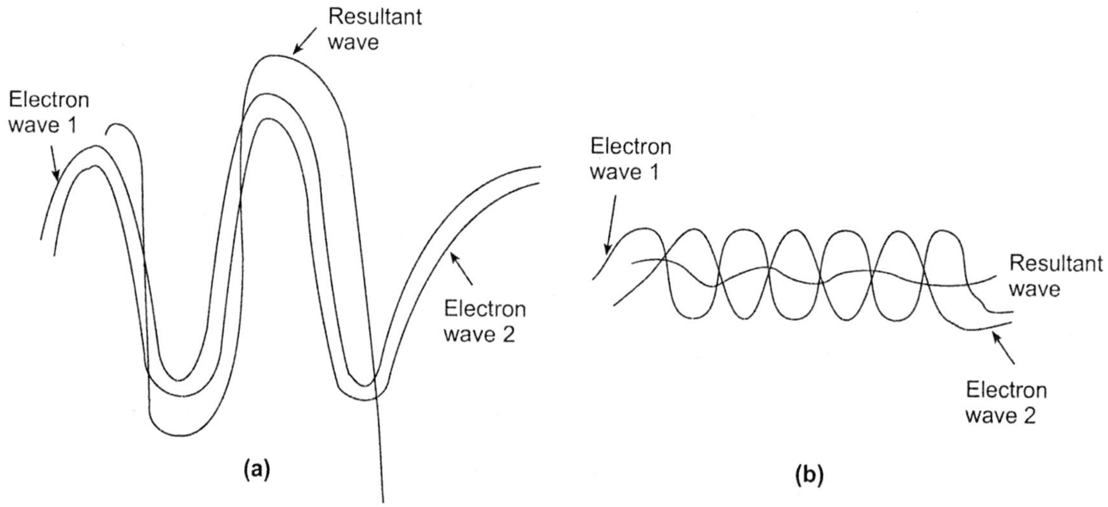


*Figure 3.18 : (a) Constructive interference where the wave functions combine with same sign and the resultant wave has enhanced amplitude, (b) Destructive interference where the two wave functions have opposite sign and the resulting superposition has a reduced amplitude.*

### 3.4.2 The necessary conditions of LCAO to form MO

In order that there may be effective combination between $\Psi_A$ and $\Psi_B$ the following conditions are to be followed :

(a) That the energies of $\Psi_A$ and $\Psi_B$ in their respective atoms should be of comparable magnitude i.e. combining atomic orbital wave functions should have comparable energy. This is known as **energy condition** which means that best combination will occur between 1s and 1s, 5s and 5s, 3p and 3p etc. but less probable combination is 1s and 5s, 2s and 5p etc.

(b) That the charge clouds of $\Psi_A$ and $\Psi_B$ should overlap one another as much as possible i.e. the overlap integral $S_{AB} = \int \Psi_A \Psi_B d\tau$ and resonance integral $H_{AB}^* = \int \Psi_A \hat{H} \Psi_B d\tau$ will have maximum value. If these integrals correspond to zero value, which means that there is no combination. This is known as **overlap condition.**

(c) $\Psi_A$ and $\Psi_B$ should have the same symmetry relation to the molecular axis AB. Let $\Psi_A$ is an s type atomic orbital and $\Psi_B$ is a $p_z$ type orbital and let z axis is directed perpendicular to the molecular axis AB (Fig. 3.19). It is evident that,

$\Psi_A$ in the upper region = $\Psi_A$ in the lower region, but

$\Psi_B$ in the upper region = $-\Psi_B$ in the lower region

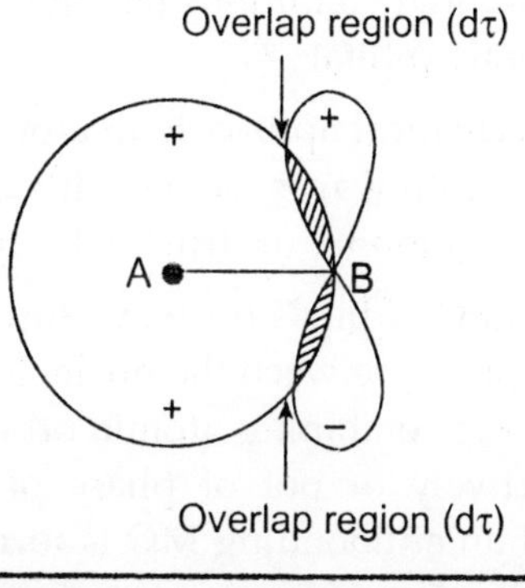


***Figure 3.19 : Representation of overlap of s and $p_z$ atomic orbitals. Combination not allowed.***

Therefore overlap integral in the upper region = $\int \Psi_A \Psi_B d\tau$ and the overlap integral in the lower region = $-\int \Psi_A \Psi_B d\tau$, so that net overlap integral = $\int \Psi_A \Psi_B d\tau - \int \Psi_A \Psi_B d\tau = 0$. Similarly net resonance integral $H_{AB} = \int \Psi_A \hat{H} \Psi_B d\tau - \int \Psi_A \hat{H} \Psi_B d\tau = 0$.

So s type atomic orbital will not combine with $p_z$ type orbital due to lack of symmetry. This is known as **symmetry condition** of the combining atomic orbitals. Thus without going details of the symmetry condition we can summarise the fact in the following tabular form (Table 3.8).

* $H_{AB}$ also represented as J.

**Table 3.8 : Permitted and disallowed atomic orbitals in LCAO**

| $\Psi_A$ | $\Psi_B$ | | Type of M.O. |
|---|---|---|---|
| | Permitted (x-bond axis) | Disallowed (x-bond axis) | |
| s | s, $p_x$, $d_{x^2-y^2}$, $d_{z^2}$ | $p_y$, $p_z$, $d_{xy}$, $d_{yz}$, $d_{xz}$ | σ |
| $p_x$ | s, $p_x$, $d_{x^2-y^2}$, $d_{z^2}$ | $p_y$, $p_z$, $d_{xy}$, $d_{yz}$, $d_{xz}$ | σ |
| $p_y$ | $p_y$, $d_{xy}$ | s, $p_x$, $p_z$, $d_{x^2-y^2}$, $d_{yz}$, $d_{xz}$, $d_{z^2}$ | π |
| $p_z$ | $p_z$, $d_{xz}$ | s, $p_x$, $p_y$, $d_{x^2-y^2}$, $d_{yz}$, $d_{xy}$ | π |
| $d_{x^2-y^2}$ | s, $p_x$, $d_{x^2-y^2}$, $d_{z^2}$ | $p_y$, $p_z$, $d_{xy}$, $d_{yz}$, $d_{xz}$ | σ |
| $d_{xy}$ | $p_y$, $d_{xy}$ | s, $p_x$, $p_z$, $d_{x^2-y^2}$, $d_{yz}$, $d_{xz}$, $d_{z^2}$ | π |
| $d_{z^2}$ | s, $p_x$, $d_{x^2-y^2}$, $d_{z^2}$ | $p_y$, $p_z$, $d_{xy}$, $d_{yz}$, $d_{xz}$ | σ |
| $d_{xz}$ | $p_z$, $d_{xz}$ | s, $p_x$, $p_y$, $d_{x^2-y^2}$, $d_{yz}$, $d_{xy}$ | π |

### 3.4.3 Essential features of molecular orbital theory

1. Atomic orbitals of individual atoms combine to form molecular orbitals according to LCAO (Sec. 3.4.2). The number of molecular orbitals will be equal to the participating atomic orbitals. As stated earlier three essential conditions (energy, overlap and symmetry) must have to be followed for combining atomic orbitals.

2. According to quantum mechanical approach an atomic orbital is represented by a wave function. Just like water waves, electron waves have positive (crest) or negative (trough) phases or amplitude. The two waves may combine destructively or constructively (Sec. 3.4.1).

In the same way when two atomic orbitals (or wave functions) undergo in phase or similar phase addition, their electron density in between the nuclei increases and a **bonding MO results** which has an energy lower than the combining atomic orbitals. On the other hand when two wave functions combine destructively or out of phase of addition, the electron density in between the nuclei decreases and an antibonding MO is formed, the energy of which is higher than the combining atomic orbitals. Thus two atomic orbitals combine to form two molecular orbitals—one is bonding and the other is antibonding. The two 1s orbitals of different atoms combine to give a bonding MO designated as $\sigma_{1s}$ and an antibonding MO represented as $\sigma^*_{1s}$ (Fig. 3.20).

In a similar way the combination of two 2s or 3s orbitals give $\sigma_{2s}$ and $\sigma^*_{2s}$ or $\sigma_{3s}$ and $\sigma^*_{3s}$ molecular orbitals respectively. The antibonding molecular orbitals have a nodal plane which divides the internuclear axis at right angles and at this plane the probability of finding the electron is zero.

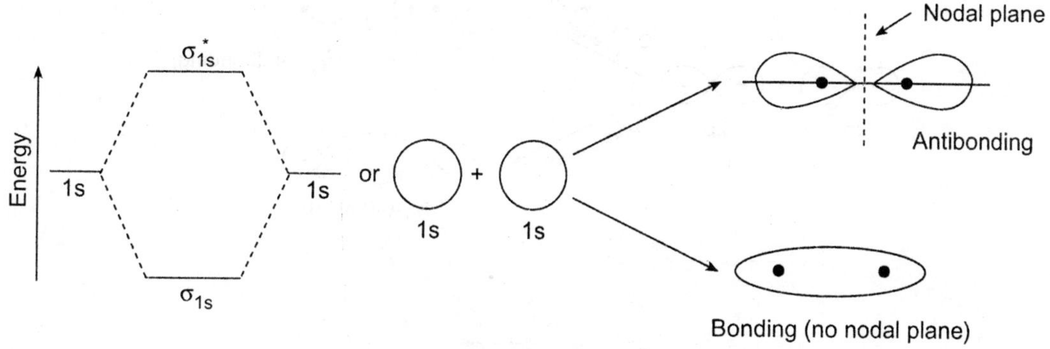


***Figure 3.20** : Formation of molecular orbitals from two 1s atomic orbitals.*

It is evident from the Table 3.8 that s and $p_x$ combination is permitted (Fig. 3.21).

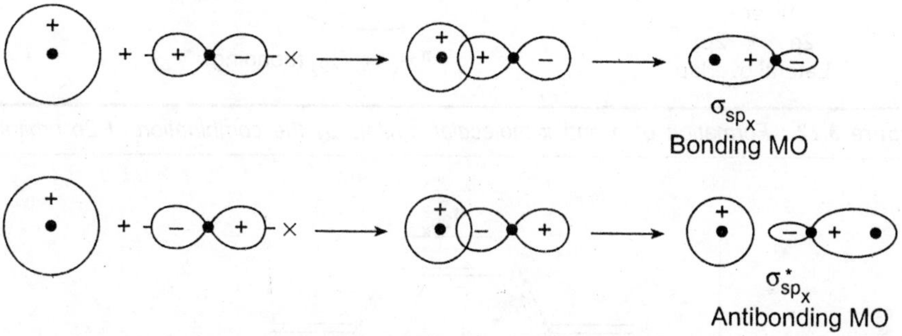


***Figure 3.21** : Formation of $\sigma_{sp_x}$ and $\sigma^*_{sp_x}$ molecular orbitals from s and $p_x$ combination.*

Combination of two sets of 2p orbitals result six molecular orbitals of which three will be bonding and three will be antibonding. The $2p_x$ orbitals overlap along the bond axis (x-axis) to give $\sigma_{2p_x}$ (bonding) and $\sigma^*_{2p_x}$ (antibonding) molecular orbitals (Fig. 3.22). On the other hand the $2p_y$ or $2p_z$ orbitals overlap laterally to form $\pi$ and $\pi^*$ molecular orbitals (Fig. 3.22).

Now with the above considerations we are now in a position to construct the molecular orbital diagram for simple homonuclear diatomic molecules upto second period elements (Fig. 3.23). In this diagram no interaction has been made between 2s and $2p_x$ orbital, though from symmetry consideration they have the same symmetry and have energies close to each other. *A consideration of these criteria is that if two molecular orbitals falling in the same energies close to each other (i.e. constructed from atomic orbitals with energies close to each other) they will interact each other i.e. a type of orbital mixing will occur* ($\sigma_{2s}$ interact with $\sigma_{2p_x}$ as also $\sigma^*_{2s}$ with $\sigma^*_{2p_x}$). Therefore we have to modify the above MO diagram with s-p mixing (Fig. 3.24).

3. Inner orbital electrons will not take part in the formation of molecular orbitals. For example in case of $O_2$ molecule only 2s and 2p electrons of the two oxygen atoms take part in

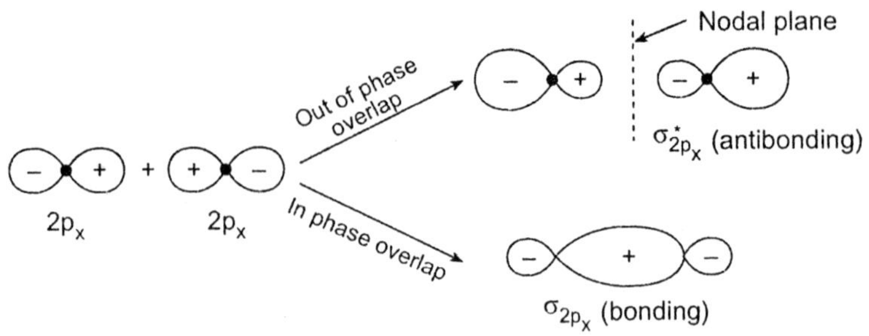


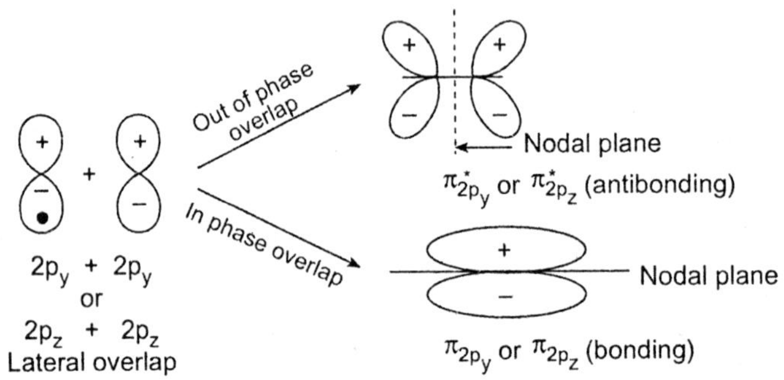


***Figure 3.22** : Formation of σ and π molecular orbitals by the combination of 2p orbitals.*

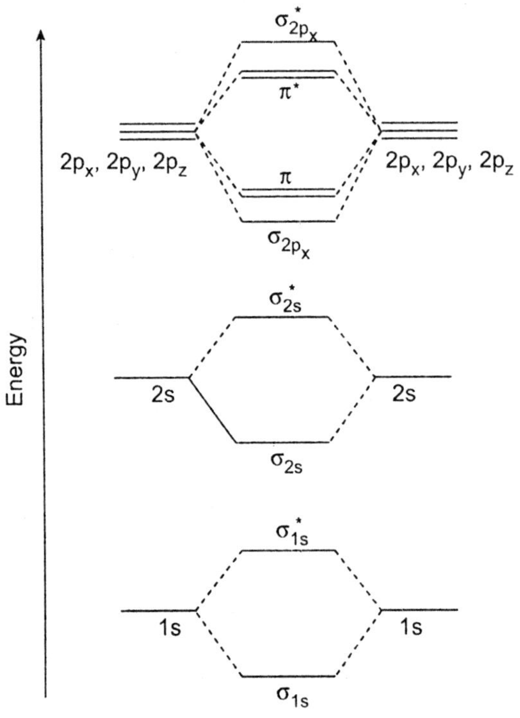


***Figure 3.23** : MO diagram for homonuclear diatomic molecules or ion of the second period.*

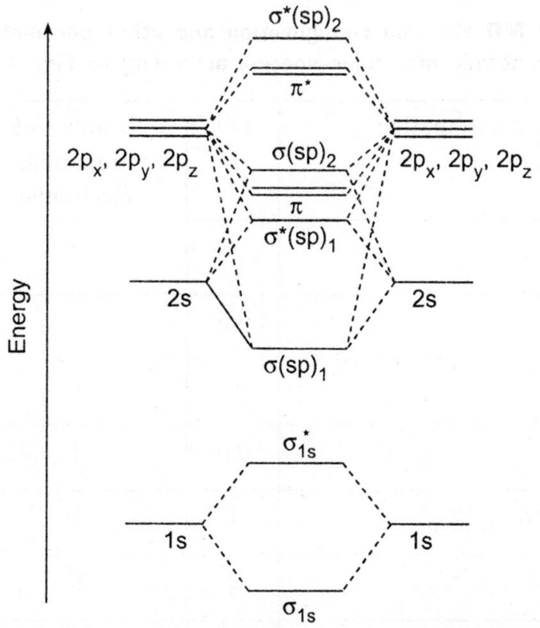


***Figure 3.24** : s-p mixing MO diagram for homonuclear diatomic molecular species of the second period (x axis is molecular axis).*

MO formation. Inner orbitals generally have substantially smaller radii than valence orbitals and do not overlap well with orbitals of adjacent atoms.

4. In using the MO diagram the same rules are to be followed as electrons are filled in atomic orbitals i.e.

(a) Each MO will be filled up with a maximum of two electrons (Pauli's exclusion principle).

(b) The lower energy MO will be filled prior to that of higher energy (Aufbau principle). The order of energy according to Fig. 3.24 is :

$$\sigma_{1s} < \sigma^*_{1s} < \sigma(sp)_1 < \sigma^*(sp)_1 < \pi_{2p_y}/\pi_{2p_z} < \sigma(sp)_2 < \pi^*_{2p_y}/\pi^*_{2p_z} < \sigma^*(sp)_2$$

(c) In case of degenrate MOs the electrons remain unpaired as far as possible (Hund's rule of maximum multiplicity).

The MO electron configurations of some molecular species are given in Table 3.9. In this context a pertinent term is associated in MO theory which is bond order (BO). It is defined as one-half of the difference between the number of electrons in bonding molecular orbital (BMO) and the number of electrons in antibonding molecular orbital (ABMO) i.e.

$$BO = \frac{\text{Number of electrons in BMO} - \text{Number of electrons in ABMO}}{2} \qquad ...(3.31)$$

Higher BO indicates higher stability of molecular species i.e. the dissociation energy of the species is higher conversely lower BO indicates low stability of the species.

**Table 3.9 : MO electron configuration and other parameters of simple diatomic molecular species according to Fig. 3.24**

| Molecular species | MO electron configuration | BO | Number of unpaired electron(s) | Bond length (pm) | Bond energy kJ mol$^{-1}$ |
|---|---|---|---|---|---|
| $H_2$ | $\sigma_{1s}^{2}$ | 1 | 0 | 74 | 432 |
| $H_2^+$ | $\sigma_{1s}^{1}$ | 0.5 | 1 | 106 | 255 |
| $He_2$ | $\sigma_{1s}^{2}\sigma_{1s}^{*2}$ | 0 | 0 | — | — |
| $He_2^+$ | $\sigma_{1s}^{2}\sigma_{1s}^{*1}$ | 0.5 | 1 | 108 | 230 |
| $B_2$ | $\sigma(sp)_1^2\, \sigma^*(sp)_1^2\, \pi_{2p_y}^{1}\pi_{2p_z}^{1}$ | 1 | 2 | 159 | 274 |
| $C_2$ | $\sigma(sp)_1^2\, \sigma^*(sp)_1^2\, \pi_{2p_y}^{2}\pi_{2p_z}^{2}$ | 2 | 0 | 124 | 602 |
| $N_2$ | $\sigma(sp)_1^2\, \sigma^*(sp)_1^2\, \pi_{2p_y}^{2}\pi_{2p_z}^{2}\, \sigma(sp)_2^2$ | 3 | 0 | 110 | 942 |
| $N_2^+$ | $\sigma(sp)_1^2\, \sigma^*(sp)_1^2\, \pi_{2p_y}^{2}\pi_{2p_z}^{2}\, \sigma(sp)_2^1$ | 2.5 | 1 | 112 | 842 |
| $O_2$ | $\sigma(sp)_1^2\, \sigma^*(sp)_1^2\, \pi_{2p_y}^{2}\pi_{2p_z}^{2}\, \sigma(sp)_2^2$ $\pi^{*}{}_{2p_y}^{1}\, \pi^{*}{}_{2p_z}^{1}$ | 2 | 2 | 121 | 494 |
| $O_2^+$ | $\sigma(sp)_1^2\, \sigma^*(sp)_1^2\, \pi_{2p_y}^{2}\pi_{2p_z}^{2}\, \sigma(sp)_2^2$ $\pi^{*}{}_{2p_y}^{1} / \pi^{*}{}_{2p_z}^{1}$ | 2.5 | 1 | 112 | — |
| $O_2^-$ | $\sigma(sp)_1^2\, \sigma^*(sp)_1^2\, \pi_{2p_y}^{2}\pi_{2p_z}^{2}\, \sigma(sp)_2^2$ $\pi^{*}{}_{2p_y}^{2}\, \pi^{*}{}_{2p_z}^{1}$ | 1.5 | 1 | 126 | 393 |
| $O_2^{2-}$ | $\sigma(sp)_1^2\, \sigma^*(sp)_1^2\, \pi_{2p_y}^{2}\, \pi_{2p_z}^{2}\, \sigma(sp)_2^2$ $\pi^{*}{}_{2p_y}^{2}\, \pi^{*}{}_{2p_z}^{2}$ | 1 | 0 | 149 | — |
| $F_2$ | $\sigma(sp)_1^2\, \sigma^*(sp)_1^2\, \pi_{2p_y}^{2}\pi_{2p_z}^{2}\, \sigma(sp)_2^2$ $\pi^{*}{}_{2p_y}^{2}\, \pi^{*}{}_{2p_z}^{2}$ | 1 | 0 | 142 | 155 |
| $Ne_2$ | $\sigma(sp)_1^2\, \sigma^*(sp)_1^2\, \pi_{2p_y}^{2}\, \pi_{2p_z}^{2}\, \sigma(sp)_2^2$ $\pi^{*}{}_{2p_y}^{2}\, \pi^{*}{}_{2p_z}^{2}\, \sigma^*(sp)_2^2$ | 0 | 0 | — | — |

**Problem 3.31** Why two atomic orbitals combine to form two MOs.

**Ans. :** As the atomic orbitals combine either by constructive interference or by destructive interference (Sec. 3.4.1) thereby two MOs are formed. In other words when two electron waves combine **in phase** bonding MO is formed whereby when they combine in **out of phase** or opposite phase antiboding MO results.

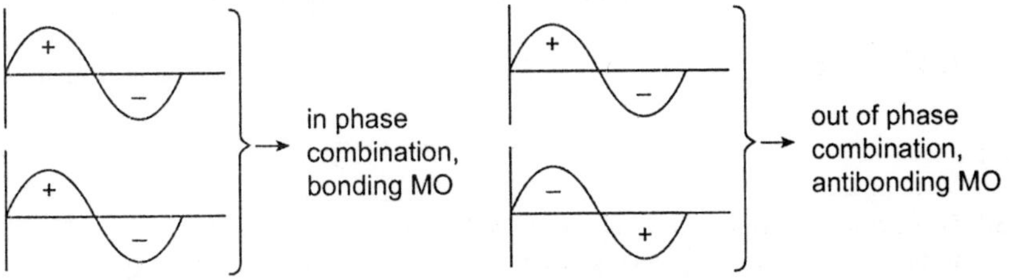


Thus combination of two atomic orbitals results the formation of two MOs.

**Problem 3.32** Distinguish between BMO and ABMO.

**Ans. :** (a) BMO results due to constructive interference (++ sign or – – sign overlap) whereas ABMO results due to destructive interference (+ – sign or – + sign overlap) of two atomic orbitals.

(b) The energy of BMO is lower than the combining atomic orbitals whereby ABMO has higher energy than the two atomic orbitals.

(c) The electron density in BMO is greater than that of ABMO.

(d) Each electron in BMO contributes to attraction between the nuclei whereas contribution of electron in ABMO results in dissociation of a molecule.

(e) Addition of electron in BMO increases BO whereas the process decreases BO when added to ABMO.

**Problem 3.33** (a) Mention the conditions for linear combination of atomic orbitals relating to the formation of molecular orbitals.

(b) Show pictorially the formation of every possible bonding and antibonding molecular orbitals from linear combinations of s and p orbitals.

(c) Compare and contrast the concept of hybrid orbitals and molecular orbitals with respect to (i) the number of atoms involved and (ii) the energies of the resulting orbitals with respect to one another. **[C.U. 1997]**

**Hint :** (a) Sec. 3.4.2, (b) Sec. 3.4.3, (c) Sec. 3.4.

**Problem 3.34** Do you expect the structures of $PCl_3F_2$ and $PF_3Cl_2$ to be different? If so why? **[NET 2000]**

**Ans. :** The two molecules has the total coordination number 5 and have a trigonal bipyramidal structure. In $PCl_3F_2$ the two fluorine atoms preferably have positions in the axial directions. The axially directed hybrid orbitals (i.e. pd; the combination of $p_z$ and $d_z^2$) having no s character is attached with more electronegative fluorine (Bent's rule).

In $PF_3Cl_2$, the more electronegative fluorines will prefer the axial sites, but as there are three fluorines and only two axial sites, the third fluorine will have to go to the equatorial site.

**Problem 3.35** The combination of atomic orbitals centered on A and B yielding π-orbitals in the molecule AB (molecular axis is the z axis) is—

1. $p_x$-$p_y$, 2. s-$p_x$, 3. $p_x$–$p_x$, 4. $p_x$–$p_z$. **[NET 2000]**

**Ans. :** 3.

**Problem 3.36** Among the oxides of nitrogen $N_2O$, NO and $NO_2$ molecules with unpaired electrons are :

1. $N_2O$ and NO, 2. NO and $NO_2$, 3. $N_2O$ and $NO_2$, 4. $NO_2$ and its dimer. **[NET 2000]**

**Ans. :** 2.

**Problem 3.37** The bond lengths of $H_2^+$, $H_2$ and $He_2^+$ follow the order—

1. $H_2^+ > H_2 > He_2^+$, 2. $H_2^+ > He_2^+ > H_2$, 3. $He_2^+ > H_2^+ > H_2$, 4. $H_2 > H_2^+ > He_2^+$. **[NET 2001]**

**Ans. :** 3.

### 3.4.4 σ and π molecular orbitals

In Sec. 3.2.10 the formation of σ and π atomic orbitals have been stated. We have to explain the formation of σ and π orbitals in terms of MO theory.

Atomic orbitals which have the symmetry about the bond axis or molecular axis are called **atomic σ orbitals** e.g. s, $p_x$, $d_{x^2-y^2}$. On the other hand atomic orbitals which are not symmetrical about the bond axis are called π **atomic orbitals** e.g. $p_y$, $p_z$, $d_{xz}$ etc.

Atomic σ orbital can combine with another atomic σ orbital to form $\sigma_{MO}$ due to favourable symmetry about the molecular axis. Similarly atomic π orbitals can combine with other π orbitals to form $\pi_{MO}$. Conversely atomic σ orbitals cannot combine with atomic π orbitals or vice-versa to form molecular orbital due to lack of symmetry. Furthermore there are two types of $\sigma_{MO}$ e.g. σ bonding MO designated by σ and σ antibonding MO designated by σ*. Similarly $\pi_{MO}$ can also be represented as bonding $\pi_{MO}$ as π and antibonding $\pi_{MO}$ as π*. The bonding and antibonding MO arises due to the ± sign in LCAO i.e. $\Psi_A \pm \Psi_B$. When the two atomic orbitals combine in same phase bonding orbital ($\Psi_A + \Psi_B$) results. When two atomic orbitals have negative overlap, they act as waves that are out of phase with one another. The crest of one wave coincides with the trough of the other in the region of overlap thereby cancelling the wave amplitude and in that case antibonding ($\Psi_A - \Psi_B$) interaction results.

The subscript g (gerade or even) and u (ungerade or odd) is also applied to MOs for a symmetry symbol when the molecule has a **centre of symmetry**. If a wave function remain unchanged in appearance under the operation of inversion (in which every point is interchanged with a corresponding point diametrically opposite accross the centre of symmetry) then it is of g type, if it changes sign then it is u type. For example s and d atomic orbitals are g type while p atomic orbitals are u type where the centre of inversion is the nucleus. This type of g and u classification can also be applied in case of σ and π type MOs. For example in a $\sigma_{MO}$ the bonding combination ($\Psi_A + \Psi_B$) is g type while in $\pi_{MO}$ the bonding combination is u type. Because σ bonding MO ($\Psi_A + \Psi_B$) is symmetrical about the centre of the molecular axis and is designated as $\sigma_g$. σ antibonding MO ($\sigma^*_{MO}$) is designated as $\sigma^*_u$ or $\sigma_u$ which is antisymmetric about the centre of molecular axis. In case of $\pi_{MO}$ reverse is the order. $\pi$ antibonding is symmetrical about the centre and designated as $\pi_g^*$ or $\pi_g$ whereas $\pi$ bonding MO is antisymmetrical about the centre and designated as $\pi_u$. Thus there are four types of MOs can be designated i.e. $\sigma_g$, $\sigma_u^*$ or $\sigma_u$, $\pi_u$, $\pi_g^*$ or $\pi_g$ (Fig. 3.25).

Finally, the number (1,2,3...) which precedes the symmetry symbol simply puts the MOs of that particular symmetry in ascending energy order. This is in accordance with aufbau principle

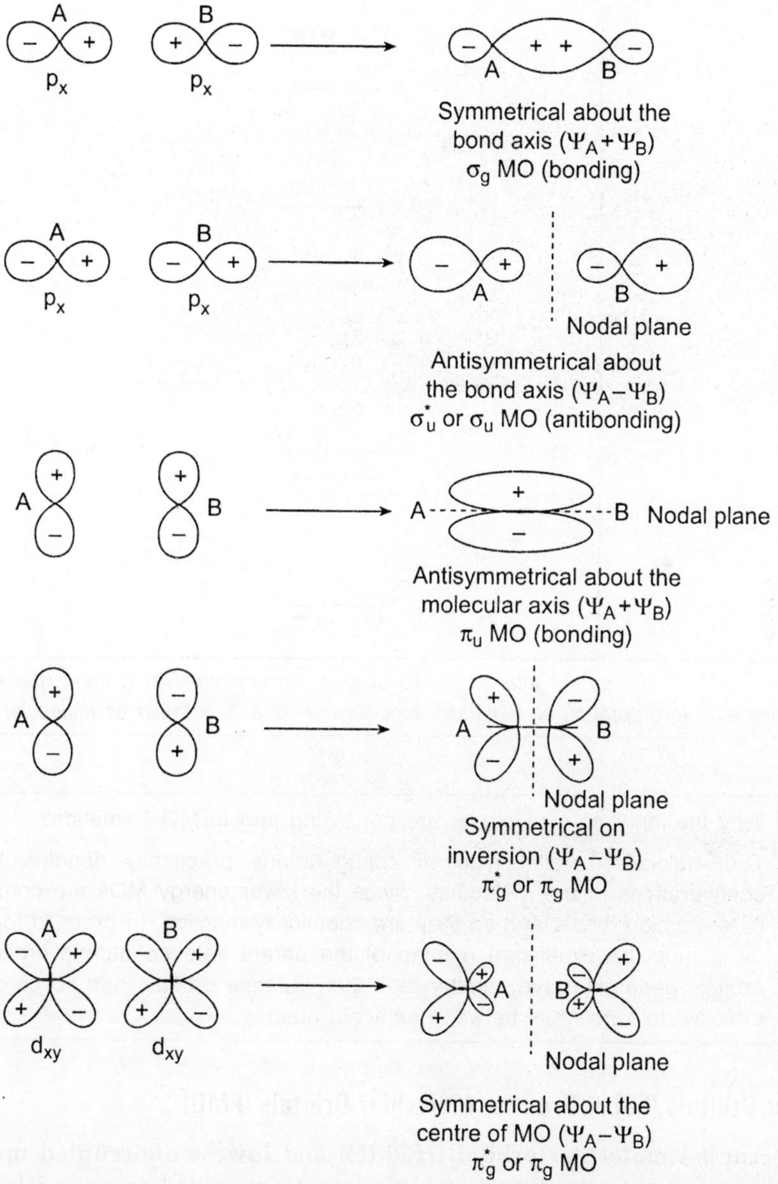


***Figure 3.25** : Formation of $\sigma$ and $\pi$ molecular orbitals by the incorporation of symmetry.*

to molecular structure, the energy order of all the available orbitals just as we did in the case of atoms. The sequence of MOs are (taking x-axis to be molecular axis) :

$$1\sigma_g < 1\sigma_u < 2\sigma_g < 2\sigma_u < 1\pi_u < 3\sigma_g < 1\pi_g < 3\sigma_u \ ......$$

With symmetry consideration of $\sigma/\sigma^*$ and $\pi/\pi^*$ molecular orbitals the molecular orbital energy diagram represented in Fig. 3.24 is modified (Fig. 3.26). It is to be noted that every $\pi$ type MO is doubly degenerate e.g. $1\pi_u$ represents the $1\pi_{yu}$ and $1\pi_{zu}$ which is constructed from $2p_y$ and $2p_z$ atomic orbitals.

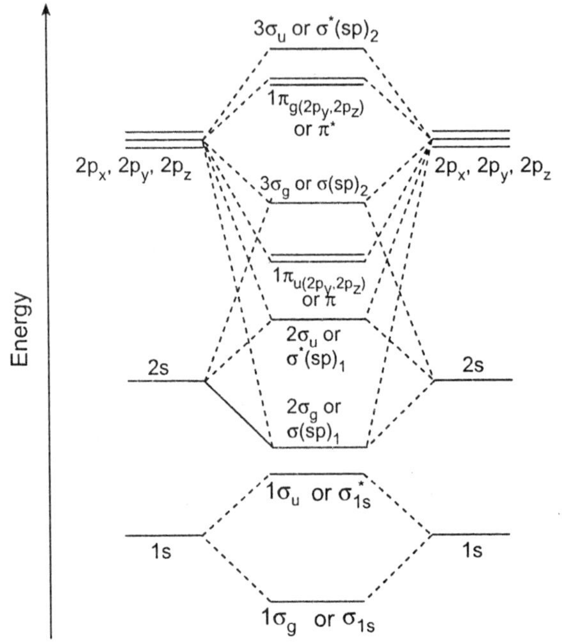


*Figure 3.26 : Modified s-p interaction MO diagram for homonuclear diatomic molecular species after incorporation of symmetry consideration (x-axis is taken as molecular axis).*

**Problem 3.38** Why the inner core electrons are not taking part in MO formation?

**Ans. :** The molecular orbital electron configurations practically denotes valence electron configurations of the molecules. Since the lower energy MOs are constructed from the core atomic orbitals and so they are chemically inactive. In order to form effective MOs there must be significant overlap of the parent valence atomic orbitals. Core atomic orbitals generally have smaller radii than valence orbitals and so core orbitals are not effective to form MOs between adjacent atoms.

### 3.4.5 Frontier Orbitals (FO) or Frontier Molecular Orbitals (FMO)

**Highest occupied molecular orbital** (HOMO) and **lowest unoccupied molecular orbital (LUMO)** for a particular molecular species are sometimes called frontier orbitals. Sometimes **singly occupied molecular orbital/orbitals** are called SOMO/SOMOs. In case of $O_2$ molecule

HOMO and LUMO are $1\pi$ or $\pi_g^*$ and $3\sigma_u$ or $\sigma^*(sp)_2$ respectively. $1\pi$ or $\pi_g^*$ is also called two SOMOs for $O_2$ molecule. HOMO, LUMO and SOMO are collectively known as FMO.

The FO or FMO has application in redox chemistry and they focus much of chemical reactivity of molecules. Hardness or softness can be explained on the basis of separation of frontier orbitals. They have application in establishing concerted mechanism and kinetic studies of organic molecules. HOMO orbital acting as **donor** whereas LUMO acting as **acceptor**. From orbital symmetry rule it can be stated that as the reactant molecules approach each other in a chemical reaction electron density will flow from the HOMO of the donor to the LUMO of the acceptor.

---

**Problem 3.39** Explain why dissociation energy of $O_2$ is less than that of $O_2^+$ but dissociation energy of $N_2$ is greater than that of $N_2^+$. **[B.U. 1984]**

**Ans. :** According to Table 3.9 bond order increases from $O_2$ to $O_2^+$ so bond dissociation energy increases from $O_2$ to $O_2^+$ while bond order decreases from $N_2$ to $N_2^+$, so bond dissociation energy decreases from $N_2$ to $N_2^+$.

---

**Problem 3.40** Explain why $SO_3$ is planar but $SO_3^{2-}$ is pyramidal? **[B.U. 1992, 2016]**

**Ans. :** According to Table 3.7 in $SO_3$ and $SO_3^{2-}$ the central S is $sp^2$ and $sp^3$ hybridised respectively.

In $SO_3$ central S: ↑↓ 3($sp^2$), ↑↓ 3($sp^2$), ↑ 3($sp^2$), ↑ $3p_z$

Two $sp^2$ hybrid orbitals having electron pairs form two coordinate link to two oxygen atoms while third $sp^2$ having a single unpaired electron forms a $\sigma$ bond with third oxygen and the remaining one electron in $3p_z$ orbital overlap with $2p_z$ orbital of third oxygen to form a $\pi$ bond and thus a double bond between S and the third oxygen is formed. Through resonance all the S–O distance are same and thus $SO_3$ molecule has a triangular planar geometry.

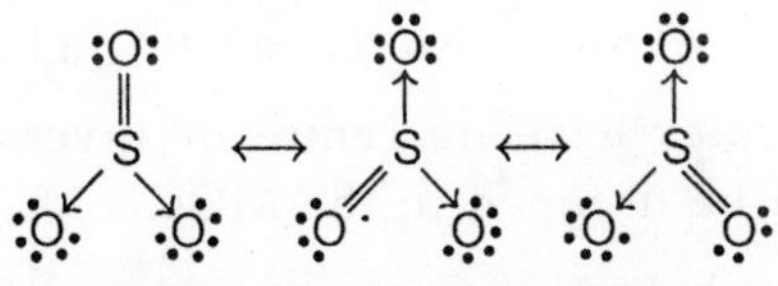

In $SO_3^{2-}$ the central S has the valence shell of 6+2 = 8 electrons :

↑↓ 3($sp^3$), ↑↓ 3($sp^3$), ↑↓ 3($sp^3$), ↑↓ 3($sp^3$)

Three $sp^3$ hybrid orbitals form three coordinate link to three oxygens and one $sp^3$ hybrid orbital has a lone pair of electron. According to VSEPR theory there is a lone pair-bond pair repulsion so that the shape of the $SO_3^{2-}$ is pyramidal having $\angle$O—S—O = 106°.

---

### 3.4.5.1 Koopmans' Theorem

Each electron in an atom or molecule possess a characteristic **binding energy** and must absorb an amount of energy equal to or in excess of the binding energy to be ionised. The kinetic energy (K.E.) of the ejected photoelectron is the energy in excess ($E_{ex}$) of the ionisation or binding energy (I.E.), i.e.

$$K.E. = E_{ex} - I.E.$$

Since the excess energy can be measured and $E_{ex}$ is known, the binding energy of the electron can be determined. The photoelectron spectroscopy (PES) records the number of photoelectrons with a particular kinetic energy against binding energy. The quantisation of energy states leads to a photoelectron spectrum having a discrete band structure. Koopmans' theorem relates ionisaiton energy to the energy of the atomic or molecular orbital in which the electron resides before ejection. This binding energies measured from PES which gives a measure of atomic or molecular orbital energies.

According to FMO concept the HOMO represents the ionisation energy and the LUMO the electron affinity for a closed shell species as stated by Koopmans' theorem. The coulomb integrals ($H_{ii}$) represent the binding energy of an electron in atom *i*. Therefore by this Koopmans' theorem the energies of the coulomb integrals are equivalent in magnitude to the ionisation potential of an electron from those orbitals. Accordingly the values used are : H(1s), 13.6 eV; C(2s), 21.4 eV; C(2p), 11.4 eV etc.

### 3.4.6 Building of molecular orbital picture for heteroatom systems

The construction of MOs for heteroatom systems is fundamentally the same for homoatom systems yet we have to consider the following aspects while constructing the MO diagram for heteronuclear diatomic molecular species :

(1) Since the combining atoms have difference in electronegativity, therefore the bonding electron cloud in MOs is non-uniformly distributed between the two nuclei and the bonding MO leans towards the more electronegative atom while the antibonding MO leans towards less electronegative atom and the molecular system assumes some ionic character with the extent of electronegativity difference of the combining atoms.

(2) The orbitals of the two combining atoms have difference in energies so that according to LCAO consideration they overlap less effectively. The BMO will have the character of lower energy atomic orbital while ABMO is associated with higher energy atomic orbital.

(3) Heteronuclear diatomic **molecules have no centre of inversion so that the terms gerade(g) or ungerade(u) will be dropped in the MOs.**

(4) Actual location of MOs in energy level diagram may change from system to system even it may change in case of homonuclear diatomic molecules. The actual location of energy levels is established by electronic absorption spectroscopy, photoelectron spectroscopy etc.

We shall try to construct MO diagram for some simple molecular systems.

(a) **Hydrogen halides (HX) X = F, Cl, Br, I :** In this case available valence orbitals are 1s orbital of hydrogen and the ns and np orbitals of halogen and the valence electrons = 1+7 = 8. The construction of MOs will be made with the consideration that the energy separation between 1s orbital of hydrogen and the halogen atoms (F, n=2; Cl, n=3; Br, n=4 and I, n=5) increases.

Simplified MO diagram depicted in Fig. 3.27. It is to be noted that as the energy difference between the combining atomic orbitals increases effective mixing of ns and np orbitals decreases. The relative position of MOs also vary from HF to HI.

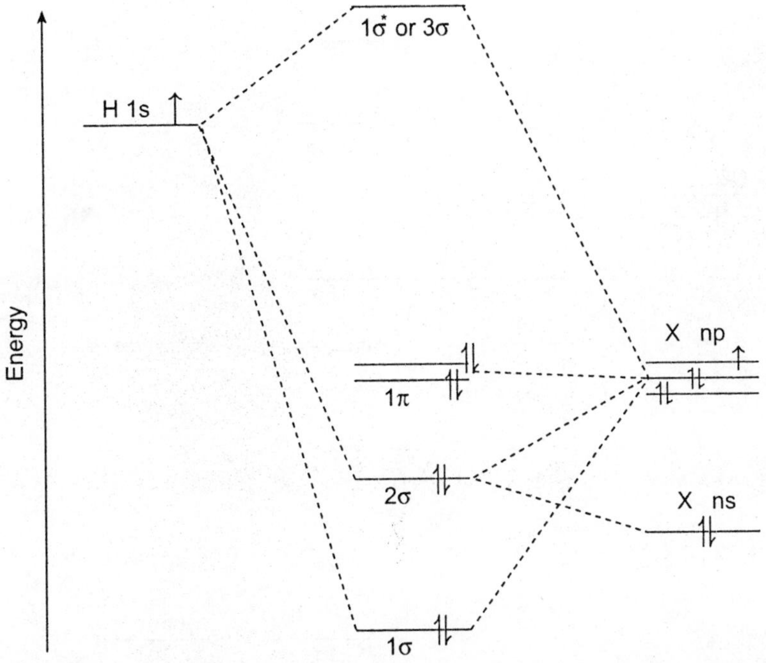


*Figure 3.27 : MO energy level diagram of HX (X=F, Cl, Br, I). The diagram is not to scale.*

(b) **Carbon monoxide :** CO is isoelectronic with $CN^-$, $N_2$ and the valence electrons are 4+6 = 10. Since C and O both have n=2 valence shell and the participating orbitals are 2s and 2p so that there is a scope of effective mixing of 2s and 2p orbitals to form sp hybrid orbitals. Moreover oxygen is more electronegative than carbon so that atomic orbitals of oxygen are of lower energy (Fig. 3.28). The bond order of CO is 3 which is in accord with the VBT structure **:C⇚O:** with the formation of a coordinate linkage from oxygen to carbon.

(c) **Nitric oxide :** The NO molecule is an odd electron one (5+6 = 11 electrons). The MO picture is comparable to that of CO having one unpaired electron in $1\pi^*$ orbital which is reponsible for paramagnetism of the molecule. The position of MOs are not similar to that of CO due to the electronegativity difference between N and O with that of C and O. Bond order for NO is 2.5. This fractional BO indicates the resonating structure for NO according to VBT $\left(:\dot{N}=\ddot{O}: \longleftrightarrow :\ddot{N}=\dot{O}:\right)$. The nitrosyl cation ($NO^+$) is a good ligand, has the bond order 3 which is more stable than NO.

### 3.4.7 Molecular orbitals for triatomic molecules or ions

(a) **Homonuclear triatomic molecule $H_3$ and molecule ion $H_3^+$, $H_3^-$ :** The electrons in $H_3$, $H_3^+$ and $H_3^-$ are 3, 2 and 4 respectively and in all the cases three 1s orbitals of three hydrogen atoms combine to form three MOs (Fig. 3.29). Among these species $H_3^+$ is resonably stable having $\sigma_g^2$ configuration, $H_3$ and $H_3^-$ has configuration $\sigma_g^2(\sigma_u^{nb})^1$ and $\sigma_g^2(\sigma_u^{nb})^2$ respectively.

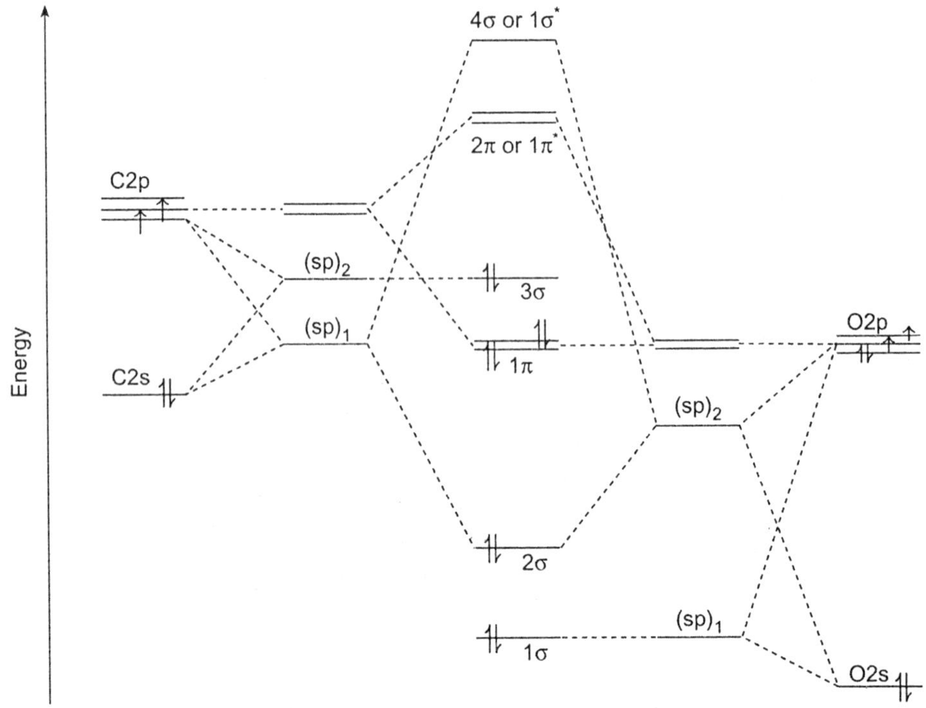


*Figure 3.28 : MO diagram (not to scale) of CO with s-p interaction. LUMO and HOMO are 2π and 3σ respectively.*

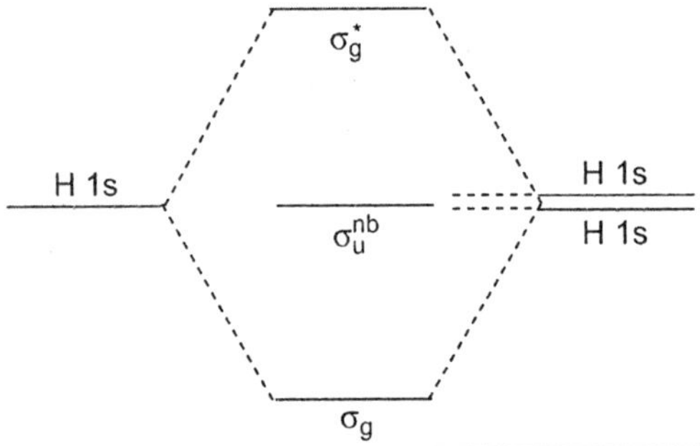


*Figure 3.29 : MO energy level diagram for triatomic molecule $H_3$ and ionic species $H_3^+$ and $H_3^-$. $\sigma_u^{nb}$ represents non bonding MO.*

In all cases the bond order is one. $H_3^+$ species is considered to be linear **three centre two electron bond (3c-2e)**. In these species there is H–H–H link so that bond order per link = $\frac{1}{2}$ = 0.5 whereas in $H_2$ the BO is one which indicates the stability sequence $H_2 > H_3^+ > H_3 > H_3^-$.

Although $H_3^+$ is considered to be unstable having 3c–2e bond but the molecule $B_2H_6$ remarkably stable where 3c–2e bond exist. The molecule has 12 valence electrons and there are two B–H–B three centre two electron bonds having four terminal hydrogen atoms ($H_t$) and two bridging hydrogens ($H_b$).

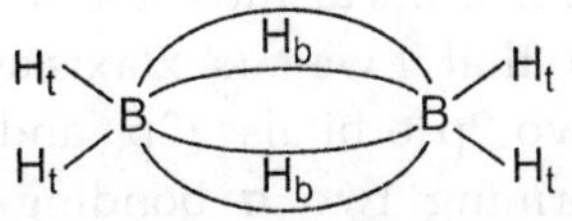


Here each boron is $sp^3$ hybridised. Two $sp^3$ hybrid orbitals of each B forms four B–$H_t$ links. One $sp^3$ hybrid orbital having one unpaired electron overlap with 1s orbital of bridging hydrogen ($H_b$) and overlap with an empty $sp^3$ hybrid orbital of another B to form B–$H_b$–B. In a similar way another B–$H_b$–B link is formed. The MO picture of B–$H_b$–B and $B_2H_6$ is given in Fig. 3.30.

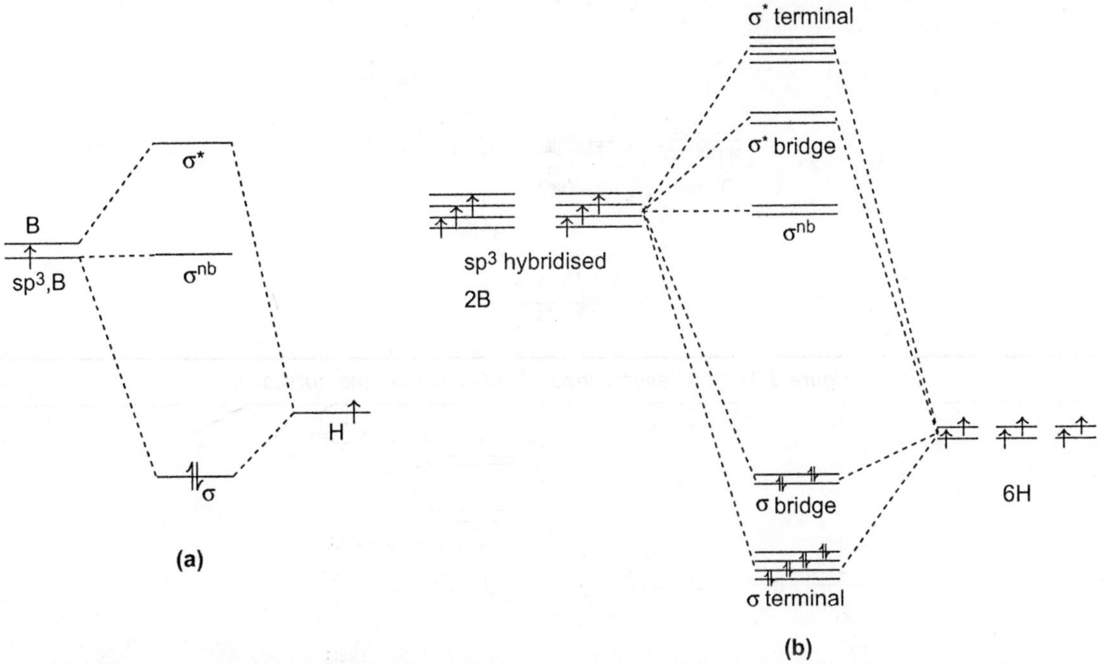


***Figure 3.30*** *: MO picture of (a) B–H–B bond, (b) $B_2H_6$ molecule.*

**Problem 3.41** Draw the energy level diagram for the MOs of CO molecule and hence explain its low polarity and ligational behaviour towards a metal ion. **[C.U. 2001]**

**Ans. :** For MO diagram Sec. 3.4.6(b). For low polarity we can say that there is a dative bond from O to C i.e. :C$\Leftarrow$O: which is responsible for low polarity or low dipole moment of CO. There are two lone pairs residing each on C and O and oxygen is more electronegative than C, so carbon lone pair will function in order to stabilise the low oxidation state of metals.

(b) **MO picture of $H_2O$ :** According to VSEPR model this molecule is angular, the central oxygen is $sp^3$ hybridised having the valence electron = 2+6 = 8. The MO energy level is drawn with the consideration of the overlap of four $sp^3$ hybrid orbitals of oxygen and two 1s orbitals of hydrogen (Fig. 3.31). Two $\sigma^{nb}_{sp^3}$ molecular orbitals are occupied by lone pairs and the two $\sigma^{*}_{sp^3}$ antibonding orbitals remain unoccupied.

(c) **MO diagram of $CO_2$ :** This is a linear molecule having valence electrons = 4 + 2 × 6 = 16. The central C is sp hybridised so that if we take x-axis as bond axis, the hybrid orbital is $sp_x$. The two $sp_x$ hybrid orbitals and two 2p orbitals ($2p_y$ and $2p_z$) of carbon overlap with six 2p orbitals of two oxygen atoms forming two σ bonding MO, two π bonding MO, two π nonbonding MO, two π* antibonding MO and two σ* antibonding MO in order of increasing energy (Fig. 3.32). The two 2s orbitals of two oxygens will not take part in MO and remain as non bonding pair.

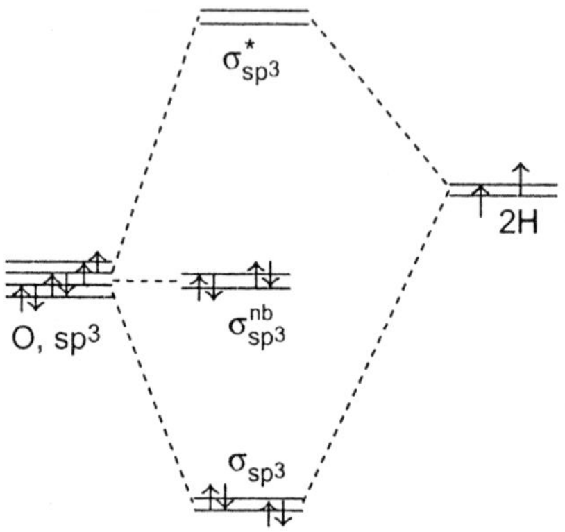


***Figure 3.31** : MO energy level of $H_2O$ molecule (not to scale).*

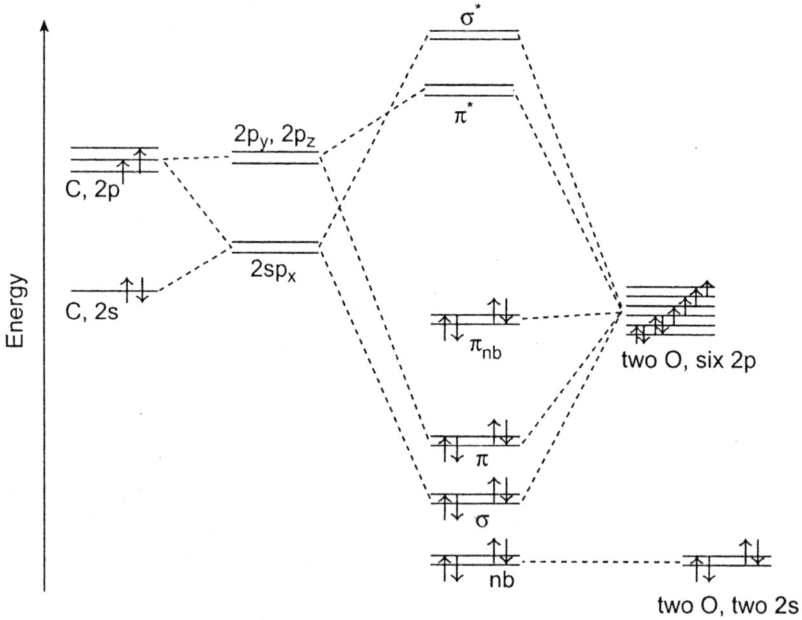


***Figure 3.32** : MO energy level diagram of $CO_2$.*

(d) **MO energy level diagram of $BeH_2$ and $BeF_2$ :** Both the molecules are linear with the central Be $sp_x$ hybridised (taking x axis as molecular axis). $BeH_2$ and $BeF_2$ have the valence electrons 4 and 16 respectively. It is to be noted that fluorine and hydrogen atomic orbitals are lower in energy than beryllium (Fig. 3.33). MO diagram of $BeCl_2$ can also be represented with the consideration of 3s and 3p orbitals of chlorine.

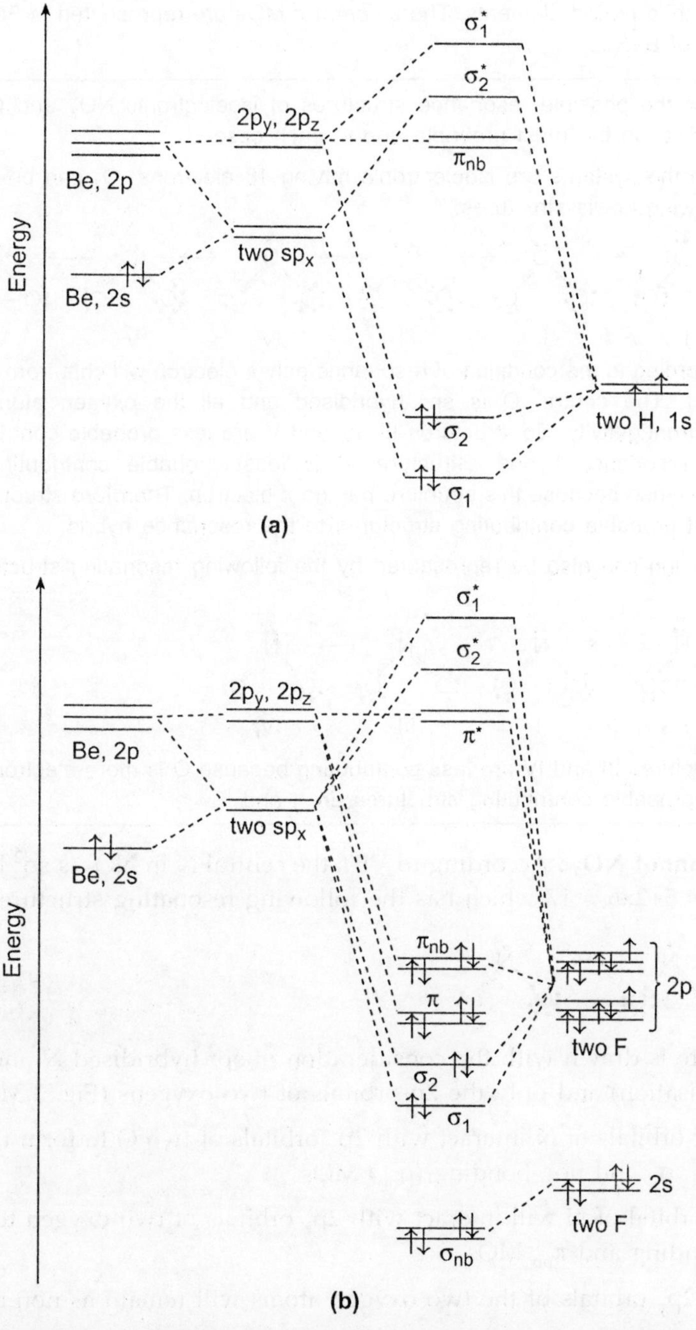


*Figure 3.33 : MO picture of (a) $BeH_2$, (b) $BeF_2$ (not to the scale). In case of $BeF_2$ two fluorine 2s orbitals forms two non-bonding MO.*

**Problem 3.42** X and Y are atoms of the second and third period of the periodic table respectively. Assuming x as the bonding axis show the different $\pi$ MOs X can form with Y. **[C.U. 2002]**

**Ans. :** This is considered to be as $BeCl_2$ molecule where Be and Cl are the members of second and third period elements. The different $\pi$ MOs are represented in Fig. 3.33(b) similar to that of $BeF_2$.

**Problem 3.43** Draw the possible resonance structures of isoelectronic $NO_2^-$ and $O_3$ and idicate with justification the most probable one in each case. **[C.U. 2000]**

**Ans. :** Both the systems are isoelectronic having 18 electrons. $O_3$ can be represented by the following Lewis structures.

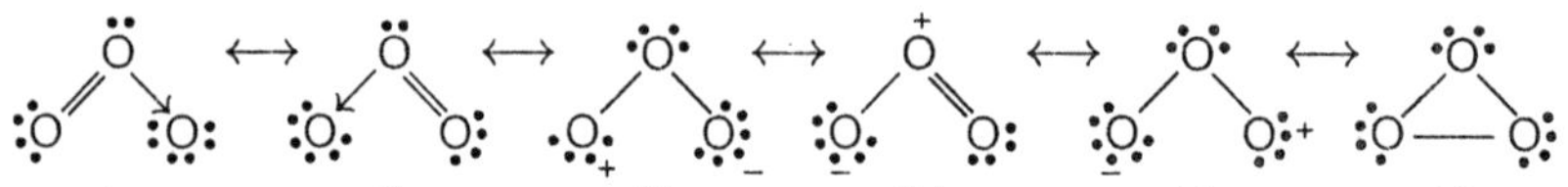

I II III IV V VI

According to the condition of resonance only $\pi$ electron will shift from one structure to the other. The central O is $sp^2$ hybridised and all the oxygen atoms have the same electronegativity. So structures III, IV and V are less probable contributing structures to the resonance hybrid. Structure VI is least probable contributing structure to the resonance because this structure has no $\pi$ electron. Therefore structures I and II are the most probable contributing structures to the resonance hybrid.

$NO_2^-$ ion can also be represented by the following resonating structures :

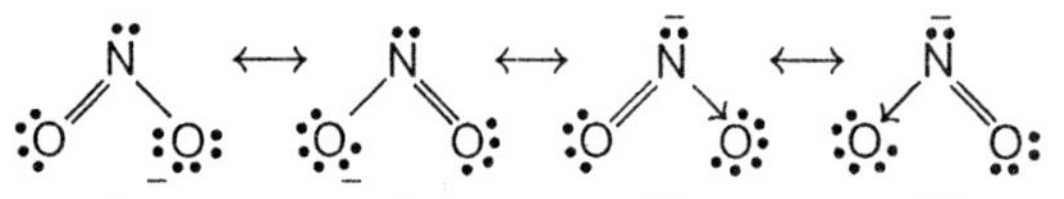

I II III IV

Structures III and IV are less contributing because O is more electronegative than N. So the probable contributing structures are I and II.

(e) **MO diagram of $NO_2$ :** According to VBT the central N in $NO_2$ is $sp^2$ hybridised having valence electrons = 5+2×6 = 17 which has the following resonating structures :

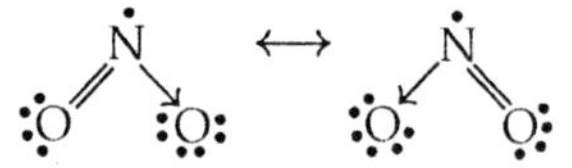

The MO picture is drawn with the consideration of $sp^2$ hybridised N and $2p_z$ orbital of N (left out of hybridisation) and only the 2p orbitals of two oxygens (Fig. 3.34).

(i) Three $sp^2$ orbitals of N interact with $2p_x$ orbitals of two O to form two degenerate $\sigma_1$, $\sigma_2$ and $\sigma_1^*$, $\sigma_2^*$ and non-bonding ($\sigma_{nb}$) MOs.

(ii) The $2p_z$ orbital of N will interact with $2p_z$ orbitals of two oxygen to form $\pi$ bonding, $\pi^*$ antibonding and $\pi_{nb}$ MOs.

(iii) The two $2p_y$ orbitals of the two oxygen atoms will remain as non-bonding pairs.

(iv) Note that the 2s orbital of each oxygen will not take part in MO formation and remain as non-bonding pair.

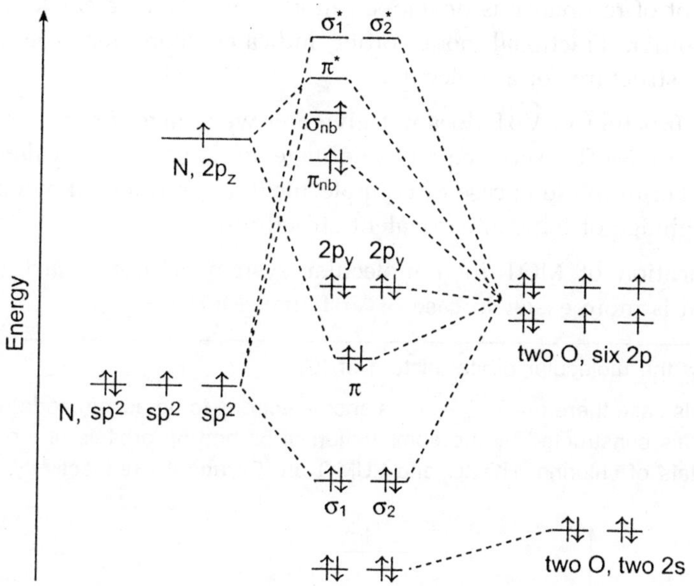


***Figure 3.34** : MO diagram of $NO_2$ (x-axis is taken as molecular axis) is not to scale.*

Fig. 3.34 can also be used to draw the MO picture of isoelectronic $NO_2^-$ and $O_3$. This is left for exercise.

### 3.4.8 Comparison of VBT and MOT

**Similarities :**

(a) Both the theories can be applied to form 2c-2e, 3c-2e as also in case of partial ionic character of a chemical bond.

(b) The symmetry and energy matching of atomic orbitals is of prime importance in both the theories.

(c) Both the theories predict the enhancement of electron density in between the nuclei of a chemical bond.

(d) The directional property of a covalent bond is established in both the theories.

**Dissimilarities :**

(a) According to VBT the concept of orbitals in a molecule is the overlap of atomic orbitals to form directional hybrid orbitals which are essentially atomic orbitals, but MOT demands the formation of molecular orbitals by LCAO which is more accurate than VBT because in a molecule the orbitals should be molecular orbitals and not atomic orbitals.

(b) VBT fails to explain the paramagnetic character of $O_2$ and $B_2$ molecules whereas this can be explained nicely by MOT.

(c) The concept of bonding, antibonding and non-bonding orbitals are introduced in case of MOT. Only σ and π bonds are formed in VBT.

(d) The concept of resonance is developed in case of VBT whereas MOT has the concept of bond order. Fractional bond order indicates there may be the possibility of resonating structures of a molecule.

(e) The wave function in VBT does not give any weightage for ionic contribution in the structure but MOT exerts equal weightage of ionic and covalent structures of a molecule. For example in case of $H_2$ molecule $\Psi_{MO}$ represented by equation (3.15) give equal weightage of ionic and covalent structures.

(f) The application of MOT in a molecular system is more exact than VBT though calculation is more easier in case of VBT than MOT.

---

**Problem 3.44** Draw the molecular orbital picture of ICl.

**Ans. :** In this case there are 7+7 = 14 valence electrons to be accommodated into the molecular orbitals constructed by the combination of 5s and 5p orbitals of iondine and 3s plus 3p orbitals of chlorine. HOMO and LUMO are $2\pi$ and $4\sigma$ respectively.

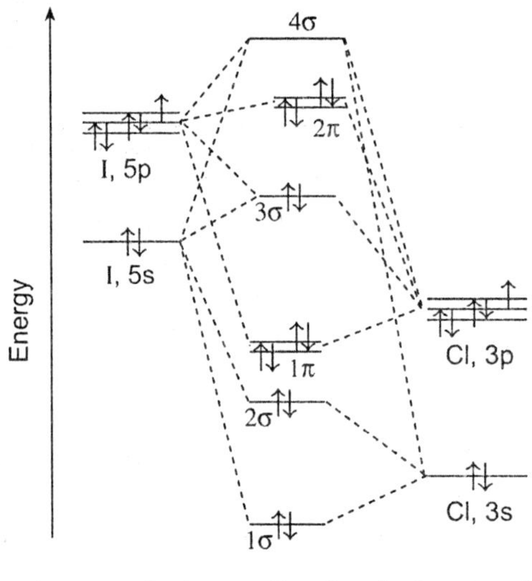


As chlorine is more electronegative than iodine, bonding orbitals are chlorine dominated while antibonding orbitals are iodine dominated.

---

### 3.4.9 Some application of molecular orbital theory

(a) $O_2$ molecule is friendly to skin but $Cl_2$ is more toxic though it is expected that $O_2$ is more reactive than $Cl_2$ in terms of of availability of electrons (unpaired electrons). On the other hand transition metals are generally reactive with elemental oxygen but less reactive to $Cl_2$.

The answer of these problems can be explained on the **spin conservation principle** which states that in any chemical reaction the total change in spin is zero i.e. $\Delta S = 0$, where

$$\Delta S = S_{product} - S_{reactant}$$

Skin has no unpaired electron, so that it does not react with $O_2$ or we may say skin is friendly with $O_2$ molecule. As $Cl_2$ molecule is diamagnetic, so spin conservation principle satisfies that diamagnetic compounds react quickly with other diamagnetic compound maintaining $\Delta S = 0$ and thereby $Cl_2$ is more toxic to skin.

Transition metals have unpaired electrons and $O_2$ has also unpaired electrons. So spin conservation principle suggests the favourable reaction between them.

(b) Photons can interact with halogen acids in a different way :

$$HCl \xrightarrow{h\nu} HCl^+ \qquad HBr \xrightarrow{h\nu} \tfrac{1}{2}H_2 + \tfrac{1}{2}Br_2$$

$$HI \xrightarrow{h\nu} \tfrac{1}{2}H_2 + \tfrac{1}{2}I_2$$

Then a question arises why there is a difference.

According to Fig. 3.27 the energy difference between 1σ and 2σ MOs is sufficient enough in case of HCl because electronegativity difference between hydrogen and the halogens (Cl → Br → I) decreases as also the population of participating atomic orbitals increases from Cl → I. σ and π orbitals have different symmetry, but photons can interact with σ orbitals only and so electron is ejected from 2σ orbital, since 2σ orbital is chlorine dominated MO in case of HCl. Therefore on light irridiation it ionises to form molecule cation ($HCl^+$).

But in case of HBr and HI, the 1σ and 2σ orbitals are populated by H and Br electrons both (or with H and I electrons both). As the energy difference between the 1σ and 2σ orbitals is small, the electrons may eject either from 1σ or 2σ orbitals and thereby the bond will be dissociated to form $\tfrac{1}{2}H_2$ and $\tfrac{1}{2}Br_2$ (or $\tfrac{1}{2}H_2$ and $\tfrac{1}{2}I_2$).

(c) Reactivity sequence of isoelectronic $C_2H_2$, CO and $N_2$ is $C_2H_2 > CO > N_2$. Molecular orbital picture of these molecules are shown below (only MOs are shown) :

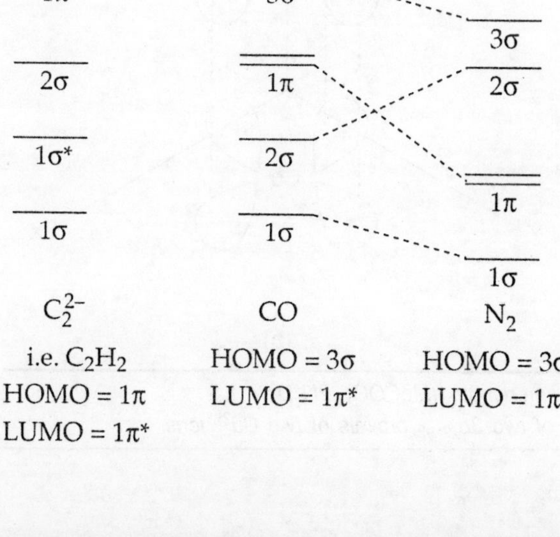


In case of $C_2^{2-}$ the exterior orbital i.e. HOMO is 1π but in case of CO and $N_2$ the exterior orbitals are 3σ. Orbitals of same symmetry will not cross each other upon interaction but σ and π MOs are heterosymmetric, so they will cross each other thereby MO picture of CO and $N_2$ are different. Reactivity of a molecule depends on the exterior π orbitals. As π orbitals are under the control of σ orbitals the reactivity will be less. In $C_2H_2$ the exterior 1π is not under the influence of σ orbital, so reactivity will be highest for $C_2H_2$. In CO, 1π orbitals are under the control of 3σ and in $N_2$, the 1π orbitals are further deep seated under the care of 2σ, 3σ orbitals, so reactivity will be decreased from CO to $N_2$.

## 3.5 SOME SPECIAL TYPES OF BOND

### (a) Delta (δ) and quadruple bond

We have seem that bonding MOs that result from the positive overlap of two atomic orbitals are σ and π molecular orbitals. The third category of MO which is formed is konwn as δ-bond. σ bond is formed by the combination of two s orbitals, which has **no nodal plane.** σ and π bonds may be formed by the combination of p orbitals and **π bond has a nodal plane.** On the other hand **when two atomic orbitals overlap to form molecular orbitals which has two nodal planes passing through both the nuclei is called δ-bond. Only d and f block elements can form δ-bonding but s and p orbitals are unable to do this bonding.** Thus transition elements and inner transition elements are the best members to form the δ-bond. Two δ-bonds may be formed per pair of atoms, one using two $d_{xy}$ and other using two $d_{x^2-y^2}$ orbitals that overlap in four regions (Fig. 3.35(b)). The formation of δ-bond is represented in cupric (II) acetate dihydrate (Fig. 3.35(a)). Here Cu-Cu distance is longer than can be expected in the metal. Partial quenching of magnetic moment occurs due to overlap of orbitals and the magnetic moment recorded 1.4 BM (BM = Bohr magneton) (Expected value for $3d^9$, Cu(II) = 1.73 BM). **δ-bonds are weaker than π bonds and the tendency to form δ-bonds increases down the d-block elements.**

Another allied behaviour of the overlap of orbitals to form an unusual bond which is known as **quadruple bond.** This is a combination of σ, π and δ bond i.e. σ+π+δ. One example of this bond is $Re_2Cl_8^{2-}$ where $ReCl_4$ unit is $dsp^2$ hybridised.

**It is to be noted that in case of δ-bond there will be no formation of σ and π bonds but quadruple bond is associated with the formation of σ, π and δ-bonds.** A quadruple bond is some what stronger than δ-bond.

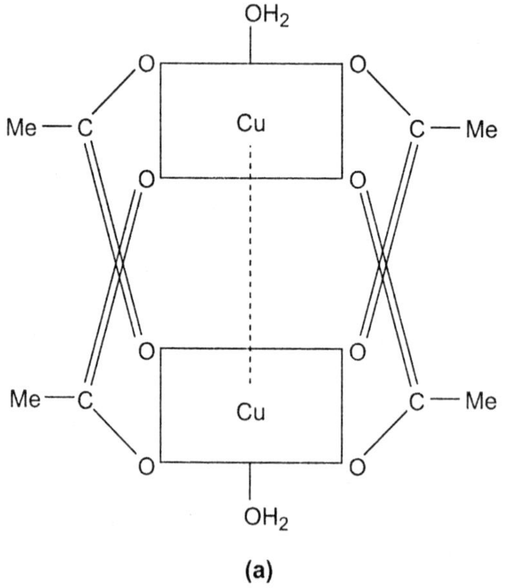


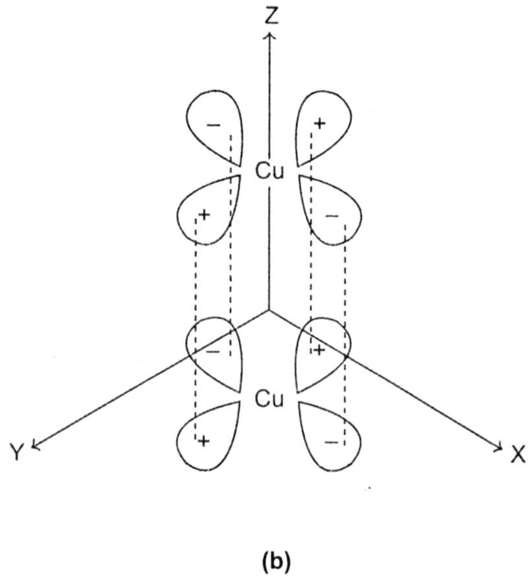


***Figure 3.35** : (a) Structure of dimeric $Cu_2\ (MeCOO)_4.2H_2O$,*
*(b) Formation of δ-bond by lateral overlap of two $3d_{x2-y2}$ orbitals of two $Cu^{2+}$ ions.*

### (b) Dative, portray dative and retrodative bond

Lewis stated that a different type of covalent bond may be formed when both the electrons for sharing between the atoms are contributed by one atom only. This is sometimes called **coordinate covalent, coordinate, dative or portray dative bond.** For example $H_3N \rightarrow BF_3$, $H_2\dot{\ddot{O}} + H^+ \rightarrow H_3O^+$ etc. This is also called **donor-acceptor bond.**

We have seen that in CO the HOMO is 3σ orbital which is populated by two electrons (Fig. 3.28). CO is a potential ligand for the stabilisation of low oxidation states of transition metals in which the electron pair of HOMO is donated to the low valent metal through σ route and in order to relieve the electron density, the low valent metal back donates the electron density to the π* antibonding (LUMO) of CO through the formation of π bond i.e. $M \underset{\pi}{\overset{\sigma}{\rightleftarrows}} CO$. This type of σ-π interaction is called **synergic effect** and this type of bond is sometimes called **retrodative bond.** Here CO acting as σ-basic and π-acidic character. Isosteric species* e.g. $CN^-$, CO, $NO^+$ can act as retrodative bond formation. Retrodative bond more stronger than dative bond. Stable complexes of transition metals are formed by retrodative bond. Dative bond breaks on simple heating but not for retrodative bond as the later is accompanied with the former.

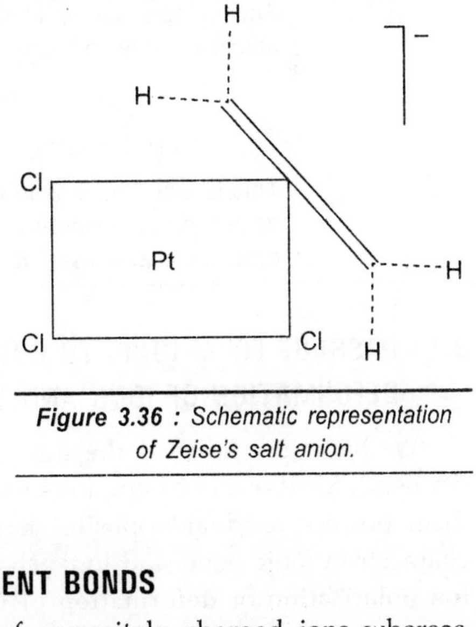


*Figure 3.36 : Schematic representation of Zeise's salt anion.*

### (c) μ-bond

Ethylene is a π donor ligand can form a special type of bond which is known as μ-bond e.g. Zeise's salt, $K[PtCl_3(C_2H_4)]$. Here central Pt(II) is $dsp^2$ having square planar geometry (Fig. 3.36). Here one of the coordination site is occupied by C=C i.e. the bonding is attributed by the interaction between π-electron of ethylene and a hybrid orbital of the metal. This type of bond is known as μ-bond in alkene and alkyne complexes of Pt(II).

## 3.6 PROPERTIES ASSOCIATED WITH IONIC AND COVALENT BONDS

1. Ionic bond is formed by electrostatic attraction of oppositely charged ions whereas covalent bond is formed by mutual sharing of electrons.

2. Since electrostatic attractions operate between cations and anions so discrete ionic compound formation is impossible–ionic compounds always exist as a giant molecule. On the other hand covalent compounds may exist as a discrete molecule or as a giant molecule e.g. CO exist as a discrete molecule while $SiO_2$ is a giant molecule.

3. Ionic compounds are hard but brittle having high melting points and boiling points. Covalent compounds are generally soft having low melting points and boiling points (There may be exception).

4. Generally ionic compounds are soluble in polar solvents while covalent compounds are soluble in non-polar solvents (There are several exception of this statement).

* The molecular species having similar outer electronic configurations are called isosteric species.

5. Ionic compounds are conductors of electricity in aqueous solution as also in fused state. But the covalent compounds are non-conductors.

6. **Ionic bond has no directional property but covalent bond must have a directional property.**

7. The reactions of ionic compounds are the reactions of their own ions but the reactions of covalent compounds are the reactions of the molecule as a whole.

---

**Problem 3.45** $N_3^-$ is more resonance stabilised than $HN_3$—comment. **[C.U. 2003]**
$NaN_3$ is more stable than $HN_3$—explain. **[B.U. 2014]**

**Ans. :** This difference is due to the fact that the N–N distances in $HN_3$ are different whereas in $N_3^-$ the distances are same.

These facts suggest that both $HN_3$ and $N_3^-$ are resonance stabilised but $N_3^-$ is more stabilised due to more number of canonical forms.

$$H-N=\overset{+}{N}=\overset{-}{N} \longleftrightarrow H-\overset{-}{N}-\overset{+}{N}\equiv N$$

$$\overset{-}{N}=\overset{+}{N}=\overset{-}{N} \longleftrightarrow \overset{2-}{N}-\overset{+}{N}\equiv N \longleftrightarrow N\equiv\overset{+}{N}-\overset{2-}{N}$$

This is also the reason why ionic azides are more stable than covalent azides. Covalent azides are considered to have the structure of undissociated $HN_3$. Due to the greater resonance stabilisation ionic azides are more stable than covalent azides.

---

## 3.7 PASSAGE FROM IONIC TO COVALENT CHARACTER. ION POLARISATION OR DEFORMATION OF IONS AND THEIR CONSEQUENCES. FAJAN'S RULE

We have just finished the discussion of two extreme type of chemical bonds (ionic and covalent). Most of the compounds neither cent percent ionic nor cent percent covalent—both of them possess residual opposite character. In order to understand the existance of covalent character in ionic bond and ionic character in a covalent bond we have to develop the idea of **ion polarisation** or **deformation of ions.**

Let us consider that two oppositely charged ions $C^+$ and $A^-$ are approaching each other from a large distance of separation (Fig. 3.37(a)). When they approach each other closely the net charge of $C^+$ will attract the electron cloud of $A^-$ towards itself due to the electrostatic force operated between them. As a consequence the electron cloud of $A^-$ becomes elongated along the direction of $C^+$ (Fig. 3.37(b)). In other words we can say that the electron shell of $A^-$ is said to **deformed** or **polarised** by the electron shell of $C^+$. Similarly on approaching each other the electron shell of $C^+$ is said to deform by the electron shell of $A^-$ i.e. both the electron shell of $C^+$ and $A^-$ is deformed (Fig. 3.37(c)). But as $C^+$ has small size than $A^-$ so that electron density in $C^+$ is more tightly bound thereby polarisation caused to $C^+$ by $A^-$ is negligibly small. The net effect of this deformation or ion polarisation is that the electron density instead of remaining either $C^+$ or $A^-$ are shared by both the nuclei. This leads to partial polar character (Fig. 3.37(d)) i.e. predominant covalent character with residual ionic character. When the ion polarisation occurs to a larger

*Concept of ion polarisation*

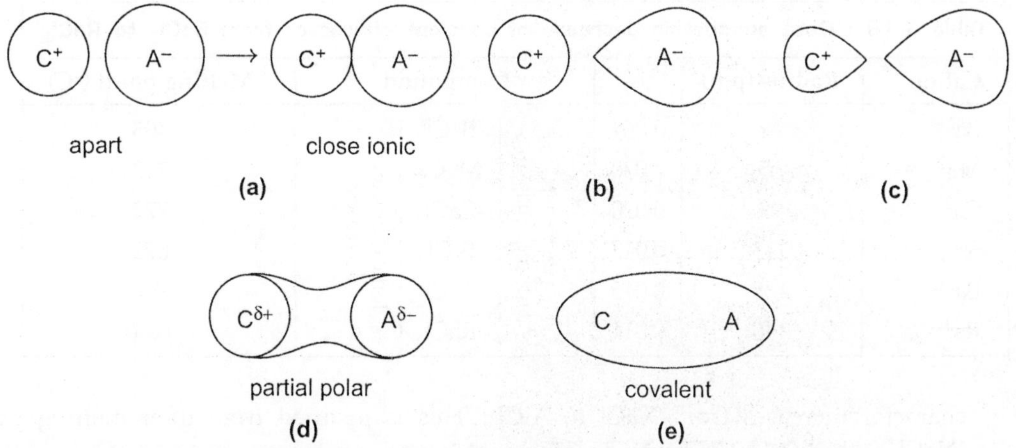


*Figure 3.37 : Representation of ion polarisation and passage from ionic to covalent character.*

extent, greater degree of covalency arises (Fig. 3.37(e)) and thus the term arises 'passage from ionic to covalent character'.

Then a question arises what characteristic features of cations ($C^+$) and anions ($A^-$) lead us this transformations from ionic to covalent character i.e. the extent of polarisation is governed by Fajan's rule. The statement of Fajan's rule are :

(a) **Either the cation or anion should have large charge :** As the charge of cation or anion increases, the extent of polarisation increases and the covalent character increases. For example among $SnCl_2$ and $SnCl_4$, the anion ($Cl^-$) is same but the cations are $Sn^{2+}$ and $Sn^{4+}$. As the charge of $Sn^{4+}$ is more compared to that of $Sn^{2+}$, so $Sn^{4+}$ will polarise the $Cl^-$ ion more strongly so that $SnCl_4$ will be more covalent than $SnCl_2$. This is justified from their melting points ($SnCl_2$, 606°; $SnCl_4$, 114°).

On the other hand as the anionic charge increases from $F^-$ to $O^{2-}$, the covalent character decreases from CaO to $CaF_2$. $Ca^{2+}$ will exert greater polarising power on the larger anion thereby CaO is more covalent than $CaF_2$.

(b) **The cation should have small size :** As charge of the cations increases the size decreases and as mentioned earlier a cation will take a pivotal role in ion polarisation so that for cations a parameter is associated for deformation of ions which is known as **ionic potential** (φ), which is denoted by charge/radius ratio. The small size cation has robust electron shell, and this will exert greater polarising power on the anion thereby covalent character predominating. The parameter φ is also an index of ion polarisation and as **the value of φ increases, covalent character increases.** The sequence of polarisation can be demonstrated by alkaline earth metal cations as $Be^{2+} > Mg^{2+} > Ca^{2+} > Sr^{2+} > Ba^{2+} > Ra^{2+}$ so that in case of alkaline earth metal halides covalent character decreases from $BeCl_2$ to $RaCl_2$ (Table 3.10).

A similar situation arises in case of chlorides of $Na^+$, $Mg^{2+}$ and $Al^{3+}$. In this case size decreases from $Na^+$ to $Al^{3+}$ ($Na^+$, 95 pm; $Mg^{2+}$, 65 pm; $Al^{3+}$, 50 pm) so that covalent

**Table 3.10 : Data supporting decrease of covalent character from $BeCl_2$ to $RaCl_2$**

| Cation | Radius (pm) | $\phi$ | Compound | Melting point (°C) |
|---|---|---|---|---|
| $Be^{2+}$ | 31 | 0.064 | $BeCl_2$ | 405 |
| $Mg^{2+}$ | 65 | 0.030 | $MgCl_2$ | 712 |
| $Ca^{2+}$ | 99 | 0.020 | $CaCl_2$ | 772 |
| $Sr^{2+}$ | 113 | 0.018 | $SrCl_2$ | 872 |
| $Ba^{2+}$ | 135 | 0.015 | $BaCl_2$ | 960 |
| $Ra^{2+}$ | 140 | 0.014 | $RaCl_2$ | 1000 |

Covalent character decreases ↓ (from $BeCl_2$ to $RaCl_2$); Ionic character decreases ↑ (from $RaCl_2$ to $BeCl_2$)

character increases from NaCl to $AlCl_3$. This is justified from their melting points (NaCl, 800°; $MgCl_2$, 712°; $AlCl_3$, 573°).

(c) **The anion will have large size :** As the charge of the anion increases, the size increases. The electron cloud of large size anion is loosely held by the nucleus so that this anion will be strongly polarised by cation and covalent character increases. This is exemplified by the calcium halides. The radius of $F^-$, $Cl^-$, $Br^-$ and $I^-$ are 136, 181, 195 and 216 pm respectively so that $Ca^{2+}$ ion will exert greater polarising power on $I^-$ ion thereby $CaI_2$ will be primarily covalent and $CaF_2$ is mostly ionic. This is justified by their melting points ($CaF_2$, 1392°; $CaCl_2$ 772°; $CaBr_2$, 730° and $CaI_2$, 575°).

(d) **The cation should have a non-inert gas configuration or pseudo inert gas configuration :** In considering the polarising power of a cation we should have to consider the periphery electron and the effective nuclear charge of the nucleus. s and p electrons are more effective to screen the nuclear charge than d electrons so that a cation having $ns^2np^6$ configuration will experience less effective nuclear charge than a configuration of $ns^2np^6(n-1)d^{10}$ which is known as **pseudo noble gas configuration**. This $ns^2np^6(n-1)d^{10}$ configuration having 18 electrons will have a loosely held periphery electron than $ns^2np^6$ configuration so that the cations having $ns^2np^6(n-1)d^{10}$ configuration is more polarising than a cation having $ns^2np^6$ configuration. This is demonstrated by the example of CuCl and NaCl. The size of $Cu^+$ and $Na^+$ are almost same (radius of $Na^+$ = 95 pm and that of $Cu^+$ = 96 pm) but as $Cu^+$ has the $3s^23p^63d^{10}$ configuration which is more deformable than $Na^+$ having $2s^22p^6$ configuration, therefore CuCl will be more covalent than NaCl which is also justified from the lesser solubility of CuCl than NaCl. Similarly AgCl is more covalent than KCl though $K^+$ and $Ag^+$ has approximately equal ionic radius.

**Problem 3.46** The shapes of nitrite and nitrile ions respectively are—
1. Linear and angular, 2. Angular and linear, 3. Both angular, 4. Both linear.
**[NET 2001]**

**Ans.** : 2.

**Problem 3.47** Compare the geometry of $NO_2$ and $NO_2^+$ from VSEPR point of view. Compare the bond angle of $NO_2$, $NO_2^+$ and $NO_2^-$. **[C.U. 2003]**

**Ans.** : The ∠O–N–O in $NO_2^+$, $NO_2$ and $NO_2^-$ is 180°, 135° and 115° respectively.

In $NO_2$ central N is $sp^2$ hybridised $\underset{sp^2}{\uparrow\downarrow}\ \underset{sp^2}{\uparrow}\ \underset{sp^2}{\uparrow}\ \underset{2p_z}{\uparrow}$. There is an odd electron in one of the $sp^2$ hybrid orbital of nitrogen and due to the repulsion of the electron with bond pair the bond angle is changed from ideal 120° to 135° and the molecule is **angular**.

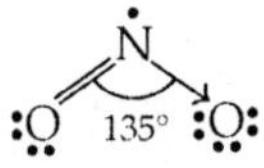

In $NO_2^+$ the valence shell nitrogen has 5+6+6–1 = 16 electrons. Here central $N^+$ is sp hybridised $\underset{sp}{\uparrow}\ \underset{sp}{\uparrow}\ \underset{2p_y}{\uparrow}\ \underset{2p_z}{\uparrow}$.

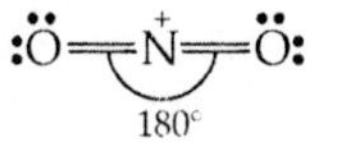

Since π bonds does not influence the geometry of the molecule, the molecule ion ($NO_2^+$) assumes **linear** geometry.

In $NO_2^-$ the central $N^-$ is $sp^2$ hybridised $\underset{sp^2}{\uparrow\downarrow}\ \underset{sp^2}{\uparrow\downarrow}\ \underset{sp^2}{\uparrow}\ \underset{2p_z}{\uparrow}$ having the valence shell of central N = 5+6+6+1 = 18 electrons. Here one of the $sp^2$ hybrid orbital of N has a lone pair of electron and due to the lone pair-bond pair repulsion the bond angle changed from 120° to 115°.

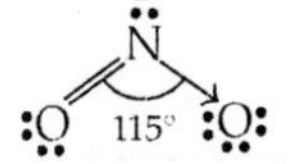

---

**Problem 3.48** Explain the structure of $TeCl_6^{2-}$ in terms of VBT. **[C.U. 2003]**

**Ans.** : Te (IV) has the electronic configuration $1s^2 2s^2 2p^6 3s^2 3p^6 4s^2 3d^{10} 4p^6 4d^{10} 5p^2$. $TeCl_6^{2-}$ has a regular octahedral geometry according to VBT but has some exception from VSEPR model in having 7 electron pairs in its valence shell. Six electron pairs form the regular octahedral geometry and one lone pair residing in spherical s orbital. This lone pair is called stereochemically inactive lone pair and this lone pair does not influence the geometry of the molecule ($:TeCl_6^{2-}$).

---

### 3.7.1 Diagonal relationship

One of the manifestation of ion polarisation is the diagonal relationship. This is also one of the application of ionic potential(ϕ).

On moving left to right in a period the size of the element decreases while on moving down the group the size increases. **When an element shows similarities with an element in the next higher group in the next period is called diagonal relationship.** This is due to Fajan's rule. Accordingly the polarising power of an ion increases with increase in charge and decreases with the size. When moving from left to right in a period, the ionic charge increases while ionic size decreases. Both these factors tend to increase polarising power. On the other hand, on moving down in a group, the ionic charge remain the same, but ionic size increases and therefore the polarising power decreases. As a result of these two opposing effects, if we move in the diagonal direction these two effects will be balanced each other partly at least i.e. there will be a balance between size increase and decrease of elements as also their ions and also there will be a balance between polarising power of ions. Hence appreciable similarity in properties between the elements exist which are placed in the diagonal position of the periodic table [Fig. 3.38(a)]. This **similarity between the elements in the diagonal direction is known as diagonal relationship.** In the first two short periods there are some similarities between the pairs of elements e.g. Li/Mg, Be/Al and B/Si. These are called diagonal pairs [Fig. 3.38(b)] and among these **the best diagonal pair is the Be/Al pair**. This is due to the fact that Be/Al pair has close ionic potential

*Cause of diagonal relationship*

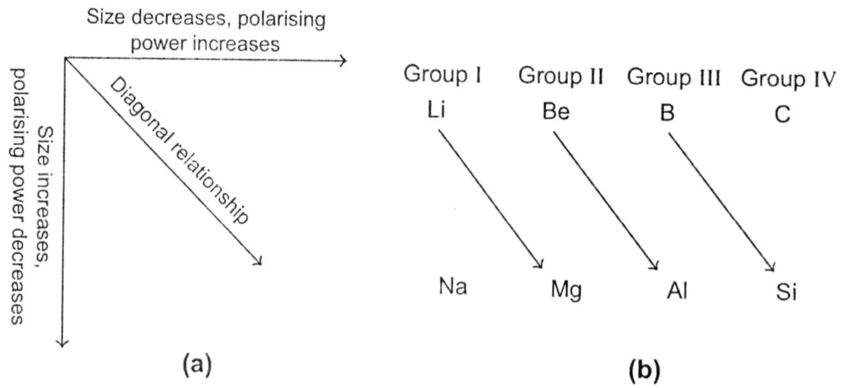


*Figure 3.38 : (a) Schematic representation of diagonal relationship, (b) Diagonal pairs in the periodic table.*

values (Table 3.11). One point should have to be mentioned here that why **diagonal relationship exist between two short periods**. This is due to the fact that when atomic number of the elements are low i.e. in those cases where the lower lying participating atomic orbitals are reduced, the variation of size increase and decrease (in a group and in a period), the polarising power etc. will occur in a more or less regular manner so that in these cases the validity of diagonal relationship exist.

**Table 3.11 : Some parameters for diagonal pairs**

| Element | Li | Mg | Be | Al | B | Si |
|---|---|---|---|---|---|---|
| Charge | +1 | +2 | +2 | +3 | +3 | +4 |
| Radius (pm) | 60 | 65 | 31 | 50 | 20 | 41 |
| Ionic potential ($\phi$) | 0.017 | 0.030 | 0.064 | 0.060 | 0.150 | 0.098 |
| Electronegativity (Pauling) | 1.0 | 1.2 | 1.5 | 1.5 | 2.0 | 1.8 |

But when the atomic number of the elements increases, the lower lying participating atomic orbitals also increases (repulsive force between the lower lying inner shell electrons increases) and the validity of the diagonal relationship vitiated so that for long period elements having lower lying d or f orbitals, the diagonal relationship does not hold good.

**Problem 3.49** $MgCO_3$ is thermally less stable than $CaCO_3$—explain. **[C.U. 2007, 2014, B.U. 2012]**

**Ans. :** According to Fajan's rule, a cation having small size will exert greater polarising power on the anion and the compound will be more covalent. Between $MgCO_3$ and $CaCO_3$, the anion ($CO_3^{2-}$) is the same but the cations $Mg^{2+}$ and $Ca^{2+}$ respectively. $Mg^{2+}$ has a small size than $Ca^{2+}$ and so $Mg^{2+}$ will exert greater polarising power on $CO_3^{2-}$ and so $MgCO_3$ will be more covalent than $CaCO_3$. We know that thermal stability of covalent compounds are less than that of ionic compounds so that $MgCO_3$ will be decomposed thermally more easily than that of $CaCO_3$.

### 3.7.2 The polar covalent bond

The concept of ion polarisation leads to the transformation of ionic character to covalent character of a chemical bond. A true covalent bond having cent percent covalent character is impossible even in case of homonuclear diatomic molecules e.g. $H_2$, $Cl_2$ etc. **Intermediate polarisation of ions leads to the formation of partial ionic and partial covalent character of a chemical bond i.e. a polar covalent bond is formed** [Fig. 3.37(d)]. Polar covalent molecule is one in which there is a separation of the centre of gravity of positive and negative charges. Polarity of a molecule is determined by the term **dipole moment which is equal to the product of the charge and distance of separation of charges.** Since the electronic charge = $4.8 \times 10^{-10}$ esu and distance is of the order of angstrom ($10^{-8}$ cm), therefore the dipole moment (μ) = $4.8 \times 10^{-10}$ esu × $10^{-8}$ cm = $4.8 \times 10^{-18}$ esu cm = 4.8 Debye or 4.8D where 1 Debye or 1D = $10^{-18}$ esu cm. Therefore Debye is defined as the value of dipole moment where the unit positive and negative charges separated by a distance ≈ 0.209Å. The current SI unit of μ is Coulomb metre (Cm). Therefore 1D = $1.6 \times 10^{-19}$ coulomb × $0.209 \times 10^{-8}$ cm = $1.6 \times 10^{-19}$ Coulomb × $0.209 \times 10^{-10}$ metre = $3.34 \times 10^{-30}$ Cm.

In order to predict the polar character of a molecule we have to consider two aspects simultaneously—**the electronegativity difference between the combining atoms and the geometry of the molecule.** For example in $BeF_2$, $BeCl_2$ molecules, though there is an electronegativity difference between Be and halogen atom yet due to the linear geometry of the molecules, the centre of gravity of partial positive and negative charges coincide on Be ($Cl^{\delta-}$—$Be^{\delta++}$—$Cl^{\delta-}$) so that the molecules are non-polar. On the other hand due to the angular nature of $H_2O$ molecule, there will be a separation of the centre of gravity of positive and negative charges so that $H_2O$ is a polar molecule 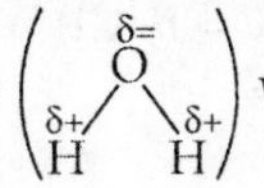

which is established from its dipole moment value (1.84D).

The dipole moment of a molecule can be obtained by the vectorial addition of different bond moments present in a molecule. Thus μ is zero for $CH_4$ and $CCl_4$ i.e. non-polar molecules whereas $CHCl_3$ (μ = 1.15D) and $CH_3Cl$ (μ = 1.85D) are polar molecules (Fig. 3.39).

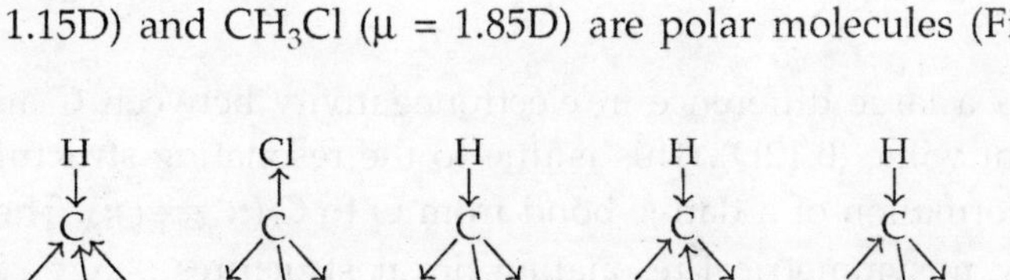


*Figure 3.39 : Representation of non-polar and polar molecules of carbon.*

In all the molecules central carbon is $sp^3$ hybridised. In $CH_4$ and $CCl_4$ molecules where central carbon is symmetrically arranged by four hydrogens and four chlorine atoms respectively thereby the centre of gravity of positive and negative charges coincide on carbon, so that the molecules are non-polar. For other molecules the sequence of dipole moment is $CH_3Cl > CH_2Cl_2 > CHCl_3$. For these molecules vector sum of the individual bond moments will give us erroneous result. In $CCl_4$, vector sum of individual bond moments cancel each other

($\mu = 0$). As we go from $CCl_4$ to $CHCl_3$, one chlorine of $CCl_4$ is substituted by hydrogen thereby symmetrical structure of $CCl_4$ becomes unsymmetrical so that $\mu$ of $CHCl_3$ is 1.15D. In this way when more electronegative chlorine atoms are substituted by less electronegative hydrogen atoms, the structure becomes more and more unsymmetrical and the separation of the centre of gravity of positive and negative charges are more wide, so that $\mu$ increases from $CH_2Cl_2$ to $CH_3Cl$ (Fig. 3.39). However the variation of dipole moment of these molecules can also be explained by variation of bond angles along with individual bond moments. But the results not always tally with the experimental value.

Effect of lone pair largely influence the polarity of a molecule. The $\mu$ of $NH_3$ (1.46D) and $NF_3$ (0.2D). In both the molecules central nitrogen is $sp^3$ hybridised and one of the $sp^3$ hybrid orbital is occupied by the lone pair. The electronegativity difference between N and F is almost the same as that between N and H. But the direction of individual bond moments are opposite to each other In $NH_3$, the N—H bond moments are along the direction of lone pair as N is more electronegative than H(Fig. 3.40). While in $NF_3$, the N—F bond moments are opposite to lone pair because F is more electronegative than N. So that charge separation in case of $NH_3$ will be more compared to that of $NF_3$ leading to higher dipole moment of $NH_3$ than that of $NF_3$. The $\angle HNH = 107°$ while $\angle FNF = 102°$ in $NH_3$ and $NF_3$ respectively. These bond angles can be utilised to calculate the individual bond moments.

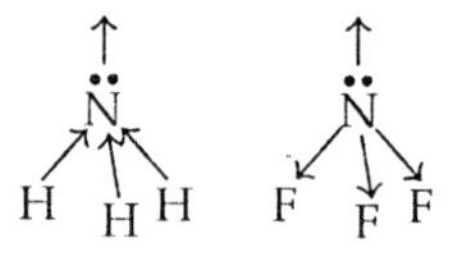


***Figure 3.40** : Effect of lone pair in the variation of $\mu$ of $NH_3$ and $NF_3$.*

The $\mu$ is zero for $CO_2$ due to the linear structure (O=C=O) while the corresponding angular $SO_2$ molecule has $\mu = 1.63D$. $BX_3(X = F, Cl)$ molecules has a poor dipole moment value due to the following resonating structures :

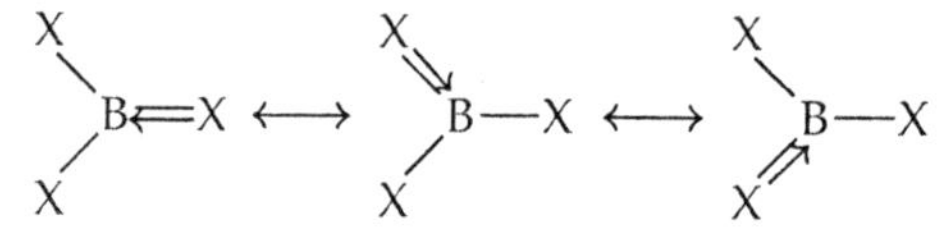

Though there is a large difference in electronegativity between C and O yet CO registers poor dipole moment value (0.12D). This is due to the resonating structures of CO (Sec. 3.2.12) as also due to the formation of a dative bond from O to C ($:C \Leftarrow O:$). The $\mu = 0.17D$ for $N_2O$ is due to its following unsymmetrical resonating linear structures :

$$:\overset{-}{\ddot{N}}=\overset{+}{N}=\ddot{O}: \longleftrightarrow :N\equiv\overset{+}{N}-\overset{-}{\ddot{\underset{\cdot\cdot}{O}}}:$$

Dipole moment determination can be applied to calculate **the percentage ionic character** of a molecule. Theoretical calculation of percent ionic character(P) is related to the electronegativity difference between the atom $A(X_A)$ and $B(X_B)$ as :

$$P = 16(X_A - X_B) + 3.5(X_A - X_B)^2 \quad \ldots(3.32)$$

For HCl,
$$P = 16(X_{Cl} - X_H) + 3.5(X_{Cl} - X_H)^2$$
$$= 16(3.0 - 2.1) + 3.5(3.0 - 2.1)^2$$
$$= 16 \times 0.09 + 3.5(0.9)^2 = 17.23\%$$

The equation (3.32) is known as **Hannay-Smith equation** The dipole moment of HCl is 1.03D and its bond length is 1.27Å. Therefore the partial charge ($\delta^+$ or $\delta^-$) that accumulated on

$$\text{H and Cl} = \frac{1.03\times10^{-18}\text{esu cm}}{1.27\times10^{-8}\text{ cm}} = 0.81\times10^{-10}\text{ esu.}$$

If HCl would have been 100% ionic then in that case there will be accumulation of +1 and –1 charge on H and Cl respectively = $4.8 \times 10^{-10}$ esu. Therefore, the percentage ionic character of

$$\text{HCl} = \frac{0.81\times10^{-10}\text{esu}}{4.8\times10^{-10}\text{esu}}\times 100 = 16.9 \approx 17\%.$$

In this way by measuring the dipole moment value and bond length we are able to calculate the percent ionic character of a molecule and this may be corrborated with equation (3.32). The percent ionic character calculated for HF, HBr and HI in this way are 45%, 12% and 5% respectively.

*Application of dipole moment*

The dipole moment values can also be applied to determine the **cis** and **trans** isomers. The cis isomer will have larger μ value compared to that of trans which has zero or almost zero μ value e.g.

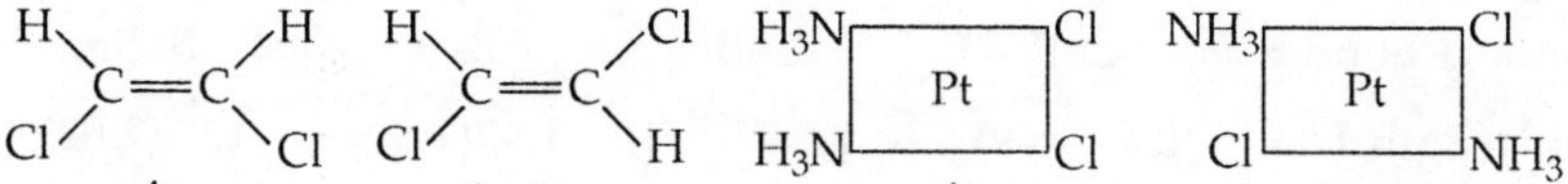


cis $\mu \neq 0$ | trans $\mu = 0$ | cis $\mu \neq 0$ | trans $\mu = 0$

However in case of $Pt(Et_2S)_2Cl_2$ ($Et_2S$ = diethylthioether) the trans isomer will have some dipole moment value as :

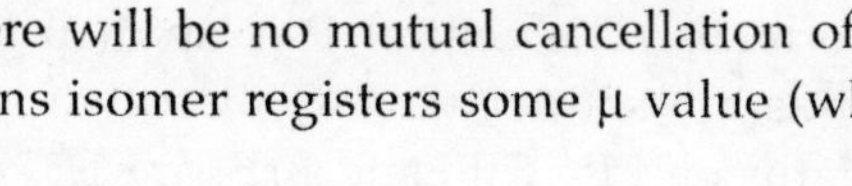


cis $\mu$ = 9.5D | trans $\mu$ = 2.4D

In this case the μ value of trans isomer is due to the contribution of lone pair on S atom of each Pt–$SEt_2$ bond. This can be explained on the fact that lone pair contribution is not in the direction of Pt–S bond axis thereby there will be no mutual cancellation of lone pair effect even in case of trans isomer so that the trans isomer registers some μ value (which is lower than cis isomer)

## 3.8 BOND LENGTH

It is defined **as the average distance between the nuclei of two covalently bonded atoms in a molecule.** Generally the bond length in a covalent molecule is the sum of the atomic radii of the combining atoms. But in some cases the bond may be stretched (due to repulsive force) or may be compressed (due to the combination of bulkier atom). For example the atomic radius of F is 0.64Å so that F–F bond length should be = 1.28Å, but the experimental value = 1.42Å. This is due to the repulsive force operated between two F atoms (Sec. 2.7). On the other hand the bond length in HCl = 1.28Å though the atomic radius of H and Cl is 0.37Å and 0.99Å respectively. This shrinkage of bond length is due to the bulkier effect of chlorine (Sec. 2.7).

The determination of bond length can be made by X-ray diffraction technique, neutron diffraction, nuclear magnetic resonance etc. With reference to Sec. 2.7 we shall consider the following aspects in brief regarding variation of bond length.

(a) **Bond multiplicity :** As bond multiplicity increases bond length decreases e.g. carbon–carbon bond length varies as :

$$-C\equiv C- \; < \; >C=C< \; < \; -\overset{|}{\underset{|}{C}}-\overset{|}{\underset{|}{C}}-$$

(b) **Effect of lone pairs :** Due to the lone pair—lone pair repulsion bond length decreases. For example C–C bond length (1.54Å) in $C_2H_6$ is higher than N–N distance (1.47Å) in $N_2H_4$. This reduction is due to the lone pair respulsion of two N atoms of $N_2H_4$ ( $H_2\ddot{N}-\ddot{N}H_2$ ).

(c) **Electronegativity difference :** As the electronegativity difference between the combining atoms increases, the bond length decreases. For example :

| Chemical bond : | C–F | C–Br | B–F | B–Br |
|---|---|---|---|---|
| Bond length (Å) | 1.36(1.49) | 1.88(1.91) | 1.42(1.60) | 1.95(2.02) |

In first bracket the calculated values are given.

(d) **Effect of π-bonding :** Due to **synergic effect** (Sec. 2.7) M–C bond length is considerably short. The C–F bond distance in $CF_4$ is 1.36Å whereas Si–F distance in $SiF_4$ is 1.56Å. This Si–F distance would be expected to be much larger than 1.56Å. This shrinkage of bond length can be explained on the basis of overlap of Si and F orbitals to form $p_\pi$–$d_\pi$ interaction. In general formation of π-bond decreases the bond length.

(e) **Effect of s-character of hybrid orbitals :** As the s-character of the hybrid orbitals increases generally the bond length decreases. Greater s-character means greater overlap thereby shorter bond length. For example C—H bond length in $C_2H_2$, $C_2H_4$ and $C_2H_6$ are 1.05, 1.07 and 1.09Å respectively where the carbon is hybridised as sp, $sp^2$ and $sp^3$.

(f) **Non-equivalent hybrid orbitals :** In Sec. 3.2.6 it has been stated that $PCl_5$ forms non-equivalent hybrid orbitals. The central P is $sp^3d(sp^2 + pd)$ hybridised where three equatorial P–Cl bonds are shorter (2.04Å) than two axial bonds (2.19Å). The axial/equatorial bond lengths on $PF_5$, $ClF_3$ and $BrF_3$ are 1.57/1.53, 1.71/1.62 and 1.81/1.72Å respectively.

## 3.9 BOND ANGLE

This is an important physical parameter of a molecular species by which we determine the symmetry and the shape of the molecule/ion. **This is defined as the angle operated by two adjacent bonds in a molecular system.** The factors governing the variation of bond angles are the rules followed in VSEPR theory (Sec. 3.3 and 3.3.1).

## 3.10 INTERMOLECULAR FORCES : VAN DER WAALS FORCES

Besides the extreme types of chemical forces there are several types of weak forces operated between the molecules which are known as intermolecular forces or van der Waals forces. It has been established that such forces are also present in the solid as also in the liquid states of several substances. The forces which hold the atoms of noble gases together in liquid or solid state are called **London forces** and collectively all van der Waals forces are also called **London forces.** All types of forces are quite weak (2-20 kJ $mol^{-1}$) compared to that of ionic or covalent forces. Covalent force is of the order of 200-400 kJ $mol^{-1}$. There are several types of intermolecular forces operative between the molecules.

(a) **Dipole-dipole interaction** : A dipolar molecule will attract another dipolar molecule causing an alignment of oppositely charged centres of the molecules. This results a weak interaction between the molecules which is known as **dipole-dipole interaction.** For example in BrF and ICl the following type of molecular association occurs :

$$\overset{\delta^+}{Br}-\overset{\delta^-}{F}\cdots\cdots\overset{\delta^+}{Br}-\overset{\delta^-}{F}\cdots\cdots\overset{\delta^+}{Br}-\overset{\delta^-}{F} \qquad \overset{\delta^+}{I}-\overset{\delta^-}{Cl}\cdots\cdots\overset{\delta^+}{I}-\overset{\delta^-}{Cl}\cdots\cdots\overset{\delta^+}{I}-\overset{\delta^-}{Cl}$$

Due to this weak interaction the melting point and boiling point is somewhat higher than can be expected for single molecule. These weak forces still operative even in liquid and solid phases.

A extreme case of dipole-dipole interaction is the hydrogen bonding (will be discussed later) operative between the molecules of some hydrides. The interaction energy ($E_{D\text{-}D}$) between two dipolar molecules having dipole moments $\mu_D$ and $\mu_D$ separated by a distance r is given by—

$$E_{D-D} = \frac{-2\mu_D^2\mu_D^2}{(4\pi\varepsilon_0)^2}\left(\frac{1}{3kT}\right)\left(\frac{1}{r^6}\right) \qquad \ldots(3.33)$$

$4\pi\varepsilon_0$ is the permittivity of the medium, k is Boltzmann constant and T is absolute temperature. $E_{D\text{-}D}$ is called **Keesom energy** which is inversely proportional to temperature. At higher temperature, kinetic energy of the molecules increases thereby the molecules are randomly oriented so that dipole-dipole interaction decreases.

These forces are effective only at very short distances and are directional in nature.

(b) **Dipole-induced dipole interaction** : A permanent dipolar molecule will induce a dipolar nature to a molecule and a weak force operated between the molecules which is known as dipole-induced dipole interaction. This is a consequence of laws of electrostatics where a charged condenser (+ or –) is in nearest neighbour of an uncharged condenser, the nearest position of the uncharged condenser is induced by opposite charge and the farthest position is induced by same charge.

The formation of inert gas hydrates ($Xe.6H_2O$) and deutarates ($Xe.6D_2O$) are the examples of this type of dipole-induced dipole interaction. The inert gas having $ns^2np^6$ configuration is considered to be a uncharged condenser and water is a dipolar molecule having a permanent dipole moment. When this dipolar water molecules come closer to the inert gas shell induces dipolar nature to the inert gas and there will be a weak attractive force operated to form $Xe.6H_2O$. Due to the greater polarisability of larger inert gases, the solubility of larger inert gases increases in water. That a weak force operated between Xe and water molecules is due to the fact that on slight heating the inert gas hydrate breaks up. Similar mechanism of formation may be followed for certain compounds of phenol e.g. $Kr(phenol)_2$, $Xe(phenol)_2$ etc.

The interaction energy ($E_{D\text{-}iD}$) is given by the expression :

$$E_{D-iD} = \frac{-2\mu_D^2\alpha}{(4\pi\varepsilon_0)^2}\left(\frac{1}{r^6}\right) \qquad ...(3.34)$$

where $\mu_D$ is the dipole moment of the dipolar molecule, $\alpha$ is the polarisability of the molecule where dipolar nature is induced, r and $4\pi\varepsilon_0$ as usual is the distance and permittivity.

(c) **Instantaneous dipoles interaction :** This type of London forces operated between the real gas molecules. These forces are also the manifestation of the attractive force exist between the real gas molecules e.g. $H_2$, $Cl_2$, $N_2$ etc. These molecules have no permanent dipolar character but at a particular moment they execute instantaneous polar nature. For example in $H_2$ molecule the two hydrogen atoms are identical (H–H) and the two electrons is midway between two H atoms. But for an instantaneous moment let the two electrons are very close to one of the hydrogen atom and instantaneous dipolar nature created e.g. ( $\overset{\delta^-}{H_A}\!:\,\overset{\delta^+}{H_B}$ ). These instantaneous dipoles will induce the neighbouring molecules to induce an electrostatic attraction among the instantaneous dipoles ( $\overset{\delta^-}{H_A}\!:\,\overset{\delta^+}{H_B} \longleftrightarrow \overset{\delta^-}{H_A}\!:\,\overset{\delta^+}{H_B}$ ). This interaction energy is known as instantaneous dipole induced dipole interaction or London dispersion forces.

The interaction energy per pair ($E_L$) of same kind of molecules is given by,

$$E_L = \frac{-3I\alpha^2}{4(4\pi\varepsilon_0)}\left(\frac{1}{r^6}\right) \qquad ...(3.35)$$

where I is the ionisation energy of the molecule and $\alpha$ as before is the polarisability. This expression (Equation 3.35) is independent of temperature which indicates that London forces operate at all temperature in case of real gas molecules. These London forces responsible for the variation of boiling points of inert gases.

This type of weak forces increases with increasing polarisability of the molecules. The ease of polarisation also depends on the molecular size which is also related to molecular weight. The polarisability increases with increase of molecular weight. Larger size molecules and larger molecular weight are more effective for polarisation than the smaller one. In case of covalent molecules boiling points/melting points increases with increasing molecular weight/increasing polarisability e.g. boiling points of C, Si, Ge and Sn hydrides is of the order $CH_4 < SiH_4 < GeH_4 < SnH_4$.

(d) **Ion-dipole interaction :** During the dissolution of an ionic compound in polar solvent, this type of interaction occurs [Sec. 3.1.4(a)]. Negative and positive ends of dipolar molecules are attracted by cations and anions respectively of ionic compounds and thereby there will be the rupture of electrostatic attractive force in ionic compounds and dissolution occurs due to **ion-dipole interaction**. The dissolution of NaCl in water is an example of this type of weak interaction force.

Similarly this type of force is responsible for the formation of hydrates and ammoniates e.g. $CuSO_4.5H_2O$ and $CoCl_3.6NH_3$. $NH_3$ is a dipolar molecule and the negative nitrogen ends of $NH_3$ molecules are attracted by $Co^{3+}$ to form ammoniate i.e. $[Co(NH_3)_6]Cl_3$. However a question arises whether ion-dipole interaction is the sole criteria for the formation of ammoniates or not because $Co^{3+}$–N bond is extremely strong, even conc. $H_2SO_4$ cannot be able to knock off the $NH_3$ molecules from the sphere of influence of $Co^{3+}$. The explanation of the extra stability of metal-ligand bonds is beyond the scope of ion-dipole interaction (very weak force) and may be explained by theory of metal-ligand bonding (Sec. 7.14).

The weak ion-dipole interaction energy ($E_{I-D}$) is represented as :

$$E_{I-D} = \frac{-Z\mu_D}{4\pi\varepsilon_0}\left(\frac{1}{r^2}\right) \quad ...(3.36)$$

where Z = charge of the ion (positive or negative), $\mu_D$ = dipole moment of the dipolar molecule, other terms as usual.

(e) **Ion-induced dipole interaction :** A charged ion will distort the electron density of a large polarisable molecule and induce a dipole moment due to distortion polarisation. This induced dipole further interact with the ion to form a charged species. The interaction of halide ion ($I^-$) with halogen molecule ($I_2$) to form a polyhalide ($I_3^-$) ion ($I^- + I_2 \rightarrow I_3^-$) is one of the well known example of this type of weak interaction. That there is a weak interaction operated between $I^-$ and $I_2$ is revealed from the following two equations :

$$I_2 + 2Na_2S_2O_3 \longrightarrow Na_2S_4O_6 + 2NaI$$

i.e. one mole $I_2$ reacts with two moles of $Na_2S_2O_3$.

$$KI_3 + 2Na_2S_2O_3 \longrightarrow KI + Na_2S_4O_6 + 2NaI$$

This reaction indicates that one mole $KI_3$ reacts with two moles $Na_2S_2O_3$ and simply $KI_3$ breaks up into KI + $I_2$, and practically $I_2$ reacts with $Na_2S_2O_3$.

The stability sequence of polyhalide ions is of the order $I_3^- > Br_3^- > Cl_3^-$. This is due to the fact that larger halogens are easily polarisable and polyhalides of larger halogens are more stable than lower halogens. This is justified from their stability constant values : $I_3^-$ (725.0), $Br_3^-$ (17.8) and $Cl_3^-$ (0.01).

The interaction energy ($E_{I-iD}$) can be represented by,

$$E_{I-iD} = -\frac{1}{2}\frac{Z^2\alpha}{r^4} \quad ...(3.37)$$

where Z is the charge of the ion, $\alpha$ = polarisability of the molecule, r denotes the distance of separation of the ion and polarisable molecule.

### 3.10.1 The Hydrogen Bond

As mentioned this is a kind of weak intermolecular forces operated between the dipolar molecules (Sec. 3.10). But the strength of the hydrogen bond (10-40 kJ $mol^{-1}$) is somewhat larger compared to other intermolecular forces (2-20 kJ $mol^{-1}$). That's why we can say that hydrogen bond more stronger than ordinary dipole-dipole interaction.

When hydrogen is joined to a strong electronegative atom e.g. fluorine, nitrogen or oxygen, the bond will be polar and hydrogen forms a bridge between two highly electronegative atoms as :

$$\cdots X—H\cdots X—H\cdots X—H\cdots X$$

where X = strong electronegative atom. The dotted line represents the hydrogen bond. The weak electrostatic bond between hydrogen and strong electronegative atom is known as hydrogen bond e.g.

$$\cdots F—H\cdots F—H\cdots F—H\cdots F$$

The cluster of HF molecules may be described as $(HF)_n$. Therefore it is defined as **the attractive force which binds hydrogen of one molecule with electronegative atom (such as fluorine) of another molecule of the same substance is known as hydrogen bond.**

### 3.10.2 Effects of Hydorgen Bonding

(a) **State of aggregation :** A molecule capable of executing hydrogen bonding is associated with several other molecules so that state of aggregation may change from gaseous to the liquid state. One important example is the liquid state of aggregation of water is due to the formation of hydrogen bonding between $H_2O$ molecules while $H_2S$ is a gas at ordinary temperature though oxygen and sulphur are the members of the same group of the periodic table. Hydrogen bonding is so important that our very existance on earth depends because in absence of hydrogen bond $H_2O$ molecules exist in gaseous state.

(b) **Abnormal melting points and boiling points :** In general for similar type of compounds the melting points and boiling points are the function of their molecular weight i.e. melting points and boiling points increase with increasing molecular weight of the compounds. But exception of boiling point occurs for the hydrides of first element of groups 15, 16 and 17 i.e. for $NH_3$, $H_2O$ and HF respectively (Fig. 3.41). This abnormal behaviour is due to the fact that the electronegativity difference between N, O and F with that of H is highest than the post elements of the same group and so hydrogen bonding will be most prominant for $NH_3$, $H_2O$ and HF. Due to this a special molecular association exist between the molecules causing increase of effective molecular weight as also increase of boiling point. It is to be noted that boiling points of $CH_4$ to $SnH_4$ rise steadily with increase of molecular weight.

(c) **Molar entropy change of vaporisation :** In case of normal liquids having no hydrogen bonding, register lower molar latent heat of vaporisation compared to those liquids where hydrogen bond exist. The molar latent heat of vaporisation of $H_2O$ at its boiling point is 40.6 kJ $mol^{-1}$ which is far larger than that of $H_2S$. The molar entropy change of vaporisation* for normal

* Molar entropy change of vaporisation = $\dfrac{\text{molar latent heat of vaporisation}}{\text{boiling point in K}}$.

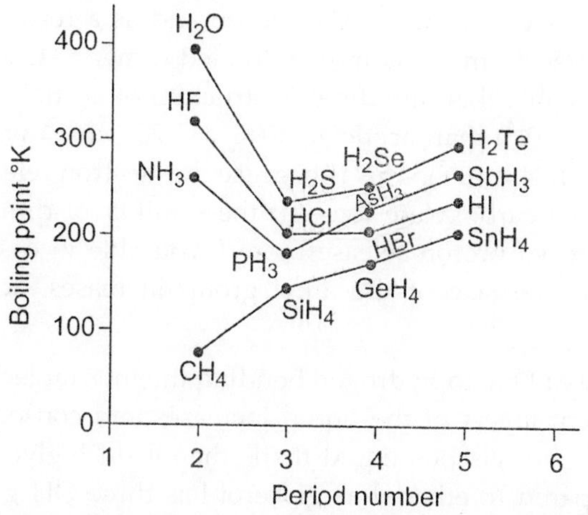


*Figure 3.41 : Variation of boiling points of the hydrides of group 15, 16 and 17 elements.*

liquids is approximately 90 $JK^{-1}$ $mol^{-1}$, the corresponding data for $H_2O$ and $NH_3$ are 109 and 97 $JK^{-1}$ $mol^{-1}$ respectively.

(d) **Unusual solubility :** A general solubility rule is followed in section 3.1.4. However there are several exception to this rule. Alcohols are soluble in water due to hydrogen bonding. Generally organic compounds having strong electronegative atom (e.g. oxygen, nitrogen) present in functional group are more soluble in $CHCl_3$, $CH_2Cl_2$ etc. than $CCl_4$. This is due to the weak interaction operated by the hydorgen atom of the solvent with that of the strong electronegative atom of the organic compound.

(e) **Variation of strength of acids and bases :** Acetic acid is a weak acid due to the hydrogen bonded dimeric structure [Fig. 3.42(a)]. Trimethylammonium hydroxide [Fig. 3.42(b)] is weaker base than tetramethylammonium hydroxide [Fig. 3.42(c)] because in the former the

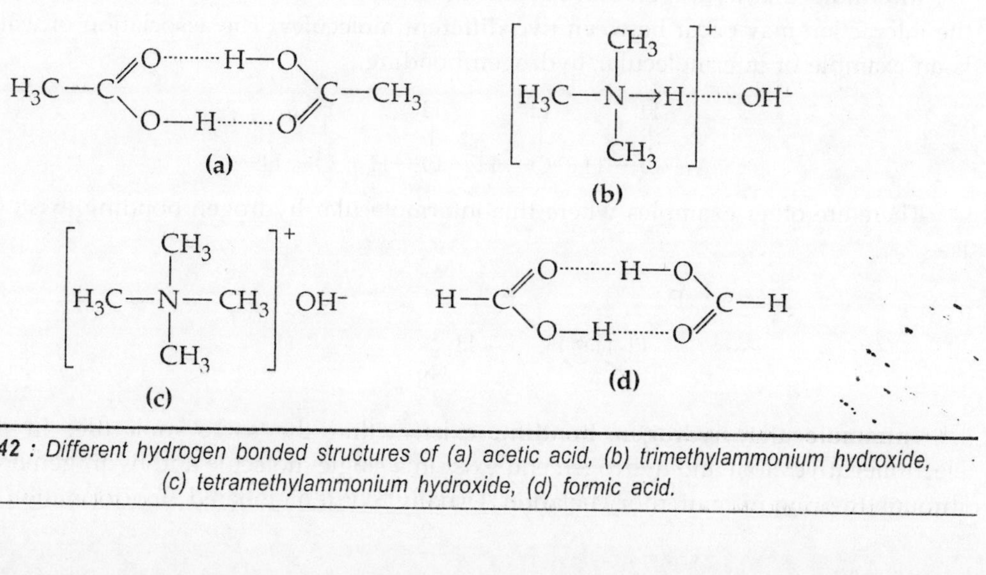

*Figure 3.42 : Different hydrogen bonded structures of (a) acetic acid, (b) trimethylammonium hydroxide, (c) tetramethylammonium hydroxide, (d) formic acid.*

$OH^-$ group is hydrogen bonded to the amino group and as a result the ionisation of $OH^-$ ion is difficult. Formic acid also forms a hydrogen bonded dimeric structure [Fig. 3.42(d)] but this dimeric structure is less stable than the dimeric structure of acetic acid. So formic acid is more stronger acid ($K_a = 17.7 \times 10^{-5}$) than acetic acid ($K_a = 1.76 \times 10^{-5}$) which in turn more stronger than propionic acid ($K_a = 1.34 \times 10^{-5}$). This is because the electron releasing alkyl groups increase the negative charge on the carboxylate ion and the stability of dimer in $CH_3COOH$ increases than that of HCOOH and so proton release more favourable in case of HCOOH than that of $CH_3COOH$. Moreover as the size of the alkyl group increases, acidity decreases in case of propionic acid.

(f) **Effect of viscosity :** Due to hydrogen bonding the intermolecular attraction between the molecules as also different layers of the liquid increases and consequently viscosity increases. For example glycerol is more viscous liquid than ethanol since glycerol forms more number of hydrogen bonding compared to ethanol as glycerol has three OH groups whereas ethanol has only one OH group. Viscosity of water is 10.05 millipoise while methanol has the value 5.97 millipoise.

(g) **Effect on dielectric constant :** The dielectric constant of liquid solvents are generally a function of hydrogen bonded structure of associated liquids. Generally there will be an increased value of dielectric constant as the number of hydrogen bond increases. The enhancement of hydrogen bonded structure depends on the dipole moment of the molecule as also the orientation of the monomeric units. For example the dielectric constants of $H_2O$ and $CH_3OH$ are 78.5 and 32.6 respectively but the dielectric constant value of HF is 83.0. Though fluorine is more electronegative than oxygen yet in $(HF)_n$ the monomeric units are angularly oriented by which the resultant dipole moment not so prominant than that of $H_2O$ causing very small increase of dielectric constant value for HF.

### 3.10.3 Types of Hydrogen Bond

Hydrogen bonds generally classified into two main groups—**intermolecular** and **intramolecular.**

**Intermolecular hydrogen bond** is one that exist between two or more molecules. In this case the interaction may occur between two different molecules. The association of water molecules is an example of intermolecular hydrogen bonding.

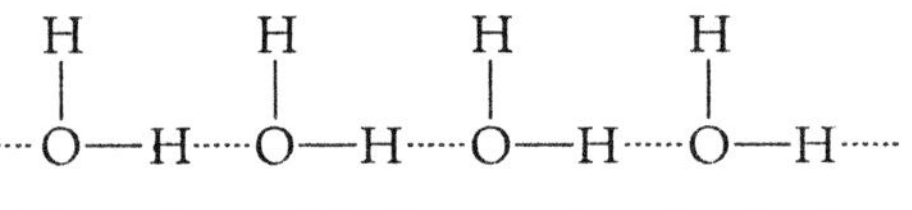

There are other examples where this intermolecular hydrogen bonding exist. One example is—

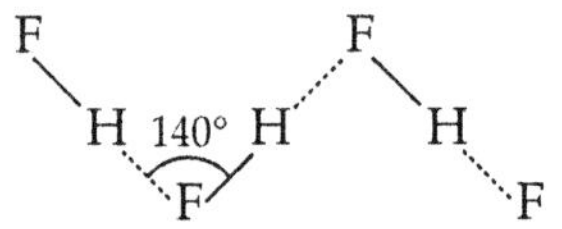

**Intramolecular hydrogen bonding** exist within the same molecule. In this case the electronegative atom and hydrogen end exist in a single molecule and hydrogen bonding occurs through the ring formation or chelation. Usually 5 or 6 membered ring formation enhances the intramolecular hydrogen bond. Ortho substituted molecules undergo intramolecular whereas para substituted molecules undergo intermolecular hydrogen bonding. Intramolecular hydrogen bonded molecules register lower boiling point/melting point compared to that of the corresponding intermolecular partner because the effective molecular weight increases in case of intermolecular hydrogen bonding compared to that of intramolecular one. For example,

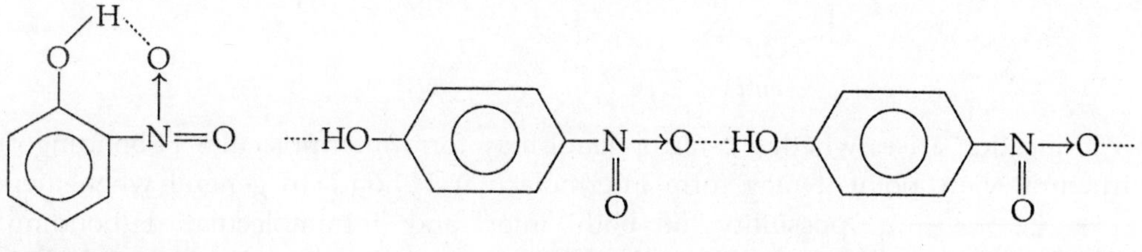

*o-nitro phenol (m.p=45°C) intramolecular hydrogen bond*

*p-nitrophenol, intermolecular hydrogen bond m.p=114°C*

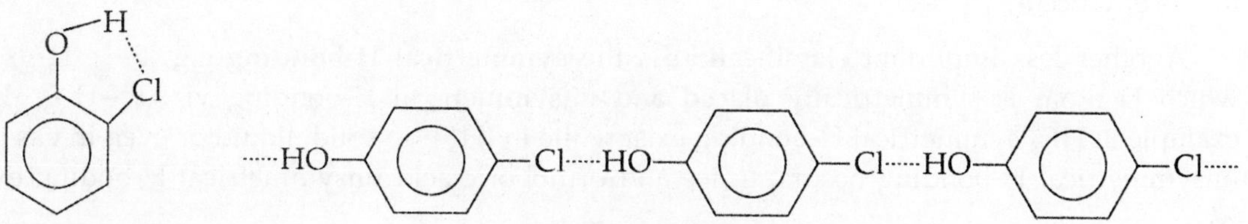

*o-chlorophenol intramolecular H-bonding*

*p-chlorophenol, intermolecular H-bonding*

The **cis** isomer (maleic acid) undergoes intramolecular H bonding while the corresponding trans isomer (fumaric acid) conforms intermolecular H-bonding as depicted below :

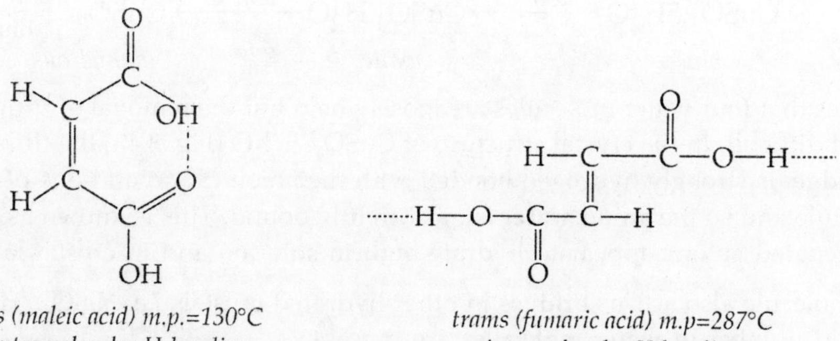

*cis (maleic acid) m.p.=130°C intramolecular H-bonding*

*trans (fumaric acid) m.p=287°C intermolecular H-bonding*

It is to be noted that for intramolecular H-bonding (a) the 5 or 6 membered ring formed should be planar (b) interacting atoms should be oriented in such a way that there is minimum strain during ring formation. The **anti** form of pyridine 2-carboxaldoxime undergoes **intramolecular H-bonding** while the **syn** form is unable to do so as the lone pair on pyridine nitrogen is not directed towards the OH group of the oxime as shown below :

Condition for intramolecular H-bonding

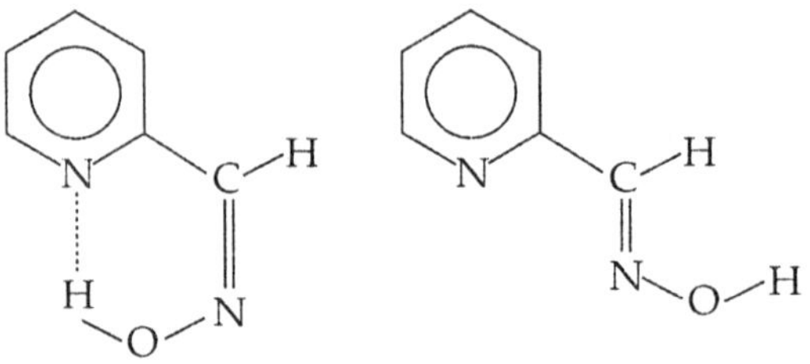


*anti* *syn*

Now a question arises whether o-nitrophenol may form intermolecular H-bonding or not. From structural view point it may form intermolecular H-bond. In general when there is a possbility of both inter and intramolecular H-bonding, the preference is the formation of latter because intramolecular H-bonding possess more thermodynamic stability. During the formation of intramolcular H-bonding there will be the ring closure i.e. chelation process which causes disorderness in the system by which greater positive change in entropy occurs.

*Comparative stability of inter and intramolecular H-bonding*

Another less important classification is the **symmetrical H-bonding** e.g. F······H······F in which H atom is symmetrically placed and unsymmetrical H-bonding viz. F—H······F. For example in $HF_2^-$ symmetrical H-bonding exist while in $(HF)_n$ in solid, liquid or even in gas phase unsymmetrical H-bonding occurs. In ice and orthoboric acid unsymmetrical H-boding exist.

### 3.10.4 Some typical example of hydrogen bonded structures

(a) **Crystal structure of $CuSO_4.5H_2O$** : Binding force of all the water molecules in $CuSO_4.5H_2O$ are not identical. This is supported by the following temperature of dehydration :

$$\underset{\text{blue}}{CuSO_4.5H_2O} \xrightarrow{200°C} \underset{\text{white}}{CuSO_4.H_2O} \xrightarrow{250°C} \underset{\text{anhydrous}}{CuSO_4} \xrightarrow[\text{heat}]{\text{strong}} \underset{\text{black}}{CuO}$$

This indicates that four water molecules are loosely held but the removal of fifth water molecule is somewhat difficult. In the crystal structure of $CuSO_4$. $5H_2O$ (Fig. 3.43) the fifth water molecule acts as a bridge, is strongly hydrogen bonded with the anion ($SO_4^{2-}$) and one of the coordinated water molecule and so the **anion water** is most tightly bound. This is known as **anion solvation** where oxygenated anions appear to hydrate both in solution and in crystal lattice.

Water molecule also acts as bridges in other hydrated crystals i.e. $NiSO_4.7H_2O$, $FeSO_4.7H_2O$ in which odd number of water molecules are present.

(b) **Hydrogen bonded structure of boric acid and ice :** The crystal structure of orthoboric acid consists of layer structure of $B(OH)_3$. $BO_3$ units are joined together through hydrogen bonding to form loosely bound parallel sheets with O—H···O type of hydrogen bonds in which the hydrogen atom is **not symmetrically** located between the two bonded oxygen atoms [Fig. 3.44(a)]. The successive parallel layers are about 3.2Å apart and are held by van der Waals forces. Each B is $sp^2$ hybridised having O—H and H···O distances 0.8 and 2.7Å respectively.

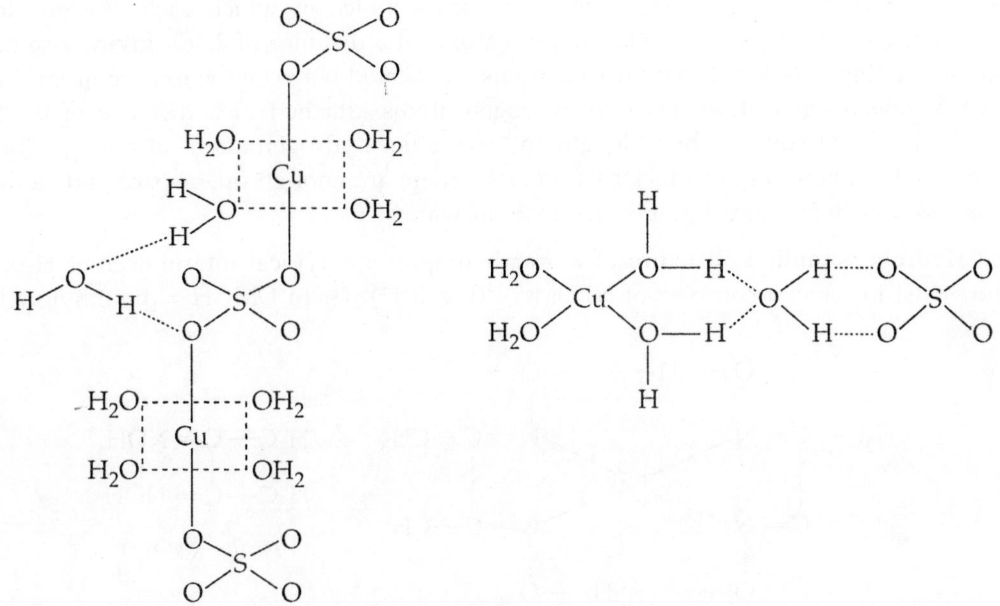

*Figure 3.43 : Crystal structure of $CuSO_4.5H_2O$*

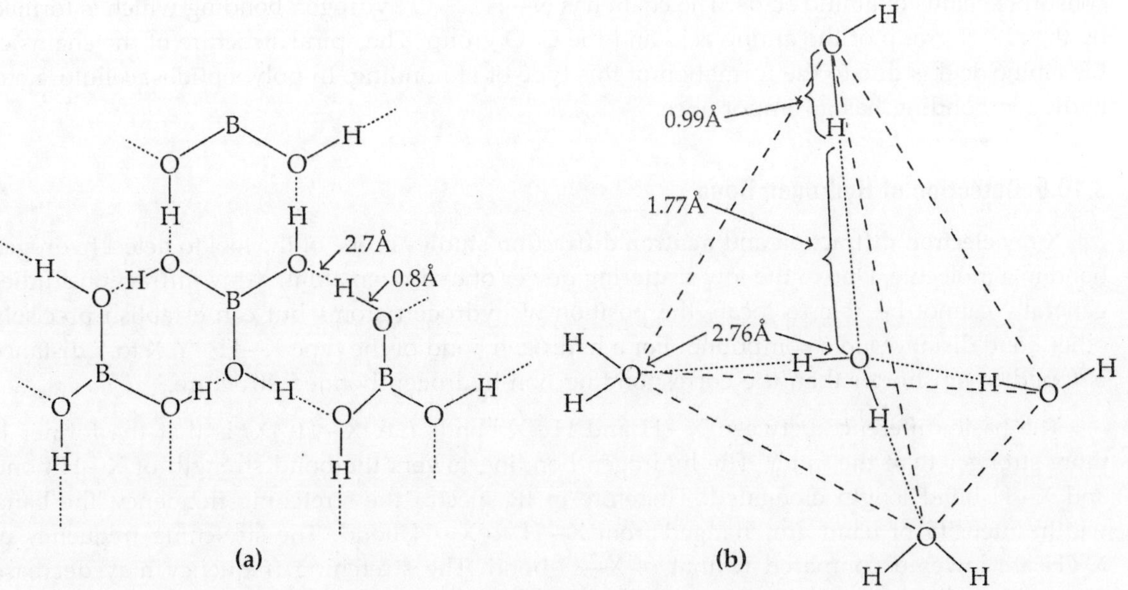


*Figure 3.44 : (a) Crystal structure of orthoboric acid, (b) Crystal structure of ice.*

Similar unsymmetrical hydrogen bonding occurs in ice in which each oxygen atom is surrounded tetrahedrally by four other oxygen atoms at a distance of 2.76Å giving rise to very open structure [Fig. 3.44(b)]. The hydrogen atoms are located between the oxygen atoms in such a way that one oxygen atom has two hydrogen atoms attached at a distance of 0.99Å i.e. practically the O—H covalent bond length and two other hydrogen atoms at a longer distance of 1.77Å due to H bonding. In this structure of ice there are enough open space and so density of ice is less than water and because ice floats in water.

(c) **Hydrogen bonded structure of a metal complex :** A typical intramolecular H-bonded structure exist in a metal complex of $Ni(DMG)_2$ (Fig. 3.45) where DMGH = dimethylglyoxime.

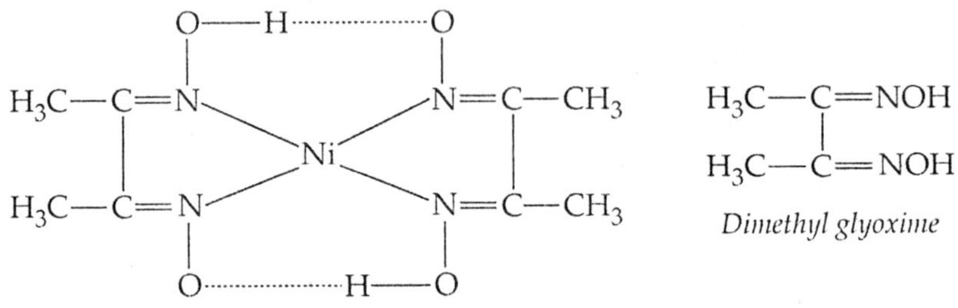

*Dimethyl glyoxime*

*Figure 3.45 : Square planar hydrogen bonded structure of $Ni(DMG)_2$.*

(d) **Hydrogen bonding in biological systems :** Hydrogen bonding has vital role in living systems. The double helix structure of DNA is stabilised through hydrogen bond. Proteins constitute chains of amino acids. The chain has N—H......O hydrogen bonding which is formed by the N—H group of the amino acid and the C=O group. The spiral structure of the chains of the amino acid is due to the formation of this type of H-bonding. In polypeptides, cellulose etc. hydrogen bonding has an importance.

### 3.10.5 Detection of Hydrogen Bond

**X-ray electron diffraction** and **neutron diffraction studies** is one of the tool to detect hydrogen bond in a molecule. Due to the low scattering power of hydrogen atom, X-ray diffraction studies generally cannot be able to locate the position of hydrogen atoms but can establish precisely other bond distances in a compound. For a hydrogen bond of the type X—H···Y, X to Y distance is found to be shorter than the corresponding non-hydrogen bonded structure.

The basic difference between X—H and H···X bonds (i.e. X—H···X) is that the former is more stronger than the latter. The hydrogen bonding lowers the bond strength of X—H bond and X···H bond length elongated. Therefore in IR spectra the stretching frequency, the band width, intensity of band are changed from X—H to X···H bond. The stretching frequency of X···H are lowered compared to that of X—H bond. The stretching frequency may decrease from 30 $cm^{-1}$ to several hundred $cm^{-1}$ depending upon the strength of H-bonding. For example, the O—H stretching frequency is observed at 3600 $cm^{-1}$ but it shifts to 3300 $cm^{-1}$ when O—H group forms intermolecular hydrogen bond. The same effect is observed for intramolecular H-bonding.

**Ultraviolet spectroscopy** or **electron absorption spectroscopy** are also employed for the presence of hydrogen bonded interaction between an organic compound and a solvent e.g. interaction between formaldehyde with hydrocarbon solvent and hydroxylic solvent.

### 3.10.6 Strength of Hydrogen Bond

The strength of the hydrogen bond depends on the electronegativity difference of hydrogen with that of the element concerned. The electronegativity of nitrogen, oxygen and fluorine is in the order N < O < F and so the strength of the H-bond is in the order $NH_3$(N—H···N) < $H_2O$(O—H···O) < HF(F—H···F). The H-bond energies in $NH_3$, $H_2O$ and HF are 8.37, 29.29 and 41.85 kJ $mol^{-1}$ respectively.

Generally strength of H-bonding decreases down the group as the electronegativity of the element decreases. Thus HF and $H_2O$ are far strongly H bonded than that of HCl and $H_2S$ respectively.

Though the electronegavity of N and Cl are same but $NH_3$ is more strongly H bonded compared to HCl. Lesser the size of the atom, the greater will be the strength of hydrogen bond. N has smaller size compared to that of Cl and so the strength of H bond order $NH_3$ > HCl.

Hydrogen bond is more stronger than intermolecular forces but weaker than covalent bond and the order is, intermolecular forces < H-bond < covalent bond < ionic bond. H-bond energy lies in the range 10-40 kJ $mol^{-1}$ while covalent bond energy is about 200-400 kJ $mol^{-1}$.

### 3.10.7 Nature of Hydrogen Bond

Hydrogen bond is not an ordinary covalent bond and hydrogen has just one electron in low energy 1s orbital. With this one electron it can be able to form one covalent bond (i.e. H—F, O—H) but it cannot be able to form simultaneous two covalent bonds (i.e. —H—), for which there will be the requirement of four electrons and two electrons may be accommodated in 1s orbital but the other two electrons will have to be accommodated to the much higher energy 2s or 2p orbital which makes the second bond energetically disfavourable. But hydrogen has a desire to expand its valency shell by the formation of a **broken hand** which is the manifestation of the formation of hydrogen bond (—H···) which is exhibited by the attachment of strong electronegative atom.

As stated H-bond arises due to the polar character of the bond and so this is a very weak electrostatic attraction i.e. $\cdots\overset{\delta^-}{X}—\overset{\delta^+}{H}\cdots\cdots\overset{\delta^-}{X}—$. Like covalent bond H-bond has directional nature.

### 3.10.8 Theories of Hydrogen Bond

There are several theories of hydrogen bonding—molecular orbital approach, valence bond approach and Coulson electrostatic theory. We shall construct a simplified MO diagram with 1s orbital of H atom and two p orbitals of two F atoms in $HF_2^-$ (Fig. 3.46). Here x-axis is taken as bond axis.

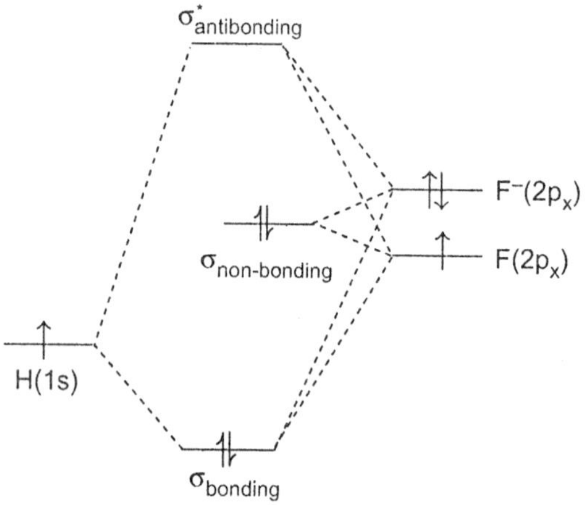


*Figure 3.46 : MO picture of $HF_2^-$.*

**Problem 3.50** Which hydrogen bond would you expect to be stronger and why : S—H······O and S······H—O? **[B.U. 2005]**

**Ans. :** Between O—H······S and O······H—S, the former will be more stronger H-bond as in that case H is linked to more electronegative element. Generally more polarity in the bond and thereby more positive character imparted on H atom.

**Problem 3.51** Show the hydrogen bonded structure of $NaHCO_3$ or $HCO_3^-$.

**Ans. :**

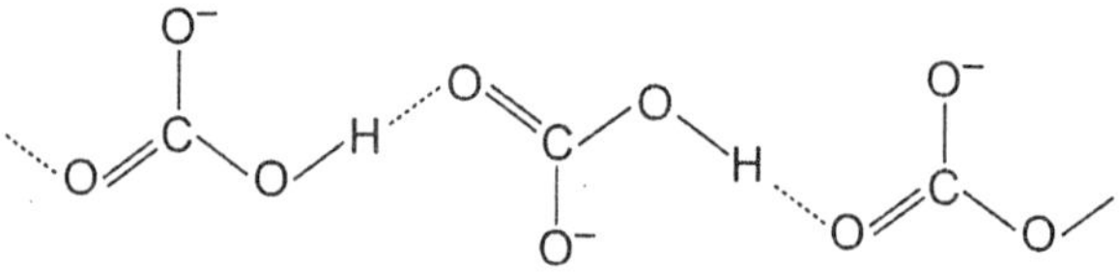

**Problem 3.52** Water has highest boiling point than HF though their hydrogen bond strengths are in reverse order. Explain. **[B.U. 2011, 2016, C.U. 2003]**

**Ans. :** The strength of the H-bond in HF (41.85 kJ $mol^{-1}$) is greater than that of $H_2O$ (29.29 kJ $mol^{-1}$). Yet the boiling point of water (100°C) is greater than that of HF (19.5°C). Though H—F is a linear molecule but association occurs through angularly oriented HF molecules and so the resultant dipole moment is relatively small. The increase of dipole moment depends on the structural orientation of the monomeric units. Water is an angular molecule but linearly associated (Sec. 3.10.3) and so that resultant dipole moment is increased.

Moreover in water statistically there are far larger number of hydrogen bonds exist than in HF. $H_2O$ has two hydrogen atoms per molecule while HF has only one. Thus the number of H-bonds arising from water molecule is much greater than that arising from HF molecule because the former has two proton donar centers and two proton acceptor centres whereas in the latter there are one proton donor center and one proton acceptor center though three lone pairs are present. Number of H-bonds control boiling point predominantly over strength of H-bond as number of bond breaking is controlling boiling point through association to a greater degree. All these explain the higher boiling point of $H_2O$ than that of HF.

### 3.10.9 Limitation of Hydrogen Bond

In hydrogen bond H forms a bridge between two electronegative atoms. This excludes the B—H—B bond in $B_2H_6$ as also W—H—W link present in $(OC)_6WHW(CO)_6$. These are not considered as H-bonds as B and W are less electronegative than hydrogen.

---

**Problem 3.53** o-nitrophenol can be easily steam distilled but p-nitrophenol cannot be. This is because of—

1. Strong intermolecular hydrogen bonding in o-nitrophenol.
2. Strong intramolecular hydrogen bonding in o-nitrophenol
3. Strong intramolecular hydrogen bonding in p-nitrophenol.
4. Dipole moment of p-nitrophenol is larger than that of o-nitrophenol. **[NET 2001]**

**Ans. :** 2.

---

**Problem 3.54** The boiling points of noble gases are illustrative of the operation of forces of the type : 1. ion-dipole, 2. dipole-induced dipole, 3. ion-induced dipole, 4. London dispersion forces. **[NET 2000]**

**Ans. :** 4.

---

## 3.11 METALLIC BOND

Last extreme type of chemical force operated between the atoms of metals is known as metallic bond. This chemical force is superior than any other chemical forces operative between atoms and molecules. Metals have some characteristic special properties which are listed below :

(a) High thermal and electrical conductivity.

(b) Generally have high melting and boiling points.

(c) Metallic lusture and also high ductility and malleability.

(d) High density and hardness.

(e) Alloy formation i.e. formation of solid solution.

(f) Photoelectric emission (Sec. 1.4.1) and thermoionic emission.

(g) Electropositive character and power to replace hydrogen from acids (have exception).

(h) Typical metallic crystalline structure e.g. face centered cubic, hexagonal close packed etc. having high coordination number.

(i) Opaque to light.

(j) Constancy of atomic heat, and

(k) Enhancement of reactivity in finely divided state than in the lump form.

All these properties of metals focus to a point that metals have free electrons in their structure. **Electron sea model, valence bond theory** supplemented with resonance concept and **molecular orbitral theory** can explain the free electron structure of metals.

### 3.11.1 Electron sea model of metallic bond

This theory also known as Drude (1900) Lorentz (1923) theory. According to this theory a metallic lattice consists of a rigid positive ions, arranged in a close packed arrangement with their valence electrons spread throughout the crystal lattice from one layer to the other. Valency electrons are weakly held to the **kernal** or **core** in metals. These valence electrons are completely delocalised and exchanged between the layers of the atoms and they move freely from one kernal to other. Hence the atoms of the metals assume a positive charge (cation) and the valence electrons form an **electron sea or pool** which binds together the positive metal kernals or cations. Hence a metallic lattice is considered as positive ions (kernals) held together by a mobile **electron gas or sea** (Fig. 3.47). Since the electrons are free from the influence of parent metal atom, these can conduct electricity easily. Mobile electrons can be easily excited by visible light and hence metal surfaces look shining.

*Explanation of conductivity and metallic lusture*

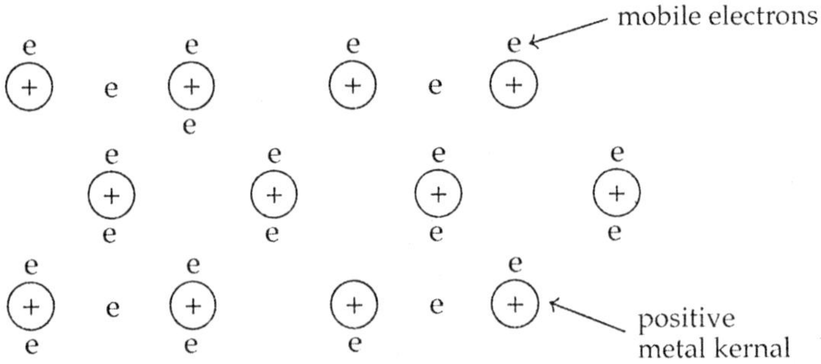


*Figure 3.47 : Schematic representation of sea of mobile electrons.*

Variation of melting point and boiling point of metals may be explained on the basis of cohesive force operated beween the two adjacent layers of metal lattice. In case of sodium only one electron per atom constitute the mobile or sea of electrons while magnesium and aluminium contributed two and three electrons/ valence electrons per atom respectively to the sea of electrons. Therefore the cohesive force between the two adjacent layers of the metals is in the order Na < Mg < Al and so the melting point and boiling point is in the order Al > Mg > Na.

*Explanation of high m.p. and b.p.*

The order of malleability and ductility is opposite to that of cohesive force. The cohesive force is less in case of sodium and highest in case of aluminium. It is therefore more easy to glide one layer over the other in case of sodium than in case of magnesium which in turn more easier than that of aluminium. **In general greater the number of valence electrons, lesser will be the malleability and ductility of metals.** Hence the order of malleability and ductility is Na > Mg > Al.

*Explanation of malleability and ductility*

This concept has certain limitations. This model can explain the conductivity of metals but the quantitative aspect cannot be explained i.e. why the conductivity decreases with rise of temperature. Why some metals show no conductivity in the solid state?

### 3.11.2 Molecular Orbital Theory—the band theory of metals

According to MO theory two atomic orbitals combine to form two molecular orbitals—one is called bonding MO (BMO) and the other is called antibonding MO (ABMO) (Sec. 3.4.3). Similarly when three atomic orbitals combine, there will be the formation of one non-bonding MO (NBMO) in addition to BMO and ABMO. NBMO is a kind of ABMO but of some lower energy.

Combination of two 3s orbitals of two Na atoms produces two molecular orbitals—one BMO and other ABMO and the two valence electrons occupy in the BMO. Similarly three Na atoms combine to form three molecular orbitals—one BMO, one ABMO and the other NBMO (Fig. 3.48). In this way there may be the formation of $Na_4$, $Na_5$ etc. atomic association and the electrons may be accommodated according to the rules of filling electrons in molecular orbitals. In a metallic lattice containing one mole sodium atoms i.e. there will be the combination of Avogadro's number (N) of 3s atomic orbitals to form N number of molecular orbitals—the half will be bonding and half will be antibonding. As there is a combination of large number of atomic orbitals so the energy gap between half BMO and half ABMO will be narrowed down (Fig. 3.48). Naturally one mole valance electrons of sodium will be occupied in the BMO to form a half filled band.

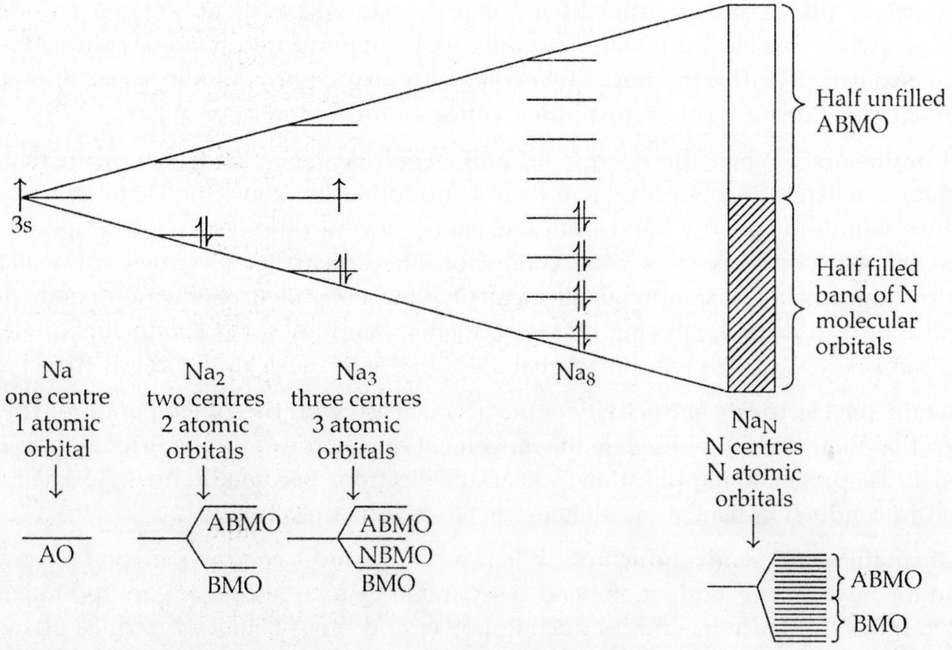


*Figure 3.48 : Schematic representation for the formation of half filled band in a crystal of sodium.*

The vacant 3p orbitals of sodium interact with 3s orbitals to form a **broad band**. The 3s and 3p orbitals have close energy so that the 3s and 3p bands overlap each other.

Even if the 3s and 3p bands will not overlap, then sodium will be a conductor of electricity since the lower half filled band transfers electrons to the upper unoccupied half filled band easily. However such type of overlap is important in case of alkaline earth metals e.g. Mg. Each

Mg atom has two electrons in 3s orbital. In a crystal of magnesium the 3s band would therefore will be completely filled. There is no scope of promoting the valence electrons to higher ABMO levels and magnesium would behave as an insulator. However, magnesium is a good conductor of electricity at room temperature due to the fact that 3p band overlap with 3s band to form an **overlap zone** and electrons can move easily from HOMO to LUMO.

### 3.11.3 Difference between metallic bond with that of the ionic and covalent bond

In metals all the atoms are same but in ionic bonds the constituent partners are different. In metals the conductance is due to the movement of free electrons while in ionic bonds the conductance is due to the movement of the ions.

The covalent bond has directional nature while in metal lattice the valence electrons are spread over the metal crystal in a non-directional character.

### 3.11.4 Classification of conducting power of substances—insulators, conductors, semiconductors and superconductors

According to band theory of metals, the highest energy electrons of metallic lattice occupy either a partially filled band or a filled band that overlap with an empty band. The electronic energy bands where the electrons can move are called the **permitted band** or **conduction band**. These are also called **Brillouin zones**. Two conduction bands or Brillouin zones are separated by empty bands which are called **forbidden zones** or **forbidden energy gap**.

(a) **Conductors** : When the energy separation between the filled band with that of the empty band is narrowed down or where there is no forbidden zone, the valence electrons can easily move within the energy band from one energy level to the other closely spaced energy level and the substance becomes good conductor. Here plenty of electrons are available for electrical conduction. For example alkali metals having half filled conduction band (Brillouin zone) and for alkaline earths having filled conduction band are good conductors of electricity. In this case uppermost filled conduction band overlap with next vacant conduction band.

Generally the electrical conductivity of metals decrease with rise of temperature. This is due to the fact that thermal energy impede the movement of electrons into the Brillouin zones when electric field is applied. On application of heat, the electrons become thermally excited and are promoted to conduction bands and thereby metals conduct electricity.

(b) **Insulators and semiconductors** : When the uppermost conduction band is completely filled and the next vacant conduction band is separated by a large energy gap (forbidden zone), [Fig. 3.49(a)] the substance has no conducting power and the substance is termed as **insulator**. On the other hand when the energy gap between filled conduction band and next vacant conduction band is not too high [Fig. 3.49(b)] the electrons can move to the vacant Brillouin zone by the application of thermal or electrical energy, then the substance is regarded as **semiconductor**. The energy gap in case of insulators is above 3 eV while in case of semiconductors this gap lies with 2-3 eV. Silicon is a semiconductor which has filled conduction band (Brillouin zone) that are slightly below the vacant conduction band. These do not conduct electricity at low temperature but a small increase in temperature, the electrons can move into the conduction band and become conductors at high temperature.

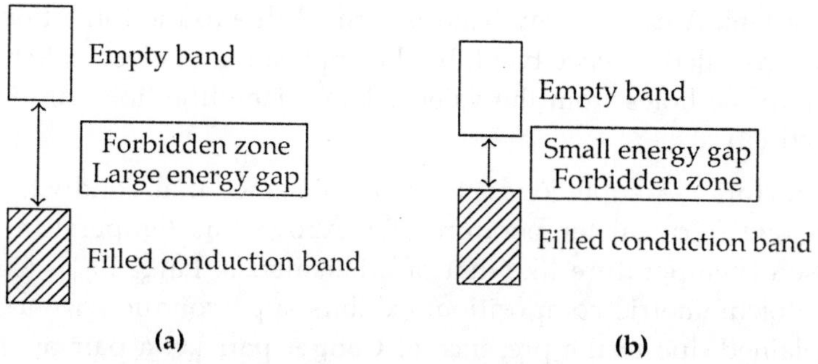


*Figure 3.49 : Energy band diagram of (a) insulators, (b) semiconductors.*

When the semiconducting property of a substance generated in a pure form then this is called **intrinsic semiconductors**. Germanium and silicon are the common examples of intrinsic semiconductors.

Sometimes an insulator can be converted to semiconductor by the application of some impurity. These are known as **extrinsic** or **impurity semiconductors**. These are classified as :

(1) **n-type semiconductors :** Here n means negatively charged electrons. This type contains an impurity which contains more valence electrons (e.g. P. As, Sb etc.) than insulators. These elements have five valence shell electrons and replacement of silicon by arsenic atoms in silicon lattice, the extra valence shell electrons form a new **negative charge** energy band. Very pure silicon is doped with arsenic, the extra valence shell electrons form a new energy band somewhat higher in energy than the filled band [Fig. 3.50(a)]. On application of electric field electrons from filled arsenic band move into the empty silicon band and conduction occurs.

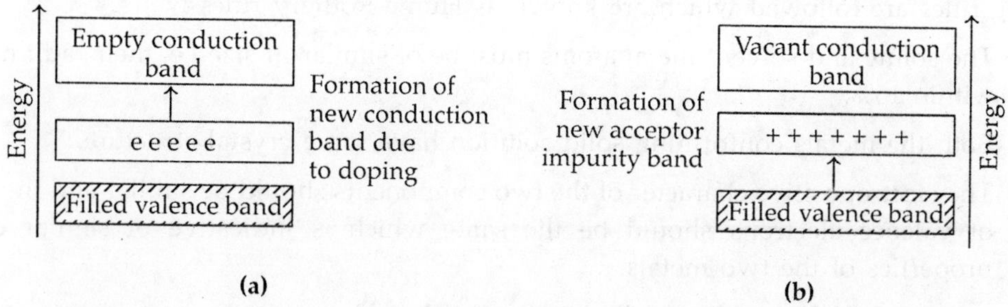


*Figure 3.50 : Conduction band of (a) n-type semiconductors, (b) p-type semiconductors.*

(2) **p-type semiconductors :** Here p means positive hole. In this type contains an impurity e.g. B, Al, Ga etc. which have less valence electrons than the insulators. These impurities belong to the one lower group than silicon so that it pulls an electron from the neighbouring silicon atom and creates a **positive charge** or **electron deficient hole** or **positive hole** on the adjacent atom [Fig. 3.50(b)]. Such holes are called **acceptors.** The number of such acceptors or positive holes is the number of impurity atoms doped

into the system. A new energy band is formed due to the formation of positive holes just above the filled valence band. By the application of electrical energy the electrons can move to the holes from the valence band. Here the holes are the prime agent to carry electricity.

(c) **Superconductors :** A superconductor is a substance that conducts electricity having zero resistance below a critical temperature ($T_c$). Above this temperature the conductivity decreases with rise of temperature like that of other metals. $La_{1.85}$ $Ba_{0.15}$ $CuO_{4-\delta}$ where $\delta = 0$ to 1, having non-stoichiometric composition, exhibits superconductivity at a temperature of 35K ($T_c$). This explained due to the presence of **Cooper pair** i.e. a pair of electrons that exist on account of their indirect interaction with one another through their interaction with the nuclei of the atoms in the lattice. Superconductivity discovered in 1911 by Kammerlingh Onnes.

### 3.11.5 Solid Solutions—Alloys

When two or more metals are melted together and the molten mixture is allowed to cool, the resulting product solidify which is known as alloy. The alloys may be considered as homogeneous solid solutions in which the atoms of one metal are randomly distributed among the atoms of the other or they may have definite composition like a chemical compound. Common examples of alloys are **brass** (upto 40% Zn in Cu), stainless steel (over 12% Cr in Fe) etc. The properties of alloys are somewhat different from those of the constituent metals. They may be classified into three types depending on their mechanism of formation.

(a) **Substitutional solid solutions or formation of intermetallic compounds :** It is a solid solution in which atoms of solute metal occupy the positions of solvent metal atom randomly in its lattice. Substitutional solid solutions having contineous miscibility can be expected if the following rules are followed which are known as Hume-Rothery rules :

(1) The solute and solvent metal atoms must be of similar in size i.e. their radii may vary within 15%.

(2) Both the metals conforming solid solution have same crystal structure.

(3) The electropositive character of the two components should be similar and the number of valence electrons should be the same which is indicative of similar chemical properties of the two metals.

Although Na and K are chemically similar and both have similar metallic structures, yet they are unable to form solid solutions because their radii vary more than 15%. Cu-Au, Ag-Au, Cu-Ni etc. systems can follow the above rules and can form solid solutions of intermetallic nature. Zn and Cu are partly miscible and can form solid solution of different composition having different phases of the following type :

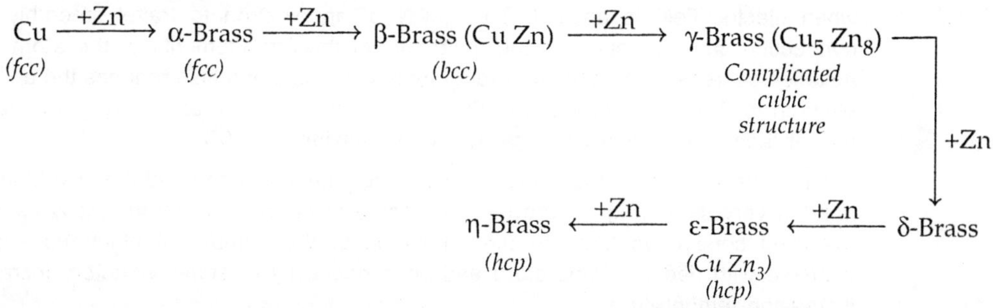


*fcc = face centered cubic, hcp = hexagonal close packed, bcc = body centered cubic.*

Among these phases β, γ ant ε phases are important. In these phases has specific ratio of valence electrons to number of atoms. This ratio is called **Hume-Rothery ratio** and has the value for β-brass (Cu Zn) = 21 : 14 = 3 : 2; for γ-brass ($Cu_5\ Zn_8$) = 21 : 13 and for ε-brass ($Cu\ Zn_3$) = 21 : 12 = 7 : 4. Generally β alloys are strong, malleable and ductile, and conducting in nature than the γ and ε alloys which are brittle and non-conducting.

(b) **Interstetial solid solutions :** This is a kind of solid solutions where a solute non-metals e.g. B, C etc. that ocupy into the interstices (holes) in the structure of the large solvent metal. The hole fitting occurs according to the radius ratio rules (Sec. 8.7.5) which is 0.22 to 0.41 for tetrahedral hole and 0.41 to 0.73 for octahedral hole. The composition may assume a definite one or variable one due to the random distribution of solute atoms into the interstices (holes) of the solvent metal. B, C, N assume definite stoichiometric composition between the host (solvent) metal and the interstitial atoms e.g. boride, carbide, nitride etc. viz. ZrB, TiB, TiC, ZrC, Tac, TiN, ZrN etc. Hardness and melting points of these significantly increased than that of the parent metals. Interstitial carbide ($Fe_3C$) is important in steel industry.

(c) **Alloys in which substituents crystallise separately :** In this case the constituent components neither mix to form a solid solution nor do they combine to form intermetallic compounds. This is considered as an intimate mixture of small crystals of the constituent metals in a relatively pure state.Zn and Cd when melted together form this type of alloy.

---

**Problem 3.55** Name the property of pure metals which make them more useful than alloys. What type of alloy would most likely be formed by a pair of metals having similar sized atoms, the same number of valence electrons and the same type of lattice when pure? **[B.U. 2006]**

**Ans. :** The electrical conductivity of pure metals is greater than that of their alloys because of more hardening (local bonding).

Substitutional solid solutions as the metals are so similar that there is little difference in bonding, no matter which sites they occupy. Hence there is random arrangement giving rise to substitutional alloy.

---

**Problem 3.56** Explain why the electrical conductivity of metal decreases with the rise in temperature but the reverse occurs with semiconductors. **[C.U. 2000]**

**Ans. :** Since thermal agitation of metal atoms impede the flow of electrons into conduction band of metals so electrical conductivity decreases with rise of temperature in case of metals when electric field is applied. The ability of an electron to travel smoothly into the conduction band depends on the uniformity of the arrangements of the atom. Thermal agitation decreases the smooth arrangements of atoms and also reduces the ability of the electron to travel from one end of the metal to other so that conductivity decreases with rise of temperature and conductivity will be maximum at 0°K.

In case of semiductors, the energy gap between the filled band and the next higher empty band is very small [Fig. 3.49(b)] so that some electrons may be thermally excited from the filled conduction band to the empty band, the number of electrons excited will increase with rise of temperature and so conductivity of semiconductors increase with increasing temperature.

---

**Problem 3.57** Suggesting suitable reason rank the following in the order of decreasing bond angles : $OF_2$, $OCl_2$, $OH_2$, $ON_2$. **[B.U. 2010, 2016]**

**Ans. :** $ON_2$ is a linear molecule having O—N—N bond angle 180°. $OF_2$, $OCl_2$ and $OH_2$ are V-shaped bent molecules. In $OCl_2$ and $OH_2$, the electronegativity of chlorine and hydrogen is less than that of oxygen, so swelled-ends of the electron density balloons at oxygen. In $OF_2$, the electronegativity of fluorine is greater than that of oxygen and reverse order is expected. Hence ∠Cl—O—Cl or ∠H—O—H is greater than that of ∠F—O—F. In $OCl_2$ and$OH_2$, ∠Cl—O—Cl is greater than ∠H—O—H due to steric reason arising due to the larger size of Cl. Hence the order of bond angle is :

$$ON_2 > OCl_2 > OH_2 > OF_2$$

bond angle : 180° 111° 104° 103°

---

**Problem 3.58** Define ionic potential and note its importance in chemistry. **[B.U. 2010]**

**Ans. :** For ionic potential (ϕ) consult Sec. 3.7(b). ϕ values correlate diagonal relationship between elements as also better complexation process for an ion having higher ϕ value.

---

**Problem 3.59** How many hydrogen bonds can a water molecule and a methanol molecule form? Identify the properties of water that stem from hydrogen bonding. **[B.U. 2010, 2016]**

**Ans. :** $H_2O$ may form four hydrogen bonds by two acidic hydrogen atoms, by hydrogen bond donors and oxygen atom hydrogen bond acceptors through its two lone pair of electrons.

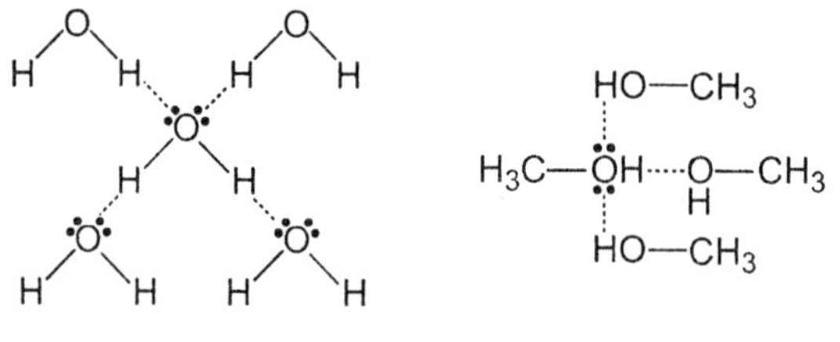

$CH_3OH$ may form three hydrogen bonds through acidic hydrogen atom of OH group, hydrogen bond donors and oxygen atom, hydrogen bond acceptors via its two lone pair of electrons.

Effect of hydrogen bonding in water are high melting point, high heat of vaporisation, lower density of ice compared to water, high specific heat, high ionic conductivity of $H_3O^+$ and $OH^-$ etc. These are the stemming from hydrogen bonding.

**Problem 3.60** State and explain Bent's rule, and apply it to predict the most probable shapes of $PCl_2F_3$ and $OSF_4$. Indicate the type of hybridisation of the central atom in each case. **[B.U. 2011]**

**Ans. :** Sec. 3.3.3.

**Problem 3.61** Mention the importance of hydrogen bonding in sustaining life. **[B.U. 2011]**

**Ans. :** Sec. 3.10.4(d).

**Problem 3.62** Draw the M.O. diagram of $SF_6$ and show how accommodation of 12 electrons around sulphur can be accounted for? **[B.U. 2013]**

**Ans. :** 12 electrons of $SF_6$ comes from 6(S) + 6(F). MO picture of $SF_6$ may be drawn from the overlap of s and p atomic orbitals of S with the $\sigma$ atomic orbitals of six terminal F atoms. We get a total of 10 molecular orbitals—4 bonding, 4 antibonding and 2 non-bonding. Out of the total 12 valence electrons 8 electrons occupy the four bonding molecular orbitals

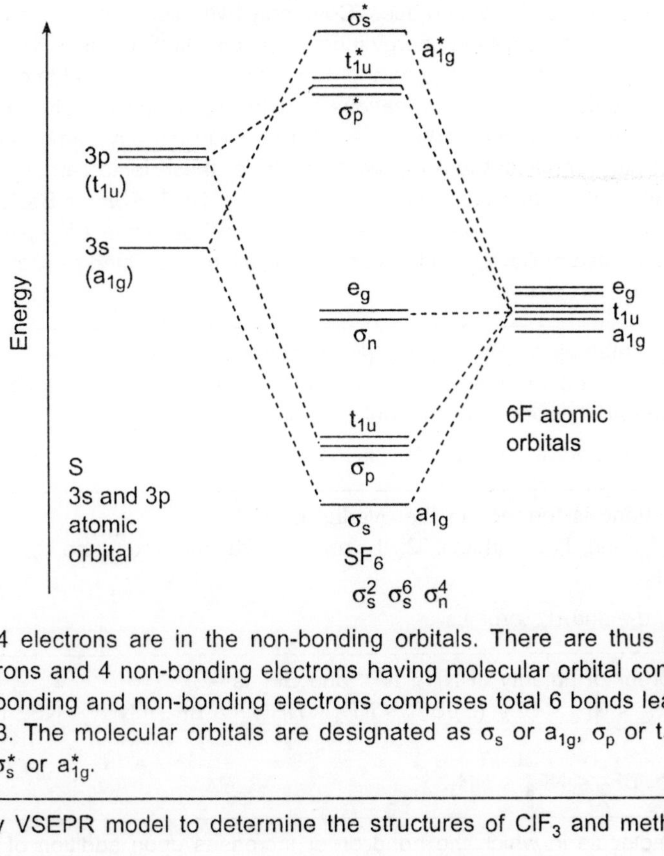


and 4 electrons are in the non-bonding orbitals. There are thus effectively 8 bonding electrons and 4 non-bonding electrons having molecular orbital configuration as $\sigma_s^2\sigma_p^6\sigma_n^4$. The bonding and non-bonding electrons comprises total 6 bonds leading to $\sigma$ bond order of 2/3. The molecular orbitals are designated as $\sigma_s$ or $a_{1g}$, $\sigma_p$ or $t_{1u}$, $\sigma_n$ or $e_g$, $\sigma_p^*$ or $t_{1u}^*$ and $\sigma_s^*$ or $a_{1g}^*$.

**Problem 3.63** Apply VSEPR model to determine the structures of $ClF_3$ and methyl carbanion. **[B.U. 2013]**

**Ans. :** $ClF_3$ has similar structure with that of $BrF_3$ (Problem 3.23). $ClF_3$ is T shaped. Central Cl has 7+3 = 10 electrons i.e. 5 electron pairs of which 3 b.p. and 2 l.p. which are arranged in TBP geometry. One F atom and 2 l.p. are in the equatorial positions and two F atoms are in the axial positions.

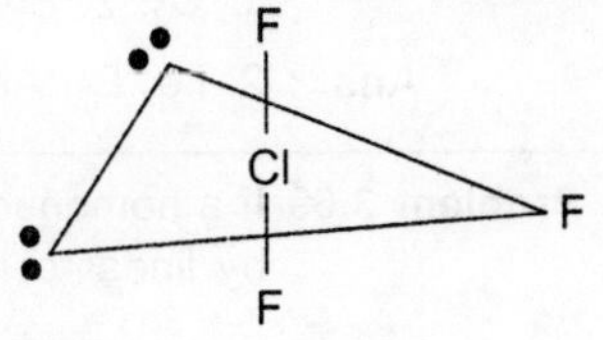

Methyl carbanion has pyramidal structure with that of isoelectronic $NH_3$. In case of $CH_3^-$ the central carbon is $sp^3$ hybridised having three b.p. and one l.p.

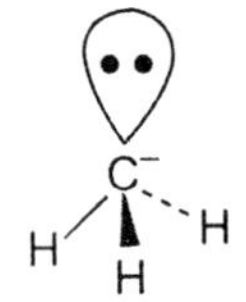

**Problem 3.64** Explain the sequence of water solubilities of the following metal carbonates : $MgCO_3 > CaCO_3 > SrCO_3 > BaCO_3$ **[B.U. 2013]**

**Ans. :** For these carbonates the anion is the same i.e. $CO_3^{2-}$ but the cations has the order of size $Mg^{2+} < Ca^2 < Sr^{2+} < Ba^{2+}$. So according to Fajan's rule $Mg^{2+}$ ion will exert greater polarising power on the anion and $Ba^{2+}$ will have least polarising power so that sequence of covalent nature of the carbonates will be in the order $MgCO_3 > CaCO_3 > SrCO_3 > BaCO_3$ and as water is a polar solvent thereby solubility order will be $BaCO_3 > SrCO_3 > CaCO_3 > MgCO_3$. But actually the water solubility sequence is just reverse of it.

This can be explained on the consideration that these carbonates are primarily ionic and in case of ionic compounds solubilities depend on two factors : (a) lattice energy and (b) hydration energy. As lattice energy increases solubility decreases or with decrease of lattice energy solubility increases. Coversely hydration energy has direct relationship with solubility i.e. as hydration energy increases solubility increases. Here the anion $CO_3^{2-}$ has large size and same charge type of the cations of Group 2 elements. Thus the interionic distance between the metal cations and the $CO_3^{2-}$ ion remain almost constant so that lattice energy of these carbonates will remain practically same and the lattice energies do not play significant role for the solubility of these ionic carbonates. But the hydration energies of the cations are in the order $Mg^{2+} > Ca^{2+} > Sr^{2+} > Ba^{2+}$. As hydration is large for $Mg^{2+}$ and least for $Ba^{2+}$ so that solubility will be highest in case of $MgCO_3$ and least will be in case of $BaCO_3$ and the sequence will be predicted as $MgCO_3 > CaCO_3 > SrCO_3 > BaCO_3$.

**Problem 3.65** The geometries of $[Br_3]^+$ and $[I_5]^+$ respectively are—
1. trigonal and tetrahedral, 2. tetrahedral and trigonal bipyramidal, 3. tetrahedral and tetrahedral, 4. linear and trigonal pyramidal. **[NET 2015]**

**Ans. :** 2. i.e. tetrahedral and trigonal bipyramidal.

**Problem 3.66** The δ-bond is formed via the overlap of—
1. $d_{x2-y2}$ and $d_{x2-y2}$ orbitals, 2. $d_{xz}$ and $d_{xz}$ orbitals, 3. $d_{xy}$ and $d_{xy}$ orbitals, 4. $d_{yz}$ and $d_{yz}$ orbitals. **[NET 2014]**

**Ans. :** 3. i.e. $d_{xy}$ and $d_{xy}$ orbitals.

**Problem 3.67** The order of polarity of $NH_3$, $NF_3$ and $BF_3$ is—
(A) $NH_3 < NF_3 < BF_3$, (B) $BF_3 < NF_3 < NH_3$, (C) $BF_3 < NH_3 < NF_3$, (D) $NF_3 < BF_3 < NH_3$. **[GATE 2012]**

**Ans. :** (B) i.e. $BF_3 < NF_3 < NH_3$.

**Problem 3.68** The molecule in which the bond order increases upon addition of an electron is—
1. $O_2$, 2. $B_2$, 3. $P_4$, 4. $N_2$ **[NET 2013]**

**Ans. :** 2. i.e. $B_2$. As per M.O. diagram (Fig. 3.24) and Table 3.9.

**Problem 3.69** If a homonuclear diatomic molecule is oriented along z-axis, the molecular orbital formed by linear combination of $p_x$ orbitals of the two atoms is—

1. $\sigma$, 2. $\sigma^*$, 3. $\pi$, 4. $\delta$ **[NET 2013]**

**Ans. :** 3. i.e. $\pi$.

---

**Problem 3.70** According to VSEPR theory, the molecule/ion having ideal tetrahedral shape is— 1. $SF_4$, 2. $SO_4^{2-}$, 3. $S_2Cl_2$, 4. $SO_2Cl_2$. **[NET 2011]**

**Ans. :** 2. i.e. $SO_4^{2-}$.

---

**Problem 3.71** Among $SF_4$, $BF_4^-$, $XeF_4$ and $ICl_4^-$ the number of species having two long pair of electrons on the central atom according to VSEPR theory is— 1. 2, 2. 3, 3. 4, 4. 0 **[NET 2011]**

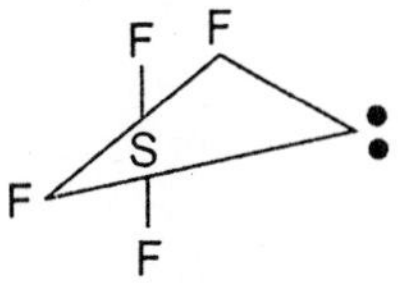


**Ans. :** 1. i.e. $XeF_4$ and $ICl_4^-$. $SF_4$ has trigonal bipyramidal geometry having one lone pair.

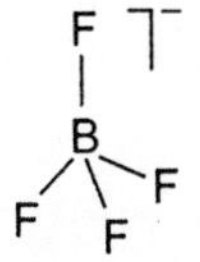


$BF_4^-$ has tetrahedral geometry having no lone pair. $XeF_4$ has square planar geometry having **two lone pairs** in axial positon i.e. the central Xe is $sp^3d^2$ hybridised i.e. $XeF_4$ has octahedral geometry.

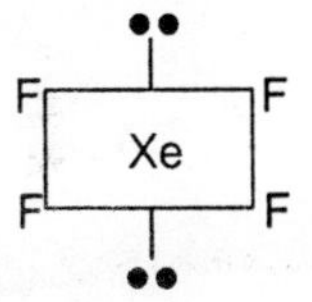


$ICl_4^-$ has similar structure with that of $XeF_4$, the central I is $sp^3d^2$ hybridised having **two lone pairs**.

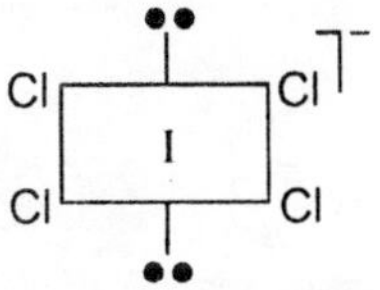


---

**Problem 3.72** Predict the trend in bond angles in $PX_3$ (X=F, Cl, Br, I) molecules with justifications. **[C.U. 2013]**

**Ans. :** Sec. 3.3.1. As per the VSEPR theory rule repulsion exerted by bond pair decreaess as the electronegativity of the B atom increases in a $AB_n$ type of molecule thereby bond angle increases as the electronegativity of B atom decreases. The bond angle of $PF_3$, $PCl_3$, $PBr_3$ and $PI_3$ are 98°, 100°, 101.5° and 102° respectively.

---

**Problem 3.73** Assuming z-axis as bond axis discuss the molecular orbitals formed from d-orbitals. Give the overlap diagram in each case.

**Ans. :** In case of first transition elements involving 3d orbitals the M.O. diagram is complicated because the 3d orbitals being contracted can give only **poor overlap with the 3d orbitals of the adjacent centres**. However the early members (e.g. Ti, V) can undergo radial extension of the 3d orbitals relatively larger compared to that of the latter members and thus overlap between the 3d orbitals is relatively favoured for the early members of the first transition series. For the latter transition series elements involving 4d and 5d orbitals, the relativistic expansion of the d-orbitals are somewhat much better to form the molecular orbitals and thus 4d and 5d orbitals are much suitable for the MOs formation.

When the z-axes are collinear two $d_{z^2}$ orbitals can combine end on for $\sigma$-bonding. The $d_{xz}$ and $d_{yz}$ orbitals form $\pi$-orbitals. When atomic orbitals meet from two parallel planes and combine side to side, as do the $d_{x^2-y^2}$ and $d_{xy}$ orbitals with collinear z-axis, they form $\delta$-orbitals. (The $\delta$ notation indicates sign changes on $C_4$ rotation about the bond axis). $\sigma$-orbitals have no nodes but $\pi$-orbitals have one node that includes the line of centres and $\delta$-orbitals have two nodes that include the line of centres. Thus the combination may be summarised as :

$d_{z^2} \pm d_{z^2}(\sigma+\sigma^*)$; $d_{xz}$, $d_{yz} \pm d_{xz}$, $d_{yz}(\pi+\pi^*)$ and $d_{x^2-y^2}$, $d_{xy} \pm d_{x^2-y^2}$, $d_{xy} \pm d_{x^2-y^2}$, $d_{xy}(\delta+\delta^*)$.

The qualitative M.O. picture of the said overlap of orbitals including the ns orbital is depicted below :

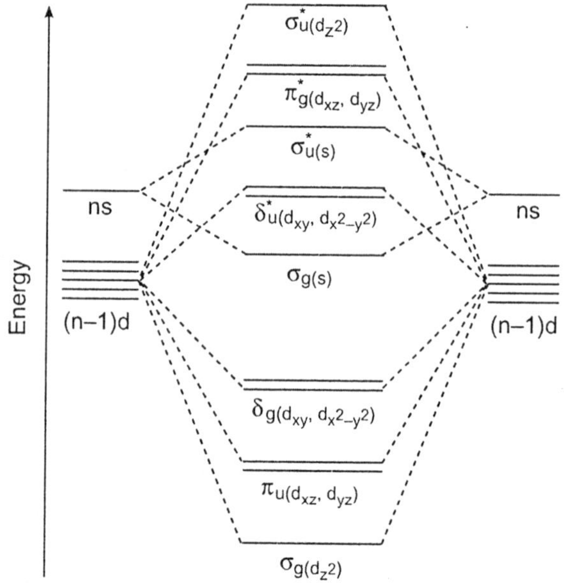


The stability sequence of the M.O.s are : $\sigma_g > \pi_g > \delta_g$.

The overlap diagram of the d orbitals to form σ, π and δ orbitals are shown below :

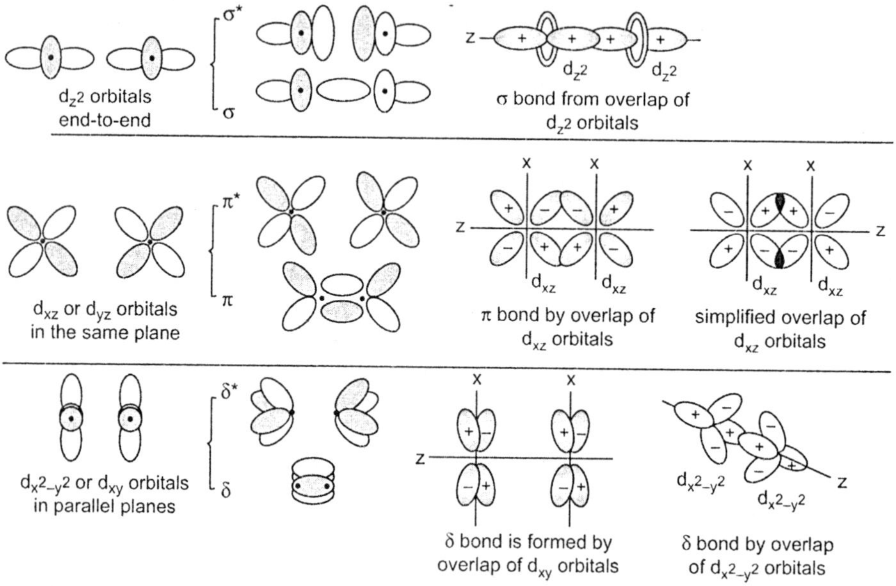


**Problem 3.74** Dipole moment of HBr is $2.60\times10^{-30}$ C.m (Coulomb metre) and interionic distance is 1.41Å. Calculate the partial charge on the bonded atoms and find out percent ionic character. [e = $1.60\times10^{-19}$ coulomb]. **[C.U. 2014]**

**Ans.** : Sec. 3.7.2. Partial charge on H and Br = $\dfrac{2.60\times10^{-30}\text{C. m}}{\dfrac{1.41\times10^{-8}}{10^2}\text{ m}}$

$$= \frac{2.60 \times 10^{-20}}{1.41} C = 1.844 \times 10^{-20} C$$

$$\text{Therefore percent ionic character} = \frac{1.844 \times 10^{-20} C}{1.60 \times 10^{-19} C} \times 100 = 11.53\%.$$

---

**Problem 3.75** Comment on $N(SiH_3)_3$ is planar while $N(CH_3)_3$ is pyramidal. **[W.B.S.U. 2019(CBCS)]**

**Ans. :** In trimethylamine $N(CH_3)_3$, the carbon atom in N–C bond is devoid of d-orbital which cannot receive electron density from nitrogen. But in trisilylamine, the silicon atom of the N–Si bond has availability of d-orbital since silicon has n=3 quantum shell. Thus the lone pair of electron on nitrogen can form dative $\pi$-bonds to the silicon atom. Because of the participation of the vacant d-orbital of silicon in $\pi$-back bonding ($3d_\pi$(si)–$2p_\pi$(O)), the structure of $N(CH_3)_2$ is totally different from $N(SiH_3)_3$. In $N(CH_3)_3$ the nitrogen is $sp^3$ hybridised and the bond angle $\angle CH_3–N–CH_3$ is about 108° while in $N(SiH_3)_3$ the nitrogen is $sp^2$ hybridised having **planar structure** with the bond angle $\angle H_3Si–N–SiH_3$ = 120°. In $N(SiH_3)_3$, the nitrogen is forced to adopt $sp^2$ hybridisation to house the lone pair in a pure p-orbital which in turn engaged in the formation of $3d_\pi$(Si)–$2p_\pi$(N) back bonding. In fact the d-orbitals on three silicon atoms can participate in the $p_\pi$–$d_\pi$ back bonding to delocalise the lone pair over the whole molecule as depicted below schematically for the structure of the two amines :

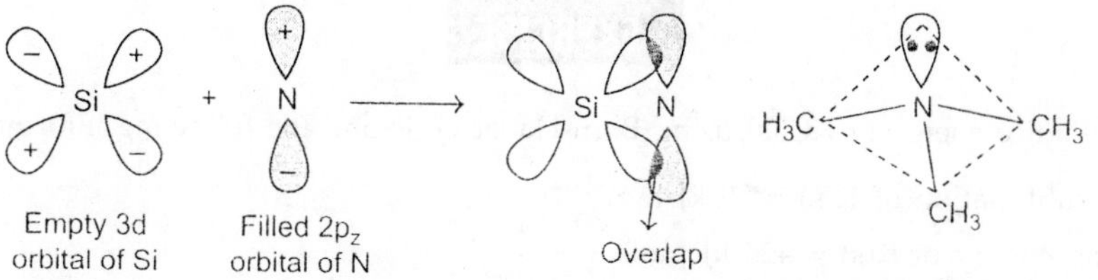


Because of this involvement of the lone pair on nitrogen, $N(SiH_3)_3$ not readily available for showing Lewis basicity. On the other hand in $N(CH_3)_3$ no such back bonding is possible and the molecule assumes a **pyramidal** shape and thus $N(CH_3)_3$ shows Lewis basicity. Due to the formation of $p_\pi$–$d_\pi$ back bonding i.e. due to the multiple bonding character between N and Si, the following resonating structures may be possible for $N(SiH_3)_3$ :

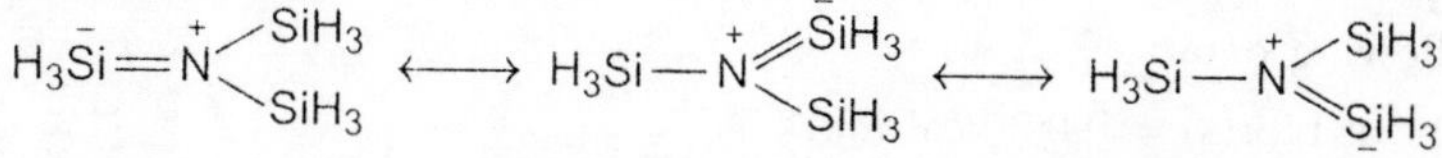

Because of this structural difference between $N(CH_3)_3$ and $N(SiH_3)_3$, they behave differently with HCl : $(CH_3)_3N + HCl \longrightarrow (CH_3)_3N^+HCl^-$ i.e. Lewis acid-base adduct formation.

$(SiH_3)_3N + HCl \longrightarrow NH_4Cl + 3SiH_3Cl$ i.e. bond cleavage occurs.

---

**Problem 3.76** Coment on the structure and basicity of $(CH_3)_2O$ and $(SiH_3)_2O$. **[B.U. 2013]**

**Ans. :** Both the structure and basicity of the two compounds differ widely. In$(CH_3)_2O$ the oxygen is $sp^3$ hybridised and two $sp^3$ hybrid orbitals houses the lone pairs. The $H_3C—O—CH_3$ bond angle ~110°. The $CH_3$ group is electron donating gorup thereby lone pair donor property facilitates so that $(CH_3)_2O$ shows substantial Lewis basicity.

In case of $(SiH_3)_2O$ the oxygen is $sp^2$ hybridised and the bond angle $H_3Si—O—SiH_3$ is expanded more compared to that of $\angle H_3C—O—CH_3$ to induce more p-character on the lone pair of oxygen to acclerate the $2p_\pi(O) \longrightarrow 3d_\pi(Si)$ back bonding. Because of the

$p_\pi$–$d_\pi$ back bonding, the availability of the lone pair on oxygen in $(SiH_3)_2O$ is significantly reduced so that $(SiH_3)_3O$ is somewhat less basic than that of $(CH_3)_2O$. Moreover in $(SiH_3)_2O$ the lone pair on oxygen is extremely delocalised and becomes less available to act as Lewis base.

$$H_3Si-\ddot{\underset{\cdot\cdot}{O}}-SiH_3 \longleftrightarrow H_3Si=\ddot{O}-SiH_3 \longleftrightarrow H_3Si-\ddot{O}=SiH_3$$

Delocalisation is possible because Si can be able to expand its octet.

**Problem 3.77** Among the following an example of a hypervalent species is—
1. $BF_3.OEt_2$, 2. $SF_4$, 3. $[PF_6]^-$, 4. $Sb_2S_3$ **[NET 2013]**

**Ans. :** 3. i.e. $[PF_6]^-$

**Problem 3.78** How many σ-bonds are present in the following molecule? $HC\equiv CCH=CHCH_3$
1. 4, 2. 6, 3. 10, 4.13 **[NET 2011]**

**Ans. :** 3. i.e. 10 σ-bonds are present.

## EXERCISE

1. Calculate the lattice energy of KCl(s) using Born-Haber cycle and the following information :

   Heat of sublimation of K(s) = 89 kJ $mol^{-1}$

   Ionisation energy of K(s) = 425 kJ $mol^{-1}$

   Bond dissociation energy of $Cl_2(g)$ = 244 kJ $mol^{-1}$

   Electron affinity of Cl(g) = –355 kJ $mol^{-1}$

   Heat of formation of KCl(s) = –438 kJ $mol^{-1}$

   **Hint :** From Sec. 3.1.3 $\Delta H_f = S+I+0.5D+E+U$ gives $U = -719$ kJ $mol^{-1}$

2. Define ionisation potential and electron affinity and hence calculate the electron affinity of bromine atom from the following data : **[B.U. 1998]**

   Lattice energy of LiBr(s) = 190.7 kcal $mol^{-1}$

   Heat of formation of LiBr(s) = 83.7 kcal $mol^{-1}$

   Sublimation energy of Li(s) = 38.4 kcal $mol^{-1}$

   Ionisation potential of Li(s) = 124.4 kcal $mol^{-1}$

   Dissociation energy of $Br_2(g)$ = 53.6 kcal $mol^{-1}$

   **Hint :** $\Delta H_f = S+I+0.5D+E+U$

   $-83.7 = 38.4+124.4+26.8+E - 109.7,$
   $E = -82.6$ kcal $= -3.58$ eV.

3. What is Born-Haber cycle? What are its applications? **[B.U. 1990]**

   **Hint :** Sec. 3.1.3 and 3.1.4.

4. Calculate the lattice energy of $MgBr_2$ from the following data :

Sublimation energy of Mg(s) = 148 kJ $mol^{-1}$

Ionisation energy of Mg(g) to $Mg^{2+}$(g) = 2187 kJ $mol^{-1}$

Vaporisation energy of $Br_2$(l) to $Br_2$(g) = 31 kJ $mol^{-1}$

Dissociation energy of $Br_2$(g) = 193 kJ $mol^{-1}$

Electron affinity of Br(g) = 331 kJ $mol^{-1}$

Heat of formation of $MgBr_2$(g) = 524 kJ $mol^{-1}$

**Hint** : The Born-Haber cycle (Sec. 3.1.3) is to be modified in case of $MgBr_2$ as :

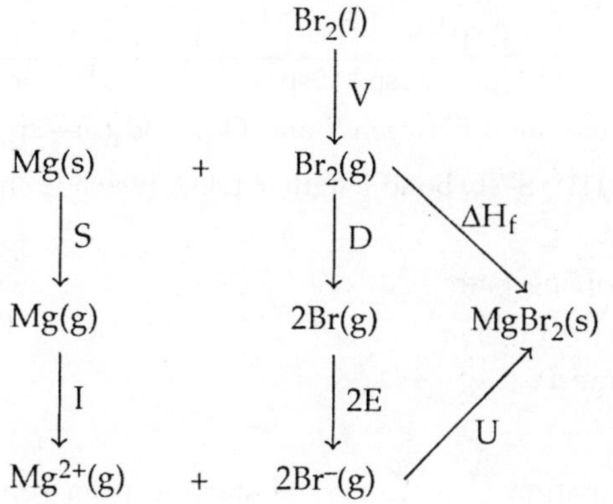


$\therefore$ $\Delta H_f$ = S+I+V+D+2E+U

where V = heat of vaporisation of $Br_2(l)$ to $Br_2$(g). Other terms are the same according to Sec. 3.1.3

i.e. –524 = 148 + 2187 + 31 + 193 – 2×331+U

$\therefore$ U = –524 – 148 – 2187 – 31 – 193 + 2×331

= –2421 kJ $mol^{-1}$.

5. What is Born-Haber cycle? How can it be applied to estimate the lattice energy of NaCl? **[B.U. 2002]**

6. Calculate the lattice energy of NaCl when a mole of crystalline NaCl is prepared from 1 gram atom of sodium and 0.5 mole of chlorine gas, 410 kJ of heat is produced. Sublimation energy of sodium metal is 108.8 kJ and the heat of dissociation of chlorine gas into atoms is 242.7 kJ. The ionisation energy of sodium is 493.7 kJ and electron affinity of chlorine is 368.2 kJ.

**Hint :** $\Delta H_f$ = S+I+0.5D+E+U

–410 = 108.8 + 493.7 + 121.35 + 368.2 + U [Ans. : U = –765.65 kJ $mol^{-1}$]

7. "The nature of hybridisation of the bonding atom has strong influence on its electronegativity"—illustrate. **[B.U. 2007, 2015]**

**Hint :** Electron attracting power of an element in a chemical bond i.e. electronegativity varies with the nature of hybridisation and as the s character of the hybrid orbital increases electronegativity increases e.g. $C_2H_6$, $C_2H_4$ and $C_2H_2$ having hybridisation $sp^3$, $sp^2$ and sp respectively which has 25%, 33% and 50% s character so that electronegativity order of carbon is sp > $sp^2$ > $sp^3$.

8. Draw the structures of iodine pentoxide and periodic acid. **[B.U. 2006]**

**Hint :** 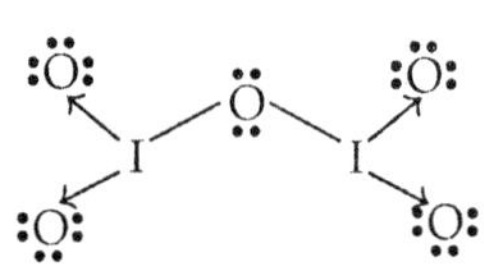

*Each iodine is $sp^2$ hybridised*

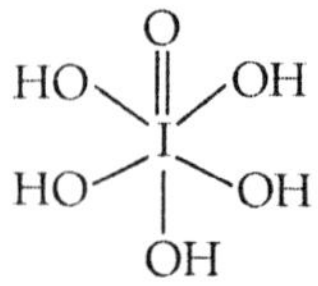

*Central iodine is $sp^3d^2$ hybridised*

9. Predict the shape of $SO_2Cl_2$.

**Hint :** 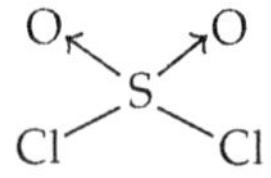 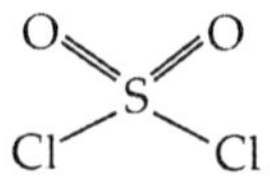

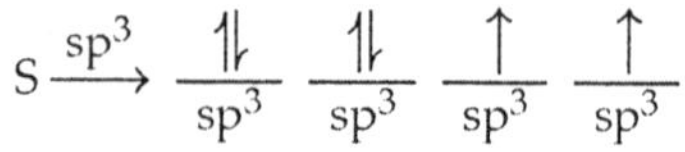 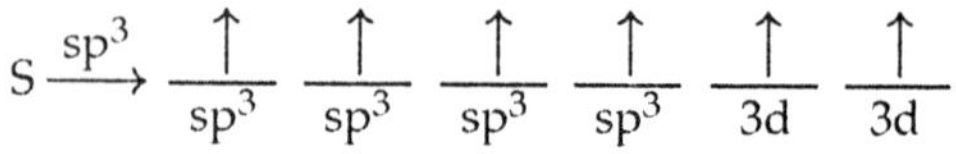

There is a scope of the formation of double bond between S and O i.e. $3d_\pi(S)—2p_\pi(O)$.

The ∠O—S—O = 120° and ∠Cl—S—Cl = 111°. S—O bond length = 1.43Å which is shorter than single bond (1.7Å).

10. Comment on the bond angles in the following pairs :
$H_2O$ (104.5°), $H_2S$ (92.2°) and $PF_3$ (104°), $PCl_3$ (100°) **[B.U. 2008, 2012]**

**Hint :** For $H_2O$ and $H_2S$, VSEPR rule I and IV

For $PF_3$ and $PCl_3$, VSEPR rule III.

11. Frame the Born-Haber cycle for the formation of NaCl crystal starting from solid sodium metal and gaseous chlorine and hence calculate the lattice energy of crystalline NaCl using the following data :

Sublimation energy of metallic Na = 108 kJ $mol^{-1}$

Bond dissociation energy of $Cl_2$ = 240 kJ $mol^{-1}$

Ionisation energy of Na = 495 kJ $mol^{-1}$

Electron affinity of Cl atom = –359 kJ $mol^{-1}$

Heat of formation of solid NaCl = –410 kJ $mol^{-1}$ **[C.U. 2005]**

12. Formation of $Mg^{2+}$ from Mg and $O^{2-}$ from $O_2$ are endothermic. In contrast $O \rightarrow O^-$ process is exothermic and $Mg^{2+}$ formation energy is about three times greater than $Mg \rightarrow Mg^+$ process. Yet MgO is always formulated as $Mg^{2+}O^{2-}$ rather than as $Mg^+O^-$. What may be the theoretical reason for the formulation $Mg^{2+}O^{2-}$? **[C.U. 2001, 2004]**

**Hint :** MgO is an ionic compound and during its formation from respective constituent ions amount of energy is evolved from the system which is known as lattice energy and this lattice energy dictates all the energetically disfavourable processes (i.e. $O_2 \rightarrow O^{2-}$ and $Mg \rightarrow Mg^{2+}$), energetically favourable so that MgO is formulated as $Mg^{2+}O^{2-}$ and not as $Mg^+O^-$.

13. For light homonuclear diatomic molecules the molecular orbital (MO) energy order may be written as—$1\sigma_g < 1\sigma_u < 2\sigma_g < 2\sigma_u < 1\pi_u < 3\sigma_g$ etc.

For the first excited, stable state of the $H_2$ molecule, the outermost occupied MO is—
1. $1\sigma_u$ 2. $2\sigma_g$ 3. $2\sigma_u$ 4. $1\pi_u$ **[NET 2001]**

**Ans. :** 1.

14. Using the concept of atomic and moelcular orbitals proceed towards formation of bonding and antibonding orbitals. What are non-bonding orbitals? Explain on the basis of molecular orbital theory the monoatomicity of inert gases. **[B.U. 2005]**

    **Hint :** Sec. 3.4.3. Non-bonding orbitals are still antibonding orbital but of lower energy than antibonding orbitals. BO is zero for inert gases.

15. Discuss the formation of a three centre bond using qualitative MO pictures and illustrate with a suitable example. **[B.U. 1999]**

    **Hint :** Sec. 3.4.7(a), Fig. 3.30.

16. Compare the bond angle of $O_3$ and $NO_2$.

    **Hint :**

Central O is $sp^2$ hybridised

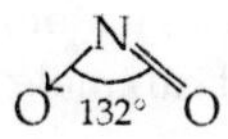


Central N is $sp^2$ hybridised

17. Draw the MO picture of $I_3^-$.

    **Hint :**

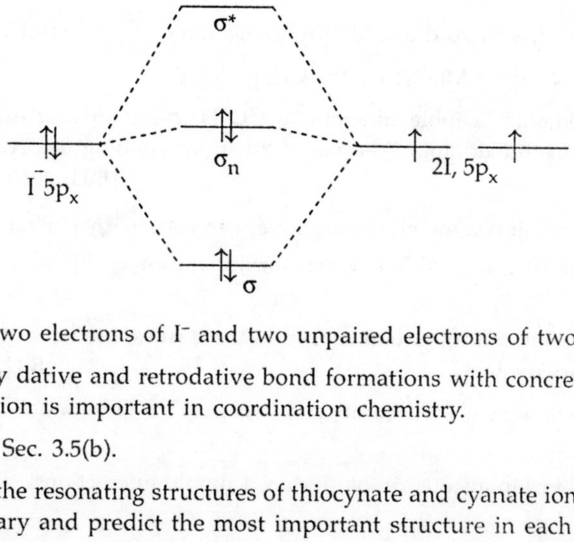


Only two electrons of $I^-$ and two unpaired electrons of two I atoms in $5p_x$ orbitals are taken.

18. Portray dative and retrodative bond formations with concrete examples. State how retrodative bond formation is important in coordination chemistry. **[B.U. 2007, 2010, 2016]**

    **Hint :** Sec. 3.5(b).

19. Draw the resonating structures of thiocynate and cyanate ions showing the formal charges whereever necessary and predict the most important structure in each case **[C.U. 2006]**

    **Hint :** Cyanate ($OCN^-$) and thiocyanate ($SCN^-$) are isosteric species have the following resonating structures :

$$\overset{+}{O}{\equiv}C{-}\overset{2-}{N} \leftrightarrow \overset{-}{O}{-}C{\equiv}N \leftrightarrow O{=}C{=}\overset{-}{N}$$

$$\overset{+}{S}{\equiv}C{-}\overset{2-}{N} \leftrightarrow \overset{-}{S}{-}C{\equiv}N \leftrightarrow S{=}C{=}\overset{-}{N}$$

(I) (II) (III)

Structure (I) is the less contributing structure as the charges are far apart.

20. Write and explain Fajan's polarisation rules. Use these rules to predict which is likely to be ionic or covalent : RbCl and AgCl. **[B.U. 2005]**

    **Hint :** Sec. 3.7. $Ag^+$ is pseudo noble gas configuration whereas $Rb^+$ is noble gas configuration.

21. Explain why o-hydroxybenzaldehyde has lower melting point than p-hydroxybenzaldehyde?

    **Hint :** In ortho form there is intramolecular whereas in para form there is intermolecular H-bonding.

22. Explain why the conductivity of Ge is enhanced manyfolds when trace amount of As is added to it. **[C.U. 2005,2002]**

    **Hint :** n-type semiconductors [Sec. 3.11.4(b)].

23. Calculate the lattice energy of NaCl crystal from the following data by the use of Born-Haber cycle.

    Sublimation energy of Na = 26 kcal/g.atom
    Dissociation energy of $Cl_2$ = 54 kcal/mole
    Ionisation for Na(g) = 117 kcal/atom
    Electron affinity for Cl(g) = 84 kcal/g.atom
    Heat of formation of NaCl = 99 kcal/mole

    **Hint :** $\Delta H_f = S + I + \frac{1}{2}D + E + U$

    $-99 = 26 + 117 + 27 - 84 + U$

    $U = -185$ kcal/mole.

24. Give example (one each) of molecules with $sp^3d$ and $d^3s$ hybridisations. **[B.U. 2010, 2015]**

    **Hint :** Table 3.4. $sp^3d$ : $PCl_5$, $XeF_2$, $I_3^-$. $d^3s$ : $MnO_4^-$, $FeO_4^{2-}$, $CrO_4^{2-}$.

25. What do you mean by dipole moment? "Dipole moment of CO is less than that expected from electronegativity difference of the constituent atoms"—account this from bonding theory. **[B.U. 2010, 2014, 2016]**

    **Hint :** Sec. 3.7.2. :C⇐O: Though oxygen is more electronegative than carbon, the VB structure of CO where a dative bond is formed from oxygen to carbon, is responsible for lower dipole moment of CO than its expected value.

26. List the Fajan's polarisation rules. Using appropriate one(s) from these rules, predict out of CsCl and AuCl, which is likely to be ionic or covalent. **[B.U. 2010, 2015]**

    **Hint :** Sec. 3.7. $Au^+$ having pseudo noble gas configuration so that AuCl will be more covalent than CsCl.

27. Explain with example the effect of instantaneous dipole induced dipole interactions. **[C.U. 2009]**

    **Hint :** Sec. 3.10.

28. Write down the shapes of $\dot{C}H_3$ and $\dot{C}F_3$ using VSEPR theory. **[B.U. 2010]**

    **Hint :** In $\dot{C}H_3$ the central carbon is $sp^2$ hybridised. The single unpaired electron in $\dot{C}H_3$ residing on $p_z$ orbital which accounts its radical nature and thus it has trigonal planar nature. Now on replacement of three hydrogen atoms by three fluorine atoms which are highly electronegative, the s character of three $sp^2$ hybrid orbitals decreases according to Bent's rule (Sec. 3.3.3). This reduction of s character of the central carbon tends to be $sp^3$ hybridised rather than that of $sp^2$ and thus bond angle of ∠F—C—F decreases and also the geometry changes from trigonal planar to pyramidal. In $\dot{C}F_3$ the unpaired electron exist in one of the $sp^3$ hybrid orbital.

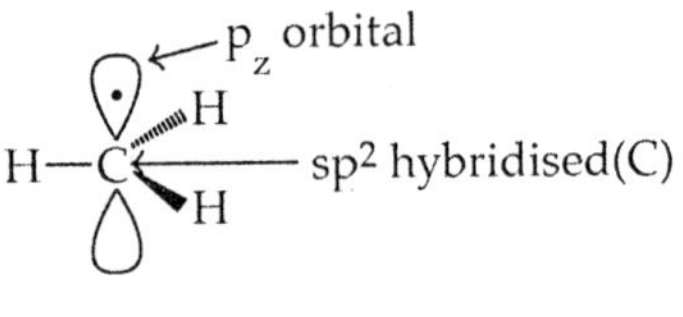


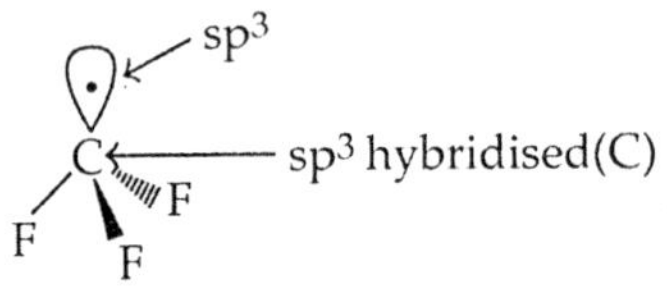

In $\dot{C}H_3$ the bp–bp repulsion operates at 120° and the unpaired electron—bond pair repulsion which is relatively less effective at 90°.

In $\dot{C}F_3$ the bp–bp repulsions are less and all the repulsions occur at about 109° since the central carbon is $sp^3$ hybridised.

29. "The more polar the bonds in a molecule, the more the value of its dipole moment".—Comment. **[B.U. 2010, 2015]**

**Hint :** Sec. 3.7.2. The statement is not always justified.

30. What is Bent's rule? Explain with an example. Predict the ∠X—C—X angle trend in $CH_2X_2$ when X=H, F, Cl. **[C.U. 2008]**

**Hint :** Sec. 3.3.3.

31. Starting with gaseous $NH_3$ and HCl, construct a Born-Haber cycle leading to the formation of solid $NH_4Cl$. **[B.U. 2009]**

**Hint :** Sec. 3.1.4(b).

32. What is the difference between equivalent and non-equivalent hybrid orbitals? Explain with examples. **[C.U. 2004, 2014]**

**Hint :** Sec. 3.2.6.

33. Predict the expected change in bond-order of the following transformation : $NO \rightarrow NO^+ + e^-$. **[B.U. 2011, 2014]**

**Hint :** Sec. 3.4.3. NO : B.O. = 2.5 $NO^+$ : B.O. = 3.

34. 'The dissociation energy of $N_2^+$ and $O_2^+$ are almost the same'.—Give two reasons. **[B.U. 2011, 2015]**

**Hint :** Sec. 3.4.3. Table 3.9. $N_2^+$ and $O_2^+$ have same B.O. (2.5) and isostructural.

35. What do you mean by hypervalence? Give examples of two hypervalent compounds and mention why hypervalence does arise in these. Count the number of total electrons around each core atom. **[B.U. 2011, 2015]**

**Hint :** Sec. 3.2.1. Examples $SF_6$, $PCl_5$ where the central core atom S and P has 12 and 10 eletrons respectively. Hypervalence arises due to the expansion of valence shell having availability of low-lying unfilled d-orbitals.

36. Draw the possible resonating structures of $O_3$ and $NO_2^-$ indicating with justification the most probable one in each case. Identify a special relationship between the molecule $O_3$ and the molecule ion $NO_2^-$. **[B.U. 2011]**

**Hint :** Sec. 3.2.12. $O_3$ and $NO_2^-$ are isoelectronic and isostructural i.e. isosteric species. See problem 3.43.

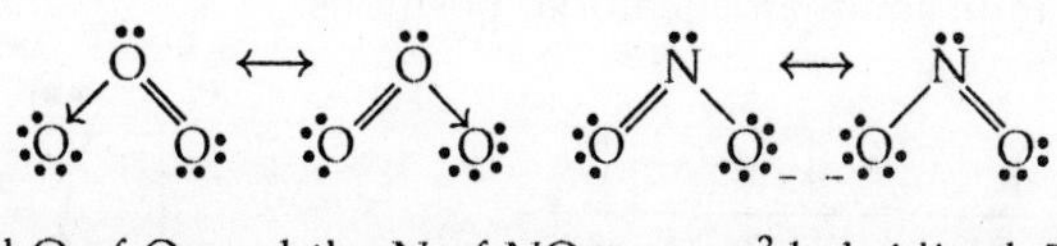

The central O of $O_3$ and the N of $NO_2^-$ are $sp^2$ hybridised. These are the most probable structures of the isosteric species.

37. Show the structure of a suitable compound having μ-bond. **[B.U. 2011]**

**Hint :** Sec. 3.5(c).

38. Distinguish between 2c–2e and 3c–2e bonds with concrete examples. **[B.U. 2011, 2013, 2015]**

**Hint :** 2c–2e bond : $NH_3$, $H_2$ 3c–2e bond : $B_2H_6$.

39. Arrange the following bonds in an expected order of increasing ionic character : B–O, Cl–O, F–O, N–O. **[B.U. 2011]**

    **Hint :** N–O, F–O, Cl–O < B–O.

40. $O_2$ molecule is paramagnetic but $O_2^{2-}$ is diamagnetic—explain. **[B.U. 2012, 2015]**

    **Hint :** Sec. 3.4.3. Fig. 3.24. $O_2$ has two unpaired electrons in the $\pi^*$ antibonding orbital while $O_2^{2-}$ has not any unpaired electron.

41. "The bond length in $N_2^+$ is greater than that in $N_2$ while bond length in $NO^+$ is less than that in NO". Explain the observation using MO theory. **[B.U. 2012]**

    **Hint :** Sec. 3.4.3. The bond order of $N_2$ and $N_2^+$ are 3 and 2.5 respectively and as bond order decreases the bond length increases from $N_2$ to $N_2^+$.

    The bond order of NO and $NO^+$ are 2.5 and 3 respectively and as bond order increases bond length decreases from NO to $NO^+$.

42. Distinguish between intramolecular and intermolecular hydrogen bonding with suitable examples. **[B.U. 2012, 2014]**

    **Hint :** Sec. 3.10.3.

43. $NH_3$ has a dipole moment of 1.49 unit against almost zero value of $NF_3$—explain. **[B.U. 2012, 2015]**

    **Hint :** Sec. 3.7.2.

44. "$He_2$ molecule does not exist". Justify. **[B.U. 2012, 2014, 2016]**

    **Hint :** Sec. 3.4.3. Bond order zero.

45. Predict the shapes of the following species and identify the hybridisations of central atoms : $XeOF_4$ and $I_5^-$. **[B.U. 2012]**

    **Hint :** In $XeOF_4$ the central Xe is $sp^3d^2$ hybridised and the molecule has octahedral geometry. Five unpaired electrons in five $sp^3d^2$ orbitals make five $\sigma$ bonds in the equatorial positions with four fluorine and one oxygen in apical position. The sixth unpaired electron and d orbital of Xe makes a double bond with the oxygen. One $sp^3d^2$ orbital occupied a lone pair in other apical site.

| Xe : | ⇅ | ↑ | ↑ | ↑ | ↑ | ↑ | ↑ |
|---|---|---|---|---|---|---|---|
| | $5(sp^3d^2)$ | $5(sp^3d^2)$ | $5(sp^3d^2)$ | $5(sp^3d^2)$ | $5(sp^3d^2)$ | $5(sp^3d^2)$ | 5d |

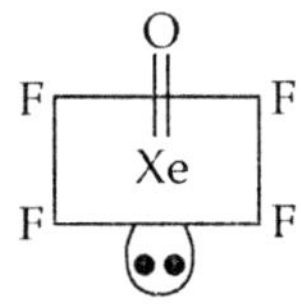


In $I_5^-$ the central $I^-$ is $sp^3d^2$ hybridised and the species $I_5^-$ has octahedral geometry. Two $sp^3d^2$ orbitals have lone pairs which occupy in axial positions and the four I atoms overlap with four $sp^3d^2$ orbitals having unpaired electrons to form four bonds in equatorial positions.

| $I^-$ : | ⇅ | ⇅ | ↑ | ↑ | ↑ | ↑ |
|---|---|---|---|---|---|---|
| | $5(sp^3d^2)$ | $5(sp^3d^2)$ | $5(sp^3d^2)$ | $5(sp^3d^2)$ | $5(sp^3d^2)$ | $5(sp^3d^2)$ |

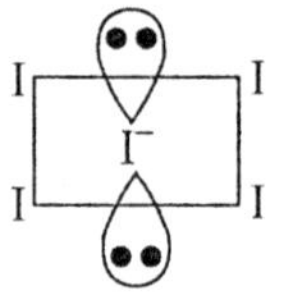


46. Between $SrSO_4$ and $MgSO_4$, which one is more soluble in water and why? **[C.U. 2012]**

    **Hint :** Sec. 3.1.4(a). $MgSO_4$ > $SrSO_4$.

47. Calculate the standard heat of formation ($\Delta H_f^\circ$) of the hypothetical $Xe^+F^-(s)$ from the following information : Lattice energy of XeF is equal to that of CsF. I.E. of Xe = 1170kJ $mol^{-1}$, of Cs = 376.5 kJ $mol^{-1}$. Heat of atomisation of Cs(s) = 79.5 kJ $mol^{-1}$, and $\Delta H_f^\circ$ of CsF = –531.4 kJ $mol^{-1}$. **[C.U. 2012]**

**Hint :** Sec. 3.1.4(c).

48. Explain the formation of sigma and pi bond in oxygen molecule. **[B.U. 2013]**

**Hint :** According to VBT $O_2$ molecule has a double bond between two oxygen atoms i.e. $:\ddot{O}=\ddot{O}:$. There is one σ-bond formed between the head-on overlap of two $2p_x$ orbitals of two oxygen atoms and one π-bond is formed by side-wise overlap of two $2p_y/2p_z$ orbitals of two oxygen atoms. In terms of hybrid orbitals concept each oxygen atom is sp hybridised. Two hybrid orbitals of two oxygen atoms overlap to form a σ-bond and the π-bond is formed as stated before.

For M.O. approach of σ and π bond see Sec. 3.4.4.

49. Obtain the bond orders of $O_2^+$, $O_2$, $O_2^-$ and $O_2^{2-}$. **[B.U. 2013]**

**Hint :** Sec. 3.4.3.

| | | | | |
|---|---|---|---|---|
| Species : | $O_2$ | $O_2^+$ | $O_2^-$ | $O_2^{2-}$ |
| Bond order : | 2 | 2.5 | 1.5 | 1 |

50. The conductivity of Ge is enhanced when trace amount of As is added to it. Explain. **[B.U. 2013]**

**Hint :** Sec. 3.11.4(b). When traces of As is added to Ge-crystal (1 in $10^9$) number of electron carriers is increased by the process of doping giving rise to n-type semiconductivity (n means negatively charged electrons).

51. Compare the bond angles between $NH_3(107°)$ and $PH_3(90°)$. **[B.U. 2013]**

**Hint :** Sec. 3.3.1 and Sec. 3.3.3.

Nitrogen has smaller size and greater electronegativity compared to phosphorus having larger size and lesser electronegativity. Acording to VSEPR model repulsion exerted in filled shell is more compared to that of repulsion exerted in incompletely filled shell. Nitrogen has filled shell (n=2) while phosphorus has incompletely filled shell (n=3). Increase in size and decrease in electronegativity of central atom in $AB_x$ molecule permit the bonding electron pairs to be drawn out further, thus decreasing the repulsion between bond pairs leading to the decrease in bond angle from $NH_3$ to $PH_3$.

52. $SnCl_4$ is low boiling liquid but $SnCl_2$ is high melting solid.—Comment. **[B.U. 2013]**

**Hint :** Sec. 3.7. According to Fajan's rule polarising power of small size cation i.e. $Sn^{4+}$ is greater than that of large size cation i.e. $Sn^{2+}$ on the $Cl^-$ so that $SnCl_4$ is more covalent than that of $SnCl_2$ i.e. ionic character of $SnCl_2$ is far greater than $SnCL_4$. We know that ionic compounds generally exist as a solid while covalent compounds exist as gaseous or liquid state. As the covalent character of $SnCl_4$ is more so that it exist as a low boiling liquid.

53. The bond angles for the hydrides of Group 15 elements are as follows : $NH_3(107.8°)$, $PH_3(93.6°)$, $AsH_3(91.8°)$, and $SbH_3(91.3°)$. Account this trend. **[C.U. 2013]**

**Hint :** Sec. 3.3 and Sec. 3.3.1.

(1) The bond angles for the hydrides of Group 15 elements decrease down the group as the electronegativity of the central element decrease. Bond pair-bond pair repulsion decreases as the electronegativity of the central atom decreases and hence bond angle decreases down the group.

(2) Repulsion exerted in the filled shell is more compared to the repulsion exerted in the incompletely filled shell and hence less repulsion and less bond angles according to VSEPR model.

54. The triiodide ion $I_3^-$ is linear but $I_3^+$ is bent. Explain. **[B.U. 2013, 2016]**

**Hint :** Valence shell electron of the central I = 8+1+1 = 10 electrons = 5 electron pairs = 2 b.p. + 3 l.p. An arrangement of five electron pairs around central iodine according to VSEPR model predicts a TBP arrangement. The three lone pairs/unshared pairs occupied in the equatorial positions and thus $I_3^-$ assumes a linear geometry i.e. $I{-}I{-}I]^-$.

$I_3^+$ : Here the valence shell of central I = 6+1+1 = 8 electrons = 4 pairs = 2 b.p. + 2 l.p. VSEPR theory predict heavy repulsion by the two non-bonding l.pl of electrons and the molecule ion will be bent

i.e. 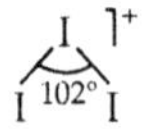 rather than linear one.

55. How Born-Haber cycle is used to get the lattice energy of an ionic crystal? Discuss with a suitable example. **[B.U. 2013]**

**Hint :** Sec. 3.1.3.

56. "The Madelung constant depends only on the crystal geometry". Explain. **[B.U. 2013]**

**Hint :** Sec. 3.1.2 and Problem 3.2.

Madelung constant (A) value arises by calculating the sum of the attractive and repulsive forces involved in the infinite series (convergent series). Thus the series converges to a particular value representing the Madelung constant for ionic crystal of a given geometry and value depends on the crystal geometry e.g.

| Crystal lattice | Madelung constant |
|---|---|
| Rock salt | 1.74756 |
| Cesium chloride | 1.76267 |
| Zinc blende (ZnS) | 1.63806 |
| Wurtzite (ZnS) | 1.64130 |

57. In spite of having sufficiently high lattice energy, CaF does not exist.—Explain. **[B.U. 2013]**

**Hint :** Sec. 3.1.4(c).

58. Metals that are malleable and ductile have cubic close packed structure.—Justify. **[B.U. 2013]**

**Hint :** The properties of a metal as noted by its malleability (ability to be beaten or rolled into sheets) and ductility (ability to be drawn into wire), depends on the crystal structure as well as on other bond strengths. These metallic characteristics are due to the presence of gliding planes having densely packed atoms, which resist breaking due to the large cohesive forces offering maximum resistance to deformation. The greater the number of such gliding planes in a structure, the more malleable and ductile will be the metal. In a cubic lattice, the plane with the largest number of atoms is the (111) plane which occurs in layers perpendicular to the four cubic body diagonals. In hcp arrangement the most dense planes (001) are the ones perpendicular to the hexagonal axis, so that the directions for deformations are limited. Hence metals with cubic lattice are expected to be more ductile than the hcp metals. It is observed that the fcc metals (Cu, Ag, Au) are more malleable and ductile than the hcp metals (Ti, Zn, Cr).

59. According to VSEPR theory, the geometry (with lone pair) around the central iodine in $I_3^+$ and $I_3^-$ ions respectively are—

1. tetrahedral and tetrahedral, 2. trigonal bipyramidal and trigonal bipyramidal,
3. tetrahedral and trigonal bipyramidal, 4. tetrahedral and octahedral. **[NET 2013]**

**Hint :** Question number 54. Answer is 3 i.e. tetrahedral and trigonal bipyramidal.

60. The reason for the chemical inertness of gaseous nitrogen at room temperature is best given by its 1. high bond energy only, 2. electronic configuration, 3. HOMO-LUMO gap only, 4. high bond energy and HOMO-LUMO gap. **[NET 2013]**

**Ans. :** 4 i.e. high bond energy and HOMO-LUMO gap.

61. In molecules $H_2O$, $NH_3$ and $CH_4$
1. the bond angles are same, 2. the bond distances are same,
3. the hybridisations are same, 4. the shapes are same. **[NET 2011]**

**Ans. :** 3 i.e. the hybridisations are same.

62. The highest occupied MO in $N_2$ and $O_2^+$ respectively are (take x axis as internuclear axis).
1. $\sigma_{2p_x}$, $\pi^*_{2p_y}$; 2. $\pi_{2p_y}$, $\pi_{2p_z}$; 3. $\sigma^*_{2p_x}$, $\sigma_{2p_x}$, 4. $\pi^*_{2p_y}$, $\pi^*_{2p_z}$ **[NET 2011]**

**Ans. :** 4. i.e. $\pi^*_{2p_y}$, $\pi^*_{2p_z}$.

63. The molecule with highest number of lone-pairs and has a linear shape based on VSEPR theory is—
1. $CO_2$, 2. $I_3^-$, 3. $NO_2^-$, 4. $NO_2^+$ **[NEt 2011]**

**Ans. :** 2 i.e. $I_3^-$ Question number 54 and Problem 3.47.

64. Between $\dot{C}H_3$ and $\dot{C}F_3$ one is pyramidal and other is planar. Which one is which and why? **[C.U. 2013]**

**Hint :** Question number 28.

65. In spite of the fact that the hypothetical $NaCl_2$ is expected to have higher lattice energy as compared to NaCl, it does not exist.—Explain. **[C.U. 2013]**

**Hint :** Sec. 3.1.4(c). The enormous amount of energy is required for the second ionisation energy of sodium (4560.5 kJ $mol^{-1}$) which should have to be released for the formation of $NaCl_2$ and thus $NaCl_2$ formation is not possible and thereby $NaCl_2$ disproportionates to the stable NaCl and elementary chlorine according to the reaction : $NaCl_2 \longrightarrow NaCl + {}^1/_2Cl_2$.

66. Draw the possible resonance structures of $ClO_4^-$ and predict the most stable one. **[C.U. 2013]**

**Hint :** $ClO_4^-$ has valence electrons = 7+4×6 + 1 = 32. The central Cl is $sp^3$ hybridised having electron distribution. $\frac{\uparrow\downarrow}{sp^3}\ \frac{\uparrow\downarrow}{sp^3}\ \frac{\uparrow\downarrow}{sp^3}\ \frac{\uparrow}{sp^3}$ or $\frac{\uparrow}{sp^3}\ \frac{\uparrow}{sp^3}\ \frac{\uparrow}{sp^3}\ \frac{\uparrow}{sp^3}\ \frac{\uparrow}{3d}\ \frac{\uparrow}{3d}\ \frac{\uparrow}{3d}$ and the $ClO_4^-$ will have the following resonating structures :

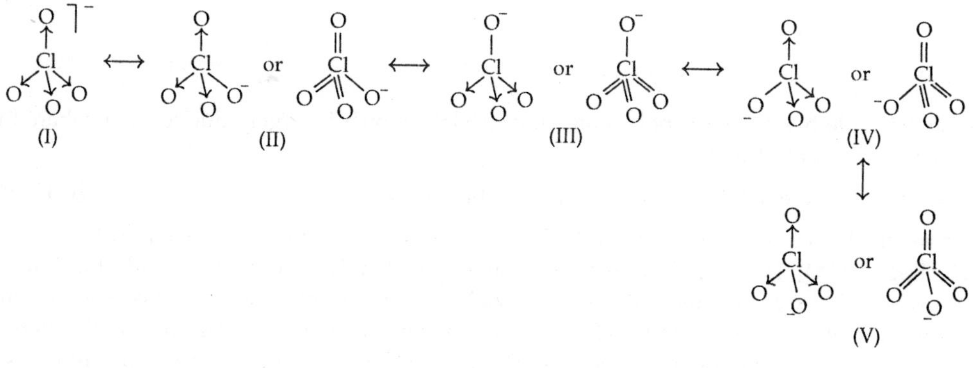


In structure (I) the central chlorine is $Cl^-$ which is $sp^3$ hybridised having electron distribution $\frac{\uparrow\downarrow}{sp^3}\ \frac{\uparrow\downarrow}{sp^3}\ \frac{\uparrow\downarrow}{sp^3}\ \frac{\uparrow\downarrow}{sp^3}$ which forms four coordinate bonds to four oxygen atoms which is the improbable structure because Cl having less electronegative than oxygen but Cl assumes the negative charge. Structures (II) to (V) are equally stable structure.

67. Why is the melting point of CuCl (422°C) much lower than that of KCl (776°C)? **[C.U. 2013]**

**Hint :** Sec. 3.7. According to Fajan's rule the pseudo noble gas configuration is more deformable than that of noble gas configuration. Though $Cu^+$ and $K^+$ has almost equal ionic radius but due to $Cu^+$ having pseudo noble gas configuration more deformable than that of $K^+$ having inert gas configuration and thus CuCl is more covalent than KCl thereby CuCl records less melting point than KCl.

68. Ozone molecule shows unusually low dipole moment.—Explain. **[C.U. 2013]**

**Hint :** Ozone is a triatomic **diamagnetic blue gas** which has according to VB theory following resonating structures having $C_{2v}$ symmetry which has 18 valence electrons :

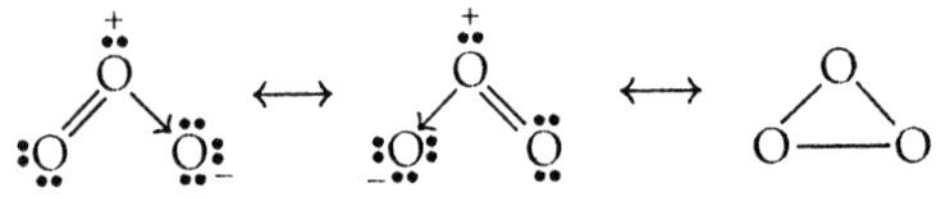

The central oxygen is $sp^2$ hybridised having electron distribution $\underset{sp^2}{\uparrow\downarrow}\ \underset{sp^2}{\uparrow\downarrow}\ \underset{sp^2}{\uparrow}\ \underset{2p}{\uparrow}$. According to MO theory each oxygen atom contributes one p orbital to yield three molecular orbitals. The orbital overlap combinations lead to the following wave functions :

$$\Psi_a = \frac{1}{2}(\Psi_1 - \sqrt{2}\Psi_2 + \Psi_3)$$

$$\Psi_n = \frac{\sqrt{2}}{2}(\Psi_1 - \Psi_3)\ (\uparrow\downarrow)$$

$$\Psi_b = (\Psi_1 + \sqrt{2}\Psi_2 + \Psi_3)\ (\uparrow\downarrow)$$

Only the $\Psi_b$ and $\Psi_n$ molecular orbitals are each doubly occupied. The orbital overlap can be shown as :

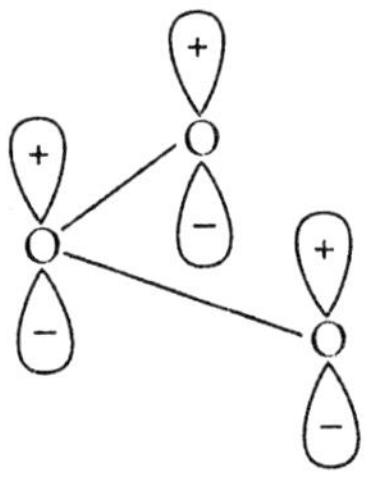

The ozone molecule has a poor dipole moment of 0.534D, which is largely due to the unshared pair of electrons on the central atom.

69. Discuss the trend in the solubility of $MClO_4$ (M = Li, Na, K). **[C.U. 2013]**

**Hint :** As per the lattice energy consideration the solubility trend would be expected $LiClO_4 < NaClO_4 < KClO_4$. But as $ClO_4^-$ is non-polarisable larger anion so that for these compounds the interionic distance ($M^+$—$ClO_4^-$) remain practically same for all the compounds thereby lattice energy remain almost same for all the compounds. Here $Li^+$ ion having smaller size registers high energy of solvation thereby $LiClO_4$ will be much more soluble than those of other alkali metal perchlorates and the solubility order will be $LiClO_4 > NaClO_4 > KClO_4$. For the same reason many simple lithium salts are normally hydrated and anhydrous salts are extremely hygroscopic and thus forms the basis of the widespread use of LiCl and LiBr as dehumidifying agent.

70. Draw the molecular orbital picture of the interhalogen ICl and write the ground state electron configuration.

**Hint :** Here the 3s and 3p valence shell orbitals of Cl and 5s and 5p orbitals of I combine to form the molecular orbitals as depicted in the figure. The bonding orbitals will be Cl dominated (because Cl is more electronegative than I) and the antibonding orbitals are predominantly I in character. The valence electrons 7+7 = 14 are accommodated in the MOs to give the ground state electron configuration as $1\sigma^2 2\sigma^2 1\pi^4 3\sigma^2 2\pi^4$.

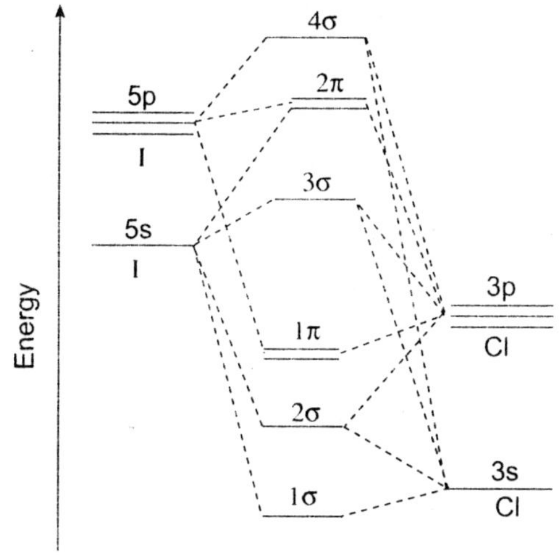


71. Explain p-type semiconductors with example. **[C.U. 2013]**

**Hint :** Sec. 3.11.4.

72. Using VSEPR theory, predict the shape and indicate the state of hybridisation of the central atom of the following : (i) $IO_2F_2^-$, (ii) $XeF_5^-$, (iii) $ClF_3$, (iv) $BrF_4^-$. **[C.U. 2013]**

**Hint :** (i) This is prepared by reaction of $KIO_3$ with 40% HF.

$$KIO_3 \xrightarrow{40\%\ HF} KIO_2F_2$$

The $IO_2F_2^-$ has trigonal bipyramidal geometry having $C_{2v}$ symmetry, the central iodine is $sp^3d$ hybridised. One equatorial position is occupied by lone pair.

$I^-$ is 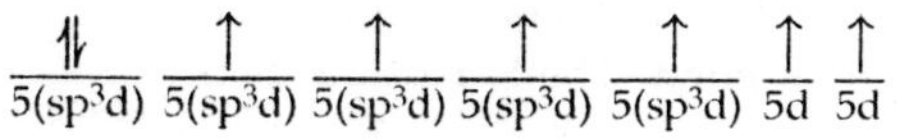


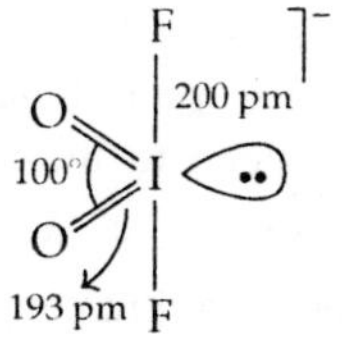


(ii) $XeF_4$ acts as a $F^-$ acceptor i.e. $XeF_4 + F^- \longrightarrow XeF_5^-$. $XeF_5^-$ has pentagonal planar structure having central Xe is $sp^2d^2$ hybridised. This has similar structure with that of $[IF_5]^{2-}$.

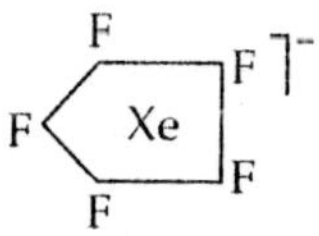


or

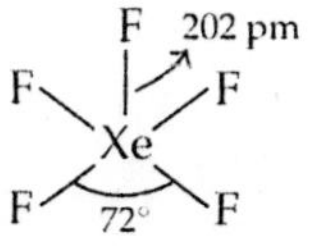

(iii) $ClF_3$ has T-shape. According to VSEPR model the central Cl has 7+1+1+1 = 10 electrons having 5 electron pairs and is $sp^3d$ hybridised $\underset{3(sp^3d)}{\uparrow\downarrow}\ \underset{3(sp^3d)}{\uparrow\downarrow}\ \underset{3(sp^3d)}{\uparrow}\ \underset{3(sp^3d)}{\uparrow}\ \underset{3(sp^3d)}{\uparrow}$.

Three $sp^3d$ hybrid orbitals having one unpaired electron in each overlap with each F to form three Cl–F bonds. There are two lone pairs which form Ψ-trigonal bipyramidal geometry. Repulsion between the lone pairs leads to a slight decrease in F–Cl–F angle from the ideal 90°.

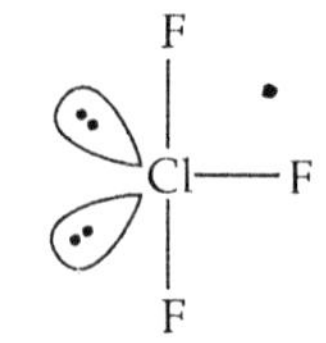


Low temperature X-ray diffraction data show that solid $ClF_3$ is a discrete T-shaped molecule.

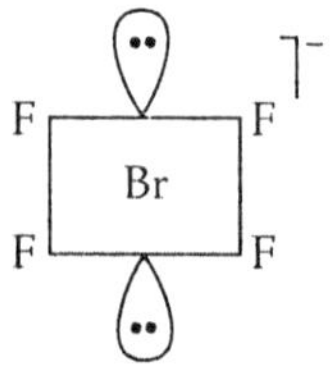


(iv) $BrF_4^-$ has octahedral geometry having central bromine = 7+1+4 = 12 electrons i.e. 6 pair of electrons. The central $Br^-$ is $sp^3d^2$ hybridised $\underset{5(sp^3d^2)}{\uparrow\downarrow}\ \underset{5(sp^3d^2)}{\uparrow\downarrow}\ \underset{5(sp^3d^2)}{\uparrow}\ \underset{5(sp^3d^2)}{\uparrow}\ \underset{5(sp^3d^2)}{\uparrow}\ \underset{5(sp^3d^2)}{\uparrow}$. The $BrF_4^-$ may be viewed as square planar arrangement of the four fluorine atoms around the bromine. The two lone pairs will point above and below the square plane. $BrF_4^-$ has $D_{4h}$ symmetry.

73. Arrange and explain the solubility of of silver halides in aqueous solution. **[C.U. 2014]**

**Hint :** Solubility sequence is AgF > AgCl > AgBr > AgI. $Ag^+$ has $4s^24p^64d^{10}$ i.e. pseudo noble gas configuration ($ns^2np^6nd^{10}$) and halide ions has the order of the radius $F^- < Cl^- < Br^- < I^-$ thereby according to Fajan's rule the AgI will be primarily covalent while AgF will be mostly ionic. Thus AgF will have highest solubility in aqueous solution and solubility decreases in the order $F^- > Cl^- > Br^- > I^-$.

74. Arrange and explain the bond angles in $OF_2$ and $OCl_2$. **[C.U. 2014]**

**Hint :** Problem 3.21 and Sec. 3.3.1.

75. Using VSEPR theory predict the shapes (indicating the state of hybridisation of the central atom) of the following : $SOF_4$, $XeOF_4$, $SbF_6^-$. **[C.U. 2014]**

**Hint :** $SOF_4$ is derived from $SF_4$ as $SF_4 \xrightarrow[NO_2 \text{ catalyst}]{O_2} SOF_4$. The $SOF_4$ has trigonal bipyramidal geometry having central S is $sp^3d$ hybridised : $\underset{sp^3d}{\uparrow}\ \underset{sp^3d}{\uparrow}\ \underset{sp^3d}{\uparrow}\ \underset{sp^3d}{\uparrow}\ \underset{sp^3d}{\uparrow}\ \underset{3d}{\uparrow}$. The four $sp^3d$ hybrid orbitals having single unpaired electron in each orbital overlap with $2p_x$ orbital of four F-atoms and the rest $sp^3d$ orbital overlaps with $2p_x$ orbital of oxygen. The single unpaired electron in 3d orbital overlaps with suitable $2p_y/2p_z$ orbital of oxygen to make the π-bond. The S–$F_{axi}$ and S–$F_{eqi}$ bond distances have close value.

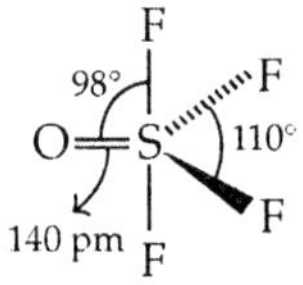


S–$F_{axi}$ = 157.5 pm
S–$F_{eqi}$ = 155 pm

$XeOF_4$ has octahedral geometry having central Xe is $sp^3d^2$ hybridised with 5 b.p. and 1 l.p. (Table 3.6 and question number 45). Actual shape of the molecule is square pyramidal.

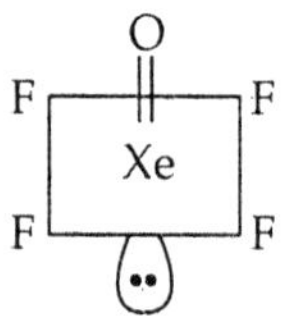


$SbF_5$ has trigonal bipyramidal geometry having no lone pair of electrons. Thus it cannot act as electron pair donors but it can be able of expanding its valence shell because of the availability of d-orbital. In fact $SbF_5$ is an extremely powerful $F^-$ acceptor :

$$NF_3 + F_2 + SbF_5 \longrightarrow [NF_4]^+[SbF_6]^-$$

The $SbF_5$ thus serves as a good Lewis acid and so $SbF_6^-$ has octahedral geometry having central Sb is $sp^3d^2$ hybridised.

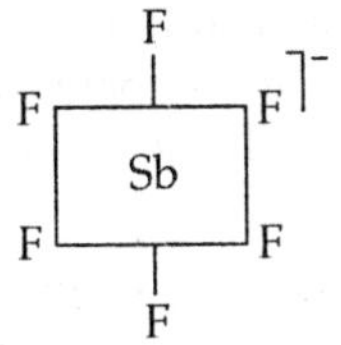

76. Derive Born-Lande expression for lattice energy of NaCl crystal. **[C.U. 2014]**

**Hint :** Sec. 3.1.2.

77. Draw resonating structures of $N_2O$. Indicate the best contributing structure with reason.**[C.U. 2014]**

**Hint :** The $N_2O$ has total valence shell electrons = 5+5+6 = 16 and has the following two resonating structures :

$$:\overset{-}{\ddot{N}} = \overset{+}{N} = \ddot{O}: \longleftrightarrow :N \equiv \overset{+}{N} - \overset{-}{\underset{\cdot\cdot}{\ddot{O}}}:$$

(a) (b)

Among those structures the contributing structure is (b) because in structure (a) the negative charge is assigned to less electronegative atom nitrogen (3.0) while oxygen is more electronegative (3.5).

78. P–O bond moment in $R_3PO$ is much less than expected. Explain. **[C.U. 2014]**

**Hint :** In the $R_3PO$ the central P is $sp^3$ hybridised. P $\xrightarrow{sp^3}$ $\underset{sp^3}{\uparrow}\ \underset{sp^3}{\uparrow}\ \underset{sp^3}{\uparrow}\ \underset{sp^3}{\uparrow}\ \underset{3d}{\uparrow}$. The $sp^3$ hybrid orbitals with each single unpaired electron is linked with three R groups and the fourth $sp^3$ orbital having an electron overlaps with suitable orbital of oxygen to form P–O σ-bond. The 3d orbital of P having single unpaired electron overlaps with suitable orbital of oxygen ($2p_y$ or $2p_z$) to form $3d_\pi(P) - 2p_\pi(O)$ π-bond. As there is a formation of the double bond in the compound i.e. $R_3P{=}O$ so that P–O bond is somewhat shorter than the expected value. R groups are electron donating groups but O is more electronegative than P, so that O will pull much electron density towards itself and thus bond polarity of P–O bond much quenched so that bond moment will be much reduced.

79. Explain why liquid $O_2$ sticks to magnetic pole? **[C.U. 2014]**

**Hint :** Oxygen condenses to a pale blue mobile paramagnetic liquid so that it will stick to magnetic pole.

80. What are intrinsic and extrinsic semiconductors? Indicate the type of semiconduction (n or p) expected in the following : (i) As doped Ge, (ii) B doped Si. **[C.U. 2014]**

**Hint :** Sec. 3.11.4. (i) n type, (ii) p type semiconductors.

81. State with reasons the expected changes in bond orders and bond distances in the following ionisation processes : (i) $O_2 \longrightarrow O_2^+ + e$, (ii) $NO + e \longrightarrow NO^-$ **[C.U. 2014, B.U. 2016]**

**Hint :** Sec. 3.4.3. For $O_2$ B.O. is 2 and for $O_2^+$ B.O. is 2.5, so from $O_2$ to $O_2^+$ B.O. increases and thus bond distance will decrease from $O_2$ to $O_2^+$.

NO has B.O. 2.5 while $NO^-$ will have B.O. = 2, so bond order decreases and thus bond distance increases from NO to $NO^-$.

82. Give reasons why $ZnCl_2$ is soluble in organic solvents while $MgCl_2$ is insoluble. **[C.U. 2015]**

**Hint :** Sec. 3.7. In both $ZnCl_2$ and $MgCl_2$ the anion $Cl^-$ is same but the cations are $Zn^{2+}$ and $Mg^{2+}$ respectively. The $Zn^{2+}$ has $ns^2np^6nd^{10}$ configuration i.e. pseudo inert gas configuration while $Mg^{2+}$ has inert gas ($ns^2np^6$) configuration. According to Fajan's rule pseudo inert gas configuration is more deformable than inert gas configuration so that $ZnCl_2$ will be more covalent compared to that of $MgCl_2$. Thus as per 'like dissolves like' rule $ZnCl_2$ will be soluble in organic solvents while $MgCl_2$ is insoluble.

83. The bond angles in $NH_3$(107°) and $NF_3$(102°) show reverse order in comparison to $PH_3$(94°) and $PF_3$(98°) pair. **[C.U. 2015]**

**Hint :** Sec. 3.8.1. Nitrogen is more electronegative than phosphorus as also phosphorus has larger size than that of nitrogen. With increasing electronegativity from H to F the bond pair electron density is pulled from the central N and charge cloud is displaced further away from the central N causing more decrease of bond angle from $NH_3$ to $NF_3$. But due to the larger size of P the charge cloud is not so displaced thereby the reverse trend is observed between $PH_3$ and $PF_3$.

84. What is resonance energy? Show the resonating structures of $OCN^-$ and compare their relative stabilities.

**Hint :** Sec. 3.2.12. Question number 19.

85. Carbon monoxide and chloroform exhibit unusually low dipole moment values.—Explain. **[C.U. 2015]**

**Hint :** Sec. 3.7.2. The poor dipole moment value of CO is due to its resonating structures.

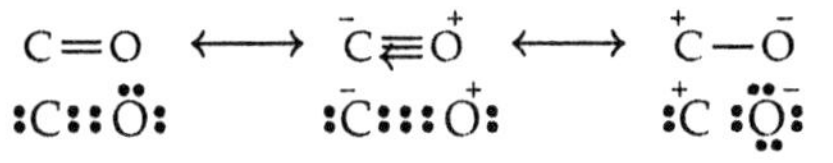

The low dipole moment of $CHCl_3$ is due to the poor unsymmetrical structure of $CHCl_3$ ($\mu$ = 1.15D) from symmetrical structure of $CCl_4$ ($\mu$ = 0).

86. Calculate the formal charge of the constituent atoms in $H{-}N=N=N$ **[C.U. 2015]**

**Hint :** Sec. 3.2.1 and Sec. 6.2.2. In this case the total number of electrons = 3×5+1 = 16 and Lewis structure can be represented as :

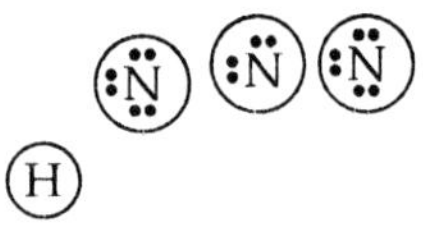

∴ Formal charge of H = 1–0 = +1

Formal charge of central N = 5–4 = +1

Formal charge of other two N = 5–6 = –1.

Alternatively, formal charge (F.C.)

= (number of valence electrons in a free atom of the element – the total number of electrons for that atom in the molecule).

The total number of electrons for that atom in the molecule = lone pair of electrons + one electron for each bond pair.

Thus in case of 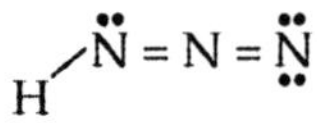

F.C. for H = 1–0 = +1

F.C. for central N = 5–4 = +1

F.C. for terminal two N = 5–6 = –1.

87. Explain the structure of $PCl_3F_2$ with a suitable theory. **[C.U. 2015]**

**Hint :** Sec. 3.3.3.1. This is a stereochemically non-rigid molecule having three Cl atoms in equatorial positions and two F atoms are in the axial positions in a TBP geometry where the central P is $sp^3d$ hybridised. NMR studies indicate that the the axial F exchange with equatorial Cl atom according to Berry pseudorotation.

88. Using VSEPR theory, establish the shape of the following : (i) $IOF_3$, (ii) $XeF_5^+$, (iii) $ICl_4^+$. **[C.U. 2015]**

**Hint :** (i) $IOF_3$ has an incomplete TBP geometry with F, O and a lone pair in the equatorial plane.

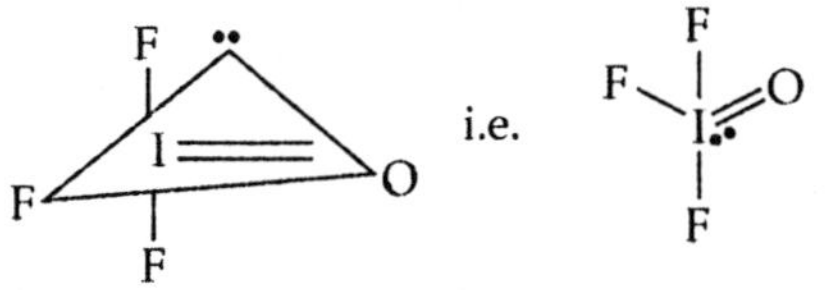

(ii) $XeF_5^+$ cation is square pyramidal as established by X-ray studies. $XeF_5^+$ is isoelectronic and isostructural with $IF_5$ with average Xe—F distance 184 pm.

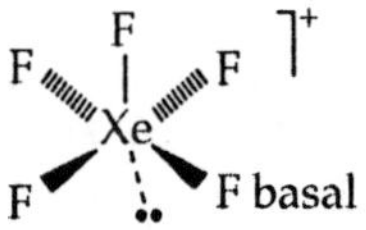


(iii) $ICl_4^+$ has structure similar to that of $SF_4$. $ICl_4^+$ has Ψ trigonal bipyramidal (see-saw) structure having central I is $sp^3d$ hybridised.

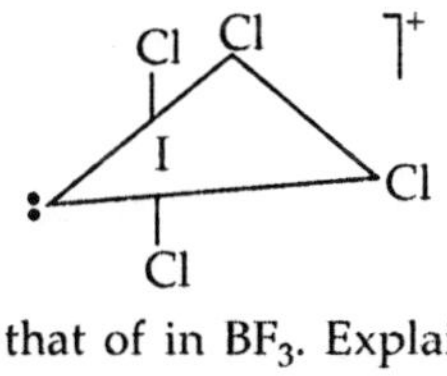

89. The B–F bond in $BF_4^-$ is longer than that of in $BF_3$. Explain. **[C.U. 2015]**

**Hint :** Reaction of $BF_3$ ($D_{3h}$ symmetry) with a Lewis base $F^-$ results in a change from triangular planar structure to tetrahedral structure.

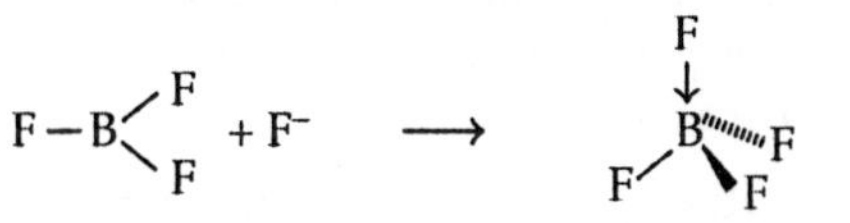

Triangular planar with $sp^2$ hybridised B centre — Tetrahedral having B is $sp^3$ hybridised

In $BF_3$, the B–F bonds contain partial π-character. Reaction with a Lewis base $F^-$ leads to a change in stereochemistry at the B centre from trigonal planar to tetrahedral and as a result the π-contributions to the B–F bonds are somewhat reduced thereby multiple bonding character of the B–F bond is lost from $BF_3$ to $BF_4^-$ and this results some lengthening of B–F bonds from $BF_3$ to $BF_4^-$. This is demonstrted by the observation that the B–F bond length increases from 130 pm in $BF_3$ to 145 pm in $BF_4^-$. This increase in bond length is also considered as steric crowding from $BF_3$ to $BF_4^-$.

90. Draw the MO of HF and find out the number of nonbonded electrons. **[C.U. 2015]**

**Hint :** Sec. 3.4.6. The ground state electronic configuration of H and F are $1s^1$ and $1s^2 2s^2 sp^5$ respectively. Let the nuclei of both the atoms lie on the z-axis. Overlap between the H 1s and F 2s orbitals is allowed by symmetry, but the energy separation is very large. Overlap between 1s orbital of H and $2p_z$ orbital of F is also symmetry allowed and there is a comparable energy matching. As F is more electronegative than H the $2\sigma$ orbital is F dominated. Because HF is non-centrosymmetric the orbitals for HF will not involve u and g notation. The two $2p_x$ and $2p_y$ atomic orbitals become non-bonding orbitals in HF since no net bonding interaction with the 1s orbital of H is possible. The

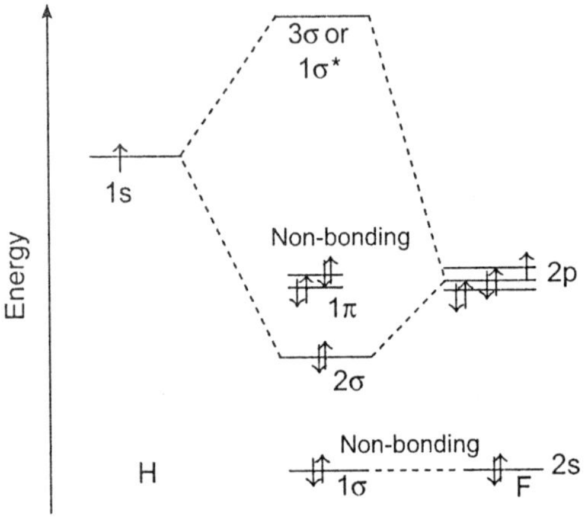


eight valence electrons are accommodated in the MO diagram giving a bond order 1 for HF. Here the molecular orbital configuration is $1\sigma^2 2\sigma^2 1\pi^4$ having HOMO is $1\pi$ level and LUMO is $3\sigma$ or $1\sigma^*$. Of the three occupied molecular orbitals only $2\sigma$ molecular orbital is bonding containing two electrons and the other six electrons remain as three non-bonding pairs. This MO picture of HF indicates polar character of HF i.e. $H^{\delta+}$–$F^{\delta-}$.

91. How and under what condition can an insulator be converted to semiconductor? **[C.U. 2015]**

**Hint :** Sec. 3.11.4(b).

92. Compare bond angles in the pair $PF_3$(104°) and $PCl_3$(100°). **[B.U. 2014]**

**Hint :** Sec. 3.3.1. Question number 10.

93. Sulphur forms $SF_6$ but not form $SH_6$ or $SCl_6$.—Explain. **[B.U. 2014]**

**Hint :** The small size and high electronegativity of F brings the central S to attain the high oxidation state and high accommodation capacity of central S atom and thus $SF_6$ is a very stable compound. Also the bond energy of $F_2$ is less than $H_2$ which favours dissociation of $SH_6$ into $H_2S + H_2$. In fact $SH_6$ is an unstable compound with respect to such decomposition.

The MO picture of $SF_6$ (Problem 3.62) indicates 4 electrons exist in non-bonding molecular orbitals localised near the F atoms and 8 electrons are in the bonding molecular orbitals meeting all 6 bonds i.e. forming σ bond order equal to 2/3. The $SCl_6$ will be unstable because 6 larger Cl atoms cannot be able to accommodate around S atom.

In case of $SH_6$ the hydrogen having low electronegativity places the non-bonding molecular orbitals comparatively higher, the non-bonding electrons are thus considerably destabilised and $SH_6$ will be unstable.

94. Using VSEPR theory, predict the shapes of $BrF_4^-$ and $XeF_2$. Indicate the state of hybridisation of the central atom in each case. **[B.U. 2014]**

**Hint :** Table 3.6 and Sec. 3.3.1.

95. In $OPF_3$ molecule, O–P bond length is shorter than that in $OPPh_3$ (Ph = Phenyl). Give reasons. **[B.U. 2014]**

**Hint :** Question number 78. In both the compounds the O–P bond has double bond character. Electron density pulling property of Ph groups much higher than that of F atoms thereby Ph will withdraw much electron density from O–P bond so that P–O bond will be somewhat longer in $OPPh_3$ compared to that of $OPF_3$.

96. Discuss the bonding in HF molecule in the light of MO theory and indicate the bonding, antibonding and non-bonding electrons on an approximate energy level diagram. **[B.U. 2014]**

**Hint :** Sec. 3.4.6 and Question number 90.

97. Give reasons for the unexpectedly high boiling point of liquid HF? **[C.U. 2014]**

**Hint :** Sec. 3.10.1. This is due to intermolecular hydrogen bonding, H–F····H–F····H–F····H–F.

98. Methyl isothiocyanate is bent while silyl isothiocyanate is linear.—Explain. **[B.U. 2014]**

**Hint :** Problem 3.75 and 3.76.

In $H_3C–N=C=S$ the bond angle $H_3C–N–C$ is 142° where the nitrogen is $sp^2$ hybridised. The $H_3Si–N–C$ bond angle in $H_3Si–N=C=S$ is 180° where the N is sp hybridised. Si has vacant 3d orbital and the lone pair resides in a pure p orbital in order to facilitate the formation of $2p_\pi(N) \longrightarrow 3d_\pi(Si)$ back bonding. The $p_\pi–d_\pi$ back bonding in $H_3Si$ NCS is given below :

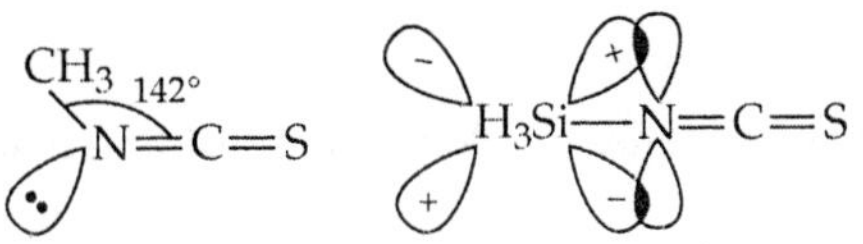


99. State and explain Bent's rule with reference to the shape and bond angles in $OSF_4$. **[B.U. 2015]**

**Hint :** Sec. 3.3.3.

100. What is ionic potential? State its significance. **[B.U. 2015]**

**Hint :** Sec. 3.7(b) and Problem 3.58.

101. '$CF_4$, $SF_4$ and $XeF_4$ have comparable molecular formula, but their shapes are different.'—Explain. **[B.U. 2015]**

**Hint :** Sec. 3.3.1. In $CF_4$ the central C is $sp^3$ hybridised and the molecule has tetrahedral structure.

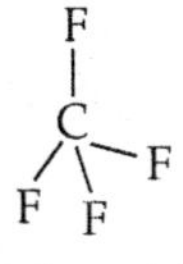


The central S of $SF_4$ is $sp^3d$ hybridised having trigonal bipyramid structure. There are 4 b.p. and 1 l.p. Actual structure is distorted tetrahedral.

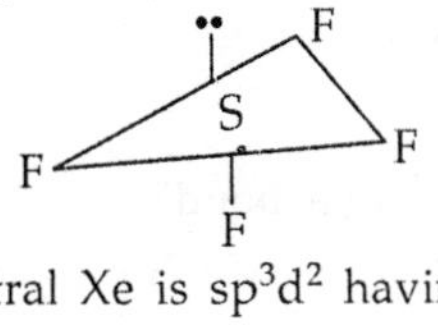


$XeF_4$ has octahedral geometry and central Xe is $sp^3d^2$ having 4 b.p. and 2 l.p. Actual geometry is square planar.

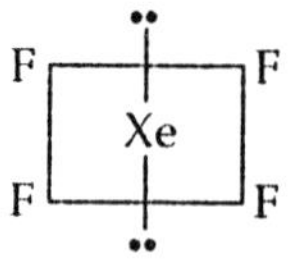


102. Cite one example of μ-bonded system. [B.U. 2015]

**Hint :** Sec. 3.5(c).

103. Compare the bond angles between $H_2O$ (104.5°) and $H_2S$ (92.2°). [B.U. 2015]

**Hint :** Sec. 3.3.1. Rule IV of VSEPR theory.

104. State the Fajan's Polarisation Rules. Applying these classify MgO and $SnCl_4$ into ionic and covalent compounds as the case may be. [B.U. 2015]

**Hint :** Sec. 3.7. MgO will be more ionic than $SnCl_4$.

105. Write down the VSEPR Rules. '$NH_3$ $BCl_3$ and $BrF_3$ have comparable molecular formula but their shapes are different.'—Explain. [B.U. 2015]

**Hint :** Sec. 3.3.1

$NH_3$ has pyramidal structure having $sp^3$ hybridised central nitrogen (3 b.p. + 1 l.p.).

The central B in $BCl_3$ is $sp^2$ hybridised having triangular planar geometry. Here all are bond pairs.

Br in $BrF_3$ is $sp^3d$ hybridised having 3 b.p. + 2 l.p. and the shape of the molecule is T-shaped.

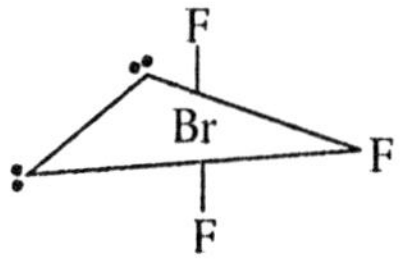


The molecule has a Ψ-trigonal bipyramidal geometry as there are two lone pairs residing in equatorial positions.

106. Define resonance. Enumerate the conditions of resonance. Show the resonating structures of carbonate and thiocyanate. [B.U. 2015]

**Hint :** Sec. 3.2.12, Problem 3.17 and Question number 19.

107. 'Metals are good conductors of electricity.'—Comment. [B.U. 2015]

**Hint :** Sec. 3.11.1 and Sec. 3.11.2.

108. Draw the structure of $BF_3$ and $BF_4^-$. Indicate the state of hybridisation of central atom and compare the B–F bond length in each compound. [B.U. 2015]

**Hint :** Question number 89.

109. Which amongst the following is diamagnetic?
(a) $O_2$, (b) $O_2^+$, (c) $O_2^-$, (d) $O_2^{2-}$.

**Hint :** Sec. 3.4.3. **Ans. :** (d) i.e. $O_2^{2-}$.

110. The hybridisation of Xe in $XeO_2F_2$ molecule is—
(a) $sp^3$, (b) $sp^3d$, (c) $sp^3d^2$, (d) $sp^3d^3$.

**Ans. :** (b) i.e. $sp^3d$.

111. Which of the following is the most polar bond?
(a) C–F, (b) C–O, (c) O–F, (d) N–F

**Ans. :** (a) i.e. C–F.

112. Amongst the following species which would show the square planar structure.
(a) $BF_4^-$, (b) $CCl_4$, (c) $NH_4^+$, (d) $XeF_4$

**Ans. :** (d) i.e. $XeF_4$.

113. The maximum hydrogen bonding occurs in—
(a) $H_2O$, (b) HF, (c) $H_2Se$, (d) $H_2S$

**Ans. :** (a) i.e. $H_2O$.

114. Which of the following molecule/ions have trigonal pyramidal structure?
(a) $H_3O^+$, (b) $NO_3^-$, (c) $CO_3^{2-}$, (d) $BF_3$

**Ans. :** (a) i.e. $H_3O^+$.

115. Which of the following has no coordinate bond?
(a) $NHO_3$, (b) $NH_4^+$, (c) $CuSO_4$, (d) $CO_3^{2-}$

**Ans. :** (d) i.e. $CO_3^{2-}$.

116. Which of the following has diagonal relationship with Al?
(a) Ga, (b) Be, (c) B, (d) Mg

**Ans. :** (b) i.e. Be.

117. Using VSEPR theory explain why $SNF_3$ is tetrahedral but $SF_4$ has a a see-saw geometry.

**Hint :** In $SNF_3$ there is no lone pair on S while in $SF_4$ has a lone pair (Table 3.6). In $SF_4$ the l.p. will occupy the equatorial site so that $SF_4$ has a distorted tetrahedral or see-saw geometry.

118. Explain why $BI_3$ is trigonal planar while $PI_3$ is trigonal pyramidal in shape.

**Hint :** In $BI_3$ there is no lone pair but in $PI_3$ there is a lone pair and hence the difference in shape.

119. Give reason why on heating $N_2F_2$ at 373K results in a change from a non-polar to polar molecule.

**Hint :** This is due to the conversion of trans isomer to cis isomer.

120. Using VSEPR theory predict the structures of—
(a) $H_2Se$, (b) $[BH_4]^-$, (c) $NF_3$, (d) $SbF_5$, (e) $H_3O^+$, (f) $IF_7$, (g) $[I_3]^-$, (h) $[I_3]^+$, (i) $SO_3$

**Ans.** (a) Bent, (b) tetrahedral, (c) trigonal pyramidal, (d) trigonal bipyramidal, (e) trigonal pyramidal, (f) pentagonal bipyramidal, (g) linear, (h) bent, (i) trigonal planar.

121. State whether you expect the following species to possess cis-trans isomers.
(a) $BF_2Cl$, (b) $POCl_3$, (c) $MePF_4$, (d) $[PF_2Cl_4]^-$

**Hint :** (a) Trigonal planar geometry—no isomer

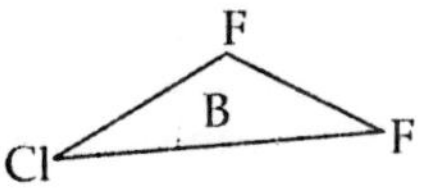


(b) Tetrahedral—no isomer

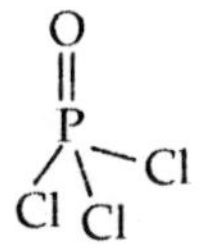


(c) Trigonal bipyramidal geometry—Me may exist axially or equatorially

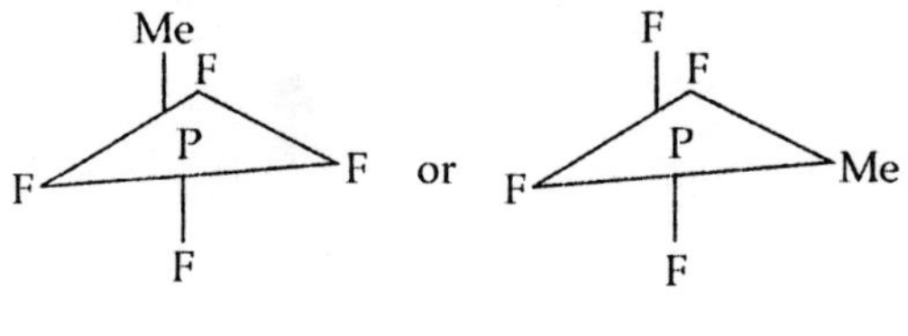

(d) Octahedral geometry—exist as cis/trans isomer

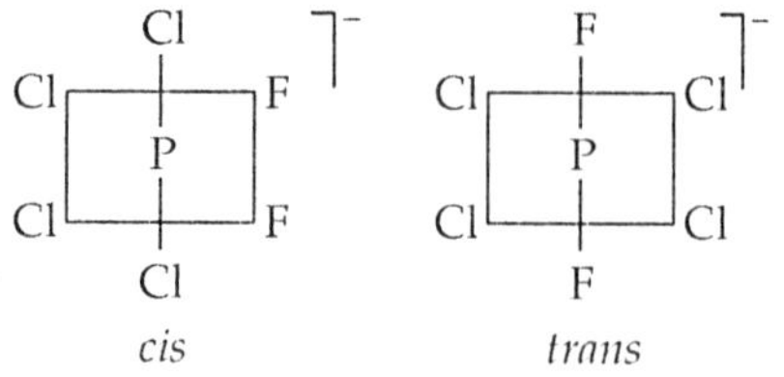


122. Write the ions that are present in the compound $[PCl_4][PCl_3F_3]$. What shape do you expect each ion to adopt? Which ion exhibits stereoisomers?

**Hint :** The ions are $[PCl_4]^+$ has tetrahedral geometry and $[PCl_3F_3]^-$ has octahedral geometry which shows stereoisomerism.

123. The structures of $XeF_2$ and $XeO_2F_2$ respectively are—
1. bent, tetrahedral, 2. linear, square planar, 3. linear, see-saw, 4. bent, see-saw.

**Ans. :** 3 i.e. linear, see-saw.

124. Among the following, species expected to show fluxional behaviour are—
A. $[NiCl_4]^{2-}$ (tetrahedral), B. $IF_7$ (pentagonal bipyramidal), C. $[CoF_6]^{3-}$ (octahedral), D. $Fe(CO)_5$ (trigonal bipyramidal).
1. B and C, 2. B and D, 3. C and D, 4. A and D **[NET 2015]**

**Ans. :** 2 i.e. $IF_7$ and $Fe(CO)_5$.

125. The correct statement among the following is—
1. $N_2$ has higher bond order than $N_2^+$ and hence has larger bond length compared to $N_2^+$.
2. $N_2^+$ has higher bond order than $N_2$ and hence has larger bond length compared to $N_2$.
3. $N_2$ has higher bond order than $N_2^+$ and hence has higher dissociation energy compared to $N_2^+$.
4. $N_2$ has lower bond order than $N_2^+$ and hence has lower dissociation energy compared to $N_2^+$ energy. **[NET 2015]**

**Hint :** Sec. 3.4.3; Table 3.9.

**Ans. :** 3 i.e. $N_2$ has higher bond order than $N_2^+$ and hence has higher dissociation energy compared to $N_2^+$.

126. The molecule $C_3O_2$ has a linear structure. This compound has—
1. 4σ and 4π bonds, 2. 3σ and 2π bonds, 3. 2σ and 3π bonds, 4. 3σ and 4π bonds. **[NET 2015]**

**Hint :** 1 i.e. 4σ and 4π bonds. O=C=C=C=O

127. The number of lone pair(s) of electrons on the central atom in $[BrF_4]^-$, $XeF_6$ and $[SbCl_6]^{3-}$ are respectively,
1. 2, 0 and 1; 2. 1, 0 and 0; 3. 2, 1 and 1; 4. 2, 1 and 0. **[NET 2015]**

**Hint :** 3 i.e. 2, 1 and 1.

128. The ground state electronic configuration of $C_2$ using all electrons is—

1. $\sigma_{1s}^2 \sigma_{1s}^{*2} \sigma_{2s}^2 \sigma_{2s}^{*2} \sigma_{2p}^2 \pi_{2p}^2$, 2. $\sigma_{1s}^2 \sigma_{1s}^{*2} \sigma_{2s}^2 \sigma_{2s}^{*2} \sigma_{2p}^2 \sigma_{2p}^{*2}$,

3. $\sigma_{1s}^2 \sigma_{1s}^{*2} \sigma_{2s}^2 \sigma_{2s}^{*2} \pi_{2p}^2 \sigma_{2p}^1 \sigma_{2p}^{*1}$, 4. $\sigma_{1s}^2 \sigma_{1s}^{*2} \sigma_{2s}^2 \sigma_{2s}^{*2} \pi_{2p}^4$ **[NET 2015]**

**Hint :** Sec. 3.4.3.; Fig. 3.24.

**Ans. :** 4 i.e. $\sigma_{1s}^2 \sigma_{1s}^{*2} \sigma_{2s}^2 \sigma_{2s}^{*2} \pi_{2p}^4$.

129. 'MgO is a stable ionic solid though large amount of energy is required to form Mg to $Mg^{2+}$ and to dissociate $O_2$ into atomic oxygen and thereafter to form $O^{2-}$ ion.'—Account for the fact.

**[B.U. 2015]**

**Hint :** Question number 12.

130. Assuming that the $CH_3$ radical is planar, the unpaired electron is in the orbital of—
(a) 1s, (b) 2s, (c) $2p_z$, (d) $2p_x$.

**Hint :** (c) i.e. $2p_z$ orbital. Question number 28.

131. Suggest the appropriate hybridisation scheme for the central atom of each of the following species :
(a) $SiF_4$, (b) $PdCl_4^{2-}$, (c) $NF_3$, (d) $F_2O$, (e) $[CoH_5]^{4-}$, (f) $[FeH_6]^{4-}$, (g) $CS_2$, (h) $BF_3$.

**Hint :** (a) $sp^3$, (b) $sp^2d$, (c) $sp^3$, (d) $sp^3$, (e) $sp^3d$, (f) $sp^3d^2$, (g) sp, (h) $sp^2$.

132. What hybridisation is expected for the O atoms in $H_2O_2$ ?

**Hint :** $sp^3$ (Table 3.7).

133. The *cis* and *trans* structure of $N_2F_2$ is the following :

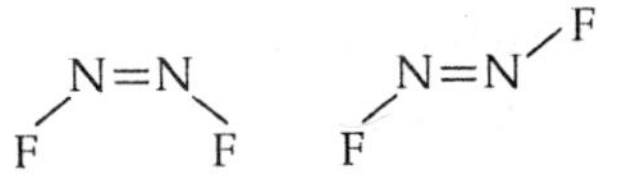

*cis* *trans*

Suggest the hybridisation of the N atoms of each isomer.

**Hint :** $sp^2$.

134. What is meant by ligand group orbital (LGO)?

**Hint :** In the molecular orbital treatment of polyatomic molecules the correct symmetry orbitals are called **ligand group orbitals (LGOs)**. In the case of linear triatomic $XH_2$ molecule in which the valence orbitals of X are the 2s and 2p atomic orbitals which combine with the two 1s orbitals of the two H atoms. Each 1s orbital has two possible phases and when the two 1s orbitals are taken as a group, there are two possible phase combinations. These are called ligand group orbitals. The number of **ligand group orbitals** formed = the number of atomic orbitals used.

135. What is called orbital hybridisation?

**Hint :** The word hybridisation means mixing and when used in connection with atomic orbitals, it describes a way of deriving **spatially directed orbitals** which may be used as per VB theory. Like all bonding theories orbital hybridisation is a **model** and should not be taken to be a **real phenomena**. Hybrid orbitals are generated by mixing the characters of atomic orbitals that are of similar energy. The character of a hybrid orbital depends on the atomic orbitals involved and their percentage contributions e.g. sp hybrid orbitals has equal percentage of s and p character. For a particular molecular species a particular set of hybrid orbitals has to be constructed e.g. for a linear molecule the sp hybrid orbital, for triangular planar species the $sp^2$ hybrid orbitals etc. An individual hybrid orbital points along a given internuclear axis within the framework of the molecule under consideration and use of a set of hybrid orbitals provides a bonding picture for a molecule in terms of **localised σ bonds.**

136. The I–I bond distance in $I_2$ (gas phase) is 267 pm and in $I_3^+$ and in $I_3^-$ is 268 pm and 290 pm respectively.
(a) Draw the Lewis structure of each of the species. Do these representations is consistent with the variation of bond distance? (b) Use MO theory to describe the bonding and deduce the I–I bond order of each species. Are your results agree with the structural data?

**Hint :** (a) Lewis structures for the species are :

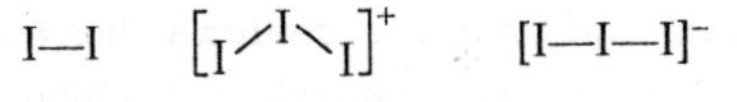

Here the effect of lone pairs are ignored. In all cases the I–I bond is 2c–2e bonds and does not represent the variation of bond lengths.

(b) From MO diagram the bond order of $I_2$=1; bond order for $I_3^+$=1 and the bond order for $I_3^-=\frac{1}{2}$. The MO diagram for $I_3^+$ and $I_3^-$ are similar to that of $H_2O$ and $XeF_2$ respectively.

137. What hybridisation would you expect for the Si atom in $SiH_4$ and what is the point group of this molecule?

**Hint :** The hybridisation of Si is $sp^3$ and the point group is $T_d$.

138. Assuming x-axis as the bonding axis predict how many π MOs would be formed by d-orbitals of two combining atoms. Give the overlap diagram in each case. **[C.U. 2013]**

**Hint :** Problem 3.73. Only combination of d-orbitals is shown excluding the ns orbital. Taking x-axis as bond axis the combinations will be :

$d_{x^2-y^2} \pm d_{x^2-y^2}$ (σ-MOs i.e. σ+σ*);

$d_{z^2} \pm d_{z^2}$ (π-MOs i.e. π+π*);

$d_{xy} \pm d_{xy}$ (π-MOs i.e. π+π*);

$d_{xz} \pm d_{xz}$ (π-MOs i.e. π+π*);

$d_{yz} \pm d_{yz}$ (δ-MOs i.e. δ+δ*).

The qualitative oribital overlap diagram is of the following type :

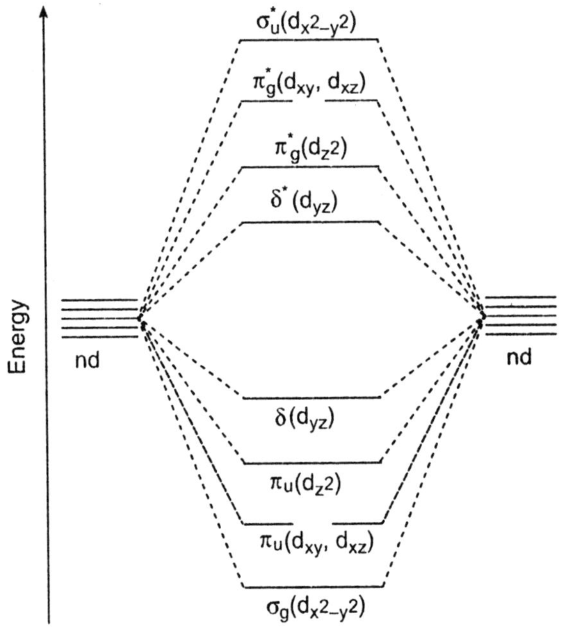


139. Adducing proper reason, draw the structures of $I_3^+$ and $I_3^-$. **[B.U. 2016]**

**Hint :** Sec. 3.3.1 (Table 3.6) and Question numbers 54, 63, 120 and 136.

$I_3^+$ has tetrahedral and $I_3^-$ has trigonal bipyramidal structure. $I_3^+$ is formed by combination of $I_2$ and $I^+$ ion held by coordinate covalent bond in which former is donor and latter is acceptor. $I^+$ has 6 electrons and so octet is formed around it by donation of lone pair of electrons from central I which has thus two bond pairs and two lone pairs around it forming a tetrahedral geometry and thus it has a bent shape (V-shaped). On the other hand $I_3^-$ is formed by combination of $I_2$ and $I^-$ ion held by coordinate

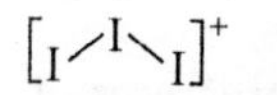

bond in which former is acceptor and latter is donor, central I has empty d-orbitals, so it can expand its octet accommodating the lone pair from $I^-$ adopting a trigonal bipyramidal geometry around it with a linear shape as the three lone pairs occupy trigonal plane and the two terminal iodine $[I–I–I]^-$ occupying the molecular axis.

140. Make a comparison of σ-bond and π-bond indicating their characteristic features. **[B.U. 2016]**

**Hint :** Sec. 3.2.11.

141. 'Gaseous sulphur is paramagnetic'. Explain using molecular orbital theory. **[B.U. 2016]**

**Hint :** Gaseous sulphur mainly exist as blue $S_2$ molecule and so like $O_2$ molecule the $S_2$ molecule will be paramagnetic to the extent of two unpaired electrons (Sec. 3.4 and Fig. 3.24).

142. State Helferich's rules. Point out the flaws. **[B.U. 2016]**

**Hint :** Sec. 3.2.1.1.

143. Note down the similarities and differences between valence bond theory and molecular orbital theory. **[B.U. 2016]**

**Hint :** Sec. 3.4.8.

144. Give an example of δ-bonded system drawing its structure. **[B.U. 2016]**

**Hint :** Sec. 3.5(a).

145. Using VSEPR rules draw the structures of $SF_4$ and $SnCl_2$. **[B.U. 2016]**

**Hint :** Sec. 3.3.1. Table 3.6 and Question number 101. $SF_4$ has TBP geometry and $SnCl_2$ has V-shaped structure.

146. Write down the favourable conditions for formation of ionic bond. **[B.U. 2016]**

**Hint :** Sec. 3.1.1.

147. 'N–O bond order in $NO_2^-$ is greater than that in $NO_3^-$. Account for the fact. **[B.U. 2016]**

**Hint :** Both $NO_2^-$ and $NO_3^-$ has the following resonating structures and the central N is $sp^2$ hybridised in each case.

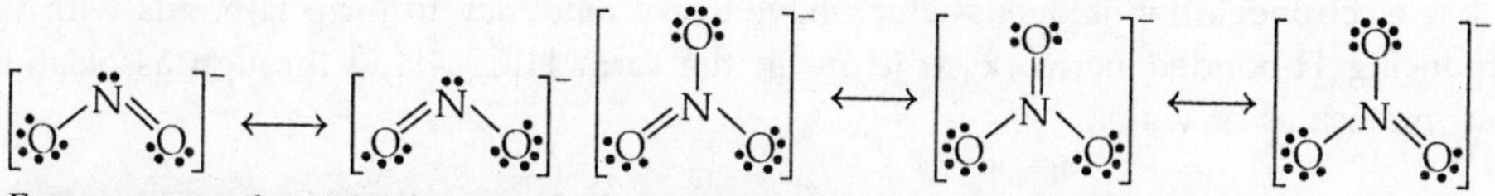

For compounds with resonance structures, to find the bond order the number of resonance structures and bonding pairs must be taken into account.

In $NO_2^-$ ion there are two equivalent resonance structures, therefore we cannot say that the bonds between N–O are single or double. Hence we consider $NO_2^-$ has two equivalent N–O bonds where three pairs of bonding electrons are distributed over two equivalent N–O bond locations. Hence to describe the bonding in resonance hybrid we consider the fractional bond order = $\frac{\text{total number of N–O bonding pairs}}{\text{number of N–O bond locations}} = \frac{3}{2} = 1.5.$

i.e. each N–O bond order in $NO_2^-$ is 1.5.

Similarly in case of $NO_3^-$ there are three equivalent resonance structures exist as shown above. Thus in $NO_3^-$ there are three equivalent N–O bonds exist where four pairs of bonding electrons are distributed over three equivalent N–O bond locations. Hence the fractional bond order =

$\frac{\text{total number of N–O bonding pairs}}{\text{number of N–O bond locations}} = \frac{4}{3} = 1.33$. Thus each N–O bond order in $NO_3^-$ = 1.33. Therefore N–O bond order in $NO_2^-$ is greater than that of $NO_3^-$.

Bond order may also be calculated by other ways. In the three resonating Lewis structures of $NO_3^-$ there are 2 single bonds and 1 double bond. The bond order is the average = $\frac{\text{total number of electron pairs}}{\text{total number of actual bonds}} = \frac{4}{3} = 1.33$. Similarly for $NO_2^-$ bond order = $\frac{3}{2} = 1.5$.

In another way bond order can be calcualted from the two informations : (1) the number of electrons that would be needed if all atoms in the molecule had their octet rule satisfied i.e. 8 electrons except hydrogen, (2) the total number of electrons the atoms in the molecule actually have i.e. total electrons in the Lewis structure. The difference between these two numbers (what they want—what they have) divided by 2 produces the number of covalent bonds i.e. the number of shared pair of electrons. This on division by actual number of bonds we get the bond order. For $NO_3^-$ the total number of wanted electrons due to octet = 4×8 = 32. Actual number of electrons = 24. Therefore bond order = $\frac{32-24}{2} \div 3 = 1.33$. Similarly for $NO_2^-$ bond order = $\frac{24-18}{2} \div 2 = 1.5$.

148. 'The thermal stability of alkaline earth metal carbonates run in the following order : $BeCO_3 < MgCO_3 < CaCO_3 < SrCO_3 < BaCO_3$'.—Explain. **[B.U. 2016]**

**Hint :** Sec. 3.7.

149. 'Compounds of the type $R_3PO$ (R = alkyl group) have lower dipole moment in comparison to $R_3NO$ compounds, though the P–O bond has a greater electronegativity difference than the N–O bond'.—State reason. **[B.U. 2016]**

**Hint :** Question number 78. $R_3NO$ has similar structure to that of $R_3PO$ i.e. 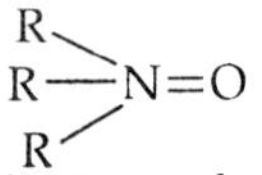 but N is more electronegative than P so that O will not be able to pull much electron density towards itself and thus bond polarity of N–O bond will be much enhanced compared to that of P–O bond so that dipole moment of N–O bond will be higher compared to that of P–O bond.

150. $HICl_4$ is not known but why its crystalline hydrate $HICl_4.4H_2O$ is known? **[B.U. 2016]**

**Hint :** $HICl_4$ has electronegative halogens which have every tendency to form H-bonds with water molecules producing H-bonded network structure in the form $HICl_4.4H_2O$ through association of $HICl_4$ and four molecules of water.

# 4

# ACIDS AND BASES

**Development of the concepts of acids and bases are based on the accumulated experimental informations. There are several theories of acids and bases. These are Arrhenius concept, Solvent System concept, Bronsted-Lowry concept, Lewis concept, SHAB principle, Usanovich concept and Lux-Flood concept. Each concept defines acids and bases in a different way and each concept has certain limitations which are modified by the next one. We shall discuss the main features of the concepts of acids and bases in a sequential order to explore the idea of acids and bases of chemical compounds.**

## 4.1 THE ARRHENIUS CONCEPT (1883)

This theory defines **an acid is a hydrogen containing compound which produces hydrogen ions** in *aqueous solution while a base is a substance which gives hydroxyl ions in aqueous solution*. Thus HCl and NaOH is an acid and a base respectively in aqueous solution.

$$\underset{acid}{HCl} \xrightarrow{H_2O} H^+ + Cl^- \quad \underset{base}{NaOH} \xrightarrow{H_2O} Na^+ + OH^-$$

Therefore **the process of neutralisation according to this concept is defined as the reaction of $H^+$ ions with that of $OH^-$ ions to form undissociated water molecules**, so that,

$$\underset{acid}{HCl} + \underset{base}{NaOH} \longrightarrow \underset{salt}{NaCl} + \underset{water}{H_2O}$$

$$\text{i.e. } H^+ + OH^- \longrightarrow H_2O \qquad ...(4.1)$$

Since the strong acids and bases are completely dissociated therefore the **heat of neutralisation** (i.e. the heat evolved for the neutralisation of one mole of an acid and a base) for the

Merit of the concept

neutralisation of one mole of a strong acid with a strong base is constant (13.7 kcal $mol^{-1}$) and this constancy of heat of neutralisation has been experimentally established. For weak acids/bases the heat of neutralisation is not constant.

Limitation of the concept

This concept lays an emphasis on aqueous solution only i.e. acid-base phenomena is restricted in aqueous solution. It does not recognise the role of non-aquesous solvents i.e. whether HCl will be an acid or a base in benzene or chloroform cannot be stated by this concept. It cannot explain the acidic character of certain compounds such as $AlCl_3$, $CO_2$, $SO_3$ etc. in aqueous solution though there is no hydrogen atom in their constitution.

A question also arises about the existance of bare $H^+$ ion in aqueous solution. $H^+$ ion is a unique ion in the sense that it has no electron. It is the smallest ion known and has a strong polarising power so that it will polarise the $H_2O$ molecules in aqueous solution and forms an hydrated species ($H_2O + H^+ \longrightarrow H_3O^+$) i.e. $H_3O^+$ which is known as **hydroxonium ion or hydronium ion**. Therefore **a nacked proton cannot survive in aqueous solution.**

## 4.2 SOLVENT SYSTEM CONCEPT

This concept developed by Franklin (1905) based on the experimental studies of the autoionisation of solvents including water. According to this concept solvents are autoionised to form solvent cations and solvent anions. If a substance in a particular solvent enhances the concentration of the solvent cation then the substance is regarded as an acid in that solvent. On the other when a species produces solvent anion in a particular solvent, then the species is considered as a base in that solvent. This generalise the Arrhenius concept in non-aqueous solvents. Thus in water solvent HCl and NaOH are considered as acid and base respectively in this solvent system concept as water is autoionised in the following way.

$$H_2O \longrightarrow H^+ + OH^-$$

$$\text{or } 2H_2O \longrightarrow \underset{\text{solvent cation}}{H_3O^+} + \underset{\text{solvent anion}}{OH^-}$$

Since HCl increases the concentration of $H^+/H_3O^+$ in water, it is considered acid in aqueous solution. Similarly NaOH produces solvent anion ($OH^-$) so that it is characterised as base in water.

The extension of this concept to non-aqueous solvents can be made by the autoionisation of liquid $NH_3$, liquid $SO_2$ and liquid $N_2O_4$ e.g.

$$2NH_3 \longrightarrow NH_4^+ + \overset{*}{N}H_2^-$$

$$2SO_2 \longrightarrow \overset{*}{S}O^{2+} + SO_3^{2-}$$

$$N_2O_4 \longrightarrow \underset{\text{solvent cation}}{\overset{*}{N}O^+} + \underset{\text{solvent anion}}{NO_3^-}$$

$NH_4Cl$ is considered as an acid in liquid $NH_3$ because it produces solvent cation ($NH_4^+$) while $NaNH_2$ is characterised as base in liquid $NH_3$ because it enhances the concentration of solvent

* $NH_2^-$ = amide ion, $SO^{2+}$ = thionyl cation, $NO^+$ = nitrosyl cation.

anion ($NH_2^-$). Thus the reaction between $NH_4Cl$ and $NaNH_2$ in liquid $NH_3$ is considered as a process of neutralisation as,

$$\underset{acid}{NH_4Cl} + \underset{base}{NaNH_2} \longrightarrow \underset{salt}{NaCl} + \underset{solvent}{2NH_3} \quad ...(4.2)$$

Thus apart from the classical concept of neutralisation in water, this theory **generalises the process of neutralisation as the reaction between an acid with a base to form salt and solvent molecules.**

Similarly thionyl chloride ($SOCl_2$) and $Na_2SO_3$ is considered as acids and bases respectively in liquid $SO_2$ and

$$SOCl_2 \longrightarrow \underset{\text{solvent cation}}{SO^{2+}} + 2Cl^- \qquad Na_2SO_3 \longrightarrow 2Na^+ + \underset{\text{solvent anion}}{SO_3^{2-}}$$

Neutralisation reaction is represented as :

$$\underset{acid}{SOCl_2} + \underset{base}{Na_2SO_3} \longrightarrow \underset{salt}{2NaCl} + \underset{solvent}{2SO_2} \quad ...(4.3)$$

In liquid $N_2O_4$, NOCl and $NaNO_3$ behaves as an acid and a base respectively.

$$NOCl \longrightarrow NO^+ + Cl^- \qquad NaNO_3 \longrightarrow Na^+ + NO_3^-$$

The neutralisation reaction is—

$$\underset{acid}{NOCl} + \underset{base}{NaNO_3} \longrightarrow \underset{salt}{NaCl} + \underset{solvent}{N_2O_4} \quad ...(4.4)$$

Some more autoionisation of solvents and neutralisation processes are given in Table 4.1. This concept emphasizes that acid behaviour is not only dependent on hydrogen containing compounds but also acid-base neuralisation process can be applied in non-protonic solvents e.g. $SO_2$, $N_2O_4$ etc.

Merit of the concept

**Table 4.1 : Autoionisation of some non-aqueous solvents and neutralisation reactions**

| Solvent autoionisation | Acid | Base | Neutralisation reaction |
|---|---|---|---|
| $2H_2SO_4 \longrightarrow H_3SO_4^+ + HSO_4^-$ | $HClO_4 + H_2SO_4$ | $KHSO_4$ | $HClO_4 + H_2SO_4 \longrightarrow H_3SO_4^+ + ClO_4^-$ |
| | | | $(HClO_4 + H_2SO_4) + KHSO_4 \longrightarrow KClO_4 + 2H_2SO_4$ |
| $COCl_2 \longrightarrow COCl^+ + Cl^-$ | $AlCl_3 + COCl_2$ | $CaCl_2$ | $AlCl_3 + COCl_2 \longrightarrow COCl^+ + AlCl_4^-$ |
| | | | $2(COCl^+ + AlCl_4^-) + CaCl_2 \longrightarrow Ca(AlCl_4)_2 + 2COCl_2$ |
| $2BrF_3 \longrightarrow BrF_2^+ + BrF_4^-$ | $SbF_5$ | KF | *$SbF_5 + KF + 2BrF_3 \longrightarrow K[SbF_6] + 2BrF_3$ |
| $POCl_3 \longrightarrow POCl_2^+ + Cl^-$ | $SbCl_5$ | NaCl | **$SbCl_5 + NaCl + POCl_3 \longrightarrow Na[SbCl_6] + POCl_3$ |

* $KF + BrF_3 \longrightarrow K^+ + BrF_4^-$, $SbF_5 + BrF_3 \longrightarrow BrF_2^+ + SbF_6^-$

** $SbCl_5 + POCl_3 \longrightarrow POCl_2^+ + SbCl_6^-$, $NaCl \longrightarrow Na^+ + Cl^-$

This concept fails to explain the inherent acid-base behaviour of chemical compounds because the concept dependent on the autoionisation of the solvents. It ignores the physical and chemical properties of solvents. Moreover the process of neutralisation of $SOCl_2$ with $Na_2SO_3$ in liquid $SO_2$ as stated previously may be erroneous. This can be explained on the basis of the autoionisation of $SO_2$ with labelled $^{35}S$.

*Limitation of the concept*

$$2\,^{35}SO_2 \longrightarrow {}^{35}SO^{2+} + {}^{35}SO_3^{2-}$$

$$SOCl_2 \longrightarrow SO^{2+} + 2Cl^-$$

$$^{35}SO^{2+} + 2Cl^- \longrightarrow {}^{35}SOCl_2$$

$$SO^{2+} + {}^{35}SO_3^{2-} \longrightarrow SO_2 + {}^{35}SO_2$$

If autoionisation of liquid $SO_2$ actually occurs, the tagged and untagged $SO^{2+}$ ions would compete with each other to combine with the $Cl^-$ ion to form $SOCl_2$. As a result a considerable amount labelled $^{35}S$ would be exchanged in $SOCl_2$. But unfortunately no exchange is noticed. The half life for the exchange of $^{35}S$ in this reaction is found to be about two years. This rules out the validity of the autoionisation of liquid $SO_2$ and the neutralisation process between $SOCl_2$ and $Na_2SO_3$.

## 4.3 BRONSTED-LOWRY CONCEPT : THE CONJUGATE ACID-BASE CONCEPT (1923)

In this concept **an acid is defined as a substance which donates proton to any other substance while a base is a substance which accepts proton from any other substance irrespective of the solvent** i.e. **acid is a proton donor and base is a proton acceptor.** Bronsted acids and bases can be classified as :

| **Bronsted acids** | **Bronsted bases** |
|---|---|
| *Molecular* : $HCl \longrightarrow H^+ + Cl^-$ | *Molecular* : $H_3N + H^+ \longrightarrow NH_4^+$ |
| *Cationic* : $[Al(H_2O)_6]^{3+} \longrightarrow [Al(H_2O)_5(OH)]^{2+} + H^+$ | *Cationic* : $[Al(H_2O)_5(OH)]^{2+} + H^+ \longrightarrow [Al(H_2O)_6]^{3+}$ |
| *Anionic* : $HCO_3^- \longrightarrow H^+ + CO_3^{2-}$ | *Anionic* : $HCO_3^- + H^+ \longrightarrow H_2CO_3$ |
| $HSO_4^- \longrightarrow H^+ + SO_4^{2-}$ | $CO_3^{2-} + H^+ \longrightarrow HCO_3^-$ |

In this concept HCl is an acid in aqueous solution according to the following equation :

$$HCl + H_2O \longrightarrow H_3O^+ + Cl^- \qquad ...(4.5)$$

Here HCl donates proton to $H_2O$ to form $Cl^-$ ion while $H_2O$ accepts proton to form $H_3O^+$. So HCl and $H_2O$ are Bronsted acid and base respectively. On the other hand $H_3O^+$ is **an acid** because it has a tendency to release a proton ($H_3O^+ \longrightarrow H_2O + H^+$) whereas $Cl^-$ is considered as **a base** because it has a tendency to accept $H^+$ ion ($Cl^- + H^+ \longrightarrow HCl$) to form undissociated HCl. $Cl^-$ ion is produced from the acid HCl, so $Cl^-$ is considered as the **conjugate base** of the acid HCl and is denoted as :

$$\underset{acid_1}{HCl} \longrightarrow \underset{base_1}{Cl^-} + H^+$$

Similarly $H_3O^+$ ion is produced from the base $H_2O$, so that $H_3O^+$ is characterised as the conjugate acid of the base $H_2O$ and is represented as :

$$\underset{base_2}{H_2O} + H^+ \longrightarrow \underset{acid_2}{H_3O^+}$$

Therefore the equation (4.5) can be represented as :

$$acid_1 + base_2 \longrightarrow acid_2 + base_1 \qquad ...(4.6)$$

This is an important **axiom** of Bronsted-Lowry concept in which the process of **neutralisation is considered as the reaction between an acid with a base to form another conjugate base and conjugate acid respectively**. For this reason this concept is called **conjugate acid base concept.** A strong acid has a weaker **conjugate base while a weak acid has a strong conjugate base.** For example HCl is a strong acid, its conjugate base $Cl^-$ is weak so that it will have no tendency to accept $H^+$ ion to form undissociated HCl so that the equilibrium of equation (4.5) will be **shifted to right.** $CH_3COOH$ is a weak acid, its conjugate base ($CH_3COO^-$) is strong so that $CH_3COO^-$ has a tendency to accept $H^+$ to form undissociated $CH_3COOH$ and the equilibrium will be **shifted to the left** i.e. a reversible process is established as :

$$\underset{\substack{acid_1 \\ weak\ acid}}{CH_3COOH} + \underset{base_2}{H_2O} \rightleftharpoons \underset{\substack{base_1 \\ strong \\ conjugate\ base}}{CH_3COO^-} + \underset{acid_2}{H_3O^+} \qquad ...(4.7)$$

Similarly a stong base has a weak conjugate acid, conversly a weak base has a strong conjugate acid. Examples of more conjugate acid-base pairs are given below :

| **$acid_1$** | + | **$base_2$** | $\rightleftharpoons$ | **$acid_2$** | + | **$base_1$** |
|---|---|---|---|---|---|---|
| $HClO_4$ | + | $H_2O$ | $\rightleftharpoons$ | $H_3O^+$ | + | $ClO_4^-$ |
| $H_2SO_4$ | + | $H_2O$ | $\rightleftharpoons$ | $H_3O^+$ | + | $HSO_4^-$ |
| $HSO_4^-$ | + | $H_2O$ | $\rightleftharpoons$ | $H_3O^+$ | + | $SO_4^{2-}$ |
| $HNO_3$ | + | $H_2O$ | $\rightleftharpoons$ | $H_3O^+$ | + | $NO_3^-$ |
| $H_2CO_3$ | + | $H_2O$ | $\rightleftharpoons$ | $H_3O^+$ | + | $HCO_3^-$ |
| $NH_4^+$ | + | $S^{2-}$ | $\rightleftharpoons$ | $HS^-$ | + | $NH_3$ |
| $NH_4^+$ | + | $OH^-$ | $\rightleftharpoons$ | $H_2O$ | + | $NH_3$ |
| $H_2O$ | + | $CO_3^{2-}$ | $\rightleftharpoons$ | $HCO_3^-$ | + | $OH^-$ |
| $H_2O$ | + | $CH_3NH_2$ | $\rightleftharpoons$ | $CH_3NH_3^+$ | + | $OH^-$ |
| $H_2O$ | + | $F^-$ | $\rightleftharpoons$ | HF | + | $OH^-$ |
| $CO(NH_2)_2$ | + | $NH_3$ | $\rightleftharpoons$ | $NH_4^+$ | + | $CO(NH_2)(NH^-)$ |
| $H_2O$ | + | $CN^-$ | $\rightleftharpoons$ | HCN | + | $OH^-$ |
| $NH_4^+$ | + | $CN^-$ | $\rightleftharpoons$ | HCN | + | $NH_3$ |
| $CH_3COOH$ | + | $NH_3$ | $\rightleftharpoons$ | $NH_4^+$ | + | $CH_3COO^-$ |
| $CH_3COOH$ | + | $HS^-$ | $\rightleftharpoons$ | $H_2S$ | + | $CH_3COO^-$ |

| | | | | | | |
|---|---|---|---|---|---|---|
| $H_2$ | + | $OD^-$ | $\rightleftharpoons$ | $DOH$ | + | $H^-$ |
| $HF$ | + | $HNO_3$ | $\rightleftharpoons$ | $H_2NO_3^+$ | + | $F^-$ |
| $HOBr$ | + | $CN^-$ | $\rightleftharpoons$ | $HCN$ | + | $OBr^-$ |
| $HI$ | + | $BrO_3^-$ | $\rightleftharpoons$ | $HBrO_3$ | + | $I^-$ |
| $H_3O^+$ | + | $IO_3^-$ | $\rightleftharpoons$ | $HIO_3$ | + | $H_2O$ |

This conjugate acid-base concept is more versatile than the former Arrhenius and Franklin concept because this represents the acid-base behaviour in terms of proton donation and acceptance property irrespective of the solvent. This concept is utilised in explaining the hydrolytic behaviour of salts. For example the aqueous solution of $FeCl_3$ is acidic due to the fact that proton donor ability of hydrated $Fe^{3+}$ ion i.e. $[Fe(H_2O)_6]^{3+}$ exceeds the proton acceptor ability of $Cl^-$ ion so that a considerable excess of $H_3O^+$ ion exist in aqueous solution making $FeCl_3$ solution acidic.

*Utility of the concept*

$$FeCl_3 + 6H_2O \longrightarrow [Fe(H_2O)_6]^{3+} + 3Cl^-$$

$$[Fe(H_2O)_6]^{3+} \longrightarrow [Fe(H_2O)_5OH]^{2+} + H^+$$

$$H^+ + H_2O \longrightarrow H_3O^+$$

Similarly aqueous solution of $Na_2CO_3$ is alkaline because the proton acceptor ability of $CO_3^{2-}$ is more than the proton donor ability of hydrated $Na^+$ ion

$$Na_2CO_3 + 2xH_2O \longrightarrow 2[Na(H_2O)_x]^+ + CO_3^{2-}$$

$$CO_3^{2-} + H_2O \longrightarrow HCO_3^- + OH^-$$

Although this concept has several utility yet there are certain drawbacks. This has an over emphasis on the proton transfer/acceptor property. Moreover there are several compounds e.g. $SO_2$, $AlCl_3$, $SOCl_2$ etc. undergo acid/base behaviour although there is no proton transfer occurs.

*Limitations of the concept*

### 4.3.1 Relative strengths of acids and bases : Levelling and differentiating solvents

In Arrhenius concept the strength of an acid HA is expressed in terms of the equilibrium constant ($K_a$) of the ionisation of HA in aqueous solution as :

$$HA \rightleftharpoons H^+ + A^- \qquad K_a = \frac{[H^+][A^-]}{[HA]} \qquad \ldots(4.8)$$

Similarly for a base we can express the strength in terms of the ionisation in aqueous solution. In Bronsted concept we can express the acid/base strength with reference to the following equilibria :

$$\underset{acid_1}{HA} + \underset{base_2}{H_2O} \rightleftharpoons \underset{acid_2}{H_3O^+} + \underset{base_1}{A^-} \qquad K_a = \frac{[H_3O^+][A^-]}{[HA]} \qquad \ldots(4.9)$$

$$\underset{base_1}{B} + \underset{acid_2}{H_2O} \rightleftharpoons \underset{acid_1}{BH^+} + \underset{base_2}{OH^-} \qquad K_b = \frac{[BH^+][OH^-]}{[B]} \qquad \ldots(4.10)$$

where $K_b$ is the equilibrium constant for the ionisation of the base(B). Here the concentration of $H_2O$ remains practically constant before and after ionisation.

The strength of an acid according to Bronsted concept is measured by its tendency to donate proton. Thus a strong acid has a strong tendency to donate proton and if we take water as solvent then the equilibrium will be shifted to the right so that the $K_a$ value of equation (4.9) will assume the value **infinity**. Similarly the value of $K_b$ in equation (4.10) will have the value **infinity** for strong base.

To compare the relative proton donating capability of a given acid generally two methods are used :

(1) *Proton donating tendencies of different acids towards the same base.* For weak acids, $H_2O$ is taken as the base and $K_a$ values are measured. For example for $CH_3COOH$ and HCN the following equilibrium experimentally studied at room temperature.

$$CH_3COOH + H_2O \rightleftharpoons H_3O^+ + CH_3COO^- \qquad K_a = 1.8 \times 10^{-5}$$

$$HCN + H_2O \rightleftharpoons H_3O^+ + CN^- \qquad K_a = 4.0 \times 10^{-10}$$

Thus $CH_3COOH$ is a stronger acid than HCN and $CN^-$ ion is a stronger base than $CH_3COO^-$ ion.

(2) *The competitive protolysis method.* In this method one acid is added to the conjugate base of another and the equilibrium concentrations are determined experimentally e.g. when $NaOC_2H_5$ is added to water, it is verified that $OC_2H_5^-$ which is the conjugate base of $C_2H_5OH$ reacts completely with $H_2O$ to form $C_2H_5OH$ and $OH^-$ ion.

$$H_2O + C_2H_5O^- \rightleftharpoons C_2H_5OH + OH^-$$

Thus $C_2H_5O^-$ ion is a stronger base than $OH^-$ and $H_2O$ is a stronger acid than $C_2H_5OH$.

The strength of protonic acids is dependent on the solvent in which the acids are dissolved. If the solvent is a good proton acceptor (base) then the different acids will donate the protons completely (100%) to the reference base, so that all acids will be equally strong. For example, $HClO_4$, $H_2SO_4$, HCl, $HNO_3$ are levelled to the strength of $H_3O^+$ in solvent $H_2O$ i.e.

$$\left.\begin{array}{l} HClO_4 + H_2O \longrightarrow H_3O^+ + ClO_4^- \\ H_2SO_4 + H_2O \longrightarrow H_3O^+ + HSO_4^- \\ HCl + H_2O \longrightarrow H_3O^+ + Cl^- \\ HNO_3 + H_2O \longrightarrow H_3O^+ + NO_3^- \end{array}\right\} \qquad ...(4.11)$$

In all cases the acid dissociation constant ($K_a$) has the value infinity and water acting as a **levelling solvent**. *A solvent in which 100% proton transfer may occur is called a levelling solvent for acids.* Similarly liquid $NH_3$ is acting as a levelling solvent for the acids HF, HBr, $CH_3COOH$ etc. where proton transfer is about cent percent, so that

$$HX + NH_3 \longrightarrow NH_4^+ + X^- \ (X = F, Br, CH_3COO)$$

If we take $CH_3COOH$ as a solvent then the acid strength of the above strong acids e.g. $HClO_4$, $H_2SO_4$, HCl and $HNO_3$ can be differentiated, i.e.

$$HX + CH_3COOH \rightleftharpoons CH_3COOH_2^+ + X^- \ (X = ClO_4, HSO_4, Cl \text{ and } NO_3) \qquad ...(4.12)$$

Here **$CH_3COOH$ is acting as a base,** therefore in order to differentiate the acid strength of strong acids we have to take a much weaker acid as a **differentiating solvent**. Therefore *the concept of acids and bases depends on the solvent* e.g. $CH_3COOH$ is *an acid in water but it is acting as a base in presence of strong acids represented by equation* (4.12).

Though it is apparent from equation (4.12) that the strengths of all strong acids are levelled to the strength of $CH_3COOH_2^+$, this is not correct because the proton donation capability of strong acids towards $CH_3COOH$ are different, so that concentration of $CH_3COOH_2^+$will be different for different acids and the measurement of acid dissociation constant will give us the order of acid strength as $HClO_4 > HBr > H_2SO_4 > HCl > HNO_3$ in anhydrous $CH_3COOH$ medium. Here $CH_3COOH$ is acting as a differentiating solvent. Generally *a differentiating solvent is a weak proton acceptor (a weak acid) which distinguish the relative strengths of strong acids.* In fact a particular solvent may act as a differentiating as well as levelling solvent. For example water acting as a levelling solvent for strong acids as represented by equation (4.11) while it acts as a differentiating solvent for weak acids such as $CH_3COOH$, HCN, HCOOH etc. This is due to the fact that $H_2O$ is an **amphiprotic** solvent i.e. acting as an acid and a base according to Bronsted concept as :

$$\underset{\text{acid}_1}{H_2O} + \underset{\text{base}_2}{H_2O} \longrightarrow \underset{\text{acid}_2}{H_3O^+} + \underset{\text{base}_1}{OH^-}$$

In a similar way $H_2O$ may acts as a levelling solvent for strong bases i.e.

$$HaH + H_2O \longrightarrow NaOH + H_2$$

$$HaNH_2 + H_2O \longrightarrow NaOH + NH_3$$

$$NaOC_2H_5 + H_2O \longrightarrow NaOH + C_2H_5OH$$

### 4.3.2 Cosolvating agents

A cosolvating agent is one which enhances the acidity of an acid HA by reaction with its conjuugate base ($A^-$) or enhance the basicity of a base B by reaction with its conjugate acid ($BH^+$). For example $Cu^{2+}$ ion acting as a cosolvating agent for $NH_4^+$ or glycine.

$$\underset{\textit{acid}}{NH_4^+} + H_2O \longrightarrow \underset{\textit{conjugate base}}{NH_3} + H_3O^+$$

$$Cu^{2+} + 4NH_3 \longrightarrow [Cu(NH_3)_4]^{2+}$$

$$\underset{\textit{acid}}{H_2NCH_2COOH} + H_2O \longrightarrow \underset{\textit{conjugate base}}{H_2NCH_2COO^-} + H_3O^+$$

$$Cu^{2+} + 2NH_2CH_2COO^- \longrightarrow Cu(NH_2CH_2COO)_2$$

## 4.4 LEWIS CONCEPT (1923)

The limitation of Bronsted concept is somewhat rectified by Lewis theory because this concept based on the electronic structure of the acids and bases. According to Lewis *an acid is defined as a species (molecule, radical or ion) which accepts electron pair and a base is a molecular system which donates electron pair.* Thus an acid is an electron deficient species (electrophile) while bases are electron rich molecular system (nucleophile). The *process of neutralisation then considered as the formation of a donor-acceptor bond (coordinate bond) from a donor base (:B) to the acceptor acid (A) as to form an adduct (AB) as :*

$$A + :B \longrightarrow A \leftarrow :B \qquad ...(4.13)$$

In terms of MO theory the adduct formation (AB) is considered as the interaction between HOMO of Lewis base (:B) and LUMO of Lewis acid (A) to form molecular orbitals as depicted below :

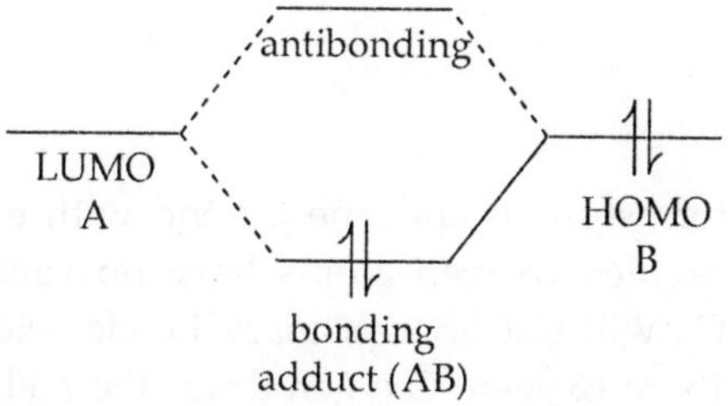


Let us consider a reaction :

$$H_3N: + H^+ \longrightarrow NH_4^+ \qquad ...(4.14)$$

Evidently the $H^+$ ion accepts one electron pair from $NH_3$, therefore $H^+$ is a Lewis acid while $NH_3$ donates an electron pair, so it is a Lewis base. Lewis bases are infact the same as that of Bronsted bases since any species which accepts proton is an electron pair donor. In the equation (4.14) $NH_3$ is a proton acceptor (Bronsted base) which is also an electron pair donor (Lewis base). $NH_3$, $H_2O$, $OH^-$, $Cl^-$ etc. are bases according to both the theories.

The Lewis acids are classified as :

(a) **A species containing a central atom having incomplete octet :** In the molecules viz. $BF_3$, $BCl_3$, $AlCl_3$, $B(CH_3)_3$ etc., the central atom is of sextet and can accept an electron pair to complete the octet.

$$F_3B + :NH_3 \longrightarrow F_3B \leftarrow NH_3$$

$$(CH_3)_3B + :NH_3 \longrightarrow (CH_3)_3B \leftarrow NH_3$$

*acid* *base* *adduct*

(b) **All metal cations are acting as Lewis acids to form coordination compounds** e.g.

$$Co^{2+} + 6H_2O \longrightarrow [Co(H_2O)_6]^{2+}$$

$$Co^{3+} + 6NH_3 \longrightarrow [Co(NH_3)_6]^{3+}$$

$$Ag^+ + 2NH_3 \longrightarrow [Ag(NH_3)_2]^+$$

*acid* *base* *adduct*

Transition metal ions have strong power of complex formation.

(c) **Molecules having central atom with vacant d-orbitals :** In the case of halides e.g. $SiCl_4$, $GeCl_4$, $TiCl_4$, $SnCl_4$, $TeCl_4$ etc., the central atom has vacant d-orbital which receive the electron pair from the Lewis base.

$TiCl_4 + 2Cl^- \longrightarrow [TiCl_6]^{2-}$

$SiF_4 + 2F^- \longrightarrow [SiF_6]^{2-}$

*acid* *base* *adduct*

(d) **Molecules containing multiple bonds between atoms of different electronegativity :** For example $CO_2$ acting as a Lewis acid when it forms $HCO_3^-$ or $H_2CO_3$ by accepting the electron pair from the oxygen atom of $OH^-$ or $H_2O$ .

$CO_2 + :\ddot{\underset{..}{O}}H^- \longrightarrow HCO_3^-$

$CO_2 + :\ddot{O}H_2 \longrightarrow H_2CO_3$

In $CO_2$ the central C is sp hybridised and forms one σ and one π bond with each oxygen i.e. O=C=O. Thus in $CO_2$ both central C and bonded oxygen atoms have no vacant orbitals for further electron pair accommodation. So $CO_2$ will not be a primary Lewis acid but its Lewis acidity arises due to the high electronegativity of oxygen. Oxygen drags the π electron cloud of C=O towards itself making carbon as electron deficient. Now carbon changes its state of hybridisation from sp to $sp^2$. In presence of a base (say $OH^-$ and $H_2O$), the $sp^2$ hybridised electron deficient central carbon accepts the electron pair donated by a base. Hence it may be stated that $CO_2$ can function as a **secondary Lewis acid** and not as a **primary acid** as :

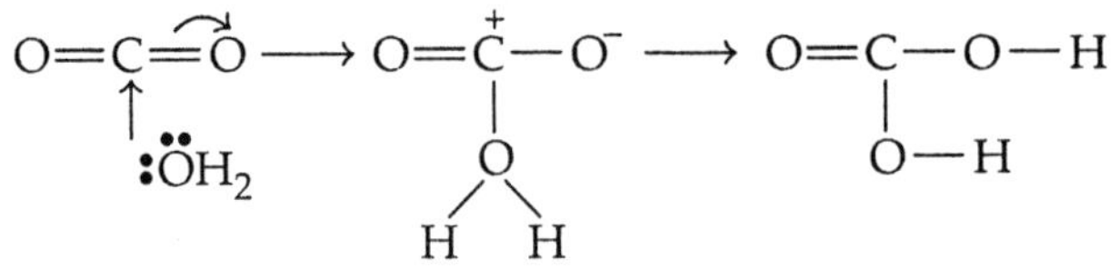

Similarly $SO_3$ can act as a secondary Lewis acid as :

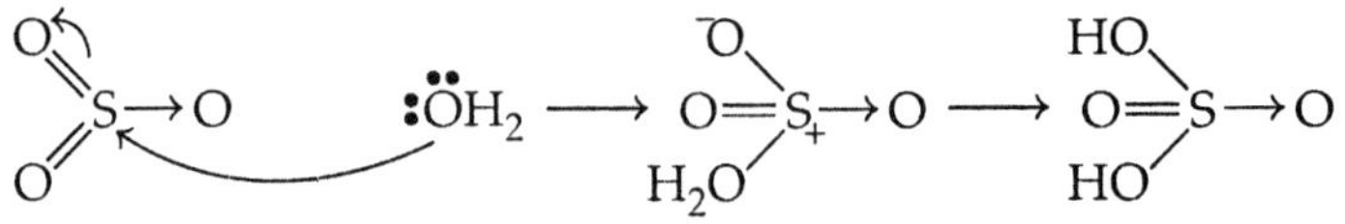

Lewis acid-base reactions are classified as :

(a) **Displacement of one Lewis base by another :** This type of reaction can be—

B–A + :B′ $\longrightarrow$ :B + A–B′

For example all Bronsted proton transfer reactions are of this type,

$HS^- + H_2O \longrightarrow H_3O^+ + S^{2-}$

Here Lewis base $H_2O$ displaces the Lewis base $S^{2-}$ from its adduct with the acid $H^+$ i.e. $HS^-$. Other example is—

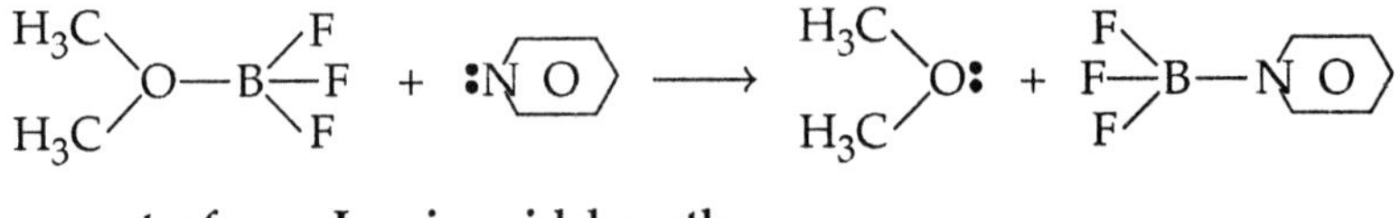

(b) **Displacement of one Lewis acid by other**

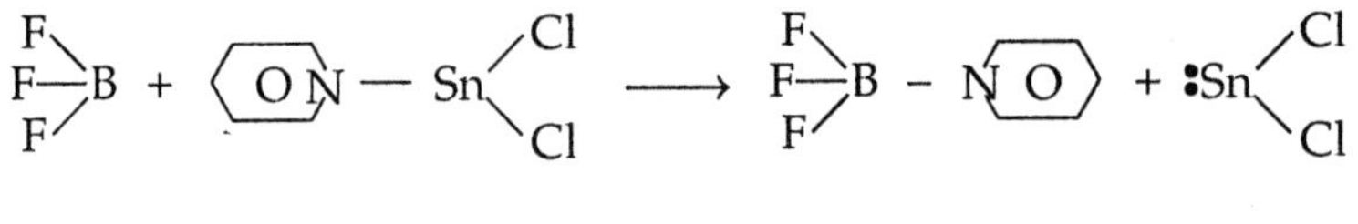

These reactions (a) and (b) are sometimes called **substitution reaction.**

(c) **Double displacement reaction or interchange reaction :** This is represented as :

$$A—B + A'—B' \longrightarrow A—B' + A'—B$$

For example,

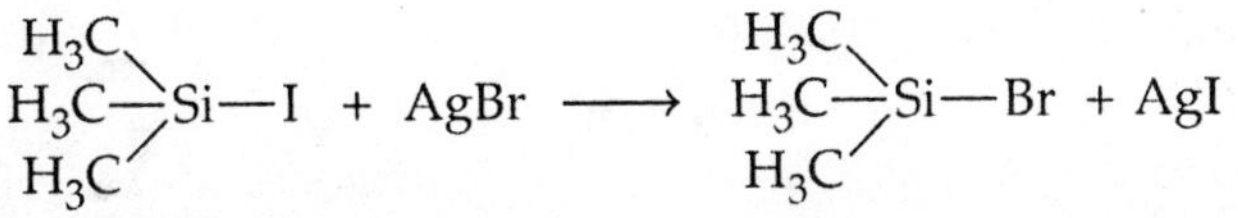

(d) **Oxidation reduction reactions are generally considered as Lewis acid-base reaction :** For example,

$$\underset{\text{Lewis acid}}{Cu^{2+}} + \underset{\text{Lewis base}}{Zn} \longrightarrow Cu + Zn^{2+}$$

The versatility of Lewis concept over other preceding concept is that it can explain the acid-base criteria for protonic and non-protonic substances. However this concept has certain **limitations :**

(a) It cannot explain the relative acid-base strength of different acids and bases. For example $Cu^{2+}$ and $Co^{3+}$ forms complexes with $NH_3$ as $Cu(NH_3)_4^{2+}$ and $Co(NH_3)_6^{3+}$. Therefore a question arises whether $NH_3$ (Lewis base) will form strong complex with $Cu^{2+}$ (Lewis acid) or $Co^{3+}$ (Lewis acid).

(b) Relative strength of protonic acids viz. HCl, $HNO_3$ or $H_2SO_4$ cannot be explained by this concept. These acids on dissociation produces $H^+$ ions which accept the electron pair from $H_2O$ to form $H_3O^+$ ion ($H_2\ddot{O}: + H^+ \longrightarrow H_3O^+$). Therefore the protonic acids are not able to accept an electron pair directly from a base so that they are not a primary acid but secondary acid like $CO_2$ or $SO_3$ etc.

---

**Problem 4.1** In what situation $H_2SO_4$ behaves as a base?

**Ans. :** In aqueous solution $H_2SO_4$ behaves as a strong acid

$$\underset{acid_1}{H_2SO_4} + \underset{base_2}{H_2O} \longrightarrow \underset{acid_2}{H_3O^+} + \underset{base_1}{HSO_4^-}$$

But $H_2SO_4$ behaves as a base in presence of a more stronger acid viz, $H[B(HSO_4)_4]$

$$\underset{acid_1}{H[B(HSO_4)_4]} + \underset{base_2}{H_2SO_4} \longrightarrow \underset{acid_2}{H_3SO_4^+} + \underset{base_1}{[B(HSO_4)_4]^-}$$

---

**Problem 4.2** Acidity of anhydrous HF is substantially increased in presence of $SbF_5$—explain.

**[B.U. 1992,1995, C.U. 2002]**

**Ans. :** According to solvent system concept anhydrous HF is autoionised as :

$$3HF \rightleftharpoons \underset{\text{solvent cation}}{H_2F^+} + \underset{\text{solvent anion}}{HF_2^-}$$

In presence of $SbF_5$, the following reaction occurs

$SbF_5 + 2HF \longrightarrow H_2F^+ + SbF_6^-$

As the concentration of solvent cation ($H_2F^+$) increases in presence of $SbF_5$, so acidity increases.

**Problem 4.3** Classify the following species into acids and bases and give the conjugate bases or acids corresponding to each.

$HSO_4^-$, $H_3PO_2$, $Al(H_2O)_6^{3+}$, $NH_4^+$, $HS^-$, $PH_3$ **[B.U. 1996,2008]**

**Ans. :**

$HSO_4^- \longrightarrow SO_4^{2-} + H^+$ (acid → conjugate base)

$HSO_4^- + H^+ \longrightarrow H_2SO_4$ (base → conjugate acid)

$H_3PO_2 \longrightarrow H_2PO_2^- + H^+$ (acid → conjugate base)

$Al(H_2O)_6^{3+} \longrightarrow [Al(H_2O)_5OH]^{2+} + H^+$ (acid → conjugate base)

$NH_4^+ \longrightarrow NH_3 + H^+$ (acid → conjugate base)

$HS^- \longrightarrow H^+ + S^{2-}$ (acid → conjugate base)

$HS^- + H^+ \longrightarrow H_2S$ (base → conjugate acid)

$PH_3 + H^+ \longrightarrow PH_4^+$ (base → conjugate acid)

**Problem 4.4** Identify the Lewis acids and bases in the reactions :

(a) $FeCl_3 + Cl^- \longrightarrow [FeCl_4]^-$ (b) $I^- + I_2 \longrightarrow I_3^-$

(c) $[SnCl_3]^- + (CO)_5MnCl \longrightarrow (CO)_5Mn - SnCl_3 + Cl^-$

(d) $BrF_3 + F^- \longrightarrow [BrF_4]^-$ (e) $(CH_3)_2CO + I_2 \longrightarrow (CH_3)_2CO - I_2$

(f) $KH + H_2O \longrightarrow KOH + H_2$

**Ans. :** (a) Acid $FeCl_3$ adds the base $Cl^-$ to form $[FeCl_4]^-$.

(b) Acid $I_2$ and base is $I^-$.

(c) $(CO)_5MnCl$ is acid and $[SnCl_3]^-$ is base.

(d) $BrF_3$ is acid and $F^-$ is base.

(e) $(CH_3)_2CO$ is a base and $I_2$ is acid.

(f) The ionic hydride KH provides $H^-$ which is a base and $H_2O$ provides $H^+$ ion which is an acid.

**Problem 4.5** Identify the acids and bases involved in the following reactions :

(i) $SO_3 + H_2O \longrightarrow HSO_4^- + H^+$

(ii) $2CH_3COCl + Al_2Cl_6 \longrightarrow 2CH_3CO^+ + 2AlCl_4^-$ **[B.U. 1995]**

**Ans. :** (i) Lewis acid $SO_3$ and Lewis base $H_2O$.

(ii) Lewis acid $AlCl_3$ and Lewis base $CH_3COCl$.

**Problem 4.6** Identify the acid, base, conjugate acid and conjugate base in :

(a) $HNO_3 + H_2O \longrightarrow H_3O^+ + NO_3^-$ (b) $CO_3^{2-} + H_2O \rightleftharpoons HCO_3^- + OH^-$

(c) $NH_3 + H_2S \longrightarrow NH_4^+ + HS^-$ (d) $HSO_4^- + OH^- \rightleftharpoons SO_4^{2-} + H_2O$

(e) $PO_4^{3-} + H_2O \rightleftharpoons HPO_4^{2-} + OH^-$

(f) $H_2Fe(CO)_4 + CH_3OH \rightleftharpoons [HFe(CO)_4]^- + CH_3OH_2^+$

**Ans.** : (a) Acid $HNO_3$ and conjugate base $NO_3^-$, base is $H_2O$ and its conjugate acid is $H_3O^+$.

(b) Acid is $H_2O$ and conjugate base is $OH^-$ while $CO_3^{2-}$ is base and its conjugate acid is $HCO_3^-$.

(c) Acid is $H_2S$ and conjugate base is $HS^-$ while $NH_3$ is a base and its conjugate acid is $NH_4^+$.

(d) $HSO_4^-$ is acid and conjugate base is $SO_4^{2-}$ while $OH^-$ is a base and its conjugate acid is $H_2O$.

(e) $H_2O$ is an acid and its conjugate base is $OH^-$ while $PO_4^{3-}$ is a base and its conjugate acid is $HPO_4^{2-}$.

(f) $H_2Fe(CO)_4$ acting as an acid that transfers a proton to form the conjugate base $[HFe(CO)_4]^-$ while $CH_3OH$ is a base which accepts the proton to form the conjugate acid $CH_3OH_2^+$.

---

## 4.5 SOFT AND HARD ACIDS AND BASES (SHAB PRINCIPLE) (1963)

In order to compare the donor properties of different bases and acceptor properties of different acids, the concept of soft and hard acids and bases developed i.e. drawbacks of Lewis theory is somewhat modified in this concept.

In order to determine the donor properties of different bases and acceptor properties of acids, the preference of a base B to bind with $H^+$ and methyl mercury (II) cation ($CH_3Hg^+$) was estimated by measuring the equilibrium constant K of the reaction :

$$BH^+ + CH_3Hg^+ \rightleftharpoons CH_3HgB^+ + H^+ \qquad ...(4.15)$$

$$\text{and} \quad K = \frac{[CH_3HgB^+][H^+]}{[BH^+][CH_3Hg^+]} \qquad ...(4.16)$$

$$= \frac{[CH_3HgB^+]}{[CH_3Hg^+][B]} \cdot \frac{[H^+][B]}{[BH^+]} = K_sK_a \qquad ...(4.17)$$

where $K_s = \dfrac{[CH_3HgB^+]}{[CH_3Hg^+][B]}$ for the reaction $CH_3Hg^+ + B \longrightarrow CH_3HgB^+$

and $K_a = \dfrac{[H^+][B]}{[BH^+]}$ for the reaction $BH^+ \longrightarrow B + H^+$

Therefore $\log K = \log K_s + \log K_a$

$$\text{or} \quad pK = pK_s + pK_a \qquad ...(4.18)$$

In the equilibruim represented in equation (4.15) if the donor atom of the base B is O, N or F, then these donors prefer to combine with proton and equilibrium will be shifted to the left while if the donor atom of the base is P, S Cl etc. then these will combine with $CH_3Hg^+$ and the equilibrium will be shifted to the right. The donor atoms of the first category are called **hard donor** or **hard bases** while the donor atoms of the second category are called **soft donors** or **soft bases.** For hard bases $pK_a$ is more than $\log K_s$ while for a soft base $pK_a$ is less than $\log K_s$.

Similarly the metal cations are also classified as hard acceptor or **hard acids** or **soft acceptor** or **soft acids**. In both the cases, the classification can be made as **class (a) acids** or **bases** as **hard acids** and **hard bases while class (b) acids** or **bases** are **classed as soft acids and soft bases.**

Now what are the characteristic features of those acids and bases. A *typical hard acid will have* (a) *high positive charge i.e. high ionic potential,* (b) *a small size,* (c) *high polarising power,* (d) *have no unshared pair of electrons in their valence shell.* The criteria for *soft acids are* (a) *small positive or zero charge,* (b) *a large size,* (c) *high polarisability and low electronegativity,* (d) *have unshared pair of electrons in their valence shell.* On the other hard there are typical acceptor atoms which have intermediate characteristic feature of class (a) and class (b) acids which are naturally called **borderline acids.** A list of **hard**, **soft** and **borderline** acids are given in Table 4.2.

The characteristic features of **hard bases** i.e. **class (a) donors** are (a) **the donor atom will have high electronegativity**, (b) **small size**, (c) **low polarisability**, (d) **hard to oxidise while the soft bases** i.e. **class (b) donors will have** (a) **the donor atom have low electronegativity**, (b) **large size**, (c) **high polarisability**, (d) **easy to oxidise**. Intermediate property of the donor atoms are classified as **borderline bases**. Table 4.3 represents the classification of hard, soft and borderline bases.

**Table 4.2 : Classification of Lewis acids as hard, soft and borderline**

| | |
|---|---|
| **Hard acids :** | $H^+$, $Li^+$, $Na^+$, $K^+$, $Be^{2+}$, $Mg^{2+}$, $Ca^{2+}$, $Sr^{2+}$, $Sc^{3+}$, $Ce^{4+}$, $Mn^{2+}$, $Al^{3+}$, $Ga^{3+}$, $In^{3+}$, $La^{3+}$, $Lu^{3+}$, $Cr^{3+}$, $Co^{3+}$, $Fe^{3+}$, $As^{3+}$, $Si^{4+}$, $Ti^{4+}$, $Th^{4+}$, $Zr^{4+}$, $UO_2{}^{2+}$, $Pu^{4+}$, $BF_3$, $Sn^{4+}$, $VO^{2+}$ $Cr^{6+}$, $AlCl_3$, $SO_3$, $BCl_3$, $CO_2$, $RCO^+$ |
| **Borderline acids :** | $Fe^{2+}$, $Co^{2+}$, $Ni^{2+}$, $Cu^{2+}$, $Zn^{2+}$, $Pb^{2+}$, $Sn^{2+}$, $Sb^{3+}$, $Rh^{3+}$, $Bi^{3+}$, $Ru^{3+}$, $BMe_3$, $SO_2$, $NO^+$ |
| **Soft acids :** | $Cu^+$, $Ag^+$, $Cd^{2+}$, $Pt^{2+}$, $Hg^{2+}$, $Hg^+$, $Pd^{2+}$, $CH_3Hg^+$, $Tl^{3+}$, $BH_3$, $I_2$, $Br_2$, zero valent metal, carbenes |

**Table 4.3 : List of Lewis bases as hard, soft and borderline (R = alkyl group)**

| | |
|---|---|
| **Hard bases :** | $H_2O$, $OH^-$, $F^-$, $Cl^-$, $CH_3COO^-$, $PO_4^{3-}$, $SO_4^{2-}$, $CO_3^{2-}$, $ClO_4^-$, $NO_3^-$, ROH, $R_2O$, $NH_3$, $N_2H_4$, $RNH_2$, $RO^-$, $O^{2-}$ |
| **Borderline bases :** | $C_6H_5NH_2$, pyridine, $N_3^-$, $N_2$, $Br^-$, $SO_3^{2-}$, $NO_2^-$ |
| **Soft bases :** | $H^-$, $I^-$, CO, $C_2H_4$, $C_6H_6$, $SCN^-$, $CN^-$, $S_2O_3^{2-}$, $R_2S$, RSH, $RS^-$, $R_3P$, $R_3As$, RNC, $HS^-$, $(RO)_3P$, $S^{2-}$ |

A qualitative correlation between the various acids and bases has been done experimentally on various metal complexes and an approximate rule has been framed by R. G. Pearson (1963), known as **Pearson rule** which states that **hard acids prefer to combine with hard bases and soft acids prefer to combine with soft bases**. The classification hard-soft acids and bases was given by Pearson. This rule classify the preference of combination of Lewis acids and bases in the following sequence :

Class (a) acids : N > P > As > Sb; O > S > Se > Te
F > Cl > Br > I

Class (b) acids : N < P < As < Sb; O < S < Se ≈ Te
F < Cl < Br < I

### 4.5.1 Basis for SHAB principle

Experimentally the **hard-hard interactions are ionic,** since the formation of a strong ionic bond is accompanied by a gain in lattice energy, the smaller the ions greater is the stability. As the size of the donor atom of hard base and acceptor atom of a hard acid is small, the ionic bond formation by their interaction is accompanied by significant release of energy.

The **soft-soft interactions are generally covalent** which is supplemented by π-bonding. The soft donor atoms and acceptor atoms have large size. The soft donor atoms are polarisable and tend to form covalent bond with soft acids e.g. $Cu^+$, $Ag^+$, $Pd^{2+}$ etc. form compounds with appreciable covalent character. In addition to conventional covalent bonding there are two factors which favour π-bonding or back bonding. The base should possess vacant orbitals for accepting π-electrons and then the acid should be able to donate electrons for π-bonding. The acids are generally the metals in low oxidation states having great tendencies to form π bond and the good π-bonding ligands e.g. CO, $PR_3$ etc. are all soft bases.

Finally the basis for Pearson's rule i.e. stability of hard-hard and soft-soft interactions can be explained by Chatt theory with reference to the following four types of interactions :

(a) **Hard-hard interaction :** In this case there is a formation of σ bond from the hard base to hard acid. In addition to this bond there is a formation of π bond by the overlapping of filled p-orbital of the hard base with that of the vacant d-orbital of the hard acid [Fig. 4.1(a)]. Thus due to the formation of this additional bond, the hard-hard interaction is strengthen and hence favourable.

(b) **Soft-soft interaction :** There is a formation of σ bond from the soft base to soft acid. In addition to this bond there is a formation of π bond by the overlapping of the filled d-orbital of soft acid with that of the vacant orbital of the soft base [Fig. 4.1(b)]. Due to the formation of this additional bond the soft-soft interaction is stable and favourable.

(c) **Soft-hard interaction :** This interaction will form unstable adduct. In this case of soft-hard interaction a σ bond is formed from the hard base to soft acid [Fig. 4.1(c)]. There is no scope of the formation of π bond because in this case there is no vacant orbital of the soft acid to receive the electron density from the hard base. So this interaction is less favourable than hard-hard and soft-soft interaction.

(d) **Hard-soft interaction :** In this case also the resulting adduct will be unstable. There is a formation of σ bond from the soft base to hard acid. But there is no formation of π bond because there is no filled d-orbital to shift electron density from one to other [Fig. 4.1(d)]. So this is less favourable than (a) and (b).

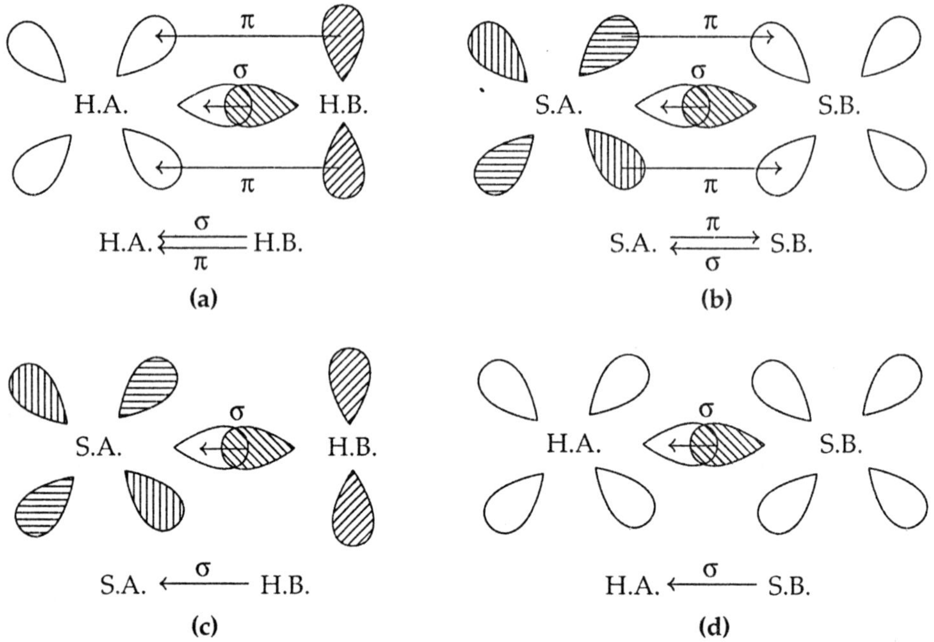


*Figure 4.1 : Representation of (a) hard acid-hard base, (b) soft acid-soft base, (c) soft acid-hard base, (d) hard acid-soft base interactions.*

### 4.5.2 Application of SHAB principle

(i) **Stability of compounds** : $AgI_2^-$ is stable while $AgF_2^-$ is unstable. Similarly $CoF_6^{3-}$ is more stable than $CoI_6^{3-}$. These are explained due to the following :

$$\underset{\textit{soft acid}}{Ag^+} + \underset{\textit{hard base}}{2F^-} \longrightarrow \underset{\textit{unstable}}{AgF_2^-} \qquad \underset{\textit{soft acid}}{Ag^+} + \underset{\textit{soft base}}{2I^-} \longrightarrow \underset{\textit{stable}}{AgI_2^-}$$

$$\underset{\textit{hard acid}}{Co^{3+}} + \underset{\textit{hard base}}{6F^-} \longrightarrow \underset{\textit{stable}}{CoF_6^{3-}} \qquad \underset{\textit{hard acid}}{Co^{3+}} + \underset{\textit{soft base}}{6I^-} \longrightarrow \underset{\textit{unstable}}{CoI_6^{3-}}$$

HgS (soft acid + soft base) is more stable than $Hg(OH)_2$ (soft acid + hard base). More stability of HgS compared to that of $Hg(OH)_2$ is suggested by the fact that $Hg(OH)_2$ dissolves readily in acidic aqueous solution but HgS is not.

The existance of natural source of metals can also be explained by SHAB principle. Alkali and alkaline earth metals are hard acids, prefer to combine with hard donors e.g. O sites so that they exist in nature as oxides, carbonates and silicates. Soft acids such as $Cu^+$, $Ag^+$ and $Hg^{2+}$. occur in nature as sulphides. The borderline metal ions e.g. $Ni^{2+}$, $Cu^{2+}$ and $Pb^{2+}$ occur in nature both as carbonates and sulphides.

If we arrange the donor atoms of most common Lewis bases in the order of increasing electronegativity, the following sequence is obtained :

As, P < C, Se, S, I < Br < N, Cl < O < F

2.0 2.1 < 2.5 2.4 2.5 2.5 < 2.8 < 3.0 3.0 < 3.5 < 4.0

Soft acids form stable complexes with left hand members of this series and hard acids will form stable complexes with right hand members of the above series.

Irving-Williams series of stability constants ($Mn^{2+} < Fe^{2+} < Co^{2+} < Ni^{2+} < Cu^{2+} > Zn^{2+}$) [(Sec. 7.14.5(A)] can also be explained by SHAB principle.

(ii) **Prediction of feasibility of reactions :** A reaction will be viable or not can be predicted by SHAB principle. The reaction between LiI and CsF to produce LiF and CsI can be explained by the following :

$$\underset{\textit{hard-soft}}{LiI} + \underset{\textit{soft-hard}}{CsF} \longrightarrow \underset{\textit{hard-hard}}{LiF} + \underset{\textit{soft-soft}}{CsI}$$

Similarly we can predict the following reaction

$$\underset{\textit{soft-hard}}{HgF_2} + \underset{\textit{hard-soft}}{BeI_2} \longrightarrow \underset{\textit{hard-hard}}{BeF_2} + \underset{\textit{soft-soft}}{HgI_2}$$

(iii) **Stability of complexes :** Jorgensen stated that in a complex having different ligands, the stability will increase if all the ligands are of same nature. On the other hand the stability of the complex will decrease if the ligands are of different nature i.e. hard-soft or soft-hard type. Thus $[Co(NH_3)_5F]^{2+}$ (where both the ligands are hard) and $[Co(CN)_5I]^{3-}$ (where both the ligands are soft) are expected to be more stable than $[Co(NH_3)_5I]^{2+}$ and $[Co(CN)_5F]^{3-}$ because $NH_3$ and $F^-$ are hard ligands while $CN^-$ and $I^-$ are soft ligands.

(iv) **Hardening and softening with reference to symbiosis :** *Hardening of hard centre by attachment with hard species and softening of a soft centre by combination with soft species is known as symbiosis.*

Let us consider that a mixture of $F^-$ and $H^-$ are added to $BF_3$. Which one prefer to combine with $BF_3$? $B^{3+}$ in $BF_3$ is a hard centre, prefer to combine with hard base $F^-$ and becomes more harder i.e. $BF_3 + F^- \longrightarrow BF_4^-$. On the other hand when a mixture of $H^-$ and $F^-$ are added to $B_2H_6$, the soft centre $B^{3+}$ in $B_2H_6$ becomes more softer by attachment of soft base $H^-$ so that the reaction $B_2H_6 + 2H^- \longrightarrow 2BH_4^-$ will occur.

Symbiosis can explain the stability of some **linkage isomers**. Thus $[Co(NH_3)_5(SCN)]^{2+}$ readily converted to $[Co(NH_3)_5(NCS)]^{2+}$. By attachment of five hard $NH_3$ ligands the $Co^{3+}$ becomes more hard, thereby it will prefer to combine with hard donor i.e. prefer to combine with hard N donor than that of soft S donor. Similarly $[Co(CN)_5(NCS)]^{3-}$ converted to $[Co(CN)_5(SCN)]^{3-}$ as $Co^{3+}$ centre becomes softer by the attachment of five soft $CN^-$ ligands so it will perfer to combine with soft centre S than the hard centre N and the conversion $[Co(CN)_5(NCS)]^{3-} \longrightarrow [Co(CN)_5(SCN)]^{3-}$ occurs. Thus we say that like ligands will prefer the like ligands.

Hard-soft parameters may often change due to substitution of suitable groups e.g. $Co^{3+}$ becomes more harder in $Co(NH_3)_5(H_2O)^{3+}$ due to the presence of five hard $NH_3$ and one hard $H_2O$ ligands. The substitution of $H_2O$ molecule in this complex by halide ligands will have the sequence $Cl^- > Br^- > I^-$. This is experimentally established by studying the equilibrium constant ($K_d$) for the reaction.

$$Co(NH_3)_5(H_2O)^{3+} + X^- \underset{}{\overset{K_d}{\rightleftharpoons}} Co(NH_3)_5X^{2+} + H_2O \qquad \text{...(4.19)}$$

$$\text{where } K_d = \frac{[Co(NH_3)_5X^{2+}]}{[Co(NH_3)_5H_2O^{3+}][X^-]} \qquad ...(4.20)$$

For $X^-$ = $Cl^-$ $Br^-$ $I^-$

$\log K_d$ = 1.25 0.48 0.21

But the reverse will occur i.e. $I^- > Br^- > Cl^-$ for the substitution of $H_2O$ in $Co(CN)_5(H_2O)^{2-}$. Here $Co^{3+}$ becomes softer by the attachment of five soft $CN^-$ ligands so that substitution will be easier for soft $I^-$ and difficult for hard $Cl^-$. Such softening of a hard and hardening of a soft reactant is also called **symbiosis**.

(v) **Analytical application :** In acid medium both $Cu^{2+}$ and $Cd^{2+}$ react with $H_2S$ to form the precipitate of CuS (black) and CdS (Yellow). The black colour of CuS will subside the yellow colour of CdS, so detection of $Cd^{2+}$ in presence of $Cu^{2+}$ is sometimes difficult. In order to obviate this difficulty before passing $H_2S$, a mixture of $Cu^{2+}$ and $Cd^{2+}$ is treated with KCN whereby the following reactions occur :

$$Cu^{2+} + 4CN^- \longrightarrow [Cu^{I}(CN)_4]^{3-}$$

$$Cd^{2+} + 4CN^- \longrightarrow [Cd^{II}(CN)_4]^{2-}$$

$Cu^{2+}$ is borderline acid and $Cu^+$ is soft acid, $CN^-$ is a soft ligand, so combination will occur by the reduction of Cu(II) to Cu(I). Both $Cu^+$ and $Cd^{2+}$ are soft acids, but $Cu^+$ is more softer than $Cd^{2+}$ and so the complex $[Cu(CN)_4]^{3-}$ is more stable than $[Cd(CN)_4]^{2-}$. Therefore on passing $H_2S$ in this solution only yellow precipitate of CdS will occur but not CuS.

### 4.5.3 Limitations of SHAB concept

$Al^{3+}$ is hard acid occurs in nature as the oxide, $Al_2O_3$ (bauxite) by combination with hard base $O^{2-}$ ion. But it does not occur in nature as the carbonate though $CO_3^{2-}$ ion is also a hard base.

The deviation of the hard-hard or soft-soft interaction is noticed in the following reaction :

$$\underset{\textit{borderline base}}{SO_3^{2-}} + \underset{\textit{hard acid + hard base}}{HF} \longrightarrow HSO_3^- + F^-$$

Here the borderline base $SO_3^{2-}$ replaces the hard base $F^-$ in HF and itself combine with hard acid $H^+$ to form $HSO_3^-$.

The esterification reaction between ethanol and acetic acid can be represented by breaking the reactants into the following ways :

$$\underset{\textit{hard-hard}}{CH_3COO^-H^+} + \underset{\textit{borderline-hard}}{C_2H_5^+OH^-} \quad \text{or} \quad \underset{\textit{hard-hard}}{CH_3CO^+OH^-} + \underset{\textit{hard-hard}}{C_2H_5O^-H^+}$$

Both the breakings are justified by the hard-hard combination of $H^+$ with $OH^-$. Though the second breaking explains the observed behaviour, the esterification reaction mechanism may also follow the first one.

Lastly the reaction,

$$CH_3^+(g) + H_2(g) \longrightarrow CH_4(g) + H^+(g)$$

proceed to completion by the combination of soft-soft combinations i.e. $CH_3^+ + H^- \longrightarrow CH_4$. But the process is endothermic which is verified by measurement of heat change by about

+359 $kJmol^{-1}$ so that the process is said to be unfavourable. The completion of the reaction is explained by the greater acidity of the proton relative to the $CH_3^+$ cation.

Finally there are several paradox of the Pearson rule and therefore SHAB principle is not a theory but considered as a statement of experimental facts.

---

**Problem 4.7** Why methyl mercury ion is taken as the standard for the SHAB principle? **[C.U. 2010]**

**Ans. :** In fact in the SHAB principle the reference acids $H^+$ and $CH_3Hg^+$ are arbitrarily taken to define the hard and soft bases.

The selection of $H^+$ ion as the reference hard acid is due to fact that the ion has highest charge density making it hardest.

The selection of $CH_3Hg^+$ as a typical soft acid is justified as :

(a) Hg being a member of sixth period of the periodic table, has large size and has electronic configuration $[Xe]4f^{14}5d^{10}6s^2$. Thus Hg(II) has $5d^{10}$ configuration with inner twenty d electrons and fourteen f electrons. Thus Hg(II) cation has low positive charge, large size and non-inert gas configuration and like $H^+$ it can form only one coordinate bond.

(b) The additional d and f inner electrons make the metal more softer.

(c) Hg(II) when attached to $CH_3$ group, its acceptor property reduces to minimum because of hyperconjugation of the σ electron of the C—H bond with the partially filled d-orbitals of mercury, the +I effect of $CH_3$ group is highest i.e. +I effect will be in the following order :

$$H_3C- < CH_3-CH_2- < (CH_3)_2CH- < (CH_3)_3C-$$

Attachment of $CH_3$ group to Hg(II) makes the Hg(II) centre more softer and thus the acceptor property of $CH_3Hg^+$ is minimum.

(d) Finally $CH_3Hg^+$ cation is a selected cation, because it is a typical soft acid and is a univalent ion like proton, making the comparison more plausable.

---

**Problem 4.8** Would the following increase, decrease or have no effect on the acidity oif the solution? Give arguments.

(i) Addition of HgO to an aqueous KI solution. **[C.U. 2002]**

(ii) Addition of $CuSO_4$ to an aqueous $(NH_4)_2SO_4$ solution.

(iii) Addition of $CH_3COONa$ to dilute $CH_3COOH$. **[B.U. 1992]**

(iv) Addition of $Al(OH)_3$ to aqueous NaOH solution.

(v) Addition of $KHSO_4$ to sulphuric acid.

**Ans. :** (i) $H_2O$ dissociates to form solvent cation $H^+$ or $H_3O^+$ and solvent anion $OH^-$

$$H_2O \longrightarrow H^+ + OH^- \quad \text{or} \quad 2H_2O \longrightarrow H_3O^+ + OH^-$$

KI reacts with HgO to form the stable complex species $HgI_4^{2-}$ and the concentration of solvent anion i.e. $OH^-$ ion increases and so acidity of the solution decreases.

$$HgO + 4KI + H_2O \longrightarrow K_2[HgI_4] + 2K^+ + 2OH^-$$

(ii) $(NH_4)_2SO_4$ on hydrolysis produces a weak base $NH_4OH$ and strong acid $H_2SO_4$

$$(NH_4)_2SO_4 + 2H_2O \longrightarrow H_2SO_4 + 2NH_4OH$$

$CuSO_4$ also hydrolysed to form strong acid $H_2SO_4$ and weak base $Cu(OH)_2$,

$$CuSO_4 + H_2O \longrightarrow Cu(OH)_2 + H_2SO_4$$

So acidity of the solution increases. Moreover according to the following reaction the concentration of $H^+$ ion increases and so acidity increases.

$$Cu^{2+}(aq) + 4NH_4^+(aq) \longrightarrow Cu(NH_3)_4^{2+} + 4H^+(aq)$$

(iii) $CH_3COONa$ is a strong electrolyte which on dissociation produces the conjugate base of $CH_3COOH$

$$CH_3COONa \longrightarrow CH_3COO^- + Na^+$$

$$CH_3COOH + H_2O \longrightarrow CH_3COO^- + H_3O^+$$

So due to the greater concentration of conjugate base $CH_3COO^-$, the dissociation of $CH_3COOH$ is favoured to the left and acidity decreases.

(iv) NaOH produces solvent anion $OH^-$. When $Al(OH)_3$ is added the concentration of $OH^-$ ion in the solution is decreased due to the formation of the complex species $Al(OH)_4^-$ and so acidity increases.

$$Al(OH)_3 + OH^- \longrightarrow Al(OH)_4^-$$

(v) $KHSO_4$ on dissociation produces $HSO_4^-$ which is the conjugate base of $H_2SO_4$. More concentration of $HSO_4^-$ inhibit the dissociation of $H_2SO_4$ and so acidity of the solution decreases.

### 4.5.4 Hardness and Softness in terms of Frontier Orbitals

We have mentioned the R.G. Pearson definition of **absolute hardness (η)** in terms of ionisation energy and electron affinity (Sec. 2.5.5). Now the hardness and softness parameter can also be explained by the energy separation of the frontier orbitals (Sec. 3.4.5). FO comprises of HOMO and LUMO and ionisation energy (I.E.) is the amount of energy required to remove electron from HOMO while electron affinity (E.A.) is associated with the energy when an electron is added to LUMO from infinity [Fig. 4.2(a)]. Thus the energy separation (ΔE) between HOMO and LUMO is a measure of chemical hardness and softness of a species.

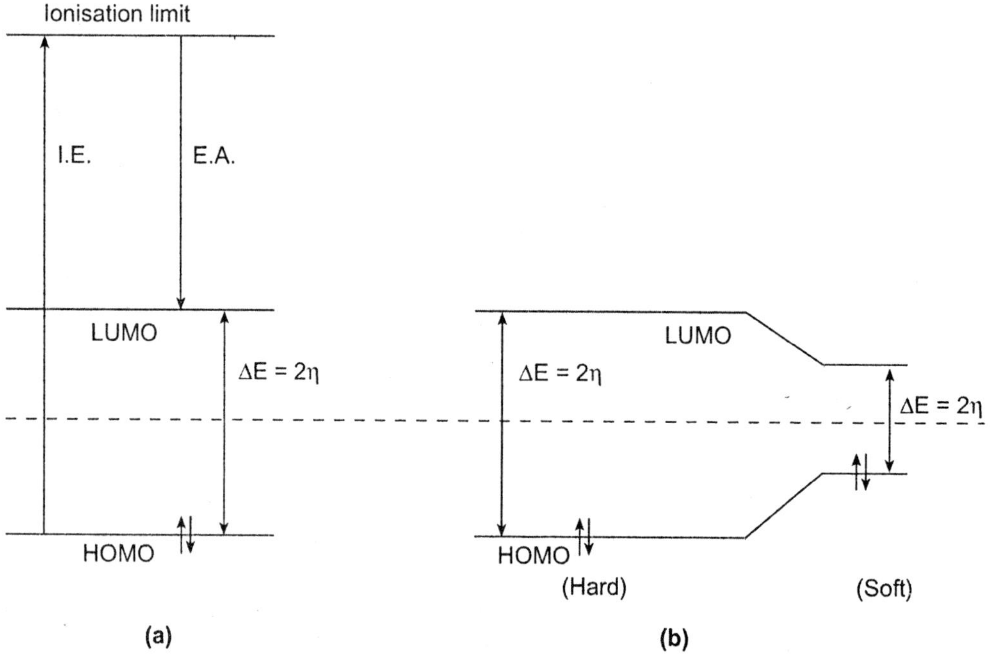


*Figure 4.2 : (a) FO with reference to I.E. and E.A., (b) Molecular hardness (η) in relation to HOMO and LUMO.*

Soft acids and bases are more polarisable than hard acids and bases. Hard acids and bases are those which have large energy separation between HOMO and LUMO while small energy separation between FO coresponds to soft character of acids and bases [Fig. 4.2(b)]. Small atoms or ions are hardest e.g. $Li^+$, $H^+$, $Be^{2+}$ etc. while heavier alkali metals or heavier halogens e.g. $Rb^+$, $Cs^+$, $I_2$, $I^+$ etc. are soft. Softest atoms or ions have low I.E. and low E.A. When the HOMO/ LUMO separation is small the electron distribution is easily perturbed and the species is termed soft. On the other hand when FO separation is large which resist the external perturbation and the species is regarded as hard. A hard acid does not have low energy LUMO and a hard base have a strongly bound HOMO of low energy. Thus the hard-hard interaction is said to be **ionic (electrostatic) as their electronic structures are rigid and cannot be perturbed externally.** Conversely **the HOMO-LUMO of soft acids and bases have electronic structures which are easily perturbed to give a covalent bond and so soft-soft interactions are called covalent.**

---

**Problem 4.9** Predict whether reactants or products are favoured in the following equilibria.

(a) $ZnI_2 + HgCl_2 \rightleftharpoons ZnCl_2 + HgI_2$

(b) $CuI_2 + Cu_2O \rightleftharpoons 2CuI + CuO$

(c) $2CH_3MgF + HgF_2 \rightleftharpoons (CH_3)_2Hg + 2MgF_2$

(d) $Nb_2S_5 + 5HgO \rightleftharpoons Nb_2O_5 + 5HgS$

(e) $(CH_3)_2OBF_3 + (CH_3)_2SBH_3 \rightleftharpoons (CH_3)_2SBF_3 + (CH_3)_2OBH_3$

(f) $La_2(CO_3)_3 + Tl_2S_3 \rightleftharpoons La_2S_3 + Tl_2(CO_3)_3$

**Ans. :** (a) In $HgCl_2$ there is soft-hard interaction, $Zn^{2+}$ is a borderline acid and $I^-$ is a soft acid thereby the equilibrium is shifted to the right to form $ZnCl_2$ (borderline acid + hard base) and $HgI_2$ (soft acid + soft base).

(b) $CuI_2$ (borderline acid + soft base), $Cu_2O$ (soft acid + hard base), CuI (soft acid + soft base), CuO (hard acid + hard base). So equilibrium is shifted to right.

(c) In $CH_3MgF$ the $Mg^{2+}$ is hard acid, $F^-$ is hard base and $CH_3^-$ with a carbon donor is a soft base. So $CH_3MgF$ is a combination of soft base-hard acid-hard base. $HgF_2$ (soft acid - hard base) while the products $(CH_3)_2Hg$ (soft base - soft acid) and $MgF_2$ (hard acid - hard base). So equilibrium will be shifted to right to give the products.

(d) The reactant cations $Nb^{5+}$ and $Hg^{2+}$ are hard acid and soft acid respectively while the anions $S^{2-}$ and $O^{2-}$ are soft and hard bases respectively. So the reaction is favoured to right.

(e) $(CH_3)_2O$ and $(CH_3)_2S$ are hard and soft bases respectively while $BF_3$ and $BH_3$ are hard and soft acids respectively. Hence equilibrium is shifted to left and reactants are preferred.

(f) $La^{3+}$ is a hard acid whereas $Tl^{3+}$ is soft acid. $CO_3^{2-}$ and $S^{2-}$ are hard and soft bases respectively. Therefore reactants are favoured and equilibrium is shifted to left.

---

### 4.5.5 The Geochemical classification of the terrestrial distribution of elements

This is the extension of the application of SHAB principle (Sec. 4.5.2) regarding the natural occurance of elements. Pearson rule of soft acids prefer soft bases while hard acids prefer hard bases, which explain certain aspects of the geochemical existance of the elements known as

**Goldschmidt classification.** Presently geochemists classify the elements into four types according to their geochemical behaviour :

(a) **Lithophiles :** The elements which occur primarily on the surface of the earth (the lithosphere) in oxides, silicates or carbonate minerals, which include lithium, magnesium, titanium, aluminium and chromium as cations. **These cations are all hard acids and the anions** all possess oxygen as a **donor atom which is essentially hard base.** In lithophiles the nonmetals that are classified are inherently hard bases or have been oxidised by atmospheric oxygen to oxo anions, which are essentially hard bases.

(b) **Chalcophiles :** The elements are commonly found in association with **sulphide** (less commonly with other soft bases such as selenide, telluride, arsenide etc. since sulphide is the most common soft base among others) minerals which include the cations of cadmium, lead, antimony and bismuth. The cations are either soft or borderline acids are found in associaton with soft base $S^{2-}$. The chalcophiles are mostly the borderline or soft acids.

(c) **Atmophiles :** These include the elements in the atmosphere which are unreactive and exist as elemental form e.g. nitrogen and the noble gases.

(d) **Siderophiles :** This is a subdivision of chalcophiles where the metals occur in nature as native i.e. in the elemental form. These metals are of soft acid category.

It is to be noted that last two classifications (c and d) are less important from the standpoint of SHAB concept.

---

**Problem 4.10** Geochemical classifications distinguish metals as lithophile if they tend to occur as oxides and chalcophile if they tend to occur as sulphide. What properties of the metals determine this distinction? Give examples. **[B.U. 2009, 2013]**

**Ans. :** Sec. 4.5.5 (a and b).

---

**Problem 4.11** Some new elements have been discovered by students which have the following symbols and properties :

| Symbol | Ax | Bx | Cx | Dx | Ex | Fx |
|---|---|---|---|---|---|---|
| Electronegativity | 2.22 | 1.33 | 2.08 | 2.60 | 3.33 | 1.90 |
| Ion formed | $Ax^{2-}$ | $Bx^{3+}$ | $Cx^{+}$ | $Dx^{2-}$ | $Ex^{2-}$ | $Fx^{2+}$ |
| Ionic radius (pm) | 188 | 63 | 125 | 155 | 117 | 102 |

Based on the informations given above (a) classify each of the ions as hard/soft acids and bases (b) arrange them according to soft acids and soft bases in order of increasing softness.

**Ans. :** (a) The anions ($Ax^{2-}$, $Dx^{2-}$ and $Ex^{2-}$) are Lewis bases and the electronegativity of Ex is highest which is smallest in size, so $Ex^{2-}$ is a hard base. Both $Ax^{2-}$ and $Dx^{2-}$ are softer than $Ex^{2-}$. Since electronegativity of Dx is higher than that of Ax while the size of $Dx^{2-}$ is small than $Ax^{2-}$, so $Dx^{2-}$ will be less soft than $Ax^{2-}$.

The cations ($Bx^{3+}$, $Cx^{+}$ and $Fx^{2+}$) are Lewis acids. $Bx^{3+}$ having high charge, small size and low electronegativity suggests that $Bx^{3+}$ is a hard acid. $Cx^{+}$ and $Fx^{2+}$ are considered to be soft compared to that of $Bx^{3+}$. $Cx^{+}$ has a low charge, larger radius and higher electronegativity than $Fx^{2+}$, so $Cx^{+}$ will be more softer than $Fx^{2+}$.

(b) Soft acids : $Fx^{2+} < Cx^{+}$, Soft bases : $Dx^{2-} < Ax^{2-}$.

## 4.6 USANOVICH CONCEPT—THE ULTIMATE ACID-BASE CONCEPT

Russian chemist M. Usanovich (1934) extended the Lewis concept in a more generalised way removing the electron pair donation/acceptance concept. According to this concept **an acid is any chemical species which reacts with a base, gives up cations or accepts anions or electrons while a base is a substance which reacts with acids, gives up anions or electrons and accepts cations.**

*Merits of the concept*

The scope of this concept is thus very wide covering all the previous definitions. It includes all acids and bases defined by Lewis modifying the idea of electron pair acceptance for acids and donation for bases. Thus this concept includes all the oxidising agents as acids and reducing agents as bases. Practically this concept includes all the reactions as acid-base category e.g.

**acid** **base**

$SO_2 + Na_2O \longrightarrow Na_2SO_3$, here $SO_2$ accepts the anion $O^{2-}$ which is given by $Na_2O$.

$Fe^{2+} + 6CN^- \longrightarrow Fe(CN)_6^{4-}$, here $Fe^{2+}$ accepts the anion $CN^-$ and $CN^-$ accepts the cation $Fe^{2+}$

$Cl_2 + 2Na \longrightarrow 2NaCl$, here Cl accepts an electron while Na gives up electron. This reaction is also an oxidation-reduction type

$2Fe^{3+} + Sn^{2+} \longrightarrow 2Fe^{2+} + Sn^{4+}$, here $Fe^{3+}$ accepts electron, so it is an acid and $Sn^{2+}$gives up electrons so it is a base.

Lewis and Usanovich concepts is more or less identical as can be demonstrated by the formation of pyridine N-oxide from pyridine ($C_5H_5N: + \ddot{O}: \longrightarrow C_5H_5N^+–O^-$). The pyridine – N– oxide is considered as Lewis adduct ($C_5H_5N$ is Lewis base and oxygen is Lewis acid) or as Usanovich acid (oxygen) and base (pyridine). The reaction also considered as oxidation-reduction reaction.

The versatility of Usanovich concept can be considered (a) acid-base character of some compounds in the periodic table and (b) coordination unsaturation. Regarding the first one we can generalise (i) acidic character increases from left to right in the periodic table e.g. $Na_2O$ strongly basic while $Cl_2O_7$ is strongly acidic (ii) acidic character decreases from top to bottom e.g. $N_2O_5$ more acidic than $P_2O_5$ (iii) the higher oxidation state of an element indicates greater acidic character e.g. $SnCl_4$ more acidic than $SnCl_2$.

Coordination unsaturation means to raise the covalency. $SO_2$ is coordinatively unsaturated and so it will accept an anion ($O^{2-}$) to form $SO_3^{2-}$ and acting as an acid.

In Usanovich concept a base is defined as a negative ion donor or a positive ion acceptor whereas an acid is positive ion donor or negative ion acceptor. Therefore we can conclude that a species with **highest positive charge density will be the strongest acid**. The strongest acid is therefore a **bare proton** because it is the smallest ion which carry the positive charge. Though bare proton cannot survive and tend to be solvated in solution, in the gaseous state it exists and have a strong tendency to be attached with any compound containing **electrons and so proton is considered as ultimate acid**. On the other hand a species having **highest negative charge**

**density** is considered to be as **strongest base.** In this game **the electron** which is the conjugate of proton, is the **ultimate base** because electron has very minute size. It exists in solution as solvated electron and the solution is considered as highly basic.

The only drawback of this concept is its extreme generality that all chemical reactions are under the perview of acid-base reactions.

## 4.7 LUX-FLOOD CONCEPT

This concept was proposed by Lux (1939) and extended by Flood (1947) which considers the acid-base reactions in terms of oxide ion ($O^{2-}$) transfer and is used mainly for molten oxides at high temperature. It states that **an acid is an oxide ion acceptor** and **a base is oxide ion donor** i.e. Base $\rightleftharpoons$ Acid + $O^{2-}$. For example,

| **Acid** | | **Base** | | **Salt** |
|---|---|---|---|---|
| $SiO_2$ | + | $CaO$ | $\longrightarrow$ | $CaSiO_3$ |
| $SO_3$ | + | $PbO$ | $\longrightarrow$ | $PbSO_4$ |
| $SiO_2$ | + | $CaSO_4$ | $\longrightarrow$ | $CaSiO_3 + SO_3$ |
| $6SiO_2$ | + | $2Ca_3(PO_4)_2$ | $\longrightarrow$ | $6CaSiO_3 + P_4O_{10}$ |
| $Na_2S_2O_7$ | + | $TiO_2$ | $\longrightarrow$ | $(TiO)SO_4 + Na_2SO_4$ |

Some oxides show amphoteric behaviour i.e. they have tendency to accept or to donate oxide ion e.g.

$Na_2O + ZnO \longrightarrow Na_2ZnO_2$, here ZnO accepts $O^{2-}$, so it is an acid.

$ZnO \longrightarrow Zn^{2+} + O^{2-}$ or $ZnO + S_2O_7^{2-} \longrightarrow Zn^{2+} + 2SO_4^{2-}$, here ZnO donates $O^{2-}$ ion, so it is a base.

Similarly $Al_2O_3 + O^{2-} \longrightarrow 2AlO_2^-$, here $Al_2O_3$ is an acid and

$Al_2O_3 \longrightarrow 2Al^{3+} + 3O^{2-}$, here $Al_2O_3$ is a base.

This concept can be extended to any other negative ion acceptor/donor processes in a generalised case to cover acid-base reactions

| e.g. | Acid | | Base | | Salt |
|---|---|---|---|---|---|
| | $CS_2$ | + | $Na_2S$ | $\longrightarrow$ | $Na_2CS_3$ |
| | $AlF_3$ | + | $3NaF$ | $\longrightarrow$ | $Na_3[AlF_6]$ |

**Problem 4.12** State the theory by which the reaction **[C.U. 2007, 2010, 2014]**
$6CaO + P_4O_{10} \longrightarrow 2Ca_3(PO_4)_2$ may be regarded as acid-base reaction.

**Ans. :** According to Lux-Flood concept, CaO is oxide ion donor so it is a base while $P_4O_{10}$ is oxide ion acceptor, so it is an acid.

**Problem 4.13** Common ore of nickel is sulphide but that of aluminium is oxide—explain. **[B.U. 1995]**

**Ans. :** According to SHAB principle $Ni^{2+}$ is borderline acid (tendency to be soft acid) and $S^{2-}$ is a soft base, so soft-soft interaction pointed the existance of common ore of nickel as sulphide. $Al^{3+}$ is a hard acid and $O^{2-}$ is a hard base, so common ore of Al will be oxide.

## 4.8 THERMODYNAMICAL ASPECT OF ADDUCT FORMATION IN ACIDS AND BASES

There are several heat terms associated with different processes e.g. **gas phase proton affinity, enthalpy change in adduct formation, donor number, acceptor number** etc. Gas phase proton afinity has been discussed in Sec. 3.1.4(b). We shall consider other thermodynamical parameters.

### 4.8.1 Thermodynamic acidity parameters of the adducts : Drago-Wayland Equation : Ionic and covalent contributions of the adducts

Attempts have been made to evaluate the stability of the acid-base adducts in terms of covalent and electrostatic interactions. The gas phase adduct formation is represented by—

$$A(g) + :B(g) \longrightarrow A\text{–}B(g) \qquad ...(4.21)$$

The standard enthalpy ($\Delta H°$) of the reaction is expressed by **Drago-Wayland equation** as :

$$-\Delta H° = E_A E_B + C_A C_B \qquad ...(4.22)$$

where E terms represent the electrostatic interactions while C terms represent the covalent contribution, A and B subscripts the acids and bases respectively. Values of E and C parameters of different acids and bases are given in Table 4.4. The standard enthalpy of adduct formation can be claculated by the E and C values. Thus the standard enthalpy of adduct formation ($\Delta H°$) between $NH_3$ and $B(CH_3)_3$ can be calculated by taking the values of E and C from Table 4.4 as :

$$-\Delta H° = E_A E_B + C_A C_B = 12.6 \times 2.78 + 3.48 \times 7.08 = 59.7 \text{ kJmol}^{-1}$$

Experimental value of $-\Delta H° = 57.5 \text{ kJmol}^{-1}$

**Table 4.4 : Drago-Wayland parameters for some acids and bases (kJmol$^{-1}$)**

| | Acids | | | | | | | Bases | | | | |
|---|---|---|---|---|---|---|---|---|---|---|---|---|
| | $B(CH_3)_3$ | $SO_2$ | $C_6H_5OH$ | $I_2$ | $BF_3$ | $SbCl_5$ | ICl | $NH_3$ | $C_5H_5N$ | $CH_3NH_2$ | $C_6H_6$ | $CH_3COCH_3$ |
| E | 12.6 | 1.88 | 8.66 | 2.05 | 20.2 | 15.1 | 10.4 | 2.78 | 2.39 | 2.66 | 0.57 | 2.02 |
| C | 3.48 | 1.65 | 0.904 | 2.05 | 3.31 | 10.5 | 1.70 | 7.08 | 13.10 | 12.0 | 2.9 | 4.67 |

Although the agreement between calculated and experimental values for enthalpy of reaction are close, the equation (4.22) is valid for molecular species. However when one of the species is ionic or both are ionic, the calculated and experimental values of $\Delta H°$ will differ and in that case equation (4.22) modified by Drago et. al. They suggested that for ionic species some charge is transferred from one species to other. Accordingly the modified form of equation (4.22) is :

$$-\Delta H° = E_A E_B + C_A C_B + T_A T_B \qquad ...(4.23)$$

where $T_A$ and $T_B$ represents the energy of the charge transfer of acids and bases respectively. One application of equation (4.23) is to calculate $\Delta H°$ for the reaction

$$H_2O + H^+ \longrightarrow H_3O^+$$

$-\Delta H° = 45 \times 2.3 + 13 \times 0.1 + 130 \times 0.43 = 160.7 \text{ kcal mol}^{-1}$ (The values of the parameters are obtained from standard books). The experimental value of $-\Delta H° = 166 \text{ kcal mol}^{-1}$.

Drago-Wayland equation (4.22) is useful by which the $\Delta H^\circ$ values for over 1500 acid-base pairs can be successfully compiled with reasonable accuracy in gas phase as well as in nonpolar, nondonor solvents. The equation is not applicable for highly exothermic reactions where the acids and bases undergo extensive interaction.

### 4.8.2 Solvents as acids and bases : Donor number (DN) and Acceptor number (AN)

Most common solvents are either Lewis acids and bases. **Solvents acidity and basicity** are of considerable interest for they help to account for the differences between reactions in aqueous and nonaqueous medium. Nonaqueous basic solvents are dimethyl sulfoxide (DMSO)($Me_2SO$), dimethyl formamide (DMF) ($HCONMe_2$) and acetonitrile (MeCN). DMSO may act as ambidentate solvent, coordination either through O or S. A good nonaqueous acidic solvent is HF. Acidity or basicity of solvents is influenced by hydrogen bonding. For example HCl is a stronger acid in $CH_3OH$ than in DMF. In $CH_3OH$ the $Cl^-$ is stabilised through hydrogen bonded species like $Cl^-...H{-}OCH_3$. The stability of HF to form a strong hydrogen bond to $F^-$ means that the conjugate base is better considered as $HF_2^-$ ion.

Solvation of a solute can be regarded as a Lewis acid base interaction followed by either displacements or double displacement reactions. Thus when $SbF_5$ dissolves in $BrF_3$, the following displacement reaction occurs :

$$SbF_5 + BrF_3 \longrightarrow [BrF_2]^+ + [SbF_6]^-$$

Here the strong Lewis acid $SbF_5$ displaces the weaker Lewis acid $BrF_2^+$ from its complex with the Lewis base $F^-$. Thus the solvation power of a species is measured as a function of the strength of Lewis basicity of the solvent which can be quantitatively expressed in terms of the reaction enthalpy for the formation of an adduct with a reference Lewis acid. Victor Gutmann taken the strong Lewis acid $SbCl_5$ as reference one in non-coordinating solvent like 1, 2 dichloroethane for the reaction :

$$SbCl_5 + \underset{solvent}{:B} \longrightarrow Cl_5Sb \leftarrow B \qquad ...(4.24)$$

and calculated the change in enthalpy. The change in enthalpy in this process is ascribed as **donor number (DN)** by Gutmann which is a measure of the basicity or donor ability of a solvent. **The DN is defined as the negative enthalpy of the reaction represented by equation (4.24).** A higher donor number indicates a higher basicity of a solvent.

*Solvent basicity*

The corresponding parameter to measure solvent acidity is the **acceptor number (AN)** has been introduced by Gutmann. This is defined as the characteristic of a solvent to form an adduct with a reference base i.e. triethyl phosphine oxide $(C_2H_5)_3PO$. When the base is dissolved in pure solvent and NMR chemical shift of $^{31}P$ is measured which gives an indication of donor-acceptor interaction. Numerical values of AN is arbitrarily taken as zero for the base in hexane and 100 for the base in $SbCl_5$. The DN and AN of some reprsentative solvents are given in Table 4.5.

*Solvent acidity*

**Table 4.5 : DN and AN of some representative solvents**

| Solvents | DN | AN |
|---|---|---|
| Water | 18.0 | 55.0 |
| Pyridine | 33.1 | 14.2 |
| DMSO | 29.8 | 19.3 |
| Acetone | 17.0 | 12.5 |
| Benzene | 0.1 | 8.2 |
| Ethanol | 19.0 | 37.1 |

The higher the acceptor number, the stronger is Lewis acidity.

### 4.8.3 Change of bond length in acid-base adduct formation : Gutmann's Rule

There will be a considerable change of bond lengths during Lewis acid base adduct formation. Thus the B–F distance in $BF_3$ is 130 pm while the same in $H_3NBF_3$ and $Me_3NBF_3$ is 138 and 139 pm respectively. The B–F distance increases in $H_3NBF_3$ and also further increases in $Me_3NBF_3$ due to stronger adduct formation. As stronger adduct is formed from $H_3NBF_3$ to $Me_3NBF_3$, the B–N distance decreases from 160 pm to 158 pm for $H_3NBF_3$ to $Me_3NBF_3$ respectively. This phenomena can be stated by a general emperical rule known as Gutmann's rule : **the stronger the adduct formed between Lewis acid and a Lewis base the shorter will be the bond length and greater will be the lengthening of the adjacent bonds in both the donor and acceptor molecules.** Strong adduct means displacement of electron density from the donor atom to the acceptor atom so that donor atom will have less partial negative than before. On the other hand the acceptor atom will assume less positive than before, thus leading to decrease in polarity of bonds and ultimately lengthening of the adjacent bonds e.g.

| Species : | $NH_3$ | $NH_4^+$ | $SiF_4$ | $SiF_6^{2-}$ | $SO_2$ | $SO_3^{2-}$ | $SbCl_5$ | $SbCl_6^-$ | $I_2$ | $I_3^-$ |
|---|---|---|---|---|---|---|---|---|---|---|
| Bond length (pm) | 103 | 105 | 154 | 171 | 143 | 150 | 231 | 247 | 266 | 283 |

---

**Problem 4.14** Complete the reaction : $BF_3 + B_2H_6 + (CH_3)_2O + (CH_3)_2S \longrightarrow ?$

**[B.U. 2002, 2010, 2015]**

**Ans. :** $BF_3$ is a hard acid and $(CH_3)_2O$ is a hard base while $(CH_3)_2S$ is a soft base and $B_2H_6$ is soft acid. So according to Pearson rule the product will be :

$$BF_3 + B_2H_6 + (CH_3)_2O + (CH_3)_2S \longrightarrow (CH_3)_2O \longrightarrow BF_3 + (CH_3)_2S \longrightarrow 2BH_3$$

---

**Problem 4.15** Predict the direction of the following equilibria :

(i) $2CH_3MgF + HgF_2 \rightleftharpoons (CH_3)_2Hg + 2MgF_2$

(ii) $La_2(CO_3)_3 + Bi_2S_3 \rightleftharpoons La_2S_3 + Bi_2(CO_3)_3$ **[C.U. 2003]**

**Ans. :** For (i) See Problem 4.9(c).

For (ii) $La^{3+}$ and $Bi^{3+}$ are hard acid and borderline soft acid respectively. $S^{2-}$ and $CO_3^{2-}$ are soft and hard base respectively and so equilibrium will be shifted to the left.

## 4.9 IONIC PRODUCT OF WATER ($K_W$) AND DISSOCIATION CONSTANT (K) OF WATER

Pure water is not a non-conductor but a bad conductor of electricity. The specific conductivity of a sample of water purified with utmost care was found to be $0.04 \times 10^{-6}$ ohm$^{-1}$ at 18°C. This indicates that even the purest quality of water feebly dissociated to form $H^+$ and $OH^-$ ions. According to Bronsted-Lowry concept (Sec. 4.3.1) water acting as an acid and a base and dissociates as :

$$2H_2O \longrightarrow H_3O^+ + OH^- \quad \text{or} \quad H_2O \longrightarrow H^+ + OH^-$$

so that the equilibrium constant or dissociation constant (K) is represented as :

$$K = \frac{[H_3O^+][OH^-]}{[H_2O]^2} \qquad ...(4.25)$$

Since the concentration of $H_3O^+$ and $OH^-$ is small in comparison to the concentration of undissociated water, so we may write,

$$K[H_2O]^2 = K_w = [H_3O^+][OH^-] \qquad ...(4.26)$$

where $K_w$ is called ionic product of water. Concentration is expressed in g-ion lit$^{-1}$ or in SI unit expressed as mol dm$^{-3}$. The concentration of $H^+$ or $H_3O^+$ ion has been determined for pure water as $10^{-7}$ M lit$^{-1}$ so that,

$$K_w = [H_3O^+][OH^-] = 10^{-7} \times 10^{-7}$$

$$= 1 \times 10^{-14}\ M^2 \text{ at } 25°C \qquad ...(4.27)$$

The values of $K_w$ varies with temperature e.g. at 0°C, 50°C and 100°C the values are $0.114 \times 10^{-14}$, $5.47 \times 10^{-14}$ and $49.0 \times 10^{-14}$ respectively. The equation (4.27) indicates that in aqueous solution at a particular temperature the concentration of $H^+$ or $H_3O^+$ and $OH^-$ are inversely proportional to each other. If the concentration of one species increases, the other one decreases proportionally so that $K_w$ remain constant.

According to equation (4.25), the dissociation constant (K) is represented as :

$$K = \frac{K_W}{[H_2O]^2} \qquad ...(4.28)$$

Assuming the density of water equal to unity so that $[H_2O] = \frac{1000}{18} = 55.56$ mol dm$^{-3}$ at 25°, so that

$$K = \frac{1 \times 10^{-14}}{(55.56)^2} = 3.26 \times 10^{-18} \qquad ...(4.29)$$

Therefore dissociation constant and ionic product of water are not the same thing.

## 4.10 THE CONCEPT OF pH AND pH SCALE

We have seen that in aqueous solution the concentration of $H^+/H_3O^+$ and $OH^-$ ions are small and are expressed by negative exponent of base 10. Such representation is troublesome so that Sorensen (1909) proposed a technique for expressing such exponential term in some simplified form which is denoted by the **term pH* of a solution** and expressed as the negative logarithm of $H^+/H_3O^+$ ion concentration i.e.

$$pH = -\log_{10}[H^+] \quad \text{or} \quad pH = -\log_{10}[H_3O^+] \quad \ldots(4.30)$$

where the concentration is expressed in mol $lit^{-1}$ or mol $dm^{-3}$ (in SI unit). The **operational definition** of pH replaces concentration term with activities so that pH appropriately defined as :

$$pH = -\log a_{H_3O^+} \quad \ldots(4.31)$$

In our common practice pH is calculated by equation (4.30). Similarly a term pOH is denoted in order to express the concentration of $OH^-$ ions and is expressed as :

$$pOH = -\log[OH^-] \quad \ldots(4.32)$$

According to equation (4.27)

$$[H^+][OH^-] = 10^{-14}$$

$$\text{so that } pH + pOH = 14 \quad \ldots(4.33)$$

$$\text{or} \quad pH + pOH = pK_w, \text{ where } pK_w = -\log K_w \quad \ldots(4.34)$$

It is evident from equation (4.30) that lower the pH value, more acidic the solution is. For example a solution having pH1 is more acidic than a solution having pH3 and the former solution has $H^+$ ion concentration 100 times greater than the latter solution. Therefore **as pH increases the solution has a tendency to transform from acidic to basic nature**. Consequently **as pOH increases the solution is converted from basic to acidic nature.**

For an **acidic solution** $[H^+] > [OH^-]$ or $[OH^-] < [H^+]$

i.e. $[H^+] > 10^{-7}$ mol $dm^{-3}$ or $[OH^-] < 10^{-7}$ mol $dm^{-3}$

On the other hand for **basic solution** $[OH^-] > [H^+]$ or $[OH^-] > 10^{-7}$ mol $dm^{-3}$

or $[H^+] < 10^{-7}$ mol $dm^{-3}$

In neutral solution $[H^+] = [OH^-] = 10^{-7}$ mol $dm^{-3}$.

In terms of pH we can say that, for neutral solution pH = 7.0; for acidic solution pH < 7 and for basic solution pH > 7.

According to equation (4.33) a scale of pH can be made (Table 4.6) by considering that as concentration of $H^+$ ion increases, the concentration of $OH^-$ ion decreases so that the product of the concentration of the two ions is equal to $10^{-14}$.

**Table 4.6 : The pH scale**

| | | | | | | | | | | | | | | | |
|---|---|---|---|---|---|---|---|---|---|---|---|---|---|---|---|
| $[H^+]$ | $10^0$ | $10^{-1}$ | $10^{-2}$ | $10^{-3}$ | $10^{-4}$ | $10^{-5}$ | $10^{-6}$ | $10^{-7}$ | $10^{-8}$ | $10^{-9}$ | $10^{-10}$ | $10^{-11}$ | $10^{-12}$ | $10^{-13}$ | $10^{-14}$ |
| $[OH^-]$ | $10^{-14}$ | $10^{-13}$ | $10^{-12}$ | $10^{-11}$ | $10^{-10}$ | $10^{-9}$ | $10^{-8}$ | $10^{-7}$ | $10^{-6}$ | $10^{-5}$ | $10^{-4}$ | $10^{-3}$ | $10^{-2}$ | $10^{-1}$ | $10^0$ |
| pH | 0 | 1 | 2 | 3 | 4 | 5 | 6 | 7 | 8 | 9 | 10 | 11 | 12 | 13 | 14 |
| pOH | 14 | 13 | 12 | 11 | 10 | 9 | 8 | 7 | 6 | 5 | 4 | 3 | 2 | 1 | 0 |

← acidic solution → neutral ← basic solution →

* pH abbreviated as puissance d' hydrogen which means the pressure of hydrogen.

A question arises whether pH of a solution may be negative. A negative pH means the solution is too much acidic e.g. if a solution has pH = –1 which means that $[H^+] = 10^1$ so that $H^+$ ion concentration has much larger value than the conventional $[H^+]$ in the pH scale.

---

**Problem 4.16** What will be the concentration of $H^+/H_3O^+$ of a solution having pH = 5.6.

**Ans. :** $pH = -\log[H^+]$ $\therefore \log[H^+] = -5.6$ or

$[H^+] = \text{antilog}(-5.6) = \text{antilog}(-6 + 6 - 5.6)$

$= \text{antilog}(\bar{6}.4) = 2.512 \times 10^{-6}$ mol dm$^{-3}$

---

**Problem 4.17** Calculate $H^+$ concentration and pH of (a) $10^{-4}$ M HCl, (b) $10^{-3}$ M $H_2SO_4$ and (c) $10^{-3}$ M NaOH solutions.

**Ans. :** (a) $[H^+] = 10^{-4}$ mol dm$^{-3}$

$\therefore pH = -\log 10^{-4} = 4.$

(b) $H_2SO_4$ produces 2 moles of $H^+$, so that

$[H^+] = 2 \times [H_2SO_4] = 2 \times 10^{-3}$ mol dm$^{-3}$

$\therefore pH = -\log(2 \times 10^{-3}) = 2.699.$

(c) $[OH^-] = 10^{-3}$ mol dm$^{-3}$

$\therefore pOH = -\log[OH^-] = -\log 10^{-3} = 3$

$\therefore pH = 14 - pOH = 14 - 3 = 11.0$

---

**Problem 4.18** Calculate the pH of (a) $10^{-8}$ M or $10^{-8}$ N HCl, (b) $10^{-7}$ M HCl solution, (c) $10^{-9}$ M NaOH and (d) $10^{-7}$ M NaOH.

**Ans. :** (a) $\frac{N}{10}$ HCl has pH = 1 or $\frac{N}{100}$ HCl, pH = 2 but the $10^{-8}$ M or $10^{-8}$ N HCl has not the pH = 8 because pH = 8 means a basic solution. In this case we have to take the concentraiton of $H^+$ ion of water, so that the total concentration of $H^+$ is—

$[H^+] = (10^{-8} + 10^{-7}) = 10^{-7}(10^{-1} + 1) = 1.1 \times 10^{-7}$ mol dm$^{-3}$

$\therefore pH = -\log(1.1 \times 10^{-7}) = 7 - \log 1.1 = 7 - 0.0414 = 6.9586.$

(b) Here $[H^+] = 10^{-7} + 10^{-7} = 2 \times 10^{-7}$ mol dm$^{-3}$

$\therefore pH = -\log(2 \times 10^{-7}) = 7 - \log 2.0 = 7 - 0.3010 = 6.6990.$

(c) Here $[OH^-[ = 10^{-9} + 10^{-7} = 10^{-7}(10^{-2} + 1) = 10^{-7} \times 1.01$ mol dm$^{-3}$

$\therefore pOH = -\log[OH^-] = -\log(1.01 \times 10^{-7})$

$= 7 - \log 1.01 = 7 - 0.0043 = 6.9957$

$\therefore pH = 14 - pOH = 14 - 6.9957 = 7.0043.$

(d) Here $[OH^-] = 10^{-7} + 10^{-7} = 2 \times 10^{-7}$ mol dm$^{-3}$

$\therefore pOH = -\log(2 \times 10^{-7}) = 7 - \log 2 = 7 - 0.3010 = 6.6990$

$\therefore pH = 14 - 6.6990 = 7.3010.$

---

### 4.10.1 Equilibirum in aqueous solutions of acids and bases

In Sec. 4.3.1 we have seen that all strong acids and bases are 100% dissociated in aqueous solution. Therefore for those solutions the calculation of pH is done by known concentration of acids and bases like that of Problem 4.17. In some cases due consideration of the cencentration of $H^+$ and $OH^-$ ions of water have been made in order to calculate the total concentration of $H^+$ and $OH^-$ ions with reference to the Problem 4.18.

For weak acids and bases which are not completely dissociated and there will be a significant amount of undissociated species. Let an weak and (HA) is disociated in the following way having the analytical concentration of HA is $C_a$ mol $dm^{-3}$ and also let the degree of dissociation is $\alpha$.

$$\underset{(1-\alpha)C_a}{HA} \rightleftharpoons \underset{\alpha C_a}{H^+} + \underset{\alpha C_a}{A^-}$$

Therefore the dissociation constant ($K_a$) of the acid can be represented as :

$$K_a = \frac{[H^+][A^-]}{[HA]} = \frac{\alpha C_a \times \alpha C_a}{(1-\alpha)C_a} = \frac{\alpha^2 C_a}{1-\alpha} \qquad \text{...(4.35)}$$

As the acid is weak so that $\alpha << 1$, then $1 - \alpha \approx 1$, we get

$$K_a = \alpha^2 C_a \quad \text{i.e.} \quad \alpha = (K_a/C_a)^{1/2} \qquad \text{...(4.36)}$$

Here $$[H^+] = \alpha C_a \qquad \text{...(4.37)}$$

so that by knowing the value of $\alpha$ and $C_a$ we are able to calculate the pH of an aqueous solution of weak acid. In a similar way we are able to calculate the pH and pOH of a weak base.

---

**Problem 4.19** Establish the equation for the determination of pH of a solution of $CH_3COOH$ and hence find the pH of 0.1(M) $CH_3COOH$ solution. (Given $K_a$ of $CH_3COOH$ = $1.8 \times 10^{-5}$ mol/lit. **[C.U. 2005]**

**Ans.** : Establish the equation $\alpha = \left(\frac{K_a}{C_a}\right)^{\frac{1}{2}}$ according to Sec. 4.10.1

$$= \left(\frac{1.8 \times 10^{-5}}{0.1}\right)^{\frac{1}{2}} = \left(1.8 \times 10^{-4}\right)^{\frac{1}{2}} = 1.34 \times 10^{-2}$$

According to equation (4.37)

$$[H^+] = \alpha C_a = 1.34 \times 10^{-2} \times 0.1 = 1.34 \times 10^{-3} \text{ mol/lit}$$

$$\therefore \text{pH} = -\log(1.34 \times 10^{-3}) = 3 - 0.1271 = 2.8729.$$

---

### 4.10.2 Hydrolysis of Salt

**Hydrolysis* is a process by which a salt reacts with water to form the acid and the base from which the salt is produced.** Therefore it is considered as the reverse process of neutralisation reaction. Depending on the nature of acids and bases, the solution may be either acidic, basic or neutral. Hydrolysis may be of the following four categories :

(a) **Hydrolysis of salts of strong acids and bases :** In this case the solution is neutral because both the acids and bases are strong e.g. NaCl, $CaCl_2$ etc. In this case the cations

* The general term for hydrolysis is solvolysis.

and anions produced from the salts will have equal tendency to donate or to accept proton towards water furnishing equal concentration of $H^+/H_3O^+$ and $OH^-$ ions so that the solution will be neutral (pH = 7). This type of salts has less importance in the hydrolysis of salts.

(b) **Hydrolysis of salts of weak acids and strong bases :** In this case as the acid is weak while the base is strong, naturally the aqueous solution of the salt will be basic e.g. NaF, $CH_3COONa$ etc. i.e.

$$\underset{}{NaF} + H_2O \longrightarrow \underset{strong}{NaOH} + \underset{weak}{HF}$$

$$CH_3COONa + H_2O \longrightarrow \underset{weak}{CH_3COOH} + \underset{strong}{NaOH}$$

Therefore we can say that aqueous solution of NaCl is neutral whereas aqueous solution of NaF is alkaline.

Let us consider that a salt MA is of this category which have the concentration c mol $dm^{-3}$. The hydrolytic equation can be represented as :

$$MA + H_2O \longrightarrow \underset{strong}{MOH} + \underset{weak}{HA}$$

$$\text{or } M^+ + A^- + H_2O \longrightarrow M^+ + OH^- + HA$$

$$\text{i.e. } A^- + H_2O \longrightarrow HA + OH^- \qquad \ldots(4.38)$$

Applying law of mass action, we get,

$$K = \frac{[HA][OH^-]}{[A^-][H_2O]} \quad \text{or} \quad K[H_2O] = \frac{[HA][OH^-]}{[A^-]}$$

$$\text{i.e. } K_h = \frac{[HA][OH^-]}{[A^-]} \qquad \ldots(4.39)$$

where $K_h$ = hydrolysis constant = $K[H_2O]$, because the concentration of $H_2O$ remain practically constant before and after hydrolysis.

Multiplying the numerator and denominator of equation (4.39) by $[H^+]$ wet get

$$K_h = \frac{K_w}{K_a} \qquad \ldots(4.40)$$

where $K_a$ = dissociation constant of the weak acid. Let the degree of hydrolysis is x, so that from equation (4.38) we get,

$$\underset{(1-x)c}{A^-} + H_2O \longrightarrow \underset{xc}{HA} + \underset{xc}{OH^-}$$

so that $[OH^-] = xc$ ...(4.41)

From equation (4.39) we get,

$$K_h = \frac{[HA][OH^-]}{[A^-]} = \frac{xc \times xc}{(1-x)c} = \frac{x^2c}{1-x}$$

As $x$ is small so that $1 - x \approx 1$, we get,

$$K_h = x^2c \text{ i.e. } x = \sqrt{\frac{K_h}{c}} \quad ...(4.42)$$

As $[OH^-] = xc$ so that $[H^+] = \frac{K_w}{[OH^-]} = \frac{K_w}{xc} = \frac{K_w}{c\sqrt{\frac{K_h}{c}}} = \frac{K_w}{\sqrt{K_h c}} = \frac{K_w}{\sqrt{\frac{K_w}{K_a}c}} = \sqrt{\frac{K_w K_a}{c}}$

(From equation 4.40)

Therefore $\log[H^+] = \frac{1}{2}\log K_w + \frac{1}{2}\log K_a - \frac{1}{2}\log c$

i.e. $-\log[H^+] = -\frac{1}{2}\log K_w - \frac{1}{2}\log K_a + \frac{1}{2}\log c$

or $$pH = \frac{1}{2}pK_w + \frac{1}{2}pK_a + \frac{1}{2}\log c \quad ...(4.43)$$

or $$pH = 7 + \frac{1}{2}pK_a + \frac{1}{2}\log c \quad ...(4.44)$$

(c) **Hydrolysis of salts of strong acids and weak bases :** In this case the aqueous solution will be acidic e.g. $FeCl_3$, $NH_4Cl$ etc. Let us consider the general expression for the hydrolysis of this type of salt (MA)

$$MA + H_2O \longrightarrow \underset{weak}{MOH} + \underset{strong}{HA}$$

or $M^+ + A^- + H_2O \longrightarrow MOH + H^+ + A^-$

i.e. $$M^+ + H_2O \longrightarrow MOH + H^+ \quad ...(4.45)$$

Proceeding in a similar way like that for weak acids and strong bases, we get,

$$K_h = \frac{K_w}{K_b}$$

where $K_b$ is the dissociation constant of the base MOH

and $$pH = \frac{1}{2}pK_w - \frac{1}{2}pK_b - \frac{1}{2}\log c \quad ...(4.46)$$

or $$pH = 7 - \frac{1}{2}pK_b - \frac{1}{2}\log c \quad ...(4.47)$$

(d) **Hydrolysis of salts of weak acids and weak bases :** In this case the aqueous solution should be neutral but from the practical sense all weak acids and bases are not equally weak so that the solution may be either slightly acidic if the weak acid slightly stronger than the weak base or may be slightly alkaline if the base is little bit stronger than that of the acid. In some cases the solution may also be neutral e.g. ammonium acetate solution is neutral but ammonium formate solution is slightly acidic while ammonium carbonate solution is slightly alkaline.

In this case we can represent the hydrolysis of the salt (MA) as :

$$MA + H_2O \longrightarrow \underset{weak}{MOH} + \underset{weak}{HA}$$

Here both the processes represented by equations (4.38) and (4.45) occur simultaneously.

In this case we may deduce the following expressions :

$$K_h = \frac{K_w}{K_a K_b} \qquad ...(4.48)$$

$$\text{and } pH = \frac{1}{2} pK_w + \frac{1}{2} pK_a - \frac{1}{2} pK_b \qquad ...(4.49)$$

$$\text{or } pH = 7 + \frac{1}{2} pK_a - \frac{1}{2} pK_b \qquad ...(4.50)$$

where the symbols have their usual significance.

---

**Problem 4.20** Calculate the pH and degree of hydrolysis of a 0.1 molar solution of $Na_2CO_3$ (For $H_2CO_3$, $K_1 = 4 \times 10^{-7}$, $K_2 = 6 \times 10^{-11}$).

**Ans. :** $Na_2CO_3 + 2H_2O \longrightarrow 2NaOH + H_2CO_3$

*strong base* *weak acid*

Therefore aqueous solution of $Na_2CO_3$ will be alkaline and the relevant formula of hydrolysis constant

$$K_h = \frac{K_w}{K_a}$$

Now what will be the value of $K_a$, because $H_2CO_3$ is a weak acid and have two dissociation constant values.

$$H_2CO_3 \longrightarrow H^+ + HCO_3^- \qquad K_1 = 4 \times 10^{-7}$$

$$HCO_3^- \longrightarrow H^+ + CO_3^{2-} \qquad K_2 = 6 \times 10^{-11}$$

According to Bronsted-Lowry concept both $HCO_3^-$ and $CO_3^{2-}$ are conjugate bases of $H_2CO_3$. But a question arises which one is more weak? Here $K_2 < K_1$. $K_2$ is the dissociation constant of $HCO_3^-$ as an acid. So $HCO_3^-$ is a much more weak acid than $H_2CO_3$ because $HCO_3^-$ is the conjugate base of the weak acid $H_2CO_3$. The conjugate base of $HCO_3^-$ is $CO_3^{2-}$ while $HCO_3^-$ is the conjugate base of $H_2CO_3$. So it is reasonable to assume that $CO_3^{2-}$ is a more stronger base than $HCO_3^-$.

Now the hydrolysis of $CO_3^{2-}$ takes place in two sptes :

$$CO_3^{2-} + H_2O \rightleftharpoons HCO_3^- + OH^- \quad \text{and} \quad HCO_3^- + H_2O \rightleftharpoons H_2CO_3 + OH^-$$

In both the cases $OH^-$ ions are produced, but the $OH^-$ ions from the first step of hydrolysis will practically repress the second step and thus the $OH^-$ ions produced by the second step may be ignored. The hydrolysis constant then corresponds to the first step and so

$$K_h = \frac{[HCO_3^-][OH^-]}{[CO_3^{2-}]}$$, the concentration of water remain practically constant before and after hydrolysis.

$$= \frac{[HCO_3^-]}{[CO_3^{2-}][H^+]}[OH^-][H^+] = \frac{K_w}{K_2} = \frac{1 \times 10^{-14}}{6 \times 10^{-11}} = 1.67 \times 10^{-4}$$

Now according to equation (4.42) the degree of hydrolysis

$$x = \sqrt{\frac{K_h}{c}} = \sqrt{\frac{1.67 \times 10^{-4}}{0.1}} = 0.0408 = 4.08 \text{ percent}$$

According to equation (4.41)

$$[OH^-] = xc \qquad x = \sqrt{\frac{K_h}{c}}$$

$$= \sqrt{\frac{K_h}{c}}c = \sqrt{K_h c} = \sqrt{1.67 \times 10^{-4} \times 0.1} = 4.08 \times 10^{-3}$$

$pOH = -\log(4.08 \times 10^{-3}) = 2.39$

$\therefore$ $pH = 14 - 2.39 = 11.61$

pH can also be calculated from equation (4.44) by taking $K_a = K_2$ of $H_2CO_3$.

**Problem 4.21** Calculate the pH of a 0.1 (M) $K_3PO_4$ solution. The third dissociation constant of $H_3PO_4$ is $1.3 \times 10^{-12}$. Assume that the hydrolysis proceeds only in the first step.

**Ans. :** Here hydrolysis occurs in the first step, so

$$K_3PO_4 + H_2O \rightleftharpoons K_2HPO_4 + KOH$$

$$\text{or} \quad \underset{0.1(1-x)}{PO_4^{3-}} + H_2O \rightleftharpoons \underset{0.1x}{HPO_4^{2-}} + \underset{0.1x}{OH^-}$$

Here $K_h = \frac{K_w}{K_a} = \frac{K_w}{K_3} = \frac{1 \times 10^{-14}}{1.3 \times 10^{-12}} = 7.7 \times 10^{-3}$

Now degree of hydrolysis $x = \sqrt{\frac{K_h}{c}} = \sqrt{\frac{7.7 \times 10^{-3}}{0.1}} = 2.8 \times 10^{-1}$

$[OH^-] = xc = 2.8 \times 10^{-1} \times 0.1 = 2.8 \times 10^{-2}$

$\therefore$ $pOH = -\log(2.8 \times 10^{-2}) = 1.5528$

$\therefore$ $pH = 14 - 1.5528 = 12.4472.$

### 4.10.3 Buffer Solutions

A buffer solution is one which have the reverse power of maintaining a constant pH by the addition of a small amount of acids and bases. Constitution of the buffer solution is that the action of small amounts of $H^+$ and $OH^-$ ions are cancelled by the formation of weak acids or bases within the solution.

A buffer solution usually consist of a mixture of (a) weak acid and its salt, (b) weak base and its salt, (c) sometimes a solution of a salt of weak acid and a base, (d) a solution of an ampholyte e.g. glycine (Table 4.7). Thus buffer solution may be considered as a solution of an acid and its conjugate base or a base and its conjugate acid.

**Table 4.7 : Some buffer solutions and their pH at room temperature**

| Buffers | pH range |
|---|---|
| $CH_3COOH + CH_3COONa$ | 3.7–5.6 |
| $Na_2HPO_4 + Na_3PO_4$ | 11–12 |
| $Na_2HPO_4 + NaH_2PO_4$ | 5.8–8.0 |
| Saturated solution of potassium hydrogen tartarate | 3.57 |
| $NH_4Cl + NH_4OH$ | 11.0 |
| $NH_2CH_2COOH + HCl$ | 1.0–3.7 |
| $NH_2CH_2COOH + NaOH$ | 8.2–10.1 |

### 4.10.4 Buffer action and Buffer capacity

Let us consider that a buffer solution is composed of a weak acid (e.g. $CH_3COOH$) and its salt MA (e.g. $CH_3COONa$). The salt MA is completely dissociated to form $M^+$ and $A^-$ whereby the HA is partially dissociated i.e.

$$MA \longrightarrow M^+ + A^-, \quad HA \rightleftharpoons H^+ + A^- \quad \text{and} \quad K_a = \frac{[H^+][A^-]}{[HA]}$$

Now in this solution when a small amount of **acid is added**, the $H^+$ ions combine with the $A^-$ ions in the buffer solution to form the undissociated acid HA and thereby the pH of the solution remain constant.

$$\underset{\text{externally added}}{H^+} + \underset{\text{solution}}{A^-} \rightleftharpoons HA$$

On the other hand when a small amount of alkali is added to a buffer solution, the added $OH^-$ ions react with $H^+$ ions present in the solution to form undissociated water molecules i.e.

$$\underset{\text{externally added}}{OH^-} + \underset{\text{solution}}{H^+} \rightleftharpoons H_2O$$

The $H^+$ ions supplied by the dissociation of the acid HA as :

$$HA \rightleftharpoons H^+ + A^-$$

The **buffer action** of a buffer comprising a weak base and its corresponding salt (e.g. $NH_4OH + NH_4Cl$) can also be similarly explained.

A term **buffer capacity** is defined for a buffer solution which denotes the power to resist the change in pH by the addition of acids and bases. **This is defined as the number of moles of a strong acid or a strong base that causes one litre of a buffer solution to undergo one unit pH change.** This is expressed in the differential form :

$$\text{Buffer capacity} = \frac{db}{d(pH)} = -\frac{da}{d(pH)} \qquad ...(4.51)$$

where db moles strong base or da moles strong acid causes a change in pH by an amount dpH. The negative sign in equation (4.51) arises due to the fact that addition of acid decreases the pH and so the negative sign arises.

A buffer solution having higher buffer capacity indicates higher capacity to resist the change in pH by the application of more amount of acids and bases.

### 4.10.5 The pH of a buffer solution : Henderson Equation

Let a buffer solution is composed of $C_A$ moles of weak acid HA and that of $C_S$ moles of its salt MA. The salt MA has a tendency to be dissociated completely while the weak acid HA is dissociated reversibly as :

$$HA \rightleftharpoons H^+ + A^- \quad \text{and} \quad K_a = \frac{[H^+][A^-]}{[HA]}$$

$$\therefore\ [H^+] = \frac{[HA]}{[A^-]} \cdot K_a \qquad \text{...(4.52)}$$

Now in a solution HA and MA having the said concentrations of the two, we can say that :

Concentration of undissociated acid = [HA] = initial concentration of the acid – concentration of $H^+$ = $C_A - [H^+]$

Concentration of $A^-$ in solution = $[A^-]$ = concentration of $A^-$ produced by the dissociation of MA (which is completely dissociated) + concentration of $A^-$ produced due to the dissociation of the acid HA = $C_S + [H^+]$

$$\text{So that, } [H^+] = \frac{C_A - [H^+]}{C_S + [H^+]} \cdot K_a \qquad \text{...(4.53)}$$

HA is a weak acid, in presence of MA this will be more weaker, so it is reasonable to assume that $[H^+]$ is small compared to $C_A$ and $C_S$ and therefore,

$$[H^+] = \frac{C_A}{C_S} K_a = K_a \frac{[\text{Acid}]}{[\text{Salt}]}$$

$$\text{and so,} \quad pH = pK_a + \log \frac{[\text{Salt}]}{[\text{Acid}]} \qquad \text{...(4.54)}$$

$$\text{or} \quad pH = pK_a + \log \frac{[\text{Conjugate base}]}{[\text{Conjugate acid}]} \qquad \text{...(4.55)}$$

Similarly for a weak base and its corresponding salts we can write,

$$pOH = pK_b + \log \frac{[\text{Salt}]}{[\text{Base}]} \qquad \text{...(4.56)}$$

$$\text{or} \quad pOH = pK_b + \log \frac{[\text{Conjugate acid}]}{[\text{Conjugate base}]} \qquad \text{...(4.57)}$$

The equations (4.54) to (4.57) are known as different forms of **Henderson equation** or Henderson-Hasselbalch equation.

**Problem 4.22** When 0.05 mole of NaOH was added to one litre of a buffer solution, its pH changed from 5.70 to 5.85. Find the buffer capacity.

**Ans.** : Buffer capacity $= \dfrac{\text{mole of alkali added per litre}}{\text{change in pH}}$

$$= \frac{0.05}{5.85 - 5.70} = 0.333.$$

**Problem 4.23** What do you mean by 'buffer capacity'? A buffer of pH 3.85 is prepared by mixing 0.1(M) $CH_3COOH$ and 0.1(M) NaOH. Calculate the ratio of volumes of required acid and alkali to prepare the buffer. **[C.U. 2004]**

**Ans.** : For buffer capacity see Sec. 4.10.4

Sum : Let $x$ ml 0.1 M $CH_3COOH$ and y ml 0.1 M NaOH is mixed together to have the pH of the buffer solution 3.85. Therefore this is an acid range buffer.

$x$ ml 0.1 M $CH_3COOH = x \times 0.1$ m mol of $CH_3COOH$ and

y ml 0.1 M NaOH $= y \times 0.1$ m mol of NaOH

| | $CH_3COOH$ | + NaOH ⟶ | $CH_3COONa$ | + $H_2O$ |
|---|---|---|---|---|
| original (m mol) | $x\times0.1$ | $y\times0.1$ | – | – |
| change (m mol) | $-y\times0.1$ | $-y\times0.1$ | $+y\times0.1$ | |
| Equilibrium (m mol) | $(x-y)\times0.1$ | – | $y\times0.1$ | |

Therefore, $[CH_3COOH] = \dfrac{(x-y)\times0.1}{x+y}$ and $[CH_3COONa] = \dfrac{y\times0.1}{x+y}$

According to Henderson equation for acid range buffer,

$$pH = pK_a + \log\frac{[\text{Salt}]}{[\text{Acid}]}$$

$K_a$ for $CH_3COOH = 1.8\times10^{-5}$ so that $pK_a = 5 - \log1.8 = 4.7447 \approx 5$

$$= 5 + \log\frac{\dfrac{y\times0.1}{x+y}}{\dfrac{(x-y)\times0.1}{x+y}} = 5 + \log\frac{y}{x-y}$$

$$\therefore\ 3.85 = 5 + \log\frac{y}{x-y} \quad \text{i.e.} \ \log\frac{x-y}{y} = 1.15$$

$$\therefore\ \frac{x-y}{y} = 14.13, \text{ so that } \frac{x}{y} = 15.13.$$

**Problem 4.24** Calculate the pH of the solution produced by adding 10 $cm^3$ of 1 M HCl to 1 $dm^3$ of a solution which is 0.1 M in $CH_3COOH$ and 0.1 M in $CH_3COONa$ ($K_a = 1.82 \times 10^{-5}$ mol $dm^{-3}$).

**Ans.** : The pH of $CH_3COOH + CH_3COONa$ buffer solution will be—

$$pH = pK_a + \log\frac{[\text{Salt}]}{[\text{Acid}]} = -\log(1.82 \times 10^{-5}) + \log\frac{[\text{Salt}]}{[\text{Acid}]}$$

$$= 4.74 + 0.0 = 4.74.$$

The $H^+$ ions from HCl react with $CH_3COO^-$ and forms undissociated $CH_3COOH$. If we neglect the change in volume from 1000 $cm^3$ to 1010 $cm^3$, we may consider :

$$[CH_3COO^-] = 0.1 - 0.01 = 0.09$$

$$\text{and} \quad [CH_3COOH] = 0.1 + 0.01 = 0.11$$

So that $pH = 4.74 + \log\frac{0.09}{0.11} = 4.74 - 0.09 = 4.65.$

---

## 4.11 OSTWALD'S DILUTION LAW

We know that weak electrolytes are incompletely dissociated in aqueous solution and the dissociation is favoured with dilution.

Let an weak electrolyte BA is dissolved in water having concentration c mol $lit^{-1}$. If $\alpha$ is the degree of dissociation of the electrolyte, then,

$$BA \rightleftharpoons B^+ + A^-$$

| | BA | $B^+$ | $A^-$ |
|---|---|---|---|
| Molar concentration before dissociation | c | o | o |
| Molar concentration at equilibrium | $(1-\alpha)c$ | $c\alpha$ | $c\alpha$ |

Applying law of mass action, the dissociation constant (K) is given by,

$$K = \frac{[B^+][A^-]}{[BA]} = \frac{\alpha c \times \alpha c}{(1-\alpha)c} = \frac{\alpha^2 c}{1-\alpha} \qquad ...(4.58)$$

This equation (4.58) is known as Ostwald's dilution law **which denotes the change in the degree of dissociation of weak electrolyte with the change in dilution.** $\alpha$ is small for weak electrolytes, so $1-\alpha \approx 1$ and equation (4.58) can be written as—

$$K = \alpha^2 c \qquad \text{or} \qquad \alpha = \sqrt{\frac{K}{c}} \qquad ...(4.59)$$

The equation (4.59) indicates that the **degree of dissociation of a weak electrolyte is inversely proportional to the square root of the concentration of the solution.** So at a fixed temperature, decrease in concentration of the solution causes the increase in the dgree of dissociation and conversely rise in concentration of the solution brings about the decrease in the degree of dissociation.

If 1 gram equivalent weak electrolyte is dissolved in V litres of solution then $V = \frac{1}{c}$ where c is the concentration in gram equivalent per litre, so that equation (4.59) transformed to—

$$\alpha = \sqrt{KV} \qquad ...(4.60)$$

This indicates that as V increases (i.e. dilution increases) $\alpha$ increases and so it is called **dilution law.**

## 4.12 ACIDITY FUNCTION

In Sec. 4.3.1 we have seen that $K_a$ value is a measure of the acid strength for weak acids in aqueous solutions but all strong acids are levelled to the strength of $H_3O^+$ so that $K_a$ value is not the index to measure the acid strength of strong acids. Hammett has introduced a function known as **Hammett acidity function** in order to measure the acidity of strong acids in aqueous solutions as also in nonaqueous media. Actually pH scale (0–14) provides an information regarding the $H^+$ ion concentration in aqueous solution. Hammett provides an **acidity scale** for solvents having dielectric constants ranging from 50 (anhydrous HCOOH) to 110 (anhydrous $H_2SO_4$).

Hammett utilises the protonation-deprotonation equilibrium of an indicator base B (e.g. p-amino azobezene) in strong acids i.e.

$$BH^+ \rightleftharpoons B + H^+$$

$$\text{so taht } K_a = \frac{[B][H^+]}{[BH^+]} \cdot \frac{\gamma_B \gamma_{H^+}}{\gamma_{BH^+}} \quad \ldots(4.61)$$

where γ's represent activity coefficients. For dilute solution γ is unity but for concentrated solution the value of γ cannot be predicted. There are several indicator bases (e.g. 2, 4-nitrochloroaniline; 2,4,6 trinitroaniline etc.) are used by Hammett to construct the acidity scale by these indicators. The $pK_a$ values of these indicators were determined colorimetrically. Let us rewrite the equation (4.61) as,

$$[H^+] = K_a \frac{[BH^+]}{[B]} \cdot \frac{\gamma_{BH^+}}{\gamma_B \gamma_{H^+}} \quad \ldots(4.62)$$

In a solution containing measurable amounts of the two forms of an indicator base whose constant is known. However, since $BH^+$ and B differ only by a $H^+$, the ratio $\frac{[BH^+]}{[B]}$, to a good approximation is independent of the indicator base used but is characteristic of the solvent. Then a quantity $h_0 = [H^+]\frac{\gamma_B}{\gamma_{BH^+}}$ is a property of the solution which is the same as that of $[H^+]$. Now we can write $h_0$ as :

$$h_0 = [H^+]\frac{\gamma_B}{\gamma_{BH^+}} = K_a \frac{[BH^+]}{[B]} \quad \ldots(4.63)$$

This equation (4.63) indicates that $h_0$ is a measure of the ability of a solution to convert a neutral basic molecule (B) to its conjugate acid ($BH^+$) since $h_0$ is proportional to the ratio of concentrations of the acidic and basic forms of the indicator. It is to be noted that $h_0$ becomes identical with $[H^+]$ in very dilute solution as the γ values aproaches to unity. The quantity $h_0$ thus makes a quantitative measure of the relative proton donating abilities of solutions of high acidity and high dielectric constant.

Hammett introduced a quantity $H_0$ (called acidity function) defined as :

$$H_0 = -\log h_0 = -\log K_a - \log \frac{[BH^+]}{[B]} \quad \text{(from equation 4.63)}$$

$$\text{i.e. } H_0 = -\log h_0 = pK_a + \log \frac{[B]}{[BH^+]} \qquad \text{...(4.64)}$$

$H_0$ is similar to that of pH of a solution. In dilute soliution $\gamma$'s approaches unity so that $h_0 = [H^+]$ and $H_0$ = pH. In concentrated solutions $\gamma$ is not equal to unity and so $H_0$ values of different acids having same concentration show considerable difference in the values of $H_0$ thereby $H_0$ values can be considered as acidity scale of strong acids. Typical $H_0$ values are :

*Utility of acidity function*

| **Solvent system** | $H_0$ |
|---|---|
| Water | +7.0 |
| 6 molal $HNO_3$ | –1.5 |
| 6.9 molal HCl | –2.0 |
| 8 molal $HClO_4$ | –2.63 |
| 70% $H_2SO_4$ | –5.54 |
| 100% $H_2SO_4$ | –10.60 |
| HF + $SbF_5$ (0.6 mol percent) | –21.1 |
| HF (100%) | –11.0 |
| HF + NaF (1M) | –8.4 |
| HCOOH | –2.2 |
| $H_3PO_4$ | –5.0 |
| $HSO_3F$ + $SbF_5$ (14 mol percent) | –26.5 |

## 4.12.1 Superacids

**The acid medium in which the $-H_0$ values are higher than 6 are frequently termed as spuperacids.** A mixture of $SbF_5 + HSO_3F$ has the highest value of $-H_0$ so that this strong acid mixture is called superacid.

To know the increased acidity in case of superacids we have to consider the autoionisation process of solvents e.g.

$$HF + HF \rightleftharpoons H_2F^+ + F^- \qquad SbF_5 + F^- \rightleftharpoons SbF_6^-$$

$$HSO_3F + HSO_3F \rightleftharpoons H_2SO_3F^+ + SO_3F^-$$

$$SbF_5 + SO_3F^- \rightleftharpoons FSO_3SbF_5^-$$

The addition of $F^-$ ion acceptor i.e. $SbF_5$ in HF, the acidity as also $H_0$ increases due to the increased concentration of solvent cation as :

$$2HF + SbF_5 \rightleftharpoons H_2F^+ + SbF_6^-$$

The ultimate reaction of a mixture of $SbF_5 + HSO_3F$ is—

$$SbF_5 + 2HSO_3F \rightleftharpoons H_2SO_3F^+ + FSO_3SbF_5^-$$

The protonated fluoro sulphuric acid acts as a powerful Bronsted acid. The addition of $SbF_5$ in HF or $HSO_3F$ increases the acidity of the system due to the removal of the anionic species of the autoionisation of HF or $HSO_3F$. This is also considered as effect of cosolvating agents (Sec. 4.3.2).

Fluroantimonic acid HF – $SbF_5$ has $-H_0$ value in the range 21 to 28. Both $SbF_5 + HSO_3F$ and $SO_3 + SbF_5$ (3 : 1 molar ratio) mixtures are considered as powerful superacids. In these superacids it is considered that $C_6H_3(NO_2)_3$ acting as a strong base.

Superacids can be able to protonate almost all organic compounds which are not considered as bases. For example the superacid prepared by mixing HF and $HSO_3F$ are effective protonating agent for alkanes to produce carbonium ion—

$$C(Me)_4 + \text{Superacid} \rightleftharpoons Me_3C(C^+H_4) \longrightarrow Me_3C^+ + CH_4$$

*Uses of superacids*

The superacids medium are useful for the stabilisation of polycationic species e.g. $I_2^+$, $S_4^{2+}$, $Se_4^{2+}$ etc. which are readily attacked by bases and they can only survive in superacids medium.

*Utility*

All strong acids are levelled to the strength of $H_3O^+$ in aqueous solution (Sec. 4.3.1) so that the concept of superacids exist only in nonaqueous medium. Generally superacids are far more stronger than the concentrated aqueous solutions of strong acids.

---

**Problem 4.25** Calculate the pH of the mixed solution obtained by mixing 100 ml 0.1 (M) $CH_3COOH$ to 50 ml 0.1(M) NaOH solution. $K_a$ for $CH_3COOH = 1.8 \times 10^{-5}$.

**Ans.** : Number of moles of $CH_3COOH$ present in 1000 ml 0.1(M) $CH_3COOH$ = 0.1

" " " " " 100 " " $= \frac{0.1 \times 100}{1000} = 0.01$

Similarly number of moles of NaOH in 50 ml 0.1(M) NaOH $= \frac{0.1 \times 50}{1000} = 0.005$

Total volume of the mixture = 100 + 50 = 150 ml.

0.005 mole of NaOH reacts with 0.005 mole of $CH_3COOH$ to produce 0.005 mole $CH_3COONa$.

∴ the no of mole of $CH_3COOH$ unreacted = 0.01 – 0.005 = 0.005

∴ the concentration of $CH_3COOH$ in the mixed solution $= \frac{0.005 \times 1000}{150} = 0.033(M)$

Similarly concentration of $CH_3COONa$ = 0.033(M)

According to Henderson equation—

$$pH = pK_a + \log\frac{[\text{Salt}]}{[\text{Acid}]} = -\log(1.8 \times 10^{-5}) + \log\frac{0.033}{0.033} = 4.74.$$

---

**Problem 4.26** Among the following mixtures, the one that would result in a buffer is—
1. NaCl and NaOH, 2. $CH_3COONa$ and $CH_3COOH$, 3. $NH_4Cl$ and HCl, 4. $NH_4Cl$ and $NH_4NO_3$. **[NET 2001]**

**Ans.** : 2.

**Problem 4.27** The $pK_a$ of a weak acid HA in water is 5. The $pK_b$ of the conjugate base in water is ($pK_w$ = 14). 1. 2.8, 2. 4, 3. 4, 4. 9. **[NET 2001]**

**Ans. :** 4.

## 4.13 ACID-BASE TITRATIONS—CHOICE OF INDICATORS

Acid-base titrations is an important tool in analytical chemistry for quantitative estimation of acids and bases in volumetric analysis. The end point in these titrations is detected by a substance which is known as **acid-base indicator** e.g. methyl orange, phenolphthalein etc.

*Theory of acid base indicators*

Generally acid-base indicators are weak organic acids or bases. In solution of acids and bases these indicators exist in two tautomeric forms in a state of dynamic equilibrium. **These two forms of indicator must differ in colour**—one form is the dissociated form exist in basic solution while the undissociated form exist in acid solution. Let $HI_n$ is the undissociated form (acid colour) and $I_n^-$ is the dissociated form (basic colour) of an indicator exist in equilibrium as :

$$HI_n \rightleftharpoons H^+ + I_n^- \quad ...(4.65)$$

$$\text{so that } K_{In} = \frac{[H^+][I_n^-]}{[HI_n]} \quad ...(4.66)$$

where $K_{In}$ is the indicator dissociation constant.

$$\therefore [H^+] = K_{In}\frac{[HI_n]}{[I_n^-]}$$

$$\text{so that } -\log[H^+] = -\log K_{In} + \log\frac{[I_n^-]}{[HI_n]}$$

$$\text{or } pH = pK_{In} + \log\frac{[I_n^-]}{[HI_n]} \quad ...(4.67)$$

Equation (4.67) suggests that ratio of the concentration of $I_n^-$ and $HI_n$ depends on pH of the solution. When the concentration of one form becomes at least ten times greater than that of the other form, then the colour of that form with higher concentration is distinctly visible to the eye.

In acid solution the indicator will exist in undissociated form ($HI_n$) so that

$$\frac{[HI_n]}{[I_n^-]} \geq 10 \quad ...(4.68)$$

Let $\frac{[HI_n]}{[I_n^-]} = 10$, so that from equation (4.67)

$$pH = pK_{In} - 1 \quad ...(4.69)$$

In basic solution the indicator will exist in dissociated form ($I_n^-$) and so,

$$\frac{[I_n^-]}{[HI_n]} \geq 10 \quad ...(4.70)$$

Let $\frac{[I_n^-]}{[HI_n]} = 10$, so that we get, $pH = pK_{In} + 1$ ...(4.71)

Therefore the pH interval limit for an indicator to change colour is $pK_{In} \pm 1$. For example, the $pK_{In}$ values of methyl orange and phenolphthalein are 3.7 and 9.4 repectively so that pH interval limit will be 2.7 to 4.7 and 8.4 to 10.4 respectively. The pH interval limit and colours of some indicators are given in Table 4.8. It is to be noted that pH interval limit have been determined experimentally by colorimetric procedure.

**Table 4.8 : The pH interval limit and colour of some acid-base indicators**

| Indicator | $pK_{In}$ | pH interval range | Acid colour | Base colour |
|---|---|---|---|---|
| Methyl orange | 3.7 | 3.1-4.4 | Red | Yellow |
| Methyl red | 5.0 | 4.2-6.3 | Red | Yellow |
| Phenolphthalein | 9.4 | 8.3-10.0 | Colourless | Pink |
| Litmus | 7.0 | 4.5-8.3 | Red | Blue |
| Bromocresol green | 4.7 | 3.8-5.4 | Yellow | Blue |
| Bromothymol blue | 7.1 | 6.0-7.6 | Yellow | Blue |
| Alizarin yellow R | – | 10.1-12.0 | Yellow | Oragne red |
| Thymolphthalein | 9.3 | 8.3-10.5 | Colourless | Blue |
| Thymol blue (base) | 8.9 | 8.0-9.6 | Yellow | Blue |
| Phenol red | 7.8 | 6.8-8.4 | Yellow | Red |
| Congo red | – | 3.0-5.0 | Blue | Red |
| Methyl yellow | 3.3 | 2.9-4.0 | Red | Yellow |

*Choice of indicators*

The selection of a suitable indicator for a particular acid-base titration can be made by the pH interval limit of an indicator. **For a specific acid-base titration there is a pH value at the equivalence point. If an indicator has pH interval limit lie within the pH range at equivalence point, then in that titration that particular indicator is used—**

(a) When a solution of strong base is added to a solution of strong acid, the pH of the solution gradually increases, but near the equivalence point there is a sudden jump in pH from 4 to 10. The pH interval limit of almost all indicators lies within this said equivalence point, so that any indicator can be used for this titration.

(b) For weak acid and strong base titration the pH at the equivalence point lies at 8.7 to 10.5 so that phenolphthalein may be used in this case.

(c) When a strong acid is added to a solution of a weak base, the pH value gradually decreases and at equivalence point there is a sudden change in pH from 3.0 to 6.5 so that methyl orange may be used in this titration.

(d) When a solution of a weak base is added to a solution of weak acid, the pH of the solution increases very slowly and near the end point and near the equivalence point the solution does not show any sudden change in pH and hence in the titration curve there is no sharp end point can be found with any simple indicators. However by the use of mixed indicators the end point in such titration can be detected. All the results are summarised in Table 4.9.

**Table 4.9 : Selection of indicators in acid-base titrations**

| Titration | pH change at the equivalence point | Indicators used |
|---|---|---|
| Strong acid and strong base e.g. HCl + NaOH | 4.0–10.0 | Any indicator |
| Weak acid and strong base e.g. $CH_3COOH$ + NaOH | 8.7–10.5 | Phenolphthalein, thymol blue, thymolphthalein |
| Weak base and strong acid e.g. $NH_4OH$ + HCl | 3.0–6.5 | Methyl red, methyl orange, bromophenol blue, bromocresol green |
| Weak acid and weak base e.g. $CH_3COOH$ + $NH_4OH$ | — | No suitable indicator |

Suitable mixing of several indicators in different proportions led to the formation of **universal or multiple range indicators** which show change in colours over a considerable pH range. Universal indicator is used for a rough idea of the pH of the solution. One such indicator is prepared by mixing 0.05g of methyl orange, 0.15 g of methyl red, 0.3 g of bromothymol blue and 0.35g of phenolphthalein in 1 litre 66% ethanol. The colour changes of this mixed indicator are : red upto pH3, orange-red at pH4, orange at pH5, yellow at pH6, yellowish-green at pH7, greenish-blue at pH8, blue at pH9, violet at pH10 and reddish-violet at pH11. Papers impregnated with these mixed indicators are widely used in the laboratory under the name pH papers.

### 4.13.1 Titrations of polyprotic acids with strong base

Polyprotic acids means **diprotic acids** $H_2B$ e.g. $H_2SO_3$, $H_2CO_3$, **triprotic acids** $H_3B$ e.g. $H_3PO_4$, **tetraprotic acid** $H_4B$ e.g. ethylenediaminetetraacetic acid ($EDTAH_4$) etc.

In general **for polyprotic acids when the successive dissociation constants differ by a factor of at least $10^4$ or the $pK_a$ values differ by 4 units, then the steps in the titration by a strong base will be distinct.** For a **diprotic acid when** $K_1/K_2$ ratio is at least $10^4$ it is treated as a mixture of two monoprotic acids. The choice of indicators depends on the value of $K_1$ and $K_2$ e.g. in $H_2SO_3$, $K_1 = 1.7 \times 10^{-2}$, $K_2 = 1.0 \times 10^{-7}$ i.e. $K_1/K_2 \approx 10^5$ and in this case there will be a sharp change in pH in acid range at the first equivalence point and so methyl orange or methyl red may be used. The $K_2$ value indicates that the change in pH will occur in alkali range and in this case thymolphthalein may be used. In case of $H_2CO_3$, $pK_1 = 6.34$, ($K_1 = 4.3 \times 10^{-7}$) and $pK_2 = 10.36$ ($K_2 = 5.6 \times 10^{-11}$), the difference of pK value is only 4.02. Therefore we can expect that the titration curve will show distinct break at the two equivalence points but as the **$K_1$ value is very small, only the first stage of neutralisation can be detected by phenolphthalein or thymolphthalein indicator.** The second equivalence point is difficult to determine by any indicator.

Suppose 50 ml 0.10 M **diprotic acid** $H_2B$ ($K_1 = 1.0 \times 10^{-3}$ and $K_2 = 1.0 \times 10^{-7}$) is titrated with 0.10 M NaOH solution. We have to calculate pH of the solution at different stages and also the concentration of $H_2B$, $HB^-$ and $B^{2-}$.

Here we have to consider the following two dissociations :

$$H_2B + H_2O \rightleftharpoons H_3O^+ + HB^- \qquad K_1 = 1.0 \times 10^{-3}$$

$$HB^- + H_2O \rightleftharpoons H_3O^+ + B^{2-} \qquad K_2 = 1.0 \times 10^{-7}$$

$HB^-$ is a much weaker acid than $H_2B$, so we can neglect the concentration of $H_3O^+$ produced from the dissociation of $HB^-$. Therefore we can consider $[H_3O^+] = [HB^-]$ and $[H_2B] = 0.10 - [H_3O^+]$. Now,

$$K_1 = \frac{[H_3O^+][HB^-]}{[H_2B]} = \frac{[H_3O^+]^2}{0.10-[H_3O^+]} = 1 \times 10^{-3} \qquad ...(4.72)$$

By solving this quadratic equation (4.72) we get $[H_3O^+] = 0.0095$

So that, $[HB^-] = 0.0095$ ...(4.73)

and $[H_2B] = 0.10 - 0.0095 = 0.09$ ...(4.74)

Now from $K_2$,

$$K_2 = \frac{[H_3O^+][B^{2-}]}{[HB^-]} = 1 \times 10^{-7}$$

As $[H_3O^+] = [HB^-]$, so that $[B^{2-}] = 1.0 \times 10^{-7}$ ...(4.75)

Now we have to calculate the pH at different stages of titrations.

(a) **Initial pH :** Solving the quadratic equation (4.72) we found

$[H_3O^+] = 9.5 \times 10^{-3}$ so that pH = 2.0223 ...(4.76)

(b) **pH after addition of 10 ml base :** At the start there is 50 ml 0.10 M $H_2B$ = 50 × 0.10 = 5.00 m mol of $H_2B$ and 10 ml 0.10 M base = 10 × 0.10 = 1.00 m mol of base. The reaction is—

$$H_2B + OH^- \longrightarrow HB^- + H_2O$$

1.00 m mol base will neutralise 1.00 m mol $H_2B$ and 4.00 m mol $H_2B$ will remain unreacted and there will be the formation of 1.00 m mol $HB^-$. Total volume is 50 + 10 = 60 ml.

$$\therefore [H_2B] = \frac{4.00}{60.0} \quad \text{and} \quad [HB^-] = \frac{1.00}{60.0}$$

Now according to Henderson equation (4.54)

$$pH = pK_1 + \log\frac{[HB^-]}{[H_2B]} = 3 + \log\frac{\frac{1.00}{60.0}}{\frac{4.00}{60.0}} = 2.40 \qquad ...(4.77)$$

(c) **pH at the first equivalence point :** pH at the first equivalence point for a diprotic acid is approximately as—

$$[H_3O^+] = \sqrt{K_1K_2} = \frac{1}{2}(pK_1 + pK_2) = \frac{1}{2}(3.0 + 7.0) = 5.00 \qquad ...(4.78)$$

First equivalence point is obtained by adding 50 ml 0.10 M base.

(d) **pH at the second equivalence point :** The second equivalence point is obtained by adding 100 ml 0.1 M base. The reaction is—

$$H_2B + 2OH^- \longrightarrow B^{2-} + 2H_2O$$

Initially there was 5.0 m mol $H_2B$ and base added = 100 × 0.10 = 10.0 m mol.

The pH of the solution is determined by the hydrolysis of $B^{2-}$ as—

$$B^{2-} + H_2O \longrightarrow HB^- + OH^- \qquad ...(4.79)$$

So that hydrolysis constant is given by—

$$K_h = \frac{[HB^-][OH^-]}{[B^{2-}]} = \frac{[HB^-]}{[B^{2-}][H^+]} \cdot [OH^-][H^+] = \frac{K_w}{K_2} \text{ where } K_2 = \frac{[H^+][B^{2-}]}{[HB^-]}$$

Since $[HB^-] \equiv [OH^-]$ and $[B^{2-}] = \frac{5.00}{150} = 0.0333$ [total volume = 100+50 = 150 ml]

Therefore $\frac{[HB^-][OH^-]}{[B^{2-}]} = \frac{[OH^-]^2}{0.0333}$

So that, $\frac{[OH^-]^2}{0.0333} = \frac{K_w}{K_2} = \frac{1 \times 10^{-14}}{1 \times 10^{-7}} = 10^{-7}$

$\therefore$ $[OH^-] = 5.8 \times 10^{-5}$ and so pOH = 4.24

and pH = 9.76 ...(4.80)

The pH beyond the second equivalence point may be calculated by the amount of excess base added. The titration curve for the diprotic acid is given in Fig. 4.3.

For a tiprotic acid $H_3B$ e.g. $H_3PO_4$, $K_1 = 7.5 \times 10^{-3}$ ($pK_1 = 2.12$), $K_2 = 6.2 \times 10^{-8}$ ($pK_2 = 7.20$) and $K_3 = 5.0 \times 10^{-13}$ ($pK_3 = 12.32$) where the pK values differ by more than four units, so that the three equivalence points may be detected. In case of citric acid the three dissociation constants are close, so that in this case the three neutralisation points cannot be differentiated easily.

Suppose 100 ml 0.1 M $H_3PO_4$ is titrated with 0.1 M NaOH solution. We have to calculate the pH at various stages of titrations. $H_3PO_4$ may be considered as a mixture of three monopritic acids.

(a) **pH at the first equivalence point :**

$$H_3PO_4 + NaOH \longrightarrow NaH_2PO_4 + H_2O$$

and the pH of the solution $= \frac{1}{2}(pK_1 + pK_2) = \frac{1}{2}(2.12 + 7.21) = 4.66$ ...(4.81)

Here the $K_1$ value indicates that pH will be in the acid range so that methyl orange or methyl red may be used in this stage.

(b) **pH at second equivalence point :**

$$NaH_2PO_4 + NaOH \longrightarrow Na_2HPO_4 + H_2O$$

As the $K_2$ value of $H_3PO_4$ is less than $10^{-7}$ so the change in pH at the second equivalence point will be in the alkaline range and the pH value calculated from $K_2$ and $K_3$ i.e.

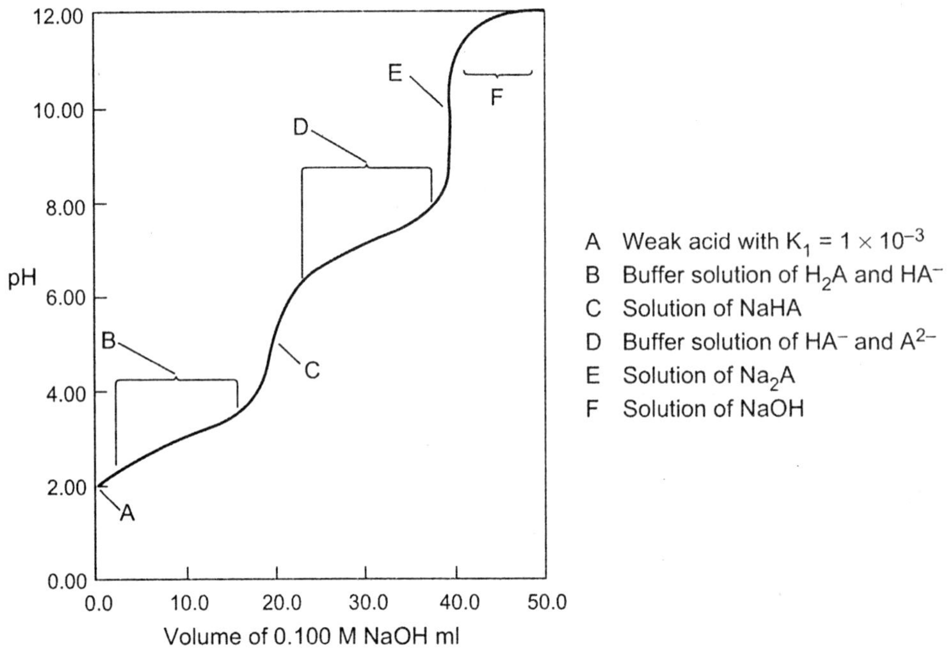


***Figure 4.3** : Titration of 50.0 ml of 0.100 M $H_2A$ with 0.100 M NaOH. For $H_2A$, $K_1$ = $1.00 \times 10^{-3}$ and $K_2$ = $1.00 \times 10^{-7}$.*

$$pH = \frac{1}{2}(pK_2 + pK_3) = \frac{1}{2}(7.21 + 12.32) = 9.76 \quad ...(4.82)$$

Phenolphthalein or thymolphthalein may be used to detect the second state.

(c) **pH at the third equivalence point :**

$Na_2HPO_4 + NaOH \longrightarrow Na_3PO_4 + H_2O$

The salt $Na_3PO_4$ is hydrolysed here according to the following way

$Na_3PO_4 + H_2O \longrightarrow Na_2HPO_4 + NaOH$

and here the hydrolysis constant $K_h = \frac{K_W}{K_3}$ and the subsequent hydrolysis may be ignored and the pH of the solution can be calculated by the consideration of the hydrolysis of weak monoprotic acid and strong base [equation (4.44)] i.e.

$$pH = 7 + \frac{1}{2}pK_3 + \frac{1}{2}\log c = 7 + 6.16 + \frac{1}{2}\log\frac{1^*}{20} = 13.16 - 0.6505 = 12.5095 \quad ...(4.83)$$

In this third stage $K_3 = 5.0 \times 10^{-13}$ indicates that the change in pH is very slow and the titration curve (Fig. 4.4) is very flat and so no suitable indicator is available in this case. For this reason detection of third-equivalence point is difficult for $H_3PO_4$.

* Here $c = \frac{100 \times 0.1}{200} = \frac{1}{20}$

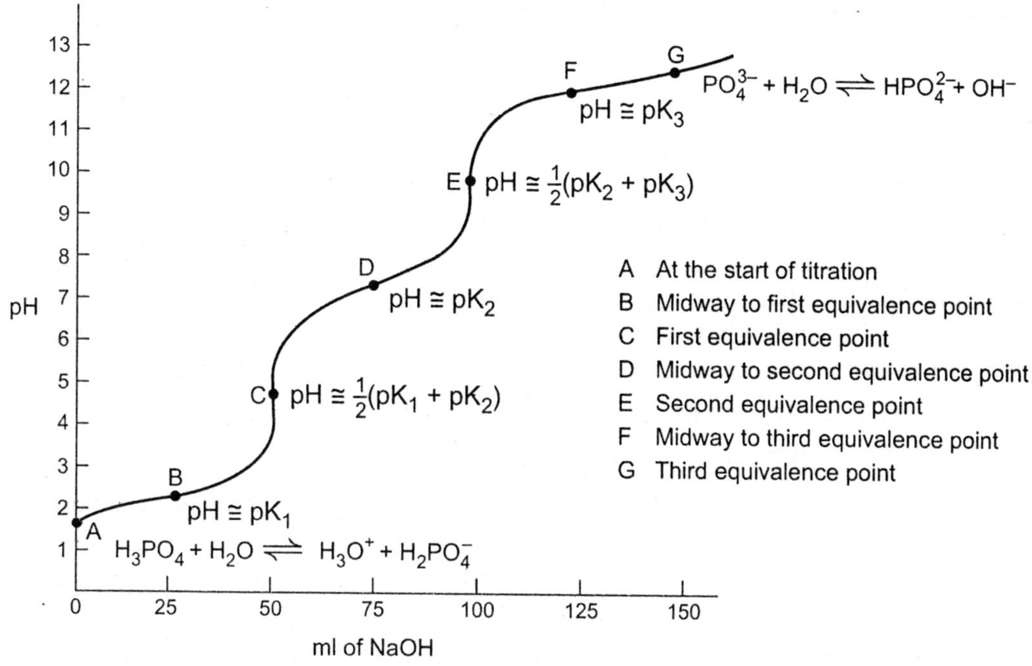


***Figure 4.4** : Titration curve of phosphoric acid. 50 ml of 0.10 M acid titrated with 0.10 M base.*

In the titration cuve (Fig. 4.4) various points have different compositions which are given below :

| Points | Approximate composition |
|---|---|
| A | $[H_3PO_4] \cong [H_3O^+]$ |
| B | $[H_3PO_4] \cong [H_2PO_4^-]$ |
| C | $[H_3PO_4] \cong [HPO_4^{2-}]$ |
| D | $[H_2PO_4^-] \cong [HPO_4^{2-}]$ |
| E | $[H_2PO_4^-] \cong [PO_4^{3-}]$ |
| F | $[HPO_4^{2-}] \cong [PO_4^{3-}]$ |
| G | $[HPO_4^{2-}] \cong [OH^-]$ |

**Problem 4.28** Calculate the pH and pOH of a 0.050 M solution of HCl.

**Ans. :** We know HCl is a strong acid so it will completely dissociated.

Therefore, $[H_3O^+] = 0.050 = 5 \times 10^{-2}$

$\therefore$ pH = 2 – log5 = 1.30

As pH + pOH = 14, pOH = 14 – 1.30 = 12.70.

## 4.14 PERIODIC VARIATION OF ACIDITY AND BASICITY

Here we shall discuss some correlation of the variation of acidic and basic properties of chemical compounds or ionic species with reference to the existing concept of acids and bases.

### 4.14.1 Variation of acidic and basic properties of hydrides

Any binary compound containing hydrogen which forms a chemical bond to a more electronegative element than itself can, in principle, behave as an acid. In aqueous solution, acid properties are associated chiefly with the hydrides of the non metallic elements of group 16 and 17 as evident from the $pK_a$ values :

| | $H_2Te$ > | $H_2Se$ > | $H_2S$ > | $H_2O$ | HI > | HBr > | HCl > | HF |
|---|---|---|---|---|---|---|---|---|
| $pK_a$ | 2.64 | 3.72 | 7.06 | 15.69 | –10.0 | –9.0 | –7.0 | 3.17 |

In a period acidity increases from left to right e.g. the acid strength increases in the order $H_3N < H_2O < HF$ or $PH_3 < H_2S < HCl$.

The variation of acidity along a period may be explained by the increasing electronegativity of the element linked to hydrogen. Here the size of the elements remain practically same, so that with increasing electronegativity of the element, proton release will be more easier making increase of acidity.

Along a group from top to bottom acidity increases, here electronegativity alone can not be able to explain the acidity. It appears acidity decreases with the increase in the electronegativity of the elements. Here opposite trend observed from what is expected from electronegativity consideration. As the size of the atoms increases, the bonding with small hydrogen atom becomes weaker so that acidity increases with increasing bond length/ decreasing bond strengh as :

| | HF < | HCl < | HBr < | HI | |
|---|---|---|---|---|---|
| Bond energy | 565 | 431 | 368 | 297 | $kJ\ mol^{-1}$ |

Critically we can say that,

(1) Charge density on the conjugate base is in the order
$F^- > Cl^- > Br^- > I^-$ and $O^{2-} > S^{2-} > Se^{2-} > Te^{2-}$

Greater charge density indicates greater affinity for proton by the conjugate base so that the corresponding acid will be weaker.

(2) Smaller the anions bonded with the hydrogen increases hydrogen bonding, making decrease in acid strength.

(3) The variation of acidity of a binary hydracid HX can be explained by the following type of Born-Haber thermodynamic cycle :

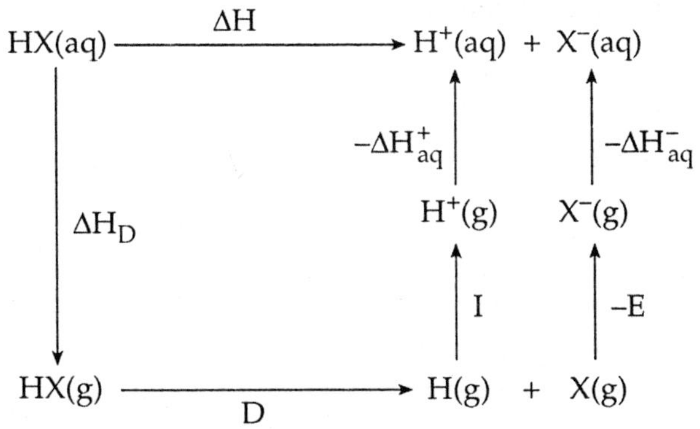


$\Delta H_D$ = heat of dehydration
I = ionisation energy of H(g)
−E = electron affinity of X(g)
ΔH = dissociation energy of the acid HX(aq)

D = heat of dissociation
$-\Delta H^+_{aq}$ = heat of hydraton of $H^+(g)$
$-\Delta H^-_{aq}$ = heat of hydration of $X^-(g)$

Therefore in order to estimate the acid strength of HX we can write :

$$\Delta H = \Delta H_D + D + I - \Delta H^+_{aq} - E - \Delta H^-_{aq} \qquad ...(4.84)$$

From the values of the thermochemical terms, ΔH values calculated as : −14, −60, −64 and −58 kJ $mol^{-1}$ for HF, HCl, HBr and HI respectively. ΔH value for HF is largely effected **by heat of dissociation** (D) and **heat of dehydration** ($\Delta H_D$). The high value of heat of dissociation (ΔH) is due to the short H–F bond compared to that of longer H–I bond.

The acid strength of HX can be related to the free energy term (ΔG) for the process $HX(aq) \longrightarrow H^+(aq) + X^-(aq)$ and equilibrium constant (K) of the process is given by :

$$\left.\begin{aligned} \Delta G &= \Delta H - T\Delta S \\ \text{and } \Delta G &= -RT\ln K \end{aligned}\right\} \qquad ...(4.85)$$

From these equations we can calculate the K values of the hydracids (HX) which revealed that HF is the weakest among them (Table 4.10).

It is interesting to consider that $NH_3$ is the only simple hydride which has basic properties in aqueous solution. Both $NH_3$ and $H_2O$ molecules have a tendency to accept protons from donor molecules. $NH_3$ is more basic than $H_2O$, in fact $NH_3$ accepts proton from water ($NH_3 + H_2O \rightleftharpoons NH_4^+ + OH^-$, $K_b = 1.8 \times 10^{-5}$). The other hydrides (e.g. $PH_3$, $AsH_3$ etc.) are less basic and have a little tendency to accept protons from water.

**Table 4.10 : Thermochemical data for HX hydraacids (energy terms given in kJ $mol^{-1}$)**

| | ΔH | TΔS | ΔG | $pK_a$ |
|---|---|---|---|---|
| HF | −14 | −29 | +15 | 3 |
| HCl | −60 | −13 | −47 | −8 |
| HBr | −64 | −4 | −60 | −10 |
| HI | −58 | +4 | −62 | −11 |

### 4.14.2 The properties of hydroxides and oxoacids

The hydroxide MOH where M represents a particular element, can dissociate in one of the following two ways in ionising solvent :

$$MOH \longrightarrow MO^- + H^+ \qquad ...(4.86)$$

$$\text{or } MOH \longrightarrow M^+ + OH^- \qquad ...(4.87)$$

MOH composed of $M^+$, $O^{2-}$ and $H^+$ ions and if $M^+$ cation is a strong polarising species then MOH dissociates according to equation (4.86) i.e. MOH suffers acidic dissociation while if $H^+$ ion will exert greater polarising power on $O^{2-}$ then MOH suffers basic dissociation according to equation (4.87). If MOH has amphoteric behaviour then it will undergo both the dissociation equally. The increase in acid character of the hydroxides along a period from left to right is shown in the following :

| Group | I | II | III | IV | V | VI | VII |
|---|---|---|---|---|---|---|---|
| Period-3 | NaOH | $Mg(OH)_2$ | $Al(OH)_3$ | $Si(OH)_4$ | $H_3PO_4$ | $H_2SO_4$ | $HClO_4$ |

There is a gradual transition of strongly basic character (NaOH) to strongly acidic character ($HClO_4$), the middle members e.g. $Al(OH)_3$ and $Si(OH)_4$ are acting as amphoteric.

Variation of the strength of oxoacids can be considered in the following aspects :

(1) The acid strength increases with the electronegativity of the element. The strongly electronegative element will pull the electron density from hydrogen due to inductive effect and thus proton release will be easier e.g.

| | HOH | < HOI | < HOBr | < HOCl |
|---|---|---|---|---|
| $pK_a$ | 15.69 | 11.0 | 8.7 | 7.4 |

(2) An oxyacid having more electronegative central element will be stronger acid. The more electronegative character is associated with the increasing oxidation state of the central element. This is known as **electronegativity acidity rule** and is demonstrated with the chlorine oxyacids :

| | $HClO_4$ > | $HClO_3$ > | $HClO_2$ > | HOCl |
|---|---|---|---|---|
| oxidation number of chlorine | +7 | +5 | +3 | +1 |
| $K_a$ | very large | large | $1\times10^{-2}$ | $3.2\times10^{-8}$ |

The same is true for sulphur and nitrogen oxyacids e.g. $H_2SO_4 > H_2SO_3$; $HNO_3 > HNO_2$

(3) The electronegativity-acidity rule will not hold good in case of phosphorus oxyacids e.g. $H_3PO_4$, $H_3PO_3$ and $H_3PO_2$. Expected trend according to electronegativity concept is $H_3PO_4 > H_3PO_3 > H_3PO_2$. But the actual order is just reversed :

| | $H_3PO_2$ > | $H_3PO_3$ > | $H_3PO_4$ |
|---|---|---|---|
| $K_a$ | $6\times10^{-2}$ | $1.6\times10^{-2}$ | $7.5\times10^{-3}$ |

This reversal of the order of acidity may be explained from the structure of the oxyacids as :

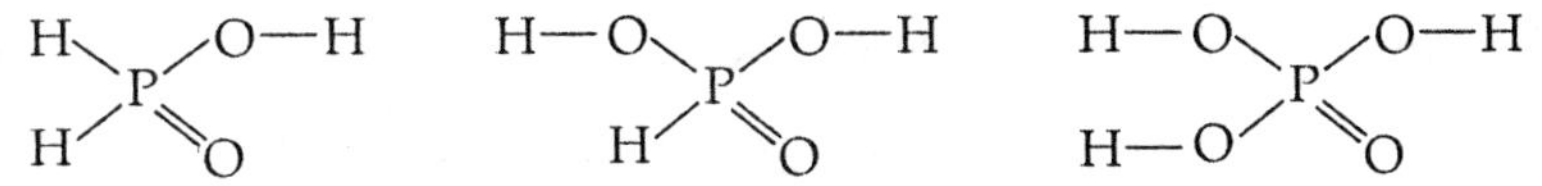

In all the oxyacids there is one =O group. In $H_3PO_2$, $H_3PO_3$ and $H_3PO_4$ there are one, two and three protonated oxygen respectively. These protonated oxygens are the vehicles to develop the acidity of oxyacids. There are two hydrogen atoms directly link to phosphorous atom in $H_3PO_2$, one hydrogen atom directly to $H_3PO_3$ and in $H_3PO_4$ there is no hydrogen atom directly link to phosphorous. Due to the inductive effect of =O, this will pull electron density from the central phosphorous whereby this will directly effect one protonated oxygen of $H_3PO_2$, this pull will be bifurcated in two protonated oxygen of $H_3PO_3$ and the same pull will be tripartite in three protonated oxygens in $H_3PO_4$. Therefore proton release will be more easier in $H_3PO_2$, this will be less easier in case of $H_3PO_3$ and this will be still more less easier in case of $H_3PO_4$ and hence the order of acidity is $H_3PO_2 > H_3PO_3 > H_3PO_4$.

(4) For oxyacids having the same composition and oxidation number, the acidity decreases with the size of the central element. Thus,

$$HClO_4 > HBrO_4 > HIO_4,\ H_3PO_4 > H_3AsO_4 > H_3SbO_4$$

$$HNO_3 > HPO_3 > HAsO_3,\ H_2SO_4 > H_2SeO_4 > H_2TeO_4$$

These can also be explained from electronegativity-acidity rule.

*Pauling's rule for oxoacids*

(5) Pauling enunciated two emperical rules which are useful in predicting the strength of mononuclear oxyacids :

(a) In oxyacids of the type $H_nXO_m$ all the oxygen atoms are directly linked to central X atom, some of the oxygen atom exist as protonated oxygen. So from the structural viewpoint the oxyacid is represented as $O_{m-n}X(OH)_n$ and emperical rule proposed for the $pK_1$ value as :

$$pK_1 = 8 - 5(m-n) \qquad ...(4.88)$$

The strength of the acid is thus related to the number of non-protonated oxygen atoms.

(b) For polyprotic acid (n>1) the successive dissociation constants, $K_1$, $K_2$, $K_3$ etc. are in the ratio $1 : 10^{-5} : 10^{-10}$ i.e. successive pK values differ approximately by 5 units.

These rules can be applied in order to predict the structure and strength of oxyacids (Experimental pK values are given in parentheses)

When m–n = 0, $pK_1 = 8$ $HOCl(pK_1 = 7.5)$, $H_3AsO_3(pK_1 = 9.2)$, $H_6TeO_6(pK_1 = 8.8)$, $H_3BO_3(pK_1 = 9.2)$

When m–n = 1 $pK_1 = 3$ $H_3PO_4(pK_1 = 2.1, pK_2 = 7.2, pK_3 = 12.5)$

$H_3PO_2(pK_1 = 2.0)$, $HNO_2(pK_1 = 3.3)$

$H_2SO_3(pK_1 = 1.9, pK_2 = 7.3)$

$H_2CO_3(pK_1 = 3.6, pK_2 = 10.3)$

$H_3PO_3(pK_1 = 1.8, pK_2 = 6.15)$

When m–n = 2 $pK_1$ = –2 $HNO_3(pK_1 = -1.4)$

$H_2SO_4(pK_1 = -2.0,\ pK_2 = 1.9)$

$HClO_3(pK_1 = -1.0)$

When m–n = 3 $pK_1$ = –7 $HClO_4(pK_1 = -10)$

$HMnO_4$($pK_1$ = highly negative)

It is interesting to note that phosphorous oxyacids are moderately weak and they have the structures of the following :

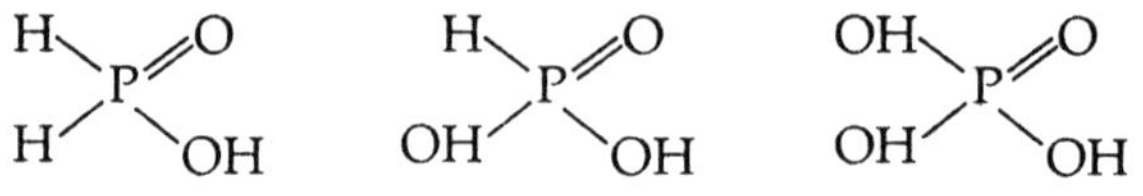

| | | |
|---|---|---|
| m–n = 1 | m–n = 1 | m–n = 1 |
| $pK_1$ = 2.0 | $pK_1$ = 1.8 | $pK_1$ = 2.1 |

Thus $H_3PO_2$, $H_3PO_3$ and $H_3PO_4$ may be formulated as $H_2P(O)(OH)$, $HP(O)(OH)_2$ and $P(O)(OH)_3$. At first sight it does not appear possible to classify them according to Pauling's first rule. However their chemical properties and structures show that P—H bonds are present and that their formulas are more accurately written as $H_2PO(OH)$, $HP(O)(OH)_2$ and $P(O)(OH)_3$. These typical oxo-acids have m – n = 1 and pK values correspond to their structures as $pK_1$ = 2.0 for $H_3PO_2$, $pK_1$ = 1.8 for $H_3PO_3$ and $pK_1$ = 2.1 for $H_3PO_4$.

**Problem 4.29** Using Pauling's first rule identify the structural formula that are nearly consistent with the actual $pK_a$ values (i) 2.12 for $H_3PO_4$ and (ii) 2.0 for $H_3PO_2$. **[C.U. 2005, 2012]**

**Ans.** : Sec. 4.14.2(5).

## 4.14.3 Hydration of cations and their acidity

In Sec. 3.1.4 we have seen that simple metal cations will not exist in aqueous solution–all exist as an hydrated species. Here we shall discuss some aspects of periodic variation of acidity of hydrated metal cations.

The species which in aqueous solution, furnish a proton by dissociation of water molecules bound to a metal ion are called **hydrated cationic acids**. We may represent this as—

$$[M(H_2O)_6]^{Z+} + H_2O \rightleftharpoons [M(H_2O)_5(OH)]^{(Z-1)+} + H_3O^+$$

$$\text{or} \quad M^{Z+}(aq) + H_2O \rightleftharpoons M(OH)^{(Z-1)+} + H^+(aq)$$

The proton is lost from the hydrated cation since the $M^{Z+}$ ion tends to displace the lone pairs on oxygen of attached water molecules towards itself as :

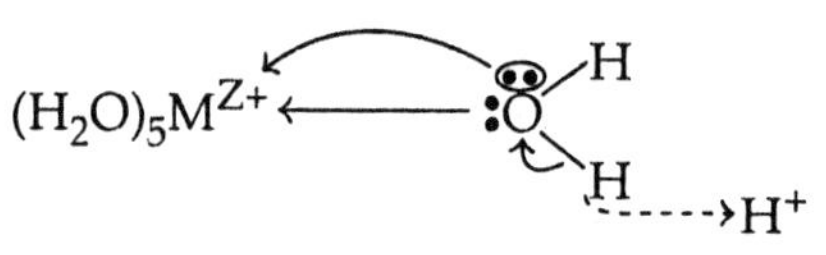

This influences the oxygen to withdraw the electron pair which shares with hydrogen towards itself. As a result the O—H bond is weakened and the proton release occurs.

Let us consider with specific example with that of $Cr^{3+}$—

$$Cr(H_2O)_6^{3+} + H_2O \rightleftharpoons Cr(H_2O)_5OH^{2+} + H_3O^+$$

$$\text{or} \quad Cr^{3+}(aq) + H_2O \rightleftharpoons Cr(OH)^{2+}(aq) + H^+(aq)$$

These acid-producing reactions may be commonly called **hydrolysis of metal cations** for which we can write :

$$K_a = \frac{[Cr(OH)^{2+}(aq)][H^+(aq)]}{[Cr^{3+}(aq)]} \quad \text{...(4.89)}$$

where $K_a$ is the equilibrium constant of the reaction, may sometimes be called $K_h$ (hydrolysis constant). We may represent this as :

$$pK_a = -\log K_a = -\log \frac{[Cr(OH)^{2+}(aq)]}{[Cr^{3+}(aq)]} + pH \quad \text{...(4.90)}$$

It is important to note that **lower values of $pK_a$ and pH correspond to higher degree of hydrolysis and higher acidities of the aqueous solution.** About 150 metal cations have been studied to correlate their reactivity and acidity. **Smaller the size and/or higher the charge on the hydrated metal cation increases its extent of hydrolysis as also increasing acidity.** The extent of acidity may be represented by an emperical relationship.

$$pK_a = 15.14 - 88.16\ Z^2/r \quad \text{...(4.91)}$$

where r is the ionic radius of the cation. Now we shall try to classify the hydrated cations according to their acidity :

*Classification of cations into acidity order*

(1) If the $Z^2/r$ value of the cation < 0.01 then it is **nonacidic** and the $pK_a$ value is in the range 14–15. These cations undergo negligible hydrolysis reactions to show appreciable acidity. Most +1 ions of the **s-block elements** e.g. $Cs^+$ and $Rb^+$ are of this category.

(2) When the $Z^2/r$ ratio is in the range 0.01 – 0.04, then the cation will produce **feeble acidity**. In this case $pK_a$ value is in the range 11.5 – 14.0 and measurement may be done by sensitive measurement. Examples are $Ca^{2+}$, $Sr^{2+}$, $Tl^+$ etc. and most +2 ions of s and f-block elements.

(3) If the value of $Z^2/r$ lie in the limit 0.04 – 0.10, the cation is **weakly acidic**. The $pK_a$ value lie in the range 6.0 – 11.5. $Mg^{2+}$ ion and +2 ions of d-block elements are of this type.

(4) When the $Z^2/r$ value lie in the range 0.10 – 0.16, the cation is termed **moderately acidic** and the $pK_a$ value of the solution lie in the range 1.0 – 6.0. Most +3 ions of d-block or most +4 ions of the f-block are of this type. $Al^{3+}$ ion is under this category.

(5) When the ratio $Z^2/r$ of a cation lie in the range 0.16-0.22, that cation will produce **strongly acidic solution** on hydrolysis. The $pK_a$ value is in the range –4 to 1. These ions (e.g. $Tl^{4+}$ and most +4 ions of d-block) react violently with water and produce strongly acidic solution.

(6) If $Z^2/r$ ratio greater than 0.22, the cation will produce **very strongly acidic solution** with $pK_a$ value below –4. These ions will react with water irreversibly with water producing an oxide or hydroxide and high concentration of $H^+$ ions in aqueous solution. Some metal ions in high oxidation states (+5 and +6) are in this category.

Equation (4.91) may be modified by introducing Pauling electronegativity ($\chi_p$) which is also an important parameter to produce acidity of the hydrated cations.

$$pK_a = 15.14 - 88.16\,[Z^2/r + 0.096(\chi_p - 1.50)] \quad \ldots(4.92)$$

This equation (4.92) is only used in those cases where the Pauling electronegativity of the cation exceeds 1.50.

### 4.14.4 Formation of Oxo cations

Sometimes two or four hydroxide groups are attached to a highly charged cations, they may lose molecules of water and form oxo cations e.g. Zr(IV) forms the oxo cation $ZrO^{2+}$. Such species are exist as +5 and +6 oxidation states of the f block elements U, Np, Pu etc., one common oxo cation is uranyl cation, $UO_2^{2+}$. Mo (V and VI) and V (IV and V) also form similar oxo species. Many nonmetal halides undergo partial hydrolysis to form oxo compound e.g. $POCl_3$. In our familiar analytical chemistry, the precipitates formed by antimony and bismuth are not the oxides but the oxo species e.g. SbOCl and BiOCl.

**Problem 4.30** Arrange the following hydrated cations in order of increasing acidity : $[Na(H_2O)_n]^+$, $[Al(H_2O)_6]^{3+}$, $[Mn(H_2O)_6]^{2+}$ and $[Ni(H_2O)_6]^{2+}$.

**Ans. :** Acidity of the hydrated cation depends on the polarising power of the cation and hence acid strength increases as : $[Na(H_2O)_n]^+ < [Mn(H_2O)_6]^{2+} < [Ni(H_2O)_6]^{2+} < [Al(H_2O)_6]^{3+}$.

## 4.15 STRUCTURAL AND STERIC FACTORS IN LEWIS ACID–BASE ADDUCT

In Sec. 4.8.1 and 4.8.3 we have discussed some aspects of Lewis acid-base adducts. Generally adduct formation is accompanied by structural changes e.g. $BF_3$ is planar while $BF_4^-$ is tetrahedral, $SiF_4$ is tetrahedral whereas $SiF_6^{2-}$ is octahedral. The stability of the adducts depends largely on the steric factors like **Front strain or F strain and Back strain or B strain.**

**When both the acids and bases have bulkier substituents in their structures, the adduct formation is hindered due to steric hindrance which is known as F-strain.** Trimethyl aluminium shows F-strain relative to $AlCl_3$. From the inductive effect of methyl group it is expected that 2-methylpyridine forms strong adduct with proton. But it acts as a weaker base towards the Lewis acid $B(CH_3)_3$ than that of pyridine while 3-methylpyridine forms stronger adduct with $B(CH_3)_3$ which are authenticated by enthalpy change ($\Delta H$) of formation of adduct : –71 kJ $mol^{-1}$ for pyridine –$B(CH_3)_3$, –42kJ$mol^{-1}$ for 2-methylpyridine –$B(CH_3)_3$ and –74 kJ$mol^{-1}$ for 3-methylpyridine –$B(CH_3)_3$. It is evident that 2,6 dimethyl pyridine will not form any adduct with $B(CH_3)_3$. The shifting of the methyl group from 2 position to 3 position of the pyridine ring reduces the F-strain from 2-methylpyridine to 3-methylpyridine to some extent and the acidity of 3-methylpyridine is more than that of 2-methyl analogue so that stronger adduct is formed in case of 3-methylpyridine. The F-strain is maximum in case of 2,6-dimethylpyridine. The F-strain is so stronger that 2-t-butylpyridine will not form any adduct with $B(CH_3)_3$.

The **B-strain arises when the substituents on the Lewis base are bulky and due to steric over-crowding, the electron donation to the Lewis acid is restricted.** This will require the reorganisation of energy from one geometry to other. Some π-bonding interaction and hybridisation energy is also considered here. For example triethylamine shows B-strain relative to $NH_3$.

We have already mentioned that some structural changes may occur during adduct formation. Thus when boron changes its geometry from triangular planar to pyramidal geometry, the substituents come closer and if there branching occurs which cause some steric crowding and the adduct formation is hindered due to B-strain. Thus the Lewis acidity order is $B(n\text{–}Bu)_3 > B(t\text{–}Bu)_3$ where the state of hybridisation of B changes from $sp^2$ to $sp^3$. In $B(t\text{–}Bu)_3$ the B atom will experience more B-strain for branching of alkyl group so that $B(t\text{–}Bu)_3$ will form weaker adduct. Similarly $B(CH_3)_3$ forms stronger adduct with $NH_3$ compared to that of $B(i\text{–}Pr)_3$.

There may occur another type of strain known as **internal strain or I-strain.** The term is applied to Lewis bases having exocyclic structure. As the size of the ring is reduced, the internal bond angles are reduced and the hybrid orbitals of cyclic atoms have less s character so that electronegativity decreases causing less acid character. As a result of I-strain, the base strength of lactams is of the following order :

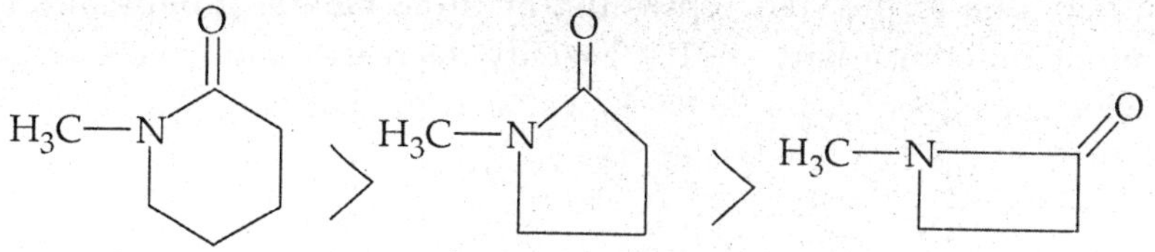


## 4.16 INFLUENCE OF INDUCTIVE EFFECT ON ACIDITY AND BASICITY

**Electron withdrawing substituents on an atom increases its electronegativity** and **hence increases acidity**. Thus the acid strength of substituted acetic acid are in the order :

$$FCH_2COOH(pK_a = 2.66) > ClCH_2COOH(pK_a = 2.86) > HCH_2COOH(pK_a = 4.76)$$

$$MeOCH_2COOH(pK_a = 3.53) > MeSCH_2COOH(pK_a = 3.72) > HCH_2COOH(pK_a = 4.76) > Me_3CCH_2COOH(pK_a = 5.05) > Me_3SiCH_2COOH(pK_a = 5.22)$$

If X is a more electronegative atom than carbon in a C—X bond, then X will pull the electron density of the bond and there will be a slightly positive charge on the carbon atom which in turn will effect the electron distribution on the adjacent carbon atoms. As this bond polarisation decreases from one atom to the next, this effect is called **inductive effect** and the effect is very poor beyond two or three atoms i.e. $\cdots \overset{\delta_4^+}{C_4} — \overset{\delta_3^+}{C_3} — \overset{\delta_2^+}{C_2} — \overset{\delta_1^+}{C_1} — \overset{\delta^-}{X}$ where $\delta_1^+ > \delta_2^+ > \delta_3^+ \ldots$ Groups that increase the positive charge on the adjacent carbon atoms are called the electron withdrawing groups having –I effect e.g.

$$R_3N^+ > R_2S^+ > NO_2 > CN > F > Cl > Br > I > OR > OH$$

On the other hand **electron donating substituents lower the electronegativity of an atom and hence lower the acidity.** Electron donating groups increase the electron density on the carbon atom and such effects are expressed as +I e.g.

$$COO^- > (CH_3)_3C > (CH_3)_2CH > CH_3CH_2 > CH_3$$

HCOOH is more stronger than $CH_3COOH$. This due to the greater +I effect of $CH_3$ group compared to that of H so that O–H bond is more stronger in $CH_3COOH$ than HCOOH. Therefore proton release is more easier in case of HCOOH than $CH_3COOH$. $FCH_2COOH$ is more stronger than $ClCH_2COOH$ is due to the greater –I effect of F than that of Cl.

Generally electron withdrawing group increases acidity so that it will reduce the base strength. Thus $NF_3$ is a weaker base than $NH_3$. Other examples of base strength order are :

$$NH_3 > H_2N - NH_2 > HO - NH_2;\ Me_3N > NH_3 > NF_3$$

Aliphatic amines are stronger base than ammonia and the base strength order is $(CH_3)_2NH > CH_3NH_2 > (CH_3)_3N > NH_3$. When one hydrogen atom of $NH_3$ is replaced by $CH_3$ group, due to the +I effect of $CH_3$ group the basicity of $CH_3NH_2 > NH_3$. When another $CH_3$ group is introduced, due to the greater number of relectron releasing groups, the availability of lone pair on nitrogen atom increased so that $(CH_3)_2NH > CH_3NH_2$. Now on further introduction of $CH_3$ group it is expected that base strength will increase, i.e. $(CH_3)_3N$ will be a stronger base than $(CH_3)_2NH$. But in fact due to the **steric crowding of three methyl groups, the lone pair on nitrogen atom cannot function easity so that basicity decreases** and hence we get the above order of basicity which is authenticated by the $K_b$ values : $NH_3(1.8 \times 10^{-5})$, $CH_3NH_2(37.0 \times 10^{-5})$, $(CH_3)_2NH(54.0 \times 10^{-5})$ and $(CH_3)_3N(6.5 \times 10^{-5})$.

## 4.17 RESONANCE AND HYPERCONJUGATION EFFECT ON ACIDITY/BASICITY

According to the inductive effect the acid strength of boron halides should be $BF_3 > BCl_3 > BBr_3 > BI_3$. But actually the sequence of acid strength is reverse of what is expected i.e. $BI_3 > BBr_3 > BCl_3 > BF_3$. This is due to the following **resonating structures** of $BX_3$ i.e.

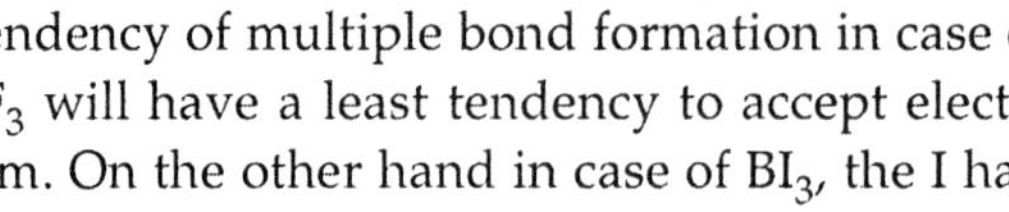

In all boron halides the central boron has sextet. As F is small compared to that of other halogens so that $p_\pi$-$p_\pi$ back bonding i.e. multiple bond formation is more prominant in case of F and this multiple bond formation tendency decreases from F→Cl→Br→I. Because multiple bond will be stronger in shorter internuclear distance compared to that of longer internuclear distance. Due to the greater tendency of multiple bond formation in case of $BF_3$, the central B here will achieve octet so that $BF_3$ will have a least tendency to accept electron pair i.e. the Lewis acidity of $BF_3$ will be minimum. On the other hand in case of $BI_3$, the I has a larger size, so that $p_\pi$-$p_\pi$ multiple bond formation is minimum and thereby in $BI_3$ the central B will not be able to achieve the octet. Therefore $BI_3$ will have a greater tendency to accept the electron pair and the Lewis acidity will be maximum. Thus the actual Lewis acidity sequence is $BI_3 > BBr_3 > BCl_3 > BF_3$.

The phenols are more acidic than the corresponding aliphatic alcohols but less so than carboxylic acids or even carbonic acid. This is evident from the following $K_a$ values :

| | Phenols | Alcohols | Carboxylic acids | Carbonic acid |
|---|---|---|---|---|
| | Ar–OH | R–OH | R–COOH | $H_2CO_3$ |
| $K_a$ | $\sim10^{-10}$ | $\sim10^{-18}$ | $\sim10^{-5}$ | $\sim10^{-7}$ |

Phenols are acidic due to the formation of stable phenoxide ion i.e.

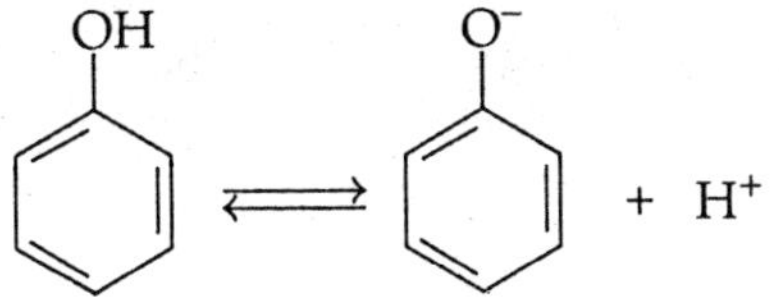

*Phenoxide ion*

The stability of the phenoxide ion is explained due to the delocalisation of negative charge on oxygen through the **resonance hybrid** of the following :

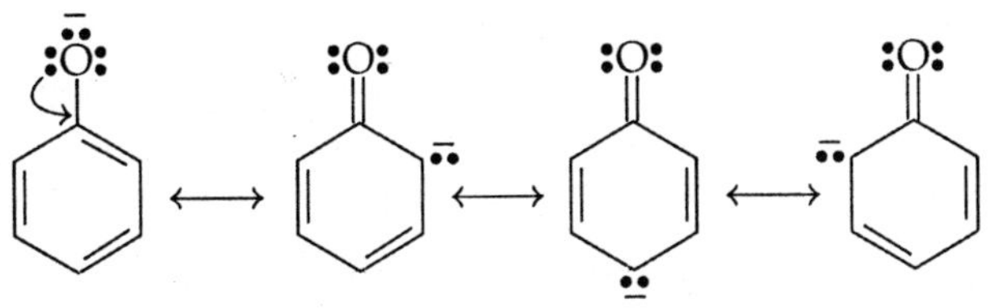

This type of **resonance stabilisation** of alkoxide ion is not possible so that alcohols are far less acidic than phenols.

Another aspect of the variation of acidity/basicity is due to the **hyperconjugation effect** which may change the order of acid/base strength. The sequence of acid strength is $BF_3 > B_2H_6 > Me_3B$. Thus $BF_3$ is a stronger Lewis acid than $B_2H_6$ or $(BH_3)_2$ but $BF_3$ does not form any compound with CO whereas $B_2H_6$ does. In a similar way $Me_2S$ can function as stronger base to $B_2H_6$ than $Me_2O$ though $BF_3$ forms strong adduct with $Me_2O$ than that of $Me_2S$. These are explained by the resonance stabilisation in the adduct through hyperconjugation which involves the protons in $B_2H_6$ for stabilisation in the adduct ($H_3B{\leftarrow}SMe_2$) which involves the use of vacant d-orbital of sulfur. But the adduct $H_3B{\leftarrow}OMe_2$ cannot be formed in this way. Pictorially these can be represented in the following way :

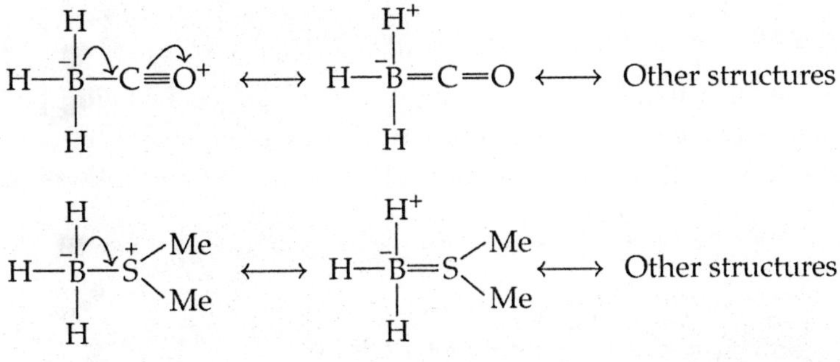

**Problem 4.31** Arrange the following : (a) Acidic nature : $Ag_2O$, $V_2O_5$, CO, $N_2O_5$; (b) Stronger base toward a proton : $NH_2^-$, $PH_2^-$.

**Ans. :** (a) The oxide with higher positive charge (high electronegativity) on the element should be most acidic. In both $V_2O_5$ and $N_2O_5$ the respective elements are in +5 oxidation state, but the electronegativity of +5 nitrogen is higher than that of +5 vanadium so that $N_2O_5$ will be the most acidic and the acidity order will be : $N_2O_5 > V_2O_5 > CO > Ag_2O$.

(b) $NH_2^-$, $PH_2^-$ are the bases of the conjugate acid $NH_3$ and $PH_3$ respectively. Actually $NH_3$ and $PH_3$ are bases and $NH_3$ is a stronger base than $PH_3$ i.e. $NH_3$ is a weaker acid than $PH_3$ and so the conjugate base of $NH_3$ is $NH_2^-$ which will be more stronger than the conjugate base of $PH_3$ i.e. $PH_2^-$.

## 4.18 HETEROGENEOUS ACID-BASE REACTIONS–SURFACE ACIDITY

There are several inorganic substances which exhibit Lewis and Bronsted acidity at solid surface so that they are called **surface acids.** These materials are solid and have large surface area.

*Solid acid*

Silicate materials or silica surfaces do not produce Lewis acidity but may show Bronsted acidity due to the prsence of dissociable protons of the –OH groups which are firmly held at the surface of the silicate materials. These Bronsted surface acidity of silica surfaces is comparable to that of $CH_3COOH$.

Extensive type of surface reactions may be carried out using Bronsted acid sites with $HOSiR_3$ or $ClSiR_3$ to produce a thin coatings at the surface of a wide variety of organic groups as of the following :

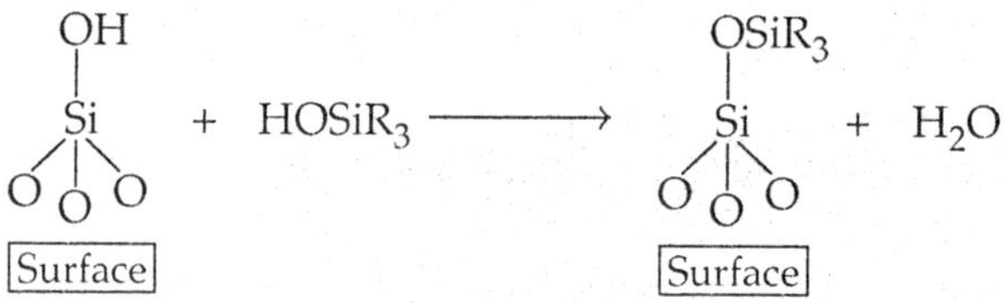


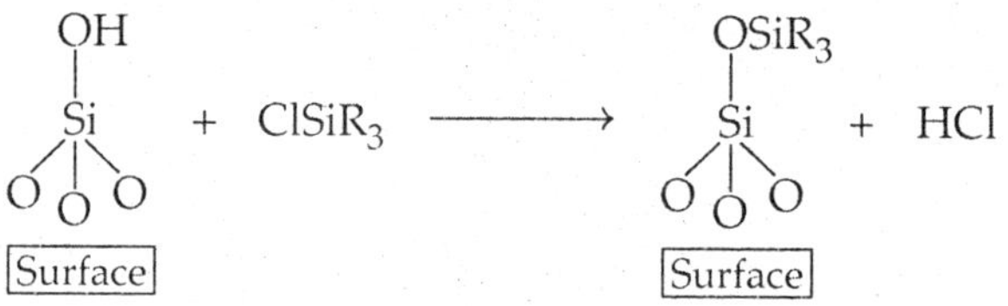


γ-alumina surface can also exhibit surface acidity. The surface OH groups on glass can be modified in a similar way e.g. on reduction of $CrO_2Cl_2$ by methanol or ethanol in the presence of a carboxylic acid there will be the formation of a solution of the following type :

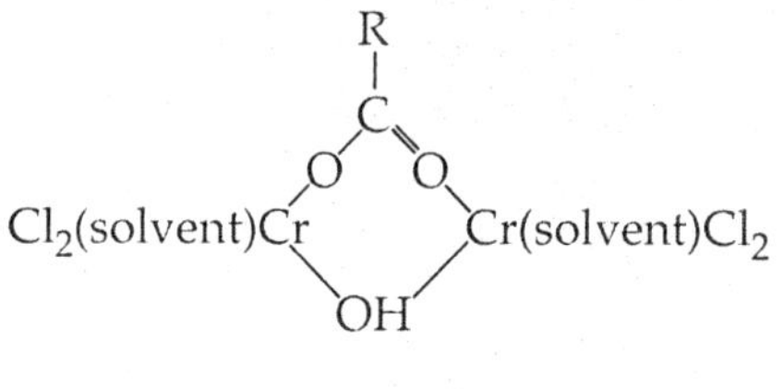

If a glass surface containing functional groups such as –OH, $-NH_2$, –COOH etc. is now treated with the above solution and heated, polymerisation occurs giving an insoluble coating which is attached firmly to the surface.

These solid acids are used as catalysts in petrochemicals and are sometimes used in chromatography. The above surface –OH groups on glass is used in the laboratory to study proton-sensitive compounds.

---

**Problem 4.32** Calculate the pH of a solution made by mixing 50 ml of 0.10M $NH_3$ and 50 ml 0.040M HCl. The $K_b$ of $NH_3 = 1.8 \times 10^{-5}$.

**Ans. :** 50 ml 0.10M $NH_3$ = 50 × 0.10 = 5.00 mmol of $NH_3$

50 ml 0.040M HCl = 50 × 0.040 = 2.00 mmol of HCl

| | $NH_3$ | + HCl | ⟶ $NH_4^+$ | + $Cl^-$ |
|---|---|---|---|---|
| Initial (mmol) | 5.00 | 2.00 | — | — |
| Change (mmol) | –2.00 | –2.00 | +2.00 | |
| Equilibrium (mmol) | 3.00 | — | 2.00 | |

$$\therefore [NH_3] = \frac{3.00 \text{ mmol}}{100} = 0.030M \text{ and } [NH_4^+] = \frac{2.00 \text{ mmol}}{100} = 0.020M$$

Now according to Henderson-Hasselbalch equation (4.57)

$$pOH = pK_b + \log\frac{[\text{Conjugate acid}]}{[\text{Conjugate base}]} = pK_b + \log\frac{[NH_4^+]}{[NH_3]}$$

$$= -\log(1.8 \times 10^{-5}) + \log\frac{0.020}{0.030} = 4.74 + \log\frac{0.020}{0.030} = 4.56$$

Since, pH + pOH = 14, so pH = 9.44.

---

**Problem 4.33** Calculate the concentration of $H^+$, $H_2PO_4^-$, $HPO_4^{2-}$, $PO_4^{3-}$ and $H_3PO_4$ in a 0.10 M solution of $H_3PO_4$. Given $K_1 = 7.5 \times 10^{-3}$ M, $K_2 = 6.2 \times 10^{-8}$ M, $K_3 = 1.0 \times 10^{-12}$ M.

**Ans. :** $H_3PO_4$ is an weak acid so that the source of $H^+$ ion is the dissociation of primary, secondary and tertiary steps of $H_3PO_4$ as also from the dissociation of $H_2O$ which is negligible. Let the degree of dissociation of $H_3PO_4$ is α and we can say,

$$\underset{0.1M-\alpha}{H_3PO_4} \rightleftharpoons \underset{\alpha}{H^+} + \underset{\alpha}{H_2PO_4^-} \qquad \text{and } K_1 = \frac{[H^+][H_2PO_4^-]}{[H_3PO_4]} = \frac{\alpha^2}{0.1M-\alpha}$$

$$\therefore \frac{\alpha^2}{0.1M-\alpha} = 7.5 \times 10^{-3}$$

Solving this quadratic equation we get,

$$\alpha = [H^+] = [H_2PO_4^-] = 2.4 \times 10^{-2} \text{ M}$$

$$\therefore [H_3PO_4] = 0.1M - \alpha = (0.1 - 2.4 \times 10^{-2})M = 7.6 \times 10^{-2} \text{ M}$$

In the secondary step we can write,

$$\underset{2.4\times10^{-2}M}{H_2PO_4^-} \rightleftharpoons \underset{2.4\times10^{-2}\text{ M}}{H^+} + HPO_4^{2-} \qquad \text{and } K_2 = \frac{[H^+][HPO_4^{2-}]}{[H_2PO_4^-]} = [HPO_4^{2-}]$$

$$\therefore K_2 = [HPO_4^{2-}] = 6.2 \times 10^{-8} \text{ so that } [HPO_4^{2-}] = 6.2 \times 10^{-8} \text{ M}$$

In the tertiary step, it may be written as,

$$HPO_4^{2-} \rightleftharpoons H^+ + PO_4^{3-} \qquad \text{and } K_3 = \frac{[H^+][PO_4^{3-}]}{[HPO_4^{2-}]}$$

$6.2\times10^{-8}M \qquad 2.4\times10^{-2}M$

$$\therefore [PO_4^{3-}] = K_3\frac{[HPO_4^{2-}]}{[H^+]} = 1.0 \times 10^{-12}\ M\ \frac{6.2\times10^{-8}M}{2.4\times10^{-2}M} = 2.58 \times 10^{-18}\ M.$$

**Problem 4.34** Calculate the pH of 0.1M solution of—(i) $Na_2HPO_4$, (ii) $NaH_2PO_4$ and (iii) $NaHCO_3$. Given that, **[C.U. 2003]**

$CO_2 + H_2O \rightleftharpoons H^+ + HCO_3^- \qquad K_1 = 4.2 \times 10^{-7}$ M

$HCO_3^- \rightleftharpoons H^+ + CO_3^{2-} \qquad K_2 = 4.8 \times 10^{-11}$ M

and for $H_3PO_4$, $K_1$, $K_2$, $K_3$ values given in Problem 4.33.

**Ans. :** (i) According to equation (4.82) for $Na_2HPO_4$ the

$$pH = \frac{1}{2}(pK_2 + pK_3) = \frac{1}{2}(7.2 + 12.0) = 9.60$$

(ii) For $NaH_2PO_4$, the pH of the solution will be according to equation (4.81) i.e.

$$pH = \frac{1}{2}(pK_1 + pK_2) = \frac{1}{2}(2.1 + 7.2) = 4.65$$

(iii) pH of $NaHCO_3$ solution will be,

$$pH = \frac{1}{2}(pK_1 + pK_2) = \frac{1}{2}[-\log(4.2 \times 10^{-7}) - \log(4.8 \times 10^{-11})] = 8.35.$$

**Problem 4.35** Calculate the pH of an aqueous solution of 0.015M pyridine (Given $K_b = 1.70 \times 10^{-9}$) write the necessary proton transfer reaction. **[B.U. 1991]**

**Ans. :** $py + H_2O \longrightarrow pyH^+ + OH^-$

$$K_b = \frac{[pyH^+][OH^-]}{[py]}$$

Let $\alpha$ is the fraction of pyridine that associate with the $H^+$ ion so that,

$$py + H_2O \longrightarrow pyH^+ + OH^-$$

$$(1-\alpha)C_b \qquad \alpha C_b \qquad \alpha C_b$$

where $C_b$ is the concentration of the base.

$$\therefore K_b = \frac{\alpha C_b \alpha C_b}{(1-\alpha)C_b}$$

When $\alpha << 1$, then $K_b = \dfrac{\alpha^2 C_b^2}{C_b} = \alpha^2 C_b$

so that $\alpha = \sqrt{\dfrac{K_b}{C_b}}$ [correct value of $\alpha$ can be obtained by solving the quadratic equation]

Therefore, $[OH^-] = \alpha C_b = \sqrt{\dfrac{K_b}{C_b}}C_b = \sqrt{K_b C_b}$ [For weak base this is the general expression for concentration of $OH^-$ ion]

$$= \sqrt{1.70\times10^{-9}\times0.015}\ M$$

$= 0.05 \times 10^{-4}$ M

$\therefore$ pOH = $-\log(0.05 \times 10^{-4}) = 5.3010$

$\therefore$ pH = 14 − 5.3010 = 8.6990.

---

**Problem 4.36** What is the pH of $10^{-3}$ M aqueous solution of $NH_4OH$? Given $K_b = 1.85 \times 10^{-5}$ M at 25°C.

**Ans. :** $\alpha = \sqrt{\frac{K_b}{C_b}} = \sqrt{\frac{1.85 \times 10^{-5} M}{0.001 M}}$ = 0.136 when $\alpha << 1$

More exact value of $\alpha$ can be obtained by solving the quadratic equation

$C_b\alpha^2 + K_b\alpha - K_b = 0$, when $\alpha = 0.127$

$\therefore$ $[OH^-] = \alpha C_b = 0.127 \times 10^{-3}$ M $= 1.27 \times 10^{-4}$ M

$\therefore$ pOH = $-\log[OH^-] = -\log(1.27 \times 10^{-4}) = 3.90$

$\therefore$ pH = 14 − 3.90 = 10.10.

---

**Problem 4.37** Calculate the pH of 0.01 M solution of monochloroacetic acid $K_a = 1.4 \times 10^{-3}$.

**Ans. :** If $\alpha$ is the degree of dissociation of the weak acid HA, then

$$HA \rightleftharpoons H^+ + A^-$$

$(1-\alpha)C_a$ $\quad \alpha C_a \quad \alpha C_a$ where $C_a$ is the concentration of the acid

Therefore when $\alpha << 1$, $\alpha = \sqrt{\frac{K_a}{C_a}}$

Now $[H^+] = \alpha C_a = \sqrt{\frac{K_a}{C_a}} C_a = \sqrt{K_a C_a} = \sqrt{1.4 \times 10^{-3} \times 0.1} = 1.8 \times 10^{-2}$ M

$\therefore$ pH = $-\log 1.18 \times 10^{-2} = 1.93$

More exact value of $\alpha$ can be obtained by solving the quadratic equation $C_a\alpha^2 + K_a\alpha - K_a = 0$.

---

**Problem 4.38** What will be the pH of the solution obtained by mixing 10 ml of 0.2(N) KOH with 30 ml of 0.1(N) $CH_3COOH$. $K_a = 2 \times 10^{-5}$.

**Ans. :** 10 ml 0.2(N) KOH will react with 20 ml 0.1(N) $CH_3COOH$ to give 20 ml 0.1(N) $CH_3COOK$. Here the total volume = 30 + 10 = 40 ml

$\therefore$ $[CH_3COOK] = \frac{20 \times 0.1}{40} = 0.05(N)$

Excess amount of $CH_3COOH$ = 30 − 20 = 10 ml

$\therefore$ $[CH_3COOH] = \frac{10 \times 0.1}{40} = 0.025(N)$

According to Henderson equation, $pH = pK_a + \log\frac{[Salt]}{[Acid]}$

$= -\log(2 \times 10^{-5}) + \log\frac{0.05}{0.025} = 4.699 + 0.3010 = 5.$

---

**Problem 4.39** 100 ml of a buffer solution of pH = 9 is to be prepared by mixing 0.1(N) HCl and 0.1(N)$NH_4OH$. Calculate the volume of each solution required to prepare the buffer. $K_b = 2 \times 10^{-5}$.

**Ans. :** Let the volume of HCl = *x* ml so volume of $NH_4OH$ = (100 – *x*) ml. So *x* ml. 0.1(N) HCl ≡ *x* ml 0.1(N) $NH_4OH$. Therefore (100 – *x*) – *x* i.e. (100 – 2*x*) ml $NH_4OH$ in 100 ml solution as also there will be *x* ml 0.1(N) $NH_4Cl$ in 100 ml solution.

$$\therefore\ [NH_4OH] = \frac{(100-2x)0.1}{100}$$

$$\text{and } [NH_4Cl] = \frac{x \times 0.1}{100}$$

Now from Henderson equation,

$$pH = pK_w - \left(pK_b + \log\frac{[Salt]}{[Base]}\right)$$

$$\text{i.e. } 9 = 14 - \left(-\log 2\times10^{-5} + \log\frac{\frac{x\times0.1}{100}}{\frac{(100-2x)0.1}{100}}\right)$$

$$\text{or, } 9 = 14 - \left(5 - \log 2 + \log\frac{x}{100-2x}\right)$$

$$\therefore\ \frac{x}{100-2x} = 2 \text{ i.e. } x = 40$$

∴ Volume of HCl = 40 ml and volume of $NH_4OH$ = 60 ml.

**Problem 4.40** 4 g of NaOH is dissolved in 1 litre of a solution containing one mole of $CH_3COOH$ and one mole of $CH_3COONa$. Calculate the pH of the final solution. $K_a$ for $CH_3COOH$ = $1.8 \times 10^{-5}$.

**Ans. :** $[NaOH] = \frac{4}{40}$ mol $lit^{-1}$ = 0.1 M

As 0.1 M NaOH is added to 1 litre of the buffer solution containing 1 mole $CH_3COOH$ and $CH_3COONa$, 0.1M NaOH will neutralise 0.1 M $CH_3COOH$ and thus reducing the concentration of $CH_3COOH$ from 1 M to 0.9 M and thereby the concentration of $CH_3COONa$ increases from 1 M to 1.1 M.

Now from Henderson equation,

$$pH = pK_a + \log\frac{[Salt]}{[Acid]} = -\log(1.8 \times 10^{-5}) + \log\frac{1.1}{0.9}$$

$$= 4.832.$$

**Problem 4.41** What will be the change in pH of water when 0.01 mole of NaOH is added to 10 litres of water?

**Ans. :** $[NaOH] = \frac{0.01}{10} = 10^{-3}$ M

$$\therefore\ [OH^-] = 10^{-3} \text{ M, so that } [H^+] = \frac{1\times10^{-14}}{10^{-3}} = 10^{-11} \text{ M}$$

$\therefore$ pH = $-\log[H^+] = -\log 10^{-11} = 11$

As pH of water is 7, its pH increases by 4 units.

---

**Problem 4.42** Predict the stronger acid and explain your answer. **[C.U. 2003]**

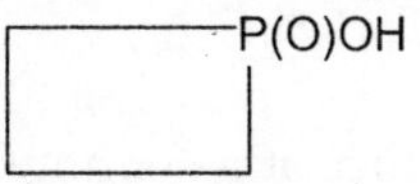

and $(CH_3)_2P(O)OH$

**Ans. :**

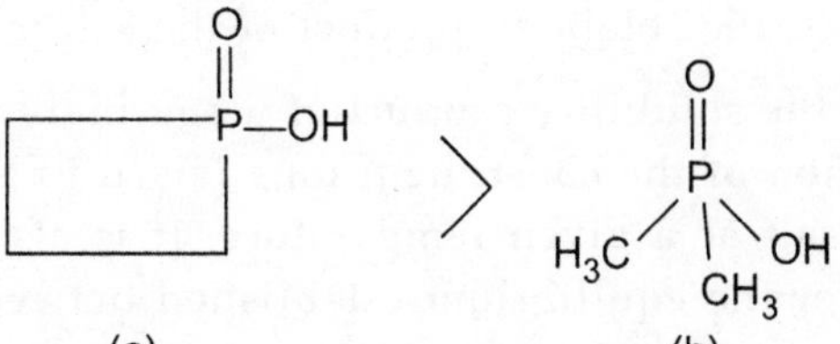

(a) (b)

In structure (a) the phosphorus utilises its two pure p-orbitals for bonding to the adjacent carbon atoms and in the residual bonds the s-orbital and another p-orbital are distributed equally having the s-character about 50%. In structure (b) the central phosphorous is roughly tetrahedrally hybridised having s-character almost 25%. As s-character increases in the cyclic structure so cyclic compound is more acidic than the non-cyclic one.

---

## 4.19 SOLUBILITY AND SOLUBILITY PRODUCT

The salts having solubility less than 0.01 gram mole per litre at ordinary temperature are commonly known as sparingly soluble salts e.g. AgCl, $BaSO_4$, $CaSO_4$ etc. In a saturated solution of a sparingly soluble salt, a part of the salt is solvated and so there will be a dynamic equilibrium is established between the dissolved and undissolved species of a sparingly soluble salt, AB such as :

$$\underset{(solid)}{\text{AB (undissolved)}} \rightleftharpoons \text{AB (dissolved)} \rightleftharpoons A^+(aq) + B^-(aq) \qquad ...(4.93)$$

Since the dissolved part of the salt is completely dissociated, so that by applying law of mass action, we can write the equilibrium constant $K = \dfrac{[A^+(aq)][B^-(aq)]}{[AB(dissolved)]}$

and since the concentration of AB (dissolved) is constant for a sparingly soluble salt, then K[AB(dissolved] = constant, which is called the **solubility product** of a sparingly soluble salt and is denoted by $K_{sp}$ so that,

$$K_{sp} = [A^+(aq)][B^-(aq)]$$

$$\text{or,} \quad K_{sp} = [A^+][B^-]$$

$K_{sp}$ is constant at a given temperature for any particular salt. The value of $K_{sp}$ changes with variation of temperature.

Let us consider a sparingly soluble salt $A_pB_q$ which has the solubility s mol $lit^{-1}$. Then,

$$A_pB_q \text{ (dissolved)} \rightleftharpoons pA^+ + qB^-$$

Therefore the concentration of $A^+$ and $B^-$ ions are ps and qs respectively so that

$$K_{sp} = [A^+]^p[B^-]^q = (ps)^p(qs)^q = p^pq^qs^{p+q} \quad ...(4.94)$$

$$\text{Thus } s = \left(\frac{K_{sp}}{p^pq^q}\right)^{\frac{1}{p+q}} \quad ...(4.95)$$

This relation connect the solubility and solubility product of a sparingly soluble salt. It is also evident that as solubility increases, the solubility product of the salt increases.

According to equation (4.94) **the solubility product of a sparingly soluble salt is defined as the product of the concentration of the constituent ions raised to the appropriate powers in a saturated solution of the salt at a given temperature.** It is also defined according to equation (4.93) **as the constant for the equilibrium established between a sparingly soluble salt and its ions in solution.**

### 4.19.1 Solubility product and precipitation of a salt in solution

In solution at a given temperature the product of the molar concentration of the constituent ions raised to the appropriate powers of a salt i.e. the ionic product can never exceed the value of $K_{sp}$ since $K_{sp}$ is the highest limit of the ionic product at a particular temperature. Therefore if in a solution the product of the concentration of constituent cations and anions of a salt at a given temperature exceed the value of $K_{sp}$, then the salt will be precipitated. Thus for a salt of the type $A_pB_q$ the condition of precipitation is :

$$[A^+]^p[B^-]^q > K_{sp} \quad ...(4.96)$$

This condition is valid for a saturated solution in equilibrium with excess solid. The solubility product values of some of the sparingly soluble salts are given in Table 4.11.

### 4.19.2 Common ion effect and solubility product

If in a saturated solution of a sparingly soluble salt, a strong electrolyte having an ion in common is added, the solublity of the sparingly soluble salt decreases e.g. the solubility of AgCl in a solution containing NaCl is less than that of water.

Let the solubility of a sparingly soluble salt (AB) is s mol/lit then the solublity product,

$$K_{sp} = [A^+][B^-] = s \times s = s^2$$

Now at a fixed temperature a strong electrolyte (AX) is added in which $A^+$ ion is common to the sparingly soluble salt (AB). Let the solubility of AX is $x$ mol/lit and let the decreased solublity of AB in presence of common ion $A^+$ is $s_1$ mol/lit. Therefore in the mixed solution the concentration of $A^+$ ion = $(s_1 + x)$ gram-ion/lit and concentration of $B^-$ = $s_1$ gram-ion/lit. Since at a particular temperature the solubility product is constant, so that,

$$K_{sp} = [A^+][B^-] = (s_1 + x) \times s_1$$

This expression suggests that in order to maintain the constancy of $K_{sp}$ value $s_1$ must be less than that of s i.e. the solubility of a sparingly soluble salt will decrease in presence of a common ion.

**Table 4.11 : Solubility products of some sparingly soluble salts at 25°C**

| Salt | Dissociation reaction | Solubility product expression | $K_{sp}$ |
|---|---|---|---|
| $AgCl$ | $AgCl(s) \rightleftharpoons Ag^+ + Cl^-$ | $K_{sp} = [Ag^+][Cl^-]$ | $1.1 \times 10^{-10}$ |
| $PbCl_2$ | $PbCl_2(s) \rightleftharpoons Pb^{2+} + 2Cl^-$ | $K_{sp} = [Pb^{2+}][Cl^-]^2$ | $1.6 \times 10^{-5}$ |
| $Hg_2Cl_2$ | $Hg_2Cl_2(s) \rightleftharpoons Hg_2^{2+} + 2Cl^-$ | $K_{sp} = [Hg_2^{2+}][Cl^-]^2$ | $1.3 \times 10^{-18}$ |
| $HgS$ | $HgS(s) \rightleftharpoons Hg^{2+} + S^{2-}$ | $K_{sp} = [Hg^{2+}][S^{2-}]$ | $3.0 \times 10^{-52}$ |
| $CuS$ | $CuS(s) \rightleftharpoons Cu^{2+} + S^{2-}$ | $K_{sp} = [Cu^{2+}][S^{2-}]$ | $4.0 \times 10^{-38}$ |
| $Bi_2S_3$ | $Bi_2S_3(s) \rightleftharpoons 2Bi^{3+} + 3S^{2-}$ | $K_{sp} = [Bi^{3+}][S^{2-}]^3$ | $1.6 \times 10^{-72}$ |
| $CdS$ | $CdS(s) \rightleftharpoons Cd^{2+} + S^{2-}$ | $K_{sp} = [Cd^{2+}][S^{2-}]$ | $1.0 \times 10^{-28}$ |
| $PbS$ | $PbS(s) \rightleftharpoons Pb^{2+} + S^{2-}$ | $K_{sp} = [Pb^{2+}][S^{2-}]$ | $3.0 \times 10^{-29}$ |
| $As_2S_3$ | $As_2S_3(s) \rightleftharpoons 2As^{3+} + 3S^{2-}$ | $K_{sp} = [As^{3+}]^2[S^{2-}]^3$ | $1.0 \times 10^{-16}$ |
| $Sb_2S_3$ | $Sb_2S_3(s) \rightleftharpoons 2Sb^{3+} + 3S^{2-}$ | $K_{sp} = [Sb^{3+}]^2[S^{2-}]^3$ | $1.6 \times 10^{-93}$ |
| $SnS$ | $SnS(s) \rightleftharpoons Sn^{2+} + S^{2-}$ | $K_{sp} = [Sn^{2+}][S^{2-}]$ | $8.0 \times 10^{-29}$ |
| $Fe(OH)_3$ | $Fe(OH)_3(s) \rightleftharpoons Fe^{3+} + 3OH^-$ | $K_{sp} = [Fe^{3+}][OH^-]^3$ | $6.0 \times 10^{-38}$ |
| $Al(OH)_3$ | $Al(OH)_3(s) \rightleftharpoons Al^{3+} + 3OH^-$ | $K_{sp} = [Al^{3+}][OH^-]^3$ | $5.0 \times 10^{-33}$ |
| $Cr(OH)_3$ | $Cr(OH)_3(s) \rightleftharpoons Cr^{3+} + 3OH^-$ | $K_{sp} = [Cr^{3+}][OH^-]^3$ | $6.7 \times 10^{-31}$ |
| $NiS$ | $NiS(s) \rightleftharpoons Ni^{2+} + S^{2-}$ | $K_{sp} = [Ni^{2+}][S^{2-}]$ | $1.0 \times 10^{-25}$ |
| $CoS$ | $CoS(s) \rightleftharpoons Co^{2+} + S^{2-}$ | $K_{sp} = [Co^{2+}][S^{2-}]$ | $7.0 \times 10^{-23}$ |
| $MnS$ | $MnS(s) \rightleftharpoons Mn^{2+} + S^{2-}$ | $K_{sp} = [Mn^{2+}][S^{2-}]$ | $5.6 \times 10^{-16}$ |
| $ZnS$ | $ZnS(s) \rightleftharpoons Zn^{2+} + S^{2-}$ | $K_{sp} = [Zn^{2+}][S^{2-}]$ | $1.0 \times 10^{-24}$ |
| $MgCO_3$ | $MgCO_3(s) \rightleftharpoons Mg^{2+} + CO_3^{2-}$ | $K_{sp} = [Mg^{2+}][CO_3^{2-}]$ | $3.0 \times 10^{-5}$ |
| $CaCO_3$ | $CaCO_3(s) \rightleftharpoons Ca^{2+} + CO_3^{2-}$ | $K_{sp} = [Ca^{2+}][CO_3^{2-}]$ | $4.7 \times 10^{-9}$ |
| $SrCO_3$ | $SrCO_3(s) \rightleftharpoons Sr^{2+} + CO_3^{2-}$ | $K_{sp} = [Sr^{2+}][CO_3^{2-}]$ | $2.0 \times 10^{-10}$ |
| $BaCO_3$ | $BaCO_3(s) \rightleftharpoons Ba^{2+} + CO_3^{2-}$ | $K_{sp} = [Ba^{2+}][CO_3^{2-}]$ | $1.6 \times 10^{-9}$ |
| $CaC_2O_4$ | $CaC_2O_4(s) \rightleftharpoons Ca^{2+} + C_2O_4^{2-}$ | $K_{sp} = [Ca^{2+}][C_2O_4^{2-}]$ | $1.3 \times 10^{-9}$ |
| $SrC_2O_4$ | $SrC_2O_4(s) \rightleftharpoons Sr^{2+} + C_2O_4^{2-}$ | $K_{sp} = [Sr^{2+}][C_2O_4^{2-}]$ | $5.6 \times 10^{-8}$ |
| $BaSO_4$ | $BaSO_4(s) \rightleftharpoons Ba^{2+} + SO_4^{2-}$ | $K_{sp} = [Ba^{2+}][SO_4^{2-}]$ | $1.0 \times 10^{-10}$ |
| $SrSO_4$ | $SrSO_4(s) \rightleftharpoons Sr^{2+} + SO_4^{2-}$ | $K_{sp} = [Sr^{2+}][SO_4^{2-}]$ | $7.6 \times 10^{-7}$ |
| $CaF_2$ | $CaF_2(s) \rightleftharpoons Ca^{2+} + 2F^-$ | $K_{sp} = [Ca^{2+}][F^-]^2$ | $3.9 \times 10^{-11}$ |
| $Ca_3(PO_4)_2$ | $Ca_3(PO_4)_2(s) \rightleftharpoons 3Ca^{2+} + 2PO_4^{3-}$ | $K_{sp} = [Ca^{2+}]^3[PO_4^{3-}]^2$ | $1.3 \times 10^{-32}$ |
| $Sr_3(PO_4)_2$ | $Sr_3(PO_4)_2(s) \rightleftharpoons 3Sr^{2+} + 2PO_4^{3-}$ | $K_{sp} = [Sr^{2+}]^3[PO_4^{3-}]^2$ | $1.0 \times 10^{-31}$ |

| Salt | Dissociation reaction | Solubility product expression | $K_{sp}$ |
|---|---|---|---|
| $Ba_3(PO_4)_2$ | $Ba_3(PO_4)_2(s) \rightleftharpoons 3Ba^{2+} + 2PO_4^{3-}$ | $K_{sp} = [Ba^{2+}]^3[PO_4^{3-}]^2$ | $6.0 \times 10^{-39}$ |
| $MgNH_4PO_4$ | $MgNH_4PO_4(s) \rightleftharpoons Mg^{2+} + NH_4^+ + PO_4^{3-}$ | $K_{sp} = [Mg^{2+}][NH_4^+][PO_4^{3-}]$ | $3.0 \times 10^{-13}$ |
| $PbCrO_4$ | $PbCrO_4(s) \rightleftharpoons Pb^{2+} + CrO_4^{2-}$ | $K_{sp} = [Pb^{2+}][CrO_4^{2-}]$ | $2.0 \times 10^{-16}$ |
| AgCN | $AgCN(s) \rightleftharpoons Ag^+ + CN^-$ | $K_{sp} = [Ag^+][CN^-]$ | $1.6 \times 10^{-14}$ |

Thus we can summarise the result as :

If ionic product > $K_{sp}$, the salt will be precipitated

ionic product = $K_{sp}$, the solution is just saturated

ionic product < $K_{sp}$, the solution is not saturated.

**Problem 4.43** What will be the solubility of AgCl in a solution of 0.1(M) NaCl solution? ($K_{sp}$ of AgCl = $1 \times 10^{-10}$)

**Ans. :** Let the solubility x mol/lit. So the concentration of $Ag^+$ and $Cl^-$ ions are x gram-ion/lit. As NaCl is a strong electrolyte, therefore, the total concentration of

$Cl^-$ i.e. $[Cl^-] = x + 0.1$

and $[Ag^+] = x$

According to solubility product principle,

$$K_{sp} = [Ag^+][Cl^-] = x(x + 0.1)$$

so that $x(x + 0.1) = 1 \times 10^{-10}$

i.e. $x^2 \times 0.1x = 1 \times 10^{-10}$

As the solubility of AgCl is very poor, so $x^2$ is very small in comparison to 0.1.

Therefore $0.1x = 1 \times 10^{-10}$

so that $x = 1 \times 10^{-9}$

Therefore the solubility of AgCl in 0.1(M) NaCl is $1 \times 10^{-9}$ mol/lit.

**Problem 4.44** The solubility of $CaF_2$ in water at 18°C is $2.05 \times 10^{-4}$ mol/lit. Calculate its solubility product.

**Ans. :** $CaF_2 \rightleftharpoons Ca^{2+} + 2F^-$

Therefore, $[Ca^{2+}] = 2.05 \times 10^{-4}$ gram-ion/lit

$[F^-] = 2 \times 2.05 \times 10^{-4}$ gram-ion/lit

$$K_{sp} = [Ca^{2+}][F^-]^2 = (2.05 \times 10^{-4})(2 \times 2.05 \times 10^{-4})^2$$
$$= 3.45 \times 10^{-11} \text{ mol}^3$$

**Problem 4.45** The solubility of $CaF_2$ in water at 18°C is $2.04 \times 10^{-4}$ mol/lit. Calculate (i) the solubility product and (ii) the solubility of $CaF_2$ in 0.01(M) NaF solution. **[C.U. 2007]**

**Ans. :** (i) $CaF_2 \longrightarrow Ca^{2+} + 2F^-$

$$K_{sp} = [Ca^{2+}][F^-]^2 = (2.04 \times 10^{-4})(2 \times 2.04 \times 10^{-4})^2$$
$$= 3.4 \times 10^{-11} \text{ mol}^3$$

(ii) Let the solubility of $CaF_2$ in NaF is $x$ mol/lit,

so that $F^- = (2x + 0.01)M$

and $[Ca^{2+}] = xM$

$\therefore K_{sp} = [Ca^{2+}][F^-]^2 = [(xM)(2x + 0.01)M]^2$

Neglecting $x^2$ and $x^3$ terms as $x$ is very small, we get

$0.0001xM = 3.4 \times 10^{-11}$

$\therefore x = 3.4 \times 10^{-7}$ mol/lit.

---

### 4.19.3 Applications of solubility product principle

A. **Precipitation of NaCl :** Pure NaCl can be prepared from impure solution of NaCl by the application of solubility product principle. When dry HCl gas is passed through a saturated solution of impure NaCl, the concentration of $Cl^-$ ions increase to a larger extent and the ionic product of $Na^+$ and $Cl^-$ ions i.e. $[Na^+][Cl^-]$ exceeds the solubility product of NaCl, thereby NaCl is separated in a pure state. The solubility product of contaminated impurities like $MgCl_2$, $CaCl_2$ etc. will not reach so that they remain in the solution.

B. **Application in analytical chemistry :** Solubility product principle is extremely important in qualitative detection of metal ions. Qualitative detection of cations is based on the solubility product principle and according to the solubility product of the precipitate they are separated into different analytical groups :

Group I    Cations : $Ag^+$, $Pb^{2+}$, $Hg_2^{2+}$

Group reagent : Dilute HCl

Precipitated as : AgCl, $PbCl_2$, $Hg_2Cl_2$

According to Table 4.11 the solubility products of the chlorides of Group I cations are sufficiently low so that on addition of dilute HCl, the ionic products of these metal chlorides exceed their solubility products and thereby they are precipitated. On the other hand the metallic chlorides of other groups have higher solubility product and thus they will not be precipitated by the application of HCl.

Group II    Cations : $Cu^{2+}$, $Pb^{2+}$, $Cd^{2+}$, $Hg^{2+}$, $Bi^{3+}$, $As^{3+}$, $Sb^{3+}$, $Sn^{2+}$

Group reagent : $H_2S$ gas is passed in presence of dilute HCl

Precipitated as : CuS, PbS, CdS, HgS, $Bi_2S_3$, $As_2S_3$, $Sb_2S_3$, SnS

$H_2S$ is feebly dissociated in aqueous solution as :

$$H_2S \rightleftharpoons H^+ + HS^- \quad K_1 = \frac{[H^+][HS^-]}{[H_2S]} = 9.6 \times 10^{-8}\ M$$

$$HS^- \rightleftharpoons H^+ + S^{2-} \quad K_2 = \frac{[H^+][S^{2-}]}{[HS^-]} = 1.3 \times 10^{-14}\ M$$

Therefore the $S^{2-}$ ion concentration in a saturated solution of $H_2S$ is given by,

$$[S^{2-}] = \frac{K_1K_2[H_2S]}{[H^+]^2} = \frac{1.25\times 10^{-21} M^2[H_2S]}{[H^+]^2}$$

In acidic solution the $[H^+]$ is large so that due to the common ion effect $H_2S$ will be less dissociated and $[S^{2-}]$ will be low. According to Table 4.11 the solubility products of the sulphides of $Cu^{2+}$, $Pb^{2+}$, $Cd^{2+}$, $Hg^{2+}$, $Bi^{3+}$, $As^{3+}$, $Sb^{3+}$ and $Sn^{2+}$ are low compared to $Fe^{3+}$, $Co^{2+}$, $Ni^{2+}$, $Zn^{2+}$ and $Mn^{2+}$ so that CuS, PbS, CdS, HgS, $Bi_2S_3$, $As_2S_3$, $Sb_2S_3$ and SnS will be precipitated from a solution of about 0.3M in $H^+$ ion concentration while the ionic products of other suphides are smaller than their respective solubility products and thus they are not precipitated from the solution. It is to be noted that among the sulfides of Group II metal ions CdS has higher value of solubility product. It is a common practice in the laboratory that yellow precipitate of CdS is obtained when the solution is diluted by water. This is due to the fact that on dilution $H^+$ ion concentration decreases thereby $S^{2-}$ ion concentration increases and CdS will be precipitated in this situation.

Group IIIA Cations : $Fe^{3+}$, $Al^{3+}$ $Cr^{3+}$

Group reagent : $NH_4OH$ in presence of $NH_4Cl$

Precipitated as : $Fe(OH)_3$, $Al(OH)_3$, $Cr(OH)_3$

Due to the common ion effect of $NH_4^+$ ion which is produced by the dissociation of the strong electrolyte $NH_4Cl$, the weak base $NH_4OH$ becomes more weaker and the concentration of $OH^-$ ions in the solution decreases. In this situation the ionic products of the least soluble hydroxides of $Fe^{3+}$, $Al^{3+}$ and $Cr^{3+}$ exceed the solubility products and they are precipitated. The solubility products of other hydroxides e.g. $Co^{2+}$, $Ni^{2+}$, $Zn^{2+}$, $Ca^{2+}$ etc. are higher and consequently higher concentration of $OH^-$ ions are required to precipitate these hydroxides from the solution than those required for the precipitation of group IIIA metal ions and thus they are not precipitated from the solution.

Group IIIB Cations : $Co^{2+}$, $Ni^{2+}$, $Mn^{2+}$, $Zn^{2+}$

Group reagent : $H_2S$ in presence of $NH_4OH$

Precipitated as : CoS, NiS, MnS, ZnS

Table 4.11 shows that the solubility products of group IIIB metal sulphides are higher than those of the sulphides of group II cations. In ammonical solution the dissociation of $H_2S$ more favoured to the right hand direction and so the concentration of $S^{2-}$ ions becomes larger and this large $S^{2-}$ ion concentration is favourable for the precipitation of the sulphides of group IIIB cations.

Group IV Cations : $Ba^{2+}$, $Ca^{2+}$, $Sr^{2+}$

Group reagents : $(NH_4)_2CO_3$ in presence of $NH_4OH + NH_4Cl$

Precipitated as : $BaCO_3$, $CaCO_3$, $SrCO_3$

The solubility product of $MgCO_3$ is somewhat higher than that of $BaCO_3$, $CaCO_3$ and $SrCO_3$ (Table 4.11) so that selective precipitation of $Ba^{2+}$, $Ca^{2+}$ and $Sr^{2+}$ ions can be done by controlling the $CO_3^{2-}$ ion concentration. $Na_2CO_3$ is not used here instead a saturated solution of $(NH_4)_2CO_3$ in presence of $NH_4Cl$, when due to the common ion effect of $NH_4^+$ ion, the concentration of $CO_3^{2-}$ ions just sufficient to precipitate group IV cations but not $Mg^{2+}$.

**Problem 4.46** At 25°C, a saturated solution of $BaSO_4$ is $3.9 \times 10^{-5}$ M. Calculate its solubility in 0.1M $Na_2SO_4$ solution.

**Ans. :** $BaSO_4 \rightleftharpoons Ba^{2+} + SO_4^{2-}$

$\therefore [Ba^{2+}] = [SO_4^{2-}] = 3.9 \times 10^{-5}$ M

Thus $K_{sp}$ of $BaSO_4 = [Ba^{2+}][SO_4^{2-}] = (3.9 \times 10^{-5}\ M)^2 = 1.52 \times 10^{-9}\ M^2$

The concentration of $SO_4^{2-}$ ion in 0.1M $Na_2SO_4$ solution = 0.1M.

$$\text{Therefore } [Ba^{2+}] = \frac{K_{sp} \text{ of } BaSO_4}{[SO_4^{2-}]} = \frac{1.52 \times 10^{-9} M^2}{0.1 M} = 1.5 \times 10^{-8}\ M$$

Thus the solubility of $BaSO_4$ is reduced from $3.9 \times 10^{-5}$ M to $1.5 \times 10^{-8}$ M due to the common ion effect of $SO_4^{2-}$.

---

**Problem 4.47** The solubility of $Mg(OH)_2$ in pure water is $9.57 \times 10^{-3}$ g $lit^{-1}$. Calculate its solubility (in gram per litre) in 0.02 M $Mg(NO_3)_2$ solution.

**Ans. :** Solubility of $Mg(OH)_2$ in pure water in mol $lit^{-1}$

$$= \frac{9.57 \times 10^{-3}}{58} = 1.65 \times 10^{-4} \quad [\text{Mol. wt. of } Mg(OH)_2 = 58]$$

$\therefore [Mg^{2+}] = 1.65 \times 10^{-4}$ mol $lit^{-1}$

$[OH^-] = 2 \times 1.65 \times 10^{-4}$ mol $lit^{-1}$

$K_{sp}$ of $Mg(OH)_2 = [Mg^{2+}][OH^-]^2 = [1.65 \times 10^{-4}][2 \times 1.65 \times 10^{-4}]^2$ $mol^3\ lit^{-3}$

$= 1.797 \times 10^{-11}$ $mol^3\ lit^{-3}$

Let the solubility of $Mg(OH)_2$ in 0.02M $Mg(NO_3)_2$ solution = s mol $lit^{-1}$

$\therefore [Mg^{2+}] = (s + 0.02)$ mol $lit^{-1}$

and $[OH^-] = 2s$ mol $lit^{-1}$

Hence $K_{sp} = (s + 0.02)(2s)^2$

$\therefore (s + 0.02)(2s)^2 = 1.797 \times 10^{-11}$

$4s^3 + 0.08s^2 = 1.797 \times 10^{-11}$

So that $0.08s^2 = 1.797 \times 10^{-11}$ (neglecting $4s^3$ as s is small)

i.e. $s = 1.5 \times 10^{-5}$ mol $lit^{-1}$ $= 1.5 \times 10^{-5} \times 58 = 8.7 \times 10^{-4}$ g $lit^{-1}$.

---

**Problem 4.48** Calculate the solubility of AgCl in 3M $NH_3$ solution. $K_{sp}$ of AgCl = $1.5 \times 10^{-10}$ $mol^2\ lit^{-2}$. Instability constant of $Ag(NH_3)_2^+ = 6.8 \times 10^{-8}$ $mol^2\ lit^{-2}$.

**Ans. :** Two equilibrium of the following types involved in the process :

$$AgCl(s) \rightleftharpoons Ag^+ + Cl^-$$

$$\text{and} \quad Ag(NH_3)_2^+ \rightleftharpoons Ag^+ + 2NH_3$$

At equilibrium the concentration of $Ag^+$ will be the same for both the processes provided solid and $Ag(NH_3)_2^+$ ions are present. For each mole of AgCl dissolved, 1 mole of $Ag(NH_3)_2^+$ and 1 mole of $Cl^-$ are formed. As the instability constant of the complex is small, so that the complex is feebly dissociated and we may assume that,

$$[Ag(NH_3)_2^+] = [Cl^-] = x \text{ mol lit}^{-1} \text{ (say)}$$

$$K_{sp} = [Ag^+][Cl^-] = 1.5 \times 10^{-10}$$

$$\therefore\ [Ag^+] = \frac{1.5\times 10^{-10}}{x}$$

The instability constant of the complex is $\dfrac{[Ag^+][NH_3]^2}{[Ag(NH_3)_2^+]}$

$$\therefore\ \frac{[Ag^+][NH_3]^2}{[Ag(NH_3)_2^+]} = 6.8 \times 10^{-8}$$

$$\text{i.e.}\quad \frac{\dfrac{1.5\times 10^{-10}}{x}\times 3^2}{x} = 6.8 \times 10^{-8}$$

$$\text{so that } x^2 = \frac{13.5\times 10^{-10}}{6.8\times 10^{-8}}$$

where $x$ = 0.141 mol lit$^{-1}$ = 0.141 × 143.5 = 20.23 g lit$^{-1}$

---

**Problem 4.49** The solubility product ($K_{sp}$) of chalk is 9.3 × $10^{-8}$ mol$^2$ lit$^{-2}$. Calculate its solubility in gram per litre.

**Ans. :** $CaCO_3 \rightleftharpoons Ca^{2+} + CO_3^{2-}$

Let the solubility of $CaCO_3$ = s mol lit$^{-1}$

$$K_{sp} = [Ca^{2+}][CO_3^{2-}] = s \times s$$

and so, $s = \sqrt{K_{sp}} = \sqrt{9.3\times 10^{-8}}$ = 0.000304 mol lit$^{-1}$

∴ Solubility in gram per litre = 0.000304 × 100 = 0.0304 g lit$^{-1}$

---

**Problem 4.50** Find the solubility of AgCl in 1(M) ammonia solution and also the order of increase in solubility of AgCl in the above mentioned solution in comparison to aqueous solution.

[Given $K_{sp}$ of AgCl = 1.8 × $10^{-10}$ mol$^2$ lit$^{-2}$ **[C.U. 2004]**

$K_f$ of $[Ag(NH_3)_2]^+$ = 1.66 × $10^7$ mol$^{-2}$ lit$^2$]

**Ans. :** Let the solubility of AgCl in water = s mol lit$^{-1}$

$AgCl(s) \rightleftharpoons Ag^+ + Cl^-$, $\quad K_{sp} = [Ag^+][Cl^-] = s \times s$

and so $s = \sqrt{K_{sp}} = \sqrt{1.8\times 10^{-10}}$ = 1.34 × $10^{-5}$ mol lit$^{-1}$

$Ag^+ + 2NH_3 \longrightarrow Ag(NH_3)_2^+$, $\quad K_f = \dfrac{[Ag(NH_3)_2^+]}{[Ag^+][NH_3]^2}$

For each mole AgCl dissolved, 1 mole of $Ag(NH_3)_2^+$ and 1 mole of $Cl^-$ are formed.

Let $[Ag(NH_3)_2^+] = [Cl^-] = x$ mol lit$^{-1}$

$\therefore\ K_f = \dfrac{[Ag(NH_3)_2^+]}{[Ag^+][NH_3]^2}$ $\quad [Ag^+][Cl^-] = K_{sp} = 1.8 \times 10^{-10}$

$$\therefore\ [Ag^+] = \frac{1.8\times 10^{-10}}{x}$$

$$\text{So that } 1.66 \times 10^7 = \frac{x}{\dfrac{1.8\times 10^{-10}}{x}\times 1}$$

whence $x^2 = 2.988 \times 10^{-3}$, so that $x = 5.46 \times 10^{-2}$ mol lit$^{-1}$.

Therefore the solublity of AgCl in 1(M) ammonia solution is $5.46 \times 10^{-2}$ mol lit$^{-1}$.

---

**Problem 4.51** What will be the minimum concentration of $S^{2-}$ ion to precipitate a sulphide of a metal ion $M^{2+}$ having solubility product of the sulphide is $K_{sp}$?

**Ans. :** Metal sulphide MS dissociated as :

$$MS(s) \rightleftharpoons M^{2+} + S^{2-}$$

$$K_{sp} = [M^{2+}][S^{2-}]$$

$$\therefore [S^{2-}] = \frac{K_{sp}}{[M^{2+}]}$$

This is the required concentration of a sulphide ion.

---

**Problem 4.52** What are the limitations of the application of solubility product principle?

**Ans. :** The following points denote the limitations of solubility product principle :

(a) It is defined as an equilibrium constant established between a sparingly soluble salt and its ions in solution. This constant does not say anything about the rate at which the equilibrium is established.

(b) Complete precipitation for a sparingly soluble salt is not possible because there is always present a residual amount of ions in the solution.

(c) The solubility product of a substance may depends upon the physical structure of the precipitate e.g. $\alpha$ and $\beta$ CoS has the $K_{sp}$ values $5 \times 10^{-23}$ and $3 \times 10^{-26}$ respectively, freshly precipitated $\alpha$ NiS has $K_{sp} = 4 \times 10^{-20}$ while aged $\beta$ NiS has $K_{sp} = 1.3 \times 10^{-25}$, pink MnS has $K_{sp} = 3 \times 10^{-11}$ while green MnS has $K_{sp} = 3 \times 10^{-16}$ etc.

(d) The process of supersaturation must be eliminated for effective precipitation of a substance.

---

**Problem 4.53** What will be the maximum concentration of $Cd^{2+}$ and $Mn^{2+}$ ions which will remain in solution after precipitation by excess of $H_2S$ in 0.25 M HCl. Given $K_{sp}$ of CdS = $5.5 \times 10^{-25}$ mol$^2$ lit$^{-2}$ and $K_{sp}$ of MnS = $1.4 \times 10^{-15}$ mol$^2$ lit$^{-2}$?

**Ans. :** The $S^{2-}$ ion concentration in a saturated solution of $H_2S$ is given by,

$$[S^{2-}] = \frac{1.25 \times 10^{-21} M^2 [H_2S]}{[H^+]^2}$$

In a saturated solution of $H_2S$, $[H_2S] = 0.1$ M

So, $[H^+]^2[S^{2-}] = 1.25 \times 10^{-21}\ M^2 \times 0.1$ M

$= 1.25 \times 10^{-22}\ M^3$

For each $Cd^{2+}$ ion precipitated, two $H^+$ ions are added to the solution according to the following reaction

$$Cd^{2+}(aq) + H_2S \rightleftharpoons CdS(s) + 2H^+(aq)$$

Hence the precipitation of all $Cd^{2+}$ ions produces 0.20 mol lit$^{-1}$ $H^+$ concentration and the final $[H^+] = 0.45$ M.

$$\therefore [S^{2-}] = \frac{1.25 \times 10^{-22} M^3}{(0.45M)^2} = 6.17 \times 10^{-22}\ M$$

Now $[Cd^{2+}][S^{2-}] = K_{sp}$ of CdS = $5.5 \times 10^{-25}$ mol$^2$ lit$^{-2}$

$$\therefore [Cd^{2+}] = \frac{5.5 \times 10^{-25} \text{ mol}^2 \text{lit}^{-2}}{6.17 \times 10^{-22} \text{ mol lit}^{-1}} = 0.89 \times 10^{-3} \text{ mol lit}^{-1}$$

So concentration of $Cd^{2+}$ in the solution after precipitation of CdS = 0.89 mol lit$^{-1}$. The solubility product of MnS indicates that MnS will not be precipitated in 0.25 M HCl solution.

---

**Problem 4.54** What will be the concentration of $H_3O^+$, $HS^-$ and $S^{2-}$ at equilibrium in a 0.1 M aqueous solution of $H_2S$. $K_1 = 9.6 \times 10^{-8}$ M; $K_2 = 1.3 \times 10^{-14}$ M.

**Ans. :** $H_2S$ is dissociated in the following way as :

$$H_2S + H_2O \rightleftharpoons H_3O^+ + HS^- \quad K_1 = \frac{[H_3O^+][HS^-]}{[H_2S]} = 9.6 \times 10^{-8} \text{ M}$$

$$HS^- + H_2O \rightleftharpoons H_3O^+ + S^{2-} \quad K_2 = \frac{[H_3O^+][S^{2-}]}{[HS^-]} = 1.3 \times 10^{-14} \text{ M}$$

$K_2$ is small compared to $K_1$ so that dissociation of $HS^-$ is practically negligible and we consider that $HS^-$ formed in the first stage may be considered to be present as it is and so the $H_3O^+$ formed in the second stage is also negligible. Therefore from the first dissociation alone, $[H_3O^+] = [HS^-]$. The concentration of undissociated $H_2S$ i.e. $[H_2S]$ = initial concentration – the amount dissociated = $0.1 - [H_3O^+]$.

$$\text{Thus, } K_1 = \frac{[H_3O^+][HS^-]}{[H_2S]} = \frac{[H_3O^+]^2}{0.1 - [H_3O^+]} = 9.6 \times 10^{-8} \text{ M}$$

As the value of $K_1$ is small, we may consider that $[H_3O^+] << [H_2S]$.

So we get, $K_1 \times 0.1 = [H_3O^+]^2$

$\therefore 9.6 \times 10^{-8} \text{ M} \times 0.1 = [H_3O^+]^2$

i.e. $[H_3O^+] = 9.8 \times 10^{-5}$ mol lit$^{-1}$ or mol dm$^{-3}$

and so $[HS^-] = 9.8 \times 10^{-5}$ mol dm$^{-3}$

We have considered $[H_3O^+] = [HS^-]$ and so from the second dissociation constant,

$[S^{2-}] = 1.3 \times 10^{-14}$ mol lit$^{-1}$ or mol dm$^{-3}$.

---

**Problem 4.55** Using electronic theory classify the following into acids, bases or neutral species : $H_2C = CH_2$, $H_3O^+$, HCl and $Me_2SO$. **[B.U. 2010]**

**Ans. :** $H_2C = CH_2$ which is $\pi$-electron donor forming $\mu$-bond and so a $\mu$-base and can also accept electrons in its $\pi^*$ orbital and so also acting as $\pi$-acid.

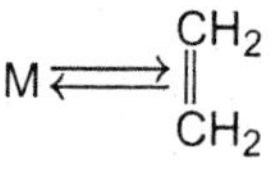

$H_3O^+$ in which a lone pair residing on central oxygen may be donated and thus a $\sigma$ base.

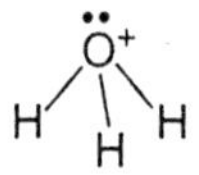

In HCl, lone pairs on Cl may be donated and thus acting as a base.

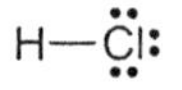

The $Me_2SO$ can behave as base through S and O donors ($\sigma$-base) as lone pair of electrons exist on S and O. Again it can function as $\pi$-acid because S has availability of d-orbital i.e. function as **$\sigma$-base** and **$\pi$-acid**.

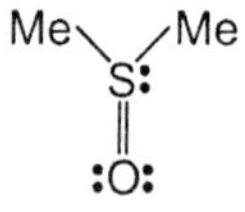

---

## EXERCISE

1. List all Bronsted acids in the following reactions in order of decreasing strength :

$HOBr^- + CN^- \rightleftharpoons HCN + OBr^-$ **[B.U. 1991]**

$$HI + BrO_3^- \rightleftharpoons HBrO_3 + I^-$$

$$H_3O^+ + IO_3^- \rightleftharpoons HIO_3 + H_2O$$

**Hint :** Bronstead acids : HOBr, HCN, HI, $HBrO_3$, $H_3O^+$ and $HIO_3$ and acidity order, $H_3O^+ > HI > HBrO_3 > HIO_3 > HOBr > HCN$.

2. (a) State the conjugate acids corresponding to the following :

$$O^{2-}, CH_3COOH, CN^-, PO_4^{3-}$$

(b) Explain why $NH_3$ is a Lewis base but $NF_3$ has practically no basic character. **[B.U.1996]**

**Hint :** (a) $OH^-$, $CH_3COOH_2^+$, HCN, $HPO_4^{2-}$

(b) F is more electronegative than H, so lone pair on N cannot function as that of $NH_3$.

3. "The amino acid, glycine exists predominantly in the form $NH_3^+CH_2COO^-$". Write down the formula for (i) conjugate base, (ii) conjugate acid of glycine. **[B.U. 2006, 2016]**

**Hint :** (i) $NH_2CH_2COO^-$, (ii) $\overset{+}{N}H_3CH_2COOH$.

4. Find out (adducing proper reason) the Bronsted acid(s) in the following reaction : $HOBr + CN^- \longrightarrow HCN + OBr^-$ **[B.U. 2007]**

**Hint :** HOBr, HCN.

5. Classify the following species into acids and bases and write their conjugate bases and conjugate acids : **[B.U. 2008, 2014]**

$$HSO_4^-, Al(H_2O)_6^{3+}, PH_3, CH_3OH$$

**Hint :** Sec. 4.3 and 4.3.1 $\underset{\textit{base}}{PH_3} + H^+ \longrightarrow \underset{\textit{conjugate acid}}{PH_4^+}$

$$\underset{\textit{acid}}{CH_3OH} + H_2O \longrightarrow \underset{\textit{conjugate base}}{CH_3O^-} + H_3O^+$$

6. Explain why acetic acid behaves as an acid in aqueous medium but as a base in nitric acid medium. **[B.U. 1984]**

**Hint :** $HNO_3 + CH_3COOH \longrightarrow CH_3COOH_2^+ + NO_3^-$

7. $HNO_3$ and HCl have equal strength in aqueous medium but their strengths differ in $CH_3COOH$ medium. **[B.U. 1988]**

**Hint :** Sec. 4.3.1

8. Explain why $CO_2$ is not a primary acid. **[B.U. 1987]**

**Hint :** Sec. 4.4(d).

9. Adducing proper reason classify the following according to electronic theory into acids, bases or neutral species : $NH_4^+$, $Me_2SO$, $BCl_3$, $H^-$ and $N_2$ **[B.U. 2005]**

**Hint :**

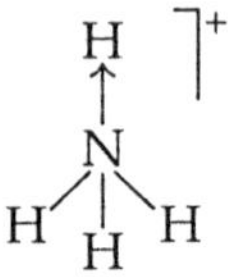

It is a neutral species neither accept nor donate lone pair.

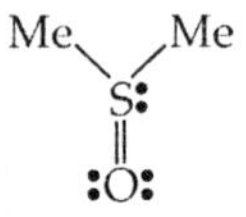

Can behave as base by donation of electron pair through S or O end. It can also acts as an acid through S centre as it has availability of d orbital for acceptance of electron pair.

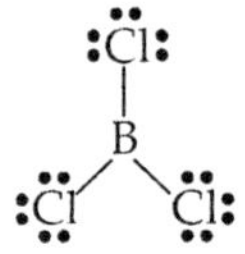

Central B is electron deficient, it can accept an electron pair to achieve octet and behave as acid. Again Cl centres have lone pairs, so these can function as basic centres.

$H:^-$

It can funciton as a base since it has a lone pair but cannot function as an acid as hydrogen is the smallest atom.

$:N\equiv N:$

It can behave as a base through each lone pair of nitrogen. It also acts as π-electron donor as also π-acid as its LUMO is stable.

10. Point out the criteria of a Lux-Flood base. **[B.U. 2006]**

    **Hint :** Sec. 4.7.

11. State with reason whether there will be any increase/decrease in acidity if anhydrous $AlCl_3$ is added to liquid phosgene. **[C.U. 2002]**

    **Hint :** $COCl_2 \rightleftharpoons COCl^+ + Cl^-$ $\qquad AlCl_3 \longrightarrow Al^{3+} + 3Cl^-$

    Acidity will decrease

12. Explain the role of solvents in the determination of strengths of acids. **[B.U. 2003, 2008]**

    **Hint :** Sec. 4.3.1.

13. Which of the following reaction actually takes place? **[B.U. 2006]**

    $CsI + LiF \longrightarrow LiI + CsF$ or $LiI + CsF \longrightarrow LiF + CsI$

    **Hint :** The second reaction $LiI + CsF \longrightarrow LiF + CsI$ will occur but the first one will not occur.

14. Indicate with reasons the direction of the following reactions—(i) $BF_3H^- + BH_3F^- \longrightarrow BF_4^- + BH_4^-$ (ii) $HgS + 2HCl \longrightarrow HgCl_2 + H_2S$. **[C.U. 2005]**

    **Hint :** (i) In $BF_3H^-$, hard acid-soft base, in $BH_3F^-$, soft acid-hard base interaction. In $BF_4^-$ and $BH_4^-$ there are hard-hard and soft-soft interaction respectively.

    (ii) $H_2S$ is a gas at ordinary temperature so that equilibrium is shifted to the right and the reaction occurs. Moreover the total stabilising contribution of the products are greater than that of reactants, so that the reaction will occur. This is sometimes referred to as Pearson paradox.

15. Predict which way the reactions will go in the gas phase with explanation : (i) $HI + NaF \longrightarrow HF + NaI$, (ii) $TiF_4 + 2TiI_2 \longrightarrow TiI_4 + 2TiF_2$. **[C.U. 2004, 2012]**

16. A buffer solution is prepared by mixing 8.0 ml 0.1(M) $CH_3COOH$ with 4.0 ml 0.1(M) NaOH solution. Find the pH of the solution (pKa = 4.75). **[C.U. 2001]**

    **Ans. :** 4.75.

17. Which quantity do we use in place of pH to measure the acidity of a concentrated aqueous solution? Show that it reduces to pH in dilute solution. **[B.U. 2000]**

    **Hint :** Sec. 4.12.

18. What is acidity function? What is its utility? [B.U. 2002, 2011, 2013]

    **Hint :** Sec. 4.12.

19. What are superacids? Indicate the parameter used to have a quantitative measure of superacid strength. Mention the basis to set up this parameter, and the condition when this becomes pH of the medium. [B.U. 2006]

    **Hint :** Sec. 4.12 and Sec. 4.12.1.

20. Write Pauling's rules for oxo-acid strengths. [B.U. 2007, 2016]

    **Hint :** Sec. 4.14.2(5).

21. What are superacids? How the acidity of such solutions is measured? Explain with an example. [C.U. 2006, 2012]

    **Hint :** Sec. 4.12.1.

22. pK of $H_3AsO_3$ is 9.2, predict the structure of the acid molecule. [C.U. 2002]

    **Hint :** Sec. 4.14.2(5), $As(OH)_3$, $m - n = 0$.

23. Arrange the following in the increasing order of given properties with reasonable explanation :

    (i) $Al(H_2O)_6^{3+}$, $Fe(H_2O)_6^{2+}$, $Fe(H_2O)_6^{3+}$ .........acidity

    (ii) $H_2O$, $H_2S$, $H_2Se$ ......... basicity towards proton. [B.U. 1999, 2012]

    **Hint :** (i) Transition metals possess greater polarising power than non-transition metal cations and hence acidity order :

    $$\begin{array}{lccccc} & Fe(H_2O)_6^{2+} & < & Al(H_2O)_6^{3+} & < & Fe(H_2O)_6^{3+} \\ pK_a & 9.5 & & 5.0 & & 2.2 \end{array}$$

    (ii) $H_2Se < H_2S < H_2O$.

24. How will you demonstrate that $H_3PO_4$ is a tribasic acid by alkalimetry. [B.U. 2000]

    **Hint :** Sec. 4.13.1.

25. A 100 ml buffer of pH 9.50 is to be prepared by mixing 0.30 M $NH_3$ and 0.10 M HCl. How many milliliters of each solution should be taken assuming volumes are additive.

    **Ans. :** 53 ml HCl, 47 ml $NH_3$.

26. What is the pH of a 0.1 M solution of methylamine $K_b = 5 \times 10^{-14}$.

    **Hint :** $[OH^-] = \sqrt{K_b C_b}$, pOH = 2.15, pH = 11.85.

27. 10 ml of 0.1 M aqueous solution of acetic acid are added to 10 ml of 0.05 M aquous solution of NaOH. Find the pH of the resulting solution. $K_a$ for acetic acid $= 1.75 \times 10^{-4}$. [B.U. 1984]

    **Hint :** See solved problem 4.23 and 4.32.

28. 5.0 lit of a solution containing 2.0 moles of $CH_3COONa$ (a fully ionised salt) is mixed with 5.0 lit of a solution containing 8.0 moles of $CH_3COOH$. Calculate $[OH^-]$ of the resulting solution. ($K_a = 1.8 \times 10^{-5}$). [B.U. 1983]

    **Hint :** Solved Problem 4.40.

29. (a) Derive suitable expressions for calculation of concentrations of all the species of a polyprotic acid $H_3A$ at a given pH.

(b) Calculate the concentrations of $HPO_4^-$ in a 0.1 M phosphate buffer of pH = 7.0, given $pK_1$ = 2.16, $pK_2$ = 7.13 and $pK_3$ = 12.30. **[B.U. 1995]**

**Hint :** Sec. 4.13.1 and Problem 4.33.

30. Indicate with reason and direction of the following reactions :
(i) $AgF_2^- + 2I^- \longrightarrow AgI_2^- + 2F^-$, (ii) $CH_4 + CF_4 = CH_3F + CHF_3$ **[C.U. 1999]**

31. 30.0 ml 0.0900 M HCl is required to titrate a sample of a weak base hydroxylamine. After the addition of 10.0 ml of HCl the pH is 6.26. Calculate the $pK_b$ of hydroxylamine.

**Ans. :** 8.04.

32. Justify the directions of the following reactions :

(i) $CuSO_4(l) + H_2S(g) \longrightarrow CuS(g) + H_2SO_4(l)$

(ii) $H_2SO_4(l) + FeS(s) \longrightarrow H_2S(g) + FeSO_4(l)$ **[C.U. 1996]**

33. Which member between the pair given below, is stronger acid—(i) $H_2CrO_4$ and $HMnO_4$, (ii) $Fe(H_2O)_6^{3+}$ / $Fe(H_2O)_6^{2+}$ **[C.U. 2004]**

**Hint :** (i) $HMnO_4 > H_2CrO_4$, (ii) $Fe(H_2O)_6^{3+} > Fe(H_2O)_6^{2+}$. This is explained by electronegativity-acidity rule (Sec. 4.14.2).

34. Illustrate with suitable example the phenomena of 'symbiosis' with explanation. **[C.U. 2004]**

**Hint :** Sec. 4.5.2(iv).

35. Apply solubility product principle and common ion effect in separations of group-II and group-IIIB metal sulphides in qualitative inorganic analysis. **[C.U. 2006, 2002]**

36. State the principle of solubility product. What is common ion effect? Explain the significance of these two aspects in the precipitation of iron, aluminium as hydroxides in conventional qualitative analysis. **[C.U. 2000]**

37. Calculate the solubility of AgCl in 0.20 M ammonia solution. Given $K_{sp}$ of AgCl = $1.7 \times 10^{-10}$ mol$^2$ lit$^{-2}$. Instability constant of $Ag(NH_3)_2^+ = 6.0 \times 10^{-8}$ mol$^2$ lit$^{-2}$.

**Ans. :** 0.0106 mol lit$^{-1}$.

38. The metal ions $X^{2+}$ and $Y^{2+}$ belong to analytical groups IIA and IIIB respectively. Which sulphide has lower solubility product, XS or YS? How can you get XS from a mixture containing both $X^{2+}$ and $Y^{2+}$ ions? **[B.U. 2001]**

**Hint :** $K_{sp}$ of XS < $K_{sp}$ of YS. If we pass $H_2S$ in HCl medium, XS will be precipitated from a mixture of $X^{2+}$ and $Y^{2+}$ ions.

39. The solubility product of $MgF_2$ is $7 \times 10^{-4}$. Find its solubility in water and in 0.01 M aqueous NaF solution approximately. **[B.U. 2002]**

**Hint :** Similar to Problem 4.45.

40. Write down the Usanovich concept of acids and bases. Show amphiprotic property of water using two equations. What is the levelling effect of acids and bases in water? **[B.U. 2010, 2016]**

**Hint :** Sec. 4.6 and Sec. 4.3.1.

Amphiprotic property of water :

$$\underset{acid}{H_2O(l)} + NH_3(aq) \longrightarrow NH_4^+(aq) + OH^-(aq)$$

$$\underset{base}{H_2O(l)} + HCl(g) \longrightarrow Cl^-(aq) + H_3O^+(aq)$$

In water strongest acid is $H^+$ and the strongest base is $OH^-$. Thus in aqueous solution strong acids cannot be stronger than $H^+$ and bases cannot be stronger than $OH^-$. Thus acids are levelled to $H^+$ and bases to $OH^-$.

41. Arrange $BF_3$, $BCl_3$, $BBr_3$ and $BI_3$ in order of their Lewis acidity with justification. **[C.U. 2009, B.U. 2013]**

**Hint :** Sec. 4.17.

42. In the light of Usanovich concept all the reactions are considered as acid-base reactions. Justify this statement. **[C.U. 2009]**

**Hint :** Sec. 4.6.

43. Find out the pH of $10^{-8}$(M) HCl solution. **[C.U. 2009]**

**Hint :** Problem 4.18.

44. $Me_3P$ acts as stronger base than $Me_3N$ in their reactions with $B_2H_6$.—Explain. **[C.U. 2009, B.U. 2012]**

**Hint :** $B_2H_6$ is soft acid and $Me_3P$ is a soft base while $Me_3N$ is hard base. So soft-soft interaction is more stronger than hard-soft interaction.

45. Draw the acid base neutralisation curves for—

(i) Strong acid by strong base and (ii) Weak acid by strong base. What would be your choice of indicators in each case. **[C.U. 2009]**

**Hint** Sec. 4.13.

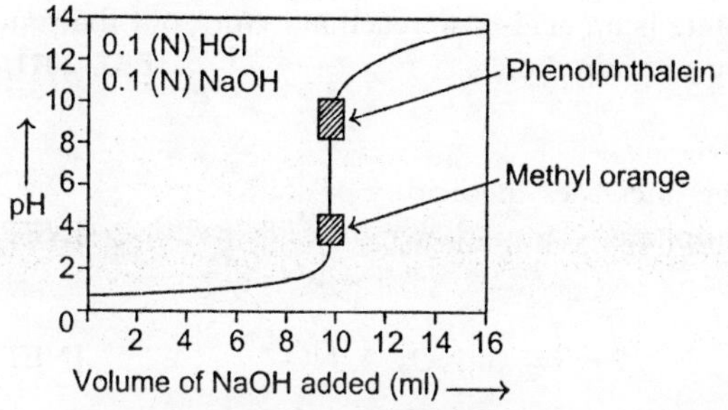


(a) Titration curve for neutralisation of strong acid and strong base

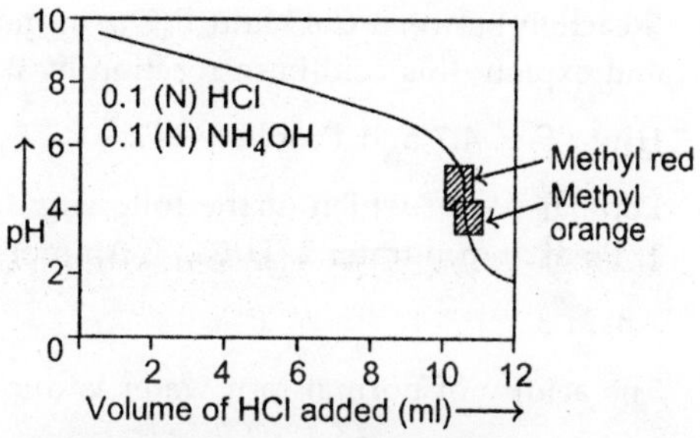


(c) Titration curve for neutralisation of strong acid and weak base

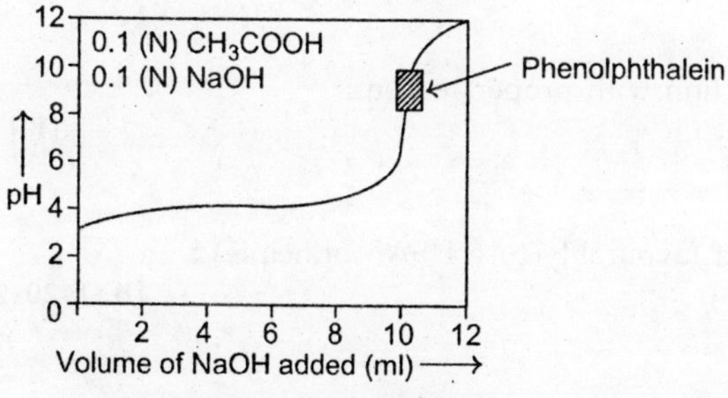


(b) Titration curve for neutralisation of weak acid and strong base

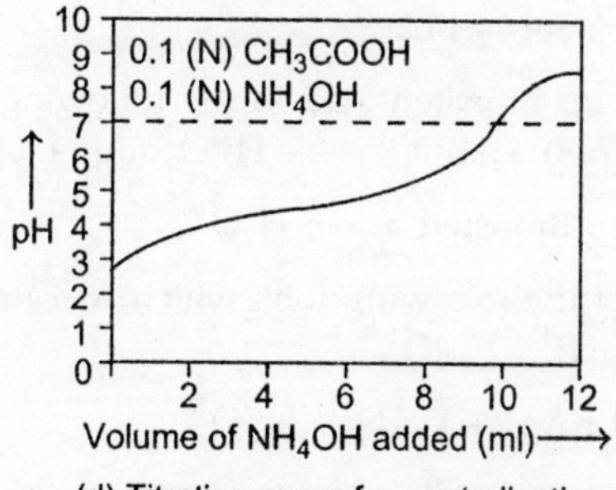


(d) Titration curve for neutralisation of weak acid and base

46. Establish the equation for the determination of pH of a solution of $CH_3COOH$ and hence find the pH of 0.1(M) $CH_3COOH$. ($K_{CH_3COOH} = 1.8 \times 10^{-5}$ mol/litre). **[C.U. 2012]**

**Hint :** Problem 4.37.

47. Write down Pauling's rules for predicting acid strength of oxoacids. Using this/these rule/rules explain the first $pK_a$ values of $H_3PO_3$ (~ 2.0) and $H_3AsO_3$ (~ 9.0) with respect to their structural difference, if any. **[C.U. 2008, 2010; B.U. 2011]**

**Hint :** 4.14.2(5). For an oxoacid of type $O_nX(OH)_m$, $pK_a \approx 8-5n$. The acid strength is related to the number of non-protonated oxygen atoms.

For polyprotic acids (m > 1) the successive $pK_a$ values increase by about 5 units.

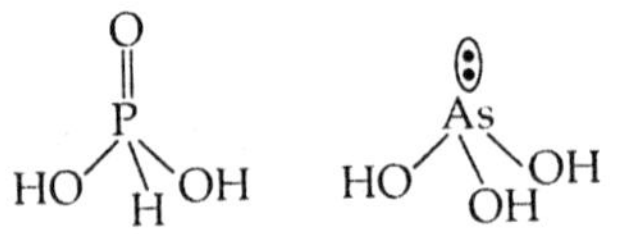

$OP(OH)_2$ : $pK_a = 8 - (5\times1) = 3$

$As(OH)_3$ : $pK_a = 8 - (5\times0) = 8$

Thus $H_3PO_3$ contains one non-protonated O but $H_3AsO_3$ does not contain such O according to Pauling's rule.

48. Point out the main reason why Ca, Al and Ni are respectively found as carbonate, oxide and sulphide in nature. **[B.U. 2011]**

**Hint :** Sec. 4.5.2. $Ca^{2+}$ and $Al^{3+}$ both are hard acids and both are non-transition element. The hardness of $Al^{3+} > Ca^{2+}$.

$Ni^{2+}$ is a soft acid being a member of transition element.

49. 'Reaction between CaO and $P_4O_{10}$ in molten state is an acid-base reaction'. Work out the product(s) and explain this acid-base reaction by a theory of your choice. **[B.U. 2011, 2014]**

**Hint :** Sec. 4.7 and Problem 4.12.

50. Popular use of which of the following fertilisers increases the acidity of soil?
1. Potassium nitrate, 2. Urea, 3. Ammonium sulphate, 4. Superphosphate of lime. **[NET 2011]**

**Ans. :** 3.

51. The acidity of normal rain water is due to—1. $SO_2$, 2. $CO_2$, 3. $NO_2$, 4. NO. **[NET 2011]**

**Ans. :** 2.

52. Justify the acid strength of $H_3PO_2$, $H_3PO_3$ and $H_3PO_4$. **[B.U. 2012]**

**Hint :** Sec. 4.14.2.

53. Pick up Bronsted acid in the following reaction with proper reason :
$PO_4^{3-}(aq) + H_2O(l) \longrightarrow HPO_4^{2-}(aq) + OH^-(aq)$ **[B.U. 2012]**

**Hint :** Bronsted acid : $H_2O$.

54. Select the following ions which will interact favourably (use HSAB principle) :
$Ag^+$, $Ni^{4+}$, $I^-$, $IO_6^{5-}$ **[B.U. 2012, 2014]**

**Hint :** $Ag^+ + I^-$ ; $Ni^{4+} + IO_6^{5-}$.

55. Among the following, the correct acid strength is represented by—
1. $[Al(H_2O)_6]^{3+} < [Fe(H_2O)_6]^{3+} < [Fe(H_2O)_6]^{2+}$
2. $[Fe(H_2O)_6]^{3+} < [Al(H_2O)_6]^{3+} < [Fe(H_2O)_6]^{2+}$
3. $[Fe(H_2O)_6]^{2+} < Fe(H_2O)_6]^{3+} < [Al(H_2O)_6]^{3+}$
4. $[Fe(H_2O)_6]^{2+} < [Al(H_2O)_6]^{3+} < [Fe(H_2O)]_6]^{3+}$ **[NET 2013]**

**Hint :** Small size cation will exert polarising power on $H_2O$ molecule and so acidity increases. The size of the cations are in the order $Fe^{2+} > Fe^{3+} > Al^{3+}$ and so the acidity order will be $[Fe(H_2O)_6]^{2+} < Fe(H_2O)_6]^{3+} < [Al(H_2O)_6]^{3+}$ i.e. 3.

56. Two tautomeric forms of phosphorus acid are—

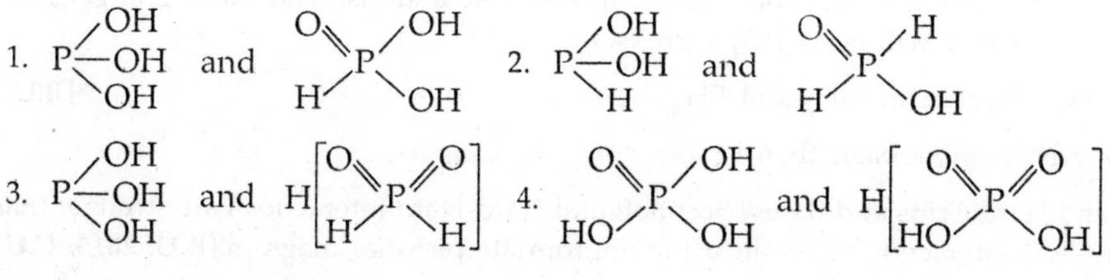

[NET 2013]

**Ans. :** 1. i.e. 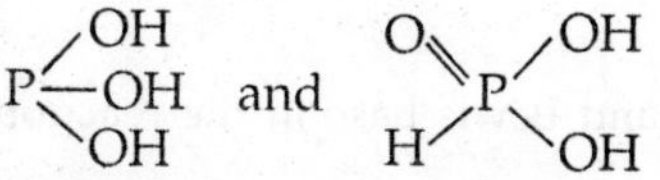

57. Xenon forms several fluorides and oxofluorides which exhibit acidic behaviour. The correct sequence of descending Lewis acidity among the given species is represented by—
1. $XeF_6 > XeOF_4 > XeF_4 > XeO_2F_2$ 2. $XeOF_4 > XeO_2F_2 > XeOF_4 > XeF_6$
3. $XeF_4 > XeO_2F_2 > XeOF_4 > XeF_6$ 4. $XeF_4 > XeF_6 > XeOF_4 > XeO_2F_2$ **[NET 2013]**

**Ans. :** 1. i.e. $XeF_6 > XeOF_4 > XeF_4 > XeO_2F_2$

58. Successive $pk_a$ values of $H_3PO_4$ are 2.1, 7.4 and 12.7 respectively. Explain the data using Pauling's rule. **[B.U. 2013]**

**Hint :** Sec. 4.14.2(5).

59. Compare the Lewis basicity between $(CH_3)_2O$ and $(SiH_3)_2O$. **[B.U. 2013]**

**Hint :** Sec. 4.16 and Problem 3.76

$(CH_3)_2O$ is more basic than $(SiH_3)_2O$.

60. Arrange the compounds with increasing proton affinity : $HS^-$, $F^-$, $NH_2^-$, $I^-$. **[B.U. 2013]**

**Hint :** $HS^- + H^+ \longrightarrow H_2S$, $F^- + H^+ \longrightarrow HF$, $NH_2^- + H^+ \longrightarrow NH_3$, $I^- + H^+ \longrightarrow HI$. Proton affinity depends on conjugate acid/base property. Thus the proton affinity order is $I^- < F^- < HS^- < NH_2^-$.

61. The molecule $(CH_3)_2N$–$PF_2$ has two basic atoms, P and N. One is bound to B in a complex with $BH_3$ while the other is bound to B in a complex with $BF_3$. Decide which is which. **[B.U. 2013]**

**Hint :** As per SHAB principle $BH_3$ is a soft acid prefer to combine with soft donor P while $BF_3$ having hard acid character prefer to combine with hard donor N.

62. Give one example each of superacid and solid acid. **[B.U. 2014]**

**Hint :** Sec. 4.12.1 and Sec. 4.18.

63. $AgClO_4$ is significantly more soluble in benzene than in n-hexane. Account in terms of Lewis acid-base theory. **[B.U. 2014, 2016]**

**Hint :** Question number 162 of chapter 7.

Here we have to consider solvent solute interaction. The π electrons of benzene, a soft base, are available for complex formation with the empty orbitals of the cation $Ag^+$, a soft acid. The $Ag^+$ ion is thus solvated favourably in benzene to form the complex $[Ag–C_6H_6]^+$. In this complex $Ag^+$ is a soft Lewis acid and benzene is a weak soft base.

64. Note down the Lux-Flood concept of acids and bases with suitable example. [B.U. 2015]

**Hint :** Sec. 4.7.

65. Which one is stronger acid and why? $H_3PO_4$ and $H_3AsO_4$? [B.U. 2015]

**Hint :** Sec. 4.14.2. According to electronegativity acidity rule and also considering larger size of As compared to P the acidity will be $H_3PO_4 > H_3AsO_4$.

66. Compare the Lewis basicity of $NH_3$ and $PH_3$ [B.U. 2015]

**Hint :** Sec. 4.4. $NH_3$ is more basic than $PH_3$.

67. What are Soft and Hard Acids and Bases? Soft-Soft and Hard-Hard interactions are stronger than Soft-Hard and Hard-Soft interaction? Explain using appropriate pictorial diagram. [B.U. 2015, C.U. 2014]

**Hint :** Sec. 4.5 and Sec. 4.5.1.

68. Adducing proper reason identify the Lewis acid and Lewis base in the reaction $I^- + I_2 \longrightarrow I_3^-$. [B.U. 2015]

**Hint :** Sec. 4.5. $I_2$ is soft Lewis acid and $I^-$ is a soft Lewis base.

69. '$(CH_3)_3P$ acts as stronger base than $(CH_3)_3N$ towards $B_2H_6$'.—Explain. [B.U. 2014]

**Hint :** Sec. 4.5.2. According to Pearson's rule $(CH_3)_3P$ is a soft base combine with the soft acid $B_2H_6$ i.e. $BH_3$ so that $(CH_3)_3P$ acting as a stronger base towards $B_2H_6$ compared to $(CH_3)_3N$. $B_2H_6$ has hydrogen bridged structure where B is electron deficient which makes $B_2H_6$ or $BH_3$ a soft acid.

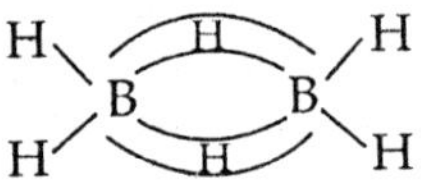


70. Identify the most basic one—
(i) $NH_2^-$ and $PH_2^-$, (ii) $ClO_2^-$ and $ClO^-$, (iii) $[Fe(CN)_6]^{3-}$ and $[Fe(CN)_6]^{4-}$ [C.U. 2013]

**Hint :** (i) $NH_2^-$ more basic than $PH_2^-$ because $NH_3$ is a sronger base than $PH_3$.

(ii) According to electronegativity acidity rule $ClO^-$ will be more basic than $ClO_2^-$.

(iii) As the charge on the metal cation increases, the acidity increases so that $H_3[Fe(CN)_6]$ is stronger acid than that of $H_4[Fe(CN)_6]$ thereby $[Fe(CN)_6]^{4-}$ will be a stronger base compared to that of $[Fe(CN)_6]^{3-}$.

71. Arrange the following complexes in order of their increasing acidity : $[Al(H_2O)_6]^{3+}$, $[Na(H_2O)_n]^+$, $[Mn(H_2O)_6]^{2+}$, $[Co(H_2O)_6]^{2+}$. [C.U. 2013]

**Hint :** Problem 4.30. Acidity of the hydrated metal cations depend on the polarising power of the cation which depends on the size of the cations. The small size will produce stronger acidity, however other factors may also take place (Sec. 4.14.3). Considering all these the acidity order will be $[Na(H_2O)_n]^+ < [Mn(H_2O)_6]^{2+} < [Co(H_2O)_6]^{2+} < [Al(H_2O)_6]^{3+}$.

72. Give the order of acidity of the following and rationalise the trend : $[Na(H_2O)_x]^+$, $[Fe(H_2O)_6]^{3+}$, $[Al(H_2O)_6]^{3+}$, $[Fe(H_2O)_6]^{2+}$. [C.U. 2013]

**Hint :** Sec. 4.14.3. Acid strength increases as : $[Na(H_2O)_x]^+ < [Fe(H_2O)_6]^{2+} < [Al(H_2O)_6]^{3+} < [Fe(H_2O)_6]^{3+}$. The polarising power of $Fe^{3+}$ is greater than that of $Al^{3+}$ ion.

73. Predict the direction of the following equilibria with explanation—
(i) $2CH_3MgF + HgF_2 \rightleftharpoons (CH_3)_2Hg + 2MgF_2$
(ii) $BF_3H^- + BH_3F^- \rightleftharpoons BF_4^- + BH_4^-$ [C.U. 2013]

**Hint :** Sec. 4.5.2. (i) Problem 4.9(c).

(ii) For $BF_3H^-$ the combination is hard acid-soft base while for $BH_3F^-$ the combination is soft acid-hard base thus the reaction proceed to the right and the equilibrium will be shifted to the right (Question number 14).

74. What is the principle of colour change of indicators in the acid-base titrations? How can you choose appropriate indicators in the following titrations? (i) $CH_3COOH$ vs NaOH, (ii) $NH_4OH$ vs HCl. **[C.U. 2013]**

**Hint :** Sec. 4.13.

75. $CH_3COOH$ is a weak acid in water but strong in liquid $NH_3$.—Explain. **[C.U. 2013]**

**Hint :** Sec. 4.3.1. Water acting as a differentiating solvent for weak acids but acting as a levelling solvent for strong acids while liquid $NH_3$ acting as a levelling solvent for weak acids. Liquid $NH_3$ will extract all $H^+$ ions from weak acid, $CH_3COOH$ to form $NH_4^+$ and $CH_3COO^-$.

$$CH_3COOH + NH_3 \longrightarrow CH_3COO^- + NH_4^+$$

But water cannot be able to extract all the $H^+$ ions from $CH_3COOH$ so $CH_3COOH$ acting as a weak acid in water but strong acid in liquid $NH_3$.

76. $SnCl_2$ can act both as Lewis acid and Lewis base.—Justify. **[C.U. 2013]**

**Hint :** Lewis acids are electron pair acceptor while Lewis bases are electron pair donor.

$SnCl_2$ acts as Lewis acid by accepting electron pair from $Cl^-$ to form $SnCl_3^-$ and this $SnCl_3^-$ acts as a ligand e.g. $[RuCl_2(SnCl_3)_2]^{2-}$, cis/trans $[PtCl_2(SnCl_3)_2]^{2-}$ etc.

$$SnCl_2 + Cl^- \longrightarrow SnCl_3^-$$

$SnCl_2$ can also acts as Lewis base in the formation of the neutral complex $[F_3B \longleftarrow SnCl_2(NMe_3)]$ (X = Cl, Br, I).

77. Comment (with reason) on the change in acidity by adding (i) $SbF_5$ in anhydrous HF, (ii) BiN in liquid $NH_3$, (iii) $CuSO_4$ in aqueous solution of $(NH_4)_2SO_4$. **[C.U. 2014]**

**Hint :** (i) Problem 4.2, acidity increases.

(ii) Liquid $NH_3$ antoionised as $2NH_3 \longrightarrow NH_4^+ + NH_2^-$

$N^{3-} + NH_3 \longrightarrow NH_2^- + NH^{2-}$ As the concentration of solvent anion increases, so acidity decreases.

(iii) Problem 4.8(ii), acidity increases due to the increase of $H^+$ ion concentration.

78. Calculate approximate $pK_a$ values of $H_3PO_4$ and $HClO_4$ using Pauling rules. **[C.U. 2014]**

**Hint :** Sec. 4.14.2(5) $pK_a$ for $H_3PO_4$ = 3 and for $HClO_4$, $pK_a$ = –7.

79. Acid strength of HF, HCl, HBr and $CH_3COOH$ differ in aqueous solution but all are equally strong in liquid $NH_3$—Explain. **[C.U. 2014]**

**Hint :** Sec. 4.3.1. Water acting as a differentiating solvent for HF, HCl, HBr and $CH_3COOH$ i.e. the concentration of $H_3O^+$ will be different for these acids in aqueous solution and the order of acid strength is given as :

$$CH_3COOH < HF < HCl < HBr$$

But liquid $NH_3$ is too basic and thereby liquid $NH_3$ will extract all $H^+$ ions equally from HF, HCl, HBr and $CH_3COOH$ and levelled to the strength of $NH_4^+$. In this case liquid $NH_3$ acting as a levelling solvent.

$$HF + NH_3 \longrightarrow NH_4^+ + F^-$$

$$HCl + NH_3 \longrightarrow NH_4^+ + Cl^-$$

$$HBr + NH_3 \longrightarrow NH_4^+ + Br^-$$

$$CH_3COOH + NH_3 \longrightarrow NH_4^+ + CH_3COO^-$$

80. Arrange the following according to increasing Lewis acid character :
(a) $B(n\text{–}Bu)_3$, $B(t\text{–}Bu)_3$, (b) $SiF_4$, $SiCl_4$, $SiBr_4$, $SiI_4$.

**Hint :** (a) Sec. 4.15. Due to B strain in case of t–Bu group because branching causes some steric crowding and the adduct formation is hindered. Thus the Lewis acidity order is $B(n\text{–}Bu)_3 > B(t\text{–}Bu)_3$ where the state of hybridisation of B changes from $sp^2$ to $sp^3$. Similarly $B(CH_3)_3$ forms stronger adduct with $NH_3$ compared to that of $B(i\text{–}Pr)_3$.

(b) In this case π-conjugation from the halogen p-orbital to the silicon d-orbital is not so prominant as that of $BX_3$. The order of acidity is in sequence of electron withdrawing capability of halogens from iodine to fluorine and the order is $SiI_4 < SiBr_4 < SiCl_4 < SiF_4$.

81. $Me_3N(pK_b = 4.2)$ and $C_5H_5N(pK_b = 8.8)$ are bases in aqueous solution, but MeCN is not.—Explain. **[C.U. 2014]**

**Hint :** We know that as s character of the hybrid orbitals increases acidity increases conversely basicity decreases.

In $Me_3N$ ($pK_b = 4.2$) where N is $sp^3$ having 25% s character.

In $C_5H_5N$ ($pK_b = 8.8$) where N is $sp^2$ having 33% s character and in MeCN the N is sp hybridised having 50% s character. Therefore acidity order will be MeCN > $C_5H_5N$ > $Me_3N$. Conversely MeCN will be least basic while $Me_3N$ and $C_5H_5N$ will have considerable basic character.

82. Show the Lewis acidity order of the compounds $BX_3$ and $SiX_4$ (X = F, Cl, Br). Give reasons. **[C.U. 2015]**

**Hint :** Sec. 4.17 and Question number 80. Lewis acidity sequence : $BI_3 > BBr_3 > BCl_3 > BF_3$ and $SiF_4 > SiCl_4 > SiBr_4 > SiI_4$.

83. Why do Ca, Al and Ni are found as carbonate, oxide and sulphide respectively in nature? **[C.U. 2015]**

**Hint :** Sec. 4.5.2 and Problem 4.13. According to Pearson rule : $Ca^{2+} - CO_3^{2-}$ (hard-hard interaction); $Al^{3+} - O^{2-}$ (hard-hard interaction); $Ni^{2+}$ is borderline acid (tendency to be soft acid) and $S^{2-}$ is a soft base.

84. Arrange the given ions in order of increasing acidity in aqueous medium with justification.
$[Ni(H_2O)_6]^{2+}$, $[Fe(H_2O)_6]^{3+}$, $[Al(H_2O)_6]^{3+}$ and $[Mn(H_2O)_6]^{2+}$. **[C.U. 2015]**

**Hint :** Sec. 4.14.3., Question number 71, 72 and Problem 4.30.

The order is $[Mn(H_2O)_6]^{2+} < [Ni(H_2O)_6]^{2+} < [Al(H_2O)_6]^{3+} < [Fe(H_2O)_6]^{3+}$ considering the effective nuclear charge on the metal cation and polarising power of metal cation.

85. Arrange and explain the proton affinity order of the following : $CH_3^-$, $NH_2^-$, $OH^-$ and $F^-$. **[C.U. 2015]**

**Hint :** Question number 60. Proton affinity depends on the conjugate acid/base property. $CH_4$ is the weakest acid so proton affinity will be highest for its conjugate base $CH_3^-$ and thus the proton affinity sequence will be $CH_3^- > NH_2^- > OH^- > F^-$. The gas phase proton affinity values for $CH_3^-$, $NH_2^-$, $OH^-$ and $F^-$ are > 1695, 1695, 1632 and 1548 kJ $mol^{-1}$ respectively.

86. The most basic species among the following is—
(A) $F^-$, (B) $OH^-$, (C) $NH_2^-$, (D) $CH_3^-$.

**Hint :** (D) i.e. $CH_3^-$. The most basic species is one which attract the proton strongly. The protonation of a strong base yields a weak acid and vice-versa. Among HF, $H_2O$, $NH_3$ and $CH_4$, the $CH_4$ is the weakest acid. This is due to the fact that carbon is not electronegative enough to polarise the C–H bond and produce a proton. Therefore among the species, $CH_3^-$ is the most basic species.

87. The acid dissociation constant of $H_2S$ and $HS^-$ are $10^{-7}$ and $10^{-13}$ respectively. THe pH value in a 0.1 molar solution of $H_2S$ is— (A) $10^{-2}$, (B) 2, (C) 3, (4) 4.

**Hint :** D i.e. 4.

For the equilibrium $H_2S \longrightarrow H^+ + HS^-$

$$K = \frac{[H^+][HS^-]}{[H_2S]}$$

Initially $[H_2S] = 0.1M$.

After equilibrium, $[H^+] = [HS^-] = x$(Say) and $[H_2S] = 0.1 - x$.

So, $\frac{x \times x}{0.1 - x} = K = 10^{-7}$

Since the dissociation of $HS^-$ is almost negligible and x is small, so that,

$$\frac{x^2}{0.1} = 10^{-7} \text{ i.e. } x^2 = 10^{-8} \text{ and}$$

so $x = 10^{-4}$ i.e. pH = 4.

88. The pH of a solution obtained by mixing 50 ml(N) HCl and 30 ml(N) NaOH is—
(A) 6.02, (B) 0.602, (C) 12.04, (D) 1.204.

**Ans. :** (B) i.e. 0.602. 20 ml(N) HCl left after neutralisaiton.

20 ml(N) HCl = 0.73 gm HCl

Total volume = 50+30 = 80 ml, therefore,

80 ml solution contains 0.73 gm HCl

1000 ml solution contains = $\frac{0.73 \times 1000}{80}$ = 9.125 gm HCl = $\frac{9.125}{36.5}$ = 0.25 gm equivalent

$\therefore$ $[H^+]$ = 0.25 g ion per litre, so that

$pH = -\log[H^+] = -\log 0.25 = 0.602$.

89. Draw the acid-base neutralisation curves for the titration of (i) HCl vs NaOH, (ii) $CH_3COOH$ vs NaOH. Explain your choice of indicators in each case. **[C.U. 2015]**

**Hint :** Sec. 4.13 and Question number 45.

90. The compound whose water solution has the highest pH is—
(A) NaCl, (B) $Na_2CO_3$, (C) $NH_4Cl$, (D) $NaHCO_3$.

**Hint :** (B) i.e. water solution of $Na_2CO_3$ which is a salt of strong base NaOH and weak acid $H_2CO_3$.

91. Which of the following molecule is the strongest protonic acid—
(A) $H_6TeO_6$, (B) $SbH_3$, (C) $PH_3$, (D) $B_2H_6$.

**Ans. :** A i.e. $H_6TeO_6$.

92. Which of the following is a soft base?
(A) $CH_3COO^-$, (B) $H^-$, (C) $NO_3^-$, (D) $CO_3^{2-}$.

**Ans. :** (B) i.e. $H^-$.

93. Sulphanilic acid is a/an—
(A) Arrhenius acid, (B) Lewis base, (C) Neither (A) nor (B), (D) Both (A) and (B).

**Hint :** D i.e. both Arrhenius acid and Lewis base.

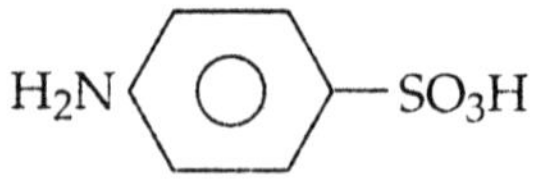


In aqueous solution it can give $H^+$ ions and also $H_2N$ group can donate electron pair.

94. Using the values $K_a$ for $NH_4^+$ = $5.6 \times 10^{-10}$ and $K_w$ = $1.00 \times 10^{-14}$, determine a value of K for equilibrium of the following :

$$NH_3(aq) + H_2O(l) \rightleftharpoons NH_4^+(aq) + OH^-(aq)$$

**Hint :** For $K_a$ we can write

$$NH_4^+(aq) + H_2O(l) \rightleftharpoons NH_3(aq) + H_3O^+(aq)$$

so that, $K_a = 5.6 \times 10^{-10} = \dfrac{[NH_3(aq)][H_3O^+(aq)]}{[NH_4^+(aq)]}$

and $K_w = 1.00 \times 10^{-14} = [H_3O^+][OH^-]$

$$K = \frac{[NH_4^+(aq)][OH^-(aq)]}{[NH_3(aq)]} = \frac{[H_3O^+][OH^-]}{\dfrac{[NH_3(aq)]H_3O^+(aq)]}{[NH_4^+(aq)]}} = \frac{1.00\times10^{-14}}{5.6\times10^{-10}} = 1.8 \times 10^{-5}.$$

Here $[OH^-(aq)]$ and $[OH^-]$ are same.

95. The graph that represents the titration of 0.1M butanoic acid (20 ml) with 0.1M NaOH is—

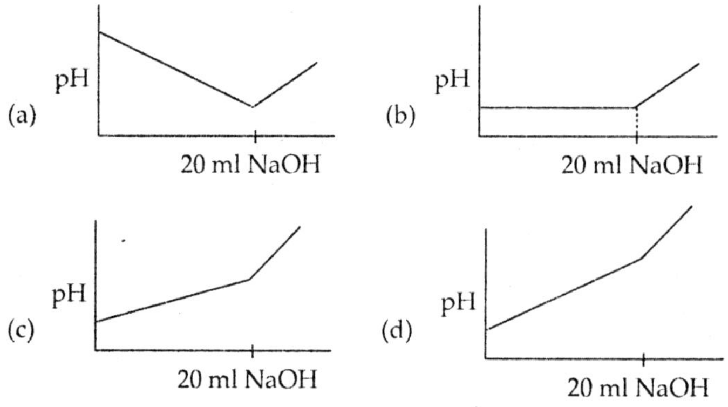


**Hint :** (c). Sec. 4.13 and Question number 45.

96. Which of the following is a hard base—
(a) $H^-$, (b) $NO_3^-$, (c) $I^-$, (d) $CN^-$

**Hint :** (b) i.e. $NO_3^-$.

97. $N(CH_3)_3$ and $N(SiH_3)_3$ gives different products on reaction with HCl.—Explain. [C.U. 2015]

**Hint :** Problem 3.75 of chapter 3.

98. Justify that CuS is precipitated in acidic medium (HCl) when $H_2S$ gas is passed through $Cu^{2+}$ solution while ZnS is precipitated in basic medium ($NH_4OH$), [$K_s$(CuS) = $8.0\times10^{-36}$, $K_s$(ZnS) = $3\times10^{-32}$ at 25°C] [C.U. 2015]

**Hint :** Sec. 4.19.3. $H_2S$ is an weak acid. In acid medium due to the common ion effect the dissociation

$$H_2S \rightleftharpoons 2H^+ + S^{2-}$$

of $H_2S$ is not favoured to the right-hand side and so the concentration of $S^{2-}$ will be low. As the solubility product of CuS is less compared to that of ZnS, so low concentration of $S^{2-}$ ion in acidic medium is sufficient to reach the solubility product of CuS and thereby CuS will be precipitated from acidic medium. But this concentration of $S^{2-}$ is not sufficient to reach the solubility product of ZnS and ZnS will not be precipitated.

In basic medium ($NH_4OH$) the dissociation of $H_2S$ will be more favoured to the right thereby the concentration of $S^{2-}$ is increased and as the solubility product of ZnS is higher compared to that of CuS, the increased concentration of $S^{2-}$ will precipitate ZnS from basic solution.

99. Show that the degree of hydrolysis of a salt of weak acid and weak base is independent of concentration of the solution. [C.U. 2015]

**Hint :** Sec. 4.10.2(d). Here $x = \sqrt{K_h}$ which indicates that degree of hydrolysis is independent of concentration of solution.

100. In group IIIA for precipitation of $Fe^{3+}$, $Al^{3+}$, $Cr^{3+}$ mixture of $NH_4Cl + NH_4OH$ is used but not NaOH. Explain. [C.U. 2014]

**Hint :** Sec. 4.19.3. $NH_4OH$ is a weak base and in presence of $NH_4Cl$ due to the common ion effect of $NH_4^+$ ion, the $NH_4OH$ will be less basic and the dissociation of $NH_4OH$ will be favoured to the left thereby the concentration of $OH^-$ will be poor. This low cencentration of $OH^-$ ions is sufficient to reach the solubility products of $Fe(OH)_3$, $Al(OH)_3$ and $Cr(OH)_3$ so that $Fe^{3+}$, $Al^{3+}$ and $Cr^{3+}$ will be precipitated as hydroxides. If NaOH is used in this case then there will be greater concentration of $OH^-$ ions because NaOH is a strong base. This larger concentration of $OH^-$ will be able to precipitate the hydroxides of succeeding groups. In order to stop this in the precipitation of Group IIIA metals a mixture of $NH_4Cl + NH_4OH$ is used.

101. Calculate the pH at which $Mg(OH)_2$ will just begin to precipitate from a solution of 0.100 M $Mg(NO_3)_2$ by addition of NaOH. $K_{sp}Mg(OH)_2 = 1.2 \times 10^{-11}$. [C.U. 2014]

**Hint :** $[Mg^{2+}] = 0.100$ M

$K_{sp}Mg(OH)_2 = [Mg^{2+}][OH^-]^2 = 1.2 \times 10^{-11}$

$$\therefore [OH^-]^2 = \frac{1.2\times10^{-11}}{[Mg^{2+}]} = \frac{1.2\times10^{-11}}{10^{-1}} = 1.2 \times 10^{-10} \text{ M}$$

$\therefore [OH^-] = 1.09 \times 10^{-5}$ M

$\therefore$ pOH = $-\log[OH^-] = -\log 1.09 \times 10^{-5} = 5 - \log 1.09 = 5 - 0.0374 = 4.9626$

$\therefore$ pH = 14 – 4.9626 = 9.0374.

102. The ionic strength of 0.1N $CH_3COOH$ solution is 0.001 at 20°C. Calculate the pH of the solution. [C.U. 2015]

**Hint :** In 0.1N $CH_3COOH$ let $[H^+] = 0.1x$ and $[CH_3COO^-] = 0.1x$

where x is the degree of dissociation of $CH_3COOH$.

We know ionic strength, $\mu = \frac{1}{2}\sum C_i Z_i^2$

where $C_i$ and $Z_i$ are the concentration and valence of the ith ion.

$\therefore \mu = \frac{1}{2}[0.1x(1)^2 + 0.1x(1)^2] = \frac{1}{2} \times 0.2x$

so, $\frac{1}{2} \times 0.2x = 0.001$. Thus, $x = 0.01$

$\therefore [H^+] = 0.1 \times 0.01 = 0.001 = 10^{-3}$

$\therefore pH = -\log[H^+] = -\log 10^{-3} = 3.$

103. What pH must be maintained to precipitate CdS but not ZnS from an aqueous solution of 0.1M each of $Cd^{2+}$ and $Zn^{2+}$ saturated with $H_2S$(0.1 M)?

Given $K_{sp}(ZnS) = 1.0 \times 10^{-21}$; $K_{sp}(CdS) = 8.0 \times 10^{-27}$

$K_{a_1}$ and $K_{a_2}$ of $H_2S = 1.1 \times 10^{-7}$ and $1.0 \times 10^{-14}$ respectively. **[C.U. 2013]**

**Hint :** In order to prevent precipitation of ZnS, $[Zn^{2+}][S^{2-}] < K_{sp}(ZnS) = 1.0 \times 10^{-21}$

Here $[Zn^{2+}] = [Cd^{2+}] = 0.1M$

$\therefore (0.1)[S^{2-}] < 1.0 \times 10^{-21}$

Or, $[S^{2-}] < 1 \times 10^{-20}$.

This is the maximum value of $[S^{2-}]$ before ZnS will precipitate.

Let $[H^+]$ to maitain this $[S^{2-}]$ be x

$$H_2S \rightleftharpoons H^+ + HS^- \quad K_{a_1} = 1.1 \times 10^{-7}$$

$$HS^- \rightleftharpoons H^+ + S^{2-} \quad K_{a_2} = 1.0 \times 10^{-14}$$

$$H_2S \rightleftharpoons 2H^2 + S^{2-}$$

$$K_a = K_{a_1} + K_{a_2} = \frac{[H^+]^2[S^{2-}]}{[H_2S]} = 1.1 \times 10^{-7} \times 1.0 \times 10^{-14} = 1.1 \times 10^{-21}$$

$$\therefore \frac{x^2(1\times10^{-20})}{0.1} = 1.1 \times 10^{-21}$$

So that $x = [H^+] = 0.1M$

$\therefore$ No ZnS will prcipitate at a concentration of $H^+$ greater than 0.1M or pH = 1.

104. Explain the significance of the principle of solubility product and common ion effect for the precipitation of iron, aluminium and chromium as hydroxides in qualitative analysis. **[C.U. 2012]**

**Hint :** Sec. 4.19.3.

105. The solubility product of ferric hydroxide is $1.1 \times 10^{-36}$ at 25°C. Calculate the solubility of ferric hydroxide in gram per litre (Fe = 56, O = 16). **[C.U. 2012]**

**Hint :** $Fe(OH)_3 \rightleftharpoons Fe^{3+} + 3OH^-$

Let the solubility of $Fe(OH)_3$ = s mol $lit^{-1}$

$\therefore K_{sp} = [Fe^{3+}][OH^-]^3 = s \times (3s)^3 = s \times 27s^3 = 27s^4$

$\therefore 27s^4 = 1.1 \times 10^{-36}$ so that,

$s = 0.4493 \times 10^{-9}$ mol $lit^{-1} = 0.4493 \times 10^{-9} \times 107$ g $lit^{-1}$

$= 48.075 \times 10^{-9}$ g $lit^{-1}$

Molecular weight of $Fe(OH)_3 = 107$.

106. The species having the strongest gas phase proton affinity among the following :
1. $N^{3-}$, 2. $NF_3$, 3. $NH_3$, 4. $N(CH_3)_3$ **[NET 2015]**

**Ans. :** 1 i.e. $N^{3-}$.

107. 10 ml 0.02M NaOH is added to 10 ml of 0.02M acetic acid ($pK_a = 4.75$). The pH of the solution will be closest to— 1. 7.0, 2. 8.4, 3. 5.6, 4. 9.6 **[NET 2014]**

**Hint :** 1 i.e. pH = 7.0

10 ml of 0.2M NaOH will neutralise the 10 ml of 0.02M acetic acid.

108. Write down the Born equation and explain the terms involved. **[C.U. 2013]**

**Hint :** Born equation is similar to that of Born-Lande equation in case of lattice energy (equation (3.8)). But Born equation is applied in case of acidity of a species in a solvent. The acidity is largely influnced by the solvation of the ions in solution. The Gibbs free energy (ΔG°) of solvation of an ion of radius r and charge Ze in a solvent of relative permittivity $\varepsilon_r$ is expressed by the Born equation :

$$\Delta G^\circ = -\frac{NZ^2e^2}{8\pi\varepsilon_0 r}\left(1-\frac{1}{\varepsilon_r}\right)$$

$Z_2/r$ is sometimes called the electrostatic parameter. Small ions of high charge will be highly stabilised in a polar solvent. Also larger the value of $\varepsilon_r$, the more negative will be ΔG°, favouring the formation of ions in solution.

109. List the merits and demerits of Usanovich's concept of acids and bases. **[B.U. 2016]**

**Hint :** Sec. 4.8. Merits of this concept is the most general of the different acid-base definitions. It includes all acids and bases defined by the Lewis concept, at the same time it states that the donation or acceptance of electrons need not take place as shared pairs. Accordingly oxidation-reduction reactions may be classified as acid-base reactions.

Demerits of this concept is its extreme generality—its scope is so large that almost all chemical reactions may be considered as acid-base reactions. The convenience of treating a particular type of compound as acids and bases is thereby lost.

110. Are there any change in the hybridisation of B and N atoms, B–F bond length, F–B–F angle and H–N–H angle as a result of the following reaction :

$$F_3B + NH_3 \longrightarrow F_3BNH_3\,?$$ **[B.U. 2016]**

**Hint :** Sec. 4.8.3. (a) The hybridisation of B is changed from $sp^2$ to $sp^3$ in the adduct formation thereby the ∠F–B–F changed from 120° in $BF_3$ ($sp^2$ hybridised B) to 109° in the adduct $NH_3BF_3$ ($sp^3$ hybridised B). The hybridisation of N in $NH_3$ and in the adduct remains the same thereby there will be change of H–N–H bond angle only by 2° (107° in $NH_3$ to 109° in the adduct). This small change in bond angle is due to the small required reorganisation of energy as $NH_3$ is a strong donor molecule.

(b) Due to the strong adduct formation there will be increase of B–F distance in the adduct. Thus the B–F distance in $BF_3$ is 130 pm while the same in $F_3BNH_3$ is 138 pm.

111. What is Hammett acidity function? Show how it compares the strengths of strong acids. State its utility. How does it behave in dilute solutions? **[B.U. 2016]**

**Hint :** Sec. 4.12.

112. Explain whether the following reaction will have equilibrium constant greater or less than unity.

$$R_3PBBr_3 + R_3NBF_3 \rightleftharpoons R_3PBF_3 + R_3NBBr_3$$ **[B.U. 2016]**

**Hint :** Both $PR_3$ and $NR_3$ are Lewis bases of which $PR_3$ has soft donor center P and $NR_3$ has hard donor center N. $BBr_3$ and $BF_3$ both acting as Lewis acids but the former has soft B center as it is linked with three relatively soft Br centers and the latter has hard B center as it is linked to three hard F atoms. According to Pearson's rule hard-hard/soft-soft interaction is more stable than hard-soft/soft-hard interaction, so the equilibrium will be shifted towards left with equilibrium constant value less than unity.

$$\underset{\text{soft-soft}}{R_3PBBr_3} + \underset{\text{hard-hard}}{R_3NBF_3} \rightleftharpoons R_3PBF_3 + R_3NBBr_3$$

113. Write down Drago-Wayland equation indicating the significance of each term involved therein. **[B.U. 2016]**

**Hint :** Sec. 8.1.

114. Using Pauling's rules obtained $pK_a$ of $H_3AsO_4$. **[B.U. 2016]**

**Hint :** Sec. 4.14.2. $H_3AsO_4$ i.e. $As(=O)(OH)_3$ is a tribasic acid. So from structural view point it can be represented as $O_{m-n}X(OH)_n$ and emperical rule proposed for the $pK_{a_1}$ value as, $pK_{a_1} = 8-5(m-n)$.

When $m-n = 0$, $pK_{a_1} = 8$, when $m-n = 1$, $pK_{a_1} = 3$ which is the case for arsenic acid.

115. Adducing proper reason arrange in descending series the acidity order of HCl, $PH_3$ and $H_2S$. **[B.U. 2016]**

**Hint :** Descending series of acidity order is $HCl > H_2S > PH_3$. In a period from left to right electronegativity increases, so sequence of electronegativity is $Cl > S > P$. Thus withdrawal of electron from H is pronounced from left to right resulting in release of $H^+$ is in the order $HCl > H_2S > PH_3$.

116. Exemplify a superacid and a solid acid. **[B.U. 2016]**

**Hint :** Sec. 4.12.1 and Sec. 4.18.

Both $SbF_5 + HSO_3F$ and $SO_3 + SbF_5$ (3 : 1 molar ratio) mixtures are acting as powerful superacids.

Silica surfaces and $\gamma$-$Al_2O_3$ are examples of solid acids.

# 5

# NON-AQUEOUS SOLVENTS

For a long time water has been designated as an universal solvent due to the easy availability, non-toxic nature, has a wide range of temperature to exist as liquid, easy to handle and above all many substances dissolve in water. Almost all laboratory chemical reactions occur in water medium. However there might be some special reactions which are not occur in aqueous solutions but may occur in some non-aqueous solvents. So the need to study reactions in non-aqueous solvents arises. Here we shall discuss special properties of non-aqueous solvents with reference to classification of solvents.

## 5.1 TYPES OF SOLVENTS

Solvents are classified with their acid-base properties or with their neutral behaviour as :

1. **Protic or protonic**–this is subdivided in two categories :

   (a) **Acidic** or **protogenic** i.e. the tendency to donate proton e.g. $H_2O$, HF, $H_2SO_4$, HCN, $CH_3COOH$ etc.

   (b) **Basic** or **protophilic** i.e. the tendency to accept proton e.g. liquid $NH_3$, $N_2H_4$, amines etc.

2. **Amphiprotic** or **amphoteric** i.e. tendency either to release or to accept proton e.g. ROH, $H_2O$.

3. **Aprotic** or **nonprotonic** i.e. no tendency either to accept or to release proton e.g. $CCl_4$, $CHCl_3$, liquid $SO_2$, $C_6H_6$ etc.

Solvents may also be classified according to their ***autoionisation property*** to form solvent cation and solvent anion (Sec. 4.2). Another classification is made due to their ionising property and classified as ***ionic*** or ***ionising solvents*** e.g. $H_2O$, liquid $NH_3$ etc. and ***non ionising*** solvents e.g. $CCl_4$, $C_6H_6$ etc.

## 5.2 CHARACTERISTICS OF IONISING SOLVENTS

The dissolution of ionic compounds in water is discussed in Sec. 3.1.4 e.g. when NaCl is dissolved in water, the $Na^+$ ions are attracted towards the negative ends i.e. oxygen ends and the $Cl^-$ ions are attracted to the positive ends i.e. the hydrogen ends of water molecules. The *force of attraction between $Na^+$ and $Cl^-$ ions are knocked off* and there will be the formation of hydrated $Na^+$ and $Cl^-$ species in solution.

The electrostatic force of attraction (F) between two oppositely charged ions $Z^+$ and $Z^-$ separated by a distance r in a medium of dielectric constant D is represented as :

$$F = \frac{1}{D} \cdot \frac{Z^+Z^-}{r^2} \text{ in CGS system} \qquad \ldots(5.1)$$

Therefore ***dielectric constant (D) of the medium is that property which reduces the interionic force between the cation and anion*** and easy solvation occurs. A solvent having high dielectric constant is therefore most suitable for the dissolution of an ionic compound.

In S.I unit $F = \frac{Z^+Z^-}{4\pi\varepsilon r^2}$ where $\varepsilon$ is the permittivity of the medium which is related with the permittivity in vacuum $\varepsilon_0$ (= $8.854 \times 10^{-12}$ Farad per metre) as $\frac{\varepsilon}{\varepsilon_0} = \varepsilon_r$ where $\varepsilon_r$ is the relative permittivity of the medium and therefore we can write,

$$F = \frac{Z^+Z^-}{4\pi\varepsilon_0\varepsilon_r r^2} \qquad \ldots(5.2)$$

The $\varepsilon_r$ is a dimensionless quantity and is also known as dielectric constant of the medium i.e.

$$\frac{\varepsilon}{\varepsilon_0} = \varepsilon_r \approx D \qquad \ldots(5.3)$$

Solvents having high dipole moments (Sec. 3.7.2) are good ionising solvents and of course have high dielectric constant. Values of dipole moment and dielectric constant of some representative solvents are given in Table 5.1. There are various factors which play part in solvent-solute interactions e.g. $\pi$ bonding effect, hard-hard or soft-soft interactions etc.

**Table 5.1 : Physical properties of some solvents**

| Solvents | Dipole moment (Debye) | Dielectric constant | Viscosity (millipoise) | Autoionisation | Ionic product |
|---|---|---|---|---|---|
| Water | 1.84 | 78.5 (298 K) | 10.1 | $2H_2O \rightarrow H_3O^+ + OH^-$ | $1 \times 10^{-14}$ |
| Ammonia | 1.46 | 22 (240 K) | 2.54 | $2NH_3 \rightarrow NH_4^+ + NH_2^-$ | $1 \times 10^{-27}$ |
| Sulphur dioxide | 1.61 | 17.3 (256 K) | 4.28 | $2SO_2 \rightarrow SO^{2+} + SO_3^{2-}$ | $1 \times 10^{-13}$ |
| Hydrogen fluoride | 1.90 | 83.6 (273 K) | 2.56 | $3HF \rightarrow H_2F^+ + HF_2^-$ | $> 1 \times 10^{-14}$ |
| Hydrogen cyanide | 2.93 | 106.8 (298 K) | 2.0 | $2HCN \rightarrow H_2CN^+ + CN^-$ | $1 \times 10^{-19}$ |
| Sulphuric acid | — | 110.0 (293 K) | 245.4 | $2H_2SO_4 \rightarrow H_3SO_4^+ + HSO_4^-$ | $1 \times 10^{-4}$ |
| Dinitrogen tetroxide | — | 2.42 (273 K) | 5.27 | $N_2O_4 \rightarrow NO^+ + NO_3^-$ | — |

We shall discuss below the reactions of some important non-aqueous solvents.

## 5.3 LIQUID AMMONIA AS SOLVENT

It is the most extensively studied non-aqueous solvent. Some important physical properties are : Boiling point–33.35°C, freezing point –77.7°C, density 0.68 g $ml^{-1}$ at –33.5°C, specific conductance $1 \times 10^{-11}$ $ohm^{-1}$ $cm^{-1}$ at –33.35°C. Compared to water liquid $NH_3$ is less ionising solvent as its dielectric constant value less than that of water.

### 5.3.1 Solubility of different substances

Ionic compounds will be less soluble than that of water. Salts having large lattice energy are practically insoluble in liquid $NH_3$. The solubility of metal halides are in the order $I^- > Br^- > Cl^- > F^-$ which is just reversed of what is expected in water. This can be explained due to the lesser amount of lattice energy of iodide compared to fluoride as also due to polarisability of the anions.

### 5.3.2 Solubility of metals in liquid $NH_3$

Alkali and alkaline earth metals dissolve in liquid $NH_3$. The solubility order of alkali metals are Li < Na < K at the b.p. of liquid $NH_3$. In the presence of catalyst e.g. iron, alkali metals react with liquid $NH_3$ to form metal amide and hydrogen i.e. $M + NH_3 \rightarrow MNH_2 + \frac{1}{2}H_2$. But in absence of any impurity alkali metals dissolve in liquid $NH_3$ to form a ***blue solution (without evolution of hydrogen)*** from which the alkali metals recovered unchanged on evaporation of the solvent. Therefore it is a purely ***physical process***. The colour of the solution changes to ***bronze like*** by increasing the concentration of the metal. It has been suggested that the blue coloured solution is composed of ***ammoniated cation*** and ***ammoniated electron***, represented as :

$$\underset{dispersed}{M(s)} \longrightarrow M\ (solution) \longrightarrow M^+ + e$$

$$M^+ + xNH_3 \longrightarrow \underset{ammoniated\ cation}{[M(NH_3)_x]^+}$$

$$e + yNH_3 \longrightarrow \underset{ammoniated\ electron}{[e(NH_3)_y]^-}$$

The transport and conductance behaviour of the solution indicates that the ***mobility of the anion*** i.e. ***ammoniated electron* is almost ~280 times that of the cation.*** Therefore the solution is ***highly conducting*** and have ***paramagnetic nature*** indicating the presence of unpaired electron. The blue colour is due to the absorption band in the visible region at ~15000 nm owing to transitions among quantised levels in their solvent cages. The blue solution has a strong ***reducing property*** and have ***low density*** compared to that of liquid $NH_3$.

* ammoniated electron also called solvated electron.

### 5.3.3 Reactions in liquid $NH_3$

(a) **Acid-base reactions :** From the solvent system concept (Sec. 4.2) some solvents are autoionised to form solvent cation and anion. Accordingly all ammonium salts are acids in liquid $NH_3$ while amides, imides etc. are acting as a base in liquid $NH_3$. Thus the process of neutralisation is given by,

$$\underset{acid}{NH_4NO_3} + \underset{base}{KNH_2} \longrightarrow \underset{salt}{KNO_3} + \underset{solvent}{2NH_3}$$

$$2NH_4I + PbNH \longrightarrow PbI_2 + 3NH_3$$

$$3NH_4I + BiN \longrightarrow BiI_3 + 4NH_3$$

Water and liquid $NH_3$ are related closely to their solvent property due to the fact that amides and imides in liquid $NH_3$ are equivalent to hydroxides in water while nitrides in liquid $NH_3$ are similar to that of oxides in water. It is to be noted that like acid-base titrations in water, $NH_4Cl$ (which is an acid in liquid $NH_3$) can be titrated with $KNH_2$ (which is a base in liquid $NH_3$) using phenolphthalein as indicator in liquid $NH_3$ solvent. The solution turns pink in excess $KNH_2$.

Because of the formation of solvent cation urea is acting as an weak acid in liquid $NH_3$ due to the following reaction :

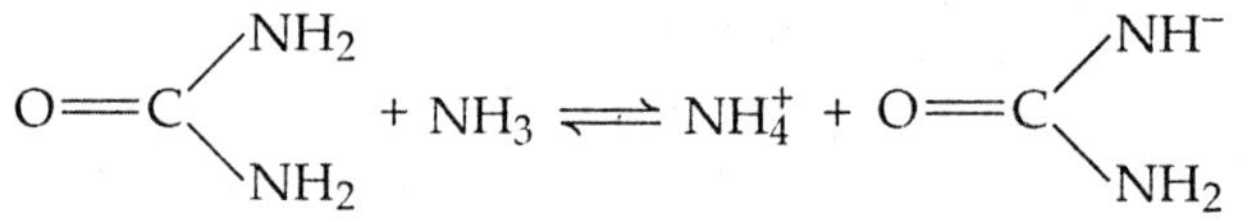

Similarly acetamide and sulphamide behave as a weak monobasic acid.

$$H_3CCONH_2 + NH_3 \longrightarrow H_3CCONH^- + NH_4^+$$

$$H_2NSO_2NH_2 + NH_3 \longrightarrow H_2NSO_2NH^- + NH_4^+$$

However sulphamic acid behaves as a dibasic acid in liquid $NH_3$ while it is monobasic acid in water.

$$H_2NSO_2OH + 2NH_3 \longrightarrow \bar{O}SO_2\bar{N}H + 2NH_4^+$$

$$H_2NSO_2OH + H_2O \longrightarrow \bar{O}SO_2NH_2 + H_3O^+$$

In a similar way amides, imides and nitrides behave as a base in liquid $NH_3$ because they produce solvent anion.

$$NH^{2-} + NH_3 \longrightarrow NH_2^- + NH_2^-$$

$$N^{3-} + NH_3 \longrightarrow NH_2^- + NH^{2-}$$

(b) **Amphoteric behaviour :** Some compounds e.g. $Zn(NH_2)_2$ or $Mg(NH_2)_2$ can behave as amphoteric compound in liquid $NH_3$ like that of water as solvent :

In water : $Zn^{2+} + 2OH^- \longrightarrow Zn(OH)_2\downarrow \underset{acid}{\overset{excess\ NaOH}{\rightleftharpoons}} Zn(OH)_4^{2-}$

In liquid $NH_3$ : $Zn^{2+} + 2NH_2^- \longrightarrow Zn(NH_2)_2\downarrow \underset{NH_4^+\ salt}{\overset{excess\ KNH_2}{\rightleftharpoons}} Zn(NH_2)_4^{2-}$

Similarly for $Mg(NH_2)_2$ we can state the reactions as :

$$\underset{acid}{2NH_4Cl} + \underset{base}{Mg(NH_2)_2} \longrightarrow \underset{salt}{MgCl_2} + \underset{solvent}{4NH_3}$$

$$\underset{base}{2NaNH_2} + \underset{acid}{Mg(NH_2)_2} \longrightarrow Na_2[Mg(NH_2)_4]$$

(c) **Solvolysis reactions :** This reaction is of similar nature as that of ***hydrolysis*** reactions e.g.

$$OPCl_3 + 6H_2O \longrightarrow OP(OH)_3 + 3H_3O^+ + 3Cl^-$$

$$OPCl_3 + 6NH_3 \longrightarrow OP(NH_2)_3 + 3NH_4^+ + 3Cl^-$$

$$NOCl + NH_3 \longrightarrow NO(NH_2) + HCl$$

$$Hg_2Cl_2 + 2NH_3 \longrightarrow Hg + \underset{\text{amidochloro mercury(II)}}{[Hg(NH_2)Cl]} + NH_4Cl$$

$$AuCl + 12NH_3 \longrightarrow AuCl{\cdot}12NH_3$$

$$SiCl_4 + 8NH_3 \longrightarrow Si(NH_2)_4 + 4NH_4Cl$$

$$AlI_3 + 6NH_3 \longrightarrow AlI_3{\cdot}6NH_3$$

$$HgCl_2 + NH_3 \longrightarrow HgNH_2Cl + HCl$$

$$HCl + NH_3 \longrightarrow NH_4Cl$$

These reactions are sometimes referred to as ***ammonolysis*** reactions.

In water medium different salts undergo hydrolysis and the aqueous solutions may be acidic or akaline i.e. the pH range may vary in aqueous solution from 0 to 14. But in liquid $NH_3$ solvent the pH may vary from 0 to 27 because like $K_w$ the $K_{NH_3} = 10^{-27}$. Here pH = 0 indicates $[NH_4^+] = 1$ mol dm$^{-3}$, pH = 13.5 indicates $[NH_4^+] = [NH_2^-]$ and pH = 27 means $[NH_2^-] = 1$ mol dm$^{-3}$.

(d) **Precipitation reactions :** The following precipitation reaction occurs in water medium e.g.

$$KCl + AgNO_3 \longrightarrow AgCl\downarrow + KNO_3$$

But the above precipitation reaction will not occur in liquid $NH_3$. In $NH_3$ solvent AgCl will form the amine complex viz. $[Ag(NH_3)_2]Cl$. Therefore a compound which is insoluble in water may be soluble in liquid $NH_3$ or vice-versa.

We can depict the following precipitation reactions in liquid $NH_3$.

$$AgCl + KNO_3 \longrightarrow KCl\downarrow + AgNO_3$$

$$(NH_4)_2S + Cd(NO_3)_2 \longrightarrow CdS\downarrow + 2NH_4NO_3$$

$$NH_4Cl + NaI \longrightarrow NaCl\downarrow + NH_4I$$

$$PbI_2 + KNH_2 \longrightarrow KI + HI + Pb(NH)\downarrow$$

$$AgCl + Ba(NO_3)_2 \longrightarrow BaCl_2\downarrow + 2AgNO_3$$

$$(NH_4)_2S + 2AgNO_3 \longrightarrow Ag_2S\downarrow + 2NH_4NO_3$$

$$3HgI_2 + 6KNH_2 \longrightarrow Hg_3N_2\downarrow + 6KI + 4NH_3$$

$$Sr(NO_3)_2 + 2NH_4Br \longrightarrow SrBr_2\downarrow + 2NH_4NO_3$$

$$BI_3 + KNH_2 \longrightarrow KI + 2HI + BiN\downarrow$$

$$Zn(NO_3)_2 + 2NH_4I \longrightarrow ZnI_2\downarrow + 2NH_4NO_3$$

$$AgNO_3 + KNH_2 \longrightarrow KNO_3 + AgNH_2\downarrow$$

$$NaCl + KI \longrightarrow KCl\downarrow + NaI$$

$$2NaCl + Ca(NO_3)_2 + 4NH_3 \longrightarrow [Ca(NH_3)_4]Cl_2\downarrow + 2NaNO_3$$

All the above reactions are of ***mutual exchange, double decomposition*** or ***metathetical reactions*** because one of the reaction product is insoluble in liquid $NH_3$ and so it is precipitated.

(e) **Redox reactions :** Liquid $NH_3$ medium is very rich in electrons so that strong oxidising agents cannot survive in this solvent. Thus $HNO_3$ is non oxidising in liquid $NH_3$. $KMnO_4$, $K_2Cr_2O_7$ are weak oxidant. Solutions of alkali and alkaline earth metals in liquid $NH_3$ are acting as a strong reductant due to the presence of ammoniated electron. Some typical reactions are given below :

$$6KMnO_4 + 6KNH_2 \longrightarrow 6K_2MnO_4 + 4NH_3 + N_2$$

$$\underset{\textit{Porassium hexa amido stannate(II)}}{K_4[Sn(NH_2)_6]} + \underset{\textit{weak oxidant}}{I_2} \xrightarrow{\text{liquid } NH_3} \underset{\textit{Potassium hexa amido stannate(IV)}}{K_2[Sn(NH_2)_6]} + 2KI$$

$$CuI + Na \xrightarrow{\text{liquid } NH_3} Cu + NaI$$

$$2Na + 2NH_4Br \longrightarrow 2NaBr + 2NH_3 + H_2$$

$$Fe(CO)_5 \xrightarrow{\text{Na in liquid } NH_3} Na_2[Fe(CO)_4] + CO$$

$$S + 2Na \xrightarrow[\text{Na}]{\text{liquid } NH_3} Na_2S$$

$$K_2MnO_4 + 4K + 3NH_3 \longrightarrow MnO + 3KNH_2 + 3KOH$$

(f) **Complex formation reactions :** There are several complex formation reactions occur like that of water as solvent e.g.

$$Zn(NO_3)_2 + 2NaOH \xrightarrow{H_2O} Zn(OH)_2\downarrow + 2NaNO_3$$

$$\downarrow \textit{excess NaOH}$$

$$Zn(OH)_2 + 2NaOH \xrightarrow{H_2O} \underset{\textit{Soluble hydroxo complex}}{Na_2[Zn(OH)_4]}$$

$$Zn(NO_3)_2 + 2KNH_2 \xrightarrow{\text{liq. } NH_3} Zn(NH_2)_2\downarrow + 2KNO_3$$

$$\downarrow \textit{excess } KNH_2$$

$$Zn(NH_2)_2 + 2KNH_2 \xrightarrow{\text{liq. } NH_3} \underset{\textit{Soluble amide complex}}{K_2[Zn(NH_2)_4]}$$

$$AgNH_2 + 2KNH_2 \xrightarrow{\text{liq. } NH_3} K[Ag(NH_2)_2]$$

*Soluble amide complex*

$$AlCl_3 + 3KNH_2 \xrightarrow{\text{liq. } NH_3} Al(NH_2)_3\downarrow + 3KCl$$

$$\downarrow \text{excess } KNH_2$$

$$Al(NH_2)_3 + KNH_2 \xrightarrow{\text{liq. } NH_3} K[Al(NH_2)_4]$$

*Soluble amide complex*

### 5.3.4 Advantages of using liquid $NH_3$ as solvent

(a) Alkali metals dissolve in liquid $NH_3$ and forms a blue solution which acts as a strong reductant and may be employed for reducing many compounds which are not possible in aqueous solutions.

(b) Solutions of alkali metals in liquid $NH_3$ are widely used in preparative inorganic chemistry.

(c) The tendency of solvolysis or rather ammonolysis is less in liquid $NH_3$ than in water.

### 5.3.5 Limitations of using liquid $NH_3$ as solvent

(a) Liquid $NH_3$ is hygroscopic in nature so that all reactions have to be carried out in moisture free conditions.

(b) As mentioned the boiling point of liquid $NH_3$ is –33.35°C, therefore all chemical reactions must be performed with maintainance of low temperature device system.

---

**Problem 5.1.** What will be the effect in respect of acidity by the addition of (i) $KNH_2$ (ii) $Li_3N$ and (iii) $CH_3COOK$ in liquid $NH_3$.

**Ans.** (i) $KNH_2$ will provide $NH_2^-$ i.e. solvent anion i.e. $KNH_2$ will be a base in liquid $NH_3$ solvent, so acidity will decrease.

(ii) Nitride ion reacts with liquid $NH_3$ to form the solvent anion according to the following reaction thereby acidity decreases.

$$N^{3-} + NH_3 \longrightarrow NH_2^- + NH^{2-}$$

$$\text{or } Li_3N + NH_3 \longrightarrow LiNH_2 + Li_2NH$$

(iii) $CH_3COOK$ will not produce any solvent cation or solvent anion and so $CH_3COOK$ will not effect the acidity in liquid $NH_3$.

---

**Problem 5.2.** (a) Why is liquid $NH_3$ called water-like solvent? (b) Give one example of precipitation reaction in liquid $NH_3$. (c) Ionic product of liquid $NH_3$ at –50°C is $10^{-30}$. If you set a pH scale for this solvent, just like that in water, what would be the readings at the two extremes and at the middle point of that scale at this temperature? **[B.U. 1997, 2002]**

**Ans.** Ammonia molecules are less strongly associated than $H_2O$. The freezing and boiling points of liquid $NH_3$ are lower than those of water. $NH_3$ is less polar than that of $H_2O$ but both have pyramidal structures. Both liquid $NH_3$ and water show similar type of autoionisation e.g.

$$\underset{}{2H_2O} \longrightarrow \underset{\text{Solvent cation}}{H_3O^+} + \underset{\text{Solvent anion}}{OH^-} \qquad 2NH_3 \longrightarrow \underset{\text{Solvent cation}}{NH_4^+} + \underset{\text{Solvent anion}}{NH_2^-}$$

There are several reactions occuring in liquid $NH_3$ system which are like that of water e.g.

$$KCl + AgNO_3 \xrightarrow{H_2O} AgCl\downarrow\ KNO_3$$

$$AgCl + KNO_3 \xrightarrow{\text{liquid } NH_3} KCl\downarrow + AgNO_3$$

For this reason liquid $NH_3$ sometimes called water like solvent.

(b) See sec. 5.3.3(d)

(c) See sec. 5.3.3(c)

---

**Problem 5.3** Complete the following reactions with explanation.

(a) $Zn(NH_2)_2 + 2KNH_2 \xrightarrow{\text{liquid } NH_3}$ **[B.U. 2005, 2008]**

(b) $NH_4Cl + Mg(NH_2)_2 \xrightarrow{\text{liquid } NH_3}$ **[B.U. 2005]**

(c) $HgCl_2 + NH_3 \xrightarrow{\text{liquid } NH_3}$ **[B.U. 2000, 1998]**

(d) $NH_4Cl + KNH_2 \xrightarrow{\text{liquid } NH_3}$ **[B.U. 1998]**

**Ans.** (a) $Zn(NH_2)_2 + 2KNH_2 \xrightarrow{\text{liquid } NH_3} K_2Zn(NH)_2 + 2NH_3$

$Zn(NH_2)_2$ shows amphoteric behaviour. Here acting as an acid with excess $KNH_2$, the reaction is $Zn(NH_2)_2 + 2KNH_2 \longrightarrow K_2[Zn(NH_2)_4]$

(b) $$\underset{\text{acid}}{2NH_4Cl} + \underset{\text{base}}{Mg(NH_2)_2} \xrightarrow{\text{liquid } NH_3} \underset{\text{salt}}{MgCl_2} + \underset{\text{solvent}}{4NH_3}$$

This is an acid-base or neutralisation reaction.

(c) $HgCl_2 + NH_3 \xrightarrow{\text{liquid } NH_3}$ This reaction is a solvolysis reaction but the products depend on the reaction conditions and the following reactions occur :

$$HgCl_2 + 2NH_3 \longrightarrow HgCl_2(NH_3)_2\downarrow$$

$$HgCl_2 + 2NH_3 \longrightarrow HgCl(NH_2)\downarrow + NH_4Cl$$

(d) $$\underset{\text{acid}}{NH_4Cl} + \underset{\text{base}}{KNH_2} \xrightarrow{\text{liquid } NH_3} \underset{\text{salt}}{KCl} + \underset{\text{solvent}}{2NH_3}$$

This is an acid-base or neutralisation reaction in liquid $NH_3$.

## 5.4 LIQUID SULPHUR DIOXIDE AS SOLVENT

Its dielectric constant is 17.4 at ~20°C which indicates that it is not a good solvent for ionic compounds but is a good solvent for covalent or organic compounds. The sequence of solubility of alkali metal halide is MI > MBr > MCl > MF. It has a long liquid range (–10.0°C to –75.5°C), but not so large compared to that of water.

### 5.4.1 Solubility

As mentioned liquid $SO_2$ is an excellent solvent for organic compounds. Among inorganic compounds iodides and thiocyanates are the most soluble. Ammonium, thallium and mercuric salts are soluble while metal sulphates, sulphides, oxides and hydroxides are practically insoluble. Dissolution of metal is not possible in this solvent. Compounds e.g. IBr, $PBr_3$, $CCl_4$, $SiCl_4$, $SnCl_4$, $(CH_3)_2CO$, $C_6H_5Cl$, alcohols, ketones, aldehydes are either more soluble or moderately soluble.

### 5.4.2 Reactions in liquid $SO_2$

(a) **Acid-base reactions or neutralisation reactions :** According to solvent system concept (Sec. 4.2) liquid $SO_2$ is autoionised to form solvent cation ($SO^{2+}$) and solvent anion ($SO_3^{2-}$). Thus, $2SO_2 \longrightarrow SO^{2+} + SO_3^{2-}$

So compounds which produce $SO^{2+}$ ion is acting as an acid while the compounds which furnish $SO_3^{2-}$ ion will be termed as a base in liquid $SO_2$ e.g.

$$\underset{acid}{SOCl_2} + \underset{base}{Cs_2SO_3} \longrightarrow \underset{salt}{2CsCl} + \underset{solvent}{2SO_2}$$

$$\underset{acid}{SO(SCN)_2} + \underset{base}{K_2SO_3} \longrightarrow \underset{salt}{2KSCN} + \underset{solvent}{2SO_2}$$

$$\underset{acid}{SOBr_2} + \underset{base}{[N(CH_3)_4]_2SO_3} \longrightarrow \underset{salt}{2[N(CH_3)_4]Br} + \underset{solvent}{2SO_2}$$

(b) **Amphoteric behaviour :** Like that in water solvent amphoteric behaviour can also be observed in liquid $SO_2$. For example $AlCl_3$ can show amphoteric behaviour both in aqueous and in liquid $SO_2$ medium.

$$*\begin{cases} AlCl_3 + 3NH_4OH \xrightarrow{H_2O} Al(OH)_3\downarrow + 3NH_4Cl \\ AlCl_3 + 3NaOH \xrightarrow{H_2O} Al(OH)_3\downarrow + 3NaCl \\ Al(OH)_3 + NaOH \xrightarrow{H_2O} \underset{Soluble}{NaAlO_2} + 2H_2O \\ NaAlO_2 + NH_4Cl + H_2O \xrightarrow{H_2O} Al(OH)_3\downarrow + NaCl + NH_3\uparrow \end{cases}$$

* These reactions are employed for qualitative detection of $Al^{3+}$.

$$\text{or } NaAlO_2 + HCl + H_2O \xrightarrow{H_2O} Al(OH)_3\downarrow + NaCl$$

$$2AlCl_3 + \underset{\textit{trimethyl ammonium sulphite}}{3[N(CH_3)_4]_2SO_3} \xrightarrow{\text{liquid } SO_2} 6[N(CH_3)_4]Cl + Al_2(SO_3)_3\downarrow$$

$$Al_2(SO_3)_3 + 3[N(CH_3)_4]_2SO_3 \xrightarrow{\text{liquid } SO_2} \underset{\textit{Soluble}}{2[N(CH_3)_4]_3Al(SO_3)_3}$$

From the soluble species i.e. $[N(CH_3)_4]_3Al(SO_3)_3$ the $Al_2(SO_3)_3$ can be reprecipitated by the addition of the acid $SOCl_2$.

$$2[N(CH_3)_4]_3Al(SO_3)_3 + 3SOCl_2 \xrightarrow{\text{liquid } SO_2} Al_2(SO_3)_3\downarrow + 6[N(CH_3)_4]Cl + 6SO_2$$

*Limitation of the reactions*

All these reactions are explained by the autoionisation of liquid $SO_2$ to form solvent cation $SO^{2+}$ and solvent anion $SO_3^{2-}$. But there is no convincing evidence for the existance of $SO^{2+}$ ion in liquid $SO_2$.

Similar type of amphoteric character is expected for $GaCl_3$ like that of $AlCl_3$.

(c) **Solvolysis reactions :** Many solvolysis reactions in liquid $SO_2$ medium occurs like that of hydrolysis in aqueous medium e.g.

$$2CH_3COONH_4 + 2SO_2 \longrightarrow (NH_4)_2SO_3 + (CH_3COO)_2SO$$

$$PCl_5 + SO_2 \longrightarrow POCl_3 + SOCl_2$$

$$UCl_6 + 2SO_2 \longrightarrow UO_2Cl_2 + 2SOCl_2$$

$$NbCl_5 + SO_2 \longrightarrow NbOCl_3 + SOCl_2$$

$$WCl_6 + SO_2 \longrightarrow WOCl_4 + SOCl_2$$

Typical ***solvates*** like that of hydrates can also formed in liquid $SO_2$ e.g. $KBr.4SO_2$; $BaI_2.4SO_2$; $NaI.4SO_2$ etc.

(d) **Precipitation reactions :** Ma iy precipitation reactions occur in liquid $SO_2$ e.g.

$$2CH_3COOAg + SOCl_2 \xrightarrow{\text{liquid } SO_2} 2AgCl\downarrow + \underset{\textit{thionyl acetate}}{SO(CH_3COO)_2}$$

$$BaI_2 + Zn(CNS)_2 \xrightarrow{\text{liquid } SO_2} Ba(CNS)_2\downarrow + ZnI_2$$

$$AlCl_3 + 3NaI \xrightarrow{\text{liquid } SO_2} 3NaCl\downarrow + AlI_3$$

$$2KI + SOCl_2 \xrightarrow{\text{liquid } SO_2} 2KCl\downarrow + SOI_2$$

$$SbCl_3 + 3LiI \xrightarrow{\text{liquid } SO_2} SbI_3\downarrow + 3LiCl$$

$$PbF_2 + Li_2SO_4 \xrightarrow{\text{liquid } SO_2} PbSO_4\downarrow + 2LiF$$

$$2NH_4SCN + SOCl_2 \xrightarrow{\text{liquid } SO_2} 2NH_4Cl\downarrow + SO(SCN)_2$$

These are sometimes called ***mutual exchange*** or ***double decomposition*** or ***metathetical*** reactions.

(e) **Redox reactions :** $SO_2$ shows oxidising and reducing behaviour in aqueous medium but liquid $SO_2$ does not show prominant redox property, it merely servs as a solvent for redox reactions :

$$6KI + 3SbCl_5 \xrightarrow{\text{liquid } SO_2} 2K_3[SbCl_6] + 3I_2 + SbCl_3$$

$$I_2 + 2R_2SO_3 \xrightarrow{\text{liquid } SO_2} R_2SO_4 + 2RI + SO_2$$

(f) **Complex formation reactions :** Large number of complex formation reactions occur in aqueous medium. But the process of complex formation is limited in liquid $SO_2$. Some typical examples are :

$$2AlCl_3 + 3Cs_2SO_3 \xrightarrow{\text{liquid } SO_2} Al_2(SO_3)_3\downarrow + 6CsCl$$

$$Al_2(SO_3)_3 + 3Cs_2SO_3 \xrightarrow{\text{liquid } SO_2} 2Cs_3[Al(SO_3)_3]$$

Zn(II) salt also produce $K_2[Zn(SO_3)_2]$ with $K_2SO_3$ in liquid $SO_2$ medium. Other examples are :

$$RbI + I_2 \xrightarrow{\text{liquid } SO_2} RbI_3$$

$$HgI_2 + 2KI \xrightarrow{\text{liquid } SO_2} K_2[HgI_4]$$

$$SbCl_5 + NOCl \xrightarrow{\text{liquid } SO_2} [NO][SbCl_6]$$

$$SbCl_5 + KCl \longrightarrow K[SbCl_6]$$

$$SbCl_3 + 3KCl \longrightarrow K_3[SbCl_6]$$

$$SbCl_5 + CH_3COCl \longrightarrow [CH_3CO]\ [SbCl_6]$$

As mentioned earlier liquid $SO_2$ is a good solvent for organic compounds, therefore we state some typical examples of organic reactions here :

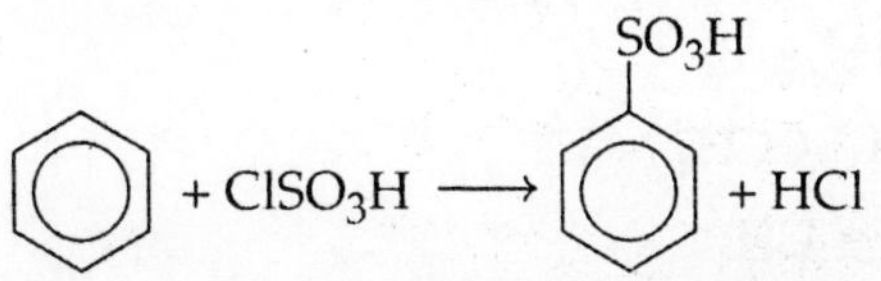


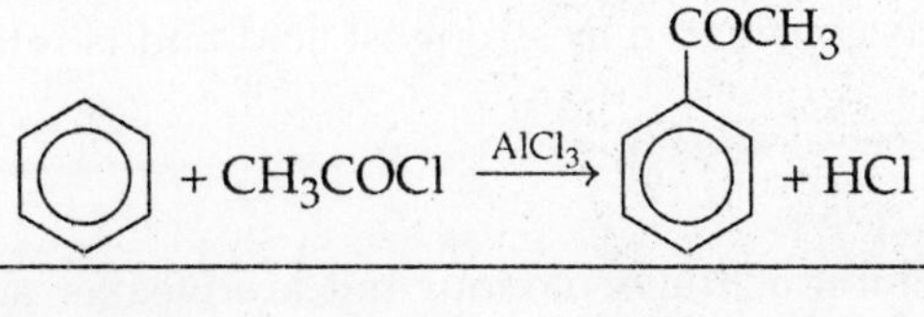


**Problem 5.4** Complete and explain the following reaction— $SOCl_2 + Cs_2SO_3 \xrightarrow{\text{liquid } SO_2}$

**Ans.** Sec. 5.4.2(a). **[B.U. 2000, 1998]**

## 5.5 LIQUID HYDROGEN FLUORIDE AS SOLVENT

Liquid HF is a superior water like ionising solvent since it has a dielectric constant value (83.6 at 0°C) higher than that of water (78.5 at 25°C). Its boiling point (19.5°C) and freezing point (–83°C) indicates that it has a wide liquid range. The use of this hydrogen bonded solvent is restricted due to its poisonous nature as also due to its extreme reactivity.

### 5.5.1 Reactions in liquid HF

(a) **Acid-base reactions :** The autoionisation of liquid HF can be depicted as :

$$3HF \longrightarrow \underset{\textit{solvent cation}}{H_2F^+} + \underset{\textit{solvent anion}}{HF_2^-}$$

The solvent cation ($H_2F^+$) is equivalent to $H^+$. HF and the solvent anion is equivalent to $F^-$.HF. Therefore a substance which produces solvent cation ($H_2F^+$) or $H^+$ ion is termed as an acid while a substance which produces solvent anion ($HF_2^-$) or $F^-$ ion is called a base. Strong mineral acids e.g. $HNO_3$, $H_2SO_4$ and $CH_3COOH$ acting as a base in this solvent i.e. HF forces to accept proton to $HNO_3$, $H_2SO_4$ and $CH_3COOH$ according to the equation :

$$HNO_3 + HF \longrightarrow H_2NO_3^+ + F^-$$

$$H_2SO_4 + HF \longrightarrow H_3SO_4^+ + F^-$$

$$CH_3COOH + HF \longrightarrow CH_3COOH_2^+ + F^-$$

However $HClO_4$ is the strongest acid in aqueous medium can behave as amphoteric one in liquid HF.

$$\underset{\textit{(base)}}{HClO_4} + HF \longrightarrow H_2ClO_4^+ + F^-$$

$$\underset{\textit{(acid)}}{HClO_4} + HF \longrightarrow ClO_4^- + H_2F^+$$

Organic compounds such as ketones, aldehydes, benzene, ethanol etc. accept proton from HF and thus behave as bases e.g.

$$C_6H_6 + HF \longrightarrow C_6H_7^+ + F^-$$

$$C_2H_5OH + HF \longrightarrow C_2H_5OH_2^+ + F^-$$

$$(CH_3)_3CCH_3 + HF \longrightarrow (CH_3)_3CCH_4^+ + F^-$$

$$(CH_3)_3CCH_4^+ \longrightarrow (CH_3)_3C^+ + CH_4$$

A mixture of $SbF_5$ and liquid HF is considered as strongest acid and is referred as super acid (Sec. 4.12.1).

$$2HF + SbF_5 \rightleftharpoons SbF_6^- + H_2F^+$$

$GeF_4$, $SnF_4$, $BF_3$, $AsF_5$, $PF_5$ etc. exhibit acidic behaviour but are weaker acid than $SbF_5$. $AlF_3$ shows amphoteric properties in liquid HF.

As acid : $AlF_3 + 3HF_2^- \longrightarrow AlF_6^{3-} + 3HF$

As base : $Na_3AlF_6 + 3HBF_4 \longrightarrow AlF_3 + 3NaBF_4 + 3HF$

The following neutralisation reaction will occur in liquid HF

$$\underset{\textit{acid}}{HBF_4} + \underset{\textit{base}}{KF} \longrightarrow \underset{\textit{salt}}{KBF_4} + \underset{\textit{solvent}}{HF}$$

(b) **Double decomposition or metathetical reactions and precipitation reactions :** Alkali metal sulphate, nitrate and perchlorate etc. salts are sufficiently soluble in liquid HF and they undergo double decomposition reaction with silver and nickel fluorides or other compounds whereby the heavy metal salt is precipitated.

$$KIO_4 + AgF \longrightarrow AgIO_4\downarrow + KF$$

$$NaClO_4 + AgF \longrightarrow AgClO_4\downarrow + NaF$$

$$Na_2SO_4 + 2AgF \longrightarrow Ag_2SO_4\downarrow + 2NaF$$

$$NaClO_4 + TlF \longrightarrow TlClO_4\downarrow + NaF$$

$$Na_2SO_4 + NiF_2 \longrightarrow NiSO_4\downarrow + 2NaF$$

$$AgNO_3 + BF_3 + 2HF \longrightarrow AgBF_4\downarrow + H_2NO_3^+F^-$$

$AlF_3$ dissolves in HF due to the formation of fluoro complex but again reprecipitated by $BF_3$ which acts as a better $F^-$ acceptor.

$$AlF_3 + HF \rightleftharpoons \underset{solution}{HAlF_4} \xrightarrow{BF_3} HBF_4 + AlF_3\downarrow$$

(c) **Solvolysis reactions :** There is a formation of covalent metal fluoride by the solvolysis reaction e.g.

$$TiCl_4 + 4HF \longrightarrow TiF_4 + 4HCl$$

(d) **Formation of addition compounds :** Like that of hydrates, there will be the formation of addition compounds with some fluorides viz. KF.HF, KF.2HF, KF.3HF, $NH_4$F.HF etc.

## 5.6 SULPHURIC ACID AS SOLVENT

Inspite of its high dielectric constant (110 at 20°C), it is not used as a good ionising solvent compared to water because its viscosity is about 25 times greater than that of water. Ionic compounds very slowly dissolve in this solvent. Its boiling point and freezing point are 300°C and 10.4°C respectively. Its density is 1.84 g $cm^{-3}$ at 25°C.

### 5.6.1 Reactions in sulphuric acid

**Acid-base reactions :** Sulphuric acid is autoionised in the following way :

$$2H_2SO_4 \longrightarrow \underset{\substack{solvent\\cation}}{H_3SO_4^+} + \underset{\substack{solvent\\anion}}{HSO_4^-}$$

The equilibrium constant of this autoionisation is about $10^{-4}$

$KHSO_4$ acts as a base in this solvent.

$$KHSO_4 \longrightarrow K^+ + HSO_4^-$$

All species which are bases in water are also bases in $H_2SO_4$.

$$OH^- + 2H_2SO_4 \longrightarrow 2HSO_4^- + H_3O^+ \text{ or } OH^- + H_2SO_4 \longrightarrow HSO_4^- + H_2O$$

$$NH_3 + H_2SO_4 \longrightarrow HSO_4^- + NH_4^+$$

Even water acting as a base in this solvent e.g.

$$H_2O + H_2SO_4 \longrightarrow HSO_4^- + H_3O^+$$

Similarly the species containing lone pair of electrons easily accept a proton from $H_2SO_4$ and acts as a base. Thus,

$$H_2NCONH_2 + H_2SO_4 \longrightarrow HSO_4^- + H_2NCONH_3^+$$

$$C_2H_5OH + H_2SO_4 \longrightarrow HSO_4^- + C_2H_5OH_2^+$$

$CH_3COOH$ and even strong acids in water acting as a base (i.e. proton acceptor) in this solvent.

$$H_3CCOOH + H_2SO_4 \longrightarrow CH_3COOH_2^+ + HSO_4^-$$

$$HNO_3 + H_2SO_4 \longrightarrow H_2NO_3^+ + HSO_4^-$$

or $$HNO_3 + 2H_2SO_4 \longrightarrow NO_2^+ + H_3O^+ + 2HSO_4^-$$

$HClO_4$ is a very strong acid which acts as a weak acid in $H_2SO_4$ because of the very strong tendency to donate protons. Similarly pyrosulphuric acid $H_2S_2O_7$ acting as a weak acid

$$HClO_4 + H_2SO_4 \longrightarrow H_3SO_4^+ + ClO_4^-$$

$$SO_3 + H_2SO_4 \longrightarrow H_2S_2O_7$$

$$H_2S_2O_7 + H_2SO_4 \longrightarrow H_3SO_4^+ + HS_2O_7^-$$

Boric acid and sulphuric acid mixture acts as a **very strong acid**. Hydrogen tetrakis (hydrogen sulfato) borate, $HB(HSO_4)_4$ forms in solution. To this solution addition of $SO_3$ removes $H_3O^+$ and $HSO_4^-$

$$H_3BO_3 + 6H_2SO_4 \longrightarrow B(HSO_4)_4^- + 3H_3O^+ + 2HSO_4^-$$

$$B(HSO_4)_4^- + 3H_3O^+ + 2HSO_4^- + 3SO_3 \longrightarrow H_3SO_4^+ + B(HSO_4)_4^- + 4H_2SO_4$$

$$H_3BO_3 + 2H_2SO_4 + 3SO_3 \longrightarrow H_3SO_4^+ + B(HSO_4)_4^-$$

In fact this reaction is considered as solvolysis of $H_3BO_3$.

**Problem 5.5** Complete the reaction $N_2O_5 + H_2SO_4 \longrightarrow$ **[B.U. 2000]**

**Ans.** $N_2O_5 + H_2SO_4 \longrightarrow H_3O^+ + 3HSO_4^- + 2NO_2^+$

In this reaction $N_2O_5$ acting as a base because it produces the solvent anion.

## 5.7 LIQUID DINITROGEN TETROXIDE AS SOLVENT

This is one of the important non-aqueous solvent. Liquid $N_2O_4$ undergoes autoionisation in the following ways :

$$N_2O_4 \rightleftharpoons NO^+ + NO_3^- \qquad K = 7 \times 10^{-8} \text{ mol dm}^{-3} \text{ at } 30°C$$

or $$N_2O_4 \rightleftharpoons NO_2^+ + NO_2^- \qquad K = 10^{-22} \text{ mol dm}^{-3} \text{ at } 30°C$$

Therefore the major species is $NO^+$ and $NO_3^-$ which are solvent cation and solvent anion respectively. Its dielectric constant (2.42) indicates that it is a poor ionising solvent. Freezing point (–11.2°C) and boiling point (21.1°C) indicates that working with this solvent requires low temperature. The autoionisation of this solvent into $NO^+$ and $NO_3^-$ is almost complete in anhydrous $HNO_3$.

Typical acid-base (neutralisation) reaction in $N_2O_4$ is

$$\underset{acid}{NOCl} + \underset{base}{AgNO_3} \longrightarrow \underset{salt}{AgCl} + \underset{solvent}{N_2O_4}$$

In liquid $N_2O_4$, NOCl is considered as ***acid*** and a reaction is represented as :

$$2NOCl + Sn \longrightarrow 2NO + SnCl_2$$

This reaction is of similar type of reaction by which Zn reacts with HCl to form $ZnCl_2$ and $H_2$.

Some ***solvolysis*** reactions depicted as :

$$CaO + 2N_2O_4 \longrightarrow Ca(NO_3)_2 + N_2O_3$$

$$Fe(CO)_5 + 3N_2O_4 \longrightarrow Fe(NO_3)_3 + 5CO + 3NO$$

In presence of $MeNO_2$ or MeCOOEt the following ***complex formation*** reaction may occur.

$$Cu + 3N_2O_4 \longrightarrow NO[Cu(NO_3)_3] + 2NO$$

Solvated species like that of hydrates may be formed by liquid $N_2O_4$ e.g. $2Fe(NO_3)_3{\cdot}3N_2O_4$; $Sc(NO_3)_3{\cdot}2N_2O_4$ etc.

## 5.8 SOME OTHER NON-AQUEOUS SOLVENTS AND THEIR AUTOIONISATION PROCESS WITH TYPICAL REACTIONS

(a) $2HCN \rightleftharpoons H_2CN^+ + CN^-$ or $HCN \rightleftharpoons H^+ + CN^-$

$$\underset{acid}{HCl} + \underset{base}{NaCN} \longrightarrow \underset{salt}{NaCl} + \underset{solvent}{HCN}$$

**Solvolysis reaction :** $Ag_2SO_4 + 2HCN \longrightarrow 2AgCN + H_2SO_4$

**Complex formation :** $FeCl_3 + 3Et_3NHCN \longrightarrow Fe(CN)_3\downarrow + 3Et_3NHCl$

$$Fe(CN)_3 + 3Et_3NHCN \longrightarrow \underset{soluble}{(Et_3NH)_3[Fe(CN)_6]}$$

**Metathetical reaction :** $2NaCl + K_2SO_4 \xrightarrow{HCN} Na_2SO_4\downarrow + 2KCl$

$$AgI + NaCl \xrightarrow{HCN} NaI\downarrow + AgCl$$

(b) $2CH_3COOH \longrightarrow CH_3COOH_2^+ + CH_3COO^-$

or $CH_3COOH \longrightarrow \underset{\substack{solvent\\cation}}{H^+} + \underset{\substack{solvent\\anion}}{CH_3COO^-}$

**Acid-base reaction :** $\underset{acid}{HCl} + \underset{base}{CH_3COONa} \longrightarrow \underset{salt}{NaCl} + \underset{solvent}{CH_3COOH}$

**Solvolysis reaction :** $SO_2Cl_2 + 4CH_3COOH \rightleftharpoons SO_2(CH_3COO)_2 + 2CH_3COOH_2^+ + 2Cl^-$

**Metathetical reaction :** $BaI_2 + 2NaNO_3 \longrightarrow Ba(NO_3)_2\downarrow + 2NaI$

$$2AgNO_3 + ZnI_2 \longrightarrow 2AgI\downarrow + Zn(NO_3)_2$$

**Complex formation reaction :** $Fe^{3+} + CNS^- \longrightarrow \underset{red}{Fe(SCN)^{2+}}$

$$Co^{2+} + 4NH_4CNS \longrightarrow \underset{blue}{(NH_4)_2[Co(SCN)_4]}$$

(c) $\underset{liquid}{H_2S} \longrightarrow \underset{\text{solvent cation}}{H^+} + \underset{\text{solvent anion}}{HS^-}$

**Acid-base reaction :** $\underset{acid}{HCl} + \underset{base}{(Et_3NH)^+HS^-} \longrightarrow \underset{salt}{(Et_3NH)^+Cl^-} + \underset{solvent}{H_2S}$

**Solvolysis reactions :** $2AgNO_3 + H_2S \longrightarrow Ag_2S + 2HNO_3$

$$Hg_2Cl_2 + 2H_2S \longrightarrow Hg_2(HS)_2 + 2HCl$$

$$2AsCl_3 + 3H_2S \longrightarrow As_2S_3 + 6HCl$$

**Redox reactions :** $Cu + H_2S \longrightarrow CuS + H_2$

$$SO_3 + H_2S \longrightarrow H_2O + SO_2 + S$$

**Metathetical reaction :** $Hg_2Cl_2 + 2(Et_3NH)^+HS^- \longrightarrow Hg_2(HS)_2 + 2(Et_3NH)^+Cl^-$

(d) $\underset{liquid}{2BrF_3} \longrightarrow \underset{\text{solvent cation}}{BrF_2^+} + \underset{\text{solvent anion}}{BrF_4^-}$

**Acid-base reaction :** $\underset{acid}{SbF_5} + BrF_3 \longrightarrow BrF_2^+ + SbF_6^-$

$$\underset{base}{KF} + BrF_3 \longrightarrow K^+ + BrF_4^-$$

$\therefore \underset{acid}{SbF_5} + \underset{base}{KF} + 2BrF_3 \longrightarrow \underset{salt}{KSbF_6} + \underset{solvent}{2BrF_3}$

**Redox reaction :** $6Nb + 16BrF_3 \longrightarrow 6BrF_2[NbF_6] + 5Br_2$

(e) $COCl_2 \longrightarrow \underset{\text{solvent cation}}{COCl^+} + \underset{\text{solvent anion}}{Cl^-}$

$COCl_2$ has a low dielectric constant and is a poor ionising solvent. The following ***solvolysis reactions*** occur :

$$AlCl_3 + COCl_2 \longrightarrow COCl^+ + AlCl_4^-$$

$$SbCl_3 + COCl_2 \longrightarrow COCl^+ + SbCl_4^-$$

$$SbCl_5 + COCl_2 \longrightarrow COCl^+ + SbCl_6^-$$

**Problem 5.6** Why boric acid acting as a strong acid in anhydrous $H_2SO_4$.

**Ans.** Sec. 5.6.1.

**Problem 5.7** What will be the effect of adding $KHSO_4$ to liquid $H_2SO_4$.

**Ans.** $KHSO_4$ produces the solvent anion ($HSO_4^-$) in liquid $H_2SO_4$. So acidity of liquid $H_2SO_4$ will reduce.

## 5.9 TITRATIONS IN NON-AQUEOUS SOLVENTS

We have seen that water acting as a levelling solvent (Sec. 4.3.1) for strong acids and bases, even the strengths of weak acids and bases may not be distinguished in this amphiprotic (water) solvent. Titration in non-aqueous solvents permits the determination of the strengths of various acids and bases which cannot be titrated in water. Except Arrhenius concept (which is restricted to water solvent) the other concepts of acids and bases may be applicable in non-aqueous solvents.

Let us consider an acid HB which will be titrated with a base say NaOH. The feasibility of the titration is to be judged by the acid dissociation constant $K_a$ of the acid. This $K_a$ value is a measure of the tendency to transfer $H^+$ ions to the solvent e.g. water, i.e.

$$HB + H_2O \rightleftharpoons H_3O^+ + B^-$$

The $K_a$ value cannot give us the idea of the absolute acid strength of HB since the same acid may dissociate more in some organic basic solvent e.g. an amine, so that,

$$HB + RNH_2 \rightleftharpoons RNH_3^+ + B^-$$

Therefore there will be a greater concentration of solvated protons in the organic amine solvent.

*Principle of acid-base titrations in nonaqueous solvents*

Thus the feasibility of the titration is to be increased by selecting a solvent more basic than water. Apart from the amine solvent, excellent titrations have been carried out in non polar solvents like benzene and chloroform. However titration of very weak acid is carried out by amine solvent better than water. In water solvent the titration reaction is represented as :

$$HB + OH^- \rightleftharpoons H_2O + B^-$$

$$\text{so that } K = \frac{[H_2O][B^-]}{[HB][OH^-]} \qquad HB \rightleftharpoons H^+ + B^- \quad K_a = \frac{[H^+][B^-]}{[HB]}$$

$$= \frac{K_a}{[H^+]} \cdot \frac{[H_2O]}{[OH^-]} \qquad \therefore \frac{[B^-]}{[HB]} = \frac{K_a}{[H^+]}$$

$$= \frac{K_a}{K_w} \qquad \text{... (5.4)}$$

since $[H_2O] = 1$ and $[H^+][OH^-] = K_\omega$

Thus equation (5.4) suggests that the titration reaction will be complete unless $K_a$ is large i.e. the acid HB is sufficiently strong. Larger K values are achieved by the smaller autoprotolysis constant ($K_s$) of the solvent which can be represented in a general way other than water as :

$$HB + S^- \rightleftharpoons HS + B^-$$

For which we can write :

$$K = \frac{K_a}{K_s} \quad \ldots(5.5)$$

where HS is the solvent, $S^-$ is the conjugate base. If HS is a weaker acid than water K for equation (5.5) will be larger than K for equation (5.4). Thus we may say that many titrations of weak acids and bases may be carried out in non aqueous solvents which are not feasible in water solvent.

*Advantage and limitations of non-aqueous titrations*

The non-aqueous titrations may be applicable to any compound which behave either as an acid or as a base in a suitable solvent. Compounds having insufficient acidic or basic strength which do not exhibit sharp end points in aqueous solutions, can be titrated in a non-aqueous levelling solvent which increase the acidic and basic character of the compounds. In order to depress the solvolysis at the end point an aprotic solvent is used to improve the end point in the titrations. Potentiometric titrations may also be employed for this purpose. Many visual indicators are used for compounds which behave as sufficiently strong acids and bases in non-aqueous solvents. The suitability of a particular indicator is judged by potentiometric titration and observing the colour change of the indicator.

Non-aqueous titrations are not widely used as normal volumetric procedure in the laboratory since many compounds are insoluble in non-aqueous solvents.

### 5.9.1 Titration of bases with non-aqueous acids i.e. titration of aniline with glacial acetic acid

Let us consider the titration of a base B with an acid in a solvent(S) according to the following equilibria :

$$B + SH^+ \rightleftharpoons BH^+ + S$$

where $SH^+$ is the solvated proton produced from the titrant. Sharp end point is obtained when the base is a strong one thereby the equilibrium will be shifted to the right. Excellent end points are obtained with glacial acetic acid as solvent and perchloric acid as titrant. It is to be noted that even aromatic amines which behave as weak bases in water, may be considered as strong bases and give satisfactory end points in acetic acid medium. Dioxan may be used for the preparation of perchloric acid solution as titrant for the titration of aliphatic and aromatic amines e.g. $C_6H_5NH_2$. Perchloric acid and acetic acid mixture solution is standardised against pure

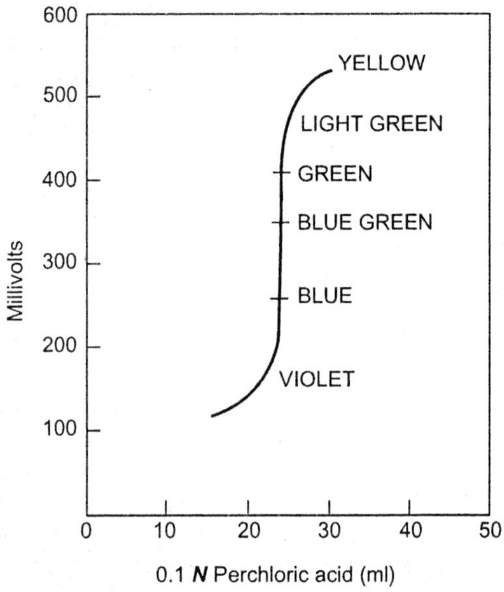


***Figure 5.1 :** Titrations of aniline with perchloric acid with acetic acid as solvent (crystal violet as indicator).*

potassium hydrogen phthalate which acts as a base in acetic acid solvent. This solution is then used to titrate the amine e.g. $C_6H_5NH_2$. Crystal violet or methyl violet may be used as indicator and the end point may be detected potentiometrically using glass electrode and calomel electrode. The colour changes may be recorded in the titration curve (Fig. 5.1). The colour at the end point found by noting that which corresponds to the inflection point on the curve using crystal violet as indicator. This is important since the colour change at the end point may vary with the compound being tirated. The colour changes for cyrstal violet in acetic acid solution is violet to blue then through several shades finally converted to yellow (Fig. 5.1). Methyl violet changes from violet to blue. A sample of aniline is dissolved in about 50 ml of acetic acid and titrated with standard perchloric acid to the first appearance of the violet colour using two drops of methyl violet indicator.

The general procedure for the titration of aniline as a base in non-aqueous solvents with the titrant of perchloric acid in glacial acetic acid. Here the solvent is used frequently as glacial acetic acid since this will increase the basicity of aniline or other amines by levelling effect. The end point of the titration may be established either by visual indicator e.g. crystal violet or methyl violet or by potentiometrically. Sometimes sharp end points may be obtained if the solvent acetic acid is diluted with another solvent having low dielectric constant such as benzene, nitrobenzene etc. For potentiometric titration glass and calomel electrode system is used if the solvent is only acetic acid. For other mixed solvents it is better to use glass and Ag/AgCl electrode assembly.

Accurately weighed about 0.3 g aniline or other amine e.g. p-toluidine, diethylaniline is dissolved in 50 ml glacial acetic acid in a 250 ml conical flask. About three drops of indicator methyl violet (colour change from violet to green) or crystal violet (colour change from blue to blue green) is added and the solution titrated with the acetic acid-perchloric acid mixture solution taking in a burette. The correct colour change of the indicator may be detected by potentiometric titration.

A blank titration may be made with 50 ml of glacial acetic acid in the presence of the indicator and a correction of the titre value is made accordingly if required.

### 5.9.2 Titration of acids with non-aqueous bases i.e. titration of phenol with quaternary ammonium compounds

Very weak acids e.g. phenols usually require a more strongly basic solvent such as anhydrous ethylenediamine, or n-butylamine which exhibit a levelling effect and increase the acid strength of weak acids sufficiently and hence the titratability enhances with sodium methoxide in benzene-methanol solution using thymol blue or azo violet indicator. Solutions of tetra-n-butyl ammonium hydroxide in benzene-methanol or in isopropanol etc. are used as valuable titrants for a large variety of organic acids. Potentiometric titrations may be employed with glass and calomel electrode. The colour change for thymol blue is from ***yellow to blue*** while for azo violet the colour change is from ***red to blue***. The azo violet gives sharp end point with ethylenediamine and n-butyl amine.

## EXERCISE

1. Why is liquid $NH_3$ called water-like solvent? 'Zinc amide is sparingly soluble in liquid $NH_3$ but dissolves in presence of potassium amide'–explain. **[B.U. 2001, C.U. 2005]**

   **Hint :** Problem 5.2(a) and 5.3(a).

2. What will be the effect of adding $KNH_2$ to liquid $NH_3$ in respect of acidity? **[B.U. 2000]**

   **Hint :** Problem 5.1(i).

3. Depict by proper equations the autoionisation of liquid $SO_2$ and hence complete the following reaction :

   $$PCl_5 + SO_2 \xrightarrow{\text{liquid } SO_2} ?$$ **[B.U. 2005, 2001, 1998]**

4. Discuss the principles of acid-base titrations in non-aqueous solvents. Outline a procedure for the titration of aniline in glacial acetic acid. **[B.U. 2003, 2005]**

5. Explain why zinc amide is insoluble in liquid ammonia, but dissolves when ammonium acetate is added.

   **Hint :** In liquid $NH_3$, $CH_3COONH_4$ acting as an acid. $Zn(NH_2)_2$ forms $Zn(NH_3)_4^{2+}$ and so dissolves in this condition.

6. Explain why alkali metals do not survive in aqueous solution but in liquid $NH_3$.

7. Explain why liquid $SO_2$ is a good solvent for carrying out organic synthesis.

   **Hint :** Because of high solubility of many organic compounds in liquid $SO_2$ e.g. sulphonation reaction of benzene.

8. Complete the following reactions in liquid ammonia medium :

   (i) $NH_4I + PbNH \longrightarrow$ (ii) $CO(NH_2)_2 + NH_3 \longrightarrow$ **[C.U. 2005]**

9. Complete the reactions occuring in liquid ammonia medium :

   (i) $PH_3 + NaNH_2 \longrightarrow$ (ii) $BiI_3 + KNH_2 \longrightarrow$ (iii) $ZnI_2 + KNH_2$ (excess) $\longrightarrow$ **[C.U. 2003]**

   **Hint :** (iii) $ZnI_2 + 2KNH_2 \longrightarrow Zn(NH_2)_2\downarrow + 2KI$

   $$Zn(NH_2)_2 + 2KNH_2 \longrightarrow K_2[Zn(NH_2)_4$$

10. (a) Solution of Na in liquid $NH_3$ is paramagnetic—the magnetic susceptibility decreases with increase in metal concentration–explain.

    (b) State with chemical equations, what happens when liquid $NH_3$ is treated with (i) powdered rhombic sulphur (ii) silicon tetrachloride. **[C.U. 1997]**

    **Hint :** (a) Metallic Na dissolves in liquid $NH_3$ to form solvated electron or ammoniaed electron

    $$e + xNH_3 \longrightarrow \underset{\substack{\textit{ammoniated}\\ \textit{electron}}}{[e(NH_3)_x]^-}$$

    Actually the ammoniated electron is trapped in the cavities of the solvent molecules and the solution is paramagnetic. On increasing the concentration of the dissolved metal, the electrons in the cavity is paired i.e. $2e \rightleftharpoons e_2$ or $[e(NH_3)_x]_2$ and the solution turns diamagnetic. The formation of ion pairs like $[Na(NH_3)_y]^+[e(NH_3)_x]^-$ etc. is also one of the cause to the reduced paramagnetism of the solution with increase in metal concentration.

    (b) (i) $10S + \underset{\textit{liquid}}{4NH_3} \longrightarrow \underset{\substack{\textit{sulphur}\\ \textit{nitride}}}{S_4N_4} + 6H_2S$ (ii) $3SiCl_4 + \underset{\textit{liquid}}{4NH_3} \longrightarrow \underset{\substack{\textit{silicon}\\ \textit{nitride}}}{Si_3N_4} + 12HCl$

11. Give examples of one acid and one base in liquid $NH_3$ solvent. **[SSC 2009]**

**Hint :** See sec. 5.3.3.

12. In the light of solvent system theory, justify the nature of BiN in liquid $NH_3$. **[C.U. 2009, 2012]**

**Hint :** Sec. 5.3.3.(a). BiN is a base in liquid $NH_3$.

13. What will be the effect of adding (i) $KNH_2$ and (ii) $LiN_3$ to liquid ammonia? **[SSC 2010]**

**Hint :** Sec. 5.3.3.

14. The reaction between $SbF_5$ and two equivalents of HF leads to the formation of—
1. $H_2SbF_3 + 2F_2$, 2. $HSbF_2 + 3F_2$, 3. $SbF_3 + H_2 + 2F_2$, 4. $[SbF_6]^-[H_2F]^+$ **[NET 2014]**

**Hint :** Sec. 5.5.1.

**Ans. :** 4 i.e. $[SbF_6]^-[H_2F]^+$

15. '$NH_4Cl$ is an acid in ammonia and $NaNH_2$ is a base in the same solvent.'—Justify or critise. **[B.U. 2015]**

**Hint :** Sec. 5.3.3. $NH_4Cl$ on dissociation produces solvent cation $NH_4^+$ so that $NH_4Cl$ is an acid in liquid $NH_3$. $NaNH_2$ will produce solvent anion $NH_2^-$ in liquid $NH_3$ so that $NaNH_2$ is a base in liquid $NH_3$

$$\underset{acid}{NH_4Cl} + \underset{base}{NaNH_2} \longrightarrow \underset{salt}{NaCl} + \underset{solvent}{2NH_3}$$

16. Can you develop a pH-like scale in liquid $NH_3$?
Given : $K_{H_2O} = 10^{-14}$ at 25°C. $K_{NH_3} = 10^{-33}$ at –50°C. **[C.U. 2013]**

**Hint :** Sec. 5.3.3(c). Problem 5.2

In liquid $NH_3$ the pH may vary from 0 to 33 because like $K_w$ the $K_{NH_3} = 10^{-33}$. Here pH = 0 indicates $[NH_4^+] = 1$ mol dm$^{-3}$, pH = 16.5 indicates $[NH_4^+] = [NH_2^-]$ and pH = 33 indicats $[NH_2^-] = 1$ mol dm$^{-3}$.

17. Careful addition of $NH_4I$ to the solution of $Zn^{2+}$ in excess $KNH_2$ in liquid $NH_3$ produces a white precipitate which dissolves if excess $NH_4I$ is added. Explain the observation. **[C.U. 2015]**

**Hint :** Sec. 5.3.3

$$Zn + 2KNH_2 + 2NH_3 \longrightarrow K_2[Zn(NH_2)_4] + H_2$$

$$[Zn(NH_2)_4]^{2-} + 2NH_4^+ \longrightarrow Zn(NH_2)_2\downarrow + 4NH_3$$

$$Zn(NH_2)_2 + 2NH_4I \longrightarrow [Zn(NH_3)_4]I_2$$

18. Suggest likely products for the following reactions in liquid $NH_3$.
(a) $ZnI_2 + 2KNH_2 \longrightarrow$, (b) $Mg_2Ge + 4NH_4Br \longrightarrow$, (c) $MeCO_2H + NH_3 \longrightarrow$,
(d) $O_2 \xrightarrow{\text{Na in liquid } NH_3}$, (e) $HC \equiv CH + KNH_2 \longrightarrow$

**Hint :** (a) $ZnI_2 + 2KNH_2 \longrightarrow 2KI + Zn(NH_2)_2$

(b) $Mg_2Ge + 2NH_4Br \longrightarrow GeH_4 + 2MgBr_2 + 4NH_3$

(c) $MeCO_2H + NH_3 \longrightarrow MeCO_2NH_4$ or $NH_4^+ + MeCO_2^-$

(d) $O_2 \xrightarrow{\text{Na in liquid } NH_3} Na_2O_2 + NaO_2$

(3) $HC \equiv CH + KNH_2 \longrightarrow K[HC \equiv C] + NH_3$

19. Addition K to $H_2O_2$ results in a vigorous reaction, addition of K to liquid $NH_3$ gives a bright blue solution which over a period of time liberates $H_2$.—Explain the observation.

**Hint :** In water : $2K + 2H_2O \longrightarrow 2KOH + H_2$

In liquid $NH_3$ : $K + NH_3 \longrightarrow \underset{\textit{ammoniated cation}}{[K(NH_3)_x]^+} + \underset{\textit{ammoniated electron}}{[e(NH_3)_y]^-}$

On standing : $2NH_3 + 2e \longrightarrow 2NH_2^- + H_2$

20. Which of the following compounds behave as acids in liquid HF : $ClF_3$, $BF_3$, $SbF_5$, $SiF_4$? Write equations to explain this behaviour.

**Hint :** Sec. 5.5.1. $BF_3$ and $SbF_5$ will behave as acids.

$$2HF + SbF_5 \longrightarrow H_2F^+ + SbF_6^-$$

$$2HF + BF_3 \longrightarrow H_2F^+ + BF_4^-$$

21. The salt $[S_8][AsF_6]_2$ may be isolated by the following reaction :

$$S_8 + 3AsF_5 \xrightarrow{\text{liquid HF}} [S_8][AsF_6]_2 + AsF_3$$

What is the role of $AsF_5$ in this reaction?

**Hint :** Here $AsF_5$ is acting as oxidising agent and $F^-$ acceptor.

22. By considering similarity of reaction between Na and $H_2O$, suggest how Na reacts with liquid $N_2O_4$.

**Hint :** Sec. 5.7. $Na + N_2O_4 \longrightarrow NO(g) + NaNO_3$

23. Gallium is dissolved in a solution of KOH in liquid $NH_3$ whereby a salt(I) is formed which is the amido salt of Ga(III). Heating one equivalent of (I) at 570K in vacuum produces an imido complex of Ga(III)(II) with evolution of two equivalents of $NH_3$. Partial neutralisation of (I) with $NH_4Cl$ produces $Ga(NH_2)_3$. Suggest the identities for the salts (I) and (II) and write equations for the thermal decomposition and partial neutralisation reaction of(I).

**Hint :** (I) is $[Ga(NH_2)_4]^-$ and (II) is $[Ga(NH)_2]^-$

$$Ga + 4NH_3 \xrightarrow{\text{KOH}} \underset{\text{(I)}}{[Ga(NH_2)_4]^-} + 2H_2$$

$$[Ga(NH_2)_4]^- \xrightarrow[\text{570K}]{\Delta} \underset{\text{(II)}}{[Ga(NH)_2]^-} + 2NH_3$$

$$[Ga(NH_2)_4]^- + NH_4Cl \longrightarrow Ga(NH_2)_3 + 2NH_3 + Cl^-$$

Amido complex(I) contains $NH_2^-$ and imido complex (II) contains $NH^{2-}$

24. $SbCl_3$ may acts as a non-aqueous solvent. Suggest the plausable autoionisation process.

**Hint :** $2SbCl_3 \longrightarrow [SbCl_2]^+ + [SbCl_4]^-$

25. Why the reaction between NOCl and $AgNO_3$ is considered as neutralisation reaction in liquid $N_2O_4$?

**Hint :** Sec. 5.7.

$$\underset{\textit{acid}}{NOCl} + \underset{\textit{base}}{AgNO_3} \longrightarrow \underset{\textit{salt}}{AgCl} + \underset{\textit{solvent}}{N_2O_4}$$

26. Explain the following observation—

(a) $AlF_3$ has only a low solubility in liquid HF but a combination of NaF and $AlF_3$ leads to dissolution of the two.

(b) When $BF_3$ is added to the solution, a precipitate forms.

**Hint :** $AlF_3 + NaF \longrightarrow \underset{\textit{which is soluble in liquid HF}}{Na[AlF_4]}$

$$Na[AlF_4] + BF_3 \longrightarrow AlF_3\downarrow + NaBF_4$$

27. "A large number of compounds get dissolved in a solvent due to complex formation."—Give one example to establish the above statement. **[C.U. 2013]**

**Hint :** $Zn(NO_3)_2$ dissolved in liquid $NH_3$ due to the formation of the complex.

$$Zn(NO_3)_2 + 2KNH_2 \xrightarrow{\text{liquid } NH_3} Zn(NH_2)_2\downarrow + 2KNO_3$$

$$\downarrow \text{excess } KNH_2 \text{ in liquid } NH_3$$

$$K_2[Zn(NH_2)_4]$$

# 6

# OXIDATION AND REDUCTION

There are various types of chemical reactions viz. substitution reaction, elimination reaction, direct synthesis, mutual exchange or double decomposition reaction, hydrolysis reaction, oxidation-reduction etc. Here we shall discuss the red-ox reactions.

## 6.1 OXIDATION AND REDUCTION PROCESSES ARE DEFINED AS

| Oxidation | Reduction |
|---|---|
| (a) Addition of oxygen e.g.<br>$2Cu + O_2 \longrightarrow 2CuO$ | (a) Addition of hydrogen e.g.<br>$N_2 + 3H_2 \longrightarrow 2NH_3$ |
| (b) Removal of hydrogen e.g.<br>$2NH_3 \longrightarrow N_2 + 3H_2$ | (b) Removal of oxygen e.g.<br>$CuO + H_2 \longrightarrow Cu + H_2O$ |
| (c) Addition of any electronegative element e.g.<br>$SnCl_2 + Cl_2 \longrightarrow SnCl_4$ | (c) Addition of any electropositive element e.g.<br>$HgCl_2 + Hg \longrightarrow Hg_2Cl_2$ |
| (d) Removal of any eletropositive element e.g.<br>$Hg_2Cl_2 \longrightarrow HgCl_2 + Hg$ | (d) Removal of any eletronegative element e.g.<br>$SnCl_4 \longrightarrow SnCl_2 + Cl_2$ |
| (e) Loss of electron(s) e.g.<br>$Fe - 2e \longrightarrow Fe^{2+}$<br>$Fe^{2+} - e \longrightarrow Fe^{3+}$ | (e) Gain of electron(s) e.g.<br>$Sn^{4+} + 2e \longrightarrow Sn^{2+}$<br>$Fe^{3+} + e \longrightarrow Fe^{2+}$ |

**Oxidation and reduction takes place simultaneously in a chemical reaction,**

$$2FeCl_3 + SnCl_2 \longrightarrow 2FeCl_2 + SnCl_4$$

Here $SnCl_2$ is oxidised to form $SnCl_4$ while $FeCl_3$ is reduced to $FeCl_2$.

An **oxidising agent** or **oxidant** is one that gains electron(s) and is thereby reduced to the lower oxidation state. On the other hand a **reducing agent** or **reductant** is one that looses electron(s) and is oxidised to the higher oxidation state. In the above reaction $FeCl_3$ is **oxidant** while $SnCl_2$ is **reductant.**

## 6.2 OXIDATION NUMBER CONCEPT OF OXIDATION AND REDUCTION

In the modern concept oxidation and reduction is the loss and gain of electron/electrons respectively. Such definition may be applied to ionic compounds where complete transfer of electron/electrons occur from one atomic orbital to other. But there are several compounds exist where such type of transfer of electron not occurs but still their formation may be considered under the perview of oxidation and reduction. For example in HCl which is primarily a covalent compound, we can say that hydrogen is oxidised and chlorine is reduced though there is no transfer of electron from hydrogen to chlorine orbital.

In order to cover such reactions e.g. the formation of $H_2O$, HCl etc. oxidation number (O.N.) concept was developed. **Oxidation number of an element in a compound is the formal charge (positive or negative) that should be assigned to that element if all the bonds are ionic.** The commission on the Nomenclature of Inorganic Chemistry (CNIC) 2005 recommended 'the oxidation number of an element is indicated by a Roman numeral placed in parenthesis immediately following the name of the element to which it refers. The oxidation number may be positive, negative or zero (represented by the number 0). The positive sign is never used. An oxidation number is always positive unless the minus sign is, explicitly used. Non-integral oxidation numbers are not used for nomenclature purposes'. In order to assign the positive and negative oxidation number, electronegativity concept has been utilised. The **more electronegative element assigned the negative O.N. and the less electronegative partner assigned the positive O.N.** For example in HF, the hydrogen is assigned +1 (I) and fluorine is assigned –1 (–I) oxidation number. The following rules are followed in the assignment of O.N. of an element :

(a) The sum of the O.N. of all the elements in a compound is equal to zero.

(b) Free elements e.g. Fe, Co, Ar etc. or for diatomic molecules e.g. $H_2$, $Cl_2$ etc. are assigned zero O.N.

(c) Hydrogen has normally +1 O.N. (e.g. in HCl, hydrogen has + 1 O.N.) but when hydrogen is joined to strong electropositive alkali and alkaline earth metals, it has assigned –1 O.N. e.g. in NaH, $BeH_2$ etc. hydrogen is assigned –1 O.N.

(d) Oxygen has normally –2 O.N. (e.g. in $H_2O$) but in peroxides (e.g. $Na_2O_2$) or superoxides (e.g. $CsO_2$) oxygen has –1 and –1/2 O.N. respectively. When oxygen is joined to more eletronegative element e.g. $F_2O$, in which oxygen has assigned +2 O.N.

The assignment of O.N. in accordance with the CNIC recommendation is to be followed for the above rules e.g. in $F_2O$ the O.N. of oxygen is II while in $H_2O_2$ in O.N. of hydrogen and oxygen are I and –I respectively.

O.N. is a fictitious number assigned to that element because O.N. depends on the bonding partner of the element e.g. O.N. of oxygen in $H_2O$ is –2 while in $F_2O$ the O.N. of oxygen is +2, in $H_2O_2$ the O.N. of oxygen is –1 etc.

In the process of oxidation the positive oxidation number is increased while in reduction the positive oxidation number is reduced. For example in the following reaction the O.N. is assigned as :

$$\overset{+3}{Fe}Cl_3 + \overset{+2}{Sn}Cl_2 \longrightarrow \overset{+2}{Fe}Cl_2 + \overset{+4}{Sn}Cl_4$$

In balancing the above redox reaction by O.N. method we have to balance the increase and decrease of O.N. as :

*decrease of O.N. = –1*

$$\overset{+3}{Fe}Cl_3 + \overset{+2}{Sn}Cl_2 \longrightarrow \overset{+2}{Fe}Cl_2 + \overset{+4}{Sn}Cl_4$$

*increase of O.N. = +2*

Now balancing the increase and decrease of O.N., the balanced reaction will be,

$$2FeCl_3 + SnCl_2 \longrightarrow 2FeCl_2 + SnCl_4$$

In some cases after balancing the increase and decrease of O.N., the number of atoms on both sides of the equation may not be balanced. In those cases simple multiplication by suitable factors (trial and error method), the number of atoms on both sides of the equation is balanced.

Oxidation number and valency may or may not be similar. Valency of an element represents the combining capacity with other elements. It is numerically equal to the number of hydrogen atoms or chlorine atoms or twice the number of oxygen atoms that combine with one atom of that element. For example in $H_2O$ and $H_2O_2$ the valency of oxygen is two but the O.N. of oxygen is –2 in $H_2O$ while in $H_2O_2$ the value is –1. Therefore in order to distinguish O.N. and valency we have to note the following :

| Valency | Oxidation number |
|---|---|
| 1. It represents the combining power of an element. It has only a numerical value, no positive or negative sign is applicable. | 1. It denotes the formal charge of the element in a compound. It always associated with positive or negative sign. O.N. may also be zero. |
| 2. Valency is always a whole number | 2. It may be whole number or fractional. |
| 3. It is generally a fixed number. | 3. It is a fictitious number assigned to that element in a compound. |

### 6.2.1 Evaluation of oxidation number

(a) **Calculation of O.N. of Mn in $KMnO_4$ :**

Let the O.N. of Mn = $x$

$\therefore$ $+1 + x + 4(-2) = 0$ i.e. $x = +7$

so that O.N. of Mn in $KMnO_4$ is +7.

(b) **O.N. of S in $H_2SO_5$ :**

Here O.N. will be calculated from the structure of the compound i.e.

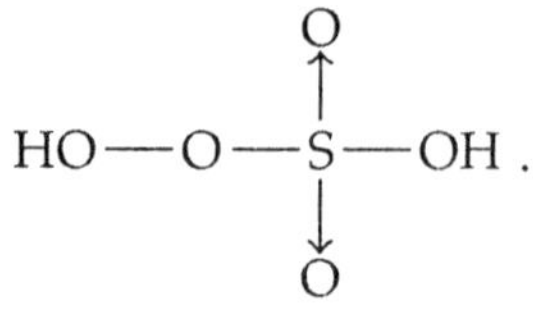

. One peroxo linkage exist in this compound.

Let the O.N. of S = $x$, so that $2\times(+1) + 2\times(-1) + 3\times(-2) + x = 0$

*(for H)* *(for 0-0)* *(for 0)*

i.e. $x = +6$, so O.N. of S in $H_2SO_5$ is +6.

(c) **O.N. of Fe in $Fe(CO)_5$ :**

Let the O.N. of Fe = $x$

$\therefore$ $x + 5\times0 = 0$ i.e. $x = 0$, so O.N. of Fe in $Fe(CO)_5 = 0$.

(d) **Evaluation of O.N. of Fe in $Fe_3O_4$ :** $Fe_3O_4$ is a mixed oxide i.e. $FeO.Fe_2O_3$. In these two oxides the O.N. of Fe is +2 and +3 respectively. But in $Fe_3O_4$ the O.N. of Fe is calculated as : $3x + 4(-2) = 0$ i.e. $x = +\frac{8}{3}$ where $x$ is the O.N. of Fe.

(e) **O.N. of carbon in H–C≡N :** In evaluating the O.N. of this type of compounds we have to follow the following rules :

(a) Each covalent bond contributes one unit value of O.N.

(b) Covalently bonded atom with less electronegativity will assume positive O.N. while the more negative one will have negative O.N.

(c) In case of dative bond +2 O.N. is assigned for donor atom while –2 value assigned to acceptor atom. Generally dative bond directed from less electronegative atom to more electronegative atom.

(d) If dative bond arises from more electronegative atom to less electronegative atom, then contribution is to be neglected.

Let the O.N. of carbon in H–C≡N is = $x$.

$\therefore$ $1 + x + 3 \times (-1) = 0$ i.e. $x = +2$

(f) **O.N. of C in H–N≡C :** Let the O.N. of C = $x$. Nitrogen has three covalent bonds and more electronegative than carbon. Therefore O.N. of N = –3, so that $1 + (-3) + x = 0$ i.e. $x = +2$.

(g) **O.N. of two chlorine atoms in Ca(OCl)Cl :** In $OCl^-$ the O.N. of chlorine is +1 and the end chlorine has the O.N. = –1.

(h) **Calculation of oxidation number of carbon in $C_6H_{12}O_6$ :** Let the O.N. of carbon in $C_6H_{12}O_6$ is $x$, so that $6x + 12 + 6(-2) = 0$ i.e. $x = 0$.

Similarly the O.N. of C in HCHO, $CH_2Cl_2$ etc. is also zero.

(i) **Evaluation of O.N. of S in $Na_2S_4O_6$ :**

Let the O.N. of S = $x$.

$\therefore$ $2 + 4x + 6(-2) = 0$ i.e. $x = +\frac{5}{2}$.

Similarly O.N. of S in $Na_2S_2O_3$ is +2. But from the structure of $Na_2S_2O_3$, the central S has O.N. +4 and the O.N. of terminal S is zero.

(j) **Calculation of O.N. of Cr in $CrO_5$ :** This O.N. is evaluated from the structure of the compound i.e. 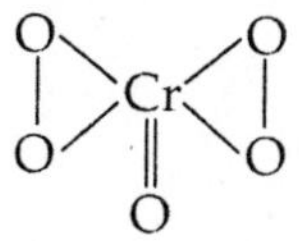. Here two peroxo linkage exist. Taking $x$ as the O.N. of Cr we can write,

$4 \times (-1) + x + 1 \times (-2) = 0$ i.e. $x = +6$

(k) **O.N. of nitrogen in NOCl :** Here NOCl is $NO^+Cl^-$ so that O.N. of nitrogen is +3.

(l) **Calculation of O.N. of I in $H_2IO_6^{3-}$ :**

Let the O.N. of I = $x$, so that $2 + x - 12 = -3$ i.e. $x = +7$.

Therefore the O.N. of I in $H_2IO_6^{3-}$ is +7

(m) **Evaluation of O.N. of S in $S_3O_6^{2-}$ :** Let the O.N. of S is $x$ so that $3x - 12 = -2$ i.e. $x = +3\frac{1}{3}$.

## 6.2.2 Evaluation of O.N. as per CNIC recommendations with reference to Lewis electron dot structures and Pauling's electronegativity

In Sec. 3.2.1 we have stated the rules for drawing the Lewis electron dot structure. In evaluating the O.N. by this procedure the following rules are to be followed :

(a) Draw the Lewis dot structure of the molecule or ion. For resonating structures only one structure will be considered.

(b) Move the electrons of each bond onto the more electronegative atom of that bond. If the two atoms are of extremely different electronegativities, the electron pair may reside completely on the more electronegative atom giving a positive and negative ion. If the two atoms are identical or have similar electronegativity, the electrons will be equally shared like a covalent bond. In most cases partial positive charge assigned to less electronegative atom and partial negative charge assigned to more electronegative atom.

(c) Depict the non-bonding electrons which exist on the atom concerned.

(d) Compare the remaining number of valence electrons on each atom with the number of valence electrons found on a free atom of that element which is indicated by its group number. The difference indicates the O.N. of the element. Generally valence electrons are the sum of the ns and np electrons.

(e) Finally we have to check the total number of electrons shown in the Lewis structure agrees with the number of valence electrons calculated for each atom and also to consider generally that each atom has an octet of electrons except hydrogen.

(I) **Evaluation of O.N. of S and O in $S_2O_3^{2-}$ :** There will be total of 32 electrons for $S_2O_3^{2-}$ and have the following Lewis structure :

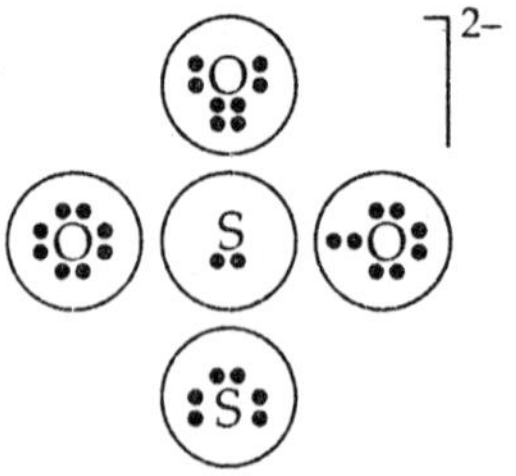

O.N. of central S = 6 – 2 = +4 or according to CNIC = IV

O.N. of terminal S = 6 – 6 = 0

O.N. of oxygens = 6 – 8 = –2 or according to CNIC = –II.

If we take the O.N. of S is $x$ then for $S_2O_3^{2-}$ we can write, $2x + 3\times(-2) = -2$ i.e. $x = +2$

Here the average O.N. of S = $\frac{4+0}{2}$ = +2.

(II) **Calculation of O.N. of S and O in $S_3O_6^{2-}$ :** Total number of electrons for this system is 18+36+2 = 56 and the Lewis structure can be represented as :

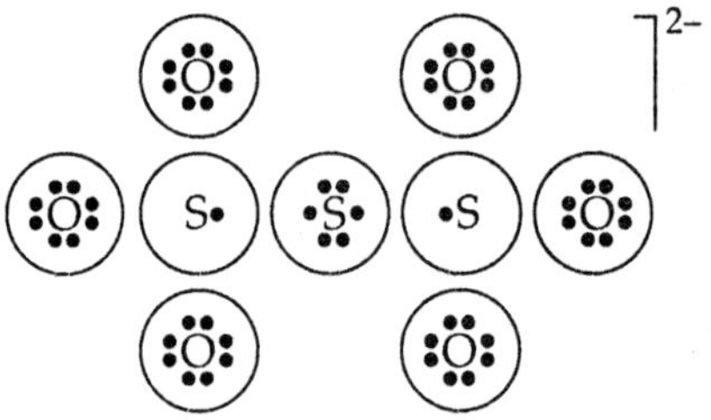

i.e.

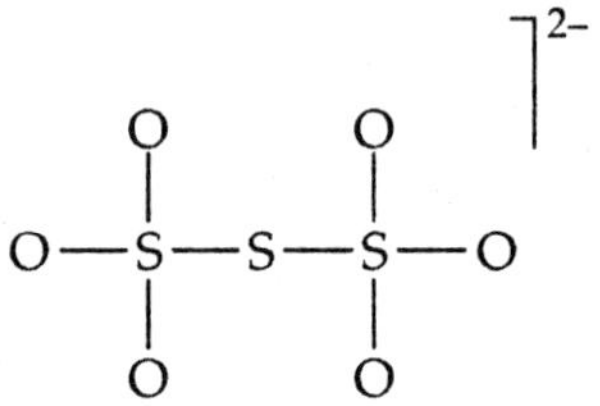

∴ O.N. of central S = 6 – 6 = 0

O.N. of terminal each S = 6 – 1 = +5 and according to CNIC is V.

O.N. of oxygen = 6 – 8 = –2 i.e. –II

Earlier the average O.N. of S has been calculated as $\frac{10}{3}$ [Sec. 6.2.1(m)].

(III) **Evaluation of O.N. of S, O and H in $H_2SO_4$ :** Total number of electrons = 2+6+24 = 32 where the structure can be represented in the following way with the consideration that oxygen is more electronegative than sulphur.

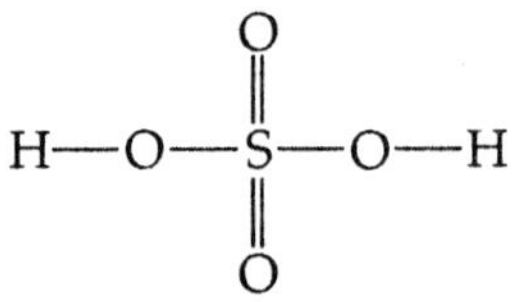

or

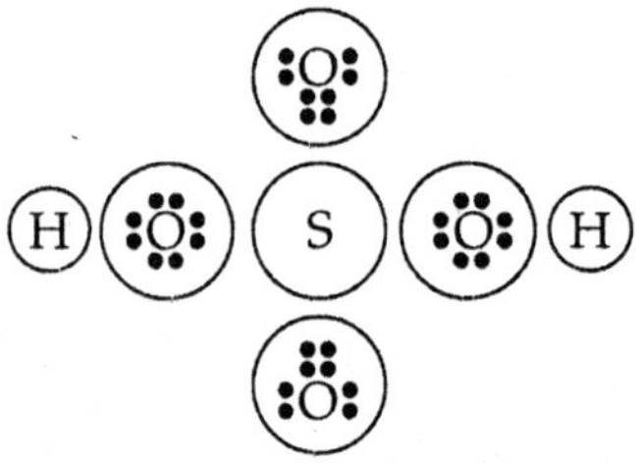

The O.N. of central S = 6 – 0 = +6 or VI

The O.N. of non-hydroxyl oxygen = 6 – 8 = –2 or –II

The O.N. of hydroxyl oxygen = 6 – 8 = –2 or –II

The O.N. of hydrogen = 1 – 0 = +1 or I.

---

**Problem 6.1** Indicate the oxidation number of all the P atoms in the following compounds :
(i) $(HO)_2(O=)$ P–P(=O)$(OH)_2$, (ii) (H)(HO)(O=)P–O–P(=O)(OH)(H). **[C.U. 2000,1995]**

**Ans. :** (i) The compound is represented as :

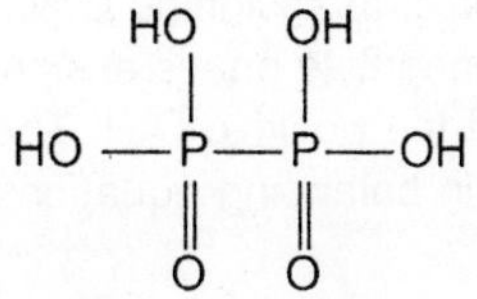

or

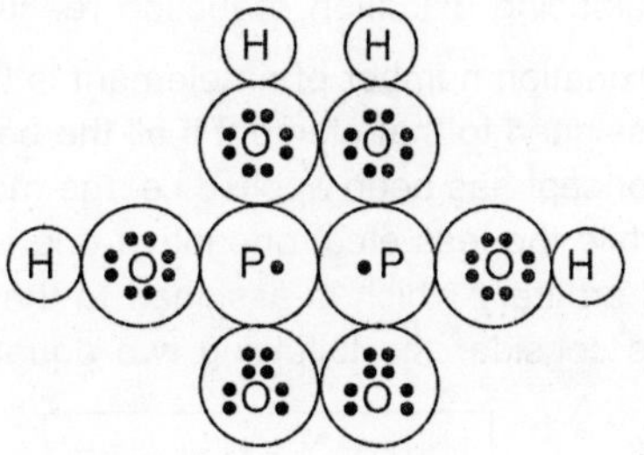

The O.N. of each P = 5–1 = +4 or IV

By simple rules the O.N. of P, $2x + 4\times(+1) + 6\times(-2) = 0$ or $x = +4$ or IV

(ii) The compound may be represented as :

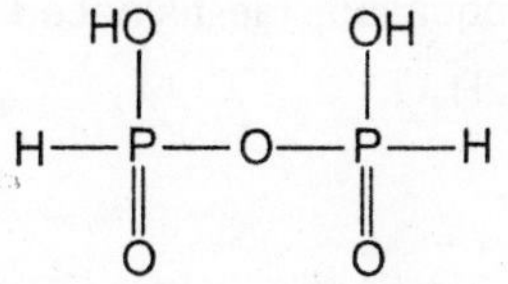

or

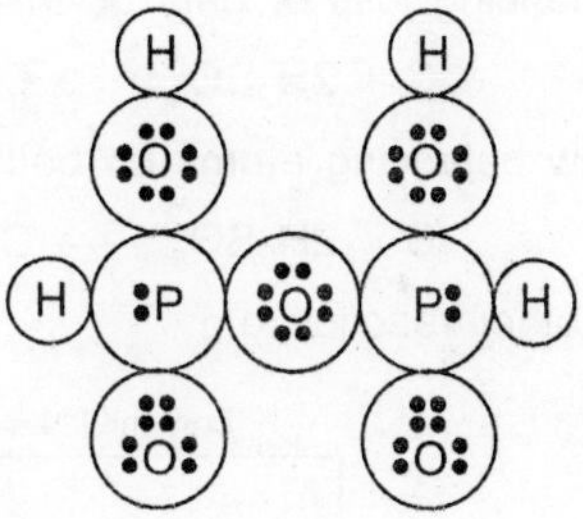

The oxidation number of each P = 5 – 2 = +3 or III

By simple rules the O.N. of P is $2x + 4(+1) + 5(-2) = 0$ or $x = +3$.

---

(IV) **Calculation of O.N. of hydrogen and oxygen of problem 6.1 :**

(i) O.N. of hydrogen = 1 – 0 = +1 or I

O.N. of double bonded oxygen = 6 – 8 = –2 or –II

O.N. of oxygen in OH = 6 – 8 = –2 or –II

(ii) O.N. of hydrogen = 1 – 0 = +1 or I

O.N. of double bonded oxygen = 6 – 8 = –2 or –II

O.N. of central oxygen = 6 – 8 = –2 or –II

O.N. of oxygen in OH = 6 – 8 = –2 or –II

### 6.2.3 Periodic variation in oxidation number

Generally the highest oxidation number of an element is equal to its group number in the periodic table. In fact no O.N. beyond +8 is known.

The s-block alkali and alkaline earth metals has the highest O.N. +1 and +2 respectively. For p-block elements it is likely that during cation formation only np electrons are lost remaining the $ns^2$ valence electron pair. In that case the O.N. of p-block cations will be two less than the group number. The group O.N. is common for less than one-half of the d-block elements which are being placed at the left of the d-block. The group O.N. for f-block elements being found at the far left of this block elements. For most of the f-block elements the common O.N. is +3.

---

**Problem 6.2** "Oxidation number is a fictitious number assigned to elements for the purpose of balancing oxidation reduction reactions"—comment. **[B.U. 1984]**

**Ans. :** Oxidation number of an element is the formal charge (positve or negative) that should be assigned to that element if all the bonds are ionic. In assigning O.N., the electronegativity concept has been applied i.e. the more electronegative one is assigned the negative O.N. while the less electronegative one is assigned the positive O.N. The O.N. of an element is arbitrary which is assigned to that element in balancing equation by O.N. method. Let us consider the following two equations.

Loss in O.N. = +4 – (+6) = –2

$$\overset{0}{C} + H_2\overset{+6}{S}O_4 \longrightarrow \overset{+4}{C}O_2 + \overset{+4}{S}O_2 + H_2O \quad \text{...(A)}$$

Gain in O.N. = +4 – (0) = +4

Balancing loss or gain of O.N., the balanced reaction will be—

$$C + 2H_2SO_4 \longrightarrow CO_2 + 2SO_2 + H_2O$$

Now balancing atoms on both sides of the equation, the balanced reaction will be—

$$C + 2H_2SO_4 \longrightarrow CO_2 + 2SO_2 + 2H_2O$$

In other reaction e.g.

Loss in O.N. = –2 – (0) = –2

$$\overset{+2-1}{FeS_2} + \overset{0}{O_2} \longrightarrow \overset{+3\ -2}{Fe_2O_3} + \overset{+4\ -2}{SO_2} \quad \text{...(B)}$$

Gain in O.N. = +4 – (–1) = +5

Gain in O.N. = +3 – (+2) = +1

Here both iron and sulphur is oxidised. Balancing loss or gain of O.N., the balanced reaction will be—

$$2FeS_2 + O_2 \longrightarrow 2Fe_2O_3 + SO_2$$

$$\text{or } 2FeS_2 + 5O_2 \longrightarrow 2Fe_2O_3 + 5SO_2$$

Now balancing atoms on both sides of the equation, the balanced reaction will be—

$$4FeS_2 + 11O_2 \longrightarrow 2Fe_2O_3 + 8SO_2$$

Therefore in the above two reactions (represented by A and B) the O.N. of S in the reactant side is +6 and –1 respectively, while in the product side the O.N. of S is +4 in both the cases. Therefore O.N. of an element is a fictitious number which is assigned to that element for the purpose of balancing equation by O.N. method.

## 6.3 BALANCING OF REDOX REACTIONS BY ION-ELECTRON METHOD

The modern concept of oxidation and reduction is the loss or gain of electrons respectively. Therefore redox reactions may be balanced by the loss or gain of electrons which is known as **ion-electron method of balancing** redox reactions. In order to apply this method the following rules are to be followed :

(a) Ascertain the reactants and products with their chemical formulas.

(b) Set up partial equation for each of the reduction of the oxidant and oxidation of the reductant.

(c) If the reaction occurs in acid medium, use requisite number of $H^+$ ions, in order to balance atoms on both sides of the equation. For alkali medium use $OH^-$ ions. Balance the charges on both sides of the partial equations by adding suitable number of electrons.

(d) Multiply each of the partial equations by suitable factors so that each of the two equations has the same number of electrons.

(e) Add the two partial equations and cancel out the species which are common on both sides of the equation.

The application of these rules may be illustrated in the following examples :

(1) **The reduction of ferric chloride with stanous chloride in hydrochloric acid medium :**

The reaction is $FeCl_3 + SnCl_2 \longrightarrow FeCl_2 + SnCl_4$.

The partial equation for oxidation is $Sn^{2+} - 2e \longrightarrow Sn^{4+}$ ...(a)

and the partial equation for reduction is $Fe^{3+} + e \longrightarrow Fe^{2+}$ ...(b)

Multiplying equation (a) by 1 and equation (b) by 2 and then adding the two equations we get the balanced reaction as :

$$Sn^{2+} - 2e \longrightarrow Sn^{4+}$$

$$2Fe^{3+} + 2e \longrightarrow 2Fe^{2+}$$

$$2Fe^{3+} + Sn^{2+} \longrightarrow 2Fe^{2+} + Sn^{4+}$$

(2) **The reaction of ferrous sulphate (Mohr salt) with postassium dichromate in acid medium :**

The reaction is $FeSO_4 + K_2Cr_2O_7 + H_2SO_4 \longrightarrow Fe_2(SO_4)_3 + Cr_2(SO_4)_3 + K_2SO_4 + H_2O$

The partial equation for oxidation is $Fe^{2+} - e \longrightarrow Fe^{3+}$ ...(a)

and the partial equation for reduction is

$$Cr_2O_7^{2-} + 14H^+ + 6e \longrightarrow 2Cr^{3+} + 7H_2O \quad ...(b)$$

Multiplying equation (a) by 6 and equation (b) by 1 and adding we get the balance reaction as :

$$6Fe^{2+} - 6e \longrightarrow 6Fe^{3+}$$

$$Cr_2O_7^{2-} + 14H^+ + 6e \longrightarrow 2Cr^{3+}\ 7H_2O$$

$$6Fe^{2+} + Cr_2O_7^{2-} + 14H^+ \longrightarrow 6Fe^{3+} + 2Cr^{3+} + 7H_2O$$

**(3) The oxidation of ferrous sulphate with potassium permanganate in acid medium :** **[B.U. 1997, 1995]**

The reaction is $FeSO_4 + KMnO_4 + H_2SO_4 \longrightarrow K_2SO_4 + MnSO_4 + Fe_2(SO_4)_3 + H_2O$

Oxidation reaction is $Fe^{2+} - e \longrightarrow Fe^{3+}$ ...(a)

and reduction reaction is $MnO_4^- + 8H^+ + 5e \longrightarrow Mn^{2+} + 4H_2O$ ...(b)

By making equation (a)×5 + equation (b)×1, the balanced reaction is—

$$5Fe^{2+} - 5e \longrightarrow 5Fe^{3+}$$

$$MnO_4^- + 8H^+ + 5e \longrightarrow Mn^{2+} + 4H_2O$$

$$5Fe^{2+} + MnO_4^- + 8H^+ \longrightarrow 5Fe^{3+} + Mn^{2+} + 4H_2O$$

**(4) The reaction of oxalic acid with potassium permanganate in dilute sulphuric acid medium :**

Representative reaction is $H_2C_2O_4 + KMnO_4 + H_2SO_4 \longrightarrow K_2SO_4 + MnSO_4 + CO_2 + H_2O$

Oxidation : $C_2O_4^{2-} - 2e \longrightarrow 2CO_2$ ...(a)

Reduction : $MnO_4^- + 8H^+ + 5e \longrightarrow Mn^{2+} + 4H_2O$ ...(b)

Making (a)×5 + (b)×2 we get the balanced reaction as,

$$5C_2O_4^{2-} - 10e \longrightarrow 10CO_2$$

$$2MnO_4^- + 16H^+ + 10e \longrightarrow 2Mn^{2+} + 8H_2O$$

$$2MnO_4^- + 5C_2O_4^{2-} + 16H^+ \longrightarrow 10CO_2 + 2Mn^{2+} + 8H_2O$$

**(5) The reaction of potassium iodide with cupric sulphate in acid medium :**

Representative reaction is $KI + CuSO_4 \longrightarrow Cu_2I_2 + K_2SO_4 + I_2$

The partial equation for oxidation : $2I^- - 2e \longrightarrow I_2$ ...(a)

and the partial equation for reduction : $Cu^{2+} + e \longrightarrow Cu^+$ ...(b)

By the process of equation (a)×1 + equation (b)×2 we get the balanced equation as:

$$2I^- - 2e \longrightarrow I_2$$

$$2Cu^{2+} + 2e \longrightarrow 2Cu^+$$

$$2Cu^{2+} + 2I^- \longrightarrow I_2 + 2Cu^+$$

**(6) The reaction of sodium thiosulphate with iodine :**

The reaction is $I_2 + Na_2S_2O_3 \longrightarrow NaI + Na_2S_4O_6$

Oxidation : $2S_2O_3^{2-} - 2e \longrightarrow S_4O_6^{2-}$ ...(a)

Reduction : $I_2 + 2e \longrightarrow 2I^-$ ...(b)

(a) + (b) i.e. $2S_2O_3^{2-} + I_2 \longrightarrow S_4O_6^{2-} + 2I^-$

This is the balanced reaction.

**(7) The reaction of potassium dichromate with potassium iodide in acid medium :**

Representative reaction is $K_2Cr_2O_7 + KI + H^+ \longrightarrow K^+ + Cr^{3+} + I_2 + H_2O$

Oxidation : $2I^- - 2e \longrightarrow I_2$ ...(a)

Reduction : $Cr_2O_7^{2-} + 14H^+ + 6e \longrightarrow 2Cr^{3+} + 7H_2O$ ...(b)

(a)×3 + (b)×1 leads to the balanced equation as :

$$Cr_2O_7^{2-} + 6I^- + 14H^+ \longrightarrow 2Cr^{3+} + 3I_2 + 7H_2O$$

(8) $P + H_2O \xrightarrow{NaOH} PH_3 + H_2PO_2^-$ **[B.U. 1984]**

This is a disproportionation reaction.

The partial equation for oxidation is $P + 2OH^- - e \longrightarrow H_2PO_2^-$ ...(a)

and the partial equation for reduction is $P + 3H_2O + 3e \longrightarrow PH_3 + 3OH^-$ ...(b)

(a)×3 + (b)×1 leads to—

$$3P + 6OH^- - 3e \longrightarrow 3H_2PO_2^-$$
$$P + 3H_2O + 3e \longrightarrow PH_3 + 3OH^-$$
$$\overline{4P + 3H_2O + 3OH^- \longrightarrow PH_3 + 3H_2PO_2^-}$$

(9) **Chlorine is passed through hot and concentrated solution of sodium hydroxide :**

$3Cl_2 + 6NaOH = 5\,NaCl + NaClO_3 + 3H_2O$

This is also a disproportionation reaction.

Oxidation : $Cl_2 + 12OH^- - 10e \longrightarrow 2ClO_3^- + 6H_2O$ ...(a)

Reduction : $Cl_2 + 2e \longrightarrow 2Cl^-$ ...(b)

(a)×1 + (b)×5 leads to—

$$Cl_2 + 12OH^- - 10e \longrightarrow 2ClO_3^- + 6H_2O$$
$$5Cl_2 + 10e \longrightarrow 10Cl^-$$
$$\overline{6Cl_2 + 12OH^- \longrightarrow 2ClO_3^- + 10Cl^- + 6H_2O}$$

(10) **The reaction of chlorine with cold dilute sodium hydroxide :**

This reaction is an example of a disproportionation reaction.

$$Cl_2 + NaOH \longrightarrow NaCl + NaOCl + H_2O$$

Oxidation : $Cl_2 + 4OH^- - 2e \longrightarrow 2OCl^- + 2H_2O$ ...(a)

Reduction : $Cl_2 + 2e \longrightarrow 2Cl^-$ ...(b)

(a)×1 + (b)×1 : $2Cl_2 + 4OH^- \longrightarrow 2Cl^- + 2OCl^- + 2H_2O$

(11) **Bismuth salt reacts with alkaline solution of sodium stannite :**

$2Bi(NO_3)_3 + 3Na_2SnO_2 + 6NaOH \longrightarrow 2Bi + 3Na_2SnO_3 + 6NaNO_3 + 3H_2O$

$SnCl_2 + 4NaOH \longrightarrow Na_2SnO_2 + 2NaCl + 2H_2O$

or **Bismuth hydroxide is treated with alkaline sodium stannite solution :** **[C.U. 2003,2007]**

$2Bi(OH)_3 + 3Na_2SnO_2 = 2Bi + 3Na_2SnO_3 + 3H_2O$

This reaction is employed for qualitative detection of bismuth.

In both the cases $Bi^{3+}$ ion is reduced to metallic Bi and stannite ion is oxidised to stannate.

Reduction : $Bi^{3+} + 3e \longrightarrow Bi$ ...(a)

Oxidation : $SnO_2^{2-} + 2OH^- - 2e \longrightarrow SnO_3^{2-} + H_2O$ ...(b)

(a)×2 + (b)×3 leads to—

$$2Bi^{3+} + 6e \longrightarrow 2Bi$$

$$3SnO_2^{2-} + 6OH^- - 6e \longrightarrow 3SnO_3^{2-} + 3H_2O$$

$$2Bi^{3+} + 3SnO_2^{2-} + 6OH^- \longrightarrow 2Bi + 3SnO_3^{2-} + 3H_2O$$

(12) **Urea is treated with nitrous acid :** This reaction is employed for the removal of nitrous acid from a reaction medium.

$$CO(NH_2)_2 + 2HNO_2 \longrightarrow 2N_2 + 3H_2O + CO_2$$

Oxidation : $CO(NH_2)_2 + H_2O - 6e \longrightarrow CO_2 + N_2 + 6H^+$ ...(a)

Reduction : $2NO_2^- + 8H^+ + 6e \longrightarrow N_2 + 4H_2O$ ...(b)

(a) + (b) $CO(NH_2)_2 + 2NO_2^- + 2H^+ \longrightarrow CO_2 + 2N_2 + 3H_2O$

(13) **The reaction between zinc metal and sodium vanadate in acid medium :**

$$Zn + NaVO_3 + HCl \longrightarrow \underset{\textit{vanadyl chloride}}{VOCl_2} + ZnCl_2 + NaCl + H_2O$$

Oxidation : $Zn - 2e \longrightarrow Zn^{2+}$ ...(a)

Reduction : $VO_3^- + 4H^+ + e \longrightarrow VO^{2+} + 2H_2O$ ...(b)

(a)×1 + (b)×2 leads to the balanced reaction as :

$$Zn + 2VO_3^- + 8H^+ \longrightarrow Zn^{2+} + 2VO^{2+} + 4H_2O$$

(14) **Action of perdisulphuric acid with $Mn^{2+}$ salt :**

$$Mn^{2+} + H_2S_2O_8 \longrightarrow HMnO_4 + H_2SO_4$$

Oxidation : $Mn^{2+} + 4H_2O - 5e \longrightarrow MnO_4^- + 8H^+$ ...(a)

Reduction $S_2O_8^{2-} + 2e \longrightarrow 2SO_4^{2-}$ ...(b)

(a)×2 + (b)×5 leads to the balanced reaction as :

$$2Mn^{2+} + 5S_2O_8^{2-} + 8H_2O \longrightarrow 2MnO_4^- + 10SO_4^{2-} + 16H^+$$

Henceforth in the following reactions we shall only mention the partial equation for oxidation and reduction.

(15) **The reaction of sodium hyposulphite with potassium ferricyanide :**

$$K_3Fe(CN)_6 + Na_2S_2O_4 \longrightarrow K_4Fe(CN)_6 + SO_2 + NaCN + KCN$$

Oxidation : $S_2O_4^{2-} - 2e \longrightarrow 2SO_2$

Reduction : $[Fe(CN)_6]^{3-} + e \longrightarrow [Fe(CN)_6]^{4-}$

(16) **The reaction of sodium hyposulphite with hydrogen sulphide :**

$Na_2S_2O_4 + H_2S \longrightarrow Na_2S_2O_3 + H_2O + S$

Oxidation : $S^{2-} - 2e \longrightarrow S$

Reduction : $S_2O_4^{2-} + 2H^+ + 2e \longrightarrow S_2O_3^{2-} + H_2O$

(17) **The reaction of silver nitrate with phosphorous acid to produce a black precipitate of metallic silver :**

$H_3PO_3 + 2AgNO_3 + H_2O \longrightarrow H_3PO_4 + 2HNO_3 + 2Ag$

Oxidation : $HPO_3^{2-} + H_2O - 2e \longrightarrow PO_4^{3-} + 3H^+$

Reduction : $Ag^+ + e \longrightarrow Ag$

(18) **The reaction of phosphorous with sodium iodate in slightly acidic medium :**

$P_4 + NaIO_3 + H_2O \longrightarrow NaI + NaH_2PO_4$

Oxidation : $P_4 + 16H_2O - 20e \longrightarrow 4H_2PO_4^- + 24H^+$

Reduction : $IO_3^- + 6H^+ + 6e \longrightarrow I^- + 3H_2O$

(19) **Chlorine reacts with alkaline solution of hydrogen peroxide :**

$Cl_2 + H_2O_2 + NaOH \longrightarrow NaCl + H_2O + O_2$

Oxidation : $H_2O_2 + 2OH^- - 2e \longrightarrow 2H_2O + O_2$

Reduction : $Cl_2 + 2e \longrightarrow 2Cl^-$

(20) **Manganese dioxide is heated with concentrated hydrochloric acid :**

$MnO_2 + HCl \longrightarrow MnCl_2 + Cl_2 + H_2O$

Oxidation : $2Cl^- - 2e \longrightarrow Cl_2$

Reduction : $MnO_2 + 4H^+ + 2e \longrightarrow Mn^{2+} + 2H_2O$

(21) **Arsenious acid is boiled with chlorine water :**

$H_3AsO_3 + Cl_2 + H_2O \longrightarrow H_3AsO_4 + HCl$

Oxidation : $AsO_3^{3-} + H_2O - 2e \longrightarrow AsO_4^{3-} + 2H^+$

Reduction : $Cl_2 + 2e \longrightarrow 2Cl^-$

(22) **Oxidation of sulphur with hot concentrated nitric acid :**

$S + HNO_3 \longrightarrow H_2SO_4 + NO$

Oxidation : $S + 4H_2O - 6e \longrightarrow SO_4^{2-} + 8H^+$

Reduction : $NO_3^- + 4H^+ + 3e \longrightarrow NO + 2H_2O$

(23) **The reaction of potassium iodate with hydrazine :**

$KIO_3 + N_2H_4 \longrightarrow KI + N_2 + H_2O$

Oxidation : $N_2H_4 - 4e \longrightarrow N_2 + 4H^+$

Reduction : $IO_3^- + 6H^+ + 6e \longrightarrow I^- + 3H_2O$

(24) **Action of sodium bismuthate on manganous sulphate in nitric acid medium :** This reaction is employed for qualitative detection as also quantitative estimation of manganese.

$MnSO_4 + NaBiO_3 + HNO_3 \longrightarrow HMnO_4 + Bi(NO_3)_3 + NaNO_3 + H_2O$

Oxidation : $Mn^{2+} + 4H_2O - 5e \longrightarrow MnO_4^- + 8H^+$

Reduction : $BiO_3^- + 6H^+ + 2e \longrightarrow Bi^{3+} + 3H_2O$

(25) **The interaction of potassium permanganate with hydrogen peroxide in presence of dilute acid :**

$KMnO_4 + H_2SO_4 + H_2O_2 \longrightarrow K_2SO_4 + MnSO_4 + H_2O + O_2$

Oxidation : $H_2O_2 - 2e \longrightarrow 2H^+ + O_2$

Reduction : $MnO_4^- + 8H^+ + 5e \longrightarrow Mn^{2+} + 4H_2O$

(26) **Reaction of sodium stannite with potassium permanganate in alkaline medium :**

$KMnO_4 + Na_2SnO_2 + NaOH \longrightarrow MnO_2 + Na_2SnO_3 + H_2O$

Oxidation : $SnO_2^{2-} + 2OH^- - 2e \longrightarrow SnO_3^{2-} + H_2O$

Reduction : $MnO_4^- + 2H_2O + 3e \longrightarrow MnO_2 + 4OH^-$

(27) **The reduction of sodium nitrate with zinc dust and sodium hydroxide : [C.U. 2005]**

$NaNO_3 + Zn + NaOH \longrightarrow NH_3 + Na_2ZnO_2 + H_2O$

Oxidation : $Zn + 4OH^- - 2e \longrightarrow ZnO_2^{2-} + 2H_2O$

Reduction : $NO_3^- + 6H_2O + 8e \longrightarrow NH_3 + 9OH^-$

This reaction may also be performed with aluminium powder. **[C.U. 2000]**

(28) **Potassium nitrite is boiled with aluminium powder and sodium hydroxide :**

$KNO_2 + 2Al + NaOH + H_2O \longrightarrow NaAlO_2 + KAlO_2 + NH_3$

Oxidation : $Al + 4OH^- - 3e \longrightarrow AlO_2^- + 2H_2O$

Reduction : $NO_2^- + 5H_2O + 6e \longrightarrow NH_3 + 7OH^-$

This reaction may also be carried out with zinc dust also.

(29) **A solution containing potassium bromate and potassium bromide is acidified with hydrochloric acid :**

$KBrO_3 + 5KBr\ 6HCl \longrightarrow 6KCl + 3Br_2 + 3H_2O$

Oxidation : $2Br^- - 2e \longrightarrow Br_2$

Reduction : $2BrO_3^- + 12H^+ + 10e \longrightarrow Br_2 + 6H_2O$

(30) **Zinc is treated with sodium hydroxide solution :**

$Zn + NaOH \longrightarrow Na_2ZnO_2 + H_2$

Oxidation : $Zn + 4OH^- - 2e \longrightarrow ZnO_2^{2-} + 2H_2O$

Reduction : $OH^- + 2H_2O + 2e \longrightarrow H_2 + 3OH^-$

(31) $\mathbf{MnO_4^- + NO_2^- \xrightarrow{\text{acid}}}$ **[B.U. 1984]**

$KMnO_4 + NaNO_2 + H_2SO_4 \longrightarrow K_2SO_4 + MnSO_4 + NaNO_3 + H_2O$

Oxidation : $NO_2^- + H_2O - 2e \longrightarrow NO_3^- + 2H^+$

Reduction : $MnO_4^- + 8H^+ + 5e \longrightarrow Mn^{2+} + 4H_2O$

(32) **$Cu + HNO_3$ (conc.) $\longrightarrow Cu^{2+} + NO_2$** **[B.U. 1984]**

$Cu + HNO_3 \longrightarrow Cu(NO_3)_2 + NO_2 + H_2O$

Oxidation : $Cu - 2e \longrightarrow Cu^{2+}$

Reduction : $2NO_3^- + 4H^+ + 2e \longrightarrow 2NO_2 + 2H_2O$

(33) **Reaction of barium peroxide with ozone :**

$BaO_2 + O_3 \longrightarrow BaO + 2O_2$

Oxidation : $BaO_2 - 2e \longrightarrow Ba^{2+} + O_2$

Reduction : $O_3 + 2e \longrightarrow O^{2-} + O_2$

(34) **Oxidation of sodium chromite with sodium hypochlorite in basic solution :**

$2NaCrO_2 + 3NaOCl + 2NaOH \longrightarrow 3NaCl + 2Na_2CrO_4 + H_2O$

Oxidation : $CrO_2^- + 4OH^- - 3e \longrightarrow CrO_4^{2-} + 2H_2O$

Reduction : $OCl^- + H_2O + 2e \longrightarrow Cl^- + 2OH^-$

(35) **Cupric sulphide dissolved in hot dilute nitric acid :**

$3CuS + 8HNO_3 \longrightarrow 3Cu(NO_3)_2 + 2NO + 3S + 4H_2O$

Oxidation : $CuS - 2e \longrightarrow Cu^{2+} + S$

Reduction : $NO_3^- + 4H^+ + 3e \longrightarrow NO + 2H_2O$

(36) **$Cr_2O_7^{2-} + NO_2^- \longrightarrow Cr^{3+} + NO_3^-$ (in acidic medium)** **[C.U. 2013]**

Oxidation : $NO_2^- + H_2O - 2e \longrightarrow NO_3^- + 2H^+$

Reduction : $Cr_2O_7^{2-} + 14H^+ + 6e \longrightarrow 2Cr^{3+} + 7H_2O$

## 6.4 EQUIVALENT WEIGHT OF OXIDANT AND REDUCTANT

The best way to calculate the equivalent weight of the oxidant or reductant is to follow the partial equations for the oxidation and reduction of reductant and oxidant respectively. The formula weight of the oxidant or reductant (M) is divided by the number of electrons lost or gained in the partial equations will give the value of the equivalent weight of the substance. In this calculation if there is a water of crystallisation, then the weight of the water molecules will be added in order to determine the formula weight. Some common examples are :

$Cr_2O_7^{2-} + 14H^+ + 6e \longrightarrow 2Cr^{3+} + 7H_2O$

$$\text{Equivalent weight of } K_2Cr_2O_7 = \frac{M}{6}$$

$C_2O_4^{2-} - 2e \longrightarrow 2CO_2$

$$\text{Equivalent weight of } H_2C_2O_4 = \frac{M}{2}$$

Crystalline oxalic acid is $H_2C_2O_4{\cdot}2H_2O$, therefore $M = 90+36 = 126$.

$MnO_4^- + 8H^+ + 5e \longrightarrow Mn^{2+} + 4H_2O$

Equivalent weight of $KMnO_4$ in acid medium $= \frac{M}{5}$

$MnO_4^- + 2H_2O + 3e \longrightarrow MnO_2 + 4OH^-$

In alkali medium the equivalent weight of $KMnO_4 = \frac{M}{3}$

$MnO_4^- + e \longrightarrow MnO_4^{2-}$

In strong alkali medium the equivalent weight of $KMnO_4 = \frac{M}{1}$

## 6.5 COMPLEMENTARY AND NON-COMPLEMENTARY REDOX REACTIONS

Reactions in which the oxidant and reductant change oxidation state by the same number of electron are termed **complementary** e.g.

$$Cr^{2+} + Ag^+ \longrightarrow Ag + Cr^{3+}$$

$$Ce^{4+} + Fe^{2+} \longrightarrow Ce^{3+} + Fe^{3+}$$

But when the oxidant and reductant will change the oxidation state by different number of electrons are termed **non-complementary electron transfer reactions** e.g.

$$Sn^{2+} + 2Fe^{3+} \longrightarrow Sn^{4+} + 2Fe^{2+}$$

$$2Fe^{2+} + Tl^{3+} \longrightarrow 2Fe^{3+} + Tl^+$$

## 6.6 TYPES OF CELLS

Two types of cells are of operative—**electrolytic cell** and **electrochemical cell.**

**An electrolytic cell is one in which a non-spontaneous chemical reaction occurs by electrical energy from an external source. Here a chemical change occurs at the expense of electrical energy**. The process is simply called electrolysis e.g. the electrolysis of water. The electrolytic cell is the reverse of electrochemical cell.

**An electrochemical cell is one in which the system of chemicals so arranged as to produce electrical energy at the expense of chemical energy.** Here a chemical energy is converted to clectrical energy. The Daniell cell is an example of electrochemical cell which is also known as **galvanic cell** or **volatic cell.** In the galvanic cell we get electrical work through the electromotive force (EMF) established in the external circuit. There may be another type of **chemical cell** where electromotive force is operative through a **dilution process** or **concentration change**. This type of cell is called **concentration cell** which are of two types— a **cell with transference** and **a cell without transference**. A cell may also be classified as **reversible** and **irreversible** one. When a cell reaction occurs in both the directions the process is known as reversible one. **The Daniell cell is an example of reversible cell**. There are several conditions exist for the reversibility of a cell but an irreversible cell is one where the cell reaction generally occurs in an unidirectional way. Irreversibility of a cell may also arise by other several factors. However electrochemical cell may be classified as :

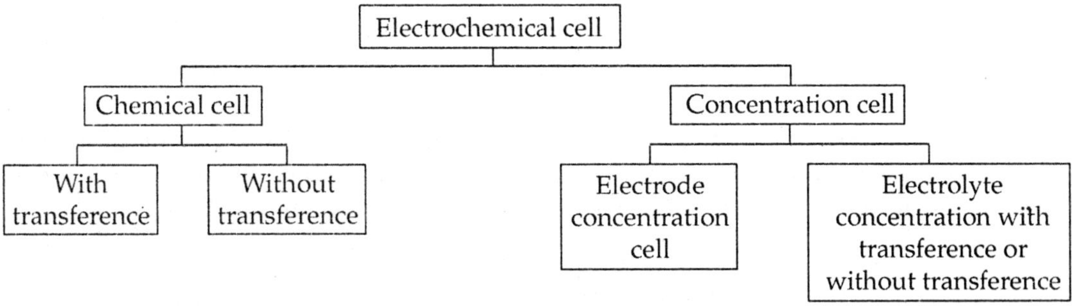


## 6.7 ELECTRODE, ELECTRODE POTENTIAL AND STANDARD ELECTRODE POTENTIAL

**When a metal (or any species) is in contact with a solution of its own ions comprises an electrode.** Each reversible cell e.g. Daniell cell has two reversible electrodes i.e. $Zn/Zn^{2+}$ and $Cu/Cu^{2+}$ electrodes.

**When a metal (or any species) is in contact with a solution of its own ions a potential difference is developed between the metal surface and the solution which is known as electrode potential (E).** In the metal (M) – metal ion ($M^{n+}$) electrode there exists a separation of charges between the metal and its salt solution. This separation of charges results a net positive or negative charge is developed on the metal with respect to its solution. The negative or positive charge developed on the metal creates an electrical potential difference between the metal and its solution which is also referred to as **electrode potential.** The electrode potential is a measure of the tendency of an electrode to lose or gain electrons i.e. a tendency to be oxidised or reduced. Electrode potential arises due to the formation of electrical double layer of oppositely charged metal plate and the solution containing its ions. It is measured in **volt** unit. **Volt is defined as the potential required to flow one coulomb of electric current per second through a uniform thread of mercury 106.3 cm long containing 14.4521 g of mercury at 0°C.**

Standard electrode potential(E°) is the electrode potential (E) when the system is in the **standard state.** For solid substances they are in the standard state. For gases 1 atmospheric pressure and for ions in solution one unit activity is referred to as standard state.

Till now there is no absolute method for the determination of standard electrode potentials of a system. Standard electrode potentials (S.E.P.) are measured by comparison with the **standard hydrogen electrode** (S.H.E.) or any standard electrode such as **standard calomel electrode.** S.E.P. of various systems are measured with reference to S.H.E. whose potential is arbitrarily fixed as **zero** at all temperatures. The selected electrode against which the measurement of S.E.P. of other system is done is called the **standard electrode.**

A S.H.E. consists of a platinum sheet or foil coated with platinum black* immersed in contact with an aqueous solution of an acid at unit activity i.e. concentration of $H^+$ ion is 1 M

* Platinum black is deposited on a bright platinum cathode—a sheet or foil—when $H_2PtCl_6$ solution is electrolysed. Bright platinum foil or sheet acting as an anode. This is requird to avoid the unimolecular layer of $H_2$ gas around the platinum electrode. In presence of platinum black surface area of platinum foil is increased which acclerates the reaction and helps to bring the equilibrium quickly.

through which hydrogen gas ($H_2$) is bubbled over the surface of the metal at 1 atmospheric pressure at a temperature of 25°C (Fig. 6.1). When two S.H.E. are electrically joined through a potentiometer and connected with a salt bridge, the potentiometer will record 'zero' potential difference.

*Determination of S.E.P.*

When one of the S.H.E. is replaced by a standard electrode of the metal whose electrode potential is to be measured is formed by immersing a metal rod in a solution of the same metal ion of 1 M concentration to form the electrochemical cell (Fig. 6.1).

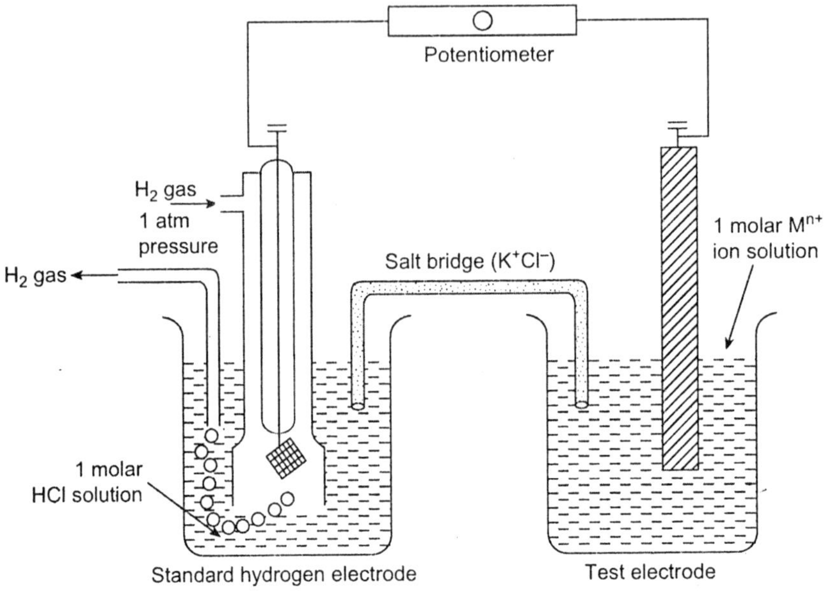


***Figure 6.1** : Determination of standard electrode potential (E°).*

The electromotive force of the cell is recorded with the help of the potentiometer. The value of the electromotive force of this electrochemical cell is the **standard electrode potential (E°)** of the metal under consideration with reference to the S.H.E. which is represented by the half cell reaction as :

$$H^+(aq) + e \longrightarrow \frac{1}{2}H_2(g) \quad \text{or} \quad \frac{1}{2}H_2(g) \longrightarrow H^+(aq) + e$$

Depending on the nature of the oxidation or reduction the sign of E° will be **positive** or **negative**. S.H.E. is difficult to handle so that it will be replaced by the more practicable electrode which is the **standard saturated calomel electrode** in which there is mercury in contact with solid $Hg_2Cl_2$ and a solution of KCl saturated with $Hg_2Cl_2$. The half cell reaction may be represented as :

$$Hg_2^{2+} + 2e \rightleftharpoons 2Hg \quad \text{or} \quad 2Hg - 2e \rightleftharpoons Hg_2^{2+}$$

With reference to the S.H.E. the potential of this saturated calomel electrode is denoted as +0.2458 volt or −0.2458 volt at 25°C.

### 6.7.1 The writing of an electrochemical cell

As mentioned earlier that in an electrochemical cell e.g. in the Daniell cell zinc rod immersed in $ZnSO_4$ solution and copper rod dipped in $CuSO_4$ solution(Fig. 6.2). The solutions in the two beakers are connected by an inverted U-tube known as salt bridge (Concentrated solutions of an electrolyte which has the same cation and anion velocity e.g. KCl, $KNO_3$, $NH_4NO_3$ etc. mixed with small amount of agar-agar or gelatin and heated. The solution is poured into the U-tube and cooled to form a semi-solid mass). Then the two metallic rods which act as electrodes are connected by a conducting wire through a voltmeter and a key. The solution in each beaker along with the metallic rod is called **half cell.** Combinatioin of two half cells forms the **complete electrochemical or volatic cell.** When the electrical circuit is closed with the help of the key, electric current will flow through the conducting wire as indicated by the deflection of the voltmeter and **electrons are flowing from zinc atoms to $Cu^{2+}$ ions.** Therefore in the Daniell cell the following reaction occurs :

$$Zn(s) + Cu^{2+}(aq) \longrightarrow Zn^{2+}(aq) + Cu(s)$$

so that at **zinc electrode oxidation** ($Zn - 2e \longrightarrow Zn^{2+}$) occurs while at **copper electrode reduction** ($Cu^{2+} + 2e \longrightarrow Cu$) occurs.

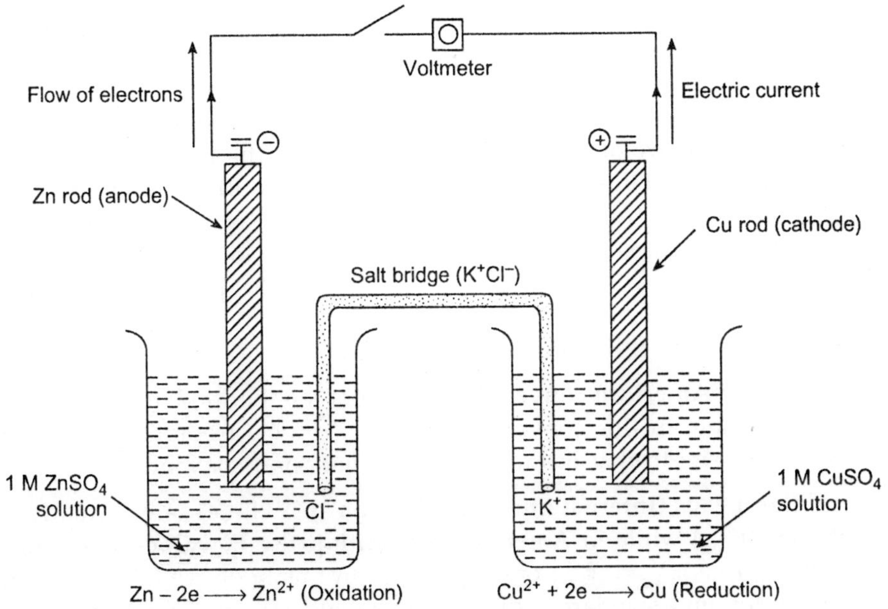


*Figure 6.2 : Representation of Daniell cell.*

According to I.U.P.A.C. system the electrodes are written from left. The electrolytes of the two electrodes if written by sign '||' which means that they are separated by the elimination of liquid junction potential, if the sign is '⋮' the liquid junction potential is not eliminated. The conducting material is written at the extreme left and right and a line '|' which indicates phase separation. Thus if platinum is the conducting material then the Daniell cell is represented as :

$$\text{Pt, Zn} \mid Zn^{2+} \parallel Cu^{2+} \mid \text{Cu, Pt} \quad \text{or} \quad \text{Pt, Zn} \mid Zn^{2+} \vdots\, Cu^{2+} \mid \text{Cu, Pt}$$

In the Daniell cell zinc electrode acting as **anode** while copper electrode works as **cathode.** Here zinc atom loses two electrons and as a result abundance of electrons is found to occur in the zinc rod and the zinc rod becomes negatively charged. On the other hand $Cu^{2+}$ ions in solution accepting two electrons from copper rod are reduced and so there will be deficit of electrons and so copper rod becomes positively charged. In general **the electrode where oxidation occurs called anode and the electrode where reduction occurs is called the cathode of an electrochemical cell. In electrochemical/galvanic cell the anode is negative and the cathode is positive.**

### 6.7.2 The Sign covention and the standard electrode potential of metals

In the Daniell cell if the concentration of the two electrolyte solutions is 1M or 1 unit activity and if the two electrodes i.e. $Zn/Zn^{2+}$ or $Cu/Cu^{2+}$ are connected with S.H.E. we get the value of standard electrode potential (E°) of the $Zn/Zn^{2+}$ or $Cu/Cu^{2+}$ systems. As mentioned earlier regarding sign convention of E°, two systems have been adopted—the **European convention (International convention)** and **American convention**. The value of E° of any particular electrode is numerically equal in both the conventions but opposite in sign. **In European convention regarding sign convention of E° states that when an electrode is coupled with a S.H.E., depending on the nature of the charge acquired by it, the potential of that electrode is either positive or negative. Thus in this convention the sign of electrode potential is the electrostatic charge on that electrode.** Thus $Zn/Zn^{2+}$ electrode has E° value –0.763 volt which means that if we make a cell of standard $Zn/Zn^{2+}$ electrodde and a S.H.E., it is the $Zn/Zn^{2+}$ electrode that will be connected to the negative terminal of the potentiometer i.e. zinc rod becomes negatively charged which means that compared to that of hydrogen, the Zn metal has a greater tendency to pass into the solution as $Zn^{2+}$ ions.

On the other hand the **American convention** considers the thermodynamic property of the respective electrode reaction. In this convention the $Zn/Zn^{2+}$ electrode has the value of E° = +0.763 volt which is opposite to that of European (International) convention.

**According to European (International) convention E° vlaues are actually standard reduction potentional ($E^\circ_{red}$) while in American convention the E° values refers to standard oxidation potential ($E^\circ_{ox}$)** i.e. $E^\circ_{ox} = -E^\circ_{red}$ or $E^\circ_{red} = -E^\circ_{ox}$. In this book we follow only European (International) convention and standard reduction potentials are simply written as E° and the electrode reaction depicted as :

$$M^{n+} + ne \longrightarrow M \quad \textit{or} \quad \text{Oxidant} + ne \rightleftharpoons \text{Reductant}$$

When metals are arranged according to standard reduction potentials (E°) we get the electrochemical series of metals (Table 6.1). According to this convention **a system having more negative reduction potential has a greater tendency to pass into the solution as a form of its own ions i.e. a system have a greater tendency to be oxidised than to be reduced while a system having more positive reduction potential has a least tendency to pass into the solution as a form of its own ions. i.e. the system has a greater tendency to be reduced than to be oxidised.**

**Table 6.1 : Standard reduction potentials of Metals at 25°C**

| Electrode | Electrode reaction | E°(volt) |
|---|---|---|
| $Li^+/Li$ | $Li^+ + e \rightleftharpoons Li$ | –3.045 |
| $K^+/K$ | $K^+ + e \rightleftharpoons K$ | –2.925 |
| $Rb^+/Rb$ | $Rb^+ + e \rightleftharpoons Rb$ | –2.925 |
| $Cs^+/Cs$ | $Cs^+ + e \rightleftharpoons Cs$ | –2.923 |
| $Ba^{2+}/Ba$ | $Ba^{2+} + 2e \rightleftharpoons Ba$ | –2.90 |
| $Sr^{2+}/Sr$ | $Sr^{2+} + 2e \rightleftharpoons Sr$ | –2.89 |
| $Ca^{2+}/Ca$ | $Ca^{2+} + 2e \rightleftharpoons Ca$ | –2.87 |
| $Na^+/Na$ | $Na^+ + e \rightleftharpoons Na$ | –2.741 |
| $Mg^{2+}/Mg$ | $Mg^{2+} + 2e \rightleftharpoons Mg$ | –2.37 |
| $Lu^{3+}/Lu$ | $Lu^{3+} + 3e \rightleftharpoons Lu$ | –2.25 |
| $Sc^{3+}/Sc$ | $Sc^{3+} + 3e \rightleftharpoons Sc$ | –2.08 |
| $Be^{2+}/Be$ | $Be^{2+} + 2e \rightleftharpoons Be$ | –1.85 |
| $Al^{3+}/Al$ | $Al^{3+} + 3e \rightleftharpoons Al$ | –1.66 |
| $Ti^{2+}/Ti$ | $Ti^{2+} + 2e \rightleftharpoons Ti$ | –1.63 |
| $Mn^{2+}/Mn$ | $Mn^{2+} + 2e \rightleftharpoons Mn$ | –1.18 |
| $Zn^{2+}/Zn$ | $Zn^{2+} + 2e \rightleftharpoons Zn$ | –0.763 |
| $Cr^{3+}/Cr$ | $Cr^{3+} + 3e \rightleftharpoons Cr$ | –0.74 |
| $Fe^{2+}/Fe$ | $Fe^{2+} + 2e \rightleftharpoons Fe$ | –0.44 |
| $Cd^{2+}/Cd$ | $Cd^{2+} + 2e \rightleftharpoons Cd$ | –0.403 |
| $Co^{2+}/Co$ | $Co^{2+} + 2e \rightleftharpoons Co$ | –0.277 |
| $Ni^{2+}/Ni$ | $Ni^{2+} + 2e \rightleftharpoons Ni$ | –0.24 |
| $Sn^{2+}/Sn$ | $Sn^{2+} + 2e \rightleftharpoons Sn$ | –0.14 |
| $Pb^{2+}/Pb$ | $Pb^{2+} + 2e \rightleftharpoons Pb$ | –0.126 |
| $H^+/\frac{1}{2}H_2$ | $H^+ + e \rightleftharpoons H$ | 0.000 |
| $Cu^{2+}/Cu$ | $Cu^{2+} + 2e \rightleftharpoons Cu$ | +0.337 |
| $Cu^+/Cu$ | $Cu^+ + e \rightleftharpoons Cu$ | +0.521 |
| $Hg^{2+}/Hg$ | $Hg^{2+} + 2e \rightleftharpoons Hg$ | +0.789 |
| $Hg_2^{2+}/Hg$ | $Hg_2^{2+} + 2e \rightleftharpoons 2Hg$ | +0.789 |
| $Ag^+/Ag$ | $Ag^+ + e \rightleftharpoons Ag$ | +0.80 |
| $Pd^{2+}/Pd$ | $Pd^{2+} + 2e \rightleftharpoons Pd$ | +0.987 |
| $Pt^{2+}/Pt$ | $Pt^{2+} + 2e \rightleftharpoons Pt$ | +1.200 |
| $Au^{3+}/Au$ | $Au^{3+} + 3e \rightleftharpoons Au$ | +1.500 |
| $Au^+/Au$ | $Au^+ + e \rightleftharpoons Au$ | +1.68 |

↓ Reduction of metal ions to metals are easier/increasing oxidising property

↑ Reduction of metal ions to metals are difficult/increasing reducing ability

It is evident from Table 6.1 that lithium is acting as **strong reductant** i.e. it has a greater tendency to be oxidised while gold having higher positive potential acting as a **strong oxidant** i.e. it will have a greater tendency to be reduced so that platinum, gold etc. are considered as noble metals.

### 6.7.3 The Nernst Equation : Relation between electrode protential (E) with that of standard electrode potential (E°)

In an electrochemical cell two half cells are connected by a conducting wire whereby oxidation and reduction occurs spontaneously at the respective electrodes and electrons flow from the positive to negative electrode e.g. in the Daniell cell electrons are flowing from zinc electrode to copper electrode and current is flowing in the reverse direction. In this type of cells the electrical energy is obtained at the expense of free energy change of the system. The reactants attain a more stable state to the products with a resulting decrease in free energy. This decrease in free energy of the cell provides the electrical energy available from an electrochemical cell.

Let us consider a spontaneous reversible redox reaction occuring in an electrochemical cell comprising of two electrodes are :

$$aA + bB \rightleftharpoons cC + dD \qquad \ldots(6.1)$$

The equilibrium constant (K) of this reaction is

$$K = \frac{a_C^c \; a_D^d}{a_A^a \; a_B^b} \qquad \ldots(6.2)$$

where 'a' represents the activity of the respective species. The degree of spontaneity of a reaction is attended by negative free energy change. **From thermodynamic consideration the electromotive force (EMF) of a cell will be positive when the cell reaction is spontaneous** and for spontaneous reaction K will be high which is related to the standard free energy change ($\Delta G°$) for the process when the activities (a) of products and reactants are all unity i.e. in the standard state :

$$\Delta G° = -RT\ln K = -2.303\ RT\log K \qquad \ldots(6.3)$$

For the above reaction (6.1) if $\Delta G$ represents the free energy change when the reactants and products are not in their standard state which is related to $\Delta G°$ by the following thermodynamic relation :

$$\Delta G = \Delta G° + RT\ln K \qquad \ldots(6.4)$$

If E is the EMF of the cell producing a quantity of electricity Q, then the electrical energy = EQ. With each equivalent of reactant involved is associated with one Faraday, F of electricity. If n electrons are to be used for balancing and writing the total cell reaction, then n Faradays are passing through the cell for one unit of reaction so that Q = nF, therefore electrical energy = nFE. This electrical energy is obtained at the expense of free energy, therefore,

$$\Delta G = -nFE \quad \text{and} \quad \Delta G° = -nFE° \qquad \ldots(6.5)$$

where E° represents the EMF of the cell when the electrodes are in the standard state.

Then we can write equation (6.4) as—

$$-nFE = -nFE^\circ + RT\ln K$$

so that $E = E^\circ - \frac{RT}{nF}\ln K$ ...(6.6)

In this equation (6.6) n represents the number of electrons involved. The electrode reaction according to European (International) convention is written as :

$$\text{Oxidant} + ne \rightleftharpoons \text{Reductant}$$

so that we can write equation (6.6) with reference to equation (6.2) as :

$$E = E^\circ - \frac{RT}{nF}\ln\frac{a_{Red}}{a_{Ox}} \quad ...(6.7)$$

where $a_{Red}$ and $a_{Ox}$ represents the activity of the reductant and oxidant respectively. The equation (6.7) is known as **Nernst equation**. Converting 'ln' to $\log_{10}$ we get,

$$E = E^\circ - \frac{2.303RT}{nF}\log\frac{a_{Red}}{a_{Ox}} \quad ...(6.8)$$

In actual practice activities are replaced by concentration in dilute solutions and inserting the value of molar gas constant R = 8.314 $JK^{-1}$ $mol^{-1}$, temperature T = 298K, Faraday F = 96500 C $mol^{-1}$ we get,

$$E = E^\circ - \frac{2.303 \times 8.314 JK^{-1}mol^{-1} \times 298K}{n \times 96500\ C\ mol^{-1}}\log\frac{[Red]}{[Ox]}$$

Again 1J = Coulomb × volt = C × V

$$\text{i.e. } E = E^\circ - \frac{2.303 \times 8.314 CVK^{-1}mol^{-1} \times 298K}{n \times 96500\ C\ mol^{-1}}\log\frac{[Red]}{[Ox]}$$

So that $E = E^\circ - \frac{0.0591^*}{n}\log\frac{[Red]}{[Ox]}$ ...(6.9)

or $E = E^\circ + \frac{0.0591}{n}\log\frac{[Ox]}{[Red]}$ ...(6.10)

Equations (6.9) and (6.10) represent the workable forms of the Nernst equation.

If we write the electrode reaction according to **American convention** i.e. Reductant – ne $\rightleftharpoons$ Oxidant, then the Nernst equation will have the form :

$$E = E^\circ - \frac{0.0591}{n}\log\frac{[Ox]}{[Red]} \quad ...(6.11)$$

The standard redox potential of non-metallic systems and for metals in different oxidation states are given in Table 6.2. Let us exemplify the two conventions :

---

* $\frac{2.303 \times 8.314 CVK^{-1}mol^{-1} \times 298K}{n \times 96500\ C\ mol^{-1}} = \frac{0.0591}{n} V$

**Table 6.2 : Standard reduction potentials at 25°C**

| Couple | E°(Volt) |
|---|---|
| $\frac{1}{2}H_2 + e \rightleftharpoons H^-$ | –2.25 |
| $AsO_4^{3-} + 3H_2O + 2e \rightleftharpoons H_2AsO_3^- + 4OH^-$ | –0.67 |
| $Fe(OH)_3 + e \rightleftharpoons Fe(OH)_2 + OH^-$ | –0.56 |
| $H_3PO_3 + 2H^+ + 2e \rightleftharpoons H_3PO_2 + H_2O$ | –0.50 |
| $Cr^{3+} + e \rightleftharpoons Cr^{2+}$ | –0.41 |
| $Ti^{3+} + e \rightleftharpoons Ti^{2+}$ | –0.37 |
| $H_3PO_4 + 2H^+ + 2e \rightleftharpoons H_3PO_3 + H_2O$ | –0.28 |
| $V^{3+} + e \rightleftharpoons V^{2+}$ | –0.26 |
| $CuI + e \rightleftharpoons Cu + I^-$ | –0.19 |
| $AgI + e \rightleftharpoons Ag + I^-$ | –0.15 |
| $2H^+ + 2e \rightleftharpoons H_2$ | 0.00 |
| $CuBr + e \rightleftharpoons Cu + Br^-$ | +0.03 |
| $S_4O_6^{2-} + 2e \rightleftharpoons 2S_2O_3^{2-}$ | +0.08 |
| $CuCl + e \rightleftharpoons Cu + Cl^-$ | +0.14 |
| $Sn^{4+} + 2e \rightleftharpoons Sn^{2+}$ | +0.15 |
| $Cu^{2+} + e \rightleftharpoons Cu^+$ | +0.153 |
| $SO_4^{2-} + 4H^+ + 2e \rightleftharpoons H_2SO_3 + H_2O$ | +0.17 |
| $Fe(CN)_6^{3-} + e \rightleftharpoons Fe(CN)_6^{4-}$ | +0.36 |
| $2H_2SO_3 + 2H^+ + 4e \rightleftharpoons S_2O_3^{2-} + 3H_2O$ | +0.40 |
| $4H_2SO_3 + 4H^+ + 6e \rightleftharpoons S_4O_6^{2-} + 6H_2O$ | +0.51 |
| $I_2 + 2e \rightleftharpoons 2I^-$ | +0.54 |
| $Cu^{2+} + Cl^- + e \rightleftharpoons CuCl$ | +0.539 |
| $MnO_4^- + e \rightleftharpoons MnO_4^{2-}$ | +0.564 |
| $O_2 + 2H^+ + 2e \rightleftharpoons H_2O_2$ | +0.68 |
| $Fe^{3+} + e \rightleftharpoons Fe^{2+}$ | +0.77 |
| $O_2 + 2H_2O + 4e \rightleftharpoons 4OH^-$ | +0.82 |
| $Cu^{2+} + I^- + e \rightleftharpoons CuI$ | +0.86 |
| $2Hg^{2+} + 2e \rightleftharpoons Hg_2^{2+}$ | +0.92 |
| $NO_3^- + 4H^+ + 3e \rightleftharpoons NO + 2H_2O$ | +0.96 |
| $Br_2 + 2e \rightleftharpoons 2Br^-$ | +1.0652 |

Reducing power increases (↑) — Oxidising power increases (↓)

| Couple | E°(Volt) |
|---|---|
| $IO_3^- + 6H^+ + 5e \rightleftharpoons \frac{1}{2}I_2 + 3H_2O$ | +1.195 |
| $O_2 + 4H^+ + 4e \rightleftharpoons 2H_2O$ | +1.229 |
| $Tl^{3+} + 2e \rightleftharpoons Tl^+$ | +1.25 |
| $Cr_2O_7^= + 14H^+ + 6e \rightleftharpoons 2Cr^{3+} + 7H_2O$ | +1.33 |
| $Cl_2 + 2e \rightleftharpoons 2Cl^-$ | +1.36 |
| $MnO_4^- + 8H^+ + 5e \rightleftharpoons Mn^{2+} + 4H_2O$ | +1.52 |
| $BrO_3^- + 6H^+ + 5e \rightleftharpoons \frac{1}{2}Br_2 + 3H_2O$ | +1.52 |
| $Ce^{4+} + e \rightleftharpoons Ce^{3+}$ | +1.68 |
| $MnO_4^- + 4H^+ + 3e \rightleftharpoons MnO_2 + 3H_2O$ | +1.695 |
| $H_2O_2 + 2H^+ + 2e \rightleftharpoons 2H_2O$ | +1.77 |
| $Co^{3+} + e \rightleftharpoons Co^{2+}$ | +1.82 |
| $S_2O_8^{2-} + 2e \rightleftharpoons 2SO_4^{2-}$ | +2.01 |
| $O_3 + 2H^+ + 2e \rightleftharpoons H_2O + O_2$ | +2.07 |
| $F_2O + 2H^+ + 4e \rightleftharpoons H_2O + 2F^-$ | +2.10 |
| $O + 2H^+ + 2e \rightleftharpoons H_2O$ | +2.20 |
| $XeO_6^{4-} + 6H^+ + 2e \rightleftharpoons XeO_3 + 3H_2O$ | +2.30 |
| $F_2 + 2e \rightleftharpoons 2F^-$ | +2.87 |
| $3N_2 + 2e \rightleftharpoons 2N_3^-$ | +3.09 |

Reducing power increases (↑) — Oxidising power increases (↓)

**European (International) convention**

$Fe^{3+} + e \rightleftharpoons Fe^{2+}$ $E^\circ = +0.77V$

$Sn^{4+} + 2e \rightleftharpoons Sn^{2+}$ $E^\circ = +0.15V$

**American convention**

$Fe^{2+} - e \rightleftharpoons Fe^{3+}$ $E^\circ = -0.77V$

$Sn^{2+} - 2e \rightleftharpoons Sn^{4+}$ $E^\circ = -0.15V$

The two conventions has the same significance. According to European (International) convention a system **having more positive reduction potential value will act as a strong oxidant than a system having less positive reduction potential value.** Thus in the above example $Fe^{3+}$ is a stronger oxidant than $Sn^{4+}$ while $Sn^{2+}$ is a stronger reductant than $Fe^{2+}$. On the other hand according to American convention **a system having more negative oxidation potential has the greater tendency to act as a stronger oxidant than a system hving less negative oxidation potential.**

An oxidant having higher positive standard redox potential (reduction potential) will generally oxidise the reduced form of a redox couple having less positive or more negative standard redox potential while a reductant having less positive E° value will reduce the oxidised form of a redox couple having more positive E° value. Thus fluorine is the most virulent oxidant known even it is stronger than the atomic oxygen, ozone or perxenate as is evident from their reduction potential values (Table 6.2).

The value of redox potential dependes on the following factors :

(a) The nature of metal or non-metal participating in the electrode reaction and the ion involved.

(b) The concentration of the electrolyte solution in which the electrode is immersed.

(c) The temperature of the electrode system.

### 6.7.4 Critique on EMF

From equation (6.5) we can write, $E = -\frac{\Delta G}{nF}$ and thus effect of amount is eliminated and so E is an **intensive property**. Therefore E of a cell or an electrode system is independent of how we write the electrode reaction as oxidised or reduced form. Let us consider the $H^+/H_2$ electrode and if we write the reaction as $H^+ + e \rightleftharpoons \frac{1}{2}H_2$ and assuming $H_2$ as an ideal gas, we can write the electrode potential according to Nernst equation as :

*E is a sign invariant quantity*

$$E_{H^+/H_2} = E^\circ_{H^+/H_2} + \frac{RT}{F} \ln \frac{a_{H^+}}{(p_{H_2})^{1/2}} \quad ...(6.12)$$

where $p_{H_2}$ = pressure of $H_2$ gas and $a_{H^+}$ = activity of $H^+$ ion in solution.

Now if we write the electrode reaction as :

$$2H^+ + 2e \rightleftharpoons H_2$$

then
$$E_{H^+/H_2} = E^\circ_{H^+/H_2} + \frac{RT}{2F} \ln \frac{a^2_{H^+}}{p_{H_2}}$$

$$= E^\circ_{H^+/H_2} + \frac{RT}{F} \ln \frac{a_{H^+}}{(p_{H_2})^{1/2}} \quad ...(6.13)$$

Thus equations (6.12) and (6.13) are identical and hence E and E° independent of how we write the electrode reaction. The IUPAC has recommended the electrode potential which according to European (International) convention as reduction potential refers to electrostatic charge on that electrode. In opposite convention i.e. in American convention which favours in writing the electrode reaction as that of oxidation and refers to as oxidation potential. Thus the two conventions are conceptually different. In the European (International) convention the sign of the electrode potential is **invariant** while in American convention the electrode potential would change its sign when reduction is involved.

**Problem 6.3** Establish Nernst equation for the following couples : (i) $Cr_2O_7^{2-}/Cr^{3+}$ in acid medium, (ii) $BrO_3^-/Br^-$ in alkaline medium. **[C.U. 2007,2000]**

**Ans. :** (i) $Cr_2O_7^{2-} + 14H^+ + 6e \longrightarrow 2Cr^{3+} + 7H_2O$

Therefore according to equation (6.10) the form of Nernst equation will be—

$$E = E^\circ + \frac{0.0591}{6} \log \frac{[Cr_2O_7^{2-}][H^+]^{14}}{[Cr^{3+}]^2}$$

(ii) $BrO_3^- + 3H_2O + 6e \longrightarrow Br^- + 6OH^-$ [B.U. 2011]

$$\therefore E = E^\circ + \frac{0.0591}{6}\log\frac{[BrO_3^-]}{[Br^-][OH^-]^6}$$

$$= E^\circ - \frac{0.0591}{6}\log[OH^-]^6 + \frac{0.0591}{6}\log\frac{[BrO_3^-]}{[Br^-]}$$

$$= E^\circ + 0.0591\ pOH + \frac{0.0591}{6}\log\frac{[BrO_3^-]}{[Br^-]}$$

### 6.7.5 Formal Potential

Standard reduction (redox) potential (E°) of a particular electrode system refers to unit activity/concentration of the species concerned. In actual practice the solutions may be quite concentrated and the activities of the species concerned may differ from concentrations. Moreover in solution there may occur several factors e.g. the ions may not be in the simple form, there may be complex formation, change in pH of the solution etc. and under this situation E° value stated above will give us misleading information. For these reasons E° value is replaced by the more practicable potential which is known as **formal potential**. Formal potential refers to the actual potential of the system under actual experimental conditions where the **formal concentrations** of the oxidised and reduced forms are the same. Formal concentrations or formality F is the number of formula weights of a substance per litre of the solution.

According to Nernst equation (6.10), the E° value of $Fe^{3+}/Fe^{2+}$ is +0.77 volt but the formal potential values are : +0.65 volt in 1M HCl, +0.73 volt in 1M $HClO_4$, +0.68 volt in 1M $H_2SO_4$, +0.70 volt in 1M $HNO_3$ and +0.61 volt in 1M $H_2SO_4$+0.5M $H_3PO_4$. The $Ce^{4+}/Ce^{3+}$ system has E° = +1.68 volt, the E° value changes to +1.44 volt in 1M $H_2SO_4$, +1.28 volt in 1M HCl etc. Formal potential covers the following effects :

(a) Effect of complex formation.

(b) Effect of evolution of a gas.

(c) Effect of pH.

(d) Effect of insolubility of a substance.

(e) Effect of addition of a foreign substance etc.

## 6.8 APPLICATION OF REDOX POTENTIAL (E°)

### (a) Direction of the course of reactions

Let us consider the following E° values of redox couples :

$$Fe^{3+} + e \rightleftharpoons Fe^{2+} \qquad E^\circ = +0.77 \text{ volt}$$

$$Sn^{4+} + 2e \rightleftharpoons Sn^{2+} \qquad E^\circ = +0.15 \text{ volt}$$

According to the convention (Sec. 6.7.3) the system having higher positive potential will be acting as a stronger oxidant while the system having less positive potential acts a reductant. In other words a system having higher positive E° value (i.e. less negative E°), the reduced form of which will be in the right hand side while the oxidised form will be in the right hand side having the system less positive or more negative E° value so that the predicted reaction will be

$$Fe^{3+} + Sn^{2+} \longrightarrow Fe^{2+} + Sn^{4+}$$

This also occurs in actual laboratory condition. Let us consider another system :

$$Cu^{2+} + 2e \rightleftharpoons Cu \qquad E^\circ = +0.337 \text{ V}$$

$$Zn^{2+} + 2e \rightleftharpoons Zn \qquad E^\circ = -0.763 \text{ V}$$

Therefore the predicted direction of the reaction will be—

$$Zn + Cu^{2+} \longrightarrow Zn^{2+} + Cu$$

In actual practice if we immersed a zinc rod in copper sulphate solution, the blue colour of the solution will be slowly converted to colourless due to the dissolution of zinc atom as $Zn^{2+}$ ion and conversion of $Cu^{2+}$ ion to metallic copper.

### (b) Evaluation of EMF of a cell

Let us consider an electrochemical cell e.g. Daniell cell (Sec. 6.7.1) which is represented as :

$$\text{Pt, Zn}\,|\,Zn^{2+}\,||\,Cu^{2+}\,|\,\text{Cu, Pt}$$

and the respective E° values are represented above. Here oxidation occurs at zinc electrode and reduction occurs at copper electrode. The **cell potential of an electrochemical cell is the algebric sum of the two E° values making negative the E° value where oxidation occurs.** Thus the cell potential of a Daniell cell = –(–0.763) + 0.337 = +1.10 V

The EMF of a cell may also be evaluated by the two half-cell reactions as :

$$\left.\begin{aligned} E_{Cell} &= E_{Ox}(\text{left}) + E_{Red}(\text{right}) \\ E^\circ_{Cell} &= E^\circ_{Ox}(\text{left}) + E^\circ_{Red}(\text{right}) \\ \text{or} \quad E_{Cell} &= E_{Ox}(\text{left}) - E_{Ox}(\text{right}) \\ E^\circ_{Cell} &= E^\circ_{Ox}(\text{left}) - E^\circ_{Ox}(\text{right}) \end{aligned}\right] \qquad \text{...(6.14)}$$

If $E^\circ_{Cell}$ is positive then cell reaction will be spontaneous.

Let us consider another galvanic cell—

$$Cu\,|\,Cu^{2+}\,||\,Ag^{+}\,|\,Ag$$

The E° value for $Ag^+/Ag$ electrode, $Ag^+ + e \longrightarrow Ag$ $\quad E^\circ = +0.80$ V

So that the cell reaction is $Cu + 2Ag^+ \longrightarrow Cu^{2+} + 2Ag$

and the cell potential, $E^\circ_{Cell} = -(+0.337) + 0.80 = +0.463$ V

### (c) Calculation of equilibrium constants

The redox reactions are of reversible type so that laws of chemical equilibrium may be applied to those reactions. Let us consider a general form of a redox reaction :

$$aOx_A + bRed_B \rightleftharpoons aRed_A + bOx_B \qquad \text{...(6.15)}$$

Then according to Nernst equation (6.10) for the two systems, we can write :

$$E_A = E_A^\circ + \frac{0.0591}{n}\log\frac{[Ox_A]^a}{[Red_A]^a} \quad ...(6.16)$$

and $$E_B = E_B^\circ + \frac{0.0591}{n}\log\frac{[Ox_B]^b}{[Red_B]^b} \quad ...(6.17)$$

When equilibrium will be reached at 25°C then $E_A = E_B$, so that

$$E_A^\circ - E_B^\circ = \frac{0.0591}{n}\log\frac{[Ox_B]^b[Red_A]^a}{[Ox_A]^a[Red_B]^b} = \frac{0.0591}{n}\log K \quad ...(6.18)$$

where K is the equilibrium constant of the reaction represented by equation (6.15)

$$\text{Hence } \log K = \frac{n(E_A^\circ - E_B^\circ)}{0.0591} \quad ...(6.19)$$

Equation (6.19) can also be established in different way. We know, $\Delta G^\circ = -nFE^\circ$ and $\Delta G^\circ = -RT \ln K$ according to equations (6.5) and (6.3). Therefore,

$$nFE^\circ = RT \ln K = 2.303\ RT \log K$$

$$\therefore \log K = \frac{nFE^\circ}{2.303RT} = \frac{nE^\circ}{0.0591} = \frac{n(E_A^\circ - E_B^\circ)}{0.0591} \quad ...(6.20)$$

where $E^\circ = E_A^\circ - E_B^\circ$ i.e. the cell potential/EMF.

We know that equilibrium constant (K) is a measure of the completeness of a reaction under experimental conditions. If K is large then the reaction proceed to completion i.e. reactants converted to products. Very small value of K indicates that the reaction takes place from right to left while when $K \approx 1$ indicates the reaction will not be complete in either direction. Let us consider an example.

$$Fe^{3+} + e \rightleftharpoons Fe^{2+} \qquad E^\circ = +0.77 \text{ volt}$$

$$Sn^{4+} + 2e \rightleftharpoons Sn^{2+} \qquad E^\circ = +0.15 \text{ volt}$$

Hence the reaction is $Sn^{2+} + 2Fe^{3+} \longrightarrow Sn^{4+} + 2Fe^{2+}$

So that cell potential $E^\circ = -(+0.15) + 0.77 = +0.62$ volt

According to equation (6.19), $\log K = \frac{2(0.77 - 0.15)}{0.0591} = \frac{2 \times 0.62}{0.0591} = 21$

so that $K = 10^{21}$.

The high value of K indicates that the reduction of $Fe^{3+}$ by $Sn^{2+}$ proceeds to completion in our quantitative analysis.

Let us consider the following more examples where K is evaluated from equation (6.19) ($E^\circ$ values taken from Table 6.2) :

$$MnO_4^- + 8H^+ + 5Fe^{2+} \longrightarrow Mn^{2+} + 5Fe^{3+} + 4H_2O;$$

$$\log K = \frac{5 \times (+1.52 - 0.77)}{0.0591} = 63.45,\ K \sim 10^{63}$$

$$Sn^{4+} + 2Ce^{3+} \longrightarrow Sn^{2+} + 2Ce^{4+};$$

$$\log K = \frac{2 \times (-1.68 - 0.15)}{0.0591} = -62, K = 10^{-62}$$

From the E° values this reaction will occur in opposite direction. So K value will be poor.

$$Ce^{4+} + Fe^{2+} \longrightarrow Ce^{3+} + Fe^{3+} \quad \log K = \frac{1.68 - 0.77}{0.0591} = 15.39, K \sim 10^{15}.$$

---

**Problem 6.4** Will metallic tin reduce $Pb^{2+}$ if :

(i) Concentration of $Pb^{2+}$ is 0.50M and that of $Sn^{2+}$ being 0.01M?

(ii) Concentration of $Pb^{2+}$ is 0.01M and that of $Sn^{2+}$ being 0.50M?

[The cell set up is $Sn|Sn^{2+}||Pb^{2+}|Pb$; E° values for $Sn^{2+}/Sn$ and $Pb^{2+}/Pb$ are respectively –0.136 V and –0.126 V] **[B.U. 1992,1999]**

**Ans. :** From the given E° values the expected reaction will be :

$Sn + Pb^{2+} \longrightarrow Sn^{2+} + Pb$ so that metallic tin will reduce $Pb^{2+}$. The equilibrium constant of the reaction is—

$$K = \frac{[Sn^{2+}][Pb]}{[Sn][Pb^{2+}]} \text{ i.e. } K = \frac{[Sn^{2+}]}{[Pb^{2+}]},$$ as the concentrations of metallic tin and lead are unity.

Hence in (i) $K = \frac{0.01}{0.50} = \frac{1}{50}$ and in (ii) $K = \frac{0.50}{0.01} = 50.$

Therefore metallic tin will reduce $Pb^{2+}$ in condition (ii) and not in condition (i) as the equilibrium constant K has larger value in condition (ii) than that of condition (i). Therefore concentration of ions is also a factor in order to proceed the reaction from left to right which is predicted from E° values.

---

**Problem 6.5** A cell is set up as : $Fe|\,Fe^{2+}(a = 0.1)\,||\,Cd^{2+}(a = 0.001)\,|\,Cd$.

(a) What will be the cell reaction? (b) Evaluate the cell potential and the polarity of the electrodes (c) Show the direction of spontaneous reaction, (d) Calculate the equilibrium constant of the cell reaction. Given $Cd^{2+} + 2e \rightleftharpoons Cd$ E° = –0.40V and $Fe^{2+} + 2e \rightleftharpoons Fe$ E° = –0.44 V.

**Ans. :** (a) From the E° values the cell reaction will be—

$$Fe + Cd^{2+} \longrightarrow Fe^{2+} + Cd$$

and $E^\circ_{Cell} = -(-0.44) - 0.40 = +0.04$ V

(b) The cell potential can be evaluated from the two single electrodes

$$E_{Cd^{2+}/Cd} = -0.40 + \frac{0.0591}{2}\log\frac{0.001}{1} = -0.49 \text{ V}$$

$$E_{Fe^{2+}/Fe} = -0.44 + \frac{0.0591}{2}\log\frac{0.1}{1} = -0.47 \text{ V}$$

$$\therefore E_{Cell} = E_{Cd^{2+}/Cd} - E_{Fe^{2+}/Fe} = -0.49 + 0.47 = -0.02 \text{ V}$$

or according to Nernst equation, $E_{Cell} = E^\circ_{Cell} + \frac{0.0591}{2}\log\frac{a_{Cd^{2+}}}{a_{Fe^{2+}}}$

$$= 0.04 + \frac{0.0591}{2}\log\frac{0.001}{0.1}$$

$$= -0.02 \text{ V}$$

As the cell potential is negative so the reverse reaction will occur as represented above and the cadmium electrode is anode and the iron electrode is cathode. If both the ions are at unit activity then $E^\circ_{Cell}$ = +0.04 V and in this case the above reaction will occur and the iron electrode will be anode and the cadmium electrode will be cathode.

(c) As the cell potential $E_{Cell}$ is negative then the reverse reaction as stated above will be spontaneous i.e. $Fe^{2+} + Cd \longrightarrow Fe + Cd^{2+}$. But when the activity of the ions at unity then the above reaction as predicted from the E° values will be spontaneous i.e. $Fe + Cd^{2+} \longrightarrow Fe^{2+} + Cd$.

(d) The equilibrium constant (K) as calculated from $E^\circ_{Cell}$ is given by—

$$\log K = \frac{nE^\circ}{0.0591} = \frac{2 \times 0.04}{0.0591} = 1.3536 \text{ so that } K = 22.6.$$

## 6.9 APPLICATION OF FORMAL POTENTIAL (E°′)

### (A) Redox Titrations

Like that of acid-base titrations (Sec. 4.13) redox titrations can be performed with redox reactions. One example is the titration of $Fe^{2+}$ with $Ce^{4+}$ solution in $H_2SO_4$ medium (Fig. 6.3). Formal potential values for $Ce^{4+}/Ce^{3+}$ and $Fe^{3+}/Fe^{2+}$ systems in 1M $H_2SO_4$ medium are :

$$Ce^{4+} + e \rightleftharpoons Ce^{3+} \qquad E^{\circ\prime} = +1.44 \text{ V}$$

$$Fe^{3+} + e \rightleftharpoons Fe^{2+} \qquad E^{\circ\prime} = +0.68 \text{ V}$$

We perform the titration by placing the two redox systems in the same container and allowing them to reach the equilibrium as :

$$Ce^{4+} + Fe^{2+} \rightleftharpoons Ce^{3+} + Fe^{3+}$$

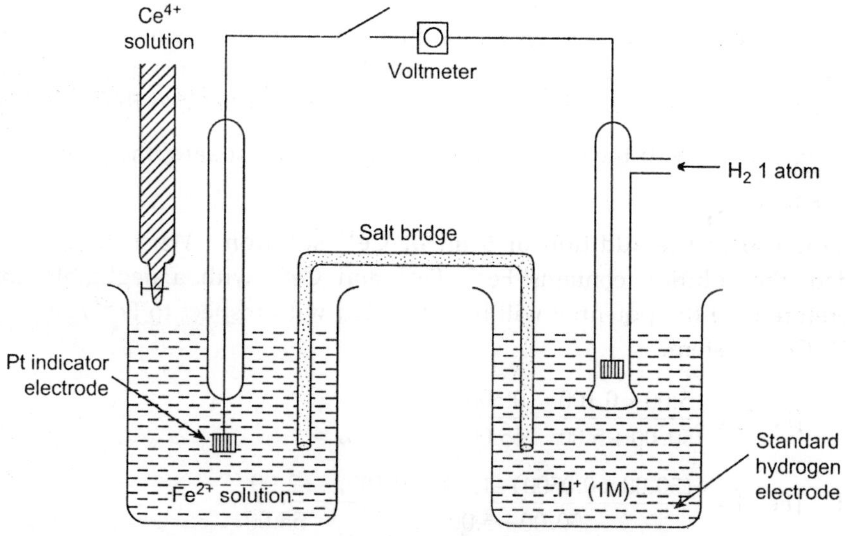


*Figure 6.3 : Titration of $Fe^{2+}$ with $Ce^{4+}$ solution.*

and the equilibrium constant of the reaction is given by—

$$K = \frac{[Ce^{3+}][Fe^{3+}]}{[Ce^{4+}][Fe^{2+}]}$$

and according to equation (6.19)

$$\log K = \frac{1 \times (1.44 - 0.68)}{0.0591} = 12.88 \text{ i.e. } K \sim 10^{13}$$

Which indicates that the reaction proceed to completion.

In the titration vessel there are two redox systems namely $Fe^{3+}/Fe^{2+}$ and $Ce^{4+}/Ce^{3+}$ for which we can apply the Nernst equation (6.10) as :

$$E_1 = E^{o'}_{Fe^{3+}/Fe^{2+}} + 0.0591 \log \frac{[Fe^{3+}]}{[Fe^{2+}]} = 0.68 + 0.0591 \log \frac{[Fe^{3+}]}{[Fe^{2+}]}$$

and $$E_2 = E^{o'}_{Ce^{4+}/Ce^{3+}} + 0.0591 \log \frac{[Ce^{4+}]}{[Ce^{3+}]} = 1.44 + 0.0591 \log \frac{[Ce^{4+}]}{[Ce^{3+}]}$$

The potential at the various stages of titration can be calculated in the titration vessel from the above Nernst expression for any one redox system by choosing one which is more convenient for our purposes. The values of potential which will be calculated at each stage of titration would be measured experimentally as shown in the Fig. 6.3.

Let us consider for convenience 50 ml 0.05M $Fe^{2+}$ is titrated with 0.10 M $Ce^{4+}$ solution in a medium which is 1M in $H_2SO_4$ at all stages of titration. Let us calculate the potential values at various stages :

(i) **At the start of titration :** In this stage it is likely that there is no cerium species present in the solution and the potential is to be calculated by the $Fe^{3+}/Fe^{2+}$ system i.e.

$$E = 0.68 + 0.0591 \log \frac{[Fe^{3+}]}{[Fe^{2+}]}$$

It is expected that due to air oxidation there might be very small concentration of $Fe^{3+}$ ion, the amount of which depends on how the $Fe^{2+}$ solution is prepared and the extent of exposure in air. Let only 0.1% iron is oxidised to $Fe^{3+}$ so that, $E = 0.68 + 0.0591 \log \frac{1}{1000} = 0.503$ V. If the concentration of $Fe^{3+}$ is almost **zero**, the potential is of course very small.

(ii) **Potential after the addition of 5 ml of $Ce^{4+}$ solution :** When first $Ce^{4+}$ solution is added, the solution contains $Fe^{2+}$, $Fe^{3+}$ and $Ce^{3+}$ with a negligible amount $Ce^{4+}$. Therefore here the potential will be calculated with respect to $Fe^{3+}/Fe^{2+}$ and not with $Ce^{4+}/Ce^{3+}$ system.

$$[Fe^{3+}] = \frac{5.00 \times 0.100}{50.00 + 5.00} = \frac{0.500}{55.00}$$

and $$[Fe^{2+}] = \frac{(50.00 \times 0.0500) - (5.00 \times 0.100)}{50.00 + 5.00} = \frac{2.00}{55.00}$$

and the potential value is,

$$E = 0.68 + \frac{0.0591}{1} \log \frac{\frac{0.500}{55.00}}{\frac{2.00}{55.00}} = 0.644 \text{ V}$$

In this way the potential for the addition of different volumes of $Ce^{4+}$ solution before equivalence point are given in Table 6.3.

(iii) **Equivalence point potential i.e. when 25 ml $Ce^{4+}$ is added :** The equivalence point potential for any redox system can be given by the expression

$$E_{eq} = \frac{n_1E_1 + n_2E_2}{n_1 + n_2} \quad \text{...(6.21)}$$

where $n_1$ and $n_2$ are the number of electrons involved in the respective half cell reduction reaction having standard reduction potential values $E_1^\circ$ and $E_2^\circ$ and $E_1$ and $E_2$ are the respective electrode potential. For any reaction in which the number of electrons lost by the reductant is equal to the number of electrons gained by the oxidant, the potential at the equivalence point is simply the arithmetic mean of the $E_1^\circ$ and $E_2^\circ$. Thus,

$$E_{eq} = \frac{1.44 + 0.68}{2} = 1.06 \text{ V}$$

(iv) **Potential beyond equivalence point i.e. by the addition of 25.10 ml $Ce^{4+}$ solution :** In this stage the concentrations of $Ce^{4+}$, $Ce^{3+}$ and $Fe^{3+}$ are to be considered and not the concentration of $Fe^{2+}$. Therefore the potential value are to be considered by $Ce^{4+}/Ce^{3+}$ system and not with $Fe^{3+}/Fe^{2+}$ system. The concentration of $Ce^{3+}$ and $Ce^{4+}$ ions are :

$$[Ce^{3+}] = \frac{25 \times 0.100}{50 + 25.10} = \frac{2.500}{75.10}$$

and $$[Ce^{4+}] = \frac{(25.10 \times 0.100) - (50.00 \times 0.0500)}{50 + 25.10} = \frac{0.010}{75.10}$$

The potential at this stage is,

$$E = 1.44 + \frac{0.0591}{1} \log \frac{\frac{0.010}{75.10}}{\frac{2.500}{75.10}} = 1.298 \text{ V}$$

The titration curve for titration of $Fe^{2+}$ with $Ce^{4+}$ is plotted in Fig. 6.4 which is similar to that of neutralisation curve of acid-base reaction.

**Table 6.3 : Potential values in the titration of $Fe^{2+}$ with $Ce^{4+}$ solution with reference to S.H.E. as shown in Fig. 6.3. 50 ml 0.050M $Fe^{2+}$ is taken**

| Volume in ml of 0.100M $Ce^{4+}$ solution | Potential in volt |
|---|---|
| 5.0 | 0.644 |
| 10.0 | 0.669 |
| 15.0 | 0.690 |
| 20.0 | 0.716 |
| 24.0 | 0.761 |
| 24.9 | 0.822 |

| Volume in ml of 0.100M $Ce^{4+}$ solution | Potential in volt |
|---|---|
| 25.0 | 1.060 ← Equivalence point |
| 25.1 | 1.298 |
| 26.0 | 1.357 |
| 30.0 | 1.399 |

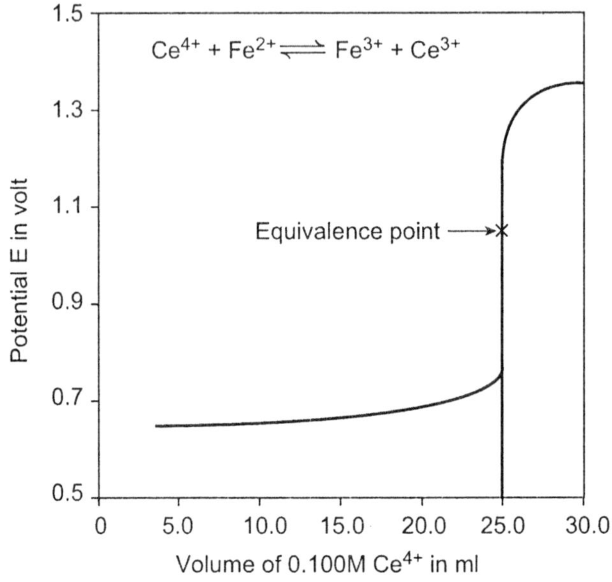


***Figure 6.4** : Titration curve for the titration of 50.0 ml 0.050M $Fe^{2+}$ with 0.100M $Ce^{4+}$ solution.*

**Problem 6.6** $\left(\frac{N}{10}\right)$ $Fe^{2+}$(aq) is titrated with $\left(\frac{N}{10}\right)$ $MnO_4^-$(aq) at (N) acidity. Derive the expression for the equivalence point potential of $Fe^{3+}/Fe^{2+}$ electrode at 25°C. **[C.U. 2007]**

Given $E^\circ_{Fe^{3+}/Fe^{2+}} = 0.76$ V and $E^\circ_{MnO_4^-/Mn^{2+}} = 1.51$ V

**Ans. :** At the equivalence point traces of $Fe^{2+}$ will still exist and an exactly equivalent amount of $MnO_4^-$ will also remain unreacted. Here the half-cell reactions are :

$$MnO_4^- + 8H^+ + 5e \longrightarrow Mn^{2+} + 4H_2O$$

$$Fe^{3+} + e \longrightarrow Fe^{2+}$$

The mole ratio of $Fe^{2+}$ and $MnO_4^-$ ions in the reaction are 5 : 1 so that at equivalence point,

$$[Fe^{2+}] = 5[MnO_4^-], \text{ similarly } [Fe^{3+}] = 5[Mn^{2+}]$$

so that, $\frac{[Fe^{2+}]}{[Fe^{3+}]} = \frac{[MnO_4^-]}{[Mn^{2+}]}$ i.e. $\frac{[Fe^{3+}][MnO_4^-]}{[Fe^{2+}][Mn^{2+}]} = 1$

Let E is the equivalence point potential, then,

$$E = E^\circ_{Fe^{3+}/Fe^{2+}} + 0.0591 \log \frac{[Fe^{3+}]}{[Fe^{2+}]}$$

and $$E = E^\circ_{MnO_4^-/Mn^{2+}} + \frac{0.0591}{5} \log \frac{[MnO_4^-][H^+]^8}{[Mn^{2+}]}$$

$$= E^\circ_{MnO_4^-/Mn^{2+}} + \frac{0.0591}{5} \log \frac{[MnO_4^-]}{[Mn^{2+}]} \quad [\because [H^+] = 1]$$

$$\therefore 5E = 5E^\circ_{MnO_4^-/Mn^{2+}} + 0..0591 \log \frac{[MnO_4^-]}{[Mn^{2+}]}$$

By adding we get,

$$6E = E^\circ_{Fe^{3+}/Fe^{2+}} + 5E^\circ_{MnO_4^-/Mn^{2+}} + 0.0591 \log \frac{[Fe^{3+}][MnO_4^-]}{[Fe^{2+}][Mn^{2+}]}$$

$$= 0.76 + 5 \times 1.51 + 0.0591 \log 1$$

$$= 8.31, \text{ so that } E = 1.385 \text{ V}$$

According equation (6.21) the equivalence point potential can be calculated as,

$$E = \frac{n_1 E_1^\circ + n_2 E_2^\circ}{n_1 + n_2} = \frac{1 \times 0.76 + 5 \times 1.51}{1 + 5} = 1.385 \text{ V}$$

---

**Problem 6.7** In an aqueous solution containing $Fe^{3+}$ and $Fe^{2+}$ ions, the redox potential is +0.70 volt at 25°C. Taking E° value of $Fe^{3+}/Fe^{2+}$ couple as +0.77 volt, calculate $Fe^{3+}/Fe^{2+}$ ratio in the solution. **[B.U. 2001, C.U. 2006]**

**Ans.** : According to Nernst equation

$$E = E^\circ + 0.0591 \log \frac{[Ox]}{[Red]}$$

$$\therefore 0.70 = 0.77 + 0.0591 \log \frac{[Fe^{3+}]}{[Fe^{2+}]}$$

$$\therefore \log \frac{[Fe^{3+}]}{[Fe^{2+}]} = \frac{0.70 - 0.77}{0.0591} = -1.184 = -2 + 2 - 1.184 = \bar{2}.816$$

so that $$\frac{[Fe^{3+}]}{[Fe^{2+}]} = 0.06546.$$

---

## (B) Redox indicator in redox titrations

In Sec. 6.9(A) we have mentioned only one redox titration i.e. titration of $Fe^{2+}$ with $Ce^{4+}$. There are also two common redox titrations very often carried out in laboratory e.g. titration of $Fe^{2+}$ with $KMnO_4$ and titration of $Fe^{2+}$ with $K_2Cr_2O_7$ solution. In the case of $Fe^{2+}$ with $KMnO_4$ titration no indicator is used because $KMnO_4$ solution is self indicating—one drop of which develops a pink colour at the equivalence point. In the titration of $Fe^{2+}$ with $K_2Cr_2O_7$ solution an indicator commonly used is **barium diphenylamine sulphonate**.

Like that of acid-base indicators which exist as dissociated and undissociated forms in basic and acidic solutions respectively having different colours, the redox indicators also exist as oxidised and reduced forms having diffrent colours.

Redox indicators are generally organic dyes which can undergo reversible oxidation or reduction in a typical redox titration and impart different colours to detect the equivalence point. The oxidised ($In_{Ox}$) and reduced ($In_{Red}$) forms of a redox indicator forms a redox electrode with a characteristic electrode potential (E) and having standard reduction potential ($E^\circ_{In}$). The redox couple of the indicator can be written as :

$$In_{Ox} + ne \rightleftharpoons In_{Red}$$

Applying Nernst equation (6.10), we can write,

$$E = E^\circ_{In} + \frac{0.0591}{n}\log\frac{[In_{Ox}]}{[In_{Red}]} \qquad ...(6.22)$$

The indicator changes colour of the oxidised form to that of the reduced form when the concentration ratio of $\frac{[In_{Ox}]}{[In_{Red}]}$ is 1 to 10 or greater i.e. a colour change is seen when

*Potential interval for the colour change of a redox indicator*

$$\frac{[In_{Ox}]}{[In_{Red}]} \geq \frac{1}{10} \quad \text{or} \quad \frac{[In_{Ox}]}{[In_{Red}]} \leq 10$$

Substituting these two values in equation (6.22) we get,

$$E = E^\circ_{In} \pm \frac{0.0591}{n} \qquad ...(6.23)$$

Thus the range of potential values across which an indicator exhibits a sharp colour change have the values $E^\circ_{In} \pm \frac{0.0591}{n}$. This equation (6.23) is of similar type of equation (4.69 or 4.71) which is applied for pH interval limit for acid-base indicator. However protons are involved in the reduction of many indicators, so the range of potentials over which a colour change occurs may often be pH dependent. Let us simplify the condition more clearly. Let colour X can be observed by the eye when $\frac{[In_{Ox}]}{[In_{Red}]} = 10$ and colour Y is observed when $\frac{[In_{Ox}]}{[In_{Red}]} = \frac{1}{10}$, so that at

$$\text{Colour X : } E = E^\circ_{In} + \frac{0.0591}{n}\log 10 = E^\circ_{In} + \frac{0.0591}{n}$$

$$\text{Colour Y : } E = E^\circ_{In} + \frac{0.0591}{n}\log\frac{1}{10} = E^\circ_{In} - \frac{0.0591}{n}$$

$$\text{Subtracting, } \Delta E = \pm 2 \times \frac{0.0591}{n} \qquad ...(6.24)$$

Therefore when one electron is involved (n=1) the change of potential is ±0.118 V while for two electrons involvement the change of ±0.0591 volt is indicated. Some typical redox indicators with their $E^\circ_{In}$ values are given in Table 6.4. It is to be noted that many redox indicators are colourless in reduced form and oxidised form is distinctly coloured.

**Table 6.4 : Some typical redox indicators**

| Indicator | Colour of | | Transition potential (volt) | Medium |
|---|---|---|---|---|
| | Oxidised form | Reduced form | | |
| Methylene blue | Blue | Colourless | +0.53 | 1M acid |
| Diphenylamine | Violet | Colourless | +0.76 | 1M $H_2SO_4$ |
| Diphenylaminesulfonic acid | Red-violet | Colourless | +0.85 | Dilute acid |
| Diphenylbenzidine | Violet | Colourless | +0.76 | 1M $H_2SO_4$ |
| Ferroin (1,10-phenanthroline iron (II) complex) | Pale-blue | Red | +1.11 | 1M $H_2SO_4$ |
| Nitroferroin (5-nitro 1, 10-phenanthroline iron (II) complex) | Pale-blue | Red-violet | +1.25 | 1M $H_2SO_4$ |
| Starch $I_3^-$, KI | Blue | Clourless | +0.53 | Dilute acid |

Let us consider some specific cases for the **choice of indicator in redox titration :**

Selection of redox indicators

(I) In the titration of $Fe^{2+}$ with $Ce^{4+}$ [Sec. 6.9(A)] what will be the indicator to be used. The equivalence point potential which is calculated as 1.06 V. According to Table 6.4 it is evident that **ferroin** with a transition potential 1.11 volt is suitable for this titration since ferroin has $E^\circ_{In}$ = 1.06 V but the formal potential where the indicator undergoes colour change is 1.11 V.

(II) Another common example is the titration of $Fe^{2+}$ with standard $K_2Cr_2O_7$ solution in which what indicator will be used? The E° values for the two couples are :

$$Fe^{3+} + e \longrightarrow Fe^{2+} \qquad E^\circ = +0.77 \text{ V}$$

$$Cr_2O_7^{2-} + 14H^+ + 6e \longrightarrow 2Cr^{3+} + 6H_2O \qquad E^\circ = +1.33 \text{ V}$$

The reaction will be—

$$6Fe^{2+} + Cr_2O_7^{2-} + 14H^+ \longrightarrow 2Cr^{3+} + 6Fe^{3+} + 7H_2O$$

According to equation (6.21) the equivalence point potential ($E_{eq}$) in this titration is

$$\frac{1 \times 0.77 + 6 \times 1.33}{1+6} = 1.25 \text{ V.}$$

Hence the indicators to be taken which have transition potential in the range of $E_{eq}$. In fact in this titration diphenylamine (+0.76V) or diphenylamine sulfonic acid (+0.85V) may be used but from potential data it is evident that dichromate will simultaneously oxidise $Fe^{2+}$ to $Fe^{3+}$ as also indicator reduced form to oxidised form so that we get the colour change of the indicator much before the equivalence point. In order to obviate this difficulty we add $H_2SO_4$ + $H_3PO_4$ mixture in the titration flask containing $Fe^{2+}$ solution before the addition of $K_2Cr_2O_7$. We have seen that the formal potential of $Fe^{3+}/Fe^{2+}$ system is reduced to +0.61V in 1M $H_2SO_4$ + 0.5M $H_3PO_4$ medium. So that in this situation $K_2Cr_2O_7$ first oxidise all $Fe^{2+}$ ion to $Fe^{3+}$ and at the end point the indicator reduced form will be oxidised and we get the colour change (violet) at the equivalence point.

In fact in this titration there is a potential break in the equivalence point in the range 0.94V to 1.30 V. For the choice of a redox indicator the following two conditions will be satisfied simultaneously :

(i) The $E^\circ_{In}$ value of an indicator will be less than the E° value of the titrant and will be greater than the E° value of the sample to be titrated i.e.

$$E^\circ_{titrant} > E^\circ_{In} > E^\circ_{sample} \quad ...(6.25)$$

e.g. $E^\circ_{Cr_2O_7^{2-}/Cr^{3+}} > E^\circ_{diphenylamine\text{-}sulfonic\ acid} > E^\circ_{Fe^{3+}/Fe^{2+}}$

(ii) The equivalence point potential ($E_{eq}$) will be in the vicinity of the transition potential ($E_{tp}$) of the indicatoir i.e.

$$E_{eq} \approx E_{tp} \quad ...(6.26)$$

However in actual situation in a redox titration this condition does not hold good due to the complicacy arises by the gradual change of potential of the titration system as titration proceeds.

Structural chemistry of redox indicators

As mentioned diphenylamine* is used in the titration of $Fe^{2+}$ with $K_2Cr_2O_7$. The indicators mentioned in Table 6.4 are organic molecules that undergo structural changes during oxidation and reduction. Since diphenylamine is difficultly soluble in water and as tungstate ion and $HgCl_2$ interfere with its action, the sodium or barium salt of diphenylaminesulfonic acid is more widely used in practice. Diphenylamine is believed to undergo the following structural changes whereby the oxidised form is deep violet and the reduced form is colourless (Fig. 6.5).

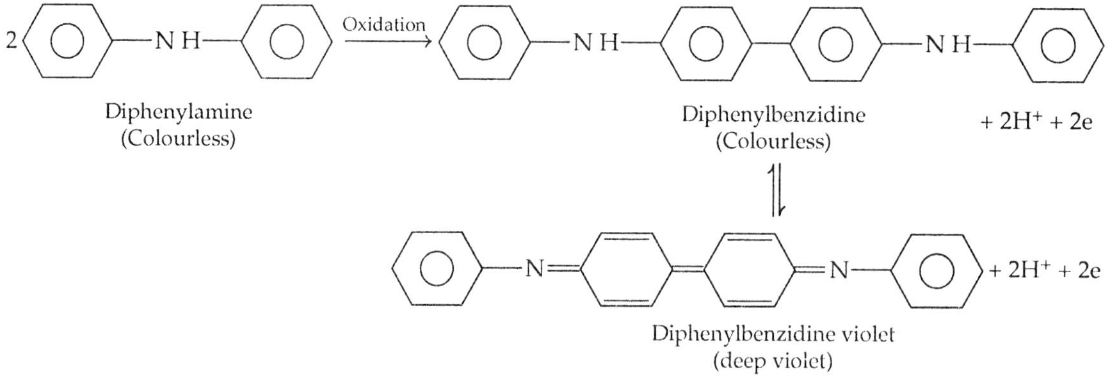


*Figure 6.5 : Structural changes of diphenylamine indicator.*

As mentioned earlier ferroin indicator which is the phenanthroline (ophen) complex of $Fe^{2+}$ having blood-red colour.

* N-phenylanthranilic acid may be used in the titration of $Fe^{2+}$ by dichromate even without adding $H_3PO_4$.

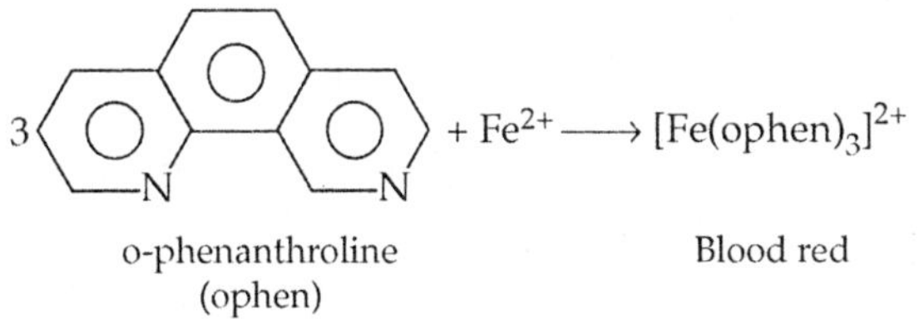

The $Fe^{2+}$ complex undergoes oxidation to $Fe^{3+}$ complex which has a light-blue colour.

$$\underset{\text{Light blue}}{[Fe(ophen)_3]^{3+}} + e \rightleftharpoons \underset{\text{Blood red}}{[Fe(ophen)_3]^{2+}} \quad E^\circ_{In} = 1.06\ V$$

The indicator is prepared by mixing equivalent amount of $FeSO_4$ and o-phenanthroline.

(III) Another indicator of common use in laboratory is the starch $-I_3^-$ indicator. This is used in oxidation or reduction reactions involving iodine as an oxidant or iodide ion as reductant. In presence of excess oxidising agent the indicator imparts a blue colour to the solution while in presence of excess reducing agent the indicator is colourless. Hence the indicator changes from **colourless to blue** in the titration of various reducing agents with many oxidising agents e.g. in the titration of iodine with $Na_2S_2O_3$ solution.

---

**Problem 6.8** Give the name and structure of a redox indicator. Show its mode of action with equation. **[C.U. 2006,1996; B.U. 2009]**

**Ans. :** Sec. 6.9(B)(II).

---

**Problem 6.9** 100 ml $\left(\frac{N}{10}\right)$ Mohr salt solution is titrated with $\left(\frac{N}{10}\right)$ $KMnO_4$ solution in 1(M) acid. Calculate the potential of the solution when (i) 90 ml of $KMnO_4$ solution has been added (ii) the equivalence point is reached (iii) 101 ml of $KMnO_4$ solution has been added. Given $E^\circ_{Fe^{3+}/Fe^{2+}} = +0.76$ V and $E^\circ_{MnO_4^-/Mn^{2+}} = 1.51$ Volt. **[C.U. 1997]**

**Ans. :** (i) $E = E^\circ_{Fe^{3+}/Fe^{2+}} + 0.0591 \log\frac{[Fe^{3+}]}{[Fe^{2+}]} = +0.76 + 0.0591 \log\frac{90}{10} = 0.82$ Volt.

(ii) At equivalence point, $E = \frac{n_1E^\circ_1 + n_2E^\circ_2}{n_1 + n_2} = \frac{1\times 0.76 + 5\times 1.51}{1+5} = 1.385$ Volt.

(iii) $E = E^\circ_{MnO_4^-/Mn^{2+}} + \frac{0.0591}{5}\log\frac{[MnO_4^-]}{[Mn^{2+}]}$

$= 1.51 + \frac{0.0591}{5}\log\frac{1}{100} = 1.474$ Volt.

---

## (C) Influence of pH

This can be exemplified in the following cases :

(I) The E° values of $Fe(CN)_6^{3-}/Fe(CN)_6^{4-}$ and $I_2/I^-$ couples are :

$[Fe(CN)_6]^{3-} + e \rightleftharpoons [Fe(CN)_6]^{4-}$ $\quad E^\circ = +0.36$ V

$I_2 + 2e \rightleftharpoons 2I^-$ $\quad E^\circ = +0.54$ V

From E° values it is evident that the predicted reaction will be :

$$2[Fe(CN)_6]^{4-} + I_2 \longrightarrow 2[Fe(CN)_6]^{3-} + 2I^-$$

i.e. $I_2$ will oxidise $[Fe(CN)_6]^{4-}$ to $[Fe(CN)_6]^{3-}$ which means that $I_2$ is a better oxidant than $[Fe(CN)_6]^{3-}$. In practice the reaction will go to the reverse direction in concentrated acidic medium of low pH (1M $H^+$). In acidic solutions both $[Fe(CN)_6]^{4-}$ and $[Fe(CN)_6]^{3-}$ combine with $H^+$ ions to form $H_4[Fe(CN)_6]$* and $H_3[Fe(CN)_6]$** respectively. We know increasing charge on a metal ion increases its electronegativity as also acidity so that protonation process will be more easier in case of $[Fe(CN)_6]^{4-}$ compared to that of $[Fe(CN)_6]^{3-}$. Thus the concentration of $[Fe(CN)_6]^{4-}$ is appreciably diminished in acid solution and according to Nernst equation (6.10) the E° value of $[Fe(CN)_6]^{3-}/[Fe(CN)_6]^{4-}$ system is raised from +0.36 V to +0.71 V in 1M HCl or $HClO_4$ medium. Thus the formal potential value is somewhat raised from E° value of $I_2/I^-$ couple so that the following reaction takes place in reality :

*Utility of formal potential*

$$2[Fe(CN)_6]^{3-} + 2I^- \longrightarrow 2[Fe(CN)_6]^{4-} + I_2$$

(II) The $AsO_4^{3-}/AsO_3^{3-}$ and $I_2/I^-$ couples have the following E° values :

*pH effect on E° value*

$$AsO_4^{3-} + 2H^+ + 2e \rightleftharpoons AsO_3^{3-} + H_2O \quad E^\circ = +0.56\text{ V}$$

$$I_2 + 2e \rightleftharpoons 2I^- \quad E^\circ = +0.54\text{ V}$$

Hence the reaction should occur in acid medium as :

$$AsO_4^{3-} + 2H^+ + 2I^- \longrightarrow I_2 + AsO_3^{3-} + H_2O$$

The E° value of $AsO_4^{3-}/AsO_3^{3-}$ system is pH dependent as is evident from the Nernst equation as :

$$E = 0.56 + \frac{0.0591}{2}\log\frac{[AsO_4^{3-}][H^+]^2}{[AsO_3^{3-}]}$$

$$= 0.56 + 0.03\log[H^+]^2 + 0.03\log\frac{[AsO_4^{3-}]}{[AsO_3^{3-}]}$$

$$= 0.56 + 0.06\log[H^+] + 0.03\log\frac{[AsO_4^{3-}]}{[AsO_3^{3-}]}$$

$$= 0.56 - 0.06pH + 0.03\log\frac{[AsO_4^{3-}]}{[AsO_3^{3-}]} \qquad ...(6.27)$$

$$= E^{\circ\prime} + 0.03\log\frac{[AsO_4^{3-}]}{[AsO_3^{3-}]} \qquad ...(6.28)$$

where $E^{\circ\prime} = 0.56 - 0.06pH$ is the formal potential of the system. The equation (6.27) denotes the pH effect on the E° value of a system.

It is evident from equation (6.27) that on lowering the $[H^+]$ from the standard state i.e. 1 to $10^{-8}$, the course of the reaction will change. When $[H^+] = 10^{-8}$ i.e. pH = 8 then,

* $H_4[Fe(CN)_6]$ is called hydroferrocyanic acid.

** $H_3[Fe(CN)_6]$ is called hydroferricyanic acid.

$$E = 0.56 - 0.06 \times 8 + 0.03 \log \frac{[AsO_4^{3-}]}{[AsO_3^{3-}]}$$

$$= 0.08 + 0.03 \log \frac{[AsO_4^{3-}]}{[AsO_3^{3-}]}$$

Thus the $E^{o\prime}$ value of the system decreases with increase in pH thus making $AsO_4^{3-}$ a weaker oxidant than $I_2$ and hence the reaction takes place as :

$$AsO_3^{3-} + I_2 + H_2O \longrightarrow AsO_4^{3-} + 2H^+ + 2I^-$$

---

**Problem 6.10** Briefly explain the nature of variation of redox potential due to change in $H^+$ ion concentration. **[C.U. 2000]**

**Ans. :** Sec. 6.9(C)(II).

---

**Problem 6.11** Establish Nernst equation for the couple $VO_3^-/VO^{2+}$. Standard potential of this couple at 25°C in 1(M) acid medium is 0.92 volt. Find its formal potential at pH = 7. **[C.U.1998]**

**Ans. :** $VO_3^- + 4H^+ + e \longrightarrow VO^{2+} + 2H_2O \quad E° = +0.92$ V

Therefore the form of Nernst equation will be,

$$E = 0.92 + \frac{0.0591}{1} \log \frac{[VO_3^-][H^+]^4}{[VO^{2+}]}$$

When pH = 7, then $[H^+] = 10^{-7}$, so that

$$E = 0.92 + 0.0591 \log[10^{-7}]^4 + 0.0591 \log \frac{[VO_3^-]}{[VO^{2+}]}$$

$$= 0.92 - 28 \times 0.0591 \log 10 + 0.0591 \log \frac{[VO_3^-]}{[VO^{2+}]}$$

$$= -0.74 + 0.0591 \log \frac{[VO_3^-]}{[VO^{2+}]}$$

Therefore the formal potential value at pH = 7 is –0.74 V.

---

**Problem 6.12** "1 N $KMnO_4$ may mean either 0.33 M or 0.20 M $KMnO_4$."—Comment.
**[B.U. 2009, 2012, 2016]**

**Ans. :** In acid solution $KMnO_4$ undergoes the following reaction i.e.

$$MnO_4^- + 8H^+ + 5e \rightleftharpoons Mn^{2+} + 4H_2O$$

$\therefore$ Equivalent weight of $KMnO_4 = \frac{M}{5}$

where M is the molecular weight of $KMnO_4$.

Hence 1N $KMnO_4$ = 0.20 M $KMnO_4$.

Another reaction of $KMnO_4$ which occurs in alkali medium is

$$MnO_4^- + 2H_2O + 3e \rightleftharpoons MnO_2 + 4OH^-$$

So that the equivalent weight = $\frac{M}{3}$

In this case 1N $KMnO_4$ = 0.33M $KMnO_4$ solution.

Therefore, we can say that 1N $KMnO_4$ may be either 0.33M or 0.20M solution.

**Problem 6.13** Explain the effect of pH on the redox potential of the system,

$$MnO_4^- + 8H^+ + 5e \rightleftharpoons Mn^{2+} + 4H_2O$$

Explain why chloride ion is oxidised by this system at low pH (<1.5) but not in neutral medium. Given E° values are $MnO_4^- / Mn^{2+}$ = 1.52 volt, $Cl_2 / Cl^-$ = 1.36 volt. **[B.U. 2001]**

**Ans. :** According to Nernst equation,

$$E = E^\circ + \frac{0.0591}{5} \log \frac{[MnO_4^-][H^+]^8}{[Mn^{2+}]}$$

$$= 1.52 + \frac{0.0591}{5} \log \frac{[MnO_4^-]}{[Mn^{2+}]} + \frac{0.0591}{5} \log[H^+]^8$$

Let pH changed from acidic to alkaline range, say pH = 8 i.e. $[H^+] = 10^{-8}$, so that

$$E = 1.52 + \frac{0.0591}{5} \log[10^{-8}]^8 + \frac{0.0591}{5} \log \frac{[MnO_4^-]}{[Mn^{2+}]}$$

$$= 0.764 + 0.012 \log \frac{[MnO_4^-]}{[Mn^{2+}]}$$

Thus the formal potential value changes from 1.52 V to 0.764 V as we change pH from acidic to alkaline range. This denotes the effect of pH on the redox potential of the system.

Now it is evident from the E° values of $MnO_4^-/ Mn^{2+}$ and $Cl_2/Cl^-$ systems that at low pH chloride ion is oxidised by $MnO_4^-$ ion i.e. the reaction will be,

$$2MnO_4^- + 16H^+ + 10Cl^- \longrightarrow 2Mn^{2+} + 5Cl_2 + 8H_2O$$

But in neutral medium i.e. when pH = 7 so that $[H^+] = 10^{-7}$, the E° value of $MnO_4^-/Mn^{2+}$ system will change as :

$$E = 1.52 + \frac{0.0591}{5} \log[10^{-7}]^8 + \frac{0.0591}{5} \log \frac{[MnO_4^-]}{[Mn^{2+}]}$$

$$= 0.858 + 0.012 \log \frac{[MnO_4^-]}{[Mn^{2+}]}$$

It is evident that the formal potential value of $MnO_4^-/ Mn^{2+}$ far lower than the E° value of $Cl_2/Cl^-$ system so that in neutral medium the reaction will go in the opposite direction i.e. in neutral medium $Cl^-$ will not be oxidised by $MnO_4^-$.

## (D) Effect of complex formation

The following examples illustrate the effect of complex formation by which the E° values change to E°′—

(I) In the titration of $Fe^{2+}$ with $K_2Cr_2O_7$ using barium diphenyl amine sulphonate indicator a question arises how the E° value of $Fe^{3+}/Fe^{2+}$ changes from +0.77V to 0.61V in 1M $H_2SO_4$ + 0.5M $H_3PO_4$ medium. If we apply Nernst equation for $Fe^{3+}/Fe^{2+}$ system then,

$$E = 0.77 + 0.0591 \log \frac{[Fe^{3+}]}{[Fe^{2+}]}$$

In the said medium $H_3PO_4$ forms a complex $[Fe(HPO_4)]^+$ with $Fe^{3+}$ ion and thus the concentration of $Fe^{3+}$ ions are removed so that the potential value is reduced.

(II) $2Hg^{2+} + 2e \rightleftharpoons Hg_2^{2+}$ $E° = +0.92$ V

and $Fe^{3+} + e \rightleftharpoons Fe^{2+}$ $E° = +0.77$ V

Therefore under normal condition $Fe^{2+}$ will reduce $Hg^{2+}$ to $Hg_2^{2+}$. But in the presence of $SCN^-$ ion $Hg^{2+}$ forms a stable complex as $[Hg(SCN)_4]^{2-}$. Then according to Nernst equation for $Hg^{2+}/Hg_2^{2+}$ system we can write,

$$E = 0.92 + \frac{0.0591}{2} \log \frac{[Hg^{2+}]^2}{[Hg_2^{2+}]}$$

As the concentration of $Hg^{2+}$ ions are removed due to the formation of the above complex, so the potential value will be reduced. Let the concentration of $Hg^{2+}$ is reduced from 1 to $10^{-3}$M. Therefore

$$E = 0.92 + \frac{0.0591}{2} \log \frac{[10^{-3}]^2}{[Hg_2^{2+}]}$$

$$= 0.743 + \frac{0.0591}{2} \log \frac{1}{[Hg_2^{2+}]}$$

As the formal potential value is substantially lowered so that in presence of $SCN^-$ the reduction of $Fe^{3+}$ with $Hg_2^{2+}$ ion occurs i.e. the reaction will go in reverse direction as we expect from simple E° values.

(III) The change in E° value has been brought about by complexation. Thus $Co^{3+}/Co^{2+}$ has E° value of +1.82 V and simple $Co^{3+}$ is easily reduced to $Co^{2+}$ by KI ($I_2 + 2e \longrightarrow 2I^-$ $E° = +0.54$ V). But in presence of $CN^-$ ion or $NH_3$ the reduction of $Co^{3+}$ to $Co^{2+}$ will not occur due to the formation of the complex and the E°′ values of $[Co(CN)_6]^{3-}/[Co(CN)_6]^{4-}$ and $[Co(NH_3)_6]^{3+}/[Co(NH_3)_6]^{2+}$ are –0.83V and +0.1V respectively.

## (E) Effect of insolubility of a substance or evolution of a gas

This effect may be illustrated by the following examples :

(I) $Cu^{2+} + e \longrightarrow Cu^+$ $E° = +0.153$ V

$I_2 + 2e \longrightarrow 2I^-$ $E° = +0.54$ V

It is evident from the above E° values that $Cu^{2+}$ will not oxidise $I^-$ to $I_2$ but common laboratory practice for the **iodometric estimation** of $Cu^{2+}$ is to titrate the liberated $I_2$ by reaction of $Cu^{2+}$ with KI i.e.

$$2Cu^{2+} + 4I^- \longrightarrow Cu_2I_2\downarrow + I_2$$

This is due to the insolubility of cuprous iodide whereby the E° value of $Cu^{2+}/Cu^+$ is changed in the following way :

$$E = 0.153 + 0.0591 \log \frac{[Cu^{2+}]}{[Cu^+]}$$

The solbulity product of CuI is $10^{-12}$ so that,

$$[Cu^+][I^-] = 10^{-12}$$

Let the $[I^-]$ = 0.1M, therefore $[Cu^+] = \frac{10^{-12}}{10^{-1}} = 10^{-11}$ M

Hence, $E = 0.153 + 0.0591 \log \frac{[Cu^{2+}]}{10^{-11}}$

$= 0.803 + 0.0591 \log[Cu^{2+}]$

As the E°′ value is changed so that idometric estimation of $Cu^{2+}$ becomes viable in practice.

(II) Let us consider the following E° values :

$Fe(CN)_6^{3-} + e \rightleftharpoons Fe(CN)_6^{4-}$ E° = +0.36 V

$I_2 + 2e \rightleftharpoons 2I^-$ E° = +0.54 V

From the potential values it is evident that under normal condition ferricyanide will not oxidise iodide to iodine. However in presence of $Zn^{2+}$ ion ferricyanide will oxidise iodide to iodine. $Zn^{2+}$ ions in neutral or acid medium react with $K_4[Fe(CN)_6]$ to form the **very sparingly soluble** precipitate of potassium zinc ferrocyanide i.e.

$3Zn^{2+} + 2K_4[Fe(CN)_6] \longrightarrow K_2Zn_3[Fe(CN)_6]_2 + 6K^+$

Therefore as long as excess $Zn^{2+}$ ions present in solution, the concentration of $[Fe(CN)_6]^{4-}$ ion is very small and according to Nernst equation,

$$E = 0.36 + 0.0591 \log \frac{[Fe(CN)_6^{3-}]}{[Fe(CN)_6^{4-}]}$$

the formal potential value exceeds from the value of 0.36V and will be raised over the value of $I_2/I^-$ couple so that in this situation ferricyanide will oxidise iodide ion. In fact $Zn^{2+}$ can be estimated by iodometric estimation of liberated $I_2$ in this way.

(III) In the quantitative estimation of $Fe^{3+}$, it is a common practice to reduce $Fe^{3+}$ to $Fe^{2+}$ in 6(N)HCl medium by $SnCl_2$ and then $Fe^{2+}$ is titrated with strong oxidants like $KMnO_4/K_2Cr_2O_7$ solution. If we take $KMnO_4$ solution as titrant then the following potential values involved in the system :

$MnO_4^- + 8H^+ + 5e \rightleftharpoons Mn^{2+} + 4H_2O$ E° = +1.52 volt

$Cl_2 + 2e \rightleftharpoons 2Cl^-$ E° = +1.36 volt

$Fe^{3+} + e \rightleftharpoons Fe^{2+}$ E° = +0.77 volt

It is evident from the above E° values that $KMnO_4$ will simultaneously oxidise $Cl^-$ to $Cl_2$ as also $Fe^{2+}$ to $Fe^{3+}$. If it is so happen that $Cl_2$ formed remains in solution and would oxidise equivalent amount of $Fe^{2+}$ to $Fe^{3+}$ then in this titration there will be no error. Unfortunately **$Cl_2$ has low solubility in aqueous acidic solution so that most of the $Cl_2$ will be escaped from the system causing an error in the titration**. To prevent this side reaction we use **Zimmerman-Reinhardt solution** (also known as **preventive solution**) which contains $MnSO_4$, $H_2SO_4$ and $H_3PO_4$. $H_2SO_4$ maintains the acidity of the solution and $H_3PO_4$ forms complex with $Fe^{3+}$ to form colourless $[Fe(HPO_4)]^+$

thereby the yellow colour of $Fe^{3+}$ is masked in solution and the end point is detected clearly. Moreover in this way the formal potential value of the $Fe^{3+}/Fe^{2+}$ is substantially reduced from the E° value (+0.77V) and the oxidation of $Fe^{2+}$ with $MnO_4^-$ is facilitated.

*Function of Zimmerman-Reinhardt solution*

On the other hand $MnSO_4$ added in the preventive solution lowers the E° value of $MnO_4^-/Mn^{2+}$ system as :

$$E = E^\circ + \frac{0.0591}{5}\log\frac{[MnO_4^-][H^+]^8}{[Mn^{2+}]}$$

$$= 1.52 + \frac{0.0591}{5}\log[H^+]^8 + \frac{0.0591}{5}\log\frac{[MnO_4^-]}{[Mn^{2+}]}$$

$$= 1.52 + \frac{0.0591\times 8}{5}\log[H^+] + \frac{0.0591}{5}\log\frac{[MnO_4^-]}{[Mn^{2+}]}$$

In the standard condition the concentration of $MnO_4^-$ and $Mn^{2+}$ are unity but due to the addition of $Mn^{2+}$ by preventive solution the ratio $[MnO_4^-]/[Mn^{2+}]$ is substantially reduced and the formal potential of the $MnO_4^-/Mn^{2+}$ system decreases and even it will be reduced to the E° value of $Cl_2/2Cl^-$ system (E° = +1.36V). In this situation $MnO_4^-$ will only oxidise $Fe^{2+}$ to $Fe^{3+}$ but not be able to oxidise $Cl^-$ to $Cl_2$.

**Problem 6.14** What happens when $I_2$ is added to excess aqueous acidic $HClO_3$? Write the half reactions. Predict which net reaction will occur? Given E° values for $I_2/I^-$ and $ClO_3^-/ClO_4^-$ couples are 0.54 V and –1.20 V respectively. **[C.U. 2006]**

**Ans. :** $I_2 + 2e \longrightarrow 2I^-$   E° = +0.54 V

$ClO_4^- + 2H^+ + 2e \longrightarrow ClO_3^- + H_2O$   E° = +1.20 V

The predicted reaction will be

$$ClO_4^- + 2I^- + 2H^+ \longrightarrow I_2 + ClO_3^- + H_2O$$

But in actual practice the reaction will go in the opposite direction. $HIO_3$ is prepared by evaporating iodine with chloric acid. Also $KIO_3$ is prepared by reaction of $KClO_3$ with $I_2$.

$$2HClO_3 + I_2 \longrightarrow 2HIO_3 + Cl_2$$

$$2KClO_3 + I_2 \longrightarrow 2KIO_3 + Cl_2$$

This is due to the metallic character of iodine than that of chlorine as also the stability factor of $KClO_3$ and $KIO_3$. Moreover we have to consider the following E° values :

$$ClO_3^-/\tfrac{1}{2}Cl_2 \quad E^\circ = 1.47\text{ V and } IO_3^-/\tfrac{1}{2}I_2 \quad E^\circ = 1.2\text{ V}$$

which suggest the actual reaction will be,

$$2ClO_3^- + I_2 \longrightarrow IO_3^- + Cl_2$$

**Problem 6.15** The overall formation constant for $Zn(CN)_4^{2-}$ is $10^{21}$. What will be the E° value for the following process : $Zn(CN)_4^{2-} + 2e \rightleftharpoons Zn + 4CN^-$. Given E° for $Zn^{2+}/Zn$ is –0.76V.

**Ans. :** $Zn^{2+} + 4CN^- \rightleftharpoons Zn(CN)_4^{2-}$   and $K_{eff} = 10^{21}$ where $K_{eff}$ is overall formation constant

$Zn^{2+} + 2e \rightleftharpoons Zn$   E° = –0.76V

Therefore for these two reactions we can write for 25°C

| | |
|---|---|
| $Zn(CN)_4^{2-} \rightleftharpoons Zn^{2+} + 4CN^-$ | $\Delta G° = -2.303RT\log 10^{-21}$ |
| $Zn^{2+} + 2e \rightleftharpoons Zn$ | $\Delta G° = -nFE° = -2F(-0.76)$ |
| $Zn(CN)_4^{2-} + 2e \rightleftharpoons Zn + 4CN^-$ | $\Delta G° = 1.24F + 1.52F = 2.76F$ |

Now $\Delta G° = -nFE°$

i.e. $2.76F = -2 \times FE°$ so that $E° = -1.88V$.

---

**Problem 6.16** From the following E° values calculate the equilibrium constant for the reaction—

$Au^+(aq) + 2CN^-(aq) \longrightarrow [Au(CN)_2]^-(aq)$

$Au^+(aq) + e \longrightarrow Au(s) \quad E° = +1.69V$

$[Au(CN)_2]^-(aq) + e \longrightarrow Au(s) + 2CN^-(aq) \quad E° = -0.60V$

**Ans. :** The predicted reaction from the above E° values may be—

$Au(s) + 2CN^-(aq) + Au^+(aq) \longrightarrow [Au(CN)_2]^-(aq) + Au(s)$

Cell potential = 1.69 + 0.60 = 2.29V

$$\text{so that } \log K = \frac{nE°}{0.0591} = \frac{2.29}{0.0591} = 38.7479$$

i.e. $K = 5.6 \times 10^{38}$.

---

## 6.10 DIGRAMMATIC REPRESENTATION OF REDOX POTENTIAL DATA

When an element exist in different oxidation states, the relative thermodynamic stability of different species can be represented in a diagrammatic way indicating the reduction potential of each species which is known as EMF diagram. There are several EMF diagrams namely **Latimer diagram, Frost diagram, Pourbaix diagram** and **Ellingham diagram** etc. Most useful one is the Latimer diagram.

### 6.10.1 Latimer diagrams

Let us consider the following reduction potential (E°) for the different oxidation states of manganese in acidic medium :

| | |
|---|---|
| $Mn^{2+} + 2e \rightleftharpoons Mn$ | $E° = -1.18V$ |
| $Mn^{3+} + e \rightleftharpoons Mn^{2+}$ | $E° = +1.51V$ |
| $MnO_2 + 4H^+ + e \rightleftharpoons Mn^{3+} + 2H_2O$ | $E° = +0.95V$ |
| $MnO_4^{2-} + 4H^+ + 2e \rightleftharpoons MnO_2 + 2H_2O$ | $E° = +2.26V$ |
| $MnO_4^- + e \rightleftharpoons MnO_4^{2-}$ | $E° = +0.564V$ |
| $MnO_4^- + 8H^+ + 5e \rightleftharpoons Mn^{2+} + 4H_2O$ | $E° = +1.52V$ |
| $MnO_4^- + 4H^+ + 3e \rightleftharpoons MnO_2 + 2H_2O$ | $E° = +1.695V$ |
| $MnO_2 + 4H^+ + 2e \rightleftharpoons Mn^{2+} + 2H_2O$ | $E° = +1.23V$ |

Wendell Latimer introduced a method connecting the E° values in volts is written over a horizontal arrow line with the species of an element in different oxidation states. The highest oxidation state is written in left and the successive lower oxidation states are depicted to the right. For example the above E° values for manganese system can be written as :

$$\underset{(+7)}{MnO_4^-} \xrightarrow{0.56} \underset{(+6)}{MnO_4^{2-}} \xrightarrow{2.26} \underset{(+4)}{MnO_2} \xrightarrow{0.95} \underset{(+3)}{Mn^{3+}} \xrightarrow{1.51} \underset{(+2)}{Mn^{2+}} \xrightarrow{-1.18} \underset{(0)}{Mn}$$

$$MnO_4^- \xrightarrow{1.51} Mn^{2+}$$

$$MnO_4^- \xrightarrow{1.70} MnO_2$$

$$MnO_2 \xrightarrow{1.23} Mn^{2+}$$

Similarly the different species of chlorine is depicted in the Latimer diagram as :

$$ClO_4^- \xrightarrow{1.20} ClO_3^- \xrightarrow{1.18} HClO_2 \xrightarrow{1.65} HClO \xrightarrow{1.67} Cl_2 \xrightarrow{1.36} Cl^-$$

**Utility of the Latimer diagram**

These diagrams can be used to calculate the E° values for any nonadjacent species. This can be derived by using the thermodynamic equation $\Delta G° = -nFE°$ where the free energy ($\Delta G°$) not depends on the path but depends only on the initial and final states so that $\Delta G°$ is a function of state and not a path function. To calculate the E° value of any composite process, we have to convert the individual E° values to $\Delta G°$ values by multiplication with –nF factor and add them together and the sum is converted to E° by dividing with –nF. Let us consider the following case :

$$MnO_4^{2-} \xrightarrow[n=2]{2.26} MnO_2 \xrightarrow[n=1]{0.95} Mn^{3+} \xrightarrow[n=1]{1.51} Mn^{2+}$$

Then what will be the E° value for $MnO_4^{2-} \longrightarrow Mn^{2+}$ i.e.

$$MnO_4^{2-} + 8H^+ + 4e \longrightarrow Mn^{2+} + 4H_2O \qquad E° = ?$$

The overall $\Delta G°$ for two successive steps is the sum of the individual values i.e.

$$\Delta G° = \Delta G_1° + \Delta G_2°$$

So that $\quad -nFE° = -nFE_1° - nFE_2°$

or $\quad -4FE° = -2F \times 2.26 - 1F \times 0.95 - 1F \times 1.51$

So that $\quad E° = \dfrac{2 \times 2.26 + 1 \times 0.95 + 1 \times 1.51}{4} = 1.745V$

In general for any composite system we can write

$$E° = \frac{n_1E_1° + n_2E_2° + n_3E_3° + ......}{n_1 + n_2 + n_3 + ......} \qquad ...(6.29)$$

In basic aqueous solution the Latimer diagram for $Cl_2$ system is :

$$ClO_4^- \xrightarrow{0.37} ClO_3^- \xrightarrow{0.30} ClO_2^- \xrightarrow{0.68} ClO^- \xrightarrow{0.42} Cl_2 \xrightarrow{1.35} Cl^-$$

What will be the E° value for $ClO^- \longrightarrow Cl^-$ i.e.

$$ClO^- + H_2O + 2e \longrightarrow Cl^- + 2OH^- \qquad E° = ?$$

The intermediate steps are :

$$2ClO^- + 2H_2O + 2e \longrightarrow Cl_2 + 4OH^- \qquad E° = +0.42V$$

$$Cl_2 + 2e \longrightarrow 2Cl^- \qquad E° = +1.35V$$

So according to equation (6.29) we get

$$E^\circ = \frac{2 \times 0.42 + 2 \times 1.35}{2+2} = 0.89 \text{ V}$$

This Latimer diagram gives us information in a compact form about the E° values for the conversion of different oxidation states of an element. This diagram can be used to judge whether **disproportionation** for a species will occur or not. **A species has a tendency to disproportionate to its nearest neighbours if the potential on the right of the species in the diagram is higher than the potential on the left**. Thus $Cl_2$ has a tendency to disproportionate in alkaline medium due to the following :

*Uses of Latimer diagram*

$$ClO^- \xrightarrow{0.42} Cl_2 \xrightarrow{1.35} Cl^-$$

**Problem 6.17** Using the following Latimer diagram, calculate the E° value for reduction of HClO to $Cl^-$ in aqueous acid medium.

$$HClO \xrightarrow{1.67} Cl_2 \xrightarrow{1.36} Cl^-$$

**Ans.** : $HClO + H^+ + e \longrightarrow \frac{1}{2}Cl_2 + H_2O \quad E^\circ = +1.67 \text{ V}$

$\frac{1}{2}Cl_2 + e \longrightarrow Cl^- \quad E^\circ = +1.36 \text{ V}$

$$\therefore E^\circ = \frac{n_1E_1^\circ + n_2E_2^\circ}{n_1+n_2} = \frac{1 \times 1.67 + 1 \times 1.36}{1+1} = 1.515 \text{ V}$$

## 6.10.2 Frost diagrams

Let an element X exist in different oxidation number as $X^n$ to $X^0$ where $X^n$ and $X^0$ forms a redox couple which indicates that the element has oxidation number n in oxidised state and zero in reduced state. **The Frost diagram* of the element X is a plot of nE° versus the various oxidation number of the element**. According to equation (6.5) we can write $\Delta G^\circ = -nFE^\circ$ so that $nE^\circ = -\Delta G^\circ/F$ and hence the nE° is proportional to the standard free energy change for the formation of the species of the element concerned in the n oxidation number. Therefore the plot of nE° against n can also be considered as the plot of ΔG° against n. Therefore the most stable oxidation state of the element which have the position lowest to the Frost diagram as shown in Fig. 6.6.

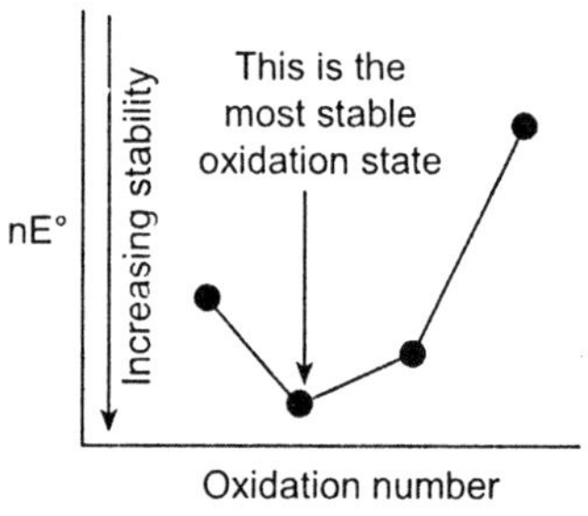


*Figure 6.6 : Frost diagram i.e. plot of nE° vs. n.*

* also known as Ebsworth diagram.

*Usefulness of Frost diagram*

The Frost diagram is a ready record for quick qualitative idea about the variation of redox properties of an element in different oxidation states. Although numerical calculations are more easier in Latimer diagram, the following features of the Frost diagram indicates some special utility :

(A) The Frost diagram for oxygen is shown in Fig. 6.7. In this diagram the slope of the line joining any two points indicates the E° value of the redox couple of the two points concerned. From the curve the value of nE° for $H_2O_2$ (oxidation number –1) is –0.70V while at oxidation number –2 (for $H_2O$) is –2.46V so that the difference of the values is –1.76V and the change in oxidation number of the two systems is –1. Therefore the slope of the line is $\frac{-1.76}{-1}$ = +1.76V which is in agreement with that of the value for the $H_2O_2/H_2O$ couple in Latimer diagram as :

$$O_2 \xrightarrow{0.70} H_2O_2 \xrightarrow{1.76} H_2O \quad (O_2 \rightarrow H_2O: 1.23)$$

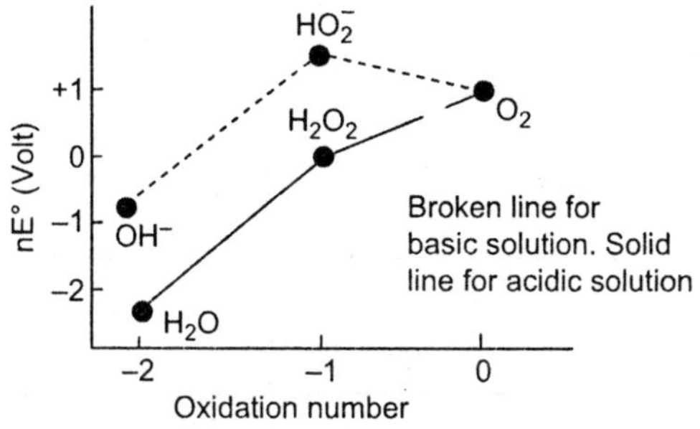


***Figure 6.7** : Frost diagram for oxygen.*

(B) **In this diagram the steeper the line joining two points, the higher is the potential of the corresponding couple (Fig. 6.8) :** The CD line is more steeper than the BC line and thus the couple corresponding to points C and D will have higher reduction potential compared to that of the couple formed by B and C points. The spontaneity of the redox reaction between the two points can be judged by comparing the slopes of the corresponding lines. Thus the species represented by the point D would undergo reduction while the species represented by the point B

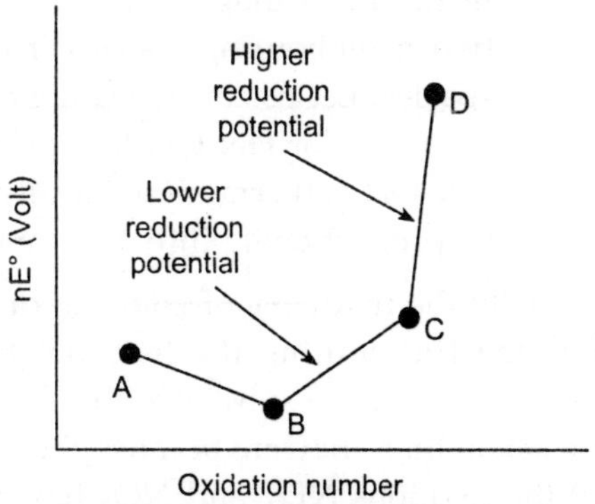


*Figure 6.8 : Representation of Frost diagram with the variation of higher and lower reduction potential of a species.*

would be liable to spontaneous oxidation. This diagram can be utilised to predict whether a particular oxidant will oxidise a particular species or not. The judgement can be made by observing that **an oxidant with the more positive slope in nE° vs. oxidation number plot would easily oxidise a species occupying the position in the said plot having less positive slope.** Thus in the diagram (Fig. 6.9) nE° vs. oxidation number plot of $HNO_3$—NO line has more greater positive slope than that of Cu—$Cu^{2+}$ line and hence $HNO_3$ would easily oxidise Cu to $Cu^{2+}$ and itself would be reduced to NO i.e. $3Cu + 8HNO_3 \longrightarrow 3Cu(NO_3)_2 + 2NO + 4H_2O$.

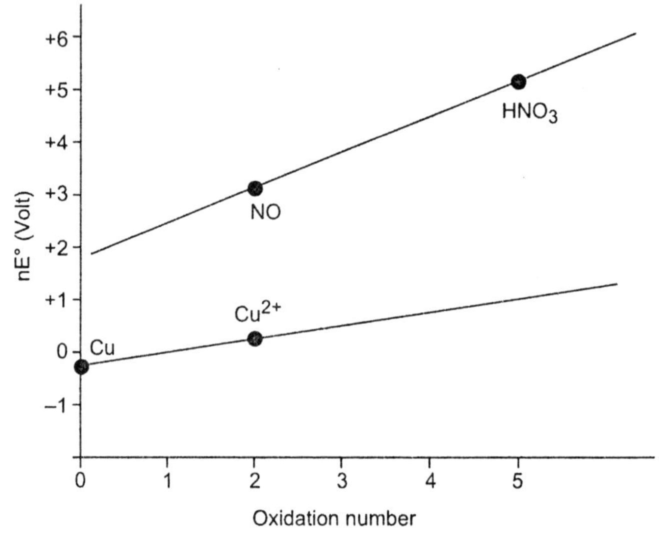


***Figure 6.9** : Frost diagram of $HNO_3$—NO and $Cu^{2+}$—Cu systems.*

(C) Another application of this diagram is that **an ion or a molecule is unstable with respect to disproportionation if it lies above the line connecting two adjacent species in the Frost diagram** i.e. the species will be thermodynamically more stable than the two neighbouring species if the said species lies below the line connecting the two species. Let us consider the Frost diagram for nitrogen compounds (Fig. 6.10) in which the point for $HN_2OH$ lies above the line connecting points for $NH_3$ and $N_2$ and hence $HN_2OH$ is thermodynamically less stable than $NH_3$ and $N_2$ so that $HN_2OH$ is liable to disproportionate into $NH_3$ and $N_2$ ($3NH_2OH \longrightarrow NH_3 + N_2 + 3H_2O$).

Finally the **tendency of comproportionation is judged when one intermediate species lies below the line joining the terminal two species.** Comproportionation is thermodynamically favourable when the substance that lies below the line joining its neighbours in the diagram is more stable than they are because their average free energy is higher. Thus in the diagram (Fig. 6.10) the two ions $NH_4^+$ and $NO_3^-$ having oxidation number of nitrogen –3 and +5 respectively comproportionate spontaneously in the following way to form the intermediation oxidation number of nitrogen i.e. +1 in $N_2O$.

$$NH_4^+ + NO_3^- \longrightarrow N_2O + 2H_2O$$

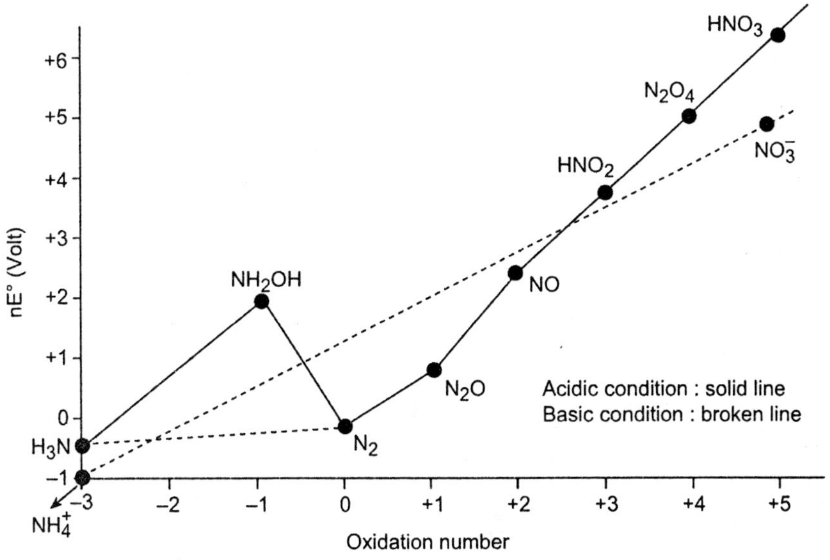


***Figure 6.10** : Frost diagram for nitrogen compounds.*

This reaction will not occur in aqueous phase but violently occurs in solid state and so $NH_4NO_3$ is sometimes used as explosive material.

Let us correlate Latimer diagram with that of Frost (Ebsworth) diagram for manganese system. The Latimer diagram for manganese system is—

$$MnO_4^- \xrightarrow{0.56} MnO_4^{2-} \xrightarrow{2.26} MnO_2 \xrightarrow{0.95} Mn^{3+} \xrightarrow{1.51} Mn^{2+} \xrightarrow{-1.18} Mn$$

which indicates $Mn^{3+}$ will disproportionate to $Mn^{2+}$ and $MnO_2$., The Frost diagram for manganese system may be constructed in the following way :

$\Delta G° = -nFE°$ so that $nE° = -\Delta G°/F$, which has to be plotted against the oxidation number (n) (Fig. 6.11).

Now for n = 2, E° for $Mn^{2+}/Mn = -1.18V$, so $2E° = -2.36$ V

for n = 3, E° for $Mn^{3+}/Mn^{2+} = 1.51$ V

$$\therefore nE° = -\frac{\Delta G°}{F} = (2\times1.18 - 1.51) \text{ V}$$

i.e. $3E° = -0.85$ V

for n = 4, E° for $MnO_2/Mn^{3+} = 0.95$ V

and so $4E° = (-0.95 + 0.85)$ V = 0.10 V

Since $Mn^{3+}$ lies above the line joining the points of $Mn^{2+}$ and $MnO_2$ so $Mn^{3+}$ will disproportionate into the species $Mn^{2+}$ and $MnO_2$ and thus Latimer and Frost diagram have the equivalent redox interpretation. Similarly $Mn^{6+}$ will disproportionate into $Mn^{5+}$ and $Mn^{7+}$. It can also be inferred that Mn and $Mn^{3+}$ comproportionate to form $Mn^{2+}$.

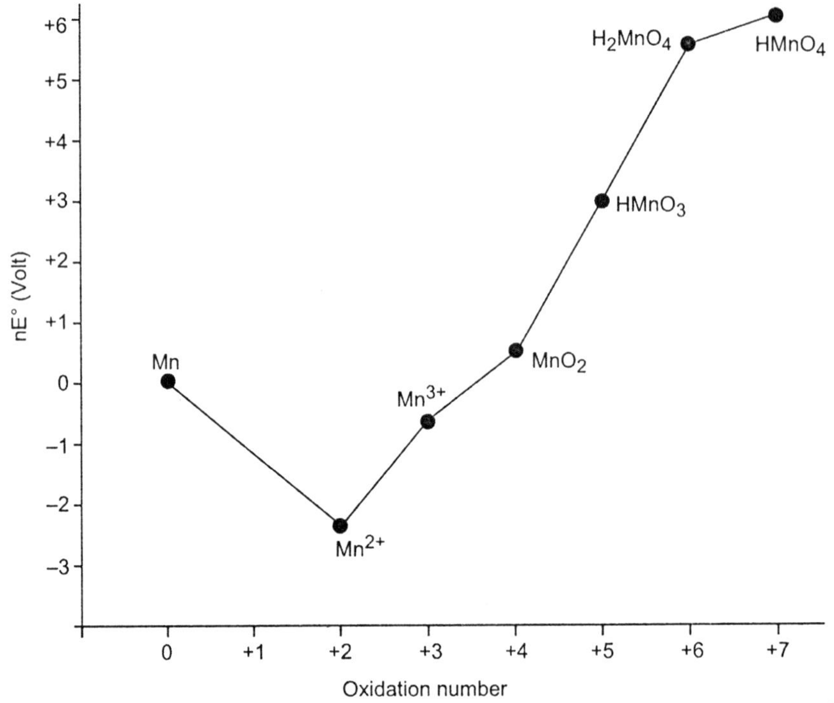


*Figure 6.11 : Frost diagram for manganese system in acidic medium.*

**Problem 6.18** Draw the Frost diagram for oxygen compounds from the following Latimer diagram :

$$O_2 \xrightarrow{0.70} H_2O_2 \xrightarrow{1.76} H_2O$$
$$O_2 \xrightarrow{1.23} H_2O$$

**Ans. :** The oxidation number of oxygen in the said diagram are 0, –1 and –2 respectively. For the conversion of $O_2$ to $H_2O_2$ the change in oxidation number(n) is from 0 to –1, here E° = +0.70 V and n = –1, so that nE° = –0.70 V. The change in oxidation number from $O_2$ to $H_2O$ is from 0 to –2 and E° = +1.23 V and hence nE° = –2 × 1.23 = –2.46 V. These nE° values are plotted against oxidation number to construct the Frost diagram (Fig. 6.7, shown by solid line.)

**Problem 6.19** Construct the Frost diagram from the following Latimer diagram :

$$Tl^{3+} \xrightarrow[n=2]{1.25} Tl^{+} \xrightarrow[n=1]{-0.34} Tl$$
$$Tl^{3+} \xrightarrow{0.72} Tl$$

Show the calculation for evaluation of coordinate of each point of the plot.

**[C.U. 2002, B.U. 2015]**

**Ans. :** n = 1 $Tl^+/Tl$ E° = –0.34 Volt 1E° = –0.34 V

n = 3 $Tl^{3+}/Tl^+$ E° = +1.25 Volt $Tl^+/Tl$ E° = –0.34 Volt

$$E^\circ = \frac{n_1E_1 + n_2E_2}{n_1 + n_2} = \frac{2 \times 1.25 - 1 \times 0.34}{2 + 1} = \frac{2.16}{3} \text{ Volt}$$

∴ 3E° = 2.16 Volt.

These results are plotted in the following figure to form the Frost diagram.

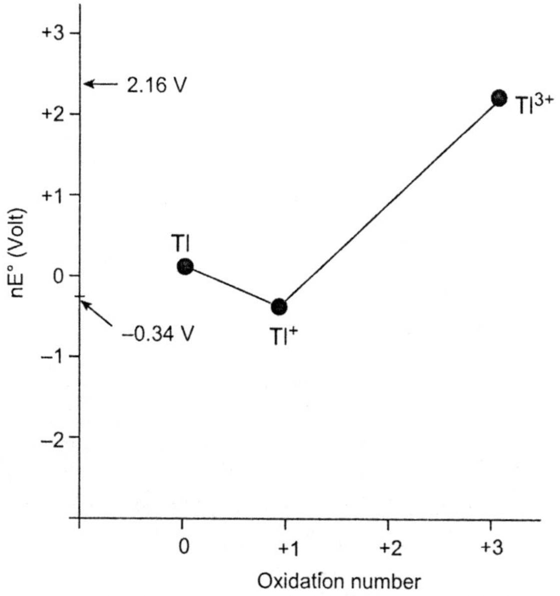


**Problem 6.20** Draw a Frost diagram for mercury in acid solution from the given Latimer diagram :

$Hg^{2+} \xrightarrow{0.911V} Hg_2^{2+} \xrightarrow{0.796\ V} Hg$ . Comment on the tendency of any of the species to undergo disproportionation. **[C.U. 2007]**

**Ans. :** What will be the potential value for the conversion of $Hg^{2+}$ to Hg i.e.

$$E° = \frac{2\times0.911+2\times0.796}{2+2} = \frac{3.414}{4} = 0.854\ V$$

So that Latimer diagram is modified as :

$$Hg^{2+} \xrightarrow{0.911} Hg_2^{2+} \xrightarrow{0.796} Hg \quad (Hg^{2+} \to Hg:\ 0.854)$$

For the change in oxidation number from +2 to +1 ($Hg^{2+}$ to $Hg_2^{2+}$) E° = +0.911 V and n = 1 so nE° = 0.911 V and for change in oxidation number +2 to 0 ($Hg^{2+}$ to Hg), E° = +0.854 V, n = 2 and so nE° = 2 × 0.854 = 1.708 V.

Therefore at the point corresponding to oxidation number +1 (for $Hg_2^{2+}$) nE° = 0.911 V and at oxidation number +2 (for $Hg^{2+}$) nE° = 1.708 V so that the difference = 1.708 – 0.911 = 0.797 V. The change of oxidation number of Hg on going from $Hg_2^{2+}$ to Hg is 1.

The slope of a line = $\frac{\text{Change in nE° value}}{\text{Change in oxidation number}}$. Therefore the slope of the line between $Hg_2^{2+}$ and $Hg^{2+}$ = $\frac{1.708 - 0.796}{1}$ = 0.912 V and the slope of the line between Hg and $Hg_2^{2+}$ = $\frac{0.796}{1}$ = 0.796 V. The results are drawn in the following Frost diagram :

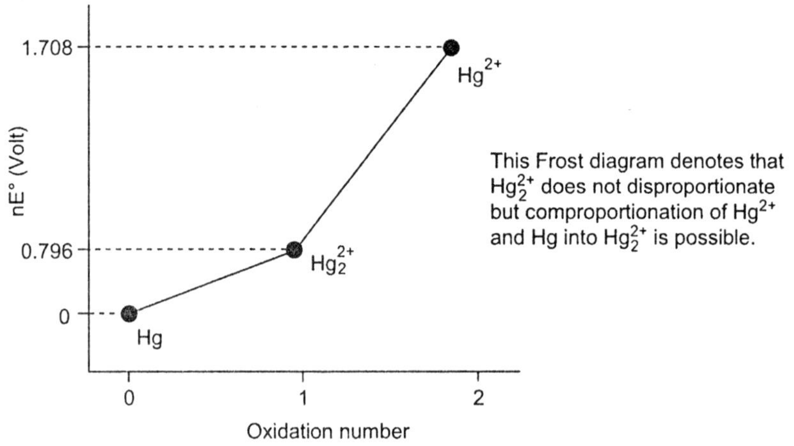


From the above Latimer diagram of mercury it is evident that disproportionation is not thermodynamically possible because reduction potential decreases from higher to the lower oxidation states. But if the formal potential value of $Hg^{2+}/Hg_2^{2+}$ couple can be lowered by the addition of alkali then disproportionation may occur.

---

**Problem 6.21** Calculate the reduction potential value for reduction of $O_2$ to $H_2O$ from the following Latimer diagram.

$$O_2 \xrightarrow{0.70} H_2O_2 \xrightarrow{1.76} H_2O$$ **[C.U. 2000]**

**Ans. :** The half reaction for the reduction of $O_2$ to $H_2O_2$ is

$$O_2 + 2H^+ + 2e \longrightarrow H_2O_2 \qquad E° = +0.70 \text{ Volt}$$

Therefore the free energy change in the process for the involment of 2 electrons = $2 \times 0.70 = 1.40$ V.

Similarly the half reaction for the reduction of $H_2O_2$ to $H_2O$ is $H_2O_2 + 2e \longrightarrow H_2O + O^{2-}$ and the free energy for this process = $2 \times 1.76 = 3.52$ V.

Therefore the total free energy change for the reduction of $O_2$ to $H_2O$ = 1.40 + 3.52 = 4.92 V.

Since the reduction involves 4 electrons change so that reduction potential = $\frac{4.92}{4}$ = 1.23 V.

Alternatively from equation (6.29) the reduction potential (E°) may be evaluated for the reduction of $O_2$ to $H_2O$ as :

$$E° = \frac{n_1E_1^\circ + n_2E_2^\circ}{n_1 + n_2} = \frac{2 \times 0.70 + 2 \times 1.76}{2 + 2} = 1.23 \text{ V.}$$

---

### 6.10.3 Pourbaix Diagrams

Acid-base and redox equilibria are the two broad types of chemical equilibria intimately related. The predominant form of an element which exist in nature (in soil or natural water) will be governed by the both types of equilibria so that in practical field we have a visual representation of the effects of both redox potential and pH on the predominant form of an element. Such a correlation diagram of these two variables as its two axes are known as Pourbaix diagrams which can also be named as E°–pH diagrams or pE–pH diagrams. Simply a **Pourbaix diagram is a representation depicting the conditions of potential and pH under which a given metal species is thermodynamically stable towards oxidation or reduction.** In fact this diagram is a merger of acid-base and redox predominance diagram having wide applicability in nature.

Here we shall illustrate a simplified Pourbaix diagram for iron species which exist in natural waters. In order to construct this we have to consider the following steps :

(a) The reaction, $Fe^{3+}(aq) + e \longrightarrow Fe^{2+}(aq)$ $E° = +0.77$ V is independent of pH since the reaction does not involve $H^+$ ions. This is represented by the horizontal line PQ (Fig. 6.12). If environment e.g. natural water contains a redox system having reduction potential above the line PQ, then the oxidised species i.e. $Fe^{3+}$ will survive. There will be no $Fe^{2+}$ ions in solution and thus the line PQ servs as a boundary that separates the regions between $Fe^{3+}$ and $Fe^{2+}$ ions domination.

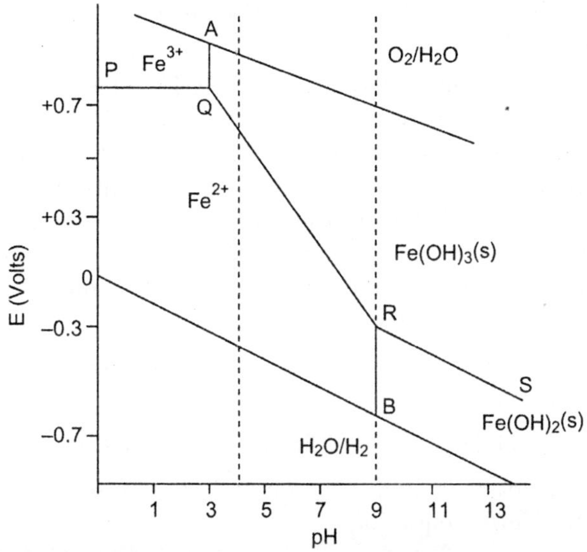


***Figure 6.12** : Pourbaix diagram for iron species in natural waters.*
*The broken lines denote the normal pH range of natural waters in lakes (pH = 4 and pH = 9).*

(b) Another reaction, $Fe^{3+}(aq) + 3H_2O(l) \longrightarrow Fe(OH)_3(s) + 3H^+(aq)$ is to be considered which is not a redox reaction (since no change of oxidation number of any element occurs) but rather an acid-base reaction i.e.

$$Fe^{3+}(aq) + 3OH^-(aq) \longrightarrow Fe(OH)_3(s)$$

The above main reaction is dependent on pH since $H^+$ ions are liberated. According to the Le Chatelier's principle the formation of $Fe(OH)_3(s)$ is favoured at high pH while $Fe^{3+}(aq)$ formation favoured at low pH. The equilibrium concentration of $Fe^{3+}(aq)$ varies with pH and the vertical boundary line AQ represents the pH where the $Fe^{3+}$ ions are dominant.

(c) Thirdly we need to consider the following reaction :

$$Fe(OH)_3(s) + 3H^+(aq) + e \longrightarrow Fe^{2+}(aq) + 3H_2O(l)$$

As $H^+$ ions involved in the half reaction, so this depends on pH. Applying Nernst equation (6.10) we can write,

$$E = E^\circ - \frac{0.0591}{1} \log \frac{[Fe^{2+}(aq)]}{[H^+(aq)]^3}$$

$$= E^\circ - 0.0591 \log[Fe^{2+}(aq)] + 3 \times 0.0591 \log[H^+(aq)]$$

$$= E^\circ - 0.0591 \log[Fe^{2+}(aq)] - 0.1773 \text{ pH}$$

This expression indicates that E decreases with increase of pH as shown by the QR line in the Fig. 6.12. The QR denotes the boundary line above which corresponds to the conditions in which the oxidised species $Fe(OH)_3(s)$ is stable and below the line corresponds to the conditions in which the reduced species i.e. $Fe^{2+}$ is stable.

Many of the boundary lines in Pourbaix diagram are diagonal and during change of pH with potential the lines crosses which indicates that the reactions are **both an acid-base and redox reactions.**

(d) Another reaction is $Fe^{2+}(aq) + 2H_2O(l) \longrightarrow Fe(OH)_2(s) + 2H^+(aq)$ which is not a redox one but pH dependent. At low pH the formation of $Fe^{2+}$ ions is favoured while the higher pH acclerates the formation of $Fe(OH)_2$. The vertical line at pH = 9 (denoted by broken line) divides the regions in which the reactants [i.e. $Fe^{2+}(aq) + H_2O(l)$] and products [i.e. $Fe(OH)_2(s) + 2H^+(aq)$] are dominating in the above reaction. Generally the presence of $H^+$ ions on the left of the reduction couple implies a slope down from left to right of a boundary in the diagram (Fig. 6.12) while $H^+$ ions on the right made an upward slope of a boundary.

(e) We have to consider another boundary line RS that separates $Fe(OH)_3$ and $Fe(OH)_2$ with the following reaction :

$$Fe(OH)_3(s) + H^+(aq) + e \longrightarrow Fe(OH)_2(s) + H_2O(l)$$

Since this reaction involves both electron and $H^+$ ion, the reduction process depends on pH. Applying Nernst equation

$$E = E^\circ - 0.0591 \log \frac{1}{[H^+]}$$

$$= E^\circ - 0.0591 \text{ pH}$$

(f) Another boundary in the predominance diagram (not shown in the Fig. 6.12) is to be considered for the existance of ferrate ion :

$$FeO_4^{2-} + 8H^+ + 3e \longrightarrow Fe^{3+} + 4H_2O \quad E^\circ = +2.20 \text{ V}$$

Applying Nernst equation we can write for the electron and pH dependent half reaction as—

$$E = 2.20 - \frac{0.0591}{3} \log \frac{[Fe^{3+}]}{[FeO_4^{2-}][H^+]^8}$$

$$= 2.20 - 0.1576 \text{ pH} - 0.0197 \log \frac{[Fe^{3+}]}{[FeO_4^{2-}]}$$

The very high half cell potential indicates that $FeO_4^{2-}$ cannot survive in acidic condition. Removal of $H^+$ ions will shift the equilibrium to the left and the stability of $FeO_4^{2-}$ increases so that $FeO_4^{2-}$ will survive in alkaline natural waters.

(g) Finally in the diagram (Fig. 6.12) the two sloping lines indicates the stability field of water. The region between these two lines is sometimes called the **window of redox stability of water.** Any species present with potentials more negative than the lower line will reduce water to $H_2$ while the species present with potentials more positive than the upper line will oxidise it to $O_2$.

*Utility of the diagram*

Biological ecosystem of natural waters can be explained by Pourbaix diagram. Thus when fresh water is in contact with atmospheric $O_2$, the dissolved $O_2$ may oxidise many species present in natural waters due to the following half-reaction having E° value of +1.23 volt.

$$O_2(g) + 4H^+(aq) + 4e \longrightarrow 2H_2O(l) \qquad E^\circ = +1.23 \text{ V}$$

Fig. 6.12 denotes $Fe^{3+}$ will exist in water if the environment is oxidising i.e. water is saturated with plenty of oxygen and the pH of water is low (< 4). If the condition of water as represented by the regions falling below the sloping boundary line QR, the iron present as insoluble $Fe_2O_3$, will enter into the solution by reduction as $Fe^{2+}$ ions. At higher pH, $Fe_2O_3$ would change into $Fe^{2+}$ ions, only possible if strong reducing agents present in natural waters so that survival of $Fe^{2+}$ ions is very unlikely in oxygen-rich waters. Organic pollutants consume a lot of oxygen for its bio-degradation which means that they are acting as a strong reducing agents so that organic pollutants rich waterlogged soils with pH around 4.5, the insoluble $Fe_2O_3$ will be reduced to soluble $Fe^{2+}$ ions. Thus the Pourbaix diagram can be used to understand the physical processes that occur in natural waters of lakes and ponds where temperature gradient exist between the surface and at greater depth of water. Completely oxygenated water is present at the surface and iron will be present as fine particles of $Fe(OH)_3$. At greater depth of water oxygen content is low (reducing environment) where the insoluble $Fe(OH)_3$ or $Fe_2O_3$ is reduced to soluble $Fe^{2+}$ ions which diffuse towards the surface and will be oxidised to form insoluble $Fe(OH)_3$ so that a diffusion process occurs.

Mention should be made here for wide use of Pourbaix diagram are : (1) strong oxidising agents are confined to the top of the diagram, (2) strong reducing agents are placed to the bottom, (3) strong acids are placed to left while strong bases are confined to the right of the diagram.

---

**Problem 6.22** Establish Nernst equation for $O_2(g)/2H_2O(l)$ couple at 25°C for which E° = 1.23 volts. Find the redox potential of the couple at pH7 when the partial pressure of $O_2(g)$ is 0.2 bar at 25°C. **[C.U. 1996, 2014]**

**Ans.** : The half-reaction for $O_2(g)/2H_2O(l)$ is represented as :

$$O_2(g) + 4H^+(aq) + 4e \longrightarrow 2H_2O(l) \qquad E^\circ = +1.23 \text{ volts}$$

According to Nernst equation,

$$E = E^\circ + \frac{0.0591}{n} \log \frac{[Ox]}{[Red]}$$

so that $E = 1.23 + \frac{0.0591}{4} \log \frac{P_{O_2}[H^+]^4}{[H_2O]^2}$

As the concentration of $H_2O$ is unity, so

$$E = 1.23 + \frac{0.0591}{4} \log P_{O_2}[H^+]^4$$

For partial pressure of oxygen is 0.2 bar at 25°C

$$E = 1.23 - 0.0591 \times 0.2 \text{ pH}$$

When the pH of the solution = 7, then

$$E = 1.23 - 0.0591 \times 0.2 \times 7 = 1.147 \text{ Volts.}$$

## 6.10.4 Ellingham diagram : Thermodynamics of metallurgical processes

The choice of a reducing agent for the conversion of a metal oxide to metal during the metallurgical operations is determined by thermodynamical factors. The spontaneity of the reduction process is attended by negative free energy change ($\Delta G°$) and the equilibrium constant (K) of the reaciton is related to the standard free energy change ($\Delta G°$) as :

$$\Delta G° = -RT\ln K$$

The free energy change on the other hand related to the change in entropy ($\Delta S$) and change in enthalpy ($\Delta H$) of the reaction occuring at temperature T by the relation as :

$$\Delta G = \Delta H - T\Delta S$$

and for the standard state we can write,

$$\Delta G° = \Delta H° - T\Delta S°$$

At a specific temperature, higher the positive value of change in entropy, greater is the decrease in free energy change and more easy it is to carry out the reaction.

In order to achieve the decrease in free energy ($\Delta G°$) for the reduction of metal oxide with carbon we have to consider the following reactions :

$C(s) + \frac{1}{2}O_2(g) \longrightarrow CO(g)$ Let free energy change = $\Delta G_1^\circ$

$\frac{1}{2}C(s) + \frac{1}{2}O_2(g) \longrightarrow \frac{1}{2}CO_2(g)$ Let free energy change = $\Delta G_2^\circ$

$CO(g) + \frac{1}{2}O_2(g) \longrightarrow CO_2(g)$ Let free energy change = $\Delta G_3^\circ$

and let the free energy change for the formation of the oxide is $\Delta G_4^\circ$ for the following reaction :

$$xM(s) + \frac{1}{2}O_2(g) \longrightarrow M_xO(s)$$

Then the free energy change for the reduction of metal oxide by the above reducing agents can be calculated by the following reactions :

$M_xO(s) + C(s) \longrightarrow xM(s) + CO(g),$ $\Delta G° = \Delta G_1^\circ - \Delta G_4^\circ$

$M_xO(s) + \frac{1}{2}C(s) \longrightarrow xM(s) + \frac{1}{2}CO_2(g),$ $\Delta G° = \Delta G_2^\circ - \Delta G_4^\circ$

$M_xO(s) + CO(g) \longrightarrow xM(s) + CO_2(g),$ $\Delta G° = \Delta G_3^\circ - \Delta G_4^\circ$

Among the reduction processes of metal oxide which have more negative free energy change and thus be spontaneous. The necessary information is obtained from the **Ellingham diagram which is a graph showing the free energy change (ΔG°) of formation for metal and carbon oxides as a function of temperature (Fig. 6.13).** Most of the moderately active metals and non-metals are reduced from their oxides using common reducing agent carbon and heat. The activity of carbon as a reducing agent increases at high temperature.

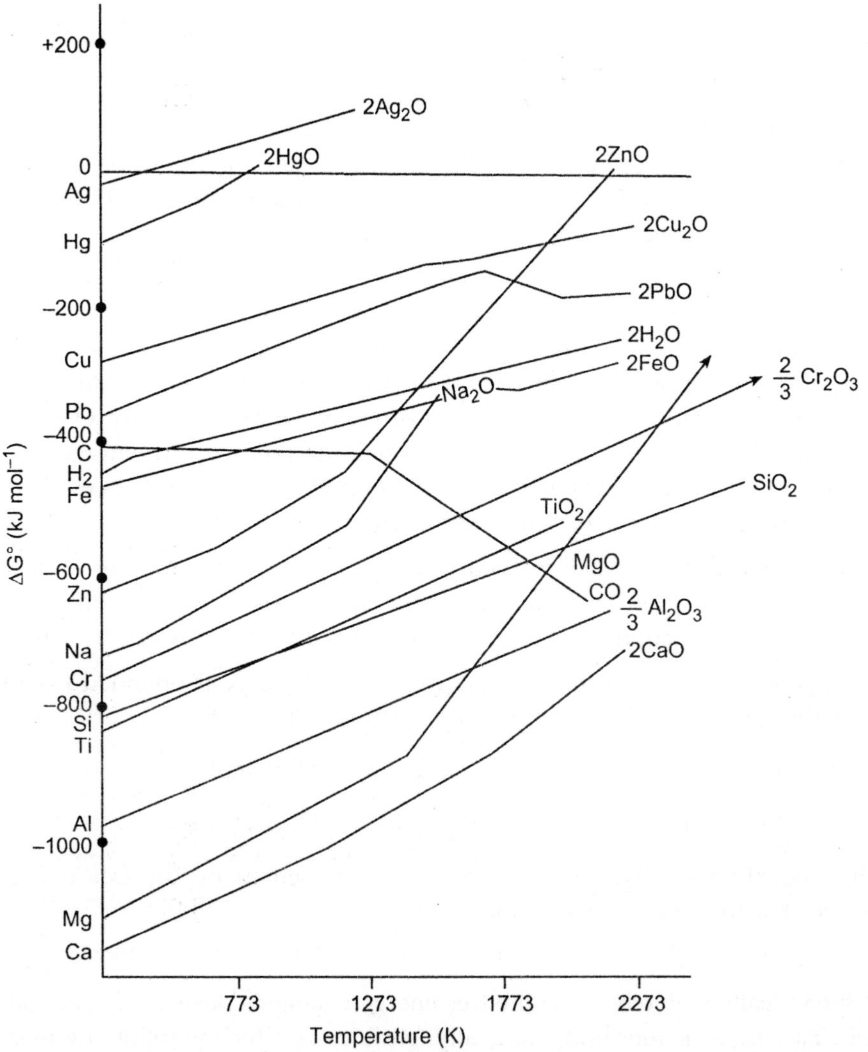


***Figure 6.13** : Ellingham diagram for the reduction of metal oxides.*

Differentiating the equation $\Delta G = \Delta H - T\Delta S$ with respect to T (considering that ΔH does not vary with change in temperature) we get, $d(\Delta G)/dT = -\Delta S$. Thus for a reaction with negative $\Delta S°$, the slope of the line obtained by plotting $\Delta G°$ vs. T will be **positive** while the slope of a similar line for a reaction with positive $\Delta S°$ will be negative. Since oxygen is a gas which has a higher entropy than the metal and the metal oxide, so that there will be a decrease in entropy on going to metal oxide from metal and oxygen and hence the slope of the lines obtained by plotting $\Delta G°$ vs. T will be positive. As $\Delta S° = 0$ for the reaction $C(s) + O_2(g) \longrightarrow CO_2(g)$, the slope of the line is zero and the plot of $\Delta G°$ vs. T will be parallel to the temperature axis. The slope of the line will be negative for the reaction $C(s) + \frac{1}{2}O_2(g) \longrightarrow CO(g)$ for which $\Delta S°$ is naturally positive.

The following points to be noted for Ellingham diagram :

(1) All the $\Delta G°$ vs. T lines have slope upwards as $\Delta G°$ becomes less negative with increase of temperature.

(2) The plots are straight line upto the temperature at which the metals remain in solid state. Large change in entropy occurs on vaporisation and fusion of metals and thus there will be a change in slope of the lines. For example the slope of the lines of Hg–HgO, Mg–MgO and Ca–CaO curves changed appreciably at 629, 1380 and 1713 K which are respectively the boiling points of Hg, Mg and Ca.

(3) With increase in temperature a situation is reached when the lines cross the $\Delta G° = 0$ line. Below the $\Delta G° = 0$ line, the $\Delta G°$ is negative which means that the heats of formation of the oxides are negative and hence the oxides are stable. Above these temperatures the $\Delta G°$ is positive and the oxides will be unstable and will decompose to form metal and oxygen. This is the situation for HgO and $Ag_2O$ (Fig. 6.13) which are unstable and spontaneously decomposed into their constituent elements. For these oxides no reducing agents are required to get the metal which is evident for the self reduction in air of cinnabar to obtain mercury.

(4) Quite frequently one metal is used to reduce the oxide of another metal. Therefore a question arises which of the following is thermodynamically feasible : (a) reduction of $Cr_2O_3$ with Al, (b) reduction of $Al_2O_3$ by Cr.

*Utility of Ellingham diagram*

At a certain temperature say at 1773 K as obtained from the diagram (Fig. 6.13) for the formation of $Al_2O_3$ and $Cr_2O_3$ can be represented as :

$$2Al + \frac{3}{2}O_2 \longrightarrow Al_2O_3 \qquad \Delta G° = -900 \text{ kJ mol}^{-1}$$

$$2Cr + \frac{3}{2}O_2 \longrightarrow Cr_2O_3 \qquad \Delta G° = -500 \text{ kJ mol}^{-1}$$

Therefore the reduction of $Al_2O_3$ with Cr may be represented by the free energy change of 400 kJ $mol^{-1}$ so that the following reaction is :

$$Al_2O_3 + 2Cr \longrightarrow Cr_2O_3 + 2Al \quad \Delta G° = -500 - (-900) = 400 \text{ kJ mol}^{-1}$$

not thermodynamically viable because the free energy change is positive. On the other hand the reduction of $Cr_2O_3$ by Al is thermodynamically feasible since for the following reaction the free energy change is negative.

$$Cr_2O_3 + 2Al \longrightarrow Al_2O_3 + 2Cr \quad \Delta G° = -900 - (-500) = -400 \text{ kJ mol}^{-1}$$

It is to be noted that in the diagram (Fig. 6.13) **the position of the curve for chromium is higher than that for aluminium so that $Cr_2O_3$ will be reduced by Al but the reduction of $Al_2O_3$ by Cr is not possible. In general an oxide can be reduced by the metal which has the position lower in the Ellingham diagram (Fig. 6.13).**

Though in the **pyrometallurgical** processes the Ellingham diagram predicts the feasibility of a reduction reaction, yet it gives no information about the rate of the reaction.

**Problem 6.23** Applying Ellingham diagram predict whether rutile ($TiO_2$) will be reduced by carbon?

**Ans. :** In the Ellingham diagram the curve of carbon is below the curve of titanium at about 1900K. At this high temperature of course $TiO_2$ is reduced but instead of titanium metal, titanium carbide will be formed.

$$TiO_2 + 3C \longrightarrow TiC + 2CO$$

In order to obviate this difficulty rutile is treated with carbon and chlorine at this temperature whereby titanium (IV) chloride is formed in the liquid state.

$$TiO_2(s) + C(s) + 2Cl_2(g) \longrightarrow TiCl_4(l) + CO_2(g)$$

Now $TiCl_4$ is purified by fractional distillation and then reduced by molten magnesium in argon atmosphere at 1000 – 1150°C.

$$TiCl_4 + 2Mg \longrightarrow Ti + 2MgCl_2$$

This is known as Kroll process for the extraction of titanium.

## 6.11 DISPROPORTIONATION AND COMPROPORTIONATION REACTIONS

When a particular species simultaneously acts as an oxidising as well as reducing agent in a redox reaction, the reaction is termed as disproportionation reaction. On the other hand the reverse reaction of disproportionation is called comproportionation i.e. in this case the higher and lower oxidation states of a species combine to form the intermediate oxidation state, e.g.

$$Cl_2 + 2NaOH \underset{\text{comproportionation}}{\overset{\text{disproportionation}}{\rightleftharpoons}} NaCl + NaOCl + H_2O$$

Now a question arises under what situation disproportionation will occur or not.

We have seen in the Latimer diagram (Sec. 6.10.1) that a species will disproportionate when the potential on the right is higher than the potential on the left. Disproportionation can also be judged having the high value of equilibrium constant (K). Let us consider the following cases where disproportionation occurs :

(A) **The Latimer diagram for manganese is of the following :**

$$MnO_4^- \xrightarrow{0.56} MnO_4^{2-} \xrightarrow{2.26} MnO_2 \xrightarrow{0.95} Mn^{3+} \xrightarrow{1.51} Mn^{2+} \xrightarrow{-1.18} Mn$$

It is evident from the above diagram the following disproportionation reactions will occur :

Case I $Mn^{3+} \longrightarrow Mn^{2+} + MnO_2$

Case II $MnO_4^{2-} \longrightarrow MnO_2 + MnO_4^-$

For Case I

| | | |
|---|---|---|
| $Mn^{3+} + e \longrightarrow Mn^{2+}$ | $E^\circ_{red} = +1.51$ V | $\Delta G^\circ = -1.51$ F |
| $Mn^{3+} - e \longrightarrow MnO_2$ | $E^\circ_{ox} = -0.95$ V | $\Delta G^\circ = +0.95$ F |
| $2Mn^{3+} \longrightarrow Mn^{2+} + MnO_2$ | | $\Delta G^\circ = -0.56$ F |

We know that $\Delta G^\circ = -nFE^\circ = -2.303\ RT\ \log K_{disprop}$

$$\therefore \log K_{disprop} = -\frac{\Delta G^\circ}{2.303\ RT} \quad ...(6.30)$$

$$= -\frac{-0.56 \times 96500\ C\ mol^{-1}}{2.303 \times 8.314\ JK^{-1} mol^{-1} \times 298K} = 9.47$$

so that $K_{disprop} = 2.95 \times 10^9$.

For Case II

| | | |
|---|---|---|
| $MnO_4^{2-} + 2e \longrightarrow MnO_2$ | $E^\circ_{red} = +2.26$ V | $\Delta G^\circ = -4.52$ F |
| $2MnO_4^{2-} - 2e \longrightarrow 2MnO_4^-$ | $E^\circ_{ox} = -0.56$ V | $\Delta G^\circ = +1.12$ F |
| $3MnO_4^{2-} \longrightarrow MnO_2 + 2MnO_4^-$ | | $\Delta G^\circ = -3.40$ F |

According to equation (6.30)

$$\log K_{disprop} = \frac{3.40 \times 96500\ C\ mol^{-1}}{2.303 \times 8.314\ JK^{-1} mol^{-1} \times 298K} = 57.50$$

In both the cases disproportionation will be almost complete but in the Case II it is better than that of Case I.

(B) **In basic aqueous solution the Latimer diagram for chlorine system can be represented partly as :**

$$ClO_3^- \xrightarrow{0.30} ClO_2^- \xrightarrow{0.68} ClO^- \xrightarrow{0.42} Cl_2 \xrightarrow{1.35} Cl^-$$

which indicates the following disproportionation reactions will occur :

$$2ClO_2^- \longrightarrow ClO_3^- + ClO^-$$

and $$Cl_2 + 2OH^- \longrightarrow ClO^- + Cl^- + H_2O$$

The relevant $\Delta G^\circ$ values and the equilibrium constant values may be calculated for the two reactions.

(C) **The related Latimer diagram for copper is—**

$$Cu^{2+} \xrightarrow{0.15} Cu^+ \xrightarrow{0.50} Cu$$

which shows that $Cu^+$ disproportionate in the following way as :

$$2Cu^+ \longrightarrow Cu^{2+} + Cu$$

for which $\Delta G^\circ = -0.35$ F and the equilibrium constant K can be calculated as :

$$\log K = \frac{0.35 \times 96500}{2.303 \times 8.314 \times 298} = 5.92$$

(D) **The EMF diagram for oxygen system is depicted as :**

$$O_2 \xrightarrow{0.70} H_2O_2 \xrightarrow{1.76} H_2O$$

so that $H_2O_2$ disproportionate in the following way :

$$2H_2O_2 \longrightarrow 2H_2O + O_2$$

$$H_2O_2 - 2e \longrightarrow O_2 + 2H^+ \qquad E^\circ_{ox} = -0.70\ V \qquad \Delta G^\circ = 0.70\ F$$
$$H_2O_2 + 2e \longrightarrow H_2O + O^{2-} \qquad E^\circ_{red} = 1.76\ V \qquad \Delta G^\circ = -1.76\ F$$
$$2H_2O_2 \longrightarrow H_2O + O_2 + 2H^+ + O^{2-} \qquad \Delta G^\circ = -1.06\ F$$

According to equation (6.30) we get the $K_{disprop}$ as :

$$\log K_{disprop} = \frac{1.06 \times 96500}{2.303 \times 8.314 \times 298} = 17.93$$

The high value of $K_{disprop}$ will suggest that the reaction will proceed to completion.

(E) **The Latimer diagram for gold is of the following :**

$$Au^{3+} \xrightarrow{1.41} Au^+ \xrightarrow{1.68} Au$$

which indicates $Au^+$ will disproportionate as $2Au^+ \longrightarrow Au^{3+} + Au$ and the corresponding equilibrium constant may be calculated.

(F) **The related EMF diagram for silver is :**

$$Ag^{2+} \xrightarrow{1.98} Ag^+ \xrightarrow{0.80} Ag$$

This diagram suggests that there will be no disproportionation occurs for $Ag^+$ to form $Ag^{2+}$ and Ag instead the reverse reaction i.e. the comproportionation reaction will occur ($Ag^{2+} + Ag \longrightarrow 2Ag^+$).

Therefore disproportionation will occur when the net $\Delta G^\circ$ is negative i.e. in this case a species will disproportionate to its nearest neighbours while comproportionation occurs when the net $\Delta G^\circ$ is positive and in this case the two species of different oxidation states combine to form the species having intermediate oxidation state.

---

**Problem 6.24** "Disproportionation and comproportionation reactions are actually redox cell reaction". Justify the statement with suitable examples. **[C.U. 2006, B.U. 2016]**

**Ans. :** Let the partial Latimer diagram for the disproportionation of $MnO_4^{2-}$ species is—

$$MnO_4^- \xrightarrow{0.56} MnO_4^{2-} \xrightarrow{2.26} MnO_2$$

so that $MnO_4^{2-}$ will disproportionate as :

$$3MnO_4^{2-} + 4H^+ \longrightarrow MnO_2 + 2MnO_4^- + 2H_2O$$

The relevant oxidation and reduction reactions are :

$$MnO_4^{2-} - e \longrightarrow MnO_4^- \qquad E^\circ = -0.56\ V$$
$$MnO_4^{2-} + 4H^+ + 2e \longrightarrow MnO_2 + 2H_2O \qquad E^\circ = +2.26\ V$$

so that $E^\circ_{Cell}$ for the disproportionation reaction = 2.26 – 0.56 = 1.70 V and $K_{disprop}$ can be calculated from equation (6.20) as :

$$\log K_{disprop} = \frac{nE^\circ}{0.0591} = \frac{2 \times 1.70}{0.0591} = 57.52$$

The Latimer diagram for silver is :

$$Ag^{2+} \xrightarrow{1.98} Ag^+ \xrightarrow{0.80} Ag$$

If the disproportionation reaction of $Ag^+$ will occur i.e. $2Ag^+ \rightarrow Ag + Ag^{2+}$ then $E°_{Cell}$ = $E°_{Ag^+/Ag} - E°_{Ag^{2+}/Ag^+}$ = 0.80 – 1.98 = –1.18 V and $\Delta G°$ = $-nFE°_{Cell}$ > 0 for which $\log K_{disprop}$ will be negative. It indicates that disproportionation will not be feasible for $Ag^+$ while the reverse reaction i.e. $Ag + Ag^{2+} \longrightarrow 2Ag^+$ will occur in practice which is actually a comproportionation reaction. For this comproportionation reaction $E°_{Cell}$ = 1.98 – 0.80 = 1.18 V so that the $\log K_{comp} = \frac{nE°}{0.0591} = \frac{1 \times 1.18}{0.0591} = 19.97$.

This indicates comproportionation will proceed to completion. Thus we say that the disproportionation and comproportionation reactions are actually termed redox cell reactions.

**Problem 6.25** The standard potentials of Fe(III)/Fe(II) and Sn(IV)/Sn(II) are 0.77 V and 0.15 volt respectively. The emf value for the reaction 2Fe(III) + Sn(II) $\longrightarrow$ 2Fe(II) + Sn(IV) is : 1. 0.47 V 2. 0.92 V 3. 0.62 V 4. 1.07 V. **[NET 2001]**

**Ans. :** 3.

## 6.12 OVERVOLTAGE OR OVERPOTENTIAL

When current passed through an electrolytic cell the decomposition products formed at electrodes convert the electrodes and the electrolyte into a chemical cell. If a substance dissolves as a result of electrolysis then the potential is called **dissolution potential** while if ions are discharged as a result of electrolysis then the potential is called **discharge potential**. When electrode reactions involve a gaseous substance e.g. $H_2$, $O_2$, $Cl_2$ etc., then very often an **irreversible electrode reaction** occur as evolution of gaseous molecules is kinetically delayed. If E′ is reversible electrode potential and E is the applied potential required to cause the electrode reaction, then the difference btween E and E′ is called the **overvoltage or overpotential** (**$E_0$**). The $E_0$ depends on the nature and surface of the electrode. Overvoltage ($E_0$) is a surface phenomena and the magnitude of $E_0$ depends on how ions are discharged and also the rate at which they are discharged. $E_0$ can be represented by the following emperical **Tafel equation :**

$$E_0 = C + B \ln j$$

where j = current density = i/A, i = current and A = area of the electrode.

$B \approx \frac{2RT}{nF}$ and C is of the order of 1 volt for $H_2$ produced on Ag or Hg, and of the order of 0.5 volt for $O_2$ discharged on Pb, Ag and other metals. The **hydrogen overvoltage** is defined as the difference between the potential of the metal electrode surface at which $H_2$ gas is actually evolved and the theoretical i.e. the reversible potential of hydrogen is called the hydrogen over-voltage. Similarly **oxygen overvoltage** may also be defined. The hydrogen overvoltage on Ni, Pb and Hg is 0.22 V, 0.64 V and 0.80 V respectively. The oxygen overvoltage on Fe, Pt and Pb is 0.24 V, 0.43 V and 0.32 V respectively.

## 6.13 REDOX STABILITY IN WATER

Water is a unique solvent which may act as an oxidising agent as also reducing agent. Water may be oxidised to $O_2$ as also reduced to $H_2$ and the relevant potential for these two processes may give us some idea about the thermodynamic stability of a redox species in water.

(a) **Water as an oxidant** : By definition the standard reduction potential of hydrogen electrode is zero volt. Hence in pure water where $[H^+] = 1 \times 10^{-7}$(M), the reduction potential for the process $2H^+(aq) + 2e \rightleftharpoons H_2(g)$ may be calculated by applying Nernst equation at 298K as :

$$E = E^\circ + \frac{0.0591}{n} \log \frac{[Ox]}{[Red]}$$

$$= 0 + \frac{0.0591}{2} \log \frac{[H^+]^2}{[H_2]}$$

$$= 0 + \frac{0.0591}{2} \log \frac{(1 \times 10^{-7})^2}{1} = -0.414 \text{ Volt.}$$

Similarly in alkaline solution having $[H^+] = 10^{-14}$(M) the E value will be –0.827 volt. Therefore any redox couple with E° values lies less than these values at the specified pH may be expected to be oxidised by $H^+$ ion with the evolution of $H_2$. Thus it is evident that the formal reduction potential change from –0.414 volt to –0.827 volt with change in pH from 7 to 14. But due to the hydrogen overvoltage and other kinetic factors a difference of 0.5 to 0.6 volt is often necessary for evolution of hydrogen. The metals like Zn and Fe is unable to evolve hydrogen from water at pH = 7 though they have more negative reduction potential values (more negative than –0.414 volt). This is due to the hydrogen overvoltage as the potential difference is not sufficient enough to overcome the hydrogen overvoltage. In more acidic medium the potential value exceed the overpotential of hydrogen and hydrogen evolution occurs. Metallic sodium has the reduction potential value :

$$Na^+ + e \rightleftharpoons Na \qquad E^\circ = -2.741 \text{ V}$$

so that Na has a pronounced tendency to react with water to evolve $H_2$ :

$$Na + H_2O \longrightarrow Na^+ + OH^- + \tfrac{1}{2}H_2$$

This is an example where water acts as an oxidant or we may write a generalised equation as :

$$M(s) + H^+(aq) \longrightarrow M^+(aq) + \tfrac{1}{2}H_2(g)$$

(b) **Water as a reducing agent** : When water acts as a reductant, it is oxidised to oxygen according to the reaction :

$$O_2(g) + 4H^+ + 4e \longrightarrow 2H_2O(l) \qquad E^\circ = +1.23 \text{ V}$$

so that $E = 1.23 + \frac{0.0591}{n} \log \frac{[H^+]^4}{P_{O_2}(g)}$

At $P_{O_2} = 1$ atm. and at pH = 7,

$$E = 1.23 + \frac{0.0591}{4} \log \frac{[10^{-7}]^4}{1} = 0.816 \text{ V}$$

At $P_{O_2} = 1$ atm and at pH = 14 $\qquad E^\circ = 0.41$ V.

In general we can write E = (1.23 – 0.0591 pH) volt. Thus any redox couple having formal reduction potential value less than 1.23 – 0.0591 pH will be oxidised by $H_2O$ while for any redox couple having formal reduction potential greater than 1.23 – 0.0591 pH will oxidise $H_2O$. This is the thermodynamic criteria of the redox stability of water. But in actual practice due to the oxygen overvoltage this potential value is about 0.3 volt higher. Thus the stability field of water can be demonstrated in terms of E and pH with consideration of hydrogen and oxygen overvoltage (Fig. 6.14).

The positive E° value of +1.23 V in strong acid medium suggest that it is a poor reducing agent. Some strong oxidising agents like $Co^{3+}$ (aq) (E° for $Co^{3+}/Co^{2+}$ = 1.82 V) and $F_2$(g) (E° for $F_2/2F^-$ = 2.87 V) are reduced by water with the evolution of oxygen :

$$4Co^{3+}(aq) + 2H_2O(l) \longrightarrow 4Co^{2+}(aq) + O_2(g) + 4H^+(aq) \qquad E^\circ = 0.59\ V$$

$$2H_2O + 2F_2 \longrightarrow 4F^- + 4H^+ + O_2 \qquad E^\circ = 1.64\ V$$

These are the examples of the reactions where water acts as a reductant.

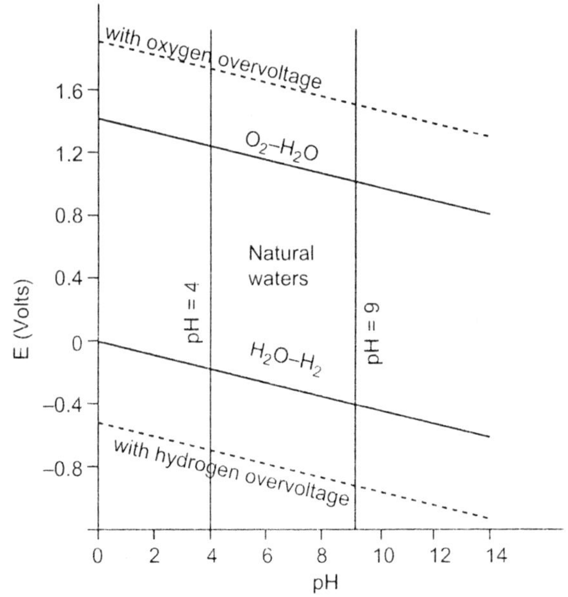


***Figure 6.14*** *: The stability field of water.*

There are several redox couples e.g. $Cr_2O_7^{2-}/Cr^{3+}$ (E° = +1.33 V), $Ce^{4+}/Ce^{3+}$ (E° = +1.68 V) and $MnO_4^-/Mn^{2+}$ (E° = +1.52 V), which have reduction potential values greater than +1.23 V can survive in aqueous solution due to the overvoltage factor. The high overvoltage is attributed to the activation barrier to the transfer of four electrons for the formation of an $O_2$ molecule and thereby an oxygen-oxygen double bond is formed. It is a challange to coordination chemist to search for a good catalyst for oxygen evolution. Some ruthenium complexes have a pivotal role in this field. Nature has its own efficient machine to overcome this dificulty with some mysterious enzyme system in the plant photosynthetic centre established by coordination biochemists.

The stability field of water shown by dots in the figure (Fig. 6.14) indicates the pH range and potential intervals for which water is thermodynamically stable towards both oxidation and reduction. This indicates that redox couples whose pH and potentials lie within this stability field of water are stable and are not destroyed by oxidising and reducing action of water.

## 6.14 CALCULATION OF STABILITY CONSTANT OF A COMPLEX FROM E° VALUES

Let us consider the following E° values :

$$Sn^{4+} + 4e \rightleftharpoons Sn \qquad E^\circ = +0.014\ V$$

$$SnF_6^{2-} + 4e \rightleftharpoons Sn + 6F^- \qquad E^\circ = -0.254\ V$$

The overall stability constant (β) can be calculated as :

$$Sn^{4+} + 6F^- \longrightarrow SnF_6^{2-} \qquad \beta = \frac{[SnF_6^{2-}]}{[Sn^{4+}][F^-]^6}$$

For the above half cell reactions, E° refers to the standard state, i.e. $a_{F^-} = a_{SnF_6^{2-}} = 1$ (M)

$$\text{Therefore, } [Sn^{4+}] = \frac{[SnF_6^{2-}]}{[F^-]^6\beta} = \beta^{-1}$$

Now applying Nernst equation :

$$E^\circ_{complex} = E^\circ_{simple\ ion} + \frac{0.0591}{n}\log[ion]$$

$$\text{i.e. } E^\circ_{SnF_6^{2-},Sn,F^-} = E^\circ_{Sn^{4+},Sn} + \frac{0.0591}{4}\log[Sn^{4+}]$$

$$\text{So that, } -0.254 = 0.014 + \frac{0.0591}{4}\log\beta^{-1}$$

$$= 0.014 - 0.014775\log\beta$$

$$\text{Therefore, } \log\beta = \frac{0.014 + 0.254}{0.014775} = 18.13$$

## 6.15 MASKING AND DEMASKING

It is a routine practice to determine $Cu^{2+}$ iodometrically in a mixture of $Cu^{2+}$ and $Fe^{3+}$. The relevant E° values are :

$$Cu^{2+} + e \rightleftharpoons Cu^+ \qquad E^\circ = +0.153\ V$$

$$Fe^{3+} + e \rightleftharpoons Fe^{2+} \qquad E^\circ = +0.77\ V$$

$$I_2 + 2e \rightleftharpoons 2I^- \qquad E^\circ = +0.54\ V$$

It is evident from the above E° values that $Fe^{3+}$ will oxidise $I^-$ to $I_2$ so that iodometric estimation will not be feasible. In order to obviate this difficulty $NH_4HF_2$ is added prior to the addition of KI. $NH_4HF_2$ maintains the correct pH for iodometric estimation of $Cu^{2+}$ as also $Fe^{3+}$ ions are **masked** by the formation of a soluble strong complex $[FeF_6]^{3-}$ and thus the E° value of

$Fe^{3+}/Fe^{2+}$ is substantially reduced from 0.77 volt so that $Fe^{3+}$ ions are unable to oxidise $I^-$ to $I_2$. But $Cu^{2+}$ will not form any fluoro complex so that it can liberate $I^-$ to $I_2$ due to the insolubility of CuI [Sec. 6.9(E)(I)]. Therefore **masking may be defined as a process in which a species in solution, without physical separation of it or its reaction products, is so transformed that it does not enter into a particular reaction. Demasking is the reverse of masking, in which the masked substance again revert back to enter into a particular reaction**. One of the effective masking agent is $CN^-$ ion. In a mixture of $Mg^{2+}$ and $Zn^{2+}$ if KCN is added, then $Zn^{2+}$ ions are masked by the formation of a cyanide complex :

$$Zn^{2+} + 4CN^- \longrightarrow [Zn(CN)_4]^{2-}$$

But $Mg^{2+}$ will not form any cyanide complex, which can be titrated directly with ethylene diamine tetraacetic acid (EDTA) using proper metal indicator. Then cyanide complex of $Zn^{2+}$ is demasked by formaldehyde-acetic acid solution and then titrated with EDTA solution :

$$[Zn(CN)_4]^{2-} + 4H^+ + 4HCHO \longrightarrow Zn^{2+} + 4HOCH_2CN$$

For more understanding of masking and demasking processes it is better to consult any analytical chemistry book.

---

**Problem 6.26** The reduction potential (E°) for silver is

$Ag^+ + e \rightleftharpoons Ag(s)$ E° = +0.80 V

$Ag^{2+} + e \rightleftharpoons Ag^+$ E° = +1.98 V

(a) Is $Ag^+$ ion will act as a good oxidant or reductant?

(b) Among $F_2$, $F^-$, $I_2$ or $I^-$ which is most suitable for converting $Ag^+$ to $Ag^{2+}$

**Ans.** : (a) $Ag^+$ is a good oxidant.

(b) $F_2$ is strong oxidant which is suitable for converting $Ag^+$ to $Ag^{2+}$.

---

**Problem 6.27** The unfavourable electrochemical reaction among the following is : **[NET 2000]**

1. $Zn + H_2SO_4 \longrightarrow ZnSO_4 + H_2$
2. $KI + Cl_2 \longrightarrow KCl + ICl$
3. $KCl + I_2 \longrightarrow KI + ICl$
4. $Al + 3HCl \longrightarrow AlCl_3 + \frac{3}{2}H_2$

**Ans.** : 3 (because $I^- + I^+ \longrightarrow I_2$ i.e. comproportionation will occur).

---

**Problem 6.28** Among the following, the redox indicator is

1. Bromocresol green 2. Ferroin 3. Phenol red 4. Phenolphthalein. **[NET 2000]**

**Ans.** : 2. Ferroin.

---

**Problem 6.29** The element which is recovered through an electrolytic process is

1. Iron 2. Lead 3. Aluminium 4. Zinc. **[NET 2001]**

**Ans.** : 3. Aluminium.

---

**Problem 6.30** Calculate the solubility product of $Mn(OH)_2$ from the following data :

$E^\circ_{Mn^{2+}/Mn} = -1.18$ V $E^\circ_{Mn(OH)_2,Mn,OH^-} = -1.59$ V

**Ans.** : The solubility product ($K_{sp}$) of $Mn(OH)_2$ is represented as :

$$Mn(OH)_2 \rightleftharpoons Mn^{2+} + 2OH^-$$

i.e. $K_{sp} = [Mn^{2+}][OH^-]^2$

The two half reactions when substracted we get the required equation for solubility product expression.

$$Mn(OH)_2 + 2e \rightleftharpoons Mn + 2OH^- \qquad E° = -1.59\ V$$

$$Mn^{2+} + 2e \rightleftharpoons Mn \qquad E° = -1.18\ V$$

(Subtracting) $Mn(OH)_2 \rightleftharpoons Mn^{2+} + 2OH^-$ $E^{\circ}_{cell} = -0.41\ V$

Then applying Nernst equation,

$$E = -0.41 - \frac{0.0591}{2}\log[Mn^{2+}][OH^-]^2$$

At equilibrium E = 0, so that,

$$\log[Mn^{2+}][OH^-]^2 = \frac{-0.41 \times 2}{0.0591}$$

i.e. $\log K_{sp} = \frac{-0.41 \times 2}{0.0591} = -13.87$

So that $K_{sp} = 1.4 \times 10^{-14}$.

---

**Problem 6.31** For the redox reaction $MnO_4^- + 5Fe^{2+} + 8H^+ \longrightarrow Mn^{2+} + 5Fe^{3+} + 4H_2O$. Calculate the equilibrium constant value and hence comment on whether such reaction can find application in quantitative estimation of iron. [Given $E°_{MnO_4^-/Mn^{2+}}$ = 1.52 volt and $E°_{Fe^{3+}/Fe^{2+}}$ = 0.77 volt] **[C.U. 2007]**

**Ans. :** According to Sec. 6.8(c) the equilibrium constant of the reaction is :

$$\log K = \frac{5(1.52-0.77)}{0.0591} = 63.45$$

Therefore $K \approx 10^{63}$, this high value of K indicates that this reaction may be employed for quantitative estimation of iron.

---

**Problem 6.32** Define comproportionation reaction with example. What is the driving force behind it? **[B.U. 2010]**

**Ans. :** Sec. 6.11. A reaction with two species containing an element in two different oxidation states combined to a product with an intermediate oxidation state of the element is called comproportionation reaction e.g. $Ag^{2+}(aq) + Ag(s) \longrightarrow Ag^+(aq)$

In the Latimer diagram when the E° vlaue in the right hand side of the intermediate species is less than that of the left, then the comproportionation will occur viz.

$M^{2+} \xrightarrow{x} M^+ \xrightarrow{y} M$ where y < x then the net E° = (x–y) V is positve, so that comproportionation reaction $M^{2+} + M \longrightarrow 2M^+$ will occur. For example,

$Ag^{2+} \xrightarrow{1.98} Ag^+ \xrightarrow{0.80} Ag$, so that E° = 1.98 – 0.80 = 1.18 V and $\log K_{comp}$ = $\frac{nE°}{0.0591} = \frac{1 \times 1.18}{0.0591} = 19.97$.

For a comproportionation reaction the net ΔG° value is positive while for a disoporoportionation the same is negative.

---

**Problem 6.33** The $E^\circ_{cell}$ (at 298K) for the Daniell Cell is 1.10V. Calculate the corresponding values of $\Delta G^\circ$ and K and comment on the thermodynamic feasibility of the cell reaction (F = 96500 C $mol^{-1}$; R = $8.314\times10^{-3}$ $kJK^{-1}$ $mol^{-1}$).

**Ans. :** In the Daniell cell the cell reaction is

$Zn(s) + Cu^{2+}(aq) \longrightarrow Zn^{2+}(aq) + Cu(s)$ where $E^\circ_{cell}$ = 1.10 V.

As per equation (6.5)

$\Delta G^\circ = -nFE^\circ_{cell} = -2\times96500\times1.10 = -212300\ Jmol^{-1}$

$\approx -212\ kJ\ mol^{-1}$

Here n = 2 for the cell reaction.

Now according to equation (6.20)

$$\log K = \frac{nE^\circ}{0.0591} = \frac{2\times1.10}{0.0591} = 37.23$$

so that $K = 1.698\times10^{37}$

The large negative value of $\Delta G^\circ$ and the large value of the equilibrium constant suggests that the reaction is thermodynamically viable and virtually goes to completion.

**Problem 6.34** The correct statement for coulometry is—

1. it is based on Faraday's law of electrolysis
2. it is a type of voltammetry
3. it is based on Ohm's law
4. it uses ion selective electrode. **[NET 2015]**

**Hint :** Sec. 1.1.1.

**Ans. :** 1 i.e. coulometry based on Faraday's law of electrolysis.

**Problem 6.35** Given that $E^\circ(Cl_2/Cl^-)$ = 1.35V and $K_{sp}(AgCl) = 10^{-10}$ at 25°C, E° corresponding to the electrode reaction $^1/_2Cl_2(g) + Ag^+(soln.) + e^- \longrightarrow AgCl(s)$ is—

1. 0.75V, 2. 1.05V, 3. 1.65V, 4. 1.95V $\left[\frac{2.303RT}{F} = 0.06V\right]$ **[NET 2015]**

**Ans. :** 4 i.e. 1.95V. Similar to Problem 6.30 and Question number 32.

The solubility of sparingly soluble salt is expressed by an equilibrium constant known as the solubility product, $K_{sp}$.

For metal ions $M^{n+}$ forming a precipitate $MX_n$ with anions $X^-$,

$$M^{n+}(aq) + nX^-(aq) \rightleftharpoons MX_n(s) \quad K_{sp} = [M^{n+}][X^-]^n$$

To generate the overall (non-redox) solubility reaction we use the difference of the two reduction half-reactions :

$$M^{n+}(aq) + ne^- \longrightarrow M(s) \qquad E^\circ(M^{n+}/M)$$

$$MX_n(s) + ne^- \longrightarrow M(s) + nX^-(aq) \qquad E^\circ(MX_n/M, X^-)$$

From which it follows that, $\ln K_{sp} = \frac{nF\{E^\circ(MX_n/M, X^-) - E^\circ(M^{n+}/M)\}}{RT}$

Therefore for the present problem,

$$Ag^+(soln.) + Cl^-(soln.) \rightleftharpoons AgCl(s),\ K_{sp} = [Ag^+][Cl^-] = 10^{-10}$$

Now for this the two reduction half-reactions :

$Ag^+(soln.) + e^- \longrightarrow Ag(s)$ $E° = xV(say)$

$AgCl(s) + e^- \longrightarrow Ag(s) + Cl^-(soln.)$ $E° = yV(say)$

Therefore $\ln K_{sp} = \frac{1F(y-x)}{RT}$, thus $2.303 \log K_{sp} = \frac{F(y-x)}{RT}$

$\therefore \log K_{sp} = \frac{F(y-x)}{2.303RT} = \frac{y-x}{0.06}$ as $\frac{2.303RT}{F} = 0.06$ (given)

$\therefore y - x = \log K_{sp} \times 0.06 = \log 10^{-10} \times 0.06 = -10 \times 0.06 = -0.6$

so that $x - y = 0.6$

Now as stated before,

$Ag^+(soln.) + e^- \longrightarrow Ag(s)$ $E° = x$

$AgCl(s) + e^- \longrightarrow Ag(s) + Cl^-(soln.)$ $E° = y$

---

(Subtracting) : $Ag^+(soln.) - AgCl(s) \longrightarrow -Cl^-(soln.)$ $E° = x - y$

From given data : $\frac{1}{2}Cl_2(g) + e^- \longrightarrow Cl^-(soln.)$ $E° = 1.35$

---

(Adding) $\frac{1}{2}Cl_2(g) + Ag^+(soln.) + e^- \longrightarrow AgCl(s)$ $E° = x - y + 1.35$

$= 0.6 + 1.35 = 1.95V$ (Ans.)

---

**Problem 6.36** For the cell reaction, $Sn(s) + Sn^{4+}(aq) \rightleftharpoons 2Sn^{2+}(aq)$, separate electrode reactions could be written with the respective standard electrode potential data at 25°C as

$Sn^{4+}(aq) + 2e \longrightarrow Sn^{2+}(aq)$, $E° = +0.15V$

$Sn^{2+}(aq) + 2e \longrightarrow Sn(s)$ $E° = -0.14V$

When RT/F is given as 25.7 mV, logarithm of the equilibrium constant (lnK) is obtained as—

1. 22.6, 2. 226, 3. 2.26, 4. $2.26 \times 10^{-1}$ **[NET 2014]**

**Ans. :** 1 i.e. 22.6. Sec. 6.8(c)

Here $Sn^{4+}(aq) + 2e \longrightarrow Sn^{2+}(aq)$, $E° = +0.15V$

$Sn^{2+}(aq) + 2e \longrightarrow Sn(s)$ $E° = -0.14V$

$\therefore$ Predicted reaction is $Sn^{4+}(aq) + Sn(s) \longrightarrow Sn^{2+}(aq) + Sn^{2+}(aq)$

Cell potential $E° = 0.15 - (-0.14) = 0.29V$

We know that $\ln K = \frac{nFE°}{RT}$ here $n = 2$ and $\frac{RT}{F} = 25.7$ mV $= 0.0257V$

$\therefore \ln K = \frac{2 \times 0.29}{0.0257} = 22.568$ (Ans.)

---

**Problem 6.37** What is the potential of hydrogen electrode at pH = 10? **[B.U. 2014]**

**Ans. :** According to Nernst equation

$$E = E° = \frac{0.0591}{n} \log \frac{[OX]}{[Red]}$$

For hydrogen electrode the reaction is $H^+(aq) + e \longrightarrow \frac{1}{2}H_2(g)$

$$\therefore E = 0 + \frac{0.0591}{1} \log \frac{[H^+]}{[H_2(g)]^{1/2}}.$$

Here pH = 10 i.e. $10 = -\log[H^+]$ ∴ $[H^+] = 10^{-10}$ and $[H_2(g)] = 1$.

Thus $E = 0 + 0.0591\log 10^{-10} = -0.0591 \times 10 = -0.591V$

---

**Problem 6.38** Given that E° for the reaction $MnO_4^-(aq) + 8H^+(aq) + 5e \rightleftharpoons Mn^{2+}(aq) + 4H_2O(l)$ is 1.51V, calculate the reduction potential E, in a solution of pH 2.5 and in which the ratio $Mn^{2+} : MnO_4^- = 1 : 100$.

**Ans. :** We know that $pH = -\log[H^+]$

∴ $[H^+] = 10^{-2.5} = 3.162 \times 10^{-3}$ mol dm$^{-3}$

Now according to Nernst equation

$$E = E^\circ + \frac{0.0591}{n}\log\frac{[OX]}{[Red]}$$

$$= 1.51 + \frac{0.0591}{5}\log\frac{[MnO_4^-][H^+]^8}{[Mn^{2+}]}$$

$$= 1.51 + \frac{0.0591}{5}\log\left\{\frac{[MnO_4^-]}{[Mn^{2+}]}[H^+]^8\right\}$$

$$= 1.51 + \frac{0.0591}{5}\log\left\{\frac{100}{1}(3.162\times10^{-3})^8\right\}$$

$$= 1.51 + \frac{0.0591}{5}\log\left\{100(3.162)^8(10^{-3})^8\right\}$$

$$= 1.51 - 0.21 = 1.30V$$

---

**Problem 6.39** Using the following data calculate the equilibrium constant K for the reaction at 298K,

$2Cu^+(aq) \rightleftharpoons Cu^{2+}(aq) + Cu(s)$

$Cu^{2+}(aq) + e \rightleftharpoons Cu^+(aq)$ E° = +0.15V

$Cu^+(aq) + e \rightleftharpoons Cu(s)$ E° = +0.52V

**Ans. :** Predicted reaction is $2Cu^+(aq) \longrightarrow Cu^{2+}(aq) + Cu(s)$

Cell potential is = 0.52 – 0.15 = 0.37V

$$\log K = \frac{n \times \text{cell potential}}{0.0591} = \frac{0.37}{0.0591} = 6.2605$$

so that $K = 1.82 \times 10^6$.

This K value indicates that disproportionation reaction is thermodynamically favourable.

---

**Problem 6.40** Calculate pK for the reaciton $HFe(CN)_6^{3-} \rightleftharpoons H^+ + Fe(CN)_6^{4-}$

Given : $Fe(CN)_6^{3-} + e \rightleftharpoons Fe(CN)_6^{4-}$, E° = 0.365V

$Fe(CN)_6^{3-} + H^+ + e \rightleftharpoons HFe(CN)_6^{3-}$, E° = 0.619V **[B.U. 2014]**

**Ans. :** The equilibrium constant for the reaction $K = \dfrac{[Fe(CN)_6^{4-}][H^+]}{[HFe(CN)_6^{3-}]}$

The predicted reaction from the E° values is :

$Fe(CN)_6^{4-} + Fe(CN)_6^{3-} + H^+ \longrightarrow Fe(CN)_6^{3-} + HFe(CN)_6^{3-}$

i.e. $Fe(CN)_6^{4-} + H^+ \longrightarrow HFe(CN)_6^{3-}$

Cell potential = –0.365 + 0.619 = 0.254V

so that $\log K = \frac{nE^\circ}{0.0591} = \frac{0.254}{0.0591} = 4.30$ (as n = 1)

$\therefore$ K = 19950

Therefore the equilibrium constant for the reaction

$HFe(CN)_6^{3-} \rightleftharpoons H^+ + Fe(CN)_6^{4-}$ will be $K = \frac{1}{19950} = 5.0 \times 10^{-5}$

and the $pK = -\log K = -\log(5.0 \times 10^{-5}) = 5 - 0.6989 = 4.3011$.

---

**Problem 6.41** The following reduction potential diagram depicts the redox chemistry of iron in aqueous solution. Calculate the value of E° for the reduction of $Fe^{3+}$(aq) to Fe°.

$$Fe^{3+}(aq) \xrightarrow{+0.77} Fe^{2+}(aq) \xrightarrow{-0.44} Fe(s)$$

$$\underbrace{Fe^{3+}(aq) \longrightarrow Fe(s)}_{E^\circ}$$

**Ans. : $Fe^{3+}$ to $Fe^{2+}$ is a one electron reduction**

$\Delta G_1^0 = -nFE^\circ = -1 \times 96500 \times 10^{-3} \times 0.77 = -74.3$ kJ per mole of $Fe^{3+}$

**$Fe^{2+}$ to Fe is a two electron reduction**

$\Delta G_2^0 = -nFE^\circ = -2 \times 96500 \times 10^{-3} \times (-0.44) = +84.9$ kJ per mole of $Fe^{2+}$

Therefore the ΔG° for the reduction of $Fe^{3+}$ to Fe is :

$\Delta G^\circ = \Delta G_1^0 + \Delta G_2^0 = -74.3 + 84.9 = +10.6$ kJ per mole of $Fe^{3+}$

**$Fe^{3+}$ to Fe is a three electron reduction.** The standard reduction potential is obtained from the corresponding value of ΔG° as :

$$E^\circ = -\frac{\Delta G^\circ}{nF} = -\frac{10.6}{3 \times 96500 \times 10^{-3}} = -0.0366V.$$

---

**Problem 6.42** From the following Frost-Ebsworth diagram comment on the relative stabilities of the different oxidation states of phosphorus in aqueous media at pH 0.

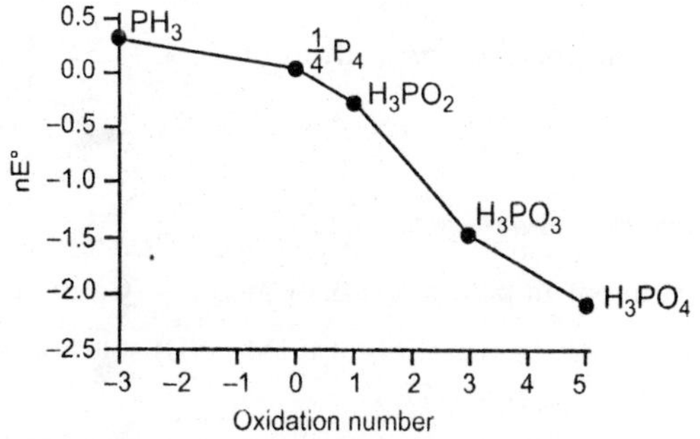


**Ans. :** (a) The most thermodynamically viable state is $H_3PO_4$ having +5 oxidation number of P.

(b) $PH_3$ having phophorus in −3 oxidation number is the least thermodynamically stable species.

(c) At pH 0 in aqueous solution phosphorus will disproportionate to $PH_3$ and $H_3PO_2$. By joining the lines between the points for $PH_3$ and $H_3PO_3$ and between $PH_3$ and $H_3PO_4$ it is evident that $H_3PO_2$ is unstable with respect to disproportionation either to $PH_3$ and $H_3PO_3$, or to $PH_3$ and $H_3PO_4$.

(d) $H_3PO_3$ is stable with respect to disproportionation.

---

## EXERCISE

1. Balance by ion-electron method
$NaNO_3 + NaOH + Al \longrightarrow NH_3 + NaAlO_2 + H_2O$ **[C.U. 2000]**

2. Balance ion-electronically the chemical reaction when potassium iodide is added to an acidified solution of sodium nitrite. **[B.U. 2000]**

**Hint :** $NO_2^- + 2H^+ + e \longrightarrow NO + H_2O \quad 2I^- - 2e \longrightarrow I_2$

3. Show that equilibrium constant K of a redox process is given by the equation $\log K = \frac{n}{0.0591}(E°_1 - E°_2)$ where the terms have usual meaning and n is the number of transferable electrons.

**Hint :** Sec. 6.8(c). Equation (6.19). **[C.U. 2004]**

4. Calculate the redox potentials at the following three stages of titration of 100 ml of $\left(\frac{M}{10}\right)$ $Fe^{2+}$ with $\left(\frac{M}{10}\right)$ $Ce^{4+}$ at 1(N) acidity— (i) 100 ml $Fe^{2+}$ + 99 ml of $Ce^{4+}$, (ii) 100 ml $Fe^{2+}$ + 100 ml of $Ce^{4+}$, (iii) 100 ml $Fe^{2+}$ + 101 ml of $Ce^{4+}$. Given $E^\circ_{Fe^{3+}/Fe^{2+}} = 0.76$ V and $E^\circ_{Ce^{4+}/Ce^{3+}} = 1.76$ V. **[C.U. 2003]**

**Hint :** Sec. 6.9(A).

**Ans.** (i) 0.878 V, (ii) 1.26 V, (iii) 1.642 V.

5. In 1(N) $H_2SO_4$ medium, calculate the redox potential values at the following three stages of titration of 0.1(N) $Fe^{2+}$ with 0.1(N) $KMnO_4$.

(i) 25 ml $Fe^{2+}$ + 24.95 ml $KMnO_4$

(ii) 25 ml $Fe^{2+}$ + 25 ml $KMnO_4$

(iii) 25 ml $Fe^{2+}$ + 25.05 ml $KMnO_4$.

Given $E^\circ_{Fe^{3+}/Fe^{2+}} = 0.77$ volt and $E^\circ_{MnO_4^-/Mn^{2+}} = 1.51$ volt. **[C.U. 2005]**

**Hint :** Sec. 6.9(A).

6. Explain why pH has a larger effect when arsenate reacts with iodine.

**Hint :** Sec. 6.9(C)II.

7. Discuss the equilibrium :

$AsO_3^{3-} + I_2 + H_2O \rightleftharpoons AsO_4^{3-} + 2I^- + 2H^+$ **[B.U. 1986]**

8. "The equivalent weight of $KMnO_4$ is different in acid and alkali media"—Comment. **[B.U. 2008]**

9. Find out the equivalence point potential during the titration of 0.1M Fe(II) solution with 0.1M Ce(IV) solution. Given $E^\circ_{Ce^{4+}/Ce^{3+}} = 1.57$ V and $E^\circ_{Fe^{3+}/Fe^{2+}} = 0.77$ V.

Name one suitable indicator for the above titration. **[B.U. 2009, C.U. 2001]**

**Ans. :** 1.17 V.

10. What do you mean by standard potential of a redox couple? How does it differ from formal potential? **[C.U. 2001]**

**Hint :** Sec. 6.7, 6.7.2 and 6.7.5.

11. What are redox indicators? Give one example with structure both in oxidised and reduced states. **[C.U. 2004]**

**Hint :** Sec. 6.9(B).

12. Explain $Cu^{2+}$ ion readily liberates iodine from iodide but in presence of ethylenediamine it does not.

Given $E^\circ_{Cu^{2+}/Cu^+} = 0.15$ Volt, $E^\circ_{Cu^{2+}/CuI} = 0.87$ Volt, $E^\circ_{\frac{1}{2}I_2/I^-} = 0.54$ Volt.

**[C.U. 2005, B.U. 2008, 2012, 2014, 2016]**

**Hint :** In presence of ethylenediamine $Cu^{2+}$ forms complex i.e.

$$Cu^{2+} + 2en \longrightarrow Cu(en)_2^{2+}$$

so that there will be no free $Cu^{2+}$ ion and $I_2$ liberation is not possible.

13. Reinhardt solution is used in the titration of $Fe^{2+}$ ion with $KMnO_4$ solution in HCl medium—explain.

**[C.U. 2002, B.U. 2008]**

**Hint :** Sec. 6.9(E)(III).

14. A solution of $K_3[Fe(CN)_6]$ cannot oxidise iodide to iodine but it can do so in presence of zinc ion—explain.

$E^\circ_{Fe(CN)_6^{3-}/Fe(CN)_6^{4-}} = +0.36$ Volt, $E^\circ_{\frac{1}{2}I_2/I^-} = +0.54$ Volt **[C.U. 1999, 2002; B.U. 1995, 2006, 2008, 2013]**

**Hint :** Sec. 6.9(E)(II).

15. "Addition of phosphoric acid is essential in the titration of $Fe^{2+}$ ion with dichromate"—Comment.

Given E° for $Cr_2O_7^{=}$ – Cr(III) = 1.33 V, $Fe^{3+}$ – $Fe^{2+}$ = 0.77 V, $I_{nd(ox)} - I_{nd(red)} = 0.76$ V. **[B.U. 2009, 2007]**

**Hint :** Sec. 6.9(B)(II).

16. Give reason why $Cl^-$ ion is oxidised to $Cl_2$ by $KMnO_4$ solution only at low pH.

[Given $MnO_4^-/Mn^{2+} = +1.51V$ and $\frac{1}{2}Cl_2/Cl^- = +1.36V$] **[C.U. 1999]**

**Hint :** Sec. 6.9(E)(III).

17. From the following Latimer diagram, calculate the reduction potential of $ClO_4^-$—HClO.

$$ClO_4^- \xrightarrow{+1.20V} ClO_3^- \xrightarrow{+1.18V} ClO_2^- \xrightarrow{+1.70V} HClO \xrightarrow{+1.63V} Cl_2 \xrightarrow{+1.36V} Cl^-$$

**[C.U. 1997]**

**Hint :**

$$ClO_4^- + 2H^+ + 2e \longrightarrow ClO_3^- + H_2O$$
$$ClO_3^- + 2H^+ + 2e \longrightarrow ClO_2^- + H_2O$$
$$ClO_2^- + 2H^+ + 2e \longrightarrow ClO^- + H_2O$$

$$E^\circ = \frac{n_1E_1^\circ + n_2E_2^\circ + n_3E_3^\circ}{n_1 + n_2 + n_3} = 1.36V$$

18. Use the Latimer diagram to calculate E° for the reduction of $ClO_3^-$ to HClO in aqueous acidic solution.

**Ans. :** 1.44V.

19. Discuss the difference between the ionisation potential of an element and the oxidation potential of it. **[C.U. 1996]**

**Hint :** Ionisation potential is a single step process while oxidation potential is a multistep process.

20. $Cu^+(aq)$ does not exist but CuI is stable—explain using the following data.

$E^\circ_{Cu^{2+}/Cu^+} = 0.153V$; $E^\circ_{Cu^+/Cu} = 0.521V$. **[C.U. 2002, B.U. 2011]**

**Hint :** From the above E° values $Cu^+$ will disproportionate to Cu and $Cu^{2+}$. But CuI is stable due to its insolubility.

21. Standard potential of $Fe^{III}/Fe^{II}$ couple is 0.77 volt. On treatment with sufficient excess of $NH_4HF_2$ both $Fe^{III}$ and $Fe^{II}$ form hexafluoro complex ions. If the $Fe^{III}$ complex is $10^{10}$ times more stable than the $Fe^{II}$ complex, calculate the formal potential of $Fe^{III}/Fe^{II}$ couple under this condition and rationalise quantitatively what would happen on addition of KI into $Fe^{III}$ solution (i) in the absence and (ii) in the presence of $NH_4HF_2$ (Given : Standard potential of $I_2/2I^-$ couple is 0.54 volt). **[C.U. 1998]**

**Hint :** $Fe^{3+} + 6F^- \longrightarrow FeF_6^{3-}$ $\quad K_1 = \dfrac{[FeF_6^{3-}]}{[Fe^{3+}][F^-]^6}$ $\quad \therefore [Fe^{3+}] = \dfrac{[FeF_6^{3-}]}{K_1[F^-]^6}$

$Fe^{2+} + 6F^- \longrightarrow FeF_6^{4-}$ $\quad K_2 = \dfrac{[FeF_6^{4-}]}{[Fe^{2+}][F^-]^6}$ $\quad \therefore [Fe^{2+}] = \dfrac{[FeF_6^{4-}]}{K_2[F^-]^6}$

For $Fe^{III}/Fe^{II}$ system According to condition $K_1 = 10^{10}K_2$

$$E = 0.77 + 0.0591 \log \frac{[Fe^{3+}]}{[Fe^{2+}]}$$

$$= 0.77 + 0.0591 \log \frac{[FeF_6^{3-}]}{K_1[F^-]^6} \bigg/ \frac{[FeF_6^{4-}]}{K_2[F^-]^6}$$

$$= 0.179 + 0.0591 \log \frac{[FeF_6^{3-}]}{[FeF_6^{4-}]}$$

Since the formal potential (0.179 volt) decreases than the E° value of $Fe^{III}/Fe^{II}$ system, the oxidising power of $Fe^{3+}$ decreases and $Fe^{3+}$ ion will not oxidise iodide ion to $I_2$, but in absence of $NH_4HF_2$, $Fe^{3+}$ ion will oxidise iodide to $I_2$ as E° values of the two systems are permitted :

$$Fe^{3+} + 2I^- \longrightarrow I_2 + Fe^{2+}$$

22. What do you know about Ellingham diagram? Illustrate with example. **[C.U. 2007]**

**Ans. :** See Sec. 6.10.4.

23. Calculate the potential of the cell $Zn/ZnCl_2(aq)(1.5M)/AgCl(s)/Ag$ at 25°C. $E^\circ_{Zn^{2+}/Zn} = -0.76$ volt, $E^\circ_{AgCl(s)/Ag} = 0.24$ volt; $\dfrac{2.303\,RT}{F} = 0.059$ at 298K. **[NET 2000]**

24. Indicate the direction in which the following reactions proceed spontaneously and assign them with appropriate name : (i) $Cu^+ + Cu^+ \longrightarrow Cu^{2+} + Cu^0$, (ii) $Ag^+ + Ag^+ \longrightarrow Ag^{2+} + Ag^0$.

$E^\circ_{Cu^{2+}/Cu^+} = 0.16V$, $E^\circ_{Cu^+/Cu^0} = 0.52V$, $E^\circ_{Ag^{2+}/Ag^+} = 1.98V$, $E^\circ_{Ag^+/Ag^0} = 0.80V$. **[C.U. 2004,1995]**

**Hint :** (i) Sec. 6.11(C). Here disproportionation will occur i.e. $2Cu^+ \longrightarrow Cu^{2+} + Cu$.

(ii) Sec. 6.11(F). Here comproportionation will occur i.e. $Ag^{2+} + Ag \longrightarrow 2Ag^+$

25. From the following Latimer diagram, establish that $H_2O_2$ is unstable toward disproportionation.

$$O \xrightarrow{+0.70V} H_2O_2 \xrightarrow{+1.76V} H_2O$$ **[B.U. 2009, 2013]**

**Hint :** Sec. 6.11(D).

26. Cite one example in each case to show that water can behave as oxidising as well as reducing agents. **[C.U. 1997]**

**Hint :** Sec. 6.13.

27. Consider the given portion of the Latimer diagram and give answer of the following :

$$MnO_4^- \xrightarrow{0.564V} MnO_4^{2-} \xrightarrow{1.44V} MnO_2 \xrightarrow{-0.70V} Mn^{2+}$$

(i) Check whether $MnO_4^{2-}$ would disproportionate in solution.

(ii) Would $MnO_4^{2-}$ oxidise $H_2O$ to $O_2$ in acidic solution $[H^+ = 1.0(M)]$

$[E^\circ_{O_2(g)/H_2O(l)} = 1.23V$ and oxygen overvoltage = 0.40V] **[C.U. 2001]**

**Hint :** (i) Sec. 6.11(A). Disproportionation will occur.

(ii) Sec. 6.13(b). Not oxidise $H_2O$ to $O_2$.

28. E° values of $MnO_4^-/Mn^{2+}$ and $Cr_2O_7^{2-}/2Cr^{3+}$ couples are 1.51 and 1.33 volts. At what acidity these two couples will have the same redox potential. **[C.U. 1991]**

29. Explain with examples masking and demasking reactions. **[C.U. 2002]**

30. Based on periodic trends, which two of the following six anions would be the weakest reductant? $CH_3^-$, $Ge^{4-}$, $O^{2-}$, $Se^{2-}$ $F^-$ and $H^-$.

**Hint :** $O^{2-}$ and $F^-$.

31. Calculate the solubility product of $Cu(OH)_2$ from the following reduction potential data.

$E^\circ_{Cu^{2+}/Cu} = 0.337V$, $E^\circ_{Cu(OH)_2/Cu,OH^-} = -0.224V$.

**Ans. :** $1.05 \times 10^{-19}$.

32. Calculate the value of E° for the half reaction— $CuCl(s) + e \rightleftharpoons Cu + Cl^-$

Given $K_{sp}$ of $CuCl(s) = 3 \times 10^{-7}$, $E^\circ_{Cu^+/Cu} = 0.52V$, $E^\circ_{CuCl(s),Cu,Cl^-} = 0.14V$.

**Ans. :** 0.14V.

33. Calculate the solubility product of $Fe(OH)_3$ from the given data :

$E^\circ_{Fe^{3+}/Fe} = -0.036V$, $E^\circ_{Fe(OH)_3,Fe,OH^-} = -0.77V$

**Ans. :** $5.6 \times 10^{-38}$.

34. Consider the given E° values :

$A^{3+} + e \rightleftharpoons A^{2+}$ $E^\circ = +1.42V$

$B^{4+} + 2e \rightleftharpoons B^{2+}$ $E^\circ = +0.40V$

In the titration of $B^{2+}$ with $A^{3+}$, the equivalence point potential will be (a) 0.61V, (b) 0.74V, (c) 0.91V, (d) 1.08V.

**Ans. :** (b).

35. Show that, $E = (5E^\circ_{MnO_4^-/Mn^{2+}} + E^\circ_{Fe^{3+}/Fe^{2+}})/6 - 0.08$ pH where E is the equivalence point potential and $Fe^{2+}$ is oxidised by $MnO_4^-$ and itself is reduced to $Mn^{2+}$.

**Hint :** Problem 6.6.

36. Calculate the equilibrium constant for the reaction of $KMnO_4$ and Fe(II) in acid medium. Given E° for $MnO_4^-/Mn^{2+} = 1.5V$, $Fe^{3+}/Fe^{2+} = 0.77V$. **[B.U. 2010]**

**Hint :** Problem 6.31. $\log K = \frac{nE^\circ}{0.0591}$ (Equation 6.20).

37. Write concisely with example the influence of pH, complexation and precipitation on redox potential. **[B.U. 2010]**

**Hint :** Sec. 6.9(C), (D) and (E).

38. The equilibrium constant for the disproportionation reaction :

$2Cu^+(aq) \rightleftharpoons Cu + Cu^{2+}(aq)$ at 298K is $1.66 \times 10^6 M^{-1}$. If the standard reduction potential of $Cu^{2+}/Cu$ system is 0.337 volt, what is the standard reduction potential for $Cu^+/Cu$ system? **[C.U. 2009]**

**Hint :** $Cu^{2+} \xrightarrow{x \text{ volt}} Cu^{+} \xrightarrow{y \text{ volt}} Cu$ (overall $Cu^{2+} \rightarrow Cu$: 0.337 volt) For disproportionation y > $x$.

$$2Cu^{+} \xrightarrow{K_{disprop}} Cu + Cu^{2+}$$

Now $E^\circ_{Cell} = E^\circ_{Cu^+/Cu} - E^\circ_{Cu^{2+}/Cu^+}$ Anodic reaction : $Cu^+ - e \longrightarrow Cu^{2+}$

$= y - x$ Cathodic reaction : $Cu^+ + e \longrightarrow Cu$

Now we can write, $2F \times 0.337 = 1F \times x + 1F \times y$

i.e. $0.674 = x + y$, so that $x = 0.674 - y$

Therefore $E^\circ_{Cell} = y - (0.674 - y) = 2y - 0.674$.

Now we know that, $\log K_{disprop} = \frac{nE^\circ}{0.0591} = \frac{1 \times (2y - 0.674)}{0.0591}$

i.e. $\log(1.66 \times 10^6) = \frac{2y - 0.674}{0.0591}$

or $6.2201 = \frac{2y - 0.674}{0.0591}$, so that y = 0.52

Therefore the standard reduction potential for $Cu^+/Cu$ is +0.52 volt.

39. From the following Latimer potential diagram draw the Frost diagram for manganese in acid solution. $MnO_4^- \xrightarrow{0.56} MnO_4^{2-} \xrightarrow{2.27} MnO_2 \xrightarrow{0.95} Mn^{3+} \xrightarrow{1.51} Mn^{2+} \xrightarrow{-1.18} Mn$

Then answer the following questions with reasons :

(i) Which is the most thermodynamically (from a redox perspective) stable state?

(ii) Which species potentially disproportionate? **[C.U. 2009]**

**Hint :** The Frost diagram (Fig. 6.11) is given below which is drawn according to Sec. 6.10.2.

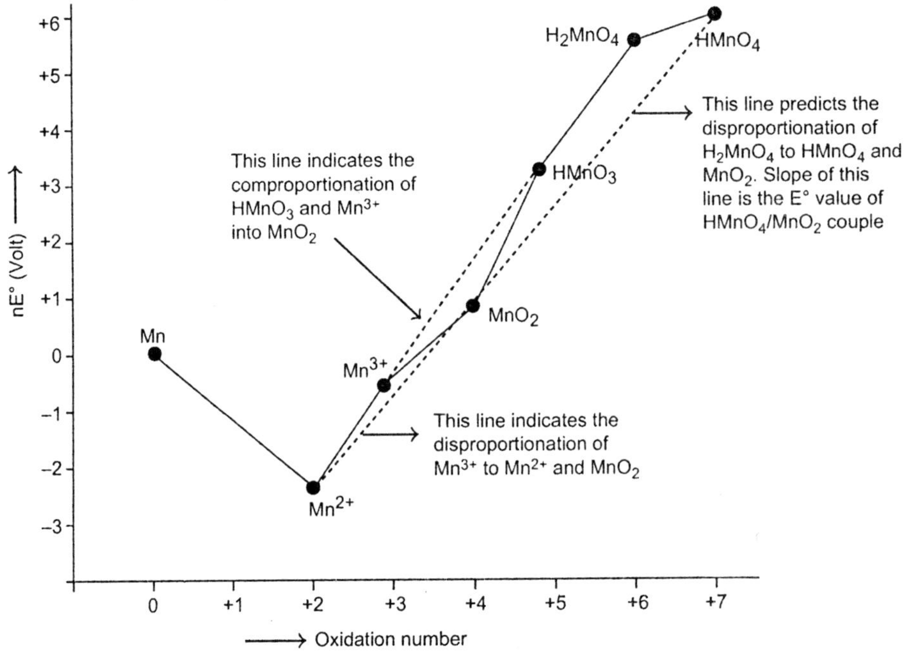

(i) In the Frost diagram the steeper the line joining the two points, the higher is the potential of the corresponding couple. The curve indicates that the line joining $Mn^{2+}$ and $MnO_2$ is more steeper than that of the line joining between $MnO_4^{2-}$ and $MnO_4^-$ which means that $MnO_2/Mn^{2+}$ couple is a better oxidant than that of $MnO_4^-/MnO_4^{2-}$ system. The minima of the curve at $Mn^{2+}$ indicates that it is the most stable species in acid solution.

(ii) Since $Mn^{3+}$ lies above the line joining the points of $Mn^{2+}$ and $MnO_2$, so $Mn^{3+}$ will disproportionate to $Mn^{2+}$ and $MnO_2$. Similarly $Mn^{6+}$ i.e. $H_2MnO_4$ will disproportionate to form $HMnO_3$ and $HMnO_4$.

40. What is Latimer diagram? What information do you get from it? **[B.U. 2011, 2015]**

**Hint :** Sec. 6.10.1.

41. Establish Nernst equation for $MnO_4^-/Mn^{2+}$ in acid medium. **[B.U. 2011, C.U. 2000]**

**Hint :** The relevant equation is $MnO_4^- + 8H^+ + 5e \longrightarrow Mn^{2+} + 4H_2O$

Therefore according to Nernst equation

$$E = E^\circ + \frac{0.0591}{5}\log\frac{[MnO_4^-][H^+]^8}{[Mn^{2+}]}$$

$$= E^\circ + \frac{0.0591}{5}\log\frac{[MnO_4^-]}{[Mn^{2+}]} + \frac{0.591}{5}\log[H^+]^8$$

$$= E^\circ + \frac{0.0591}{5}\log\frac{[MnO_4^-]}{[Mn^{2+}]} - \frac{0.591\times 8}{5}\left(-\log[H^+]\right)$$

$$= E^\circ + \frac{0.0591}{5}\log\frac{[MnO_4^-]}{[Mn^{2+}]} - 0.09456\text{ pH}$$

42. If the ratio of composition of oxidised and reduced species in electrochemical cell is given as $\frac{[O]}{[R]} = e^2$, the correct potential difference will be

1. $E - E^{\circ\prime} = +\frac{2RT}{nF}$ 2. $E - E^{\circ\prime} = -\frac{2RT}{nF}$ 3. $E - E^{\circ\prime} = \frac{RT}{nF}$ 4. $E - E^{\circ\prime} = -\frac{RT}{nF}$ **[NET 2011]**

**Ans. :** 1. (Sec. 6.7.3).

43. What are Frost diagrams? Write down its salient features. **[C.U. 2012]**

**Hint :** Sec. 6.10.2.

44. Find out the E° value of the following half reaction, $Co(CN)_6^{3-} + e \longrightarrow Co(CN)_6^{4-}$.

Given that the overall formation constant of $Co(CN)_6^{4-}$ and $Co(CN)_6^{3-}$ ions are $1\times10^{19}$ and $1\times10^{64}$ respectively. $E^\circ_{Co^{3+}/Co^{2+}} = 1.82V$. **[C.U. 2012]**

**Hint :** Problem 6.15.

45. The standard EMF of the cell Pt, $H_2(g)$ | HCl (soln.) | AgCl(s), Ag(s)
1. increases with T, 2. decreases with T,
3. remains unchanged with T, 4. decreases with [HCl] **[NET 2015]**

**Ans. :** 2 i.e. decreases with T

46. For the oxidation state(s) of sulphur atoms in $S_2O$, consider the following : **A** : –2 and +4, **B** : 0 and +2, **C** : +4 and 0.

The correct answer(s) is/(are)
1. A and B, 2. A and C, 3. B and C, 4. C only. **[NET 2015]**

**Ans. :** 1 i.e. A and B.

47. The correct ΔG for the cell reaction involving steps
$Zn(s) \longrightarrow Zn^{2+}(aq) + 2e^-$
$Cu^{2+}(aq) + 2e^- \longrightarrow Cu(s)$ is—
1. $\Delta G^\circ - RT\ln \frac{a_{Zn^{2+}}}{a_{Cu^{2+}}}$, 2. $\Delta G^\circ + RT\ln \frac{a_{Zn^{2+}}}{a_{Cu(s)}}$,
3. $\Delta G^\circ - RT\ln \frac{a_{Zn(s)}}{a_{Cu^{2+}}}$, 4. $\Delta G^\circ + RT\ln \frac{a_{Zn^{2+}}}{a_{Cu^{2+}}}$

**Ans. :** 4. i.e. $\Delta G^\circ + RT\ln \frac{a_{Zn^{2+}}}{a_{Cu^{2+}}}$. Sec. 6.7.3.

Reaction steps indicate the reaction as

$$Zn(s) + Cu^{2+} \longrightarrow Zn^{2+} + Cu(s)$$

For which the equilibrium constant $K = \frac{a_{Zn^{2+}}a_{Cu(s)}}{a_{Zn(s)}a_{Cu^{2+}}}$

For solid substances activity = 1, so that $K = \frac{a_{Zn^{2+}}}{a_{Cu^{2+}}}$

The free energy change for a spontaneous process, the K is related as $\Delta G = \Delta G^\circ + RT\ln K = \Delta G^\circ + RT\ln \frac{a_{Zn^{2+}}}{a_{Cu^{2+}}}$

48. The oxoacid of phosphorus having P atoms in +4, +3 and +4 oxidation states respectively, is—
1. $H_5P_3O_{10}$, 2. $H_5P_3O_7$, 3. $H_5P_3O_8$, 4. $H_5P_3O_9$ **[NET 2015]**

**Ans. :** 3 i.e. $H_5P_3O_8$.

Sum of the total oxidation states = 0 i.e. for H(+5) + for P(+4+3+4) + for O(−2×8) = 0.

49. Explain the functioning of a redox indicator (avoid detail structure). **[B.U. 2013]**

**Hint :** Sec. 6.9(B). Problem 6.8.

50. $K_2Cr_2O_7$ is an oxidant in acid medium. Express its equivalent weight in terms of its molecular weight. **[B.U. 2013]**

**Hint :** Sec. 6.4. Equivalent weight = $\frac{M}{6}$, where M = molecular weight.

51. Which of the following gas-phase reaction(s) can proceed spontaneously?
$Kr + He^+ \longrightarrow Kr^+ + He$
$Si + Cl^+ \longrightarrow Si^+ + Cl$
$Cl^- + I \longrightarrow Cl + I^-$ **[B.U. 2014]**

**Hint :** $Si + Cl^+ \longrightarrow Si^+ + Cl$ may proceed spontaneously.

52. What do you mean by complementary and non-complementary redox reactions? Give example. **[B.U. 2015]**

**Hint :** Sec. 6.5.

53. 'Chloride is oxidised by permanganate ion in acid medium but not in neutral condition'."Explain. **[B.U. 2015]**

**Hint :** Sec. 6.9(E)(III), Question Number 16.

54. At 298K, $E^{\circ}_{Cell}$ for the Daniell cell is 1.10V. Determine the equilibrium ratio $[Cu^{2+}]/[Zn^{2+}]$.

**Ans. :** $6.90 \times 10^{-38}$.

55. Solutions containing cupic ions readily oxidise potassium iodide to iodine though the standard oxidation potential of the $Cu^{2+} - Cu^{+}$ system (–0.15V) is higher than that of the $2I^{-} - I_2$ system (–0.53V). Justify. **[C.U. 2015]**

**Hint :** Sec. 6.9(E). This is the due to the insolubility of CuI.

56. From Nernst equation derive the expression for equivalence point potential during the titration of $Fe^{2+}$ with standard $KMnO_4$ solution at 1(N) acidity. Hence calculate its magnitude.

Given : $E^{\circ}_{Fe^{3+}/Fe^{2+}} = 0.77V$ and $E^{\circ}_{MnO_4^-/Mn^{2+}} = 1.51V$. **[C.U. 2013]**

**Hint :** Sec. 6.9(A) and Problem 6.6.

$$\text{Equivalent point potential} = \frac{n_1E_1^0 + n_2E_2^0}{n_1 + n_2} = \frac{1\times 0.77 + 5\times 1.51}{1+5} = 1.386V.$$

57. Which one is a stronger oxidant–acidic $K_2Cr_2O_7$ or neutral $K_2Cr_2O_7$? Comment. **[C.U. 2013]**

**Hint :** In acidic medium the E° value for $Cr_2O_7^{2-}/Cr^{3+}$ is $Cr_2O_7^{2-} + 14H^+ + 6e \rightleftharpoons 2Cr^{3+} + 7H_2O$, E° = +1.33V.

The E° value of the system is highly dependent on pH. As per Nernst equation we can write,

$$E = 1.33 - \frac{0.059\times 14}{6}pH - \frac{0.591}{6}\log\frac{[Cr^{3+}]^2}{[Cr_2O_7^{2-}]} \text{ (at 25°C)}.$$

At pH = 7 i.e. in neutral medium the potential falls below 0.5V so that it acts as a weaker oxidant i.e. in acidic medium $K_2Cr_2O_7$ acts as stronger oxidant compared to that of neutral medium.

58. Balance the following reaction by ion-electron method :
$Zn + NaNO_3 + KOH \longrightarrow Na_2ZnO_2 + K_2ZnO_2 + NH_3 + H_2O$. **[C.U. 2014]**

**Hint :** Sec. 6.3. Reaction number (27).

59. Indicate the thermodynamically feasible reduction step(s) from the following reduction potential diagram :

$$FeO_4^{2-} \xrightarrow{+2.30V} Fe^{3+} \xrightarrow{+0.77V} Fe^{2+} \xrightarrow{-0.47V} Fe^0$$

Calculate E° for the reaction $Fe^{3+} \longrightarrow Fe^0$. **[C.U. 2014]**

**Hint :** Sec. 6.10.3.

(a) Applying Nernst equation for the reduction of $FeO_4^{2-}$ to $Fe^{3+}$ we can establish the relation :

$FeO_4^{2-} + 8H^+ + 3e \longrightarrow Fe^{3+} + 4H_2O$ E° = +2.20V

$$E = 2.20 - 0.1576pH - 0.0197\log\frac{[Fe^{3+}]}{[FeO_4^{2-}]}$$

This indicates that $FeO_4^{2-}$ cannot survive in acidic condition and will be reduced to $Fe^{3+}$. Removal of $H^+$ ions will shift the equilibrium to the left and the stability of $FeO_4^{2-}$ increases so that $FeO_4^{2-}$ will survive in alkaline condition. Therefore thermodynamically feasible reduction is possible at low pH but not at high pH.

(b) The reduction of $Fe^{3+} + e \longrightarrow Fe^{2+}$, E° = +0.77V is independent of pH since the reaction does not involve $H^+$ ions. The reduction of $Fe^{3+}$ to $Fe^{2+}$ is not thermodynamically viable.

(c) The reduction potential for the reaction, $Fe^{2+} + 2e \longrightarrow Fe$, $E° = -0.47V$ indicates that the reduction of $Fe^{2+}$ to Fe is also not thermodynamically possible. Calculation of E° value for $Fe^{3+}$ to Fe as per Problem 6.41.

60. For the disproportionation of Cu(I) to Cu and Cu(II), the equilibrium constant value $K = 1.81 \times 10^6$ at 298K. Calculate $\Delta G°$ for the reaction per mole of Cu(I).

**Ans. :** $-17.8$ kJ $mol^{-1}$.

61. For the redox reaction—

$MnO_4^- + 5Fe^{2+} + 8H^+ \rightleftharpoons Mn^{2+} + 5Fe^{3+} + 4H_2O$

Calculate the equilibrium constant value.

[Given E° $MnO_4^-/Mn^{2+} = +1.52$ volt

E° $Fe^{3+}/Fe^{2+} = +0.77$ volt] **[C.U. 2015]**

**Hint :** Sec. 6.8(c). $K \sim 10^{63}$.

62. What will be the effect on the potential of the $Fe^{3+}/Fe^{2+}$ system on adding—
(i) $NH_4HF_2$, (ii) 2, 2′ bipyridyl. **[C.U. 2015]**

**Hint :** (i) Sec. 6.9(D) and Question number 21.

Due to the formation of the complex, $FeF_6^{3-}$ the E° value of $Fe^{3+}/Fe^{2+}$ system is reduced from 0.77 volt to about 0.40 volt.

(ii) 2, 2′ bipyridyl, o-phenanthroline etc. are π acid ligands and can stabilise $Fe^{2+}$ more than $Fe^{3+}$ as the metal to ligand π-back bonding is more favoured for the metal ions of less positive charge. Due to the more favourable complex formation with $Fe^{2+}$ ion here the potential value is substantially increased from 0.77 volt.

$Fe(bpy)_3^{3+} + e \rightleftharpoons Fe(bpy)_3^{2+}$ $E°' = 1.14V$.

63. According to reduction potential value of $Cu^{2+}/Cu^+$ ($E° = +0.15V$) and $\frac{1}{2}I_2/I^-$ ($E° = +0.54V$) system, $Cu^{2+}$ should not oxidise $I^-$. Explain how can iodometric titration of $Cu^{2+}$ is possible? [$K_s$(CuI) ≈ $1 \times 10^{-12}$ at 25°C] **[C.U. 2015]**

**Hint :** Sec. 6.9(E). This is due to the insolubility of CuI thereby E° value of $Cu^{2+}/Cu^+$ is changed from 0.15V to ~ 0.803V and thus iodometric estimation becomes viable in practice.

64. From Frost diagram compare the stability of $Cu^+$ and $Ag^+$ in solution. Given—

(i) $Cu^{2+} \xrightarrow{+0.15V} Cu^+ \xrightarrow{+0.52V} Cu^0$

(ii) $Ag^{2+} \xrightarrow{+1.99V} Ag^+ \xrightarrow{+0.80V} Ag^0$ **[C.U. 2015]**

**Hint :** Sec. 6.10.1 and Sec. 6.10.2.

$Cu^+$ will not be stable in solution and will disproportionate to Cu and $Cu^{2+}$ but $Ag^+$ will be stable in solution.

65. Balance the following equation by ion-electron method :
$Br_2 + NaOH \longrightarrow NaBrO_3 + NaBr + H_2O$ **[C.U. 2015]**

**Hint :** Sec. 6.3. Similar to reaction number (9).

66. Write down the spontaneous cell process and calculate $E^°_{Cell}$ and $\Delta G°$ for the following combination of half-cells :

(a) $Ag^+(aq) + e \rightleftharpoons Ag(s)$ $E° = +0.80V$
$Zn^{2+}(aq) + 2e \rightleftharpoons Zn(s)$ $E° = -0.76V$

(b) $Br_2(aq) + 2e \rightleftharpoons 2Br^-(aq)$ $E° = +1.07V$
$Cl_2(aq) + 2e \rightleftharpoons 2Cl^-(aq)$ $E° = +1.36V$

(c) $Cr_2O_7^{2-}(aq) + 14H^+ + 6e \longrightarrow 2Cr^{3+}(aq) + 7H_2O$ $E° = +1.33V$
$Fe^{3+}(aq) + e \longrightarrow Fe^{2+}(aq)$ $E° = +0.77V$

**Hint :** (a) $2Ag^+(aq) + Zn(s) \longrightarrow 2Ag(s) + Zn^{2+}(aq)$

$E^0_{Cell} = 1.56V$ $\Delta G° = -301$ kJ per mole of reaction.

(b) $Cl_2(aq) + 2Br^-(aq) \longrightarrow Br_2(aq) + 2Cl^-(aq)$

$E^0_{Cell} = 0.29V$ $\Delta G° = -nFE° = -2 \times 96{,}500 \times 10^{-3} \times 0.29 = -55.97$ kJ per mole of reaction.

(c) $Cr_2O_7^{2-}(aq) + 14H^+(aq) + 6Fe^{2+}(aq) \longrightarrow 2Cr^{3+}(aq) + 6Fe^{3+}(aq) + 7H_2O$

$E^0_{Cell} = 0.56V$ $\Delta G° = -nFE° = -6 \times 96{,}500 \times 10^{-3} \times 0.56 = -324.24$ kJ per mole of reaction.

67. Using the following potential diagram at pH = 14, calculate $E^{\circ}_{O_3^-/O_2}$.

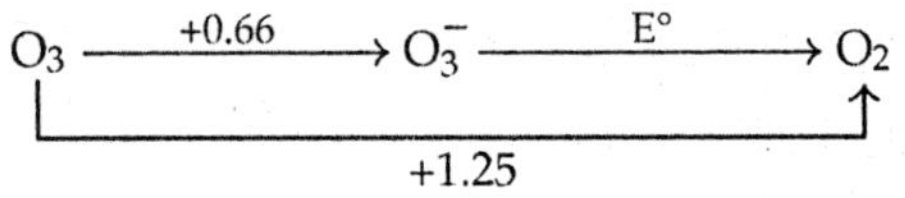


**Hint :** $E^{\circ}_{O_3^-/O_2}$ is 1.84V.

(1) $O_3$ to $O_3^-$ is a one electron reduction

$\therefore \Delta G_1^{\circ} = -nFE° = -1 \times 0.66 \times 96500 \times 10^{-3}$ kJ mol$^{-1}$

(2) $O_3^-$ to $O_2$ is a one electron reduction

$O_3^- + H_2O + e \longrightarrow O_2 + 2OH^-$

$\therefore \Delta G_2^{\circ} = -1 \times E° \times 96500 \times 10^{-3}$ kJ mol$^{-1}$

(3) $O_3$ to $O_2$ is a two electron reduction

$O_2 + H_2O + 2e \longrightarrow O_2 + 2OH^-$

$\therefore \Delta G° = -2 \times 1.25 \times 96500 \times 10^{-3}$ kJ mol$^{-1}$

Thus $\Delta G° = \Delta G_1^{\circ} + \Delta G_2^{\circ}$

$-1 \times 0.66 \times 96500 \times 10^{-3} - 1 \times E° \times 96500 \times 10^{-3} = -2 \times 1.25 \times 96500 \times 10^{-3}$

i.e. $0.66 + E° = 2 \times 1.25$, so that $E° = 1.84V$.

68. Half reactions of a cell are—
$Au(s) + 4Cl^- \longrightarrow AuCl_4^- + 3e$ $Au^{3+} + 3e \longrightarrow Au(s)$

E° for the cell is +0.50V. The equilibrium constant for the reaction $Au^{3+} + 4Cl^- \rightleftharpoons AuCl_4^-$ will be

1. $22 \times 10^{25}$, 2. $2.2 \times 10^{25}$, 3. $25 \times 10^{22}$, 4. $2.5 \times 10^{22}$

**Ans. :** 2 i.e. $2.2 \times 10^{25}$

Here the reduction potential values are :

$AuCl_4^- + 3e \longrightarrow Au(s) + 4Cl^-$ $E° = +1.0V$

$Au^{3+} + 3e \longrightarrow Au(s)$ $E° = 1.50V$

so that the predicted reaction is

$Au(s) + 4Cl^- + Au^{3+} \longrightarrow AuCl_4^- + Au(s)$

i.e. $Au^{3+} + 4Cl^- \longrightarrow AuCl_4^-$

The cell potential is 1.50 – 1.0 = 0.5V

$$\log K = \frac{nE^\circ}{0.0591} = \frac{3 \times 0.5}{0.0591} = 25.38$$

so that $K = 2.291 \times 10^{25}$.

69. Calculate $\Delta G^\circ$ at 298K for the reaction

$2CuCl(s) \rightleftharpoons Cu^{2+}(aq) + 2Cl^-(aq) + Cu(s)$

Given : $2Cu^+(aq) \rightleftharpoons Cu^{2+}(aq) + Cu(s)$ $K = 1.81 \times 10^6$

$CuCl(s) \rightleftharpoons Cu^+(aq) + Cl^-(aq)$ $K_{sp} = 1.72 \times 10^{-7}$

What information you will get from the $\Delta G^\circ$ value about the tendency of the precipitated CuCl to disproportionate?

**Hint :** $\Delta G^\circ = 41.57$ kJ $mol^{-1}$. In this case disproportionation of precipitated CuCl is not thermodynamically favourable.

70. Calculate the reduction potential for the following reaction if the ratio of the concentration of $VO^{2+} : V^{3+}$ is 1 : 2 and the pH of the solution is 2.2.

$VO^{2+}(aq) + 2H^+(aq) + e \rightleftharpoons V^{3+}(aq) + H_2O(l)$

The E° value for the said reaction is 0.34V.

**Hint :** E = 0.07V.

$$E = 0.34 + 0.0591 \log \frac{[VO^{2+}][H^+]^2}{[V^{3+}]}$$

Here pH = 2.2 so that $[H^+] = 10^{-2.2}$ So that, $E = 0.34 + 0.0591\left\{\log\frac{[VO^{2+}]}{[V^{3+}]} + \log[H^+]^2\right\}$

$= 0.34 + 0.0591\{\log(\frac{1}{2}) + \log(10^{-2.2})^2\}$

$= 0.34 + 0.0591\{-0.3010 - 4.4\} = 0.07V.$

71. Comment on the following data :

$Cd^{2+}(aq) + 2e \rightleftharpoons Cd(s)$ $E^\circ = -0.40V$

$[Cd(CN)_4]^{2-}(aq) + 2e \rightleftharpoons Cd(s) + 4CN^-$ $E^\circ = -1.03V.$

**Hint :** From the reduction potential values the predicted reaction will be

$Cd^{2+} + 4CN^- \longrightarrow [Cd(CN)_4]^{2-}$

For which the stability constant of the complex species can be determined as per Problem 6.16.

72. Calculate the overall formation constant for $Fe(CN)_6^{3-}$, given that the overall formation constant for $Fe(CN)_6^{4-}$ is ~$10^{32}$ and that :

$Fe^{3+}(aq) + e \rightleftharpoons Fe^{2+}(aq)$ $E^\circ = +0.77V$

$Fe(CN)_6^{3-}(aq) + e \rightleftharpoons Fe(CN)_6^{4-}(aq)$ $E^\circ = +0.36V$

**Hint :** Formation constant K ~ $10^{39}$.

73. Calculate the ratio of the overall stability constants of the complexes $Fe(ophen)_3^{2+}$ and $Fe(ophen)_3^{3+}$ at 298K.

**Hint :** $\dfrac{\text{Overall stability constant } (\beta) \text{ of } Fe(ophen)_3^{3+}}{\text{Overall stability constant}(\beta) \text{ of } Fe(ophen)_3^{2+}} = 1.2\times10^{-6}$.

74. Prove that disproportionation reaction in aqueous solution is thermodynamically favourable for the system given below :

$Cu^+(aq) \rightleftharpoons Cu^\circ + Cu^{2+}(aq)$

$E^\circ_{Cu^{2+}(aq)/Cu^+(aq)} = 0.153V, E^\circ_{Cu^+(aq)/Cu^\circ} = 0.521V)$

Would you expect similar disproportionation in ammonia solution. **[C.U. 2014]**

**Hint :** Sec. 6.11. The disproportionation reaction $2Cu^+(aq) \longrightarrow Cu^{2+}(aq) + Cu^\circ$ is thermodynamically favourable.

In ammonia solution due to the formation of cupro ammine complex the disproportionation reaction is not viable.

75. Consider the cell :
$Zn \mid Zn^{2+}(a = 0.01) \mid\mid Fe^{2+}(a = 0.001), Fe^{3+}(a = 0.01) \mid Pt$
$E_{Cell} = 1.71V$ at 25°C for the above cell. The equilibrium constant for the reaction $Zn + 2Fe^{3+} \rightleftharpoons Zn^{2+} + 2Fe^{2+}$ at 25°C would be close to
1. $10^{27}$, 2. $10^{54}$, 3. $10^{31}$, 4. $10^{40}$ **[NET 2013]**

**Hint :** The answer is 2 i.e. close to the value of $10^{54}$.

We know $\log K = \dfrac{nE^\circ}{0.0591} = \dfrac{2\times1.71}{0.0591} = 57.868$ i.e. $K = 10^{57}$.

76. Identify the complex ions in sequential order when ferroin is used as an indicator in the titration of iron(II) with potassium dichromate. (phen = 1, 10-phenanthroline)
1. $[Fe(phen)_3]^{2+}$ and $[Fe(phen)_3]^{3+}$, 2. $[Fe(phen)_3]^{3+}$ and $[Fe(phen)_3]^{2+}$
3. $[Fe(CN)_6]^{4-}$ and $[Fe(CN)_6]^{3-}$, 4. $[Fe(CN)_6]^{3-}$ and $[Fe(CN)_6]^{4-}$ **[NET 2015]**

**Ans. :** 1. i.e. $[Fe(phen)_3]^{2+}$ and $[Fe(phen)_3]^{3+}$.

77. What do you mean by Pourbaix diagram? Note down the information obtainable from it. **[B.U. 2016]**

**Hint :** Sec. 6.10.3. Pourbaix diagram is a plot of potential (E) against pH for different species of an element in aqueous solution with electrode potentials from negative potentials at lower end to positive potentials at upper end (Fig. 6.12).

Since positive potentials are the hallmark of strong oxidants and negative potentials of strong reducing agents, such diagrams have strong oxidants at the top and strong reducing agents at the bottom. For example from this diagram the stability ranges in terms of potential and pH of different species of iron such as $FeO_4^{2-}$, Fe(III), Fe(II), $Fe(OH)_3$, $Fe(OH)_2$, Fe etc. can be identified.

78. Define formal potential. In this context explain the statement that 'in neutral solution ferricyanide cannot liberate iodine but in presence of $ZnSO_4$, it can do so'. **[B.U. 2016]**

**Hint :** Sec. 6.7.5 and Sec. 6.9(E)(II). Question number 14.

79. $Cu^+$ disproportionates in aqueous solution but $Ag^+$ does not.—Explain. **[B.U. 2016]**

**Hint :** Sec. 6.11. Question number 24 and 64 and Problem 6.24.

# 7

# COORDINATION CHEMISTRY

The transition metals exist in the central portion of the periodic table which have enormous capability of forming various types of complexes. Therefore a natural question arises what is complex compounds and how do they formed. We shall define complex compounds with reference to simple and double salts.

## 7.1 SIMPLE SALT, DOUBLE SALT AND COMPLEX SALT

A **simple salt** is formed by the interaction of an acid with a base e.g.

$$NaOH + HCl \longrightarrow \underset{\textit{simple salt}}{NaCl} + H_2O$$

$$NaCl \xrightarrow{\text{water}} \underset{\textit{ions}}{\underline{Na^+ + Cl^-}}$$

The existance of simple ions in aqueous solution is verified in case of simple salt.

When two or more simple salts are mixed in requisite proportion and allowed to crystallise together, there will be the formation of a **double salt.** Thus,

$$FeSO_4 + (NH_4)_2SO_4 + \text{Water} \longrightarrow \underset{\textit{double salt (Mohr salt)}}{FeSO_4{\cdot}(NH_4)_2SO_4{\cdot}6H_2O}$$

Aqueous solution of a double salt shows the existance of simple ions e.g. aqueous solution of Mohr salt shows the presence of $Fe^{2+}$, $NH_4^+$ and $SO_4^{2-}$ ions. This type of double salts only exist in the crystalline state.

It has been observed that when two simple salts e.g. $FeSO_4$ and KCN are separately dissolved in water and then mixed, the mixed solution then concentrated and allowed to crystallise, we get a compound which is potassium ferrocyanide, $K_4[Fe(CN)_6]$. The aqueous solution of $K_4[Fe(CN)_6]$ will not give tests for the existance of simple ions viz. $Fe^{2+}$, $CN^-$ instead there will be the existance of $K^+$ and $[Fe(CN)_6]^{4-}$ ions. The $[Fe(CN)_6]^{4-}$ ion is formed by the combination of two simple $Fe^{2+}$ and $CN^-$ ions and so $[Fe(CN)_6]^{4-}$ ion is called a **complex ion.** The salts in which one complex ion exist is called **complex salts** or **complex compounds.** They are also called **complex, coordination compounds** or **Werner complex.** Complex salts retain different identity both in the solid state as also in aqueous solution from simple salts and double salts.

### 7.1.1 Classification of Complexes

Classification of complexes have been made by the reversible dissociation of complex ions. Those complexes which are reversibly dissociated in solution are called **normal** or **imperfect complexes** e.g.

$$[Cd(CN)_4]^{2-} \rightleftharpoons Cd^{2+} + 4CN^-$$

$$[Co(NH_3)_6]^{2+} \rightleftharpoons Co^{2+} + 6NH_3$$

While those complexes which possess sufficient stabilities to retain their identity and are not reversibly dissociated in solution are called **penetration** or **perfect complexes** e.g. $[Cu(CN)_4]^{3-}$, $[Fe(CN)_6]^{4-}$ etc.

Practical application of this classification have been made in qualitative detection of $Cd^{2+}$ in presence of $Cu^{2+}$ ion. Both $Cu^{2+}$ and $Cd^{2+}$ ions are precipitated as sulphide from HCl solution—CuS is black while CdS is yellow. Hence if a small amount of $Cd^{2+}$ ions present in solution, then black colour of CuS will subside the yellow colour of CdS and we are not able to detect $Cd^{2+}$ in presence of $Cu^{2+}$. In order to obviate this difficulty before passing $H_2S$, the mixture of the solution of $Cu^{2+}$ and $Cd^{2+}$ is treated with KCN solution drop by drop so that the blue colour of the solution is discharged due to the formation of the following complexes :

$$Cu^{2+} + 4KCN \longrightarrow [Cu(CN)_4]^{3-} + 4K^+$$

$$Cd^{2+} + 4KCN \longrightarrow [Cd(CN)_4]^{2-} + 4K^+$$

$[Cu(CN)_4]^{3-}$ is of penetration or perfect type while $[Cd(CN)_4]^{2-}$ is of normal or imperfect category so that it will further dissociate in aqueous solution by the above way and so when $H_2S$ is passed to this solution, there will be the precipitation of only yellow CdS but not CuS.

### 7.1.2 Detection of complex formation in solution

What are the evidences by which we can prove that there might be the formation of coordination complexes? The following points are to be noted in order to detect complex formation in solution :

(a) **Sudden change in colour and solubility :** If aqueous ammonia solution is added to a solution of copper (II) sulphate, there will be the precipitation of greenish-blue basic $CuSO_4$ i.e. $CuSO_4.Cu(OH)_2$. If addition of ammonia solution is continued, this precipitate is dissolved due to the formation of the cupro ammine complex and the solution turns deep blue.

$$2CuSO_4 + 2NH_3 + 2H_2O \longrightarrow \underset{\text{greenish-blue precipitate of basic } CuSO_4}{CuSO_4\cdot Cu(OH)_2\downarrow} + (NH_4)_2SO_4$$

$$CuSO_4 \cdot Cu(OH)_2 + (NH_4)_2SO_4 + 6NH_3 \longrightarrow \underset{\text{deep blue solution, cupro ammine complex}}{2[Cu(NH_3)_4]SO_4} + 2H_2O$$

This sudden change in colour and solubility indicates the formation of a complex in aqueous solution.

AgCN is a practically insoluble substance, this will go into the solution by the addition of KCN due to the formation of the following potassium argentocyanide complex.

$$\underset{insoluble}{AgCN} + KCN \longrightarrow \underset{soluble}{K[Ag(CN)_2]}$$

In qualitative analysis of chloride ion, there will be the formation of white precipitate of AgCl when $AgNO_3$ solution is added to a chloride salt solution. AgCl is practically an insoluble substance (solubility 0.0002% only). When excess ammonia solution is added to this precipitate, AgCl goes into solution due to the formation of argento ammine complex.

$$AgCl + 2NH_3 \longrightarrow [Ag(NH_3)_2]Cl$$

AgCl also dissolves in $Na_2S_2O_3$ solution due to the formation of the following thiosulphato complex.

$$AgCl + 2Na_2S_2O_3 \longrightarrow Na_3[Ag(S_2O_3)_2] + NaCl$$

The solubility of AgCl may also occur in other reagents also. All these indicate the formation of complexes.

(b) **Drop in conductivity :** Conductivity of a solution depends upon the nature of the ions as also the number of the ions. As mentioned earlier if we measure the conductivity of the mixed solution of $FeSO_4$ and KCN and after a certain lapse of time if we measure the conductivity of the solution again, then there will be the drop in conductivity due to the formation of the complex, $K_4[Fe(CN)_6]$. Because $Fe(CN)_6^{4-}$ ion is formed by the combination of $Fe^{2+}$ and $CN^-$ ions and so the number of ions decreases, thereby conductivity decreases. Therefore conductivity measurement is one of the important diagnostic tool for the detection and formulation of complexes. Molar conductivity of some of the complexes are given in Table 7.1. This indicates that molar conductivity approximately increases with the number of ions in solution.

(c) **Absorption spectral study :** Transition metals are good complex former and they absorb light in visible and ultraviolet regions mainly for electronic transitions between various energy levels. When the absorption mainly occurs in the visible region, there is a colour change indicating complex formation. In most of the cases the intensity of absorption increases many times due to the replacement of coordinated water molecules by more basic ligands and the intensity of the colour also increases e.g.

$$\underset{\text{light blue}}{[Cu(H_2O)_6]^{2+}} + 4NH_3 \rightleftharpoons \underset{\text{deep blue}}{[Cu(NH_3)_4(H_2O)_2]^{2+}} + 4H_2O$$

$$\underset{\text{light blue}}{[Cu(H_2O)_6]^{2+}} + 3en \rightleftharpoons \underset{\text{deep blue}}{[Cu(en)_3]^{2+}} + 6H_2O \qquad \text{[en = ethylenediamine]}$$

**Table 7.1 : Molar conductivity (0.001M solution) of some of the complexes**

| Compound | Molar conductivity ($ohm^{-1}$ $cm^2$) | No. of ions | Formulation of the complex |
|---|---|---|---|
| $CoCl_3.6NH_3$ | 431.6 | 4 | $[Co(NH_3)_6]Cl_3$ |
| $CoCl_3.5NH_3$ | 261.3 | 3 | $[Co(NH_3)_5Cl]Cl_2$ |
| $PtCl_4.6NH_3$ | 523.0 | 5 | $[Pt(NH_3)_6]Cl_4$ |
| $PtCl_4.5NH_3$ | 404.0 | 4 | $[Pt(NH_3)_5Cl]Cl_3$ |
| $PtCl_4.4NH_3$ | 229.0 | 3 | $[Pt(NH_3)_4Cl_2]Cl_2$ |
| $PtCl_4.3NH_3$ | 97.0 | 2 | $[Pt(NH_3)_3Cl_3]Cl$ |
| $PtCl_4.2NH_3$ | 7.0 | 0 | $[Pt(NH_3)_2Cl_4]$ |
| $K_2PtCl_6$ | 256.0 | 3 | $K_2[PtCl_6]$ |

(d) **Change in pH :** Generally compounds containing acidic functional groups can form complexes with metal ions and such groups release proton in solution and hence hydrogen ion concentration of a solution can be utilised for the detection of complex ion. For example 8-hydroxyquinoline, salicylaldehyde etc. can form complexes with metal ions viz.,

$$Al^{3+} + 3oxinH \longrightarrow Al(oxin)_3 + 3H^+$$

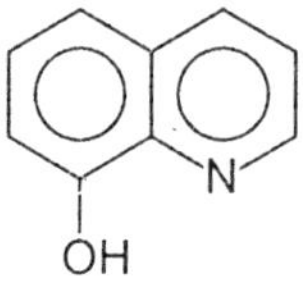

*8-hydroxyquinoline (oxinH)*

Thus measuring the variation of pH before and after the formation of a complex, we are able to detect the existance of complex ions in solution.

Besides above mentioned methods, several others e.g. polarographic method, emf measurement, ion exchange process etc. are available for the detection of complex formation in solution.

## 7.2 THEORY OF COORDINATION COMPOUNDS—WERNER'S COORDINATION THEORY

The early observation by Tassaert was the accidental isolation of the orange coloured cobalt--amine compound, $CoCl_3.6NH_3$ in 1798 was the landmark in the understanding of coordination compounds. Practically it had to wait for about 100 years to have an proper understanding of Tassaert's observation because in 1893 Werner proposed his famous coordination theory.

Within about 50 years of the discovery of Tassaert compound, $CoCl_3.6NH_3$ serveral different cobalt-amine compounds were isolated by different chemists as also different other compounds apart from amine complexes were synthesised.

In order to explain the formation of such compounds Blomstrand put forward his **chain theory** in analogy with the **catenating property of tetracovalent carbon** having chain structure in higher hydrocarbons. This idea was extended by **J. M. Jorgensen** to explain the cobalt-amine and platinum-amine compounds. Comparing the chain structure like $CH_3$–$CH_2$–$CH_2$–$CH_2$–, they suggested the following chain structure of those amine compounds considering the valence of nitrogen is 5 and that of cobalt is 3 :

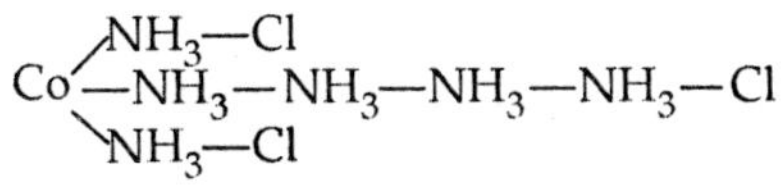

($CoCl_3.6NH_3$)

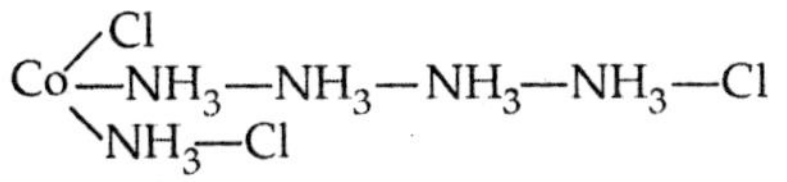

($CoCl_3.5NH_3$)

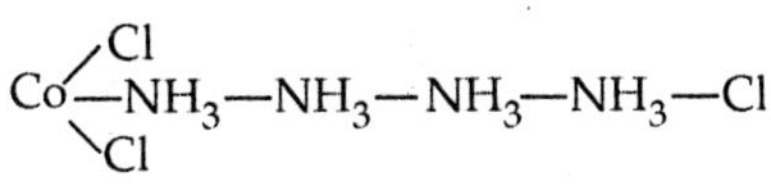

($CoCl_3.4NH_3$)

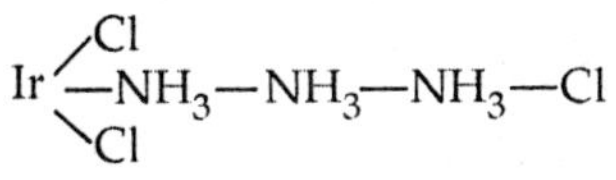

($IrCl_3.3NH_3$)

But according to **Arrhenius** the above compounds having chain structure undergo ionisation and will precipitate as AgCl by the addition of $AgNO_3$. Unfortunately several above compounds will not give any precipitate of AgCl by the addition of $AgNO_3$. It was early suggested that Co–Cl or Ir–Cl bonds will not be ionised but according to Arrhenius ionisation theory the Cl atom which is joined in the $(NH_3)_n$ chain must ionise and give precipitate of AgCl. In fact in case of $IrCl_3.3NH_3$ no precipitate of AgCl was obtained by the addition of $AgNO_3$ and thus the chain theory failed to explain the properties of those compounds.

The influence of conception of organic chemistry on catenation property misdirected Jorgensen. He followed the structure of ethylenediamine dichloride, ClH–$NH_2$–$CH_2$–$CH_2$–$NH_2$–HCl and ascribed different valences of cobalt mainly and other metal ions. He was unable to think of anything outside the chain theory to explain the then complicated compounds, which Werner solved with some bold propositions.

In order to explain the properties of those compounds Alfred Werner put forward his famous theory of coordination compounds in 1893. In recognition of his pioneering work in the field of complexes he was awarded the Nobel prize in chemistry in 1913. This was the first example of Nobel prize in inorganic chemistry. In his attempt to explain the experimental facts gathered with cobaltammines and other similar compounds, Werner not only suggested the formulation of those compounds but also developed his theory of the formation of complexes which is still applicable today.

His theory based on the following postulates :

1. Every complex species has a central metal ion.
2. Metals possess two types of valencies :
   (a) Primary, principal or ionisable valency,
   (b) Secondary, auxiliary or non-ionisable valency.

   Essentially the **primary valency** corresponds to the **oxidation state** of the metal ion while **secondary valency** corresponds to **coordination number. The maximum number of atoms, groups or ions that may be accommodated around the central metal is known as the coordination number (C.N.) of the central metal which is a characteristic of a particular metal and the groups or ions are called ligands.** The

C.N. depends on the nature of the metal as also the nature of the ligands. The common C.N. of some of the representative metal ions are given below :

| Metal ion | Coordination number |
|---|---|
| $Cu^{2+}$ | 4 |
| $Fe^{3+}$ | 6 |
| $Co^{3+}$ | 6 |
| $Pt^{2+}$ | 4 |
| $Pt^{4+}$ | 6 |
| $Ni^{2+}$ | 4 |
| $Cd^{2+}$ | 4 |

The sphere enclosing the metal and the coordinated groups is called the **coordination sphere** which is denoted by square bracket. Werner represented $CoCl_3.6NH_3$ as $[Co(NH_3)_6]Cl_3$ where $[Co(NH_3)_6]^{3+}$ denotes the coordination sphere.

3. Primary valencies may be satisfied either by positive or negative ions while secondary valencies may be satisfied by either positive or negative ions or neutral molecules.
4. Primary valencies have no directional property but secondary valencies must have directional property. Due to the directional property of secondary valencies the phenomena of isomerism arises in case of coordination compounds. Thus in case of $CoCl_3.6NH_3$, the six ammonia molecules are directed towards the six corners of an octahedron and the three chloride ions are in anywhere under the influence of $Co^{3+}$ ion.
5. The net charge on the complex ion is the resultant of the charges borne by the central metal ion and the ligands e.g.

$$Fe^{3+} + 6CN^- \longrightarrow [Fe(CN)_6]^{3-}$$

Net charge on the complex = –6 + 3 = –3

$$Co^{3+} + 6NH_3 \longrightarrow [Co(NH_3)_6]^{3+}$$

So that net charge on the complex ion = +3 + 6 × 0 = +3.

The significant aspect of Werner's theory is the statement of the directional nature of the secondary valencies which results the existance of stereoisomerism of inorganic compounds. Werner was the first who established optical isomerism in coordination compounds. At the time of Werner there was no sophisticated instrument available to elucidate the structure of coordination complexes. He had to establish the property and stereochemistry of the complexes on the basis of some synthetic routes.

### 7.2.1 Verification of Werner's theory in case of cobalt ammines

Tassaert made the begining of coordination chemistry in 1798 by preparing an orange coloured compound $CoCl_3.6NH_3$. The formation of this compound by Tassaert was quite intriguing in terms of valency considerations because at that time it was difficult to establish the structure of this type of compound. Later on several compounds of this type were synthesized and Werner finally elucidated the structure of the following cobaltammine compounds :

I. **Luteocobaltic chloride, $CoCl_3.6NH_3$ (Orange yellow)**

II. **Roseocobaltic choride, $CoCl_3.5NH_3.H_2O$ (Pink)**

III. **Purpureocobaltic chloride, $CoCl_3.5HN_3$ (Violet)**

IV. **Praseocobaltic chloride and violeocobaltic chloride, $CoCl_3.4NH_3$ (Praseo is green while violeo is violet)**

All the above compounds were treated with conc. $H_2SO_4$ and $AgNO_3$ solution and also conductivity of the aqueous solution were measured and following observations were obtained :

| Compound | Treatment of conc. $H_2SO_4$ | Treatment of $AgNO_3$ | Molar conductivity $ohm^{-1}cm^2$ | No. of ions | Formulation |
|---|---|---|---|---|---|
| I | Compound obtained as $Co_2(SO_4)_3.12NH_3$ where Co : $NH_3$ = 2 : 12 = 1 : 6 | All the $Cl^-$ ions are precipitated as AgCl | 431.6 | 4 | $[Co(NH_3)_6]Cl_3$ |
| II | Compound obtained as $Co_2(SO_4)_3.10NH_3.5H_2O$ which on slight heating converted to $Co_2(SO_4)_3.10NH_3.2H_2O$ where Co : $NH_3$ : $H_2O$ = 2 : 10 : 2 = 1 : 5 : 1 | All the $Cl^-$ ions are precipitated as AgCl | — | 4 | $[Co(NH_3)_5H_2O]Cl_3$ |
| III | Compound obtained as $CoClSO_4.5NH_3$ where Co : $NH_3$ : Cl = 1 : 5 : 1 | Only two $Cl^-$ ions are precipitated as AgCl | 261.3 | 3 | $[Co(NH_3)_5Cl]Cl_2$ |
| IV | Compound obtained as $Co_2Cl_4SO_4.8NH_3$ where Co : $NH_3$ : Cl = 2 : 8 : 4 = 1 : 4 : 2 | Only one $Cl^-$ ion is precipitated as AgCl | 102.0 | 2 | $[Co(NH_3)_4Cl_2]Cl$ |

In all these compounds the C.N. of the central cobalt (III) is 6 and in compound (I) all the 6 secondary valencies are directed towards the six corners of an octahedron and is satisfied by six $NH_3$ molecules. The primary valencies of central cobalt (III) is satisfied by three $Cl^-$ ions (Fig. 7.1). The atoms, molecules or ions satisfying the secondary valencies of the central metal ion conventionally written in a square bracket e.g. $[Co(NH_3)_6]^{3+}$. In compound (II) the secondary valencies of the metal ion is satisfied by five $NH_3$ and one $H_2O$ molecules and the primary valencies are satisfied by three $Cl^-$ ions (Fig. 7.1). In compounds (III) and (IV) the secondary valencies are satisfied by five $NH_3$ plus one $Cl^-$ ion and four $NH_3$ plus two $Cl^-$ ions respectively (Fig. 7.1). All the complexes have octahedral geometry.

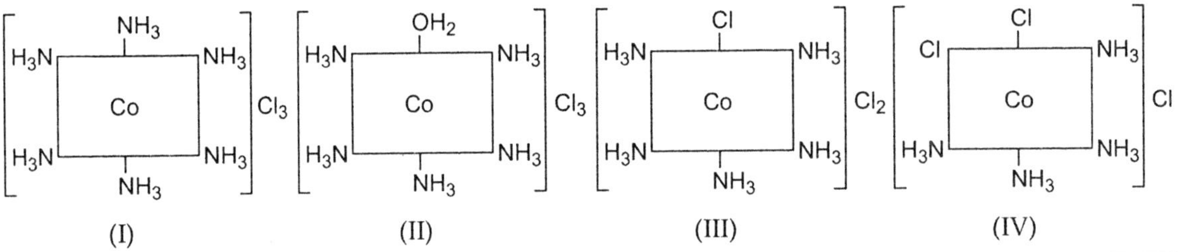


*Figure 7.1 : Structure of different cobaltammine complexes.*

Most important aspect of Werner's theory is the establishment of directional property of secondary valencies of central metal ions which forms the **coordination sphere** and denoted by square bracket which is also called **first sphere of attraction**. This is also called as **coordination zone**. Thus the concept of primary and secondary valencies developed by Werner brought a new horizon in the field of inorganic chemistry.

## 7.3 SIDGWICK ELECTRONIC INTERPRETATION OF COORDINATION

The formation of a complex is regarded as the formation of the donor-acceptor bond of the Lewis acid-base type where the acceptor is the central metal ion while the donors are the ligands. Electronic theory of valency was not known when Werner put forward his coordination theory. Sidgwick made notable contribution in this field. According to Sidgwick all molecules or ions capable of being coordinated to the central metal ion have at least one electron pair i.e. ligands must have a donor atom which possess at least a lone pair of electron. The secondary valency of the central metal ion arises by the donation of electron pairs by such atoms to the central metal ion which is deficient in electron. Therefore ligands are considered as **donors** while metal ions are considered as **acceptors**.

In cobaltammine complexes e.g. $CoCl_3.6NH_3$, the six $NH_3$ molecules will donate six pairs of electron to the central $Co^{3+}$ ion. Naturally according to Sidgwick's view the $Co^{3+}$ ion will be negatively charged with respect to six $NH_3$ molecules. Therefore Sidgwick electronic interpretation of coordination is unable to explain the stability of complexes. In order to explain the stability of complexes a new theory is developed which is called **theory of metal-ligand bonding**. This will be discussed at the appropriate place of this chapter.

## 7.4 CLASSIFICATION OF LIGANDS

The classification may be made in several ways according to coordination motif of ligands. The number of donor atom(s) is sometimes called the **denticity** of the ligand. Thus when a ligand has one donor atom it is called **unidentate** or **monodentate** i.e. a monodentate or unidentate ligand will occupy one stereochemical position of a metal ion. When a ligand will occupy two positions of a metal ion it is called **bidentate** or **didentate** and so on tridentate, tetradentate, pentadentate, hexadentate ligands have 3, 4, 5, 6 donor centres respectively to the metal ion (Table 7.2). Ligands having more than one donor centres are called **polydentate ligands** and thus bidentate to hexadentate all ligands are collectively known as polydentate ligands. Polydentate ligands may be homo or hetero atom donors.

**Table 7.2 : Some important coordinating ligands**

| Type | Name | Abbreviation | Structure showing linkage sites to metal by → from the donor atom /atoms (abbreviation) | Character |
|---|---|---|---|---|
| **Monodentate** | Water | aquo | 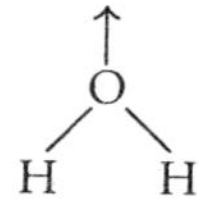 | Neutral O donor |
| | Ammonia | ammine | 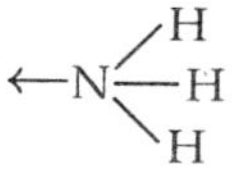 | Neutral N donor |
| | Pyridine | py | 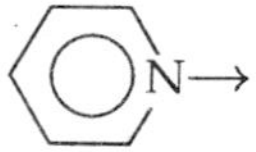 | Neutral N donor |
| | Triphenylphosphine | $Ph_3P$ | $(C_6H_5)_3P$ | Neutral P donor |
| **Bidentate** | Ethylenediamine $H_2NCH_2CH_2NH_2$ | en | $\underset{\downarrow}{H_2N}CH_2CH_2\underset{\downarrow}{NH_2}$ | Neutral NN donor |
| | Biguanide 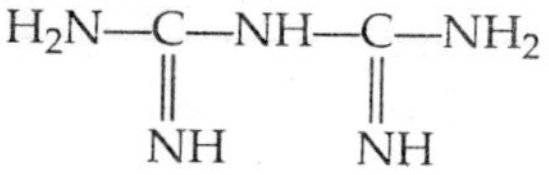 | BigH | 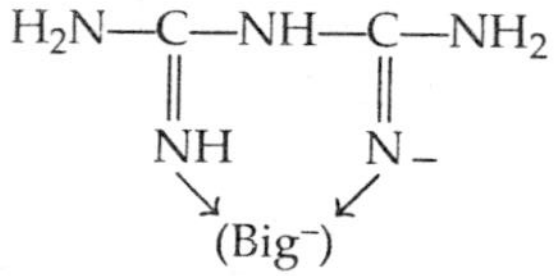 | Monobasic acid NN donor |
| | Acetylacetone $CH_3COCH_2COCH_3$ | acacH | 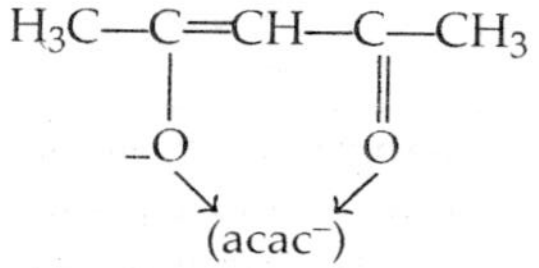 | Keto form converted to enol form and function as mono basic acid OO donor |

*Table contd. next page*

| Type | Name | Abbreviation | Structure showing linkage sites to metal by → from the donor atom /atoms (abbreviation) | Character |
|---|---|---|---|---|
| | O-phenanthroline | ophen | 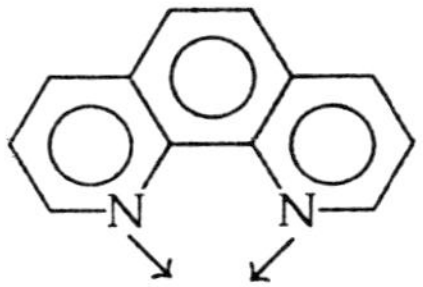 | Neutral NN donor |
| | 2,2′-dipyridyl<br>or<br>2,2′-bipyridine | dipy<br>or<br>bpy | 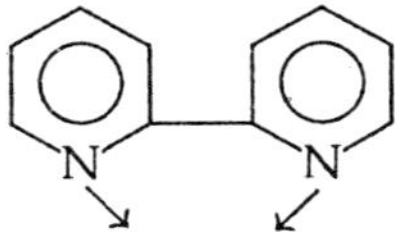 | Neutral NN donor |
| | Dimethylglyoxime<br>$H_3C-C=NOH$<br>$\vert$<br>$H_3C-C=NOH$ | DMGH | 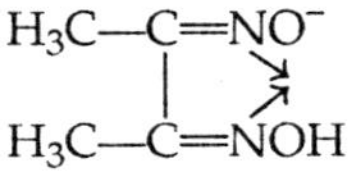<br>(DMG⁻) | Monobasic acid<br>NN donor |
| | Oxalic acid<br>$H_2C_2O_4$ | $oxH_2$ | 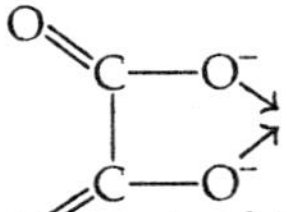<br>($ox^{2-}$) | Dibasic acid<br>OO donor |
| | Salicylaldehyde<br>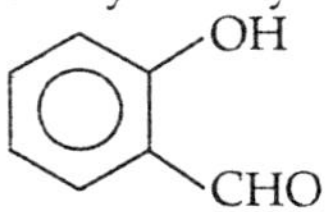 | salH | 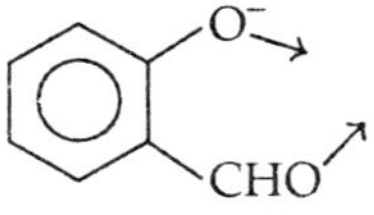<br>($sal^-$) | Monobasic acid<br>OO donor |
| | Glycine<br>$H_2NCH_2COOH$ | glyH | 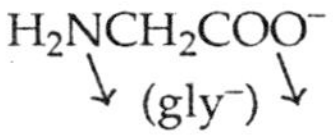<br>($gly^-$) | Monobasic acid<br>NO donor |

*Table contd. next page*

| Type | Name | Abbreviation | Structure showing linkage sites to metal by → from the donor atom /atoms (abbreviation) | Character |
|---|---|---|---|---|
| | 8-hydroxyquinoline 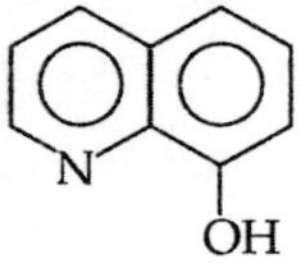 | oxinH | 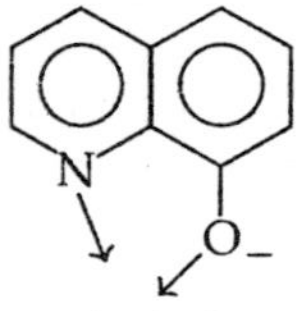 $(oxin^-)$ | Monobasic acid NO donor |
| | Quinaldinic acid 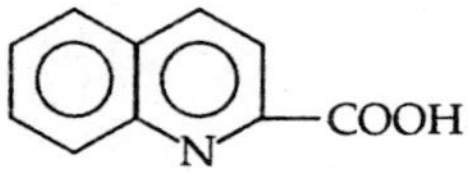 | QuiH | 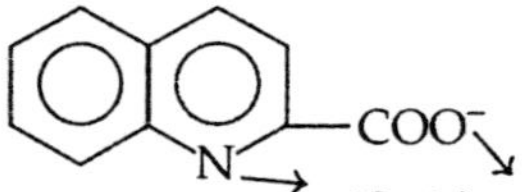 $(Qui^-)$ | Monobasic acid NO donor |
| | Salicylaldioxime 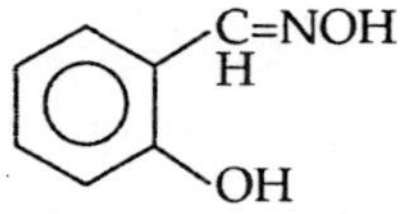 | SaloxH | 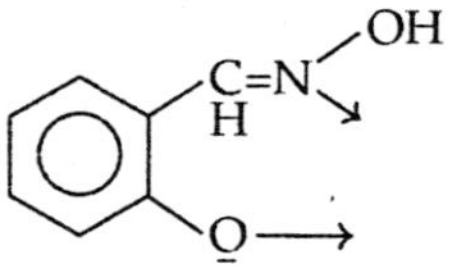 $(Salox^-)$ | Monobasic acid NO donor |
| | Catechol 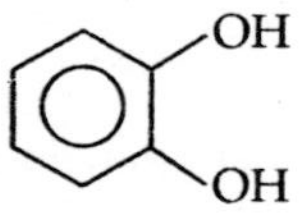 | $CatH_2$ | 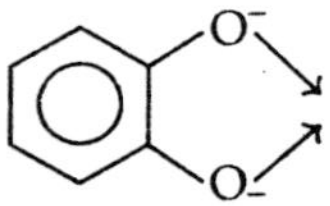 $(Cat^{2-})$ | Dibasic acid O O donor |
| | Orthophenylenediamine | Opn | 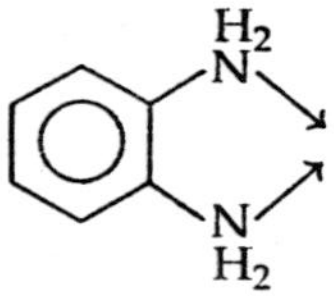 | Neutral NN donor |

*Table contd. next page*

| Type | Name | Abbreviation | Structure showing linkage sites to metal by → from the donor atom /atoms (abbreviation) | Character |
|---|---|---|---|---|
| | Orthophenylene bis dimethyl arsine | diars | 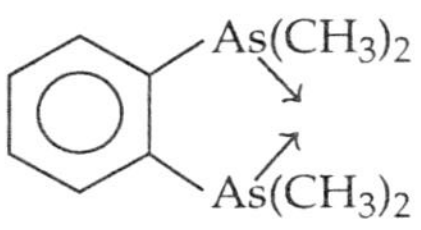  | Neutral As As donor |
| | Picolinic acid or pyridine 2-carboxylic acid 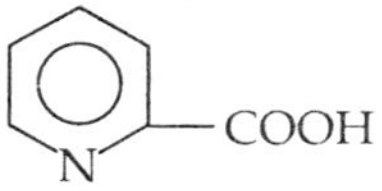  | PicH | 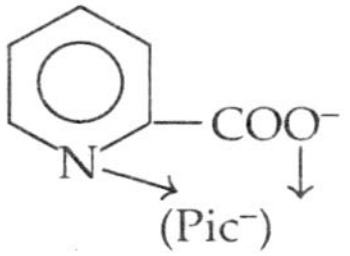  | Monobasic acid NO donor |
| | Nicotinic acid or Pyridine 3-carboxylic acid 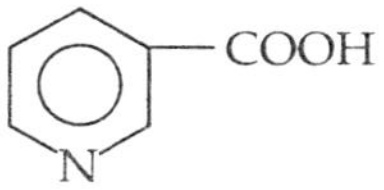  | NicH | 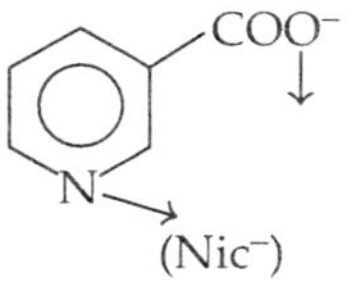  | Monobasic acid NO donor |
| | 2-benzoylpyridine oxime 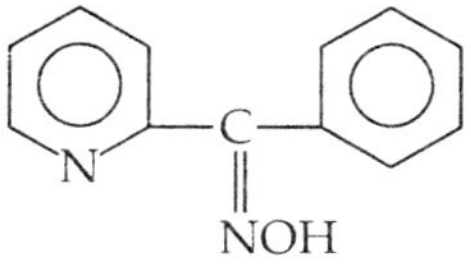  | BPOH | 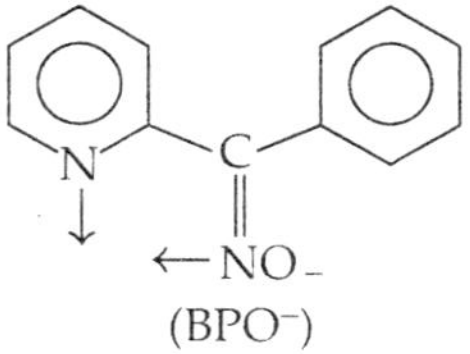  | Monobasic acid NN donor |

*Table contd. next page*

| Type | Name | Abbreviation | Structure showing linkage sites to metal by → from the donor atom /atoms (abbreviation) | Character |
|---|---|---|---|---|
| | α-nitroso β-napthol 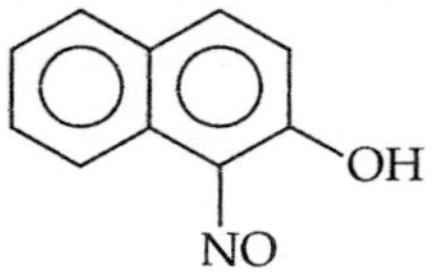 | NTAOH | 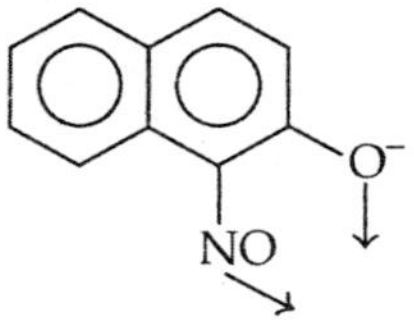 (NTAO⁻) | Monobasic acid NO donor |
| | Carbonic acid $H_2CO_3$ | — | 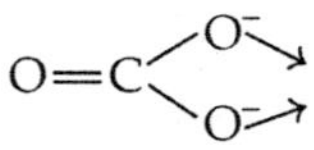 | Dibasic acid OO donor |
| | Ethylene diphosphine | diphos | $H_2P—CH_2—CH_2—PH_2$ | Neutral PP donor |
| | Dimethylglycol | glyme | $H_3CO—CH_2—CH_2—OCH_3$ | Neutral OO donor |
| **Tridentate** | Diethylenetriamine | dien | $H_2NCH_2CH_2NHCH_2CH_2NH_2$ | Neutral NNN donor |
| | Iminodiacetic acid $HOOCH_2C—N(H)—CH_2COOH$ | $IDAH_2$ | $^-OOCH_2C—N(H)—CH_2COO^-$ ($IDA^{2-}$) | Dibasic acid NOO donor |
| | Dipicolinic acid or pyridine 2, 6 dicarboxylic acid or lutidinic acid 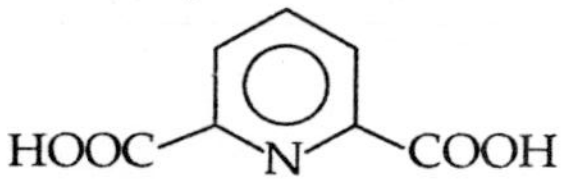 | $LutH_2$ | 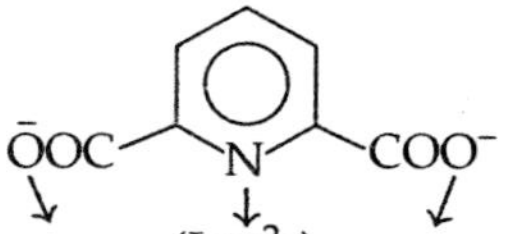 ($Lut^{2-}$) | Dibasic acid NOO donor |

*Table contd. next page*

| Type | Name | Abbreviation | Structure showing linkage sites to metal by → from the donor atom /atoms (abbreviation) | Character |
|---|---|---|---|---|
| | 2, 2′, 2″-terpyridine | terpy | 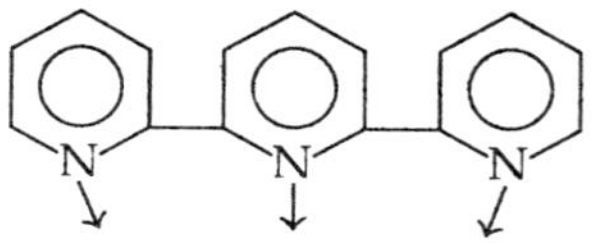 | Neutral NNN donor |
| | Cysteine<br>$HS-CH_2-CH(NH_2)-COOH$ | $CysH_2$ | 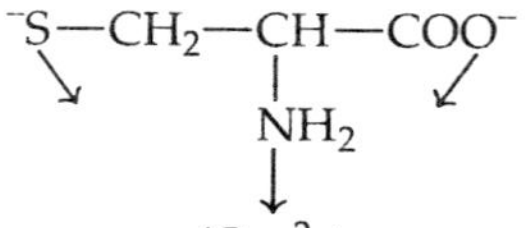<br>($Cys^{2-}$) | Dibasic acid<br>SNO donor |
| | Histidine<br>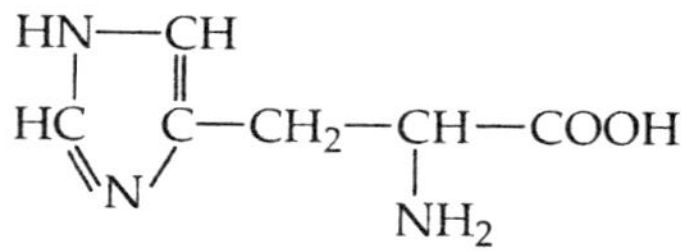 | HistH | 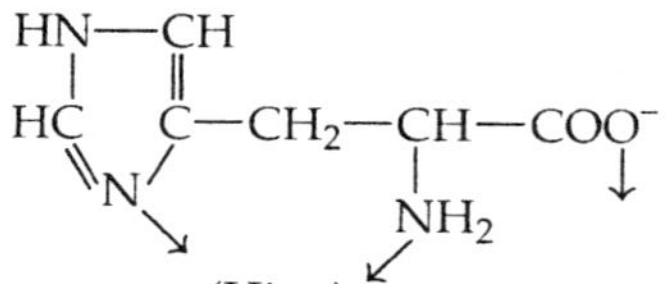<br>($Hist^-$) | Monobasic acid<br>NNO donor |
| **Tetradentate** | Triethylenetetraamine | trien | 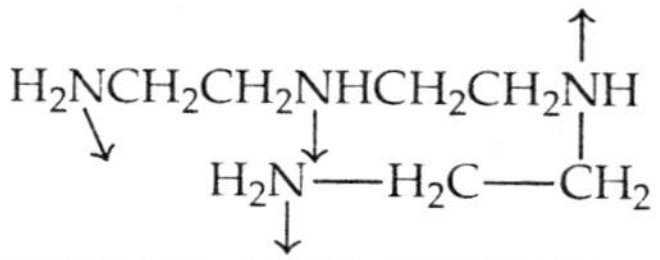 | Neutral<br>NNNN donor |
| | Nitrilotriacetic acid<br>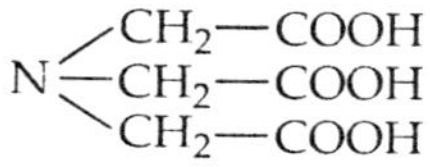 | $NTAH_3$ | 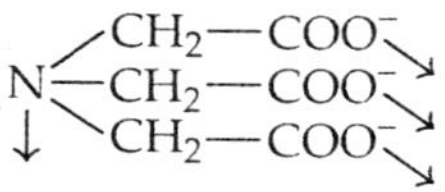<br>($NTA^{3-}$) | Tribasic acid<br>NOOO donor |

*Table contd. next page*

| Type | Name | Abbreviation | Structure showing linkage sites to metal by → from the donor atom /atoms (abbreviation) | Character |
|---|---|---|---|---|
| | Ethylenedibiguanide | $Et(BigH)_2$ | 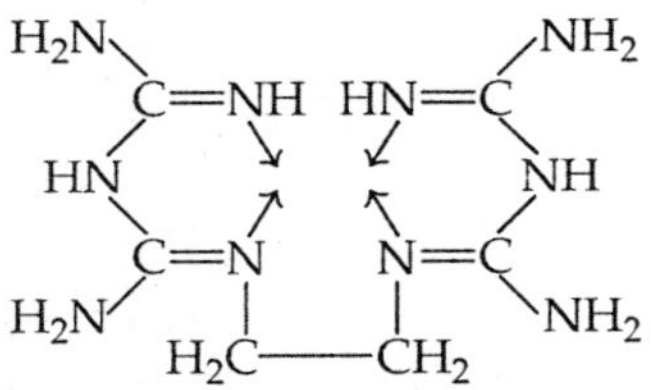 | Neutral<br>NNNN donor |
| **Pentadentate** | Tetraethylenepentaamine | tetraen | 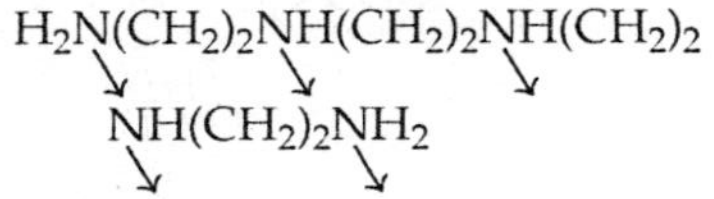 | Neutral<br>NNNNN donor |
| | Ethylenediaminetriacetic acid<br>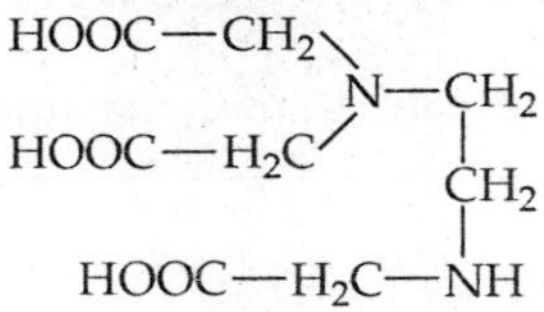 | $EDTIH_3$ | 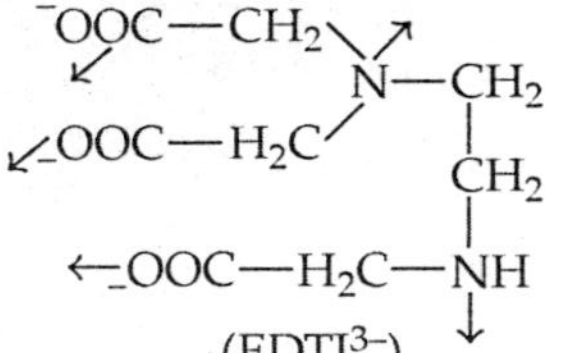<br>$(EDTI^{3-})$ | Tribasic acid<br>NNOOO donor |
| **Hexadentate** | Ethylenediamine tetraacetic acid<br>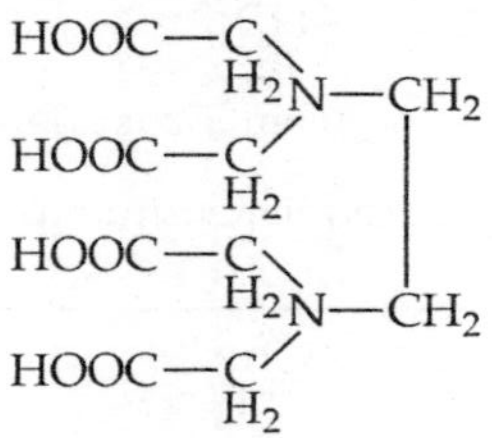 | $EDTAH_4$ | 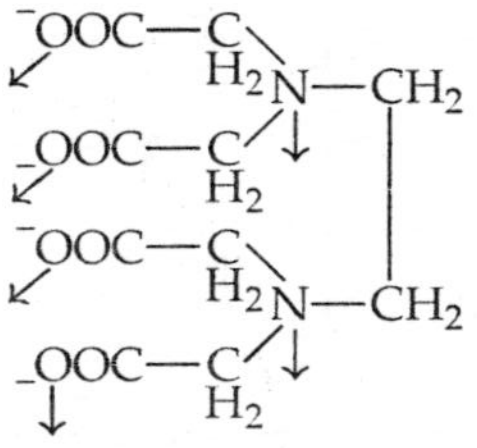<br>$(EDTA^{4-})$ | Tetrabasic acid<br>NNOOOO donor |

*Table contd. next page*

| Type | Name | Abbreviation | Structure showing linkage sites to metal by → from the donor atom /atoms (abbreviation) | Character |
|---|---|---|---|---|
| | 1,2 diaminocyclohexane N,N,N′, N′-tetraacetic acid 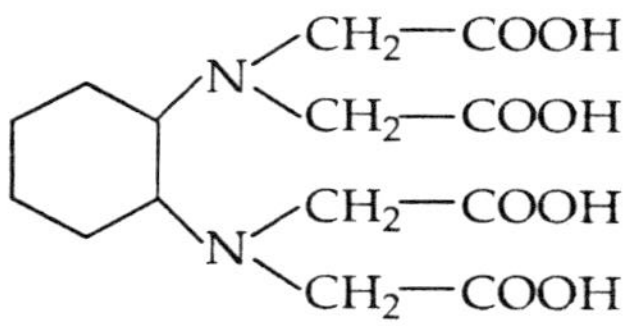 | $DCTAH_4$ | 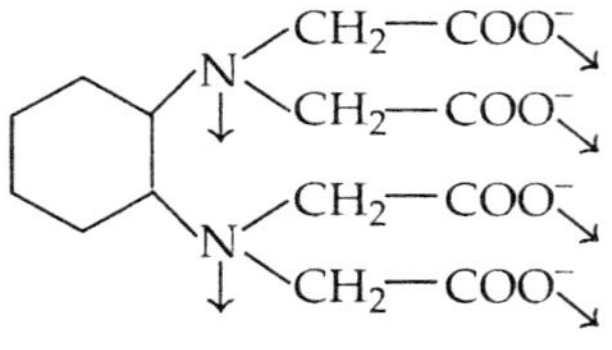 ($DCTA^{4-}$) | Tetrabasic acid<br>NNOOOO donor |
| **Octadentate** | Diethylenetriamine pentaacetic acid 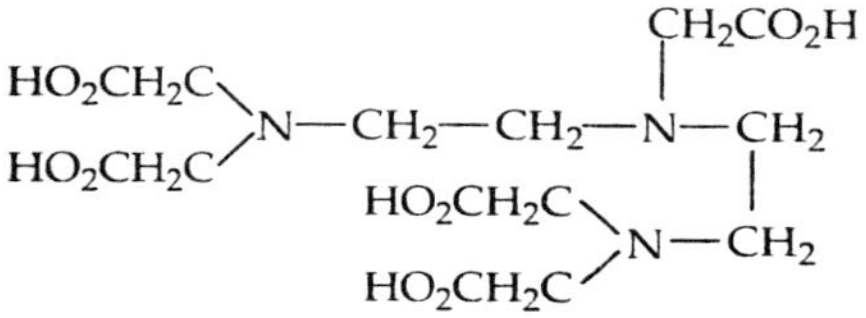 | $DTPAH_5$ | 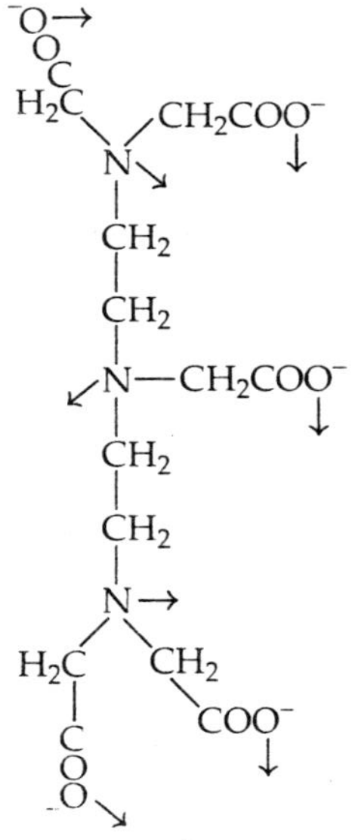 ($DTPA^{5-}$) | Pentabasic acid<br>NNNOOOOO donor |

### 7.4.1 Schiff-base ligands

These are the condensation products of amine with a carbonyl compound. These ligands are potentially versatile and have an immense interest for the formation of complexes with metals in various oxidation states.

A tridentate diabasic acid schiff base ligand may be prepared by the reaction of benzoyl hydrazide with salicylaldehyde i.e.

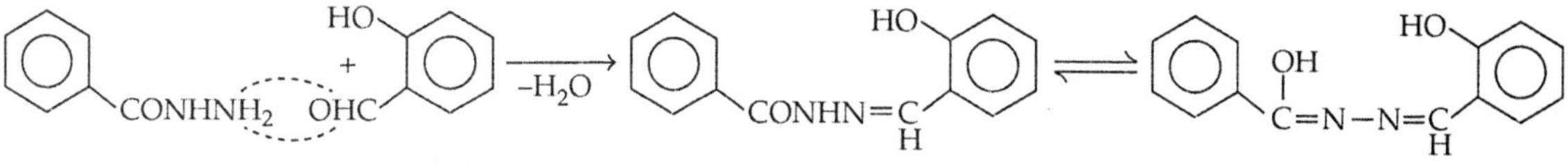


Keto form
(A)

Enol form
(B)

The keto form is converted to enol form and then function as dibasic acid (OON) donor in the following way :

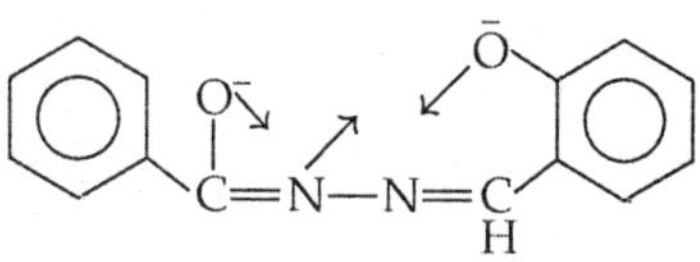


It is to be noted that due to the close proximity of the two nitrogen donors in the enol form (Structure B) there will be a steric repulsion and so only one of the nitrogen atom will act as a donor.

Another important schiff base ligand is synthesized by the condensation of ethylenediamine with salicylaldehyde in the following way :

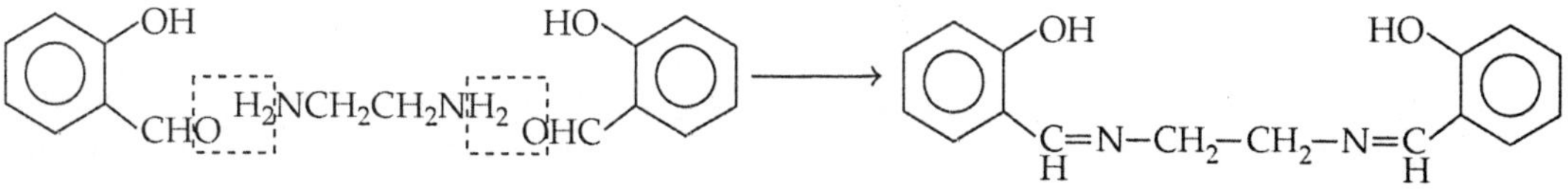


(C)

This ligand (C) can function as tetradentate dibasic acid as :

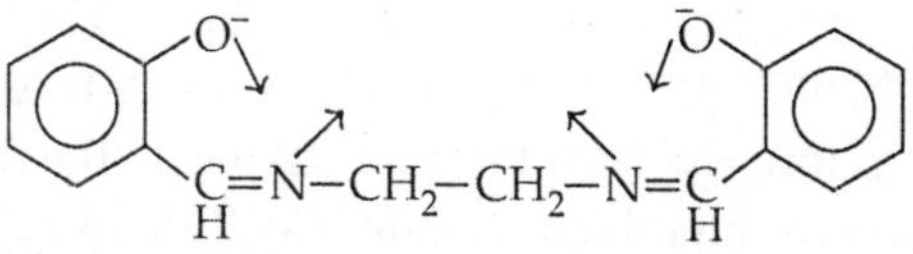


In this ligand the two nitrogen atoms are wide apart and so repulsion is minimised and thereby the two nitrogen atoms can function as donors.

### 7.4.2 Chelating ligand, bridging ligand, flexidentate ligand and ambidentate ligand

All bidentate to octadentate ligands are collectively known as **polydentate ligands** and all such ligands are sometimes called **chelating ligands**. Therefore **chelating ligands are those**

**which have two or more donor atoms attached to the same metal ion in the first sphere of coordination to form a ring structure.** The resulting complexes are called **metal chelates or** simply **chelates.** Thus all chelates are metal complexes but all metal complexes are not chelates. Ethylenediamine forms the following chelate complex (7-I). Usually strain free rings are formed around the metal ion and 5 or 6 membered rings are more stable than 2,3 or 4 membered rings. A chelating ligand having anionic coordination site may simultaneously satisfy the **secondary valency** as also **primary valency** or charge of the metal ion. Such type of complexes are called **inner metallic complexes** (will be discussed later).

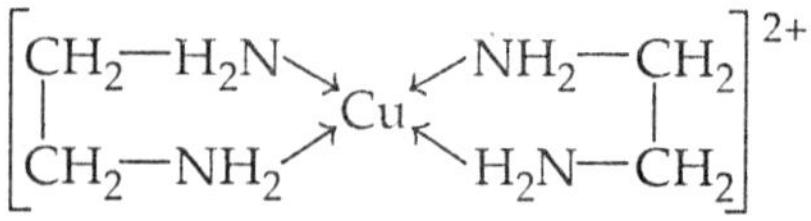

(7-I)

**Bridging ligands** are essentially monodentate but those monodentate ligands have more than one lone pairs can function as bridging ligands. In many cases they serve as monodentate bridging ligands e.g. $OH^-$, $Cl^-$, $NH_2^-$. Essentially they have more than one lone pair but cannot serve as chelating ligands because there will be the formation of three membered ring which is unstable, so that these ligands are not be able to span the two coordination sites of a metal ion. Due to this reason though hydrazine ($\ddot{N}H_2—\ddot{N}H_2$) has two donor atoms, yet it can function only as a monodentate ligand because it is unable to satisfy the two adjacent coordination sites of a metal ion. Examples of complexes acting $Cl^-$ and $OH^-$ as bridging ligands are :

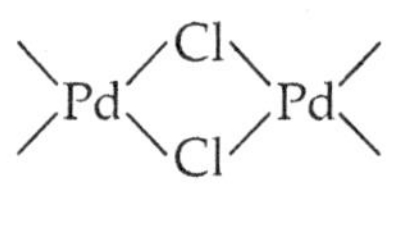

(7-II)

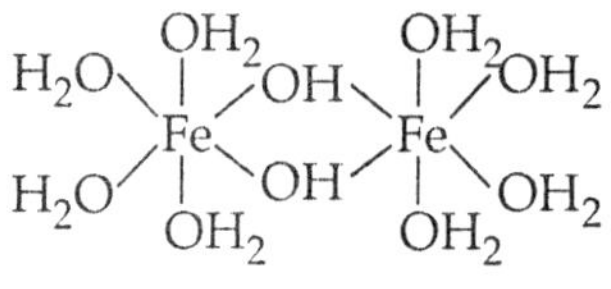

(7-III)

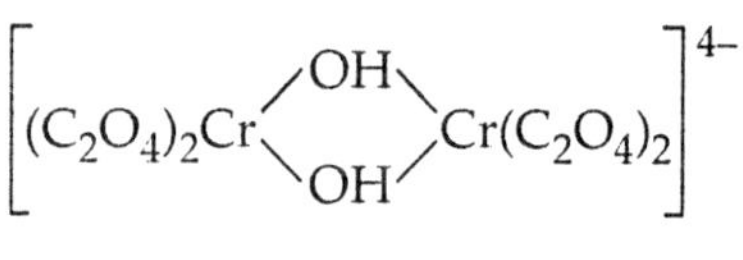

(7-IV)

A ligand which shows different denticity i.e. flexible mode to attachment to a metal ion is called **flexidentate ligand.** Several polydentate ligands will change their modous operandi to attach a metal ion due to several factors e.g. pH at the time of complex formation, competition to achieve a particular coordination number to form a mixed ligand complex with two or more ligands, the nature of the metal ion etc. Thus $EDTAH_4$ is usually a sexadentate ligand but function as pentadentate in green $[Ni(H_2O)(H_2EDTA)]$, blue $[Cu(H_2O)(H_2EDTA)]$, $[Cr(OH)(EDTA)]^{2-}$, $[CoBr(EDTA)]^{2-}$, $Na_2[Co(CN)(EDTA)]$; function as **tetradentate** in $[Pd/Pt(H_2EDTA)]$; function as **bidentate** behaviour in $[Pd/Pt(H_4EDTA)Cl_2]$. $IDAH_2$ essentially a tridentate ligand function as a bidentate one in the mixed chelate $[Co(IDA)(BigH)_2]^+$. Dipicolinic acid or lutidinic acid $LutH_2$ is potentially a tridentate ligand function as bidentate in $[Co(Lut)(BigH)_2]^+$.

One of the important tool to characterise the manifestations of flexidentate behaviour of polydentate ligands is to study the **infrared spectra** both in the solid state as also in solution phase. Electronic spectra is also one of the method. X-ray crystallographic study gives us the information regarding the donor atoms around the metal ion as also structural studies can also be made.

When ligands change their donor atom keeping the same denticity, the ligands are termed **ambidentate.** Thus in $NO_2^-$ the donor atom is nitrogen while in $ONO^-$ the donor atom is oxygen. Similarly thiocyanate can coordinate through nitrogen as in $[Co(NCS)_4]^{2-}$ while with $Hg^{2+}$ it coordinates through sulphur as in the case $[Hg(SCN)_4]^{2-}$. Cyanide can also function as an ambidentate ligand. Such ambidentate behaviour gives rise to structural, salt or linkage isomerism in coordination compounds (will be discussed later).

## 7.4.3 Other classification of ligands

(a) Ligands in which the donor atoms have one or more nonbonded electron pairs are $H_2O$, $NH_3$, $F^-$, $H^-$ etc. These ligands are called **classical ligands** because they have **σ-donor property.** On the other hand **π-donor ligands** are called **non-classical ligands** e.g. $C_2H_4$, $C_6H_6$, $C_5H_5^-$ (cyclopentadienide) etc. Since in these compounds have carbon donor atoms to metal, the metal compounds are sometimes called organometallic compounds.

(b) Ligands having a σ bonding pair of electrons and low energy empty π-antibonding orbitals that can accept suitably oriented d-orbital electrons from the low valent metal are called **σ-donor and π-acceptor ligands** e.g. CO, $R_3P$, $R_3As$, $CN^-$, acacH etc. In $Ni(CO)_4$, the ligand CO acts as a σ-donor and π acceptor through the linkage of carbon with the metal. This is called **synergic effect** i.e.

$$Ni \underset{\pi}{\overset{\sigma}{\rightleftarrows}} CO$$

Moreover CO can also function as **ambidentate ligand** coordination through oxygen in the isocarbonyl of gold i.e. COAuCO.

(c) $F^-$ is an **innocent ligand** because it has a fixed oxidation number –1 while NO can function as NO, $NO^+$ and $NO^-$. So NO is a **non-innocent ligand.** In $Cr(NO)_4$ and $Fe(CO)_2(NO)_2$ the ligand NO function as neutral molecule. Nitrosyl cation ($NO^+$) exist in $[Fe(H_2O)_5NO]SO_4$ and $[RhCl(NO)(CO)(PPr_3)_2]^+$. $NO^-$ exist in $[Ir(CO)Cl(PPh_3)_2NO]^+$ and $[Co(en)_2Cl(NO)]^+$.

(d) A special type of chelation occurs with **macrocyclic ligands.** Macrocyclic ligands are essentially polydentate ligands in which several donor atoms form a large ring and form complexes with transition metal ions. The donor atoms in these ligands may be nitrogen, sulphur or phosphorus. These ligands encapsulate the metal ion and form stable complexes when the metal ion exactly fit into the cavity formed by the donor atoms e.g. porphyrin ring (7-V). Due to the encapsulation properties they are termed as **cryptands** and the complexes are called **cryptates** or **sepulchrates.** These ligands also termed as **crown-ethers** becuase of their crown like structures. Large number of synthetic macrocyclic ligands have nitrogen and oxygen donors. However mixed (N, O), (N, S) or (N, O, P) etc. donors have also been synthesized. There are several macrocyclic ligands which have conjugated π-systems.

*Template effect and template synthesis*

A metal ion can be used to orient a group of ligands which then undergo a condensation reaction to form a macrocyclic ligand. Here metal ion has a controlling influence to form the macrocycles which is called **template effect.** Here metal ion is considered as a template agent to form the macrocyclic metal complex and any reaction of this type is called **template synthesis.** One of the template synthesis to form a macrocyclic complex (7-VI) is schemated in the following way :

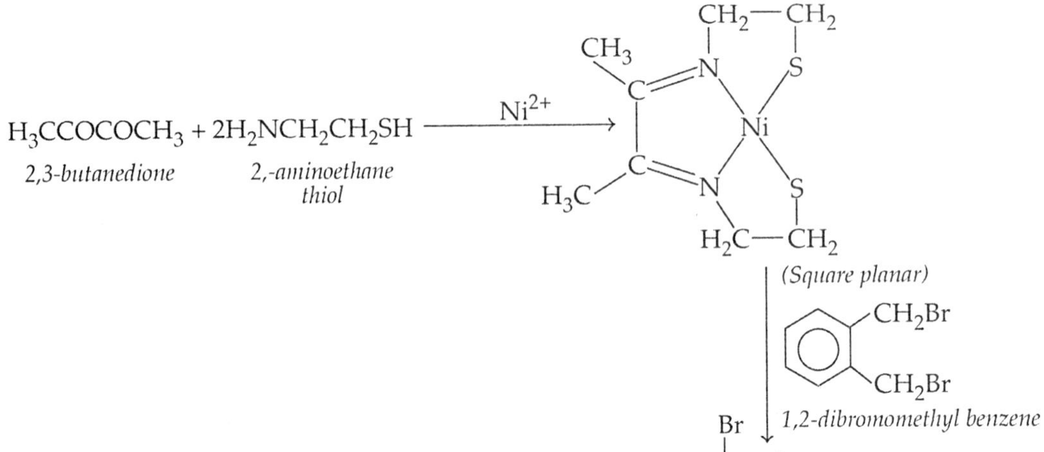


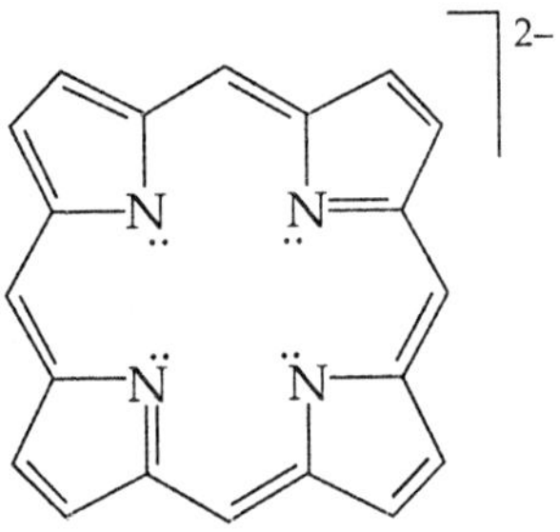

(7-V)

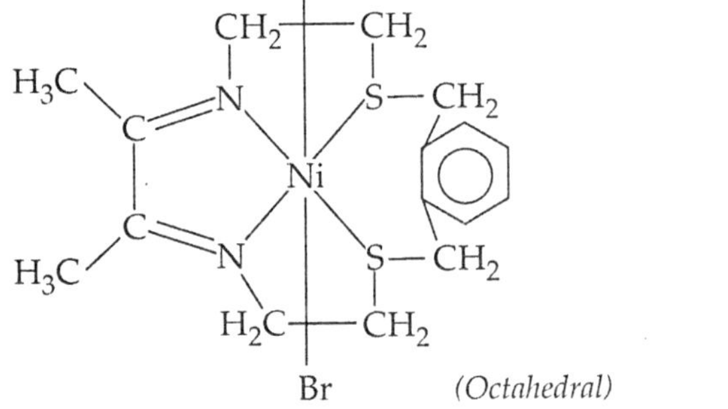

(7-VI)

Template effect may be considered in two aspects—**kinetic template effect** which denotes the course of the rate of the reaction and **thermodynamic template effect** which denotes the extra stability of the macrocyclic product i.e. the equilibrium will be shifted to the formation of macrocyclic metal complex. In absence of the metal ion the simple macrocycles are unstable but the stability is enhanced by the incorporation of the metal ion and thus the term thermodynamic template effect is applied in the synthesis of these types of complexes. One of the self condensation reaction in presence of $Ni^{2+}$ to form a tridentate (7-VII) and tetradentate (7-VIII) macrocyclic complexes are shown below. This is an example of **thermodynamic template effect** in presence of a metal ion.

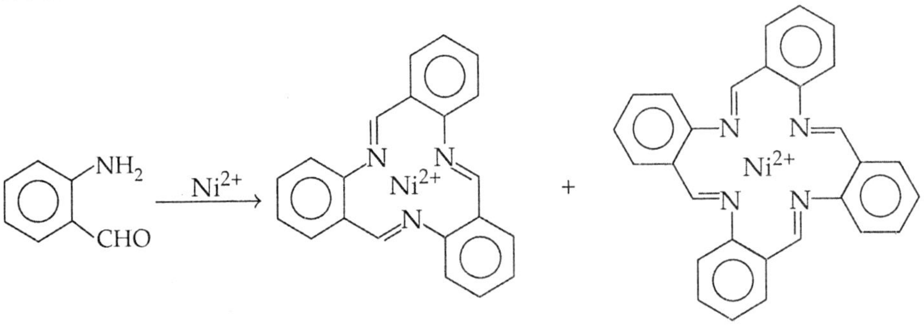

(7-VII) (7-VIII)

As mentioned earlier a new type of macrocyclic ligands containing oxygen donors are the crown ethers or cryptands were synthesized. C.J. Pederson (1967) synthesized several macrocyclic polyethers which are the cyclic polymers of ethylene glycol, $(OCH_2CH_2)_n$. Their actual naming is complicated but they usually named as α-crown-β where α represents the total number of atoms in the ring and β is the number of oxygen atoms. Therefore the cyclic hexamer of ethyleneglycol is called 18-crown-6 (7-IX) and its benzo derivative is called dibenzo-18-crown-6 (7-X).

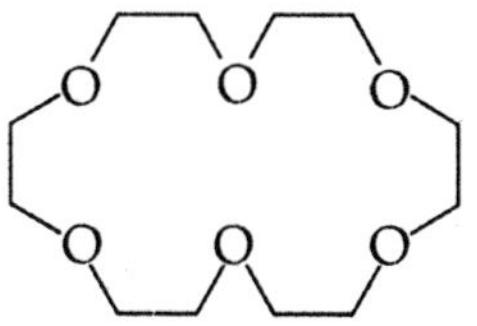

18-crown-6
(7-IX)

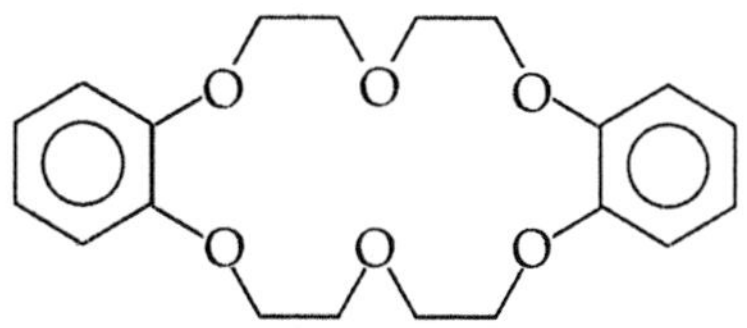

dibenzo-18-crown-6
(7-X)

The above two compounds were prepared by Pederson by the following ways :

$$C_6H_4(OH)_2 + (ClCH_2CH_2)_2O \xrightarrow[\text{n-butanol}]{\text{NaOH}} \text{dibenzo-18-crown-6}$$

*catechol* *dichloroethyl ether*

$$(CH_2OCH_2CH_2Cl)_2 + (CH_2OCH_2CH_2OH)_2 \xrightarrow{\text{aqKOH}} \text{18-crown-6}$$

*triethylene dichloroethyl ether* *triethylene glycol*

These crown ethers form very stable complexes with alkali metals.

Overwhelming number of macrocyclic and crown ether ligands have been synthesised in recent years and these form complexes with various transition and non-transition metals. Sometimes **in situ** metal promoted reactions has also been carried out e.g.

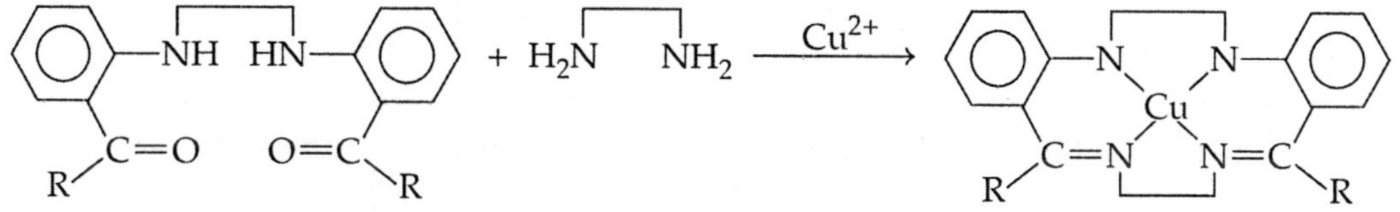

(7-XI)

In absence of the metal ion (i.e. $Cu^{2+}$) only a polymeric ligand is obtained.

One example of the schiff base is of the following :

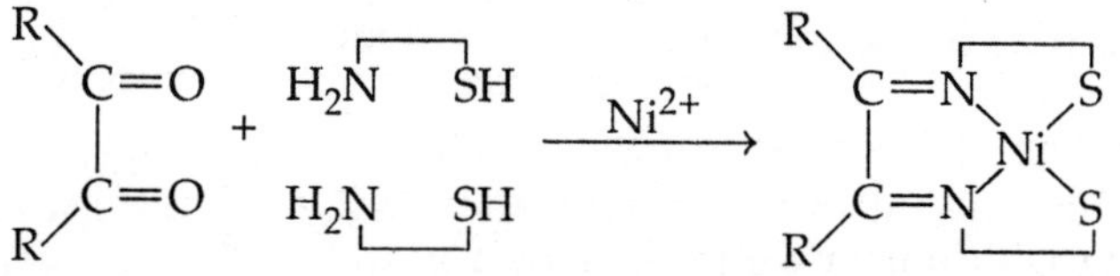

α-diketone 2-aminoethanethiol (7-XII)

The (7-XII) complex of Ni(II) is converted to macrocyclic complex (7-XIII) through **kinetic template effect** leading to ring closure in the following sequence :

(7-XII) +

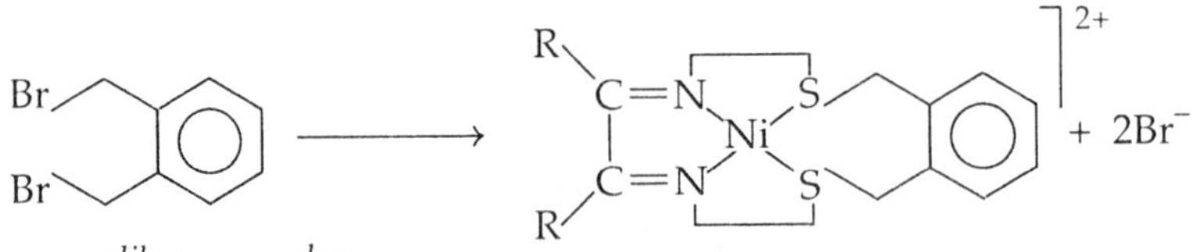

*dibromo-o-xylene*

(7-XIII)

(e) **Some specific coordination mode of ligands :**

(i) $N_3^-$ $[:\ddot{N}=N=\ddot{N}:]^-$

End on (EO) coordination : $N{=}N{=}N\begin{smallmatrix}\nearrow M \\ \searrow M'\end{smallmatrix}$

End to end (EE) coordination : $M \leftarrow N{=}N{=}N \rightarrow M'$

It can also function as bridging ligand.

(ii) $NCS^-$ $[:\ddot{N}=C=\ddot{S}:]^-$

Monodentate anionic ambidentate behaviour showing structural, salt or linkage isomers : Coordination through S i.e. $NCS \rightarrow M$ and coordination through N i.e. $SCN \rightarrow M$.

End to end (EE) coordination : $M \leftarrow NCS \rightarrow M'$

End on (EO) monobridged coordination : $N{=}C{=}S\begin{smallmatrix}\nearrow M \\ \searrow M'\end{smallmatrix}$

or

$S{=}C{=}N\begin{smallmatrix}\nearrow M \\ \searrow M'\end{smallmatrix}$

End on (EO) bibridged motif : $S{=}C{=}N\begin{smallmatrix}\nearrow M \nwarrow \\ \searrow M' \nearrow\end{smallmatrix}N{=}C{=}S$

or

$N{=}C{=}S\begin{smallmatrix}\nearrow M \nwarrow \\ \searrow M' \nearrow\end{smallmatrix}S{=}C{=}N$

End to end (EE) bibridged (symmetrical) mode : $M\begin{smallmatrix}\swarrow N{=}C{=}S \searrow \\ \nwarrow N{=}C{=}S \nearrow\end{smallmatrix}M'$

or the same with unsymmetrical mode : $M\begin{smallmatrix}\swarrow N{=}C{=}S \searrow \\ \nwarrow S{=}C{=}N \nearrow\end{smallmatrix}M'$

The Lewis electron dot structure can also be represented as : $[:N \equiv C - \ddot{\underset{..}{S}}:]^-$ or $[:S \equiv C - \ddot{\underset{..}{N}}:]^-$.

Therefore it can function as classical σ donor behaviour : 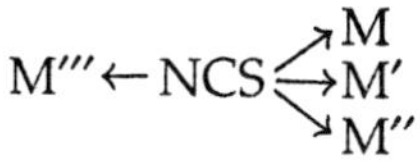 or

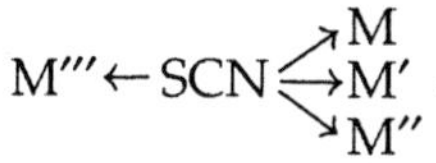 .

Moreover in principle it can function as non-classical ligand i.e. π-cloud donor.

(iii) 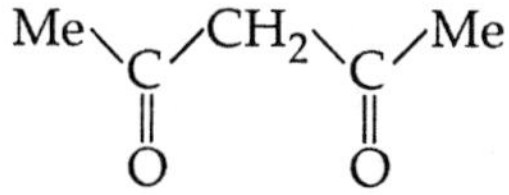

acacH

In neutral form it can form bidentate (O, O) chelate i.e. 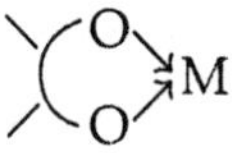

Function as monobasic C-donor i.e. organometallic : 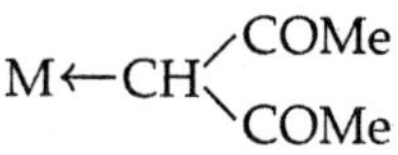

Function as monobasic (O, O) inner metallic form* as : 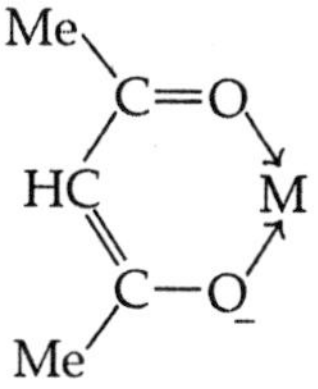

Acetylacetone can function (C, O, O) donor as : 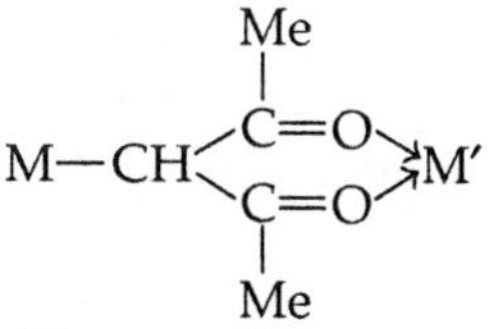

Here M is a soft metal ion and M′ is a hard metal ion.

Again the ligand can function as non-classical (π-electron donor) ligand as :

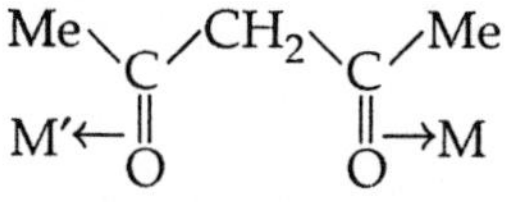

(f) **Compartmental ligands** : There are some open-ended polydentate ligands and macrocyclic ligands which can accommodate two metal centres side by side. Those ligands are called compartmental ligands e.g.

* will be discussed later.

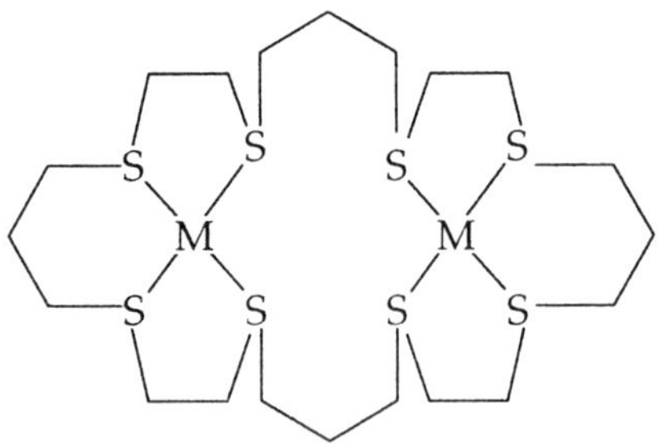


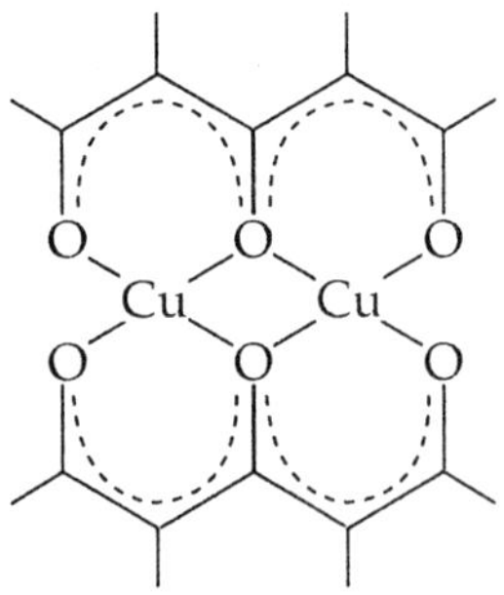


(triketonate)

(g) **σ donor + π donor ligands :** Ligands having more than one lone pairs can function as σ+π donor ligands. In that case the lower energy lone pair can form σ donation while the higher energy electron pair may participate in π-donation e.g. $OH^-$, $F^-$, $I^-$ etc. Such ligands are weak field ligands. The pure σ donor ligands and the ligands with the π-donor properties can stabilise the higher oxidation state of metal. $O^{2-}$, $OH^-$, $NH^{2-}$, $F^-$, $Br^-$, $I^-$, $S^{2-}$ are **σ donor + good π donor** ligands while $H_2O$, $NH_2^-$, $R_2O$, $R_2S$, $R_3PO$ (O-donor), $R_3AsO$ (O-donor) are **σ donor + weak π donor** ligands.

(h) **Tripod ligands :** These are generally tetradendate ligands where the central atom itself acting as a donor and there are three branches from the central donor atom and each branch also acting as a donor centre to the metal e.g.

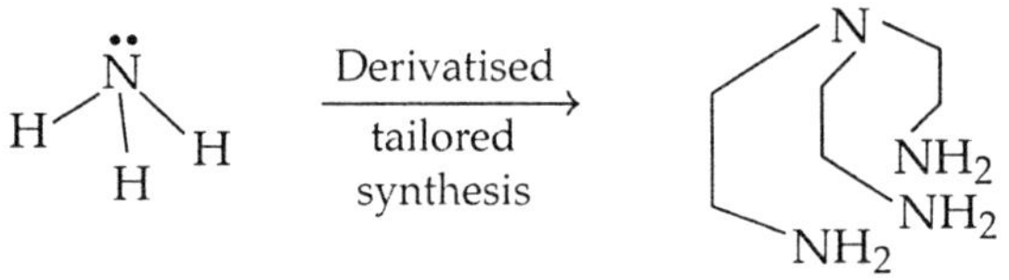


tris(2-aminoethyl)amine ⇒ tren, tripod ligand

Depending on the flexibility of the branching these types of ligands can form tetrahedral, trigonal bipyramidal and octahedral geometry with transition metals e.g.

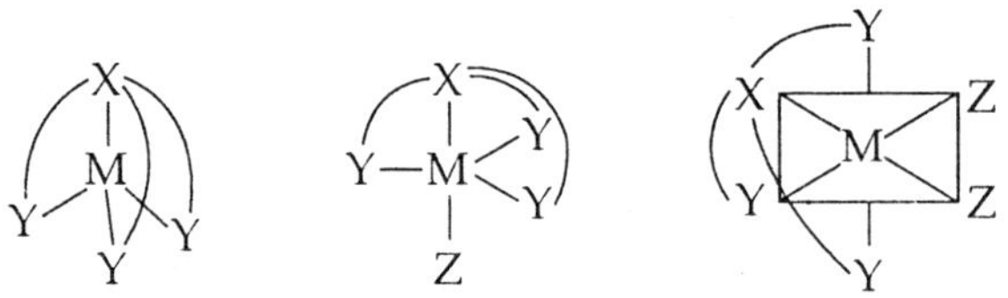


Tetrahedral    Trigonal bipyramid    Octahedral

In all the above figures X = tertiary N and Z is a monodentate ligand. Other examples of tripod ligands are :

$$:N(CH_2CH_2PH_2)_3, \quad :N(CH_2COO^-)_3$$

$$NTA^{3-}$$

These ligands behave umbrella like tetradentate ligands and represented as X(⁓Y)$_3$ where X = N, P or As; Y = $NR_2$, $AsR_2$, $NH_2$ etc. and the segment ⁓ denotes $-CH_2CH_2-$, $-CH_2-CH_2-CH_2-$ or $o-C_6H_4$ etc.

In the trigonal bipyramidal geometry the tetradentate umbrella like tripodal ligand function symmetrically to occupy three equatorial sites and one axial position is occupied by the monodentate ligand(Z). For example in [M(tripodal ligand)Z]$^+$ where tripodal ligand = tren, M = $Mn^{2+}$, $Fe^{2+}$, $Ni^{2+}$, $Cu^{2+}$ etc. and Z = $Cl^-$, $Br^-$, $CN^-$ etc. as depicted above. Bis complex of tripod ligand is the model complex for the DNA or biological system.

(i) The **Chlorin ring or corrin ring** : The porphyrin related **macrocylic ligand** (7-V) are used in chlorophyll and vitamin $B_{12}$. In chlorophyll the ring coordinates with Mg(II) and the ring is called **chlorin ring**. The similar type of ring coordinates with cobalt metal centre in vitamin $B_{12}$ and the ring is called **corrin ring** as depicted below :

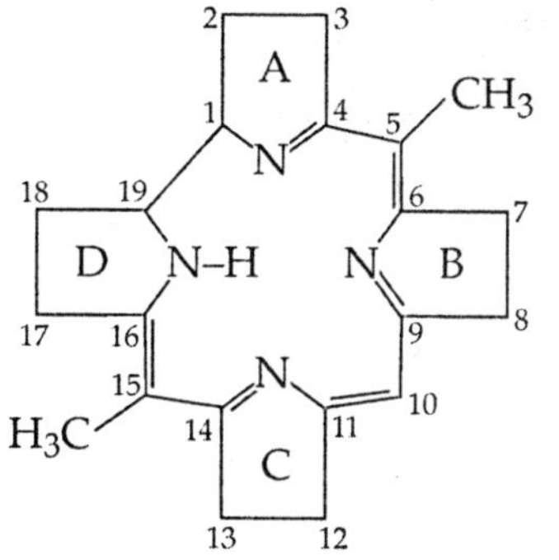


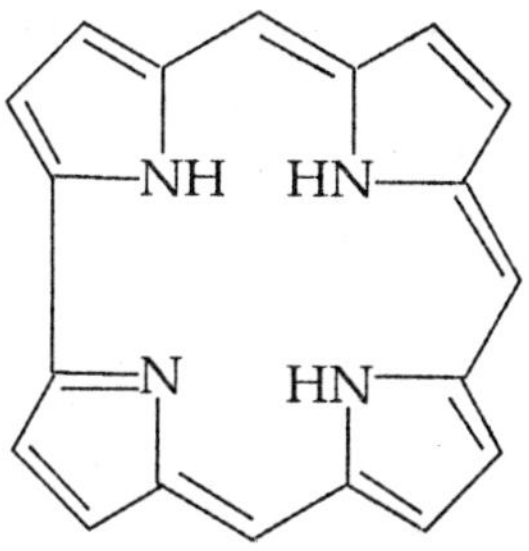


Corrin ring of vitamin $B_{12}$ Corrole ring

**Corrole ring** is intermediate between porphyrin and corrin ring as shown above.

(j) **Clathro-Chelate** : When a macrocylic ligand is sufficiently large and flexible it can encapsulate the metal centre in a three-dimensional cage like structure. This encapsulation effect is sometimes called **wrapping effect**. The flexibility of the macrocycle is acclerated when the chain of the ligand is largely saturated and the resulting complex is called **clathro-chelate** or **cryptate**. One of the clathro-chelate is derived by reaction with dimethylglyoxime, $BF_3$ and cobalt (III). The resulting complex has the following formula :

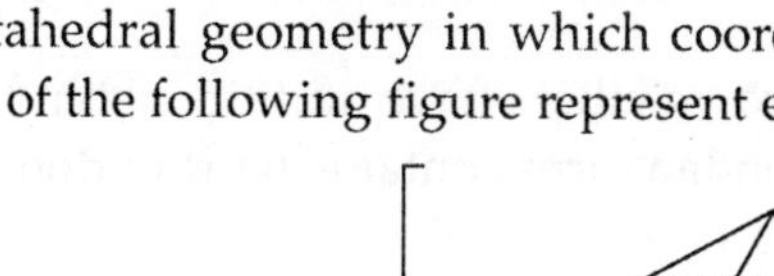


and the structure is depicted below wherein boron atoms form apexes on lower left and upper right in an octahedral geometry in which coordinate positions are ocupied by nitrogen atoms. The solid lines of the following figure represent edges of the polyhedron spanned by chelate rings.

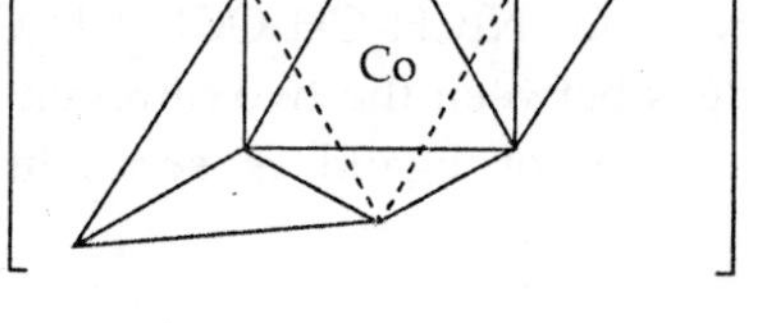

A distinction is to be noted for **clathrate compounds** (associated with the trapping of inert gases) and clathro-chelate. In case of former there occurs no chemical reaction and this is purely a physical process but in case of the latter the metal ion is encapsulated within the hole of the macrocycle through chemical process.

(k) **Corands :** According to the terminology of **supramolecular chemistry** the monocyclic macrocycles e.g. cyclic polymers of ethylene glycol $(OCH_2CH_2)_6$ are termed as corands. These are also called crown ethers. 18-crown-6 is a corand.

(l) **Podands :** The synthetic macrocyclic ligands like the crown ethers and cryptands are used as the molecular **receptors** in the supramolecular interaction for **molecular recognition**. These macrocycles and their related ligands are classified into three heads such as :

(i) **Podands** which are open chain macrocycles e.g. polyglymes, $CH_3(OCH_2CH_2)_xOCH_3$,

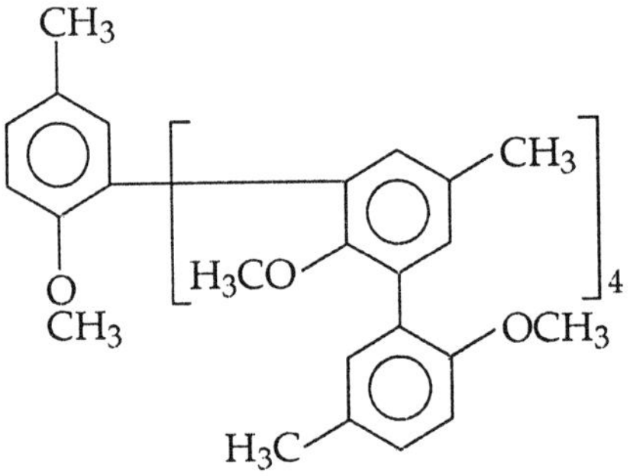


(ii) **Corands** has already been discussed e.g. crown ether.

(iii) **Cryptands** which is a multicyclic macrocycle giving rise to a spherical shape e.g.

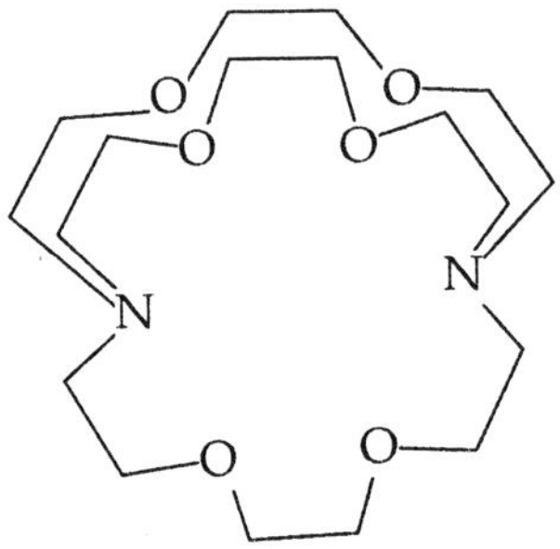


The corresponding metal complexes are termed as **podates, corates** and cryptates respectively.

(m) **Lariat crown ether :** These macrocyclic crown ethers have additional binding sites in the side chain which is called **pendant** arm. If the pendant arm contains hard O-donor site then the lariat crown ethers can bind the hard metal ions more strongly.

(n) **Football ligand :** The cryptand, $N(CH_2CH_2OCH_2CH_2OCH_2CH_2)_3N$ is called football ligand because the polyether bridges between the two nitrogen atoms resemble the seams of a football. Other example of this class of ligand is **sepulchrate** which is formed by the condensation of formaldehyde and ammonia onto the nitrogen atoms of $[Co(en)_3]^{3+}$. Football ligands form bicyclic cryptands or may form multicyclic macrocycles.

(o) **Cyclometallated ligand :** This is organometallic + chelating ligand that have another donor other than C-atom e.g. bipyridine, azobenzene i.e. 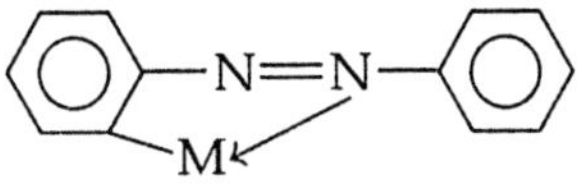

represented as 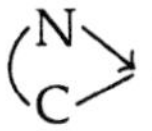

.

(p) **cis chelating and trans chelating ligand :** Cis chelators are those whose donor centres occupy different axes as depicted below :

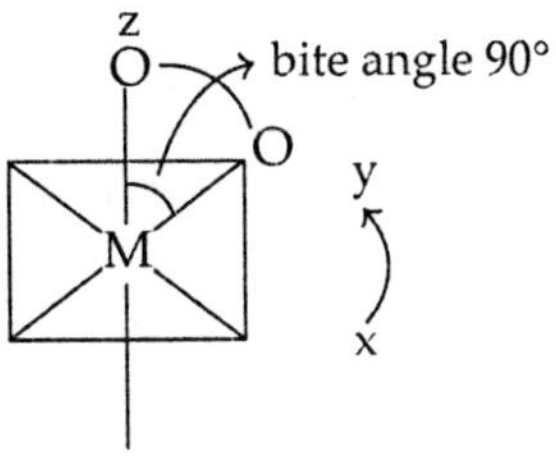


Most of the bidentate ligands are cis chelating e.g. acetylacetone.

In case of trans chelator the donor centres occupy two different opposite points on the same axis is noted below :

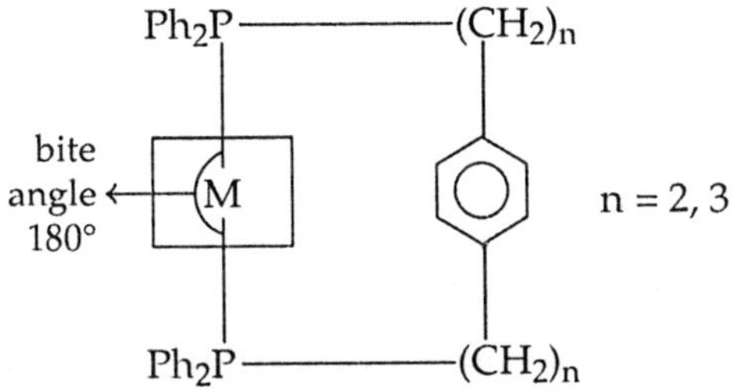


cis chelator has smaller arm whereas the trans chelator should have very larger tailored arm.

(q) **Metalloorganic ligand :** Organometallic ligands are essentially metalloorganic but the reverse is not true. For organometallic ligand the origin is organic i.e. C-donor centre but in case of metalloorganic ligand the origin is organic but any donor centre other than C-donor. For example phthalocyanine, the origin is organic having four N donors is only metalloorganic but bipyridine is both metalloorganic as well as organometallic ligand.

(r) **Metalloligand :** Metal complex behaving as ligand is called metalloligand. Actually the ligand which was originally used must have pendant arm (unused arm) thus giving rise to **duplex behaviour**. The following metal complex cannot be used as metalloligand as all donor centres are used here and there is no **pendant arm** :

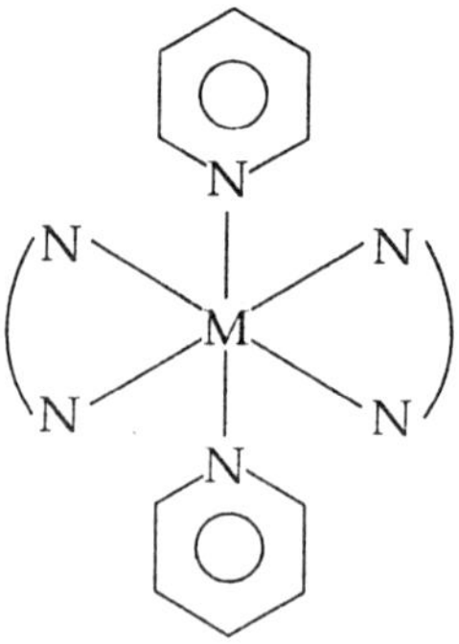

Here N–N is bpy

The duplex behaviour is to be considered in those cases of metalloligand where aparat from the orginal coordination mode to the central metal, the metalloligand have pendant arm (unused arm) which may be used as additional donor centre to other metal e.g. the complex $[M(bpy)_2(pyz)_2]^{z+}$ where pyz is pyrazine, . The said complex may be represented as :

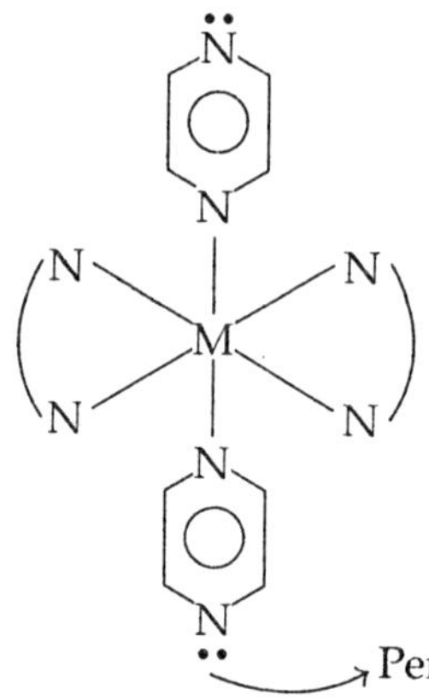


The above complex having pendant arm can be used as **metalloligand** comprising duplex behaviour.

The above metalloligand may form **molecular wire** in the following fashion :

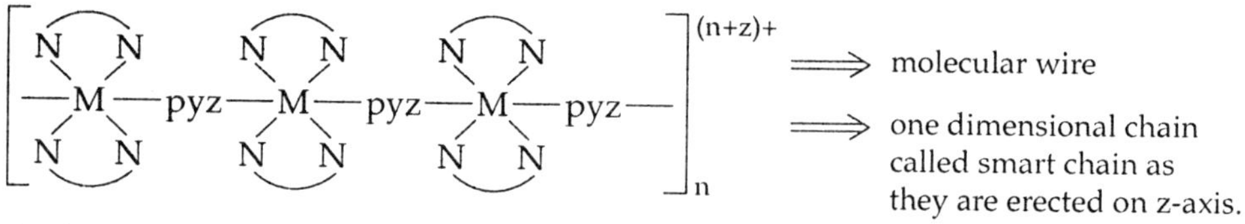


As the metalloligand at first forms a stable complex through its donor centres and afterwards forms a new complex acting as a bridge, so it has dual nature which is called **duplex behaviour.** For example $[Ru(pyrazine)_2(bpy)_2]^{2+}$ complex can behave as a metalloligand with its two pendant pyrazine N atoms even after coordination to Ru(II) but $[Ru(pyridine)_2(bpy)_2]^{2+}$ complex cannot behave as a metalloligand as pyridine has no such pendant N atom.

(s) **Chelator and congregator :** Chelator is a ligand containing two or more donor centres so disposed in space as these may simultaneously occupy positions around a single core metal ion forming a five or six membered chelate loop around the metal ion in a particular coordination complex with definite geometry e.g. ethylediamine (en) is a chelator which forms a chelate complex $[Co(en)_3]^{3+}$ in which there are formation of three 5-membered chelate rings around $Co^{3+}$ in the following way :

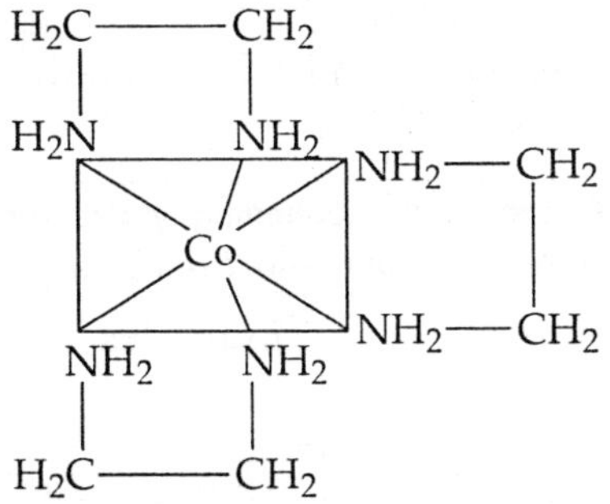

**Congregator** is an organic ligand which incorporates more than one metal ion forming a coordination complex with considerable stability. Modification of ligand framework of a classical chelator judiciously needed to convert a chelator into a congregator. The tetradentate (N, N, N, N) schiff base formed by condensation of butane-1, 4-diamine and pyridine 2-carboxaldehyde in the ratio 1 : 2 is a good example of a congregator forming dinuclear complex encapsulating two metal ions in the following way :

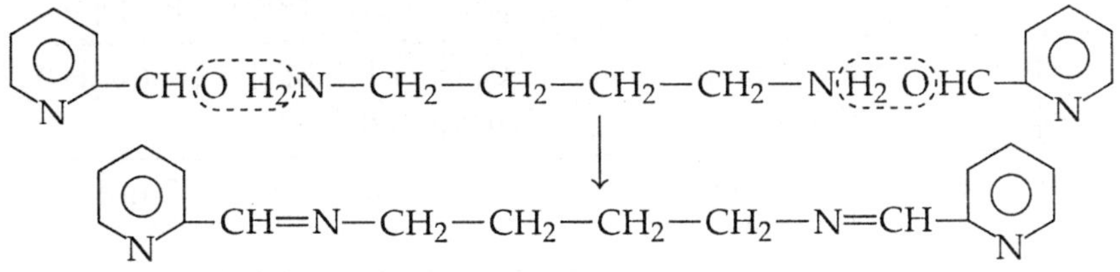

Schiff base ligand

and the coordination mode of the ligand represented as :

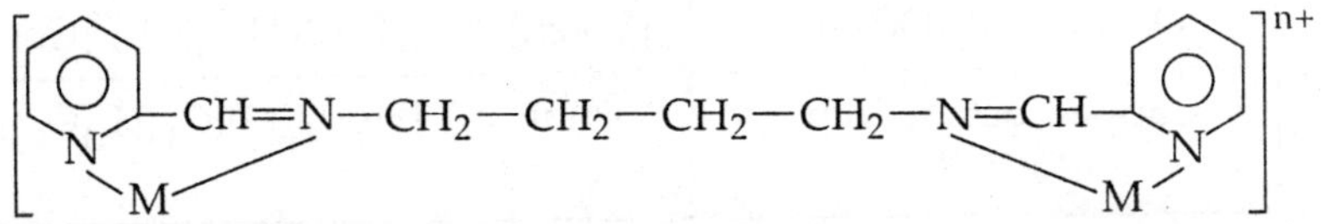

## 7.5 STEREOCHEMISTRY AND COORDINATION NUMBER OF COMPLEXES

As covalent molecules have definite geometry (Sec. 3.2.5) and since coordinate bond is a special type of covalent bond and therefore interaction of metal ions with ligands to form coordination complexes which also have definite geometry in space. The arrangement of ligands around a central metal ion depends not only on the nature of the metal ion but also on the nature of the ligands i.e. the **coordination number** of a central metal ion depends on the specific interaction of the metal ion with that of a ligand and the geometry of the complex entirely depends on the coordination number. Therefore there will be an intimate connection between coordination number and geometry/stereochemistry of the complex. Since during the formation of a complex a specific interaction between ligands and metal ion occurs, therefore the arrangement of ligands around a specific metal ion minimises electrostatic attraction so that the coordination number (C.N.) will be maximum.

The interaction of a metal ion with ligands may assume C.N. from 2 to 9 having geometry linear, triangular planar, tetrahedral, square planar, square pyramid, trigonal bipyramid, octahedral, pentagonal bipyramid, capped octahedral etc. Earlier (Sec. 3.2.5, Fig. 3.8) we have shown the disposition of different geometry. Examples of complexes having C.N. from 2 to 9 with their stereochemistry are recorded in Table 7.3. As already mentioned (Sec. 3.2.5) the total coordination number upto 14 are known but the examples are not common and their stereochemistry are not known properly.

**Table 7.3 : Different stereochemistry (Polyhedral symbol) with coordination number and examples of complexes**

| Coordination number | Stereochemistry | Examples |
|---|---|---|
| 2 | Linear (L-2) | $[CuCl_2]^-$, $[Ag(NH_3)_2]^+$, $[Au(CN)_2]^-$ etc. |
| 3 | Equilateral triangle, may be distorted (TP-3) | $[((C_6H_5)_3P)_3Pt]$, $[HgI_3]^-$, $[CuCl(tu)_2]$ where tu = thiourea |
| 4 | Tetrahedral (T-4) | $[Cd(CN)_4]^{2-}$, $[CoCl_4]^{2-}$, $[ZnCl_4]^{2-}$, $[BF_4]^-$ etc. |
| 4 | Square planar (SP-4) | $[Ni(DMG)_2]$, $[Ni(CN)_4]^{2-}$, $[Cu(en)_2]^{2+}$ $[Pt(NH_3)_2Cl_2]$, $[PtCl_2(PPh_3)_2]$ etc. |
| 5 | Square pyramid (SPY-5) | $[VO(acac)_2]$, $[SbF_5]^{2-}$, $[Ni((C_2H_5)_3P)_2Br_3]$, $[NiBr(diars)_2]^+$ etc. |
| 5 | Trigonal bipyramid (TBP-5) | $Fe(CO)_5$, $[Ni(CN)_5]^{3-}$, $[CuI(dipy)_2]I$ etc. |
| 6 | Trigonal prismatic (TPR-6) | $[Re(S_3(CH_2)_3CMe)_2]$, $[Re(S_2C_2Ph_2)_3]$ $[V(S_2C_2Ph_2)_3]$ etc. |
| 6 | Octahedral (OC-6) | $[Co(en)_3]^{3+}$, $[Co(NH_3)_6]Cl_3$, $[Co(CN)_6]^{3-}$ etc. |
| 7 | Capped octahedron (OCF-7) | $[NbOF_6]^{3-}$, $[Ho(H_2O)(C_{15}H_{11}O_2)_3]$ |
| 7 | Capped trigonal prism (TPRS-7) | $[NbF_7]^{2-}$, $[TaF_7]^{2-}$ |
| 7 | Pentagonal bipyramid (PBPY-7) | $[UO_2F_5]^{3-}$, $[V(CN)_7]^{4-}$, $[ZrF_7]^{3-}$ |
| 8 | Cubic (CU-8) | $[UF_8]^{3-}$, $[PaF_8]^{3-}$ |
| 8 | Square antiprism (SAPR-8) | $[ReF_8]^{2-}$, $[Mo(CN)_8]^{3-}$, $[M(acac)_4]$ ($M=Zr^{4+}$, $Ce^{4+}$, $Th^{4+}$) |
| 8 | Dodecahedral (DD-8) | $K_4[Mo(CN)_8].2H_2O$ |
| 9 | Tricapped trigonal prismatic (TPRS-9) | $[Nd(H_2O)_9]^{3+}$, $[TcH_9]^{2-}$ etc. |

Among these the well studied coordination numbers are 4, 5 and 6. For clarity we illustrate the stereochemistries in these coordination numbers (Fig. 7.2). The following points are to be noted for these coordination numbers :

(a) Among the two geometries for C.N. 4, the tetrahedral geometry (Fig. 7.2(a)) has four triangular faces, four vertices and six edges.

(b) The C.N. 5 has two stereochemistries–trigonal bipyramidal (Fig. 7.2(b)) which has five vertices, six triangular faces and nine edges while square pyramidal geometry (Fig. 7.2(c)) has one square face and four triangular faces, five vertices and eight edges.

(c) For C.N. 6 corresponds to two stereochemistries–the trigonal prismatic (Fig. 7.2(d)) one has six vertices, three reactangular faces and two triangular faces and nine edges. The extensive studied octahedral geometry (Fig. 7.2(e)) has eight faces, six vertices and twelve edges.

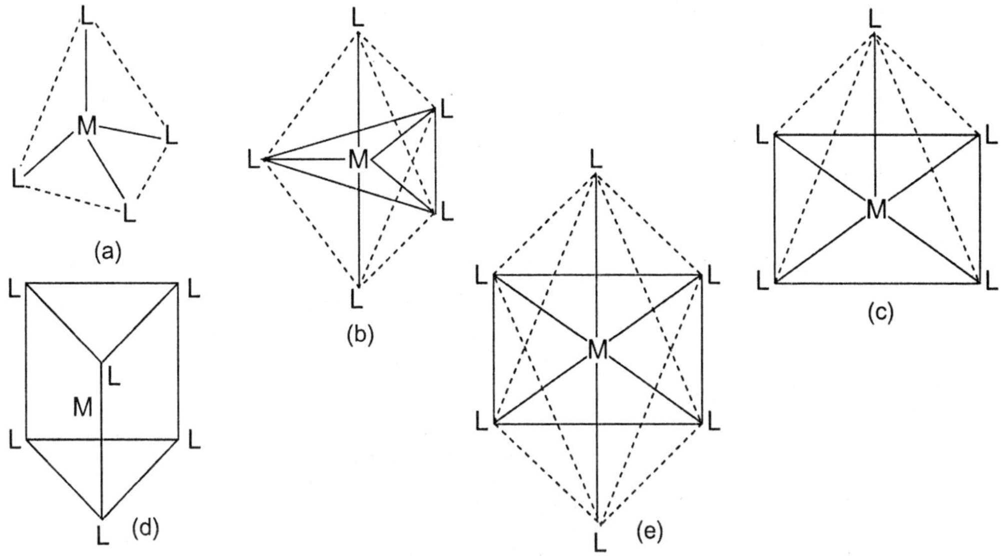


***Figure 7.2** : Disposition of (a) tetrahedral, (b) trigonal bipyramidal, (c) square pyramidal, (d) trigonal prismatic and (e) octahedral stereochemistries.*

It is to be noted that C.N. of a complex depends on the size of the metal ion, the steric interactions between the ligands and the electronic interactions between the ligands and the metal ions. For d-block metal ion complexes there occur some distortions and in several cases distorted geometry is obtained.

## 7.6 NOMENCLATURE OF COORDINATION COMPOUNDS

The International Union of Pure and Applied Chemistry (IUPAC) was formed in 1921 and many meetings took place at different places of the world and finally first published rules for systematic naming of coordination compounds in 1957 by the commission on the Nomenclature of Inorganic chemistry (CNIC) which is the apex body of IUPAC. The CNIC is responsible for the nomenclature of entire inorganic compounds. With the accumulation of large number of compounds the editorial committee modified the rules for nomenclature in 1970, in 1990, in 2001 and finally in 2005. In 1990 recommendations the CNIC introduced configuration index of coordination compounds which is discussed in the 'red book'*. General restructuring of IUPAC was made in the recommendations 2000** and finally 2005 recommendations*** is more or less is the same as that of 1990 recommendations except in case of organometallic nomenclature. The detailed discussion of the rules of nomenclature of 1990 and 2005 recommendations is beyond the scope of this book. We shall discuss here the following broad outline of recommendations of 2005.

In order to describe the constitution of coordination compounds the following sequential steps are to be considered :

(a) **Identification of central atom(s).** For complicated structures the name is easier to form if more central atoms are chosen.

(b) **Identification of ligands.**

(c) **Naming of ligands.** Anionic ligands require special endings.

(d) **Specification of coordination mode for each ligand.** The κ(kappa) convention may be applicable.

(e) **Order of ligands and central atom(s).** Ligand names are ordered alphabetically. Central atom(s) names are ordered accordingly.

(f) **Identification of coordination geometry and selection of polyhedral symbol.** Most structures will deviate from ideal polyhedra. The closest should be chosen.

(g) **Description of relative configuration.** Configuraton idex priority rule (CIP) is used.

(h) **Absolute configuration is determined.**

The systematic naming of coordination compounds follow the following steps :

(1) **Sequences of ligands and central atoms within names :**

(a) Ligand names are listed before the name(s) of the central atom(s).

(b) No spaces are left between parts of the name that refer to the same coordination entity.

(c) Ligand names are listed in alphabetical order (multiplicative prefixes indicating the number of ligands are not considered in determining that order).

(d) The use of abbreviations of ligands are not applicable.

(2) **Number of ligands in a coordination entity :**

(a) For simple ligands e.g. for monodentate ligands the prefixes di, tri, tetra, etc. are used for 2, 3, 4 etc. number of ligands. No enclosing marks are required.

---

* Nomenclature of Inorganic Chemistry, IUPAC recommendations 1990 (Red Book I), Edited by G. J. Leigh.

** Nomenclature of Inorganic Chemistry, IUPAC recommendations 2000 (Red Book II), Edited by J.A. Mc Cleverty and N.G. Connelly.

*** Nomenclature of Inorganic Chemistry, IUPAC recommendations 2005. Prepared for publication by Neil G. Connelly, RichardM. Hartshorn, Ture Damhus, Alan T Hutton.

(b) For complicated ligands such as for polydentate or ambidentate ligands etc. the prefixes bis, tris, tetrakis, etc. are used instead of di, tri, tetra, etc. Here enclosing marks are to be used around the multiplicand.

For example the naming for $(NH_3)_2$ will be diamine while for $(CH_3NH_2)_2$ will be bis (methylamine) to make a distinction from dimethylamine. No hyphen will be used e.g. tetra-ammine a wrong statement, correct one is tetraammine, petaammine etc.

(3) **Representation of the names of ligands :**

(a) For inorganic or organic anionic ligands if the name of the anions ends with 'ide', 'ite' or 'ate', the final 'e' is replaced by 'o' producing 'ido', 'ito' and 'ato' respectively. e.g. the halide ligands are named **fluorido, chlorido, bromido** and **iodido** and coordinated cyanide, hydride ($H^-$) are named as **cyanido, hydrido** respectively.

(b) Naming of neutral and cationic ligands including organic ligands are done without modification even if they carry the endings with 'ide', 'ite' or 'ate' e.g.

$CH_3CONH_2$ named as **acetamide** not as acetamido,

$CH_3OS(O)OH$ named as **methyl hydrogen sulfite**

(c) Enclosing marks are used for (i) naming of cationic and neutral ligands, (ii) naming of inorganic anionic ligands including multiplicative prefixes (such as triphosphato), (iii) naming of compositional names e.g. carbon disulfide, (iv) naming of substituted organic ligands and wherever necessary to avoid ambiguity. But for common ligand names such as aqua, ammine, carbonyl, nitrosyl, methyl, ethyl etc. no enclosing marks are required.

(4) **Charge numbers, oxidation numbers in a complex entity :**

(a) The oxidation number of the central atom in a coordination entity is shown by Roman numeral in brackets immediately following its name. If the complex entity is negative then the ending will be with the 'ate'. When necessary a negative sign is placed before the number. Arabic zero indicates the oxidation number zero.

(b) Alternatively, the charge on a complex entity may be indicated. The net charge may be indicated by arabic number which is preceding with the charge sign and denoted in brackets. It is to be noted that the name of the central atom including the ending 'ate' must be written without the intervention of a space e.g. $K_4[Fe(CN)_6]$ named as potassium hexacyanidoferrate(II), or potassium hexacyanidoferrate (4–), or tetrapotassium hexacyanidoferrate.

(5) **Specification of donor atoms–the kappa (κ) convention :** In cases where coordination occurs through one of several possible donor atoms of a ligand, the specific indication of the donor atom is achieved through the use of kappa convention in which the Greek lower case kappa (κ) is used to indicate the donor atom e.g. if glycinate anion ($gly^-$) coordinates through the nitrogen atom, the abbreviation of the would be gly-κN as in the complex $[M(gly\text{-}\kappa N)_3X_3]$ where X = halide ligand.

The cases where specification of the donor atom is not necessary for a ligand that can bind the central atom in more then one way are :

(a) monodentate O-bonded carboxylate groups
(b) monodentate C-bonded cyanide ligand (name of the ligand is 'cyanido')
(c) monodentate C-bonded carbon monoxide which is named as 'carbonyl'
(d) monodentate N-bonded nitrogen monoxide which is named as 'nitrosyl'.

It is evident that in these cases the ligand names imply the binding modes as indicated above.

Single ligating atoms are indicated by a Greek kappa (κ). κ symbols are placed after the portion of the ligand name that represents the ring, chain or substituent group in which the ligating atom is found. Multiplicative prefixes which apply to a ligand or portions of a ligand also apply to the donor atom symbols.

Simply the nitrogen bonded NCS is denoted by **thiocyanato-κN** while sulfur bonded NCS is named as **thiocyanato-κS.** Similarly nitrogen bonded nitrite is denoted by **nitrito-κN** and oxygen bonded nitrite is named as **nitrito-κO**. For example,

$[Co(NH_3)_5ONO]Cl_2$ named as pentaamminenitrito-κOcobalt(III) chloride and $[Co(NH_3)_5NO_2]Cl_2$ is named as pentamminenitrito-κNcobalt(III) chloride.

Donor atoms of a particular element may be distinguished by adding a right superscript numerical locant to the italicized element symbol or in simple cases a prime or primes e.g.

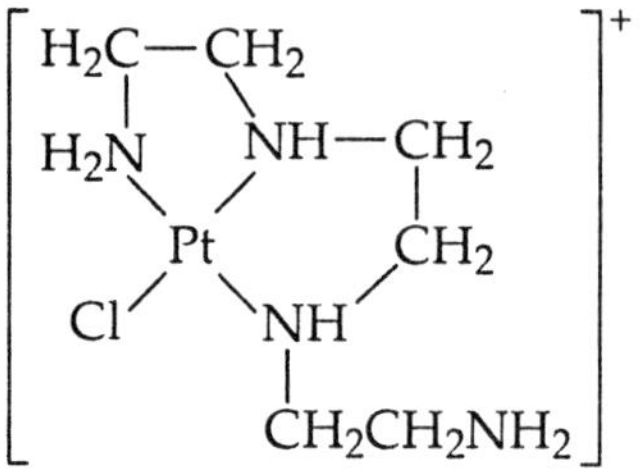


[N-(2-amino-κN ethyl)-N'-(2-aminoethyl) ethane-1,2-diamine-$\kappa^2$ N, N'] chloridoplatinum (II)

In cases where two or more identical ligands are involved then a superscript is used on κ to indicate the number of such ligational motif. Any multiplicative prefixes for complex entities may be operated on the κ symbol. Thus we can use the partial name '......bis(2-amino-κN ethyl)......' and not '......bis(2-amino-$\kappa^2$N ethyl)......' in the following example :

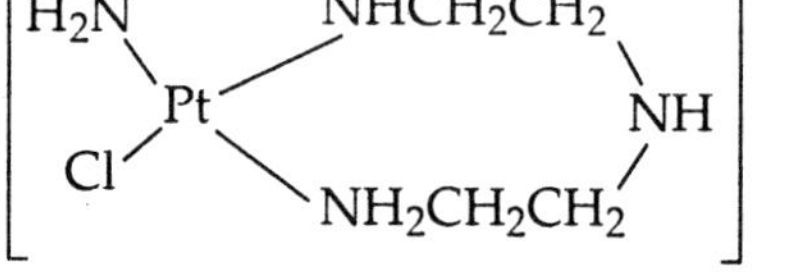


[N,N'-bis(2-amino-κNethyl)ethane-1,2-diamine-κN]chloridoplatinum(II)

$EDTAH_4$ or (ethane-1,2 diyldinitrilo) tetraacetato can function as hexadentate but may function as flexidentate behaviour namely bidentate, tetradentate and pentadentate as illustrated in the following examples :

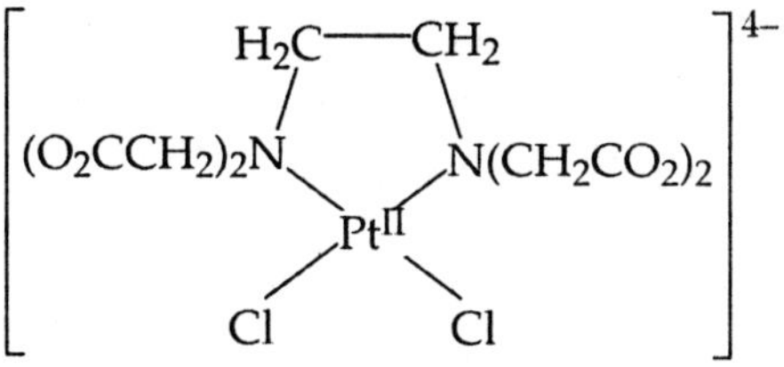

dichlorido[(ethane-1,2-diyldinitrilo-κ$^2$N,N′)tetraacetato]platinate(4–)

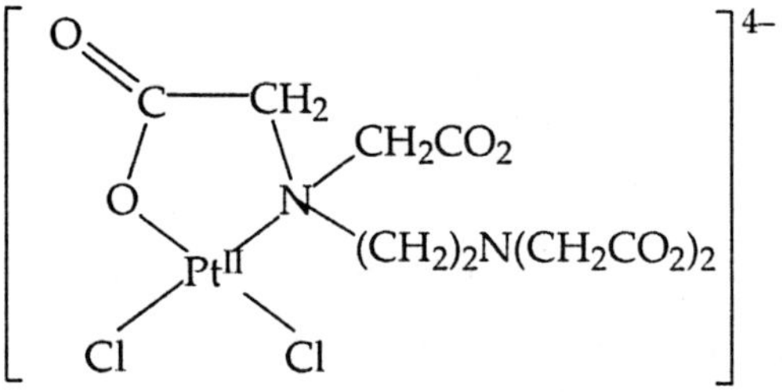

dichlorido[(ethane-1,2-diyldinitrilo-κN)(acetato-κO)triacetato]platinate(4–)

or

dichlorido[(ethane-1,2-diyldinitrilo-κN)(acetato-κO)triacetato]platinate(II)

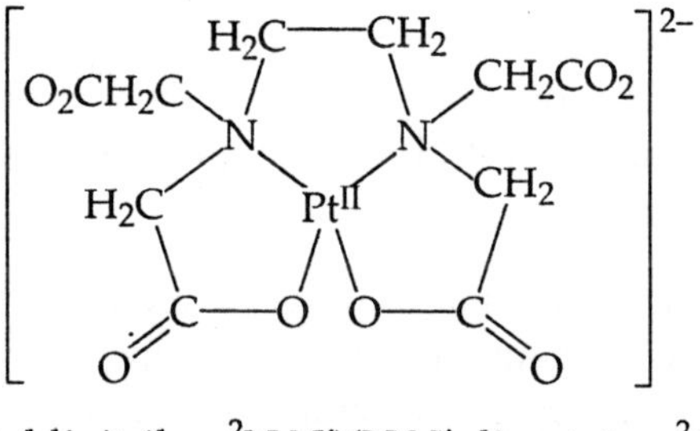

[(ethane-1,2-diyldinitrilo-κ$^2$N,N′)(N,N′-diacetato-κ$^2$O,O′)(N,N′-diacetato)platinate(II)

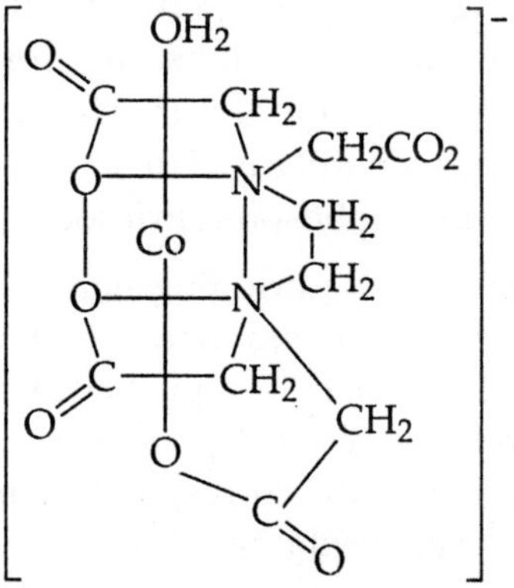

aqua[(ethane-1,2-diyldinitrilo-κ$^2$N,N′)tris(acetato-κO)acetato]cobaltate(III) *or*

aqua[(ethane-1,2-diyldinitrilo-$\kappa^2$N,N′)tris(acetato-κO)acetato]cobaltate(1–).

(6) **The eta (η) convention :** This convention is used in those cases where the central atom is bonded to contiguous donor atoms within one ligand. In most of the case this type of convention is used in organometallic compounds. One example of coordination compound is the following whose naming given below :

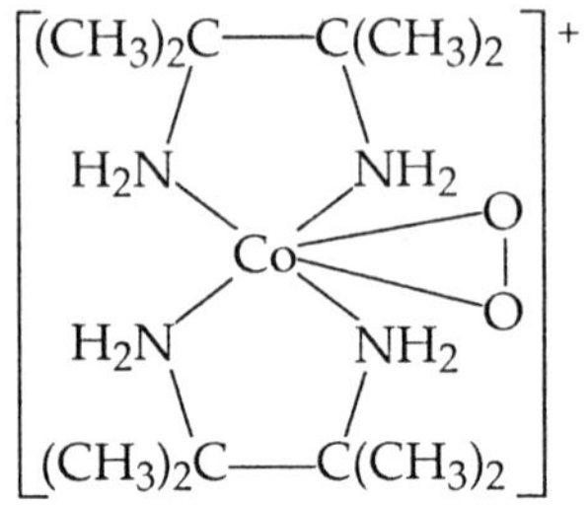

bis(2,3-dimethylbutane-2,3-diamine)($\eta^2$-peroxido)cobalt(1+)

*Comparison between κ and η convention*

Thus this is used only when there is more than one ligating atom and the contiguous atoms are often the same element. The κ convention is used to specify bonding from isolated donor atoms to one or more central atoms. In those cases where two or more identical ligands are attached to the central metal atom, a superscript is used on κ to indicate the number of donor atom to central metal atom bonds. In some cases the κ convention may be simplified. Donor atoms of a ligand may be denoted by adding only the italicized symbol(s) for the donor atom(s) to the end of the name of the ligand e.g. **thiocyanato-N** and **thiocyanato-S, nitrito-N** and **nitrito-O** etc.

(7) **Naming of polynuclear complexes :**

(a) If in a complex there is more than one metal atom exist, the complex entity is termed polynuclear. As a general rule in writing the formula or naming of polynuclear complexes as much structural information as possible should be presented. Ligands in polynuclear complexes are indicated in alphabetical order both in the formula and naming of the complex. The naming of each ligand is specified by subscript numerical multipliers in formula and by proper multiplicative prefixes in names. The number of central metal atoms of a given kind if greater than one, is written similarly. For example $[Rh_3H_3\{P(OMe)_3\}_6]$ is named as trihydridohexakis(trimethylphosphite)trirhodium(III).

(b) If in a polynuclear complex there exist a bridging ligand which is denoted by the greek letter μ appearing before the ligand name and separated from it by a hyphen(–). Bridging ligands are denoted in alphabetical order together with the other ligands. Multiplicative prefixes are employed e.g. di-μ-chlorido, tri-μ-chlorido etc. Bridging ligands are named before the corresponding non-bridging ligand as in di-μ-chlorido-tetrachlorido etc. The bridging index η which is the number of coordination centres connected by a bridging ligand is placed as a right subscript. The bridging index 2 is not generally indicated. Multiple bridging is listed in descending order of e.g. $\mu_3$-oxido-di-μ-oxido-trioxido. If required enclosing marks for ligand naming then μ is to be indicated within those marks.

The κ convention is used along with μ whenever necessary to denote which central atoms are bridged and through which donor atoms. The κ descriptor counts all donor atom-to-central atom bonds.

When single ligating atoms bind to two or more central atoms, the central atom locants are separated by a colon e.g. tri-μ-chlorido-1:2$\kappa^2$Cl; 1:3$\kappa^2$Cl; 2:3 $\kappa^2$Cl indicates that there are three bridging chloride ligands and they bridge between central atoms 1 and 2, 1 and 3, and 2 and 3. It is to be noted that because of the use of the colon, sets of bridge locants are separated here by semicolons rather than commas. For example,

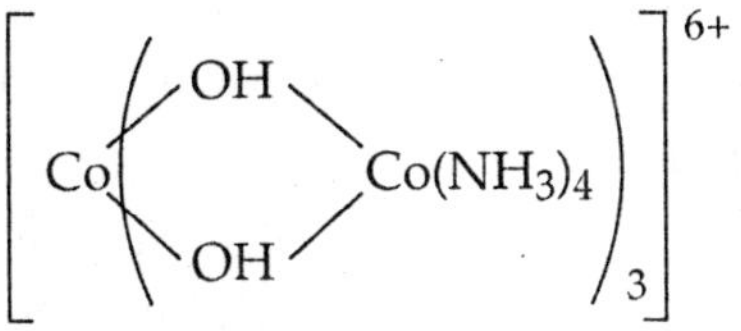

Formula $[Co\{(\mu\text{-}OH)_2Co(NH_3)_4\}_3]^{6+}$

Naming : dodecaammine-1$\kappa^4$N, 2$\kappa^4$N, 3$\kappa^4$N-hexa-μ-hydroxido-1:4$\kappa^4$O; 2:4$\kappa^4$O; 3:4$\kappa^4$O-tetracobalt (6+) or

dodecaammine-1$\kappa^4$N, 2$\kappa^4$N, 3$\kappa^4$N-hexa-μ-hydroxido-1:4$\kappa^4$O; 2:4$\kappa^4$O; 3:4$\kappa^4$O-tetracobalt (III)

(c) **Metal-metal bonding** or more generally, bonding between central atoms in complexes, may be indicated in names by placing italicized atomic symbols of the appropriate central atoms, separated by an 'em' dash and enclosed in parentheses, after the list of central atom names and before the ionic charge. The central atom element symbols are placed in the same order as the central atoms appear in the name. The number of such bonds are denoted by an arabic numeral placed before the first element symbol and separated from it by a space. For example,

*Naming of polynuclear complexes with metal-metal bonds*

$[Cl_4\overset{1}{Re}\overset{2}{Re}Cl_4]^{2-}$ which is named as :
octachlorido-1$\kappa^4$Cl, 2$\kappa^4$Cl-dirhenate(Re-Re)(2–)

The **symmetrical dinuclear complexes** are named on simplification by employing multiplicative prefixes e.g.

1. $[Re_2Br_8]^{2-}$ bis(tetrabromidorhenate)(Re-Re)(2–)
2. $Cs_3[Re_3Cl_{12}]$ caesium dodecachlorido-triangulo-trirhenate (3Re-Re)(3–)
3. $Mn_2(CO)_{10}$ bis(pentacarbonylmanganese)(Mn-Mn)
4. $[\{Cr(NH_3)_5\}_2(\mu\text{-}OH)]^{5+}$ μ-hydroxido-bis(pentaamminechromium)(5+)
5. $[\{Cu(py)\}_2(\mu\text{-}O_2CMe)_4]$ tetrakis(μ-acetato-κO : κO′) bis[(pyridine)copper(II)]

6. $[\{PtCl(PPh_3)\}_2(\mu\text{-}Cl)_2]$ — di-μ-chlorido-bis[chlorido(triphenylphosphane)platinum]
7. $[Pb_2(C_2H_5)_6]$ or $[(C_2H_5)_3PbPb(C_2H_5)_3]$ — hexaethyldilead (Pb-Pb) or bis(triethyllead)(Pb-Pb)

The naming of the unsymmetrical dinuclear coordination entity results from the rules governed by rules of (a), (b) and (c) of naming of polynuclear complexes above. For the complexes of this type—(1) the central atom with greater number of ligands is numbered 1 and the other is numbered 2. If for a complex where the same number of ligands are attached to both the central atoms, then the lower number is applied to the central atom which has greater number of alphabetically preferred ligands but for complexes where dissimilar central atoms exist, the lower number is assigned to more metallic central atom. For example the following complexes are named as :

1. $[ClHgIr(CO)Cl_2(PPh_3)_2]$ — carbonyl-1κC-trichlorido-1κ²Cl, 2κCl-bis(triphenylphosphane-1κP) iridiummercury (Ir-Hg)
2. $[(H_3N)_3\overset{1}{C}o(\mu\text{-}NO_2)(\mu\text{-}OH)_2\overset{2}{C}o(NH_3)_2(py)]^{3+}$ — pentaammine-1κ³N,2κ²N-di-μ-hydroxido-μ-nitrito-1κN:2κO-(pyridine-2κN)dicobalt(3+)
3. $[(H_3N)_5\overset{1}{C}r(\mu\text{-}OH)\overset{2}{C}r(NH_2Me)(NH_3)_4]^{5+}$ — nonaammine-1κ⁵N,2κ⁴N-μ-hydroxido-(methanamine-2κN)dichromium(5+)
4. $[\{Co(NH_3)_3\}_2(\mu\text{-}NO_2)(\mu\text{-}OH)_2]^{3+}$ — di-μ-hydroxido-μ-nitrito-κN:κO-bis(triamminecobalt)(3+)

For symmetrical species it is not necessary to denote the number explicitly for two central atoms.

In summary the naming of polynuclear complexes is as follows :

(i) Identify the central atoms and ligands.

(ii) Name the ligands with κ, η and μ designations. It is to be noted that ligand names may be modified if κ, η or μ designations applied only to some portions of the ligand that are otherwise equivalent and described by multiplicative prefix such as **tri** or **tris**.

(iii) Naming of ligands will be done according to alphabetical order.

(iv) Numbering of central atom(s) is done by following rules :

(a) Generally the element sequence are followed according to the group sequence of the periodic table.

(b) Within each class of identical central atoms, lower locant numbers are assigned to central atoms with higher coordination numbers.

(c) Specify the ligating atoms with κ or η designation whereever necessary.

(d) Any central atoms that are inequivalent and have not yet been assigned distinct locant numbers will differ only in the other central atoms to which they are directly bonded. The locant numbers of these directly bonded neighbouring central atoms are compared and the central atom with the lowest-locant neighbouring atoms is given the lowest of the remaining possible locants e.g.

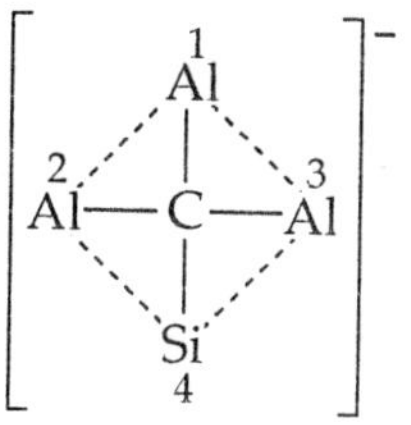


$\mu_4$-carbido-quadro-(trialuminiumsilicon)ate(Al[1]-Al[2])(Al[1]-Al[3])(Al[2]-Si)(Al[3]-Si)(1–)

**As per recommendations of 2005 the chemical formula and nomenclature of following coordination complexes are given below :**

1. $[CoCl(NH_3)_5]Cl_2$
   pentaamminechloridocobalt(2+) chloride *or*
   pentaamminechloridocobalt(III) chloride
2. $[Co(NH_3)_6]Cl_3$
   hexaamminecobalt(III) chloride
3. $[CoCl(NH_3)_4(NO_2)]Cl$
   tetraamminechloridonitrito-κN-cobalt(III) chloride
4. $K_2[PdCl_4]$
   potassium tetrachloridopalladate(II)
5. $K_2[OsCl_5N]$
   potassium pentachloridonitridoosmate(2–)
6. $[Co(en)_3]Cl_3$
   tris(ethane-1, 2-diamine)cobalt(III) chloride
7. $[ReCo(CO)_9]$
   nonacarbonylrheniumcobalt
8. $[(OC)_5\overset{1}{Re}\overset{2}{Co}(CO)_4]$
   nonacarbonyl-1$\kappa^5$C, 2$\kappa^4$C-rheniumcobalt *or*
   nonacarbonyl-1$\kappa^5$C, 2$\kappa^4$C-rheniumcobalt (Re-Co)
9. $[Ni(OH_2)_6]^{2+}$
   hexaaquanickel(II) or hexaaquanickel(2+)

10. 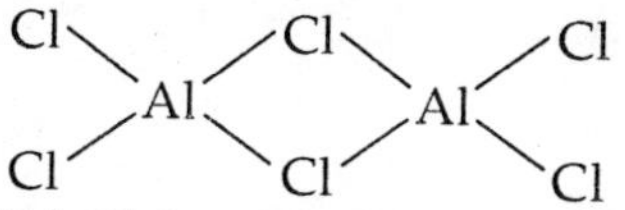


$[Al_2Cl_4(\mu\text{-}Cl)_2]$ or $[Cl_2Al(\mu\text{-}Cl)_2AlCl_2]$
di-μ-chlorido-tetrachlorido-1$\kappa^2$Cl, 2$\kappa^2$Cl dialuminium

11. 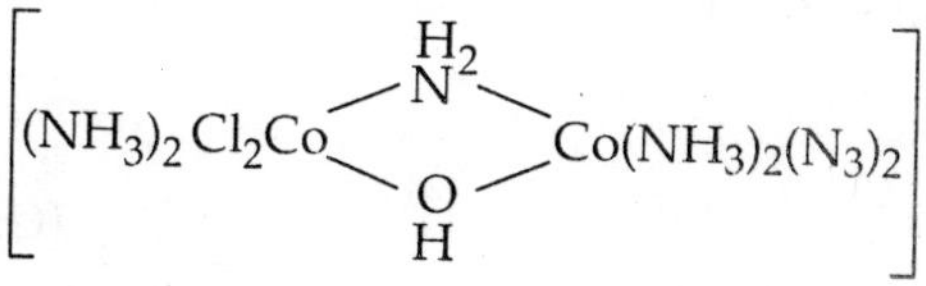 [B.U.2012]


diamminedichloridocobalt(III)-μ-amido-μ-hydroxidodiamminediazidocobalt(III)

12. $[Ru(NH_3)_5N_2]Cl_2$ [B.U.2009]
pentaammine(dinitrogen)ruthenium(II) chloride *or*
pentaammine(dinitrogen)ruthenium(2+) chloride

13. $[Ni(NH_3)_6][Co(NO_2)_6]$ [B.U.2009, C.U.2004]
hexaamminenickel(III)hexakis(nitrito-κN)cobaltate(III) *or*
hexaamminenickel(III)nitrito-$\kappa^6$Ncobaltate(III)

14. $[Cr(en)_3][Co(CN)_6]$ [C.U.2007]
tris(ethane-1,2-diamine)chromium(III)hexacyanidocobaltate(III)

15. 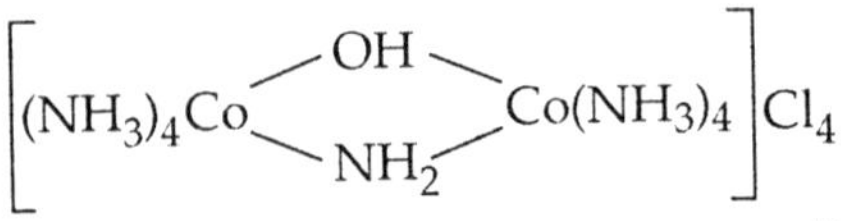

[B.U.2013,2008,2005,2000,1992; C.U.2003]
μ-amido-μ-hydroxidooctaamminedicobalt(III) chloride

16. $Rb_2[Os(CN)_5NO]$ [B.U.2006]
rubidium pentacyanidonitrosyl-κNosmate(2–)

17. $[Co(en)_3]^{3+}$ [B.U.2006]
tris(ethane-1,2-diamine)cobalt(III)

18. $[Cr(C_2O_4)_3]^{3-}$ [B.U.2006]
tris(oxalato)chromate(III)

19. $[Co(NH_3)_5ONO]Cl_2$ [C.U.2006, B.U.2012]
pentaamminenitrito-κOcobalt(III) chloride

20. 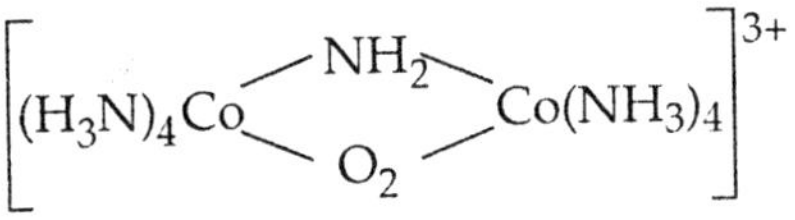

[C.U.2007]
μ-amido-μ-peroxidooctaamminedicobalt(III)

21. $[(NH_3)_4Co(OH)(NH_2)Co(NH_3)_4]^{4+}$ [C.U.2006]
μ-amido-μ-hydroxidoctaaminedicobalt(III) *or*
μ-amido-μ-hydroxidooctaamminedicobalt(4+)

22. $NH_4[Cr(NCS)_4(NH_3)_2]$ [B.U.2004,2002]
ammonium diamminetetrakis(thiocyanato-κN)chromate(III) *or*
ammonium diammine thiocyanato-$\kappa^4$Nchromate(III)

23. $[(NH_3)_5Cr(OH)Cr(NH_3)_5]Cl_5$ [C.U.2004]
μ-hydroxidodecaamminedichromium(III) chloride *or*
μ-hydroxido bis(pentaamminechromium(III)) chloride

24. $[Cr(NH_3)_4Cl_2]Cl$ [B.U.2001]
tetraamminedichloridochromium(III) chloride

25. 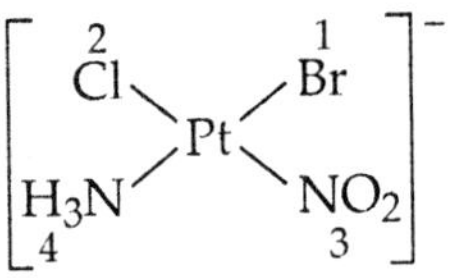

[C.U.2003]

[SP-4-4]amminebromidochloridonitrito-κNplatinate(II)(Priority number rule is applied)
or
amminebromidochloridonitrito-κNplatinate(1–)

26. $NH_4[Cr(SCN)_4(NO)_2]$ [B.U.1999]
ammonium bis(nitrosyl-κN)tetrakis(thiocyanato-κS)chromate(III) *or*
ammonium nitrosyl-$\kappa^2$Nthiocyanato-$\kappa^4$Schromate(III)

27. $[Pt(NH_3)_4Cl_2][PtCl_4]$ [B.U.1999]
tetraamminedichloridoplatinum(IV)tetrachloridoplatinate(II)

28. $[(CO)_3\overset{1}{Fe}(CO)_3\overset{2}{Fe}(CO)_3]$ [B.U.1999,1992]
hexacarbonyl-1$\kappa^3$C, 2$\kappa^3$Ctri-μ-carbonyldiiron *or*
tri-μ-carbonyl-bis(tricarbonyliron)(Fe-Fe)

29. $Na[Pt(Cl)(Br)(NO_2)(H_2O)]$ [B.U.1998]
sodium aquabromidochloridonitrito-κNplatinate(II)

30. $[Cr(NH_3)_6][Mo(CN)_6]$ [B.U.1998]
hexaamminechromium(III)hexacyanidomolybdate(III)

31. 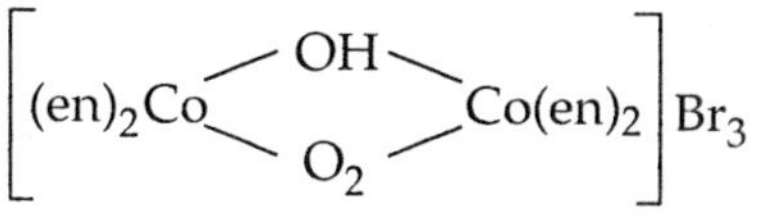

[B.U.1996]
μ-hydroxido-μ-peroxidotetrakis(ethane-1,2-diamine)dicobalt(III) bromide

32. $[Co(NH_3)_5SO_4]Br$ [B.U.1996,1995]
pentraamminesulphatocobalt(III) bromide

33. $[Co(en)_2SO_4]Br$ [B.U.1996]
bis(ethane-1,2-diamine)sulphatocobalt(III) bromide

34. $[Co(NO_3)_5Br]SO_4$ [B.U.1996]
bromidopentanitratocobalt(2+) sulphate

35. $[(H_3N)_5Co(O_2^{2-})Co(NH_3)_5]^{4+}$ [B.U.1996]
μ-peroxidodecaamminedicobalt(III)

36. $[AlCl_4]^-$
tetrachloridoaluminate (1–) *or* tetrachloridoaluminate(III)

37. $LiAlH_4$ [B.U.1989]
lithium tetrahydridoaluminate(1–) *or* lithium tetrahydridoaluminate(III)

38. 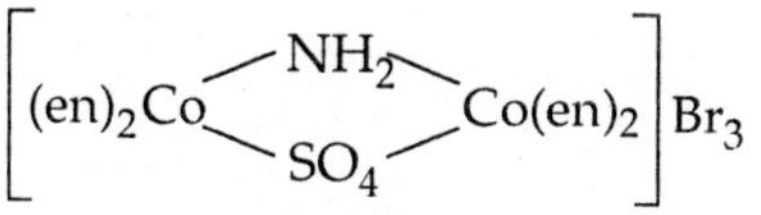

[B.U.1995]
μ-amido-μ-sulphatotetrakis(ethane-1,2-diamine)dicobalt(III) bromide

39. $[Co(NH_3)_5Br]SO_4$ [B.U.1995]
pentaamminebromidocobalt(III) sulphate

40. $[Co(NH_3)_4ClNO_2]Cl$ [B.U.1993]
tetraamminechloridonitrito-κNcobalt(III) chloride

41. $[Cr(en)_2(H_2O)_2]Br_3$ [B.U.1993]
diaquabis(ethane-1,2-diamine)chromium(III) bromide

42. $[Cr(en)_2(H_2O)Br]Br_2 \cdot H_2O$ [B.U.1993]
aquabromidobis(ethane-1,2-diamine)chromium(III) bromide-water

43. K[Co(EDTA)] [B.U.1989]
potassium (2,2′,2″,2‴-(ethane-1,2-diyldinitrilo)tetracetato)cobaltate(III) *or*
potassium (ethylenediaminetetraacetato)cobaltate(III)

44. $[Co(NH_3)_5N_3]Cl_2$ [B.U.1987]
pentaammineazidocobalt(III) chloride

45. $Na_2[Fe(CN)_5NO]$ [B.U.1989]
sodium pentacyanidonitrosyl-κN ferrate(2–)

46. $K[PtCl_3(C_2H_4)]$ [B.U.1980]
potassium trichlorido(ethylene)platinate(II)

47. $K_2[OsNCl_5]$ [B.U.1980]
potassium pentachloridonitridoosmate(VI)

48. $[Pt(NH_3)_2Cl_2]$ [B.U.1984]
diaminedichloridoplatinum(II)

49. $Na[Co(CO)_4]$
sodium tetracarbonyl-$\kappa^4$C cobaltate(1–)

50. $K[CrOF_4]$
potassium tetrafloridoxidochromate(1–)

51. $[(CO)_5\overset{1}{Mn}\text{-}\overset{2}{Mn}(CO)_5]$
decacarbonyl-1$\kappa^5$C, 2$\kappa^5$C-dimanganese *or*
decacarbonyl-1$\kappa^5$C, 2$\kappa^5$C-dimanganese(Mn-Mn) *or*
bis(pentacarbonylmanganese)(Mn-Mn)

52. $Na_4[Fe(CN)_5NOS]$
sodium pentacyanidonitrosyl-κNsulphidoferrate(4–) *or*
sodium pentacyanidonitrosyl-κNsulphidoferrate(II)

53. $[Fe(H_2O)_5NO]SO_4$
pentaaquanitrosyl-κNiron(2+) sulphate

54. $Na_3[Ag(S_2O_3)_2]$
sodium bis(thiosulphato-$\kappa^2$S)argentate(3–) *or*
sodium bis(thiosulphato-$\kappa^2$S)argentate(I)

55. $Ba[BrF_4]_2$
barium tetrafluoridobromate(2–)

56. $H_2[PtCl_6]$
hydrogen hexachloridoplatinate(2–) *or*
hydrogen hexachloridoplatinate(IV)

57. $[Pd(bpy)(NCS)_2]$
(2,2′-bipyridine)bis(thiocyanato-$\kappa^2$N)palladium(II)

58. $[Pd(bpy)(SCN)_2]$
(2,2′-bipyridine)bis(thiocyanato-$\kappa^2$S)palladium(II)

59. $[Pt(NH_3)_2Cl(NH_2CH_3)]Cl$
diamminechlorido(methylamine)platinum(II) chloride *or*
diamminechlorido(methanamine)platinum(II) chloride

60. $[Co(en)_2(H_2O)(OH)]^{2+}$ [B.U.1985]
aquahydroxidobis(ethane-1,2-diamine)cobalt(2+)

61. $[CuCl_2\{O{=}C(NH)_2\}_2]$
dichloridobis(urea)copper(II)

62. $[Fe(CNMe)_6]Br_2$
hexakis(methylisocyanide)iron(II) bromide

63. $Na[PtBrCl(NH_3)(NO_2)]$
sodium amminebromidochloridonitrito-κN-platinate(1–) *or*
sodium amminebromidochloridonitrito-κN-platinate(II)

64. $[Rh_3H_3\{P(OMe)_3\}_6]$
trihydridohexakis(trimethyl phosphite)trirhodium

65. $[Cl_4\overset{1}{Re}\overset{2}{Re}Cl_4]^{2-}$
octachlorido-1$\kappa^4$Cl, 2$\kappa^4$Cl-dirhenate(2–) *or*
octachlorido-1$\kappa^4$Cl, 2$\kappa^4$Cl-dirhenate(Re-Re)(2–)

66. $Na[B(NO_3)_4]$
sodium tetranitratoborate(III) *or*
sodium tetranitratoborate(1–)

67. $[CuCl_2(NH_2Me)_2]$
dichloridobis(methylamine)copper(II)

68. $[Pt(\eta^2\text{-}C_2H_4)Cl_3]^-$
trichlorido ($\eta^2$-ethene)platinate(II)

69. $[\{Cr(NH_3)_5\}_2(\mu\text{-}OH)]^{5+}$
μ-hydroxido-bis(pentaamminechromium)(5+)

70. $[Co(CO)_4]^-$
tetracarbonylcobaltate(1–)

71. $[\{Co(NH_3)_3\}_2(\mu\text{-}NO_2)(\mu\text{-}OH)_2]^{3+}$
di-μ-hydroxido-nitrito-κN :κO-bis(triamminecobalt)(3+)

72. $[Cr_2(\mu\text{-}O)(OH)_8(\mu\text{-}OH)]^{5-}$
μ-hydroxido-octahydroxido-μ-oxido-dichromate(5–)

**Problem 7.1** Molar conductances at a dilution of 1024 litres of $PtCl_4 \cdot 2NH_3$, $PtCl_4 \cdot 3NH_3$ and $PtCl_4 \cdot 6NH_3$ are 7, 97 and 520 $ohm^{-1}cm^2$ respectively. Rationalise these data in the light of Werner's theory. **[C.U. 2007, B.U. 2012]**

**Ans. :** The compound $PtCl_4 \cdot 2NH_3$ is formulated as $[Pt(NH_3)_2Cl_4]$ which shows poor conductance at a dilution of 1024 litres because the complex is not dissociated to form ions.

Compound $PtCl_4 \cdot 3NH_3$ is formulated as $[Pt(NH_3)_3Cl_3]Cl$ which in solution dissociated to form $[Pt(NH_3)_3Cl_3]^+$ and $Cl^-$ ions for which the conductance 97 $ohm^{-1}$ $cm^2$ is observed because in this case there will be the formation of two ions.

More greater conductance value is to be expected for the compound $PtCl_4 \cdot 6NH_3$ which is formulated as $[Pt(NH_3)_6]Cl_4$ which on dissociation produces $[Pt(NH_3)_6]^{4+}$ and $4Cl^-$ ions i.e. a total of five ions are produced and so a conductance value of 520 $ohm^{-1}$ $cm^2$ is registered for this complex.

All the complexes has coordination number six according to Werner's theory and all have octahedral geometry.

**Problem 7.2** A solution of 0.1332g of $CrCl_3 \cdot 6H_2O$ was passed through a cation exchange resin in the acid form. 8.0 ml of a standard caustic soda solution (containing 5.0g NaOH/litre) was required for the titration of the liberated acid. Determine the correct formulation of the complex and give its IUPAC name. Given atomic weight of Cr is 52. **[C.U. 1999]**

**Ans. :** Strength of NaOH solution = $\frac{5}{40}$(N) = 0.125(N)

8 ml 0.125(N) NaOH $\equiv$ 1.0 ml(N) NaOH $\equiv$ 1 ml(N) HCl

1 ml(N) HCl $\equiv$ 0.0355 g$Cl^-$ ion.

Let $x$ is the number of $Cl^-$ ions outside the coordination entity of $CrCl_3 \cdot 6H_2O$

Molecular weight of $CrCl_3 \cdot 6H_20$ = 266.5

$\therefore$ 266.5g $CrCl_3 \cdot 6H_2O$ has $x \times 35.5$ g exchangable $Cl^-$ ions, so that 0.1332 g has

exchangable $Cl^-$ ion = $\frac{x \times 35.5 \times 0.1332}{266.5}$ g.

According to condition, $\frac{x \times 35.5 \times 0.1332}{266.5}$ = 0.0355, so that $x$ = 2

Therefore the correct formulation of the complex is, $[Cr(H_2O)_5Cl]Cl_2.H_2O$

The IUPAC name of the compound is pentaaquachloridochromium(III) chloride-water.

## 7.6.1 Classical names of some complexes

Based on the names of discoverers some of the complexes named as :

| Complex | Name |
|---|---|
| cis-$Pt(NH_3)_2Cl_4$ | Cleve's salt |
| $K[Co(NH_3)_2(NO_2)_4]$ | Erdmann's salt |
| $NH_4[Cr(NH_3)_2(NCS)_4]$ | Reinecke's salt |
| $K[PtCl_3(C_2H_4)]$ | Zeise's salt |
| $[Pt(NH_3)_3Cl]Cl$ | Cleve's triammine |

| | |
|---|---|
| $K[Pt(NH_3)Cl_3]$ | Cossa's first salt |
| $K[Pt(NH_3)Cl_5]$ | Cossa's second salt |
| $[Pt(NH_3)_4][PtCl_4]$ | Magnus's green salt |
| $[Pt(NH_3)_3Cl]_2[PtCl_4]$ | Magnus's pink salt |
| $K_3[Co(NO_2)_6]$ | Fischer's salt |
| $[Co(NH_3)_3(NO_2)_3]$ | Gibb's salt |
| $[Ir(CO)(PPh_3)_2Cl]$ | Vaska's complex |
| $[Rh(PPh_3)_3Cl]$ | Wilkinson's catalyst |
| $Pt(EtNH_2)_4Cl_3.2H_2O$ | Wolfram's red salt |
| $[Ni(en)_2]X_2$ | Lifschitz salt (en=substituted ethylenediamine) |

## 7.7 SYNTHETIC STRATEGIES OF COORDINATION COMPLEXES

There are several methods available for the synthesis of complexes. We shall discuss here some broad outline of the methods for the preparation of homo and hetero ligand complexes.

### 1. Substitution of coordinated ligands by other ligands

Generally water is taken as the solvent for the synthesis of coordination complexes and transition metal ions will form hydrated metal ion in aqueous solution. The existance of these hydrated species is verified by several experimental studies. Hydration is associated to form primary and secondary hydration sphere (Fig. 3.3) e.g. $Li^+$ is attached with four water molecules in primary hydration sphere and eight water molecules in the secondary sphere. Sometimes hydrated crystal contains aquated ion e.g. $NiSO_4 \cdot 7H_2O$ actually have $[Ni(H_2O_6]^{2+}$ ion, one molecule of water is held by hydrogen bonds to an oxygen atom of sulphate and the hydrogen atom of a coordinated water molecule (7-XIV).

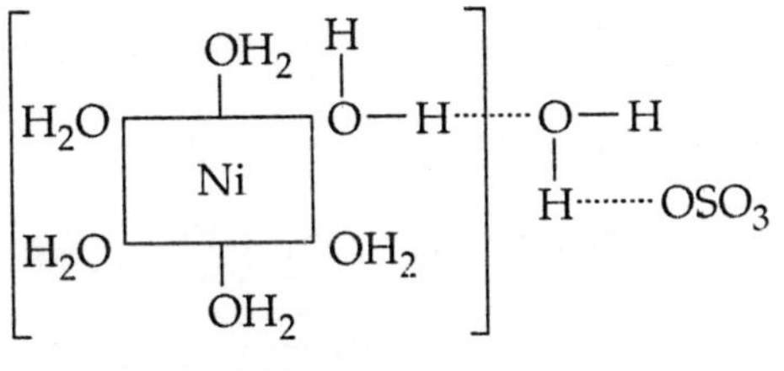

(7-XIV)

Synthesis of coordination complexes is carried out by substitution of coordinated water molecules by other ligands. Thus,

$$[Ni(H_2O)_6]^{2+} \xrightarrow[NH_4Cl\ /\ NH_3]{25\%\ \text{aqueous}\ NH_3} \text{isolation of } [Ni(NH_3)_6]Cl_2$$

$$CuSO_4 \cdot 5H_2O \longrightarrow [Cu(H_2O)_4]SO_4 + H_2O$$

$$[Cu(H_2O)_4]SO_4 \xrightarrow{\text{aqueous}\ NH_3} [Cu(NH_3)_4]SO_4 + 4H_2O$$

In certain cases the substitution is extremely slow e.g.

$$[Cr(H_2O)_6]^{3+} + 6NH_3 \longrightarrow [Cr(NH_3)_6]^{3+} + 6H_2O$$

Thus to carry out substitution reaction both thermodynamic stability* as also kinetic aspects of the reaction have to be considered.

Other examples are : $\underset{wine\ red}{[RhCl_6]^{3-}} + 3C_2O_4^{2-} \xrightarrow[2\ hr]{100°C} \underset{yellow}{[Rh(C_2O_4)_3]^{3-}} \xrightarrow[evaporation]{K^+} K_3[Rh(C_2O_4)_3](s)$

$$[Fe(H_2O)_6]^{2+} \xrightarrow[\text{dissolved in alcohol}]{3bpy} [Fe(bpy)_3]^{2+} + 6H_2O$$

$$[Co(NO_2)_6]^{3-} + 2en \longrightarrow [Co(NO_2)_2(en)_2]^+$$

$$[CoCl_2(en)_2]^+ + C_2O_4^{2-} \longrightarrow [Co(en)_2C_2O_4]^+ + 2Cl^-$$

In general we can say that interaction of a ligand L with a metal $M^{n+}$(aq) results in successive replacement of water molecules from the coordination sphere of the metal. Thus,

$$[M(H_2O)_6]^{n+} + L \rightleftharpoons [M(H_2O)_5L]^{n+} + H_2O$$

$$[M(H_2O)_5L]^{n+} + L \rightleftharpoons [M(H_2O)_4L_2]^{n+} + H_2O \text{ and so on.}$$

This is the most common technique of synthesis and is usually employed excess ligand to ensure complete substitution and to avoid the formation of intermediate mixed ligand complexes.

## 2. Redox reactions

The stabilisation of a particular oxidation state of a metal is influenced by judicious choice of ligands. Thus CO will stabilise Ni° while $NH_3$ or DMGH will stabilise $Ni^{2+}$. Cobalt is present in aqueous solution as $Co^{2+}$, but in presence of $NH_3$ this is oxidised to $Co^{3+}$. Thus $[Co(NH_3)_6]Cl_3$ is prepared by the oxidation of an aqueous solution of $CoCl_2$ containing $NH_4Cl$ and $NH_3$ with $H_2O_2$.

$$2CoCl_2 + 2NH_4Cl + 10NH_3 + H_2O_2 \xrightarrow[catalyst]{charcoal} 2[Co(NH_3)_6]Cl_3 + 2H_2O$$

or by bubling air through aqueous solution of $Co^{2+}$

$$4[Co(H_2O)_6]Cl_2 + 4NH_4Cl + 20NH_3 + O_2 \xrightarrow[catalyst]{charcoal} 4[Co(NH_3)_6]Cl_3 + 26H_2O$$

or $2CoCl_2 + 2NH_4Cl + 10NH_3 + \frac{1}{2}O_2 \xrightarrow[catalyst]{charcoal} 2[Co(NH_3)_6]Cl_3 + H_2O$

In absence of the charcoal catalyst there will be the formation of pentaammineaqua complex.

$$2[Co(H_2O)_6]Cl_2 + 2NH_4Cl + 8NH_3 + H_2O_2 \longrightarrow 2[Co(NH_3)_5(H_2O)]Cl_3 + 12H_2O$$

Similarly $[Co(NH_3)_5NO_3](NO_3)_2$ can be prepared by oxidising $[Co(H_2O)_6]^{2+}$ containing $NH_4NO_3$ and $NH_3$ with $H_2O_2$.

$K_3[Cr(C_2O_4)_3]$ can be prepared by reducing $K_2Cr_2O_7$ with $H_2C_2O_4$ and $K_2C_2O_4$.

$$K_2Cr_2O_7 + 7H_2C_2O_4 + 2K_2C_2O_4 \longrightarrow 2K_3[Cr(C_2O_4)_3] + 6CO_2 + 7H_2O$$

* will be discussed later.

Other examples are :

$$K_2[Ni(CN)_4] \xrightarrow[\text{liquid } NH_3]{\text{excess K}} K_4[Ni(CN)_4]$$

$$2CuCl_2 + 8N_2H_4 \longrightarrow 2[Cu^I(N_2H_4)_2]Cl + 2N_2H_4{\cdot}HCl + 2NH_3 + N_2$$

$$[Pt^{II}(NH_3)_4]Br_2 + 2K \xrightarrow[\text{reduction}]{\text{liquid } NH_3} [Pt(NH_3)_4] + 2KBr$$

$$[Fe(CN)_6]^{4-} + [IrCl_6]^{2-} \longrightarrow [Fe(CN)_6]^{3-} + [IrCl_6]^{3-}$$

In $[Pt(NH_3)_4]$, low oxidation state of metal is stabilised by ammonia though it has no capacity to accept π-back bonding.

### 3. Direct combination

Direct reaction between metal/metal salts with ligands in nonaqueous medium led to the formation of coordination complexes. For example,

$$\underset{\textit{anhydrous}}{CrCl_3} + 3en \xrightarrow{\text{ether}} \underset{\textit{yellow}}{[Cr(en)_3]Cl_3}$$

$$VCl_4 + 2py \xrightarrow{\text{toluene}} [VCl_4(py)_2]$$

$$SnCl_4 + 2Et_2O \xrightarrow{\text{benzene}} [SnCl_4(Et_2O)_2]$$

Many metal ammines are prepared by direct combination of metal salt with liquid $NH_3$ and then evaporating the excess $NH_3$. For example,

$$NiCl_2 + 6NH_3 \longrightarrow [Ni(NH_3)_6]Cl_2$$

$$\underset{\textit{anhydrous}}{CrCl_3} + 6NH_3(l) \xrightarrow[\text{liquid } NH_3 \text{ is evaporated}]{\text{excess}} \underset{\textit{pink}}{[CrCl(NH_3)_5]Cl_2} + \underset{\textit{yellow}}{[Cr(NH_3)_6]Cl_3}$$

$$FeCl_2 + 6NH_3 \longrightarrow [Fe(NH_3)_6]Cl_2$$

$$CdCl_2 + 4NH_3 \longrightarrow [Cd(NH_3)_4]Cl_2$$

These reactions are particularly suitable in those cases where the metal is precipitated as hydroxide in aqueous $NH_3$ medium.

Besides ammine complexes, other complexes have been prepared by direct combination method e.g.

$$Ni + 4CO \xrightarrow[\text{and temperature}]{\text{ordinary pressure}} Ni(CO)_4$$

$$Fe + 5CO \xrightarrow{200°C/100 \text{ atm. pressure}} Fe(CO)_5$$

$$CrCl_3 + 6CO \xrightarrow{Et_3Al \text{ in ether}} Cr(CO)_6$$

$$SnCl_4 + 2NMe_3 \xrightarrow[\text{petroleum ether}]{40\text{-}60°C} [SnCl_4(NMe_3)_2]$$

$$2CoCO_3 + 8CO \xrightarrow[150°C/250 \text{ atm. pressure}]{2H_2} Co_2(CO)_8 + 2CO_2 + 2H_2O$$

$$NiCl_2 + 3en \longrightarrow [Ni(en)_3]Cl_2$$

### 4. Template synthesis

This is discussed in Sec. 7.4.3(d).

### 5. Ligand transformation or destruction by reactions of coordinated ligands

In case of inert complexes or complexes of chelating ligands, the coordinated ligand is transformed to other ligand remaining the similar metal-donor environment. The O–C bond of $CO_3^{2-}$ is readily broken by acids liberating $CO_2$ and transforming the carbonato complex to aqua complex :

$$\underset{pink}{[(NH_3)_5Co(O{-}CO_2)]^+} + 2H^+ \longrightarrow \underset{pink}{[(NH_3)_5Co(OH_2)]^{3+}} + CO_2(g)$$

The $NO_2$ group of a Pt(II) complex can be converted to coordinated $NH_3$ molecule by reduction with Zn/HCl :

$$[(NH_3)_3Pt - NO_2]^+ \xrightarrow{Zn/HCl} [(NH_3)_3Pt - NH_3]^{2+}$$

This reaction is of similar type in organic chemistry where aromatic nitro compounds are reduced to aromatic amines.

Sometimes ultraviolet or visible radiations are used to promote reactions of coordinated ligands. Thus in the following chelate the $C_2O_4^{2-}$ is eliminated by irradiation to form two mole of $CO_2$ and two mole of electrons and a complex is formed having lower coordination number and lower oxidation state of the metal than the starting complex. The complex formed is coordinatively unsaturated, adds up with other ligands to form a new complex :

$$[Pt^{II}(C_2O_4)(PR_3)_2] \xrightarrow{h\nu} [Pt(PR_3)_2] + 2CO_2(g)$$

$$\downarrow +2L$$

$$[Pt(PR_3)_2L_2]$$

This reaction may be considered as redox reaction.

### 6. Synthesis of mixed ligand complexes

Mixed ligand complexes are also called heteroleptic complexes while homoligand complexes are called homoleptic complexes. Synthesis of mixed ligand complexes of transition metals are governed by two important factors :

(a) favourable formation constant* over those of the component homoligand complexes

(b) favourable solubility relations.

Unless the formation constant of the mixed ligand complexes in solution is higher than those of the homoligand complexes, the mixed ligand complexes is likely to dissociate into homoligand complexes to a significant extent. Secondly it is important that none of the component homoligand complexes be sparingly soluble in the reacting solvent in which case the mixed ligand complex is likely to dissociate completely into the homoligand complexes. Examples of mixed ligand complexes from the homoligand complexes are :

* will be discussed later.

$$[Cu(en)_2](NO_3)_2 + [Cu(bpy)_2](NO_3)_2 \xrightarrow[\text{the two complexes}]{1:1 \text{ mixture of}} [Cu(en)(bpy)](NO_3)_2 \cdot 0.5H_2O$$

*grayish purple needles*

$$[Cu(gly)_2)] + [Cu(ophen)_2](ClO_4)_2 \xrightarrow[\text{of the two chelates}]{\text{equimolar mixture}} [Cu(gly)(ophen)]ClO_4 \cdot 3H_2O$$

*Blue needles*

$$[Cu(en)_2]^{2+} + [Cu(ox)_2]^{2-} \longrightarrow 2[Cu(en)(ox)]$$

Sometimes substitution reactions are carried out by other ligands e.g.

$$[Fe(CO)_2(NO)_2] + \text{ophen} \longrightarrow [Fe(NO)_2(\text{ophen})] + 2CO$$

$$Na_4[Fe(CN)_6] + NO^+ \longrightarrow Na_2[Fe(CN)_5(NO)] + NaCN + Na^+$$

$$[Ru(H_2O)(NH_3)_5]^{2+} + N_2 \longrightarrow [Ru(NH_3)_5(N_2)]^{2+}$$

$$[PtCl_4]^{2-} + \text{en} \xrightarrow{H_2O,\,\Delta} [PtCl_2(\text{en})](s) + 2Cl^-$$

## 7.8 FACTORS FAVOURING THE FORMATION OF COMPLEX COMPOUNDS

The formation of coordination complexes is influenced by the internal parameter viz. choice of metal and ligand, synergic effect etc. as also dictated by external parameter viz. temperature, pressure, reagent, aerobic and anaerobic factors etc. Some of the factors discussed below :

### 1. Environmental factors

Temperature and pressure are the two external factors which play an important part in complex formation and stability of a particular coordination complex. Coordination compounds containing volatile ligands e.g. water, ammonia, ethylenediamine etc. are less stable at elevated temperatures and transformed to other products when heated with elimination of volatile ligand e.g.

$$[Co(NH_3)_6]Cl_3 \xrightarrow{180^\circ C} [Co(NH_3)_5Cl]Cl_2 + NH_3$$

$$[Cr(en)_3]Cl_3 \xrightarrow{210^\circ C} [Cr(en)_2Cl_2]Cl + en$$

$$\underset{\textit{diamagnetic}}{2[Co(NH_3)_6]F_3} \xrightarrow{115^\circ C} \underset{\textit{paramagnetic}}{[Co(NH_3)_6][CoF_6]} + 6NH_3$$

$$\underset{\textit{white}}{[Pt(NH_3)_4]Cl_2(s)} \xrightarrow{250^\circ C} \underset{\textit{yellow}}{\text{trans}[PtCl_2(NH_3)_2]} + 2NH_3(g)$$

Pressure controls the formation of a complex is demonstrated by some authors in the **flexidentate character** of an amino acid *l*-arginine (ArgH). When l-arginine and $Cu(ClO_4)_2 \cdot 6H_2O$ (2 : 1 molar ratio) reacted in minimum amount of water, a blue microcrystalline product of composition $[Cu(ArgH)_2(ClO_4)_2]$(7-XV) separated. IR spectra of (7-XV) suggested monodentate nitrogen coordination of the amino acid and a bidentate perchlorate coordination. When (7-XV) was recrystallised from water a complex $[Cu(ArgH)_2(H_2O)](ClO_4)_2$(7-XVI) was isolated. The IR spectrum of (7-XVI) suggested bidentate (NO) coordination of the amino acid and the perchlorate group was ionic, the water molecule acting as a bridge. Interestingly when (7-XVI) was pressed in KBr and IR spectrum recorded, the spectrum compared well with that of (7-XV). Thus the following interaction was suggested by the authors :

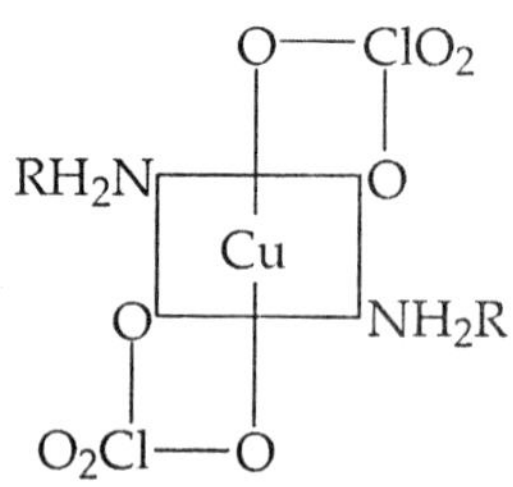

(7-XV)

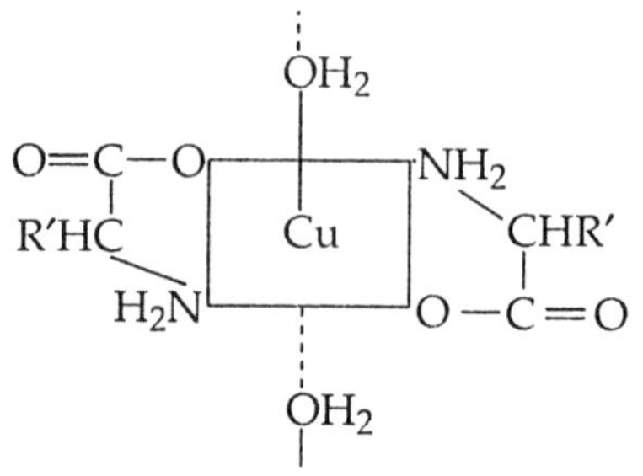

(7-XVI)

$R = H_2N{-}\underset{\underset{^{+}NH_2}{||}}{C}{-}CH_2{-}CH_2{-}CH_2{-}\underset{\underset{COO^-}{|}}{CH}{-}$ $\quad R' = H_2N{-}\underset{\underset{^{+}NH_2}{||}}{C}{-}CH_2{-}CH_2{-}CH_2{-}$

$$\underset{(7\text{-}XVI)}{[Cu(ArgH)_2(H_2O)](ClO_4)_2} \underset{H_2O}{\overset{\text{pressure}}{\rightleftharpoons}} \underset{(7\text{-}XV)}{[Cu(ArgH)_2(ClO_4)_2]} + H_2O$$

Therefore the bidentate and monodentate behaviour of the amino acid is exposed by the application of pressure.

## 2. Concentration

Stability and existance of complex ions in solution is markedly influenced by concentration changes of the coordinating ligands. Some complexes may exist in water in presence of a high concentration of a particular ligand e.g.

$$\underset{blue}{[Co(CNS)_4]^{2-}} + 6H_2O \rightleftharpoons \underset{pink}{[Co(H_2O)_6]^{2+}} + 4CNS^-$$

$$\underset{green}{[Cr(H_2O)_5Cl]^{2+}} + H_2O \rightleftharpoons \underset{violet}{[Cr(H_2O)_6]^{3+}} + Cl^-$$

$$Na_4[Fe(CN)_6] + NaNO_2 \rightleftharpoons Na_4[Fe(CN)_5NO_2] + NaCN$$

## 3. The nature of the central metal ion

Generally transition metals are good complex former. This is due to the extra stability gained by the formation of the complex by interaction of the metal ions with the ligands which is known as **crystal field stabilisation energy** (CFSE)*. We can say also **high charge** and **small size** of the metal ion as also **presence of vacant d orbitals** on metal ions are responsible for the formation of complexes. The driving force for the formation of a complex is the attraction between the lone pair of electrons with that of the metal ions. If the charge is high then the size

* will be discussed later.

will be small, the internuclear distance between the metal and the ligand is small and thereby the stability of the complex will be high. This is justified from the fact that the formation constant of $[Fe(CN)_6]^{4-}$ is ~$10^6$ while that for $[Fe(CN)_6]^{3-}$ the value is ~$10^{31}$. The order of stability of complexes with the same ligand with that of different metal ions are :

| | $Cu^{2+}$ > | $Ni^{2+}$ > | $Co^{2+}$ > | $Fe^{2+}$ > | $Mn^{2+}$ |
|---|---|---|---|---|---|
| ionic radius (Å) | 0.87 | 0.72 | 0.73 | 0.76 | 0.80 |

This order is commonly known as Irving-Williams series of stability constants*.

A complex formation is accompanied by the donation of electrons from the ligands to the metal ion. Transition metals having incompletely filled d-orbitals are able to accept the electrons from the ligands, thereby favouring complex formation.

## 4. The nature of the ligand

Ligands have a pivotal role for the stability of complexes. Greater the tendency to donate electron pair to the metal ion, the more is the basic strength of the ligand and greater will be the stability of complexes. Irrespective of the nature of the metal cation the donor ability of the ligands has been found to increase more or less in the following sequence of ligands which is known as **spectrochemical series****.

$$I^- < Br^- < S^{2-} < SCN^- < Cl^- < NO_3^- < F^- < \text{urea} \sim OH^- \sim ONO^- < N_3^- < C_2O_4^{2-} < H_2O$$
$$< NCS^- < EDTA \sim gly^- < py < NH_3 < N_2H_4 < en < SO_3^{2-} < NO_2^- < bpy < ophen < CN^-$$

This indicates $CN^-$ will form stable complex and $I^-$ forms less stable complex with a particular metal ion.

The nature of ligand is also important in stabilising the oxidation state of a metal. Thus $NH_3$ will stabilise the high oxidation state while CO is suitable for low oxidation state stabilisation of a metal. For example $[Ni(NH_3)_6]^{2+}$ and $Ni(CO)_4$. The stabilisation of low oxidation states of metal is governed by **synergic effect** (Sec. 7.4.3(b)) through the π-acid ligands.

## 5. Cyclisation or ring formation

Chelate complexes are more stable than the corresponding non-chelated complexes (Sec. 7.4.2). The formation of chelate ring (7-I) around the metal ion increases the stability of **chelates**. Thus $[Cu(en)_2]^{2+}$ is more stable than $[Cu(NH_3)_4]^{2+}$. The extra stability of chelate complexes compared to that of non-chelated complexes with similar chromophoric arrangement (i.e. similar metal donor environment) is due to chelate effect (will be discussed later).

## 6. Aerobic and anaerobic factor

The complexes in which the metal ions are in the low oxidation states are generally reactive/sensitive in presence of atmospheric ligands e.g. $O_2$, $N_2$ etc.

---

* will be discussed later. ** will be discussed later.

Ruthenium having electronic configuration [Kr]$4d^75s^1$ has low oxidation states +2, 0 etc. In an attempt to synthesise the ammine complex of Ru(II) in nitrogen atmosphere the following reaction occurs.

$$[Ru(NH_3)_5(H_2O)]^{2+} + N_2 \longrightarrow [Ru(NH_3)_5N_2]^{2+} + H_2O$$

This process is known as **anaerobic** one. In anaerobic processes the survival of the respective complex is difficult in presence of atmosphere.

In general the complexes in which the metal ions are in the high oxidation states are not susceptible to react with atmospheric ligands and the complexes are called aerobic. Most of the complexes synthesised in aerobic condition e.g. Co(III), Cr(III) complexes of different ligands. Plants take up solar energy and utilise it for the formation of complex organic molecules (carbohydrate) from simple molecules like $CO_2$ and $H_2O$. This reaction is known as photosynthesis reaction which is a redox reaction where solar energy is converted to chemical energy i.e.

$$nH_2O + nCO_2 \xrightarrow{\text{light}} (CH_2O)_n + nO_2$$

This process is initiated by chloroplasts in green plants which possess chlorophylls. Chlorophyll is the magnesium chelate of a tetrapyrrole (7-XVII) where a magnesium ion of coordination number 4 is chelated by the four electron donor nitrogen atoms of the four pyrrole rings. All **aerobic** life on earth owes its existance to chlorophyll and which is responsible for photosynthesis reaction and this reaction is our source of oxygen as well as that of the food we eat.

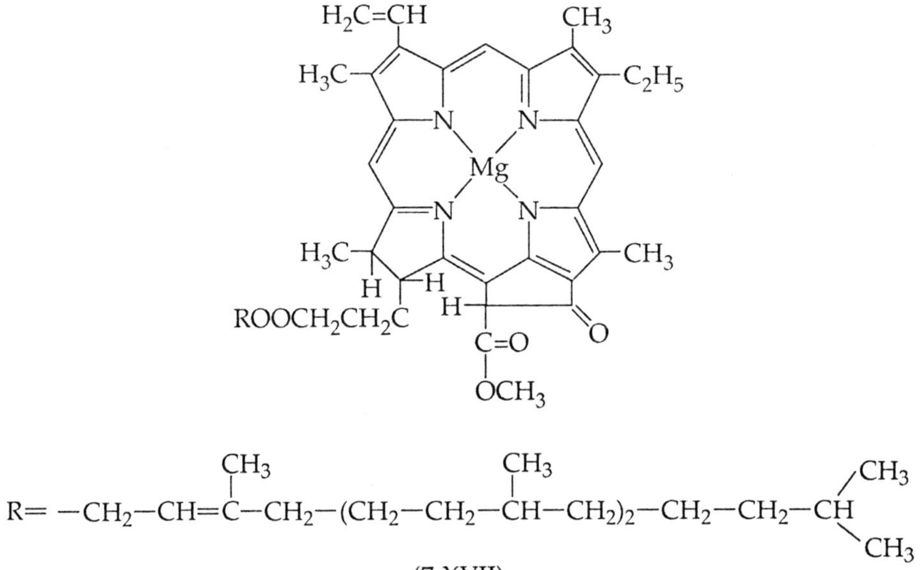


(7-XVII)

Various aerobic and anaerobic processes occur in life processes. **Aerobic** oxidation of glucose through citric acid cycle releases much energy. Formation of acetyl coenzyme is the starting point of **aerobic** oxidation of glucose. Anaerobic oxidative breakdown of glucose to ethanol or lactate produces a small amount of energy. Aerobic-anaerobic equilibrium may exist during oxygen transport by hemoglobin through heme which is the ferrous chelate of a tetrapyrrole.

The inorganic arsenic species is converted to different organoarsenic compounds through biomethylation in aerobic and anaerobic processes.

## 7.9 ISOMERISM AMONG COORDINATION COMPOUNDS

Isomerism is very common in organic chemistry e.g. ortho, meta and para substituted products, geometrical isomerism of maleic and fumeric acid, optical isomerism of lactic acid, conformational isomerism of cyclohexane etc. Since isomerism is a phenomena depending on the position and arrangement of atoms, so it cannot be limited to any particular class of compounds. A large number of inorganic complexes also exhibit the phenomena of isomerism. **Compounds having the same molecular formula but having different structural arrangements are called isomers and the phenomena is called isomerism.** (Greek, 'iso' means equal or the same and 'meros' means a part). Following are the types of isomerism exhibited by coordination compounds.

### 7.9.1 Hydrate isomerism or Solvate isomerism

This isomerism arises due to the position of water molecules inside and outside the coordination sphere. Three isomers of $CrCl_3{\cdot}6H_2O$ are known :

(1) $[Cr(H_2O)_6]Cl_3$ (violet) which has three ionic chloride ions, which can be precipitated by $AgNO_3$ solution. The compound does not lose water over conc. $H_2SO_4$ in a desiccator.

$$[Cr(H_2O)_6]Cl_3 + 3AgNO_3 \longrightarrow [Cr(H_2O)_6](NO_3)_3 + 3AgCl\downarrow$$

(2) $[Cr(H_2O)_5Cl]Cl_2{\cdot}H_2O$ (green) which loses one water molecule over conc. $H_2SO_4$ and two ionic chlorides are precipitated by $AgNO_3$ solution.

(3) $[Cr(H_2O)_4Cl_2]Cl{\cdot}2H_2O$ (dark green) which loses two water molecules over conc. $H_2SO_4$ and one ionic chloride is precipitated by $AgNO_3$ solution.

These isomers can also be characterised by thermogravimetric analysis as also by conductance measurements. Conductance measurements of compounds (1), (2) and (3) behave as tri-uni, bi-uni and uni-univalent electrolytes respectively. Commercial $CrCl_3{\cdot}6H_2O$ exist as a mixture of two green variety of isomers, the dark green dichloro variety exist as a predominating trans form. The isomers may be separated by cation exchange resin with dilute $HClO_4$ as eluting agent. Aqueous solution of commercial $CrCl_3$ is green, but on standing for a week or more the green solution turns violet from which crystals of $[Cr(H_2O)_6]Cl_3$ may be obtained.

Other examples of this type of isomers are :

(a) $[Cr(en)_2(H_2O)_2]Br_3$ and $[Cr(en)_2(H_2O)Br]Br_2{\cdot}H_2O$

(b) $[Co(NH_3)_4H_2O\,Cl]Cl_2$ and $[Co(NH_3)_4Cl_2]Cl{\cdot}H_2O$

(c) $[Co(en)_2H_2O\,Cl]Cl_2$ and $[Co(en)_2Cl_2]Cl{\cdot}H_2O$

(d) $[Cr(H_2O)_2Cl_2(py)_2]Cl$ and $[Cr(H_2O)Cl_3(py)_2]H_2O$

(e) $[VO(Qui)_2H_2O]$(olive green) and $[VO(Qui)_2]H_2O$ (cream)

where QuiH = quinaldinic acid

Thermogravimetric analysis indicates loss of water at 170-180°C for olive green variety while the cream variety loses water below 100°C.

### 7.9.2 Ionisation isomerism

This isomerism arises when the coordinating ions/ligands exchange inside and outside the coordination entity. This can also be considered as the extension of hydrate isomerism because in case of hydrate isomerism also there are several examples where apart from the exchange of water molecules, the ions inside and outside the coordination sphere may change. Examples of ionisation isomers are :

(a) $[Co(NH_3)_5Br]SO_4$ and $[Co(NH_3)_5SO_4]Br$

(b) $[Co(NH_3)_5NO_3]SO_4$ and $[Co(NH_3)_5SO_4]NO_3$

(c) $[Pt(NH_3)_4Cl_2]Br_2$ and $[Pt(NH_3)_4Br_2]Cl_2$

(d) $[Co(NH_3)_5Cl]SO_4$ and $[Co(NH_3)_5SO_4]Cl$

(e) $[Co(en)_2NO_2Cl]SCN$ and $[Co(en)_2NO_2SCN]Cl$

These isomers are easily characterised by simple conductance measurements or by IR spectral studies and sometimes by simple analytical procedure.

The following synthetic routs are followed for the preparation of violet and red variety of the isomer(a) :

$$CoBr_2 \xrightarrow[NH_4Br,\ O_2]{\text{Aqueous } NH_3} [Co(NH_3)_5H_2O]Br_3 \xrightarrow{\Delta} [Co(NH_3)_5Br]Br_2$$

$$[Co(NH_3)_5Br]Br_2 \xrightarrow{Ag_2SO_4} \underset{\textit{Violet}}{[Co(NH_3)_5Br]SO_4}$$

$$\underset{\textit{Red}}{[Co(NH_3)_5SO_4]Br} \xleftarrow{BaBr_2} [Co(NH_3)_5SO_4]HSO_4 \xleftarrow{\text{Conc. } H_2SO_4} \underset{\textit{Violet}}{[Co(NH_3)_5Br]SO_4}$$

The violet variety forms white precipitate of $BaSO_4$ on treatment with $BaCl_2$ solution while the red compound gives a pale yellow precipitate of AgBr with $AgNO_3$ solution. The violet one shows only one IR active stretching vibration while the red compound where $SO_4^{2-}$ coordinated, shows three IR stretching vibrations.

### 7.9.3 Coordination isomerism

This type of isomerism is found in compounds where the cation and anion both are complex entity and ligands will interchange between these two coordination zones e.g.

$[Co(NH_3)_6][Cr(CN)_6]$ and $[Cr(NH_3)_6][Co(CN)_6]$

$[Co(en)_3][Cr(C_2O_4)_3]$ and $[Cr(en)_3][Co(C_2O_4)_3]$

$[Pt^{II}(NH_3)_4][Pt^{IV}Cl_6]$ and $[Pt^{IV}(NH_3)_4Cl_2][Pt^{II}Cl_4]$

$[Cu(NH_3)_4][PtCl_4]$ and $[Pt(NH_3)_4][CuCl_4]$

$[Cr(NH_3)_6][Cr(SCN)_6]$ and $[Cr(SCN)_2(NH_3)_4][Cr(SCN)_4(NH_3)_2]$

$[Cr(NH_3)_6][Cr(CN)_6]$ and $[Cr(CN)_2(NH_3)_4][Cr(CN)_4(NH_3)_2]$

Characterisation of these isomers are difficult. Complete elemental analysis as also different structural studies can give us the idea of the composition of these isomers.

### 7.9.4 Coordination position isomerism

This type of isomerism arises in case of polynuclear complexes having bridging ligands. Isomers involve an exchange of ligands between the two metal atoms that are linked through one or more bridging groups e.g.

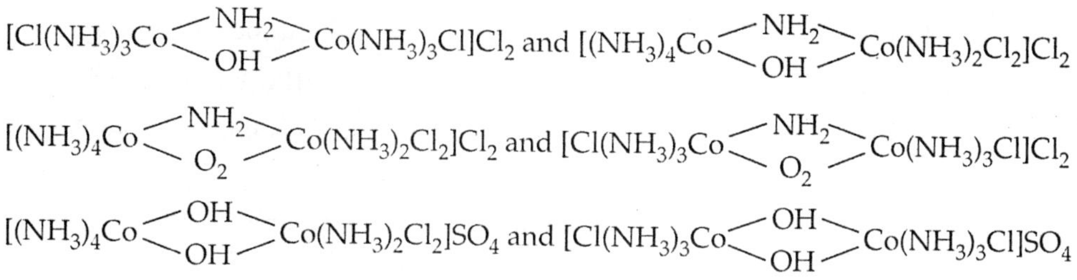


### 7.9.5 Structural, Salt or Linkage isomerism

Ambidentate ligands (Sec. 7.4.2) will exhibit this type of isomerism due to the different linking mode of the donor atoms of the ambidentate ligands. Thus the nitro group may be attached to the central atom through pentavalent nitrogen or may be attached through oxygen atom with trivalent nitrogen :

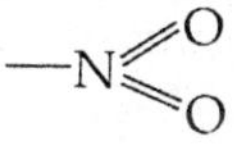

—N—O=O

$[Co(NH_3)_5(NO_2)]Cl_2$ — *Yellow brown*

pentaamminenitrito-Ncobalt(III) chloride *or* pentaamminenitrito-κNcobalt(III) chloride

$[Co(NH_3)_5(ONO)]Cl_2$ — *Red*

pentaamminenitrito-Ocobalt(III) chloride *or* pentaamminenitrito-κOcobalt(III) chloride

The above two variety of compounds may be prepared in the following schematic way :

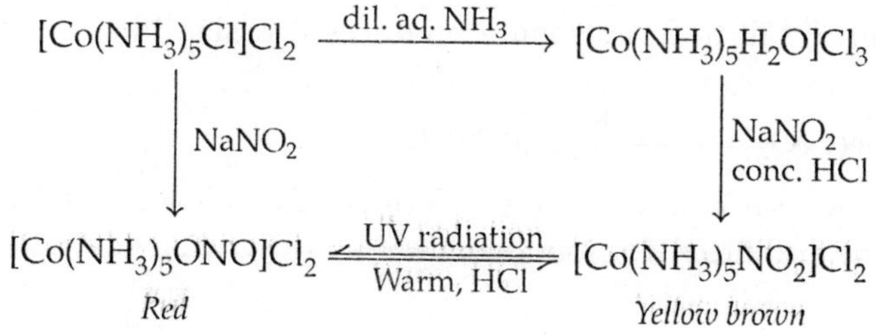

The characterisation of these isomers was made through IR study. The red compound having M-ONO bonding exhibits stretching frequency ~1060cm$^{-1}$ while yellow brown compound (M-$NO_2$ bonding) shows stretching frequency ~820 cm$^{-1}$. The red isomer is less stable than the yellow brown form. $NO_2^-$ ion can also function as **chelating** and **bridging modes** as shown below :

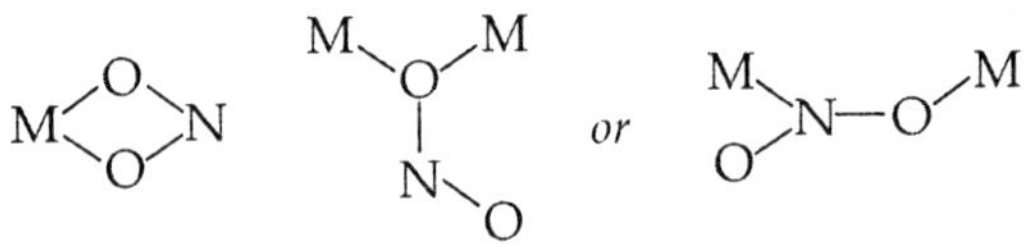

Another important ambidentate ligand is thiocyanate ion ($SCN^-$) which can function as end on (EO) mode through coordination with S or N as :

| | |
|---|---|
| M—SCN, thiocyanato-S *or* | M—NCS, thiocyanato-N *or* |
| thiocyanato-κS | thiocyanato-κN |
| $[Pd(bpy)(SCN)_2]$ | $[Pd(bpy)(NCS)_2]$ |
| (2,2′-bipyridine)bis(thiocyanato-$\kappa^2$S)palladium(II) | (2,2′-bipyridine)bis(thiocyanato-$\kappa^2$N)palladium(II) |

The two isomeric forms of thiocyanato complexes may be prepared in the following way :

$$K_2[Pd(SCN)_4] + bpy \xrightarrow{-78°C} [Pd(SCN)_2(bpy)] + 2KSCN \quad \textit{(Orange yellow)}$$

$$K_2[Pd(SCN)_4] + bpy \xrightarrow{-25°C} [Pd(NCS)_2(bpy)] + 2KSCN \quad \textit{(Light yellow)}$$

$$[Pd(SCN)_2(bpy)] \xrightarrow{\text{Solid heated to 150°C}} [Pd(NCS)_2(bpy)]$$

These two isomers are characterised by IR spectra and electronic spectra. In M—SCN and M—NCS bonding the C—S stretching frequency lies in the range 690-720 cm$^{-1}$ and 780-860 cm$^{-1}$ respectively while the C—N stretching frequency lies in the range above and below 2000 cm$^{-1}$ respectively.

Thiosulphate ion can exhibit the coordination mode through O or through S atom e.g. $[Co(NH_3)_5(OS_2O_2)]^+$ and $[Co(NH_3)_5(SSO_3)]^+$. These are also distinguished by IR studies.

Cyanide ion ($CN^-$) is also an important coordinating ligand can function through C-end or through N-end e.g. cis $[Co(trien)(CN)_2]^+$ and cis$[Co(trien)(NC)_2]^+$. In many complexes such as in Prussian blue $CN^-$ ion bridged through C-end to one metal ion and the N-end to another metal ion. Addition of ferric salt to ferrocyanide produces Prussian blue.

$$Fe^{3+} + [\overset{II}{Fe}(CN)_6]^{4-} \longrightarrow \overset{III}{Fe_4}[\overset{II}{Fe}(CN)_6]_3$$

i.e. Prussian blue is a ferric ferrocyanide and also formulated as $M^IFe^{III}[Fe(CN)_6]$. Addition of ferrous salt to ferricyanide produces Turnbull's blue

$$Fe^{2+} + [Fe(CN)_6]^{3-} \longrightarrow \overset{II}{Fe_3}[\overset{III}{Fe}(CN)_6]_2$$

i.e. Turnbull's blue is ferrous ferricyanide and may be formulated as $M^{I}Fe^{II}[Fe^{III}(CN)_6]$. Practically these two compounds are the same. Crystal structure determination shows that half of the iron atoms face **six carbon atoms** of $CN^-$ groups as nearest neighbours and another half of the iron atoms face **six nitrogen atoms** and thus exhibiting linkage isomerism.

Another exhibition of M—CN and M—NC coordination is followed in $KFe^{II}[Cr^{III}(CN)_6]$ (brick red) and $KCr^{III}[Fe^{II}(CN)_6]$ (dark green) which are prepared in the following scheme :

$$FeSO_4 + K_3Cr(CN)_6 \longrightarrow \underset{\textit{Brick red}}{KFe^{II}[Cr^{III}(CN)_6]} + K_2SO_4$$

$$\downarrow 100°C$$

$$\underset{\textit{Dark green}}{KCr^{III}[Fe^{II}(CN)_6]}$$

The brick red compound has Cr–CN coordination and a chain is formed by $Fe^{II}$ with N-end in the following way :

$$-Fe^{II}-N{\equiv}C-Cr^{III}-C{\equiv}N-Fe^{II}-N{\equiv}C-Cr^{III}-$$

and the dark green compound followed the following sequence where rearrangement of Cr–CN to Cr–NC occurs :

$$-Fe^{II}-C{\equiv}N-Cr^{III}-N{\equiv}C-Fe^{II}-C{\equiv}N-Cr^{III}-$$

$-Fe^{II}$–CN linkage complex has larger CFSE rather than –Cr–CN linkage complex and thus the transformation occurs.

Carbon monoxide a π–acid ligand normally coordinate through carbon to stabilise the low oxidation states of metals. But it may show ambidentate behaviour coordinating through oxygen. Spectroscopic data at 6-10K indicates that reaction of Au with large excess of CO exhibits the formation of isocarbonyl (carbonyl) of Au i.e. C≡O–Au–C≡O where both the carbon and oxygen mode of coordination is exhibited. On the other hand when Au and CO interact in presence of a rare gas, the normal C-linkage isomer O≡C–Au–C≡O is produced.

SHAB principle (Sec. 4.5) and symbiotic effect (Sec. 4.5.2(iv)) may be applied to explain the linkage isomerism.

### 7.9.6 Ligand isomerism

There are certain ligands which exist as isomers e.g. the diamine derivative of propane i.e. 1,2 diaminopropane or propylenediamine(pn) and 1,3 diaminopropane or trimethylenediamine (tn).

$$H_3C-CH_2-CH_3 \qquad H_2N-CH_2-\underset{\displaystyle NH_2}{\underset{|}{CH}}-CH_3 \qquad H_2N-CH_2-CH_2-CH_2-NH_2$$

*propane* *propylene diamine* *trimethylene diamine*

The pn and tn form the isomeric complexes as $[CoCl_2(pn)_2]Cl$ and $[CoCl_2(tn)_2]Cl$. Therefore **ligand isomerism arises due to the isomerism of the ligand**. Another example is $[Ag^{II}(PiC)_2]$ and $[Ag^{II}(Nic)_2]$.

### 7.9.7 Valence isomerism

This isomerism was proposed by Werner in case of complexes where the same group satisfy the primary valency in one complex and in other complex that group satisfy the secondary valency, for example,

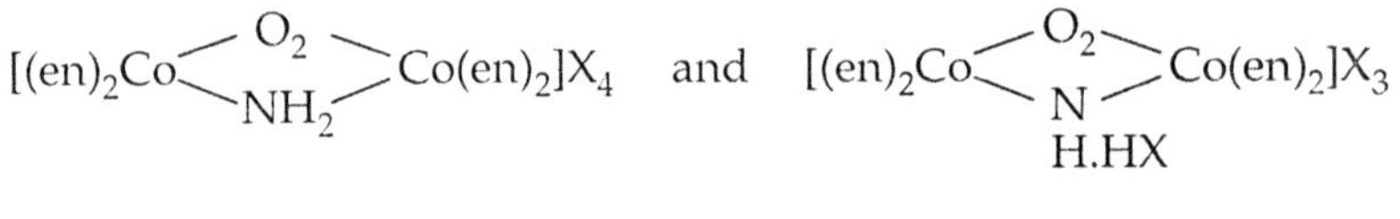

### 7.9.8 Conformation isomerism

Remaining the same donor atoms a metal complex may exist in different geometrical arrangement and in that case the complexes exhibit conformation isomerism. Thus when $[Ni(sal)_2]$ reacts with n-propylamine a square planar complex (7-XVIII) is obtained but when isopropylamine is used the geometry is changed from square planar to tetrahedral (7-XIX) due to the steric hinderence of isopropyl group.

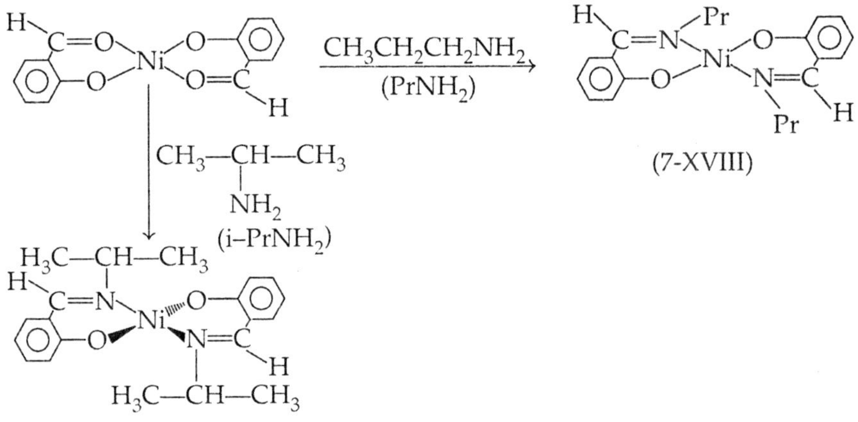

(7-XVIII)

(7-XIX)

Another example is,

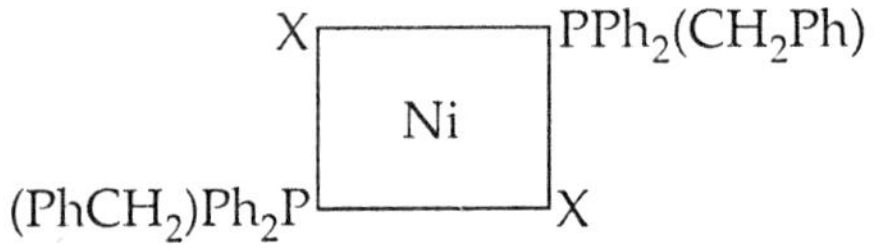

*Square planar*

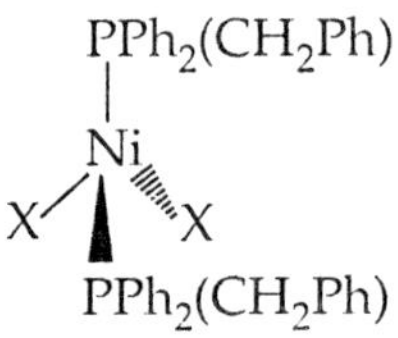

*Tetrahedral*

X = Cl, Br or I.

A typical example of this type of isomerism is of the following :

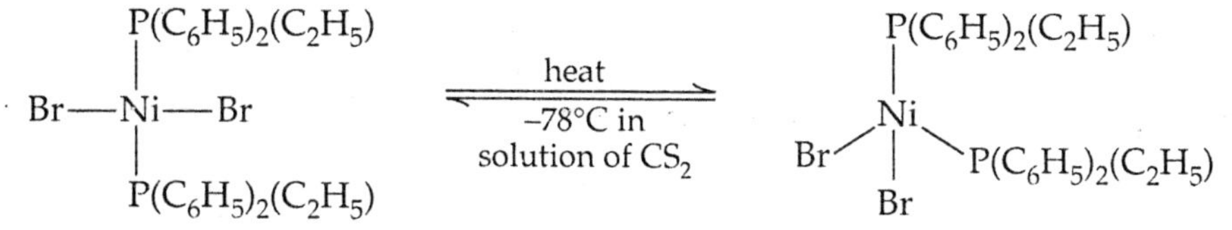


*Square planar (dark brown, diamagnetic)* *Tetrahedral (dark green, paramagnetic)*

This type of isomerism is also known as **polytopal isomerism** or **polytopism.**

## 7.9.9 Summation isomerism

Chan proposed this isomerism in the following isomeric pairs :

cis$[Co(en)_2(H_2NCH_2CH_2Br)Cl](NO_3)_2$ and cis$[Co(en)_2(H_2NCH_2CH_2Cl)Br](NO_3)_2$

These isomers discussed in Sec. 7.9.1 to 7.9.9 are sometimes called constitutional isomers.

---

**Problem 7.3** Write down the possible isomers of the compound $Co(NH_3)_4(CO_3)Br$. How would you distinguish the coordination isomers chemically? **[C.U. 2001]**

**Ans. :** For $Co(NH_3)_4(CO_3)Br$ there exist three isomers (a), (b) and (c). In (a) and (b) $CO_3^{2-}$ function as monodentate while in (c) the same function as bidentate.

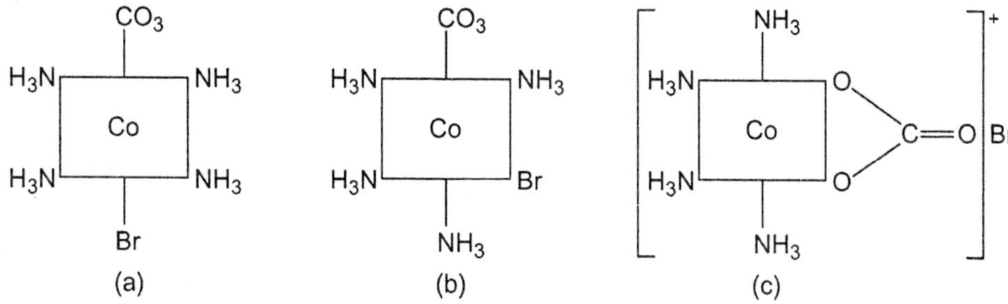


(a) (b) (c)

---

## 7.9.10 Stereoisomerism

**Stereoisomerism is the isomerism which arises on account of the different arrangement of atoms or groups in space in a molecule.** It is also known as **space isomerism.** It may be of two types : (a) Geometrical isomerism, (b) Optical isomerism.

Both geometrical and optical isomerism are encountered among the coordination complexes. In the case of complexes the common coordination numbers are four and six and therefore we shall restrict our attention to tetrahedral, square planar and octahedral complexes only. A correlation between stereoisomerism and geometry is given in Table 7.4. The most significant aspect of Werner's theory is the directional property of secondary valencies of the central metal atoms and the prediction of stereoisomers. This is the new horizon of Werner's theory.

| Table 7.4 : Prediction of stereoisomers and geometry | | | |
|---|---|---|---|
| Coordination number | Geometry | Geometrical isomers | Optical isomers |
| 4 | Tetrahedral<br>Square planar | Not possible<br>Possible | Possible, but rare<br>Not possible |
| 6 | Octahedral | Possible | Possible |

### 7.9.11 Geometrical isomerism

This isomerism is also called **cis-trans** isomerism. Cis means 'adjacent to' while trans means 'opposite to'. Thus when two similar ligands are adjacent, the isomer is named **cis** while if the ligands are in the farthest position in a specific geometry, the isomer is called **trans**. Geometrical isomerism with respect to metal has not been found among tetrahedral complexes. In organic chemistry, compound of the type $C_2X_2Y_2$ exhibits geometrical isomerism and the two isomers may be represented as :

*cis isomer* *trans isomer*

**(A) Square planar complexes having coordination number 4**

Square planar complexes of the following categories exhibit geometrical isomerism. Here M designate the central metal atom and a, b, c and d are ligands.

(i) **$Ma_2b_2$ type shows two isomers :** The $Ma_4$ type of square planar complex will not exhibit geometrical isomerism, but when two 'a' ligands are replaced by two 'b' ligands, the composition of the complex is $Ma_2b_2$ which exhibits geometrical isomerism (cis-trans) of the following type :

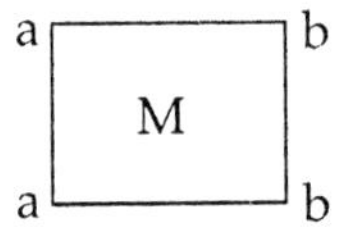


*cis*

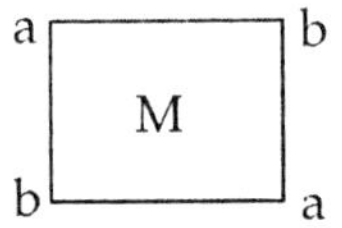


*trans*

For example $[Pt(NH_3)_2Cl_2]$

(ii) **$Ma_2bc$ shows two isomers :** In this type when the two a's are nearest, the isomer is called **cis** while the two a's are in the farthest position*, the isomer is called **trans**.

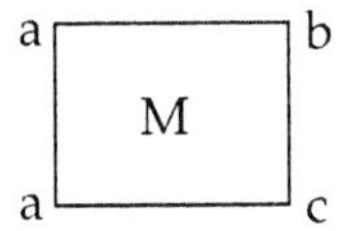


*cis*

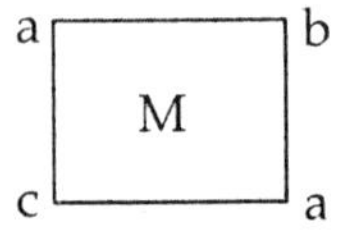


*trans*

The example of this type of complex is $[Pt(NH_3)_2ClBr]$.

(iii) **Mabcd type of complex exhibits three isomers :** The example of this type of complex is $[Pt(NH_3)(Br)(Cl)(py)]$. In this case the position of any one ligand, say 'a' is fixed while the other ligands i.e. b, c and d are successively placed at the opposite end i.e. farthest

* In square planar geometry the diagonal is the farthest position.

end, then the remaining two positions in each structure are randomly filled with rest of the two ligands.

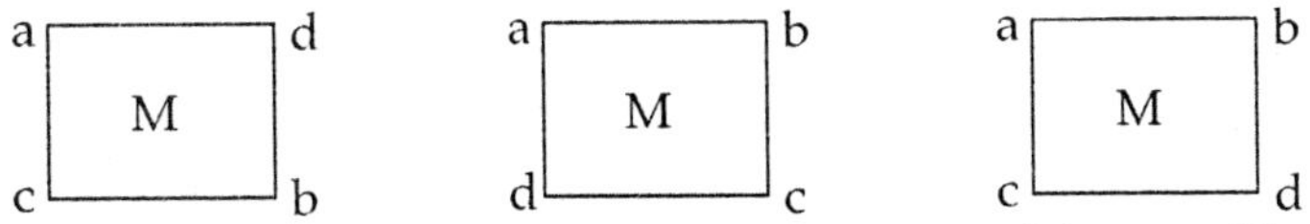


*'a' trans to 'b'* *'a' trans to 'c'* *'a' trans to 'd'*

(iv) **$Ma_4$ type chelate complex shows two isomers :** Normally $Ma_4$ type of complexes will not show geometrical isomerism. But a chelate complex of the type $Ma_4$ with substituents on the ring atoms may be expected to offer geometrical isomerism which are **cis** and **trans** with respect to the median plane of the molecule :

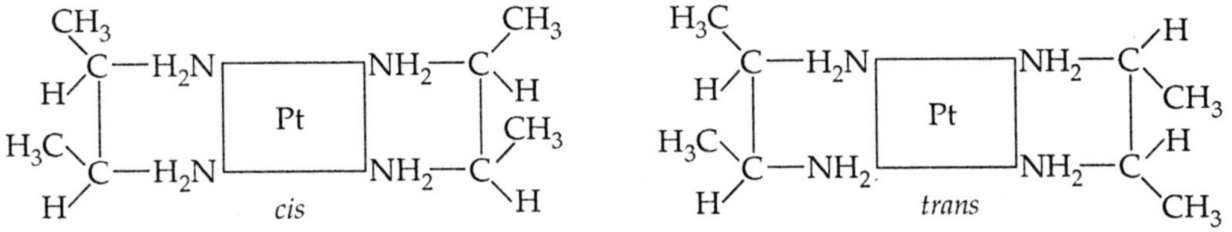

*cis* *trans*

(v) **Complexes with unsymmetrical chelating ligands show two isomers :** Ethylenediamine is a symmetrical chelating ligand. If we divide it into two halves by an imaginary line, the upper half is a mirror image of the lower half. Such ligands do not show cis-trans isomerism. However unsymmetrical ligands such as glycinate ion in which the two halves are not the mirror image of each other may form cis-trans isomers of the following :

*upper half is the mirror image of lower half* *upper half is not the mirror image of lower half*

*cis* *trans*

**(B) Octahedral complexes having coordination number 6**

Following categories of octahedral complexes exhibit geometrical isomerism.

(i) **$Ma_4b_2$ type shows two isomers :** The octahedral complexes having composition $Ma_6$ would not exhibit geometrical isomerism because all six positions of the octahedron are identical. But if the two 'a' ligands are substituted by two 'b' ligands then the composition will be $Ma_4b_2$ and in that case cis-trans isomerism arises. When both the substituents occur in the edgewise position, the compound is called **cis** isomer. If the two substituents occur in the axial position, the compound is referred to as trans isomer i.e.

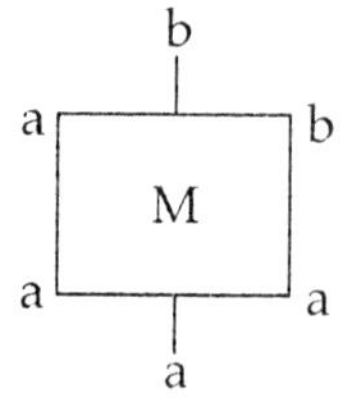


*cis*

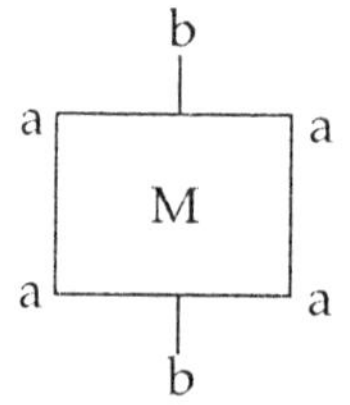


*trans*

For example $[Co(NH_3)_4Cl_2]Cl$ has two isomers. Similarly $[Co(NH_3)_4(NO_2)_2]NO_3$ has two isomers which differ in colour :

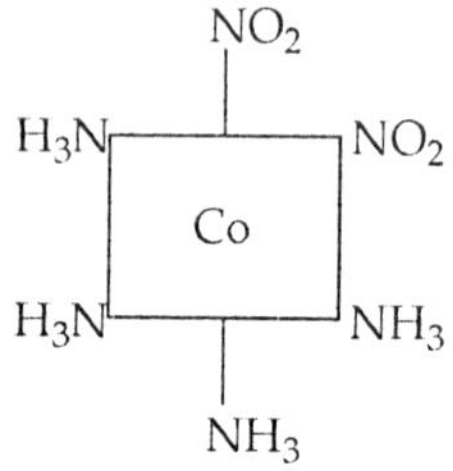


*cis (yellow)*

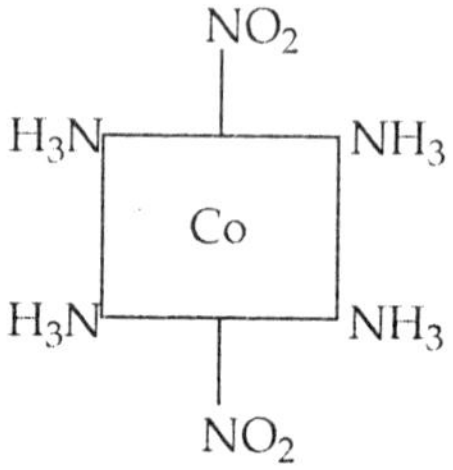


*trans (orange yellow)*

In this category large number of complexes are known e.g. cis $[Co(NH_3)_4Cl_2]^+$ (violet) and trans $[Co(NH_3)_4Cl_2]^+$ (green); cis $[CoCl_2(en)_2]Cl$ (purple) and trans $[CoCl_2(en)_2]Cl$ (green); cis $[Cr(en)_2(NCS)_2]^+$ (orange red) and trans $[Cr(en)_2(NCS)_2]^+$ (yellow orange); trans $NH_4[Cr(NCS)_4(NH_3)_2]$ etc.

(ii) **$Ma_4bc$ shows two isomers :** In this type for example $[Co(NH_3)_4H_2O\,Cl]^{2+}$ exists in two isomeric forms. When the four $NH_3$ molecules exist in the nearest position the isomer is **cis** while when the four $NH_3$ molecules are wide apart the isomer is trans i.e.

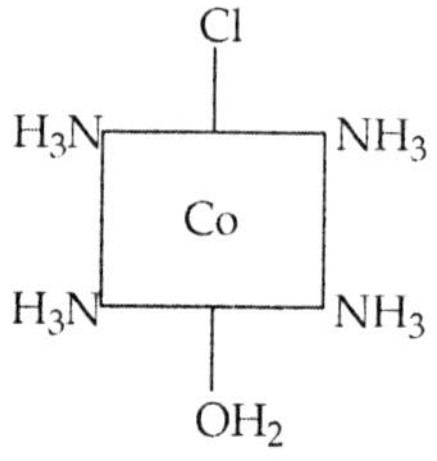


*cis*

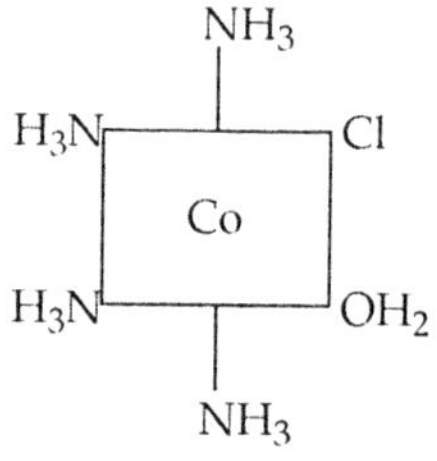


*trans*

(iii) **$Ma_3b_3$ type exists in two isomers :** This is one of the important type where the two isomers are designated as **facial** (cis) or **fac** and **meridional** (trans) or **mer**. In the **fac** form the three a's are all adjacent at one face of the ectahedron while in case of **mer** isomer the two a's are placed opposite to each other, two b's are placed opposite to each other and the remaining two positions are randomly occupied by a and b i.e.

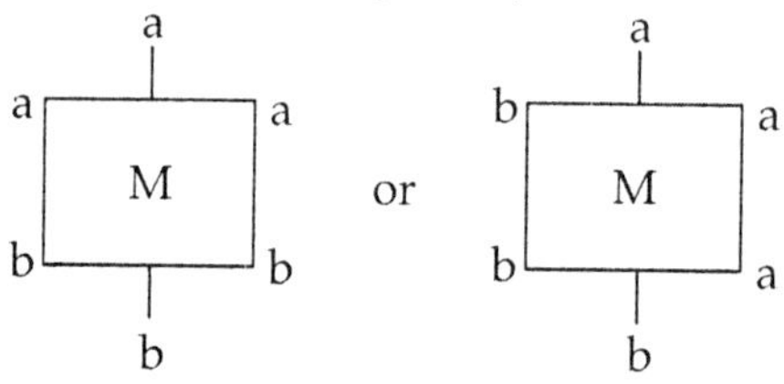


fac $Ma_3b_3$

e.g. fac $[Co(NH_3)_3(NO_2)_3]$ or fac $[Co(NH_3)_3Cl_3]$

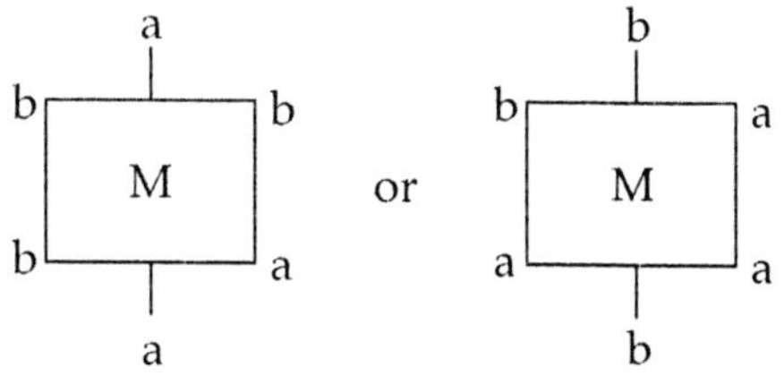

mer $Ma_3b_3$

e.g. mer $[Co(NH_3)_3(NO_2)_3]$ or mer $[Co(NH_3)_3Cl_3]$

The a and b ligands are along the meridian plane of a sphere and so this isomer is called **mer**. In the **fac** isomer the three a ligands/atoms or the three b ligands/atoms are cis to each other. Other examples are $[RhCl_3(py)_3]$, $[Co(gly)_3]$ and $[IrCl_3(PMe_3)_3]$. $[Co(gly)_3]$ exists as following two forms :

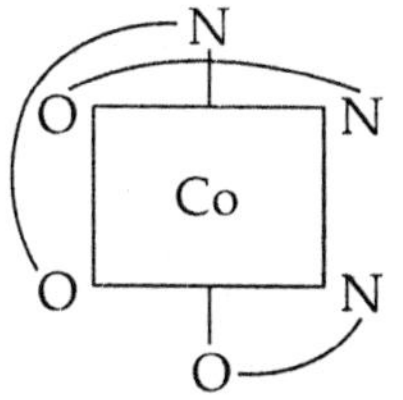

facial (fac)

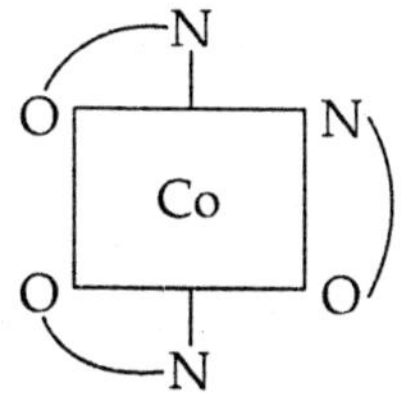

meridional (mer)

(iv) **$Ma_2b_2c_2$ type shows five isomers :** In this type e.g. $[Pt(NH_3)_2Cl_2(py)_2]^{2+}$ which may exhibit five geometrical isomers of the following type but only three isomers have been actually isolated.

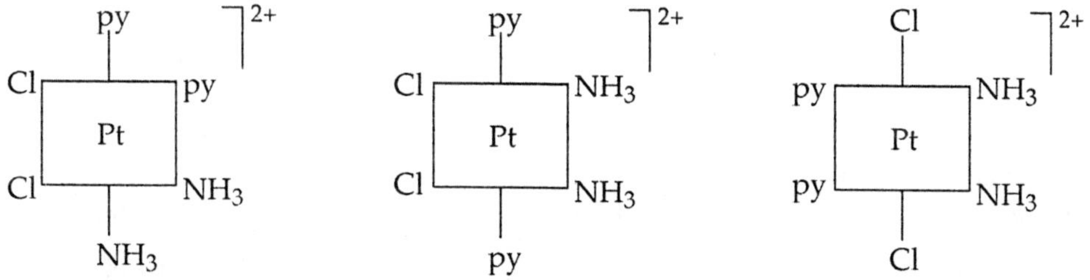

*cis isomer*

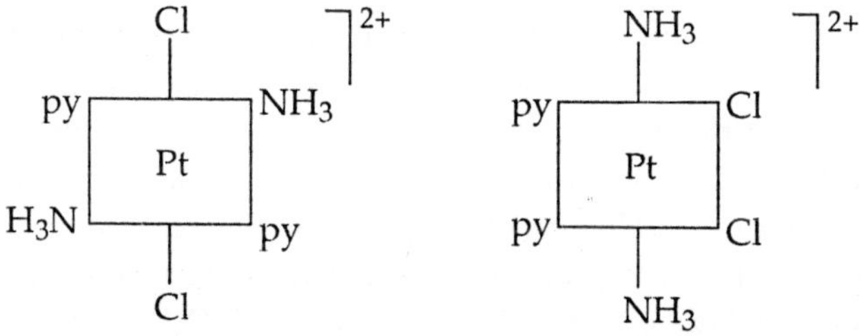

*trans isomer*

(v) **$M(AA)_2a_2$ type has two isomers :** In this type 'AA' represents a bidentate ligand and 'a' is a monodentate ligand e.g. $[Co(en)_2Cl_2]^+$, $[Co(en)_2(NH_3)_2]^{3+}$, $[Cr(en)_2Cl_2]^+$, $[Co(en)_2(NO_3)_2]^+$, $[Cr(C_2O_4)_2(H_2O)_2]^-$ etc. All these complexes exist in following **cis** and **trans** forms of which **cis** form may exhibit optical isomerism :

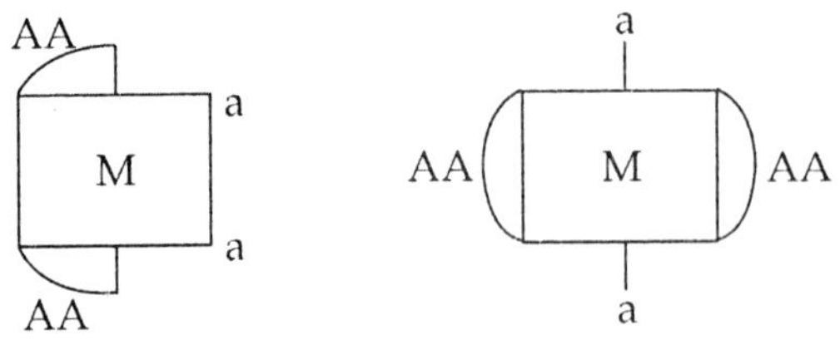


*cis isomer* *trans isomer*

(vi) **$M(AA)a_2b_2$ type shows three isomers** : The example of this type of complex is $[Co(en)(NH_3)_2Cl_2]^+$ which has the following three isomers :

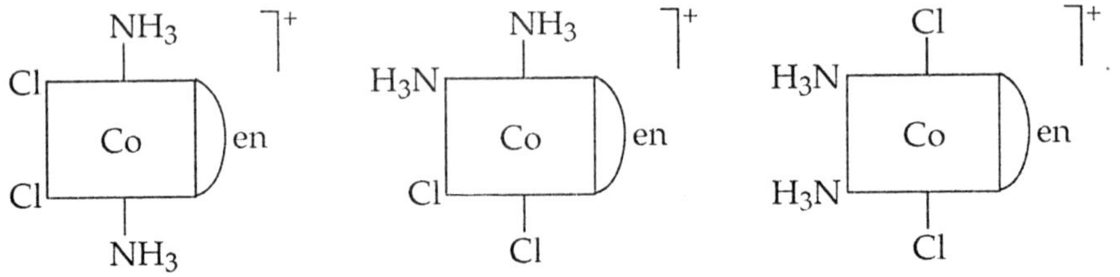


*trans two $NH_3$ and cis two $Cl^-$* *cis two $NH_3$ and cis two $Cl^-$* *trans two $Cl^-$ and cis two $NH_3$*

### 7.9.12 Determination of configuration of cis-trans isomers

There are several methods for the determination of configuration of geometrical isomers but the simplest one is based on the fact that a bifunctional chelate group e.g. carbonate, oxalate, ethylenediamine etc. can occupy only cis position i.e. 1, 2 position only from consideration of molecular dimensions. Hence assuming that no change of structures occur during substitution reaction, the isomer which reacts with a chelate group or which is formed by displacing such a group must have a **cis** structure. Thus violet $[Cr(en)_2Cl_2]Cl$ is a **cis** isomer and green $[Cr(en)_2Cl_2]Cl$ is **trans**, since the latter fails to form an oxalato compound when treated with potassium oxalate by displacing the two chlorine which are obviously in the trans positioin but the violet one gives the oxalato complex $[Cr(en)_2C_2O_4]Cl$ when similarly treated.

This conclusion is valid as long as it is supposed that the incomming group occupy the place of the group displaced. This is however not always true since molecular rearrangement may takes place and a proof is required whether molecular rearrangement is occuring or not. Sometimes in case of coordination compounds this proof is easy as **cis** and **trans** isomers may have considerable colour difference. In order to establish this point the configuration of Erdman's salt $NH_4[Co(NH_3)_2(NO_2)_4]$ was studied. This is obtained by passing air through a solution of $CoCl_2$ containing $NH_3$, $NH_4Cl$ and $NaNO_2$. It forms an oxalato complex $NH_4[Co(NH_3)_2(NO_2)_2C_2O_4]$. If the two ammonia molecules are in the trans positions (7-XX) in Erdman's salt only one oxalato compound is possible (7-XXI), while if the two ammonia molecules are in the cis positions (7-XXII) two oxalato compounds (7-XXIII) and (7-XXIV) are possible, of which the form (7-XXIV) should be resolvable into optical isomers (Scheme 7-I). The two oxalato compounds were actually obtained and it was concluded that the Erdman's salt has the cis structure (7-XXII). But the X-ray evidence shows that the anion of the Erdman's salt has a trans structure (7-XX). Hence it may be concluded that a change of structure takes place during the substitution reaction so that the **trans** structure is converted to **cis** and two oxalato products are obtained.

A **cis** complex has a much larger dipole moment than the symmetrical **trans** isomer so that dipole moment measurement is one of the important tool to distinguish cis-trans isomers. The rates of substitution reaction of **cis** isomer is greater than that of trans isomer e.g. the substitution of $Cl^-$ by water in cis$[Co(en)_2Cl_2]^+$ has the K value $1.5 \times 10^{-2}$ $min^{-1}$ at 25°C while for trans$[Co(en)_2Cl_2]^+$ the value is $2 \times 10^{-3}$ $min^{-1}$ at 25°C. A **cis** complex has a higher molar extinction coefficient than the corresponding **trans** complex and the absorption spectra of the two complexes may sometimes differ significantly*. **cis** and **trans** isomers may be separated by simple crystallisation techniques, by chromatography or by simple vacuum distillation.

The four coordinated complexes has either a tetrahedral or square planar geometry. Pt(II) complexes are some of the best known examples of the square planar geometry while $(NH_3)_2CdCl_2$, $CoCl_4^{2-}$ etc. will assume the tetrahedral geometry.

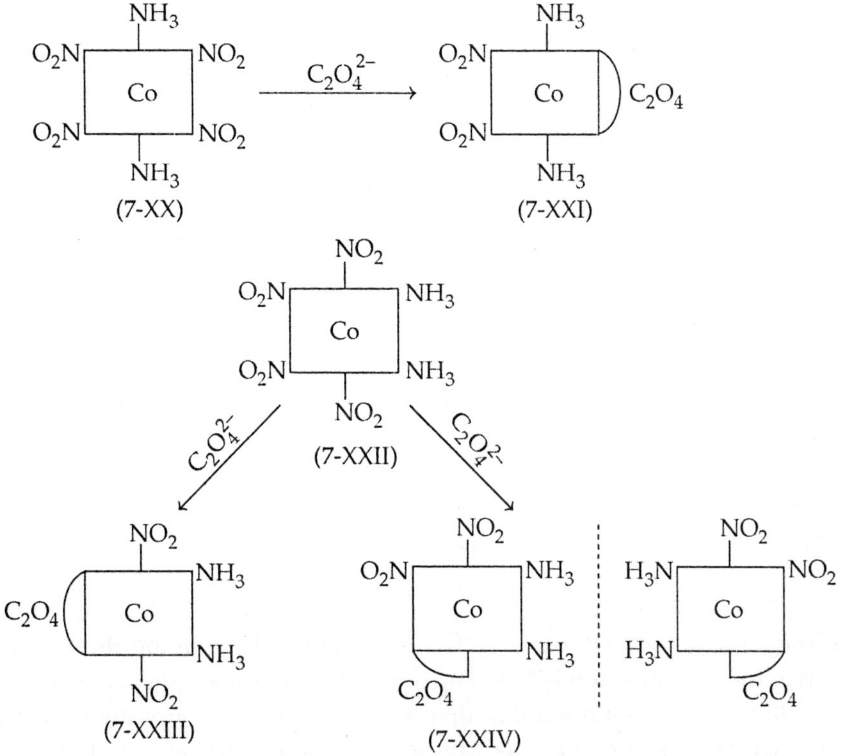


Scheme 7-I. Reactions of Erdman's salt with oxalate ion.

Existance of diaminedichloro platinum (II) in two isomeric forms led Werner (1893) to suggest a planar geometry giving rise to **cis** and **trans** isomers of the following :

* See Problem 7.68.

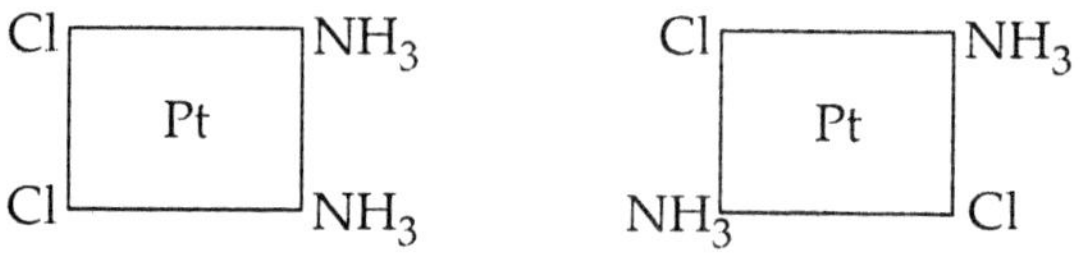

*cis* *trans*

Werner named them as α and β form.

Treatment of tetraammineplatinum(II) chloride $[Pt(NH_3)_4]Cl_2$ with concentrated HCl produces the β-form (trans form) of $[Pt(NH_3)_2Cl_2]$ while treatment of potassium tetrachloridoplatinate(II) with aqueous $NH_3$ gives the α-form (cis form) of $[Pt(NH_3)_2Cl_2]$.

We have stated that dipole moment measurement have been useful in distinguishing cis-trans isomers. The cis isomer (α-form) will register dipole moment because of lack of symmetry whereas the trans isomer will not. Therefore the β-form is the trans isomer.

*Grinberg's method*

Chemical support to distinguish these configurations came from the preparation of the corresponding isomeric dinitratodiammines. When the α and β forms were treated with oxalic acid, the nitrato groups were replaced and two complexes were formed. The α-form had the composition $Pt(NH_3)_2C_2O_4$ and the β-form had the formula $Pt(NH_3)_2(C_2O_4H)_2$. Assuming that the oxalato group acts as a bidentate ligand only if it replaces two nitrato groups in the cis position, then the reactions may be represented as :

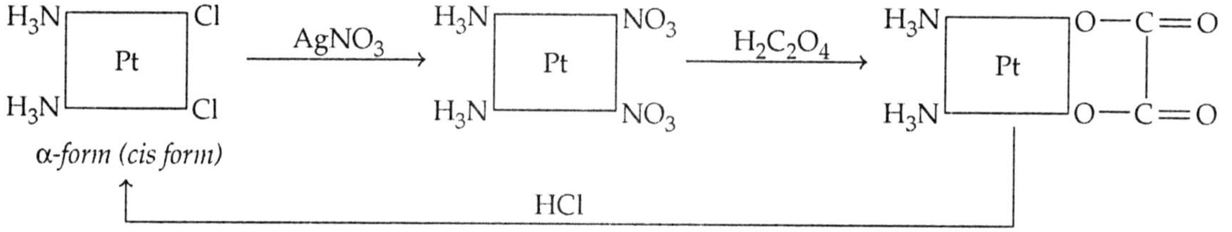


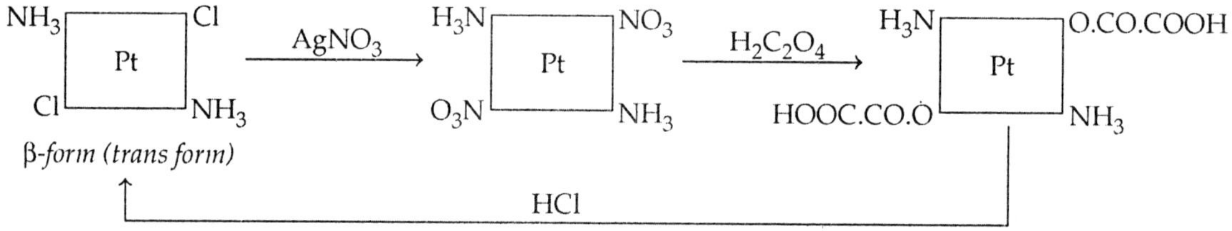


Finally Werner was able to characterise the two isomers by treating the two isomers with pyridine to have the composition $[Pt(NH_3)_2(py)_2]Cl_2$ and noted the substances that were eliminated on heating them. Assuming that thermally eliminated groups are trans to each other, elimination of two moles of $NH_3$ or pyridine would indicate a trans structure to $[Pt(NH_3)_2(py)_2]Cl_2$ and hence to the original compound $[Pt(NH_3)_2Cl_2]$, whereas the elimination of one mole of $NH_3$ or pyridine would indicate a cis structure. The α-form was thus established to have the cis structure as shown in the following way :

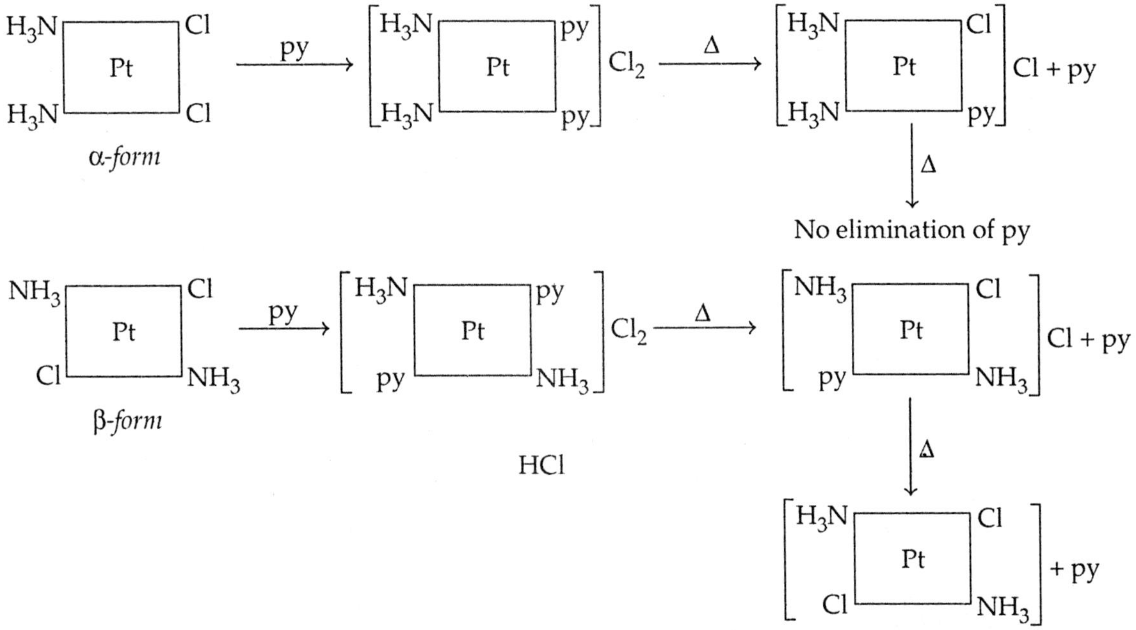


The cis and trans isomers may also be identified by the experimentally observed trans effect. According to which a negative group coordinated to a metal atom weakens the bond of a group trans to it. Thus $[Pt(NH_3)_4]Cl_2$ loses $NH_3$ on heating giving $[Pt(NH_3)_3Cl]Cl$ which in its turn, again loses $NH_3$ on further heating, giving $[Pt(NH_3)_2Cl_2]$. This $NH_3$ comes from the trans position to the coordinated chlorine atom, the product formed is trans $[Pt(NH_3)_2Cl_2]$ (β-form). When $[PtCl_4]^{2-}$ is treated with $NH_3$, a molecule of $NH_3$ enters the coordination sphere. The remaining trans chlorine atoms mutually weakens their bonds to the platinum so that the second $NH_3$ molecule enters the cis position to the first, forming the cis isomer i.e. α-form :

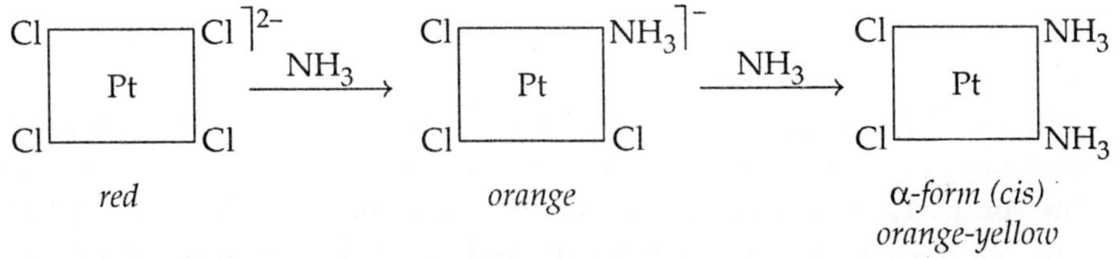


The trans isomer is made by treating $[Pt(NH_3)_4]^{2+}$ with $Cl^-$. Here the superior trans directing influence of $Cl^-$ causes the second $Cl^-$ to enter trans to the first one, producing the trans isomer (β-form) i.e.

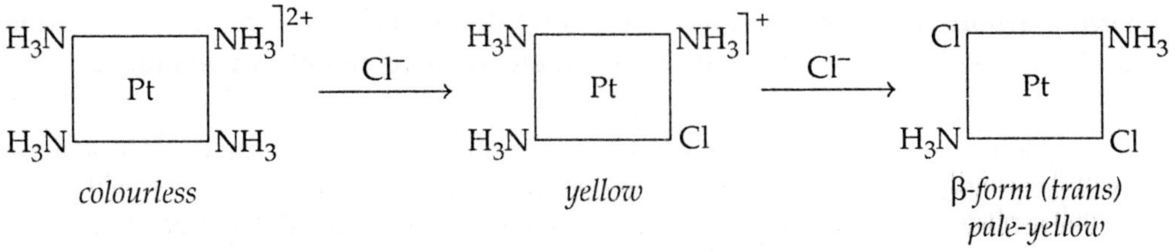

This phenomena is known as the trans effect and the order of ligands is, in part

$$HO^- < H_2O < NH_3 < py < Cl^- < Br^- < I^- \approx NO_2^- << CO \sim CN^-$$

Kurnakov utilised the phenomena of trans effect to distinguish the cis-trans pairs of $Pt(NH_3)_2Cl_2$ by treating them with thiourea (tu). The cis isomer produces $[Pt(tu)_4]^{2+}$ while the trans isomer converted to $[Pt(NH_3)_2(tu)_2]^{2+}$ as depicted below in the following reaction sequence :

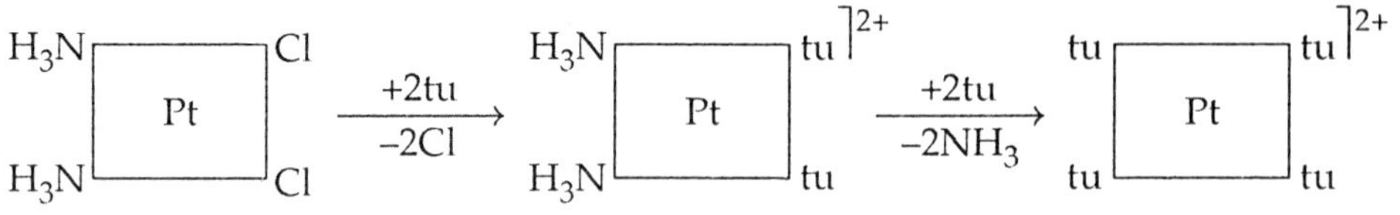


*α-form (cis)* (7-XXV)

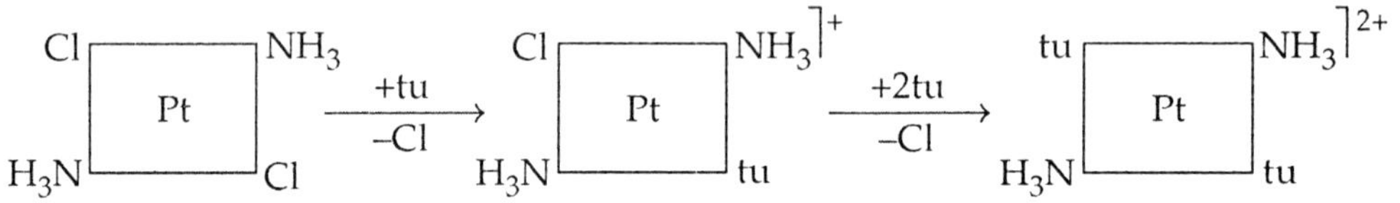


*β-form (trans)* (7-XXVI)

In the case of cis isomer (α-form) compound (7-XXV) further reacts with thiourea to give the completely substituted product $[Pt(tu)_4]^{2+}$ due to the larger trans effect of thiourea than $NH_3$. Such total substitution is not possible for trans isomer (β-form) because in compound (7-XXVI) the trans effect of $NH_3$ being smaller than that of thiourea so that $NH_3$ molecules are not sufficiently reactive to be replaced by thiourea. Thus $Cl^-$ ion trans to thiourea in (7-XXVI) is replaced further by thiourea to form $[Pt(NH_3)_2(tu)_2]^{2+}$.

### 7.9.13 Optical isomerism

In the begining it was thought that optical isomerism was associated only in case of organic compounds but it also exists in case of coordination compounds was decisively established by Werner.

When solution of a certain complex is placed in the path of a **plane polarised light***, it rotates the plane of polarisation through a certain angle θ which may be either to the left or to the right. **This property of a complex of rotating the plane of polarised light is called the optical activity** and **the complex having this property is said to be optically active.** The extent of optical activity depends on the wavelength of the polarised light, the concentration of the optically active species, the temperature, the nature of the solvent etc.

If the four valency of carbon is satisfied by four different groups then this carbon is called **asymmetric carbon**. If in an organic compound there exists an asymmetric carbon, the compound may be optically active provided there is no **elements of symmetry** (Centre of

---

* Ordinary light is an electromagnetic radiation that vibrates in all directions perpendicular to the direction of propagation. However when passed through a Nicol prism, the transmitted light vibrates in only one direction. The plane in which the light now vibrates is called the plane of polarisation and the light is called plane polarised light.

symmetry, alternating axis of symmetry etc.). But in case of coordination compounds there might not be the presence of asymmetric carbon so that in these cases the optical activity arises due to the asymmetry of the entire molecule.

Optically active substances may exist in the following forms :

(a) One which rotates the plane of polarised light towards right i.e. in clockwise direction, is said to be **dextrorotatory or d-form**. It is also represented by placing (+) sign before its name or formula.

(b) If the substance rotates the plane of polarised light towards left i.e. in anticlockwise direction, it is said to be **levorotatory or *l*-form** and may also be represented by (–) sign.

The above (a) and (b) classification is arbitrary i.e. if we take clockwise rotation as *l*-form then anticlockwise represented as d-form or in other words if in one case they are arranged clockwise, in the other they are anticlockwise. This offers a satisfactory explanation of the opposite rotatory powers of the (+) and (–) isomers.

(c) The d and l isomers are the mirror images to each other but not superimposable so that optical isomers sometimes called **mirror image isomers**. The isomeric substances which they represent are called enantiomers or antimers and the phenomena is said to be **enantiomorphism**. As stated above the asymmetry of the entire molecule means, one form is the mirror image of the other but not superimposable and the mirror image configuration is called a **chiral** or **asymmetric complex** (molecules or ions). A chiral species should not have elements of symmetry e.g. trans $[CoCl_2(en)_2]^+$ (7-XXVII) has plane of symmetry as also centre of symmetry at the metal, so that the complex gives superimposable mirror image and hence it is called a **achiral** species and would not exhibit optical isomerism.

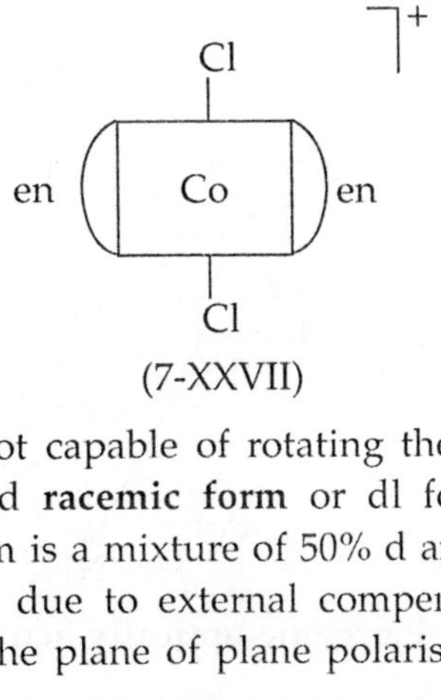


(7-XXVII)

(d) One form of isomer is not capable of rotating the plane of polarised light in either directions which is called **racemic form** or dl form or (±) form. This is optically inactive. The recemic form is a mixture of 50% d and 50% l-form. Optical inactivity of dl mixture is said to be due to external compensation. The solution of a recemic mixture does not rotate the plane of plane polarised light.

Whether an optical isomer is d or l cannot be identified by their structures. They can be identified only experimentally by measuring the rotation of the compounds in a polarimeter.

*Identificaiton of optical isomers*

The variation of the angle of rotation with the wavelength of light used is known as **optical rotatory dispersion (ORD)** and the plot of specific rotation vs wavelength is known as ORD curve. The ORD curve of the same enantiomer shows an abrupt change in the sign of the specific rotation at a wavelength near that of the absorption maximum ($\lambda_{max}$) of the complex. This phenomena is known as the **cotton effect** which may be either positive or negative depending on the nature of the specific rotation of the complex. A chiral species also displays **circular dichroism** (CD) i.e. the molecule has different absorption coefficients for right and left circularly polarised light at any given wavelength.

As mentioned earlier that optical isomerism was originally exist in case of organic compounds and so it was suggested optical isomerism in case of coordination compounds was due to the presence of organic ligands. In order to obviate this idea Werner ingeneously synthesise a complex of Co(III) (7-XXVIII) which is devoid of any organic ligand. The complex is well-known **hexol complex** where $Co(NH_3)_4(OH)_2^+$ serves as a bidentate ligand. This complex was

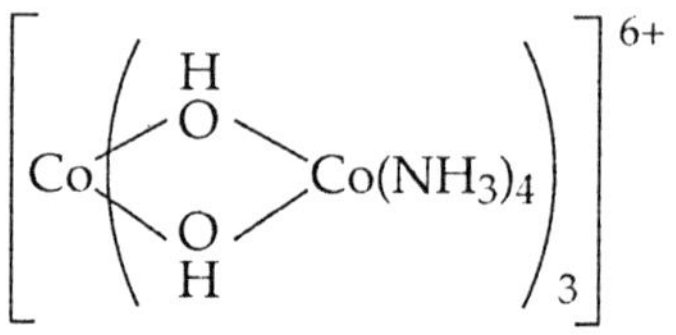


(7-XXVIII)

resolved by Werner into d and l forms, one being the mirror image of the other but not superimposable so that the complex shows optical activity. Since it is optically active, each of the $[Co(NH_3)_4(OH)_2]^+$ unit is expected to have cis structure. The structure of the compound (7-XXVIII) will be of the following :

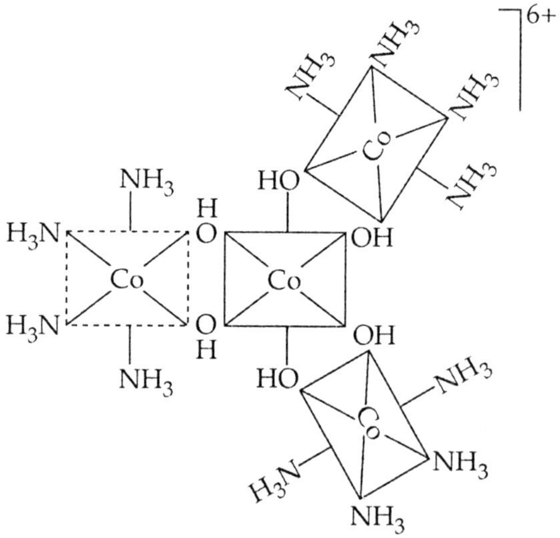


This is one example of **purely inorganic optically active compound** which has octahedral geometry.

It has been stated optical isomerism is not possible for square planar geometry (Table 7.4) but in exceptional case with asymmetric ligand this geometry will exhibit optical isomerism in the complex (7-XXIX) due to the presence of asymmetric nitrogen atom.

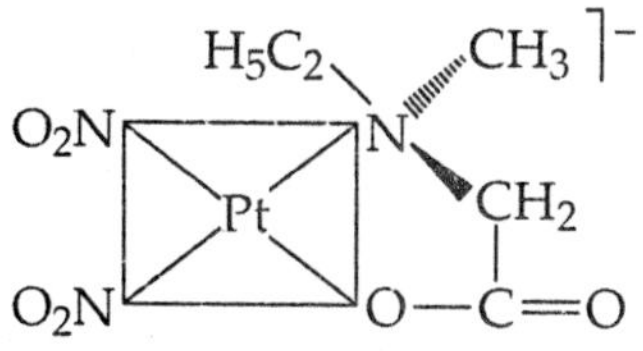


(7-XXIX)
bis(nitrito-κN)(N-methyl-N-ethylglycinato)platinate(II)

Though geometrical isomerism is not possible for tetrahedral complexes, opitcal isomerism due to the presence of four different groups/ligands are possible for Mabcd type of complex as shown below :

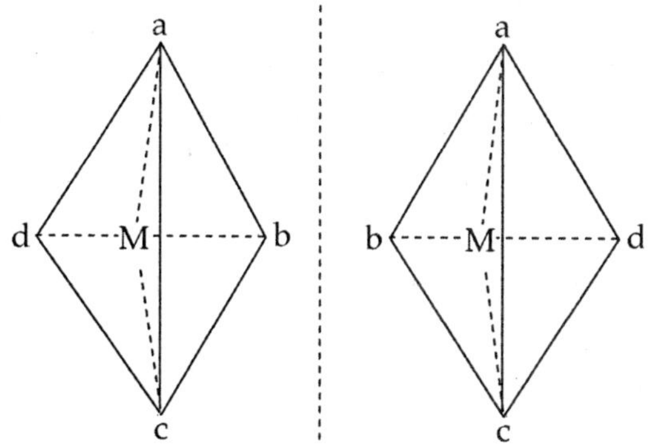


For example, ethylmethylpropylsulfidoarsenic(V) has been resolved into following two isomers :

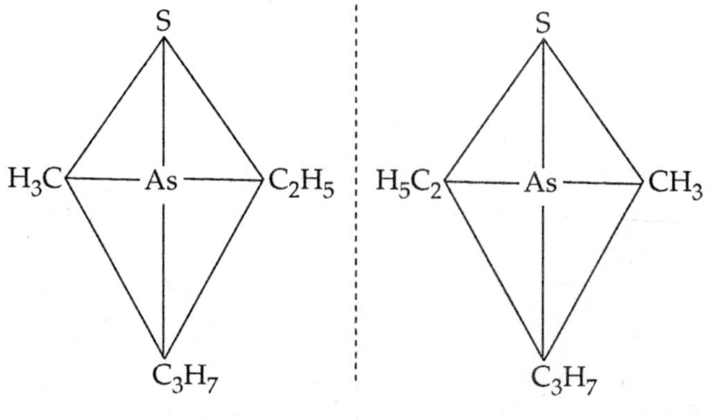


Bis(salicylaldehydato)boron(III) has also been resolved into optical isomers which has a tetrahedral geometry. Cis-diaquabis(sulfamido)rhodate(III) i.e. cis$[Rh(H_2O)_2(NH–SO_2–NH)_2]^-$ (7-XXX) is a **purely inorganic optically active complex** which has been resolved into optical isomers. The ligand sulphamide, $SO_2(NH_2)_2$ function as bidentate diprotic acid ligand. The compound (7-XXX) provided the first example of purely inorganic optically active complex.

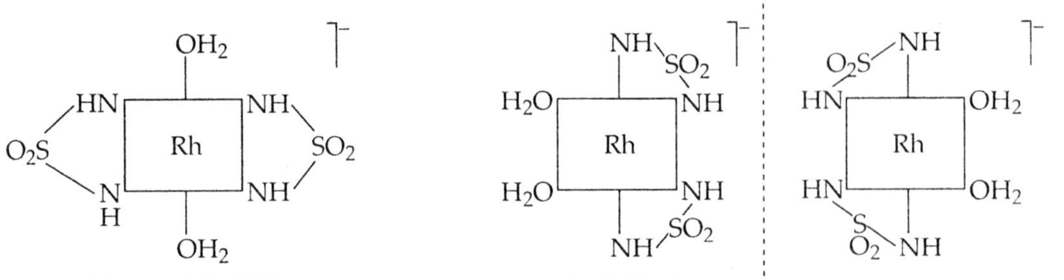

Trans of (7-XXX)
(optically inactive)

(7-XXX) cis
(optically active)

No complex in octahedral geometry with six different unidentate ligands e.g. Mabcdef may be resolved into optical isomers. Ethylenediaminetetraacetic acid complex of cobalt(III) and chromium(III) are octahedral and $[M(EDTA)]^-$ [M=Co(III) or Cr(III)] ions (7-XXXI) has no centre of inversion and no mirror plane and therefore they are **chiral** and produce non-superimposable mirror images and may be resolved into optical isomers (d and *l* form).

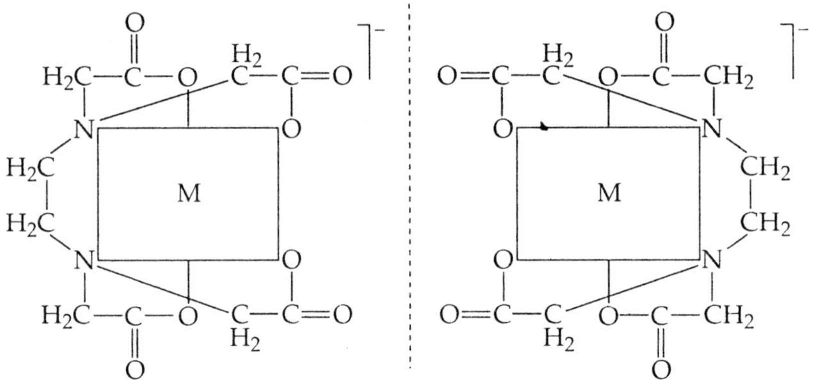

(7-XXXI)

A large number of cationic, anionic and neutral complexes in octahedral geometry exhibit optical isomerism. They are classified into the following categories :

(A) Complexes of the type $M(AA)_3$ where AA is a bidentate ligand. They show the following optical isomers.

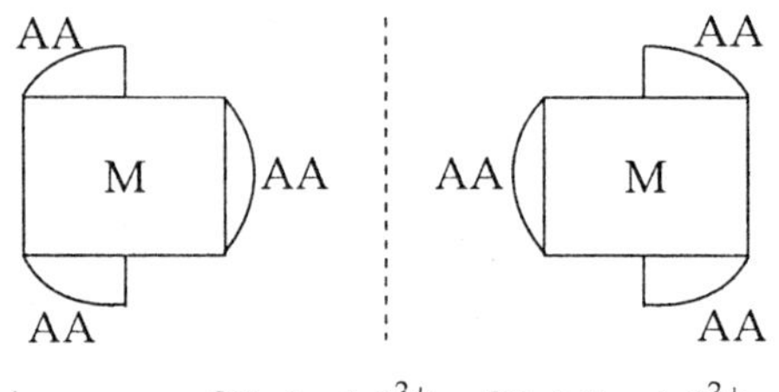

Examples are $[Co(en)_3]^{3+}$, $[Fe(dipy)_3]^{3+}$, $[Rh(en)_3]^{4+}$, $[Zn(en)_3]^{2+}$, $[Al(C_2O_4)_3]^{3-}$, $[Fe(C_2O_4)_3]^{3-}$, $[Co(C_2O_4)_3]^{3-}$, $[Co(gly)_3]$ etc.

(B) Complexes of the type $M(AA)_2BB$ where AA and BB are two different bidentate ligands e.g. $[Co(en)_2C_2O_4]^+$, $[Co(en)_2CO_3]^+$ etc. can be resolved into following d and l forms :

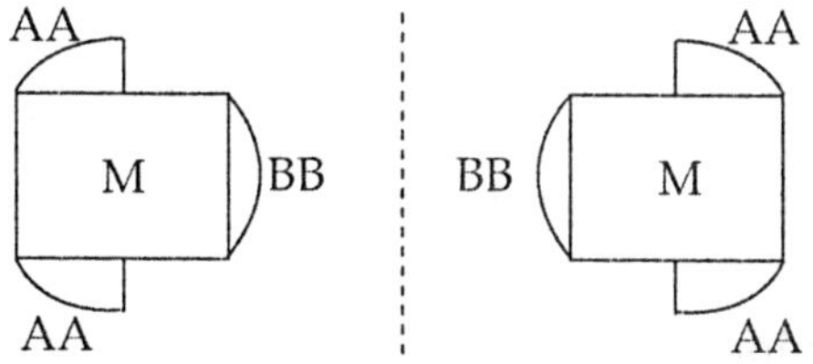


(C) Complexes of the type $M(AA)_2a_2$ where as before AA is a bidentate and 'a' is a monodentate ligand. These complexes show cis-trans isomerism with respect to 'a' ligand and the cis isomer will exhibit mirror image isomer of the following type :

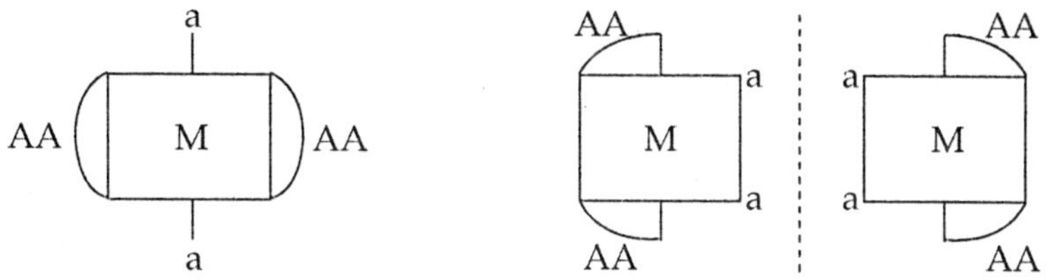


*Trans*
*(will produce super-imposable mirror image)*

*cis*
*d-form*

*l-form*

Examples are $[Co(en)_2Cl_2]^+$, $[Co(en)_2(NO_2)_2]^+$, $[Ir(C_2O_4)_2Cl_2]^{2-}$ etc.

(D) Complexes of the type $M(AA)_2ab$ where AA is a bidentate ligand and 'a', 'b' are two different monodentate ligands. Here when the two monodentate ligands are in the nearest position, the complex is optically active but when the two monodetate ligands are in the farthest position, the complex will be optically inactive and will produce superimposable mirror image which is shown below :

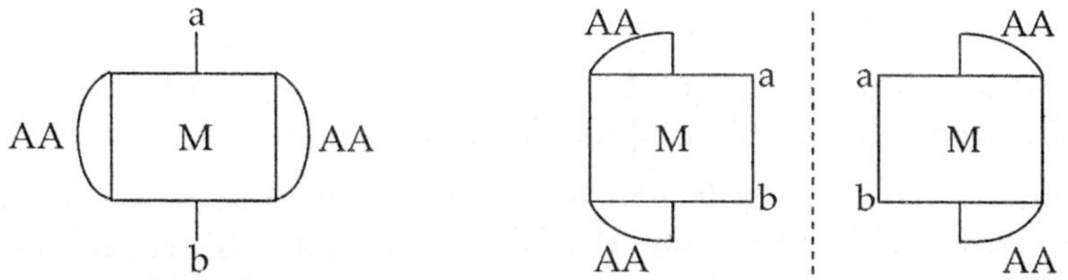


*This is*
*optically inactive*

*d-form* *l-form*
*This form is optically active*

Examples are $[Co(en)_2NH_3Cl]^{2+}$, $[Co(DMG)_2NH_3Cl]$

### 7.9.14 Resolution of optical isomers

**The separation of racemic mixture into d and l forms is known as resolution.** Since the d- and l-forms of a racemic mixture has similar chemical and physical properties, so that ordinary methods such as crystallisation/fractional crystallisation etc. are not suitable for their separation.

The most widely used method of separating a racemic mixture is to form **diastereoisomers.** The diastereoisomers are usually optically active but are not mirror images of each other. They are a pair having two optically active centres of which one is common to both while the other two are non-superimposable mirror images of each other. Therefore they are not **enantiomorphs** and thus **enantiomers** are stereoisomers with non-superimposable mirror images and hence have different solubility in water and can be separated by fractional crystallisation.

If the enantiomers of a dl mixture is an **optically active cation** they will be treated with an **optically active anion** while if it is **an anion** which will be treated with an **optically active cation** to form a **pair of diastereoisomers**. As these diastereoisomers have difference in solubility they can be separated by fractional crystallisation. The separated salts are then treated with a non optically active substance when the optically active d- and l-forms are regenerated leaving the resolving agent. Thus :

$$\begin{matrix} dA \\ + \\ lA \end{matrix} \quad + \quad dM \longrightarrow dAdM + lAdM$$

*Racemic mixture. dA and lA is the d- and l-form of the optically active anion A*

*Resolving agent. Optically active cation (dM)*

*This is a pair of diastereoisomers and separated by fractional crystallisation*

$$dAdM \xrightarrow{KI} KdA + dMI$$

*They are also separated by fractional crystallisation*

$$lAdM \xrightarrow{KI} KlA + dMI$$

The optically active anion $[Fe(C_2O_4)_3]^{3-}$ has been resolved by using an optically active cation such as strychnine and brucine. $[Co(EDTA)]^-$ was resolved with *l*-strychine by the fractional crystallisation of *l*-strychnine [Co(EDTA)] disastereoisomeric salt.

Similarly optically active cation $[Co(en)_3]^{3+}$ can be resolved by treatment with an optically active anion such as d-tart$^{2-}$*. When a solution of d*l*-$[Co(en)_3]Cl_3$ is treated with optically active d-tartaric acid, there will be the formation of d-$[Co(en)_3]Cl$ d-tart and *l*-$[Co(en)_3]Cl$ d-tart. The solubility of d-$[Co(en)_3]Cl$ d-tart is less than that of *l*-$[Co(en)_3]Cl$ d-tart so that fractional crystallisation leads to the formation of large crystals of d-$[Co(en)_3]Cl$ d-tart. The more soluble fraction *l*-$[Co(en)_3]Cl$ d-tart is then recovered from the mother liquor. Treatment of the two

---

*

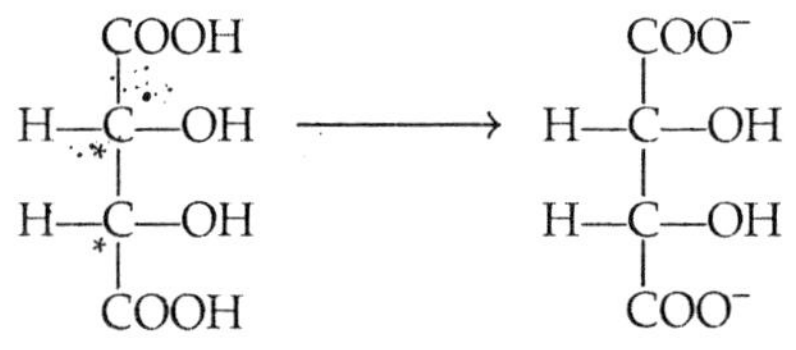

*Tartaric acid d-$H_2$tart* *d-tart$^{2-}$*

diastereoisomers with a non-optically active compound e.g. HCl we get the d- and *l*-chloride complexes of the optically active cation and the resolving agent is obtained :

$$dl\text{-}[Co(en)_3]Cl_3 + d\text{-}tart^{2-} \longrightarrow \underbrace{d\text{-}[Co(en)_3]Cl\ d\text{-}tart + l\text{-}[Co(en)_3]Cl\ d\text{-}tart}_{\text{pair of diastereoisomers}}$$

$\downarrow$ Fractional crystallisation

$$d\text{-}[Co(en)_3]Cl_3 + d\text{-}H_2tart \xleftarrow{2HCl} d\text{-}[Co(en)_3]Cl\ d\text{-}tart \text{ and } l\text{-}[Co(en)_3]Cl\ d\text{-}tart$$

*Resolving agent regenerated* — *(less soluble)* — *(more soluble)*

$$l\text{-}[Co(en)_3]Cl_3 + d\text{-}H_2tart \xleftarrow{2HCl} (\text{more soluble})$$

*Resolving agent regenerated*

Apart from d-tart$^{2-}$ a number of other resolving agents may be used e.g. antimonyl tartarate ion, SbOd-tart$^-$ and d-$\alpha$-bromo-camphor-$\pi$-sulphonate anion etc.

Resolution of the neutral complexes cannot be possible by the formation of diastereoisomeric salts in the usual way. A partial resolution of $[Co(acac)_3]$ was made by Dwyer by extraction into an organic solvent from an aqueous solution containing d-$[Co(en)_3]I_3$. The solubilities of d- and *l*-$[Co(acac)_3]$ are different in water containing an optically active ion. Very recently d- and *l*-$[Co(gly)_3]$ has been resolved by adsorption chromatography.

---

**Problem 7.4** The compound $Co(en)_2(NO_2)_2Cl$ (en = ethylenediamine) has been prepared in a number of isomeric forms. One form undergoes no reaction with either $AgNO_3$ or ethylenediamine and is optically inactive. A second form reacts with $AgNO_3$ but not with ethylenediamine and is also optically inactive. A third form reacts with both $AgNO_3$ and ethylenediamine and is optically active. Identify each of the three forms by their IUPAC names and discuss the above reactions in the light of Werner's theory. **[C.U. 1997, 2009]**

**Ans. :** $AgNO_3$ will react with the complex if the chloride ion is outside the coordination entity. Ethylenediamine will react with the complex if there are two monodentate ligands are at adjacent positions in the coordination entity.

The first isomer is optically inactive and undergoes no reaction with ethylenediamine or $AgNO_3$. Therefore $Cl^-$ ion exists in the coordination sphere and the $Cl^-$ and $NO_2^-$ groups are not in the adjacent position i.e. they exist in the **trans** position, so that this isomer is represented as :

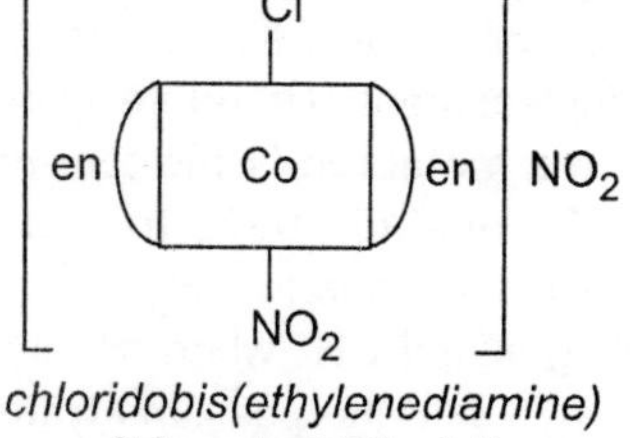


*chloridobis(ethylenediamine) nitritocobalt(III) nitrite*

The second isomer reacts with $AgNO_3$ but not with ethylenediamine which is also optically inactive so that the $Cl^-$ ion is outside the coordination entity and two $NO_2^-$ groups are not adjacent i.e. the $NO_2^-$ groups exist in **trans** position as depicted below :

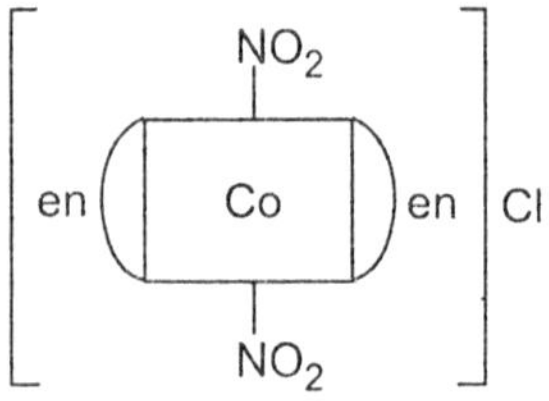


*bis(ethylenediamine)dinitrito cobalt(III) chloride*

In the case of third isomer which reacts both with $AgNO_3$ and ethylenediamine and is optically active. Therefore the $Cl^-$ ion is outside the coordination sphere and the two $NO_2^-$ groups exist in adjacent position i.e. **cis** positoin which may be represented as :

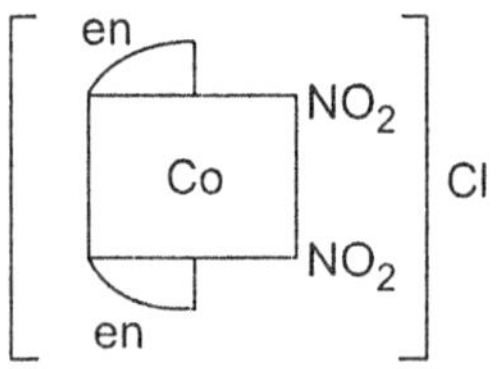


*bis(ethylenediamine)dinitrito cobalt (III) chloride*

The reaction of this isomer with ethylenediamine may be indicated as :

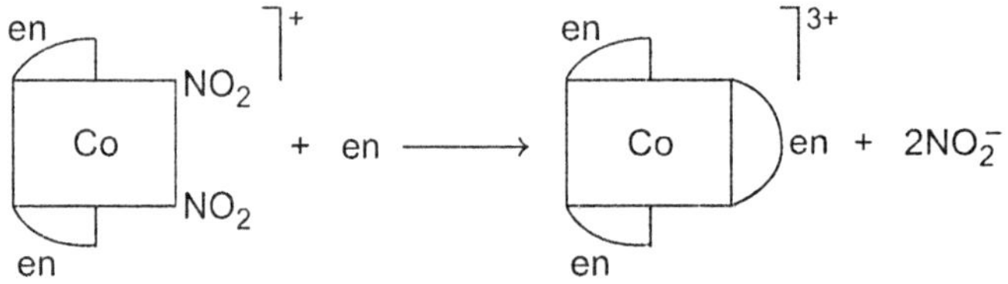


**Problem 7.5** What is trans effect? Outline a synthetic route for the synthesis of cis and trans-$[Pt(NH_3)(NO_2)Cl_2]^-$. The trans directing influence is $NH_3 < Cl^- < NO_2^-$.

**Ans. :** The phenomena of trans effect was first established by Chernyaev (1926) from observations on substitution reactions of platinum (II) complexes. He noticed that certain ligands have a trans labilising influence i.e. certain ligands facilitate the substitution of the ligand that is in a position **trans** to them. The trans directing influence of the ligands and the synthesis of cis and trans- $[Pt(NH_3)_2Cl_2]$ is explained in Sec. 7.9.12. In both the cases the greater trans directing influence of $Cl^-$ over $NH_3$ determines the geometry of the final product in the second step of replacement. Therefore **trans effect is defined as the effect of a coordinated ligand to a metal ion is the tendency of that ligand to direct an incoming ligand to occupy the position trans to that ligand**. The $Cl^-$ ion trans to each other are more labile than $Cl^-$ ion trans to $NH_3$. The reaction sequences given in Sec. 7.9.12 shows that substituting ligand $NH_3$ when advancing towards $[PtCl_4]^{2-}$ or $Cl^-$ when advancing towards $[Pt(NH_3)_4]^{2+}$ find the positions **trans** to a $Cl^-$ ligand more labile than the positions **trans** to $NH_3$.

As the trans effect arises primarily from greater rate of substitution at the trans position so that this effect is also called **kinetic trans effect** and this is sometimes called **thermodynamic effect**.

With reference to the above trans effect the synthesis of cis and trans-$[Pt(NH_3)(NO_2)Cl_2]^-$ is depicted below :

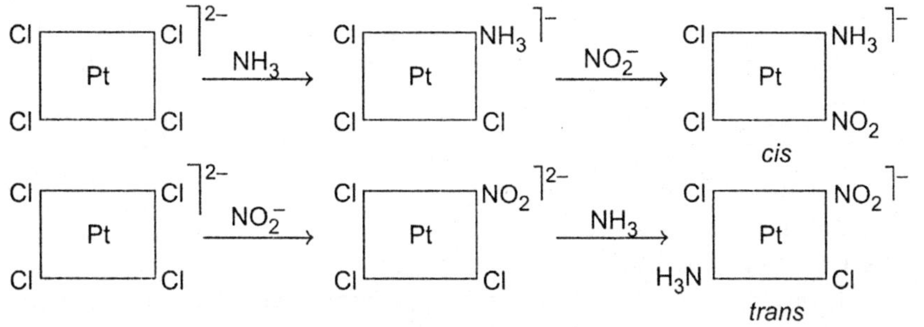


**Problem 7.6** Give the synthetic schemes for preparation of all the three isomers of $[Pt(py)(NH_3)(Br)(Cl)]$.

Given : (i) The order of trans directing ability : $NH_3 < py < Cl^- < Br^-$

(ii) Pt-N bond strength is greater than Pt-Cl bond strength. **[C.U. 2003]**

**Ans. :** The trans effect merely gives an emperical guideline on the expected geometry of the product in a substitution process provided we know the stoichiometry of the reaction.

This effect will not forecast the nature of the product i.e. which ligand is going to be replaced by other. Thus both trans effect and bond strength is to be considered for the synthesis of the three isomers of $[Pt(py)(NH_3)(Br)(Cl)]$ :

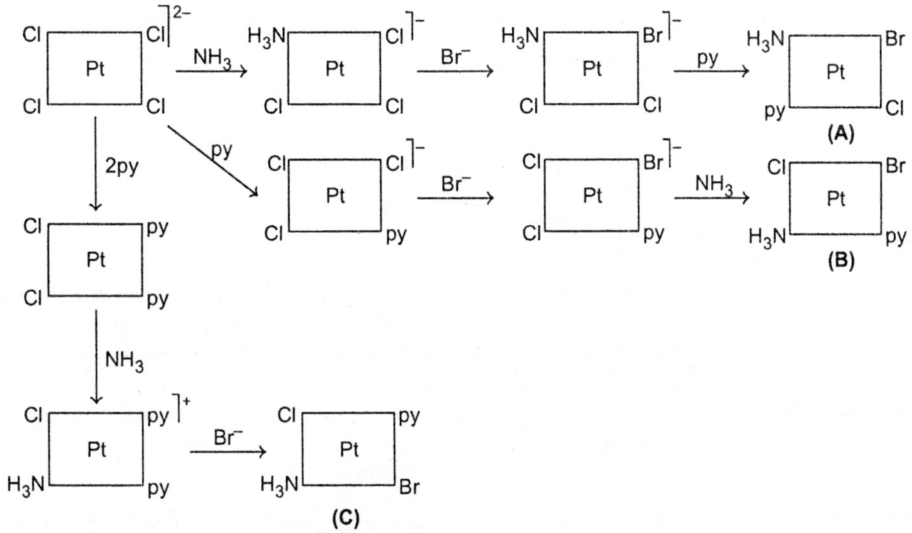


**Problem 7.7** Justify ferrioxalates are resolvable in optical isomers but ferricyanides are non resolvable. **[B.U. 1984,1990]**

**Ans. :** $C_2O_4^{2-}$ is a bidentate ligand while $CN^-$ is a monodentate ligand so that ferrioxalates has the following non superimposable mirror images so that it can be resolvable into optical isomers. But ferricyanides will have superimposable mirror images and so it will be optically inactive.

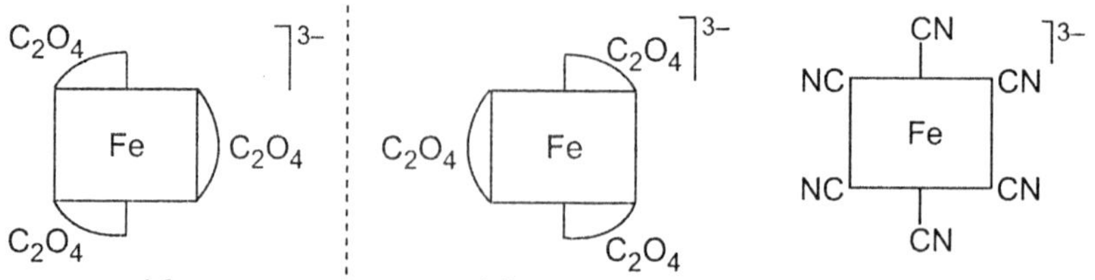

*d-form* *l-form* *optically inactive, will produce superimposable mirror image*

*non superimposable mirror image, optically active*

**Problem 7.8** On analysis, a complex compound was found to exhibit the ratio Co : $NH_3$ : Cl : Br = 1 : 4 : 2 : 1. Write down the structural formula of all possible isomers of the compound and give their respective IUPAC names. **[B.U. 1986]**

**Ans. :** According to the ratio of Co : $NH_3$ : Cl : Br = 1 : 4 : 2 : 1, the number of isomers are :

(1) $[Co(NH_3)_4Cl_2]Br$ which exists in the following cis-trans forms :

*trans* *cis*

*trans-teramminedichloridocobalt(III) bromide* *cis-teramminedichloridocobalt(III) bromide*

(2) The compound may also exist in the following ionisation isomers :

$[Co(NH_3)_4ClBr]Cl$ which also exist in the following cis-trans isomers :

*cis* *trans*

*cis-tetramminechloridobromidocobalt(III) chloride* *trans-tetramminechloridobromidocobalt(III) chloride*

**Problem 7.9** Assuming monomeric and dimeric formula state with reasons the different type of isomerism in the following cases :

(i) $[Pt(NH_3)_2Cl_2]$, (ii) $[Co(NH_3)_5(ONO)](NO_3)_2$ **[B.U. 1990]**

**Ans. :** (i) According to Sec. 7.9.11(A), $[Pt(NH_3)_2Cl_2]$ exhibits the following cis-trans isomerism in monomeric formulation :

*cis* *trans*

With dimeric formulation, the complex exhibits the following coordination isomerism :

$[Pt(NH_3)_4][PtCl_4]$ or $[PtCl_4][Pt(NH_3)_4]$

(ii) In this case with monomeric formulation the compound shows structural, salt or linkage isomerism :

$[Co(NH_3)_5(ONO)](NO_3)_2$ and $[Co(NH_3)_5(NO_2)](NO_3)_2$

---

**Problem 7.10** A compound when analysed shows 65% Pt, 23.6% Cl, 9.3% N and 2% H. The compound has a molecular weight of 600 and forms two ions in solution. No chloride can be removed from solution by the addition of $AgNO_3$ solution and no ammonia can be evolved. Draw two possible isomers of the compound and identify the type of such isomers. **[B.U. 1991]**

**Ans. :** $Pt = \frac{65.0}{195.1} = 0.333$, $Cl = \frac{23.6}{35.5} = 0.664$, $N = \frac{9.3}{14.0} = 0.664$ and $H = \frac{2}{1} = 2.$

At wt. of Pt = 195.1

So that, $Pt = \frac{0.333}{0.333} = 1$, $Cl = \frac{0.664}{0.333} = 1.99 \approx 2$, $N = 1.99 \approx 2$

and $H = \frac{2}{0.333} = 6.006 \approx 6$

Therefore the emperical formula of the compound is $PtCl_2N_2H_6$ so that $(PtCl_2N_2H_6)_n$ = 600, whereby n = 2 and the molecular formula of the compound is $Pt_2Cl_4N_4H_{12}$.

Now as the compound does not react with $AgNO_3$ and no ammonia is evolved, so that as stated in the molecular formula of the compound all nitrogen and hydrogen atoms combine to form $NH_3$ molecules and the compound may be formulated as $Pt_2(NH_3)_4Cl_4$. The compound exhibits coordination isomerism of the following type :

$[Pt(NH_3)_4][PtCl_4]$ and $[PtCl_4][Pt(NH_3)_4]$

---

**Problem 7.11** Three compounds A, B and C have formula $CrCl_3{\cdot}6H_2O$. When 1 g of each of A and B is dehydrated in presence of a dehydrating agent, a constant weight of 0.865 g and 0.932 g respectively is obtained. Compound C shows no loss of water.

(a) Find the composition of A, B and C.

(b) If an excess of aqueous $AgNO_3$ solution is added to 1 g solution of A, B and C, what amount of AgCl will be precipitated in each case?

**Ans. :** (a) Molecular weight of $CrCl_3{\cdot}6H_2O = 52 + 35.5 \times 3 + 6 \times 18 = 266.5$

From the given problem,

1 g of A gives (1 – 0.865) = 0.135 g of $H_2O$ on dehydration

$\therefore$ 266.5 g of A gives $0.135 \times 266.5 = 36.0$ g $H_2O$, which corresponds to two $H_2O$ molecules.

Therefore, the composition of A is $[Cr(H_2O)_4Cl_2]Cl{\cdot}2H_2O$

In a similar way, the composition of B is $[Cr(H_2O)_5Cl]Cl_2{\cdot}H_2O$

As C records no loss of water, so its composition is $[Cr(H_2O)_6]Cl_3$

(b) Molecular weight of AgCl = 143.5

Since A has one $Cl^-$ ion outside the coordination entity, so

$$1 \text{ g of A will produce} = \frac{143.5}{266.5} = 0.538 \text{ g AgCl}$$

Since B has two $Cl^-$ ions outside the coordination sphere, so

$$1 \text{ g of B will produce} = \frac{2 \times 143.5}{266.5} = 1.076\text{g AgCl}$$

$$\text{and C will produce} = \frac{3 \times 143.5}{266.5} = 1.614\text{g AgCl}$$

---

**Problem 7.12** Chromium (III) chloride forms three different hydrates of some composition Cr : Cl : $H_2O$ = 1 : 3 : 6. The violet salt does not lose water over conc. $H_2SO_4$ and gives 3 equivalents of AgCl on treatment with $AgNO_3$. The two other hydrates, both being green, loses 1 and 2 mol of $H_2O$ over conc. $H_2SO_4$ and respectively gives 2 and 1 equivalents of AgCl on treatment with $AgNO_3$. Write down the coordination structures of the three isomeric Cr(III) complexes and explain the type of isomerism involved. **[B.U. 1996,2001,2015]**

**Ans. :** Cr : Cl : $H_2O$ = 1 : 3 : 6

The violet form does not lose water over conc. $H_2SO_4$ and produces 3 equivalents of AgCl on treatment with $AgNO_3$ so it has the coordination structure as $[Cr(H_2O)_6]Cl_3$ i.e.

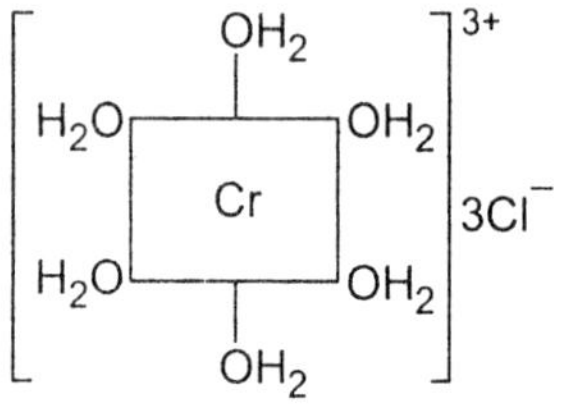


The two other hydrates have the same colour namely green. One variety loses 1 mol of $H_2O$ over conc. $H_2SO_4$ and produces 2 equivalents of AgCl on treatment with $AgNO_3$, so that this have the composition as,

$[Cr(H_2O)_5Cl]Cl_2 \cdot H_2O$ i.e.

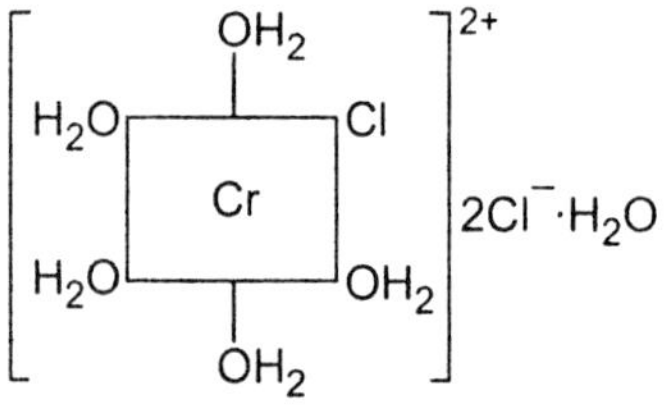


The other green variety on similar treatment produces 1 equivalent of AgCl and loses 2 mol of $H_2O$, so that it has the composition, $[Cr(H_2O)_4Cl_2]Cl \cdot 2H_2O$ i.e.

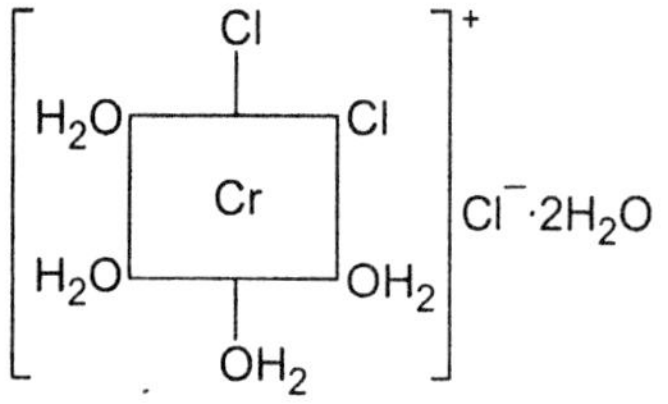

This isomer also exists in the following cis-trans or geometrical isomers :

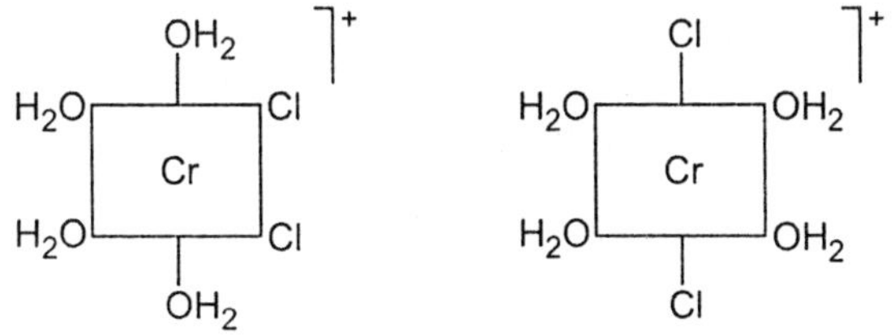

*cis* *trans*

---

**Problem 7.13** Arrange the complexes in order of their molar conductivities :
$Pt(NH_3)_6Cl_4$, $Pt(NH_3)_3Cl_4$, $K_2PtCl_6$ **[B.U. 2001,2009]**

**Ans. :** $Pt(NH_3)_6Cl_4 > K_2PtCl_6 > Pt(NH_3)_3Cl_4$

---

**Problem 7.14** Write down the exact formula of the possible polymerization isomers derived from the emperical formula $Co(NH_3)_3(NO_2)_3$. **[B.U. 2006]**

**Ans. :** This is also known as coordination isomers, viz.,

dimeric : $[Co(NH_3)_6][Co(NO_2)_6]$; $[Co(NH_3)_5(NO_2)][Co(NH_3)(NO_2)_5]$ and $[Co(NH_3)_4(NO_2)_2][Co(NH_3)_2(NO_2)_4]$

trimeric : $[Co(NH_3)_4(NO_2)_2]_2[Co(NH_3)(NO_2)_5]$; $[Co(NH_3)_5(NO_2)][Co(NH_3)_2(NO_2)_4]_2$

tetrameric : $[Co(NH_3)_4(NO_2)_2]_3[Co(NO_2)_6]$; $[Co(NH_3)_6][Co(NH_3)_2(NO_2)_4]_3$

pentameric : $[Co(NH_3)_5(NO_2)]_3[Co(NO_2)_6]_2$; $[Co(NH_3)_6]_2[Co(NH_3)(NO_2)_5]_3$

---

**Problem 7.15** Differentiate π-acidic and π-complexing ligands with concrete examples. **[B.U. 2006,2009]**

**Ans. :** σ-donor to metal ion and π-acceptor from metal ion i.e. $M \underset{\pi}{\overset{\sigma}{\leftrightarrows}} L$ is called **π-acidic ligand**. Here metal ion should be in low oxidation state and ligand should have stable LUMO e.g. CO, $PPh_3$, 2, 2′-bipyridine etc.

**π-complexing ligands** are π-electron cloud donor e.g. $CH_2{=}CH_2$, $CH{\equiv}CH$, ;  etc.; forms μ-bond and essentially a π-acidic ligand as LUMO is stable i.e.

$$M \underset{\mu}{\overset{\pi}{\rightleftarrows}} \begin{matrix} CH_2 \\ \| \\ CH_2 \end{matrix}$$

Hence a π-complexing ligand is a π-acceptor ligand but the reverse is not true. π-acidic ligand donates lone pair in a σ-bond whereas π-complexing ligand donates π-electron cloud to a metal ion in a non-classical σ-bond often called μ-bond.

---

**Problem 7.16** 'A double salt in a coordinating solvent behaves as complex salt(s)'—account for this behaviour. **[B.U. 2007]**

**Ans. :** $K_2SO_4{\cdot}Al_2(SO_4)_3{\cdot}24H_2O$ when dissolved in tetrahydrofuran (THF) solvent, both $K^+$ and $Al^{3+}$ form solvent complexes and do not exist as free ion.

---

**Problem 7.17** $Co(H_2O)_6Cl_3$ gives two water molecules on dehydration. How many equivalents of $AgNO_3$ will be required for each molecule of the complex to precipitate AgCl?

**[B.U. 2008]**

**Ans.** : $Co(H_2O)_6Cl_3$ gives two water molecules on dehydration, so that the composition of the complex species is $[Co(H_2O)_4Cl_2]Cl{\cdot}2H_2O$. Therefore in order to precipitate AgCl one equivalent of $AgNO_3$ will be required.

---

**Problem 7.18** Draw the enantiomers of the following complex : $[Cr(ox)_3]^{3-}$ and $[RhCl_2(NH_3)_4]^{1+}$. **[B.U. 2008]**

**Ans.** :

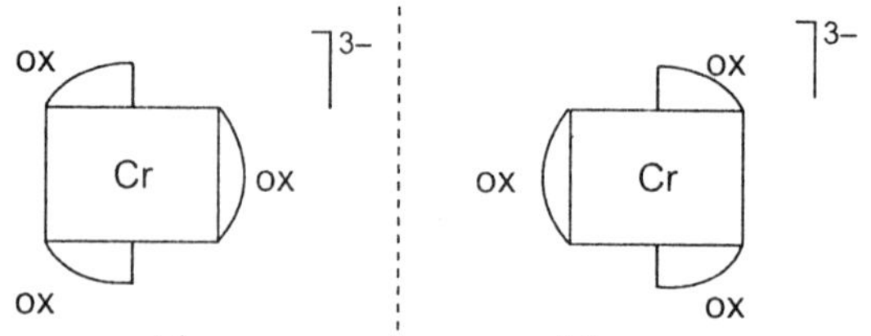


*d-form* *l-form*

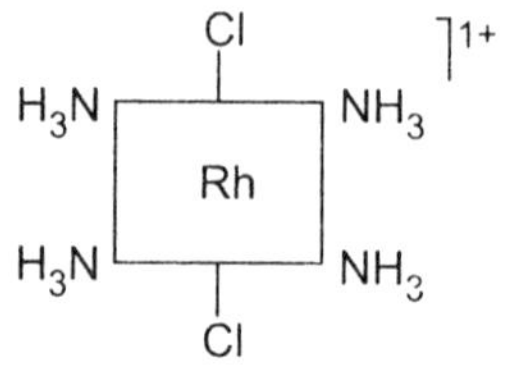


*trans*
*optically inactive*

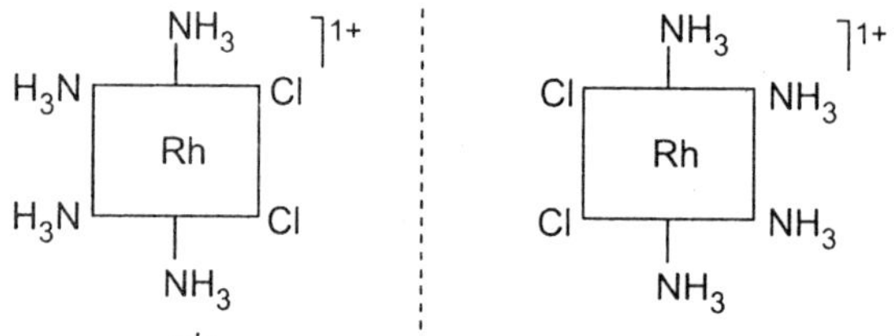


*cis*

---

**Problem 7.19** What is the formula of potassium trichloro (ethylene) platinate(II)? **[B.U. 2009]**

**Ans.** : $K[PtCl_3(C_2H_4)]$.

---

**Problem 7.20** State the stereochemistry of the following compounds with suitable drawings : $[Mg(oxin)_2]$; $[Co(NH_3)_3Cl_3]$ and $[VO(acac)_2]$. **[B.U. 2009]**

**Ans.** :

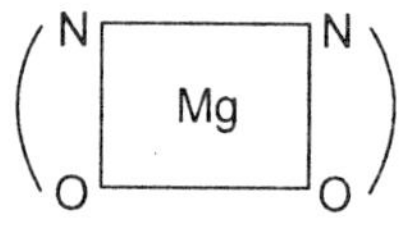


*square planar*
*$Mg(oxin)_2$*

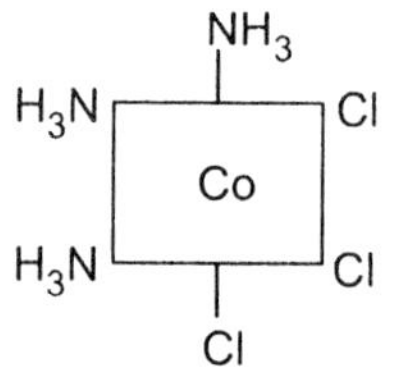


*Octahedral*
*$[Co(NH_3)_3Cl_3]$*

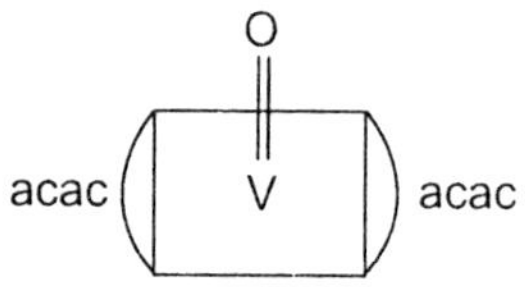


*square pyramid*
*$[VO(acac)_2]$*

---

## 7.10 INNER METALLIC COMPLEXES

Some organic ligands contain both acidic and neutral donor groups (such as $-NH_2$, $-SH_2$, $-COOH$ etc.) at suitable positions and can therefore through the formation of chelate ring may satisfy the primary valency and coordination number of a metal ion and thereby a special type of complex is formed which is known as **inner metallic complex** or **inner complex**. Inner complexes are into the following category :

(a) **First order inner complex :** In those cases where the charge and coordination number of the central metal ion is simultaneously satisfied by the ligands are called **first order**

**inner metallic complex**. Here the ligands should have salt forming groups as also neutral donor groups at suitable distance apart so that a five or six membered chelate ring is formed around the metal ion. Examples are (7-XXXII to 7-XXXVI).

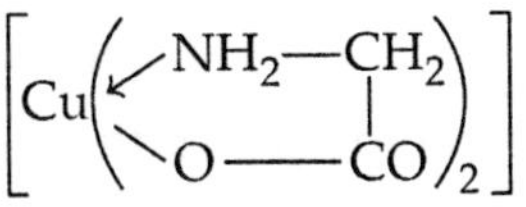

(7-XXXII)
*bis(glycinato) copper(II)*

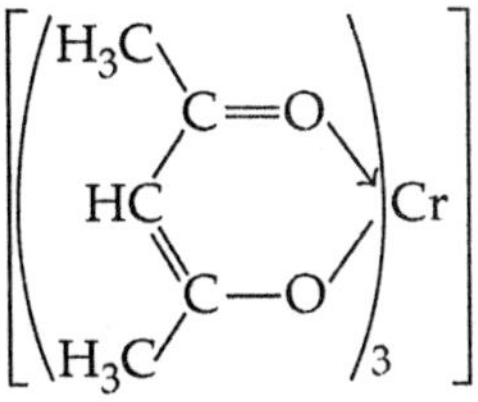

(7-XXXIII)
*tris(acetylacetonato) chromium(III)*

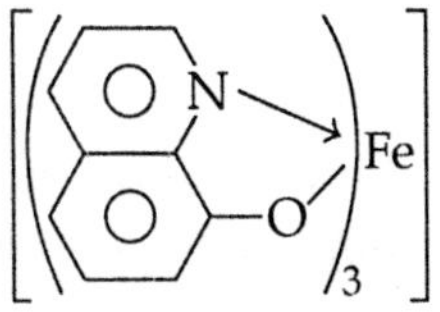

(7-XXXIV)
*tris(8-hydroxyquinolato) iron(III)*

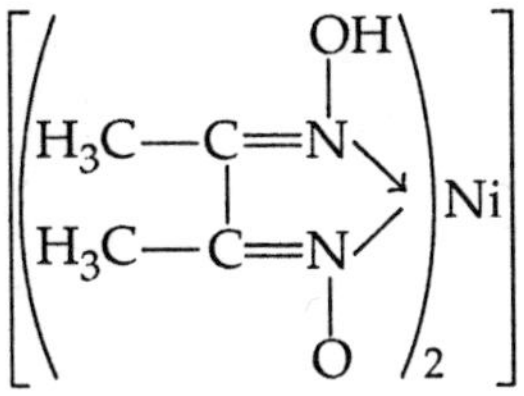

(7-XXXV)
*bis(dimethylglyoximato) nickel(II)*

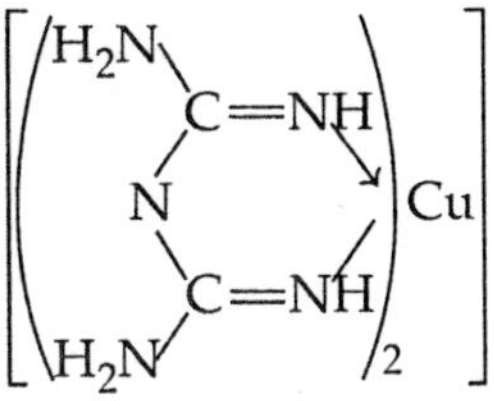

(7-XXXVI)
*bis(biguanide) copper(II)*

Other examples are $[Mg(oxin)_2]$, $[Al(oxin)_3]$, $[Co(gly)_3]$ etc. Many such first order inner complexes have characteristic colours and because of their non-electrolytic nature, they have played a significant role in the analytical chemistry for the detection and estimation of metal ions.

(b) **Second order inner complex :** In this case the coordination number of the central metal ion is satisfied but the charge of the metal ion is not neutralised. Thus there is a formation of an ionic species which is called **inner metallic complex of second order**. For example $[Si^{IV}(acac)_3]Cl$, $Na[Co^{II}(acac)_3]$, $[Ti^{IV}(acac)_3]Cl$, $[B^{III}(acac)_2]ClO_4$ etc.

(c) First order inner complexes sometimes possess suitable basic or acidic groups for salt formation and are therefore converted to ionic complexes under suitable conditions. Thus insoluble $Cu(Big)_2$(7-XXXVI) on protonation produces a soluble complex (7-XXXVII) as also 8-hydroxyquinoline-5-sulphonic acid forms first order insoluble inner complexes with divalent metal ions (7-XXXVIII) which on treatment with a base form soluble complexes (7-XXXIX) as depicted below :

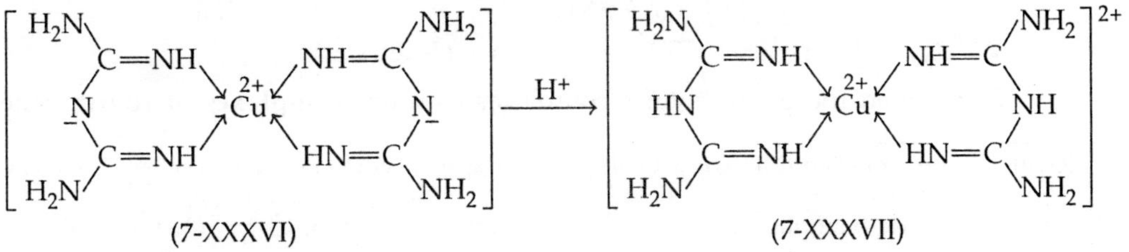

(7-XXXVI) *(insoluble)*  (7-XXXVII) *(soluble)*

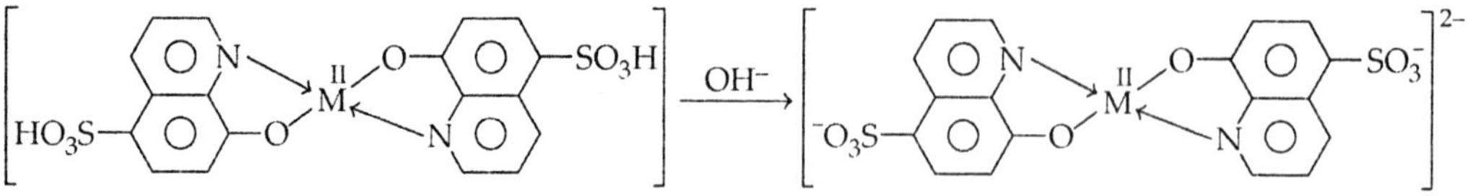

(7-XXXVIII) (*insoluble*) $M^{II} = Mg^{2+}, Ca^{2+}$ (7-XXXIX) (*soluble*)

Prof. P Ray of Calcutta University (who made immense contribution to metal-biguanide complexes) named these type of inner complexes as **third order**.

(d) Prof. P. Ray also demonstrated examples of first order inner complexes where equal number of acidic and basic groups are present in the ligand which may exist as Zwitter ion*. The copper (II) complex of phenylbiguanide-p-sulphonic acid** (7-XL) which is first order inner complex salt forms simultaneously cationic complex (7-XLI) on treatment with acid as also forms anionic complex (7-XLII) on treatment with a base :

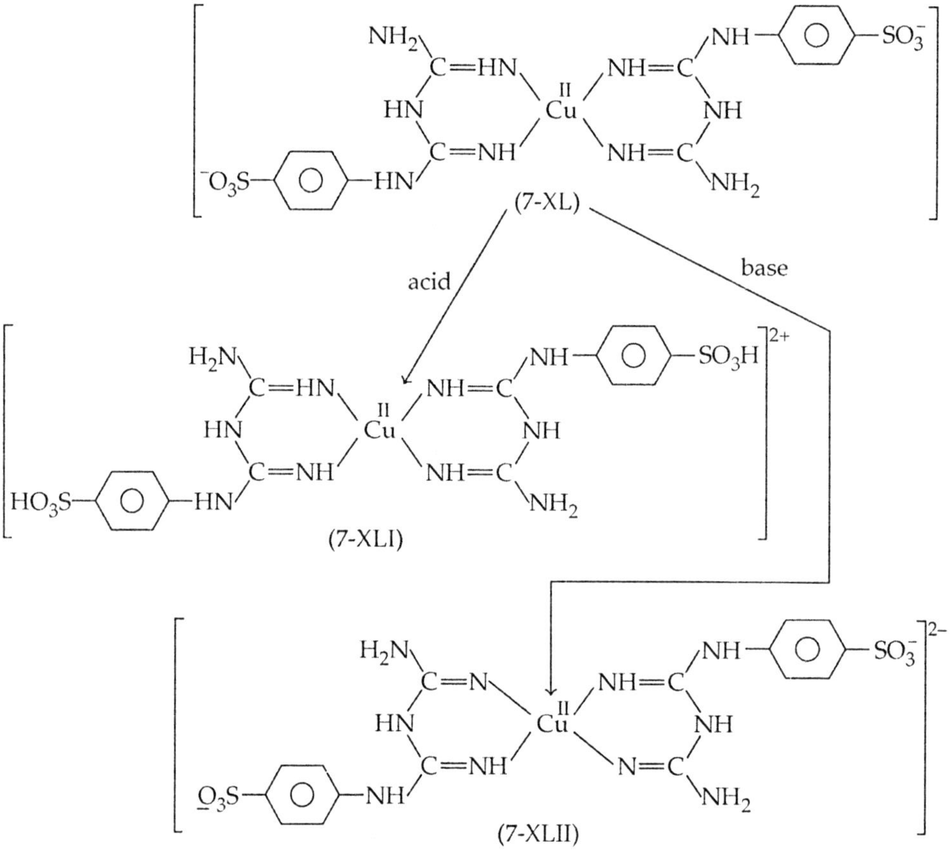

(7-XL) (7-XLI) (7-XLII)

Ray named these amphoteric complexes as **inner complexes of fourth order**.

---

* Zwitter ion e.g glycine ($NH_2CH_2COOH$) may exist as Zwitter ion structure ($N^+H_3CH_2COO^-$).

** Phenylbiguanide-p-sulphonic acid, 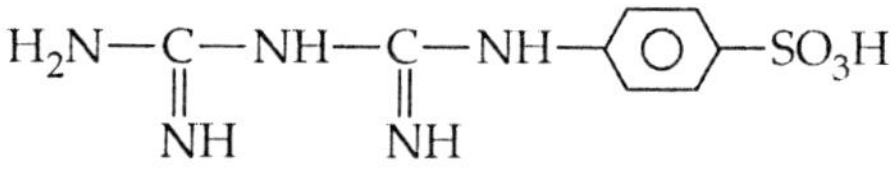

**Problem 7.21** Which among the following bidentate ligands would you expect to give innermetallic hexacoordinated complexes with transition metals $M^{3+}$ ions?
$H_2N—CH_2—COOH$, $HOOC—COOH$, $H_3C—CO—CH_2—CO—CH_3$ **[C.U. 2003]**

**Ans. :** $H_2N—CH_2—COOH$ and $H_3C—CO—CH_2—CO—CH_3$ both form first order inner complexes because they simultaneously satisfy the charge and coordination number of the central metal ion $M^{3+}$ and they also form suitable strain free five or six membered ring around the central $M^{3+}$ ion. These ligands form tris complex with $M^{3+}$. HOOC—COOH forms second order inner complex because in this case coordination number is satisfied but charge is not satisfied due to the formatiion of tris complex.

**Problem 7.22** State the essential characteristics of a ligand to form an inner metallic complex. Give an example of such a ligand. **[C.U. 1999]**

**Ans. :** Those ligands which possess both salt forming groups and neutral donor groups and can therefore simultaneously satisfy both the primary valency and coordination number of the central metal ion and which form strain free five or six membered rings around the central metal ion, are suitable for the formation of inner complex. Examples are glycine, acetylacetone etc.

## 7.11 STABILITY CONSTANTS or FORMATION CONSTANTS

In the case of labile system when the formation and dissociation of metal-ligand bonds are very rapid, the extent of formation of a metal ligand complex (ML) at equilibrium is expressed by its **thermodynamic stability constant** ($K^a$) which is activity quotient of the complex formation equilibria as :

$$M + L \longrightarrow ML \qquad K^a = \frac{{}^aML}{{}^aM\,{}^aL} \qquad ......(7.1)$$

Since the activities (a's) are not possible to determine experimentally so that for practical purposes activities are replaced by concentration and in that case we define **stoichiometic stability constant** (K) as :

$$K = \frac{[ML]}{[M][L]} \qquad ......(7.2)$$

Generally a metal cation ($M^{n+}$) exists in aqueous solution as hydrated species, $M^{n+}(aq)$ and interaction of a ligand L results in successive replacement of water molecules from the coordination sphere of the metal ion :

$$[M(H_2O)_6]^{n+} + L \rightleftharpoons [M(H_2O)_5L]^{n+} + H_2O$$

$$[M(H_2O)_5L]^{n+} + L \rightleftharpoons [M(H_2O)_4L_2]^{n+} + H_2O \text{ and so on and}$$

thereby there will be the formation of ML, $ML_2$, $ML_3$ etc. complexes.

Omitting the charge of the metal ions M in solution let us consider monodentate ligands L interact with the metal ions to form soluble mononuclear complexes, the system at equilibrium may be represented by the following equations and equilibrium constants.

$$\left.\begin{array}{lll} M + L \longrightarrow ML & K_1 = \dfrac{[ML]}{[M][L]} \\ ML + L \longrightarrow ML_2 & K_2 = \dfrac{[ML_2]}{[ML][L]} \\ ML_2 + L \longrightarrow ML_3 & K_3 = \dfrac{[ML_3]}{[ML_2][L]} \\ \vdots \qquad\qquad \vdots & \\ ML_{n-1} + L \longrightarrow ML_n & K_n = \dfrac{[ML_n]}{[ML_{n-1}][L]} \end{array}\right] \qquad ......(7.3)$$

There will be n such equilibria where n represents the maximum coordination number of the metal ion M for the ligand L. For example $Al^{3+}$ forms $AlCl_4^-$ and $AlF_6^{3-}$ and $Co^{2+}$ forms $CoCl_4^{2-}$ and $[Co(NH_3)_6]^{2+}$ as the highest complexes with the ligand indicated.

Another way of expressing the equilibrium relations is the following. In this case also the highest coordination complex $ML_n$ is formed.

$$\left.\begin{array}{lll} M + L \longrightarrow ML & \beta_1 = \dfrac{[ML]}{[M][L]} \\ M + 2L \longrightarrow ML_2 & \beta_2 = \dfrac{[ML_2]}{[M][L]^2} \\ M + 3L \longrightarrow ML_3 & \beta_3 = \dfrac{[ML_3]}{[M][L]^3} \\ \vdots \qquad\qquad \vdots & \\ M + nL \longrightarrow ML_n & \beta_n = \dfrac{[ML_n]}{[M][L]^n} \end{array}\right] \qquad ......(7.4)$$

Since there can be only n independent equilibrium in such a system, it is clear that the $K_i$'s and $\beta_i$'s must have a relation between them. Let us consider the expression for $\beta_3$ and the numerator and deonominator both multiplied by $[ML][ML_2]$ and then rearranges in the following way :

$$\beta_3 = \frac{[ML_3]}{[M][L]^3} \cdot \frac{[ML][ML_2]}{[ML][ML_2]} = \frac{[ML]}{[M][L]} \cdot \frac{[ML_2]}{[ML][L]} \cdot \frac{[ML_3]}{[ML][ML_2]} = K_1K_2K_3$$

Therefore we can say, $\beta_n = K_1K_2K_3......K_n$ ......(7.5)

The $K_i$'s are called the **stepwise formation constants (or stepwise stability constants)** and $\beta_i$'s are called the **overall formation or stability constants**. Reciprocal of these stability constants are called respecitve **instability constants**.

With few exceptions, there is generally a slow decrease in the values of $K_i$'s in any particular system. This is justified by the data for the $Cu^{2+}$–$NH_3$ and $Cd^{2+}$–$NH_3$ systems where the ligand is neutral and by $Cd^{2+}$–$CN^-$ system where the ligand is charged :

$[Cu(H_2O)_4]^{2+} + NH_3 \rightleftharpoons [Cu(H_2O)_3(NH_3)]^{2+} + H_2O$ $K_1 = 1.66\times10^4$ (1 : 1 complex)

$[Cu(H_2O)_3(NH_3)]^{2+} + NH_3 \rightleftharpoons [Cu(H_2O)_2(NH_3)_2]^{2+} + H_2O$ $K_2 = 3.10\times10^3$ (1 : 2 complex)

$[Cu(H_2O)_2(NH_3)_2]^{2+} + NH_3 \rightleftharpoons [Cu(H_2O)(NH_3)_3]^{2+} + H_2O$ $K_3 = 8.31\times10^2$ (1 : 3 complex)

$[Cu(H_2O)(NH_3)_3]^{2+} + NH_3 \rightleftharpoons [Cu(NH_3)_4]^{2+} + H_2O$ $K_4 = 1.51\times10^2$ (1 : 4 complex)

$$\beta_4 = 1.66 \times 3.10 \times 8.31 \times 1.51 \times 10^{11} = 64.57 \times 10^{11}$$

$[Cd(H_2O)_4]^{2+} + NH_3 \rightleftharpoons [Cd(H_2O)_3(NH_3)]^{2+} + H_2O$ $K_1 = 10^{2.65}$

$[Cd(H_2O)_3(NH_3)]^{2+} + NH_3 \rightleftharpoons [Cd(H_2O)_2(NH_3)_2]^{2+} + H_2O$ $K_2 = 10^{2.10}$

$[Cd(H_2O)_2(NH_3)_2]^{2+} + NH_3 \rightleftharpoons [Cd(H_2O)(NH_3)_3]^{2+} + H_2O$ $K_3 = 10^{1.44}$

$[Cd(H_2O)(NH_3)_3]^{2+} + NH_3 \rightleftharpoons [Cd(NH_3)_4]^{2+} + H_2O$ $K_4 = 10^{0.93}$

$$\beta_4 = 10^{2.65} \times 10^{2.10} \times 10^{1.44} \times 10^{0.93} = 10^{7.12}$$

$[Cd(H_2O)_4]^{2+} + CN^- \longrightarrow [Cd(H_2O)_3(CN)]^+ + H_2O$ $K_1 = 10^{5.48}$

$[Cd(H_2O)_3(CN)]^+ + CN^- \longrightarrow [Cd(H_2O)_2(CN)_2] + H_2O$ $K_2 = 10^{5.12}$

$[Cd(H_2O)_2(CN)_2] + CN^- \longrightarrow [Cd(H_2O)(CN)_3]^- + H_2O$ $K_3 = 10^{4.63}$

$[Cd(H_2O)(CN)_3]^- + CN^- \longrightarrow [Cd(CN)_4]^{2-} + H_2O$ $K_4 = 10^{3.65}$

$$\beta_4 = K_1K_2K_3K_4 = 10^{18.88}$$

Thus typically as ligand is added to the solution of the metal ion, ML is first formed more rapidly than any other complex in the series. As the ligand concentration is increased, the $ML_2$ concentration rises rapidly whereby the ML concentration drops. In the successive stages the $ML_3$ becomes dominat, ML and $ML_2$ becoming less important. In this way the highest coordination complex $ML_n$ is formed to the nearly complete exclusion of all others at very high ligand concentrations. Typical formation constant values are given in Table 7.5.

**Table 7.5 : Formation constant values for metal complex in aqueous solution at 25°C**

| Ligand | Metal ion | $\log K_1$ | $\log K_2$ | $\log K_3$ | $\log K_4$ | $\log K_5$ | $\log K_6$ |
|---|---|---|---|---|---|---|---|
| Ammonia ($NH_3$) | $Ag^+$ | 3.31 | 3.91 | | | | |
| | $Cu^{2+}$ | 4.04 | 3.43 | 2.80 | 1.48 | | |
| | $Co^{2+}$ | 2.11 | 1.63 | 1.05 | 0.76 | 0.18 | –0.62 |
| | $Ni^{2+}$ | 2.72 | 2.17 | 1.66 | 1.12 | 0.75 | 0.03 |
| | $Zn^{2+}$ | 2.37 | 2.44 | 2.50 | 2.15 | | |
| | $Hg^{2+}$ | 8.8 | 8.6 | 1.0 | 0.7 | | |
| Chloride ($Cl^-$) | $Ag^+$ | 2.85 | 1.87 | 0.32 | 0.86 | | |
| | $Fe^{3+}$ | 0.62 | 0.11 | –1.40 | –1.92 | | |
| Cyanide ($CN^-$) | $Cd^{2+}$ | 5.48 | 5.12 | 4.63 | 3.65 | | |
| | $Hg^{2+}$ | 17.00 | 15.75 | 3.56 | 2.66 | | |

| Ligand | Metal ion | $\log K_1$ | $\log K_2$ | $\log K_3$ | $\log K_4$ | $\log K_5$ | $\log K_6$ |
|---|---|---|---|---|---|---|---|
| Ethylenediamine | $Cu^{2+}$<br>$Co^{2+}$<br>$Ni^{2+}$ | 10.55<br>5.89<br>7.52 | 9.05<br>4.83<br>6.28 | −1.0<br>3.10<br>4.26 | $\log\beta = 20.2$<br>$\log\beta = 10.9$<br>$\log\beta = 14.5$ | | |
| Salicylaldehyde | $Cu^{2+}$<br>$Ni^{2+}$<br>$Co^{2+}$ | 7.40<br>5.22<br>4.67 | 5.91<br>3.97<br>3.63 | | | | |
| Ethylenediamine tetraacetic acid | $Ca^{2+}$<br>$Fe^{2+}$<br>$Fe^{3+}$<br>$Co^{2+}$<br>$Co^{3+}$<br>$Ni^{2+}$<br>$Cu^{2+}$<br>$Al^{3+}$<br>$Zn^{2+}$ | 10.70*<br>14.33<br>25.10<br>16.31<br>36.00<br>18.62<br>18.80<br>16.13<br>16.50 | | | | | |

* This also represent $\log\beta$ value.

Another formation constant is known as **conditional** or **effective stability constant** which gives the measure of stability of a complex under actual reaction conditions. This actually depends on the pH of the reaction medium and will be discussed later. Thermodyamic stability constants ($K_n^a$) depend only on temperature. These are independent of ionic strength and dielectric constant of the medium. Stoichiometric stability constant ($K_n$) on the other hand depends on temperature as also on ionic strength and dielectric constant of the medium.

### 7.11.1 Factors influencing the stability of complexes : Factors influencing the stepwise and overall formation constants

A perusal of Table 7.5 revealed that the successive formation constants ($K_1$, $K_2$, $K_3$ etc.) decreases. There are several reasons for a steady decrease in $K_i$ values as the number of ligands increases :

1. **Statistical factor :** In the absence of any valence interaction (viz. ionic, covalent, $\sigma$ or $\pi$ bonding) the successive stability constants of a series of mononuclear complexes tend to decrease in magnitude due to statistical factors. Thus, in the complex formation reaction between an aquametal ion and a ligand, the ease of substitution of a water molecule coordinated to the metal ion decreases as the number of water molecules (reaction sites) available for the reaction decreases with the successive addition of the ligand. Thus for the equilibrium

$$M(H_2O)_{N-n+1}L_{n-1} + L \underset{K_d}{\overset{K_f}{\rightleftharpoons}} M(H_2O)_{N-n}L_n + H_2O \qquad \text{......(7.6)}$$

here N = coordination number i.e. highest number of groups attached and n = the number of groups already attached.

The rate of formation of $M(H_2O)_{N-n}L_n = K_f[M(H_2O)_{N-n+1}L_{n-1}][L]$

Due to the statistical factor the forward rate constant will be proportional to the number of reaction sites i.e. the number of water molecules in the complex $M(H_2O)_{N-n+1}L_{n-1}$ and so we may write,

$$K_f = K'_f(N-n+1)$$

The forward rate ($r_1$) is given by,

$$r_1 = K'_f(N-n+1)[M(H_2O)_{N-n+1}L_{n-1}][L]$$

Similarly the rate of dissociation ($r_2$) of $M(H_2O)_{N-n}L_n$, we can write,

$$r_2 = K'_d n[M(H_2O)_{N-n}L_n]$$

where $K'_d n = K_d$

Hence, the equilibrium constant $K_n$ for the reaction (7.6) is given by,

$$K_n = \frac{[M(H_2O)_{N-n}L_n]}{[M(H_2O)_{N-n+1}L_{n-1}][L]} = \frac{K_f}{K_d} = \frac{K'_f(N-n+1)}{K'_d n}$$

Therefore $K_n \propto \frac{N-n+1}{n}$. In a similar way it can be stated that $K_{n+1} \propto \frac{N-n}{n+1}$ so that the ratio between the two successive stability constants will be—

$$\frac{K_n}{K_{n+1}} = \frac{\frac{N-n+1}{n}}{\frac{N-n}{n+1}} = \left(\frac{n+1}{n}\right)\left(\frac{N-n+1}{N-n}\right) \qquad \ldots\ldots(7.7)$$

Alternatively we may derive equation (7.7) in a more simplified way. Thus we can say, tendency to split off a group $\propto$ n i.e. the number of groups already attached, and tedency to add an addtional group $\propto$ N–n i.e. the number of unfilled positions.

Therefore the ratio of two successive stability constants will be inversely proportional to the relative number of groups already attached i.e. $\frac{n}{n+1}$ and directly proportional to the relative number of unfilled positions i.e. $\frac{N-n+1}{N-n}$ so that $K_n \propto \frac{1}{n}$, $K_{n+1} \propto \frac{1}{n+1}$ and again $K_n \propto N-n+1$, $K_{n+1} \propto N-n$ and thereby,

$$\frac{K_{n+1}}{K_n} = \left(\frac{n}{n+1}\right)\left(\frac{N-n}{N-n+1}\right)$$

For the successive association of six monodentate ligands to form an octahedral complex, $ML_6$ e.g. $Ni^{2+}$–$NH_3$ system to form $Ni(NH_3)_6^{2+}$, the statistical ratio of the stepwise formation constants are given in Table 7.6. Although the agreement between the calculated and experimental values are poor but as far as the trend is concerned the results are well accord with the statistical predictions. It is evident that experimental values are greater than the calculated statistical values indicating that the other factors play vital role to explain the difference.

For the association of three bidentate ligands (A–A) to form an octahedral complex $M(A-A)_3$, the ratio of successive stability constants will be,

$$K_1 : K_2 : K_3 = \frac{12}{1} : \frac{5}{2} : \frac{4}{15}$$

and for coordination number four (square planar or tetrahedral geometry) having composition $M(A\text{–}A)_2$ the ratio $K_1 : K_2 = 2/1 : 1/2$.

**Table 7.6 : Calculated and experimental values of the ratio of successive stability constants of $Ni^{2+}(aq)\text{–}NH_3$ system to form $Ni(NH_3)_6^{2+}$. Experimental values at 25°C are given in parenthesis.**

| Ratio | Value* |
|---|---|
| $K_1/K_2$ | 2.40 (3.57) |
| $K_2/K_3$ | 1.89 (3.23) |
| $K_3/K_4$ | 1.79 (3.45) |
| $K_4/K_5$ | 1.89 (2.78) |
| $K_5/K_6$ | 2.40 (5.26) |

* Values are calculated using equation (7.7).

Statistical values of $K_n/K_{n+1}$ are found to change under the influence of strong metal-ligand π-bonding, due to abrupt change in the coordination number and hybridisation at some stages of complex formation, steric effects, change of the spin state, electronic configuration of the metal ion at a specific stage of complex formation etc. Due to the strong π-acidic nature of the thiol (SH) group $K_2 > K_1$ for $Ni^{2+}$, $Pd^{2+}$ complexes of cysteine. For halide complexes of Hg(II), the experimental value of $K_2/K_3$ is greater than the statistical value. This is due to the **change in stereochemistry** from linear in $HgCl_2$ (sp hybridisation) to tetrahedral in $HgCl_3(H_2O)^-$ or $HgCl_4^{2-}$ ($sp^3$ hybridisation). This is also true for $Zn^{2+}$ complexes with ethylenediamine where the tetrahedral complex $Zn(en)_2^{2+}$ ($sp^3$ hybridisation) changes to octahedral one i.e. $Zn(en)_3^{2+}$ ($sp^3d^2$ hybridisation) where $K_2/K_3$ ratio is greater than the calculated statistical value. In case of $Ag^+\text{–}NH_3$ system $K_2 > K_1$. This unusual reversal value is due to the fact that $[Ag(NH_3)_2]^+$ has the most stable linear geometry for the d° metal ion but $[Ag(NH_3)(H_2O)_3]^+$ has probably tetrahedral structure. For Fe(II) complexes of o-phenanthroline or 2′, 2′-bipyridine $K_3 > K_2$ due to the **change of spin state**. $Fe(ophen)_3^{2+}$ is diamagnetic and is low spin with $t_{2g}^6$ configuration whereas the $Fe(ophen)_2(H_2O)_2^{2+}$ is high spin with $t_{2g}^4e_g^2$ configuration. This change from $t_{2g}^4e_g^2$ to $t_{2g}^6$ configuration with the change from bis complex to tris complex respectively causes an enthalpy change for the addition of third ligand to be large enough because electrons in the $e_g$ level are much higher in energy than in the $t_{2g}$ level.

*Variation of formation constants*

2. **Electronic structure of metal ions :** From an exhaustive study of literature Irving and Williams suggested that the stability constant $\beta_n$ of complexes of bivalent metals of first transition series with common coordinating ligands having N, O or S donors will follow the following sequence :

$$\begin{matrix} Cr^{2+} & > Mn^{2+} & < Fe^{2+} & < Co^{2+} & < Ni^{2+} & < Cu^{2+} & > Zn^{2+} \\ 3d^4 & 3d^5 & 3d^6 & 3d^7 & 3d^8 & 3d^9 & 3d^{10} \end{matrix}$$

This order is commonly known as Irving-Williams order of stability constants which is based on experimental determination of formation constants. This order is partly due to the decrease of ionic radius along the series but mainly due to the decrease of CFSE of the $d^n$ ions across the series.

3. **Nature of the ligand :** Hard and soft donor atoms of ligands greatly influence the formation constants of metal complexes with varied oxidation states of metals. Hard-soft parameters of donor atoms may also change and thereby the values of formation constants may change [Sec. 4.5.2(iii) and (iv)]. With the exception of very soft ligands, generally the stability constant increases with the increase of the oxidation state of the metal. With hard donor atom formation constants increase with increase of oxidation state of the metal e.g. logK value of EDTA complex of $Co^{2+}$ and $Co^{3+}$ are 16.31 and 36.00 (Table 7.5) respectively.

4. **Chelation and chelate effect :** Chelate complexes are more stable than analogous non-chelated complexes as is evident from the greater value of formation constants of chelate complexes compared to that of non-chelated complex (Table 7.5). This enhanced stability is called **chelate effect. The term refers to the enhanced stability of a complex system containing chelate rings as compared to the stability of a system which is as similar as possible but contains none or few rings.**

When a metal chelate is formed by the displacement of unidentate ligands there is an increase in the disorder of the system because of release of more molecules which are registered by the greater overall formation constants :

$$Ni(H_2O)_6^{2+} + 6NH_3 \longrightarrow Ni(NH_3)_6^{2+} + 6H_2O \qquad \log\beta_6 = 8.61$$

$$Ni(H_2O)_6^{2+} + 3en \longrightarrow Ni(en)_3^{2+} + 6H_2O \qquad \log\beta_3 = 14.50$$

$$Ni(H_2O)_6^{2+} + 2\ dien \longrightarrow Ni(dien)_2^{2+} + 6H_2O \qquad \log\beta_2 = 19.1$$

To understand the chelate effect we must consider the following well known thermodynamic expression :

$$\Delta G^0 = -RT\ln\beta \text{ and } \Delta G^0 = \Delta H^0 - T\Delta S^0$$

Thus $\beta$ increases as $\Delta G^\circ$ becomes more negative. A more negative $\Delta G^0$ can be obtained by making $\Delta H^0$ more negative or from making $\Delta S^0$ more positive. Enthalpy change ($\Delta H^0$) for the displacement of water by related ligands e.g. $NH_3$ is small. There is no net increase in the number of molecules when six monodentate $NH_3$ molecules displace six $H_2O$ molecules. But in case of substitution by three bidentate ligands there will be net increase of three unbound molecules (7 – 4 = 3) while in case of two tridentate ligands there will be net increase of four unbound molecules (7 – 3 = 4). Such increase in the number of unbound molecules in the product side recorded more randomness in the system and thus registering higher entropy value.

To clarify the entropy effect, let us consider the following reactions :

$$Cd(H_2O)_4{}^{2+} + 4CH_3NH_2 \longrightarrow [Cd(CH_3NH_2)_4]^{2+} + 4H_2O \qquad \log\beta_4 = 6.52$$

$$Cd(H_2O)_4{}^{2+} + 2en \longrightarrow [Cd(en)_2]^{2+} + 4H_2O \qquad \log\beta_2 = 10.6$$

In this case or in general enthalpy change between chelated and non-chelated complexes of non-transition metal ions ($Cd^{2+}$, $Zn^{2+}$ etc.) is small due to similar metal-donor linkages, hence chelate effect is almost entirely dependent on the entropy difference as is evident from the data (Table 7.7). Hence chelate effect is primarily an entropy effect.

**Table 7.7 : Thermodynamic data for chelate effect**

| Complex | $\Delta H^0$ (kJ mol$^{-1}$) | $-T\Delta S^0$ (kJ mol$^{-1}$) | $\Delta G^0$(kJ mol$^{-1}$) |
|---|---|---|---|
| $Cd(CH_3NH_2)_4^{2+}$ | –57.3 | 20.1 | –37.2 |
| $Cd(en)_2^{2+}$ | –56.5 | –4.2 | –60.7 |

For the complexes of transition metal ions it is in part an enthalpy effect which possibly results from the crystal field effect which makes a contribution to the stability of transition metal complexes. Here also entropy factor makes a significant role for the stability of chelate complexes.

Following Sidgwick the explanation of the chelate effect may be presented more pictorially. If one of the two metal-ligand bonds of a bidentate chelate system is broken, the remaining bond will still hold the molecule in place, so that the broken link can be reestablished. But a ligand bound by a single link will drift away if the bond is broken. Primarily a chelate complex dissociates less than a non-chelated complex and hence will record higher formation constant.

*Ring size and stability*

Finally the stability of chelate complexes increase with increase in the number and size of the chelate rings. Coordination compounds having ring sizes ranging from four to nine members are known but five and six membered rings are more stable than the other ring sizes. Five membered rings without double bonds and six membered rings with double bonds have least angular strain and the corresponding complexes are more stable e.g.

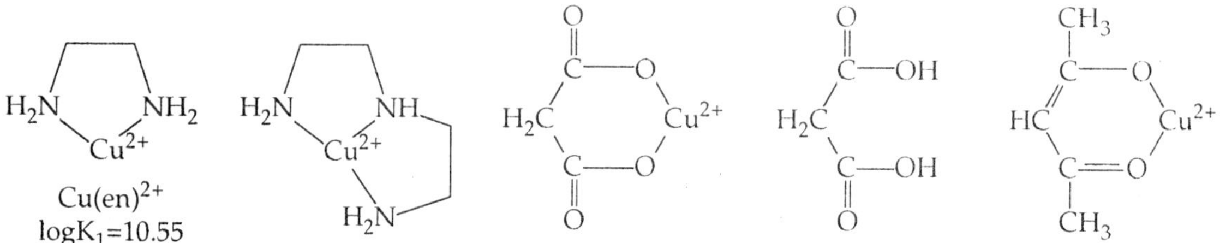


$Cu(en)^{2+}$ $\log K_1 = 10.55$ | $Cu(dien)^{2+}$ $\log K_1 = 16.02$ | Cu(mal) $\log K_1 = 6.02$ | mal $H_2$ | $Cu(acac)^+$ $\log K_1 = 8.79$

It is to be noted that ethylenediamine (en) and diethylenetriamine (dien) form five membered rings while malonic acid (mal$H_2$) and acetylacetone (acacH) form six memebred rings. The vlaues of $\log K_1$ at 25°C for the 1 : 1 complexes formed by Ni(II) with oxalate, malonate and succinate ions are 5.2, 4.1 and 2.3 respectively which also show the following decreasing order of stability with increasing ring size :

five membered > six membered > seven membered

5. **Steric effects :** When a bulky group is either attached to or present close enough to the donor centres of a chelating ligand, mutual repulsion among the donor atoms occur and consequently the metal ligand bond is weakened and thereby the formation constant value decreases. In some cases the bulky group may obstruct formation of chelate rings around the metal centre. This effect is commonly known as **steric hindrance** or **steric effect** or **steric strain.** For example 2, 2′-bipyridine forms **tris** complexes with metal ions but 6, 6′-dimethyl-2, 2′-bipyridine (7-XLIII) forms only **bis** or mono chelate complexes and in some cases no isolable complexes are formed due to steric hindrance between the methyl groups.

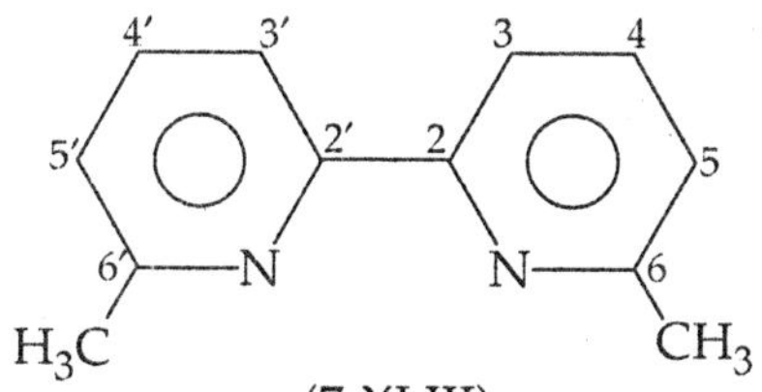

(7-XLIII)

2-and 5-methyl derivative of 8-hydroxyquinoline i.e. monomethyl derivative has higher base strength compared to that of the parent ligand. But 5-methyl derivative forms much stronger complex than the 2-methyl derivative with Cu(II) ion as the steric strain is more minimised in case of 5-methyl derivative compared to that of 2-methyl one as is evident from the formation constant values (Table 7.8). The strain effect may also be considered with the geometry of the ligand as also with the stereochemistry of the metal complexes. For example **trien** can coordinate all its four donor nitrogen atoms along the four corners of a square plane but triethyltriamine, $(NH_2–CH_2–CH_2)_3N$ (tren) cannot do so as straight chain trien may assume to form square planar geometry with $Cu^{2+}$ while tren with branched-chain is unable to form such type of complex due to steric strain.

**Table 7.8 : Formation constant values of oxine and its derivatives with Cu(II) in 50% dioxan-$H_2O$ medium**

| Ligand (HL) | $pK_B$ | $\log K_1$ $CuL^+$ | $\log K_2$ $CuL_2$ |
|---|---|---|---|
| 8-hydroxyquinoline | 10.8 | 13.0 | 12.4 |
| 5-methyl-8-hydroxyquinoline | 11.1 | 13.6 | 12.4 |
| 2-methyl-8-hydroxyquinoline | 11.0 | 10.2 | 9.3 |

6. **Macrocyclic effect** : This is the extension of chelate effect and template effect as discussed previously. When the ends of a linear polydentate ligand are joined, a macrocyclic ligand results. Complexes of macrocyclic ligands are both thermodynamically and kinetically more stable than the corresponding complexes with open chain multidentate ligands. Such

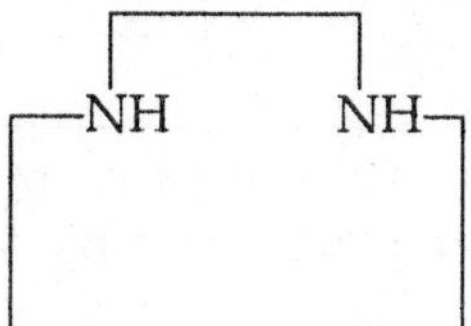

(7-XLIV)

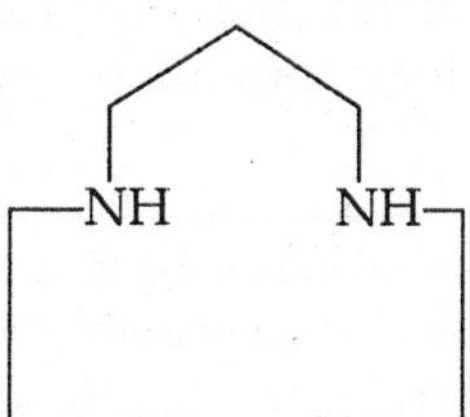

(7-XLV)

stabilising influence of macrocyclic ligands are known as macrocyclic effect. The formation constant of Cu(II) complexes of the ligand (7-XLIV) (logK = 20.1) while that for the ligand

(7-XLV) (logK = 23.9). Similar enhancement of stability constant value can also be observed in the case of following $Ni^{2+}$ complexes :

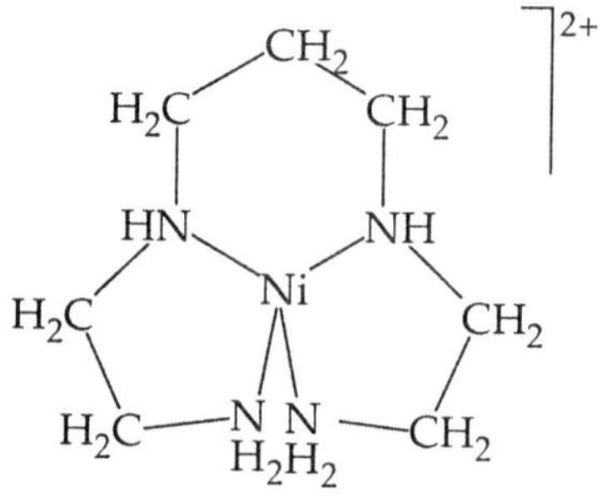


$K = 10^{11}$

(7-XLVI)

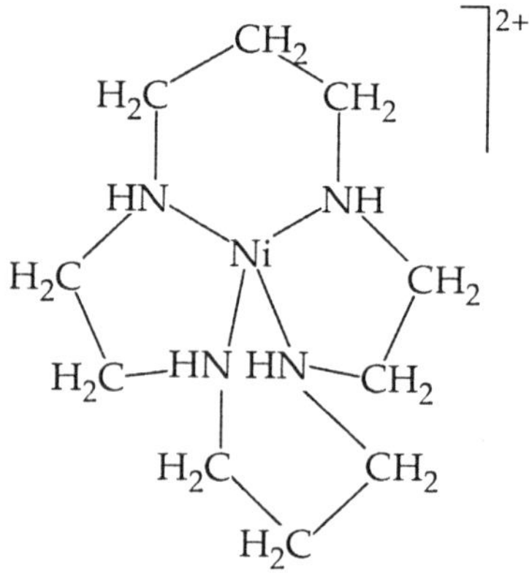


$K = 10^{16}$

(7-XLVII)

Macrocyclic complexes have the following characteristics :

(a) A marked kinetic inertness both to the formation of the complexes from the ligand or metal ion and to the reverse the dissociation of the metal ion from the ligand.

(b) They have high thermodynamic stability i.e. the formation constant for $N_4$ macrocycles is much greater than the formation constant of non-macrocyclic $N_4$ ligand. Thus for Ni(II) the formation constant for the macrocycle cyclam (7-XLVII) is many times greater than that for non-macrocycle tetradentate (7-XLVI) complex.

(c) Rates of formation and dissociation of complexes with macrocyclic ligands are slow due to the excessive geometric strain the ligand has to overcome in the simultaneous formation and dissociation of all the metal-ligand bonds.

(d) Due to the **macrocyclic effect** they stabilise the high oxidation states of metals. Macrocyclic effect is very common with biological ligands such as porphyrins, proteins, nucleic acids, chlorophyll etc.

(e) Macrocyclic effect remains the same for all metal ions of same charge so long as the macrocyclic ligand does not impose a geometry that is different from its open chain counterpart.

---

**Problem 7.23** The stepwise formation constants for complexes in general follow the order $K_1 > K_2 > K_3$ but for $Fe^{2+}$ with 1, 10-phenanthroline complex the order is $K_1 > K_2 < K_3$ – Explain.

**Ans. :** See Sec. 7.11.1(1). This is due to change of spin state. **[C.U. 2008]**

---

**Problem 7.24** A rose red precipitate is formed when $NiSO_4$ is treated with dimethylglyoxime in presence of ammonia. What is its formula? State the nature of the compound.

**[C.U. 2007]**

**Ans. :** 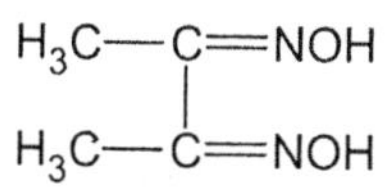

DMGH

The formula of the compound is $Ni(DMG)_2$. This is a first order innermetallic complex.

---

**Problem 7.25** Formulate an expression for the stoichiometric stability constant of the complex, $Pb(CH_3COO)^+$ ion, formed according to equilibria (I) in aqueous solution :

$$Pb^{2+} + CH_3COOH \rightleftharpoons Pb(CH_3COO)^+ + H^+ \quad \text{......(I)}$$ **[C.U. 2006]**

**Ans. :** According to Sec. 7.11 the stoichiometric stability constant of the complex $Pb(CH_3COO)^+$ is represented as :

$$K = \frac{[Pb(CH_3COO)^+][H^+]}{[Pb^{2+}][CH_3COOH]}$$

---

**Problem 7.26** Solution of the compound, $PtCl_2{\cdot}2NH_3$, on treatment with an excess of $AgNO_3$ solution, precipitates $Ag_2[PtCl_4]$, leaving the complex $[Pt(NH_3)_4](NO_3)_2$ in solution. Find the coordination formula of the parent complex and give its IUPAC name. **[C.U. 2006]**

**Ans. :** The coordination formula of the parent complex is $[Pt(NH_3)_4][PtCl_4]$ or $[PtCl_4][Pt(NH_3)_4]$. This is an example of coordination isomers which on dissociation produces two ions $[Pt(NH_3)_4]^{2+}$ and $[PtCl_4]^{2-}$. These ions in solution react with $AgNO_3$ to form a precipitate of $Ag_2[PtCl_4]$ and in solution there remains the complex $[Pt(NH_3)_4](NO_3)_2$. The name of the parent complex is tetraammineplatinum(II)tetrachloridoplatinate(II).

---

**Problem 7.27** 'An inner metallic ligand is essentially a chelating ligand'—Elucidate. **[B.U. 2013, 2007]**

**Ans. :** Characteristic of inner metallic ligands are that they have a salt forming group as also neutral donor group at suitable distance apart so that primary and secondary valency of the central metal ion is simultaneously satisfied by forming suitable strain free five or six membered rings. Therefore inner metallic ligands are essentially chelating ligands because in inner metallic complexes there always occur the formation of chelate rings i.e. glycine, acetylacetone etc.

---

**Problem 7.28** Establish the relation between stepwise formation constants and overall formation constant for a complexation reaction between $[Cu(H_2O)_6]^{2+}$ and ethylenediamine. **[C.U. 1999]**

**Ans. :** $Cu(H_2O)_6^{2+} + en \longrightarrow Cu(H_2O)_4(en)^{2+} + 2H_2O \qquad K_1 = \dfrac{[Cu(H_2O)_4(en)^{2+}]}{[Cu(H_2O)_6^{2+}][en]}$

$$Cu(H_2O)_4(en)^{2+} + en \longrightarrow Cu(H_2O)_2(en)_2^{2+} + 2H_2O \qquad K_2 = \frac{[Cu(H_2O)_2(en)_2^{2+}]}{[Cu(H_2O)_4(en)^{2+}][en]}$$

$$Cu(H_2O)_2(en)_2^{2+} + en \longrightarrow Cu(en)_3^{2+} + 2H_2O \qquad K_3 = \frac{[Cu(en)_3^{2+}]}{[Cu(H_2O)_2(en)_2^{2+}][en]}$$

where $K_1$, $K_2$ and $K_3$ are called stepwise formation constants and the overall formation constant ($\beta_3$) is :

$$Cu(H_2O)_6^{2+} + 3en \longrightarrow Cu(en)_3^{2+} + 6H_2O \qquad \beta_3 = \frac{[Cu(en)_3^{2+}]}{[Cu(H_2O)_6^{2+}][en]^3}$$

Multiplying the numerator and denominator of the expression of $\beta_3$ by

$\frac{[Cu(H_2O)_4(en)^{2+}][Cu(H_2O)_2(en)_2^{2+}]}{[Cu(H_2O)_4(en)^{2+}][Cu(H_2O)_2(en)_2^{2+}]}$ and then rearranging we get,

$$\beta_3 = \frac{[Cu(en)_3^{2+}]}{[Cu(H_2O)_6^{2+}][en]^3} \cdot \frac{[Cu(H_2O)_4(en)^{2+}][Cu(H_2O)_2(en)_2^{2+}]}{[Cu(H_2O)_4(en)^{2+}][Cu(H_2O)_2(en)_2^{2+}]}$$

$$= \frac{[Cu(en)_3^{2+}]}{[Cu(H_2O)_2(en)_2^{2+}][en]} \cdot \frac{[Cu(H_2O)_2(en)_2^{2+}]}{[Cu(H_2O)_4(en)^{2+}][en]} \cdot \frac{[Cu(H_2O)_4(en)^{2+}]}{[Cu(H_2O)_6^{2+}][en]} = K_3K_2K_1$$

$\therefore \beta_3 = K_1K_2K_3$

This is the relationship between overall and stepwise formation constants.

---

**Problem 7.29** The logarithm of stability constants values of $Cu(en)^{2+}$, $Cu(NH_3)_2^{2+}$, $Ag(en)^+$ and $Ag(NH_3)_2^+$ complexes are 10.7, 7.8, 4.7 and 7.2 respectively. Explain why is the en complex of copper is more stable than the ammine complex, whereas the reverse is true for the corresponding silver complexes. **[C.U. 1998]**

**Ans. :** According to Schwarzenbach the **chelate effect** is larger for transition metal compared to that of non-transition metal and so as $Cu^{2+}$ is a transition metal ion thereby due to the chelate effect the logK values increases from $Cu(NH_3)_2^{2+}$ to $Cu(en)^{2+}$. In case of $Ag^+$ which is a d° system i.e. a non-transition metal ion. $Ag(NH_3)_2^+$ has a linear geometry (sp hybridisation of the central $Ag^+$) but when two $NH_3$ molecules are replaced by one ethylenediamine molecule to form $Ag(en)^+$, the ethylenediamine is unable to satisfy the linear coordination without considerable distortion and hence much strain in the Ag-en chelate ring arises so that formation constant values decreases from $Ag(NH_3)_2^+$ to $Ag(en)^+$.

---

## 7.11.2 Determination of metal-ligand interacting ratio in solution and stability constants of complexes

Largely due to the efforts of Bjerrum, Sillen, Calvin, Irving and Schwarzenbach several methods namely potentiometry, spectrophotometry, polarography, solvent extraction etc. are now avilable for the evaluation of stability constant. The determination of the composition of a complex is done by the variation of any suitable property of the solution or reactants or products. In this process we shall consider the main complex forming equilibria ignoring all the additional or side reactions may occur in solution. We shall discuss here outline of few methods.

When a metal M and a ligand L react to form a complex $M_aL_b$ which has sufficient stability, then the physical properties e.g. absorbance of a solution containing M, L and the complex will alter with a change in the ratio of the molar concentrations of the M and L so that these ratio is related to the complex formation equilibria as :

$$aM + bL \longrightarrow M_aL_b$$

In a series of solutions containing the M and L in different proportions having known total concentrations of these two species i.e. $T_M/T_L$ are plotted against the physical property of the solution, a straight line is obtained which sharply changes the slope at the molar ratio (Fig. 7.2) corresponding to complex i.e. $T_M/T_L = a/b$. This is the principle adopted in **molar ratio method**.

In the **slope ratio method**, for a series of solutions having constant large excess of M and varying amounts of $L(C_L)$, the absorbance $(A_M)$ in each case will be,

$$A_M = \varepsilon[M_aL_b]t = \varepsilon\frac{C_L}{b}t$$

where ε = molecular extinction coefficient and t is the path length. Similarly for a series of solutions having a constant large excess of L and varying amounts of $M(C_M)$, the absorbance $(A_L)$ in each case is given by,

$$A_L = \varepsilon[M_aL_b]t = \varepsilon\frac{C_M}{a}t$$

Plot of $A_M$ vs. $C_L$ and $A_L$ vs. $C_M$ we get the $slope_M = \frac{\varepsilon t}{b}$ in the first case and in the second case $slope_L = \frac{\varepsilon t}{a}$, therefore,

$$\frac{slope_M}{slope_L} = \frac{a}{b}$$

from which we get the composition of the complex, $M_aL_b$.

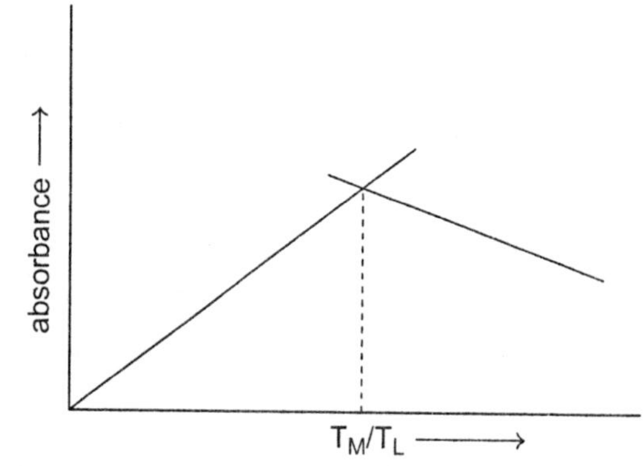


***Figure 7.2** : Plot of absorbance vs. $T_M/T_L$ of Fe(II) with ophen. A 1:3 composition is indicated i.e. $Fe(ophen)_3^{2+}$.*

Another important method is **Job's method of contineous variation** where the absorbance of a series of solutions are measured having constant sum of the total concentration of metal and ligand ($T_M + T_L$ = constant). Here absorbance is measured at suitable wavelength where the coloured complex $M_aL_b$ absorbed strongly where the metal (M) and ligand (L) practically non-absorbing. In that case the maximum absorbance will occur where L : M ratio is equal to that in the complex. By suitable variation of Job's method it is often useful to evaluate the formation constant of a complex.

## 7.12 TERNARY COMPLEXES

As mentioned earlier [Sec. 7.7(6)] that the stability of 1 : 1 : 1 ternary complexes MAB (where A and B are two different ligands and M is a metal ion) are more compared to homoligand complexes. These ternary complex formation is important in biological systems such as metalloprotein or metalloenzyme etc.

Generally in the case of homoligand complexes $K_2$ is less than that of $K_1$ (Sec. 7.11). However there might be some exceptions. But in the case of mixed ligand complexes when the metal M binds to different ligands A and B, with few exceptions $K_2$ is greater than $K_1$ because stability of MAB type complexes are influenced by intramolecular ligand-ligand interactions mediated by metal ions. The formation of ternary complex MAB from the binary complex or in some other ways with their corresponding stability constant (K) values are represented below :

$$M + A \longrightarrow MA \qquad K_{MA}^{M} = \frac{[MA]}{[M][A]} \qquad ......(7.8)$$

$$M + B \longrightarrow MB \qquad K_{MB}^{M} = \frac{[MB]}{[M][B]} \qquad ......(7.9)$$

$$MA + A \longrightarrow MA_2 \qquad K^{MA}_{MA_2} = \frac{[MA_2]}{[MA][A]} \qquad \ldots\ldots(7.10)$$

$$MB + B \longrightarrow MB_2 \qquad K^{MB}_{MB_2} = \frac{[MB_2]}{[MB][B]} \qquad \ldots\ldots(7.11)$$

$$M + A + B \longrightarrow MAB \qquad K^{M}_{MAB} = \frac{[MAB]}{[M][A][B]} \qquad \ldots\ldots(7.12)$$

$$MA + B \longrightarrow MAB \qquad K^{MA}_{MAB} = \frac{[MAB]}{[MA][B]} \qquad \ldots\ldots(7.13)$$

$$MB + A \longrightarrow MBA \qquad K^{MB}_{MBA} = \frac{[MBA]}{[MB][A]} \qquad \ldots\ldots(7.14)$$

$$MA + MB \longrightarrow MAB + M \qquad K_M = \frac{[MAB][M]}{[MA][MB]}$$

$$\text{or, } 10^{\Delta \log K_M} = \frac{[MAB][M]}{[MA][MB]} \qquad \ldots\ldots(7.15)$$

$$MA_2 + MB_2 \longrightarrow 2MAB \qquad K = \frac{[MAB]^2}{[MA_2][MB_2]} \qquad \ldots\ldots(7.16)$$

The stability of ternary complex is generally expressed in terms of $\Delta \log K_M$ and logK values as these are related to the stability constant of binary and ternary complexes in the following way :

From equation (7.15), $10^{\Delta \log K_M} = \frac{[MAB][M]}{[MA][MB]}$, so that

$$\Delta \log K_M = \log[MAB] + \log[M] - \log[MA] - \log[MB]$$

From equation (7.13), $K^{MA}_{MAB} = \frac{[MAB]}{[MA][B]}$, therefore,

$$\log[MAB] = \log K^{MA}_{MAB} + \log[MA] + \log[B]$$

so that, $\Delta \log K_M = \log K^{MA}_{MAB} + \log[MA] + \log[B] + \log[M] - \log[MA] - \log[MB]$

$$= \log K^{MA}_{MAB} + \log[B] + \log[M] - \log[MB]$$

Now from equation (7.9), $K^{M}_{MB} = \frac{[MB]}{[M][B]}$, we get

$$\log[MB] = \log K^{M}_{MB} + \log[M] + \log[B]$$

and so, $\Delta \log K_M = \log K^{MA}_{MAB} + \log[B] + \log[M] - \log K^{M}_{MB} - \log[M] - \log[B]$

$$\text{i.e. } \Delta \log K_M = \log K^{MA}_{MAB} - \log K^{M}_{MB} \qquad \ldots\ldots(7.17)$$

In a similar way by considering that the ternary complexes MAB and MBA are same, $\Delta \log K_M$ can also be represented as :

$$\Delta \log K_M = \log K^{MB}_{MBA} - \log K^{M}_{MA} \quad ......(7.18)$$

and $$\Delta \log K_M = \log K^{M}_{MAB} - \log K^{M}_{MA} - \log K^{M}_{MB} \quad ......(7.19)$$

For a binary system having monodentate ligands (A = B), the statistical values of $\Delta \log K_M$ lie in the range –0.5 to –0.8. For symmetrical bidentate ligands the value of $\Delta \log K_M$ for octahedral, square planar or tetrahedral and distorted octahedral are –0.68, –0.6 and –0.9 to –1.2 respectively.

From equations (7.17) and (7.18), it is evident that both the ligands A and B mutually influence the stability of the ternary complex MAB. In fact $\Delta \log K_M$ corresponds to logX for the process,

$$MA + MB \rightleftharpoons MAB + M \qquad X = \frac{[MAB][M]}{[MA][MB]} \quad ......(7.20)$$

Therefore a positive value for $\Delta \log K_M$ would indicate that the equilibrium as expressed in equation (7.20) is in favour of the formation of ternary complex. Positive values of $\Delta \log K_M$ indicate enhanced binding of B with MA as a result of the effects of A on the properties of the metal centre and vice versa. The idex of stability of ternary complex is represented by equation (7.12) i.e. $K^{M}_{MAB}$ is actually represents the overall formation constant ($\beta^{M}_{MAB}$). $\Delta \log K_M$ values of several ternary complexes of metal ions e.g. $Cu^{2+}$, $Ni^{2+}$, $Co^{2+}$ etc. with various bidentate-bidentate or bidentate-monodentate ligand combination were determined. Data suggest that in general $\Delta \log K_M$ is positive when either of the two ligands is a strong donor ($\sigma$ and $\pi$) and the other is a strong $\pi$-acid ligand ($M \rightarrow L$) e.g. [M(bpy)(ox)], $[M(NTA)(py)]^-$ etc.

The formation of ternary complex is also represented by equation (7.16) whereby,

$$\log K = 2\log[MAB] - \log[MA_2] - \log[MB_2]$$

$$= \log[MAB] + \log[MBA] - \log[MA_2] - \log[MB_2]$$

$$= \log K^{MA}_{MAB} + \log[MA] + \log[B] + \log K^{MB}_{MBA} + \log[MB] + \log[A] - \log K^{MA}_{MA_2}$$

$$- \log[MA] - \log[A] - \log K^{MB}_{MB_2} - \log[MB] - \log[B]$$

$$= \left(\log K^{MB}_{MBA} - \log K^{MA}_{MA_2}\right) + \left(\log K^{MA}_{MAB} - \log K^{MB}_{MB_2}\right) \quad ......(7.21)$$

As from equation (7.13), $\log[MAB] = \log K^{MA}_{MAB} + \log[MA] + \log[B]$

from equation (7.14), $\log[MBA] = \log K^{MB}_{MBA} + \log[MB] + \log[A]$

from equation (7.10), $\log[MA_2] = \log K^{MA}_{MA_2} + \log[MA] + \log[A]$

from equation (7.11), $\log[MB_2] = \log K^{MB}_{MB_2} + \log[MB] + \log[B]$

The statistical value of logK as represented by equation (7.21) is always +0.6 irrespective of geometry of the mixed ligand complex. Any value of logK greater than 0.6 indicates enhanced stability of ternary complex MAB compared to binary complexes $MA_2$ and $MB_2$.

**Problem 7.30** If $Pb^{2+}$ and acetic acid (acH) react in solution to form the lead acetate complex ($Pbac^+$) according to $Pb^{2+} + acH \rightleftharpoons Pbac^+ + H^+$ ......(I)

then show that the equilibrium constant (K) of the reaction (I) is given by $\log K = \log K^{Pb}_{Pbac} + \log K^{H}_{acH}$ where $K^{Pb}_{Pbac}$ is the stability constant by $Pbac^+$ complex and $K^{H}_{acH}$ is the deprotonation of acH respectively. **[C.U. 2001]**

**Ans. :** The equilibrium constant (K) for the reaction (I) is $K = \dfrac{[Pbac^+][H^+]}{[Pb^{2+}][acH]}$ ...(A)

Now $Pb^{2+} + ac^- \longrightarrow Pbac^+$ so that $K^{Pb}_{Pbac} = \dfrac{[Pbac^+]}{[Pb^{2+}][ac^-]}$ ...(B)

and $acH \longrightarrow H^+ + ac^-$ $K^{H}_{acH} = \dfrac{[H^+][ac^-]}{[acH]}$ ...(C)

From equation (A) we get,

$$\log K = \log[Pbac^+] + \log[H^+] - \log[Pb^{2+}] - \log[acH]$$

Now from equation (B) we get,

$$\log[Pbac^+] = \log K^{Pb}_{Pbac} + \log[Pb^{2+}] + \log[ac^-]$$

$$\therefore \log K = \log K^{Pb}_{Pbac} + \log[Pb^{2+}] + \log[ac^-] + \log[H^+] - \log[Pb^{2+}] - \log[acH]$$

$$= \log K^{Pb}_{Pbac} + \log[ac^-] + \log[H^+] - \log[acH]$$

$$= \log K^{Pb}_{Pbac} + \log\frac{[H^+][ac^-]}{[acH]}$$

$$= \log K^{Pb}_{Pbac} + \log K^{H}_{acH} \text{ [according to equation(C)]}$$

## 7.13 USE OF COORDINATION COMPLEXES IN ANALYTICAL CHEMISTRY

Analytical chemistry is a branch where analysis of various types are performed in order to know the quality and also quantity of a particular species present in a bulk sample. Here we shall discuss some aspects of analytical procedure where the formation of coordination complexes is taken as the prime aspect in order to detect and to estimate a particular species.

### 7.13.1 Use of coordination complexes for detection of metal ions or anions

As mentioned earlier (Sec. 7.1.1) the separation and qualitative detection of $Cu^{2+}$ and $Cd^{2+}$ have been made by the formation of complexes with KCN. $Cu^{2+}$ itself may be detected by the formation of deep blue cupro ammine complex, $[Cu(NH_3)_4]^{2+}$. Both $Cu^{2+}$ and $Cd^{2+}$ are in group IIA of analytical chemistry. $Cu^{2+}$ may also be detected by rubeanic acid* (dithio-oxamide) reagent which forms a black precipitate of copper rubeanate $Cu\{C(=NH)S\}_2$ from ammonical or weakly acid solution. This is a very delicate test for detection of $Cu^{2+}$ even in solution of 1 part in $10^9$ parts in solution.

* Rubeanic acid, 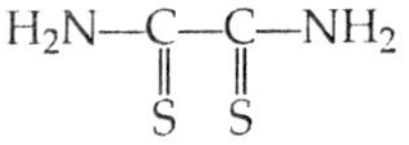

In group IIIA the blood red colour is formed by the interaction of $Fe^{3+}$ with $SCN^-$ whereby the complex $[Fe(SCN)(H_2O)_5]^{2+}$ exists in the solution. In this group both $Fe^3$ and $Al^{3+}$ react with oxine to form a pale yellow precipitate of $Al(oxin)_3$ from a weakly ammonical medium while $Fe(oxin)_3$ forms in aqueous solution over a wide pH range 2 to 9 which forms a dark coloured solution in chloroform medium.

In group IIIB the cations $Ni^{2+}$ and $Co^{2+}$ are separated and detected by the addition of KCN whereby $K_2[Ni(CN)_4]$ and $K_4[Co(CN)_6]$ are formed. On boiling $K_4[Co(CN)_6]$ is converted to $K_3[Co(CN)_6]$ but $K_2[Ni(CN)_4]$ remains unchanged.

$$4K_4[Co(CN)_6] + 2H_2O + O_2 \longrightarrow 4K_3[Co(CN)_6] + 4KOH$$

The $K_2[Ni(CN)_4]$ is decomposed by boiling with NaOH and $Br_2$ thereby a black precipitate of higher oxide of nickel is formed leaving the yellow $[Co(CN)_6]^{3-}$ in solution.

$$2NaOH + Br_2 \longrightarrow NaOBr + NaBr + H_2O$$

$$2K_2[Ni(CN)_4] + NaOBr + 4NaOH + H_2O \longrightarrow 2Ni(OH)_3 + 4KCN + 4NaCN + NaBr$$

The well known red coloured complex $Ni(DMG)_2$ is formed by $Ni^{2+}$ with DMGH in ammonical medium. $Co^{2+}$ is detected by its formation of a blue coloured complex with thiocyanate i.e. $[Co(NCS)_4]^{2-}$. The yellow sparingly soluble complex $K_3[Co(NO_2)_6]$ is formed by the reaction of $Co^{3+}$ with $KNO_2$ in acetic acid medium.

The anion $Cl^-$ is detected by the formation of insoluble white precipitate of AgCl which is soluble in excess ammonia due to the formation of the following complex :

$$AgCl + 2NH_3 \longrightarrow [Ag(NH_3)_2]Cl$$

Various complexes are employed for qualitative detection of metal ions with different spot reagents used in semimicro analysis.

## 7.13.2 Application of first order inner metallic complexes in gravimetric determination of metal ions

As mentioned earlier (Sec. 7.10) inner complexes has immense use in gravimentric estimation of metal ions due to the insolubility of the complexes in aqueous solution. Many such inner metallic ligands contain acidic functional groups and such groups release the proton while coordinating through a metal ion. Therefore formation of inner complexes require suitable pH range so that many metal ions form insoluble inner complexes in aqueous solution and after filtration of the insoluble complex, if we make proper arrangement to know the weight of the dry insoluble complex (which is done by drying arrangement in air oven) then from the compositiion of the inner complex we are able to know the concentration of metal ions in solution. Thus, $Al^{3+}$, $Fe^{3+}$ and $Ni^{2+}$ can be estimated by the formation of yellow coloured $Al(oxin)_3$ from feebly ammonical solution, dark coloured $Fe(oxin)_3$ from wide pH range 2 to 9 and red coloured $Ni(DMG)_2$ from ammonical medium respectively (Sec. 7.13.1). The $Cu^{2+}$ forms dull green $Cu(Salox)_2$ (SaloxH = salicylaldioxime) from dilute acetic acid medium as also light green $Cu(Qui)_2$ (QuiH = quinaldinic acid) from dilute $H_2SO_4$ medium.

*Inner complex salt formation depends on pH*

Apart from the above ligands cupferron (7-XLVIII) the ammonium salt of N-nitroso N-phenylhydroxylamine is used for $Fe^{3+}$ in strongly acid medium. The reagent function as bidentate ligand (OO coordination) as 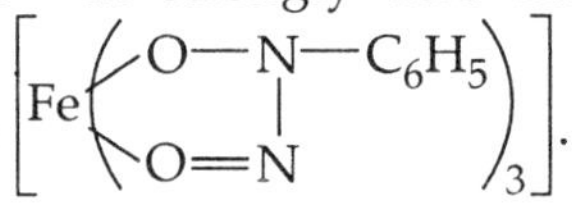. This reagent also forms selective precipitate with niobium, tantalum, tungsten and zirconium.

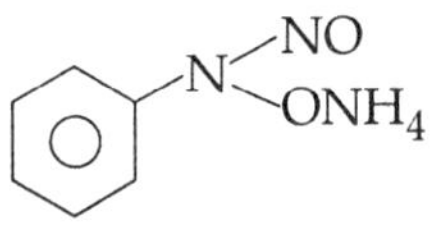

(7-XLVIII)

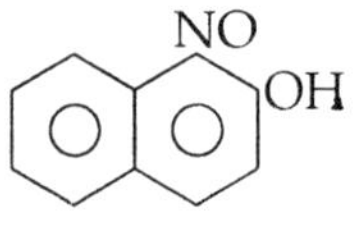

(7-XLIX)

Another important ligand is α-nitroso-β-napthol (7-XLIX) which function as bidentate monobasic one. This reagent precipitates quantitatively cobalt, iron, palladium, vanadium, zirconium etc. from slightly acidic solutions. This is widely used for the separation of cobalt and nickel.

The drying temperature of the precipitate ranging from 120 to 140°C depending on the nature of the inner complex and in some cases the precipitate is ignited to get suitable stoichiometric composition of the metal oxide which is then weighed.

### 7.13.3 Solvent extraction and spectrophotometric determination of metal ions through inner complex formation

Both $Fe^{3+}$ and $Al^{3+}$ forms inner complexes $Fe(oxin)_3$ and $Al(oxin)_3$ (Sec. 7.13.2). $Fe(oxin)_3$ forms between wide pH range 2 to 10 while $Al(oxin)_3$ forms from slightly ammonical medium. Therefore keeping the pH 2 to 3 of a mixture of aqueous solution of $Fe^{3+}$ and $Al^{3+}$, if we add 1% solution of oxine in chloroform, then $Fe(oxin)_3$ will be formed which can be extracted with chloroform in a separating funnel. The dark-coloured chloroform extract absorbs at 470 nm and from the standard curve we may know the concentration of unknown $Fe^{3+}$ solution. After separation of $Fe^{3+}$ the pH of the aqueous solution is adjusted to 9 and oxine is added whereby $Al(oxin)_3$ is formed which is filtered and dried in a air oven and weighed as $Al(oxin)_3$. Alternatively $Al^{3+}$ concentration is determined by adjusting the pH at 6 and then Eriochrome Cyanine R reagent is added to the solution. The absorbance of this solution is measured at 535 nm.

Spectrophotometric determination of $Ni^{2+}$ is carried out by the formation of red coloured $Ni(DMG)_2$ in ammonical medium which is extracted with chloroform and the absorbance of chloroform solution is measured at 445 nm against a reagent blank.

An application of solvent extraction in forensic science is employed for the estimation of trace amount of $Pb^{2+}$ in blood. The reagent used is diphenylthiocarbazone (dithizone) (7-L) which exists in solution as a tautomeric mixture (A and B) and functions as a monoprotic acid bidentate ligand upto a pH 12. It forms first order inner complex salt (7-LI) with $Pb^{2+}$ at a pH around 9.5. The complex is extracted with chloroform and absorbance is measured at 510 nm against a reagent blank in 1 cm cell. From the calibration curve of $Pb^{2+}$ the concentration of unknown $Pb^{2+}$ solution is determined.

(A) (7-L) (B)

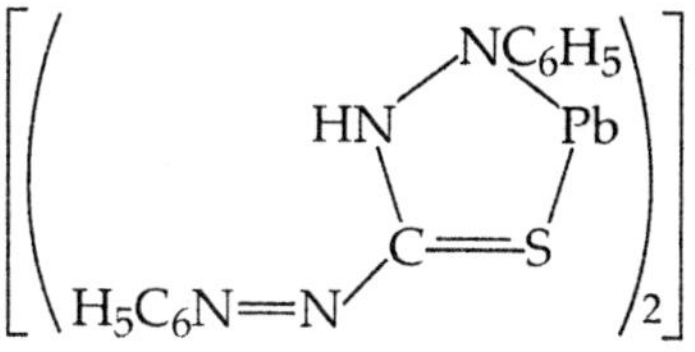

(7-LI)

In another variation only solvent extraction can be employed for the separation and estimation of a mixture of $Fe^{3+}$ and $Cu^{2+}$. Here separation be made through the formation of complex but not through inner complex formation. To the aqueous solution of $Fe^{3+}$ and $Cu^{2+}$, if conc. HCl is added both the species form $FeCl_4^-$ and $CuCl_4^{2-}$, the $FeCl_4^-$ is extracted with ether and $CuCl_4^{2-}$ remains in aqueous solution. The ether extract is treated with water and stripped by adding $NaHCO_3$ whereby $Fe^{3+}$ will go from organic layer to the aqueous layer and $Fe^{3+}$ is estimated in the usual way. The aqueous layer containing $CuCl_4^{2-}$ is evaporated to suitable volume and copper is estimated iodometrically.

### 7.13.4 Separation and estimation of metal ions by ion exchange

A metal cation normally will be retained by cation exchanger but when a metal ion forms an anionic complex it will be retained by anion exchanger. Thus several metal ions e.g. $Fe^{3+}$, $Al^{3+}$, $Zn^{2+}$, $Co^{2+}$, $Mn^{2+}$ form anionic complexes in HCl medium and can be retained by anion exchanger due to the formation of negatively charged chloro complexes. Absorption of each metal is governed by well-defined pH range and this property can be utilised as a basis for separation. The efficient separation of $Zn^{2+}$ and $Mg^{2+}$ is made by treatment of the solution with 2(M) HCl solution whereby $Zn^{2+}$ forms $ZnCl_4^{2-}$ which is strongly absorbed by anion exchanger and $Mg^{2+}$ is passed through the column of resin bed. The $Zn^{2+}$ is then eluted with dilute $HNO_3$ and both $Zn^{2+}$ and $Mg^{2+}$ is subsequently estimated by complexometric titration with ethylenediaminetetraacetic acid (EDTA).

Similarly $Co^{2+}$ and $Ni^{2+}$ is separated by an anion exchanger on the basis that $Co^{2+}$ forms a monovalent chloro complex (probably$[CoCl_3]^-$) in 9(M) HCl medium which is retained by a strongly basic anion exchanger. The $Ni^{2+}$ is passed through the exchanger. Due to the instability of the chloro complex, this can be washed out from the column with 9(M) HCl. On washing the resin bed, the cobalt complex is decomposed and passed out in the effluent as Co(II) chloride. The separated $Ni^{2+}$ and $Co^{2+}$ solutions are then evaporated to remove excess HCl and titrated with EDTA in the usual way.

### 7.13.5 Sequestering agents and selectivity of a particular metal ion

When **masking** (Sec. 6.15) is done by a chelating ligand the process is known as **sequestration** and the chelating ligand is termed as **sequestering agent**. Commonly several organic chelating agents are used and therefore these reagents are popularly known as **organic sequestering agents** used in inorganic analysis. Naturally **the term sequestration is used in a restricted sense to describe the suppression of a precipitation reaction of interfering metal ions with that of the main metal ion(s) to be determined.**

For example an effective masking agent is $CN^-$ which forms stable complexes with $Cd^{2+}$, $Zn^{2+}$, $Hg^{2+}$, $Cu^{2+}$, $Ni^{2+}$ etc. but not with $Ca^{2+}$, $Mg^{2+}$, $Pb^{2+}$ etc. It is therefore possible to determine the amount of $Ca^{2+}$, $Mg^{2+}$, $Pb^{2+}$ etc. by

$$M^{2+} + 4CN^- \longrightarrow [M(CN)_4]^{2-}$$

EDTA titrations in presence of the above mentioned interfering ions by masking with an excess of KCN or NaCN. It is to be noted that $CN^-$ is not a chelating ligand. On the other hand sequesteration is associated with precipitation processes. For example, the precipitation of soaps as their calcium and magnesium salts by hard water can be checked by forming these metals into soluble complexes with EDTA. Again the formation of scale in the boilers can be prevented by complex formation with EDTA. Here EDTA is acting as a **sequestering agent**. By the use of sequestering agent selectivity of estimation of a particular metal ion increases in presence of interfering ions. Thus benzimidazole is a specific and selective ligand for $Ag^+$ which forms a first order inner metallic complex (7-LII) in ammonical medium.

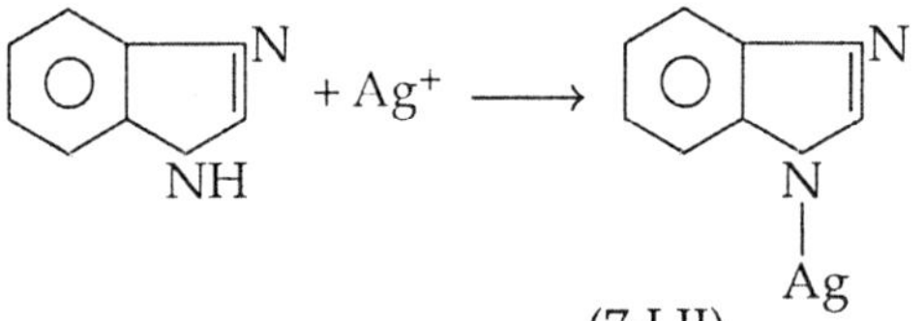

(7-LII)

This complex is not photosensitive and is employed for gravimetric estimation of $Ag^+$ in the usual way. But in presence of other ions e.g. $Cu^{2+}$, $Ni^{2+}$, $Co^{2+}$ etc. the benzimidazole also forms insoluble precipitates thus rendering the estimation of $Ag^+$ difficult. In order to obviate this difficulty, before adding the organic ligand the EDTA is added which forms the soluble complexes with interfering ions ($Cu^{2+}$, $Ni^{2+}$, $Co^{2+}$ etc.) but will not react with $Ag^+$. In this situation if we add benzimidazole which only reacts with $Ag^+$ to form insoluble precipitate but the interfering ions will remain indifferent to benzimidazole. Once again the selectivity of benzimidazole to $Ag^+$ increases by the use of EDTA as sequestering agent.

**Problem 7.31** Would substitution at 3 and 3′ positions of 2, 2′ bipyridine effect the stability of its metal complex?

**Ans. :** According to Sec. 7.11.1(5) substitution at 6, 6′ positions will lower the stability of the complex because of steric strain. However, substitution at 3, 3′ although they are far away from the coordination sites, also lowers the stability of the complex. Because substitution at 3, 3′ positions make the ring puckered and so chelation is difficult.

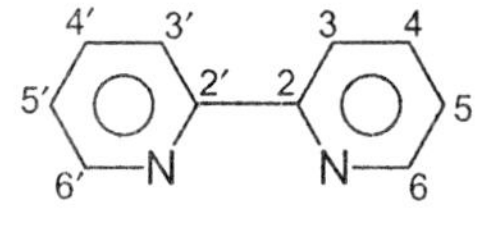

### 7.13.6 Complexometric titrations

Chelating agent having polyatomic donor centres are termed **chelon** which include polyamines such as trien, polyaminocarboxylic acids such as EDTA, and related compounds that form water soluble stable 1 : 1 complex with metal ions and hence these compounds are employed as titrants for metals. So these titrants are sometimes referred to as **complexing agents** and the titrations involving them is called **complexometric titrations.** Therefore **complexometric titrations, chelometric titrations** or **chelatometric titrations are those in which the concentration of metal ions are determined by way of complex formation with a strong multidentate chelon.**

Apart from the $NTAH_3$, tetraen, $EDTAH_4$, $DCTAH_4$ and $DTPAH_5$, 2, 2′-ethylenedioxybis {ethyliminodi(acetic acid)} (7-LIII) (EGTA) and triehtylenetetramine–N, N, N′, N″, N‴, N‴-hexa-acetic acid (7-LIV) (TTHA) are used as complexometric titrant.

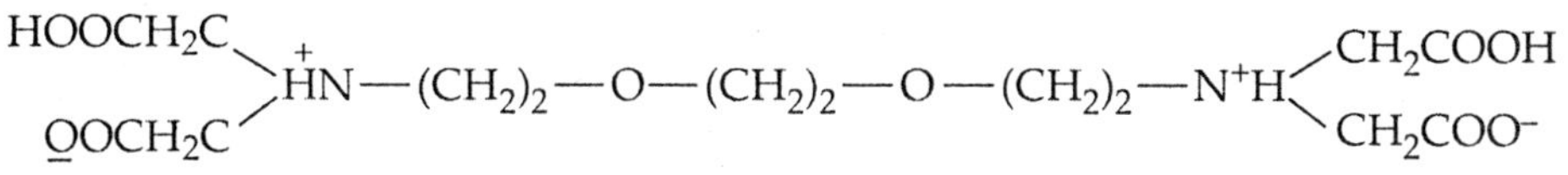

(7-LIII)

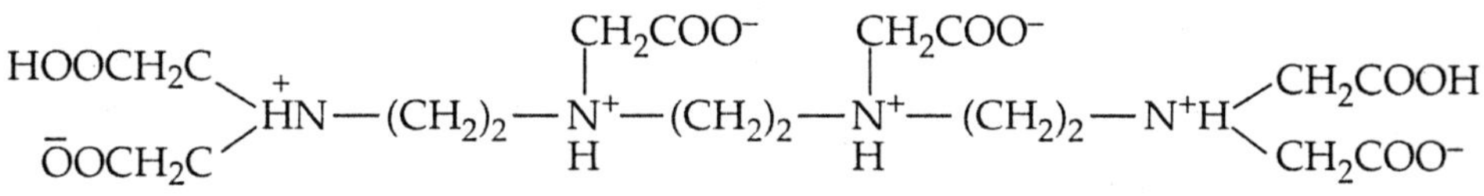

(7-LIV)

Among these chelons the most widely used complexone is $EDTAH_4$. The versatility of $EDTAH_4$ is due to its low cost (of its disodium salt), high water solubility of its chelates and it strongly forms chelates with most of the metals of periodic classification.

Like that of acid-base titration where an acid-base indicator is used, in complexometric titrations an indicator is used which is known as **metal indicator.** Metal indicators are organic dye which form metal-indicator complex with the metal ion. The metal indicator complex must have a specific colour with that of the free metal indicator so that at the end of the titration the complexometric ligand (EDTA) reacts with the metal indicator complex and the free metal indicator is produced which is detected by the sharp colour change at the end point.

To simplify the discussion of complexometric titrations, henceforth EDTA is abbreviated as $H_4Y$ and the disodium salt is denoted as $Na_2H_2Y$ which affords the complex forming ion as $H_2Y^{2-}$ in aqueous solution and it reacts with all metals in a 1 : 1 ratio and the feasibility of the titrations will be discussed in the following aspects :

### 7.13.6(A) Complex-Formation Reactions and selection of Complexometric titrant

Here the titrimetric methods involve complex formation with polydentate ligands. Therefore a natural question arises why multidentate ligands are used as titrant? Multidentate ligands particularly those having four or six donor groups have several advantages over their monodentate or bidentate counterparts. Firstly greater number of donor atoms will generate extra stability due to chelate effect and thus provide sharper end points. Secondly the multidentate ligands react with metal ions in a **single-step** to achieve the highest coordinated complex, whereas complex formation with unidentate or bidentate ligands usually involves two or more intermediate steps. Thus if we take $H_4Y$ as titrant, the six points of attachment of a metal ion is satisfied in a single step while in case of ethylenediamine or other bidentate ligands the six coordination sites are satisfied by three steps so that the resulting complex will have a greater scope of dissociation compared to that of multidentate ligand complex. The advantage of a single-step reaction can also be demonstrated by the titration curves of monodentate, bidentate, tridentate ligands which suggest that for polydentate ligands the formation constants will be highest. This is also illustrated (Table 7.5) from the fact that the logK values of en/salH complexes of $Cu^{2+}/Ni^{2+}$ are close to the logβ values of $H_4Y$ complexes but the point is that en/salH complexes are formed in three steps while the $H_4Y$ complexes will be formed in a single step. Thus the **feasibility of the titration lies in the fact that the highest coordination complex must form in a single step so that the end point will be sharp**. For this reason polydentate ligands e.g. ethylenediaminetetraacetic acid or other aminopolycarboxylic acids are preferred for complexometric titrations.

### 7.13.6(B) The absolute stability or Formation constant

The reactions with cations such as $M^{2+}$ can be written as :

$$M^{2+} + H_2Y^{2-} \rightleftharpoons MY^{2-} + 2H^+$$

For other valent metal ions, the reactions are :

$$M^{3+} + H_2Y^{2} \rightleftharpoons MY^{-} + 2H^+$$

$$M^{4+} + H_2Y^{2-} \rightleftharpoons MY + 2H^+$$

$$M^{n+} + H_2Y^{2-} \rightleftharpoons MY^{(n-4)+} + 2H^+ \quad \text{......(7.22)}$$

$$M^{n+} + Y^{4-} \rightleftharpoons MY^{(n-4)+} \quad \text{......(7.23)}$$

The **absolute** or **overall formation** or **stability constant** ($\beta_{abs}$) for the last reaction (7.23) is :

$$\beta_{abs} = \frac{[MY^{(n-4)+}]}{[M^{n+}][Y^{4-}]} \quad \text{......(7.24)}$$

### 7.13.6(C) Different Zwitter ion structures of $H_4Y$ in dissociated and undissociated forms

$H_4Y$ behaves like an amino acid such as glycine. $H_4Y$ exists as the following double zwitter ion structure and also the various dissociated forms of $H_4Y$ exist as the zwitter ion structure :

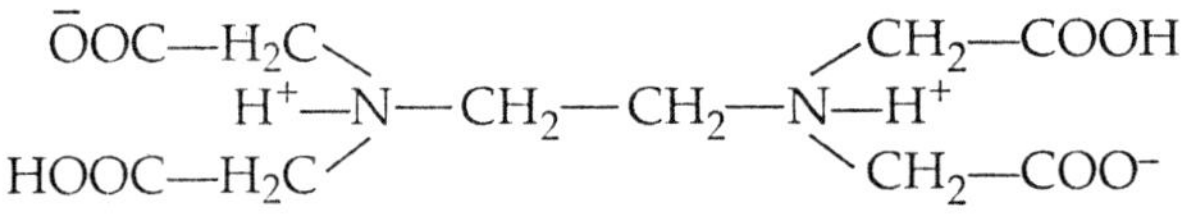

i.e. $H_4Y$

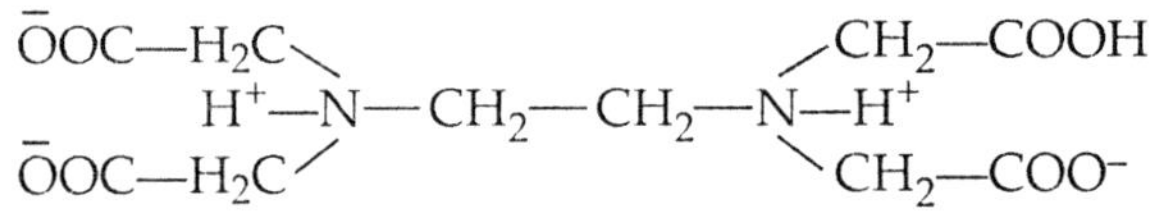

i.e. $H_3Y^-$

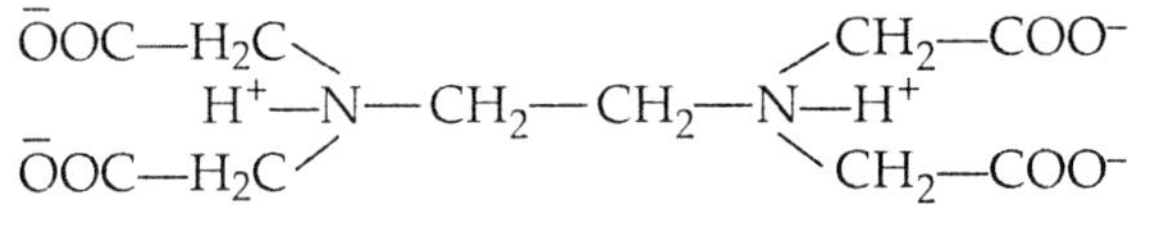

i.e. $H_2Y^{2-}$

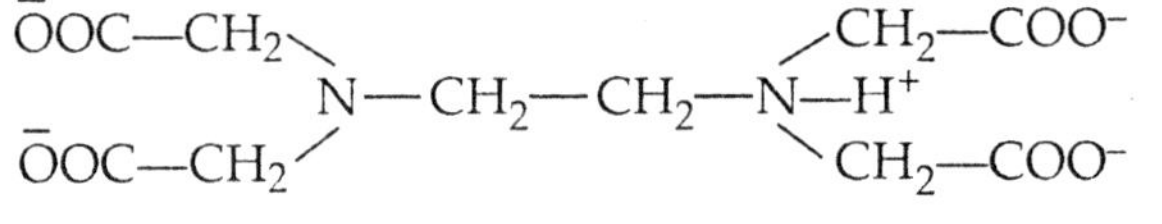

i.e. $HY^{3-}$ and

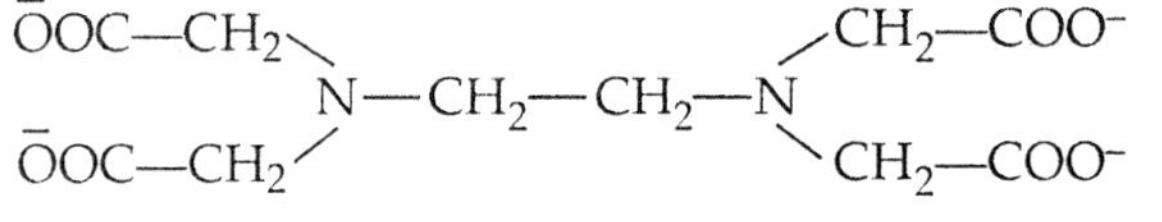

i.e. $Y^{4-}$

### 7.13.6(D) The pH dependent equilibria involved in Complexometric titration

Since $H_4Y$ has six basic sites i.e. four carboxylate oxygens and two nitrogens, there might be several ways of protonation of the molecule but we shall consider only the following four dissociation equilibria because others will be less important.

$$H_4Y + H_2O \rightleftharpoons H_3O^+ + H_3Y^- \quad K_{a_1} = \frac{[H_3O^+][H_3Y^-]}{[H_4Y]} = 1.02 \times 10^{-2} \quad ...(7.25)$$

$$H_3Y^- + H_2O \rightleftharpoons H_3O^+ + H_2Y^{2-} \quad K_{a_2} = \frac{[H_3O^+][H_2Y^{2-}]}{[H_3Y^-]} = 2.14 \times 10^{-3} \quad ...(7.26)$$

$$H_2Y^{2-} + H_2O \rightleftharpoons H_3O^+ + HY^{3-} \quad K_{a_3} = \frac{[H_3O^+][HY^{3-}]}{[H_2Y^{2-}]} = 6.92 \times 10^{-7} \quad ...(7.27)$$

$$HY^{3-} + H_2O \rightleftharpoons H_3O^+ + Y^{4-} \quad K_{a_4} = \frac{[H_3O^+][Y^{4-}]}{[HY^{3-}]} = 5.50 \times 10^{-11} \quad ...(7.28)$$

As $H_4Y$ is an aminopolycarboxylic acid, therefore the above ionisation processes will be pH dependent. The first two dissociation constants ($K_a$) values are much larger compared to that of third and fourth because the two protons in $H_2Y^{2-}$ are attached to two nitrogen atoms and are lost less readily than protons attached to oxygen. The composition of five $H_4Y$ species as a function of pH shows that maximum efficiency of this titrant is attained at and above pH 8 and above pH 12 the predominant species is $Y^{4-}$ anion. At lower pH the species $H_3Y^-$, $H_2Y^{2-}$ etc. exist. Therefore it is clear that the formation of metal-chelate at any particular pH cannot be determined by the $\beta_{abs}$ (equation 7.24) alone. For example at pH 4 the predominant species is $H_2Y^{2-}$ which binds with $Ca^{2+}$ to form the metal-chelate as :

$$Ca^{2+} + H_2Y^{2-} \rightleftharpoons CaY^{2-} + 2H^+$$

Naturally as the pH is lowered the above equilibrium will be shifted to the left and the formation of $CaY^{2-}$ is hindered so that there will be a pH value below which the titration of $Ca^{2+}$ with $H_4Y$ will not be feasible. We want to know that pH value for chelometric titrations. This calculation involves $\beta_{abs}$ and the $K_a$ expressions of $H_4Y$.

Let $C_x$ is the analytical concentration of all the species of EDTA as :

$$C_x = [Y^{4-}] + [HY^{3-}] + [H_2Y^{2-}] + [H_3Y^-] + [H_4Y] \quad ...(7.29)$$

Now from equation (7.28) we get,

$$[Y^{4-}] = \frac{[HY^{3-}]K_{a_4}}{[H_3O^+]},$$

successively by inserting the values of $[HY^{3-}]$, $[H_2Y^{2-}]$, $[H_3Y^-]$ from equations (7.27), (7.26) and (7.25) respectively we get,

$$[Y^{4-}] = \frac{[H_4Y]K_{a_1}K_{a_2}K_{a_3}K_{a_4}}{[H_3O^+]^4} \quad ...(7.30)$$

Therefore,
$$\frac{[Y^{4-}]}{C_x} = \frac{\dfrac{[H_4Y]K_{a_1}K_{a_2}K_{a_3}K_{a_4}}{[H_3O^+]^4}}{[Y^{4-}]+[HY^{3-}]+[H_2Y^{2-}]+[H_3Y^-]+[H_4Y]}$$

$$= \frac{[H_4Y]K_{a_1}K_{a_2}K_{a_3}K_{a_4}}{[H_3O^+]^4[Y^{4-}]+[H_3O^+]^4[HY^{3-}]+[H_3O^+]^4[H_2Y^{2-}]+[H_3O^+]^4[H_3Y^-]+[H_3O^+]^4[H_4Y]} \quad ...(7.31)$$

Now from equations (7.28), (7.27), (7.26) and (7.25) the values of $[Y^{4-}]$, $[HY^{3-}]$, $[H_2Y^{2-}]$ and $[H_3Y^-]$ are successively replaced in equation (7.31), which gives,

$$\frac{[Y^{4-}]}{C_x} = \frac{K_{a_1}K_{a_2}K_{a_3}K_{a_4}}{[H_3O^+]^4+[H_3O^+]^3K_{a_1}+[H_3O^+]^2K_{a_1}K_{a_2}+[H_3O^+]K_{a_1}K_{a_2}K_{a_3}+K_{a_1}K_{a_2}K_{a_3}K_{a_4}} \quad ...(7.32)$$

Let $\alpha_4 = \frac{[Y^{4-}]}{C_x}$ so that $[Y^{4-}] = \alpha_4 C_x$ ...(7.33)

where $\alpha_4$ = fraction of EDTA in the $Y^{4-}$ form.

The value of $\alpha_4$ can be calculated at any pH for $H_4Y$ and $\beta_{abs}$ (equation 7.24) yields a new constant as :

$$\beta_{abs} = \frac{[MY^{(n-4)^+}]}{[M^{n+}]\alpha_4 C_x}$$

$$\text{i.e.} \quad \beta_{abs}\alpha_4 = \frac{[MY^{(n-4)^+}]}{[M^{n+}]C_x} = K_{eff} \quad ...(7.34)$$

where $K_{eff}$ is called the **effective** or **conditional stability constant** which is pH dependent equilibrium constant because of pH dependence of $\alpha_4$ and $K_{eff}$ can be applied at a single pH only. $K_{eff}$ is more useful than $\beta_{abs}$ because $K_{eff}$ denotes the actual formation of metal-chelate complex at a particular pH in question. It may be noted that as pH is lowered $\alpha_4$ becomes smaller and hence $K_{eff}$ becomes smaller. As mentioned before, $H_4Y$ is completely dissociated at pH12 or above and in that case $\alpha_4 \approx 1$ so that $K_{eff} \approx \beta_{abs}$. $K_{eff}$ may be computed and provide a simple way to calculate equilibrium concentrations of the metal ion and the chelate at the equivalence point.

### 7.13.6(E) Indicators used in complexometric titrations

Like acid-base titrations indicators are used in complexometric titrations to detect the end point of the titration. The speciality of these indicators are that they form **metal-indicator** complex and hence these indicators are called **metal indicators** or **metallochromic indicators**. Generally these indicators are organic dyes that form coloured chelates with metal ions in a pH range that is characteristic for the particular metal and the dye. The metal-indicator complexes are often intensely coloured and are discernible to the human eye at concentrations in the range of $10^{-6}$ to $10^{-7}$ M and these indicators should function in the pH range where the chelons react with the metal ions. Examples are Eriochrome Black T (7-LV) ($H_2In^-$), Calmagite (7-LVI)($H_2In^-$), Calcon (7-LVII)($H_2In^-$) etc. The pH dependent equilibria of these indicators are given below :

Eriochrome Black T : $H_2In^- \underset{K_1 = 5\times10^{-7}}{\overset{pH5.5 - 7.2}{\rightleftharpoons}} HIn^{2-} \underset{K_2 = 2.8\times10^{-12}}{\overset{pH10.5 - 12.0}{\rightleftharpoons}} In^{3-}$ ...(7.35)

or

Solochrome Black T : *red* *blue* *orange yellow*

Calmagite : $H_2In^- \underset{pK_1 = 8.2}{\overset{pH7.1 - 9.1}{\rightleftharpoons}} HIn^{2-} \underset{pK_2 = 12.4}{\overset{pH11.4 - 13.3}{\rightleftharpoons}} In^{3-}$ ...(7.36)

*bright red* *clear blue* *reddish orange*

Calcon : $H_2In^- \underset{pK_1 = 7.4}{\overset{pH7 - 8}{\rightleftharpoons}} HIn^{2-} \underset{pK_2 = 13.5}{\overset{pH12.3 - 13.5}{\rightleftharpoons}} In^{3-}$ ...(7.37)

*pink* *blue* *pink*

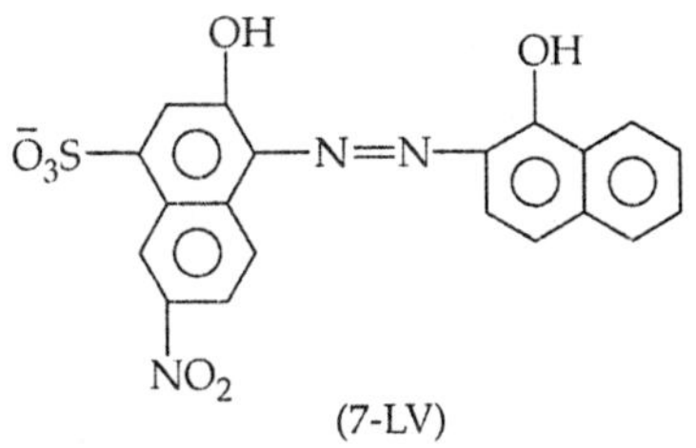

(7-LV)

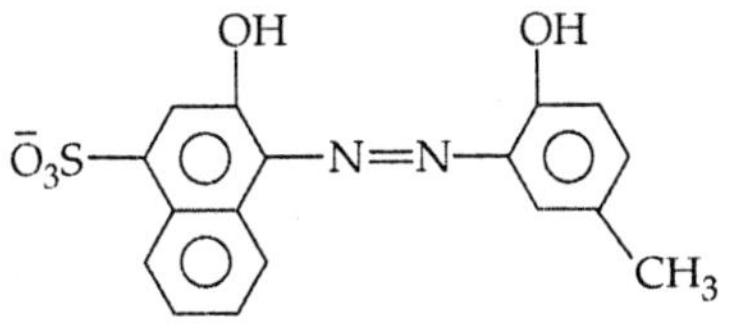

(7-LVI)

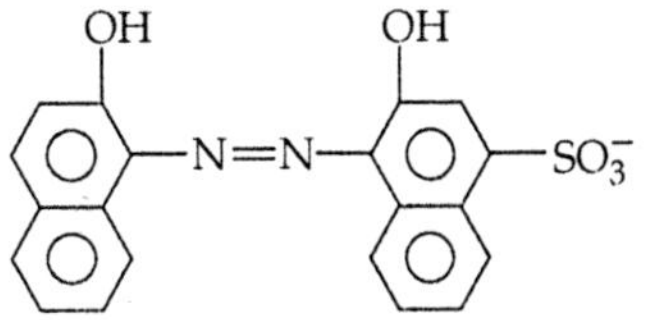

(7-LVII)

In all cases the sulphonic acid group gives up proton long before the pH range of 7-12 which is one of the important aspect of using metal indicator. Thus only the dissociation of two hydrogen atoms of the phenolic groups need to be considered.

EDTA reacts with many metal ions e.g. $Ca^{2+}$, $Mg^{2+}$ etc. at a constant pH so that a buffer used ($NH_4Cl + NH_4OH$ buffer, pH = 11) and Eriochrome Black T forms 1 : 1 complex which have a **red colour :**

$$M^{2+} + In^{3-} \longrightarrow MIn^- \qquad K_{In} = \frac{[MIn^-]}{[M^{2+}][In^{3-}]} \qquad ...(7.38)$$

At the end of the titration the metal-indicator complex ($MIn^-$) reacts with $Y^{4-}$ in the following way to free $In^{3-}$ and thus the end point is detected :

$$MIn^- + Y^{4-} \longrightarrow MY^{2-} + In^{3-} \qquad ...(7.39)$$

Thus the basic requisites of a good metal indicator are :

(a) The metal indicator should function at the same pH where the complexometric ligand reacts with the metal ion.

(b) The stability constant of the metal-indicator complex will be less than the metal-complexometric ligand complex so that the feasibility of detection of end point of tritimetric estimation is permitted as represented by equation (7.39).

(c) There must be a sharp colour contrast between the metal-indicator complex and the free metal indicator.

### 7.13.6(F) Complexometric Titration Curves

Like acid-base or redox tiration curves complexometric titration curves (Fig. 7.3) can be constructed by plotting the negative logarithm of metal ion concentration (pM = $-\log M^{n+}$) at various stages of titrations vs. milliliters of titrant.

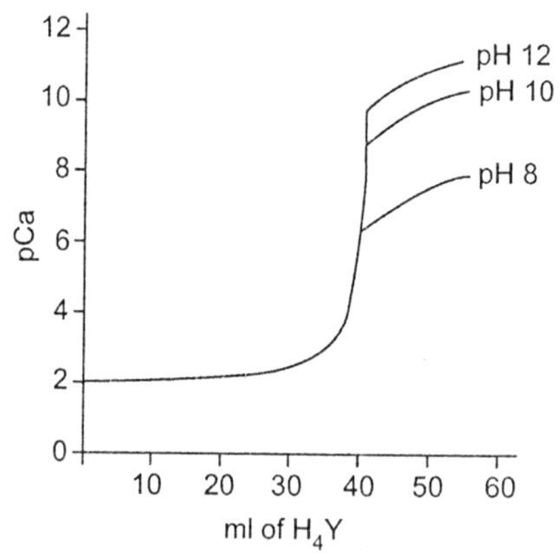


***Figure 7.3** : Titration curves for titration of 50 ml 0.01M$Ca^{2+}$ with 0.01 M $H_4Y$ at pH8, 10 and 12.*

## 7.13.6(G) Methods Employed in Complexometric Titrations

Several methods are employed in chelometric titrations.

A. **Direct titration :** This is carried out in those cases where the metal ions react with the complexometric ligand very fast for which suitable metal indicators are available. About 25 cations are titrated by this process of which $Ca^{2+}$, $Mg^{2+}$, $Zn^{2+}$, $Cu^{2+}$ are very common.

B. **Back Titration :** Back titration is useful in those cases where the cations form stable complexes with $H_4Y$ but the reaction is slow and for which no satisfactory indicator is available. A measured excess of standard $H_4Y$ solution is added to the analyte solution and the solution is heated. After the reaction is complete the excess EDTA is back tirated with standard $Mg^{2+}$ or $Zn^{2+}$ solution with suitable metal indicator. $Cr^{3+}$, $Al^{3+}$ are estimated by back titration technique.

C. **Replacement or Substitution titration :** This is useful when no suitable indicator is available for the metal ions being determined. The modus operandi is the following :

(1) If we want to determine $Me^{2+}$ then excess $MgY^{2-}$ solution is added, $Me^{2+}$ displaces $Mg^{2+}$ which is then titrated with standard EDTA solution using Calmagite as the indicator.

$$Me^{2+} + MgY^{2-} \rightleftharpoons MeY^{2-} + Mg^{2+}$$

The feasibility of the process depends on the fact that the stability of $MgY^{2-}$ complex is less compared to that of $MeY^{2-}$.

(2) In this case a suitable metal complex such as $[Ni(CN)_4]^{2-}$ is added to the solution of cation to be determined, whereby $Ni^{2+}$ is produced which is titrated with standard EDTA solution using murexide indicator. Silver and gold may be estimated in this way.

$$[Ni(CN)_4]^{2-} + 2Ag^+ \rightleftharpoons 2[Ag(CN)_2]^- + Ni^{2+} \quad ...(7.40)$$

D. **Indirect Titrations :** This is employed for estimation of anions. Sulphate may be estimated by precipitating with excess $BaCl_2$ as $BaSO_4$. The precipitate is filtered and suspended in water, dissolved in a measured excess of standard EDTA solution in presence of aqueous $NH_3$, the excess EDTA is titrated with standard $Mg^{2+}$ solution using Eriochrome Black T indicator. Similarly phosphate is precipitated as $Mg(NH_4)PO_4$, which is dissolved in dil. HCl, measured excess standard EDTA is added and the excess EDTA is titrated with standard $Mg^{2+}$ solution. Halides are estimated by employing the reaction (equation 7.40) whereby halides are precipitated as AgX (X = Cl, Br)

$$2AgX + [Ni(CN)_4]^{2-} \rightleftharpoons Ni^{2+} + 2[Ag(CN)_2]X$$

E. **Estimation of Multiple cations in a mixture : Selectivity of Complexometric Titrations by Masking and Demasking procedure.** $Pb^{2+}$, $Zn^{2+}$ and $Mg^{2+}$ in a mixture can be determined by two titrations with standard EDTA and one titration with standard $Mg^{2+}$.

An aliquot solution is first treated with excess NaCN/KCN whereby $Zn^{2+}$ is masked by the formation of $Zn(CN)_4^{2-}$.

$$Zn^{2+} + 4CN^- \longrightarrow Zn(CN)_4^{2-}$$

$Pb^{2+}$ and $Mg^{2+}$ present in the solution is titrated with standard EDTA solution. The titre value gives the corresponding $Mg^{2+}$ concentration in the solution. After the equivalence point is reached, the solution is treated with the bidentate ligand BAL* abbreviated as $R(SH)_2$ which forms complex with $Pb^{2+}$ that is much stabler than $PbY^{2-}$, so the following reaction occurs :

$$PbY^{2-} + 2R(SH)_2 \longrightarrow Pb(RS)_2^{2-} + 2H^+ + Y^{4-}$$

The liberated $Y^{4-}$ is then titrated with standard $Mg^{2+}$ solution which gives the concentration of $Pb^{2+}$ in the solution. Finally $Zn^{2+}$ is demasked by adding formaldehyde and the liberated $Zn^{2+}$ is directly titrated with standard EDTA solution using Eriochrome Black T indicator. Colour changes from wine red to blue.

$$Zn(CN)_4^{2-} + 4HCHO + 4H_2O \longrightarrow Zn^{2+} + 4HOCH_2CN + 4OH^-$$

It has been stated that EDTA forms 1 : 1 complex with divalent cations e.g. $Ca^{2+}$, $Mg^{2+}$ etc. The fact that stability of EDTA complex depends considerably on pH which makes the complexometric titration selective. For example in solution of pH 1, none of the divalent cations form complexes with EDTA whereas the trivalent cations are quantitatively complexed. Thus $Fe^{3+}$ can be estimated directly without interference of the divalent cations e.g. $Ca^{2+}$, $Mg^{2+}$, $Cd^{2+}$, $Zn^{2+}$, $Ni^{2+}$ etc. if the titration is carried out in strongly acidic solution (pH ~ 1–2). By buffering the reaction mixture at pH7, selectivity of EDTA titration increases for $Zn^{2+}$ and $Cd^{2+}$ and can be titrated in presence of $Mg^{2+}$ without interference.

### 7.13.6(H) Application of complexometric titrations

The free acid $H_4Y$ exist as the dihydrate of the sodium salt, $Na_2H_2Y.2H_2O$ which is commercially available as A.R. quality. Under normal atmospheric conditions, the dihydrate (molecular weight = 372.25) contains 0.3% moisture in excess of stoichiometric amount and therefore the pure dihydrate ($Na_2H_2Y.2H_2O$) is prepared by drying the commercial sample at 80°C for several days. This dried substance can be used as a primary standard reagent and can be stored for prolonged period. For the preparation of 250 ml 0.1 M solution 9.306 g of the dihydrate is required. Typical examples of EDTA titrations are :

* BAL is 2, 3-dimercapto-1-propanol, $CH_2SHCHSHCH_2OH$, British Anti Lawisite.

(a) **Determination of total hardness of water :** An aliquot of 50 ml water is treated with 2 ml $NH_4Cl + NH_4OH$ buffer of pH 11 followed by the addition of 3-4 drops of Eriochrome Black T whereby the solution assumes **wine red** colour. This solution is then titrated with 0.01 M EDTA solution until one drop changes its colour from **wine red** to clear **sky blue** at the end point. The hardness of water is expressed in parts of $CaCO_3$ per million of water (ppm). The equivalence condition can be established as :

$$Ca^{2+} + Y^{4-} \longrightarrow CaY^{2-}$$

so that 1000 ml 1 M EDTA $\equiv$ 40 g $Ca^{2+}$ $\equiv$ 100 g $CaCO_3$

i.e. 1 ml 0.01 M EDTA $\equiv$ 0.4 mg $Ca^{2+}$ $\equiv$ 1 mg $CaCO_3$

(b) **Estimation of $Ca^{2+}$ and $Mg^{2+}$ in a mixture :** Titration of an aliquot of solution with 0.01 M EDTA solution like that of determination of hardness of water gives the amount of total $Ca^{2+}$ and $Mg^{2+}$ in the solution. Let the titre value is $x$ ml. The same aliquot of solution (say 50 ml) is treated with 5 ml diethylamine base which precipitates $Mg^{2+}$ quantitatively as $Mg(OH)_2$ and 4 drops of Calcon indicator is added. The solution is titrated with 0.01 M EDTA until the colour changes from **pink to pure blue**. The titre value (say y ml) corresponds to $Ca^{2+}$ present in the solution. Therefore the difference of titre value $(x - y)$ ml corresponds to the amount of $Mg^{2+}$ in the solution. The equivalence conditions are :

1 ml 0.01 M EDTA $\equiv$ 0.40 mg $Ca^{2+}$

1 ml 0.01 M EDTA $\equiv$ 0.24 mg $Mg^{2+}$

### 7.13.6(I) Use of coordination complexes in various aspects

Cis-platin such as cis$[Pt(NH_3)_2X_2]$, $[Pt(en)X_2]$ (where X = $Cl^-$, $Br^-$) are widely used as anticancer drug in medicine. Vanadium (IV) chelates e.g. bis(maltolato) oxovanadium (IV) (7-LVIII) may be used as the potential insulin mimetic agents in the treatment of human diabetes. The new generation $^{99m}Tc$ radiopharmaceuticals are now widely used for cerebral blood flow imaging, renal function imaging, cancer imaging etc. e.g.

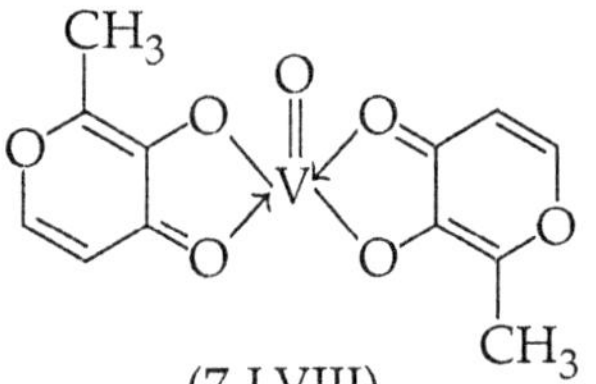

(7-LVIII)

$$^{99m}Tc\ (CNR)_6^+ \left( R = -CH_2-O-\underset{CH_3}{\overset{CH_3}{\overset{|}{\underset{|}{C}}}}-OCH_3 \right)$$ is used as a heart imaging agent. The recommended chelating antidotes for detoxification of the poisonous gas Lewisite is BAL.

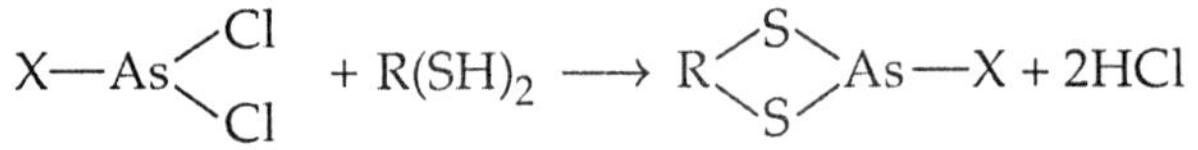

Lewisite BAL

(X=—CH=CH—Cl) R=—$CH_2$=CH—$CH_2OH$

Iron (II) porphyrin complexes are acting for oxygen transporting metalloproteins in biological systems.

In industrial application $K[Ag(CN)_2]$ and $K[Au(CN)_2]$ are employed as an electrolyte for silver and gold plating respectively. In metallurgical operation $Ni(CO)_4$ is used for the extraction of nickel by Mond carbonyl process. In fact coordination complexes have wide applicability in living systems, in medicine, in agriculture etc. Recently scientist from University of California reported* a molybdenum-oxygen complex that catalyses hydrogen production from ordinary water with high efficiency.

*Metal complex used in agriculture*

Many transition and non-transition metals such as iron, copper, zinc, manganese and molybdenum are essential micro-nutrient for plant growth. The deficiency of any of these metals caused various types of disease to plants. Deficiency of chlorophyll or rather deficiency of iron causes yellow leaves of plants and the disease is known as **iron chlorosis**. This decrease is prone to occur in plants in alkaline soil because in alkaline soil the uptake of iron by plants is hindered due to the poor solubility of $Fe(OH)_2$ ($K_{sp} \approx 2\times10^{-15}$) or $Fe(OH)_3$ ($K_{sp} \approx 6\times10^{-38}$). The addition of soluble Fe(III)-EDTA chelate to the soil eliminates this disease. Some plants such as wheat, oats etc. can grow in alkaline soil because these plants have mechanism in their roots to solubilise insoluble $Fe(OH)_3$ by producing the chelating agents such as polyaminoacids.

*Metal chelates used in food preservation*

Traces of metal ions present in food stuff like fruit juce, cake, bread etc. act as a catalyst to decompose the food stuff by the oxidation process leading to food spoilage. This is prevented by the use of chelating agents like citric acid, EDTA etc. which are acting as preservatives. They are acting as sequestering agent and can sequester the metal ions like $Cu^{2+}$, $Fe^{2+}$, $Fe^{3+}$ etc. thereby the oxidation process of food stuff are prevented.

*Use of metal complex in solar energy conservation*

Some Ru(II) complex can be used to harvest solar energy in many reactions like photochemical decomposition of water, photochemical reduction of nitrogen etc. For example the $[Ru(bpy)_3]^{2+}$ complex can catalyse the splitting of water to generate the fuel gas ($2H_2+O_2$). This fuel gas in the next step can be utilised for the production of electricity. Actually this type of reaction is carried out in space craft where solar energy is converted to electrical energy.

---

**Problem 7.32** 50 ml of a solution which is 0.01 M in $Ca^{2+}$ and buffered at pH10 is titrated with 0.01 M EDTA solution. Calculate values of pCa at various stages of the titration. $\beta_{abs}$ for $CaY^{2-}$ is $5\times10^{10}$, $\alpha_4$ at pH10 is 0.36.

**Ans. :** Here $K_{eff} = \beta_{abs}\alpha_4 = 5\times10^{10}\times0.36 = 1.8\times10^{10}$

(a) **Start of titration :** $[Ca^{2+}] = 0.01$ M  $pCa = -\log[Ca^{2+}] = -\log(0.01) = 2$

(b) **After addition of 10 ml of titrant :** There is a considerable excess of $Ca^{2+}$ at this stage. We may assume that the reaction goes to completion.

Original : 50 ml 0.01 M $Ca^{2+}$ = (50×0.01) mmol = 0.50 mmol $Ca^{2+}$

Now, 10 ml 0.01 M EDTA ≡ 10 ml 0.01 M $Ca^{2+}$ = (10×0.01) mmol = 0.10 mmol $Ca^{2+}$

$$\therefore [Ca^{2+}] = \frac{(0.50 - 0.1)\text{ mmol}}{60\text{ ml}} = 0.0067\text{ M}$$

so that pCa = 2.17.

---

* The Statesman dated 12.5.2010

(c) **At equivalence point :** $[Ca^{2+}] = C_x$, $[CaY^{2-}] = \dfrac{0.50 \text{ mmol}}{100 \text{ ml}} = 5\times10^{-3}$ M

According to equation (7.34)

$$K_{eff} = \frac{[CaY^{2-}]}{[Ca^{2+}]C_x} = \frac{5\times10^{-3}}{[Ca^{2+}]^2}$$

$$\therefore \frac{5\times10^{-3}}{[Ca^{2+}]^2} = 1.8\times10^{10}, \text{ so that } [Ca^{2+}] = 5.2\times10^{-7}, \text{ pCa} = 6.28.$$

(d) **After addition of 60 ml 0.01 M EDTA solution :** Excess EDTA = 10 ml 0.01 M = (10×0.01) mmol = 0.10 mmol.

$$\therefore C_x = \frac{0.10 \text{ mmol}}{110 \text{ ml}} = 9.1\times10^{-4} \text{ M}$$

$$[CaY^{2-}] = \frac{0.50 \text{ mmol}}{110 \text{ ml}} = 4.55\times10^{-3} \text{ M}$$

$$K_{eff} = \frac{[CaY^{2-}]}{[Ca^{2+}]C_x} = \frac{4.55\times10^{-3}}{[Ca^{2+}]\, 9.1\times10^{-4}}$$

$$\therefore \frac{4.55\times10^{-3}}{[Ca^{2+}]\, 9.1\times10^{-4}} = 1.8\times10^{10}, \text{ so that } [Ca^{2+}] = 2.8\times10^{-10}, \text{ pCa} = 9.55.$$

---

**Problem 7.33** Evaluate the conditional formation constant for the reaction $Zn^{2+} + Y^{4-} \longrightarrow ZnY^{2-}$. Given $[Y^{4-}] = 8.22\times10^{-22}$ M and $C_x = 9.05\times10^{-21}$ and $\beta_{abs} = 6.336\times10^7$.

**Ans. :** We know $\alpha_4 = \dfrac{[Y^{4-}]}{C_x} = \dfrac{8.22\times10^{-22}}{9.05\times10^{-21}} = 9.08\times10^{-2}$

$$K_{eff} = \beta_{abs}\alpha_4 = 6.336\times10^7\times9.08\times10^{-2} = 5.753\times10^6.$$

---

**Problem 7.34** 50 ml of 0.01 M solution of a metal ($M^{2+}$) is titrated with 0.01 M EDTA solution. Evaluate $K_{eff}$, when 49.95 ml of the titrant has been added. The reaction is virtually complete and the pM changes by 2 units on the addition 2 more drops (0.1 ml) of the titrant.

**Ans. :** 49.95 ml 0.01 M EDTA = (49.95×0.01) mmol = 0.4995 mmol EDTA

Initially 50 ml 0.01 M metal solution taken

$\therefore [M^{2+}]$ = (50×0.01) mmol = 0.50 mmol

49.95 ml 0.01 M $M^{2+}$ solution ≡ 49.95 ml 0.01 M EDTA solution

Hence after the titration with EDTA the $M^{2+}$ remain unreacted = 50 – 49.95 = 0.05 ml.

i.e. (0.05×0.01) mmol = 0.0005 mmol $M^{2+}$ remain unreacted.

Hence, $[M^{2+}] = \dfrac{0.0005 \text{ mmol}}{99.95 \text{ ml}} = 5\times10^{-6}$ M

so that pM = 5.30

Here ΔpM = 2 units, so that pM value changes after the addition of 2 more drops (0.1 ml) i.e. 49.95 + 0.1 = 50.05 ml titrant = 5.30 + 2 = 7.30, whereby $[M^{2+}] = 5\times10^{-8}$ M

At this stage, the concentration of uncomplexed EDTA ($C_x$)

$$= \frac{0.05 \times 0.01 \text{ mmol}}{100.05 \text{ ml}} = 5 \times 10^{-6} \text{ M}$$

$$\text{Now, } [MY^{2-}] = \frac{0.5 \text{ mmol}}{100 \text{ ml}} = 5 \times 10^{-3} \text{ M}$$

$$\text{Therefore } K_{eff} = \frac{[MY^{2-}]}{[M^{2+}]C_x} = \frac{5 \times 10^{-3}}{(5 \times 10^{-8})(5 \times 10^{-6})} = 2 \times 10^{10}.$$

---

**Problem 7.35** 10.0 ml solution of sodium salt, free from other cations is passed through a strongly acidic cation exchange resin in the Mg form and the liberated $Mg^{2+}$ is titrated with 0.0495 M EDTA using EBT indicator and 22.13 ml is consumed. Calculate the amount of $Na^+$ present (in gms) in the sample. **[C.U. 2007]**

**Ans. :** 1000 ml 1 M EDTA = 23 g $Na^+$

$$22.13 \text{ ml } 0.0495 \text{ M EDTA} = \frac{23 \times 22.13 \times 0.0495}{1000} = 0.0252 \text{ g.}$$

---

**Problem 7.36** How does metal indicators act in complexometric titration? Explain with suitable example. Outline the procedure for estimation of Cu and Zn in a mixture complexometrically. **[C.U. 2009, 2008]**

**Ans. :** For action of metal indicators see Sec. 7.13.6(E).

The use of masking and selective demasking procedure permits the successive estimation of Cu and Zn in a mixture. Thus a solution containing Cu and Zn can be titrated as follows :

(1) An aliquot portion of the solution is treated with measured excess EDTA solution and the excess EDTA is back titrated with standard $Mg^{2+}$ solution using EBT indicator. The titre value gives the sum of Cu and Zn in the mixture.

(2) The same aliquot portion of the solution is treated with excess KCN whereby both $Cu^{2+}$ and $Zn^{2+}$ forms cyano complexes as a process of masking :

$$Cu^{2} + KCN \longrightarrow K_3[Cu(CN)_4]$$

$$Zn^{2+} + KCN \longrightarrow K_2[Zn(CN)_4]$$

Then excess chloral hydrate or formaldehyde-acetic acid solution (3:1) is added whereby $[Zn(CN)_4]^{2-}$ complex will be demasked to form free $Zn^{2+}$ in the solution which is being titrated with EBT indicator until the colour changes to blue. The titre value gives the amount of $Zn^{2+}$ in the mixture. The amount of $Cu^{2+}$ is obtained by the difference of titre values.

$$[Zn(CN)_4]^{2-} + 4H^+ + 4HCHO \longrightarrow Zn^{2+} + 4HO.CH_2.CN$$

1 ml of 0.1 M EDTA $\equiv$ 6.538 mg of $Zn^{2+}$ and 1 ml of 0.05M EDTA $\equiv$ 3.177 mg of $Cu^{2+}$.

---

**Problem 7.37** Discuss the advantages of multidentate ligands (e.g. EDTA) for complexometric titrations over monodentate ligands. Write down the structural formula of Ca–EDTA complex. **[B.U. 2009]**

**Ans. :** For advantages of multidentate ligand see Sec. 7.13.6(A). Structure of Ca-EDTA complex is octahedral.

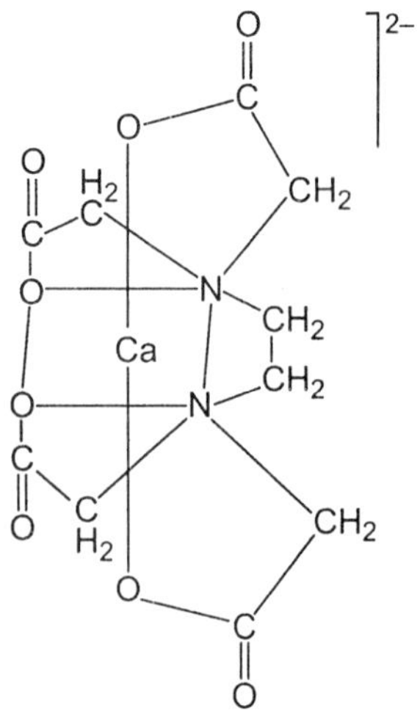

**Problem 7.38** In the complexometric estimation of $Ca^{2+}$ by disodium EDTA using EBT indicator, some $Na_2MgEDTA$ is added to the $Ca^{2+}$ solution before starting the titration. Explain its necessity showing all the reactions involved in the process. **[C.U. 2007, 2004]**

**Ans. :** In the complexometric estimation of $Ca^{2+}$ with $Na_2H_2Y$ using EBT indicator a relatively stable calcium complex is formed :

$$Ca^{2+} + H_2Y^{2-} \rightleftharpoons CaY^{2-} + 2H^+$$

With $Ca^{2+}$ ions alone, no sharp end point is obtained with EBT indicator and the transition from red to blue is not observed. In presence of $Mg^{2+}$ ions, a somewhat less stable complex with $Mg^{2+}$ is formed :

$$Mg^{2+} + H_2Y^{2-} \rightleftharpoons MgY^{2-} + 2H^+$$

and magnesium-indicator complex is more stable than the calcium-indicator complex but less stable than $MgY^{2-}$ complex. Naturally, when a solution containing $Ca^{2+}$ and $Mg^{2+}$ ions is titrated with EDTA in presence of EBT indicator, the EDTA reacts first with the free $Ca^{2+}$ ions, then with the free $Mg^{2+}$ ions and finally with magnesium-indicator complex. As the magnesium-indicator complex is wine red in colour and the free indicator is blue between pH7-11, the colour of the solution changes from wine red to blue at the end point :

$$\underset{red}{MgIn^-} + H_2Y^{2-} \longrightarrow MgY^{2-} + \underset{blue}{HIn^{2-}} + H^+$$

If $Mg^{2+}$ ions are not present in the solution containing $Ca^{2+}$ ions they must be added, as they are required for the colour change of the indicator.

**Problem 7.39** By applying complexometric titration, 6.5 ml of 0.0091(M) EDTA solution was required for titration of a 50.0 ml sample of water for total hardness. Calculate the total hardness in terms of ppm of $CaCO_3$. **[B.U. 2008]**

**Ans. :** 50 ml hard water ≡ 6.5 ml 0.0091 M EDTA.

1000 ml of 0.01 M EDTA ≡ 1000 ml of 0.01 M $CaCO_3$ ≡1 g of $CaCO_3$

(Mol. wt. of $CaCO_3$ = 100)

$$\therefore 6.5 \text{ ml } 0.0091 \text{ M EDTA} = \frac{1\times6.5\times0.0091}{1000\times0.01} = 0.00591 \text{ g } CaCO_3$$

$\therefore$ 50 parts of hard water contains 0.00591 g $CaCO_3$

$10^6$ " " " " " $\frac{0.00591\times10^6}{50} = 118.2$ ppm.

Alternatively,

Molarity of EDTA solution ($M_1$) = 0.0091

Volume of EDTA solution ($V_1$) = 6.5 ml

Volume of water sample ($V_2$) = 50.0 ml

Molarity of water sample ($M_2$) = $\frac{M_1V_1}{V_2} = \frac{0.0091\times6.5}{50}$

Total hardness of water = $M_2\times100\times1000$ ($\because$ Mol. wt. of $CaCO_3$ = 100)

$$= \frac{0.0091\times6.5}{50}\times100\times1000 = 118.2 \text{ ppm.}$$

**Problem 7.40** Establish the equivalence condition 1 ml 0.01 M EDTA = 1.0 mg $CaCO_3$. **[B.U. 1994]**

**Ans. :** We know that,

1000 ml 0.01 M EDTA $\equiv$ 1000 ml 0.01 M $CaCO_3$

$\equiv$ 1 g of $CaCO_3$ ($\because$ Mol. wt of $CaCO_3$ = 100)

$\therefore$ 1 ml 0.01 M EDTA $\equiv$ 1.0 mg of $CaCO_3$

**Problem 7.41** Calculate $\alpha_4$ and the mole percent of $Y^{4-}$ in a solution of EDTA that is buffered at pH 10.20. Given, $K_1 = 1.02\times10^{-2}$, $K_2 = 2.14\times10^{-3}$, $K_3 = 6.92\times10^{-7}$ and $K_4 = 5.50\times10^{-11}$ for $H_4Y$.

**Ans. :** $[H^+]$ = antilog of (–10.20) = $6.31\times10^{-11}$

$K_1K_2 = 2.18\times10^{-5}$, $K_1K_2K_3 = 1.51\times10^{-11}$, $K_1K_2K_3K_4 = 8.31\times10^{-22}$

$$\text{We know, } \alpha_4 = \frac{[Y^{4-}]}{C_x} = \frac{K_1K_2K_3K_4}{[H^+]^4+[H^+]^3K_1+[H^+]^2K_1K_2+[H^+]K_1K_2K_3+K_1K_2K_3K_4}$$

$$= \frac{8.31\times10^{-22}}{(6.31\times10^{-11})^4+(6.31\times10^{-11})^3 1.02\times10^{-2}+(6.31\times10^{-11})^2 2.18\times10^{-5}+6.31\times10^{-11}\times1.51\times10^{-11}+8.31\times10^{-22}}$$

$$= \frac{8.31\times10^{-22}}{1.78\times10^{-21}} = 0.466$$

Since $\alpha_4$ is the fraction of EDTA in the $Y^{4-}$ form, thus mole percent of $Y^{4-}$ = $0.466\times100$ = 46.6%.

## 7.14 THEORY OF METAL-LIGAND BONDING

As mentioned earlier (Sec. 7.3) Sidgwick electronic interpretation of coordination is insufficient to explain the stability of complexes. In order to explain the nature of coordinate linkage we outline here three theories—**valence bond theory, crystal field theory** and **molecular orbital theory**. Actually the bonding theory we used to describe molecules and ions was the **valence bond theory**, which is indeed an essential bonding theory to describe the vast majority of molecules and ions (organic and inorganic) during the year 1930-1950. This theory is based on the orbital overlap model and was developed by Linus Pauling. The inadequacies of valence bond theory arises in case of **electron deficient molecules** as also to explain the colour of coordination complexes. Thus crystal field theory (1930) gained ground and was supplemented by the **ligand field theory**. The ligand field and crystal field theory was first developed by H-Bethe. Crystal field theory assumes the metal ions and ligands are point positive and negative charges respectively and the electrostatic attraction between them force some orientation of d-orbitals. Van Vleck laid the basis of molecular orbital theory which largely considers that there is a good overlap of the orbitals of the ligands and the metal ions so that the bonding turns out to be largely covalent.

Crystal field theory strictly valid for ionic complexes. Molecular orbital theory explains the covalent bonding in complexes. The intermediate interactions between ionic and covalent bonding is explained in ligand field theory which covers the interelectronic repulsion, spin-orbit coupling etc. so that ligand field theory sometimes called **adjusted crystal field theory**.

### 7.14.1 Valence Bond Theory

This deals with the ground state electronic structure of the central metal ion and today still it is concerned primarily with the shapes and magnetic properties of complexes. The orbitals of the complexes are designated only in terms of the central atomic orbitals. Pauling has devised a very simple and satisfactory means for presenting the bonding picture which involves the following assumptions :

(a) First the requisite number of electron(s) is/are lost from the metal to form the metal ion. When the metal is in zero oxidation state this step is not applied.

(b) As every metal complex has specific geometry as also coordination number (C.N.) with a particular ligand, so metal atom or ion orbitals should be directed at the specified directions to assume the desired coordination number. Metal atom or ion orbitals e.g. s, p or d alone cannot be able to generate the desired stereochemistry such as octahedral, tetrahedral etc. Pauling therefore suggested the hybridisation of s, p, d orbitals of metal atom/ion so that the desired stereochemistry of the metal complexes are achieved. The number of hybrid orbitals are equal to the number of participating atomic orbitals. The most common hybridisation of first transition metals are $sp^3$ (tetrahedral), $dsp^2$ (square planar), $sp^3d$ (trigonal bipyramidal), $dsp^3$ (square pyramid), $d^2sp^3$ (inner orbital octahedral), $sp^3d^2$ (outer orbital octahedral) etc. (Table 3.4)

(c) The directional empty hybrid orbitals are then occupied by lone pair electrons of ligands resulting in the formation of $L \rightarrow M$ $\sigma$ bond.

(d) The non-bonding electrons of the metal atom/ion are then reorganised in the remaining metal atom/ion orbitals (s, p, d orbitals which are not taking part in hybridisation) obeying the Hund's rule of maximum multiplicity.

(e) When the metal is in the low oxidation/zero oxidation state, there will be the formation of synergic effect i.e. not only the formation of $L \rightarrow M$ $\sigma$ bond there also be the formation of $M \rightarrow L$ $\pi$ bond.

A particular complex has a specific geometry for which there will be the operation of specific hybridisation. Let us consider the formation of $[Co(NH_3)_6]^{3+}$ and $[FeF_6]^{3-}$ ions. Both of them are octahedral and $Fe^{3+}$ and $Co^{3+}$ is $3d^5$ and $3d^6$ system respectively.

$Co^{3+} \longrightarrow 3d^6$ system 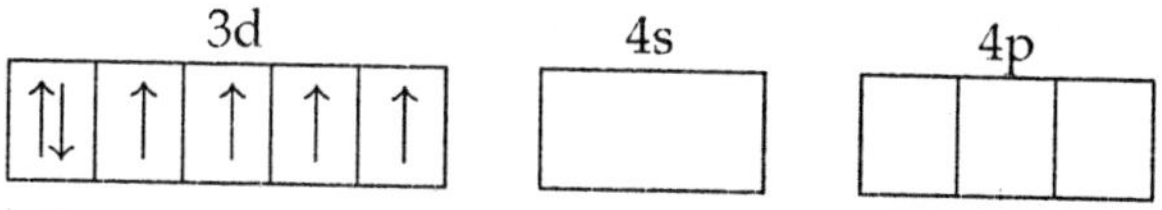


In order to accommodate the six electron pairs donated by six ammonia molecules, six vacant orbitals must be made available. A question arises how the two electrons removed from them?

(a) One possibility is that they may be promoted to the next lowest orbitals above the 4p orbital, these being either the 4d or 5s orbitals, probably the later.

(b) The other alternative is to put them into two of the other 3d orbitals.

Approximate values of energies required to arrange the two electrons in each of these two ways to indicate quite surely that the number (b) way would be more favourable. So the following diagram of electron distribution of $d^2sp^3$ hybridisation is obtained (ligand electrons are marked X) :

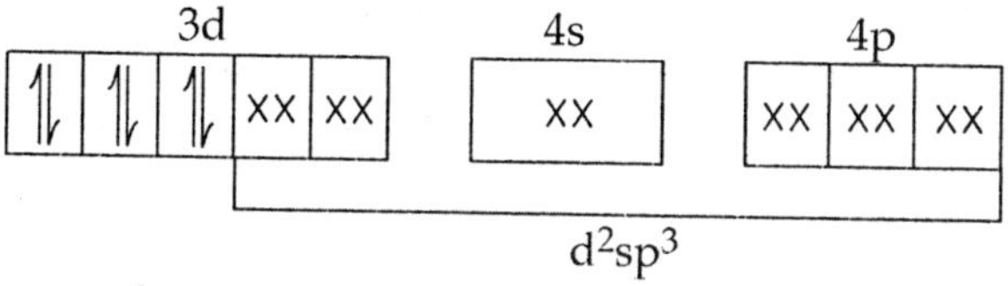


Evidently, $[Co(NH_3)_6]^{3+}$ has no unpaired electron and it should be **diamagnetic**. Magnetic moment studies indicate that the complex is diamagnetic. Here it may be noted that the d-orbitals used are inner 3d orbitals so that this is called **inner orbital complex.**

In fact several $Co^{3+}$ complexes are virtually diamagnetic, there might be some exception e.g. $[CoF_6]^{3-}$. In this case the complex has four unpaired electrons just as in the free $Co^{3+}$ ion so that following hybridisation scheme was proposed :

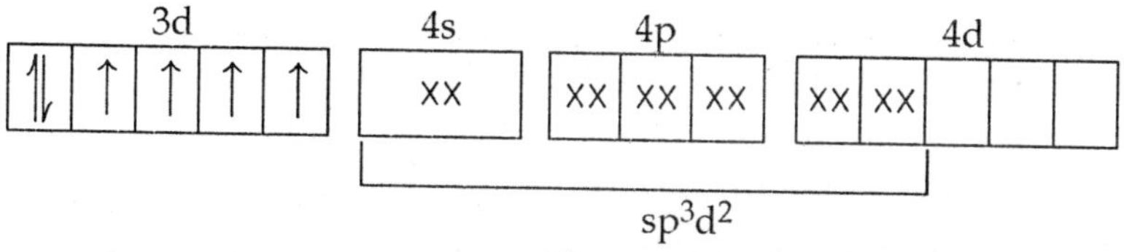


Now let us consider the formation of $[FeF_6]^{3-}$ ion. The electron distribution of the free ion is :

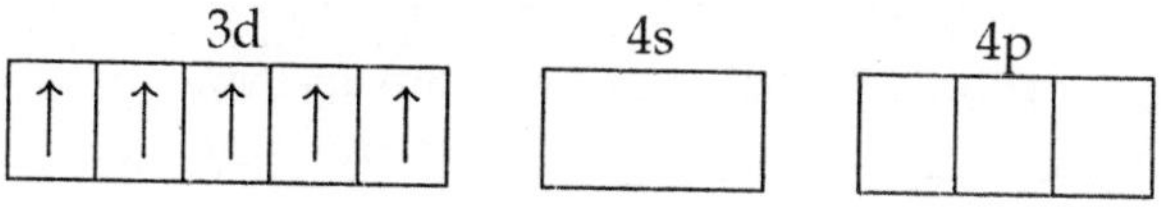


Magnetic studies indicate the presence of five unpaired electrons for $[FeF_6]^{3-}$ in 3d orbitals. Thus to get six hybrid orbitals none of the five 3d electrons should be disturbed but should be allowed to remain unpaired as they are in the free $Fe^{3+}$ ion. Pauling therefore suggested that in the complexes like $[FeF_6]^{3-}$ and $[CoF_6]^{3-}$, the metal-ligand bonds are primarily ionic and not covalent and the 3d electrons in these complex ions are therefore not disturbed but are allowed to remain unpaired just as they are in the free metal ion. Thus Pauling and also Huggin assumed that metal ion d-orbitals are left free to accommodate only the metal ion electrons and the ligand electrons will exist in separate orbitals located primarily on the ligand. According to them when electronegative atoms are involved as ligands, the electrostatic bonding takes place which does not disturb the electrons of the d-orbitals and so the following hybridisation occurs :

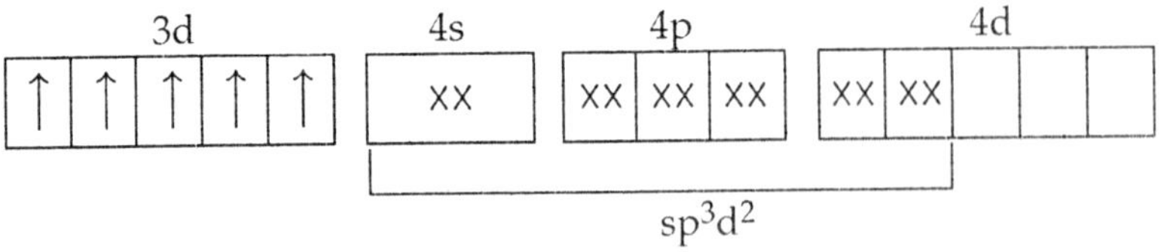


Complexes where the outer d-orbitals are involved are called **outer orbital complex.**

Criteria for the formation of outer and inner orbital octahedral complexes may be summarised as :

(a) Strongly electronegative donor atoms having smaller donor orbitals can overlap better with the expanded higher d-orbitals of the metal thereby $sp^3d^2$ hybridisation facilitated.

(b) Inner orbital $d^2sp^3$ hybridisation is facilitated by less electronegative donor atoms having expanded donor orbitals which overlap with the relatively contracted lower d-orbitals of metals. In this case metal-ligand π-bonding is of importance.

(c) Generally second and third transition elements prefer to form inner orbital $d^2sp^3$ hybridisation while the first transition elements prefer to form $sp^3d^2$ hybridisation. This is due to the fact that 4d and 5d orbitals are more spread out in space than the 3d orbitals. This leads to a smaller interelectronic repulsion and thus occurs closer approach of the ligand orbitals to the metal orbitals i.e. the higher d-orbitals are too diffused to overlap effectively with the donor ligand orbitals. This makes the higher congeners of transition elements to form inner orbital complex e.g. $[FeF_6]^{3-}$ is outer orbital complex while $[RuF_6]^{3-}$ is inner orbital.

(d) Increasing positive charge on metal ions brings closer approach of the metal ion orbitals with that of ligand orbitals thereby inner orbital ($d^2sp^3$) hybridisation is favoured e.g. $[Co(NH_3)_6]Cl_3$ is inner orbital complex ($d^2sp^3$) while $[Co(NH_3)_6]Cl_2$ is outer orbital ($sp^3d^2$) complex.

Specific hybridisation of the first transition metal complexes will correspond to differnt number of electrons (Table 7.9) which can be judged by the magnetic moment measurements. In case of octahedral geometry having $d^2sp^3$ or $sp^3d^2$ hybridisation $d_{x2-y2}$ and $d_{z2}$ orbitals are involved. The tetrahdral $d^3s$ hybridisation utilises $d_{xy}$, $d_{xz}$ and $d_{yz}$ orbitals while square planar $dsp^2$ utilises only $d_{x2-y2}$ orbital. The five coordinate trigonal bipyramid geometry having $dsp^3$ or $sp^3d$ hybrids used $d_{z2}$ orbital while square pyramid geometry ($dsp^3$) used $d_{x2-y2}$ orbital.

**Table 7.9 : Number of unpaired electrons with different $d^n$ ions of first transition series**

| $d^n$ configuration | Example | Hybridisation | Stereochemistry | Unpaired electron(s) | Magmetic Moment(μ)* B.M. |
|---|---|---|---|---|---|
| $d^1$ | $[ScF_6]^{3-}$, $[TiF_6]^{3-}$, $[Ti(H_2O)_6]^{3+}$<br>$[VO(acac)_2]$ | $d^2sp^3$<br>$dsp^3$ | Octahedral<br>Square pyramid | 1<br>1 | 1.73<br>1.73 |
| $d^2$ | $[V(acac)_3]$, $[V(C_2O_4)_3]^{3-}$, $[V(CN)_6]^{3-}$ | $d^2sp^3$ | Octahedral | 2 | 2.83 |
| $d^3$ | $[Cr(H_2O)_6]^{3+}$, $[Cr(NH_3)_6]^{3+}$, $[Cr(NH_3)_4Cl_2]^+$<br>$[MnF_6]^{2-}$ | $d^2sp^3$<br>$sp^3d^2$ | Octahedral<br>Octehedral | 3<br>3 | 3.87<br>3.87 |
| $d^4$ | $[Cr(CN)_6]^{4-}$, $[Mn(CN)_6]^{3-}$<br>$[Cr(H_2O)_6]SO_4$ | $d^2sp^3$<br>$sp^3d^2$ | Octehedral<br>Octahedral | 2<br>4 | 2.83<br>4.90 |
| $d^5$ | $[Fe(CN)_6]^{3-}$, $[Mn(CN)_6]^{4-}$<br>$[FeF_6]^{3-}$ | $d^2sp^3$<br>$sp^3d^2$ | Octahedral<br>Octahedral | 1<br>5 | 1.73<br>5.92 |
| $d^6$ | $[Co(NH_3)_6]Cl_3$, $Cr(CO)_6$, $[Fe(CN)_6]^{4-}$<br>$[FeCl_4]^{2-}$ | $d^2sp^3$<br>$sp^3$ | Octahedral<br>Tetrahedral | 0<br>4 | 0<br>4.90 |
| $d^7$ | $[Co(NO_2)_6]^{4-}$, $[Co(CN)_6]^{4-}$<br>$[Co(NH_3)_6]^{2+}$<br>$[CoCl_4]^{2-}$ | $d^2sp^3$<br>$sp^3d^2$<br>$sp^3$ | Octahedral<br>Octahedral<br>Tetrahedral | 1<br>3<br>3 | 1.70<br>3.87<br>3.87 |
| $d^8$ | $[Ni(DMG)_2]$, $[Ni(CN)_4]^{2-}$<br>$[Ni(NH_3)_6]^{2+}$ | $dsp^2$<br>$sp^3d^2$ | Square planar<br>Octahedral | 0<br>2 | 0<br>2.83 |
| $d^9$ | $[CuCl_4]^{2-}$<br>$[Cu(BigH)_2]^{2+}$<br>$[Cu(NH_3)_4]^{2+}$ | $sp^3$<br>$dsp^2$<br>$sp^2d$ | Tetrahedral<br>Square planar<br>Square planar | 1<br>1<br>1 | 1.73<br>1.73<br>1.73 |
| $d^{10}$ | $Ni(CO)_4$ | $sp^3$ | Tetrahedral | 0 | 0 |

* $\mu = \sqrt{n(n+2)}$ B.M. where n = number of unpaired electron(s), B.M. = Bohr magneton = $\frac{eh}{4\pi mc}$.

The magnetic moment of a complex will depend on the actual number of unpaired electrons which has of great practical importance in understanding the nature of the complex regarding bond type, geometry etc. The number of unpaired electrons present in a given complex depends on the geometry of the complex and the type of hybridisation executed by the central metal atom/ion. Table 7.10 shows a relationship between hybridisation and number of unpaired electrons in different $d^n$ systems. It is to be noted that inner orbital octahedral complexes formed by strong ligands while weaker ligands form outer orbital octahedral complexes.

**Table 7.10 : Relation between the number of unpaired electrons for different geometry of $d^n$ ions/atoms**

| $d^n$ ion/atom | Number of unpaired electrons for | | | |
|---|---|---|---|---|
| | Inner orbital octahedral $d^2sp^3$ | Outer orbital octahedral $sp^3d^2$ | Tetrahedral $sp^3$ | Square planar $dsp^2$ or $sp^2d$ |
| $d^1$ | 1 | 1 | 1 | 1 |
| $d^2$ | 2 | 2 | 2 | 2 |
| $d^3$ | 3 | 3 | 3 | 3 |
| $d^4$ | 2 | 4 | 4 | 4 |
| $d^5$ | 1 | 5 | 5 | 3 |
| $d^6$ | 0 | 4 | 4 | 2 |
| $d^7$ | 1 | 3 | 3 | 1 |
| $d^8$ | 0 | 2 | 2 | 0 |
| $d^9$ | 1 | 1 | 1 | 1 |

**Limitations of Valence Bond Theory (VBT)**

1. Table 7.10 shows that for octahedral ($d^2sp^3$ and $sp^3d^2$), tetrahedral ($sp^3$) and square planar ($dsp^2$) complexes of $d^1$, $d^2$, $d^3$ and $d^9$ systems have the same number of unpaired electrons, so magnetic moment would not be able to distinguish the correct geometry of the complex.

2. The square planar copper (II) complexes may be represented as—

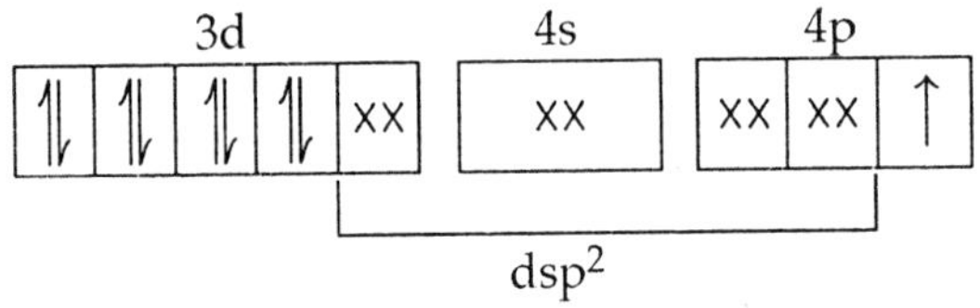


Here one d electron has been promoted to 4p orbital so that the promoted electron should have p character. But electron spin resonance (ESR) spectra of several copper (II) complexes shows that the orbital occupied by the electron does not have any p character.

3. For the formation of inner orbital octahedral complexes of $d^7$, $d^8$ and $d^9$ systems, the non-bonding electrons of central metal ion will have to be promoted to higher energy orbital (presumably to 4d or 5s orbital). Such promotion of electrons led to the oxidation of the central metal ion. This mechanism can explain the oxidation of $Co^{2+}$ complexes to $Co^{3+}$ complexes but the same mechanism is not valid for the easy oxidation of $Cu^{2+}$ to $Cu^{3+}$ complexes. In fact the synthesis of Cu(III) complexes are difficult.

4. It is firmly established that both the colour and magntic properties of complexes are due to their possessing d-orbital electrons, so that there must be close connection between the colour and magnetic properties of complexes. Unfortunately in the case of valence bond theory no such connection have been made.

5. Though valence bond theory can explain the approximate value of magnetic moment of complexes predicted from the number of unpaired electrons, yet this theory fails to explain the temperature dependence of magnetic moment as also orbital contribution to the spin only values e.g. octahedral $[Ni(NH_3)_6]Cl_2$ has $\mu$ = 2.83 B.M., but the experimental value is 3.32 B.M. This enhanced value of magnetic moment is not explained by this theory.

6. Valence bond theory cannot explain the reaction rates and mechanism of reactions.

7. This theory fails to explain the distorted geometry of complexes from regular structures.

## 7.14.2 Crystal Field Theory (CFT) : Theory of Crystalline Field (TCF)

To correlate the colour and magnetic properties of complexes and to obviate the defects of valence bond theory, the more plausable crystal field theory was developed. In crystal field theory the ligands are considered as point negative charges and metal ions are point positive charges so that there will be an electrostatic attraction between a metal ion and the ligands. Since this type of electrostatic attraction originally prevaling in case of ionic crystals e.g. KCl, NaCl etc., so that this model is termed as **electrostatic crystal field theory**. In this theory no consideration of overlap between ligand orbitals and metal ion orbitals have been made.

As transition metals are good complex former, therefore this electrostatic approach forces some perturbation or ordering of d-orbitals in different stereochemistries such as octahedral, tetrahedral square planar etc.

### 7.14.2(A) Octahedral complexes

In octahedral complexes, the metal atom/ion is at the centre of an octahedron while six ligands are existing at the six corners of the octahedron along the arbitrarily chosen x, y and z axes (Fig. 7.4). Let us consider the metal has only s-orbital. For a given n value, the s-orbital is spherically symmetrical so that on interaction with six ligands, the energy of the s-orbital will be raised compared to field free state. Instead of s-orbital if there is a p-orbital which is three fold degenerate, similarly with s-orbital, the energy of the three p-orbitals i.e. $p_x$, $p_y$ and $p_z$ will be raised, but the degeneracy will not be lifted. In place of octahedral field if we take a square planar arrangement, then degeneracy of p-orbitals will be lifted and two sets of orbitals are obtained—one having degenerate $p_x$ and $p_y$ orbitals of higher energy and the $p_z$ orbital of lower energy because a square plane may be considered by taking $x$ and y axes. These situations are less important in case of complex formation.

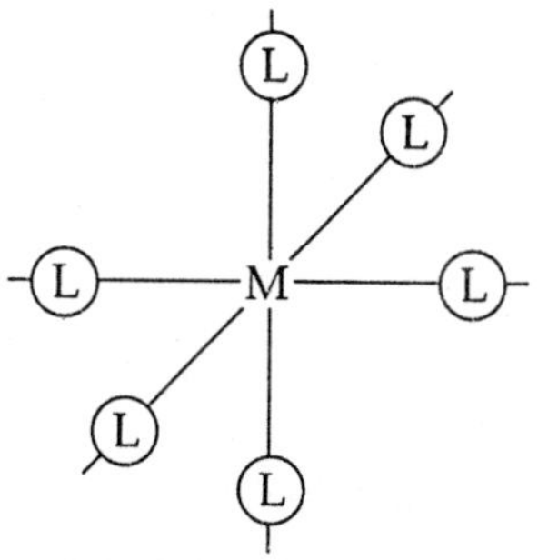


*Figure 7.4 : The octahedral complex where the six ligands are along the three cartesian axes.*

Now what happen when metal atom/ion have d-orbital which is of prime importance in our present discussion. The d-orbital is five fold degenerate and all the five d-orbitals are not alike.

Three of these orbitals, viz., $d_{xy}$, $d_{yz}$ and $d_{xz}$ are similar in all respects but their spatial orientation are different. Since they occupy spaces between the three axes, their electron density is minimum along the axes and maximum along the diagonals between the axes. On the other hand the electron density of the remaining two orbitals i.e. $d_{x^2-y^2}$ and $d_{z^2}$ orbitals, is directed along the axes.

The six ligands initially surrounded the metal will create a **spherically symmetrical field** of negative charges. On the application of octahedral crystal field such a set of five d-orbitals will be no longer degenerate. The repulsion between metal and ligand electrons raises the energy of the five d-orbitals initially, although it does not lift the degeneracy. Finally since the $d_{x^2-y^2}$ and $d_{z^2}$ orbitals are directed along the axes, there will be a strong interaction by the six ligands and their energy will be raised whereas the $d_{xy}$, $d_{yz}$ and $d_{xz}$ will experience lesser interacting force by the ligands and will be of lower energy. Thus there will be a splitting of d-orbitals between two sets of orbitals one having $d_{x^2-y^2}$ and $d_{z^2}$ orbitals designated as $e_g$ of higher energy and the other having $d_{xy}$, $d_{yz}$ and $d_{xz}$ orbitals are classed as $t_{2g}$ of lower energy (Fig. 7.5). The splitting will follow the the centre of gravity (barycentre) rule i.e. the amount of energy raised by $e_g$ orbitals is equal to the amount of energy lowered by $t_{2g}$ orbitals so that the total energy of the system is constant. If one electron of the metal ion are in the $e_g$ set of orbitals, it will be experiencing more repulsion than the electron in $t_{2g}$ set of orbitals in the case of octahedral complex than the hypothetical spherical complex. Figure 7.5 records an **energy level diagram** comparing the relative energies of d-orbitals that the d-electron(s) could occupy in these two

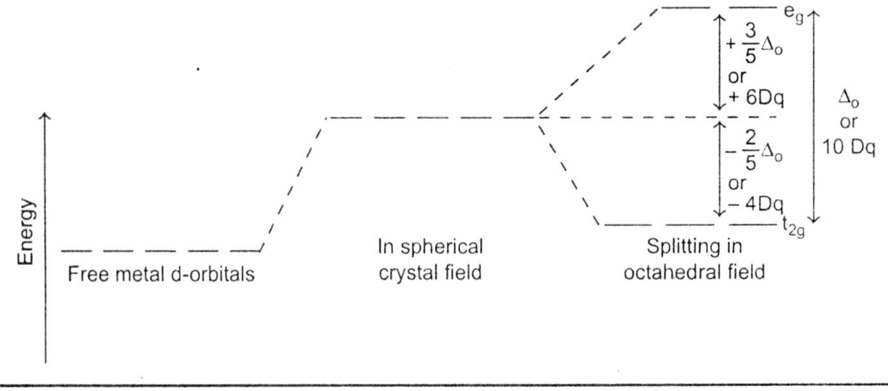


*Figure 7.5 : Splitting (qualitative) of d-orbitals in an octahedral crystal field.*

real octahedral and hypothetical spherical geometries. Any d-orbital in the spherical complex of the five degenerate d-orbitals in a uniformly smeared out field of six ligands is taken as the zero of energy while in the octahedral complex the energy is determined by the preferential filling of d-electrons in the $e_g$ and $t_{2g}$ sets of orbitals. The difference of energy between the two sets of orbitals ($e_g$ and $t_{2g}$) is called the (octahedral) **crystal field splitting** or **ligand field splitting** and is represented as 10Dq or $\Delta_o$. The subscript 'o' stands for octahedral. The terms D and q were obtained by the quantitative treatment of crystal field theory for d-orbital wave function. The potential energy and its effects on d-orbital wave function is related to the terms D and q in octahedral field as :

$$D = \frac{35Ze}{4a^5}, \quad q = \frac{2er^4}{105}$$

where Z = charge on the ligand, e=electronic charge, a = metal-ligand distance and r = distance of the electron from the nucleus. Electronic spectra of complexes give the experimental value of 10Dq.

The symbol t denotes the **triplet orbital degeneracy** while e symbolise **two fold degeneracy.** The terms $t_{2g}$ and $e_g$ has the origin in group theory–the subscript 2 indicates particular symmetry operations. The subscript g stands for **gerade** or **even** (Sec. 3.4.4). In octahedral crystal field the metal is at the centre of inversion so g term is applied for centrosymmetric octahedral field while in tetrahedral field g term is omitted because the tetrahedron has no centre of inversion.

Now a question arises how the $t_{2g}$ electrons are stabilised by the amount $-\frac{2}{5}\Delta_0$ or –4Dq while $e_g$ electrons are stabilised by $+\frac{3}{5}\Delta_0$ or +6Dq with respect to the average energy of five degenerate d-orbitals in a spherical crystal field? When an electron in a complex resides on $t_{2g}$ orbital which experiences less repulsive force by the ligands in an octahedral complex than in the hypothetical spherical complex. The ligands are able to move closer to the metal ion along the d-electron free x, y and z axis, thereby donating their electrons more effectively to the metal ion and produce a more stable complex. The extra energy evolved as a result of this non-spherical interaction is called the **crystal field stabilisation energy (CFSE)**. Since this energy term arises due to the interaction of ligand d-orbital interaction so this is also called **ligand field stabilisation energy (LFSE).** The **CFSE is also defined as the amount of energy gained by a $d^n$ system due to the preferential filling of electrons in $t_{2g}$ and $e_g$ orbitals.** Therefore the metal ion in spherical field has zero CFSE. The stabilisations resulting from the three $t_{2g}$ orbitals must just balance the repulsions from the two $e_g$ electrons. Naturally, **the repulsion per $e_g$ electron is equal to $\frac{3}{2}$ times the stabilisaiton per $t_{2g}$ electron.** This can also be stated in other way. Let each $e_g$ electron will go up by an amount of energy α and each $t_{2g}$ electron will go down by an amount of energy β and if E is energy per electron in a hypothetical spherical field, then according to barycentre rule we can say,

$$10E = 4(E+\alpha) + 6(E-\beta) \quad \text{or} \quad 2\alpha = 3\beta$$

again, $\alpha+\beta = \Delta_0$ or 10Dq, so that $\alpha = \frac{3}{5}\Delta_0$ and $\beta = \frac{2}{5}\Delta_0$.

Therefore taking the spherical field energy as zero, each $e_g$ electron will be destabilised by an amount $+\frac{3}{5}\Delta_0$ or + 6Dq or $+0.6\Delta_0$ and each $t_{2g}$ electron will be stabilised by an amount of $-\frac{2}{5}\Delta_0$ or –4Dq or $-0.4\Delta_0$.

### 7.14.2(B) Pairing energy : High and low spin electron configuration

CFSE primarily depends on the extent of crystal field splitting parameter $\Delta_0$ and the number of electrons occupying $t_{2g}$ and $e_g$ orbitals. However it, also depends on whether pairing of electron(s) takes place or not. The CFSE is the sum of the energies of all the metal d-electrons of a particular $d^n$ ion in an octahedral field compared to that of the hypothetical spherical field. Let a particular $d^n$ ion has the configuration $t_{2g}^x\ e_g^y$ where $x$ and y are the electrons occupied by $t_{2g}$ and $e_g$ orbitals, so that,

$$\text{energy, } E = x\left(-\frac{2}{5}\Delta_o\right) + y\left(+\frac{3}{5}\Delta_o\right) + ZP \qquad \text{...(7.41)}$$

where P denotes the pairing energy and Z is the number of electrons paired in the particular electron configuration.

For $d^1$, $d^2$, $d^3$, $d^8$, $d^9$ and $d^{10}$ systems, there will be no scope for alternative arrangements of electrons in $t_{2g}$ and $e_g$ orbitals (Fig. 7.6). So these systems have fixed E values.

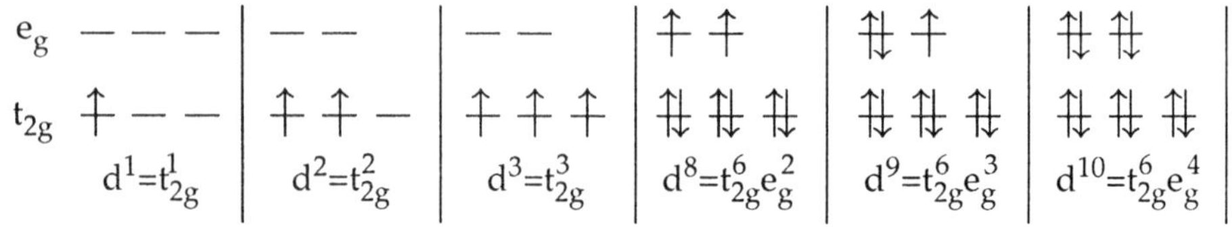


***Figure 7.6** : Configurations of $d^1$ to $d^3$ and $d^8$ to $d^{10}$ systems in octahedral field.*

For $d^4$, $d^5$, $d^6$ and $d^7$ systems alternative arrangements of the occupation of electrons in $t_{2g}$ and $e_g$ orbitals occur (Fig. 7.7). For a $d^4$ metal ion e.g. $Cr^{2+}$, the fourth electron may be paired either to the three $t_{2g}$ orbitals or may be promoted to the higher $e_g$ orbitals depending upon the strength of the crystal field. There will operate two opposing forces in these configurations ($d^4$, $d^5$, $d^6$ and $d^7$). The crystal field splitting $\Delta_o$ or 10Dq will force to pair as many electrons into the lower $t_{2g}$ orbitals while pairing energy (P) will counter spin pairing. In fact **when $\Delta_o$ > P, the electron(s)**

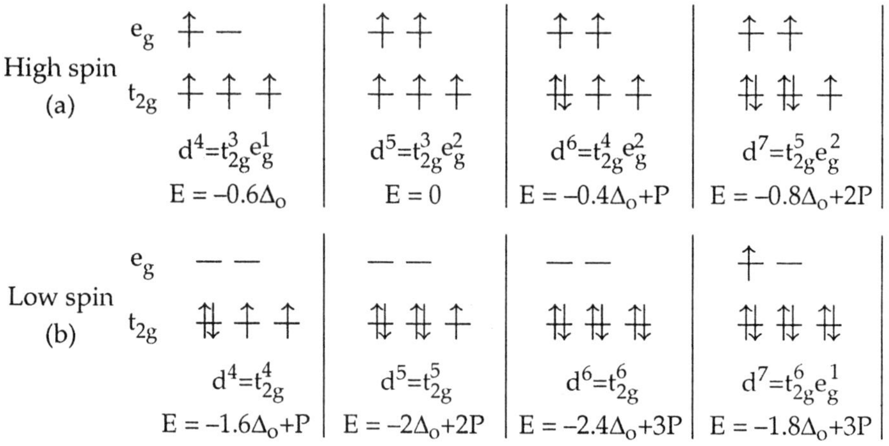


***Figure 7.7** : Energy level diagram of $d^4$, $d^5$, $d^6$ and $d^7$ metal ions (a) high spin configuration i.e. weak field ligand (b) low spin configuration i.e. strong field ligand.*

**is force to pair to the lower $t_{2g}$ orbitals and we get low spin complex i.e. strong field ligand will produce low spin complex. Complexes in which the metal d-electrons would pair than fight the strong electric field of donor atoms are known as strong field complexes i.e. low spin complexes** e.g.$[Cr(CN)_6]^{4-}$($Cr^{2+}$,$d^4$ system). On the other hand when **$\Delta_o$ <P, the electron(s)**

**is free to move from lower $t_{2g}$ to upper $e_g$ orbitals and we get high spin complexes i.e. weak field ligands formed high spin complex. Complexes in which the metal d-electrons can more readily overcome the elctric field of the ligands are known as weak field complexes i.e. high spin complexes** e.g. $[Cr(H_2O)_6]^{2+}$. It is to be noted that if we put $\Delta_o = P$, the energy for both the spin states are equal. We may generalise that in crystal field terminology the low spin and the high spin complexes correspond, respectively to the inner orbital and outer orbital complexes in valence bond theory.

It is evident that in an octahedral field $d^4$, $d^5$ $d^6$ and $d^7$ metal complexes may exist either as high spin or low spin depending upon the strength of the crystal field. Hund's rule states that pairing of electrons is an unfavourable process and so at the begining (spherical field) the high spin state is of lower energy and the hypothetical low spin state is of higher energy. But as the crystal field strength begins to increase, it is observed that the energies of both the spin states begin to fall—that of the low spin state rather more steeply than that of the high spin state. At a particular crystal field strength when $\Delta_o = P$, the energies of both the spin states become equal. When $\Delta_o = P$, it is called **critical crystal field strength** above which all ligands will produce low spin state and below which the ligands will generate high spin state. This particular critical crystal field is sometimes called **cross-over region** (Fig. 7.8). At either side of the cross over region, a thermal equilibrium will be established where the difference of energy of the high and low spin states is of the order of thermal energy kT, the population of each spin states may be determined by Blotzman's distribution law. The spin state equilibrium will be shifted either to the high or low spin state by changing the temperature. At the **spin cross-over** region change of temperature may results unusual magnetic behaviour. For example tris(dialkyl dithiocarbamato) iron (III) 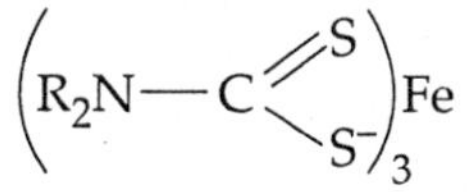 in which the magnetic moment dropped from 5.7–5.8 B.M. at 300K, to 4.7–5.0 B.M. at 150K, to 3.6–4.0 B.M. at 78K as the proportion of molecules in the high and low spin states changed with temperature. Within a narrow range of crystal field strength, represented by A–A′ in the figure (Fig. 7.8) where $\Delta_o = P$ and the actual values do not differ by more than 2000 $cm^{-1}$, there might be the possibility for an equilibrium of the type low spin state $\rightleftharpoons$ high spin state, a phenomena which is called **spin state isomerism**. Due to the varying degree of tetragonal distortion maximum cases of spin state isomerism is expected for $3d^8$ Ni(II) complexes. Co(II) and Fe(II) complexes e.g. $[Co(terpy)_2]Cl_2.3.5H_2O$, $[Fe(o\text{-}phen)_2](NCS)_2$ also exhibit this phenomena.

*Spin state isomerism*

From Fig. 7.8 one point may naturally be noted that why the energy of the low spin state falls more rapidly compared to that of the high spin state. Let us take the case of a specific case of $3d^6$. For the high spin state, $\Delta_o < P$ i.e. $P > \Delta_o$. So we can write $P = \Delta_o + x\Delta_o$. Therefore E for $t_{2g}^4e_g^2 = -\frac{2}{5}\Delta_o + P = -\frac{2}{5}\Delta_o + \Delta_o + x\Delta_o = \frac{3}{5}\Delta_o + x\Delta_o$, so that, $-\left(\frac{dE}{d\Delta_0}\right) = -\left(\frac{3}{5} + x\right)$. For low spin state ($t_{2g}^6$), $E = -\frac{12}{5}\Delta_o + 3P = -\frac{12}{5}\Delta_o + 3\Delta_o + 3x\Delta_o = \frac{3}{5}\Delta_o + 3x\Delta_o$ so that, $-\left(\frac{dE}{d\Delta_0}\right) = -\left(\frac{3}{5} + 3x\right)$, which indicates the energy of the low spin state falls rapidly compared to that of high spin state.

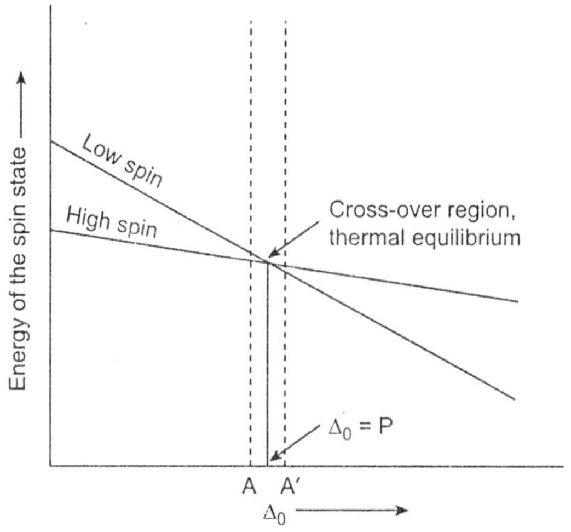


*Figure 7.8 : Spin-state cross-over region.*

### 7.14.2(C) Tetrahedral complexes

In a tetrahedral complex the splitting pattern of d-orbitals are just reverse to that of octahedral complexes. A tetrahedral arrnagement can be considered by taking alternate corners of a cube (Fig. 7.9). In a tetrahedral complex none of the d-orbitals are directed along the metal-ligand bond axes but the orientation of the $d_{xy}$, $d_{yz}$ and $d_{xz}$ orbitals are such that their lobes are much closer to the metal-ligand bond axes than the $d_{x^2-y^2}$ and $d_{z^2}$ orbitals. So the $d_{xy}$, $d_{yz}$ and $d_{xz}$ orbitals ($t_2$ set, here g subscript will be dropped) will experience greater repulsive force than the $d_{x^2-y^2}$ and $d_{z^2}$ orbitals (e set). As a result maintaing the barycentre rule the $t_2$ set will go up and e set will go down (Fig. 7.10). As the four ligands come closer to the metal, first a spherical field is operated (Fig. 7.10). Here the energy of the metal d-orbitals is raised but the degeneracy will not be lifted. Finally when the ligands are positioned at the four corners of the tetrahedron, the splitting occurs. The difference in energy between $t_2$ and e sets of orbitals is called **crystal field splitting for tetrahedral complexes** and denoted by $\Delta_t$.

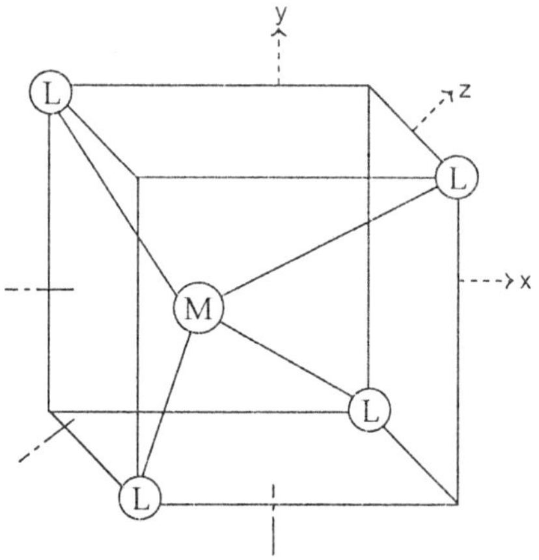


*Figure 7.9 : A tetrahedral arrangements of four ligands in a cube.*

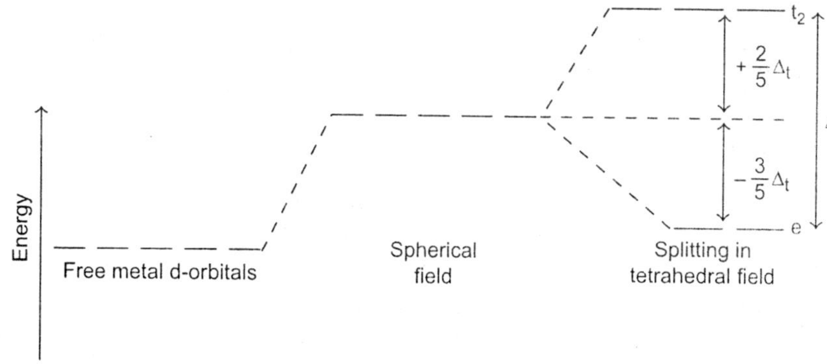


***Figure 7.10** : Crystal field splitting (qualitative) in tetrahedral geometry.*

The magnitude of crystal field splitting $\Delta_t$ is smaller than that of octahedral field ($\Delta_o$). This is due to the following two reasons :

(a) There are only four ligands in tetrahedral complex compared to that of six ligands in octahedral one so that four ligands will create more weaker crystal field compared to that of six ligands. The crystal field splitting will therefore $^4/_6$ times i.e. $^2/_3$ times less in tetrahedral geometry.

(b) Unlike the octahedral geometry, none of the metal-ligand bonds are along the coordinate axes x, y and z. This reduces the crystal field aplitting $\Delta_t$ further by $^2/_3$ times.

Therefore the crystal field splitting parameter $\Delta_t$ is $\frac{2}{3} \times \frac{2}{3}$ or $\frac{4}{9}$ times that of octahedral crystal field i.e.

$$\Delta_t = \frac{4}{9}\Delta_o \qquad \ldots(7.42)$$

This indicates that tetrahedral field always weaker than octahedral field i.e. $\Delta_t < \Delta_o$ so that all tetrahedral complexes are weak enough to initiate spin pairing. Therefore **all tetrahedral complexes exist as high spin**. There is no longer any need of the weak-field/strong-field distinction in case of tetrahedral geometry. Moreover the CFSE for tetrahedral complexes is less than that of octahedral complexes so that octahedral complexes are more stable and more numerous compared to that of tetrahedral analogs. Equation (7.42) indicates tetrahedral splitting is less than half the octahedral splitting e.g. $[Co(NH_3)_6]^{2+}$ and $[Co(NCS)_4]^{2-}$ have $\Delta_o$ and $\Delta_t$ values 10.1 and 4.9 kK respectively (1 kK = 1000 $cm^{-1}$) (kK = kilo Keyser).

### 7.14.2(D) Tetragonal distortion : Splitting of d-orbitals in square planar, trigonal-bipyramidal and square pyramidal geometry

A square planar geometry may be considered by withdrawing two ligands from octahedral geometry. Suppose the two ligands are withdrawn along the z axis from the field of the metal ion in an octahedral complex of the type $ML_6$. When the distance of these two ligands increases from metal ion as compared to the other four, the geometry of the complex becomes distorted and we get an **irregular geometry**. This formation of irregular geometry is called **tetragonal distortion** and can be represented as an intermediate stage between octahedral and square planar geometries (Fig. 7.11). Due to the tetragonal distortion the degeneracy of the $e_g$ and $t_{2g}$ orbitals will be lifted. The d-orbitals having orientation in z-direction ($d_{z^2}$ in $e_g$ set; $d_{xz}$ and $d_{yz}$ in $t_{2g}$ set) become more stable, since the metal-ligand repulsion gradually diminished along

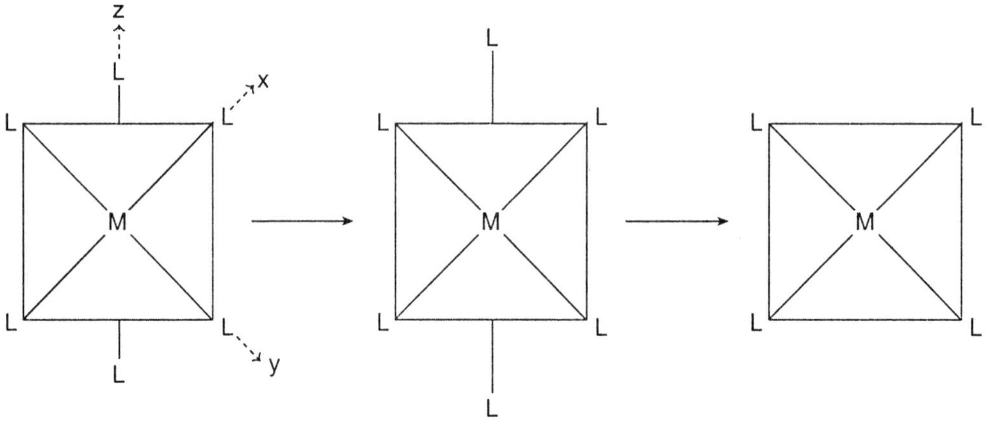


*Figure 7.11 : Representation of octahedral to square planar geometry through tetragonal distortion.*

z direction. A square plane be represented by two axes namely x and y components so that metal-ligand repulsion progressively increased along the orbitals which have x and y components ($d_{x^2-y^2}$ in $e_g$ set and $d_{xy}$ in $t_{2g}$ set). These orbitals will be destabilised. In this situation due to the tetragonal distortion we get **distorted octahedral geometry**. The $d_{z^2}$ orbital becoming more stable than $d_{x^2-y^2}$, simultaneously $d_{xy}$ becoming less stable than degenerate $d_{yz}$ and $d_{xz}$ orbitals because of spatial distribution [Fig. 7.12(d)]. This also represent **square-pyramidal** geometry.

The limiting case of tetragonal distortion is square planar complex, where the two ligands on the z axis is completely withdrawn and the complex is tetracoordinated ($ML_4$). In square planar geometry only the $d_{x^2-y^2}$ orbital will experience greater repulsive force by the four ligands and so this orbital is the most destabilised. In an ideal square planar geometry no ligand exist in the z direction, so that $d_{z^2}$ orbital will be expected mostly stabilised and will be in the lowest position of the energy level diagram. But the extent of splitting in a square planar geometry would depend on the nature of the metal ions and the ligands. Semiquantitative calculations with parameters for square planar complxes of $Cu^{2+}$, $Ni^{2+}$ and $Co^{2+}$ shows the energy level diagram [Fig. 7.12(e)] in which the $d_{z^2}$ orbital has dropped far below the $d_{xy}$ orbital but $d_{z^2}$ orbital has slight higher energy compared to that of $d_{xz}$, $d_{yz}$ pair.

In the trigonal bipyramidal (TBP) geometry [Fig. 7.12(f)] the $d_{z^2}$ will have the highest energy as it faces two ligands directly. The $d_{xz}$ and $d_{yz}$ orbitals will be the stablest as there will be least interaction with the ligands. The other degenerate doublet ($d_{x^2-y^2}$, $d_{xy}$) will experience some interaction with the ligands and so they have higher energy compared to that of the lowest doublet ($d_{xz}$, $d_{yz}$). In TBP geometry the crystal field splitting parameter $\Delta_{tb} \approx 0.98\ \Delta_0$. The energy

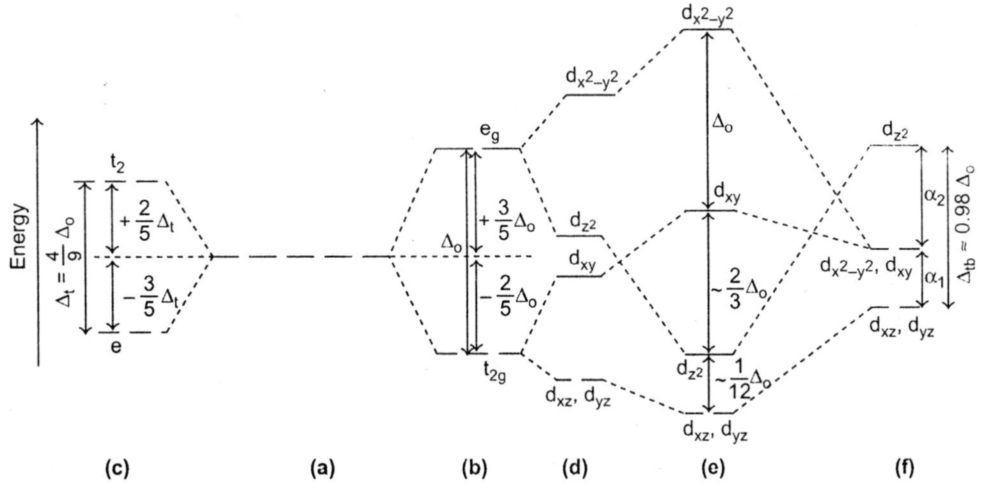


***Figure 7.12 :** Qualitative energy level diagram for the splitting of d-orbitals in different geometries : (a) Hypothetical spherical field, (b) octahedral geometry, (c) tetrahedral symmetry, (d) tetragonal distortion, square pyramidal geometry, (e) square planar stereochemistry and (f) trigonal bipyramidal geometry.*

difference ($\alpha_1$) between the lowest doublet ($d_{xz}$, $d_{yz}$) and the middle doublet ($d_{x^2-y^2}$, $d_{xy}$) is much lower compared to the energy difference ($\alpha_2$) between $d_{z^2}$ and the middle doublet ($d_{x^2-y^2}$, $d_{xy}$). $\alpha_2$ may be either less or greater than pairing energy (P) but $\alpha_1$ always less than P. Therefore in this TBP geometry $d^1$, $d^2$, $d^3$ and $d^4$ metals always give high spin complex while $d^5$, $d^6$, $d^7$ and $d^8$ metal ions may form either high spin or low spin configuration depending on the nature of the ligands. Experimentally $3d^8$ nickel (II) complexes in TBP geometry exist as high spin and low spin states. Anomolous magnetic properties may be expected when $\alpha_2 \approx P$ for $d^5$, $d^6$, $d^7$ and $d^8$ systems with suitable ligands.

Finally the above energy level diagrams (Fig. 7.12) has been established from spectroscopic data. As $E = h\nu = hc/\lambda$ where h is Planck's constant ($6.63\times10^{-27}$ erg.sec), $\nu$ is the frquency, c is the velocity of light ($3\times10^{10}$ cm.sec.$^{-1}$), $\lambda$ is the wavelenth of the absorption band, it is generally expressed the value of $\Delta_o$ or 10Dq in terms of the value of $\frac{1}{\lambda} = \bar{\nu}$ where $\bar{\nu}$ is the wave number expressed in cm$^{-1}$. Hence, E (cal.per gram mol) = $\frac{hcN}{4.2\times10^7}\times\frac{1}{\lambda}$ (1 erg = $10^{-7}$ Joule, 4.2 Joule = 1 cal, therefore 1 erg = $\frac{1}{4.2\times10^7}$ cal)

$$= \frac{6.63\times10^{-27}\,\text{cal.sec.}\times3\times10^{10}\,\text{cm.sec}^{-1}\times6.023\times10^{23}}{4.2\times10^7}\times\frac{1}{\lambda}$$

$= 2.85$ cal.cm.$\frac{1}{\lambda}$ = 2.85 cal.cm $\bar{\nu}$ . Here N is the Avogadro's number = $6.023\times10^{23}$. Thus multiplying the value of wavenumber in cm$^{-1}$ with 2.85 we get the value of energy in cal.mol$^{-1}$.

Thus 1000 cm$^{-1}$ (i.e. 1 kK) is 2.85 kcalmol$^{-1}$. The wavelength of light in nm = $10^7$ divided by the frequency in cm$^{-1}$ e.g. 493 nm = $\frac{10^7}{493}$ = 20284 cm$^{-1}$ = 243 kJ mol$^{-1}$.

### 7.14.2(E) Splitting pattern of 4f orbitals in Octahedral and Tetrahedral Crystal Field*

In order to have an idea of the splitting pattern of 4f orbitals in octahedral and tetrahedral geometry we have to consider the cubic set of f orbitals instead of general set of seven 4f orbitals (Fig. 1.19). The cubic set of f orbitals are designated as : $f_{z^3}$, $f_{x^3}$, $f_{y^3}$ having point group ($O_h$, $T_{2u}$); $f_{z(x^2-y^2)}$, $f_{x(y^2-z^2)}$, $f_{y(z^2-x^2)}$ having point group ($O_h$, $T_{1u}$); $f_{xyz}$ having point group ($O_h$, $A_{2u}$). These cubic symmetry functions are useful in order to consider the splitting pattern of 4f orbitals in octahedral and tetrahedral crystal field.

In an **octahedral crystal field** the seven cubic set of 4f orbitals split into two groups of triply degenerate sets and one degenerate orbital (Fig. 7.12(A)). In this octahedral geometry the six ligands are at the centers of the faces of the cube and it is quite plausable from the Fig. 7.12(A) that the $f_{x^3}$, $f_{y^3}$ and $f_{z^3}$ orbitals ($t_{2u}$) have lobes pointing towards the ligands directly and hence they will be maximum destabilised and have highest energy. The eight lobes of $f_{xyz}$ orbital ($a_{2u}$) pointed directly at the corners of the farthest distance from the ligands and hence this will be stabilised to maximum extent and thereby have lowest energy. The equivalent orbitals $f_{z(x^2-y^2)}$, $f_{y(z^2-x^2)}$ and $f_{x(z^2-y^2)}$($t_{1u}$) have lobes midway between the edges of the cube and thus these

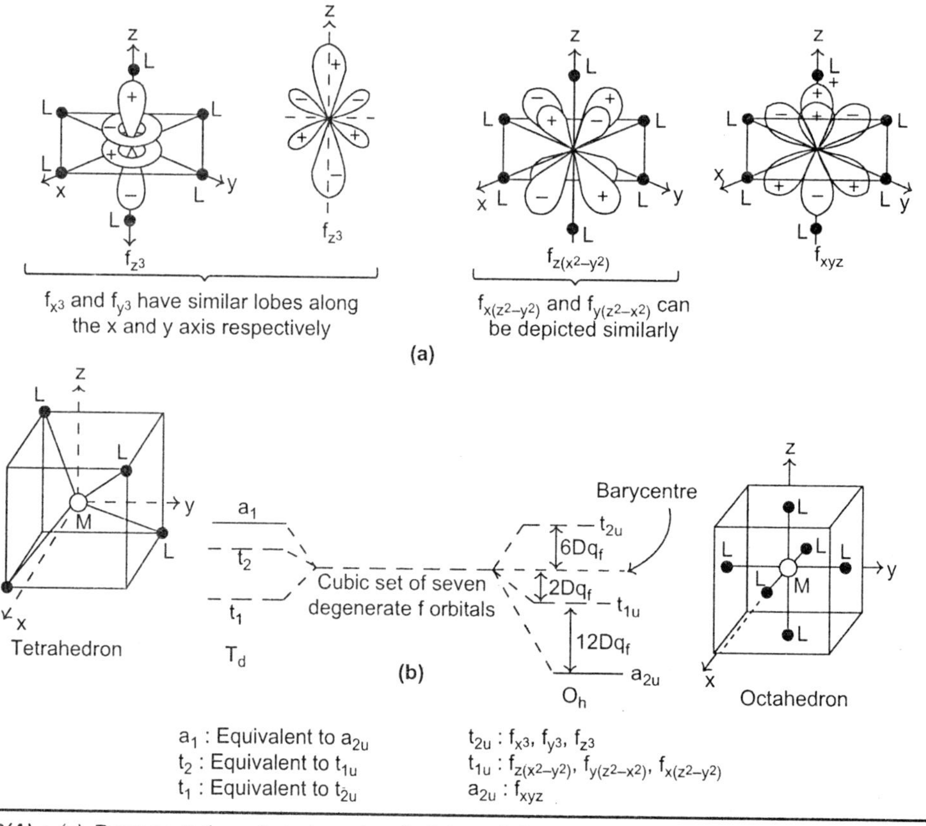


*Figure 7.12(A) : (a) Representation of electron density lobes of the seven f orbitals relative to the ligand positions in an octahedral ligand field. (b) Splitting pattern of seven f orbitals in tetrahedral and octahedral crystal field.*

* For more details about this the readers may go through the authors' book 'Comprehensive Coordination Chemistry'.

orbitals have intermediate stability between $t_{2u}$ and $a_{2u}$ orbitals (Fig. 7.12(A)). Quantitative perturbation theory suggested that the $t_{2u}$ orbitals are destabilised by the amount $6Dq_f$ while the $t_{1u}$ and $a_{2u}$ orbitals are stabilised by the amount $2Dq_f$ and $12Dq_f$ respectively (Fig. 7.12(A)). Here the parameter $Dq_f$ is not the same as represented by equation $D = \frac{35Ze}{4a^5}$, $q = \frac{2er^4}{105}$ (Sec. 7.14.2(A)) for splitting of d-orbitals in case of transition metals. Here also the barycentre rule is followed whereby the destabilisation of energy of $t_{2u}$ orbitals is balanced by the stabilisation of energy of $t_{1u}$ and $a_{2u}$ orbitals (Fig. 7.12(A)). This ordering of 4f orbitals can also be justified from the character tables of $O_h$ point group. The $f_{x^3}$, $f_{y^3}$ and $f_{z^3}$ consititute a $T_{2u}$ set, the $f_{z(x^2-y^2)}$, $f_{y(z^2-x^2)}$ and $f_{x(z^2-y^2)}$ form $T_{1u}$ set while the $f_{xyz}$ orbital has $A_{2u}$ symmetry.

The splitting pattern of cubic set of 4f orbitals in tetrahedral field is also depicted in Fig. 7.12(A). It is to be noted that the splitting pattern in a tetrahedral field is just reversed of that of the octahedral field and the same symmetry consideration can also be justified from the character table for $T_d$ group.

### 7.14.3 Crystal field stabilisation energy

This is an important term used in modern coordination chemistry. Owing to the perturbation of d-orbitals in a complex, the extra stability gained by a $d^n$ ion due to preferential filling of lower lying d-orbitals is called **crystal field stabilisation energy** or **ligand field stabilisation energy**, abbreviated as **CFSE** or **LFSE**. This can be calculated mainly for tetrahedral and octahedral complexes of different metal ions in terms of Dq values. Equation (7.41) denotes the calculation of energy in octahedral geometry where P indicates pairing energy. In the calculation of CFSE the pairing energy may be ignored but has to be considered to calculate the total energy (Fig. 7.7). Omission of P term in the expression for CFSE is justified from the fact that crystal field splitting tends to force as many electrons in the lower energy $t_{2g}$ orbital while P tends to decouple the spin pairing and thus tends to put electrons in the upper $e_g$ levels. Pairing of electrons is an unfavourable process. Pairing energy is the energy of repulsion between paired electrons in the same orbital and represents the price that we must pay in energy if we are to violate Hund's rule of maximum multiplicity. Thus P opposes the stabilisation energy due to perturbation of d-orbitals in crystal field. Appreciating this it is better to drop the term P in the

**Table 7.11 : CFSE/LFSE for tetrahedral and octahedral geometries**

| $d^n$ system | Octahedral | | Tetrahedral ($\Delta_t = {}^4/_9\Delta_0 = 4.44$ Dq) |
|---|---|---|---|
| | High spin | Low spin | |
| $d^1$ | $-0.4\Delta_0/-4$Dq | $-0.4\Delta_0/-4$Dq | $-0.6\Delta_t = -0.266\Delta_0 = -2.66$ Dq |
| $d^2$ | $-0.8\Delta_0/-8$Dq | $-0.8\Delta_0/-8$Dq | $-1.2\Delta_t = -0.532\Delta_0 = -5.32$ Dq |
| $d^3$ | $-1.2\Delta_0/-12$Dq | $-1.2\Delta_0/-12$Dq | $-0.8\Delta_t = -0.355\Delta_0 = -3.55$ Dq |
| $d^4$ | $-0.6\Delta_0/-6$Dq | $-1.6\Delta_0/-16$Dq | $-0.4\Delta_t = -0.177\Delta_0 = -1.77$ Dq |
| $d^5$ | 0 | $-2\Delta_0/-20$Dq | 0 |
| $d^6$ | $-0.4\Delta_0/-4$Dq | $-2.4\Delta_0/-24$Dq | $-0.6\Delta_t = -0.266\Delta_0 = -2.66$ Dq |

| $d^n$ system | Octahedral | | Tetrahedral ($\Delta_t = {}^4/_9\Delta_0$ = 4.44 Dq) |
|---|---|---|---|
| | High spin | Low spin | |
| $d^7$ | $-0.8\Delta_0/-8Dq$ | $-1.8\Delta_0/-18Dq$ | $-1.2\Delta_t = -0.532\Delta_0 = -5.32$ Dq |
| $d^8$ | $-1.2\Delta_0/-12Dq$ | $-1.2\Delta_0/-12Dq$ | $-0.8\Delta_t = -0.355\Delta_0 = -3.55$ Dq |
| $d^9$ | $-0.6\Delta_0/-6Dq$ | $-0.6\Delta_0/-6Dq$ | $-0.4\Delta_t = -0.177\Delta_0 = -1.77$ Dq |
| $d^{10}$ | 0 | 0 | 0 |

expression of CFSE/LFSE in all the geometries of metal complexes (Table 7.11). The single electron energies of the different d-orbitals of a central metal ion for different geometries have been calculated assuming the same overall charge density and metal-ligand distance. These values are tabulated in terms of $\Delta_0$ value for octahedral geometry (Table 7.12). These values are useful for calculating CFSE/LFSE for different types of complexes. It may be noted from Table 7.12 and Fig. 7.12 that square planar complexes have high CFSE so that square planar configuration is more stable than octahedral configuration. It should be memorised that :

**Table 7.12 : Energy of an electron ($\Delta_0$/Dq) in different d-orbitals in different crystal fields**

| Coordination number | Geometry | $d_{z^2}$ | $d_{x^2-y^2}$ | $d_{xy}$ | $d_{xz}/d_{yz}$* |
|---|---|---|---|---|---|
| 2 | Linear (along z axis) | 1.028/10.28 | –0.628/–6.28 | –0.628/–6.28 | 0.114/1.14 |
| 3 | Trigonal planar (in the xy plane) | –0.321/–3.21 | 0.546/5.46 | 0.546/5.46 | –0.386/–3.86 |
| 4 | Tetrahedral | –0.267/–2.67 | –0.267/–2.67 | 0.178/1.78 | 0.178/1.78 |
| 4 | Square planar (in the xy plane) | –0.428/–4.28 | 1.228/12.28 | 0.228/2.28 | –0.514/–5.14 |
| 5 | Trigonal bipyramid | 0.707/7.07 | –0.082/–0.82 | –0.082/–0.82 | –0.272/–2.72 |
| 5 | Square pyramid | 0.086/0.86 | 0.914/9.14 | –0.086/–0.86 | –0.457/–4.57 |
| 6 | Octahedral | 0.600/6.00 | 0.600/6.00 | –0.400/–4.00 | –0.400/–4.00 |
| 7 | Pentagonal bipyramid | 0.493/4.93 | 0.282/2.82 | 0.282/2.82 | –0.528/–5.28 |
| 8 | Cube | –0.534/–5.34 | –0.534/–5.34 | 0.356/3.56 | 0.356/3.56 |
| 9 | Tricapped trigonal prism | –0.225/–2.25 | –0.038/–0.38 | –0.038/–0.38 | 0.151/1.51 |
| 12 | Icosahedral | 0 | 0 | 0 | 0 |

* *$d_{xz}/d_{yz}$ means the two orbitals have the same $\Delta_0$/Dq values.*

(a) The 10Dq or $\Delta_0$ value is large for square planar geometry than for octahedral one.

(b) Ligand-ligand repulsion is more in octahedral geometry compared to that of square planar configuration.

(c) Total binding energy for six ligands are more compared to that of four ligands.

CFSE/LFSE value has an importance in assigning the stereochemistry and stability of complexes.

### 7.14.4 Factors influencing the crystal field splitting

The following factors influence the variation of $\Delta_0$/10Dq values.

1. **Nature of the ligands and metal ions :** As mentioned earlier [Sec. 7.14.2(B)] the high and low spin states arises due to the character of ligands. In a particular geometry the magnitude of 10Dq or $\Delta_0$ depends on the following sequence of ligands irrespective of the nature of cations :

$I^- < Br^- < S^{2-} < SCN^- < Cl^- < F^- < OH^- < C_2O_4^{2-} < H_2O < NCS^- < EDTA \sim gly$
weak field ligands

$< py < NH_3 < en < bpy < ophen < NO_2^- < CN^- < CO$
strong field ligands

This is the well-known **spectrochemical series** or Fajans-Tsuchida series which is obtained from spectroscopic data. The C-bonded $CN^-$ has the strongest crystal field strength and its Dq value is about 1.7 times greater than that of water but N-bonded $CN^-$ has the crystal field strength intermediate between that of $H_2O$ and $NH_3$. The high crystal field strength of $CN^-$ suggests that although $FeF_6^{3-}$, $CoF_6^{3-}$ are high spin, $Fe(CN)_6^{3-}$ and $Co(CN)_6^{3-}$ are low spin.

Similar with the spectrochemical series of ligands a **spectrochemical series of metals** can also be established. The approximate $\Delta_0$/10Dq can be estimated from the two independent parameters of metal and ligand i.e.

$$10Dq = f \times g$$

where f denotes the strength of the ligand relative to water (f=1). The value of f generally in the range 0.7 to 1.7 and g values lie in the range 8.0 to 36.0 kK. All these results the variation of 10Dq values of known octahedral complexes are in the range ~6000 to ~45000 $cm^{-1}$. As f and g values are valid for only octahedral geometry, so that significant variation of 10Dq values may be observed (from the value of the product fg) for other geometry of the complex. On the basis of g values metal ions are arranged which is known as **spectrochemical series of metals** as :

$Mn^{2+} < Ni^{2+} < Co^{2+} < Fe^{2+} < V^{2+} < Fe^{3+} < Cr^{3+} \sim V^{3+} < Co^{3+} < Mn^{4+} < Mo^{3+} < Rh^{3+} \sim Ru^{3+} < Pd^{4+} < Ir^{3+} < Re^{4+} < Pt^{4+}$

2. **Oxidation state of the metal ion :** Having the same ligand and same stereochemistry, the metal ion with higher oxidation state causes larger crystal field splitting than is done by the ion with lower oxidation state e.g.

$[Co(H_2O)_6]^{3+}$ $\Delta_0 = 18600\ cm^{-1}$ $[Co(H_2O)_6]^{2+}$ $\Delta_0 = 9300\ cm^{-1}$

$[Fe(H_2O)_6]^{3+}$ $\Delta_0 = 13700\ cm^{-1}$ $[Fe(H_2O)_6]^{2+}$ $\Delta_0 = 10400\ cm^{-1}$

$[Mn(H_2O)_6]^{3+}$ $\Delta_0 = 21.00$ kK $[Mn(H_2O)_6]^{2+}$ $\Delta_0 = 8.5$ kK

With increase in oxidation state the metal-ligand bond length decreases due to the closer approach of ligands to metal and so $\sigma$ bonding effect increases from ligand to metal. All these cause greater splitting of d-orbitals to a larger extent thereby larger values of 10Dq is observed with increase of oxidation state of metals.

3. **Size of d-orbitals :** Given the same ligand, the same geometry, the same oxidation state $\Delta_0$ values increases about 30-50% on going from first to second transition series and again by the same amount on going from second to third transition series i.e.

$$5d > 4d > 3d$$

This can be explained on the basis that 4d and 5d orbitals of second and third trasition series respectively are bigger in size and spread out in space than the 3d orbital of first transition series. As a result the metal d-electrons in 5d orbital will experience smaller interelectronic repulsion compared to that of 4d electrons which in turn also experience less interelectronic repulsion than that of 3d electrons with that of the ligand orbitals. Since the interelectronic repulsion reduces as also nuclear charge increases from first to third transition series, so that there will be a closer approach of metal orbitals to ligand orbitals. This leads to highest perturbation of d-orbitals in case of third transition series (5d orbital) and will be less in case of 4d and also will be still less in case of 3d orbital. Thus the third transition series elements will register highest 10Dq value and the first one will have a lowest value e.g.

| | |
|---|---|
| $[Co(NH_3)_6]^{3+}$ | 10Dq = 23000 $cm^{-1}$ |
| $[Rh(NH_3)_6]^{3+}$ | 10Dq = 34000 $cm^{-1}$ |
| $[Ir(NH_3)_6]^{3+}$ | 10Dq = 41000 $cm^{-1}$ |
| $[Co(en)_3]^{3+}$ | 10Dq = 24.0 kK |
| $[Rh(en)_3]^{3+}$ | 10Dq = 35.0 kK |
| $[Ir(en)_3]^{3+}$ | 10Dq = 41.0 kK |

This highest value of $\Delta_0$/10Dq for the later transition series leads to spin pairing. Thus $[NiCl_4]^{2-}$ is high spin while $[PdCl_4]^{2-}$ and $[PtCl_4]^{2-}$ is low spin.

4. **Geometry of the complex :** Given the same ligand and same metal ion the $\Delta_0$/10Dq varies from geometry to geometry, i.e. square planar > octahedral > tetrahedral.

Ligand field splitting energy of tetrahedral complexes is nearly half the value for octahedral complexes ($\Delta_t \approx \frac{4}{9}\Delta_0$). In other words the splitting energy for tetrahedral complexes, in general, is small as compared to the pairing energy P. The tetrahedral complexes are therefore mostly high spin.

**Problem 7.42** Explain why $K_2[NiCl_4]$ is paramagnetic, whereas $K_2[PtCl_4]$ is diamagnetic although both $Ni^{II}$ and $Pt^{II}$ are $d^8$ ions. **[C.U. 2009]**

**Ans. :** According to Sec. 7.14.4 the $K_2[NiCl_4]$ is a tetrahedral complex where nickel is a member of first transition series and the $3d^8$ configuration can be represented in tetrahedral geometry as :

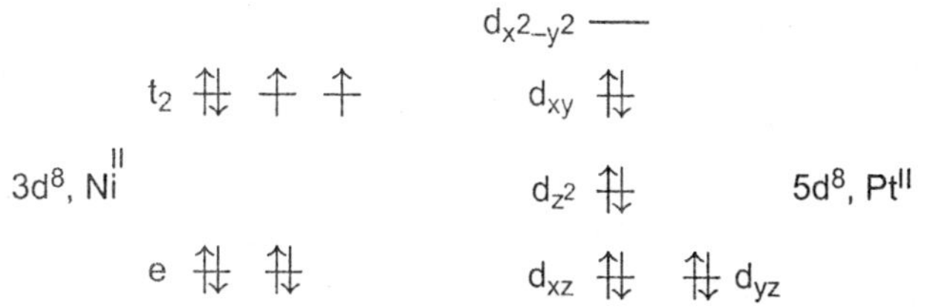


where-there are two unpaired electrons and so the complex is paramagnetic. In case of $K_2[PtCl_4]$, the platinium is a member of third transition series and the complex has a square planar geometry where all the eight electrons are paired in the lower energy orbitrals i.e. $d^2_{xz}$, $d^2_{yz}$, $d^2_{z^2}$ and $d^2_{xy}$. As $Cl^-$ is a weak field ligand and as the energy gap between $d_{xy}$ and $d_{x^2-y^2}$ is large enough, we get the diamagnetic spin paired configuration for $K_2[PtCl_4]$ as depicted above.

---

**Problem 7.43** The position of CO in the spectrochemical series is higher compared to $CN^-$—Explain. **[C.U. 2008]**

**Ans. :** The negative charge on $CN^-$ enhances its $\sigma$ donor ability but weakens $\pi$-acceptor ability. The $\sigma$ donor ability decreases in the order $CN^- > NO^+ \sim CO$ while $\pi$-acceptor effectiveness increases in the order $NO^+ > CO > CN^-$.

A correlation between the position of ligands in the spectrochemical series and the $\pi$-bonding abilities can be made. The MOT of CO indicates that the LUMO is $\pi^*$ antibonding orbital which is lying vacant. The ligands which can accept electrons from the metal ion into their $\pi^*$ orbitals have the position higher in the spectrochemical series. These ligands have strong $M \rightarrow L$ $\pi$ bonding ability. As the $\pi$ bonding ability of CO is greater than that of $CN^-$, so CO will have position higher in spectrochemical series. Such ligands are called **$\pi$-acceptor ligands** and they cause larger splitting so that they register larger $\Delta_0$/10Dq values. The ligands which donate electrons to the metal ion (i.e. the ligands having strong $L \rightarrow M$ $\pi$ bonding ability) have position lower in the series. Such ligands are called $\pi$-**donor ligands** and they cause smaller splitting. The ligands have intermediate $\pi$-bonding ability have position intermediate between the two :

| $CO>CN^->NO_2^->ophen$ | $>NH_3>H_2O$ | $>OH^->F^->Cl^-$ | $>SCN^->Br^->I^-$ |
|---|---|---|---|
| Ligands have strong $M \rightarrow L\pi$ bonding ability and have larger splitting i.e. have larger $\Delta_o$/10Dq values | Ligands have no $\pi$-bonding ability | Ligands have little $L \rightarrow M$ $\pi$-bonding ability | Ligands have strong $L \rightarrow M$ $\pi$-bonding ability. These ligands are called **$\pi$-donor ligands** and cause small perturbation of d-orbitals and have smaller $\Delta_0$/10Dq values. |
| | ⟵ These ligands have intermediate $\Delta_o$ values ⟶ | | |

⟶ $\Delta_o$/10Dq decreases ⟶

---

**Problem 7.44** Both Ni(IV) and Co(III) are $d^6$ systems but $K_2[NiF_6]$ is diamagnetic while $K_3[CoF_6]$ is paramagnetic.—Explain. **[C.U. 2008, B.U. 1999]**

**Ans. :** In$[CoF_6]^{3-}$ and $[NiF_6]^{2-}$ the $Co^{3+}$ and $Ni^{4+}$ ions respectively has the $3d^6$ configuration. Here $F^-$ is a weak field ligand therefore it is expected both the complexes will be high spin octahedral. Now since $Ni^{4+}$ ion has higher positive charge than $Co^{3+}$, the $\Delta_o$/10Dq value of $[NiF_6]^{2-}$ is greater than the $\Delta_o$/10Dq value of $[CoF_6]^{3-}$. The enhance value of $\Delta_o$

for $[NiF_6]^{2-}$ makes $\Delta_o$ > P and thus overcome the effect of interelectronic repulsion and so the complex will be low spin. On V.B. theory $[NiF_6]^{2-}$ has $d^2sp^3$ hybridisation and in CFT model it has $t_{2g}^6$ configuration making the complex diamagnetic.

On the other hand in case of $[CoF_6]^{3-}$, 10Dq < P and hence the complex will be high spin having $sp^3d^2$ hybridisation. In CFT model it has configuration $t_{2g}^4\ e_g^2$. Both the VBT and CFT indicate four unpaired electrons with a magnetic moment of 5.2 B.M. Expected value is 4.90 B.M.

---

**Problem 7.45** 10Dq of $[Mn(H_2O)_6]^{3+}$ is known from electronic spectrum to be 21000 $cm^{-1}$. The pairing energy of $Mn^{3+}$ is 28800 $cm^{-1}$. Predict whether the complex is high spin or low spin.

**Ans. :** Since pairing energy (P) > 10Dq, the given complex is high spin octahedral.

$Mn^{3+}$ is $3d^4$ system and the energy of the high spin state is,

$$E_{h,s}(t_{2g}^3\ e_g^1) = -6Dq = -6 \times \frac{21000}{10} = -12600\ cm^{-1}$$

and the energy of the low spin state is,

$$E_{l,s}(t^4{}_{2g}) = -16Dq + P = -16 \times \frac{21000}{10} + 28800 = -4800\ cm^{-1}$$

As $E_{h,s} < E_{l,s}$, the complex exist as high spin state i.e. the complex is stable in high spin state.

---

**Problem 7.46** For the $[Cr(H_2O)_6]^{2+}$ ion, the mean pairing energy 'P' is found to be 23500 $cm^{-1}$. The magnitude of $\Delta$ is 13900 $cm^{-1}$. Calculate the CFSE for the complex in configurations corresponding to high spin and low spin states. Which is more stable?

**[B.U.2014, 2008]**

**Ans. :** As pairing energy P > $\Delta$ so the complex exist in high spin octahedral.

$Cr^{2+}$ is $3d^4$ system, so the energy for high spin state,

$$E_{h,s}(t_{2g}^3\ e_g^1) = -6Dq = -6 \times \frac{13900}{10} = -8340\ cm^{-1}$$

and the energy of the low spin state,

$$E_{l,s}(t^4{}_{2g}) = -16Dq + P = -16 \times \frac{13900}{10} + 23500 = +1260\ cm^{-1}$$

Since $E_{h,s} < E_{l,s}$ so that high spin state is more stable than low spin state.

CFSE for high spin state = $-8340\ cm^{-1}$

and CFSE for low spin state = $-22240\ cm^{-1}$

---

**Problem 7.47** Draw and name the isomers of $[IrCl_3(PPh_3)_3]$ produced from the addition of $Cl_2$ in $[IrCl(PPh_3)_3]$.

**[B.U.2007]**

**Ans. :** This is an octahedral of the $Ma_3b_3$ type complex which exist as facial (cis) or fac and meridional (trans) or mer forms [Sec. 7.9.11(B)] as depicted below :

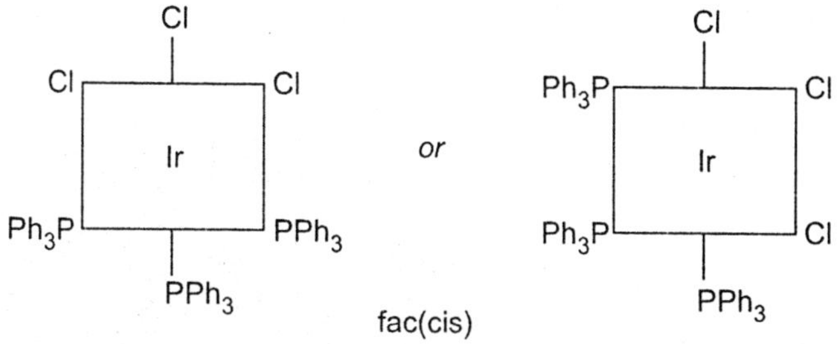


fac(cis)

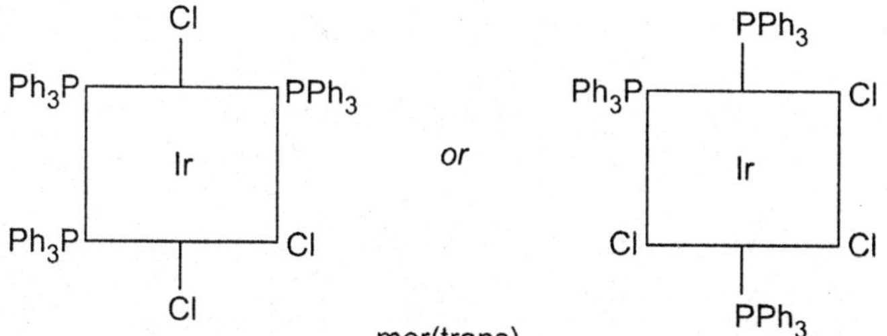


mer(trans)

In the fac form the three $Cl^-$ ions are all adjacent at one face of the octahedron while in case of mer isomer the two $Cl^-$ ions are placed opposite to each other, the two $PPh_3$ ligands are placed opposite to each other and the remaining two positions are randomly occupied by $Cl^-$ and $PPh_3$ ligands.

---

**Problem 7.48** Draw the possible isomers of $[Co(DMGH)_2(SCN)(py)]$ and comment on their stability (DMGH = dimethylglyoxime anion). **[C.U.2007]**

**Ans. :**

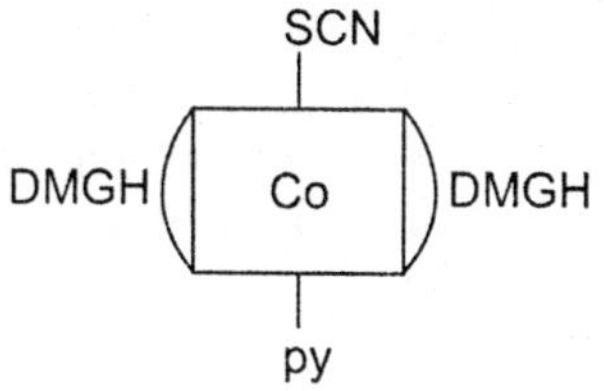


*This is optically in active*

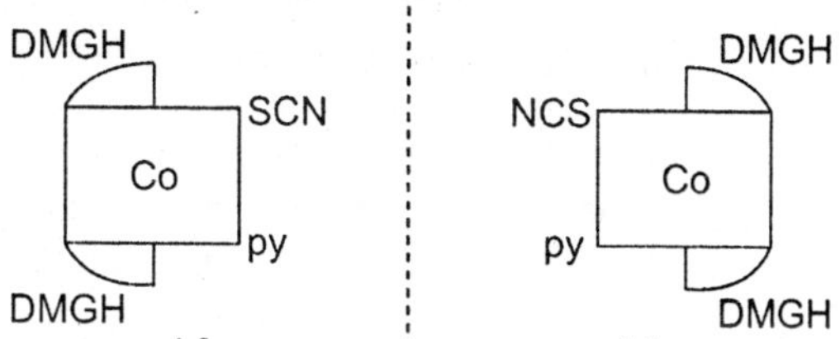


*d-form* *l-form*

*Optically active forms*

Optically inactive form generally more stable than optically active forms.

---

**Problem 7.49** Explain why $[Co(CN)_6]^{4-}$ is paramagnetic while $[Co(CN)_6]^{3-}$ is diamagnetic.

**Ans. :** In both the complexes the ligand $CN^-$ is a strong field one, therefore both the complexes will be low spin. $Co^{2+}$ in $[Co(CN)_6]^{4-}$ is a $3d^7$ system having $t_{2g}^6\,e_g^1$ configuration which gives only one unpaired electron and the complex $[Co(CN)_6]^{4-}$ is paramagnetic. In $[Co(CN)_6]^{3-}$ the $Co^{3+}$ is a $3d^6$ system having $t_{2g}^6$ configuration, so that there is no unpaired electron and the complex is diamagnetic.

---

**Problem 7.50** Explain why $Ni(CO)_4$ is tetrahedral while $[Ni(CN)_4]^{2-}$ is square planar.

**Ans. :** In $[Ni(CN)_4]^{2-}$ the central metal is $Ni^{2+}$ and the strong field ligand $CN^-$ is associated with $Ni^{2+}$ ($3d^8$ system) so that large perturbation of d-orbitals will occur and thereby square planar geometry is assumed by the complex. According to CFT it can be represented as :

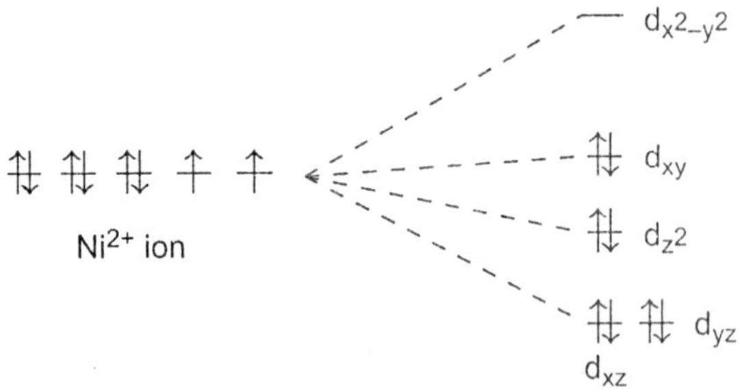


As the energy gap between $d_{x^2-y^2}$ and $d_{xy}$ orbitals are large so the electron distribution is of the above type and the complex is diamagnetic.

According to VBT the following representation of $dsp^2$ hybridisation is depicted where $d_{x^2-y^2}$ orbital is taking part in the hybridisation :

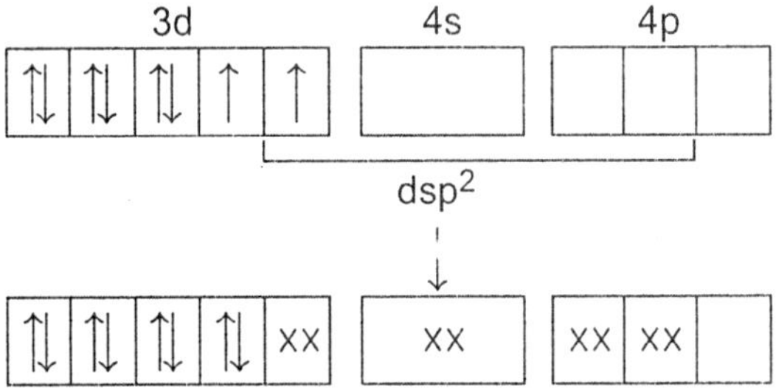


The VBT also indicates diamagnetic character of $[Ni(CN)_4]^{2-}$.

In case of $Ni(CO)_4$ the central metal is in zero oxidation state having $4s^2 3d^8$ configuration. The CO is more stronger ligand than that of $CN^-$ (Problem 7.43) so it is expected that splitting of d-orbitals will be more in case of $Ni(CO)_4$. We know that both CO and $CN^-$ are strong M $\rightarrow$ L $\pi$ bonding ligands. As Ni is in zero oxidation state and as CO has $\pi^*$ antibonding orbital lying vacant so that the two electrons present in the 4s orbital of Ni(O) shifted to 3d orbitals making 3d orbitals are completely filled and the available hybridisation will be $sp^3$ as depicted below :

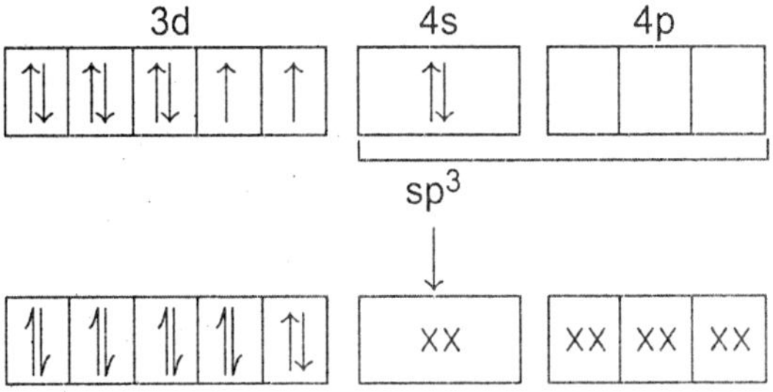


Therefore $Ni(CO)_4$ have tetrahedral geometry and the complex is diamagnetic. Here more stronger ligand CO will convert square planar geometry in $[Ni(CN)_4]^{2-}$ to tetrahedral geometry in $[Ni(CO)_4]$. In terms of CFT a tetragonal compression may be expected to convert square planar to tetrahedral geometry.

---

**Problem 7.51** Magnetic moment of brown ring compound $[Fe(H_2O)_5NO]^{2+}$ in nitrate test is 3.89 BM. Find out the oxidation state and the type of hybridisation of the central metal ion.

**Ans.** : The magnetic moment $\mu = \sqrt{n(n+2)}$ B.M. where n is the number of unpaired electrons.

$\therefore$ $3.89 = \sqrt{n(n+2)}$ which on solving we get,

n = –5 or 3, therefore the complex has three unpaired electrons. Fe has the electronic configuraiton $4s^2 3d^6$. If iron is in +1 oxidation state and the hybridisation is $sp^3d^2$ as depicted below then the complex will have three unpaired electrons.

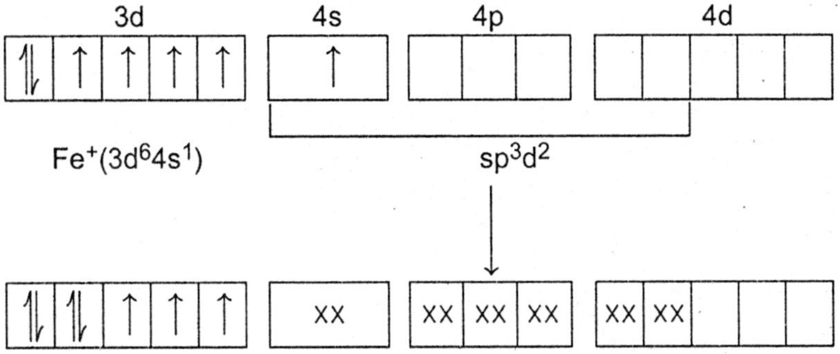


## 7.14.5 Application of Crystal Field Theory

The CFT can be employed to explain several aspects e.g. spectral, magnetic, stereochemical, thermodynamic etc. properties of complexes. We shall discuss here some of the variation in properties in brief.

### 7.14.5(A) Thermodynamic and structural consequences of CFSE

CFSE/LFSE (Sec. 7.14.3, Table 7.11) plays an important role for the stability of complex ions. The enthalpy of formation of many complexes is related to CFSE which is also related to the hydration of cations. As the value of $\Delta$ lie in the range 10-30 kK, so that CFSE will be in the range 5-40 kK i.e. 60-480 kJ $mol^{-1}$.

Hydration is a process by which a metal cation is attached with certain number of water molecules [Sec. 3.1.4(a)] and in this process an amount of energy is released which is known as hydration energy. As ionic radius has a periodic nature (Sec. 2.8) so here we have made an attempt to correlate hydration energy with CFSE and radius of divalent $3d^n$ metal cations.

A plot of hydration energies vs $3d^n$ metal ions of +2 charge shows a characteristic double-humped curve (Fig. 7.13). The hydration energy of the first $3d^0$ ion i.e. $Ca^{2+}$ is –1592 kJ $mol^{-1}$ while the end element $Zn^{2+}(3d^{10})$ has the value –2044 kJ $mol^{-1}$. If we consider the 3d metal ions are spherical then it is expected the successive ions from $Ca^{2+}$ to $Zn^{2+}$ will decrease in size and thereby hydration energy will increase.

As already stated the CFSE/LFSE is the extra stability gains by a system due to crystal field splitting. The variation of CFSE of the said metal ions has also similar trends like that of Fig. 7.13. The two humps in such a curve are due to addition of the contribution of the weak field, stabilisation energies to the hydration energies expected for hypothetical spherical metal ions of steadily decreasing radius. For a $3d^1$ system in an octahedral field the extra stability is 0.4 $\Delta_0$. As the $\Delta_0$ values for bivalent metals even with weak field $H_2O$ ligand is about 10000 $cm^{-1}$ and for trivalent metal ions is ~20000 $cm^{-1}$, the extra stability in these cases is about 12 and 24 kcal $mol^{-1}$ respectively. For strong field ligands these values are about 25% higher than $H_2O$. Thus CFSE/LFSE variation are of the same order of magnitude as the hydration energies of $3d^n$ metal ions and hence they are expected to influence the thermodynamic properties of the transition metal complexes.

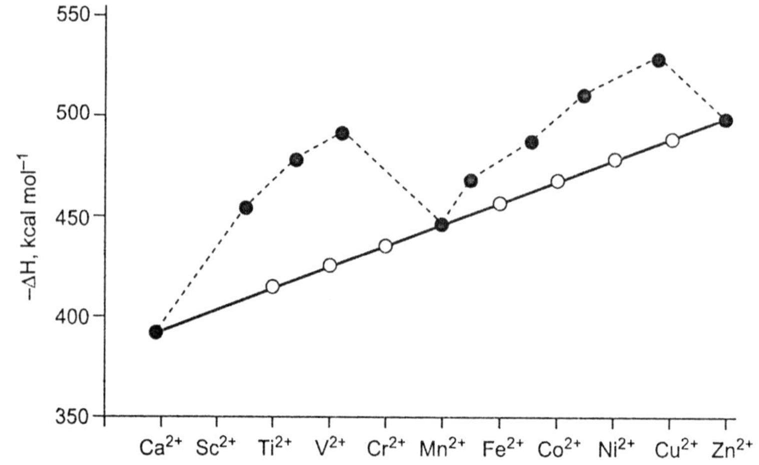


***Figure 7.13** : Hydration energies of $3d^n$ ions in +2 state.*

The full circles in the figure 7.13 denote the hydration energies estimated using thermodynamical process for the following net change :

$$M^{2+} + nH_2O \longrightarrow [M(H_2O)_n]^{2+}$$

If the CFSE is subtracted from the hydration energies, the resulting values of the corrected heats of hydration fall on the straight line denoted by the open circles in Fig. 7.13. As the CFSE is zero for $Ca^{2+}(3d^0)$, $Mn^{2+}(3d^5)$ and $Zn^{2+}(3d^{10})$ ions in weak octahedral field, in their cases there is no change in the values of hydration energies and they lie on a smooth curve. The remaining ions have higher enthalpies of hydration than these three cations. Actually CFSE values are of the order of 5-10% of the hydration energies.

The hydration energies are inversely proportional to the size of the metal ion. It is to be noted that metal ions having larger hydrated radius has greater heat of hydration. The hydration enthalpies of $d^0$, $d^1$ and $d^2$ ions are actually less than those of several other ions of the given series. Due to the contribution from the crystal field effect the hydration enthalpy for $d^3$ ion is considerable. In fact fine refinement of hydration enthalpies can be made by CFSE and thus these hydration energies can be employed to explain the consequences in magnetism, colour and other properties of complexes.

(i) **Variation of ionic radius** : Fig. 7.14 denotes the periodic variation of radii of octahedrally coordinated $M^{2+}$ ions of first transition series ($3d^n$) including $Ca^{2+}$ and $Zn^{2+}$. At first as we move from $Ca^{2+}$ to $Zn^{2+}$ the ions decrease in size which is the normal trend. There is a partial filling of $t_{2g}$ orbitals first, then partial population of $e_g$ orbitals and then again pairing occurs in $t_{2g}$ level and finally we get $t_{2g}^6e_g^4$ configuration for $Zn^{2+}$. Therefore it is expected there should be a variation of $M^{2+}$ radius from

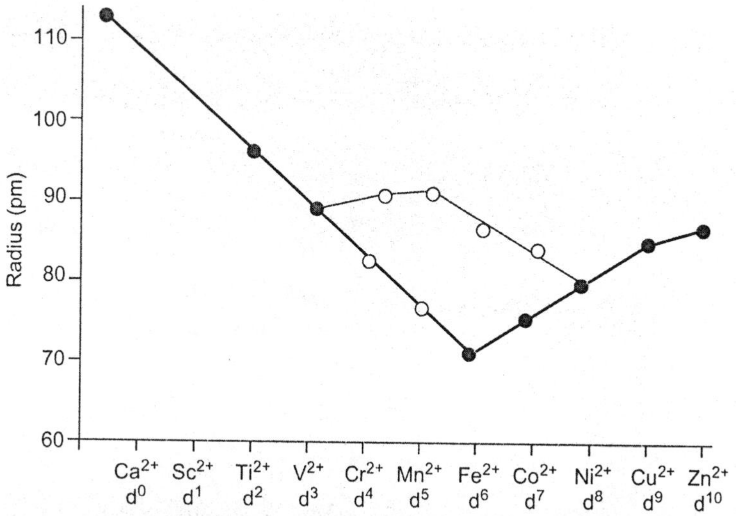


***Figure 7.14** : Variation of ionic radii of $3d^n$ ions ($M^{2+}$). Open circles (○) represent high spin and closed circles (●) respresent low spin.*

$Ca^{2+}$ to $Zn^{2+}$ (Table 2.11). In strong octahedral field, the metal electrons gradually occupied by the $t_{2g}$ level. As $t_{2g}$ orbitals are not directed along the path of the ligands, the ligands can easily approach towards the metal ion so that a steady decrease of radius occurs until $t_{2g}^6$ configuration is obtained. As the next electron enters into $e_g$ level and as $e_g$ electrons repel the ligands, resulting in an increase in radius. This explains the hump in the curve (Fig. 7.14). In a weak octahedral field (high spin) electron first enter into $e_g$ orbital in case of $d^4$ ion. Consequently the size of the ions at $d^4$ and $d^5$ and we get a double humped curve (open circles in Fig. 7.14). The radius of low spin ions (closed circles) decrease across the period until $e_g$ set are first occupied at $d^7$ configuration. The radii of $d^6$ to $d^8$ ions decrease again as electrons are occupied in $t_{2g}$ orbitals. $d^9$ and $d^{10}$ configurations again show increase in size as the electrons are populated in $e_g$ level.

(ii) **Lattice energies of metal halides :** Lattice energies (Sec. 3.1.2) of divalent metal halides $MX_2$ ($M = Ca^{2+} - Zn^{2+}$, $X = F^-, Cl^-, Br^-$) are plotted in Fig. 7.15. The curve indicates double humped nature which indicates LFSE is responsible for about 5-10% of the lattice energies of these compounds. It is expected that radius of $M^{2+}$ ions gradually should decrease so that lattice energies of $MX_2$ compounds should increase corespondingly. This is expected for $d^0$, $d^5$ and $d^{10}$ metal ions. The experimental lattice energies can be explained by the contribution of CFSE on the octahedral environment of $MX_2$ halides. The halides ($MF_2$, $MCl_2$ and $MBr_2$) particularly $MF_2$ provides weak ligand field in which lower $t_{2g}$ orbitals are occupied first in case of $d^2$ and $d^3$ metal ions which explains the steady rise of lattice energies. For $d^4$ and $d^5$ metal ions, the electron is occupied in $e_g$ level which lowers the CFSE ($\Delta_0$ or 10Dq) and also increase the size of the metal ions which in turn decrease the lattice enegy. However uncertainties in experimental values vitiate the correlation.

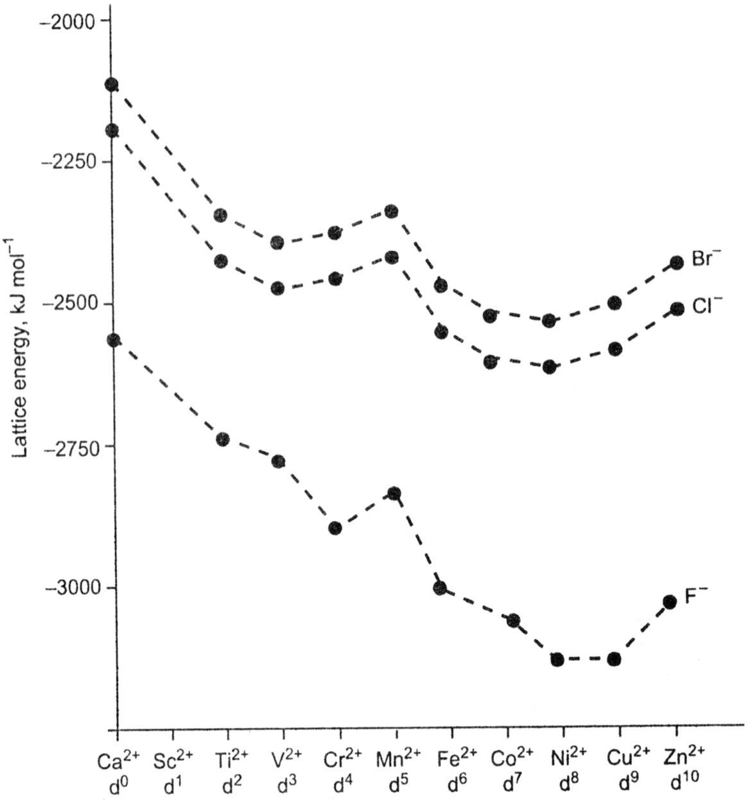


*Figure 7.15 : Lattice energy of $MX_2$ type halides ($X = F^-, Cl^-, Br^-$).*

(iii) **Variation of formation constants :** The first stepwise formation constants will follow the double humped curve (Fig. 7.16). Here some nitrogen donor ligands are used and

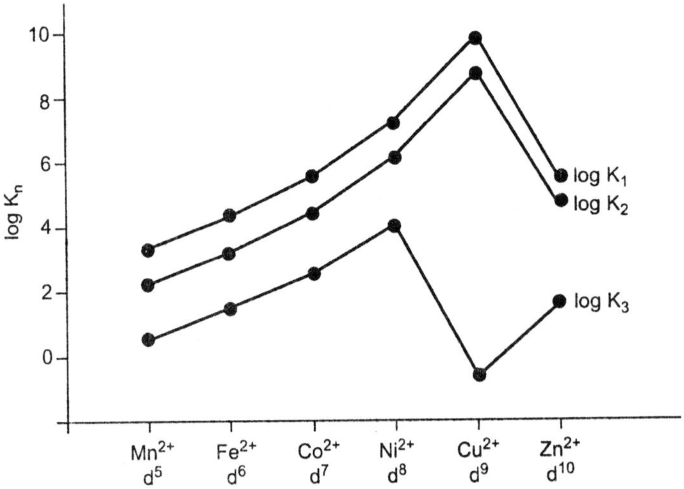


*Figure 7.16 : Variation of stepwise stability constants of ethylenediamine complexes of some $3d^n$ ions.*

as the nitrogen donor ligands are border line bases (Sec. 4.5) so that early first transition series metal cations ($M^{2+}$) will exhibit no measurable formation constant values. For weak field ligands the first stepwise formation constants ($K_1$) will follow the sequence $Mn^{2+} < Fe^{2+} < Co^{2+} < Ni^{2+} < Cu^{2+} > Zn^{2+}$ which is also knwon as **Irving Williams series of stability constants.** $K_2$ also follow similar trend like that of $K_1$ but $K_3$ in some cases [e.g. $Cu(en)_3^{2+}$] deviate from $K_1$ and $K_2$. The destabilisation of $Cu(en)_3^{2+}$ is due to distortion for $Cu^{2+}$ ($3d^9$).

(iv) **Redox potential and CFSE :** Let us consider the following reduction potential values :

| | E° (volt) |
|---|---|
| $[Co(H_2O)_6]^{3+} + e \longrightarrow [Co(H_2O)_6]^{2+}$ | +1.84 |
| $[Co(EDTA)]^{-} + e \longrightarrow [Co(EDTA)]^{2-}$ | +0.60 |
| $[Co(ox)_3]^{3-} + e \longrightarrow [Co(ox)_3]^{4-}$ | +0.57 |
| $[Co(ophen)_3]^{3+} + e \longrightarrow [Co(ophen)_3]^{2+}$ | +0.42 |
| $[Co(NH_3)_6]^{3+} + e \longrightarrow [Co(NH_3)_6]^{2+}$ | +0.10 |
| $[Co(en)_3]^{3+} + e \longrightarrow [Co(en)_3]^{2+}$ | –0.26 |
| $[Co(CN)_6]^{3-} + e \longrightarrow [Co(CN)_5]^{3-} + CN^-$ | –0.83 |

It is evident from the above reduction potential (E°) values that $Co^{3+}$ is more stable than $Co^{2+}$ which can also be correlated for the above systems having approximately the order of increasing ligand field strength of the ligands. But CFSE greatly influence the E° values due to the hydration energies, entropy effect due to the formation of chelate rings, π-bonding effect etc. so that the above order of decreasing E° values is not exactly the order of increasing ligand field strength of ligands. The stability of a complex depends on the nature of the ligand as also CFSE for the system. This can be explained in the following ways :

(a) $H_2O$ is a weak field ligand and $Co^{2+}$ and $Co^{3+}$ form complexes with $H_2O$ as $[Co(H_2O)_6]^{2+}$ and $[Co(H_2O)_6]^{3+}$ respectively. For $H_2O$ ligand, the $[Co(H_2O)_6]^{2+}$ is more stable than $[Co(H_2O)_6]^{3+}$. This is because in weak field environment $Co^{2+}$ ($3d^7$) has much higher value of CFSE ($-0.8\ \Delta_0/-8$ Dq) compared to $Co^{3+}$ ($3d^6$) which has the corresponding CFSE value is $-0.4\ \Delta_0/-4$ Dq.

(b) Both en and ophen are bidentate NN donor ligands–the en is only the σ-donor while ophen is also a σ-donor but due to the presence of benzene ring it can also function as π-acceptor. The two ligands form high spin $Co^{2+}$ complexes while the corresponding $Co^{3+}$ complexes are low spin. From Table 7.11 the CFSE for low spin $Co^{3+}$ and high spin $Co^{2+}$ complexes are –24 Dq and –8 Dq respectively so that the difference is 24 Dq – 8 Dq = 16 Dq. It has been experimentally verified that the net stabilisation energy is 358 kcal $mol^{-1}$ in case of ophen while the value is 336 kcal $mol^{-1}$ for en for $Co^{3+}$-$Co^{2+}$ system. Therefore due to crystal field effect the stabilisation of $Co^{3+}$ compared to that of $Co^{2+}$ is more in case of ophen than that of en. Due to the π-acceptor property of ophen, this has a greater preference for $Co^{2+}$ compared to that of $Co^{3+}$ (due to the synergic effect) and this effect probably more than compensate the crystal field effect and as a result en stabilises $Co^{3+}$ relative to $Co^{2+}$. Thus the E° value of $Co^{3+}/Co^{2+}$ system is more higher for ophen compared to en. The more negative E° value in case of $CN^-$ is due to the negative charge of the ligand which enhances the greater attraction of $Co^{3+}$ compared to that of $Co^{2+}$. Both $Co^{3+}/Co^{2+}$ form low spin complexes with $CN^-$, the difference of CFSE = 24 Dq – 18 Dq = 6 Dq is also one of the factor for more negative E° value.

(c) Less positive E° value is justified for strong field $NH_3$ ligand due to the fact that $Co^{3+}$ forms low spin diamagnetic complex $[Co(NH_3)_6]^{3+}$ compared to that of $Co^{2+}$. Greater CFSE value of $Co^{3+}$ (–24 Dq) compared to that of $Co^{2+}$ (–18 Dq) is also one of the cause to stabilise $Co^{3+}$ compared to that of $Co^{2+}$.

### 7.14.5 (B) Geometric site preferences

Several factors such as CFSE, ligand–metal ion interaction energy and ligand-ligand repulsion energy due to steric crowding of ligands particularly for high coordination numbers are responsible for the actual stereochemistry of a complex for a given $d^n$ metal ion with a given set of ligands. However among these factors CFSE has taken a pivotal role to predict which $3d^n$ electron configurations has a greater preference for a particular geometry compared to other geometry. In some cases a five coordinate complex exist in equilibrium between two geometries e.g. $CdCl_5^{3-}$ and $Ni(CN)_5^{3-}$ exist in equilibrium between trigonal bipyramid (TBP) and square pyramid (SP) geometry.

The two most common geometry encountered in coordination chemistry is the octahedral (C.N–6) and tetrahedral (C.N–4) (Fig. 7.12). We can calculate the CFSE value for each $d^n$ (high spin) ions and then take difference of CFSE for the two geometries which is known as the **octahedral site preference** or **stabilisation energy**, abbreviated as OSPE or OSSE, thus.

$$\text{OSPE or OSSE} = \text{CFSE}_{oct} - \text{CFSE}_{tet} \qquad \ldots(7.43)$$

High value of OSPE/OSSE indicates the stability of octahedral coordination while low value suggests the preference of tetrahedral geometry. In order to make the subtraction in equation (7.43) we must use $\text{CFSE}_{Tet}$ in units of $\Delta_o$ as shown in Table 7.11. Calculation of OSPE/OSSE (Table 7.13) indicates $d^3$ and $d^8$ configurations has a preference for octahedral coordination. Cr(III)($3d^3$) virtually never forms tetrahedral complexes as also Ni(II)($3d^8$). $3d^5$($Mn^{2+}$), $3d^6$($Fe^{2+}$) and $3d^7$($Co^{2+}$) commonly forms tetrahedral complexes with weak ligands as also with bulky ligands. Table 7.13 suggests that for Ni(II) there is a better preference of octahedral geometry than $Co^{2+}$ which has a better choice for tetrahedral geometry so that the following reaction proceed from left to right in

$$[Co(H_2O)_6]^{2+} + 4Cl^- \rightleftharpoons [CoCl_4]^{2-} + 6H_2O$$

presence of a weak $Cl^-$ ligand. $d^0$, $d^5$ and $d^{10}$ ions have no CFSE (Table 7.13) so that these ions have no stereochemical preference according to CFT model.

**Table 7.13 : Calculation of OSPE/OSSE for octahedral high spin and tetrahedral $d^n$ ions**

| $d^n$ ion | CFSE, $O_h$ | CFSE, $T_d$ | OSPE/OSSE |
|---|---|---|---|
| $d^1$ ($Ti^{3+}$) | $-0.4\ \Delta_o$ | $-0.266\ \Delta_o$ | $-0.134\ \Delta_o$ |
| $d^2$ ($V^{3+}$) | $-0.8\ \Delta_o$ | $-0.532\ \Delta_o$ | $-0.268\ \Delta_o$ |
| $d^3$ ($Cr^{3+}$) | $-1.2\ \Delta_o$ | $-0.355\ \Delta_o$ | $-0.845\ \Delta_o$ |
| $d^4$ ($Cr^{2+}$) | $-0.6\ \Delta_o$ | $-0.177\ \Delta_o$ | $-0.423\ \Delta_o$ |
| $d^5$ ($Mn^{2+}$) | 0 | 0 | 0 |
| $d^6$ ($Fe^{2+}$) | $-0.4\ \Delta_o$ | $-0.266\ \Delta_o$ | $-0.134\ \Delta_o$ |
| $d^7$ ($Co^{2+}$) | $-0.8\ \Delta_o$ | $-0.532\ \Delta_o$ | $-0.268\ \Delta_o$ |
| $d^8$ ($Ni^{2+}$) | $-1.2\ \Delta_o$ | $-0.355\ \Delta_o$ | $-0.845\ \Delta_o$ |
| $d^9$ ($Cu^{2+}$) | $-0.6\ \Delta_o$ | $-0.177\ \Delta_o$ | $-0.423\ \Delta_o$ |
| $d^{10}$ ($Zn^{2+}$) | 0 | 0 | 0 |

Another comparison of geometrical preference may be made for octahedral and square pyramidal coordination (Fig. 7.12) where the perturbation of d-orbitals are represented in terms of $\Delta_o$ as :

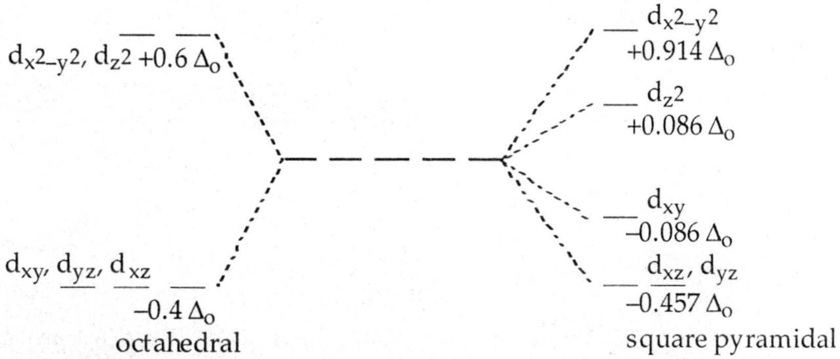


From the above figure OSPE/OSSE calculated for octahedral versus square pyramidal coordination (OSSE = $CFSE_{octahedral} - CFSE_{square\ pyramidal}$) as :

$d^1$ : OSSE = $-0.4\ \Delta_o + 0.457\ \Delta_o = +0.057\ \Delta_o$

$d^2$ : OSSE = $2(-0.4\ \Delta_o) - 2(-0.457\ \Delta_o) = +0.114\ \Delta_o$

$d^3$ : OSSE = $3(-0.4\ \Delta_o) - [2(-0.457\ \Delta_o) - 0.086\ \Delta_o] = -0.2\ \Delta_o$

$d^4$ : OSSE = $[3(-0.4\ \Delta_o) + 0.6\ \Delta_o] - [2(-0.457\ \Delta_o) - 0.086\ \Delta_o + 0.086\ \Delta_o] = +0.314\ \Delta_o$

$d^5$ : OSSE = $[3(-0.4\ \Delta_o) + 2(+0.6\ \Delta_o)] - [2(-0.457\ \Delta_o) - 0.086\ \Delta_o + 0.086\ \Delta_o + 0.914\ \Delta_o)] = 0$

It is evident from the above calculation that $d^3$ as also $d^8$ metal ions has least preference for tetrahedral coordination, which causes them to resist the dissociation of coordinated water. OSSE value for $3d^1$ ($V^{4+}$) indicates highest preference for square pyramidal coordination. $V^{4+}$ ion has high charge density and exist as an acidic cation $[V(H_2O)_6]^{4+}$, which actually exist as an oxospecies e.g. $[VO(H_2O)_5]^{2+}$ in solution. A well known square pyramidal $V^{4+}$ oxo complex is $[VO(acac)_2]$.

One important application of CFSE is to explain the structure of **spinels**. Spinels are the mixed metal oxides having composition $AB_2O_4$ where A and B are two different metals or in some cases they may be same. They are classified as **normal spinels** and **inverted spinels**. Normal spinels are the mixed metal oxides of divalent and trivlent metals, $A^{II}\ B_2^{III}O_4$ e.g. $MgAl_2O_4$. Inverted spinels have the composition $B^{III}[A^{II}B^{III}]O_4$ e.g. $Al^{III}[Ni^{II}Al^{III}]O_4$. Close packing of oxide ions ($O^{2-}$) will create two types of holes—tetrahedral hole and octahedral hole*. In *normal spinels the bivalent A metal ions occupy tetrahedral holes and the trivalent B metal ions occupy the octahedral holes. In inverted spinels the half trivalent B metal ions occupy tetrahedral holes and half B ions occupy octaheral holes while the bivalent A metal ions occupy the octahedral interstices* e.g. $Fe_3O_4$ which may be represented as $Fe^{III}[Fe^{II}Fe^{III}]O_4$. The spinels $Mn_3O_4$, $Fe_3O_4$ and $Co_3O_4$ are all intensely black and are much darker than the normal oxides of these metals, due to charge transfer transitions. The conductivity of $Fe_3O_4$ is far greater than that of $Fe_2O_3$. $Mn_3O_4$ and $Co_3O_4$ has normal spinel structure.

Structure of spinels

Whether a particular spinel will assume normal or inverted structure will depend on the CFSE values of the system. The crystal field effect of $O^{2-}$ is comparable to that of $H_2O$ i.e. $O^{2-}$ is considered to be a moderately weak field ligand. Let us consider the normal spinel $Mn_3O_4$ having the composition $Mn^{II}[Mn_2^{III}]O_4$ where Mn(II) occupied to tetrahedral site and Mn(III) occupied octahedral site. Therefore,

$$\text{for Mn(II)(3d}^5\text{) : CFSE} = 2\left(-\frac{3}{5}\Delta_t\right)+3\left(+\frac{2}{5}\Delta_t\right)=0$$

$$\text{for Mn(III)(3d}^4\text{) : CFSE} = 3\left(-\frac{2}{5}\Delta_o\right)+\frac{3}{5}\Delta_o=-\frac{3}{5}\Delta_o$$

$$\text{For two Mn(III), CFSE} = -\frac{6}{5}\Delta_o$$

$$\text{So that the total CFSE for the system} = -\frac{6}{5}\Delta_o$$

If $Mn_3O_4$ is considered to be an inverted spinel having the composition $Mn^{III}[Mn^{II}Mn^{III}]O_4$ where half Mn(III) occupied by tetrahedral site and other half Mn(III) occupied by octahedral site and Mn(II) occupied by octahedral site, then

$$\text{Mn(III)(3d}^4\text{) tetrahedral site : CFSE} = 2\left(-\frac{3}{5}\Delta_t\right)+2\left(+\frac{2}{5}\Delta_t\right)=-\frac{2}{5}\Delta_t$$

$$\text{Mn(III)(3d}^4\text{) octahedral site : CFSE} = 3\left(-\frac{2}{5}\Delta_o\right)+\frac{3}{5}\Delta_o=-\frac{3}{5}\Delta_o$$

$$\text{Mn(II)(3d}^5\text{) octahedral site : CFSE} = 3\left(-\frac{2}{5}\Delta_o\right)+2\left(+\frac{3}{5}\Delta_o\right)=0$$

$$\text{Therefore the net CFSE for the system} = -\frac{2}{5}\Delta_t-\frac{3}{5}\Delta_o+0=-\frac{2}{5}\times\frac{4}{9}\Delta_o-\frac{3}{5}\Delta_o\approx-\frac{7}{9}\Delta_o$$

As the CFSE for normal spinel structure is more negative than that of inverted structure so that $Mn_3O_4$ has normal spinel structure. Similar calculation of $Co_3O_4$ dictates the normal spinel

* will be discussed in the chapter of Inorganic Solid.

structure for it. Here we have to consider $O^{2-}$ forces spin pairing in Co(III)($3d^6$) having large CFSE = $-2.4\Delta_o$. If we consider $Fe_3O_4$ has inverted spinel structure then CFSE = $-0.4\Delta_o$ while in normal spinel structure the CFSE = $-0.3\Delta_o$ so that $Fe_3O_4$ will assume inverted spinel structure.

As mentioned earlier that for geometric site preferences the CFSE is not the sole criteria to dictate the structure of spinels. Crystal packing and the interaction of the ions has also to be considered. The overall structure consists of a cubic close packing (ccp) of oxide ions which create both octahedral and tetrahedral hole. Spinels are generally used in solid-state electronics industry for their electrical and magnetic properties.

### 7.14.5(C) Distortions from Regular geometry : The Jahn Teller Effect*

In Sec. 7.14.2(D) we have seen how tetragonal distortion of an octahedral complex ($ML_6$) leading to the formation of square planar ($ML_4$) one (Fig. 7.12). In a regular octahedral complex splitting of the d-orbitals provide greater stability by CFSE. Further removal of degeneracy of upper $e_g$ and lower $t_{2g}$ level can sometimes result additional stabilisation of the system. The additional stabilisation is executed by a theorem which is well known Jahn-Teller theorem developed in 1937. This is actually the extension of tetragonal distortion leading to the formation of additional stabilisation. This states that *any nonlinear molecular system in degenerate electronic state will undergo some kind of distortion which will lower the energy and split the degenerate state.* This is also a case of wave mechanical phenomenon. This theorem only states the distortion of degenerate states but does not give any idea of geometrical nature of distortion as also the magnitude and direction of distortion. This does not state if the undistorted configuration has a centre of symmetry, whether it will be maintained or not after distortion. However this theorem can be applied with systems having unsymmetrically filled $e_g$ level e.g.

> *Limitations of Jahn-Teller theorem*

$d^4$ (high spin) : Cr(II), Mn(III) ($t_{2g}^3e_g^1$)
$d^7$ (low spin) : Ni(III), Co(II) ($t_{2g}^6e_g^1$)
$d^9$ : Cu(II) ($t_{2g}^6e_g^3$)

The reason behind this distortion is rather easy to understand. Let us begin with $3d^9$ ($Cu^{2+}$) octahedral complex where $e_g$ orbitals ($d_{x^2-y^2}$ and $d_{z^2}$) have unequal population. The electrons in the $d_{x^2-y^2}$ orbital repel the electrons donated by four ligands along the x and y axes, the electrons in the $d_{z^2}$ orbital repel the electrons of two ligands along the z axis. If the two orbitals are equally populated then the repulsion would be same but practically the two orbitals ($d_{x^2-y^2}$ and $d_{z^2}$) are unequally populated so that we have to decide by physical reasoning which electronic cofiguration is permitted by which the repulsion is minimised :

(1) Let the electronic configuration is $d_{x^2-y^2}^2 d_{z^2}^1$. The four ligands in the xy plane will be more screened from the electrostatic attraction of the Cu(II) ion than the two ligands along the z axis. Thus the two axial ligands will be drawn more closely than the other four equatorial ligands i.e. the ligands along the most populated axis will move farther away while the ligands along the less populated axis will move closer. These opposing motions will produce **tetragonally distorted octahedral geometry** with a **contraction along z axis.**

---

* Like Jahn-Teller Effect there is Renner-Teller effect which is given in Appendix-J.

(2) Let the electronic configuration is $d^1_{x^2-y^2}d^2_{z^2}$. The four ligands along the xy plane will be less screened from the electrostatic attraction of the $Cu^{2+}$ ion than the two ligands along the z axis and thus the four ligands along the xy plane will be drawn more closer than the other two ligands on the z-axis. In this case we get **tetragonally distorted octahedron** with **elongation along z-axis.**

In either case, the unequal bond lengths lower the symmetry below that expected for a regular octahedral structure. Experimentally for Cu(II) we find that elongation occurs along z axis with two long axial bonds and four short equatorial bonds along the xy plane. Therefore the orbitals having the z component among both the $e_g$ and $t_{2g}$ orbitals are stabilised with respect to orbitals having x and y components. As a result, the degeneracy of $e_g$ orbital is lifted as also for $t_{2g}$ orbital partly (Fig. 7.17). The centre of gravity rule is followed and we get a net stabilisation of the system equal to $-{}^1/_2\delta_1$. There will be no net change of stabilisation for the $t_{2g}$ electrons. The electron configuration of $Cu^{2+}$ is thus $t^6_{2g}d^2_{z^2}d^1_{x^2-y^2}$. Due to the net gain in stability by $-{}^1/_2\delta_1$, which motivates the system to undergo distortion. Mixed ligand complexes undergo this type of distortion due to the different types of crystal field strength of ligands.

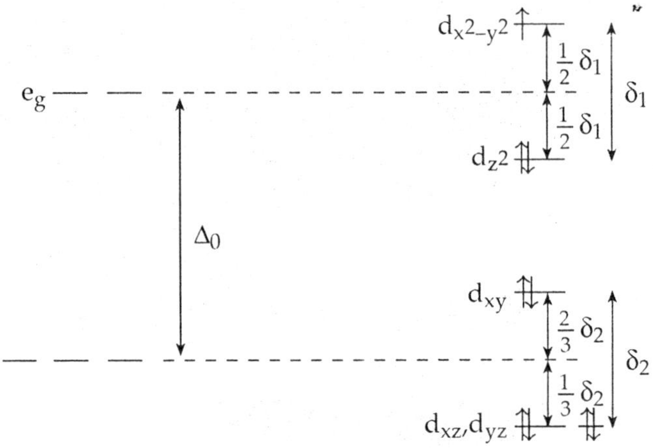


***Figure 7.17** : Jahn-Teller distortion of an octahedral $3d^9$ ($Cu^{2+}$) complex ($\delta_1$, $\delta_2$ << $\Delta_o$).*

The strong experimental evidence in favour of Jahn-Teller distortion comes from X-ray crystallographic data of several Cu(II) compounds in the solid state. Both tetragonal elongation and tetragonal compression may occur which stabilise the complex species. But experimental evidence in most of the cases revealed elongation along the z-axis leading to two long axial bonds and four short equatorial bonds e.g. in $CuF_2$ the four short Cu–F bonds (1.93Å) and two long Cu–F (2.27Å) bonds exist. Several compounds e.g. $NaCuF_3$, $CuCl_2.2py$, $MnF_3$, $CuCl_2.2H_2O$, $CrF_2$ etc. exhibit this type of difference in bond lengths. However some Cu(II) complexes are known to exist in other tetragonal geometry where the two Cu-ligand bonds are shorter along the z-axis e.g. $K_2[CuF_4]$ (7-LIX).

The existance of measurable bond length differences are the examples of **static Jahn-Teller distortion**. In such a case all the molecules exist in only one stable distorted form and in this case the energy gap between the splitted $e_g$ and $t_{2g}$ orbitals are somewhat larger. On the other hand there are some examples of complexes which exist in equilibrium between the two distorted forms i.e. tetragonal elongation and tetragonal compression forms. This is known as **dynamic Jahn-Teller distortion.** In this case the energy gap between the splitted $e_g$ and $t_{2g}$ orbitals are low compared to that of static Jahn-Teller distortion. $[Ti(H_2O)_6]^{3+}$, $K_2M[Cu(NO_2)_6]$ ($M = Ca^{2+}, Ba^{2+}, Pb^{2+}$) are the examples of dynamic Jahn-Teller distortion. These are characterised by low temperature crystallographic data as also absorption spectral studies.

*Static and dynamic Jahn-Teller distortion*

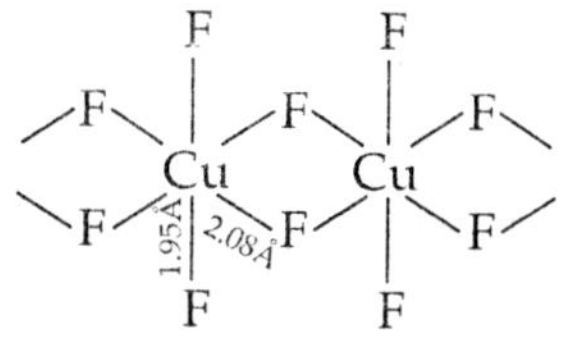


(7-LIX)

Finally Table 7.14 records Jahn-Teller effect of different $d^n$ ions. Tetrahedral complexes e.g. $Cs_2[CuCl_4]$ and $Cs_2[CuBr_4]$ also exhibit Jahn-Teller effect. The chelate complexes e.g. $[Cu(en)_3]^{2+}$, $[Cu(H_2O)_4en]^{2+}$ etc. are susceptible to Jahn-Teller distortion. Here two opposing effects occur– Jahn-Teller effect causes stabilisation while geometrical constraint for chelate ring would cause destabilisation.

**Table 7.14 : Jahn-Teller effect of different $d^n$ ions in Octahedral and Tetrahedral geometry**

| $d^n$ ion | $d^1$ | $d^2$ | $d^3$ | $d^4$ | $d^5$ | $d^6$ | $d^7$ | $d^8$ | $d^9$ | $d^{10}$ |
|---|---|---|---|---|---|---|---|---|---|---|
| Octahedral (high spin) | $t_{2g}^1$ D.P. | $t_{2g}^2$ D.P. | $t_{2g}^3$ N.P. | $t_{2g}^3e_g^1$ D.P. | $t_{2g}^3e_g^2$ N.P. | $t_{2g}^4e_g^2$ D.P. | $t_{2g}^5e_g^2$ D.P. | $t_{2g}^6e_g^2$ N.P. | $t_{2g}^6e_g^3$ D.P. | $t_{2g}^6e_g^4$ N.P. |
| Octahedral (low spin) | – | – | – | $t_{2g}^4$ D.P. | $t_{2g}^5$ D.P. | $t_{2g}^6$ N.P. | $t_{2g}^6e_g^1$ D.P. | – | – | – |
| Tetrahedral | $e^1$ D.P. | $e^2$ N.P. | $e^2t_2^1$ D.P. | $e^2t_2^2$ D.P. | $e^2t_2^3$ N.P. | $e^3t_2^3$ D.P. | $e^4t_2^3$ N.P. | $e^4t_2^4$ D.P. | $e^4t_2^5$ D.P. | $e^4t^6$ N.P. |

*D.P. = distortion permitted* *N.P.=distortion not permitted*

**Problem 7.52** 'Tetrahedral $Co^{2+}$ complexes are not uncommon whereas corresponding $Ni^{2+}$ complexes are'. Explain. **[B.U. 2009]**

**Ans. :** $Co^{2+}$ : $3d^7$ having $t_{2g}^5e_g^2$ in octahedral high spin configuration and $CFSE_{oct} = -0.8\Delta_o$.

In tetrahedral geometry ($e^4t_2^3$) the $CFSE_{tet} = -0.532\Delta_o$.

$\therefore$ OSPE / OSSE = $-0.268\Delta_o$.

$Ni^{2+}$ : $3d^8$ having $t_{2g}^6e_g^2$ in octahedral configuration the $CFSE_{oct} = -1.2\Delta_o$.

For tetrahedral geometry having $e^4t_2^4$ configuration the $CFSE_{tet} = -0.355\Delta_o$, so that OSPE/OSSE = $-0.845\Delta_o$.

High value of OSSE indicates preference for octahedral geometry while low value indicates preference for tetrahedral site. Therefore Co(II) will have a preference for tetrahedral geometry but Ni(II) will assume octahedral one.

**Problem 7.53** A chromium atom forms a neutral complex with carbon monooxide and 1, 10 phenanthroline molecule. The structure of the complex is :

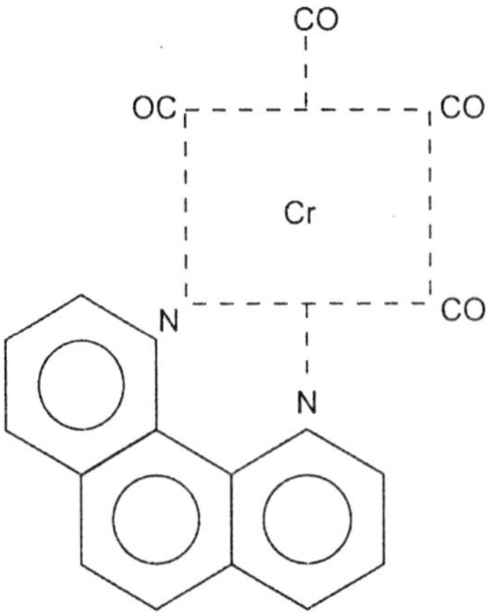


(i) Suggest the shape of the complex (ii) What feature of the structure of a nitrogen atom makes it possible for it to take part in this sort of complex? (iii) What type of ligand 1, 10 phenanthroline is in the complex? (iv) What is the oxidation state of chromium in the complex? (v) What is the coordination number of chromium in the complex? (vi) The complex has no stereoisomers—suggest reason for this. **[C.U. 1997]**

**Ans. :** (i) Octahedral ($O_h$), (ii) weak $\pi$-acidic property as a ligand, (iii) bidentate chelating ligand having soft character, (iv) zero, (v) six, (vi) 1, 10 phenanthroline occupies two cis positions so only one structure is possible.

---

**Problem 7.54** More stable state of Cu is +2 while that of Au is +3 state—Explain it from CFT. **[C.U. 1997, B.U. 2006, 2008]**

**Ans. :** Cu(II) is $3d^9$ while Au(II) is $5d^9$. We know that $\Delta_0$/10Dq value increases from the first transition series to the second and then to third transition series. It is evident from tetragonal elongation that the ninth electron of Au(II) is at much larger energy than that of Cu(II). Hence the ninth electron is susceptible to oxidation and thereby ready oxidation of Au(II) to Au(III) occurs.

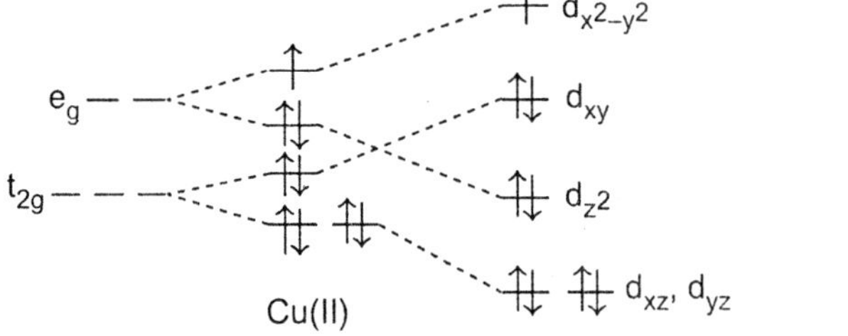


---

**Problem 7.55** Distinguish between the possibilities in complex ions of $\Delta$ = 0 and CFSE = 0. Give an example of each. **[B.U. 2009, 2000]**

**Ans. :** $\Delta$ = 0 is for free metal ion and for spherically symmetric ligand environment where no possibility of splitting of d-orbitals.

CFSE = 0 for $d^0$, $d^5$ (high spin) and $d^{10}$ systems.

**Problem 7.56** Draw a crystal field splitting diagram for a complex in a linear field. Assume that the ligands lie on the z-axis. **[B.U. 2008, 2011]**

**Ans. :** In a linear field with two negative charges (ligands) on the z-axis, the $d_{z^2}$ orbital will be highest in energy. An electron in this orbital will be located in a region of space directly adjacent to the negative charges. The $d_{yz}$ and $d_{xz}$ orbitals will be degenerate and are next highest in energy. The $d_{xy}$ and $d_{x^2-y^2}$ orbitals will be degenerate (both lying in the xy plane) and lowest in energy. In case of p-orbitals the $p_z$ orbital will be in highest positon and the two degenerate $p_x$ and $p_y$ orbitals will be in lowest position.

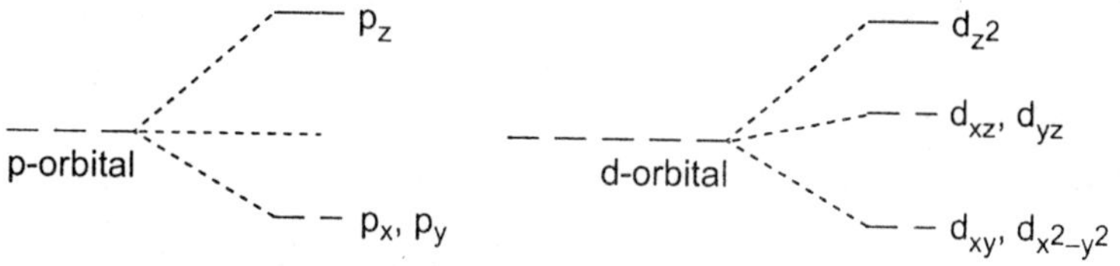


**Problem 7.57** "$OH^-$ ion is in lower position than $H_2O$ in spectrochemical series".—Explain. **[B.U. 2008, 2016]**

**Ans. :** π-donor property of the ligands reduces the positive charge on the central metal and consequently the $\Delta_o$/10Dq value decreases. In $H_2O$ out of the two lone pairs one is involved in σ-donation while the other lone pair is available for π-donation.

In $OH^-$, out of the three lone pairs, one lone pair is involved in σ-donation while the other two lone pairs are involved in π-donation. Thus $OH^-$ is a better π-donor ligand than $H_2O$. In $OH^-$, the lone pairs engaged in a π-donation bear more p-character compared to the lone pair of $H_2O$ where the lone pair engaged in a π-donation is basically residing in a $sp^3$ hybrid orbital.

Weak σ and strong π donor property produce small $\Delta_o$ while strong σ donor and weak π donor or strong π acceptor property produce large $\Delta_o$. $OH^-$ has two π donor electron pairs while $H_2O$ has only one and so larger $\Delta_o$ is expected for $H_2O$ than that of $OH^-$.

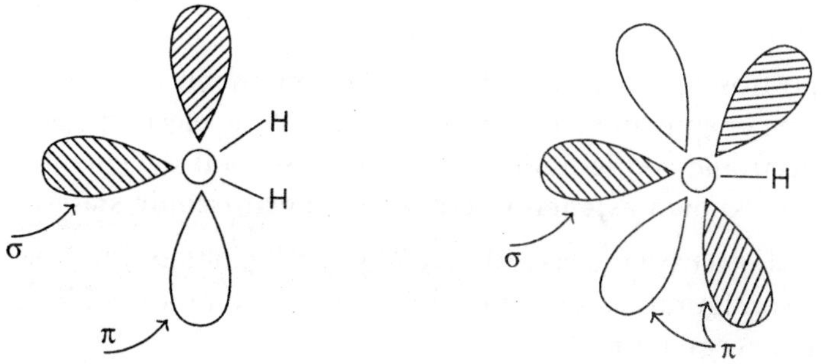


**Problem 7.58** "$Fe_3O_4$ has an inverse spinel structure while $Mn_3O_4$ has a normal spinel structure." Explain the observation from CFT. **[B.U. 2008, 2013]**

**Ans. :** CFSE calculation for $Mn_3O_4$ and $Fe_3O_4$ is to be shown according to Sec. 7.14.5(B).

For $Mn^{3+}(3d^4)$ and $Fe^{2+}$ $(3d^6)$ the CFSE values are greater for octahedral than tetrahedral sites. Thus $Mn^{3+}$ and $Fe^{2+}$ ions will preferentially occupy the octahedral sites and maximising the CFSE values. Hence in $Mn_3O_4$ all the $Mn^{3+}$ ions occupy octahedral sites and all $Mn^{2+}$ ions are in the tetrahedral sites i.e. normal spinel structure i.e. $Mn^{II}[Mn_2^{III}]O_4$.

In $Fe_3O_4$ all the $Fe^{2+}$ ions and half of the $Fe^{3+}$ ions are in the octahedral sites, while the remaining half of the $Fe^{3+}$ ions occupy tetrahedral sites. Thus $Fe_3O_4$ has an inverted spinel structure i.e. $Fe^{III}[Fe^{II}Fe^{III}]O_4$.

---

**Problem 7.59** In terms of CFT, cite an example in which the term 'strong field' does not imply 'low spin'. **[B.U. 2008]**

**Ans. :** Strong field but no possibility of spin pairing is cited by $3d^1$ system.

$e_g$ — —  $t_2$ — — —

$t_{2g}$ ↿ — —  $e$ ↿ —

octahedral  tetrahedral

---

**Problem 7.60** In $CrF_2$ the Cr-F bonds are unequal but in $MnF_2$ the Mn-F bonds are equal.—Explain.

**Ans. :** $CrF_2$ has a distorted octahedral rutile structure. $Cr^{2+}$ having $d^4$ configuraiton (high spin $t^3_{2g}e^1_g$) is susceptible to Jahn-Teller distortion thereby there will be two long Cr-F bonds (243 pm) and four short Cr-F bonds (200 pm).

In case of $MnF_2$ the $Mn^{2+}$ ($3d^5$, $t^3_{2g}e^2_g$) has symmetrically filled $t_{2g}$ and $e_g$ orbitals and for $Mn^{2+}$ Jahn-Teller distortion is not permitted. Therefore all the Mn-F bonds are equal in $MnF_2$.

---

## 7.15 ORIGIN OF COLOUR IN COORDINATION COMPLEXES

One of the important application of crystal field theory is to explain the colour of coordination complexes which arises due to two aspects (1) d-d transitions—known as d-d spectra or ligand field spectra/crystal field spectra, (2) charge transfer transitions—known as charge transfer spectra.

The crystal field spectra of coordination complexes arise due to the changes in the electronic arrangements in the d-orbitals (i.e. due to d-d transition) because of the absorption of electromagnetic radiation by the complexes. In order to understand the d-d spectra the working knowledge of the following is necessary :

(a) Atomic spectroscopy which deals with the consequences of interelectronic repulsions and coupling of spin and angular momentum of the electrons. Such interactions within a given electronic configuration give rise to several degenerate and non-degenerate energy levels known as **energy terms** and **spectroscopic states**.

(b) The effect of electrostatic crystal fields by the ligands on the energy terms of a given electronic configuration of a metal ion in its complex which very often split the degenerate energy terms.

(c) Application of selection rules to characterise the spectral lines.

### 7.15.1 Energy states and Term symbols or Spectroscopic states

The atomic spectra of many electron atoms are very complex in nature—they contain many lines. The energy states of the atom corresponding to these transitions cannot be calculated as easily like that of hydrogen like species having only one electron. In case of many electron system we have to consider the mutual interaction between the electrons. Experimental spectroscopists have developed a procedure to assign the various energy states of an atom corresponding to observed spectral transitions.

There are two ways in which the energy states of a configuration may be derived :

(1) We can write all electronic arrangements of a configuration (microstates) and calculate their energies due to actual interelectronic repulsion and magnetic interactions. It is a very laborious process and is not practically used.

(2) The second way is to apply orbital coupling, spin-spin coupling and spin-orbit coupling effects which are consequences of interelectronic repulsions and magnetic interactions of a configuration and find out the energy states.

However both the processes lead to the same result. We shall discuss here in brief the process (2) i.e. the energy states are derived by spin-spin, orbital and spin-orbit coupling effects on electrons of a configuration.

### 7.15.2 Coupling schemes

The orbital angular momentum and the spin angular momentum of the individual electrons in a multielectron system may be combined on the principle of vector addition to obtain the resultant spin and angular momentum of the system. The interactions between the spin and orbital angular momentum of many electrons system may be considered in terms of one electron quantum numbers (n, l, $m_l$ and $m_s$) of the individual electrons. For many electrons system there may be three types of interactions : (1) orbit-orbit coupling in which the orbital angular momentum of the electrons couple, (2) spin-spin coupling in which the spin angular momentum of the electrons couple, (3) spin-orbit coupling in which the spin and orbital angular momentum of the electrons couple. The coupling of the spin of one electron with the orbital angular momentum of a different electron is small and can be ignored. There are two ways to consider such interactions :

(a) **Russel-Saunders coupling scheme** which is applied to configurations of atoms or ions where interelectronic repulsions are much stronger than magnetic effects. This scheme is applicable for lighter atoms.

(b) **j-j coupling scheme** which is applied where magnetic effects acting on electrons of a configuration are stronger than interelectronic repulsion effect. This is generally applied in case of heavier atomic species.

The Russel-Saunders coupling scheme is more important than j-j coupling.

### 7.15.3 Russel-Saunders coupling or L-S coupling

This scheme is applicable for most of the transition metal ions which are of interest in inorganic chemistry. In this scheme it is assumed that :

spin-spin coupling > orbit-orbit coupling > spin-orbit coupling

This scheme involves the following steps :

(1) The first step is to find out the resultant orbital angular momenta (L) ($l_1 + l_2 + l_3$ ...... = L) and spin angular momenta (S) ($s_1 + s_2 + s_3$ ...... = S) of all electrons present in the configuration.

(2) The second step is to write all the possible terms in the configuration with the help of L and S. Some of these terms may readily exist and some may not.

(3) The third step is to determine the terms which really exist out of all the possible terms of a configuration.

(4) The fourth step is to couple the resultant orbital angular momentum (L) with resultant spin angular momentum (S) and thereby we get different values of resultant angular momentum (J).

### 7.15.4 Term symbols

Each electronic state can be designated by a symbol which is known as term symbol. For example the term symbol for ground electronic state of carbon is $3_P$, for $V^{3+}$ is $3_F$ etc. The modus operandi to establish the term symbols are the following :

(1) **The resultant orbital quantum number (L) :** The resultant orbital angular momentum ($M_L$) of the system is obtained by vector sum of individual $m_l$ values (obtained from l values) of all the electrons i.e.

$$M_L = \sum m_l$$

The total orbital angular momentum quantum number (L) which is equal to the maximum value of the sums $M_L$ of the angular momentum quantum numbers $m_l$ (e.g. for l = 2, $m_l$ = +2, +1, 0, −1, −2) of the individual orbitals being occupied i.e. L = maximum value of $M_L$. The energy states i.e. the terms of the system corresponding to different L values 0, 1, 2, 3, 4 etc. are designated as :

| | | | | | | | | | |
|---|---|---|---|---|---|---|---|---|---|
| L | = | 0 | 1 | 2 | 3 | 4 | 5 | 6 | 7 |
| State symbol | = | S | P | D | F | G | H | I | K |
| Degeneracy of L | = | 1 | 3 | 5 | 7 | 9 | 11 | 13 | 15 |

The orbital degeneracy is denoted by 2L + 1.

(2) **The resultant spin quantum number (S) :** The spin angular momentum is given by $\sqrt{s(s+1)}\frac{h}{2\pi}$ where s is the spin quantum number $\left(+\frac{1}{2}\right)$. The z component of this vector (chosen arbitrarily) is given by $m_s\left(\frac{h}{2\pi}\right)$ where $m_s$ spin quantum vector and this can be +s or −s i.e. $+\frac{1}{2}$ or $-\frac{1}{2}$ indicating either the spin up (↑) or spin down (↓). The resultant spin quantum number (S) which gives the resultant spin angular momentum as $\sqrt{S(S+1)}\frac{h}{2\pi}$. Algebric summation of $m_s$ values of individual electrons gives a resultant spin component vector $M_S$ i.e.

$$M_S = \sum m_s$$

$M_S$ has altogether (2S+1) values : S, (S−1)...0 − (S−1), −S. The quantity (2S+1) is known as **spin multiplicity** or **multiplicity of a state**. The total spin quantum number (S) is simply the maximum value of $M_S$ found among the microstates participating in that electronic state.

(3) **Resultant total angular momentum quantum number (J)** : This can be represented as :

$J = L + S = (l_1 + l_2 + l_3 + \ldots) + (s_1 + s_2 + s_3 + \ldots)$ and the total angular momentum quantum vector represented as $\sqrt{J(J+1)}\,\frac{h}{2\pi}$. J can have values L+S, L+S–1, ...... |L–S|.

The |L–S| means only the value of L–S is involved. Like j for a single electron, J can be either positive or zero. Therefore there may be the possibility of two cases :

(a) L > S, there are 2S+1 possible values of J,

(b) L < S, in this case there will be 2L+1 possible values of J.

Thus the term symbol for a polyelectronic system can be represented as :

$$2S + 1_{L_J}$$

Now we shall consider the ground state terms for few systems :

**Hydrogen atom ($1s^1$)** : Here l = 0, hence L = 0. This corresponds to an S state. For the single electron $S = \frac{1}{2}$ so that $2S + 1 = 2\times\frac{1}{2}+1 = 2$ and $J = L + S = 0 + \frac{1}{2} = \frac{1}{2}$. So the term symbol is $^2S_{\frac{1}{2}}$.

**Carbon atom ($1s^2 2s^2 2p^2$)** : For $ns^2$ or $np^6$ configurations i.e. for closed shell configurations $M_L$ and $M_S$ will be zero. So to establish the term symbol the closed shell electrons will not be considered. Therefore for carbon only two p electrons will be considered and the term symbols will be $^3P_2$, $^3P_1$ and $^3P_0$.

Similarly for $d^2$ configuration of a free metal ion ($V^{3+}$) the terms $^3S$, $^3P$, $^3D$, $^3F$, $^3G$ and $^1S$, $^1P$, $^1D$, $^1F$, $^1G$ are possible, out of which only $^3F$, $^3P$, $^1G$, $^1D$, $^1S$ actually exist.

## 7.15.5 Ground state term and Hund's rule

Once the term symbols representing electronic configurations for a given electronic configuration is determined, it is necessary to known the **ground state term** i.e. relative position of these energy states. This is done by the application of Hund's rule which states that so long orbitals are available electrons should occupy these orbitals with parallel spin as far as possible. The ground state term is determined by—

(1) The term with the greatest spin multiplicity lies lowest in energy. Thus the ground state of an atom should therefore possess the maximum number of unpaired spins.

(2) Among those of maximum spin multiplicity, the ground state term will be the one with the greatest total orbital angular momentum i.e. L value.

(3) Having the same spin multiplicity and same L value, the term having the lowest value of J is most stable.

Thus the ground state term for $d^2(V^{3+})$ is $^3F_2$ and the order of energy for carbon is $^3P_0 < {}^3P_1 < {}^3P_2$.

### 7.15.6 Ground term for $d^n$ ions

The calculation of energies of the terms of a particular electronic configuration is a complicated procedure. But we can calculate the ground term of any free metal ion configuration by applying Hund's rule of maximum multiplicity. Our interest is to calculate the ground term for $d^1$ to $d^{10}$ configurations which is related to explain the colour of transition metal complexes. Let us draw five boxes for the five d-orbitals (l=2 for d-orbitals) with their $m_l$ values in decreasing order written above the boxes. The electrons are filled in the boxes taking care that the electrons exist in the d-orbitals in such a way that these have maximum parallel spin and maximum $M_L$ (i.e. maximum $\Sigma m_l$). Since there are five degenerate d-orbitals, spin pairing does not occur upto $d^5$ configuration. The multiplicities of the ground terms from $d^1$ to $d^{10}$ are :

| $d^1$ | $d^2$ | $d^3$ | $d^4$ | $d^5$ | $d^6$ | $d^7$ | $d^8$ | $d^9$ | $d^{10}$ |
|---|---|---|---|---|---|---|---|---|---|
| 2 | 3 | 4 | 5 | 6 | 5 | 4 | 3 | 2 | 1 |

Now L has to be taken in such a way that $M_L$ obtained from the arrangement forms the highest value for this L. In this way L will have the same value as of $M_L$ obtained from the arrangement. Similarly S should be taken in such a way that $M_S$ obtained from this arrangement forms the highest value for the S. Here also S will have the same value as $M_S$ obtained from the arrangement. The spin multiplicity is calculated from 2S+1 and ground term is written from L and 2S+1. By this way the ground terms for $d^1$ to $d^{10}$ configurations of free metal ions are derived (Table 7.15). Let us take a trial for example to calculate the ground term for $d^4$ and $d^7$ metal ions :

**Table 7.15 : Ground state term for $d^n$ ions**

| $d^n$ ion | Example | Electronic arrangement $m_l$ | | | | | $M_L$ ($\Sigma m_l$) | $M_S$ ($\Sigma m_s$) | L | S | Ground term |
|---|---|---|---|---|---|---|---|---|---|---|---|
| | | +2 | +1 | 0 | –1 | –2 | | | | | |
| $d^1$ | $Ti^{3+}$ | ↑ | | | | | +2 | $+\frac{1}{2}$ | 2 | $\frac{1}{2}$ | $^2D_{3/2}$ |
| $d^2$ | $V^{3+}$ | ↑ | ↑ | | | | +3 | +1 | 3 | 1 | $^3F_2$ |
| $d^3$ | $Cr^{3+}$ | ↑ | ↑ | ↑ | | | +3 | $+\frac{3}{2}$ | 3 | $\frac{3}{2}$ | $^4F_{3/2}$ |
| $d^4$ | $Cr^{2+}$ | ↑ | ↑ | ↑ | ↑ | | +2 | +2 | 2 | 2 | $^5D_0$ |
| $d^5$ | $Mn^{2+}$ | ↑ | ↑ | ↑ | ↑ | ↑ | 0 | $+\frac{5}{2}$ | 0 | $\frac{5}{2}$ | $^6S_{5/2}$ |
| $d^6$ | $Fe^{2+}$ | ↑↓ | ↑ | ↑ | ↑ | ↑ | +2 | +2 | 2 | 2 | $^5D_4$ |
| $d^7$ | $Co^{2+}$ | ↑↓ | ↑↓ | ↑ | ↑ | ↑ | +3 | $+\frac{3}{2}$ | 3 | $\frac{3}{2}$ | $^4F_{9/2}$ |
| $d^8$ | $Ni^{2+}$ | ↑↓ | ↑↓ | ↑↓ | ↑ | ↑ | +3 | +1 | 3 | 1 | $^3F_4$ |
| $d^9$ | $Cu^{2+}$ | ↑↓ | ↑↓ | ↑↓ | ↑↓ | ↑ | +2 | $+\frac{1}{2}$ | 2 | $\frac{1}{2}$ | $^2D_{5/2}$ |
| $d^{10}$ | $Zn^{2+}$ | ↑↓ | ↑↓ | ↑↓ | ↑↓ | ↑↓ | 0 | 0 | 0 | 0 | $^1S_0$ |

| | +2 | +1 | 0 | −1 | −2 | |
|---|---|---|---|---|---|---|
| $d^4$ : | ↑ | ↑ | ↑ | ↑ | | l = 2 |

$$m_l = +2, +1, 0, -1, -2$$
$$M_L = 2 + 1 + 0 - 1 = 2, L = 2$$
$$M_S = \frac{1}{2} + \frac{1}{2} + \frac{1}{2} + \frac{1}{2} = 2, S = 2$$
$$2S + 1 = 5 \text{ and } J = L - S = 2 - 2 = 0$$

So that term symbol is $^5D_0$.

| | +2 | +1 | 0 | −1 | −2 | |
|---|---|---|---|---|---|---|
| $d^7$ : | ⇅ | ⇅ | ↑ | ↑ | ↑ | l = 2 |

$$m_l = +2, +1, 0, -1, -2$$
$$M_L = +2 + 1 + 0 - 1 - 2 + 2 + 1 = 3, L = 3$$
$$M_S = \frac{1}{2} + \frac{1}{2} + \frac{1}{2} + \frac{1}{2} + \frac{1}{2} - \frac{1}{2} - \frac{1}{2} = \frac{3}{2}, S = \frac{3}{2}$$
$$\therefore 2S + 1 = 2 \times \frac{3}{2} + 1 = 4 \text{ and } J = L + S = 3 + \frac{3}{2} = \frac{9}{2}$$

It is to be noted that J will be L–S from $d^1$ to $d^5$ and from $d^6$ onwards J = L+S.

Therefore the term symbol for $d^7$ ion is $^4F_{9/2}$. In this way for other systems the ground term may be worked out.

### 7.15.7 Terms arising from $d^n$ configurations in crystal fields: Spectroscopic terms in various crystal fields

In order to explain the spectra of complexes in which the metal ion have $d^1$ to $d^9$ configurations we have to establish various spectroscopic terms (i.e. free metal ion ground term is splitted in crystal field) between which the transitions of electrons occur to give rise absorption bands (d-d transition). We shall only consider octahedral and tetrahedral field and in this case also the Russel-Saunders ground terms are splitted into the states $E_g$ and $T_{2g}$ in octahedral geometry and the states E and $T_2$ in tetrahedral geometry (Table 7.16) where E denotes

**Table 7.16 : The splitting of Russel-Saunders terms in octahedral and tetrahedral* symmetry**

| Atomic term | Number of states | States in octahedral field ($O_h$) | Configuration | Ground term |
|---|---|---|---|---|
| S | 1 | $A_{1g}$ | $d^1$, $d^9$ | $^2D$ |
| P | 3 | $T_{1g}$ | $d^2$, $d^8$ | $^3F$ |
| D | 5 | $T_{2g}+E_g$ | $d^3$, $d^7$ | $^4F$ |
| F | 7 | $T_{1g}+T_{2g}+A_{2g}$ | $d^4$, $d^6$ | $^5D$ |
| G | 9 | $A_{1g}+E_g+T_{1g}+T_{2g}$ | $d^5$ | $^6S$ |
| H | 11 | $E_g+T_{1g}+T_{1g}+T_{2g}$ | | |
| I | 13 | $A_{1g}+A_{2g}+E_g+T_{1g}+T_{2g}+T_{2g}$ | | |

* *In tetrahedral symmetry the terms are the same but g subscript will be dropped.*

the energy level resulting from the splitting of the free ion state while e denotes a set of two orbitals (e.g. $d_{x^2-y^2}$, $d_{z^2}$). Similar is the case for $T_2$ and $t_2$. Once again it should be rembered that g subscript is applied for octahedral field while it is omitted in case of tetrahedral one. It is to be noted that—

(a) Orbital multiplicity is retained in the splitted component terms. Thus an F term (L=3, orbital multiplicity = 7) splits into one singlet term ($A_2$) and two triplet terms ($T_1$ and $T_2$), so that the total degeneracy = 1+3+3=7.

(b) S and P states are not split in the crystal field. Thus S term gives rise to the $A_1$ term and P term gives a triplet term $T_1$ in the ligand field.

The terms A, T, E etc. are symbolised as per Mulliken symbol as :

| Symbol for atomic and molecular orbitals | Explanation |
|---|---|
| a, A | Non-degenerate orbital (singlet); symmetric to principal $C_n$ axis. |
| b, B | Non-degenerate orbital (single); unsymmetric to principal $C_n$ axis. |
| e, E | Doubly degenerate (doublet) orbital. |
| t, T | Triply degenerate (triplet) orbital. |
| g | Gerade, even. |
| u | Ungerade, uneven or odd. |

### 7.15.8 Energy level diagram (Orgel Diagram)

The diagrammatic reprsentations of the effect of crystal field strength on the energies of the states arising due to free ion ground term configurations of $d^1$ and $d^9$ is shown in Fig. 7.18. The right side of the diagram ($d^1$ configuration) is the reverse of the left side ($d^9$ configuration) which is in accordance with the **hole formalism rule** which states that the splitting pattern of

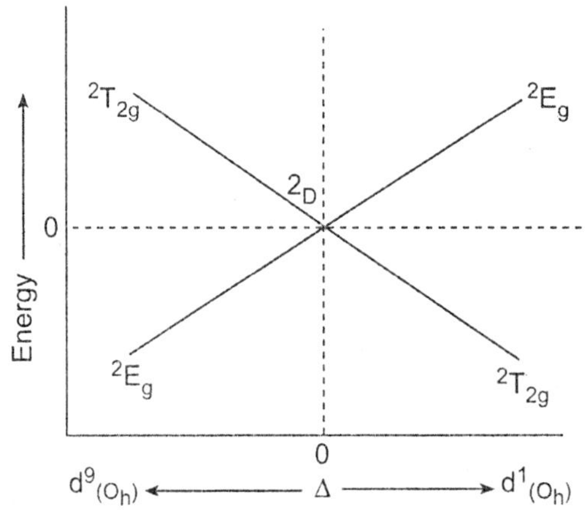


*Figure 7.18 : Energy level diagram for the splitting of the free ion ground term $^2D$ ($d^1$ and $d^9$ configuration) in $O_h$ symmetry.*

energy levels of $d^{10-n}$ configuration will be similar to that of the $d^n$ configuration under identical environment e.g. tetrahedral and octahedral, with the difference that the order of energy levels of the splitting of ground term will be reverse. This type of energy level diagram (Fig. 7.18) is known as **Orgel diagram**. Due to the Jahn-Teller distortion in $d^9$ configuration $E_g$ and $T_{2g}$ states will be further splitted (Fig. 7.17).

Now applying the hole formalism rule we can draw qualitatively the Orgel diagram for high spin $d^4$ and $d^6$ and for other $d^n$ configurations (Fig. 7.19). It is to be noted that quantitative Orgel diagram is sometimes called **Tanabe-Sugano diagram**. In this diagram (Fig. 7.19) the multiplicity of the term is ignored for simplification.

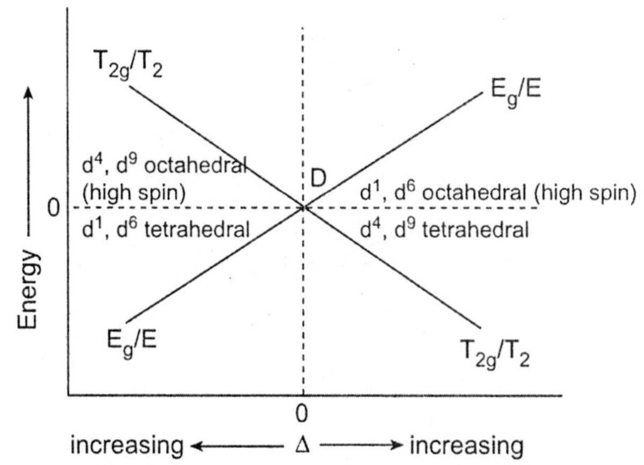


***Figure 7.19** : Orgel diagram for ground D term in octahedral and tetrahedral symmetry.*

Similarly the Orgel diagram for $d^2$ and $d^8$ system with Russel-Saunders terms $^3F$ and $^3P$ of which $^3F$ is the ground term (Table 7.16) is shown (Fig. 7.20). This diagram is useful for the interpretation of spectra of octahedral $Ni^{2+}$ complexes e.g. $[Ni(NH_3)_6]^{2+}$, $[Ni(en)_3]^{2+}$ etc.

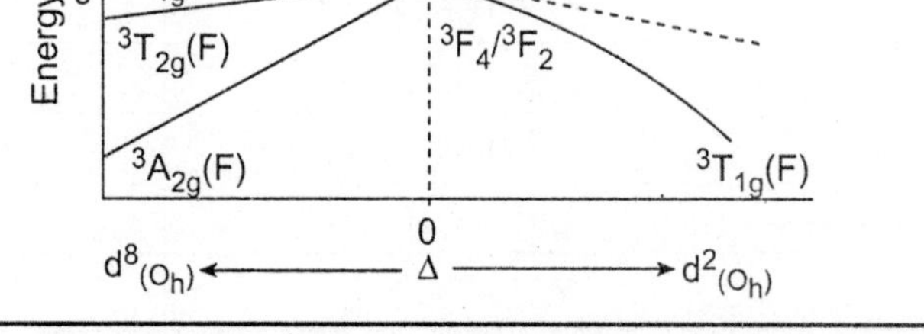


***Figure 7.20** : Energy level diagram (Orgel diagram) for the splitting of $^3F$ and $^3P$ free ion terms of $d^2$ and $d^8$ configurations.*

The following points are to be noted for Orgel diagrams :

(a) In a crystal field the spectroscopic terms have the same multiplicity as that of the free ion term from which they originate.

(b) States having identical designation i.e. having the same symmetry obey the **noncrossing rule** which states that if the increasing crystal field causes two weak field terms of the same symmetery to approach, they do not cross each other but bend apart and hence their lines will show a curvature. There had been no mixing (or crossing) of the states $T_{1g}(F)$ and $T_{1g}(P)$ in Fig. 7.20 which is indicated by the hypothetical dotted lines.

### 7.15.8.1 Colour of Transition Metal Complexes

The colours exhibited by coordination complexes can be explained by crystal field theory (CFT). In CFT the d-orbitals of such metal complexes in $O_h$ and $T_d$ symmetry split into two sets and the molecules absorb energy of visible light thereby one or more electrons may momentarily jump from the lower energy d-orbitals to the higher energy state. The difference in energy between the two sets is equal to the energy of the absorbed light and related inversely to the wavelength of the light. Because only certain wavelength of light are absorbed which are maching exactly the energy difference between the two sets and the complexes looks in the complementary colour. The higher energy light has smaller wavelength i.e. higher frequency. The gradation of different radiation can be made as :

γ-ray > X-ray > ultraviolet > visible > infrared > microwave > radio frequency

$\xrightarrow{\hspace{10cm}}$

energy decreases

$\xrightarrow{\hspace{10cm}}$

wavelength increases

In fact the absorption of light is a subtractive process i.e. a solution that appears blue is not absorbing blue light but rather light that is the complementary colour to blue. A rough idea of complementary colours can be obtained from colour wheel; those colours that are opposite to one another are complementary. Red and green are complementary colours as blue green and purple as depicted below :

| Colour absorbed | Approximate wavelength (Å)* of the absorbed colour | Observed/transmitted /complementary colour |
|---|---|---|
| Violet | 4000-45000 | Yellow-green |
| Blue | 4500-4800 | Yellow |
| Green-blue | 4800-4900 | Orange |
| Blue-green | 4900-5000 | Red |
| Green | 5000-5600 | Purple |
| Yellow-green | 5600-5750 | Violet |
| Yellow | 5750-5900 | Blue |
| Orange | 5900-6250 | Green-blue |
| Red | 6250-7500 | Blue-green |

* 10Å = 1 mμ (millimicron) = 1 nm (nanometre)

The above table shows that a solution which appears blue is actually absorbing yellow light and a solution which appears orange is actually absorbing green-blue light. Thus the complex $[Ni(H_2O)_6]^{2+}$ which is green in colour is actually absorbing purple colour. Since red/purple light has a longer wavelength (lower energy) than green light, the $NH_3$ is producing stronger ligand field splitting than $H_2O$ which is consistent with spectrochemical series.

### 7.15.9 Selection Rules

Before proceeding further it is necessary to consider the selection rule which denotes the actual transitions among the splitted crystal field terms. From detailed theoretical consideration (which is unnecessary to discuss here) two important selection rules are framed for d-d transition :

(1) **Spin selection rule :** According to which the electronic transitions from a lower to a higher electronic energy level of a molecular species will take place provided the two energy levels have the same multiplicity e.g. ${}^3T_{1g} \rightarrow {}^3T_{2g}$ transition is allowed but ${}^3T_{1g} \rightarrow {}^1T_{2g}$ is spin forbidden. For spin sllowed transition $\Delta S = 0$ and for spin forbidden transition $\Delta S \neq 0$. Due to spin-orbit coupling the spin forbidden transitions may be slightly allowed but the intensity of the absorption bands are very poor compared to spin allowed transitions.

(2) **Orbital selection rule or Laporte selection rule :** From the consideration of *transition dipole moment integral** this selection rule states that *in a centrosymmetric molecule* or *ion an allowed transition must involve orbitals having different symmetry properties with respect to inversion* i.e. gerade (g) to ungerade (u) transitions or vice-versa are allowed i.e. $g \longleftrightarrow u$ but $g \longleftrightarrow g$ or $u \longleftrightarrow u$ are not allowed. Since d-orbitals are g type and hence d-d transition i.e. $g \longleftrightarrow g$ transition is forbidden. Similarly f-f transition is also forbidden. Thus the allowed transitions for a system will be accompanied by $\Delta L = \pm 1$. Therefore $d \rightarrow p$, $s \rightarrow p$ transitions are allowed but $d \longleftrightarrow d$ and $s \longleftrightarrow d$ transitions are not.

### 7.15.10 Interpretation of Electronic Absorption Spectra of complexes

Orgel diagrams are used for the interpretation of the spin allowed absorption bands of crystal field or d-d origin in the electronic spectra of tetrahedral and octahedral transition metal complexes. Sometimes the spin forbidden transitions i.e. electronic transitions between energy levels of different spin multiplicity can also appear as absorption bands but the intensities of such bands is much lower than the intensities of the spin allowed transition. The intensities of the absorption bands in tetrahedral complexes are higher than in octahedral complexes due to d-p mixing. Due to this it is not always easy to distinguish between d-d transition and charge transfer transitions in case of tetrahedral complexes. Table 7.17 records some spectra of transition metal complexes. It is to be noted that the intensities of charge transfer bands is highest in comparison to other transitions which is registered by the following molar absorption coefficient (ε) :

* Transition dipole moment integral, $Q = \int_{-\infty}^{+\infty} \Psi_1 |\vec{M}| \Psi_2 d\tau$, $\vec{M}$ is the electric dipole moment operator.

| **Transitions** | **$\varepsilon$(L mol$^{-1}$ cm$^{-1}$)** |
|---|---|
| Spin forbidden ($\Delta S \neq 0$) | < 1 |
| Laporte forbidden ($\Delta L \neq \pm 1$) | 10–100 |
| Laporte allowed ($\Delta L = \pm 1$) | 250 |
| Charge transfer ($\Delta S = 0$, $\Delta L = \pm 1$) | $5 \times 10^3 - 35 \times 10^3$ |

**Table 7.17 : Some classification of spectra of complexes**

| Example | Type of transition | Classification | Value of $\varepsilon$ |
|---|---|---|---|
| $[Mn(H_2O)_6]^{2+}$ (pale pink) | d $\rightarrow$ d (change of multiplicity) | Spin forbidden ($\Delta S \neq 0$) Laporte forbidden ($\Delta L \neq \pm 1$) | 0.1 |
| $[Ti(H_2O)_6]^{3+}$ (purple) | d $\rightarrow$ d (no change of multiplicity) | Spin allowed, Laporte forbidden | 10 |
| $[CoCl_4]^{2-}$ (intensely blue) | d $\rightarrow$ dp (no change of multiplicity) | Spin allowed, partially Laporte allowed | $10^2$ |
| $[TiCl_6]^{2-}$ (yellow) | charge transfer | Spin allowed, ($\Delta S = 0$) Laporte allowed ($\Delta L = \pm 1$) | $10^4$ |
| $[Co(H_2O)_6]^{2+}$ (pale pink) | d $\rightarrow$ d | $\Delta S = 0$, $\Delta L \neq 0$ | 10 |

### 7.15.10(A) $d^1$ and $d^9$ case

The relevant Orgel diagrams (Fig. 7.18) indicates transition $^2T_{2g} \rightarrow {}^2E_g$ for $d^1$ (octahedral) system. Thus the observed maxima in the absorption spectra of $[Ti(H_2O)_6]^{3+}$ is assigned the said transition at around 20,400 cm$^{-1}$ and the aqueous solution of the complex is purple in colour. This transition is spin allowed but Laporte forbidden [Fig. 7.21(a)]. The broad peak is actually due to Jahn-Teller effect in the excited state.

Similarly for $d^9$ (octahedral) system there will be a single transition $^2E_g \rightarrow {}^2T_{2g}$. Thus the aqueous solution of $CuSO_4.5H_2O$ i.e. the complex species $[Cu(H_2O)_6]^{2+}$ absorbs in the visible region around 12,000 cm$^{-1}$ and the solution coloured blue. Here the Jahn-Teller distortion is

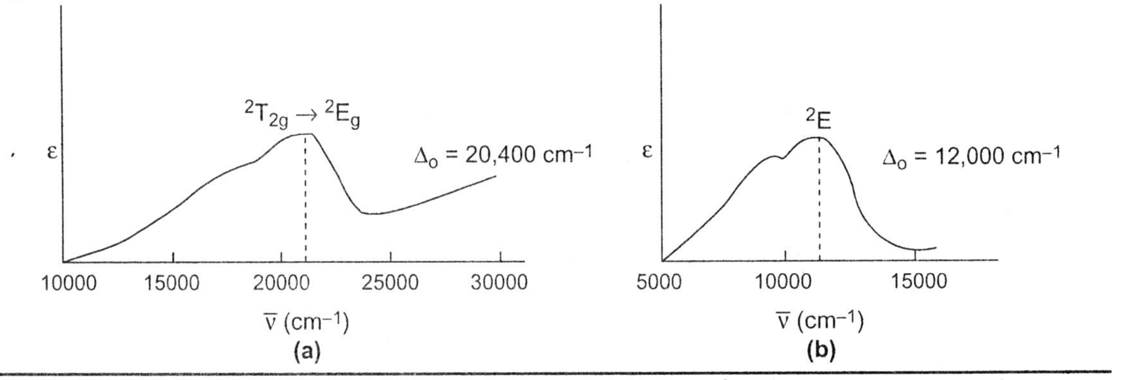


*Figure 7.21 : (a) Absorption spectra of $[Ti(H_2O)_6]^{3+}$ ($d^1$ system), (b) Electronic absorption spectra of blue aqueous solution of $[Cu(H_2O)_6]^{2+}$.*

considerably greater than that of $[Ti(H_2O)_6]^{3+}$ ($d^1$ system) and overlap of several bands gives rise to a broad asymmetric band [Fig. 7.21(b)] having a long tail into the near infrared region.

### 7.15.10(B) $d^2$ and $d^8$ systems

The Orgel diagram (Fig. 7.20) for $d^2$ system indicates three transitions $^3T_{1g(F)} \rightarrow {}^3T_{2g(F)}$, $^3T_{1g(F)} \rightarrow {}^3T_{1g(P)}$ and $^3T_{1g(F)} \rightarrow {}^3A_{2g(F)}$. But actually only two absorption bands at 17,200 $cm^{-1}$ and 25,600 $cm^{-1}$ [Fig. 7.22(a)] are observed for first two transitions for the complex species $[V(H_2O)_6]^{3+}$ and the aqueous solution of the complex is green in colour. The third transition ($^3T_{1g(F)} \rightarrow {}^3A_{2g(F)}$) is forbidden.

For $d^8$ system e.g. $[Ni(H_2O)_6]^{2+}$ which has greenish blue colour in aqueous solution. The Orgel diagram (Fig. 7.20) indicates three transitions $^3A_{2g(F)} \rightarrow {}^3T_{2g(F)}$ (8,700 $cm^{-1}$), $^3A_{2g(F)} \rightarrow {}^3T_{1g(F)}$ (14,500 $cm^{-1}$) and $^3A_{2g(F)} \rightarrow {}^3T_{1g(P)}$ (25,300 $cm^{-1}$) [Fig. 7.22(b)]. For $[Ni(NH_3)_6]^{2+}$ three transitions occur at 10,750 $cm^{-1}$, 17,500 $cm^{-1}$ and 28,200 $cm^{-1}$ respectively while for $[Ni(en)_3]^{2+}$ these transitions are at 11,200 $cm^{-1}$, 18,350 $cm^{-1}$ and 29,000 $cm^{-1}$ respectively.

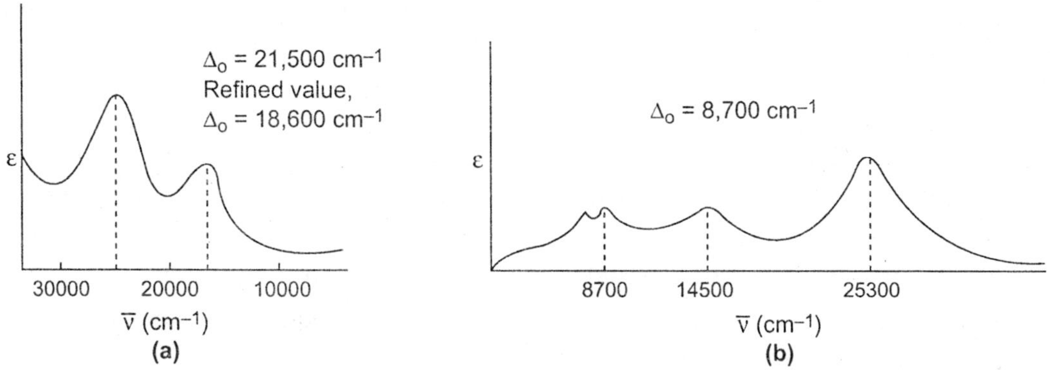


***Figure 7.22 :** (a) Ligand field spectra of perchlorate salt of vanadium compound i.e. $[V(H_2O)_6]^{3+}$, (b) Absorption spectra of $[Ni(H_2O)_6]^{2+}$.*

Similarly the energy level diagram for $d^3/d^7$ and $d^4/d^6$ systems for octahedral and tetrahedral symmetry can also be drawn and their absorption spectra can also be studied. Moreover the spectra for tetrahedral ($d^8$) $[NiCl_4]^{2-}$ has intense absorption bands compared to that of $[Ni(H_2O)_6]^{2+}$ due to the lack of centre of symmetry for tetrahedral complexes.

**Problem 7.61** Aqueous solution of Mn(II) sulfate is almost colourless, whereas aqueous solution of Cu(II) sulfate is blue.—Explain. **[C.U. 2007]**

**Ans. :** Mn(II) is a $3d^5$ system which has a $^6A_{1g}$ ground term in high spin configuration. For $Mn^{2+}$ the excited states are $^4T_{1g}$, $^4T_{2g}$, $^2A_{2g}$, $^2T_{1g}$ and all these states have different multiplicities compared to that of ground term. Hence all transitions in case of high spin $[Mn(H_2O)_6]^{2+}$ are doubly forbidden i.e. spin forbidden ($\Delta S \neq 0$) and Laporte forbidden ($\Delta L \neq \pm 1$). Thus all transitions have extermely poor intensity (very poor ε) and so the aqueous solution of Mn(II) salts is almost colourless or very pale pink in colour.

For Cu(II) solution see Sec. 7.15.10(A) $d^9$ system as also Fig. 7.21(b).

### 7.15.11 Charge Transfer Spectra

Certain $d^0$ and $d^{10}$ species like $K_2CrO_4$, $KMnO_4$ and $HgI_2$ etc. as also $FeCl_3$, $FeBr_3$ and $FeI_3$ (difficult to prepare) are intensely coloured. The intense colour of these compounds are not due to the d-d transition (as discussed earlier) but results from an electronic **charge transfer absorption**. Light energy is capable of transferring the electron from the valence orbital of the reducing ion (e.g. $Cl^-$ or $Br^-$) to the empty valence orbital of the oxidising ion i.e. $Fe^{3+}$ ion. Therefore charge transfer may be considered as a redox process. A metal having high ionisation energy which will have empty orbitals at fairly low energies will favour charge transfer from ligand to metal. Ligands having a nonmetallic donor atom with a relatively low electron affinity would be ideal which will have filled orbitals of fairly high energy and would be readily oxidisable. Since the light energy is to move in the same direction as the electron being transferred, therefore light energy has a high probability of being absorbed in the charge transfer process than that of d-d spectra. Hence in contrast to d-d transition, charge transfer absorption will produce intense colour. Due to the spin and Laporte allowed character of charge transfer absorption exhibit intense colour. Therefore we have to note the following points to distinguish d-d spectra and charge transfer spectra :

(a) In case of d-d spectra of metal complexes light absorption does not involve transfer of electron—it will reside on the same atom in the same type of orbital and hence absorption of light energy is much less compared to that of charge transfer absorption and so intensity of colour is much less in case of d-d spectra compared to that of charge transfer process.

(b) Charge transfer spectra is a redox process but d-d transition is not.

(c) Charge transfer spectra both spin and Laporte allowed but d-d spectra Laporte forbidden but may or may not be spin allowed.

(d) d-d spectra have low molar absorptivity ($\varepsilon$) compared to that of charge transfer spectra (CTS).

(e) In d-d spectra d electrons are involved while in CTS the d-electrons may not be essential.

(f) Analytical procedure for the estimation of very low concentration of metal ions is comparatively more suitable in case of CTS than that of d-d spectra.

(g) All the charge transfer bands are more intense compared to crystal field bands (d-d transition), the former sometimes obscure the later and the detection of d-d bands are somewhat difficult.

Three types of charge transfer transitions are to be considered :

(1) **Ligand to metal (L→M) charge transfer (LMCT) :** This type of spectra is very common in case of octahedral and tetrahedral complexes. Here electrons are transferred from $\sigma$ bonding orbitals or filled $\pi$ orbitals of ligands to the empty $T_{2g}$ or $E_g^*$ orbitals of metals. For LMCT the metal-ligand combination will be such that the energy difference is close enough so that transition may occur with light in the visible or near ultraviolet region. In other words LUMO of metal ion and HOMO of ligands will be close enough so that the energy difference between them is very small. Examples are :

$MnO_4^-$ (Purple) : Here four L→M transitions are observed but a transition at 17,700 $cm^{-1}$ in the visible region is responsible for its purple colour.

$FeI_3$ (Intensely black), $PbI_2$ (Yellow), $HgI_2$ (Red) : Metal ions in these cases are not good oxidant but charge transfer occurs (L→M) due to the fact that $I^-$ is easily oxidised.

CdS (Yellow), HgS (Red), PbO (Yellow), $PbCrO_4$ (Orange yellow) : Charge transfer occurs from ligands π orbital to metal respective ns, np and nd orbitals. Soft-soft interaction e.g. CdS, HgS may also have intense colour due to LMCT.

(2) **Metal to ligand (M→L) charge transfer (MLCT) :** MLCT is favourable in those cases where the ligands possess low-lying vacant orbitals and the metal can provide filled orbitals at a higher level. Here electrons from σ orbitals are excited to vacant π* orbitals localised on the ligands. MLCT is observed for octahedral complexes with CO, py, bpy, ophen etc. ligands. Here $T_{2g}$ and $E_g^*$ orbitals of metals are occupied and antibonding π* orbital of ligands are vacant so that transitions are : $E_g^* \to \pi^*$ and $T_{2g} \to \pi^*$. The deep red colour of $[Fe(SCN)_6]^{3-}$, $[Fe(ophen)_3]^{2+}$ and $[Fe(bpy)_3]^{2+}$ are examples of MLCT.

(3) **Metal to metal (M→M) charge transfer (MMCT) :** This is the less prevailage class of charge transfer spectra. MMCT occurs in those cases where a metal ion exist in two different oxidation states e.g. Prussian blue, $M^IFe^{III}[Fe^{II}(CN)_6]$, molybdenum blue or tungsten blue contains both the metals in (V) and (VI) oxidation states having non-stoichiometric composition e.g. $MoO_{2.5}(OH)_{0.5}$; $MoO_{2.88}{\cdot}H_2O$; $Mo_{2.67}(OH)_{0.33}$, black gold i.e. $Cs_2Au^IAu^{III}Cl_6$.

### 7.15.11.1 Limitations of Orgel Diagram

The Orgel diagram is only applicable in **high spin** complexes and the spin allowed transitions. In case of $d^5$ high spin configuration there is no spin allowed transitions. This diagram fails to explain the spectra for low spin octahedral complexes. Moreover in case of $d^1$, $d^9$ and high spin $d^4$ and $d^6$ systems only **one** spin allowed transition is expected. Similarly for $d^2$, $d^8$, high spin $d^7$ and $d^8$ systems only **three** spin allowed transitions are expected. But in most of the cases the actual spectra differ from what is predicted from Orgel diagram because in those cases apart from the spin allowed transitions there will be some weak spin forbidden transitions. So in order to explain the actual spectra of complexes we have to modify simple Orgel diagram to a more practicable quantitative Orgel diagram which is known as Tanabe-Sugano diagram proposed in 1954.

### 7.15.11.2 Racah Parameters

We have already discussed term symbols for different $d^n$ systems (Sec. 7.15.6). In constructing the term symbols no weightage has been given for the **repulsive energy** among the electrons. Coulombic repulsion exists among electrons of a complex, therefore different terms of a configuration have different energies. The energies of states arising from different $d^n$ configuration incorporating the interelectronic repulsion can be expressed by two parameters namely Condon-Shortley parameters and Racah parameters. Brief discussion of Racah parameters is made here because this is related to Tanabe-Sugano diagram.

The interelectronic repulsive energy of each term of a $d^n$ configuration is expressed as sums of three quantities. These three different sums of integrals are called **Racah parameters*** and these three parameters are denoted as A, B and C. Actually these Racah parameters are obtained by linear combination of **exchange** and **coulomb** integrals. Accurate evaluation of these parameters are difficult. These parameters are calculated from spectral data of free ion e.g. for a $d^2$ configuration the possible terms are given below :

$E(^1G) = A + 4B + 2C$ $\quad$ $E(^1D) = A - 3B + 2C$

$E(^3F) = A - 8B$ $\quad$ $E(^3P) = A + 7B$

$E(^1S) = A + 14B + 7C$

The evaluation of Racah parameters i.e. A, B, C can be made from the observed energies of the terms obtained from spectral data. The actual energies of the terms can be determined spectroscopically and the order is found as $^3F < {}^1D < {}^3P < {}^1G < {}^1S$ and the predicted order from Hund's rule is $^3F < {}^3P < {}^1G < {}^1D < {}^1S$. This involves the observation of transition involving both triplet and singlet states.

The following points are to be noted for Racah parameters :

(1) A term is common for all the energy terms. To consider the relative energies of the terms it is not required to know the value of A. The energy difference between the triplet terms (i.e. $^3P$ and $^3F$) is 15B for which we need only the knowledge of B term. Similarly to know the energy difference between $^1D$ and $^1G$ terms only B term is required. To know the energy difference between the terms $^3F$ and $^1D$ having different spin multiplicity we require the values of both B and C. A term is not required to know the energy difference between the same or different spin multiplicity terms.

(2) For most of the first transition metal ions $C/B \approx 4$ and B values are about 1000 $cm^{-1}$ as determined from spectral data. The interelectronic repulsive parameters B and C are decreased in the complexes compared to their values in the free ions. The B values decreased about 70% during complexation.

(3) As stated the B values are reduced during complexation. Such reduction in interelectronic repulsion can be obtained by the overlap of the metal d-orbitals and the ligand orbitals. The metal d-orbitals expand in size and penetrate into ligand orbitals so that repulsion between metal d-orbitals becomes smaller than in free ion. Moreover, the ligand electrons partially screen the metal d-orbital electrons from the nuclear charge of the metal ion. Thus the electron density of the d-orbitals will spend some time to the ligand orbitals i.e. the metal d electrons are partially delocalised into the ligand orbitals. As a result the average distance between d-orbitals thus increases. **This influence of expansion of metal d-electron clouds is known as nephelauxetic effect or cloud expansion effect.** This expansion effectively increases the separation between the d-d orbitals. As a result the interelectronic repulsion decreases. It is to be noted that for higher oxidation state of a metal ion, the lower is the size and thereby higher will be its interelectronic repulsion parameter e.g. B values of $Co^{2+}$ and $Co^{3+}$ are 971 $cm^{-1}$ and 1100 $cm^{-1}$ respectively.

* More details about Racah parameters are discussed in authors' book 'Comprehensive Coordination Chemistry'.

As a consequence of the nephelauxetic effect the average interelectronic repulsion is reduced and the reduction parameter is given by another term **nephelauxetic ratio** $\beta = B'/B$ where B′ is the interelectronic repulsion parameter for the complex while B denotes the same for free ion. The β value is always less than unity and depends on both the metal ion and the ligand. Ligands may be arranged according to the decreasing order of β for a particular metal ion and such arrangement of ligands is known as **nephelauxetic ligand series** and may be depicted as :

$F^- > H_2O >$ urea $> NH_3$ en $\approx C_2O_4^{2-}$, glyH $> NCS^- > Cl^- > CN^- > Br^- > S^{2-}, I^-$

For $F^-$ the β value ~1 and 0.3 for $I^-$

For a given ligand and a given geometry the **nephelauxetic metal ion** series can also be formed which is given below :

$Mn^{2+} \approx V^{2+} > Ni^{2+} \approx Co^{2+} > Mo^{3+} > Cr^{3+} > Fe^{3+} > Ir^{3+} > Co^{3+} > Mn^{4+} \approx Pt^{4+} > Ir^{6+} > Pt^{6+}$

The smaller the value of β, the more polarising a metal ion and the extent of covalent bonding between the metal and the ligands can be predicted.

(4) In a Tanabe-Sugano diagram, the spectral data is plotted by E/B where E is the energy and the ligand field strength is expressed as $\Delta_o/B$ or 10Dq/B.

### 7.15.11.3 Tanabe-Sugano Diagram

As mentioned earlier Orgel diagram is not sufficient to explain the quantitative electronic spectra of complexes (Sec. 7.15.11.1). The limitations of Orgel diagram are more or less removed in Tanabe-Sugano diagrams (TS diagrams) which has the following important features (Fig. 7.22(A)) :

(1) The TS diagrams were first proposed by the Japanese workers Tanabe and Sugano which is the elaborated version of Orgel diagrams. These diagrams provide an alternative way of denoting the variation of term energies with field strength.

(2) These TS diagrams are obtained by plotting E/B along y-axis against $\Delta_o/B$, Dq/B or 10Dq/B along x-axis for a particular $d^n$ configuration.

(3) These TS diagrams include the energy states of both the high and low spin complexes particularly in case of $d^4$, $d^5$, $d^6$ and $d^7$ systems i.e. these diagrams can be applied for strong and weak field cases.

(4) In these TS diagrams the abscissa i.e. x-axis represents the ground state while in the Orgel diagram the energy of the ground state decreases with the increase of the ligand field strength. Thus in this TS diagram the **lowest energy term is taken as zero irrespective of ligand field strength.** The energies of the other states are plotted relative to that of ground state energy, therefore the TS diagrams is discontineous when there is a change in the ground state i.e. when the ligand field becomes strong enough to produce spin pairing.

(5) Some lines in these diagrams are non-linear because of the mixing of terms of the same symmetry type.

(6) In these TS diagrams the energies of the states and crystal field strength are expressed in terms of B which is different for the different isoelectronic systems. Thus in these diagrams the energies of the states are plotted in the unit of B and so a particular diagram for a $d^n$ configuration can be used for all members of a particular isoelectronic systems. Therefore **one advantage of TS diagrams over Orgel diagrams is that this TS**

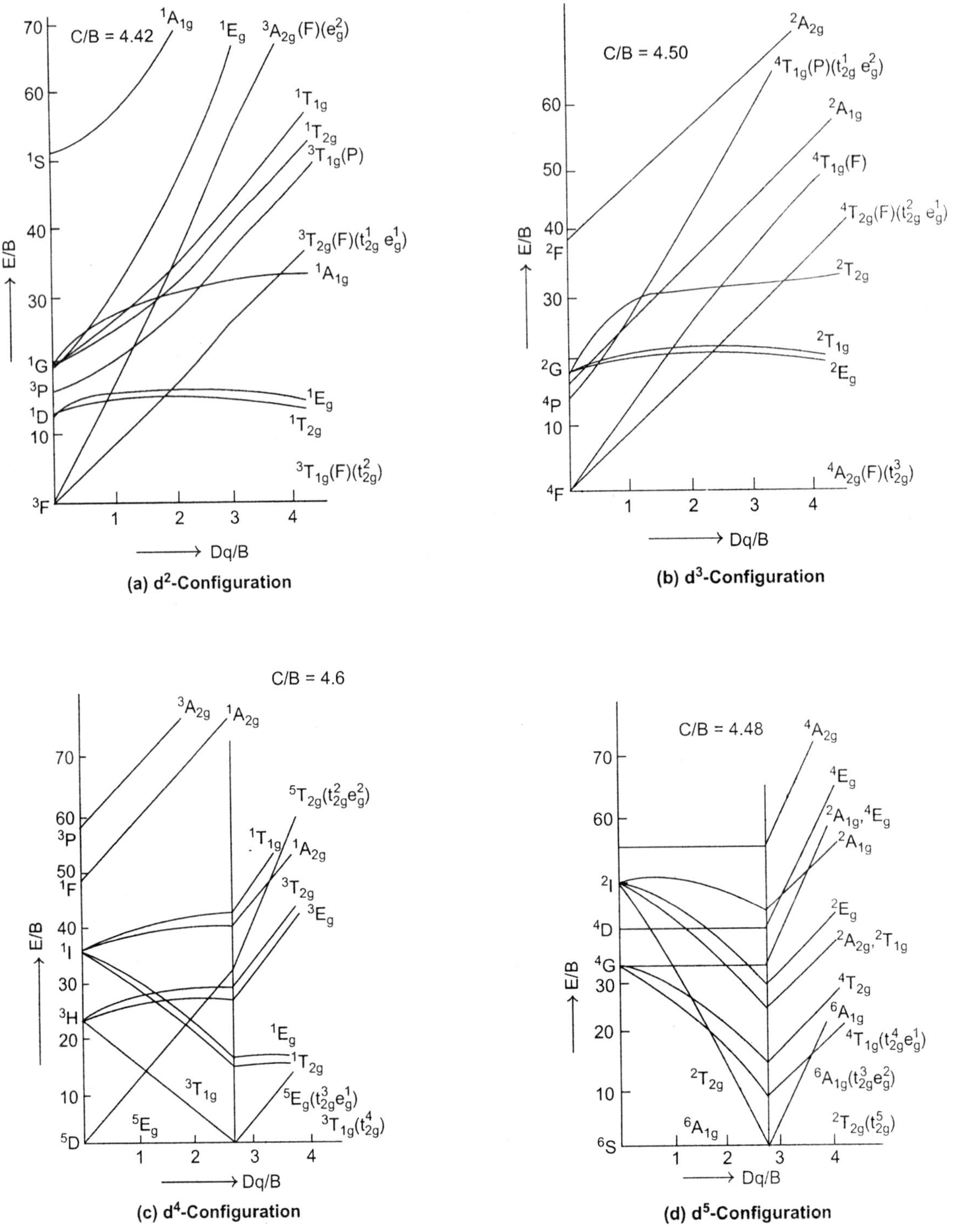


***Figure 7.22(A)** : Tanabe-Sugano diagrams for different $d^n$ configurations in octahedral field.* **(contd. next page)**

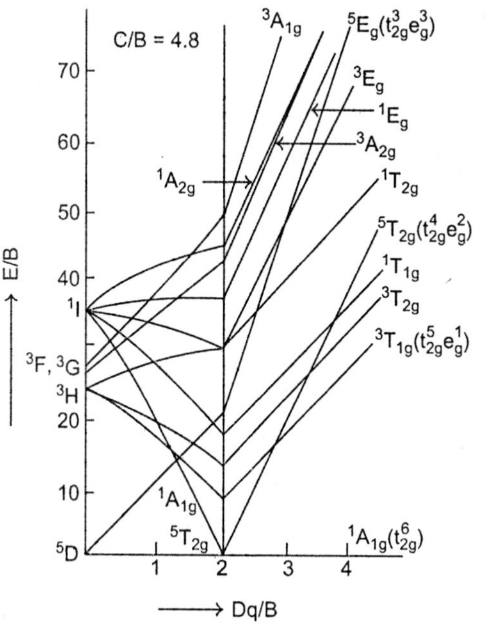


**(e) $d^6$-Configuration**

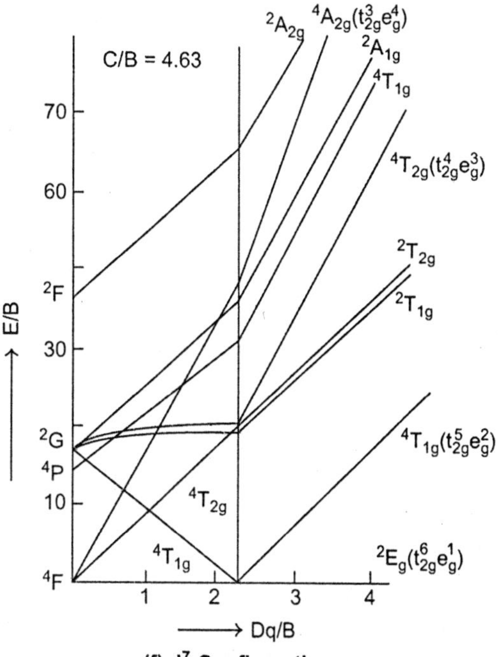


**(f) $d^7$-Configuration**

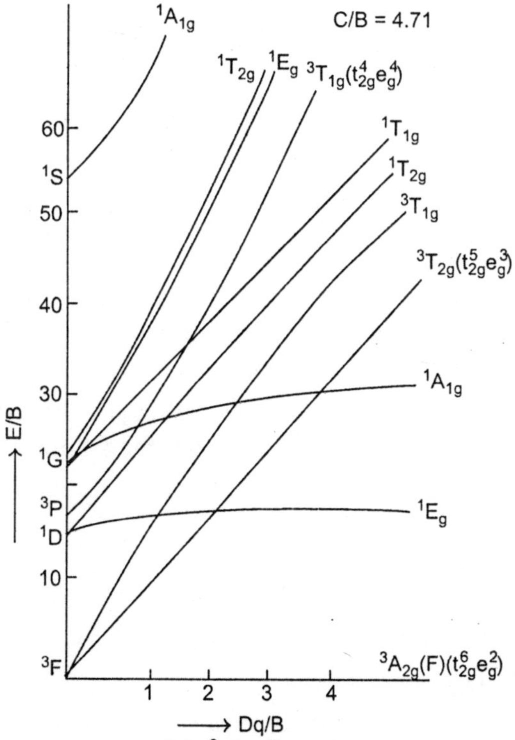


**(g) $d^8$-Configuration**

***Figure 7.22(A)** : Tanabe-Sugano diagrams for different $d^n$ configurations in octahedral field.*

**diagrams is specific for a particular group of isoelectronic ions of $d^n$ configuration while Orgel diagram is not.**

(7) These TS diagrams indicate that if the increasing ligand field causes two weak field terms of the same symmetry to approach, they do not cross but bend apart from each other. This is called **non-crossing rule according to which the energy states of same symmetry will never cross each other.**

(8) The TS diagrams can explain the spectra for cubic symmetry complexes as also these diagrams may be applied for $O_h$ and $T_d$ symmetry complexes and obey hole formalism rule.

(9) The Δ value for a complex can be determined by incorporating the spectral data in the TS diagram for a particular metal complex.

(10) These diagrams involve the states of different spin multiplicities and these can predict the spectra for both the spin allowed and spin forbidden transitions.

(11) From TS diagrams the Racah parameter B can be evaluated.

One **disadvantage** of these diagrams are the states of different spin multiplicities are involved and their energies are the functions of B and C. As stated previously (Sec. 7.15.11.2) the energy difference between the states of the same spin multiplicity is a function of B only while the difference in energy between the states of different spin multiplicities is a function of both B and C and thereby all TS diagrams are drawn for a given ratio of the C/B values. Therefore the systems having different C/B values cannot be applied for a particular C/B value. Thus for a particular $d^n$ configuration having different metal ioins, theoretically several TS diagrams may be drawn depending on the C/B values. This is one of the **serious limitation** of TS diagrams. However in most of the cases of first transition metal ions C/B value is found around 4.0.

### 7.15.11.4 Application of Tanabe-Sugano Diagram

TS diagrams has the advantage that both weak and strong field complexes are included showing the connection between the two. Here only the octahedral field considered since all tetrahedral field are weak and no consideration of strong field is expected in case of tetrahedral ligand field. The TS diagrams are depicted in Fig. 7.22(A) where $d^1$ and $d^9$ configurations excluded since they give rise only D term which is exactly the same for weak and strong octahedral field.

From this diagram we can identify the spin allowed and spin forbidden transitions although all are Laporte forbidden e.g. in case of $d^4$ configuration, the spin allowed transitions are :

$$^3T_{1g}(F) \longrightarrow {}^3T_{2g}(F)$$
$$^3T_{1g}(F) \longrightarrow {}^3A_{2g}$$
$$^3T_{1g}(F) \longrightarrow {}^3T_{1g}(P)$$

while spin forbidden transitions are :

$$^3T_{1g}(F) \longrightarrow {}^1T_{2g} \text{ and } {}^3T_{1g}(F) \longrightarrow {}^1E_g.$$

The TS diagrams provide a quantitative measure of spin pairing in different $d^n$ configurations. The $\Delta_o/B$ or $Dq/B$ values as also the Racah parameter B can be evaluated for different $d^n$ configurations as :

| $d^n$ configuration | Dq/B |
|---|---|
| $d^4$ | 2.7 |
| $d^5$ | 2.8 |
| $d^6$ | 2.0 |
| $d^7$ | 2.2 |

Since B values do not differ very much amongst the first transition series elements, spin pairing occurs most easily for $d^6(Fe^{2+}, Co^{3+})$ and $d^7(Co^{2+})$, and rather less easily for $d^4(Cr^{2+}, Mn^{3+})$ and $d^5(Mn^{2+}, Fe^{3+})$. The procedure adopted to evaluate the B values and also for Dq/B values alongwith the interpretation of the spectra of complexes can be made by using TS diagrams is to find the ratio of the energies of the second to first absorption peak and from this we can locate the position along the x-axis from which Dq/B can be determined. Having found this value, then tracing a vertical line up in the diagrams (Fig. 7.22(A)) will give the values (in E/B units) of all spin-allowed and spin-forbidden transitions.

Let us illustrate the following examples :

(1) The electronic spectra of $d^3$ complexes e.g. $[V(H_2O)_6]^{2+}$ and $[Cr(H_2O)_6]^{3+}$. Here the transitions are :

| | ${}^4A_{2g} \longrightarrow {}^4T_{2g}$ ($\nu_1$ $cm^{-1}$) | ${}^4A_{2g} \longrightarrow {}^4T_{1g}$ ($\nu_2$ $cm^{-1}$) | ${}^4A_{2g} \longrightarrow {}^4T_{1g}(P)$ ($\nu_3$ $cm^{-1}$) |
|---|---|---|---|
| $[V(H_2O)_6]^{2+}$ | 12300 | 18500 | 27900 |
| $[Cr(H_2O)_6]^{3+}$ | 17400 | 24500 | 37800 |

The Racah parameter B can be evaluated for $[Cr(H_2O)_6]^{3+}$ by fitting the ratio $\nu_1/\nu_2$ to the TS diagram for $d^3$ ion (Fig. 7.22(A)) in octahedral ligand field.

$$\frac{\nu_1}{\nu_2} = \frac{17400}{24500} = 0.710$$

This fits in the TS diagram at $\frac{Dq}{B} = 2.40$, when $\frac{\nu_1}{B} = 24$, $\frac{\nu_2}{B} = 34$ and $\frac{\nu_3}{B} = 52$.

We know that **lowest energy transition** corresponds to $\Delta_o$ so that, $\Delta_o = \nu_1 = 17400\ cm^{-1}$

$$\therefore\ B = \frac{10Dq \text{ or } \Delta_o}{2.40} = \frac{17400}{2.40} = 725\ cm^{-1}$$

The calculated B value ($725\ cm^{-1}$) is less than the free ion B value for $Cr^{3+}(1030\ cm^{-1})$ which indicates the cloud expansion i.e. nephelauxetic effect in the complex. The decreased value of B in the complex indicates reduced interelectronic repulsion.

(2) The blue-violet $[V(H_2O)_6]^{3+}$ complex i.e. $d^2$ system has the following electronic transitions :

$${}^3T_{1g}(F) \longrightarrow {}^3T_{2g}(F) \qquad \nu_1 = 17800\ cm^{-1}$$
$${}^3T_{1g}(F) \longrightarrow {}^3T_{1g}(P) \qquad \nu_2 = 25700\ cm^{-1}$$
$${}^3T_{1g}(F) \longrightarrow {}^3A_{2g}(F) \qquad \nu_3 = 35700\ cm^{-1}$$

Therefore $\frac{\nu_1}{\nu_2} = 0.693$

From TS diagram (Fig. 7.22(B)) for $d^2$ system we have to note that $\frac{\nu_1}{\nu_2}$ occurs at $\frac{Dq}{B} \approx 3.1$. From this $\frac{Dq}{B} \approx 3.1$, the corresponding $\frac{\nu_1}{B}$ and $\frac{\nu_2}{B}$ values obtained as 28.7 and 41.7 respectively. Therefore for $\frac{Dq}{B} \approx 3.1$ the values obtained, $\frac{\nu_1}{B} = 28.7$, $\frac{\nu_3}{B} = 41.7$ and $\frac{\nu_2}{B} \approx 59$.

$$\therefore B = \frac{\nu_1}{28.7} = \frac{17800 \text{ cm}^{-1}}{28.7} = 620 \text{ cm}^{-1}.$$

Again $\frac{Dq}{B} \approx 3.1$, so that $Dq = 3.1 \times 620 = 1922 \text{ cm}^{-1}$.

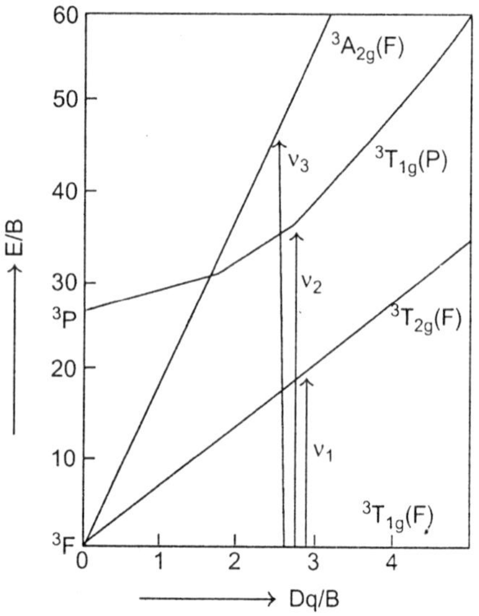


***Figure 7.22(B)** : TS digram for $d^2$ ion in $O_h$ symmetry.*

Therefore $10Dq = 19220 \text{ cm}^{-1}$.

This B value ($620 \text{ cm}^{-1}$) should be compared to that of B value ($861 \text{ cm}^{-1}$) for the free V(III) ion. The determined value ($620 \text{ cm}^{-1}$) is only 72% of the B value for free V(III) ion.

As $\frac{\nu_2}{B} \approx 59$, so that $\nu_2 = 59 \times 620 \text{ cm}^{-1} = 36580 \text{ cm}^{-1}$.

This electronic transition is not checked since this region of the spectrum of $[V(H_2O)_6]^{3+}$ is obscured by absorption arising from charge transfer. However all the three bands for $V^{3+}$ may discerned in the spectrum of $V^{3+}$ incorporated into $Al_2O_3$ lattice ($V^{3+}/Al_2O_3$).

### 7.15.11.5 Band Intensity

We have stated that band intensity of charge transfer transitions are more compared to other d-d transitions (Sec. 7.15.10) and which is also supported by higher molar absorption coefficient (Table 7.17).

Due to d-p mixing in case of non-centrosymmetric complexes the Laporte selection rule is relaxed which induces greater covalent bonding between metal and the ligand thereby crystal field transition intensity increases.

The intensities of the square planar complexes ($\varepsilon$ = 50–300 L mol$^{-1}$cm$^{-1}$) lie intermediate to those of octahedral and tetrahedral complexes. The enhancement of band intensity close to charge transfer absorption is known as **intensity stealing** and is inversely related to the energy separation between the concerned energy states. Intensity stealing may also occur from intra ligand transitions.

The centrosymmetric octahedral $d^5$ complexes have extremely low intensity ($\varepsilon = 10^{-2}$ L mol$^{-1}$ cm$^{-1}$) while the same $d^5$ non-centrosymmetric tetrahedral complexes e.g. $[MnX_4]^{2-}$ have much higher intensity e.g. $[MnCl_4]^{2-}$ ($\varepsilon$ = 0.5 L mol$^{-1}$ cm$^{-1}$) and $[MnBr_4]^{2-}$ ($\varepsilon$ = 1 L mol$^{-1}$ cm$^{-1}$). The highest crystal field transition of $[MnBr_4]^{2-}$ has intensity of about 10 due to intensity stealing from closely lying Mn–Br charge transfer band.

Generally the band intensities are measured with reference to ε values. This comparison of band intensity is only applicable in case of **sharp, symmetric** and of the **same band width cases.** But in most of the cases the bands are unsymmetric and they are of different band widths. Moreover the bandwidths may change with temperature. In those cases ε values are not the parameter to measure the band intensity. In those cases the area under the bands are measured to compare the band intensities and is expressed as :

$$I = \int_{\bar{v}_1}^{\bar{v}_2} \varepsilon_{\bar{v}} d\bar{v}$$

where I is the integrated band intensity with reference to starting wave number ( $\bar{v}_1$) and the ending wave number ( $\bar{v}_2$) expressed in $cm^{-1}$, $\varepsilon_{\bar{v}}$ is the molar absorption coefficient at the concerned wave number ( $\bar{v}$ ). In this expression I represents the area under the absorption curve which is conversely a measure of band intensity of crystal field transitions.

### 7.15.11.6 Bandwidths and Shapes : Franck-Condon Vertical Transition

The bandwidth of the spectral line for the free ion is less than 1 $cm^{-1}$ but when the same metal ion is subjected to crystal field by the ligands the half-bandwidth vary in the range 500-2000 $cm^{-1}$. Such enhancement of bandwidth is explained on the basis of the change in the vibrational energy levels along with electronic transitions. The greater the difference in the equilibrium internuclear distance between the ground state and the excited state, the broader will be the overall appearance of the spectrum.

According to the Franck-Condon principle, the electronic transitions of a molecule will not change the nuclear configuration of the same. This is due to the fact that nuclei are much more massive than electrons and the electronic transition takes place faster than the nuclei can respond. The nucleus in a molecule has coulombic forces acting on it from the electrons and other nuclei of the system. Once a molecule undergoes the electronic transition, the resulting coulombic forces serve to change the energy of the molecule. This change brings it from the ground state to an excited state and results in the nuclei changing its vibrational state. This vibrational structure of an electronic transition shows that the absorption spectrum consists of several lines instead of a single sharp electronic absorption line. Classically, the Franck-Condon principle is the approximation that an electronic transition is most liley to occur without changes in the positions of the nuclei in the molecular entity and its environment. The resulting state is called Franck-Condon state and the transition involved a **vertical transition.** Thus the electronically excited molecules are in the vibrationally excited levels having the same internuclear distance of the ground state electronic configuration to execute the Franck-Condon vertical transition (Fig. 7.22(C)).

The quantum mechanical formulation of this Franck-Condon principle is that the intensity of a vibronic transition is proportional to the square of the overlap integral between the vibrational wave functions of the two states that are involved in the transition i.e. ground state and excited state. Room temperature solid and solution bands of d-metal complexes are in general, quite wide ~ 1000 $cm^{-1}$ and only occasionally are narrow bands with widths < 100 $cm^{-1}$ found. This implies that the energy terms in the complexes do not correspond to a single energy level, but that they spread over a range of energies comparable with band widths which are found.

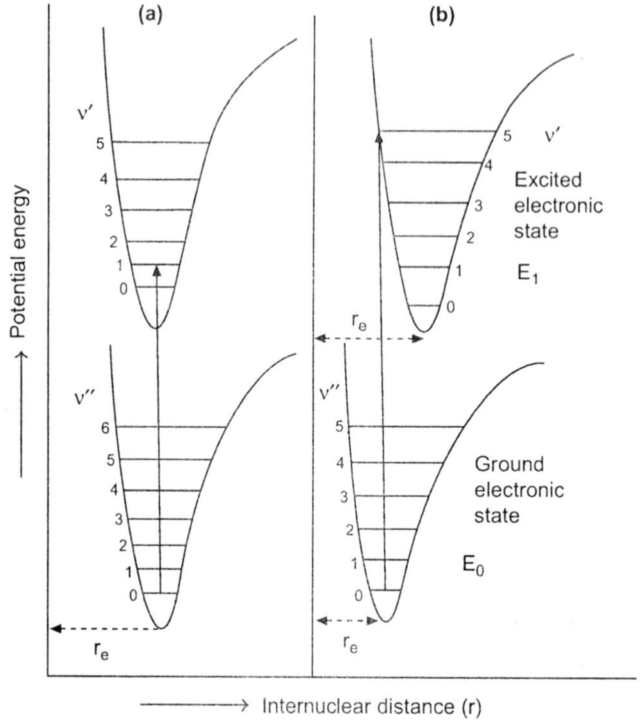


***Figure 7.22(C)*** *: Morse potential energy curve (which is a plot of potential energy versus internuclear distance) to show the Franck-Condon vertical transition. $r_e$ = equilibrium internuclear distance. (a) curve denotes $r_e$ (ground state) = $r_e$ (excited state) and (b) denotes $r_e$ (ground state) < $r_e$ (excited state). In both the (a) and (b) curves v denotes vibrational energy levels for both the ground and excited states.*

There are several factors that may serve to remove the degeneracy of the terms and thereby produce broad bands. This includes symmetry lowering by (a) molecular vibration, (b) spin-orbit coupling, (c) the Jahn-Teller effect. All of these factors lead to asymmetrically shaped bands and band broadening by the vibrational mechanism is most easily seen physically. Both spin-orbit coupling and the Jahn-Teller effect may serve to broaden bands by lifting the degeneracy. However, if the degeneracy has already been lifted by low symmetry static ligand field or by spin-orbit coupling then Jahn-Teller effect will be non-operative.

### 7.15.12 Limitations of crystal field theory (CFT)

1. CFT only consider the metal ion d-orbitals to explain the properties of complexes. This theory does not give any importance of the metal s and p orbitals and ignores their effect on bonding properties of complexes. No scope has been given for π-acidic or π-complexing ligands in CFT.

2. No explanation has been accounted for the higher positon of $H_2O$ than that of $OH^-$ in the spectrochemical series.

3. CFT considers the electrostatic attraction (ionic interaction) between point positive

charge of the metal ions and point negative charge of the ligands. It gives no explanation for covalent nature of metal-ligand bonds.

4. In CFT no provision for cloud expansion of metal ions in metal-ligand bonding is provided. Due to the nephelauxetic effect the d-electron cloud of metals overlap with ligand orbitals providing a pathway for the electrons to spread out or delocalised.

5. Antiferromagnetic behaviour (which will be discussed in the chapter of magneto-chemistry) of complexes through super-exchange of metal-metal bonds are not explained by CFT. Antiferromagnetism via super-exchange is a proof of overlap of metal-ligand orbitals.

### 7.15.13 Comparison between CFT and VBT

1. The inner orbital ($d^2sp^3$) and outer orbital ($sp^3d^2$) octahedral complexes according to VBT are the same as that of low spin and high spin terminology respectively according to CFT.

2. Orbital contribution to magnetic moment can be explained by CFT. Enhanced magnetic moment of some $d^n$ configurations cannot be explained by VBT.

3. According to VBT for inner orbital octahedral processes some electron(s) may have to be promoted in case of $d^7$, $d^8$ and $d^9$ systems thereby the system vulnerable to oxidation. In CFT no such type of mechanism will be followed because the d-electrons will be regrouped into the splitted $t_{2g}$ and $e_g$ orbitals.

4. No provision for tetragonal distortion leading to the transition from octahedral to square planar through square pyramidal geometry according to the increasing strength of the ligands, have been provided in VBT. CFT successfully explain the tetragonal elongation and tetragonal compression leading to the orientation of d-orbitals in crystal field diagram.

5. Most important aspect of CFT is the explanation of colour of coordination complexes. Quantitative Orgel diagrams/Tanabe–Sugano diagrams are the crops of CFT. These diagrams has immense importance for the explanation of colour and other magnetochemical properties of complexes. The VBT has nothing to say about the colour of coordination complexes.

## 7.16 MOLECULAR ORBITAL THEORY

CFT has many defects to interpret the spectra of complexes and the estimation of $\Delta_o$/10Dq values are not always up to the mark. A question arises about the position of CO in the spectrochemical series which causes large splitting though it has no charge and having poor dipole moment value.

Molecular orbital theory (MOT) provided some refinement of $\Delta_o$/10Dq values over the CFT values. Like MOT for simple molecules (Sec. 3.4) the essential feature of MOT for complexes is the inherent symmetry requirements for the overlap of metal ion orbitals with that of ligand orbitals. LCAO, terminal atom symmetry orbitals (TASO) concepts are applied on the metal and ligand and thereby the molecular orbitals formed are arranged according to their energy sequence. The estimated order can be verified by comparison with experimental data particularly by optical absorption and photoelectron spectroscopy.

We shall restrict our discussion only in case of octahedral complex ($ML_6$) involving $\sigma$ bonding only. MOT can also be considered for $\pi$ bonding between metals and ligands.

In octahedral symmetry the metal atom orbitals are categorised by the following four sets :

| **Metal atom orbitals** | **Symmetry terminology** |
|---|---|
| S | $a_{1g}$ (non-degenerate, symmetrical) |
| $p_x, p_y, p_z$ | $t_{1u}$ (triply degenerate) |
| $d_{xy}, d_{yz}, d_{xz}$ | $t_{2g}$ (triply degenerate) |
| $d_{x^2-y^2}, d_{z^2}$ | $e_g$ (double degenerate) |

The following symmetry operated linear combinations of the six ligands can also be formed :

$a_{1g} : \sigma_1 + \sigma_2 + \sigma_3 + \sigma_4 + \sigma_5 + \sigma_6$

$t_{1u} : \sigma_1 - \sigma_3, \sigma_2 - \sigma_4, \sigma_5 - \sigma_6$

$e_g : \sigma_1 - \sigma_2 + \sigma_3 - \sigma_4, 2\sigma_6 + 2\sigma_5 - \sigma_1 - \sigma_2 - \sigma_3 - \sigma_4$

$t_{2g}$ orbitals do not participate in σ bonding because of lack of symmetry.

Moleuclar orbitals are now formed according to LCAO approach with the same symmetry type orbitals of the metal ion and the ligands to overlap. These molecular orbitals are now arranged according to their ascending energies which is known as molecular orbital energy level diagram (Fig. 7.23).

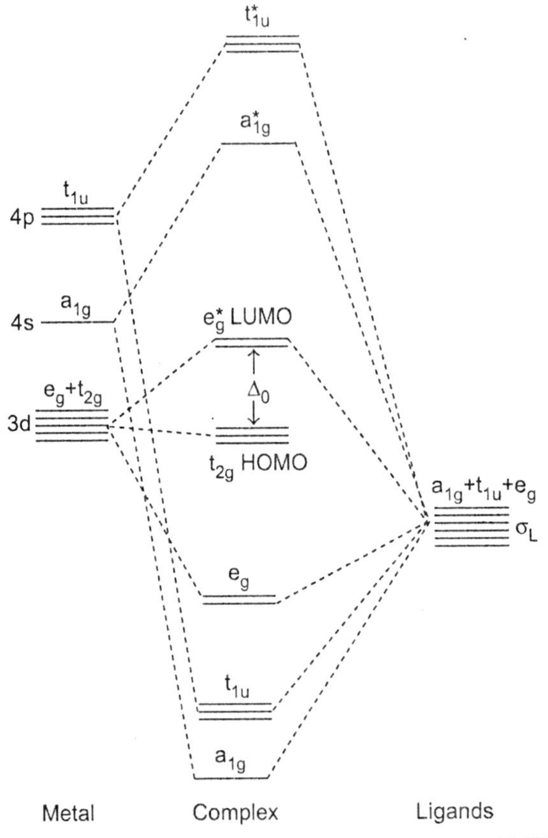


*Figure 7.23 : Molecular orbital energy level diagram for an octahedral ($ML_6$) complex involving σ bonding.*

The HOMO and LUMO molecular orbitals of the complex are the non-bonding $t_{2g}$ orbitals (essentially metal in character) and the antibonding $e^*_g$ orbitals (mainly metal in character) respectively. Here $\Delta_o$/10Dq denotes the separation between the frontier orbitals while in CFT the $\Delta_o$/10Dq denotes purely the separation between the metal d-orbitals.

Now we shall consider π-interaction on the metal-ligand bond and show how the energy level diagram is modified with the ligands having π-character. The modification depends mainly on two factors : (a) the relative energies of ligand π-orbitals and the metal $t_{2g}$ orbitals and (b) whether the ligand is π-donor or π-acceptor (Fig. 7.24).

**π-donor ligands** are those which have filled orbitals having π-symmetry around the metal-ligand bond axis. The π-symmetry orbitals are formed by the overlap of ligand p orbital perpendicular to the metal-ligand bond axis and a metal $d_{xy}$ orbital. Moreover the energies of these orbitals are similar to those of the metal d-orbitals and the ligands has no low energy vacant π-orbitals e.g. $Cl^-$, $I^-$, $H_2O$ etc., the position of which are low in spectrochemical series. As the halogens are more electronegative, the filled p orbitals will be at a lower energy than the corresponding metal 3d orbitals ($e_g$+$t_{2g}$). When they form molecular orbitals with the $t_{2g}$ orbitals of the metal ion, a bonding and an antibonding combination of the π-type will result. The bonding orbital (π($t_{2g}$)) lies lower than the ligand orbitals and the antibonding orbital (π*($t^*_{2g}$))

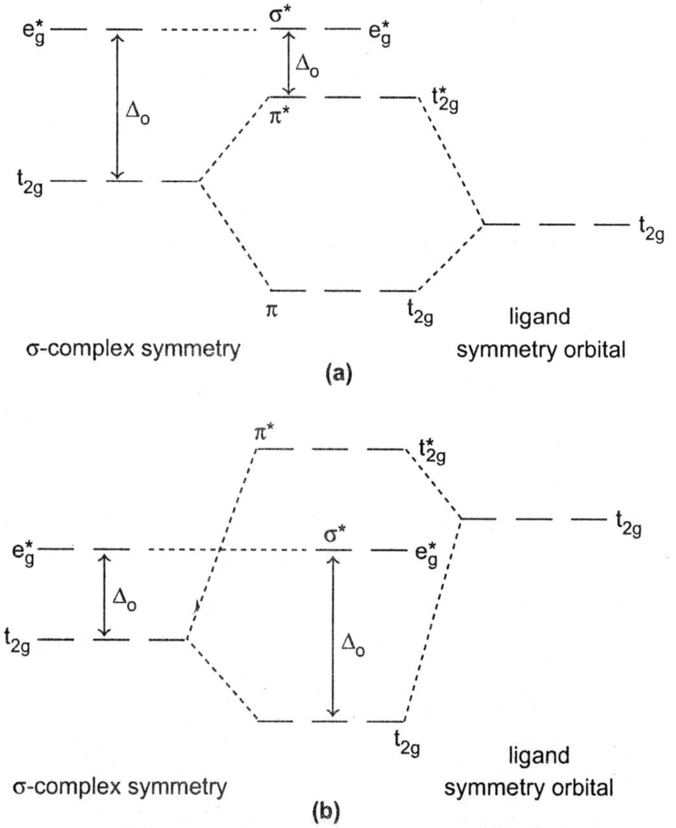


*Figure 7.24 : π bonding effect on octahedral σ complex.*
*(a) Ligands acting as π donor, (b) Ligands acting as π-acceptor.*

lies above the energy of the d-orbitals of the free metal ion [Fig. 7.24(a)]. The electrons supplied by the ligand lone pairs occupy and fill the bonding orbitals ($\pi(t_{2g})$) leaving the electrons orginally in the d-orbitals of the central metal ion to occupy the antibonding ($\pi^*(t^*_{2g})$) orbital. $\Delta_o$ now corresponds to the new separation between ($\pi^*(t^*_{2g})$) and $e^*_g$ level which is lower than the σ bonding one (Fig. 7.23). The net effect is that the non-bonding metal ion $t_{2g}$ orbitals converted to antibonding ($\pi^*(t^*_{2g})$) and thus are raised closer in energy to the mainly metal antibonding ($e^*_g$) orbital. Hence it is concluded that **strong π-donor ligands decrease $\Delta_o$/10Dq**.

Another disposition of the energy level diagram can be framed by **π-acceptor ligands** which has filled π-orbitals at lower energies than metal $t_{2g}$ orbitals but have empty π-orbitals of higher energies than the metal $t_{2g}$ orbitals available for occupation. $PH_3$, $AsH_3$, CO, $N_2$ etc. belong to this type of ligands. Typically the π-acceptor orbitals are vacant antibonding orbitals on the ligand ($\pi^*(t^*_{2g})$) and these orbitals lie above the metal d-orbitals in energy. These vacant $\pi^*(t^*_{2g})$ ligand orbitals with the correct symmetry interact with the metal ions $t_{2g}$ orbitals and the metal-ligand π-overlap is strong. These ligands are therefore called π-acceptor ligands or π-acid ligands (Lewis acid). Practically these **are σ Lewis base and π-Lewis acid ligands**. The excellent π-acceptor ligand is CO which forms metal carbonyls, the HOMO of CO (i.e. 3σ) (Fig. 3.28) which is localised on carbon first forms C→M σ donor bond. The LUMO of CO (1π* Fig. 3.28) has the correct symmetry for overlap with the $t_{2g}$ orbitals of the metal. Due to the more electronegative character of oxygen full bonding π orbital of CO is low in energy and is largely dominated on the oxygen atom and so the π donor character of CO is very poor and infact in all metal carbonyls CO function as π-acceptor ligand. Due to this π-acceptor property of these ligands, they form molecular orbitals with the $t_{2g}$ orbitals of metals which are of low energy (Fig. 7.24(b)). The net effect of a π-acceptor ligand is therefore to **increase the separation of $\Delta_o$**.

Now we shall make a ctitique on the position of CO and $OH^-$ in the spectrochemical series. The position of the ligands in the said series depend on the strength of σ donor property of the ligands to form the metal-ligand bonds. Thus $CH_3^-$, $H^-$ etc. have higher position in the spectrochemical series due to the strong σ donor property of them. But when a ligand have π-bonding property which significantly influence the $\Delta_o$/10Dq and is responsible for CO (being a strong π-acceptor and increases $\Delta_o$) being higher position in spectrochemical series than $OH^-$ (being a strong π donor ligand which decreases $\Delta_o$) being lower in the series and their position in the spectrochemical series is justified. Now the position of the ligands in the said spectro-chemical sereis is depicted with π-bonding effects in the following way (exceptions in case of $CH_3^-$ and $H^-$ which are neither π-donor nor π-acceptor ligands being only strong σ donors) :

$$\longleftarrow \quad \Delta_o \text{ increases} \quad \longrightarrow$$

$$\pi \text{ donor} < \text{weak } \pi \text{ donor} < \text{no } \pi \text{ effects} < \pi \text{ acceptor}$$

$$\text{matching ligands are : } I^- < Br^- < Cl^- < F^- < H_2O < NH_3 < PR_3 < CO$$

---

**Problem 7.62** The enthalpy of hydration of $Cr^{2+}$ is –460 kcal/mol. In the absence of crystal field stabilisation energy, the value of ΔH would be –435 kcal/mol. Estimate the value of Δ for $Cr(H_2O)_6^{2+}$. **[B.U. 2010, 2015]**

**Ans.** : For octahedral $Cr^{2+}(3d^4)$ complex the crystal field splitting diagram is ↑ — (upper level), ↑ ↑ ↑ (lower level) for which

$$\text{CFSE} = 3\times\left(-\frac{2}{5}\Delta\right)+\frac{3}{5}\Delta = -0.6\Delta$$

$\therefore -0.6\Delta = -460 - (-435) = -25$ kcal/mol.

So that $\Delta = \dfrac{25}{0.6} \times \dfrac{350\ cm^{-1}}{1\ kcal/mol}$ [$\because$ 350 $cm^{-1}$ = 1 kcal/mol]

$= 14584\ cm^{-1}$ [Actual value is 13900 $cm^{-1}$]

---

**Problem 7.63** The oxides of formula MO which all have octahedral coordination of the metal ions have the following lattice enthalpies.

| CaO | TiO | VO | MnO |
|---|---|---|---|
| 3460 | 3878 | 3913 | 3810 kJ $mol^{-1}$ |

Account for the trends in terms of the CFSE given that $O^{2-}$ is in lower position in spectrochemical series.

**Ans. :** Among these oxides $Ca^{2+}(3d^0)$, $Ti^{2+}(3d^2)$, $V^{2+}(3d^3)$ and $Mn^{2+}(3d^5)$ and their CFSE values in octahedral coordination are 0, $-0.8\Delta_0$, $-1.2\Delta_0$ and 0 respectively in high spin configuration as $O^{2-}$ is a weak field ligand. As the CFSE increases, the lattice enthalpy increases from CaO to VO. As CFSE is zero for $Mn^{2+}$ and so lattice enthalpy will decrease.

---

**Problem 7.64** In the reaction $[Co(NH_3)_4Cl_2] + Cl^- \longrightarrow [Co(NH_3)_3Cl_3] + NH_3$, only one isomer of the complex product is obtained. Is the initial complex *cis* or *trans*? **[B.U. 2010]**

**Ans. :** In the initial complex the two $Cl^-$ ions are in the trans position thereby all the other four positions which could be replaced will be identical. If the original compound have a cis structure, then the third $Cl^-$ ion could replace an $NH_3$ molecule cis to both or trans to one and so two isomers would be expected.

---

**Problem 7.65** What are pre-requisites for a coordination compound to be amenable to high spin-low spin crossover? Discuss with example. Give an example of a $d^5$ system known to exhibit this behaviour. **[B.U. 2010, 2012]**

**Ans. :** The origin of spin state cross-over is related to magnitude of orbital splitting ($\Delta_0$) and the pairing energy P. When P and $\Delta_0$ are almost equal, the high spin and low spin states will be in thermal equilibrium i.e. their energies will be almost similar (Fig. 7.8). Because of the change in the M–L bond lengths, which decrease in low spin state (by 0.1 to 0.2Å) the behaviour of the system is largely dependent on the pressure as well as temperature. For example, bis (2, 6 pyridinedialdehyde hydrazone) cobalt (II) has a moment of 1.9 BM at 80 K while the value increases to 3.7 BM at 337 K.

For $3d^5$ system, tris (dialkyl dithiocarbamato) iron (III), 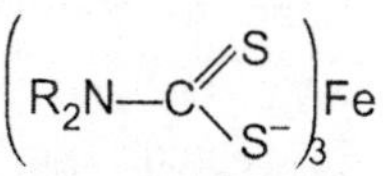 has $\mu$ = 5.7 to 5.8 B.M at 300 K, 4.7 to 5.0 B.M at 150 K, 3.6 to 4.0 B.M at 78 K as the proportion of molecules in the high and low spin states changed with temperature.

---

**Problem 7.66** When $[Ni(NH_3)_4]^{2+}$ is treated with concentrated HCl, two compounds having the formula $Ni(NH_3)_2Cl_2$ (designated I and II) are formed. (I) can be converted into (II) by boiling in dilute HCl. A solution of (I) reacts with oxalic acid to form $[Ni(NH_3)_2C_2O_4]$; (II) does not react with oxalic acid. Deduce the configurations of (I) and (II) and the geometries of nickel (II) complexes. **[B.U. 2010]**

**Ans. :** 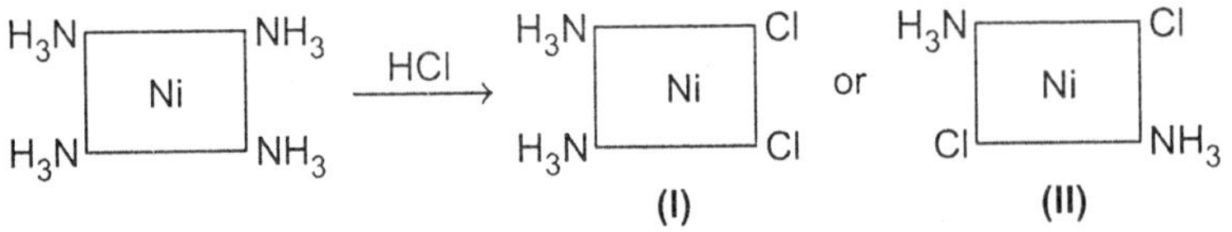

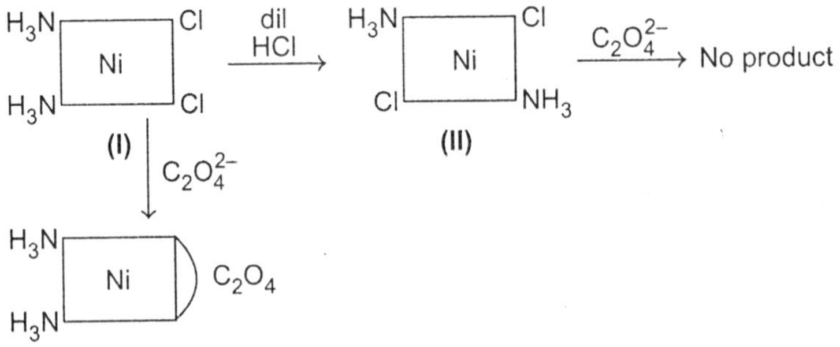

(I) is a cis isomer for which the two $Cl^-$ groups are replaced by $C_2O_4^{2-}$ group while the (II) is a trans isomer and so the two $Cl^-$ groups are in the farthest position and cannot be replaced by $C_2O_4^{2-}$. (I) and (II) have the square planar geometry.

---

**Problem 7.67** Write true or false for the following statements and give reasons : (a) Complexes of $Cu^{2+}$ usually depart considerably from $O_h$ geometry. (b) The ion $[FeF_6]^{3-}$ is colourless wehreas $[CoF_6]^{3-}$ is coloured. **[B.U. 2010]**

**Ans. :** (a) True, see Sec. 7.14.5(C). Reason is the uneven occupation of $e_g$ orbital.

(b) $[FeF_6]^{3-}$ where $Fe^{3+}$ is a $3d^5$, CFSE = 0 for high spin state. Here $\Delta S \neq 0$ and $\Delta L \neq \pm 1$ i.e. double forbidden and so $[FeF_6]^{3-}$ is colourless.

$[CoF_6]^{3-}$ where $Co^{3+}$ is a $3d^6$ system. For high spin state $\Delta S = 0$ i.e. spin allowed but Laporte forbidden transition will occur. Due to the spin allowed transition $[CoF_6]^{3-}$ is coloured $[{}^1A_{1g} \rightarrow {}^1T_{1g}$ and ${}^1A_{1g} \rightarrow {}^1T_{2g})$. Actually $[CoF_6]^{3-}$ shows a single absorption peak corresponding to the transition ${}^5T_{2g} \rightarrow {}^5E_g$ (Tanabe-Sugano diagram) at 13000 $cm^{-1}$ and the complex is **deep blue** in colour.

---

**Problem 7.68** Which one of each pair should have comparatively more intense d-d transition : trans $[Cr(H_2O)_4Cl_2]^+$ or cis $[Cr(H_2O)_4Cl_2]^+$ and $Ni(CO)_4$ or $Fe(CO)_5$? **[B.U. 2010]**

**Ans. :** As mentioned earlier (Sec. 7.9.12) a cis complex has a higher molar extinction coefficient than the corresponding trans complex.

With cis and trans isomers of the type $Ma_4b_2$ the same number of bands would be expected. The terms ${}^1E_g$ and ${}^1A_{2g}$ are inverted in the cis with respect to the trans. ${}^1E_g$ is higher in energy than ${}^1A_{2g}$ in trans isomer, but ${}^1A_{2g}$ is higher than ${}^1E_g$ in cis isomer. So the difference in frequency for the bands due to transitions ${}^1A_{1g} \rightarrow {}^1E_g$ and ${}^1A_{1g} \rightarrow {}^1A_{2g}$ should be twice as great in the trans as in the cis isomer. In practice with cis isomer the splitting only broadens the bands but with trans isomer actual splitting has been observed.

Moreover the **cis isomer has no centre of symmetry but the trans isomer has a centre of symmetry**. Therefore Laporte selection rule is better applicable to the trans isomer. As a result the d-d transition (i.e. $g \leftrightarrow g$ type) are more Laporte forbidden in trans isomer than that of cis isomer. Due to lack of symmetry in the cis isomer, the energy levels cannot be termed as g or u if we apply symmetry rules strictly. The lowering of molecular symmetry relaxes Laporte selection rule and increases the intensity of absorption in cis isomer. Thus intense absorption spectra is obtained in case of cis isomer compared to that of trans isomer.

$Ni(CO)_4$ has a tetrahedral geometry where Ni(O) is $sp^3$ configuration and the complex is diamagnetic. $Fe(CO)_5$ having Fe(O) is $dsp^3$ and the complex also diamagnetic having trigonal bipyramidal geomtry. Since in a tetrahedral complex the metal ion is not at the centre of symmetry the mixing of d and p orbitals is much favoured and as such the intensity of Laporte forbidden d-d transition bands in tetrahedral complex is much higher than in trigonal bipyramidal or octahedral complex. Thus $Ni(CO)_4$ will register more intense absorption band than that of $Fe(CO)_5$.

**Problem 7.69** Which of the following are expected to show Jahn-Teller distortion : $[CuCl_6]^{4-}$, $[Cr(acac)_3]$ and $[Co(CN)_6]^{4-}$ (low spin)? **[B.U. 2010]**

**Ans. :** $[CuCl_6]^{4-}$ where $Cu^{2+}$ is a $3d^9$ system. This undergoes J-T distortion from repulsion of ligands by the unevenly populated electrons in $e_g$ and $t_{2g}$ set.

$[Cr(acac)_3]$ where $Cr^{3+}$ is $3d^3$ system which is devoid of J-T distortion. $[Co(CN)_6]^{4-}$ where $Co^{2+}$ is $3d^7$ and in low spin state the configuration is $t_{2g}^6e_g^1$. Here distortion is permitted.

## 7.17 STABILISATION OF UNFAMILIAR OXIDATION STATES

Compared to common oxidation state(s) the high and low oxidation states are termed unfamiliar oxidation states. Table 7.18 records the unfamiliar oxidation states of first transition elements including the common oxidation states. Redox potential data is an important factor for the stabilisation of high and low oxidation states. Since increasing charge on the metal ion increases its acidity so high oxidation states are stabilised in alkaline medium. We shall discuss here briefly the stabilisation of high and low oxidation states through complex formation.

**Table 7.18 : Less familiar oxidation states of first transition elements**

| Ti($4s^23d^2$) | V($4s^23d^3$) | Cr($4s^13d^5$) | Mn($4s^23d^5$) | Fe($4s^23d^6$) | Co($4s^23d^7$) | Ni($4s^23d^8$) | Cu($4s^13d^{10}$) |
|---|---|---|---|---|---|---|---|
| [+4] | [+5] | [+6] | [+7] | +6 | +4 | +4 | +3 |
| [+3] | (+4) | +5 | +6 | +4 | [+3] | +3 | [+2] |
| +2 | +3 | +4 | +5 | [+3] | [+2] | [+2] | (+1) |
| 0 | +2 | [+3] | +4 | (+2) | +1 | +1 | |
| –1 | +1 | +2 | +3 | +1 | | | |
| | 0 | +1 | [+2] | 0 | 0 | 0 | |
| | –1 | 0 | +1 | –2 | –1 | | |
| | | | 0 | | | | |
| | | | –1 | | | | |

[ ] denotes the most common oxidation states, ( ) denotes the next in stability of common oxidation states.

### 7.17.1 High oxidation states

(a) Small and hard donor ligands having strong σ donor properties like $F^-$ and $O^{2-}$ can stablise high oxidation states e.g. $K_2MF_6$ where M = Co, Ni, Cu having +4 state, $FeO_4^{2-}$ (Fe, +6), $PtF_6$ (Pt, +6).

(b) Anions containing a central atom in a high oxidation state are able to stabilise the high oxidation states e.g. $KNiIO_6$(Ni, +4), $M^I_9[Ag(TeO_6)_2]$ (Ag, +3), $M^I_7[Ag(IO_6)_2]$ (Ag, +3) etc.

(c) Several organic ligands which can withstand the oxidation by any oxidising agent used or by the oxidising power of the high oxidation state metal ion. e.g. $[Ag^{II}(Pic)_2]$ (PicH = picolinic acid), $[Ag^{II}(dipy)_2]^{2+}$, $[Ag^{II}(ophen)_2]^{2+}$, $[Ag^{II}(py)_4]^{2+}$. In general ligands having strong σ donor properties will stabilise high oxidation states.

### 7.17.2 Low oxidation states

(a) The lower oxidation states are stabilised by the π-acid ligands e.g. CO, $CN^-$, $PF_3$, $PR_3$ etc. Examples are $Cr^{-II}(CO)_5^{2-}$, $Cr^0(CO)_6$, $Cr^0(dipy)_3$, $Cr^I(dipy)_3^+$. This low oxidation state stabilisation is operated through synergic effect.

(b) Mixed ligand complexes having molecular nitrogen as one of the ligand e.g. $[IrCl(N_2)(PPh_3)_2]$. This type of complex is less stable in comparison to the complexes of CO due to the non-polar nature of $N_2$ and greater stabilisation of HOMO and destabilisation of LUMO, thereby dinitrogen molecule both a weak σ donor and poor π-acceptor.

(c) The hydride ion, $H^-$ is a strong σ donor ligand, will form mixed ligand complex in conjunction with a π-acceptor ligand where the metal is in low oxidation state e.g. $[Co(N_2)(H)(PPh_3)_3]$, $[Pt(H)Cl(PEt_3)_2]$. On the whole, π-acceptor ligands would stabilise lower oxidation states. Intermediate oxidation states are stabilised by ligands having not too strong σ donor property and not too strong π-acceptor property.

## 7.18 COORDINATION POLYMERS

Any macromolecular materials which contain coordinate covalent bonds between the metal and the ligands is termed as **coordination polymer**. These are essentially inorganic polymers but large number of organic ligands are used for their synthesis. Coordination polymers generally classified into the following categories :

(a) Natural coordination polymers and

(b) Synthetic coordination polymers.

### 7.18.1 Natural coordination polymers

The word 'natural' does not mean that these polymers exist in nature. The term natural implies that this class of compounds are not essentially polymeric at the initial stage of their synthesis but become polymeric in course of time. Therefore these are to be formulated as monomeric one. These polymers are insoluble inorganic compounds which contain ligand bridged metal ions in a linear, planar or even three dimensional network arrangements. For example polymeric gold (I) and silver (I) cyanides are commonly written as AuCN and AgCN should be written as $(AuCN)_n$ or $(AgCN)_n$.

The example of three dimensional network polymer is the Prussian blue and Turnbull's blue. On adding Fe(III) solution to $K_4[FeCN)_6]$ a dark blue precipitate is obtained.

$$FeCl_3 + K_4[Fe(CN)_6] \longrightarrow \underset{\textit{Prussian blue}}{K[Fe^{III}Fe^{II}(CN)_6]} + 3KCl$$

Prussian blue therefore is potassium ferri ferrocyanide. Turnbull's blue is prepared by adding Fe(II) solution to $K_3[Fe(CN)_6]$ whereby a deep blue precipitate is formed so that Turnbull's blue

$$Fe^{2+} + K_3[Fe(CN)_6] \longrightarrow \underset{\textit{Turnbull's blue}}{K[Fe^{II}Fe^{III}(CN)_6]} + 2K^+$$

is potassium ferro ferricyanide. These two compounds practically have the same structure. X-ray crystallographic method suggests the structure consists of a cubic lattice (7-LX) with the Fe(II) and Fe(III) occupying the alternate corners of a cube with the $CN^-$ group present in between at the edges of the cube and bonded to both in the following way where each Fe(II) is C-bonded and each Fe(III) is N-bonded of the $CN^-$ group. Thus half of the iron

$$\ldots Fe^{2+}—CN—Fe^{3+}—NC—Fe^{2+}—CN—Fe^{3+}\ldots$$

atoms face six carbon atoms of $CN^-$ groups as nearest neighbours and another half of the iron atoms face six nitrogen atoms. Each $[Fe(CN)_6]^{n-}$ anion is surrounded by six attached cations and each cation by six $[Fe(CN)_6]^{n-}$ anions and thus an infinite chain network is built up.

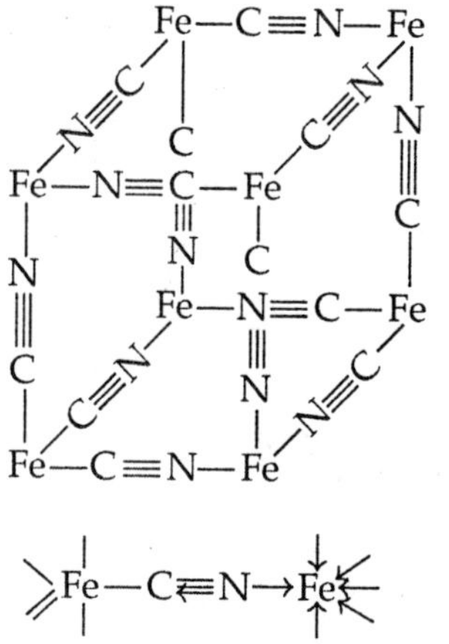

(7-LX)

Many other cyanides e.g. polymeric gold (I) and silver (I) cyanides, $M^{I}Cu^{II}Fe^{III}(CN)_6$ and $M^{I}Ru^{II}Fe^{III}(CN)_6$ have also the infinite chain network structure.

### 7.18.2 Synthetic coordination polymers

This class of coordination polymers essentially synethesised as polymeric compounds and in this class most of the ligands are organic origin.

One of the polymer of this class is Volan and Quilon polymer of chromium (7-LXI). This is synthesised by the reduction of $CrO_2Cl_2$ by methanol or ethanol in the prsence of an appropriate carboxylic acid.

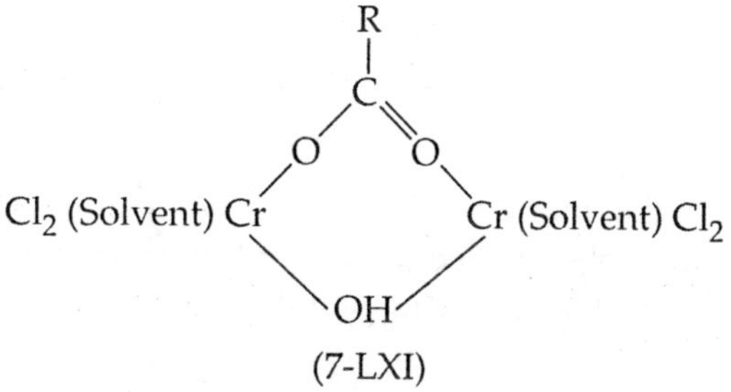

(7-LXI)

Various bis-chelating agents like bis (β-diketons), bis (schiff bases), bis (aminoacids) etc. are used for the synthesis of coordination polymers e.g. polymeric bis (β-diketone) berryllium (II) (7-LXII). One of the most thermally stable two dimensional sheet polymers of phthalocyanines (7-LXIII) have also been prepared.

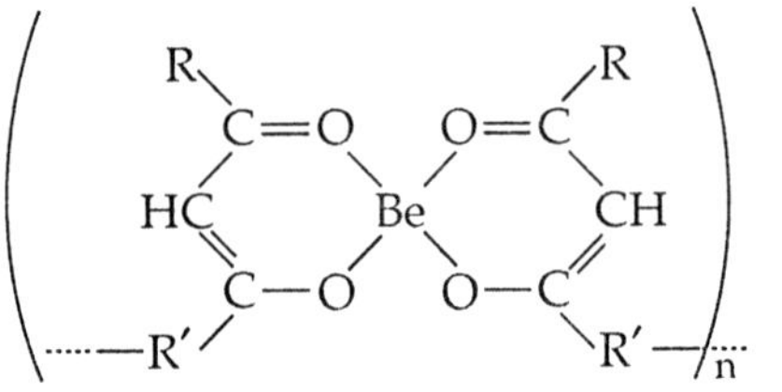

(7-LXII)

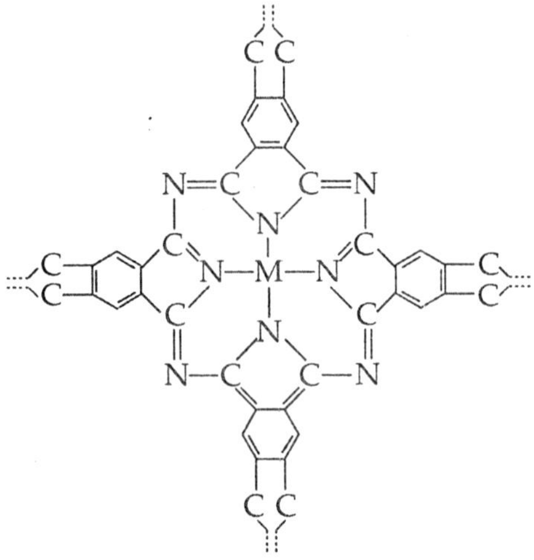

(7-LXIII)

The reaction of bis (β-diketone) Cu(II) i.e. $Cu(CH_3COCHCOCH_3)_2$ with $C_2(CN)_4$ results in a phthalocyanine like polymer of (7-LXIII) results.

Polymeric schiff bases constitute an important class of polymeric ligands. One of these ligands can be prepared in the following way :

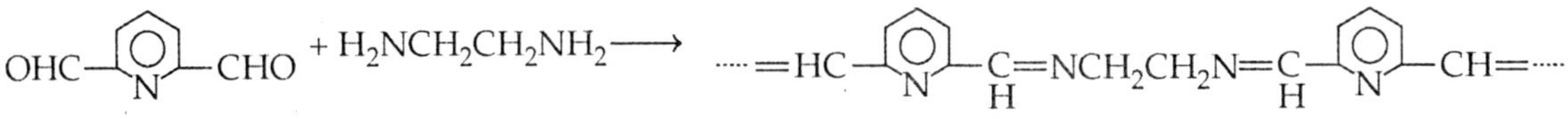

The above schiff base ligand reacts with Fe(II) salts to produce the following coordination polymer :

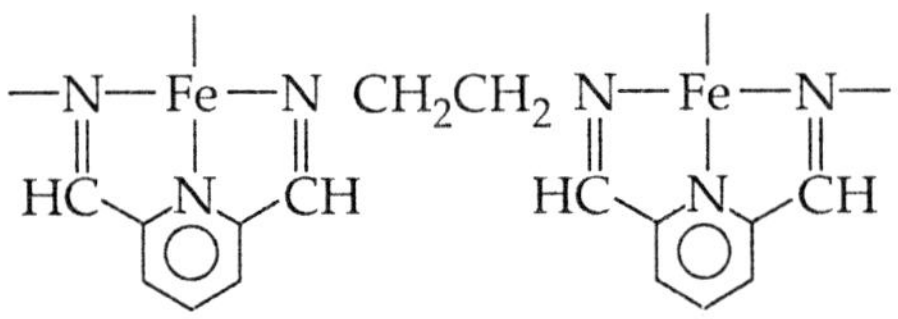

Polymeric schiff base ligands can also be prepared by condensation with salicylaldehyde and o-phenylenediamine and coordination polymers can be prepared by divalent metal ions having the following stability sequence :

$$Cu^{2+} < Co^{2+} < Zn^{2+} < Ni^{2+} < Fe^{2+} < Cd^{2+}$$

In general the thermal stability of synthetic coordination polymers are not much different from that of oragnic polymers.

**Problem 7.70** Of the two redox couples, $[Co(H_2O)_6]^{3+}/[Co(H_2O)_6]^{2+}$ and $[Co(NH_3)_6]^{3+}/[Co(NH_3)_6]^{2+}$, which one is more oxidising and why? **[C.U. 2009]**

**Ans. :** According to redox potential and CFSE [Sec. 7.14.5(A)(iv)], $[Co(H_2O)_6]^{3+}/[Co(H_2O)_6]^{2+}$ is more oxidising than $[Co(NH_3)_6]^{3+}/[Co(NH_3)_6]^{2+}$.

**Problem 7.71** Calculate the formation constant $K_f$ for $[Ag(CN)_2]^-$ if the cell $Ag(CN)_2^-$ ($7.50 \times 10^{-3}$ M), $CN^-$ (0.025 M) | Ag || SCE develops a pontential of 0.625 volt. **[C.U. 2009]**
Given : $Ag^+ + e \rightleftharpoons Ag(s)$, E° = +0.799 V, E of SCE = 0.244 V.

**Ans. :** $E_{Cell} = E_{SCE} - E_{Ag(CN)_2^-/Ag, CN^-} = 0.244 - x$ (Say)

$\therefore 0.625 = 0.244 - x$ i.e. $x = -0.381$ V

Now $Ag^+ + e \longrightarrow Ag(s)$ and $Ag^+ + 2CN^- \longrightarrow Ag(CN)_2^-$

$$\therefore K_f = \frac{[Ag(CN)_2^-]}{[Ag^+][CN^-]^2} \text{ so that } [Ag^+] = \frac{[Ag(CN)_2^-]}{K_f[CN^-]^2}$$

Now according to Nernst equation,

$$E_{Ag^+/Ag} = E°_{Ag^+/Ag} + 0.0591 \log[Ag^+] = 0.799 + 0.0591 \log[Ag^+]$$

In the presence of $CN^-$

$$x = E_{Ag(CN)_2^-/Ag, CN^-} = 0.799 + 0.0591 \log \frac{[Ag(CN)_2^-]}{K_f[CN^-]^2}$$

$$= 0.799 - 0.0591 \log K_f + 0.0591 \log \frac{[Ag(CN)_2^-]}{[CN^-]^2}$$

$$= 0.799 - 0.0591 \log K_f + 0.0591 \log \frac{7.50 \times 10^{-3}}{0.025 \times 0.025}$$

$$\therefore 0.0591 \log K_f = 0.799 - x + 0.0591 \log \frac{7.50 \times 10^{-3}}{0.025 \times 0.025}$$

$$\therefore \log K_f = \frac{0.799}{0.0591} - \frac{x}{0.0591} + \log \frac{7.50 \times 10^{-3}}{0.025 \times 0.025}$$

$$= \frac{0.799}{0.0591} + \frac{0.381}{0.0591} + \log 12$$

$$= 13.52 + 6.45 + 1.0792$$

$$= 21.04$$

## 7.19 INORGANIC REACTION MECHANISMS

By the term reaction mechanisms we first consider the transformations of d-block metal complexes to other complexes in presence of ligands and assign a suggestive mechanism for this transformations from some experimental observations. Before going into the mechanistic aspects of reactions we first consider the **stability** of d-metal complexes which is intimately related with **lability** and **inertness** terms associated with those complexes.

### 7.19.1 Kinetic versus thermodynamic stability : Lability and Inertness of complexes

The term 'stability' of complexes may be defined in two ways :

(a) **Thermodynamic stability :** This is a measure of the extent to which the complex will form or will be transformed into other complex (under certain conditions) when the system has reached equilibrium. This type of stability is related with metal-ligand bond energies, stability constants etc.

(b) **Kinetic stability :** This is related to the speed with which transformations leading to the attainment of equilibrium will occur. This stability is related to kinetic aspects of reactions such as rates and mechanisms of ligand substitution, isomerisation, racemisation and electron transfer reactions.

H. Taube classified the complexes as **inert** or **labile. Inert complexes** are thermodynamically unstable but survive for long time and undergo substitution reaction slowly. Complexes that undergo more rapid equlibrium are called **labile** and in **labile complexes** ligand substitution takes place rapidly i.e. rate of substitution reaction is high.

*Kinetic vs thermodynamic stability and inert vs labile*

As the terms **labile** and **inert** are related with the speed at which the substitution reactions occur, therefore these are related to **kinetic stability**. These terms should not be confused with **thermodynamic stability term** such as **unstable** and **stable.** Although thermodynamically stable complexes may be labile or inert and unstable complexes which are usually labile may also be inert, there is no correlation between thermodynamic and kinetic stability terms e.g. $Ni(CN)_4^{2-}$, $Cr(CN)_6^{3-}$, $Hg(CN)_4^{2-}$ which are **thermodynamically stable** but **kinetically labile** since in solution $Hg(CN)_4^{2-}$ ($K_f = 10^{42}$) exchanges $CN^-$ ligands with labelled $^{14}CN^-$ at a very fast rate :

$$Hg(CN)_4^{2-} + 4\,^{14}CN^- \rightleftharpoons Hg(^{14}CN)_4^{2-} + 4CN^-$$

Thus thermodynamic stability will not ensure the inertness of complexes. On the other hand $Co(NH_3)_6^{3+}$ which is **kinetically inert** but **thermodynamically unstable** since the complex remains unchanged in acid solution for prolonged period. Thus the kinetically inert complexes are not necessarily thermodynamically stable and conversely the kinetically labile complexes are not necessarily thermodynamically unstable. The lability of a complex depends on the activation energy while stability of a complex depends on the energy associated with the reaction. Larger activation energy means the lowering of lability i.e. the complex with larger activation energy will react slowly.

Finally as per H. Taube the labile and inert complexes are characterised by CFSE/LFSE which is intimately related with $d^n$ metal ions having high and low spin configurations (Table 7.19). In an octahedral complex with electrons in the antibonding $e^*_g$ orbitals, the ligands will be relatively weakly bound and thus may be displaced easily. Thus all high spin $d^4$, $d^5$, $d^6$ and $d^7$, $d^8$, $d^9$, $d^{10}$ configurations should be labile. Taube suggested the **activation energy** parameter for the formation of octahedral complexes as well as for the formation of complexes in other geometries. The activation energy is also related to CFSE in order to determine the **lability** and **inertness** of complexes. The octahedral complexes formed for which there is large loss of CFSE are least labile i.e. such complexes are **kinetically inert** while the complexes where there is little or no loss in CFSE are **labile** and such complexes react rapidly. Octahedral complexes of $d^0$, $d^1$, $d^2$, $d^9$, $d^{10}$ as also **high spin** complexes of $d^4$, $d^5$, $d^6$ and $d^7$ are labile. On the other hand $d^3$ and **low spin** complexes of $d^4$, $d^5$, $d^6$ and $d^7$ are generally inert. It is to be noted that the **order of inertness** of low spin complexes of $d^n$ configurations are : $d^6 > d^3 > d^4 > d^5$ while the **lability order** for high spin complexes are : $d^6 < d^3 < d^4 < d^5$.

**Table 7.19 : Examples of inert and labile complexes of $d^n$ configurations**

| $d^n$ configuration | Inert | Labile |
|---|---|---|
| $d^0$ | | $[Ca(EDTA)]^{2-}$, $[TiCl_6]^{2-}$ |
| $d^1$ | | $[Ti(H_2O)_6]^{3+}$, $[VO(H_2O)_5]^{2+}$ |
| $d^2$ | | $[V(ophen)_3]^{3+}$ |
| $d^3$ | $[V(H_2O)_6]^{2+}$, $[Cr(H_2O)_6]^{3+}$ | |
| $d^4$ | $[Cr(CN)_6]^{4-}$(l.s.), $[Mn(CN)_6]^{4-}$(l.s.) | $[Cr(H_2O)_6]^{2+}$ (h.s.) |
| $d^5$ | $[Fe(CN)_6]^{3-}$(l.s.) | $[Mn(H_2O)_6]^{2+}$(h.s.) |
| $d^6$ | $[Fe(CN)_6]^{4-}$(l.s.), $[Co(NH_3)_6]^{3+}$ $[Co(en)_2)(H_2O)_2]^{2+}$(l.s.) | $[Fe(H_2O)_6]^{2+}$(h.s.) |
| $d^7$ | | $[Co(NH_3)_6]^{2+}$, $[Co(H_2O)_6]^{2+}$ |
| $d^8$ | | $[Ni(en)_3]^{2+}$, $[Ni(CN)_4]^{2-}$ |
| $d^9$ | | $[Cu(NH_3)_4(H_2O)_2]^{2+}$ |
| $d^{10}$ | | $[Hg(CN)_4]^{2-}$ |

## 7.19.2 Ligand substitution reactions related to lability and inertness

Since the formation of complexes in solution (Sec. 7.7) is accompanied by substitution reactions by ligands, it is necessary to have an idea of the mechanisms of these reactions in order to understand the **lability** and **inertness** of complexes.

Let us consider a ligand substitution reaction :

$$M - X + Y \longrightarrow M - Y + X \qquad \text{...(7.44)}$$

where X is termed **leaving group** and Y is called **entering group** e.g.

$$[Co(H_2O)_6]^{2+} + Cl^- \longrightarrow [Co(H_2O)_5Cl]^+ + H_2O$$

The equilibrium constant of this type of substitution reaction can be used to make Lewis basicity order of ligands (Y). The activation barrier of a substitutioin reaction has two aspects. First the energy required to break the bond between the metal ion and the leaving group (X). Secondly the energy evolved when the metal ion forms a bond with the entering group (Y). The second aspect is related to kinetic phenomena of the basicity of the entering group which is called **nucleophilicity** (from the Greek meaning 'nucleus loving'). The term may be defined as :

$$\text{nucleophilicity of a ligand Y (base)} = \frac{\text{rate of attack on a complex by a given base Y}}{\text{rate of attack on the same by a standard base Y}^\circ} \qquad \text{...(7.45)}$$

**Nucleophilicity parameter** is measured by comparing rate constants for substitution reactions like that of basicity measurements by comparing equilibrium constants for acid-base reactions. Faster the reaction with an entering group, the greater will be its nucleophilicity. Increasing rates of nucleophilic substitution at a given central atom is in the order :

$$Y : H_2O < Cl^- < I^- < H^- < PR_3 < CO\ ,\ CN^- \quad ...(7.46)$$

and for a leaving group an order of increasing rates of substitution at a given central atom as :

$$X : H_2O > Cl^- > I^- > H^- > PR_3 > CO\ ,\ CN^- \quad ...(7.47)$$

The kinetic analog of acidity is **electrophilicity** (an affinity for negative centres) which is the reverse of **nucleophilicity** (an affinity for positive centres). Faster reaction of a Lewis acid with an entering group indicates greater electrophilicity of the acid.

### 7.19.3 Types of Inorganic reactions

**(1) Substitution reactions :**

(a) **Acid hydrolysis** e.g.

$$[Co(NH_3)_5Cl]^{2+} + H_2O \xrightarrow{pH \approx 1} [Co(NH_3)_5H_2O]^{3+} + Cl^-$$

$$[Pt(NH_3)_3Cl]^+ + H_2O \longrightarrow [Pt(NH_3)_3(H_2O)]^{2+} + Cl^-$$

(b) **Base hydrolysis** e.g.

$$[Co(NH_3)_5Cl]^{2+} + OH^- \xrightarrow{pH > 7} [Co(NH_3)_5(OH)]^{2+} + Cl^-$$

(c) **Aquation reaction** e.g.

$$[Cr(H_2O)_6]^{3+} + NCS^- \longrightarrow [Cr(H_2O)_5(NCS)]^{2+} + H_2O$$

(d) **Deaquation reaction** e.g.

$$[Ni(H_2O)_6]^{2+} + NH_3 \longrightarrow [Ni(H_2O)_5NH_3]^{2+} + H_2O$$

**(2) Isomerisation reaction** e.g.

$$\underset{green}{\text{Trans-}[Co(en)_2Cl_2]^+} \rightleftharpoons \underset{violet}{\text{cis}[Co(en)_2)Cl_2]^+}$$

$$[Co(NH_3)_5(ONO)]^{2+} \rightleftharpoons [Co(NH_3)_5(NO_2)]^{2+}$$

**(3) Racemisation reaction** e.g.

$$\text{dextro-}[Ni(ophen)_3]^{2+} \rightleftharpoons \text{levo}[Ni(ophen)_3]^{2+}$$

$$\text{dextro-}[Co(C_2O_4)_3]^{3-} \rightleftharpoons \text{levo}[Co(C_2O_4)_3]^{3-}$$

**(4) Oxidation-Reduction reaction (electron transfer reaction)**

(a) Where no net chemical change e.g.

$$Fe^{2+}(aq) + Fe^{*3+}(aq) \rightleftharpoons Fe^{3+}(aq) + Fe^{*2+}(aq)$$

(b) Where a net chemical change occurs e.g.

$$\underset{inert}{[Co^{III}(NH_3)_5X]^{2+}} + \underset{labile}{[Cr^{II}(H_2O)_6]^{2+}} + 5H_3O^+ \rightleftharpoons \underset{labile}{[Co(H_2O)_6]^{2+}} + \underset{inert}{[Cr(H_2O)_5X]^{2+}} + 5NH_4^+$$

In this type standard free energy change occurs.

Among the above type of reactions we shall only discuss the ligand substitution reaction in octahedral Co(III) and square planar Pt(II) complexes.

### 7.19.4 Classification of mechanisms of ligand substitution reaction or ligand displacement reactions

The term mechanism is meant all the individual collision involving molecules (atoms, radicals and ions included) or other elementary processes that take places simultaneously or consecutively in producing observed overall reaction. It is also understood that the mechanism of a reaction should give a detailed stereochemical picture of each step as it occurs. This implies a knowledge of the so called activated complex or transition state not only in terms of constitutional molecules but also in terms of the geometry such as interatomic distances and bond angles.

Ligand substitution reactions are designated as **nucleophilic substitution reactions** by organic chemists. The mechanisms of these reactions was proposed by Langford and Gray who classified in terms of **stoichiometric mechanism** and **intimate mechanism**. The task can be divided into two distinct phases. The first is the discovery of the sequence of elementary steps by which a complicated overall reaction is accomplished and the second is to understand the magnitude of rate constants for the individual steps in terms of rearrangement of atoms and bonds taking place. We shall designate the elaboration of the reactions in terms of elementary steps, the study of the stoichiometric mechanism and the analysis of the individual steps, the study of intimate mechanism.

Three types of **stoichiometric mechanism** for ligand substitution reactions are to be considered :

Dissociative D    Interchange I    and    Associative A

The first and third are two-step mechanisms in which an intermediate is formed and then transformed into the product.

$$D : MX_n \underset{+X}{\overset{-X}{\rightleftharpoons}} MX_{n-1} \underset{-Y}{\overset{+Y}{\rightleftharpoons}} MX_{n-1}Y$$

$$A : MX_n \underset{-Y}{\overset{+Y}{\rightleftharpoons}} MX_nY \underset{+X}{\overset{-X}{\rightleftharpoons}} MX_{n-1}Y$$

$$I : MX_n \underset{-Y}{\overset{+Y}{\rightleftharpoons}} MX_nY \longrightarrow MX_{n-1}Y......X$$

Let us consider a ligand substitution reaction according to equation (7.44) where X is the leaving group with all the other ligands are not taking part in the reaction (and so they will be called spectator ligands). Omitting the effect of **spectator ligands** we may consider the following mechanisms for ligand substitution reactions :

(1) **Dissociative mechanism (D) :**

This involves a step in which an intermediate is formed with a reduced coordination number. The intermediate should be stable. Let,

$$\underset{\textit{Metal complex}}{M - X} \xrightarrow{\text{slow } (K_1)} M + \underset{\substack{\textit{Leaving}\\\textit{group}}}{X} \qquad \text{...(7.48)}$$

$$M + \underset{\substack{\textit{Entering}\\\textit{group}}}{Y} \xrightarrow{\text{fast } (K_2)} M - Y \qquad \text{...(7.49)}$$

For such a process the rate of formation of M – Y can be represented by considering first order rate law as :

$$\frac{d[M\text{–}Y]}{dt} = -\frac{d[M\text{–}X]}{dt} = K_1[M\text{–}X] \qquad \text{...(7.50)}$$

where $K_1$ is the rate constant for the slow step.

The slow step is followed by the rapid step in which the entering group Y enters to form M–Y.

The possibility of step (7.48) is that there will be a rapid equilibrium reaction that has time for its reverse to be established. Actually there might be a possibility of recapture of X in step (7.48) so that

$$M - X \underset{K_{-1}}{\overset{K_1}{\rightleftharpoons}} M + X;\ M + Y \xrightarrow{K_2} M - Y \qquad \text{...(7.51)}$$

where $K_2$ is the rate constant for the fast step and $K_{-1}$ is the rate constant of the reverse process of dissociation of M – X.

Now applying steady state principle for the step (7.51) we may write :

$$\frac{d[M]}{dt} = K_1[M - X] - K_{-1}[M][X] - K_2[M][Y] \qquad \text{...(7.52)}$$

According to the **steady state approximation** for the steps (7.48) and (7.49), the concentration of M is neither increasing nor decreasing with time, so that,

$$\frac{d[M]}{dt} = 0 \qquad \text{...(7.53)}$$

Therefore, $K_1[M - X] - K_{-1}[M][X] - K_2[M][Y] = 0$

i.e. $K_1[M - X] = K_{-1}[M][X] + K_2[M][Y] = [M]\{K_{-1}[X] + K_2[Y]\}$

$$\text{so, } [M] = \frac{K_1[M\text{–}X]}{K_{-1}[X] + K_2[Y]} \qquad \text{...(7.54)}$$

Now from step (7.49), we can write,

$$\frac{d[M\text{–}Y]}{dt} = K_2[M][Y] \qquad ...(7.55)$$

From equation (7.54), we obtain,

$$\frac{d[M\text{–}Y]}{dt} = \frac{K_1K_2[M\text{–}X][Y]}{K_{-1}[X] + K_2[Y]} \qquad ...(7.56)$$

The denominator of equation (7.56) contains the sum of two expressions, either of which may be small compared to other and hence may be neglected. Thus under limiting situations :

(a) If $K_2[Y] >> K_{-1}[X]$, we can neglect $K_{-1}[X]$ w.r.t $K_2[Y]$, so that the expression (7.56) simplifies to,

rate of formation of M–Y i.e. $$\frac{d[M\text{–}Y]}{dt} \approx \frac{K_1K_2[M\text{–}X][Y]}{K_2[Y]} \approx K_1[M\text{–}X] \qquad ...(7.57)$$

which is the rate law we expect for this dissociative mechanism in which the first step (slow step) is the rate determining. A possible mechanism that can be ambiguous is one in which the first step is a slow one, the departing ligand is replaced, not by the entering ligand, but by a solvent molecule, which is subsequently rapidly replaced by the entering ligand. As the concentration of solvent molecules remain constant so that the rate law follow the apparent first order kinetics.

(b) If $K_{-1}[X] >> K_2[Y]$, then,

$$\frac{d[M\text{–}Y]}{dt} \approx \frac{K_1K_2[M\text{–}X][Y]}{K_{-1}[X]} \qquad ...(7.58)$$

In this case the second step is slower than the reverse of the first step. One example of this type of reaction is :

$$W(CO)_6 \longrightarrow W(CO)_5 + CO \qquad W(CO)_5 + PPh_3 \longrightarrow W(CO)_5PPh_3$$

(2) **Associative Mechanism (A) :** This involves a step in which a somewhat persistent intermediate is formed in which the central metal ion has expanded coordination number. Associative mechanism involves the following steps :

$$M - X + Y \underset{K_{-1}}{\overset{K_1}{\rightleftharpoons}} \underset{\textit{intermediate}}{\{Y - M - X\}} \qquad ...(7.59)$$

$$\{Y - M - X\} \xrightarrow{K_2} M - Y + X \qquad ...(7.60)$$

Applying steady state principle, we have

$$\frac{d[\{Y\text{–}M\text{–}X\}]}{dt} = K_1[M\text{–}X][Y] - K_{-1}[\{Y\text{–}M\text{–}X\}] - K_2[\{Y\text{–}M\text{–}X\}]$$

so, that $$[\{Y - M - X\}] = \frac{K_1[M\text{–}X][Y]}{K_{-1} + K_2}$$

Now, $$\frac{d[M\text{–}Y]}{dt} = K_2[\{Y - M - X\}] = \frac{K_1K_2[M\text{–}X][Y]}{K_{-1} + K_2} \qquad ...(7.61)$$

So that the rate law for this A mechanism to have the form

$$\text{rate} = \frac{K[M{-}X][Y]}{K_{-1} + K_2} \qquad \ldots(7.62)$$

One example of this type of mechaism is :

$$\underset{\textit{square planar}}{[Ni(CN)_4]^{2-}} + {}^{14}CN^- \longrightarrow [Ni(CN)_4({}^{14}CN)]^{3-} \longrightarrow [Ni(CN)_3({}^{14}CN)]^{2-} + CN^-$$

(3) **Interchange Mechanism (I) :** In this mechaism the leaving and entering groups do not form a stable intermediate instead they exchange themselves in a single step to form an **activated complex**. This activated complex almost instantaneously converted to products or revert back to the reactants. The activated complex is most difficultly achieved the point of highest energy during the passage from reactants to products. Thus in this case both bond dissociation of M–X and bond formation of M–Y occurs simultaneously in the activated complex. In fact D and A mechanism is very rare in metal complexes and almost all substitution reactions take place through I mechanism which is further subdivided (Scheme 7-II) into $I_d$ and $I_a$ according to the bond dissociation of M–X bond and bond formation of M–Y bond in the activated complex. Operationally the I mechanism may be represented as :

$$M - X + Y \xrightarrow{K_{OS}\,(\text{fast})} \underset{\substack{\textit{encounter}\\ \textit{complex}}}{\{M - X, Y\}} \xrightarrow{k\,(\text{slow})} \underset{\substack{\textit{activated complex}\\ \textit{or interchange}\\ \textit{complex } \{M-Y,\ X\}}}{\{Y\ldots M\ldots X\}} \xrightarrow{\text{fast}} M - Y + X \qquad \ldots(7.63)$$

According to the **Eigen-Wilkins mechanism** there is very fast formation and breakup of such encounter complex to reach the equilibrium. This step is followed by the slow interchange step of X and Y to form the interchange complex {M–Y, X} and finally this leaving the ligand rapidly diffuses out of the second coordination sphere to give the final product. The rate law for these simultaneous processes may be represented as :

$$\frac{d[M{-}Y]}{dt} = \frac{kK_{OS}[M{-}X][Y]}{1 + K_{OS}[Y]} \qquad \ldots(7.64)$$

In limiting cases the equation (7.64) may be transformed to—

(a) $K_{OS}[Y] << 1$, so that, $\frac{d[M{-}Y]}{dt} = kK_{OS}[M{-}X][Y]$ ...(7.65)

(b) $K_{OS}[Y] >> 1$, we obtain, $\frac{d[M{-}Y]}{dt} = k[M{-}X]$ ...(7.66)

This is exemplified by reactions of the six coordinated $3d^8$ complex of $[Ni(H_2O)_6]^{2+}$.

### 7.19.5 Intimate Mechanism for Ligand Substitution Reactions

This involves the more detailed picture of the electronic nature of the reaction in the rate determining step. The kinetic data for rate constants involve changes in the electronic nature of the leaving group X, the entering group Y, the spectator ligands and the central metal ion M. If the rate constant depends on the entering group Y, then the activated complex has strong involvement of bonding with Y while if the rate constant is less influenced by Y then the value of rate constant will be largely influenced by the rate at which the bond between the metal atom and the leaving ligand X can break. Thus in the intimate mechanism the classifications between the reactions have been made where the rate constant depends largely on the entering group (Y) or the rate constant may not be dependent on the entering group (Y) and such distinction may be operated in two extremes called **associative** (a) and **dissociative** (d) paths.

In the substitution reaction of one water molecule in $[Ti(H_2O)_6]^{3+}$ with $NCS^-$, the rate of substitution strongly depends on the nature of the entering group Y (e.g. $NCS^-$) and the nature of the M–Y bond formed is affecting the rate determining step and the mechanism was suggested as **associative** (a). Associative reactions are found among $d^8$ square planar complexes of Pt(II), Pd(II) and Au(III).

$$Ti(H_2O)_6^{3+} + NCS^- \longrightarrow Ti(H_2O)_5NCS^{2+} + H_2O \qquad K = 8.0\times10^3\ M^{-1}s^{-1}$$

In another study of the effect of entering group Y is made by varying Y on the substitution of water molecule in $[Co(NH_3)_5H_2O]^{3+}$. Here the rate constant will little effect on varying Y and we classify the rate determining step of this substitution reaciton as **dissociative** (d). One such example is :

$$[Co(NH_3)_5(H_2O)]^{3+} + Cl^- \longrightarrow [Co(NH_3)_5Cl]^{2+} + H_2O \qquad K = 2.1\times10^{-5}\ s^{-1}$$

Thus if both **stoichiometric** and **intimate** mechanism have been operated, we can select the proper choice from the overall four classifications i.e. A, D, $I_a$ and $I_d$ (Scheme 7-II). This classification was proposed by Langford and Gray (1965) and is now unequivocally used by coordination chemists and the above mechanisms are related to the conventional classical systems as :

D : $S_{N1}$ (limiting), $I_d$ : $S_{N1}$

A : $S_{N2}$ (limiting), $I_a$ : $S_{N2}$

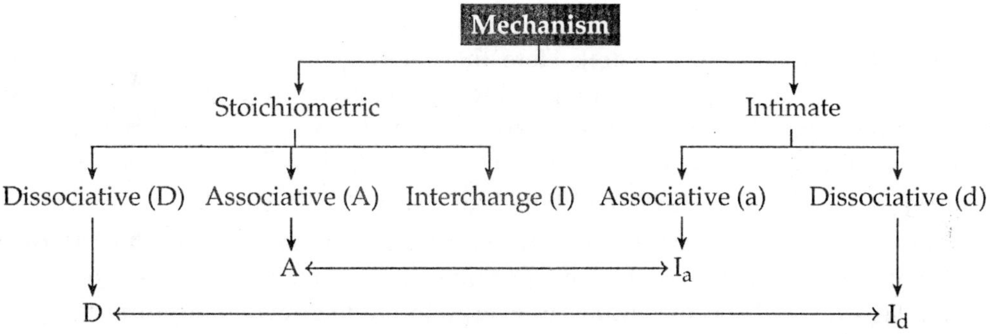


Scheme 7-II : Classification of mechanisms of ligand substitution reactions.

Associative ($I_a$ or A) mechanisms increase the total coordination number of central metal ion in the activated complex while dissociative ($I_d$ and D) mechanisms will decrease the total coordination number of the central metal ion. Associative or specially $I_a$ mechanisms are common for square planar complexes of fourth period elements. This mechanisms are less common for octahedral complexes of

*Periodic trend in mechanisms*

this period. Dissociative (D or commonly $I_d$) mechanisms are expected to be more common for octahedral complexes of fourth period d-metals than for square planar complexes.

### 7.19.6 Reaction profiles

The free energy differences between I, D and A mechanisms as the reactions proceed may be depicted by drawing a curve which is known as **reaction profiles** (Fig. 7.25). It is a qualitative graphical representation of energy as a function of reaction coordinate. The reaction coordinate represents the progress of the reaction from the reactants to achieve the products. In all cases of mechanisms the energy of the reactant(s) must increase in order to rupture the existing stable bonds until finally an **activation energy** is supplied to cross the activation energy barier thereby an **activated complex** or an **intermediate** is formed in which the existing bonds are stretched enough and that the new bonds are formed in the stable product(s).

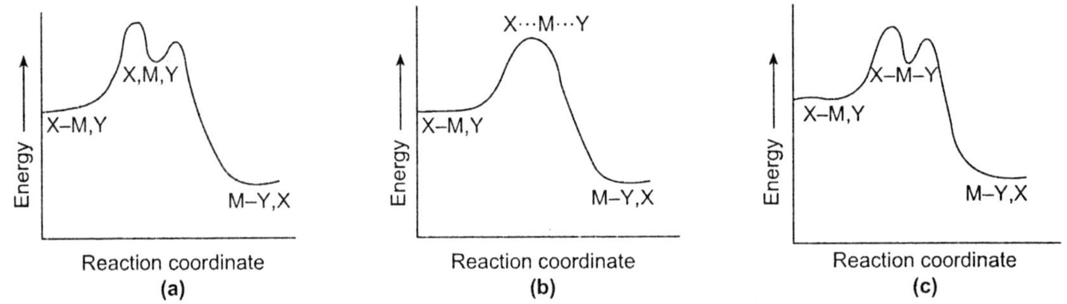


***Figure 7.25** : Reaction profiles for (a) dissociative (D), (b) interchange (I), and (c) associative (A) mechanisms.*

A **dissociative** path involves a step in which an intermediate of reduced coordination number is formed after the removal of the leaving group (X). This consists of a single intermediate and **two activated states** [Fig. 7.25(a)].

An **associative** path in which an activated complex is formed of higher coordination number than the original complex [Fig. 7.25(c)]. Here the entering group (Y) binds to the original complex (M–X) before any bond weakening of the leaving group takes place. This consists of a single intermediate similar to that of dissociative path and have two activated states—one for bond formation and other for bond breaking of the intermediate. Square planar complexes of Pt(II), Pd(II) and Ir(I) have probability to follow this type of mechanism.

One common feature of D and A mechanisms is the formation of **single true intermediate.** This single intermediate will exist in high position of the energy profile and how long the intermediate lasts depends on the supply of energy to the intermediate to form the products.

In an **interchange** path the leaving(X) and entering (Y) groups exchange in a **single step** by forming an activated complex but there is **no formation of true intermediate**. In this case the activated complex either revert back to the reactants or proceed forward to give the final products.

### 7.19.7 Acitvation parameters for A, D and I paths

The reaction profile curves (Fig. 7.25) also known as **energy profiles**. The peak of the curve represents the **transition state** (TS) and the configuration attained by the reactants at this peak is called **activated complex**. To attain the activated complex at the transition state activation energy is required which is the energy required to transform the reactants into the transition state (i.e. intermediate activated complex). Activation energy is always positive and contribution of energy is made up due to changes in metal ligand bond lengths, ligand-ligand repulsions, CFSE etc. This activation energy is also related to some thermodynamic terms i.e. $\Delta G$, $\Delta H$, $\Delta S$ etc. The difference in free energy between the activated complex and the reactants is called the **free energy of activation ($\Delta G^{\neq}$)** and the respective $\Delta H$ and $\Delta S$ terms denote the enthalpy of activation ($\Delta H^{\neq}$) and entropy of activation ($\Delta S^{\neq}$) respectively. They are related as :

$$\Delta G^{\neq} = \Delta H^{\neq} - T\Delta S^{\neq} \quad ...(7.67)$$

$$\text{and} \quad \Delta G = \Delta H - T\Delta S, \ \Delta G^0 = \Delta H^0 - T\Delta S^0 \quad ...(7.68)$$

where $\Delta G$ = free energy of reaction and $\Delta G^0$ = standard free energy of reaction and $\Delta H$, $\Delta H^0$, $\Delta S$ and $\Delta S^0$ denote the corresponding enthalpy and entropy terms. Inspite of the favourable free energy of reaction ($\Delta G$) the reaction may be kinetically not favoured if $\Delta G^{\neq}$ is very high.

These thermodynamic parameters are depicted for D, A and I pathways (Fig. 7.26). In the D process a bond rupture occurs to achieve the TS having reduced coordination number. The bond dissociation energy is a measure of activation energy for this process. While in the case of A process an activated complex is formed with higher coordination number with the release of energy due to new bond formation from the original complex. The higher coordination number requires new hybridisation of the TS thereby energy released due to the energetics of hybridisation (Sec. 3.2.6). Provided steric factor, leaving group effects are not very high the free energy of activation for dissociative pathway (D) is higher for associative pathway (A) i.e. $\Delta G_D^{\neq} > \Delta G_A^{\neq}$ (Fig. 7.26(a and b)).

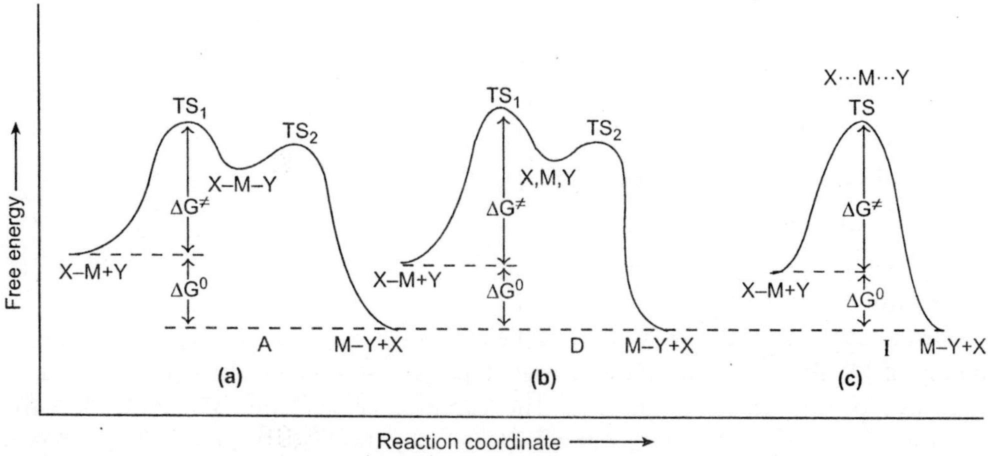


***Figure 7.26** : Activation parameter for (a) associative, (b) dissociative and (c) interchange pathways. The dip in the A and D curves indicate that the $TS_1$ (first transition state) changes to the intermediate and this further changes through $TS_2$ (second transition state) into the products.*

In case of D process only bond breaking important in TS while in A process bond breaking partially compensated by bond formation in TS and so also $\Delta H_D^{\neq} > \Delta H_A^{\neq}$.

In D pathway the number of species in the TS or activated complex is increased than that of the reactants and so **$\Delta S^{\neq}$ is positive or less negative.** As the activated state is more compact in the TS, in A process the **$\Delta S^{\neq}$ is negative or less positive.**

The activation parameter for I process is less important as in this case there is no formation of true intermediate and reaction passes to the products through the transition state of high energy (Fig. 7.26(c)).

### 7.19.8 Leaving group effect and Linear Free Energy Relationship

In nucleophillic substitution reactions a good entering group make poor leaving group. On the other hand the nature of the leaving group profoundly influence the rate of a substitution reaction. This is demonstrated in the aquation reactions of $Co(NH_3)_5X^{2+}$ and $Cr(H_2O)_5X^{2+}$. The demonstration of leaving group effect is the application of linear free energy relationship (LFER) in the following aquation reaction :

$$Co(NH_3)_5X^{2+} + H_2O \longrightarrow Co(NH_3)_5(OH_2)^{3+} + X^- \quad ...(7.69)$$

where X is different leaving groups e.g. $F^-$, $Cl^-$, $I^-$, $NCS^-$, $NO^-_3$, $SO_4^{2-}$ etc. It is found that for these complexes there is a linear relationship between the $\log k_{aq}$ and $\log K_{aq}$ where $k_{aq}$ and $K_{aq}$ is the rate constant and equilibrium constant of the reaction (7.69), which can be represented as,

$$\log k_{aq} = \log K_{aq} + C \quad ...(7.70)$$

where C is a constant and the plot of $\log k_{aq}$ vs. $\log K_{aq}$ is a straight line with a slope = 1.0 (Fig. 7.27). As we know,

$$\log k_{aq} = \frac{-\Delta G^{\neq}}{2.303RT} - \log\frac{h}{KT} \quad ...(7.71)$$

$$\text{and } \log K_{aq} = \frac{-\Delta G^0}{2.303RT} \quad ...(7.72)$$

Thus we get,

$$\frac{-\Delta G^{\neq}}{2.303RT} - \log\frac{h}{KT} = \frac{-\Delta G^0}{2.303RT} + C$$

and which on rearrangement produces,

$$\Delta G^{\neq} = \Delta G^0 + X \quad ...(7.73)$$

where X is another constant. Thus the plot (Fig. 7.27) is practically the plot of $\Delta G^{\neq}$ vs. $\Delta G^0$ which denotes the **linear free energy relation** (LFER). It shows that on changing X has the same effect on $\Delta G^{\neq}$ for the conversion of Co–X bond to the activated complex as it has on $\Delta G^0$ for complete elimination of $X^-$. The same variation of unit slope is observed for divalent $ox^{2-}$ and $SO_4^{2-}$ and a straight line parallel to $X^-$ is obtained. The variation of $\Delta G^0$ and $\Delta G^{\neq}$ with the variation of leaving group X is depicted in Fig. 7.28 which denotes the leaving group effect. This suggests that in a dissociative reaction the overall $\Delta G$ values with variations of $X^-$ is mainly the variation of Co–X bond strength and the solvation of $X^-$. This indicates a practically complete dissociation of Co–X bond in the transition state.

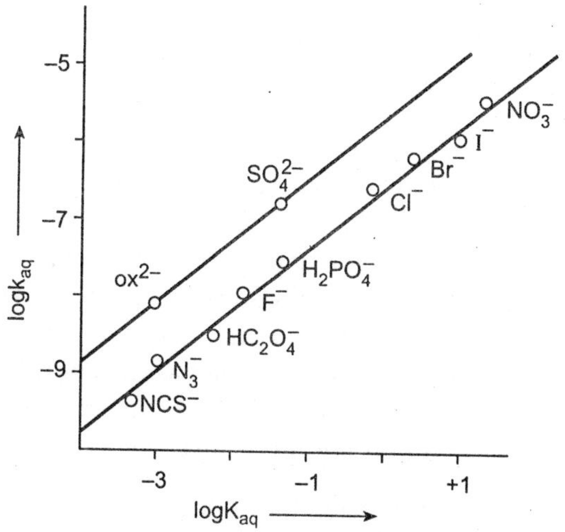


**Figure 7.27** : *Plot of $\log k_{aq}$ vs. $\log K_{aq}$ for the aquation reaction (7.69).*

As mentioned earlier for aquation reaction of $Cr(H_2O)_5X^{2+}$ similar LFER plot is obtained with a slope of 0.56 which indicates a substantial dissociation of Cr–X bond in the transition state. Similar LFER is observed for Rh(III) complex with a slope < 1 having varying leaving groups and the reaction rates are in the order $I^- < Br^- < Cl^-$.

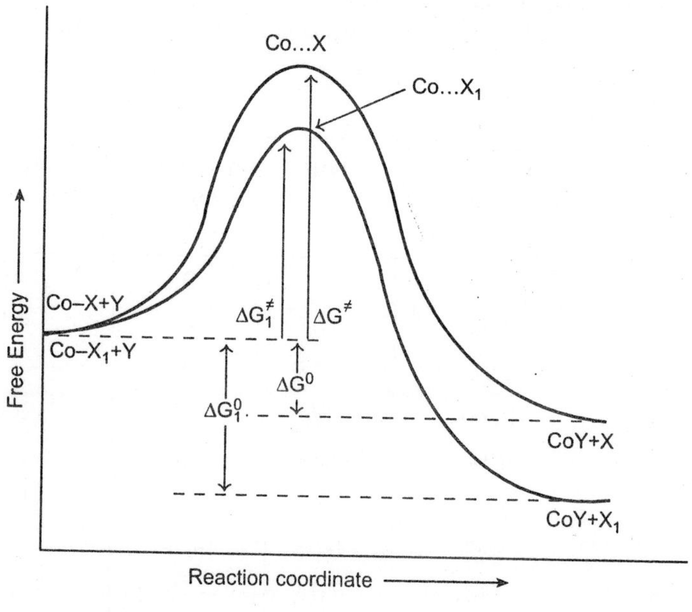


**Figure 7.28** : *LFER showing the variation of $\Delta G^{\neq}$ for changing the leaving group from X to $X_1$.*

### 7.19.9 Crystal Field Stabilisation Energy (CFSE) and Crystal Field Activation Energy (CFAE) in substitution reactions of occtahedral complexes

In **dissociative mechanism** (D) the **intermediate** is formed with reduced coordination number ($ML_5$) having square pyramidal geometry from the orginal complex ($ML_6$). A trigonal bipyramidal intermediate is less favourable for D pathway. On the other hand in **assocative pathway** the **intermediate** should have expanded coordination number 7 formed by the entering group Y having the geometry of the intermediate either pentgonal bipyramid or capped octahedron (Fig. 3.8). The difference of CFSE between the original complex and that of the intermediate formed in the transition state is called as the contribution to the total activation energy and is sometimes referred as ligand field contribution ($\Delta E_a$) or crystal field activation energy (CFAE) or ligand field activation energy (LFAE) i.e.

$$\Delta E_a/\text{CFAE}/\text{LFAE} = \text{CFSE of intermediate} - \text{CFSE of reacting complex} \quad ...(7.74)$$

Therefore CFAE depends on the CFSE/LFSE of original complex as also of the intermediate formed in TS. For simplicity's sake we shall restrict our discussion for D mechanism ($S_N1$ limiting) having square pyramidal intermediate and CFAE/LFAE is calculated for different $d^n$ systems (Table 7.20) with the help of Table 7.11 for high and low spin complexes. A positive value of CFAE indicates slowness of the rate of reaction and negative value indicates the instability of the original octahedral complex.

**Table 7.20 : CFAE/LFAE values for D path ($S_{N1}$ limiting) in octahedral to square pyramidal transition state**

| $d^n$ system | CFSE for weak field complexes of | | CFAE | CFSE for strong field complexes of | | CFAE |
|---|---|---|---|---|---|---|
| | Octahedral geometry (Dq) | Square pyramidal geometry* (Dq) | (Dq) | Octahedral geometry (Dq) | Square pyramidal geometry* (Dq) | (Dq) |
| $d^0$ | 0 | 0 | 0 | 0 | 0 | 0 |
| $d^1$ | –4 | –4.57 | –0.57 | –4 | –4.57 | –0.57 |
| $d^2$ | –8 | –9.14 | –1.14 | –8 | –9.14 | –1.14 |
| $d^3$ | –12 | –10.00 | 2.00 | –12 | –10.00 | 2.00 |
| $d^4$ | –6 | –9.14 | –3.14 | –16 | –14.57 | 1.43 |
| $d^5$ | 0 | 0 | 0 | –20 | –19.14 | 0.86 |
| $d^6$ | –4 | –4.57 | –0.57 | –24 | –20 | 4.0 |
| $d^7$ | –8 | –9.14 | –1.14 | –18 | –19.14 | –1.14 |
| $d^8$ | –12 | –10.00 | 2.00 | –12 | –10.00 | 2.00 |
| $d^9$ | –6 | –9.14 | –3.14 | –6 | –9.14 | –3.14 |
| $d^{10}$ | 0 | 0 | 0 | 0 | 0 | 0 |

* *Calculated as per Table 7.12 and Fig. 7.12(d).*

From Table 7.20 we can draw the following generalisations :

(a) For weak field octahedral complexes, those having $d^0$, $d^1$, $d^2$, $d^4$, $d^5$, $d^6$, $d^7$, $d^9$ and $d^{10}$ metal ion configurations have either negative or zero CFAE. Such complexes are therefore termed **labile**. For $d^3$ and $d^8$ configurations have positive CFAE and they would therefore be slow to react and are termed **inert**.

(b) For strong field octahedral complexes having $d^0$ $d^1$, $d^2$, $d^7$, $d^9$ and $d^{10}$ configurations have either negative or almost zero CFAE which indicates the **lability** of these complexes. Conversely $d^3$, $d^4$, $d^5$, $d^6$ and $d^8$ configurations register positve CFAE and are termed **inert** indicating slow rate of reactions by that particular path. Their reactivity sequence is : $d^5 > d^4 > d^3, d^8 > d^6$.

## 7.19.10 Spectator Ligands : Steric effect of non-leaving ligands

Entering group causes steric crowding to the formation of intermediate in the TS and influence the rate of a reaction. Steric crowding favours dissociative mechanism (D) and hindered associative mechanism (A). Favourable dissociative pathway is followed due to steric effect because of the relieve of steric strain by the formation of five coordinate activated complex but reduce the chance of association by discouraging attack of entering group. This is demonstrated by the substitution reactions of octahedral complexes of Co(III), Cr(III) where both **cis** and **trans** ligands affect the rate of substitution reaction. Ligands **cis** to a leaving group usually have non-specific effects on rates. A ligand that is present in the complex but not lost in the reaction is called **spectator ligand**. There is no important **trans** effect. An example of the study of acid hydrolysis of Co(III) complex of bidentate ligands will indicate the effect of non-leaving ligands which are acting as **spectator ligands :**

*Steric effect or steric crowding*

$$\text{trans } Co(AA)_2Cl_2^+ + H_2O \longrightarrow CoCl(AA)_2H_2O^{2+} + Cl^- \qquad ...(7.75)$$

For this hydrolytic reaction (7.75) the dissociative (first order) rate constants will vary according to the following bidentate nitrogen donor **spectator ligands** as :

| **Ligands** | **rate constant (k) at 25°C ($s^{-1}$)** |
|---|---|
| $H_2NCH_2CH_2NH_2$ | $3.2 \times 10^{-5}$ |
| $H_2N{-}CH_2{-}CH(CH_3){-}NH_2$ | $6.2 \times 10^{-5}$ |
| $H_2N{-}CH(CH_3){-}CH(CH_3){-}NH_2$ | $1.5 \times 10^{-4}$ |
| $H_2N{-}C(CH_3)_2{-}C(CH_3)_2{-}NH_2$ | Very fast (instantaneous) |

The above rate constant values revealed a correlation between rate and increasing number of methyl substituents on the spectator ligands increases the rate constant by a significant factor. This accleration of rate constant is in accordance with a dissociative mechanism in which steric strain is conducive to the formation of an activated complex (intermediate) of reduced coordination number. Similar variation of rate constants with macrocyclic **cyclam** or its derivative acting as nitrogen donor **spectator ligands**.

Another instance of hydrolysis reactions with better electron donating substituents (Spectator ligands) may be expected to acclerate the rates by dissociative pathway since they donate more electron density to compensate for the reduction of coordination number. In the following reaction—

$$NiXL_5^+ + H_2O \longrightarrow NiL_5(H_2O)^{2+} + X^- \quad ...(7.76)$$

replacement of $X^-$ by $H_2O$ is much faster when L is $NH_3$ than when it is $H_2O$. This is due to the strong σ donor ability of $NH_3$ compared to that of $H_2O$, thus increases the electron density on the central metal ion and hence acclerates the bond breaking of M–X bond to have the activated complex of reduced coordination number. The effect is enhanced if the cis ligands are good π donors while good σ donor ligands stabilise the reduced coordination number in a better way.

*Electron donating effect*

### 7.19.11 Factors affecting reaction rates and mechanism

The rate of reactions are influenced by the (a) effect of d electron structure of the central metal ion, (b) effect of net charge on the complex ion, (c) steric effect or steric crowding, (d) influence of concentration and nature of incoming ligands i.e. entering groups, (e) influence of outgoing ligands i.e. leaving groups and also spectator ligands, (f) effect of temperature, (g) influence of ionic strength, (h) effect of external pressure, (i) influence of dielectric constant of the solvent, (j) effect of pH etc.

Conclusion about the mechanism of reactions may be drawn from experimental observations.Comparing values of rate constants for formation of metal complexes we can choose whether the reaction is asociative or dissociative. For example, if the rates of the reaction is independent of the nature of entering group Y but only dependent on the leaving group and naturally the reaction is dependent on the bond breaking of M–X bond and thus the reaction will follow dissociative (D) pathway or dissociative intimate mechanism. On the other hand if the rate of a reaction depends on the entering group Y and so the raction is controlled by the rate of formation of M–Y bond and thus follow associative (A) mechanism. Sensitivity of the rate of reactions depends on the nature of entering and leaving groups.

Activation parameters may also influence to indicate specific mechanisms. Free energy of activation ($\Delta G^{\neq}$) is graeter for dissociative path than that of associative path. Entropy parameter ($\Delta S^{\neq}$) is more positive in case of dissociative than that of associative mechanism.

The rates of reactions and mechanisms depends on the **inert** and **labile** species. Generally labile complexes undergo more facile substitution reactions than that of inert complexes. Lastly CFSE and CFAE also influence associative and dissociative activation in the establishment of stoichiometric mechanism.

### 7.19.12 Experimental study of mechanisms

In the elucidation of mechanisms of inorganic reactions first we have to study the kinetics (rate) of the reaction and to determine the rate constant. This is based on the experiments in which the **concentration** of various reactants, products and intermediates are determined by various techniques. A large variety of conventional methods viz direct chemical analysis, conductometric method, potentiometric method, UV-visible spectrophotometry, polarimetric method (specific rotation measurement), isotopic tracer technique etc. are available for slow or moderate reactions. For very fast reactions of **labile** systems measurement requires in the range of nanoseconds ($10^{-9}$s) while **inert** systems undergo slow reaction. For very fast reactions sophisticated techniques such as stopped-flow spectrophotometry, NMR($^1$H), laser flash photolysis, EPR, Mossbauer spectroscopy, pulse radiolysis etc. are applied. In both the process the measurements of concentrations have been made which then subsequently helps to determine the **rate constants** and to establish plausable **rate law** for the specific reactions. Then we have to judge logically to **identify the mechanism** for the reaction by comparing it with standard established types. However ambiguity may arise which will be removed by proper scrutiny of experimental data to fit with the another alternative mechanism.

---

**Problem 7.72** Define nucleophilicity parameter $^nPt$. **[B.U. 2010, 2008]**

**Ans. :** As per equation (7.45) (Sec. 7.19.2)

$$^nP_{t(y)} = \frac{\text{rate of attack on a complex by a given base y}}{\text{rate of attack on the same by a standard base y°}}$$

$= \log\frac{K_y}{K_s}$ where $K_y$ is the second-order rate constant of the reaction,

trans $Pt(py)_2Cl_2 + Y \longrightarrow$ trans $Pt(py)_2ClY^+ + Cl^-$ and $K_s$ is the rate constant for the same reaction with the standard base, methanol, therefore

$^nP_{t(y)} = \log\frac{K_y}{K_{CH_3OH}}$ and thus $^nP_{t(y)}$ is a dimensionless quantity.

---

**Problem 7.73** How activation parameters can be evaluated from temperature coefficient of a reaction rate? **[B.U. 2015, 2009, 2003]**

**Ans. :** By the term temperature coefficient of a reaction is generally defined as the ratio of the constants at two temperatures differing by 10°C. Usually the two temperatures are chosen as 25°C and 35°C.

The activation parameters i.e. $\Delta H^{\neq}$, $\Delta G^{\neq}$ and $\Delta S^{\neq}$ are related as—

$$\Delta G^{\neq} = \Delta H^{\neq} - T\Delta S^{\neq} \text{ (Equation 7.67)}$$

Now according to Arrhenius equation the activation energy ($E_a$) is expressed by the equation,

$$k = Ae^{-E_a/RT} \quad \text{or} \quad \ln k = \text{constant} - \frac{E_a}{RT}$$

where A is the pre-exponential factor has the same unit of rate constant (k).

The temperature dependence of k may be derived from Vant Hoff equation as

$$\frac{d\ln k}{dT} = \frac{E_a}{RT^2}$$

Integrating between two temperatures $T_1$ and $T_2$ and considering that $E_a$ is independent of temperature, we get

$$\ln\left(\frac{k_1}{k_2}\right) = \frac{E_a}{R}\left(\frac{T_2 - T_1}{T_1T_2}\right)$$

This equation relates the temperature coefficient with activation energy ($E_a$).

Now from Arrhenius equation, $\ln k = -\frac{E_a}{RT}$ + constant (c)

Therefore plot of lnk vs $\frac{1}{T}$ we get a straight line with a slope = $\frac{E_a}{R}$ and thus $E_a$ is determined. Now according to transition state theory we can write,

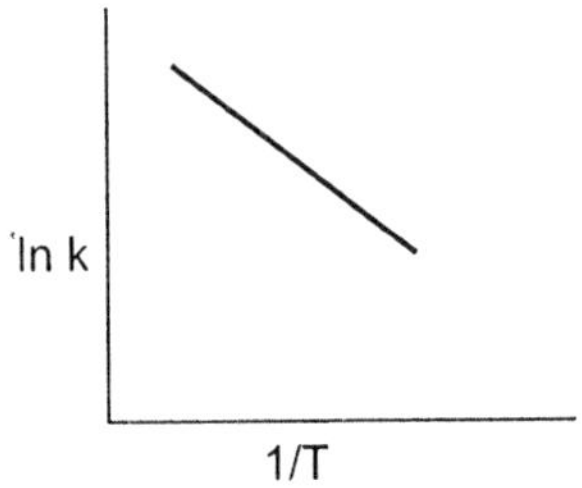


$$k = \frac{K_BT}{h} e^{\Delta S^{\neq}/R} e^{-\Delta H^{\neq}/RT}$$ where $K_B$ = Blotzmann constant

and thus $\frac{kh}{K_BT} = e^{\Delta S^{\neq}/R} e^{-\Delta H^{\neq}/RT}$

so $\ln\left(\frac{kh}{K_BT}\right) = \frac{\Delta S^{\neq}}{R} - \frac{\Delta H^{\neq}}{RT}$ and thus plot of $\ln\left(\frac{kh}{K_BT}\right)$ vs $\frac{1}{T}$ we get a straight line with a slope = $\frac{\Delta H^{\neq}}{R}$ and intercept = $\frac{\Delta S^{\neq}}{R}$. Thus from the slope and intercept we get the values of $\Delta H^{\neq}$ and $\Delta S^{\neq}$ and thus we know

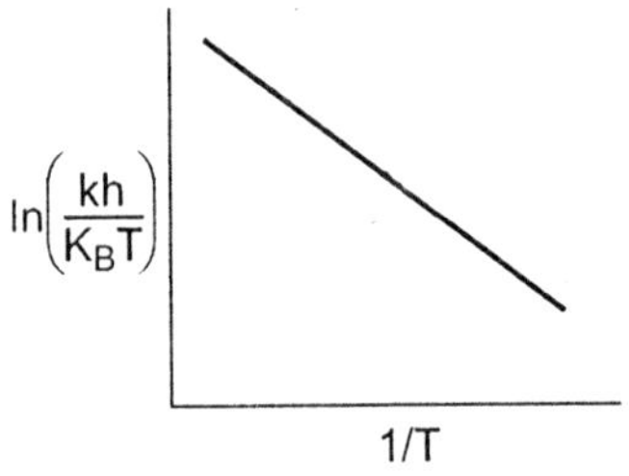


the value of $\Delta G^{\neq}$ from equation $\Delta G^{\neq} = \Delta H^{\neq} - T\Delta S^{\neq}$ and thereby all the activation parameters are determined.

---

**Problem 7.74** The second order rate constants for the reaction $[V(H_2O)_6]^{2+} + X^- \longrightarrow [V(H_2O)_5X]^+$ ($X = Cl^-$, $NCS^-$ and $N_3^-$) are in the ratio 1 : 2 : 10. What do the data suggest about the intimate mechanism of substitution? Explain with reason(s). **[B.U. 2009, 2012]**

**Ans. :** Since all the ligands are univalent anionic species having similar size, the encounter quilibrium constants are similar and second order rate constants are proportional to the first order rate constants for substitution in the encounter complex. The larger value of rate constants for $NCS^-$ compared to that of $Cl^-$ and five fold increase in case of $N_3^-$ suggests structural similarity between them as also some contribution from nucleophilic attack by $N_3^-$ compared to that of $NCS^-$. This data suggest associative intimate mechanism for the reactions. But generalisations cannot be made in this way because the same anions when reacting with Ni(II) complex is believed to follow dissociative mechanism.

---

**Problem 7.75** "A mechanistic switch-over is observed as we move along a period and along a group".— Discuss with examples. **[B.U. 2008]**

**Ans.** : In a period from left to right e.g. from Cr(III) to Co(III) i.e. from $d^3$ to $d^6$, mechanism is transformed from associative to dissociative. More space in d-orbitals invites electrons from nucleophile and thus favours associative pathway while later members have more filled d-orbital as also size decreased and thus favours dissociative pathway.

Co(III) — $3d^6$
↓
Rh(III) — $4d^6$
↓
Ir(III) — $5d^6$

In a group from top to bottom mechanistic switch over is from dissociative to associative. Here size increases and more space in d-orbital exist and thus favours associative mechanism to the later members of the group.

## 7.19.13 Ligand substitution reactions in square planar complexes

Typical square planar complexes are formed by Pt(II), Pd(II), Au(III) etc. but largely the work of square planar complexes were exhibited by Pt(II) and several square planar Pt(II) complexes in geometrical isomeric forms (cis and trans) were synthesised by Werner ($\alpha$ and $\beta$ forms) and their characterisation were made through trans effect (Sec. 7.9.12). Several Pt(II) complexes were synthesised by the application of trans effect.

Substitution reactions of square planar complexes mainly done on highly stereospecific moderately inert Pt(II) complexes e.g.

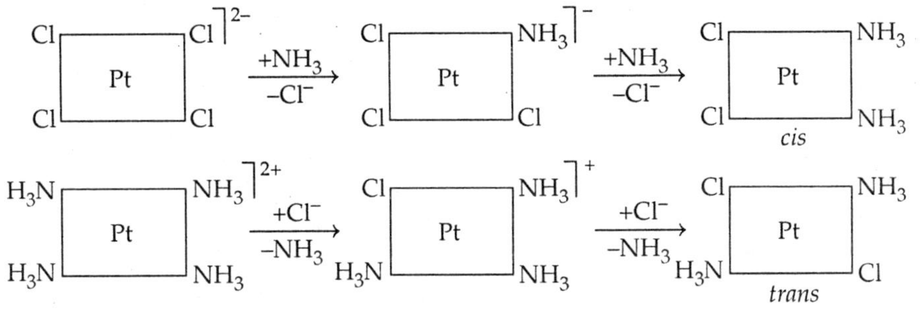


Due to the inertness of square planar Pt(II) complexes the rates of substitution reactions are slow and thus easier to study and to elucidate the mechanism of those reactions. The mechanism was suggested for those reactions as **associative** $S_N2$ rather than that of **dissociative** $S_N1$ due to the following reasons :

(a) In case of square planar complexes of Pt(II) five empty metal orbitals (one d, one s and three p) are made available for $\sigma$ bonding. Since four of the orbitals are occupied by ligands to form the $\sigma$ bonds the fifth orbital can easily be accommodated by the entering group to form the five coordinate intermediate or activated complex and thus follow associative mechanism.

(b) The rates of the reactions largely dependent on the nature of the reacting complex e.g.

$$\text{trans }[\text{MCl (o-tolyl)}(\text{PEt}_3)_2]^+ + \text{py} \longrightarrow \text{trans}[\text{Mpy(o-tolyl)}(\text{PEt}_3)_2]^{2+} + \text{Cl}^- \quad ...(7.77)$$

for which it was found that when M = Ni(II) the reaction was 50 times faster than reaction when M = Pd(II) which in turn $10^5$ times faster than the reaction when M = Pt(II). This is in good agreement of the reactivity sequence of square planar complexes of Ni(II)($3d^8$), Pd(II)($4d^8$) and Pt(II)($5d^8$) with their ease to expand their coordination number.

(c) Let us consider a general reaction

$$Pt(A{-}A{-}A)X^+ + Y^- \longrightarrow Pt(A{-}A{-}A)Y^+ + X^- \quad ...(7.78)$$

where A-A-A is a tridentate neutral ligand and $Y^-$ is entering group and $X^-$ is a leaving group. If the rate of the reaction is independent of $Y^-$ and **first order** with respect to the complex then,

$$\text{rate of the reaction} = k_1[Pt(A\text{-}A\text{-}A)X^+] \quad ...(7.79)$$

On the other if the reaction will follow another path where the rate of the overall rate is **second order** — **first order** with respect to complex and **first order** with respect to entering group $Y^-$ then,

$$\text{rate of the reaction} = k_2[Pt(A\text{-}A\text{-}A)X^+][Y^-] \quad ...(7.80)$$

where $k_1$ and $k_2$ are the first and second order rate constants respectively.

If the reaction will follow both the pathways, then,

$$\text{total rate of the reaction} = k_1[Pt(A\text{-}A\text{-}A)X^+] + k_2[Pt(A\text{-}A\text{-}A)X^+][Y^-]$$

$$= [Pt(A\text{-}A\text{-}A)X^+](k_1 + k_2[Y^-]) \quad ...(7.81)$$

$$= k_{obs}[Pt(A\text{-}A\text{-}A)X^+] \text{ where } k_{obs} = k_1 + k_2[Y^-]$$

Experimentally it was found that in presence of excess $Y^-$ the rate law will follow **pseudo first order** with the observed rate constant as,

$$k_{obs} = k_1 + k_2[Y^-] \quad ...(7.82)$$

where $k_{obs}$ is a pseudo first order rate constant which is plotted against $[Y^-]$ whereby we get a straight line whose intercept = $k_1$ and slope = $k_2$. Equation (7.82) is experimentally observed in the majority of cases of substitution reactions of square planar complexes of Pt(II), Ni(II), Pd(II) etc. which follow associative $S_{N2}$ mechanism.

(d) Another variation of substitution reaction for Pt(II) complex is the strong dependence of the rate on the nature of the incoming nucleophile Y, the leaving group X and also the non-leaving group (Spectator ligand) in the reaction :

$$PtL_2X_2 + Y \longrightarrow [PtL_2X_2Y] \longrightarrow PtL_2XY + X \quad ...(7.83)$$

*Intermediate activated complex. Five coordinated trigonal bipyramidal (TBP)*

In this reaction (7.83) a **strong reaction rate acclerating effect is observed of the ligand trans to the leaving group**. This is known as **trans effect** and the trans directing influence of ligands are shown partly in Sec. 7.9.12 and this effect also known as **kinetic effect**.

In this reaction the **associative pathway** will follow TBP intermediate (Scheme 7-III). As approaching nucleophile (entering group Y) approaches to the metal centre, there will be a minimum steric hindrance and the attachment occur at the plane perpendicular to the four ligands [Scheme 7-III(b)]. Now the leaving group X gradually weakens the Pt-X bond and bends away from the plane of the other three ligands. Thus a TBP intermediate [Scheme 7-III(c)] contains the entering group (Y), the leaving group (X) and the group originally **trans** to it i.e. group X. The groups which are originally **cis** to the leaving group (X) i.e. L are now occupy in the axial positons [Scheme 7-III(c)]. With the removal of leaving group (X), the entering group (Y) adjusts itself to form Pt-Y bond in the plane with the other three groups. The bond breaking of Pt-X bond and bond formation of Pt-Y bond has comparable contribution to the **associative**

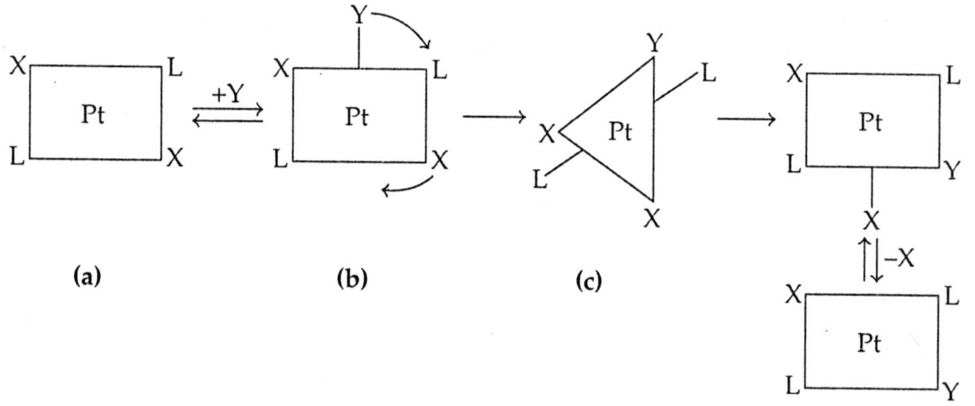


Scheme 7-III : Steric course of substitution reaction of square planar Pt(II) complex.

**mechanism** and spectral evidence has verified the existance of five coordinate TBP intermediate as activated complex. It is also evident from the observation of the reaction (Scheme 7-III) that substitution reactions of square planar geometry will retain the stereochemistry. Thus cis-$PtL_2X_2$ and trans-$PtL_2X_2$ will form cis-$PtL_2XY$ and trans-$PtL_2XY$ respectively by substitution reactions of the entering group (Y).

This type of substitution reactions of square planar complexes is the facinating example of **spectator ligand effects**. The ligand **trans** to the leaving group (X) has a substantial effect on the rate of substitution reaction while the **cis** ligands have a little effect. In the Scheme 7-III(c) the relatively less influential **cis L spectator ligands** in TBP geometry occupy the **axial positions** and the entering group (Y), leaving group (X) and the **trans** ligand (X) all lie in the equatorial positions. Here the mechanism followed symbolically represented as $I_a$.

### 7.19.14 Trans effect

As mentioned earlier the ligand **trans** to the leaving group (X) has profound influence on the rate of substitution reactions of square planar complexes and this effect is known as **trans effect**. The expanded trans-labilising influence of ligands are :

$HO^- < H_2O < NH_3 \sim RNH_2 < py < Cl^- < Br^- < SCN^- < I^- \approx NO^-_2 \approx Ph^- <$ thiourea $< H^- \approx PR_3$

← Moderate → ←——— Large ———→ ←———

$< NO \approx CO \approx C_2H_4 \approx CN^- \approx Bu$

——— Very large ———→

This trans effect series can be utilised to synthesise some typical cis/trans complexes of Pt(II) e.g.

(1) Synthesis of cis/trans $[PtCl_2(NH_3)(PPh_3)]$ from $PtCl_4^{2-}$

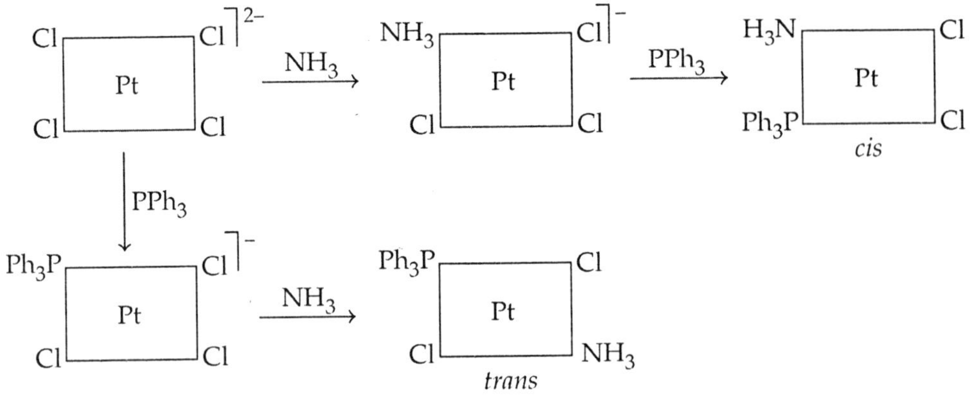


(2) Synthesis of cis/trans $[Pt(NH_3)(C_2H_4)Cl_2]$ from $K_2PtCl_4$

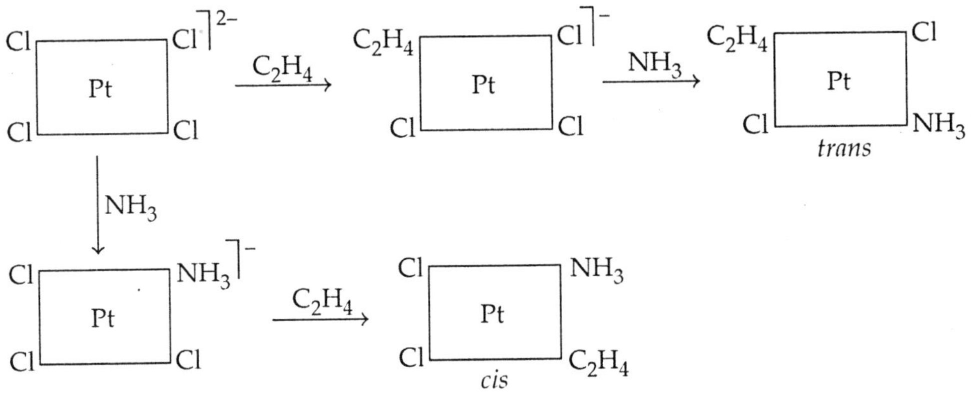


However mere consideration of trans effect will not led us to get the expected product in a substitution reaction. In order to replace one ligand by other we should have to consider trans effect as also **bond strength effect** to follow a synthetic route to synthesise a particular isomer. The sequence of **trans effect** and **bond strength** are the following :

trans effect : $NO_2^- > Br^- > Cl^- > py > NH_3$ and

bond strength : $Pt–NH_3 > Pt–py > Pt–Br > Pt–Cl$

The order of bond strength for $Pt–NO_2$ is not definitely known. However we have attempted plausable synthetic routes for the synthesis of some isomers starting from $K_2PtCl_4$ :

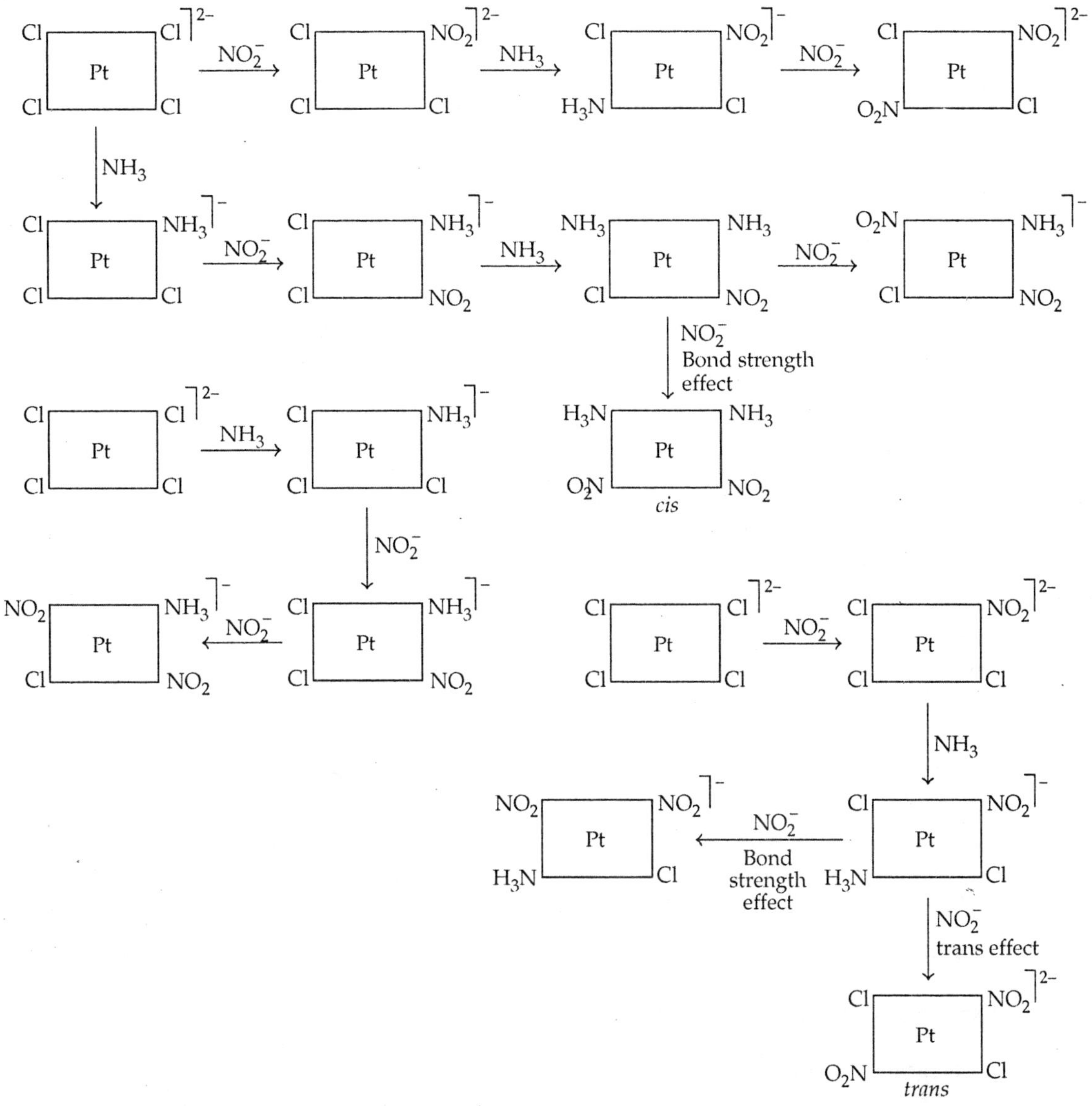


### 7.19.15 Explanation of Trans effect

An inspection of the trans effect series shows that simple σ donor ligands have widely different trans effect compared to that of π-acid ligands. The function of these ligands in acclerating the substitution reaction represented by equation (7.83) suggests that π-electron acceptor ligands are good trans directing ligands in the substitution reactions of square planar Pt(II) complexes. The acclerating rate of substitution denotes the lowering of activation energy which is established by the fact that this trans effect is a combination of **kinetic effect** as also **thermodynamic influence** by which the following two theories was established :

(A) Polarisation theory (σ-effect), (B) π-bonding theory.

A. **Polarisation theory :** This is a classical one in which we represent a symmetrical Pt(II) complex as $PtX_4$ (Fig. 7.29) where there is no net polarisation of the Pt atom. Replacement of one X by Y causes polarisation as depicted in Fig. 7.29. Mutual polarisibility of the central ion and

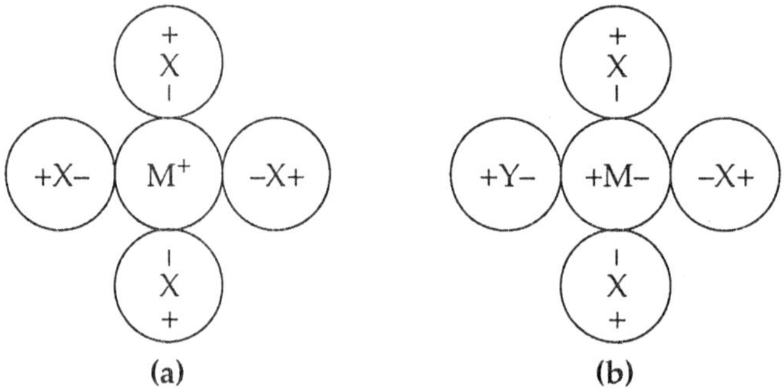


*Fig. 7.29 : Polarisation theory of trans effect (a) symmetrical arrangement, (b) induced dipole arrangement, charges of the complex species (a and b) are omitted for simplicity.*

the ligands result in induced dipoles in both. If the four ligands are identical then there is a symmetrical distribution of the induced dipole and the resultant dipole is zero [Fig. 7.29(a)]. If one ligand is replaced by other (say Y), then there is no longer a mutual compensation of the induced dipole and the result will be a dipole in the central ion as well as in the system as a whole [Fig. 7.29(b)]. On the basis of such polarisation picture, the suggestion was made that if Y is more polarisable than X, the central ion in its turn polarised in such a way that the less positive directly opposite (trans) to Y than it is adjacent to Y. It follows from the pictorial representation that **M-X bond trans to Y will be weaker than that of the bond cis to Y.** Since Pt(II) is more polarisable than Pd(II), the trans effect is only of significance in case of former. According to this theory it is thermodynamic effect of a ligand in weakening a bond trans to itself called trans influence. This is also known as **σ trans effect**.

B. **π-bonding theory :** Let us consider the following reactions :

$$(NH_3)PtCl_3^- + NH_3 \xrightarrow{\text{slow}} cis(NH_3)_2PtCl_2 + Cl^- \qquad ...(7.84)$$

$$(C_2H_4)PtCl_3^- + NH_3 \xrightarrow{\text{fast}} trans(NH_3)(C_2H_4)PtCl_2 + Cl^- \qquad ...(7.85)$$

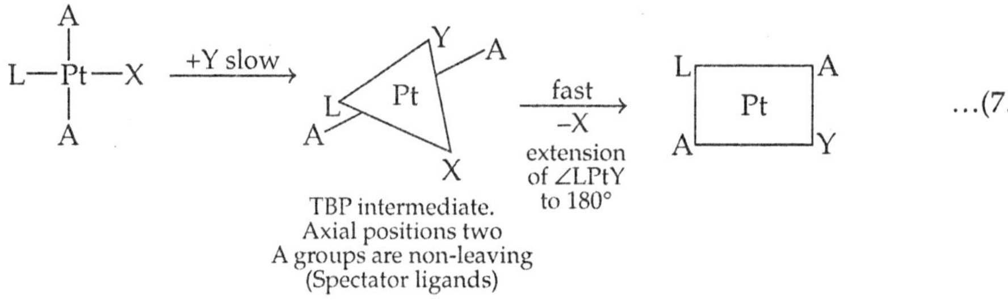


...(7.86)

It was observed that the rate of substitution in the presence of a strong trans directing group is very much faster than the corresponding reaction in the presence of a group having a low trans effect. So this effect may be treated as **kinetic effect**. As mentioned earlier that the ligand substitution in square planar Pt(II) complexes is bimolecular displacement process.

(a) **Kinetic trans effect :** It is very often observed with π-acid ligands such as amines, phosphines, olefins etc. which have a high trans effect, have suitable unfilled oribtals to overlap with the filled d-orbitals of Pt(II) to form π-bonds of M–L type. In the kinetic trans effect in which TBP intermediate is formed where three equatorial ligands (X, Y and L) overlap with the equatorial Pt $6p_x$, $6p_y$, $5d_{x2-y2}$ and $5d_{xy}$ orbitals as depicted in Fig. 7.30 with reference to the said reaction (equation 7.86). The π-acid ligand withdraws electron density at the trans position and thus acclerates the nucleophilic attack by Y at the trans position of L. J. Chatt and Orgel idependently suggested that

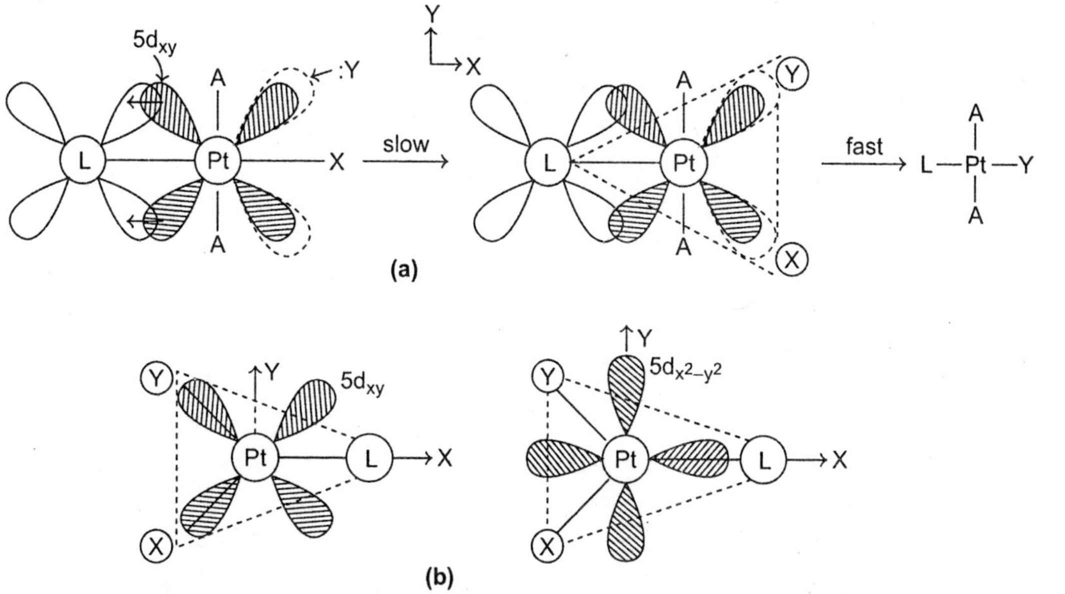


*Figure 7.30 : (a) Interaction of the nucleophile (Y) with $PtA_2LX$ where L is trans directing ligand. (b) Formation of TBP intermediate with equatorial ligands i.e. L with $Pt5d_{xy}$; X and Y with $Pt5d_{x2-y2}$.*

as a result of π-bond formation there will be considerable reduction of electron density at the **trans position** and thus activation energy is lowered due to the formation of five coordinate TBP intermediate. Conclusively the greater the π-bonding tendency of L, the greater is the trans effect because of its increased stabilising influence on the five coordinate TBP intermediate.

(b) **Thermodynamic trans influence :** This is also known as **static trans effect.** It is evident that a strong σ-bonding ligand will weaken the bond to the trans ligand as both competing for the formation of σ-bond with the same metal orbital. Thus whether a ligand will stabilise or destabilise the **trans bond** will entirely depend on the nature of the ligands participating in the process e.g. L and X (in equation 7.86) which are mutually trans to each other to the metal. The effect of a ligand on the strength of the bond in the trans position to itself is described as **trans influence.** The dominating feature among the ligand (L and X) are :

(1) strong σ-donor ligands will weaken the trans ligand

(2) better π-acceptor ligands will weaken the trans ligand more effectively.

Some octahedral and square planar complexes show both **static trans effect (trans influence)** and also **kinetic trans effect**. Trans-$Co(en)_2(H_2O)X^{n+}$ complexes where X = $S_2O_3^{2-}$, $SO_3^{2-}$ in which X is S-bonded to Co(III) and show increasing Co-O bond length trans to Co-S bond and also observed **kinetic effect** of the **trans** ligand. Similarly Pt-Cl bond distances in **cis** and **trans** isomers of square planar $[Pt(PEt_3)_2Cl_2]$ are different. Pt-Cl bond in **trans** isomer is much shorter in which the two Cl atoms are trans to other and are formed by the overlap of Pt $6p_x$ orbital. In the **cis** isomer the bonds are much longer than **trans** form. In the cis form there is a overlap of Cl atom and $PEt_3$ with the common Pt6p orbital. As $PEt_3$ is a good σ-donor ligand so that Pt-Cl bonds will be longer. Thus thermodynamic trans influence and kinetic trans effect are simultaneously applied for these complexes and they have a linear relationship between them.

In case of π-bonding ligands (L and X) stronger π-acceptor one will weaken the trans bond more effectively (Fig. 7.31).

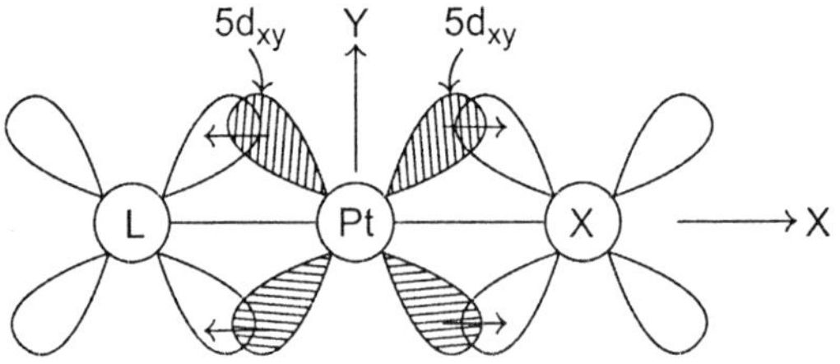


***Figure 7.31** : Competeting π-back bonding with the filled Pt $5d_{xy}$ orbitals with the trans bonded L and X ligands.*

## 7.19.16 Cis effect

This effect is also a kinetic effect and is not considered as parallel to trans effect. In substitution reactions of square planar complexes, **the labilisation effect of a group cis to a particular group is called cis effect**. This cis effect is less effective in case of square planar Pt(II) complexes but more pronounced in case of substitution reactions of octahedral complexes having **dissociative pathway**. **The reactivity of cis isomers are more compared to that of trans isomers** e.g.

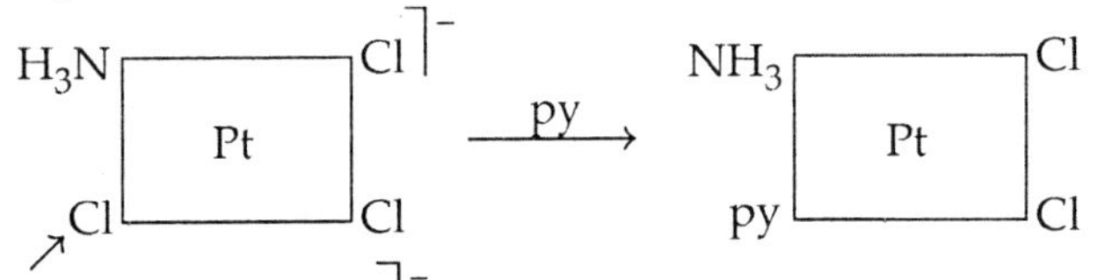


...(7.87)

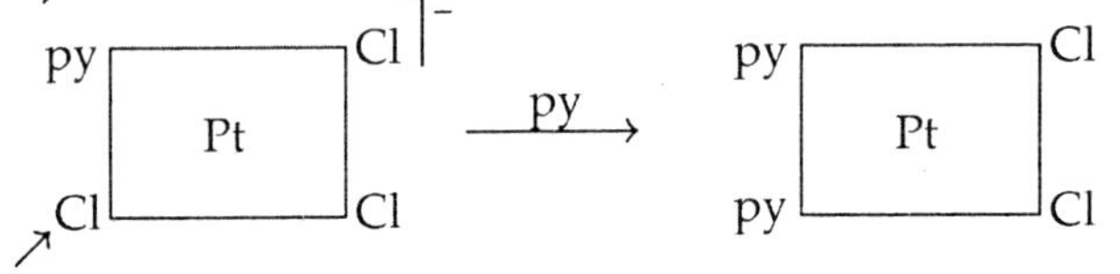


...(7.88)

Reaction (7.88) is more faster than that of reaction (7.87). This is due to the greater cis effect of pyridine compared to that of $NH_3$ to destabilise the group cis to themselves which is shown by arrow marks. Experimental kinetic studies on the substitution reactions of cis and trans isomers of $PtA_2LX$ where X = $Cl^-$ is replaced by $H_2O$ in which the following sequence of cis and trans effects are observed where the rate constants are given in parentheses :

**cis effect :** $C_2H_4(0.05) < Br^-(0.3) < Cl^-(0.4) < NH_3(1) \approx H_2O(1) < Me_2SO(5)$

**trans effect :** $H_2O(1) < NH_3(200) < Cl^-(330) < Br^-(3\times10^3) < I^-(4\times10^4)......<<C_2H_4(\sim10^{11})$

It is evident from the above that a ligand having higher trans effect will show poor cis effect and vice-versa. Once again the above rate constants indicate the poor cis effect compared to that of trans effect.

Let us consider the following aquation reaction of cis and trans Co(III) complexes for the explanation of effective cis effect :

$$\text{cisCo(en)}_2\text{(OH)Cl}^+ + H_2O \longrightarrow \text{cisCo(en)}_2\text{(OH)}H_2O^{2+} \quad ...(7.89)$$

$$k_{cis} = 1200\times10^{-5}\ s^{-1}$$

$$\text{transCo(en)}_2\text{(OH)Cl}^+ + H_2O \longrightarrow \text{cis and transCo(en)}_2\text{(OH)}H_2O^{2+} \quad ...(7.90)$$

$$k_{trans} = 160\times10^{-5}\ s^{-1}$$

It is evident that for ligands $OH^-$, $Cl^-$ and $NCS^-$, the cis isomers react more rapidly than the trans. Also **the cis isomers unlike the trans react with the retention of stereochemical configuration**. A likely explanation in terms of π-bonding has been suggested.

The expulsion of $Cl^-$ and hence the dissociation of cis $Co(en)_2ACl^+$ (where A = a π-bonding ligand e.g. –OH, $-NH_2$ etc.) via five coordinated square pyramidal intermediate has been formed readily as the p-orbital on the non-leaving π-bonding ligand. A can form π-bond with ease to a vacant $d^2sp^3$ orbital of Co(III). As in this mechanism no rearrangement is necessary, cis isomer should react with retention of configuration with dissociative (D) pathway (Fig. 7.32).

In the case of trans $Co(en)_2ACl^+$, the p-orbital of $OH^-$ cannot overlap with a vacant $d^2sp^3$ orbital without substantial rearrangement or isomerisation. The rate of aquation of such trans isomers is thus predicted to be lower than that of the analogous cis isomers. Rearrangement to a trigonal bipyramidal geometry has a vacant $3d_{x^2-y^2}$ orbital. In the trigonal plane which can accept electron by the π-donor A ligands (e.g. $OH^-$, $Cl^-$). TBP intermediate is stabilised by $A \rightarrow Co(III)$ π-bonding. In this structure entering water molecule can occupy either **cis** or **trans** positons (Fig. 7.32) leading to the cis and trans products. On the other hand for negative ligands where π-bonding of this type is not important the trans isomer will react faster than the cis.

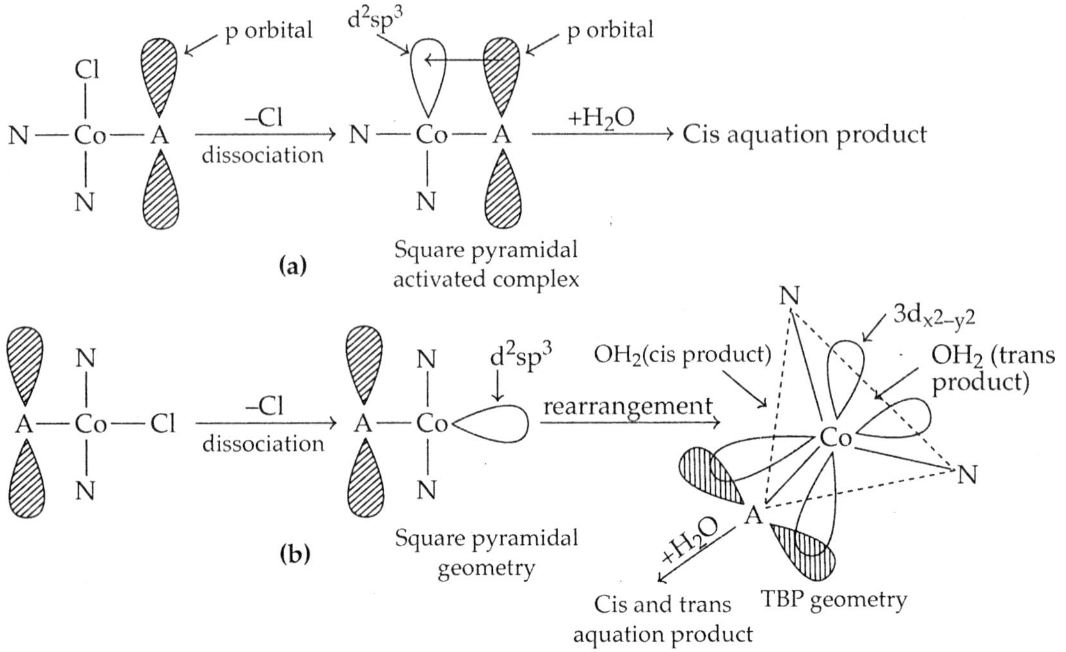


*Figure 7.32 : Aquation reaction of cis and trans Co(III) complexes (Equation 7.89 and 7.90). Two other trans axial ligands (N) are not shown. (a) A at the cis positon with respect to leaving group ($Cl^-$), (b) A at the trans position with respect to leaving group ($Cl^-$).*

**Problem 7.76** For the reaction trans $PtL_2Cl_2$ + Y → trans $PtL_2ClY$ + $Cl^-$, the rate constant k varies as follows :

| | Y | L | k($10^{-3}$ $M^-S^{-1}$) |
|---|---|---|---|
| (i) | $PPh_3$ | py | 249000 |
| (ii) | $SCN^-$ | py | 180 |
| (iii) | $I^-$ | py | 107 |

What is the mechanism of the reaction and why? **[B.U. 2010]**

**Ans. :** As the rate of the reaction is dependent on the nature of the entering nucleophile, the mechanism is associative (A) in nature. π-acid character of the ligands increases as $PPh_3$ > $SCN^-$ > $I^-$ and the rate also increases in the same order due to better metal-ligand π-bonding.

**Problem 7.77** Which of the following exhibit distorted geometry, relative to perfect octahedral? Amongst those, comment on their extent of distortion and justify your answer.
$[V(H_2O)_6]^{3+}$, $[Ru(NH_3)_6]^{2+}$, $[Co(CN)_6]^{4-}$, $[Ni(H_2O)_6]^{2+}$ **[NET 2001]**

**Hint :** Sec. 7.14.5(C). Table 7.14 indicates distortions permitted in case of $[Co(CN)_6]^{4-}$.

**Problem 7.78** With the help of an Orgel diagram, explain how many absorption bands are expected in the electronic spectra of $[Cr(H_2O)_6]^{3+}$. Label the energy levels and assign the transitions. **[NET 2001]**

**Ans. :** For $Cr^{3+}(3d^3)$ the energy terms are $^4F$ and $^4P$ of which $^4F$ is the ground term. The energy levels (Orgel diagram) obtained by splitting the energy terms in the octahedral field as :

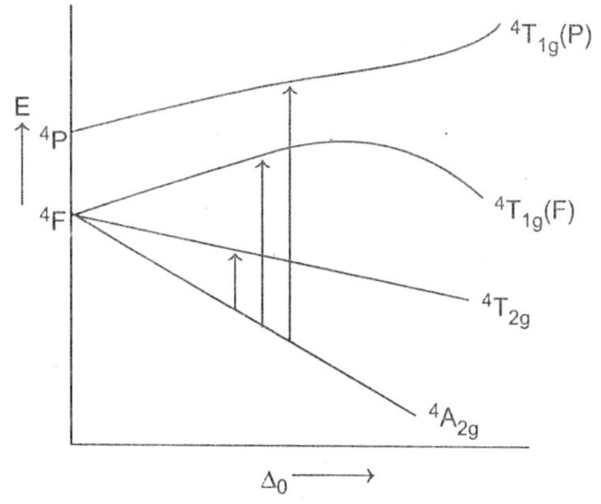


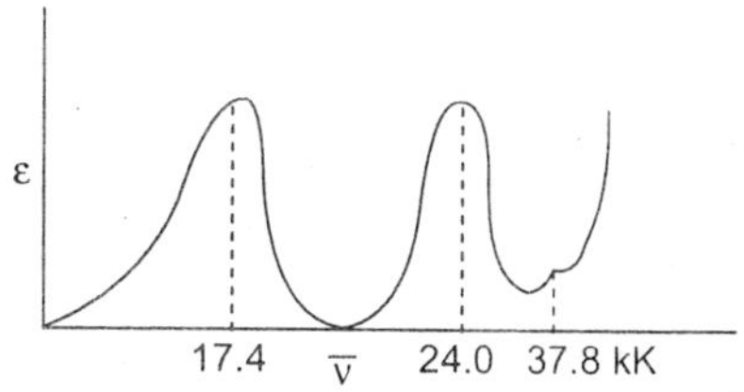


Therefore in case of $Cr^{3+}$ the three spin allowed but Laporte forbidden transtions are expected : $^4A_{2g} \longrightarrow {}^4T_{2g}$ (17.4 kK); $^4A_{2g} \longrightarrow {}^4T_{1g}$ (F) (24 kK) and $^4A_{2g} \longrightarrow {}^4T_{1g}$ (P) (37.8 kK) as shown in the figure. The lowest energy $^4A_{2g} \longrightarrow {}^4T_{2g}$ transitions (17.4 kK) gives the value for 10Dq.

---

**Problem 7.79** The $^4F$ term corresponds to—(1) $S=^1/_2$, L=2, (2) $S=^3/_2$, L=2, (3) $S=^3/_2$, L=3, (4) $S=^1/_2$, L=3. **[NET 2000]**

**Ans. :** 3. See Sec. 7.15.6.

---

**Problem 7.80** How many absorption peaks are expected in the electronic spectrum of $Ni(H_2O)_6^{2+}$. **[SSC 2007]**

**Ans. :** See Sec. 7.15.10(B). Three transitions are permitted, $^3A_{2g}(F) \rightarrow {}^3T_{2g}(F)$, $^3A_{2g}(F) \rightarrow {}^3T_{1g}(F)$ and $^3A_{2g}(F) \rightarrow {}^3T_{1g}(P)$.

---

**Problem 7.81** Consider the structural changes shown below :

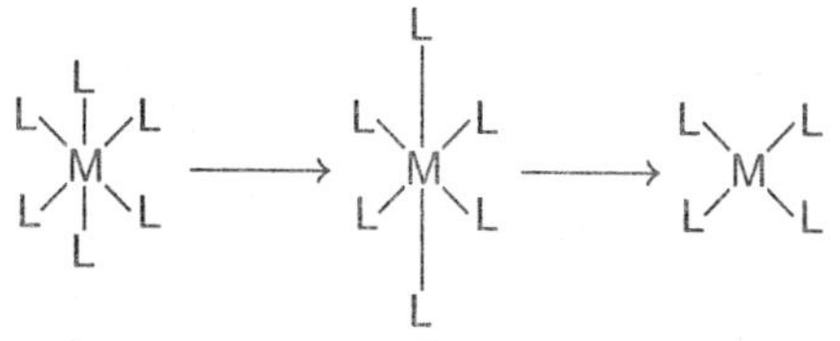


I II III

Write the d-orbital splitting pattern for I, II and III. Also indicate the point group to which each belongs. **[NET 2000]**

**Ans. :** See Sec. 7.14.2(D) and Fig. 7.12. I : Octahedral : point group $O_h$, II : Square pyramidal : point group $C_{4v}$, III : Square planar : point group $D_{4h}$.

---

**Problem 7.82** $[Ni(H_2O)_6]^{2+}$ has an absorption band at 1176 nm in near IR region and two visible absorption bands at 649 and 385 nm. Calculate $\Delta_o$.

**Ans. :** These wavelengths should be converted to frequencies by dividing them with $10^7$ (Sec. 7.14.2(D)). Therefore $v_1 = \frac{10^7}{1176} = 8503\ cm^{-1}$; $v_2 = \frac{10^7}{649} = 15408\ cm^{-1}$; $v_3 = \frac{10^7}{385} = 25974\ cm^{-1}$. Since the lowest energy transition corresponds to $\Delta_o$, therefore $\Delta_o = 8503\ cm^{-1}$.

---

**Problem 7.83** Explain why tetrahedral complexes shows intense colour than octahedral complexes? **[NET 2010]**

**Ans. :** This is due to the lack of symmetry in the tetrahedral complexes compared to that of octahedral complexes thereby Laporte selection rules are relaxed to a greater extent. Due to the lack of symmetry there is a possibility of the mixing of metal 3d orbitals with metal 4p orbitals or with ligand p-orbitals and thus pure d-character is lost and Laporte forbbidden transitions now become Laporte allowed. The intense blue colour of $CoCl_4^{2-}$ is due to the spin allowed and partially Laporte allowed transitions. Forbidden transitions for octahedral complexes have low molar extinction coefficient (10 – 200 $L mol^{-1} cm^{-1}$) compared to tetrahedral complexes ($10^2 - 10^3$ $L mol^{-1} cm^{-1}$).

**Problem 7.84** Write the d-orbital configurations and identity the ground term of (a) $[Ti(H_2O)_6]^{3+}$, (b) high spin $[Fe(H_2O)_6]^{3+}$.

**Ans. :** (a) $3d^1$, $^2T_{2g}$, (b) $3d^5$, $^6A_{1g}$.

**Problem 7.85** Explain why the difference in energy between the $d_{x^2-y^2}$ and $d_{xy}$ orbitals in a square planar field is identical to the difference between the same orbitals in the octahedral field. **[B.U. 2011]**

**Ans. :** Sec. 7.14.2(D), Fig. 7.12. The formation of a square planar complex from an octahedral one is due to the tetragonal elongation of two z-axis ligands. As these ligands are further removed, the effect will be the same on the two orbitals which are in the xy plane. When the two z-axis ligands are completely removed, the $d_{x^2-y^2}$ and $d_{xy}$ orbitals will have the same difference in energy as that of the octahedral geometry since both have been equally raised.

**Problem 7.86** The enthalpy of hydration of the $Fe^{2+}$ ion is 11.4 kcal/mol higher than would be expected if there were no crystal field stabilisation energy. Assuming the aquo complex to be high spin, estimate the magnitude of $\Delta$ for $[Fe(H_2O)_6]^{2+}$. **[B.U. 2011]**

**Ans. :** CFSE for $3d^6$ weak field ion is $-0.40\Delta_o$. The enthalpy of hydration $\Delta H_{hyd}$ is higher because the energy of the complex is lowered by that much, –11.4 kcal/mol.

$$\text{CFSE} = -11.4 \text{ kcal/mol} = -0.40\ \Delta_o$$

so that $\Delta_o$ = 28.5 kcal/mol.

**Problem 7.87** Explain why $[Cu(H_2O)_4]^{2+}$ is blue, $[Cu(NH_3)_4]^{2+}$ deep blue, while anhydrous $CuSO_4$ is white? **[B.U. 2011]**

**Hint :** The colour of a substance is due to the absorption of light in the visible region—the energy of absorbed light of appropriate wavelength is just enough to promote d-electrons from lower to higher energy levels. Hydrated copper sulphate which is $[Cu(H_2O)_4]^{2+}$ is coloured blue because it absorbs yellow light while $[Cu(NH_3)_4]^{2+}$ is deep blue due to the stronger donor $NH_3$ which shifts the transition to higher enerngy ~600 nm/mμ. Anhydrous $CuSO_4$ is white as it absorbs light in IR region (the ligand field strength of $SO_4^{2-}$ is very small) and therefore d-d transition occurs at much lower energy.

**Problem 7.88** The magnetic moments of $[Mn(CN)_6]^{3-}$ and $[MnBr_4]^{2-}$ are respectively 2.8 and 5.9 BM. What are the geometries of these complex ions? **[B.U. 2011]**

**Ans. :** In $[Mn(CN)_6]^{3-}$ the central $Mn^{3+}$ is $d^2sp^3$ hybridised whereby two unpaired electrons exist while in $[MnBr_4]^{2-}$ the central $Mn^{2+}$ is $sp^3$ hybridised producing five unpaired electrons.

$[Mn(CN)_6]^{3-}$ has therefore inner orbital octahedral geometry while $[MnBr_4]^{2-}$ has tetrahedral geometry.

---

**Problem 7.89** According to crystal field theory $Ni^{2+}$ can have two unpaired electrons in
1. Octahedral geometry only 2. Square planar geometry only 3. Tetrahedral geometry only 4. Both octahedral and tetrahedral geometry. **[NET 2011]**

**Ans, :** 4.

---

**Problem 7.90** $[Ni(CN)_4]^{2-}$ and $[NiCl_4]^{2-}$ complex ions are
1. both diamagnetic 2. both paramagnetic 3. diamagnetic and paramagnetic respectively 4. antiferromagnetic and diamagnetic respectively. **[NET 2011]**

**Ans. :** 3.

---

**Problem 7.91** Which of the following compounds show a charge-transfer band?
1. lanthanum nitrate 2. ceric ammonium nitrate 3. manganese (II) acetate 4. copper (II) sulphate pentahydrate. **[NET 2011]**

**Ans. :** 3

---

**Problem 7.92** Which of the following spectroscopic techniques will be useful to distinguish between M-SCN and M-NCS binding mode?
1. NMR 2. IR 3. EPR 4. Mass **[NET 2011]**

**Ans. :** 2.

---

**Problem 7.93** The number of possible isomers for $[Ru(bpy)_2Cl_2]$ is (bpy = 2, 2′-bipyridine)
1. 2 2. 3 3. 4 4. 5 **[NET 2011]**

**Ans. :** 2.

---

**Problem 7.94** *cis* and *trans* complexes of the type $[PtA_2X_2]$ are distinguished by
1. chromyl chloride test 2. carbylamine test 3. Kurnakov test 4. Ring test **[NET 2011]**

**Ans. :** 3.

---

**Problem 7.95** 'The aqueous solution of $[Ti(OH_2)_6]^{3+}$ shows a maximum absorption around 20300 $cm^{-1}$ in its electronic spectrum'. Express the band position in nm. Has the complex ion visible in colour? **[B.U. 2012]**

**Ans. :** Here $1/\lambda = 20300\ cm^{-1}$, so that

$\lambda = 4.92\times10^{-5}$ cm $= 4.92\times10^{-7}$ m

1 Nanometre = 1 nm = $10^{-7}$ cm = $10^{-9}$ m

$\therefore$ 1 m = $10^9$ nm

so that $\lambda = 4.92\times10^{-7}$ m = 492 nm.

Visible range is 400–800 nm, so the complex will be visible in colour.

---

**Problem 7.96** What is the formula of ammoniumdiamminetetrakis (isothiocyanato)chromate(III)? **[B.U. 2013]**

**Ans. :** $NH_4[Cr(NH_3)_2(NCS)_4]$.

**Problem 7.97** Identify (A) and (B) from the following sequence of reactions of coordination complex of cobalt in the light of Werner's theory :

$$[Co(NH_3)_4(CO_3)]^+ \xrightarrow[-CO_2]{H^+/H_2O} (A) \xrightarrow[NH_3]{Aqueous} [Co(NH_3)_4(H_2O)(OH)]^{2+} \xrightarrow[-H_2O(18\ g/mol)]{\Delta 100°C} (B)$$

The complex (B) bears a net +4 unit of charge. Write the structures of (A) and (B) and give the IUPAC name of (B). [C.U. 2013]

**Ans. :** A is $[Co(NH_3)_4(H_2O)_2]^{3+}$ and B is 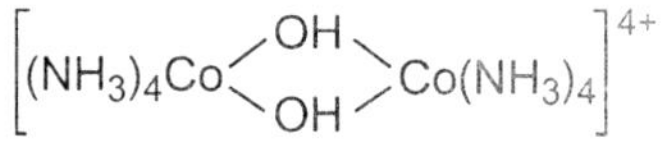 . Both the complexes has octahedral geometry and the IUPAC name of B is tetraamminecobalt(III)-μ dihydroxidotetramminecobalt (III).

or

octammine-μdihydroxidodicobalt(III).

**Problem 7.98** An octahedral metal ion $M^{2+}$ has magnetic moment of 4.0 BM. The correct combination of metal ion and d-electron configuration is given by—

1. $Co^{2+}$, $t_{2g}^5eg^2$, 2. $Cr^{2+}$, $t_{2g}^4e_g^2$, 3. $Mn^{2+}$, $t_{2g}^3e_g^1$, 4. $Fe^{2+}$, $t_{2g}^4e_g^2$. [NET 2013]

**Ans. :** 1 i.e. $Co^{2+}$, $t_{2g}^5eg^2$. The magnetic moment value corresponds to three unpaired electrons so that it will correspond to 1.

**Problem 7.99** The ground state terms of $Sm^{3+}$ and $Eu^{3+}$, respectively are—

1. $^7F_0$ and $^6H_{5/2}$, 2. $^6H_{5/2}$ and $^7F_0$, 3. $^2F_{5/2}$ and $^5I_4$, 4. $^7F_6$ and $^2F_{7/2}$ [NET 2013]

**Ans. :** 2. i.e. $^6H_{5/2}$ and $^7F_0$.

Sm(52) and Eu(63) has the electronic configurations $[Xe]4f^66s^2$ and $[Xe]4f^77s^2$ respectively. Thus $Sm^{3+}$ : $4f^5$ and $Eu^{3+}$ : $4f^6$.

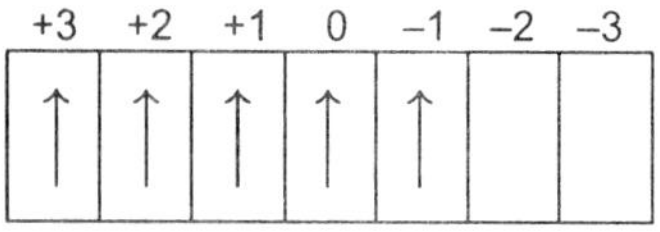


$M_L$ = +3+2+1+0−1 = 5, L = 5

$M_S = +\frac{1}{2}+\frac{1}{2}+\frac{1}{2}+\frac{1}{2}+\frac{1}{2} = \frac{5}{2}$, $S = \frac{5}{2}$

$2S+1 = 2\times\frac{5}{2}+1 = 6$

$J = L-S = 5-\frac{5}{2} = \frac{5}{2}$

∴ term symbol for $Sm^{3+}$ is $^{2S+1}L_J = {^6H_{5/2}}$

Here $J = L-S = 5-\frac{5}{2} = \frac{5}{2}$

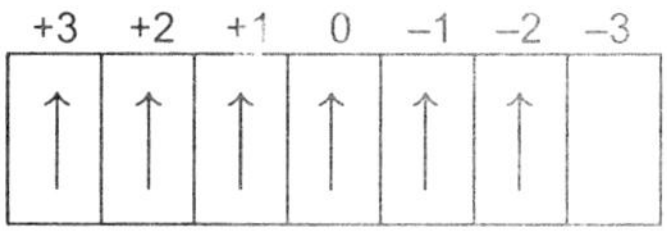


$M_L$ = +3+2+1+0−1−2 = 3, L = 3

$M_S = +\frac{1}{2}+\frac{1}{2}+\frac{1}{2}+\frac{1}{2}+\frac{1}{2}+\frac{1}{2} = \frac{6}{2} = 3$, S = 3

2S+1 = 3×2+1 = 7

J = L−S = 3−3 = 0

So term symbol for $E_u^{3+}$ is $^7F_0$

**Problem 7.100** Compounds $K_2Ba[Cu(NO_2)_6]$(A) and $Cs_2Ba[Cu(NO_2)_6]$(B) exhibit tetragonal elongation and tetragonal compression respectively. The unpaired electron in A and B are found respectively, in orbitals—

1. $d_{z^2}$ and $d_{x^2-y^2}$, 2. $d_{x^2-y^2}$ and $d_{z^2}$, 3. $d_{z^2}$ and $d_{z^2}$, 4. $d_{x^2-y^2}$ and $d_{x^2-y^2}$ [NET 2013]

**Ans. :** Sec. 7.14.5(C). The answer is 2 i.e. $d_{x^2-y^2}$ and $d_{z^2}$. In case of $Cu^{2+}$ for tetragonal elongation the configuration is $d^1_{x^2-y^2}d^2_{z^2}$ while for tetragonal contraction the configuration is $d^2_{x^2-y^2}d^1_{z^2}$.

**Problem 7.101** Adducing proper reason compare the magnitude of trans-effect of the following : $Cl^-$, $Br^-$, $NH_3$ and $H_2O$. **[B.U. 2013]**

**Ans. :** Trans-effect order is $H_2O < NH_3 < Cl^- < Br^-$. The greater the electronegativity of donor centre in ligand, the less will be reducing property and thereby covalent character will also be less, so less will be **trans** labilising effect. Electronegativity order O > N ≈ Cl > Br. $Br^-$ having least electronegative Br atom is most trans labilising and $H_2O$ has least **trans labilising** effect where O atom has highest electronegativity. Another view emphasizing the effectiveness of good σ-donor ligands at weakening the trans bond with $dsp^2$ hybrids, trans ligands shares $d_{x2-y2}$, s and one p orbital. A good σ-donor claims a larger share of Mσ orbitals, leaving a less share to the bond the trans ligand.

**Problem 7.102** Among the following the correct combination of complex and its colour is—

| Complex | Colour |
|---|---|
| 1. $[Co(CN)_4]^{2-}$ | Red |
| 2. $[CoCl_4]^{2-}$ | Orange |
| 3. $[Co(NCS)_4]^{2-}$ | Blue |
| 4. $[CoF_4]^{2-}$ | Yellow |

**[NET 2013]**

**Ans. :** 3 i.e. $[Co(NCS)_4]^{2-}$ has blue colour.

**Problem 7.103** Which type of octahedral distortion would you expect for $[Co(bpy)_3]^{3+}$ and $[Co(NH_3)_4Cl_2]^+$? (bpy = 2, 2′-bipyridine) **[B.U. 2013]**

**Ans. :** Both the complexes are octahedral having $d^2sp^3$ i.e. $t_{2g}^6e_g^0$ configuration and both have Co(III) i.e. $3d^6$ system. Therefore both are devoid of Jahn-Teller distortion (Sec. 7.14.5(C)). The only distortion in $[Co^{III}(NH_3)_4Cl_2]^+$ is tetragonal one as Co–N and Co–Cl distances are different whereas in $[Co^{III}(bpy)_3]^{3+}$ containing chelating ligand bpy, the distortion from ideal octahedral ($O_h$) geometry may arise due to the difference in bite and non-bite angle. The bite angle being much more less than the non-bite angle and both deviate from idealised 90°. Such situation does not arise in $[Co(NH_3)_4Cl_2]^+$ as both $NH_3$ and $Cl^-$ are monodentate ligands.

**Problem 7.104** What is the formula of potassium pentacyanonitrosyl ferrate(II)? **[B.U. 2014]**

**Ans. :** $K_2[Fe(CN)_5NO]$.

**Problem 7.105** Note down the conditions for MLCT and LMCT transitions with examples. **[B.U. 2014]**

**Ans. :** Sec. 7.15.11. In case of MLCT, the metal ion should be in lower oxidation state having strong reducing behaviour arising from its electron richness and the ligand should be excellent π-acceptor with strong oxidising property pertaining to it electron deficient character due to the presence of conjugated unstaturated framework. HOMO primarily residing on metal ion should not be too stable and but LUMO primarily residing on ligand must be very stable e.g. $[Fe(bpy)_3]^{2+}$.

For LMCT the metal ion should be on higher oxidation state with strong oxidising property arising from its electron deficiency and ligand should be strong σ as well as π donor coupled with strong reducing property pertaining to its high charge density. HOMO primarily residing on ligand, should not be too stable but LUMO primarily residing on metal ion must be very stable e.g. $MnO_4^-$.

**Problem 7.106** In the following reaction, $[PtCl_4]^{2-} + NO_2^- \longrightarrow A \xrightarrow{NH_3} B$

Compound B is—

1. trans-$[PtCl_2(NO_2)(NH_3)]^-$ 2. cis-$[PtCl_2(NO_2)(NH_3)]^-$
3. trans-$[PtCl_2(NH_3)_2]$ 4. cis-$[PtCl_2(NO_2)_2]^{2-}$ **[NET 2014]**

**Ans. :** 1. i.e. trans-$[PtCl_2(NO_2)(NH_3)]^-$

For explanation question number 177.

**Problem 7.107** Work out the Russel-Saunders ground state term for the iridium (IV) ion stating its total degeneracies. **[B.U. 2014]**

**Ans. :** Question number 158.

Ir (IV) has $[Xe]4f^{14}5d^5$ electron configuration.

| $m_l$ = | +2 | +1 | 0 | −1 | −2 |
|---|---|---|---|---|---|
| | ↑ | ↑ | ↑ | ↑ | ↑ |

$M_L = +2+1+0-1-2 = 0, \; L = 0$

$M_S + \frac{1}{2}+\frac{1}{2}+\frac{1}{2}+\frac{1}{2}+\frac{1}{2}, \; S = \frac{5}{2}$

$2S+1 = 2\times\frac{5}{2}+1 = 6 \quad \therefore$ term symbol is $^6S_{5/2}$

Here $J = L+S = 0+\frac{5}{2} = \frac{5}{2}$

Total degeneracies $= (2L+1)(2S+1) = (2\times0+1)(2\times\frac{5}{2}+1) = 6$

**Problem 7.108** The number of stereoisomers of trans-$[CoCl_2$(trithylenetetramine)]Br is—

1. One, 2. Two, 3. Three, 4. Four **[NET 2014]**

**Ans. :** 3 i.e. Three.

**Problem 7.109** "A and B are two complexes of nickel (II) – one of which with tetrahedral and the other with octahedral structures. The intensity in colour of B is greater than that of A". Guess the geometry of each one from such spectral findings. Measurement of which physicochemical property would exclude the possibility of either complex being square planar? **[B.U. 2014]**

**Ans. :** The colour of Ni(II) complexes is due to d–d transition which is a $3d^8$ system. In $O_h$ symmetry there is a centre of inversion but a $T_d$ symmetry has no centre of inversion. Thus d–d transition is Laporte forbidden in case of $O_h$ field since these have centre of inversion. But Laporte selection rule is relaxed in case of $T_d$ symmetry as these are not centrosymmetric. So intensity of colour is less for the octahedral complexes compared to that of tetrahedral complexes (Problem 7.83). Thus the more intensely coloured compound is expected in case of tetrahedral structure and less intensity is expected for octahedral structure.

Magnetic moment measurement may be a diagnostic tool to distinguish the geometry of the complex is either square plannar or $O_h/T_d$. In square planar geometry the $3d^8$ Ni)II) will be diamagnetic having $dsp^2$ hybridisation while for $O_h/T_d$ field the complex will be paramagnetic to the extent of two unpaired electrons.

**Problem 7.110** Determine the wavelength and frequency of 24900 $cm^{-1}$ light. **[B.U. 2014]**

**Ans. :** Problem 7.95. Here $\frac{1}{\lambda}$ = 24900 $cm^{-1}$, ∴ $\lambda = 4.01\times10^{-5}$ cm

We know $\nu\lambda = c$ i.e. $\nu = \frac{c}{\lambda} = 24900\times3\times10^{10}$ $sec^{-1}$ = $7.47\times10^{14}$ $sec^{-1}$.

**Problem 7.111** 'The existance of a series of entering groups with different rate constants is an evidence for an associative mechanism (A or $I_d$)'.—Justify or criticize. **[B.U. 2014]**

**Ans. :** In associative process both the leaving and entering groups involve in the rate determining step. This means change of an entering group will have pronounced effect on the rate of reaction. So by using a series of entering ligands it will be found that the rate of reactions will be changed. If there is no change of reaction rate, then it will not be associative. Thus it can be concluded that the existance of a series of entering groups with different rate constants is in accordance with the fact of associative (A or $I_d$) pathway.

**Problem 7.112** $Pt(NH_3)_2Cl_2$ forms two isomers—one of which gives $[Pt(NH_3)_2(tu)_2]^{2+}$ and the other, $[Pt(tu)_4]^{2+}$ on treatment with thiourea. Identify the isomers giving proper reason. **[B.U. 2014, C.U. 2015]**

**Ans. :** As the trans effect order is $NH_3 < Cl^- < tu$, so cis-$[Pt(NH_3)_2Cl_2]$ gives $[Pt(tu)_4]^{2+}$ and trans-$[Pt(NH_3)_2Cl_2]$ gives trans-$[Pt(NH_3)_2(tu)_2]^{2+}$.

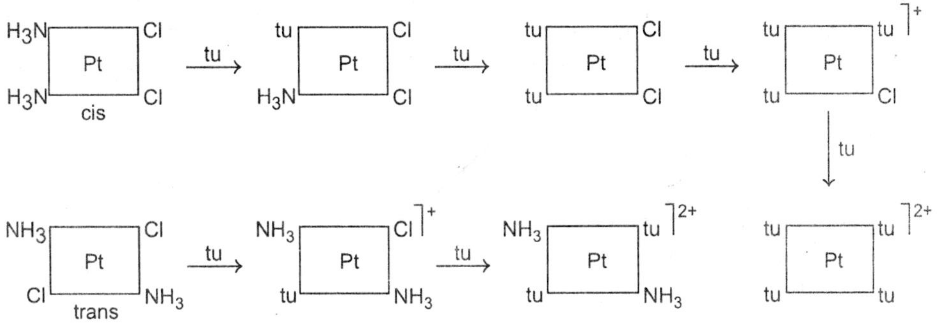


**Problem 7.113** Write the formula of tetramminechloridonitrito-κNcobalt(III) and draw the possible geometric structures. **[B.U. 2015]**

**Ans. :** $[Co(NH_3)_4ClNO_2]^+$ The geometry of the species is octahedral which may have the following geometric structures :

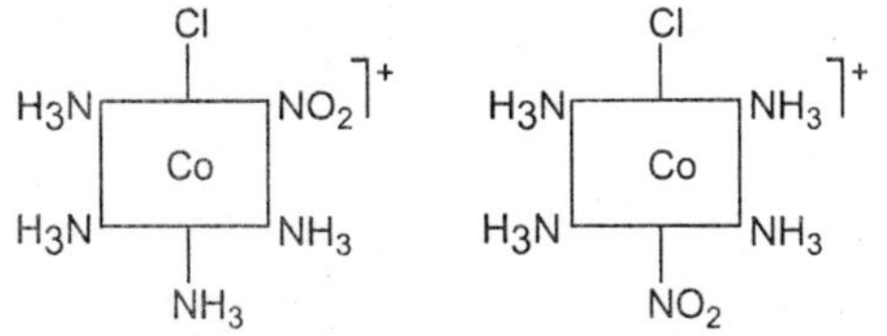


**Problem 7.114** The complex $[Fe(phen)_2(NCS)_2]$ (phen = 1, 10-phenanthroline) shows spin cross-over behaviour. CFSE and $\mu_{eff}$ at 250 and 150K, respectively are :

1. 0.4 $\Delta_o$, 4.90 BM and 2.4 $\Delta_o$, 0.00 BM; 2. 2.4 $\Delta_o$, 2.90 BM and 0.4 $\Delta_o$, 1.77 BM; 3. 2.4 $\Delta_o$, 0.00 BM and 0.4 $\Delta_o$, 4.90 BM; 4. 1.2 $\Delta_o$, 4.90 BM and 2.4 $\Delta_o$, 0.00 BM.

**[NET 2015]**

**Ans. :** In the complex Fe(II) is $3d^6$ and the complex is octahedral. As per Table 7.9, 7.10 and 7.11 in the high spin state the CFSE is 0.4 $\Delta_o$ and the number of unpaired electrons are 4 which corresponds to $\mu_{eff}$ = 4.90 BM. At 250K the complex exist in high spin state. On lowering the temperature at 150K the complex is converted to low spin thereby spin pairing occurs which corresponds to CFSE as $2.4\Delta_o$ and having no unpaired electron, so the $\mu_{eff}$ = 0. Therefore the answers is 1.

---

**Problem 7.115** Consider the following statements for $(NH_4)_2[Ce(NO_3)_6](Z)$—
A. Coordination number of Ce is 12, B. Z is paramagnetic,
C. Z is an oxidising agent, D. Reaction of $Ph_3PO$ with Z gives a complex having coordination number 10 for Ce.
The correct statements are—
1. A, B and C; 2. B, A and D; 3. B, C and D; 4. A, C and D **[NET 2015]**

**Ans. :** 4. i.e. A, C and D

$NO_3^-$ generally function as symmetrical didentate 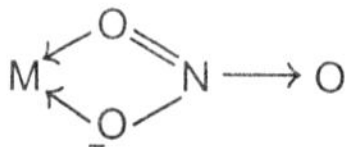

and so the coordination number of $Ce^{4+}$ is 12. Ce is a member of lanthanides which has the electronic configuration $[Pd]4f^{14}5s^25p^6$ and its common oxidation state is +3. In the Z complex the $Ce^{4+}$ therefore acting as an oxidising agent. The Z complex when reacts with a bulkier molecule like $Ph_3PO$ there will be the reduction of the coordination number from 12 to 10.

---

**Problem 7.116** Calculate spin-only magnetic moment in BM for $K_3[CuF_6]$ and $K_2[MnF_4]$. **[B.U. 2015]**

**Ans. :** In $K_3[CuF_6]$ the Cu(III) is $3d^8$ having two unpaired electrons i.e. n = 2.

$$\therefore \mu_{spin\text{-}only} = \sqrt{n(n+2)}\ \text{BM} = \sqrt{2(2+2)}\ \text{BM} = 2.82\ \text{BM}$$

In case of $K_2[MnF_4]$ the Mn(II) is $3d^5$ having n = 5

$$\therefore \mu_{spin\text{-}only} = \sqrt{5(5+2)}\ \text{BM} = 5.91\ \text{BM}.$$

---

**Problem 7.117** Find out Russel Saunders ground state terms for nickel (IV) and manganese (III). **[B.U. 2015]**

**Ans. :** Ni(IV) is $3d^6$ for which R–S ground term is $^5D_4$ and for Mn(III) the R–S ground term is $^5D_0$ (Table 7.15 and Sec. 7.15.6).

---

**Problem 7.118** Draw crystal field splitting diagram for $[Ni(dmg)_2]$; dmgH = dimethylglyoxime. **[B.U. 2015]**

**Ans. :**

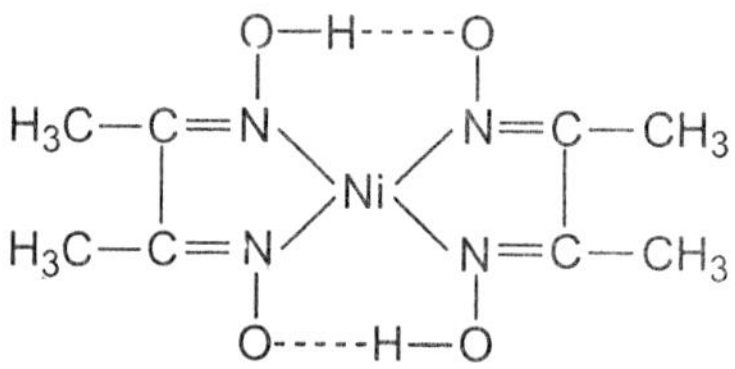


$Ni(dmg)_2$ has a square planar geometry and the splitting of d-orbitals will be as follows :

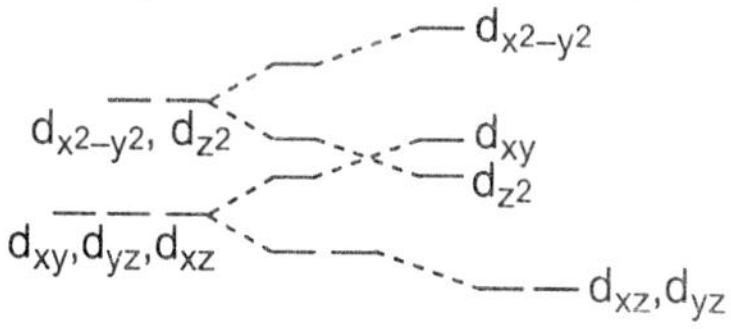

**Problem 7.119** Justify the order of ligand field splitting of the following hexammine complexes :
$[Co(NH_3)_6]^{3+} < [Rh(NH_3)_6]^{3+} < [Ir(NH_3)_6]^{3+}$ **[B.U. 2015]**

**Ans. :** Sec. 7.14.4 (size of d-orbitals). For a given ligand and same geometry, the magnitude of $\Delta_o$ increases by about 30-50% on passing from the first transition series to the second transition series and again by about the same amount from the second transition series to the third transition series. (The reason for this is that the atomic size of transition metal elements from the 1st to the 2nd to the 3rd is increasing and d-orbitals become more diffused. As a result metal d-orbitals can interact more effectively so that the d-orbital splitting increases). Thus the general trend in $\Delta_o$ for 1st transition metal < $\Delta_o$ for second transition metal < $\Delta_o$ for 3rd transition metal.

**Problem 7.120** State the basic postulates of crystal field theory (CFT). **[B.U. 2012]**

**Ans. :** (1) The central metal ion is considerd as point positive charged species and ligands are either point negativity charged or negative end of a dipolar neutral ligand molecule facing the metal ion.

(2) The electrostatic interaction between the core metal ion and periphery ligands is invoked without any orbital overlap of metal and ligands.

(3) All the ligands approach central metal ion along definite direction in a given geometry and repulsive interaction comes into play between electrons of d-orbitals of metal ion and negatively charged ligands.

(4) As the d-orbitals are differently oriented in space, so different repulsive interactions with the incoming ligands lift the degeneracy of the d-orbital set into two or more subsets.

(5) Due to preferential filling-up of the d-electrons present in the metal ion in the split subsets of d-orbitals obeying aufbau principle, Pauli exclusion principle and Hund's rule, extra stability is attainded which was otherwise absent in free metal ion.

**Problem 7.121** Three electronic transitions at 14900, 22700 and 34400 $cm^{-1}$ are observed in the absorption spectrum of $[CrF_6]^{3-}$. The $\Delta_o$ value (in $cm^{-1}$) and the corresponding transition are—

1. 7800 and $^4A_{2g} \longrightarrow {}^4T_{2g}$; 2. 14900 and $^4A_{2g} \longrightarrow {}^4T_{2g}$;
3. 14900 and $^4T_{2g} \longrightarrow {}^4T_{1g}(F)$; 4. 7800 and $^4T_{2g} \longrightarrow {}^4T_{1g}(F)$. **[NET 2015]**

**Ans. :** 2 i.e. 14900 and $^4A_{2g} \longrightarrow {}^4T_{2g}$. Because we know that lowest energy transition corresponds to $\Delta_o$. Question number 166 (Exercise).

**Problem 7.122** The symmetry allowed atomic transition among the following is—
1. $^3F \longrightarrow {}^1D$, 2. $^3F \longrightarrow {}^3D$, 3. $^3F \longrightarrow {}^1P$, 4. $^3F \longrightarrow {}^3P$. **[NET 2015]**

**Ans. :** 2 i.e. $^3F \longrightarrow {}^3D$ here selection rule is permitted.

**Problem 7.123** What is cis-effect? Give an example. **[B.U. 2016]**

**Ans. :** Certain ligands facilitate the substitution of the ligand that is in a position cis to them. Cis effect is much weaker than trans effect. Example of cis effect is the formation of $PtCl_2(NH_3)py$ from $PtCl_3(NH_3)^-$ (Question number 153).

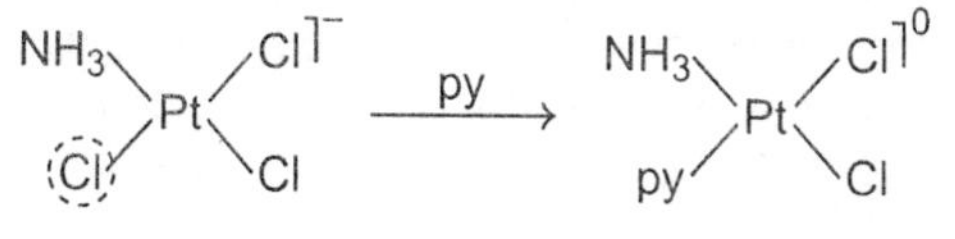


......(I)

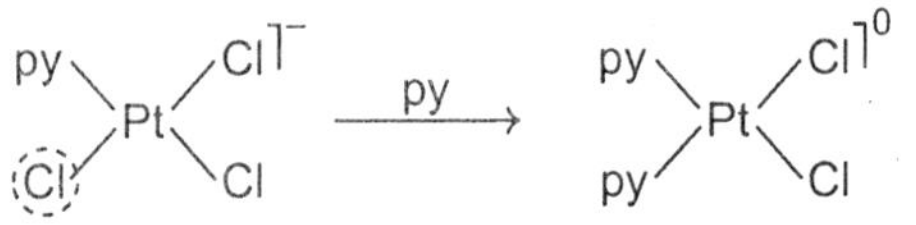

......(II)

Reaction I is about 40% faster than reaction II which indicates greater cis effect of py over $NH_3$.

---

**Problem 7.124** How do the spectator ligands effect rates of substitution reactions? Give an example. **[B.U. 2016]**

**Hint :** Sec. 7.19.10 and question number 159. Spectator ligands show steric accleration and steric retardation in substitution reaction. They also show electronic effects (i.e. electron pushing and withdrawing). If the starting complex is sterically crowded then in the dissociative process, the rate experience a steric acceleration while in the associative process the rate experiences a steric retardation.

Example is the amino acid exchange reaction at the square planar complex of Cu(II). The amino acid exchange rate at the square planar complex of Cu(II) decreases with the increase of steric crowding. The sequence for the amino acid is glycine > methyl glycine > diethyl glycine.

π-donor L : $\ddot{L}$—$Co(en)_2$—Cl (dissociative path)

π-acceptor L : L—$\ddot{C}o(en)_2$—Cl $\ddot{N}u$ (associative path).

---

## EXERCISE

1. What are the ingredients used to prepare a Schiff base. Write down the structures of a bidentate and a tridentate Schiff base. **[B.U. 2007, 2015]**

   **Hint :** Sec. 7.4.1.

2. Show the end on (EO) and end to end (EE) coordination motifs of bridging $N_3^-$. **[B.U. 2005]**

   **Hint :** Sec. 7.4.3(e)(i).

3. State the postulates of Werner. Write down the Lewis electron dot structures and show possible coordination motifs of the following : $NCS^-$ and $CH_3COCH_2COCH_3$. **[B.U. 2005, 2012]**

   **Hint :** Sec. 7.2 and 7.4.3(e).

4. What is the difference between ambidentate and polydentate ligand? **[C.U. 2004, 2015]**

   **Hint :** Sec. 7.4.2.

5. State the postulates of Werner's coordination theory. **[C.U. 1999]**

6. Molar conductances of $PtCl_4.xNH_3$ complexes at a dilution of 1024 litres are 0, 97, 229 and 523 $ohm^{-1}$ $cm^2$ for $x = 2, 3, 4$ and 6 respectively. Rationalise the modes of ionisation of these complexes according to Werner's theory of coordination. **[C.U. 1995]**

   **Hint :** Problem 7.1.

7. A solution containing 0.319 g of a hydrate isomer, with molecular formula $CrCl_3.6H_2O$ was passed through a cation exchange resin in acidic form and the acid so liberated was neutralised with 19 ml of 0.125 (N) NaOH. Identify the isomer. Molecular weight of $CrCl_3.6H_2O$ = 266.5.

   **Hint :** $[Cr(H_2O)_5Cl]Cl_2.H_2O$ Problem 7.2.

8. One red coloured coordination compound A slowly transforms into another coordination compound B, that is yellow in colour. Elemental analysis of both A and B give the same composition Co : $NH_3$ : Cl : $NO_2$ = 1 : 5 : 2 : 1. 1 m mol of each of A and B on treatment with an excess of $AgNO_3$ solution in dil $HNO_3$ medium give 2 m mol of AgCl.

   Write the possible coordination formulas of A and B, give their IUPAC names and rationalize the transformation A → B. **[C.U. 1998, B.U. 2014]**

   **Hint :** Sec. 7.9.5 and Sec. 7.6.

9. How would you identify cis and trans isomers of $[Co(NH_3)_4Cl_2]Cl$ by chemical methods?

   **Hint :** Sec. 7.9.12. **[C.U. 2004]**

10. Draw possible isomers of $[Co(NH_3)_2(H_2O)_2(CN)_2]^+$ ion. **[C.U. 2004]**

    **Hint :** Five isomers are possible. See Sec. 7.9.11(B)(iv).

11. Write down and draw the structures of the possible isomers of the coordination complex of formula $[Co(en)_2Cl_2]^+$ (where en = ethylenediamine). **[C.U. 2014, 2002, 1992, B.U. 2002, 2012]**

    **Hint :**

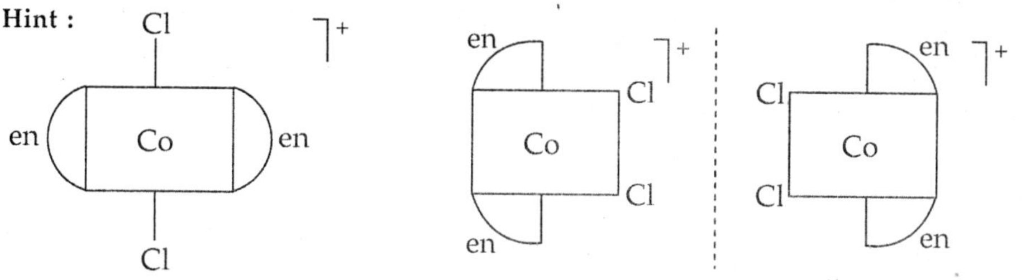


*trans* *optically inactive* *cis* *optically active* *d-form* *l-form*

For separation of enantiomers see Sec. 7.9.14.

12. $[Pt(NH_3)_2Cl_2]$ has two isomeric forms. Suggest a chemical pathway to elucidate the structures of the isomers. **[C.U. 2001]**

    **Hint :** Sec. 7.9.12.

13. Briefly discuss, four important factors, affecting complex formation. **[B.U. 1982]**

    **Hint :** Sec. 7.8.

14. Write the formula and structure of the different complex isomers of the formula $MCl_3.6H_2O$ and name the type of isomerism exhibited. **[B.U. 1982]**

    **Hint :** Sec. 7.9.1

15. In which cases the phenomena of optical isomerism expectd in tetra-coordinated and hexa-coordinated inorganic complexes. **[B.U. 1984]**

    **Hint :** Sec. 7.9.13. For hexacoordinated complexes discuss classifications (A), (B), (C) and (D).

16. Potassium sulphocyanido palladium (II) reacts with dipyridyl at –78°C to give a light orange solid (I) which when heated to 150°C gives a light yellow compound (II). Both (I) and (II) on elemental analysis give the same results. Explain the phenomenon with equation. **[B.U. 1987, 1994]**

    **Hint :** Sec. 7.9.5 (Structural, salt or linkage isomerism).

17. Mixing solutions of iron (II) salts and potassium hexacyano chromate (III), results in a brick-red precipitate which turns green on heating. Both the compounds have the same composition. Indicate the type of isomerism involved.

**Hint :** $Fe^{2+} + K^+ + [Cr(CN)_6]^{3-} \longrightarrow KFe^{II}[Cr^{III}(CN)_6] \xrightarrow{100°C} KCr^{III}[Fe(CN)_6]$
*brick-red* *dark green*

This has been interpreted in terms of linkage isomers reported by Shriver and coworkers of the following type :

$\cdots Fe^{II}–NC–Cr^{III}–CN–Fe^{II}–NC–Cr^{III} \cdots \longrightarrow \cdots Fe^{II}–CN–Cr^{III}–NC–Fe^{II}–CN–Cr^{III} \cdots$
*brick-red* *dark green*

See also Sec. 7.9.5.

18. How might one distinguish between the following :
$[Co(NH_3)_5Br]SO_4$ and $[Co(NH_3)_5SO_4]Br$ ? **[B.U. 1990, 1994, 2001, 2002, 2003]**

**Hint :** Sec. 7.9.2 for ionisation isomerism.

19. Give your comments on the statement : 'Isomerisms encountered in case of coordination complexes are more diverse than in use of organic compounds.' **[B.U. 1990]**

**Hint :** Sec. 7.9.13.

20. Bis(quinaldinato) oxovanadium (IV) forms two monohydrates–(A) cream and (B) olive green.
(i) one variety (A) looses water at 170-180°C and is synthesized in methanol medium.
(ii) the other variety (B) looses water at 100° and is synthesized in aqueous medium.
Give the formula of the two compounds and give reasons for your answer. **[B.U. 1990]**
**Hint :** Sec. 7.9.1 (Hydrate isomerism).

21. Dichlorodiamino platinum (II) has two geometrical isomers. How will you identify each of them? **[B.U. 1990, 1994, 2001, 2002, 2007]**

**Hint :** Sec. 7.9.12 (Determination of configuration of square planar complexes).

22. Give an example of linkage isomerism explaining why you call it linkage isomerism? How can you convert one form into the other?

**Hint :** Sec. 7.9.5.

$$[Co(NH_3)_5ONO]Cl_2 \underset{\text{UV ray}}{\overset{\text{let stand or warm, HCl}}{\rightleftharpoons}} [Co(NH_3)_5NO_2]Cl_2$$
*red* *yellow brown*

23. A polydentate ligand may not use all its donor centres for complexation—Comment.
**Hint :** Sec. 7.4.2 (flexidentate ligand). **[B.U. 1992]**

24. Give one physical and one chemical method which could be used to distinguish between cis and trans isomers of a complex compound. **[B.U. 1992]**

**Hint :** Sec. 7.9.12.

25. State the names and draw diagrams for all the isomers of $[Co(en)_2(NH_3)Cl]^{2+}$. **[B.U. 1994]**

**Hint :** Sec. 7.9.13(D).

26. A silver complex $Ag_4[Co(S_2O_3)(CN)_5]$ on careful manipulation with HCl gives two isomeric acids. The yellow acid on decomposition gives sulphur and $H_2SO_3$ while the red variety produces $H_2S$ and $H_2SO_4$. Give the structural formula of each isomer. What is the nature of isomerism? **[B.U. 1995]**

**Hint :** $Ag_4[Co(S_2O_3)(CN)_5] + HCl \longrightarrow H_4[Co(S_2O_3)(CN)_5]$

*This exists in two forms –*
*one is yellow and the other is red.*

The yellow variety on hydrolytic decomposition produces sulphur and $H_2SO_3$ so that in this case $S_2O_3^{2-}$ exist in oxygen bonded :

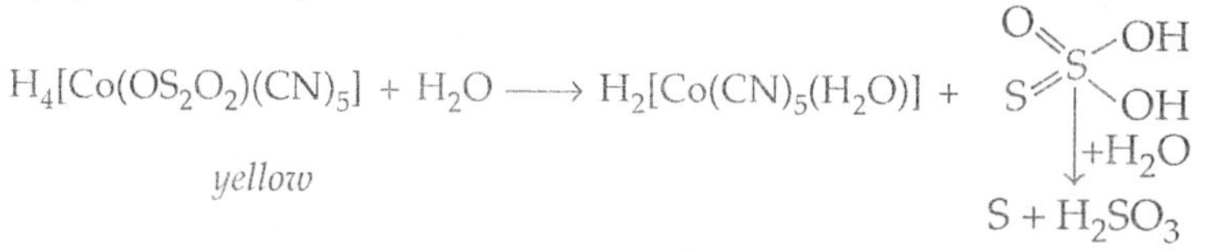


The red variety on decomposition produces $H_2S$ and $H_2SO_4$, so that here $S_2O_3^{2-}$ exist as sulphur bonded :

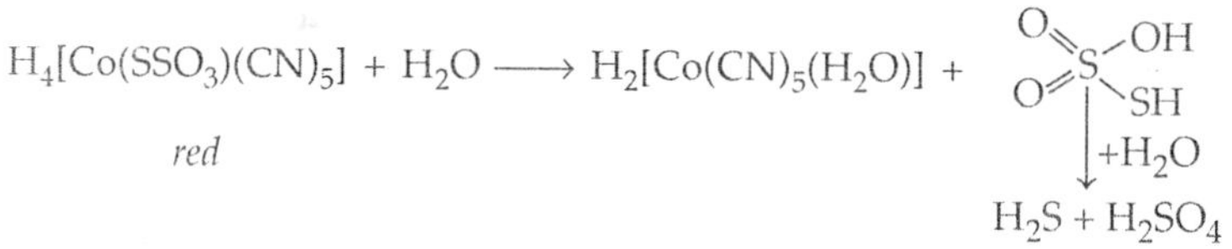


The red and yellow forms of $H_4[Co(S_2O_3)(CN)_5]$ is the example of structural, salt or linkage isomers (Sec. 7.9.5).

27. Arrange the following complexes in order to their molar conductivities in aqueous solution : $Pt(NH_3)_6Cl_4$, $Pt(NH_3)_3Cl_4$, $K_2PtCl_6$, $Co(NH_3)_5(NO_2)Cl_2$, $Co(NH_3)_3(NO_2)_3$ **[B.U. 1996]**

**Hint :** $Pt(NH_3)_6Cl_4 > K_2PtCl_6 \sim Co(NH_3)_5(NO_2)Cl_2 > Pt(NH_3)_3Cl_4 > Co(NH_3)_3(NO_2)_3$

28. Depict the strucural formula of Co(III) complex with EDTA. **[B.U. 1997]**

**Hint :** Sec. 7.9.13. Structure (7-XXXI).

29. How many isomers are possible for the following compounds? Give reasons for your answer. (i) $[Pt(NH_3)(py)(Br)(Cl)]$, (ii) $[Cr(en)_2(NCS)(Cl)]^+$ **[B.U. 1999]**

**Hint :** (i) Sec. 7.9.11. Geometrical isomers of Mabcd type complex.

(ii) Sec. 7.9.13. Optical isomerism for $M(AA)_2ab$ type complex.

30. Give the structural formula of a purely inorganic optically active compound. **[B.U. 1999, 2002]**

**Hint :** Sec. 7.9.13 Optical isomerism structure (7-XXVIII) as also structure (7-XXX).

31. Discuss the principal experimental evidences which were used by Werner for elucidating the structures of coordination complexes and their nature and behaviour. State the postulates of Werner's theory. **[B.U. 2000, 2005]**

**Hint :** Sec. 7.2 and 7.2.1.

32. "$NO_2^-$ is a versatile ligand and can coordinate in different ways"? Comment. **[B.U. 2000, 2001]**

**Hint :** Sec. 7.9.5 (Structural, salt or linkage isomerism).

33. How might one distinguish between the cis and trans isomers of $[Co(en)_2Cl_2]Cl$. **[B.U. 2003, 2015]**

**Hint :** Sec. 7.9.12 (Determination of configuration of cis-trans isomers of octahedral complex.)

34. Give examples of classical and non-classical ligand. Write down the structures of macrocycle, transchelating and organometallic compounds. **[B.U. 2004]**

**Hint :** For classical and non-classical ligand and for macrocyle see Sec. 7.4.3.

The compound

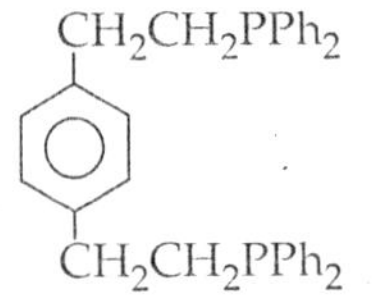


froms trans compound so that this ligand is called transchelating ligand.

Alkali metals form a number of organometallic compounds. Lithium alkyls are readily obtained by reacting lithium metal with alkyl or aryl halides in benzene or petroleum.

$$RX + 2Li \longrightarrow RLi + LiX$$

The alkalide ion $M^-$ e.g. $Na^-$ ion exist in the cryptand or macrocyclic ligand. The addition of a cryptand to ethylamine solution of Na or K greatly increases the solubility of the alkali metal by the process.

$$2M(s) + C \rightleftharpoons MC^+ + M^-$$

One example of the actual crystal is $[NaC]^+Na^-$ where C represents cryptan or macrocyclic ligand.

35. Cite an example of non-innocent ligand. Point out the criteria of a ligand to behave as a non-innocent ligand. Predict the oxidation state of iron in $[Fe(H_2O)_5NO]SO_4$. **[B.U. 2006, 2008, 2012]**

**Hint :** For non-innocent ligand see Sec. 7.4.3(c).

The oxidation state of iron is +1 when NO is $NO^+$ and the oxidation state of iron is +2 when there is neutral NO.

36. Draw and count the number of possible geometrical isomers for the complex ion $[Co(en)_2Cl_2]^+$. Which one may show optical activity? How would you resolve the optical isomers from a racemic mixture? **[B.U. 2006]**

**Hint :** For 1st part see question number 11. For resolution of racemic mixture see Sec. 7.9.14.

37. The complex $[Co(NO_2)(NH_3)_5]Cl_2$ exists in unstable red and stable yellow isomeric forms. Draw the structure of each and name the type of isomerism. **[B.U. 2007]**

**Hint :** Sec. 7.9.5

| | |
|---|---|
| $[Co(ONO)(NH_3)_5]Cl_2$ | $[Co(NO_2)(NH_3)_5]Cl_2$ |
| *red* | *yellow* |
| pentaamminenitrito-KOcobalt(III) chloride | pentaaminenitrito-KNcobalt(III) chloride |

38. Which of the two complexes $[Cr(EDTA)]^-$ and $[CrCldien]^+$ is/are chiral? **[B.U. 2007]**

**Ans. :** $[Cr(EDTA)]^-$ is chiral.

39. What is the peculiarity of the complex, $Na[Rh(NH\text{–}SO_2\text{–}NH)_2(H_2O)_2]$? **[B.U. 2008]**

**Hint :** This is purely inorganic optically active complex (Sec. 7.9.13).

40. What is an ambidentate ligand? Give examples. How does it differ from a polydentate ligand. **[C.U. 2008, B.U. 2015]**

Hint : Sec. 7.4.2.

41. $[Cr(en)_2Cl_2]Cl$ may be found in two forms, one violet and other green. On reaction with oxalate ion the violet species produces corresponding oxalato derivative, while the green one does not. Explain the result and write the IUPAC name of the oxalato derivative. **[C.U. 2009]**

**Hint :** Sec. 7.9.12 (Determination of configuration of cis-trans isomers)

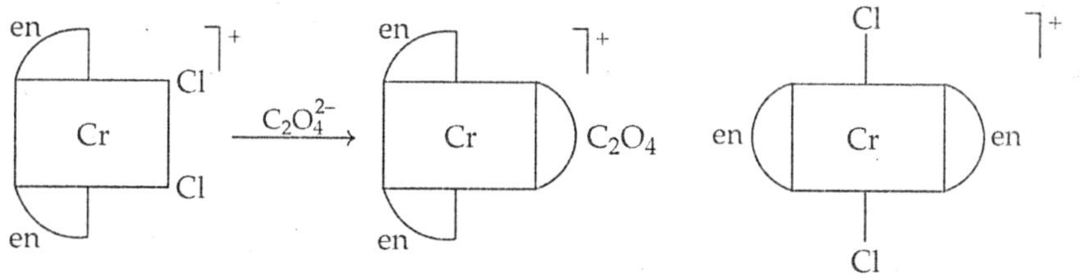


*cis (violet)* *trans (will not produce any oxalato product)*

The oxalato derivative is $[Cr(en)_2C_2O_4]Cl$ which is named as bis(ethane-1,2-diamine) (oxalato)chromium(III) chloride.

42. Give all geometrical isomers of the complexes given below : (i) $[Ru(py)_3Cl_3]$, (ii) $[Co(en)_2(NO_2)Cl]^+$. **[C.U. 2009]**

**Hint :** Sec. 7.9.11(B)(iii) for (i) and Sec. 7.9.11(B)(v) for (ii).

43. Which among the following ligands A, B and C will be able to form innermetallic complex of the first order with $M^{2+}$ ion? Justify. **[C.U. 1999]**

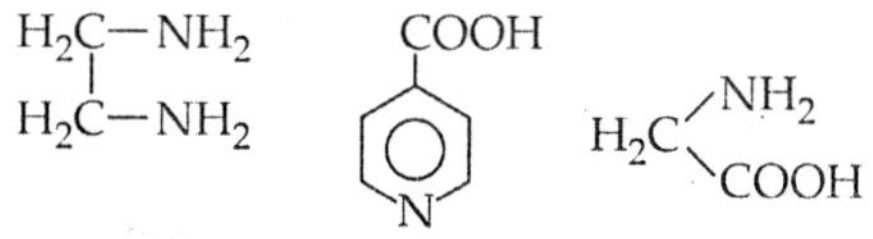


A B C

**Hint :** B and C will form first order inner complex with $M^{2+}$ ion.

44. What do you mean by trans-effect? Apply this concept to prepare cis and trans isomer of $[Pt(NH_3)_2Cl_2]$ starting with $PtCl_4^{2-}$ and $Pt(NH_3)_4^{2+}$ ion respectively? **[C.U. 2008]**

**Hint :** For trans effect see problem 7.5. For synthesis of $[Pt(NH_3)_2Cl_2]$ consult Sec. 7.9.12, 7.19.14.

45. What are crypts and cryptates? Discuss their peculiarities. **[C.U. 1997]**

**Hint :** Sec. 7.4.3(d).

46. What do you mean by chelate effect? Explain why it is called an entropy effect? **[B.U. 2015, 2012, 2007, 2003, 1993, 1991]**

**Hint :** Sec. 7.11.1(4).

47. What is a chelate complex? Give an example. Explain its stability. **[C.U. 2008]**

**Hint :** Sec. 7.8(5) and 7.11.1(4).

48. $Cu^{2+}$ ion reacts with $NH_3$ to form a series of complexes, $Cu(NH_3)^{2+}$, $Cu(NH_3)_2^{2+}$, $Cu(NH_3)_3^{2+}$ and $Cu(NH_3)_4^{2+}$. Write down the complex formation equilibria and give expressions for the stepwise and overall formation constants. **[B.U. 1998]**

**Hint :** Sec. 7.11.

49. Give the structure of dimethylglyoximato nickel(II). What type of complex is it? Indicate the different types of bonds in this compound. **[B.U. 1999]**

**Hint :** For structure see Fig. 3.45. This is a first order inner metallic complex. There are four coordinate bonds with four nitrogen atoms and two intramolecular hydrogen bonds as depicted in Fig. 3.45.

50. The stability of $Ni(en)_3^{2+}$ is much more than that of $Ni(NH_3)_6^{2+}$, although both contains Ni-N bond. Offer an explanation, showing the physical parameter mainly responsible for it. **[B.U. 2015, 1995]**

**Hint :** The stability of $Ni(en)_3^{2+}$ compared to that of $Ni(NH_3)_6^{2+}$ is due to chelate effect [Sec. 7.11.1(4)]. The value of logK at 25°C for the following reaction is 9.67 due to the enthalpy change of –2.9 kcal $mol^{-1}$ whereas $-T\Delta S°$ is –13.2 kcal $mol^{-1}$. The enthalpy change is primarily due to an increase in the crystal field stabilisation energy caused as a result of the formation of $Ni(en)_3^{2+}$ from $Ni(NH_3)_6^{2+}$.

$$Ni(NH_3)_6{}^{2+} + 3en \rightleftharpoons Ni(en)_3{}^{2+} + 6NH_3$$

51. Account for inner complex salt formation depends on pH of the solution. **[B.U. 1992]**

**Hint :** Sec. 7.10 and 7.13.2. Taking DMGH (HL) as the inner metallic ligand which is a monoprotic acid in water and pH of the solution does govern the concentration of the ligand anion ($L^-$, which is responsible for the formation of inner complex salt with $Ni^{2+}$ as $Ni(DMG)_2$ or $NiL_2$) according to the following equilibrium :

$$HL \rightleftharpoons H^+ + L^-$$

At pH below 4, the backward reaction is favourable and free $L^-$ is available only in limited quantity and hence the formation of inner complex salt will be incomplete and nonquantitative.

52. Discuss the role of formation constant of a complex in the use of a sequestering agent in a reaction. **[B.U. 1990]**

**Hint :** Sec. 7.13.5. $EDTAH_4$ is widely used as a sequestering agent and sequestering power for this is maximum at or above pH8. When a solution containing two cations M and N where N reacts with the organic ligand e.g. benzimidazole and M is an interfering ion, then the feasibility of using $EDTAH_4$ as sequestering agent will be considered in terms of stability constant values as $K_M/K_N \geq 10^6$ where $K_M$ and $K_N$ are the respective stability constants of M and N with the sequestering agent.

53. Explain the stepwise stability constants and overall stability constant of a complex with an example of your choice. How are they related? **[C.U. 2003]**

**Hint :** Sec. 7.11.

54. What do you mean by 'innermetallic complex'? Which among the following ligands A, B and C will be able to form innermetallic complex with $Fe^{3+}$ and $Cu^{2+}$?

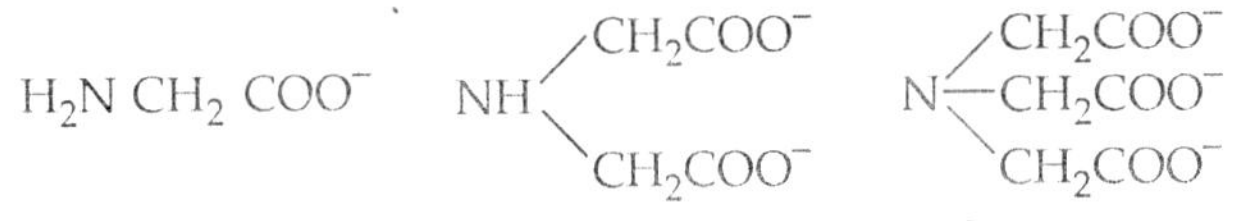

A B C

Give structural formula of these complexes. **[C.U. 1996]**

**Hint :** $Fe^{3+}$ and $Cu^{2+}$ will form first order innermetallic complexes with A ligand as $FeA_3$ and $CuA_2$.

55. Which one of 'metal-EDTA complex' and 'metal-indicator complex' is more stable for suitable direct complexometric titration of the metal? **[B.U. 2008]**

**Ans. :** The stability of metal-EDTA complex is greater than metal-indicator complex.

56. A12.60 ml of 0.008826 M EDTA solution was needed for titration of a 100.0 ml sample of water for total hardness. Calculate the total hardness in terms of ppm of $MgCO_3$. **[B.U. 2004]**

**Ans. :** 93.41 ppm.

57. Draw the structures of all possible isomers of the complex $[Co(den)_2]^{3+}$ where den = $H_2NCH_2CH_2NHCH_2CH_2NH_2$. **[C.U. 2009]**

**Hint :**

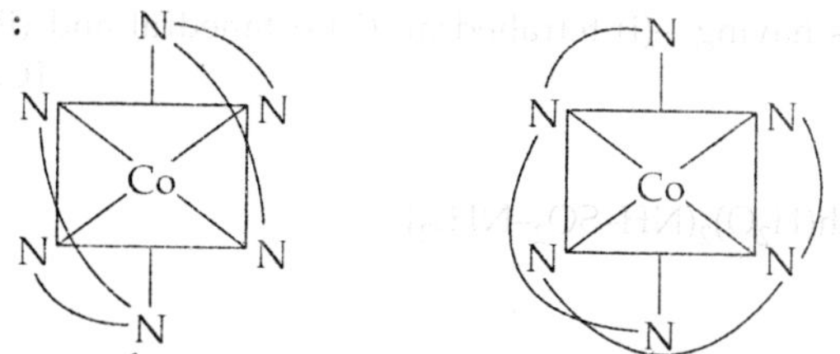


fac mer

58. Give the expressions for CFSE of $d^5$, $d^6$ and $d^7$ ions in weak and strong octahedral crystal fields in terms of Dq values and pairing energy (P). [C.U. 2009]

**Hint :** Sec. 7.14.2(B).

59. Give reasonable explanations, "while $Fe^{3+}$ (aq) ion reacts rapidly with EDTA at room temperature, $Cr^{3+}$ (aq) reacts extremely slow". [C.U. 2009]

**Hint :** Both $Fe^{3+}$ and $Cr^{3+}$ are hard acid but $Fe^{3+}(3d^5)$ is a labile system while $Cr^{3+}(3d^3)$ is an inert system. In EDTA the oxygen atoms are hard donors. The formation constant of $Fe^{3+}$ complex of EDTA at 25°C is $1.3 \times 10^{25}$ but the corresponding value for $Cr^{3+}$ is not available in the literature.

60. Compare the crystal field splitting parameter (Δ) of the following pairs with proper reason :

(i) $[Co(H_2O)_6]^{3+}$ and $[Co(H_2O)_6]^{2+}$ (ii) $[CoCl_6]^{4-}$ and $[CoCl_4]^{2-}$

(iii) $[TiF_6]^{3-}$ and $[VF_6]^{2-}$ (iv) $[W(CO)_6]$ and $[Cr(CO)_6]$ [B.U. 2007]

**Hint :** (i) $[Co(H_2O)_6]^{3+} > [Co(H_2O)_6]^{2+}$ due to greater charge of $Co^{3+}$ than $Co^{2+}$ (Sec. 7.14.4).

(ii) $[CoCl_6]^{4-} > [CoCl_4]^{2-}$ as octahedral complex will have higher Δ compared to that of tetrahedral complex. (Sec. 7.14.4).

(iii) $[VF_6]^{2-} > [TiF_6]^{3-}$ reason same as that of (i).

(iv) $W(CO)_6 > Cr(CO)_6$ as W is a member of third transition series (Sec. 7.14.4).

61. Indicate the geometries of $[NiCl_4]^{2-}$ and $[Ni(CN)_4]^{2-}$ with explanation. [B.U. 2007]

**Hint :** $[Ni(CN)_4]^{2-}$ is square planar while $[NiCl_4]^{2-}$ is tetrahedral as $CN^-$ is a strong field and $Cl^-$ is a weak field ligand.

62. State requisite properties of a metal indicator for its use in EDTA titration. [C.U. 2007]

**Hint :** Sec. 7.13.6(E).

63. Show the d-orbital splitting of the metal ions in the following complexes using crystal field theory. Calculate crystal field stabilisation energy and predict magnetic properties : (i) $[Cr(H_2O)_6]^{3+}$, (ii) $[CoF_6]^{3-}$, (iii) $[Fe(CN)_6]^{4-}$. [C.U. 2007]

**Hint :**

(i) — — $e_g$; ↑ ↑ ↑ $t_{2g}$; $Cr^{3+}(3d^3) : t_{2g}^3$; μ = 3.8 BM corresponding to three unpaired electrons

(ii) ↑ ↑ $e_g$; ↑↓ ↑ ↑ $t_{2g}$; $Co^{3+}(3d^6) : t_{2g}^4 e_g^2$; μ = 4.90 BM corresponding to four unpaird electrons

(iii) — — $e_g$; ↑↓ ↑↓ ↑↓ $t_{2g}$; $Fe^{2+}(3d^6) : t_{2g}^6$; μ = 0

For CFSE see the Table 7.11.

64. Give examples of optically active complexes having—(i) tetrahedral, (ii) octahedral and (iii) square planar geometry. [C.U. 2007]

**Hint :** See Sec. 7.9.13.

(i) ethylmethylpropylsulfidoarsenic (V), $[Rh(H_2O)_2(NH{-}SO_2{-}NH)_2]^-$

(ii) $[Co(EDTA)]^-$ or $[Cr(EDTA)]^-$

(iii) 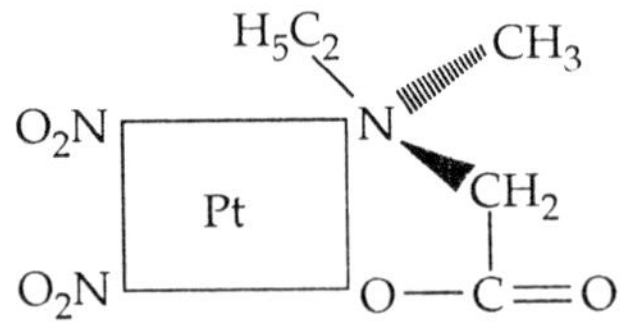


*bis(nitrito-κN)(N-methyl–N–ethyl glycinato) platinate(II)*

65. Which of the following pairs of complex has higher 10Dq value and why?

(a) $[Co(NH_3)_6]^{3+}$ and $[Rh(NH_3)_6]^{3+}$, (b) $[Fe(C_2O_4)_3]^{4-}$ and $[Fe(C_2O_4)_3]^{3-}$,

(c) $[Cr(en)_3]^{3+}$ and $[Cr(C_2O_4)_3]^{3+}$

**Hint :** (a) $[Rh(NH_3)_6]^{3+} > [Co(NH_3)_6]^{3+}$

(b) $[Fe(C_2O_4)_3]^{3-} > [Fe(C_2O_4)_3]^{4-}$

(c) $[Cr(en)_3]^{3+} > [Cr(C_2O_4)_3]^{3+}$ as en is a strong field ligand compared to that of $C_2O_4^{2-}$.

66. Explain why $[NiCl_4]^{2-}$ is paramagnetic but $[Ni(CO)_4]$ is diamagnetic though both are tetrahedral.

**Hint :** In $[NiCl_4]^{2-}$ the $Ni^{2+}$ ion is $sp^3$ hybridised making two unpaired electrons i.e. paramagnetic. Ni(O) in $Ni(CO)_4$ is $sp^3$ hybridised which corresponds to no unpaired electron and the complex is diamagnetic.

67. Guess the magnetic moments of the following : $K[AgF_4]$, $K_4[Fe(CN)_6]$, $[Ni(CO)_4]$ and $Na_2[CoCl_4]$ [B.U. 2009]

**Hint :** $K[AgF_4]$ is diamagnetic. $K_4[Fe(CN)_6]$ is diamagnetic and $[Ni(CO)_4]$ is also diamagnetic. $Na_2[CoCl_4]$ where $Co^{2+}$ ($3d^7$) has $e^4t_2^3$ configuration and the expected magnetic moment corresponds to three unpaired electrons (3.87 B.M.). Experimental value = 4.6 B.M.

68. Rationalise the fact that $\Delta_o$ increase in the order $CrCl_6^{3-}$, $Cr(NH_3)_6^{3+}$ and $Cr(CN)_6^{3-}$. [B.U. 2008, 2013]

**Hint :** Inspite of negative charge of $Cl^-$, $\Delta_o$ for $Cl^-$ is less than $\Delta_o$ for $NH_3$ because of repulsive interaction between the $p_\pi$ electrons of $Cl^-$ and the $d_\pi$ electrons of $Cr^{3+}$. $\Delta_o$ for $CN^-$ is very high because of back bonding i.e. the shift of electron density from the $d_\pi$ orbitals of $Cr^{3+}$ to the antibonding $\pi^*$ orbitals of $CN^-$. This interaction stabilises the $d_\pi$ orbitals and increases $\Delta_o$. The comparison can be made in the following way :

(1) $Cl^-$ is a good $\sigma$ donor + weak $\pi$ acceptor ligand and anionic. The overall charge on the complex $CrCl_6^{3-}$ is 3– and so $\Delta_o$ decreases to the most extent among the three.

(2) $NH_3$ is only a $\sigma$ donor ligand and neutral and so decrease in $\Delta$ not to an appreciable extent, again overall charge on the complex $Cr(NH_3)_6^{3+}$ is 3+. This will increase the $\Delta_o$ value from that of $CrCl_6^{3-}$ but will be less than that of the complex $Cr(CN)_6^{3-}$.

(3) $CN^-$ is a good $\sigma$ donor and anionic ligand. The overall charge on the complex $Cr(CN)_6^{3-}$ is 3- thereby $\Delta_o$ should decrease. But $CN^-$ has excellent $\pi$-acceptor property so that $\Delta_o$ value increases to a great extent with highest $\Delta_o$ for the complex $Cr(CN)_6^{3-}$.

69. Write down the characteristic coordination number of Cu(I) and Ag(I). [B.U. 2008]

**Hint :** Cu(I) : 2 (linear), 4 (tetrahedral), 3 (planar)

Ag(I) : 2 (linear), 3 (planar, $sp^2$), 4 (tetrahedral)

70. $[VO(acac)_2]$ has magnetic moment 1.7 B.M. while $[V(acac)_3]$ shows 2.8 B.M. as magnetic moment. How do you reconcile? [B.U. 2008]

**Hint :** $[VO(acac)_2]$ has square pyramidal geometry where $V^{4+}$ is $3d^1$ while $[V(acac)_3]$ has octahedral geometry where $V^{3+}$ has $3d^2$ configuration.

71. Why do 4d and 5d metal ions form square planar complexes than tetrahedral complexes? [B.U. 2008, 2013]

**Hint :** 4d and 5d orbitals spread out in space more than 3d orbitals. There is a reduced interelectronic repulsion thereby closer approach of metal and ligand orbitals leading to larger crystal field splitting. Thus $[NiCl_4]^{2-}$ is tetrahedral while $[PdCl_4]^{2-}$ and $[PtCl_4]^{2-}$ are square planar.

72. Differentiate between 'crystal field strength' and 'crystal field stabilisation energy'? [B.U. 2008]

**Hint :** Crystal field strength denotes interacting power and splitting power of the orbitals. CFSE denotes the gain in energy by a $d^n$ ion due to preferential filling of electrons in $t_{2g}$ and $e_g$ orbitals or $t_2$ and e orbitals.

73. Predict the number of isomers in $[Ni(NH_3)(Br)(Cl)(H_2O)]$. [B.U. 2008]

**Hint :** Number of isomers equal to three. MABCD type complex.

74. What will be the magnetic moment of Li $[Ti(bpy)_3]$ in Bohr magneton (bpy = 2, 2′-bipyridine). [B.U. 2008]

**Hint :** Ti (–I) is $3d^5$.

75. Explain the nature of Jahn-Teller distortion expected for an octahedral complex of $d^9$ metal ion. [C.U. 2007]

State Jahn-Teller theorem with a suitable example. [B.U. 2012]

**Hint :** Sec. 7.14.5(C).

76. Predict the products of the following reaction and justify your answer.

cis - $[Pt(NH_3)_2(py)_2]^{2+} + 2Cl^- \longrightarrow$ [C.U. 2003]

**Hint :**

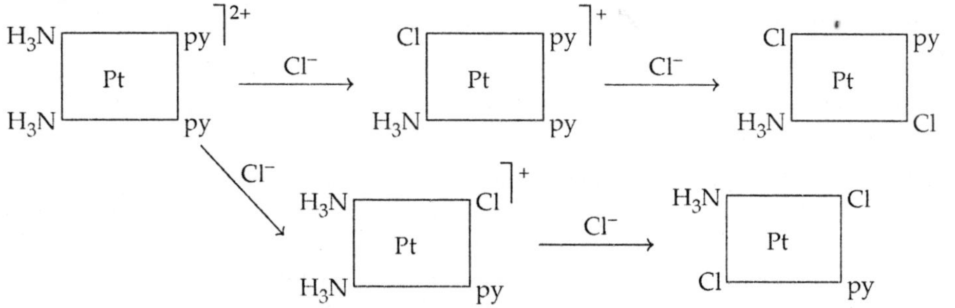


Both the first step products are of almost equal probability. Second step is governed by trans directing influence of $Cl^-$. Trans effect series : $Cl^- > NH_3 \sim py$

Some authors say the trans effect series : $Cl^- > py > NH_3$

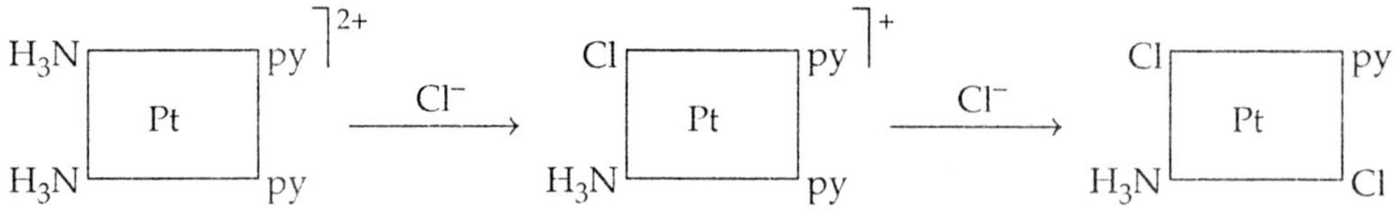

77. What do you mean by stability constant of a metal ligand complex? [C.U. 2003]

78. $K_2Ca[Cu(NO_2)_6]$ and $K_2Ba[Cu(NO_2)_6]$ exhibit static J-T distortion while in $Tl_2Pb[Cu(NO_2)_6]$ it is in the dynamic state.—Explain. [C.U. 2008]

**Hint :** Sec. 7.14.5(C).

79. Using Orgel diagram, explain the electronic transitions for $[Ti(H_2O)_6]^{3+}$ in weak octahedral field indicating the spectroscopic ground state. [C.U. 2007]

**Hint :** Fig. 7.18 and 7.21(a). Sec. 7.15.8 and Sec. 7.15.10(A).

80. Write d-orbital splitting diagrams for a square pyramid and trigonal bipyramidal complex stating the relative energies of various orbitals in Dq units. [B.U. 2010]

**Hint :** For splitting diagrams see Fig. 7.12. For relative energies of orbitals see Table 7.12.

For square pyramid : $d_{z^2}(0.86)$, $d_{x^2-y^2}(9.14)$ $d_{xy}(-0.86)$, $d_{xz}$, $d_{yz}(-4.57)$.

For trigonal bipyramidal : $d_{z^2}(7.07)$, $d_{x^2-y^2}(-0.82)$, $d_{xy}(-0.82)$, $d_{xz}$, $d_{yz}(-2.72)$.

81. Using approximate Orgel diagram explain the electronic spectrum of $Cu^{2+}$ (aq) ion. [C.U. 2009]

**Hint :** Sec. 7.15.8 (Fig. 7.18) and Sec. 7.15.10(A) (Fig. 7.21(b)).

82. Give reasonable explanation why $[FeF_6]^{3-}$ ion is colourless, whereas $[Fe(CN)_6]^{3-}$ ion is coloured? [C.U. 2009, 2006, B.U. 2015]

**Hint :** See problem 7.67. $[Fe(CN)_6]^{3-}$ is low spin which is also Laporte forbidden but partly spin allowed. Thus the partly spin allowed d-d bands of $[Fe(CN)_6]^{3-}$ are often rendered obscure by charge transfer bands in the near ultraviolet region with prominant tails in the visible region.

83. Using qualitative Orgel diagram explain the electronic spectral transitions for $3d^6$ ion in weak octahedral field. [C.U. 2008]

**Hint :** $3d^6$ ion in weak octahedral field i.e. $[CoF_6]^{3-}$. This is a high spin complex and Tanabe-Sugano diagram (modified Orgel diagram) shows single absorption peak corresponding to the transitions $^5T_{2g} \rightarrow {}^5E_g$. The deep blue colour of $[CoF_6]^{3-}$ shows a single peak at 13,000 $cm^{-1}$.

84. $Cr^{2+}$ and $V^{2+}$ give intensely coloured complexes with dipyridyl and 1, 10-phenanthroline and not with simple unsaturated π-acceptor ligand like CO, $C_2H_4$. [C.U. 2008]

**Hint :** A transfer of charge from metal to ligand (MLCT) is most commonly observed with ligands that have low lying π* orbital especially aromatic ligands (Sec. 7.15.11). When the metal ions are in the low oxidation state, the d-orbital will be relatively high in energy and the transition ($E_g^* \rightarrow \pi^*$, $T_{2g} \rightarrow \pi^*$) will occur at low energy. In this game the important ligands are dipyridyl, 1, 10 phenanthroline etc. which involved in MLCT transitions. As $Cr^{2+}$ and $V^{2+}$ are in low oxidation states thereby MLCT transitions produces intense colour with dipyridyl and 1, 10 phenanthroline complexes. For CO, $C_2H_4$ etc. having unsaturated π-acceptor character ligands are unable to induce MLCT transitions.

85. Discuss the Laporte selection rule for electronic spectral transitions. [C.U. 2008]

Write down the Laporte selection rule and spin selection rule in words and in mathematical forms. Find out their significances in explaining the colour of transition metal complexes. [B.U. 2012]

State the selection rules in electronic spectroscopy. [B.U. 2015]

**Hint** : Sec. 7.15.9.

86. The cis isomer of the complex $[Co(NH_3)_4Br_2]$ is more intensely coloured than the trans isomer.—Explain. [C.U. 2008]

**Hint** : See Problem 7.68.

87. Explain why $HgCl_2$ is white in colour but $HgI_2$ is red. [C.U. 2007]

**Hint** : The metal ion in these compounds e.g. $Hg^{2+}$ is not a good oxidising agent but despite that the ligand to metal charge transfer (LMCT) occurs because of the fact that $I^-$ is easily oxidised compared to that of $Cl^-$. So $HgCl_2$ is white but $HgI_2$ is red due to LMCT.

88. Give one example of each for MLCT and LMCT transitions. [B.U. 2009, 2017]

**Hint** : MLCT : $[Fe(open)_3]^{2+}$, $[Fe(SCN)_6]^{3-}$.

LMCT : HgS(red), $KMnO_4$, $HgI_2$.

89. Explain which one of the following pairs of complexes has higher CFSE : (a) Tetrahedral and octahedral complexes of the same $d^3$ metal ion.. (b) $[M(H_2O)_6]^{2+}$ and $[M(H_2O)_6]^{3+}$ where both have three d-electrons. [B.U. 2009]

**Hint** : (a) Octahedral complex has higher CFSE than tetrahedral one.

(b) $[M(H_2O)_6]^{3+} > [M(H_2O)_6]^{2+}$

90. "On addition of conc. HCl, $[Co(H_2O)_6]^{2+}$ transforms to $[CoCl_4]^{2-}$".—Explain. [C.U. 2015, 2005, B.U. 2009]

**Hint** : It is well known that on addition of conc. HCl to a pink coloured $[Co(H_2O)_6]^{2+}$, deep blue coloured tetrahedral complex $[CoCl_4]^{2-}$ is readily formed. This is due to the change in stereochemistry from octahedral to tetrahedral. It is to be noted that intensities of d-d absorption bands of tetrahedral complexes (e.g. $CoCl_4^{2-}$) are much greater than that of octahedral complex (e.g. $Co(H_2O)_6^{2+}$). The cause for this is the lack of symmetry in the tetrahedral complex. The spectrum of $CoCl_4^{2-}$ has bands at 3,500 and 6,300 $cm^{-1}$ and some intense bands at ~15,000 $cm^{-1}$. Using an appropriate Dq = 350 $cm^{-1}$, the first two bands are due to the transition $^2A_2(F) \rightarrow {}^4T_2(F)$ and $^4A_2(F) \rightarrow {}^4T_1(F)$. The predicted third band due to $^4A_2(F) \rightarrow {}^4T_1(P)$ at ~19,600 $cm^{-1}$ which is observed at ~15,000 $cm^{-1}$. Also see Sec. 7.14.5(B).

91. How will you differentiate the origins of colours in $[Fe(ophen)_3]^{2+}$ and $Cr_2O_7^{2-}$ ions. [B.U. 2009]

**Hint** : For $[Fe(ophen)_3]^{2+}$ it is MLCT and for $Cr_2O_7^{2-}$ it is LMCT.

92. How the stabilisation of low, medium and high oxidation states of metal ions can be made through judicious choice of ligands? [B.U. 2006]

**Hint** : Sec. 7.17.

93. What is CFAE? [B.U. 2010, 2012]

**Hint** : Sec. 7.19.9.

94. Discuss the followings in connection with the ligand substitution reaction for an octahedral complex : (i) Linear free energy relationship, (ii) Effects of spectator ligands. [B.U. 2010, 2008]

**Hint** : Sec. 7.19.8, Sec. 7.19.10 and Sec. 7.19.13.

95. Differentiate between CFSE and CFAE. [B.U. 2008]

**Hint** : Sec. 7.19.9 and Sec. 7.14.3.

96. State the factors which effect rate laws and mechanisms in inorganic reactions. **[B.U. 2009]**

**Hint :** Sec. 7.19.11.

97. Define the terms 'stoichiometric mechanism' and 'intimate mechanism'. **[B.U. 2014, 2009, 2008]**

**Hint :** Sec. 7.19.4 and Sec. 7.19.5. Stoichiometric mechanism relates to the identification of the sequence of elementary steps involved in a complicated overall reaction. Three paths for ligand substitutions are : dissociative (D), associative (A) and interchange (I).

Intimate mechanism is the understanding of the magnitude of the rate constans for the individual steps in terms of rearrangements of atoms and bonds. Two categories of intimate mechanisms are : associative activation (a) and dissociative activation (d).

98. What are labile and inert complexes? **[B.U. 2009, C.U. 2007]**

**Hint :** Sec. 7.19.1.

99. Classify the substitution reactions as A, D, $I_a$ and $I_d$ with scientific diagram.

**Hint :** Sec. 7.19.4 and Sec. 7.19.5. **[B.U. 2016, 2009, 2006]**

100. The early 3d metals prefer an associative mechanism whereas the late members prefer dissociative mechanism.—Comment. **[B.U. 2008]**

**Hint :** Early members e.g. Cr(III) ($d^3$ system) size larger, has room for accommodation of electrons from nucleophile and prefer associative mechanism. Later members e.g. Co(III) ($d^6$ system) has less space due to small size and thus prefer dissociative mechanism.

101. How do $\Delta H^{\neq}$ and $\Delta S^{\neq}$ values help us in assigning a reaction mechanism. **[B.U. 2008, 2013]**

**Hint :** Sec. 7.19.7. $\Delta H^{\neq}$ and $\Delta S^{\neq}$ are called the enthalpy and entropy of activation respectively.

**Dissociative mechanism :** $\Delta H^{\neq}$, positive as energy is required to break the bond between the leaving group and central metal ion. $\Delta S^{\neq}$, positive as number of species increases causing more randomness and in transition state (TS) bond breaking is important resulting in the formation of activated complex with reduced coordination number (CN).

**Associative mechanism :** $\Delta H^{\neq}$, less positive as bond making and bond breaking occurs simultaneously, hence bond breaking energy is partially compensated by that from bond making. $\Delta S^{\neq}$, negative as number of species decreases with increasing CN in the TS.

102. Depending on the time scale how the substitution reactions can be classified? **[B.U. 2008]**

**Hint :** Slow : Stable for at least for 1 min., half life > 1 min.

Fast : Half life < 1 min.

103. Cite the condition when the reactions of octahedral substitutions can acquire an associative path. **[B.U. 2007]**

'Depending on reaction conditions substitution reaction may acquire an associative path'.—Comment. **[B.U. 2015]**

**Hint :** This path is followed in substitution reactions of octahedral complexes when the persistent intermediate is formed in which the central metal ion has expanded coordination number.

Octahedral complexes may follow associative path if the following conditions are attained :

(a) High effective nuclear charge ($z_{eff}$) of the central metal atom/ion and high overall positive charge on the complex favours a new metal-ligand bond formation and the reaction will follow an associative path.

(b) Increase in coordination number becomes easier for large metal ions i.e. the formation of an intermediate with extended coordination number is less favoured.

(c) Presence of vacant orbitals in $t_{2g}$ level (i.e. $d^1$ and $d^2$ metal ions) favours the approach of the incoming nucleophile.

(d) Gain in CFSE/LFSE upon increase in coordination number favours associative path.

(e) Presence of bulky ligand(s) facilitates dissociative path.

(f) Electronic effects (σ-donor, π-donor and acceptance properties) of spectator ligands also play a crucial role in determining the reaction mechanism.

104. $[NiXL_5]^+ + H_2O \longrightarrow [NiL_5(H_2O)]^{2+} + X^-$ is much faster if L is $NH_3$ instead of $H_2O$.—Explain. **[B.U. 2007]**

**Hint :** This is due to strong σ donor ability of $NH_3$ compared to that of $H_2O$. See Sec. 7.19.10 (electron donating effect).

105. Which one of $d^5$ and $d^6$ ions is more inert in low spin octahedral complexes? **[B.U. 2006]**

**Hint :** $d^6 > d^5$ See Sec. 7.19.1.

106. Explain why Cr(II) is strong reducing agent but Mn(III) is strong oxidising agent, although both are $d^4$ ions. **[B.U. 2009, 2000]**

**Hint :** $Cr^{3+} + e \rightarrow Cr^{2+}$ $E° = -0.41$ volt

$Mn^{3+} + e \rightarrow Mn^{2+}$ $E° = +1.51$ volt

E° values indicate $Cr^{2+}$ is highly reducing agent while $Mn^{3+}$ is strong oxidising agent.

$Cr^{2+}$ $(3d^4)$ CFSE = –6Dq   $Cr^{3+}$ $(3d^3)$ CFSE = –12 Dq

$Mn^{2+}$ $(3d^5)$ CFSE = –20Dq   $Mn^{3+}$ $(3d^4)$ CFSE = –6Dq

As CFSE of $Cr^{3+}$ is more compared to that of $Cr^{2+}$ so that $Cr^{2+}$ will be easily oxidised to form $Cr^{3+}$ while as the CFSE of $Mn^{2+}$ is higher than that of $Mn^{3+}$ so that $Mn^{3+}$ will be easily reduced to form $Mn^{2+}$.

107. 'Although 6-coordinated complexes are formed by Cr(III) and Co(III) with cyanide ligand but most stable nickel (II) cyanide complex is tetracoordinated'.—Why? **[B.U. 2000]**

**Hint :** $Cr^{3+}(3d^3)$, $Co^{3+}(3d^6)$ and $Ni^{2+}(3d^8)$. As population of d-electron increases coordination number decreases.

108. What is 'trans effect'? Given the reactants $PPh_3$, $NH_3$ and $PtCl_4^{2-}$, propose effective routes to the synthesis of cis and trans $[PtCl_2(NH_3)(PPh_3)]$. **[B.U. 2009]**

**Hint :** Sec. 7.19.14.

109. Explain why square planar complexes prefer an associative mode of activation? **[B.U. 2009]**

**Hint :** Associative mechanism requires the accommodation of electrons as well as the ligands in the transition state. Sec. 7.19.13.

110. $[PtCl(dien)]^+ + I^- \longrightarrow [PtI(dien)]^+ + Cl^-$;

Considering the above equation predict the path of the reaction when

(I) rate = $k_1[\{PtCl(dien)\}^+]$ and

(II) rate = $k_2[\{PtCl(dien)\}^+][I^-]$

(where $k_1$ and $k_2$ are the rate constants in two paths. **[B.U. 2006]**

**Hint :** (I) first oder with respect to complex. (II) Second order—first order with respect to complex and first order with respect to $I^-$. Sec. 7.19.13(c).

111. Suggest the synthetic routes to get cis and trans $[PtCl_2(NH_3)_2]$ from the complex ions $[Pt(NH_3)_4]^{2+}$ and $[PtCl_4]^{2-}$. [B.U. 2006, NET 2001]

**Hint :** Sec. 7.9.12 and Sec. 7.19.14.

112. The colour of $[Mn(H_2O)_6]^{2+}$ is extremely faint because—

1. the d-d transitions involved are spin-forbidden but Laporte-allowed
2. the d-d transitions involved are Laporte-forbidden but spin allowed
3. the d-d transitions involved are both Laporte and spin forbidden
4. of the metal-ligand charge transfer. [NET 2001]

**Ans. :** 3.

113. cis-trans isomerism is exhibited by—

1. $[PtCl(NH_3)_3]^+$, 2. $[Pt(NH_3)_4]^{2+}$, 3. $[PtCl_4]^{2-}$, 4. $[PtCl_2(NH_3)_2]$ [NET 2001]

**Ans. :** 4.

114. The species which has four unpaired electrons is—

1. $[Co(CN)_6]^{4-}$, 2. $[Cr(H_2O)_6]^{3+}$, 3. $[FeCl_4]^{2-}$, 4. $[Fe(H_2O)_6]^{3+}$ [NET 2001]

**Ans. :** 3.

115. The species which has CFSE = –4Dq is—

1. $[CoF_6]^{3-}$, 2. $[Mn(H_2O)_6]^{3+}$, 3. $[V(H_2O)_6]^{3+}$, 4. $[Ni(H_2O)_6]^{2+}$ [NET 2001]

**Ans. :** 1.

116. The paramagnetic cobalt (III) complex is exemplified by—

1. $[Co(NH_3)_6]^{3+}$, 2. $[Co(NO_2)_6]^{3-}$, 3. $[CoF_6]^{3-}$, 4. $[Co(NH_3)_4Cl_2]^+$ [NET 2001]

**Ans. :** 3.

117. For the complexes showing the square pyramidal structure, the d-orbital involved in the hybridisation is (z axis is along the $C_4$ axis).

1. $d_{x^2-y^2}$, 2. $d_{z^2}$, 3. $d_{yz}$, 4. $d_{xz}$

**Ans. :** $d_{x^2-y^2}$.

118. Write the d-orbital splitting for $[CuCl_2]^-$ (ligands along z axis). [NET 2001]

**Hint :** $[CuCl_2]^-$ has a linear geometry and the splitting of d-orbitals will be as per problem 7.56.

119. Which isomer of $[Pt(NH_3)_2Cl_2]$ is obtained by reacting (a) 2 mol of HCl with $[Pt(NH_3)_4]^{2+}$ and (b) 2 mol of $NH_3$ with $[PtCl_4]^{2-}$.

**Hint :** Sec. 7.9.12.

120. Classify the following complexes into 'Labile' and 'Inert' giving reasons :

(i) $[VO(H_2O)_5]^{2+}$, (ii) $[Cr(H_2O)_6]^{2+}$, (iii) $[Cr(CN)_6]^{3-}$ [NET 2001]

**Hint :** Sec. 7.19.1 and Table 7.19.

121. The species with maximum CFSE is—

1. $[FeCl_4]^-$, 2. $[FeCl_4]^{2-}$, 3. $[CuCl_4]^{2-}$, 4. $[ZnCl_4]^{2-}$ **[NET 2000]**

**Ans. :** 2.

122. The species for which a change in coordination geometry occurs in acidic aqueous solution is : 1. $[Cr_2O_7]^{2-}$, 2. $[MnO_4]^-$, 3. $[Fe(H_2O)_6]^{2+}$, 4. $[CoCl_4]^{2-}$. **[NET 2000]**

**Ans. :** 4.

123. How can you estimate phosphate by EDTA solution? **[SSC 2007]**

**Hint :** 7.13.6(G).

Phosphate solution $\xrightarrow{\text{Excess } MgCl_2/NH_4Cl/NH_3}$ $NH_4(Mg)PO_4\cdot 6H_2O\downarrow$

↓ Filtered, washed free from $Mg^{2+}$

Precipitate $\xrightarrow{HCl}$ Solution

Solution → (1) Measured excess standard EDTA is added (2) Heat the solution nearly to boiling (3) pH adjusted to 10 by $NH_4Cl + NH_3$ buffer (4) EBT indicator added → Blue solution indicates excess EDTA

↓ Excess EDTA back titrated with $Mg^{2+}$ solution

Blue colour changes to wine red

$$EDTA \equiv Mg^{2+} \equiv NH_4(Mg)PO_4\cdot 6H_2O \equiv PO_4^{3-}$$

$$\therefore 1000 \text{ ml (M) } Mg^{2+} \equiv 1000 \text{ ml (M)EDTA}$$
$$\equiv 1000 \text{ ml(M)}PO_4^{3-} \equiv 95 \text{ g of } PO_4^{3-}$$

124. The species with spin only magnetic moment of $\sqrt{24}$ BM is—

1. $[CoF_3(H_2O)_3]$, 2. $[CoCl_4]^{2-}$, 3. $[NiCl_4]^{2-}$, 4. $[Ni(H_2O)_6]^{2+}$ **[NET 2000]**

**Ans. :** 1.

125. The optically active species among the following is—

1. $[Cr(NH_3)_6]^{3+}$, 2. $[Co(CN)_6]^{3-}$, 3. $[Co(gly)_3]$, 4. $[Ru(NH_3)_6]^{3+}$ **[NET 2000]**

**Ans. :** 3.

126. Predict the product, $Pt(NO_2)Cl_3^{2-} \xrightarrow{NH_3}$ ? **[SSC 2007]**

**Hint :** Since trans effect of $NO_2^- > Cl^-$ so,

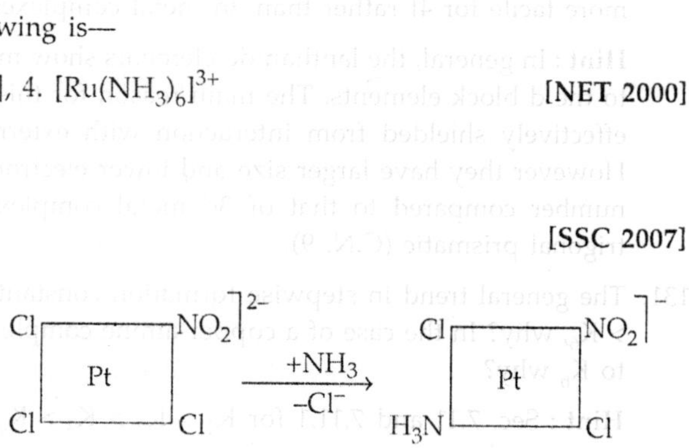


trans-$[Pt(NO_2)(NH_3)Cl_2]^-$

127. The IUPAC nomenclature of the complex $Na[PtBrCl(NO_2)(NH_3)]$ is— [NET 2000]

1. Sodium amminechlorobromonitro-N-platinum(II)
2. Sodium nitrochlorobromoammine-N-platinate(II)
3. Sodium amminobromochloronitro-N-platinate(II)
4. Sodium amminobromochloronitro-N-platinum(II)

**Ans. :** 3.

128. Show the structure of $K[(\eta^2-C_2H_4)Pt^{II}Cl_3]$ [SSC 2007]

**Hint :**

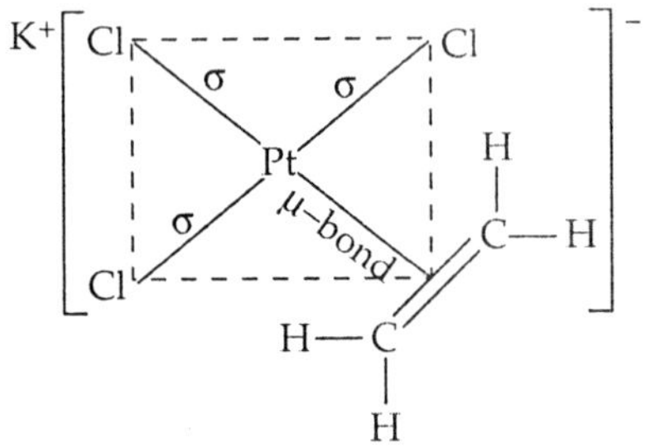


Square planar

129. In aqueous solution $[Cr(H_2O)_6]^{3+}$ and $[Ni(H_2O)_6]^{2+}$ exhibit the lowest energy electronic transitions at 17,400 $cm^{-1}$ and 8,500 $cm^{-1}$ respectively. (i) Assign the transition and obtain the 10Dq values in each case, (ii) For the former complex the effective magnetic moment ($\mu_{eff}$) is comparable to the spin only ($\mu_{s.o.}$) value but for the latter $\mu_{eff} > \mu_{s.o.}$. Explain with the help of Orgel diagram. [NET 2000]

**Hint :** (i) $[Cr(H_2O)_6]^{3+}$ : ${}^4A_{2g} \rightarrow {}^4T_{2g}$ : 10Dq = 17,400 $cm^{-1}$.

$[Ni(H_2O)_6]^{2+}$ : ${}^3A_{2g}(F) \rightarrow {}^3T_{2g}(F)$ : 10Dq = 8,500 $cm^{-1}$

Lowest energy transition denotes 10Dq.

(ii) $Ni^{2+}$ : The first excited state is ${}^3T_{2g}(F)$ and the second excited state is ${}^3T_{1g}(F)$. Both these excited states are susceptible to orbital contribution and so $\mu_{eff} > \mu_{s.o.}$.

130. The transformation of octahedral geometry to trigonal prismatic or square antiprismatic geometry is more facile for 4f rather than 3d metal complexes. Explain. [NET 2000]

**Hint :** In general, the lanthanide elements show much less tendency of complex formation compared to the d block elements. The main reason for this is that the 4f electrons or the 4f orbitals are very effectively shielded from interaction with external forces by the overlying $6s^2$ and $6p^2$ orbitals. However they have larger size and lower electronegativity so that they exhibit higher coordination number compared to that of 3d metal complexes e.g. square antiprism (C.N. 8) and tricapped trigonal prismatic (C.N. 9).

131. The general trend in stepwise formation constant of octahedral complexes $K_1 > K_2 > K_3 > K_4 > K_5 > K_6$, why? In the case of a copper-amine complex there is an abrupt decrease when we go from $K_5$ to $K_6$ why? [NET 2000]

**Hint :** Sec. 7.11 and 7.11.1 for $K_1 > K_2 > K_3 > K_4 > K_5 > K_6$.

For copper-amine complex generally we can say that unless one of the intermediate complex is extremely stable, there will be no extended range of concentration of complexing agent over which a single species is predominant. In case of copper-amine complex the highest coordination complex is $[Cu(NH_3)_4]^{2+}$ which has highest stability so that there will be an abrupt decrease from $K_5$ to $K_6$.

132. $[Cr(H_2O)_6]^{3+}$ ions are pale blue-green ($\varepsilon = 20\ mol^{-1}Lcm^{-1}$) but $CrO_4^{2-}$ ions are intense yellow ($\varepsilon = 8000\ mol^{-1}Lcm^{-1}$) why? **[NET 2000, C.U. 2015]**

**Hint :** In $[Cr(H_2O)_6]^{3+}$ there is d-d transition while in $CrO_4^{2-}$ there is charge transfer transition. (Sec. 7.15.10, 7.15.11 and problem 7.78).

133. Consider the reaction—

$$[Cu(H_2O)_6]^{2+} + en \xrightleftharpoons{K_1} [Cu(en)(H_2O)_4]^{2+}$$

$$[Cu(en)(H_2O)_4]^{2+} + en \xrightleftharpoons{K_2} [Cu(en)_2(H_2O)_2]^{2+}$$

$$[Cu(en)_2(H_2O)_2]^{2+} + en \xrightleftharpoons{K_3} [Cu(en)_3]^{2+}$$

If $K_1$, $K_2$, $K_3$ denote the formation constants, explain briefly why $K_2 >> K_3$. **[NET 2000]**

**Hint :** The span of ligation for which a multidentate ligand can be attached is sometimes related to the stability of the complex. For $Cu^{2+}$-en system when the third ethylenediamine molecule is attached, there will be a steric strain involved due to the attachment of two trans positions of the metal. Also for $Cu^{2+}$ there is a Jahn-Teller distortion and so $K_2 >> K_3$.

134. In octahedral V(III) and Cr(III) complexes the d-d transition frequencies are of the order $\nu_1 < \nu_2 < \nu_3$. Explain why 10Dq corresponds to $\nu_1$ for Cr(III) but $\nu_2 - \nu_1$ for V(III). **[NET 2000]**

**Hint :** Generally lowest energy transition corresponds to 10Dq, so $\nu_1$ corresponds to 10Dq in case of $[Cr(H_2O)_6]^{3+}$ which corresponds to $^4A_{2g} \rightarrow {}^4T_{2g}$ transition. In case of $V^{3+}$ ($3d^2$) the ground state is $^3T_{1g}(F)$ and the excited states are $^3T_{2g}$, $^3T_{1g}(P)$ and $^3A_{2g}$. Three spin allowed but Laporte forbidden transitions are permitted, $^3T_{1g}(F) \rightarrow {}^3T_{2g}(\nu_1)$, $^3T_{1g}(F) \rightarrow {}^3T_{1g}(P)(\nu_2)$ and $^3T_{1g}(F) \rightarrow {}^3A_{2g}(\nu_3)$. Actually first two transitions are obtained. Since there are two $T_{1g}$ terms which develop some configurational interaction because of similar symmetry propertries and so here 10Dq corresponds to the energy gap between $^3T_{2g}(F)$ and $^3T_{1g}(P)$ i.e. $\nu_2 - \nu_1$.

135. Guess the products of the following reactions : (a) $[Pt(PR_3)_4]^{2+} + 2Cl^-$, (b) $[PtCl_4]^{2-} + 2PR_3$, (c) cis-$[Pt(NH_3)_2(py)_2]^{2+} + 2Cl^-$.

**Hint :** (a) cis-$[PtCl_2(PR_3)_2]$, (b) trans-$[PtCl_2(PR_3)_2]$, (c) trans-$[PtCl_2(NH_3)(py)]$.

136. Starting from $K_2[PtCl_4]$ design two step synthesis of cis and trans isomers of $[Pt(NH_3)_2(NO_2)_2]$, explaining the principle involved therein. **[C.U. 2009, 2006]**

137. Write the $d^n$ configuration of the folowing complexes and also state their CFSE values. $[Ti(H_2O)_6]^{2+}$, $[Mn(H_2O)_6]^{2+}$ and $[Cu(acac)_2]$ **[NET 2010]**

**Hint :** $[Ti(H_2O)_6]^{2+}$ : $3d^2$, CFSE = –8 Dq

$[Mn(H_2O)_6]^{2+}$ : $3d^5$, CFSE = 0

$[Cu(acac)_2]$ : $3d^9$, CFSE = –12.28 Dq

138. How many microstates exist in $d^2$ configuration? Arrange $^3F$, $^3P$, $^1D$, $^1G$, $^1S$ for $d^2$ configuration. **[NET 2010]**

**Hint :** Microstates = $^{10}C_2 = \frac{L10}{L2L8} = 45$.

$$\text{Microstates} = \Sigma(2S+1)(2L+1) = \underset{^3F}{3(2\times3+1)} + \underset{^3P}{3(2\times1+1)} + \underset{^1G}{1(2\times4+1)} + \underset{^1D}{1(2\times2+1)} + \underset{^1S}{1(2\times0+1)} = 45.$$

$^3F < {}^3P < {}^1G < {}^1D < {}^1S$ :

(1) States having higher spin multiplicity lies lowest.

(2) States having the same spin multiplicity but of higher L values lies lowest.

139. "An inner metallic ligand is essentially a chelating ligand, but the reverse is not true"—Justify. [B.U. 2010, 2015]

**Hint :** Problem 7.27. Inner metallic ligand with both neutral and acidic donors satisfied coordination number and charge but chelating ligand with neutral donors satisfies coordination number only. For example $[Co(en)_3]^{3+}$ where ethylenediamine is a chelating ligand but not inner metallic ligand while in $[Co(gly)_3]$ where glycine acting as an inner metallic ligand.

140. Draw the structure of the complex formed by nickel(II) with dimethylglyoxime and state its important features. [B.U. 2010]

**Hint :** Structure of the complex : Fig. 3.45.

Important features : (1) Square planar with $NiN_4$ chromophore.(2) Nitrone N is acidic (3) Oxime N is neutral centre (4) Two oxime nitrogens bind Ni(II) through coordinate covalent bond (5) Two nitrone nitrogens bind Ni(II) through covalent bond (6) Two five membered chelate rings are formed (7) Two six membered O—H......O hydrogen bonded rings are formed (8) Two oxime nitrogens are in trans position, two nitrone nitrogens are in trans position.

141. Give the examples of a classical ligand, a trans chelating ligand and a non-classical ligand depicting the ligational motifs. [B.U. 2010]

**Hint :** Classical and non-classical ligand : Sec. 7.4.3.

$NH_3$ (Classical ligand) : $M \xleftarrow{\sigma \text{ bond}} NH_3$

$H_2C = CH_2$ (non-classical ligand) : $M \underset{\mu \text{ bond}}{\rightleftarrows} \begin{matrix} CH_2 \\ \| \\ CH_2 \end{matrix}$

According to exercise number 34 the trans chelating ligand is

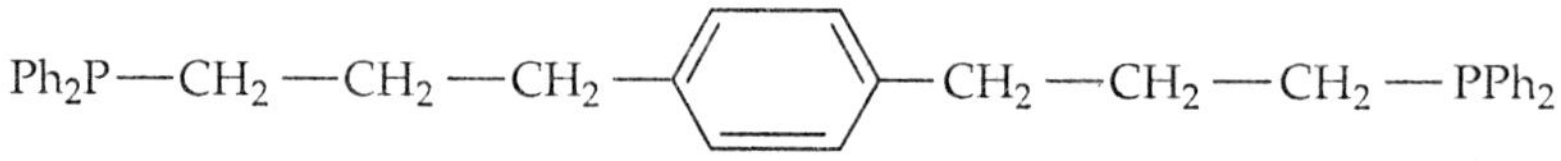


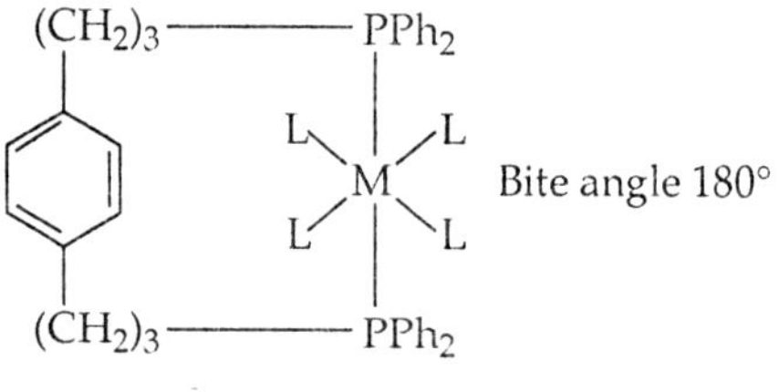


142. Show the schematic representations for preparations of a bidentate and a tetradentate schiff bases. [B.U. 2010]

**Hint :** Sec. 7.4.1.

143. Define innocent and non-innocent ligands and give examples (one each). Write the structural formula of two coordination molecules containing non-innocent ligands. [B.U. 2010, 2015, 2016]

**Hint :** Sec. 7.4.3. Innocent ligand does not involve in electron transfer reaction e.g. $Cl^-$, $OH^-$ etc. while non-innocent ligand does involve in electron transfer reaction e.g. NO, bpy, catechol etc.

$NO^+ \rightleftarrows NO \rightleftarrows NO^-$ having three oxidation states +1, 0 and –1 respectively.

$[Fe(H_2O)_5(NO)]SO_4$, here iron is +1 and NO is +1

$Na_2[Fe(CN)_5NO]$, here iron is +2 and NO is +1

Each of the complex has octahedral geometry.

144. Give the IUPAC names of the folowing

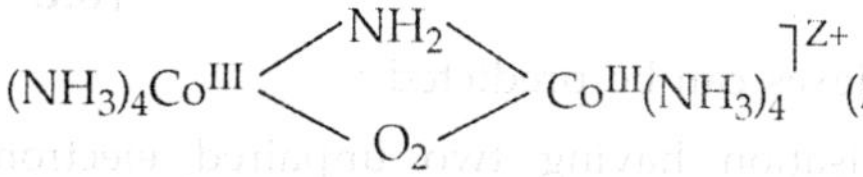

(Z = 5, 4, 3) **[B.U. 2010, 2015]**

**Hint :** When Z = 5, μ-amido-μ-dioxygenoctamminedicobalt(III)

when Z = 4, μ-amido-μ-superoxidooctamminedicobalt(III)

when Z = 3, μ-amido-μ-peroxidooctamminedicobalt(III)

145. Write the Lewis electron dot structure of cysteine and show its possible coordination motifs. **[B.U. 2010, 2016]**

**Hint :**

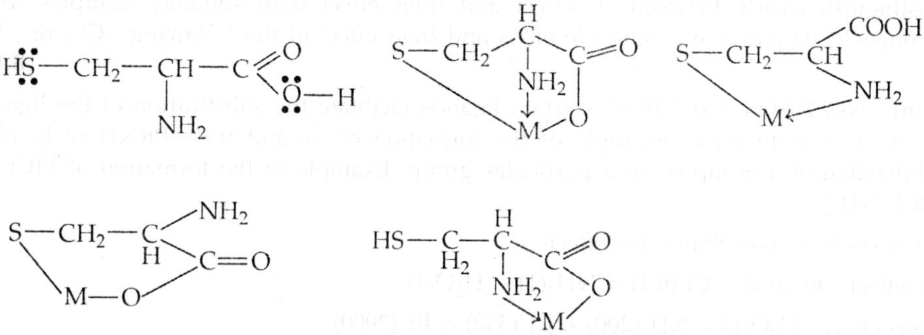

146. The different possible energy states for $d^2$ configuration are $^3F$, $^3P$, $^1G$, $^1D$ and $^1S$. Predict the ground state term with necessary arguments. **[C.U. 2007]**

**Hint :** Sec. 7.15.5 and 7.15.6.

147. Arrange the following in respect of their placement in spectrochemical series of ligands : $F^-$, $I^-$, $H_2O$, $CN^-$. **[B.U. 2011]**

**Hint :** Sec. 7.14.4. $I^- < F^- < H_2O < CN^-$.

148. What is the maximum number of unpaired electrons possessed by a high spin complex of the first transition series in the ground state? Which metal ions could show this maximum number? **[B.U. 2011]**

**Hint :** Sec. 7.14.1. Table 7.9 and 7.10. 5 unpaired electrons, $Fe^{3+}$, $Mn^{2+}$, i.e. $3d^5$ metal ions.

149. Explain the difference in colour which arises due to *cis* and *trans* form of $[Co(en)_2Cl_2]$ complex? **[B.U. 2011]**

**Hint :** Problem 7.68. The *cis* isomer will have more intense colour compared to that of *trans* isomer.

150. $Pt(NH_3)_4^{2+}$ is colourless and $PtCl_4^{2-}$ is red in the solution and the $[Pt(NH_3)_4][PtCl_4]$ named Magnus' green salt is strongly coloured.—Explain the observation. **[B.U. 2010]**

**Hint :** The square-planar units stack, alternate cation and anion, directly above each other in the crystal with a Pt-Pt distance of 3.25Å. Similar species with small ligands which allow short Pt-Pt distances are abnormally coloured, betokening a metal-metal interaction in the crystal. Such species have aroused interest as possible 'one dimensional conductors'. The short Pt-Pt distance gives rise to some innermetal interaction which shift the d-d absorption region of $PtCl_4^{2-}$ from green to red region, transmitting the green colour observed.

151. Give a concise account on charge transfer spectra involving metal to ligand and ligand to metal transitions. Use suitable examples. **[B.U. 2011]**

**Hint :** Sec. 7.15.11.

152. On the basis of valence bond theory predict different types of complexes containing $d^8$ ions. Indicate the pattern of hybridisation and magnetic properties of each type. Give suitable illustrations. **[B.U. 2011]**

**Hint :** For $d^8$ ions the following three types of complexes can be predicted :

(a) Octahedral involving $sp^3d^2(sp^3d_{x^2-y^2}d_{z^2})$ hybridisation having two unpaired electrons i.e. paramagnetic e.g. $[Ni(acac)_2(H_2O)_2]$.

(b) Tetrahedral having paramagnetic character to the extent of two unpaired electrons involving $sp^3$ hybridisation e.g. $[NiX_4]^{2-}$ (X = Cl, Br, I).

(c) Square planar configuration having $d_{x^2-y^2}sp_xsp_y$ i.e. $dsp^2$ hybridisation where the eight electrons are paired in d orbital and the complex will be diamagnetic e.g. $[Ni(CN)_4]^{2-}$.

153. Distinguish clearly between *cis* effect and *trans* effect with suitable examples. Which effect is stronger? Compare the magnitude of *cis* and *trans* effect of the following : $Cl^-$, $Br^-$, $NH_3$, $H_2O$. **[B.U. 2015, 2011]**

**Hint :** Sec. 7.19.14 and 7.19.16. Certain ligands facilitate the substitution of the ligand that is in a position trans to them. Example of the formation of *cis* and *trans* $Pt(NH_3)_2Cl_2$. *cis* effect is the labilization of a group *cis* to a particular group. Example of the formation of $PtCl_2(NH_3)py$ from $PtCl_3(NH_3)^-$.

*cis* effect is weaker than *trans* effect.

*cis* effect : $Br^-(0.3) < Cl^-(0.4) < NH_3(1) \sim H_2O(1)$

*trans* effect : $H_2O(1) < NH_3(200) < Cl^-(330) < Br^-(3000)$

In substitution reaction of square planar complexes the labilisation effect of a group cis to a particular group is called **cis effect**. It is more pronounced in octahedral complexes.

The ligand trans to the leaving group has profound influence on the rate of substitution reaction of square planar complexes and this effect is known as **trans effect**.

| **cis effect** | **trans effect** |
|---|---|
| (a) Follow dissociative pathway, more energy required for bond breaking and formation, so slower process. | (a) It is a direct process, follow associative pathway, no bond breaking is required and so faster process. |
| (b) Executed in octahedral complexes. | (b) Applied in tetrahedral complexes. |
| (c) Kinetic effect. | (c) Both kinetic and thermodynamic effect. |
| (d) It is weaker, not always influential. | (d) It is stronger and more influential. |

154. Predict the rate of base hydrolysis in case of the following complexes $[Co(NH_3)_5Cl]^{2+}$ and $[Co(py)_5Cl]^{2+}$ and explain with reason. **[B.U. 2011]**

**Hint :** Sec. 7.19.4. The mechanism suggested for the base hydrolysis reaction involves the following steps with $K_{OH^-} = 8.5 \times 10^{-1}\ M^{-1}sec.^{-1}$.

$$[Co(NH_3)_5Cl]^{2+} + OH^- \xrightleftharpoons{\text{fast}} [Co(NH_3)_4(NH_2)Cl]^+ + H_2O$$

$$[Co(NH_3)_4(NH_2)Cl]^+ \longrightarrow [Co(NH_3)_4(NH_2)]^{2+} + Cl^-$$

$$[Co(NH_3)_4(NH_2)]^{2+} + H_2O \xrightarrow{\text{fast}} [Co(NH_3)_5(OH)]^{2+}$$

A complex without an acidic proton should react with $OH^-$ much more slowly and the rate of the reaction would be expected to be independent of $OH^-$.

155. The correct order of LMCT energies is
1. $MnO_4^- < CrO_4^{2-} < VO_4^{3-}$ 2. $MnO_4^- > CrO_4^{2-} > VO_4^{3-}$ 3. $MnO_4^- > CrO_4^{2-} < VO_4^{3-}$ 4. $MnO_4^- < CrO_4^{2-} > VO_4^{3-}$. **[NET 2011]**

**Ans. :** 1. (With the increase of oxidation state the LMCT occurs at lower energy,

$\overset{VII}{Mn}O_4^- < \overset{VI}{Cr}O_4^{2-} < \overset{V}{V}O_4^{3-}$ ).

156. For the complexes (A) $[Ni(H_2O)_6]^{2+}$, (B) $[Mn(H_2O)_6]^{2+}$, (C) $[Cr(H_2O)_6]^{3+}$ and (D) $[Ti(H_2O)_6]^{3+}$, the ideal octahedral geometry will not be observed in
1. (A) and (D) 2. (C) and (D) 3. (B) only 4. (D) only. **[NET 2011]**

**Hint :** 3. (B) only (Sec. 7.14.5(C)).

157. Write the d-orbital energy splitting pattern for a $\overset{II}{Cu}L_5$ complex having the unpaired electron in the $d_{z^2}$ orbital. **[NET 2000]**

158. Find out the Russel-Saunders ground state term for $Nb^{3+}$ ion and note down the total degeneracies. **[B.U. 2012]**

**Hint :** $Nb^{3+}$ has the electronic configuration $[Kr]4d^2$

| $m_l$ = | +2 | +1 | 0 | –1 | –2 |
|---|---|---|---|---|---|
| | ↑ | ↑ | | | |

$M_L = 3$ i.e. $L = 3$, $M_S = \frac{1}{2} + \frac{1}{2} = 1$ so that $S = 1$

$$2S + 1 = 2 \times 1 + 1 = 3$$

$J = (L+S)$ to $(L–S) = (3+1)$ to $(3–1) = 4$ to $2$

As the configuration is $d^2$ i.e. less than half-filled so $J = 2$.

Therefore the ground term is $^{2S+1}L_J$ i.e. $^3F_2$.

Total degeneracies = $(2L+1)(2S+1) = (2\times3+1)(2\times1+1) = 21$.

159. Define spectator ligand. With a suitable example state how spectator ligands may effect rates of substitution reactions. **[B.U. 2012]**

**Hint :** Sec. 7.19.10. This ligand does not take part in substitution reaction but which can influence the course of the reaction through cis-trans effect to the leaving group producing selective reaction is called spectator ligand. The rate of the substitution reaction is influenced by spectator ligand in two ways (a) steric effect and (b) electronic effect. cis isomer will react more rapidly than trans through cis effect (Sec. 7.19.10).

160. Using CFSE indicate spinels to be normal or inverse : $MnCr_2O_4$, $NiFe_2O_4$. **[C.U. 2015; B.U. 2012, 2016]**

**Hint :** $MnCr_2O_4$ is normal spinel while $NiFe_2O_4$ is inverse spinel. Sec. 7.14.5(B).

**$MnCr_2O_4$ :** For normal spinel structure : $[\overset{Td}{Mn(II)}\,\overset{O_h}{Cr_2(III)}]O_4$

∴ Total CFSE = –24 Dq

For inverted spinel structure : $[\overset{Td}{Cr(III)}\,\overset{O_h}{Mn(II)}\,\overset{O_h}{Cr(III)}]O_4$

∴ Total CFSE = –15.56 Dq

As CFSE for normal spinel structure is greater than that of inverted spinel structure, therefore $MnCr_2O_4$ exist as normal spinel structure.

**$NiFe_2O_4$ :** For normal spinel structure : $[\overset{Td}{Ni(II)}\ \overset{O_h}{Fe_2(III)}]O_4$

Total CFSE = –6.56 Dq

For inverted spinel structure : $[\overset{Td}{Fe(III)}\ \overset{O_h}{Ni(II)}\ \overset{O_h}{Fe(III)}]O_4$

Total CFSE = –12 Dq

Therefore $NiFe_2O_4$ exist as inverted spinel as CFSE(IS) > CFSE(NS).

161. Depending on reaction conditions octahedral substitution reactions may acquire an associative path. **[B.U. 2012]**

**Hint :** Solved problem 7.75.

162. Silver perchlorate is significantly more soluble in benzene than in n-hexane—explain. **[B.U. 2012, 2014]**

**Hint :** $Ag^+$ is a soft acid which interacts with the π-electrons of the benzene which is a soft base to form the complex species $[Ag–C_6H_6]^+$. This type of interaction is not possible for n-hexane which is devoid of π-electrons.

163. "Tetracoordinated boron compounds are always tetrahedral but never square planar"—comment. **[B.U. 2012]**

**Hint :** B has the electronic configuration $1s^22s^22p^1$ having outer quantum shell n=2 which is devoid of d-orbital. For square planar geometry ($dsp^2$ hybridisation) the involvement of d-orbital is necessary and thus tetracoordination of boron is always exhibited by tetrahedral geometry.

164. Explain the phenomena of linkage isomerism giving examples with thiosulphate as ligand. **[C.U. 1998; B.U. 2012]**

**Hint :** Problem number 26.

165. Write the IUPAC name of the following compound 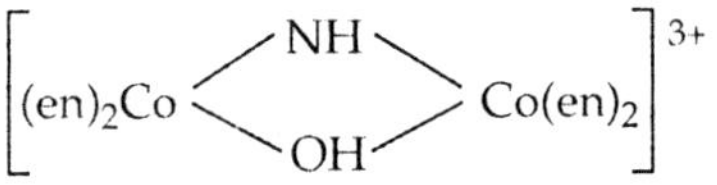 . **[C.U. 2012]**

**Ans. :** μ-imido-μ-hydroxidotetrakis(ethane-1,2-diamine)dicobalt(III)

or

μ-imido-μ-hydroxidotetrakis(ethane-1,2-diamine)dicobalt(3+)

166. The electronic spectrum of $[CrF_6]^{3-}$ shows three bands at 14900 $cm^{-1}$, 22400 $cm^{-1}$ and 34800 $cm^{-1}$. The value of $\Delta_0$ in this case is—
1. 5500 $cm^{-1}$, 2. 14900 $cm^{-1}$, 3. 22400 $cm^{-1}$, 4. 34800 $cm^{-1}$. **[NET 2011]**

**Ans. :** 2. Problem 7.78.

167. $[Pt(NH_3)_2Cl_2]$ exists in two geometrical isomeric forms. How one can distinguish them by one physical and one chemical tests? **[B.U. 2013]**

**Hint :** Sec. 7.9.12 and Question number 12.

168. Predict the sites of $SCN^-$ ligand suitable for complexing with $Fe^{3+}$ and $Ag^+$ respectively. Justify your answer.

The ligand $SCN^-$ also forms isomeric complexes with a particular metal atom using its two sites. Cite example of such a case. What is the type of isomerism? **[B.U. 2013]**

**Hint :** $Fe^{3+}$ is a hard acid complexing with (EO) mode with N atom and $Ag^+$ is a soft acid complexing with the same (EO) mode through S atom of $SCN^-$.

The isomeric complex with $SCN^-$ is $[Pd(bpy)(SCN)_2]$ and $[Pd(bpy_3)(NCS)_2]$. This is a case of structural, salt or linkage isomerism (Sec. 7.9.5).

169. Draw the possible geometrical isomers of $[Co(en)(NH_3)_2BrCl]^+$ and hence predict which of them would be optically active. **[C.U. 2013]**

**Hint :** This is a $M(AA)a_2bc$ type complex which exhibits 4 geometric isomers of which two are optically active where AA is a symmetrical bidentate ligand.

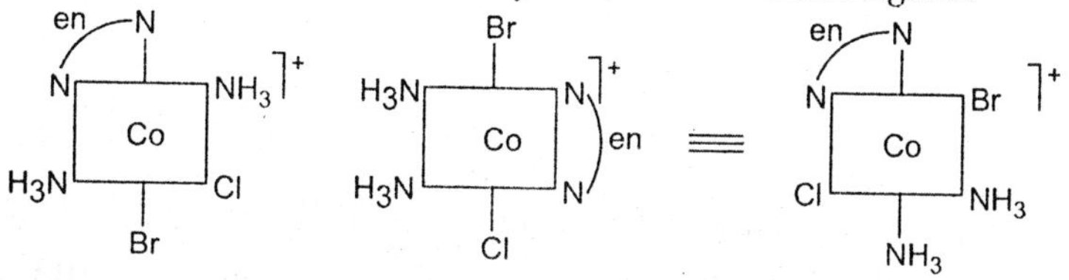

(One basal plane contains 3N, Cl), $2NH_3$ (trans), Br and Cl (cis). Basal plane containing en, Br and Cl (i.e. $H_3N$–Co–$NH_3$ axial direction) is the plane of symmetry.

(Basal plane contains 4N); $2NH_3$ (cis),, Br and Cl (trans). Vertical plane containing Br–Co–Cl and perpendicularly bisecting en is a symmetrical plane.

The opitcally active isomers place two different monodentate ligands, $2NH_3$ and en at the cis positions i.e.

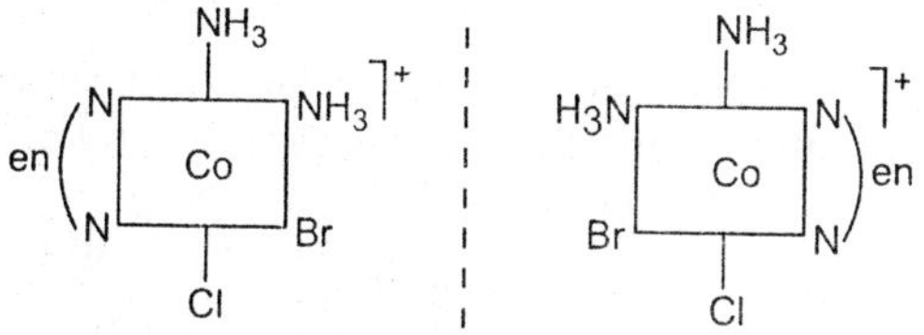

170. Rationalise the bond lengths of CO (1.128Å) and $CO^+$ (1.115Å) with the help of MO diagram. Explain the ligating behaviour of CO. **[C.U. 2013]**

**Hint :** Sec. 3.4.5 and Sec. 3.4.6.

CO has electron distribution $1\sigma^2 2\sigma^2 1\pi^2 1\pi^2 3\sigma^2$ or $\sigma(sp)_1^2\ \sigma^*(sp)_1^2\ \pi 2p_{y^2}\ \pi 2p_{z^2}\ \sigma(sp)_2^2$ and the $BO = \frac{8-2}{2} = 3$

In $CO^+$ the $BO = \frac{7-2}{2} = 2.5$ i.e. BO decreases.

So according to MO theory the BO of $CO^+$ decreases, so it is expected that bond length of $CO^+$ will increase compared to that of CO. But actually the bond length of $CO^+$ (1.115Å) is somewhat shorter compared to that of CO (1.128Å). This can be explained in the following way.

CO is not a homonuclear molecule like $C_2$, $N_2$ etc. and there is an electronegativity difference between C and O. So there is a large discrepancy in the atomic energy levels of 2s and 2p electron of C and O. As a result $2p(pi)_x$, $2p(pi)_y$ and $2p(sigma)_z$ have lower energy from 2s (sigma*). So the electron lost is from $2s(sigma^*)^2$ and not $2p(sigma)_z$ and hence the bond order increases from 3 to 3.5 and not decreases to 2.5. Thus simply during ionisation of CO the electron is lost from $\sigma^*(sp)_1$ or from $2\sigma$ thereby the $BO = \frac{8-1}{2} = 3.5$ and thus the bond length decreases in case of $CO^+$ compared to that of CO.

The CO is ligated to metal in most of the cases through C to form the carbonyls e.g. $Fe(CO)_5$, $Ni(CO)_4$ since O is more electronegative than C. Moreover CO is a $\pi$-acid ligand so CO will stabilise the low oxidation states of metals through synergic effect i.e. $\sigma/\pi$ interaction such as :

$$M \underset{\pi}{\overset{\sigma}{\rightleftarrows}} CO$$

171. Acetylacetone is a potential ligand that forms a square planar complex with Cu(II). Draw the structure of the complex and predict the formal charge on the complex. **[C.U. 2013]**

**Hint :** Acetlylacetone is a bidentate monobasic acid ligand, so it will neutralise the charge of Cu(II) when two ligands attached to Cu(II).

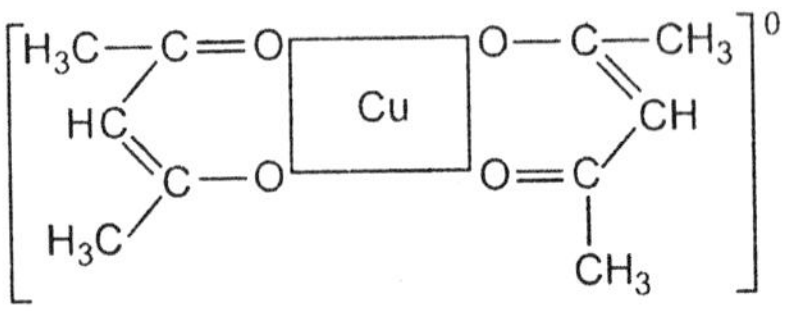

172. "In the reaction $Ma_4b_2 \xrightarrow[-a]{+b} Ma_3b_3$, two geometric isomers of the product are formed". Assign the stereochemistry of the substrate and the product. **[B.U. 2013]**

**Hint :**

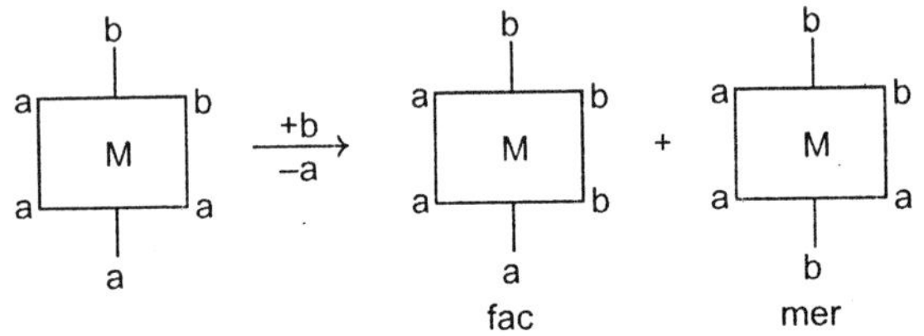

173. "It is now firmly established that colour and magnetic moment of transition metal complexes are due to their possessing of d-orbital electrons; so a quantitative connection between spectra and magnetic moment must exist". Write down the expression connecting these two molecular properties. **[B.U. 2013]**

**Hint :** The relationship between spectra and magnetic moment of a complex is given by,

$$\mu_{eff} = \mu_{spin\ only}\left[1 - \alpha\frac{\lambda}{10Dq \text{ or } \Delta_0}\right] \text{ (Sec. 10.9.3)}$$

where $\alpha$ = ground term constant. $\alpha = 2$ for E ground term, 4 for $A_2$ term and 0 for $A_1$ term; $\lambda$ = spin-orbit coupling constant for the term involved; 10D or $\Delta_0$ = energy separation between the ground term and the first excited state.

174. In its electronic spectrum $KMnO_4$ exhibits a band at 18000 $cm^{-1}$, while $K_2CrO_4$ shows its spectrum at 26000 $cm^{-1}$. Express the band in nm, assign the transition and explain the trend. **[B.U. 2013]**

**Hint :** Problem 7.95. 1 nm = $10^{-9}$ m, $\therefore$ 1 m = $10^9$ nm

For $KMnO_4$, $1/\lambda = 18000\ cm^{-1}$; $\therefore \lambda = 5.56\times10^{-5}$ cm = $5.56\times10^{-7}$ m

$= 5.56\times10^{-7}\times10^{9}$ nm = 556 nm.

For $K_2CrO_4$, $1/\lambda = 26000\ cm^{-1}$, i.e. $\lambda = 385$ nm.

The greater is the oxidising power of the central metal ion, the greater will be the electron drainage from periphery resulting in lower energy band position in LMCT transition. Mn(VII) > Cr(VI) (both $3d^{\circ}$ system) as oxidising agent so LMCT will occur at lower energy in case of former.

175. $MnCr_2O_4$ is likely to have normal spinel structure because—

1. $Mn^{2+}$ will have a LFSE in the octahedral site whereas $Cr^{3+}$ will not.
2. Mn is in +2 oxidation state and both the Cr are in +3 oxidation state.
3. Mn is in +3 oxidation state and 1 Cr is in +2 and the other is in +3 state.
4. $Cr^{3+}$ will have a LFSE in the octahedral site whereas the $Mn^{2+}$ ion will not. **[NET 2013]**

**Hint :** Sec. 7.14.5(B). The answer is 4 because in case of normal spinel the trivalent ions will occupy in the octahedral site.

176. Write down the structural formulations of sodium nitroprusside and brown ring compound and find out the oxidation states of metal ions and ligands. Comment on their magnetic properties and electrophilic/nucleophilic behaviours. **[B.U. 2013]**

**Hint :** The sodium nitroprusside has the formula $Na_2[Fe(CN)_5NO]$ which has octahedral geometry ($O_h$). The oxidation state of the metal is +2 while the ligands $CN^-$ and $NO^+$ has the oxidation states –1 and +1 respectively. This complex is diamagnetic having $d^2sp^3$ hybridisation.

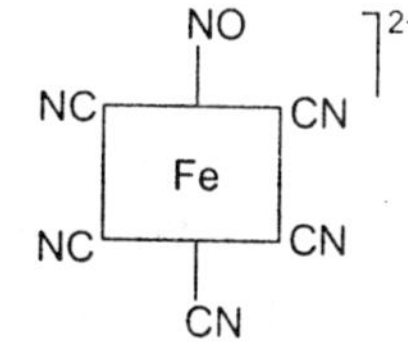


The brown ring compound has the composition $[Fe(H_2O)_5NO]SO_4$ which has also $O_h$ geometry. The oxidation state of the metal is +1 while the ligands $H_2O$ and $NO^+$ corresponds to the oxidation states 0 and +1 respectively. This complex is paramagnetic corresponding to three unpaired electrons having $sp^3d^2$ hybridisation (Problem 7.51). Both the complexes electrophilic as $NO^+$ present.

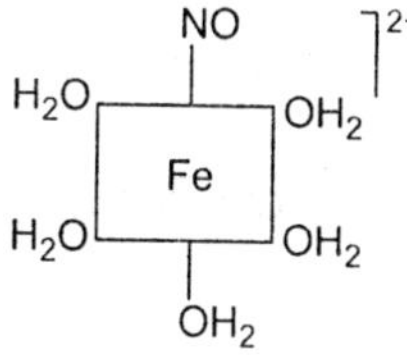


177. Define cis effect and trans effect with suitable example. Which one is stronger? **[B.U. 2013]**

**Hint :** The ability of a group (spectator ligand) to direct substitution in the position **cis** to itself is known as **cis effect.** The ability of a group (spectator ligand) to direct substitution in the **trans** position is known as **trans effect.**

cis/trans $[Co(en)_2(A)Cl]^+ + H_2O \longrightarrow [Co(en)_2(A)(OH_2)]^+ + Cl^-$

$A = OH^-, Cl^-, NCS^-$

The rate constant (k) of this reaction is measured as

| A | cis $[10^5 \times k(s^{-1})]$ |
|---|---|
| $OH^-$ | 1200 |
| $Cl^-$ | 24 |
| $NCS^-$ | 11 |

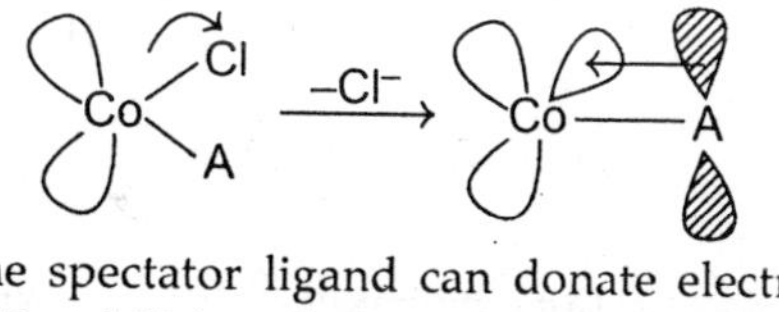


In cis form the spectator ligand can donate electron to vacated $d^2sp^3$ hybrid orbital formed after removal of $Cl^-$ stabilising the transition state (TS). $OH^-$ is best $\pi$ donor among the three ligands mentioned above, so it can donate electron effectively increasing the rate to the greater extent.

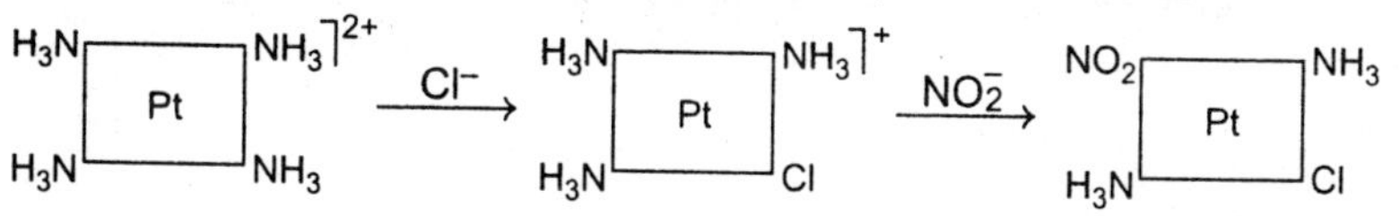


trans form

**Trans effect** is much stronger than **cis effect** because association of incoming group in position **trans** to trans directing group becomes facile due to electron deficiency in this position propounded in $\pi$-bonding theory of square planar complexes where the metal ion is coordinatively unsaturated.

178. How many isomers are possible for $[Pt(NH_3)(py)(NO_2)Br]$ (py = pyridine)? Suggest the methods for their synthesis. **[B.U. 2013]**

**Hint :** Question number 29.

Six isomers are possible–three geometric isomers with corresponding linkage isomers for $NO_2^-$/ $ONO^-$.

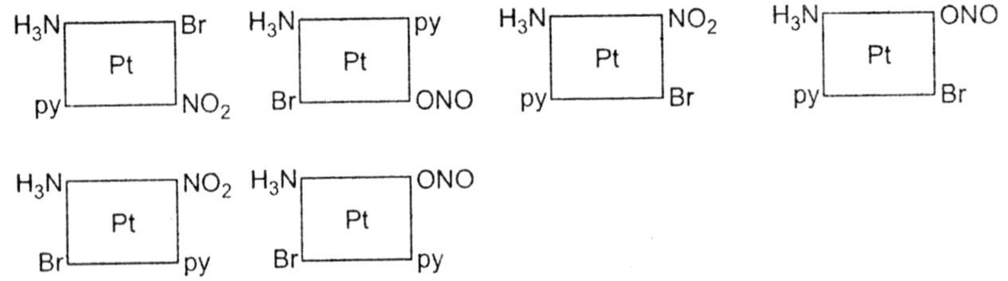

The isomers are synthesised by

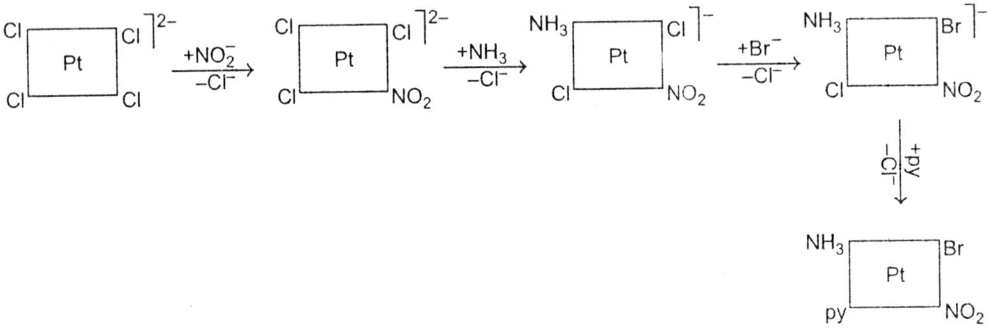

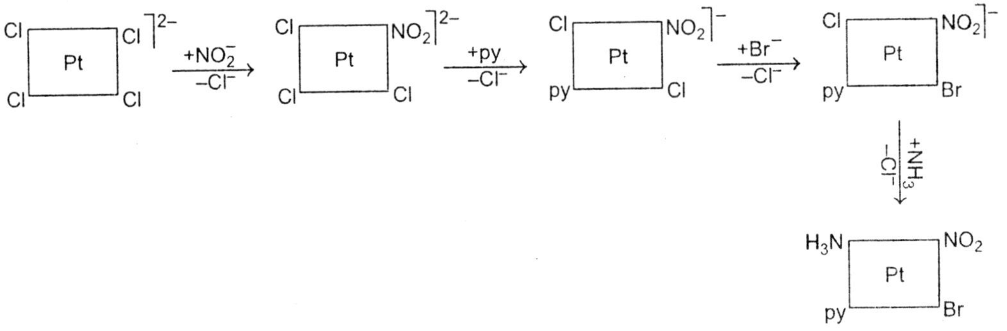

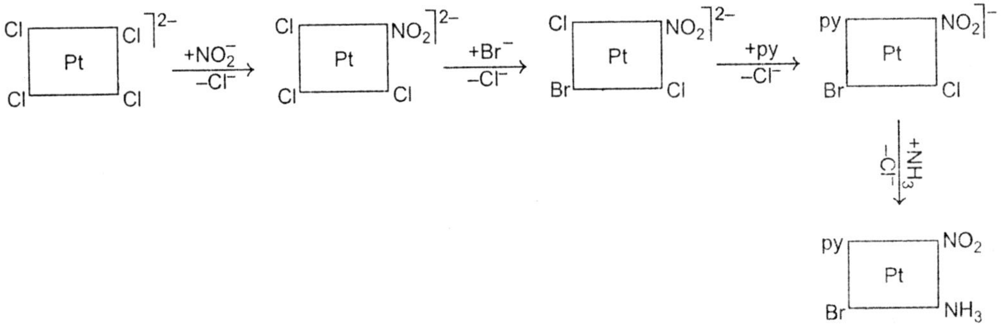

In preparing the isomers, the trans effect and bond strength order govern the reaction course :
Trans-effect order : $NO_2^- > Br^- > Cl^- >$ py

Bond strength order : Pt-$NH_3$ > Pt-py > Pt-Br > Pt-Cl

179. The most appropriate structure for the complex $[Pt_2(NH_3)_2(NCS)_2(PPh_3)_2]$

1. 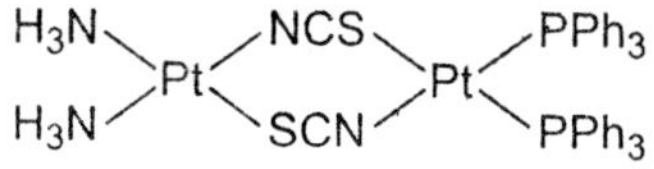


2. 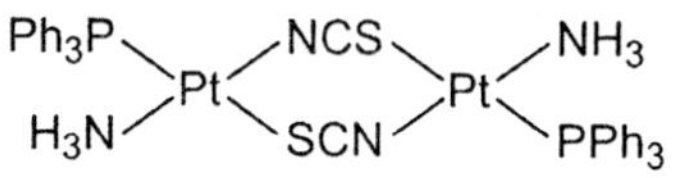


3. 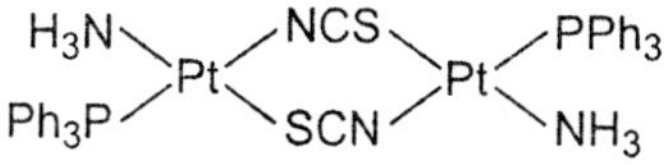


4. 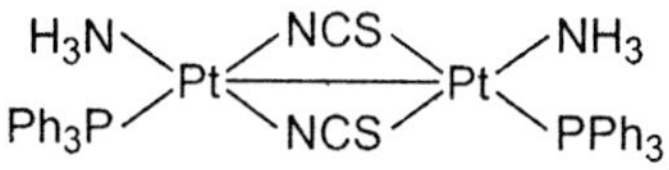


[NET 2013]

**Ans.** : 3 i.e. 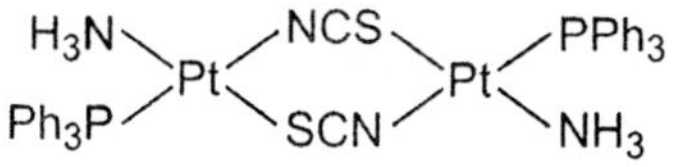


The trans structure is more stable than cis structure because in case of trans structure the repulsive force is more minimised compared to cis structure. The answer may also be 2.

180. The structure of ethylene molecule in Zeise salt is not planar. Explain. **[B.U. 2013]**

**Hint** : The Zeise's salt (Sec. 3.5(c)) has the following structure :

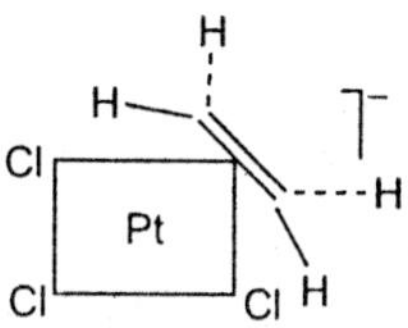


The π-electron cloud of filled π bonding MO of $C_2H_4$ is donated to a vacant $dsp^2$ hybrid orbital of Pt(II) forming a μ-bond. On the other hand Pt(II) to $C_2H_4$ π back bonding occurs from overlap of filled unhybridised d-orbital to vacant π* orbital on $C_2H_4$. This type of bonding keeps C=C bond of $C_2H_4$ perpendicular to the molecular plane. Though $C_2H_4$ is planar in free state but on complexation the molecule gets a new conformation in which the plane containing C=C does not contain the four H atoms which again constitute a new plane apart from the Pt–C bond axis so that no interaction between d-electrons and hydrogen atoms of $C_2H_4$ can decrease stability.

181. Construct a table showing salient distinctions between d–d transitions and charge transfer transition. **[B.U. 2013, 2016]**

**Hint** : Sec. 7.15.11.

182. **(a)** The electronic spectrum of $Ti(H_2O)_6^{3+}$ shows a broad band having peak at 20100 $cm^{-1}$ and a shoulder at 17400 $cm^{-1}$. Assign the ligand field transitions for both the peak and the shoulder for $H_2O$ as ligand. **(b)** Using trans effect phenomenon prepare all possible sterioisomers of $PtCl(NH_3)(py)Br$ starting from $PtCl_4^{2-}$. **[C.U. 2013]**

**Hint** : **(a)** For $d^1$ system Ti(III) the ground term is $^2D$ which gives rise to crystal field terms $^2T_{2g}$ and $^2E_g$. The $^2T_{2g}$ term is the ground state which is triply degenerate and $^2E_g$ term is doubly degenerate, thus only one transition is expected for $[Ti(H_2O)_6]^{3+}$ at ~20100 to 20400 $cm^{-1}$ (Sec. 7.15.10(A)). This absorption shows a shoulder at ~17400 to 17500 $cm^{-1}$. The only transition from $^2T_{2g} \longrightarrow {}^2E_g$ makes the aqueous Ti(III) complexes purple.

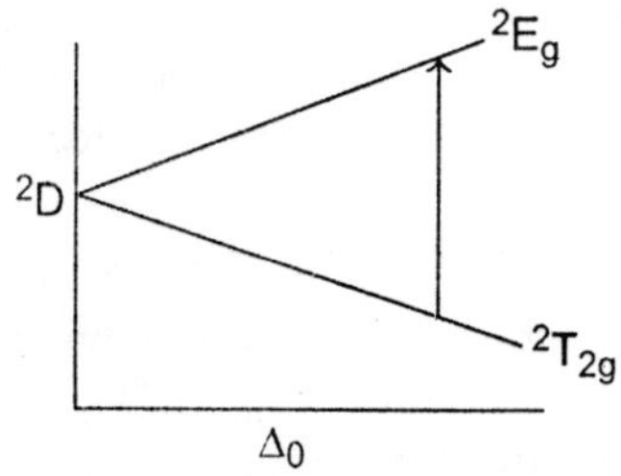


However both the ground term $^2T_{2g}$ and the excited $^2E_g$ terms are Jahn-Teller susceptible and the shoulder arises due to Jahn-Teller distortion of the excited $^2E_g$ state (arised from configuration $e_g^1$).

In general the splitting of the ground term is of the order 420-800 $cm^{-1}$ while that of the excited state is of the order of 1000-4500 $cm^{-1}$. Therefore instead of sharp single absorption band ($^2T_{2g} \longrightarrow {}^2E_g$) a broad asymmetric band covering the following close transition occurs due to tetragonal elongation :

$$^2B_{2g} \xrightarrow{\nu_1} {}^2B_{1g} \text{ i.e. } d_{xy} \longrightarrow d_{x^2-y^2} \text{ at } 19500 \text{ cm}^{-1}$$

$$^2B_{2g} \xrightarrow{\nu_2} {}^2A_{1g} \text{ i.e. } d_{xy} \longrightarrow d_{z^2} \text{ at } 21000 \text{ cm}^{-1}$$

The main peak at 21000 $cm^{-1}$ showing a shoulder overlap to produce a resultant broad peak having the average maximum at about 20100-20400 $cm^{-1}$. The splitting of the $^2T_{2g}$ and $^2E_g$ terms and the corresponding transitions are given below :

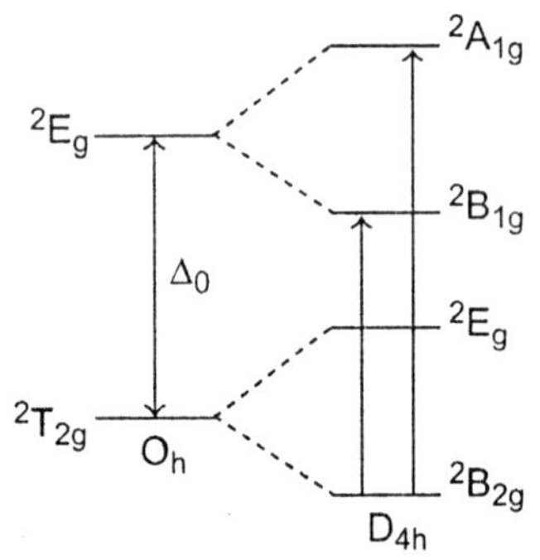


**(b)** Sec. 7.19.14 and Question number 178.

183. **(a)** How will you chemically differentiate between cis and trans isomers of $[Pt(NH_3)_2Cl_2]$? **(b)** $[FeF_6]^{3-}$ is colourless but $[Fe(SCN)_6]^{3-}$ is blood red though both are $d^5$ complexes? Explain. **[C.U. 2013]**

**Hint :** (a) Sec. 7.9.12 and Question number 12.

(b) Problem 7.67. In$[FeF_6]^{3-}$ the CFSE = 0 because $Fe^{3+}$ is $3d^5$ system and due to the weak field character of $F^-$ the $[FeF_6]^{3-}$ is a high spin complex. Here $\Delta S \neq 0$ and$\Delta L \neq \pm 1$ i.e. doubly forbidden, so that $[FeF_6]^{3-}$ is colourless.

$[Fe(SCN)_6]^{3-}$ is an example of MLCT (Question number 88) and Sec. 7.15.11. We know that charge transfer transition will produce intense colour compared to that of d–d transition. Here the deep red colour of $[Fe(SCN)_6]^{3-}$ is due to the transitions $E^*_g \rightarrow \pi^*$ and $T_{2g} \rightarrow \pi^*$.

184. **(a)** What is a tetragonal distortion? Which $d^n$ configuration would lead to weak and strong Jahn-Teller distortion in octahedral complexes? **(b)** For the complex $[Mn(H_2O)_6]^{3+}$ the mean pairing energy P is found to be 28800 $cm^{-1}$. The magnitude of 10Dq or $\Delta$ is 2100 $cm^{-1}$. Calculate crystal field stabilisation energy for the complex corresponding to high spin and low spin state. **[C.U. 2013]**

**Hint :** (a) Sec. 7.14.2(D) and Sec. 7.14.5(C).

(b) Problem 7.45 and Problem 7.46.

185. **(a)** Determine the spectroscopic ground state for $Cr^{3+}$. What is a spin forbidden electronic transition? **(b)** Hard base $F^-$ occupies the weaker end while neutral CO is located in the stronger end of the spectrochemical series. Explain. **[C.U. 2013]**

**Hint :** (a) Spectroscopic ground term for $Cr^{3+}$ is $^4F_{3/2}$ (Sec. 7.15.6). According to spectroscopic selection rule the electronic transitions occur between the terms having same spin multiplicity i.e. $^3T_{1g} \rightarrow {}^3T_{2g}$ transition but the spin forbidden transition is $^3T_{1g} \rightarrow {}^1T_{2g}$ i.e. spin forbidden transiton is exhibited by terms having different spin multiplicity.

(b) Sec. 7.14.4 and Problem 7.43.

186. (a) $[Fe(CN)_6]^{4-}$ and $[Fe(H_2O)_6]^{2+}$ are octahedral complexes of Fe(II). Comment on their (i) magnetic properties, (ii) colour feature. (b) Cu(II) acetate monohydrate shows subnormal magnetic moment. Explain. **[B.U. 2013]**

**Hint : (a)** $[Fe(CN)_6]^{4-}$, here Fe(II) is $3d^6$ and the complex is **low spin** having $t^6_{2g}$ configuration which is **diamagnetic**. The ground term is $^1A_{1g}$ and the allowed transitions are $^1A_{1g} \rightarrow {}^1T_{1g}$ (31.00 kK) and $^1A_{1g} \rightarrow {}^1T_{2g}$ (37.04 kK) occur within visible and the near UV range. The complex is intensely coloured due to charge transfer transitions which frequently obscure the two predicted above spin allowed transitions.

The complex $[Fe(H_2O)_6]^{2+}$ is high spin ($t^4_{2g}e^2_g$) which produce $^5T_{2g}$ ground term and $^5E_g$ excited state separated by 10Dq or $\Delta_0$. $[Fe(H_2O)_6]^{2+}$ ion gives pale green solution and there is weak absorption band in the red at ~10000 $cm^{-1}$ which is assigned due to the transition $^5T_{2g} \rightarrow {}^5E_g$. More corectly the absorption consists of a peak at 10400 $cm^{-1}$ and a shoulder at 8300 $cm^{-1}$. This complex undergoing Jahn-Teller distortion and is trigonally distorted having local symmetry $D_{3d}$. The $^5E_g$ term is subjected Jahn-Teller distortion resulting the splitting the term into $^5A_1 + {}^5B_1$ in $D_{4h}$ symmetry. Thus the degeneracy of the $^5E_g$ is lifted which is the only possible mechanism for the splitting of the spin allowed transition. The splitting of the Fe(II) complex is depicted below :

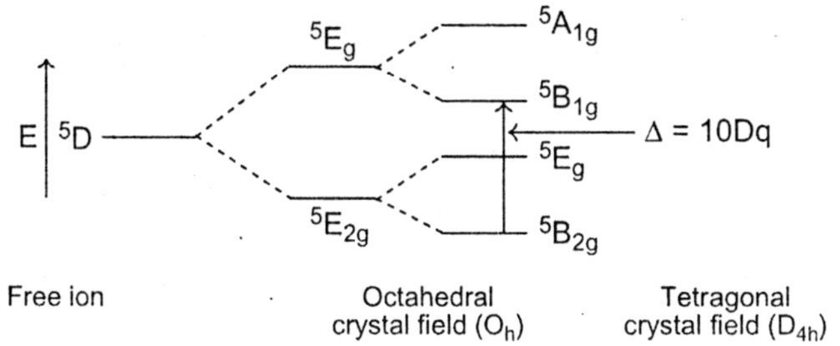


(b) Sec. 3.5 and Sec. 10.9.4 (Vol. II).

187. State the essential characteristics of a ligand to form an inner metallic complex. Show that acetylacetone will form such a complex of first order with $M^{3+}$ ion and the second order with $M^{4+}$ ion. **[B.U. 2014]**

**Hint :** Sec. 7.10 and Problem 7.22. acacH is a bidentate monobasic acid ligand so that it will form first order inner metallic complex with $M^{3+}$ ion having composition $M(acac)_3$ where charge and coordination number of $M^{3+}$ is internally copensated. In case of $M^{4+}$ ion the tris complex will satisfy the coordination number but not the charge so that the composition will be $[M(acac)_3]^+$.

188. Cite one example of an organometallic compound. **[B.U. 2014]**

**Hint :** $Ni(CO)_4$, $Fe(CO)_5$, $CH_3MgBr$.

189. Indicate the Lewis electron dot structure and possible ligational modes of 2, 2′ bipyridine. **[B.U. 2014]**

**Hint :** 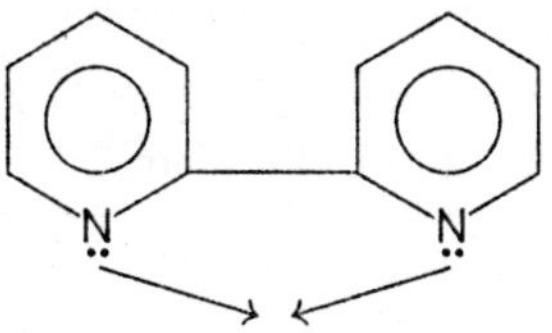 Moreover π–electron of two pyridine rings may function to metal.

190. Which of the two compounds $B(CH_3)_3$ and $B(i\text{–}C_3H_7)_3$ will form stronger complexes with $NH_3$ and why? **[B.U. 2014]**

**Hint :** The inductive effect of $i\text{–}C_3H_7$ is greater than that of $CH_3$ so that $B(CH_3)_3$ is acting as a stronger Lewis acid than that of $B(i\text{–}C_3H_7)_3$. Thus $B(CH_3)_3$ will form stronger complex with $NH_3$ compared to that of $B(i\text{–}C_3H_7)_3$.

191. '$PbCl_2$ is white while $PbI_2$ is colourd'.—Explain. **[B.U. 2014]**

**Hint :** $PbI_2$ is coloured due to LMCT (Sec. 7.15.11). $PbI_2$ is more covalent thatn $PbCl_2$ due to more polarisability of $I^-$ than that of $Cl^-$. Electronic transition i.e. LMCT occurs from $I^-$ to $Pb^{2+}$ or from HOMO to LUMO. The energy is available in the visible region while for $PbCl_2$ the required energy for the corresponding transition lies in the ultraviolet region.

192. Between **cis** and **trans** isomeric forms of $Co(en)_2Cl_2^+$ (en = ethylenediamine) which one will show optical activity? How the enantiomers may be separated? **[B.U. 2014]**

**Hint :** Question number 11 and 36. The cis isomer (d and l form) will show optical activity.

193. Draw geometrical isomers with coordination numbers 4 and 6 with specific examples. **[B.U. 2014]**

**Hint :** Sec. 7.9.10 and Sec. 7.9.11.

194. The spin-only magnetic moment and spectroscopic ground state term symbol of magnanese centre in $[MnF_6]^{3-}$ ion respectively are—
1. 4.9 BM and $^5D$, 2. 4.9 BM and $^4F$, 3. 3.9 BM and $^3D$, 4. 4.9 BM and $^3F$. **[NET 2014]**

**Hint :** Sec. 7.15.6 for ground state term.

In this complex the central metal ion is $Mn^{3+}$ which is $3d^4$ system

| $m_l =$ | +2 | +1 | 0 | −1 | −2 |
|---|---|---|---|---|---|
| | ↑ | ↑ | ↑ | ↑ | |

$$M_L = +2 + 1 + 0 - 1 = +2,\ L = 2$$

$$M_s = +\frac{1}{2} + \frac{1}{2} + \frac{1}{2} + \frac{1}{2} = \frac{4}{2} = 2,\ S = 2$$

$$2S + 1 = 2\times 2 + 1 = 5$$

$$J = L - S = 2 - 2 = 0$$

Therefore term symbol will be $^5D_0$.

$F^-$ is a weak field ligand so that hybridisation will be $sp^3d^2$ whereby there will be four unpaired electrons.

$\therefore$ Spin-only magnetic moment ($\mu$) = $\sqrt{n(n+2)}\beta = \sqrt{4(4+2)}\beta = \sqrt{24}\beta$ = 4.9 BM.

Thus the answer of the question is 1 i.e. 4.9 BM and $^5D$.

195. Tabulate the factors which affect rate laws and mechanism in inorganic substitution reactions. **[B.U. 2014]**

**Hint :** Factors affecting rate laws and mechanism are :

(a) d-electron configuration of metal ion.

(b) Mechanistic switch-over is observed along a period and along a group (Problem 7.75) i.e. periodic series of the metal ion.

(c) Overall charge of the complex ion.

(d) Concentration of the ligand.

(e) Nature of incoming, leaving and spectator ligands.

(f) Dielectric constant of the solvent leading to particular solvent structure around coordination sphere.

(g) Ionic strength of the medium.

(h) Temperature and external presure on the medium.

(i) Steric effect of the actual reaction.

(j) pH of the medium and effect of catalyst on the reaction.

196. Predict and explain the relative positions of fluoride ion, water and hydroxide ion in the spectrochemical series. **[B.U. 2014]**

**Hint :** Problem 7.57. $H_2O$ is a neutral primarily σ-donor ligand so that Δ value will be greater compared to that of $F^-/OH^-$. The two ligands $F^-/OH^-$ are anionic, strong σ as well as π-donor, so Δ decreases appreciably. Among the two ligands, $F^-$ has graeter electron density due to its smaller size so that Δ will be lowest for this ligand. Hence the order will be $F^- < OH^- < H_2O$.

197. What do you mean by spin crossover (SCO)? Give an example of a coordination compound showing this phenomenon. **[B.U. 2014, 2016]**

**Hint :** Problem 7.65. Coordination compounds of first transition metal ions with $d^4$ to $d^7$ configurations in an octahedral environment may exist either in high spin (h, s) or in low spin (l, s) state depending on crystal field strength. When the critical crystal field strength is reached, the compound may pass from h, s state to l, s state or vice-versa around the crossover region by changing external factors such as temperature as the difference in energy between these two spin states is in the range of thermal energy. This phenomenon is known as spin crossover (SCO)(Fig.7.8).

A good example is $3d^5$ tris (dialkyldithiocarbamato) iron(III), $\left(R_2N—C\begin{smallmatrix}\nearrow S\\ \searrow S^-\end{smallmatrix}\right)_3$ Fe (R = pyrrolidoyl).

Variable temperature magnetic susceptibility measurements show that μ = 5.7 to 5.8 B.M. at 300K, 4.7 to 5.0 B.M. at 150 K, 3.6 to 4.0 B.M. at 78K as the proportion of molecules in the high and low spin states changed with temperature.

198. An aqueous solution of $[Mn(H_2O)_6]^{2+}$ complex is pale pink in colour. The probable reasons for it are—

A. presence of $^6A_{1g}$ ground state, B. disallowed transition by spin selection rule, C. presence of $^2T_{2g}$ ground state, D. charge transfer transition. **[NET 2014]**

**Hint :** For $Mn^{2+}$ the ground state in octahedral field is $^6A_{1g}$ (Table 7.16) and the transition is spin disallowed and so the answer is A and B.

199. "A light pink aqueous solution of cobalt(II) chloride turns deep blue on addition of concentrated hydrochloric acid, but the blue solution again turns pink on addition of Zn(II) chloride". Account for the fact with reference to structural transformation. **[B.U. 2014]**

**Hint :** The light pink colour is due to the formation of octahedral complex, $[Co(H_2O)_6]^{2+}$ in aqueous solution. On addition of conc. HCl it is converted to blue tetrahedral complex $[CoCl_4]^{2-}$ i.e. $H_2[CoCl_4]$. This blue solution again turns pink on addition of $ZnCl_2$ because of the formation of the complex $[Co^{II}(H_2O)_6][Zn^{II}Cl_4]$. The octahedral complex cation $[Co^{II}(H_2O)_6]^{2+}$ is pink but the tetrahedral complex anion $[Zn^{II}Cl_4]^{2-}$ is colourless because of the $3d^{10}$ configuration of Zn(II).

200. What is meant by nucleophilicity parameter ($n_{pt}$) for the entering group? **[B.U. 2014]**

**Hint :** Problem 7.72. Ligand property of entering group influencing the rate of attack at the metal centre is known as nucleophilicity parameter ($n_{pt}$). Most of the square planar Pt(II) complexes undergo reactions in associative process as these are coordinatively unstaturated. The reaction

$$Pt^{II}L_2Cl_2 + Y^- \longrightarrow PtL_2(Cl)Y + Cl^-$$

follows a second order rate = $(k_2[Y^-] + k_1)[PtL_2Cl_2]$.

If the reaction is carried out with a standard base, say MeOH, then the second rate constant can be represented as $k_2^0$. Thus we can set up a nucleophilicity parameter

$$^nPt = \log(k_2[Y^-]/k_2^0) \text{ i.e. } \log k_2[Y^-] = \log k_2^0 + {}^nPt.$$

201. 'In the series nickel(II), palladium(II) and platinum(II), only platinum(II) shows significant trans-effect'. Justify. **[B.U. 2014]**

**Hint :** The group 10, congeneric elements, Ni(II), Pd(II) and Pt(II) are the members of the first (3d), second(4d) and third(5d) transition series respectively. These elements usually formed tetracoordinated (except Ni(II)) which are coordinatively unsaturated. The separations of 3d and 4p orbitals and that of 4d and 5p orbitals are very much appreciable. But the separation between 5d and 6p orbitals is sufficiently small to give rise to effective dp hybrid orbitals suitable for π-bonding with the ligands in trigonal bipyramid (TBP) transition state (TS) that is operating in Pt(II) substitution reactions. (Sec. 7.19.15). Chatt and Orgel rationalised **trans**-effect by assuming stabilisation of a TBP structure in presence of strong **trans**-directing group. The π-acceptor ligands with empty acceptor orbitals can form π-bond with filled metal $d_\pi$ orbitals stabilising the TS in which the leaving ligand (L), entering ligand(L) and *trans* -directing ligand(L) occupy equatorial positions in TBP geometry with the axial positions being occupied by the two groups that were originally **cis** to the leaving group. The π-acceptor ligands occupy the equatorial positions as these can withdraw d-electrons via $d_\pi$ orbitals. Effective hybrid dp orbitals can be involved in such π-overlap.

202. Briefly rationalise the fact that complexes of acetone are not particularly stable, but those of acetylacetone are. **[C.U. 2014]**

**Hint :** $CH_3COCH_3 \rightleftharpoons H_2C{=}C(OH){-}CH_3$ and so acetone function as monodentate i.e. $H_2C{=}C(O^{-}\rightarrow){-}CH_3$

$H_3CCOCH_2COCH_3 \rightleftharpoons H_3C{-}C(OH){=}CH{-}C(=O){-}CH_3$ and so acetylacetone function as bidentate

monobasic acid ligand i.e. $H_3C{-}C(O^{-}\rightarrow){=}CH{-}C(=O\rightarrow){-}CH_3$

As there will be the formation of chelate ring in case of acetylacetone and no such ring formation is possible for acetone, thus complexes of acetylacetone are more stable compared to that of acetone complexes.

203. 12-crown-4 binds with the alkali metal ions in the following order : $Li^+ >> Na^+ > K^+ > Cs^+$. It is due to the

1. right size of cation, 2. change in entropy being positive,

3. conformational flexibility of crown ether, 4. hydrophobicity of crown ether. **[NET 2014]**

**Ans. :** 1 i.e. right size of cation.

204. The complex $[Pt(NH_3)_2(SCN)_2]$ forms two stereoisomers, whereas the complex $[Pt(en)(SCN)_2]$ forms only one

What does this prove about the geometry of these complexes? As part of your answer sketch and name these three isomers. **[C.U. 2014]**

**Hint :** The complexes are all square planar. The $[Pt(NH_3)_2(SCN)_2]$ exists as cis and trans isomers while $[Pt(en)(SCN)_2]$ exists in only one form.

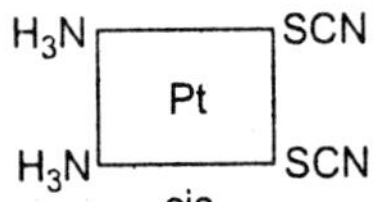


cis

cis-diaminebis(thiocyanato-κS) platinum (II)

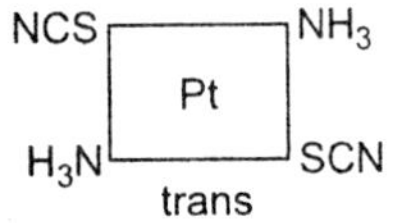


trans

trans-diaminebis(thiocyanato-κS) platinum (II)

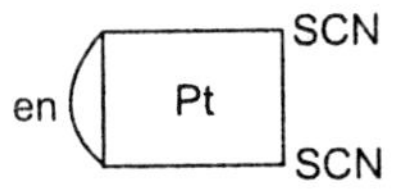


ethylenediaminebis (thiocyanato-κS) platinum (II)

205. The three dimensional structure of compound $[Co(Co(NH_3)_4(OH)_2)_3]Br_6$ has

1. twelve Co–O and twelve Co–N bonds, 2. ten Co–O and ten Co–N bonds
3. fourteen Co–O and ten Co–N bonds, 4. twelve Co–O and ten Co–N bonds. **[NET 2014]**

**Ans. :** 1 i.e. twelve Co–O and twelve Co–N bonds.

206. Predict whether the following complexes would be inert or labile, and justify your choice. The magnetic moment is given in Bohr magneton ($\mu_B$) after each complex :

| **Compound** | $\mu_B$ |
|---|---|
| Ammonium oxopentachlorochromate(V) | 1.82 |
| Potassium hexaiodomanganate(IV) | 3.82 |

**[B.U. 2014]**

**Hint :** For $(NH_4)_2[Cr^V(O)Cl_5]$, $\mu_B$ = 1.82 BM

$\therefore \sqrt{n(n+2)}$ = 1.82 and so n = 1 i.e. having one unpaired electron

In case of $K_2[Mn^{IV}I_6]$ $\mu_B$ = 3.82 Bm

$\therefore \sqrt{n(n+2)}$ = 3.82 so that n = 3 which corresponds to three unpaired electrons.

Cr(V) is $3d^1$ system which in octahedral environment has CFSE = –4Dq and Mn(IV) in the same octahedral field has CFSE = –12Dq. So for ligand substitution reaction $d^3$ system will be inert because of high CFSE value and $d^1$ system will be labile due to lower CFSE value.

207. Comment on the structure, bonding or stereochemistry as the case for the following pairs :

(i) $K_2[MCl_4]$(M = Ni and Pd); (ii) $K_3[Fe(ox)_3]$ and $[Fe(gly)_3]$;

(iii) $K[Pt(acac)_2Cl]$ and $[Me_3Pt(acac)]$ [ox = oxalate anion, gly = glycinate monoanion and acac = monoanion of pent-2, 4 dione] **[B.U. 2014]**

**Hint :** (i) As we go from the first ⟶ second ⟶ third transition series spin pairing is a common phenomena. $K_2[NiCl_4]$ has tetrahedral ($T_d$) geometry while $K_2[PdCl_4]$ has square planar geometry because Ni(II) and Pd(II) are the members of the first and second transition series respectively. The spherical surface of Pd(II) is much larger than that of Ni(II) as 4d > 3d in size thereby ligand-ligand repulsion in case of the former is more minimised in case of square planar structure with ∠LML to be 90°. But this small angle will give rise to ligand-ligand repulsion which leads to tetrahedral structure in Ni(II) complex with larger ∠LML angle value 109.5°. This is from steric factor. From view point of electronic factor, the effective nuclear charge on Pd(II) is greater, so square planar splitting is larger stabilising the system.

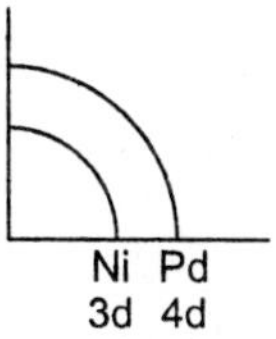


$^1/_4$th surface with ∠LML 90°.

Again $\Delta_{sq.pl} > \Delta_{Td}$, thus when interplay of steric and electronic factors are in operation, the predominating factor controls the geometry.

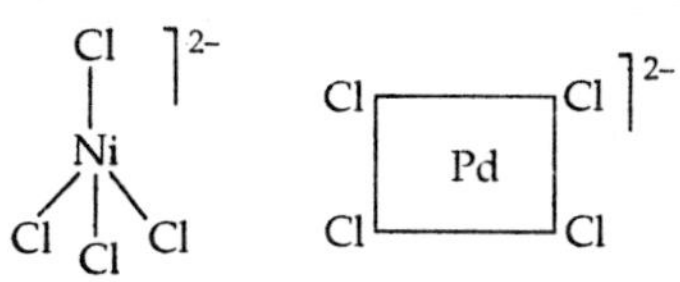


(ii) Both $K_3[Fe(ox)_3]$ and $[Fe(gly)_3]$ exist in one structural form i.e. octahedral ($O_h$). As ox is a symmetric bidentate ligand so $[Fe(ox)_3]^{3-}$ exists as d and l optical isomers. $[Fe(gly)_3]$ exists in **fac** and

**mer** isomers as gly is an unsymmetrical bidentate ligand. The complex also exists as d and l optical isomers.

$[Fe(ox)_3]^{3-}$ 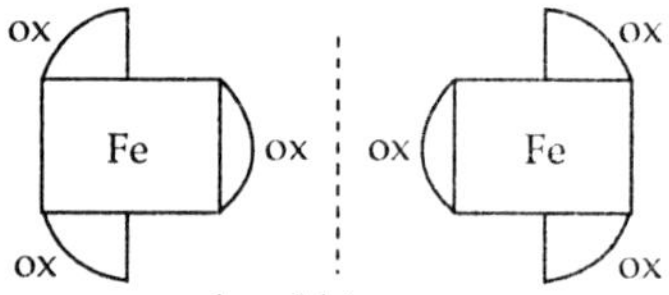


d and l isomers

$[Fe(gly)_3]$ 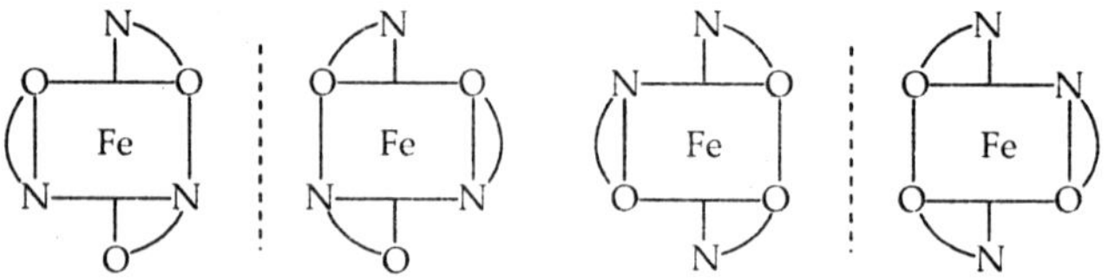


fac d and l isomers    mer d and l isomers

(iii) $K[Pt(acac)_2Cl]$, here Pt(II) has coordination number four, one acac function as bidentate monobasic (O, O) donor and other acac funciton as an organometallic, C-donor and the fourth coordination number is statisfied by $Cl^-$ ion i.e.

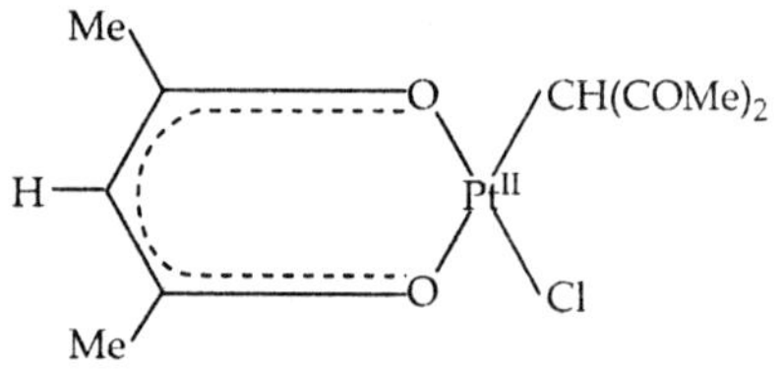


CN = 4 square planar

$[Me_3Pt^{IV}(acac)]$ is a dimer with acac behaving as a (C, O, O) bridge, each Pt(IV) is hexacoordinated.

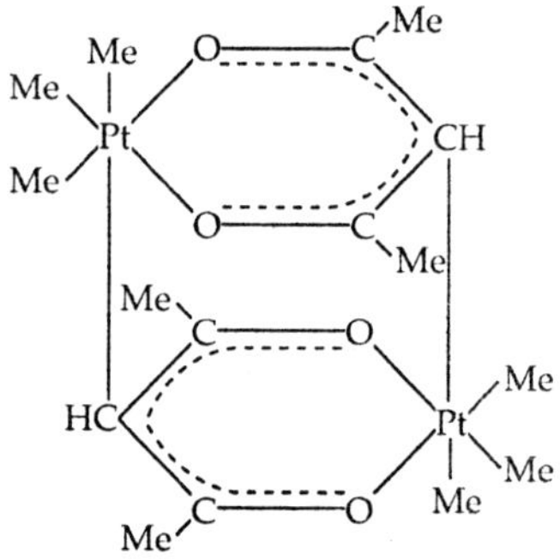


208. Arrange the ligands $H_2O$, $NH_3$, $NO_2^-$ and $Cl^-$ in order of increasing trans effect and hence design two step synthesis of the cis and trans isomers of $[Pt(NH_3)_2(NO_2)_2]$ starting from $K_2[PtCl_4]$.

**[C.U. 2014]**

**Hint :** Question number 136 and Sec. 7.19.14.

Trans effect order $H_2O < NH_3 < Cl^- < NO_2^-$

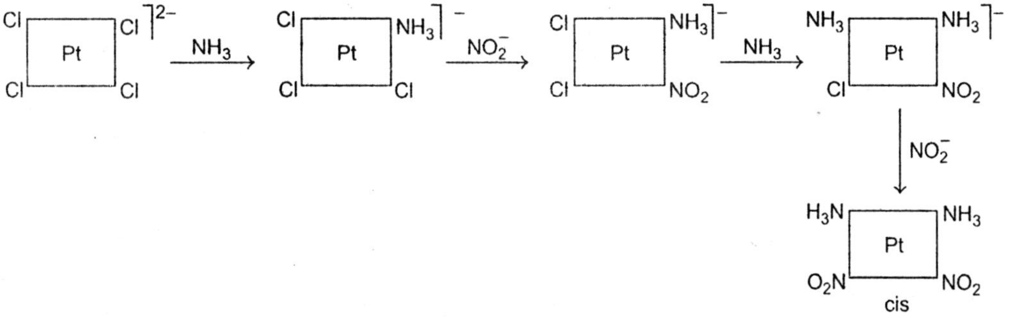


$Cl^-$ has a larger trans effect than $NH_3$ and $NO_2^-$ has a larger trans effect than either $Cl^-$ or $NH_3$. In the first step $Cl^-$ is a better leaving group than $NH_3$.

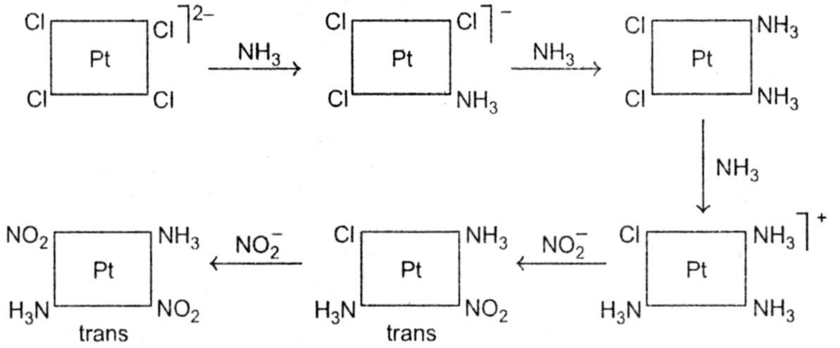


The second replacement of $NH_3$ is trans to $Cl^-$. Further replacement of $NH_3$ in the second reaction is possible.

209. Explain why a solution of $CuCl_2$ in aqueous HCl is deep greenish blue whereas similar aqueous HCl solution of CuCl is colourless. **[C.U. 2014]**

**Hint :** $CuCl_2$ in aqueous HCl forms tetrahedral $[CuCl_4]^{2-}$ complex which is intensely greenish blue. We know that in the case of octahedral complex ($O_h$) due to vibronic coupling there will be some relaxation of Laporte selection rule and thus some intensity of ligand field band is earned. In the case of noncentrosymmetric tetrahedral complex e.g. $[CuCl_4]^{2-}$, this vibronic coupling is also applicable. But in the tetrahedral symmetry ($T_d$), the more efficient mechanism is d-p mixing (mixing of the metal $t_2$ d orbitals with the ligand p orbitals) to relax the Laporte selection rule. This d-p mixing is not so efficient in the centrosymmetric octahedral complex.

Cu(II) has $^2D$ ground term which splits in tetrahedral field as $^2T_2$ and $^2E$ terms. Thus possible electronic transition for $T_d$ symmetry is $^2T_2 \longrightarrow {}^2E$ i.e. $e \longrightarrow t_2$.

Due to Jahn-Teller distortion undistorted tetrahedral complex of Cu(II) is unexpected and Cu(II) complex adopt the square planar geometry. This distortion will generate a structure of $D_{2d}$ symmetry (which is an intermediate state of the two limiting gemometries i.e. tetrahedral and square planar). For $[CuCl_4]^{2-}$ the splitting of the energy levels and d-orbitals in the $T_d$ and $D_{2d}$ symmetry are given below :

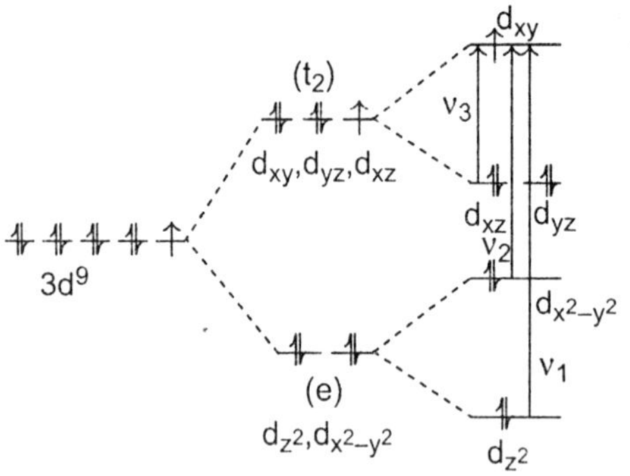


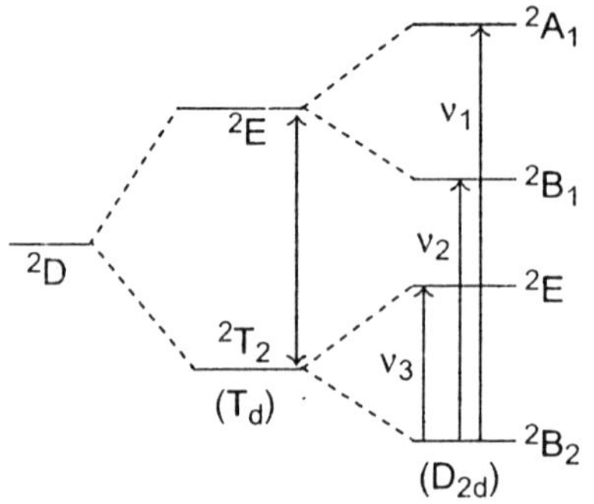


The corresponding transitions are :

$^2B_2 \longrightarrow {}^2A_1$  $\nu_1 = 9050\ cm^{-1}$

$^2B_2 \longrightarrow {}^2B_1$  $\nu_2 = 7900\ cm^{-1}$

$^2B_2 \longrightarrow {}^2E$  $\nu_3 = 5000\ cm^{-1}$

The finer resolution indicates that $\nu_3$ band consists of two components (5,500 $cm^{-1}$ and 4,800 $cm^{-1}$). It is suggested that further distortion leads to $D_{2d}$ to $C_s$ symmetry through the splitting of the $d_{xz}$ and $d_{yz}$ orbitals. This splitting leads to the following transitions :

$d_{xz} \longrightarrow d_{xy}$ (5,500 $cm^{-1}$); $d_{yz} \longrightarrow d_{xy}$ (4,800 $cm^{-1}$)

Due to the spin-orbit coupling effect the $^2E$ level further split. This can explain the possible four transitions in case of $[CuCl_4]^{2-}$. The orbitally forbidden transition $^2B_2 \longrightarrow {}^2B_1$ gains some intensity through the mixing with one component of the $^2E$ state. Thus the spectral features of $[CuCl_4]^{2-}$ may be explained by the spin orbit coupling and/or further distortion of $D_{2d}$ to $C_s$ symmetry.

In case of CuCl which forms a linear complex $[CuCl_2]^-$ with HCl and this will be colourless as Cu(I) is a $3d^{10}$ system.

210 **(a)** The two octahedral complex ions are $[CoF_6]^{3-}$ and $[Co(NH_3)_6]^{3+}$. Find out the number of unpaired electrons and the magnetic moment (μ).

**(b)** The magnetic moment of $K_3[CoF_6]$ is 4.2 BM and that of $K_3[CuF_6]$ is 2.8 BM.—Explain.

**[C.U. 2014]**

**Hint :** (a) In both the complexes the central metal ion is Co(III) which is $3d^6$ system. As $F^-$ is a weak field ligand so $[CoF_6]^{3-}$ is a high spin complex having $sp^3d^2$ hybridisation so that there will be four unpaired electrons.

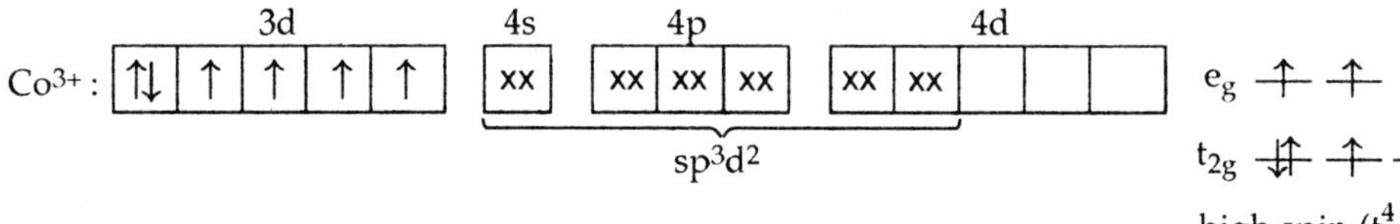


high spin ($t_{2g}^4e_g^2$)

In both the cases (VBT and CFT) the number of unpaired electrons are four having $\mu = \sqrt{n(n+2)}\beta = \sqrt{4(4+2)}\beta = 4.89$ BM.

As $NH_3$ is a strong field ligand so $[Co(NH_3)_6]^{3+}$ will be diamagnetic having no unpaired electron i.e. $\mu = 0$.

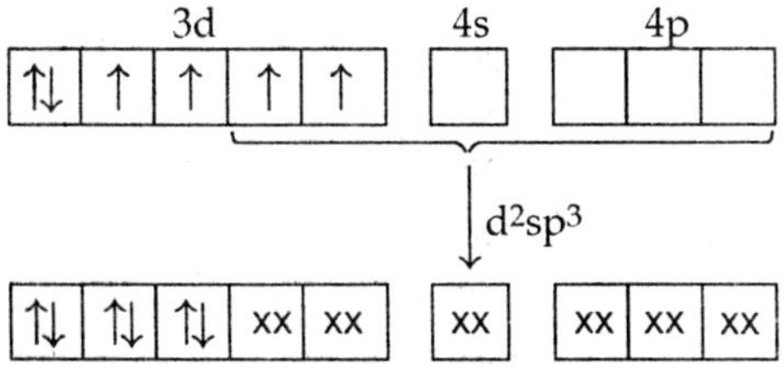


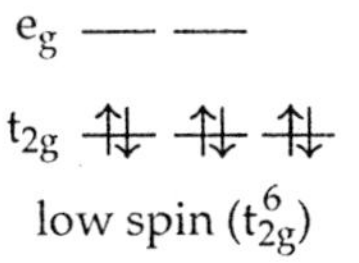


(b) In case of $K_3[CuF_6]$ the Cu(III) is a $3d^8$ system and according to VBT and CFT for weak field $F^-$ ligand corresponds to two unpaired electrons.

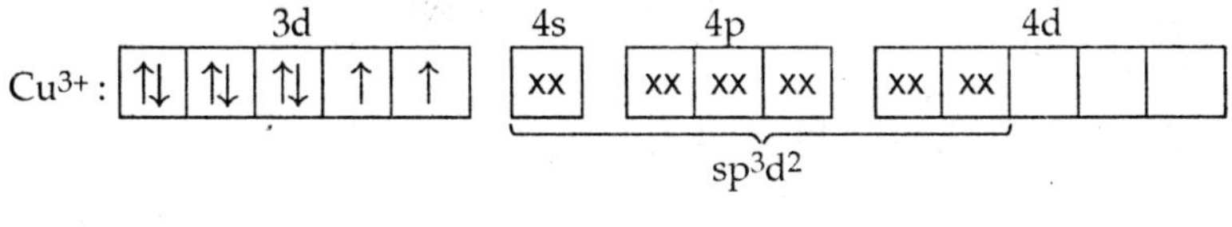


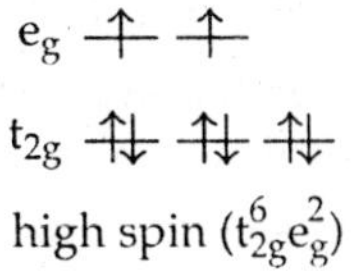


$\mu = \sqrt{n(n+2)}\beta = \sqrt{2(2+2)}\beta = 2.8$ B.M.

211. **(a)** Using an approximate Orgel diagram explain the electronic spectrum of $Ni^{2+}$(aq) ion. **(b)** Determine the spectroscopic ground state term for Co(II) ion. **[C.U. 2014]**

**Hint :** (a) Sec. 7.15.10(B); (b) Sec. 7.15.6.

212. **(a)** $F^-$ stabilises higher valent while CO stabilises lower valent state of metal ions.—Explain with the molecular orbital diagrams.

**(b)** Explain the origin of colour in $CrO_4^{2-}$ and $[Fe(open)_3]^{2+}$. **[C.U. 2014]**

**Hint :** (a) $F^-$ is a classical ligand having σ-donor property, so it will be not be able to stabilise lower valent metal ions, but CO acting simultaneously as σ-donor as also π-acceptor ligand because according to molecular orbital picture (Sec. 3.4.6; Fig. 3.28) the π* antibonding orbital lying vacant and thus CO will stabilise lower valent metal through synergic effect.

$$M \underset{\pi}{\overset{\sigma}{\rightleftarrows}} CO$$

(b) Sec. 7.15.11. The colour of $CrO_4^{2-}$ is due to LMCT while the colour of $[Fe(ophen)_3]^{2+}$ is due to MLCT.

213. Mention the formal begining of coordination chemistry made by Tassaert highlighting his observation. **[B.U. 2015]**

**Hint :** Sec. 7.2.

214. 'A π-complexing ligand is essentially a π acidic ligand but the reverse is not true.'—Justify. **[B.U. 2015]**

**Hint :** Problem 7.15.

215. Define metalloorganic and organometallic ligands with one example in each case. **[B.U. 2015]**

**Hint :** Question number 34 and Sec. 7.4.3(q).

216. 'Sulphate can behave as a flexidentate ligand.'—Comment. **[B.U. 2015]**

**Hint :** $SO_4^{2-}$ can function as monodentate oxygen donor in the complex $[Co(NH_3)_5(OSO_3)]^+$, CN = 5+1 = 6 or it may function as didentate $\eta^2 - SO_4^{2-}$ in the complex $[Co(en)_2(SO_4)]^+$, CN = 4+2 = 6.

217. What are podands and coronds? **[B.U. 2015]**

**Hint :** Sec. 7.4.3.(k and l).

218. How do $(C_6H_5)_3P$, $(C_6H_5)_3PO$ and $(C_6H_5)_2SO$ behave as coordinating ligands? **[B.U. 2015]**

**Hint :** $(C_6H_5)_3P$ behave as monodentate ligand in the heteroleptic complexes $[Cu^I(NO_3)(PPh_3)_2]$ where $NO_3^-$ function as symmetric didentate 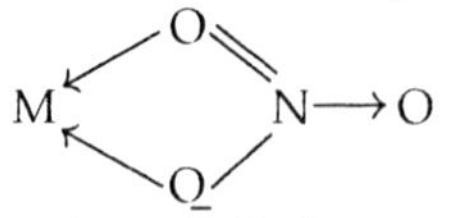 and $[Rh^ICl(CO)(PPh_3)_2(SO_2)_2]$. $(C_6H_5)_3PO$ and $(C_6H_5)_2SO$ both function as monodentate O donor to metal.

219. "Out of $F^-$ and $I^-$, $Co(NH_3)_5^{3+}$ ion prefers $F^-$ rather than $I^-$ to achieve coordination number 6."—Comment. **[B.U. 2015]**

**Hint :** $Co^{3+}$ is a hard acid, this becomes more harder on attachment of hard $NH_3$. $F^-$ is a hard base/ligand while $I^-$ is a soft base/ligand. So $Co(NH_3)_5^{3+}$ will prefer $F^-$ to achieve coordination number 6 compared to that of $I^-$ because hard-hard interaction more preferred compared to that of hard-soft interaction (Pearson rule).

220. Mention the salient features of Werner's coordination theory. Why was it a new horizon in the then inorganic chemistry? **[B.U. 2015]**

**Hint :** Sec. 7.2. The concept of secondary valency is the new horizon in the Werner's coordination theory.

221. Write down the Lewis dot structure and show the possible coordination motifs of dimethyl sulfoxide and acetylacetone. **[B.U. 2015]**

**Hint :** Sec. 7.4.3(e)(iii). $(CH_3)_2SO$ is a relatively poor nucleophile and can function as $\eta^1$-O or $\eta^1$-S mode in various complexes. It can also function as $\mu_2$-O, S or $\mu_3$–O, O, S bridging in complexes. Pd(II) and Pt(II) are frequently S-bonded as in $[Pd(en)(Me_2SO)Cl]ClO_4$, but it is O-bonded in $[Pd(bpy)(Me_2SO)Cl]BF_4$.

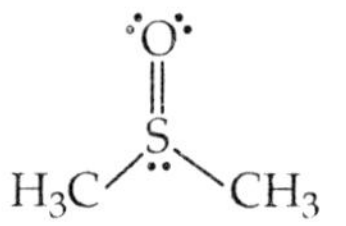

Central S = 4+2 = 6 electrons.
Coordination motif through O and S.

222. '$[Ag(NH_3)_2]^+$ is more stable than $[Ag(en)]^+$ [en = ethylenediamine]'.—Answer giving reasons. **[B.U. 2015]**

**Hint :** $[Ag(NH_3)_2]^+$ has a linear geometry $[H_3N—Ag—NH_3]^+$. As en is a bidentate ligand so that in case of $[Ag(en)]^+$ the coordination mode of en 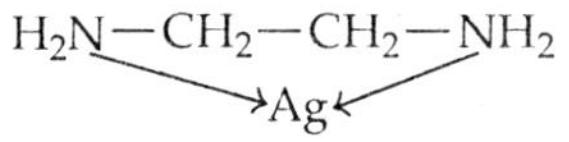 where there is a formation of three membered ring which according to Bayer's strain theory is unstable and so stability of $[Ag(NH_3)_2]^+ > [Ag(en)]^+$.

223. The reaction $[Co(CN)_5H_2O]^{2-} + X^- \longrightarrow [Co(CN)_5X]^{2-} + H_2O$ follows a/an :

1. Interchange dissociative ($I_d$) mechanism.
2. Dissociative (D) mechanism.
3. Associative (A) mechanism.
4. Interchane Associative ($I_a$) mechanism. **[NET 2015]**

**Hint :** Both 1 and 2 may be possible because dissociative (D or $I_d$) mechanisms are expected to be more common for octahedral complexes of fourth period d-metal complexes (Sec. 7.19.5).

224. The refluxing of $RhCl_3.3H_2O$ with an excess of $PPh_3$ in ethanol gives a complex A. Complex A and the valence electron count on rhodium are respectively,

1. $[RhCl(PPh_3)_3]$, 16; 2. $[RhCl(PPh_3)_5]$, 16; 3. $[RhCl(PPh_3)_3]$, 18; 4. $[RhCl(PPh_3)_5]$, 18. **[NET 2015]**

**Hint :** In all the complexes the metal is Rh(I) which has the electronic configuration $[Kr]4d^8$. Therefore electron count in case of $[RhCl(PPh_3)_3]$ = 8+2+6 = 16 i.e. 1 is answer.

225. In transition metal phosphine ($M-PR_3$) complexes, the back bonding involves donation of electrons from :

1. $M(t_{2g}) \rightarrow PR_3(\sigma^*)$, 2. $M(t_{2g}) \rightarrow PR_3(\pi^*)$, 3. $M(e_g) \rightarrow P(d)$, 4. $PR_3(\pi) \rightarrow M(t_{2g})$. **[NET 2015]**

**Hint :** $M(t_{2g}) \rightarrow PR_3(\sigma^*)$ i.e. answer 1.

226. The magnitude of the stability constants for $K^+$ ion complexes of the following supramolecular hosts follows the order,

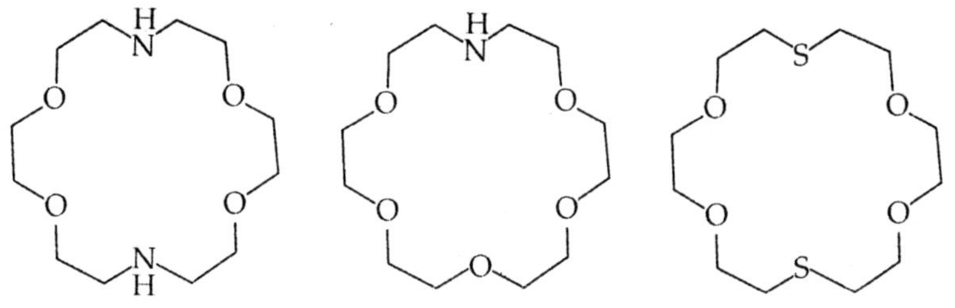


A B C

1. B > A > C, 2. C > A > B, 3. A > B > C, 4. C > B > A **[NET 2015]**

**Hint :** Donor ability O > N > S and so answer is 1 i.e. B > A > C.

227. Among the complexes, $K_4[Cr(CN)_6]$(A), $K_4[Fe(CN)_6]$(B), $K_3[Co(CN)_6]$(C) and $K_4[Mn(CN)_6]$(D), Jahn–Teiler distortion is expected in—

1. A, B and C, 2. B, C and D, 3. A and D, 4. B and C. **[NET 2015]**

**Hint :** Sec. 7.14.5(C). In $K_4[Cr(CN)_6]$ where Cr(II)($3d^4$) susceptible to J-T distortion(Table 7.14). In (B) the Fe(II) ($3d^6$) low spin is not undergoing J-T distortion. In (C) Co(III)($3d^6$) in low spin condition distortion not permitted. For(D) Mn(II)($3d^5$) in low spin condition J-T distortion permitted and so the answer is 3 i.e. A and D.

228. With respect to σ and π bonding in Pt-III in the structure given below, which of the following represent the correct bonding

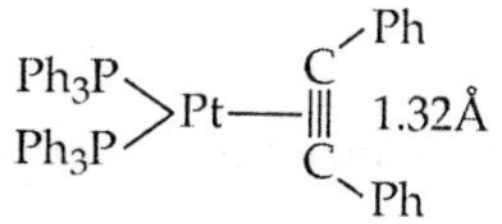


1. $M(\sigma) \rightarrow L(\sigma)$ and $M(\pi) \rightarrow L(\pi^*)$
2. $L(\sigma) \rightarrow M(\pi)$ and $L(\pi) \rightarrow M(\pi)$
3. $L(\pi) \rightarrow M(\pi)$ and $L(\sigma) \rightarrow M(\pi)$
4. $L(\pi) \rightarrow M(\sigma)$ and $M(\pi) \rightarrow L(\pi^*)$ **[NET 2015]**

**Ans. :** 4 i.e. $L(\pi) \rightarrow M(\sigma)$ and $M(\pi) \rightarrow L(\pi^*)$.

229. Tabulate the basic postulates of Valence Bond Theory. **[B.U. 2015]**

**Hint :** Sec. 7.14.1.

230. CO and $N_2$ are isoelectronic but they differ greatly in donor properties.—Explain. **[C.U. 2015]**

**Hint :** CO and $N_2$ both the molecules are isoelectronic (ten electrons) and isostructural (linear).

$$:C\equiv O: \quad :N\equiv N:$$

Being a heteroatom system, in CO, oxygen is more electronegative than carbon and so carbon lone pair will function generally. Moreover in CO the π* antibonding orbital (LUMO) is lying vacant and so it will stabilise the low valent metal through σ/π interaction known as synergic effect.

$$M \underset{\pi}{\overset{\sigma}{\rightleftarrows}} :C \equiv O:$$

This type of synergic effect is not possible for $N_2$ and infact $N_2$ is known for a longtime to have no ligational motif. The energies of the two atomic orbitals of the two nitrogen atoms are the same. Thus $N_2$ is a weaker $\sigma$ donor as well as weaker $\pi$-acceptor than CO. That $N_2$ is a weaker $\sigma$ donor than CO follows from higher electronegativity of nitrogen.

CO can form homoleptic or heteroleptic complexes while $N_2$ is known to exist as a donor in mixed ligand complexes only. CO can function as both as a terminal monodentate as well as bridging ligand in polynuclear carbonlyls where carbon is acting as a donor atom e.g.

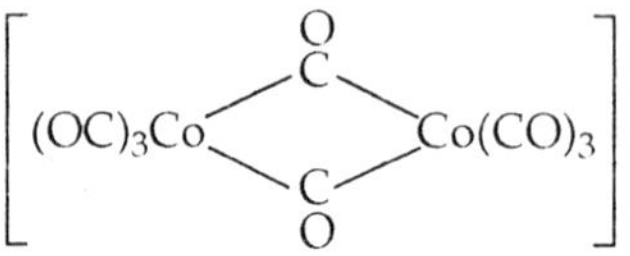

On the other hand $N_2$ functions both as a terminal ligand and as an end-on (EO) bridging ligand e.g. $[(NH_3)_5Ru^{II} \leftarrow N{\equiv}N \rightarrow Ru^{II}(NH_3)_5]^{4+}$. In fact molecular $N_2$ complexes are much less stable than the metal carbonyls.

231. 'Magnetite is expected to have invese spinel structure.'—Justify or criticise. **[B.U. 2015]**

**Hint :** Sec. 7.14.5(B) and Problem 7.58.

Normal spinel is $\overset{T_d}{A(II)}\,\overset{O_h}{B(III)}O_4$ and inverted spinel is $\overset{T_d}{B(III)}\,(\overset{O_h}{A(II)}\,\overset{O_h}{B(III)})O_4$. Magnetite has the composition $Fe_3O_4$.

CFSE for Fe(III) in both $O_h$ and $T_d$ field = 0.0.

Fe(II) in $T_d$ field is $e^3t_2^3$ and CFSE = $-0.6\Delta_t = -0.6\times{}^4/_9\Delta_o = -0.27\Delta_o$.

Fe(II) in $O_h$ field is $t_{2g}^4eg^2$ and CFSE = $-0.4\Delta_o$.

If $Fe_3O_4$ has normal spinel structure then $(\overset{T_d}{Fe(II)}\,\overset{O_h}{Fe_2(III)}O_4)$

CFSE = $-0.27\Delta_o + 0.0 = -0.27\Delta_o$

and for inverted spinel structure, $\overset{T_d}{Fe(III)}\,(\overset{O_h}{Fe(II)}\,\overset{O_h}{Fe(III)})O_4$

CFSE = $0.0 + (-0.4\Delta_o + 0.0) = -0.4\Delta_o$.

As the CFSE value for inverted spinel is more negative than that of normal structure, so $Fe_3O_4$ exist as inverted spinel structure.

232. 'Addition of ammonium thiocyanate to aqueous solution of uranyl nitrate produces deep red colouration'—State reason. **[B.U. 2015]**

**Hint :** In uranyl nitrate $UO_2(NO_3)_2$ uranium is U(VI) oxidation state having electron configuration $[Rn]5f^06d^07s^0$. So $UO_2(NO_3)_2$ will be colourless. Addition of $NH_4SCN$ to its aqueous solution produces charge transfer L → M i.e. $SCN^- \rightarrow$ U(VI)(LMCT) transition and the solution produces deep red colouration.

233. Guess the relative positions of $CN^-$ and $NH_3$ in spectrochemical series with proper reason. **[B.U. 2015]**

**Hint :** $\pi$-acceptor ligands are stronger in spectrochemical series than the ligands having no $\pi$-bonding ability ($\pi$-acceptor ligands have multiple bonds, hence it makes more perturbation with the metal ions). As $CN^-$ is a $\pi$-acceptor ligand and $NH_3$ is not, so the former exerts stronger ligand field on metal ions.

234. What are chelates? Mention one application of it each in qualitative and quantitative analysis. [C.U. 2015]

**Hint :** Sec. 7.8(5), 7.11.1(4), 7.13.1 and 7.13.2 and question number 47.

235. Draw and mention the type of isomers of the complex $[Co(NO_2)_3(NH_3)_3]$. [C.U. 2015]

**Hint :** Sec. 7.9.11(B). $Ma_3b_3$ type of complex exist as fac and mer isomer.

236. Write the IUPAC name of the complex $[Co(NH_3)_6][Co(NO_2)_6]$. [C.U. 2015]

**Hint :** Sec. 7.6. Worked out example number 13 (Page 532).

237. The oxidising strength of $[M^{VII}O_4]^-$ (M=Mn, Tc and Re) decreases and the energy of LMCT band position increases down the group. [B.U. 2012].

**Hint :** LMCT occurs from periphery to core within a molecule. The more oxidising metal is and more reducing ligand is, the lower is the energy for LMCT transition. In $[M^{VII}O_4]^-$ (M = Mn, Tc and Re) $O^{2-}$ is same for all Mn(VII) $\rightarrow$ Tc(VII) $\rightarrow$ Re(VII). As size increases and electron greediness decreaes due to decrease in electronegativity of metal. Hence the energy required for LMCT transition increases from $MnO_4^-$ to $ReO_4^-$ via $TeO_4^-$. Among the three species which one has lowest energy LMCT is the strongest oxidising agent.

238. '$[Pt(NH_3)_4][PtCl_4]$(A) has an intense green colour whereas $[Pt(EtNH_2)_4][PtCl_4]$(B) has not such green colour'.—Explain this with respect to their structures. [B.U. 2012]

**Hint :** In $[Pt(NH_3)_4][PtCl_4]$ the square planar cation i.e. $[Pt(NH_3)_4]^{2+}$ is electron deficient and colourless while the anion i.e. $[PtCl_4]^{2-}$ is electron rich and red. They stack one over another forming a linear chain. The short Pt···Pt distance gives rise to somewhat metal-metal interaction which shifts the d-d transition giving green colour due to MMCT. In $[Pt(EtNH_2)_4][PtCl_4]$, the Pt···Pt interaction is prevented due to the presence of bulky $EtNH_2$ ligand and so there will be no possibility of MMCT.

239. Distinguish between Crystal Field Splitting Energy (CFSE) and Crystal Field Stabilisation Energy (CFSE). [C.U. 2015]

**Hint :** Sec. 7.14.3. Crystal field splitting energy includes pairing energy (P) but in crystal field stabilisation energy P term is omitted.

240. Calculate the electronic state and magnetic moment of an ion with $^3F_4$ ground term. [C.U. 2015]

**Hint :** $^3F_4$ ground term metallic ion is Ni(II) which is $3d^8$ system. The electronic states will be $O_h$ field as $^3A_{2g}$, $^3T_{2g}$ and $^3T_{1g}$. In $T_d$ field the terms will be same but g subscript will be dropped. The number of unpaired electron in square planar geometry is zero and so $\mu =$ 0 and for octahedral and tetrahedral geometry the number of unpaired electrons = 2 so that $\mu = 2.83$ BM.

241. Of the two $Co^{2+}$ and $Cr^{3+}$ one is highly prone to form octahedral but the other to form tetrahedral complexes. Give reason in favour of your judgement.

**Hint :** Sec. 7.14.5(B). $Cr^{3+}(3d^3)$ is prone to form octahedral complexes as value of OSPE/OSSE is high while $Co^{2+}(3d^7)$ is commonly forms tetrahedral complexes as OSPE/OSSE value is low.

242. What do you understand by "dynamic Jahn-Teller distortion"? How does it differ from that of 'static' one? [C.U. 2015]

**Hint :** Sec. 7.14.5(C).

243. Both copper(II)($d^9$) acetate monohydrate and chromium(II)($d^4$) acetate monohydrate have the similar structure, the former is paramagnetic having lower $\mu_s$ value but the latter is completely diamagnetic at room temperature.—How will you explain the fact? [C.U. 2015]

**Hint :** Question number 16 and 18 (Exercise) of chapter 10 (Vol. II). With similar structure of $Cu_2(CH_3COO)_4.2H_2O$ it should be mentioned that δ-bonding leading to spin pairing is very weak.

Here no $\sigma$ and $\pi$ bonds are formed in contrast to quadruple bond in $Cr_2(CH_3COO)_4.2H_2O$. In $Cr_2(CH_3COO)_4.2H_2O$ the quadruple bond is quite strong enough for spin pairing and thus $Cr_2(CH_3COO)_4.2H_2O$ is diamagnetic. In Cu(II) acetate only partial spin pairing occurs which reduce the magnetic moment from spin only value from 1.73 BM to 1.4 BM.

It is to be noted that both copper(II) acetate monohydrate, $Cu(CH_3COO)_2.H_2O$ and chromium(II) acetate monohydrate, $Cr(CH_3COO)_2.H_2O$ exist in crystalline state as dimer.

244. 'Both $[Mn(CO)_4(NO)]$ and $[Fe(CO)_4CN]^-$ have trigonal bipyramidal structure., In the former NO lies in the equatorial plane while in the latter the $CN^-$ occupies the axial position.'—account the fact. [B.U. 2015]

**Hint :** Metal to ligand ($M \rightarrow L$) $\pi$ back bonding is favoured at an equatorial plane relative to axial position. Between NO and $CN^-$, NO is stronger $\pi$-acceptor (since $CN^-$ is negatively charged), hence it will be at equatorial plane.

245. The correct statement for Mn–O bond lengths in $[Mn(H_2O)_6]^{2+}$ is—

1. All bonds are equal.
2. Four bonds are longer than two others.
3. Two bonds are longer than four others.
4. They are shorter than the Mn–O bond in $[MnO_4]^-$. [NET 2015]

**Ans. :** 1 i.e. all bonds are equal.

246. $MnO_4^-$ is deep purple in colour whereas $ReO_4^-$ is colourless. This is due to greater energy required for

1. d-d transitions in Re compound compared to the Mn compound.
2. d-d transitions in the Mn compound compared to the Re compound.
3. Charge transfer from O to Re compared to O to Mn.
4. Charge transfer from O to Mn compared to O to Re. [NET 2015]

**Ans. :** 3. Re is a member of third transition series where charge transfer occurs at higher energy.

247. The correct statement about the substitution reaction of $[Co(CN)_5Cl]^{3-}$ with $OH^-$ to give $[Co(CN)_5(OH)]^{3-}$ is

1. it obeys first order kinetics
2. its rate is proportional to the concentration of both the reactants
3. it follows the $S_N^1CB$ mechanism
4. its rate is dependent only on the concentration of $[OH^-]$. [NET 2015]

**Ans.** : 1. i.e. it follws first order kinetics.

248. Aqueous $Cr^{2+}$ effects one electron reduction of $[Co(NH_3)_5Cl]^{2+}$ giving compound Y. Compound Y undergoes rapid hydrolysis Y is—

1. $[Co(NH_3)_5]^{2+}$, 2. $[Co(NH_3)_5(OH)]^+$, 3. $\{Co(NH_3)_4(OH)_2]$, 4. $[Cr(H_2O)_5Cl]^{2+}$ [NEt 2015]

**Ans. :** 1 i.e. $[Co(NH_3)_5]^{2+}$.

249. Choose the correct statements about Tanabe-Sugano diagrams : (A) E/B is plotted against $\Delta_o/B$. (B) The zero energy is taken as that of the lowest term, (C) Terms of the same symmetry cross each other, (D) Two terms of the same symmetry upon increase of ligand field strength bend apart from each other. Correct answer is—

1. A and B, 2. A and C, 3. A B and D, 4. A, B, C and D. [NET 2015]

**Ans. :** 3 i.e. A, B and D.

250. Among the following complexes— A. $[Co(ox)_3]^{3-}$, B. trans-$[CoCl_2(en)_2]^+$, C. $[Cr(EDTA)]^-$ the chiral one(s) is/are—

1. A and B, 2. C and B, 3. C only, 4 A and C. **[NET 2015]**

**Ans. :** 4 i.e. A and C. Both the complexes have mirror image isomers which are not superimposable.

251. 'Crystal field splitting of the d-orbitals is more pronounced in the octahedral field than in the tetrahedral field.'—Explain. **[B.U. 2016]**

**Hint :** (a) In tetrahedral geometry the $t_2$ orbitals, although closest to the ligands do not point directly at the ligands but in an octahedral complex there is ligands lies directly along the axes.

(b) In tetrahedral geometry there are only four ligands but in octahedral geometry there are six ligands i.e. octahedral field is more stronger than tetrahedral field.

252. Write down the crystal field splittins of the d-orbitals in square pyramidal and trigonal bipyramidal structures. **[B.U. 2016]**

**Hint :** Sec. 7.14.2(D).

Orbital ordering in square pyramidal geometry is $d_{xz}, d_{yz} < d_{xy} < d_{z2} < d_{x2-y2}$

The orbital ordering for trigonal bipyramidal geometry is $d_{xz}, d_{yz} < d_{xy}, d_{x2-y2} < d_{z2}$

253. Comment on the relative stabilities of $[Ni(CN)_4]^{2-}$ (square planar), $[Ni(CN)_5]^{3-}$ (TBP) and $[Ni(CN)_5]^{3-}$ (square pyramidal) in terms of CFT. **[B.U. 2016]**

**Hint :** Sec. 7.14.2(D). As per Fig. 7.12 the relative order of d-orbitals in different geometries are :

Square planar : $d_{xz}, d_{yz} < d_{z2} < d_{xy} < d_{x2-y2}$

TBP : $d_{xz}, d_{yz} < d_{x2-y2}, d_{xy} < d_{z2}$

Square pyramidal : $d_{xz}, d_{yz} < d_{xy} < d_{z2} < d_{x2-y2}$

As per Table 7.12 the CFSE values for those geometries for Ni(II) are :

**Square planar configuration :** $d_{xz}^2\ d_{yz}^2\ d_{z2}^2\ d_{xy}^2$

$$CFSE = [2\times(-0.514) + 2\times(-0.514) + 2\times(-0.428) + 2\times0.228]\Delta_o$$

$$= [-1.028 - 1.028 - 0.856 + 0.456]\Delta_o = -2.456\Delta_o = -24.56\ Dq$$

**TBP configuration :** $d_{xz}^2\ d_{yz}^2\ d_{x2-y2}^2\ d_{xy}^2$

$$CFSE = [2(-0.272) + 2(-0.272) + 2(-0.082) + 2(-0.082)]\Delta_o$$

$$= [-0.544 - 0.544 - 0.164 - 0.164]\Delta_o = -1.416\Delta_o = -14.16\ Dq$$

**Square pyramidal configuration :** $d_{xz}^2\ d_{yz}^2\ d_{xy}^2\ d_{z2}^2$

$$CFSE = [2(-0.457) + 2(-0.457) + 2(-0.086) + 2\times0.086]\Delta_o$$

$$= [-0.914 - 0.914]\Delta_o = -1.828\Delta_o = -18.28\ Dq.$$

$\therefore$ the relative stabilities of the three geometries are : square planar > square pyramidal > TBP.

254. 'Ni(II) forms the tetrahedral and octahedral complexes respectively with the ligands such as $Cl^-$ and $NH_3$ but Pd(II) and Pt(II) form square planar complexes with both.'—Explain. **[B.U. 2016]**

**Hint :** Sec. 7.14.5(B)

According to OSPE/OSSE = $CFSE_{oct} - CFSE_{tet}$

in case of Ni(II)($d^8$), OSPE/OSSE = $-1.2\Delta_o - (-0.355\Delta_o) = -0.845\Delta_o$.

As OSPE/OSSE is highest for Ni(II), so Ni(II) forms octahedral geometry with $NH_3$. As $Cl^-$ is a weak field ligand so with $Cl^-$, Ni(II) forms tetrahedral complex.

Ni, Pd and Pt are the members of first, second and third transition elements respectively. We known that size of d-orbitals increases in the order 3d < 4d < 5d. As size of the d-orbital increases from 3d → 4d → 5d, the interelectronic repulsion between metal ion orbitals with that of ligand orbitals decreases and this brings closer approach of metal-ligand bond thereby spin pairing is a common phenomena for second and third transition series elements making greater perturbation of d-orbitals. This makes higher CFSE i.e. Pd(II) and Pt(II) complexes will assume square planar complexes having largest separation between the d-orbitals i.e. tetragonal distortion will be maximum in case of square planar geometry.

255. Discuss the merits and demerits of CFT. **[B.U. 2016]**

**Hint :** Sec. 7.15.12.

256. Find out the Russel-Saunders ground state term for copper(III) ion and calculate its total degeneracy. **[B.U. 2016]**

**Hint :** Cu(III) is $3d^8$ system for which the ground term is $^3F_4$.

Total degeneracy = (2L+1)(2S+1) = (2×3+1)(2×1+1) = 21.

257. 'There are cursory reports of three-membered chelate rings.' Give example of two complexes having such three-membered chelate loops. **[B.U. 2016]**

**Hint :** Two complexes with three-membered chelate loop are : $[Ir(CO)Cl(PPh_3)_2(\eta^2\text{–}O_2)]$ and Zeise's salt $K[PtCl_3(C_2H_4)]$(Fig. 3.36).

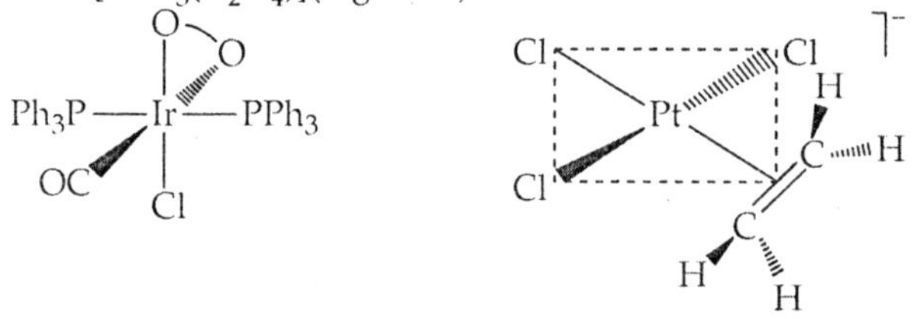


258. Define metalloligand with an example. In this context write down the meaning of duplex behaviour. **[B.U. 2016]**

**Hint :** Sec. 7.4.3(r).

259. Define chelator and congregator. How can a chelator be made a congregator by judicious change in the ligand framework? **[B.U. 2016]**

**Hint :** Sec. 7.4.3.(s).

260. Note down Jorgensen's effort to formulate different metal-ammine compounds known at that time. How was he misdirected? **[B.U. 2016]**

**Hint :** Sec. 7.2.

261. Delineate Werner's theory and explain why it was a new horizon in the then inorganic chemistry. Note down the defects of Werner's theory in the light of present knowledge on coordination chemistry. **[B.U. 2016]**

**Hint :** Sec. 7.2. The concept of secondary valency is the new horizon in the then inorganic chemistry. The secondary valencies being covalence, are directed in space around the central metal ion in the first sphere of attraction in a definite geometrical arrangement. The coordination complex is to be written with a square bracket.

Main drawback of Werner's theory is that it fails to explain the nature of metal-ligand bonding. Werner prescribed octahedral structure ignoring the other two possible structures i.e. trigonal prism and planar hexagon, and explained the optical and geometrical isomerism. So this was an adhoc conclusion made by him. Werner could not explain why most of the complexes he handled were with corodination numbers four and six. Now different coordination numbers and stereochemistries are known. Werner's theory is unable to explain spectral properties, magnetic behaviours, kinetic lability or inertness, thermodynamic stabilities and several other properties of the complexes.

262. List the factors which affect isomer population. **[B.U. 2016]**

**Hint :** Factors affecting isomer population are : (a) statistical factor, (b) coulombic factor, (c) steric factor, (d) intramolecular hydrogen bonding, (e) electronic factor ($\pi$-bonding), (f) SHAB principle etc.

263. Write down the basic requirements for stabilisation of low, medium and high oxidation states of transition metal ions in their complexes. **[B.U. 2016]**

**Hint :** Sec. 7.17, Sec. 7.17.1 and Sec. 7.17.2.

Complex formation through a judicious choice of ligands has considerable effect on the stability of a particular oxidation state for transition metals. The $\sigma$-donor ligands stabilise medium oxidation states of metal ions. The $\sigma$-donor plus $\pi$-acceptor ligands will stabilise low oxidation states of metals with synergic bonding. The ligands with $\sigma$-donor plus $\pi$-donor properties will stabilise high oxidaiton states of metals.

The reaction condition is also a deciding factor. An anaerobic condition is often needed for isolation of complexes in low oxidation states of metals. Aerobic condition and use of oxidising agent are required for isolation of complexes in high oxidation states. The role of solvent is also important. The reducing solvents like alcohols favour isolation of low oxidation states of metals. A ketonic solvent does not reduce the oxidation state of the metal ion in a starting material while isolating a compound in the same oxidation state as in starting materials, since a ketone is not easily oxidised.

264. Write the IUPAC names of (a) $[Ni(NH_3)_4(ONO)(SCN)]$, (b) $[Cr(NH_3)_6][Co(CN)_6]$. Write the formula of tetraammine dichlorocobalt(III) sulphate. **[B.U. 2016]**

**Hint :** (a) tetraamminenitrito-κ-O thiocyanato-κ-S nickel(II).

(b) hexaamminechromium(III)hexacyanidocobaltate(III).

The formula of the complex is $[Co(NH_3)_4Cl_2]_2(SO_4)$.

265. State the important differences between double salt and complex salt along with one similarity. **[B.U. 2016]**

**Hint :** Sec. 7.1. In double salt the constituent simple salts are in equimolar proportions but in complex salt, the constituent simple salts may or may not be in equimolar proportions. Double salts dissociate in aqueous solutions whereas complex ion does not dissociate in aqueous solution. Double salt contains ionic bond but complex salt may have ionic as well as covalent bonds.

266. Acetylacetone may function as an organometallic ligand, a chelating/innermetallic ligand and a bridging ligand.—Explain with respect to Lewis structure. **[B.U. 2016]**

**Hint :** Sec. 7.4.3(e)(iii).

267. 'Both $N_2$ and HC≡CH can behave as classical ligand as well as non-classical ligand'—Comment. **[B.U. 2016]**

**Hint :** Sec. 7.4.3. Classical ligand is a lone pair donor forming $\sigma$-bond whereas non-classical ligand is a $\pi$-electron cloud donor forming $\mu$-bond. $N_2$ has a lone pair on each N and one $\sigma$-bond and two $\pi$-bonds i.e. :N≡N: whereas HC≡CH has no lone pair but with one $\sigma$-bond and two $\pi$-bonds. So $N_2$ can behave both as classical and non-classical ligands, but HC≡CH can behave only as a non-classical ligand.

268. What is the criterion of a ligand to behave as a non-classical ligand? **[B.U. 2016]**

**Hint :** A non-classical ligand is a $\pi$-electron cloud donor forming $\mu$-bond with the central ion. So the criterion of a ligand to behave as a non-classical ligand is that it must contain unsaturated framework with multiple bond (double or triple bond).

269. 'Both 2, 2'-bipyridine and 1, 10-phenanthroline can act as organometallic ligands'—Justify or criticise. **[B.U. 2016]**

**Hint :** As per the structure of 2, 2'-bipyridine, the molecule has free rotation about C2–C2' bond axis, so it can behave as (N, N)chelator and (N, C) organometallic ligand as one of the two donor centres is carbon. 1, 10-phenanthroline has fused ring system, so it is having rigid framework and thus cannot rotate. Hence it can behave only (N, N) chelator to all metals.

270. What is intensity stealing? Give one example. **[B.U. 2016]**

**Hint :** Sec. 7.15.11.5. This causes a marked gain in intensity of absorption in ligand field transitions occuring close to allowed/CT transitions. The electronic wave function of forbidden excited term gets mixed with the wave functions of the allowed CT state.

In these cases the forbidden transition steals some intensity from the allowed/CT transitions because the energy levels lie closer. For example in case of $[Mn(H_2O)_6]^{2+}$ which is pale pink (both Laporte and spin forbidden despite gains some intensity because of intensity stealing).

271. What parameters are plotted against X and Y-axes of Tanabe-Sugano diagram? **[B.U. 2016]**

**Hint :** Sec. 7.15.11.3. Along X axis Dq/B and along Y-axis E/B is plotted for a particular $d^n$ configuration where Dq is crystal field splitting, B is interelectronic repulsioin parameter and E is energy of spectroscopic terms.

272. What is Jahn-Teller distortion? Point out criterion (criteria) for a coordination molecule to be Jahn-Teller active? Give an example of a Jahn-Teller active system explaining the reason for this activity. What do you mean by tetragonal elongation? **[B.U. 2016]**

**Hint :** Sec. 7.14.5(C) and Question numbers 75 and 184. JT distortion in $O_h$ geometry where the axial bonds are longer than the equatorial bonds is called **tetragonally distorted octahedron with elongation along z-axis.**

273. What is spinel? Write down the general formula stating the oxidation state of the metal ions involved. '$NiAl_2O_4$ forms inverted spinel structure'—why? **[B.U. 2016]**

**Hint :** Sec. 7.14.5(B) and Question numbers 160 and 231.

$$NiAl_2O_4 : Ni^{2+} \longrightarrow O_h \longrightarrow t^6_{2g}e^2_g,\ CFSE = -1.2\Delta_o$$
$$T_d \longrightarrow e^4t^4_2,\ CFSE = -0.8\Delta_t = -0.36\Delta_o.$$

So $Ni^{2+}$ is better stabilised in $O_h$ hole. Taking one $Al^{3+}$ from $O_h$ to $T_d$ hole make no loss in CFSE as $Al^{3+}$ is not a transition metal ion. So easily it can be stated that $NiAl_2O_4$ takes the form $Al^{Td}(NiAl)^{Oh}O_4$ (inverse spinel), not the form $Ni^{Td}(Al_2)^{Oh}O_4$ (normal spinel).

274. What is nephelauxetic effect? **[B.U. 2016]**

**Hint :** Sec. 7.15.11.2. The apparent value of **interelectronic repulsion parameter** B(B') in complexes is always smaller than that of free ion. This results from a phenomenon known as nephelauxetic (meaning 'cloud expansion') effect and is attributed to the delocalisation of metal electrons over molecular orbitals that encompass both the metal and the ligand, as a result the average interelectronic repulsion is reduced.

275. How will you estimate the total hardness of water complexometrically? By applying this method, 13.0 ml of 0.0065 M EDTA solution was required for titration of 60.0 ml of water sample for determination of total hardness. Calculate the total hardness in terms of ppm of calcium carbonate. **[B.U. 2016]**

**Hint :** Sec. 7.13.6.

60 ml of water ≡ 13.0 ml of 0.0065 M EDTA

$$\equiv \frac{100}{1000} \times 13.0 \times 0.0065 \text{ g } CaCO_3$$

$$\therefore 10^6 \text{ ml of water contain } \frac{1}{10 \times 60} \times 13.0 \times 0.0065 \times 10^6 \text{ g } CaCO_3 = 140.83 \text{ g } CaCO_3$$

So that the total hardness of water = 140.83 ppm.

276. Name one each of masking and demasking agents with suitable examples of complexometric titrations. **[B.U. 2016]**

**Hint :** Sec. 7.13.6(G).

Masking agents are those reagents which prevent interfering ion from the reaction without physical separation by forming complexes with interfering ions which are more stable than complexes formed with indicator and EDTA.

KI is acting as masking agent for $Hg^{2+}$ to form $[HgI_4]^{2-}$. KCN is also acting as same for $Ag^+$, $Cu^{2+}$, $Co^{2+}$, $Ni^{2+}$, $Cd^{2+}$, $Zn^{2+}$ etc.

Demasking agents are those which help to regain the ability of masked ion to enter into the reaction with indicator and EDTA e.g. masking by $CN^-$ can be removed by a mixture of HCHO + $CH_3COOH$ to $[Zn(CN)_4]^{2-}$, thereby free $Zn^{2+}$ goes into solution.

$$[Zn(CN)_4]^{2-} + 4(HCHO + CH_3COOH) \longrightarrow Zn^{2+} + 4CH_2\begin{matrix}\diagup CN \\ \diagdown OH\end{matrix} + 4CH_3COO^-.$$

# Appendix–A

## Names, symbols (as per IUPAC recommendations 2005) and atomic weights of the elements

| Element | Symbol | Atomic number | Atomic weight | Element | Symbol | Atomic number | Atomic weight |
|---|---|---|---|---|---|---|---|
| aluminium | Al | 13 | 26.982 | indium | In | 49 | 114.82 |
| antimony | Sb | 51 | 121.76 | iodine | I | 53 | 126.9 |
| argon | Ar | 18 | 39.948 | irridium | Ir | 77 | 192.22 |
| arsenic | As | 33 | 74.922 | iron | Fe | 26 | 55.845 |
| barium | Ba | 56 | 137.33 | krypton | Kr | 36 | 83.8 |
| beryllium | Be | 4 | 9.0122 | lanthanum | La | 57 | 138.91 |
| bismuth | Bi | 83 | 208.98 | lead | Pb | 82 | 207.2 |
| boron | B | 5 | 10.811 | lithium | Li | 3 | 6.94 |
| bromine | Br | 35 | 79.904 | lutetium | Lu | 71 | 174.97 |
| cadmium | Cd | 48 | 112.41 | magnesium | Mg | 12 | 24.305 |
| caesium | Cs | 55 | 132.91 | manganese | Mn | 25 | 54.938 |
| calcium | Ca | 20 | 40.078 | mercury | Hg | 80 | 200.59 |
| carbon | C | 6 | 12.011 | molybdenum | Mo | 42 | 95.94 |
| cerium | Ce | 58 | 140.12 | neon | Ne | 10 | 20.180 |
| chlorine | Cl | 17 | 35.453 | nickel | Ni | 28 | 58.693 |
| chromium | Cr | 24 | 51.996 | niobium | Nb | 41 | 92.906 |
| cobalt | Co | 27 | 58.933 | nitrogen | N | 7 | 14.007 |
| copper | Cu | 29 | 63.546 | osmium | Os | 76 | 190.23 |
| fluorine | F | 9 | 18.998 | oxygen | O | 8 | 15.999 |
| gallium | Ga | 31 | 69.723 | palladium | Pd | 46 | 106.42 |
| germanium | Ge | 32 | 72.61 | phosphorus | P | 15 | 30.974 |
| gold | Au | 79 | 196.97 | platinum | Pt | 78 | 195.09 |
| hafnium | Hf | 72 | 178.49 | plutonium | Pu | 94 | 244.06 |
| helium | He | 2 | 4.0026 | polonium | Po | 84 | 209.98 |
| hydrogen | H | 1 | 1.0079 | potassium | K | 19 | 39.098 |

| Element | Symbol | Atomic number | Atomic weight |
|---|---|---|---|
| radium | Ra | 88 | 226.03 |
| radon | Rn | 86 | 222.02 |
| rhenium | Re | 75 | 186.21 |
| rhodium | Rh | 45 | 102.91 |
| rubidium | Rb | 37 | 85.468 |
| ruthenium | Ru | 44 | 101.07 |
| scandium | Sc | 21 | 44.956 |
| selenium | Se | 34 | 78.96 |
| silicon | Si | 14 | 28.086 |
| silver | Ag | 47 | 107.87 |
| sodium | Na | 11 | 22.990 |
| strontium | Sr | 38 | 87.62 |
| sulfur | S | 16 | 32.066 |

| Element | Symbol | Atomic number | Atomic weight |
|---|---|---|---|
| tantalum | Ta | 73 | 180.95 |
| tellurium | Te | 52 | 127.60 |
| thallium | Tl | 81 | 204.38 |
| thorium | Th | 90 | 232.04 |
| tin | Sn | 50 | 118.71 |
| titanium | Ti | 22 | 47.867 |
| tungsten | W | 74 | 183.84 |
| uranium | U | 92 | 238.03 |
| vanadium | V | 23 | 50.942 |
| xenon | Xe | 54 | 131.29 |
| zinc | Zn | 30 | 65.39 |
| zirconium | Zr | 40 | 91.224 |

# Appendix–B
## Point groups of some common Molecular Species

| Structure | Molecular species | Point groups |
|---|---|---|
| Linear | $CO_2$, $N_2$, $H_2$, $C_2H_2$<br>CO, HCl, HCN | $D_{\infty h}$<br>$C_{\infty v}$ |
| Triangular pyramidal | $H_2O$, $SO_2$, HOCl<br>$NH_3$, $NF_3$, $PCl_3$ | $C_{2v}$<br>$C_{3v}$ |
| Triangular planar | $BF_3$, $BCl_3$ | $D_{3h}$ |
| Tetrahedral | $NH_4^+$, $CH_4$, $SiF_4$, $SiCl_4$, $BF_4^-$<br>$CH_2Cl_2$, $SO_2Cl_2$ | $T_d$<br>$C_{2v}$ |
| Square planar | $PtCl_4^{2-}$, $AuCl_4^-$, $ICl_4^-$, $Ni(CN)_4^{2-}$<br>trans-$Pt(NH_3)_2Cl_2$, $B_2H_6$, $N_2O_4$<br>cis-$Pt(NH_3)_2Cl_2$ | $D_{4h}$<br>$D_{2h}$<br>$C_{2v}$ |
| Trigonal bipyramidal | $PF_5$, $Fe(CO)_5$, $CuCl_5^{3-}$<br>$PCl_4F$<br>$FF_4Cl$ | $D_{3h}$<br>$C_{3v}$<br>$C_{2v}$ |
| Square pyramidal | $BrF_5$, $Ni(CN)_5^{3-}$<br>$[VO(acac)_2]$ | $C_{4v}$<br>$C_{2v}$ |
| Open book structure | $H_2O_2$ | $C_2$ |
| Hexagonal planar | $C_6H_6$ | $D_{6h}$ |
| Octahedral | $SF_6$, $Cr(NH_3)_6^{3+}$, $FeF_6^{3-}$, $Co(NH_3)_6^{3+}$<br>trans-$Co(en)_2Cl_2^+$<br>cis-$Co(en)_2Cl_2^+$<br>$Fe(ox)_3^{3-}$<br>trans-$Co(NH_3)_4Cl_2^+$<br>cis-$Co(NH_3)_4Cl_2^+$ | $O_h$<br>$D_{2h}$<br>$C_2$<br>$D_3$<br>$D_{4h}$<br>$C_{2v}$ |

# Appendix–C
# Electronic configurations and Ground States

| Element (Atomic number) | Electron configuration | Ground state |
|---|---|---|
| H(1) | $1s^1$ | $^2S_{\frac{1}{2}}$ |
| He(2) | $1s^2$ | $^1S_0$ |
| Li(3) | $[He]2s^1$ | $^2S_{\frac{1}{2}}$ |
| Be(4) | $[He]2s^2$ | $^1S_0$ |
| B(5) | $[He]2s^2 2p^1$ | $^2P_{\frac{1}{2}}$ |
| C(6) | $[He]2s^2 2p^2$ | $^3P_0$ |
| N(7) | $[He]2s^2 2p^3$ | $^4S_{\frac{3}{2}}$ |
| O(8) | $[He]2s^2 2p^4$ | $^3P_2$ |
| F(9) | $[He]2s^2 2p^5$ | $^2P_{\frac{3}{2}}$ |
| Ne(10) | $[He]2s^2 2p^6$ | $^1S_0$ |
| Na(11) | $[Ne]3s^1$ | $^2S_{\frac{1}{2}}$ |
| Mg(12) | $[Ne]3s^2$ | $^1S_0$ |
| Al(13) | $[Ne]3s^2 3p^1$ | $^2P_{\frac{1}{2}}$ |
| Si(14) | $[Ne]3s^2 3p^2$ | $^3P_0$ |
| P(15) | $[Ne]3s^2 3p^3$ | $^4S_{\frac{3}{2}}$ |
| S(16) | $[Ne]3s^2 3p^4$ | $^3P_2$ |
| Cl(17) | $[Ne]3s^2 3p^5$ | $^2P_{\frac{3}{2}}$ |

| Element (Atomic number) | Electron configuration | Ground state |
|---|---|---|
| Ar(18) | $[Ne]3s^2 3p^6$ | $^1S_0$ |
| K(19) | $[Ar]4s^1$ | $^2S_{\frac{1}{2}}$ |
| Ca(20) | $[Ar]4s^2$ | $^1S_0$ |
| Sc(21) | $[Ar]3d^1 4s^2$ | $^2D_{\frac{3}{2}}$ |
| Ti(22) | $[Ar]3d^2 4s^2$ | $^3F_2$ |
| V(23) | $[Ar]3d^3 4s^2$ | $^4F_{\frac{3}{2}}$ |
| Cr(24) | $[Ar]3d^5 4s^1$ | $^7S_3$ |
| Mn(25) | $[Ar]3d^5 4s^2$ | $^6S_{\frac{5}{2}}$ |
| Fe(26) | $[Ar]3d^6 4s^2$ | $^5D_4$ |
| Co(27) | $[Ar]3d^7 4s^2$ | $^4F_{\frac{9}{2}}$ |
| Ni(28) | $[Ar]3d^8 4s^2$ | $^3F_4$ |
| Cu(29) | $[Ar]3d^{10} 4s^1$ | $^2S_{\frac{1}{2}}$ |
| Zn(30) | $[Ar]3d^{10} 4s^2$ | $^1S_0$ |
| Ga(31) | $[Ar]3d^{10} 4s^2 4p^1$ | $^2P_{\frac{1}{2}}$ |
| Ge(32) | $[Ar]3d^{10} 4s^2 4p^2$ | $^3P_0$ |
| As(33) | $[Ar]3d^{10} 4s^2 4p^3$ | $^4S_{\frac{3}{2}}$ |
| Se(34) | $[Ar]3d^{10} 4s^2 4p^4$ | $^3P_2$ |

| Element (Atomic number) | Electron configuration | Ground state |
|---|---|---|
| Br(35) | $[Ar]3d^{10}4s^24p^5$ | $^2P_{\frac{3}{2}}$ |
| Kr(36) | $[Ar]3d^{10}4s^24p^6$ | $^1S_0$ |
| Rb(37) | $[Kr]5s^1$ | $^2S_{\frac{1}{2}}$ |
| Sr(38) | $[Kr]5s^2$ | $^1S_0$ |
| Y(39) | $[Kr]4d^15s^2$ | $^2D_{\frac{3}{2}}$ |
| Zr(40) | $[Kr]4d^25s^2$ | $^3F_2$ |
| Nb(41) | $[Kr]4d^45s^1$ | $^6D_{\frac{1}{2}}$ |
| Mo(42) | $[Kr]4d^55s^1$ | $^7S_3$ |
| Tc(43) | $[Kr]4d^55s^2$ | $^6S_{\frac{5}{2}}$ |
| Ru(44) | $[Kr]4d^75s^1$ | $^5F_5$ |
| Rh(45) | $[Kr]4d^85s^1$ | $^4F_{\frac{9}{2}}$ |
| Pd(46) | $[Kr]4d^{10}$ | $^1S_0$ |
| Ag(47) | $[Kr]4d^{10}5s^1$ | $^2S_{\frac{1}{2}}$ |
| Cd(48) | $[Kr]4d^{10}5s^2$ | $^1S_0$ |
| In(49) | $[Kr]4d^{10}5s^25p^1$ | $^2P_{\frac{1}{2}}$ |
| Sn(50) | $[Kr]4d^{10}5s^25p^2$ | $^3P_0$ |
| Sb(51) | $[Kr]4d^{10}5s^25p^3$ | $^4S_{\frac{3}{2}}$ |
| Te(52) | $[Kr]4d^{10}5s^25p^4$ | $^3P_2$ |
| I(53) | $[Kr]4d^{10}5s^25p^5$ | $^2P_{\frac{3}{2}}$ |
| Xe(54) | $[Kr]4d^{10}5s^25p^6$ | $^1S_0$ |
| Cs(55) | $[Xe]6s^1$ | $^2S_{\frac{1}{2}}$ |
| Ba(56) | $[Xe]6s^2$ | $^1S_0$ |
| La(57) | $[Xe]5d^16s^2$ | $^2D_{\frac{3}{2}}$ |
| Ce(58) | $[Xe]4f^26s^2$ | $^3H_5^*$ |
| Pr(59) | $[Xe]4f^36s^2$ | $^4I_{\frac{9}{2}}^*$ |
| Nd(60) | $[Xe]4f^46s^2$ | $^5I_4$ |
| Pm(61) | $[Xe]4f^56s^2$ | $^6H_{\frac{5}{2}}^*$ |
| Sm(62) | $[Xe]4f^66s^2$ | $^7F_6$ |
| Eu(63) | $[Xe]4f^76s^2$ | $^8S_{\frac{7}{2}}$ |
| Gd(64) | $[Xe]4f^25d^16s^2$ | $^9D_2$ |
| Tb(65) | $[Xe]4f^96s^2$ | $^6H_{\frac{15}{2}}^*$ |
| Dy(66) | $[Xe]4f^{10}6s^2$ | $^5I_8^*$ |
| Ho(67) | $[Xe]4f^{11}6s^2$ | $^4I_{\frac{15}{2}}^*$ |
| Er(68) | $[Xe]4f^{12}6s^2$ | $^3H_6^*$ |
| Tm(69) | $[Xe]4f^{13}6s^2$ | $^2F_{\frac{7}{2}}$ |
| Yb(70) | $[Xe]4f^{14}6s^2$ | $^1S_0$ |
| Lu(71) | $[Xe]4f^{14}5d^16s^2$ | $^2D_{\frac{3}{2}}$ |
| Hf(72) | $[Xe]4f^{14}5d^26s^2$ | $^3F_2$ |
| Ta(73) | $[Xe]4f^{14}5d^36s^2$ | $^4F_{\frac{3}{2}}$ |
| W(74) | $[Xe]4f^{14}5d^46s^2$ | $^5D_0$ |
| Re(75) | $[Xe]4f^{14}5d^56s^2$ | $^6S_{\frac{5}{2}}$ |
| Os(76) | $[Xe]4f^{14}5d^66s^2$ | $^5D_4$ |
| Ir(77) | $[Xe]4f^{14}5d^76s^2$ | $^4F_{\frac{9}{2}}$ |
| Pt(78) | $[Xe]4f^{14}5d^96s^1$ | $^3D_3$ |
| Au(79) | $[Xe]4f^{14}5d^{10}6s^1$ | $^2S_{\frac{1}{2}}$ |
| Hg(80) | $[Xe]4f^{14}5d^{10}6s^2$ | $^1S_0$ |
| Tl(81) | $[Xe]4f^{14}5d^{10}6s^26p^1$ | $^2P_{\frac{1}{2}}$ |
| Pb(82) | $[Xe]4f^{14}5d^{10}6s^26p^2$ | $^3P_0$ |
| Bi(83) | $[Xe]4f^{14}5d^{10}6s^26p^3$ | $^4S_{\frac{3}{2}}$ |
| Po(84) | $[Xe]4f^{14}5d^{10}6s^26p^4$ | $^3P_2$ |
| At(85) | $[Xe]4f^{14}5d^{10}6s^26p^5$ | $^2P_{\frac{3}{2}}^*$ |
| Rn(86) | $[Xe]4f^{14}5d^{10}6s^26p^6$ | $^1S_0$ |
| Fr(87) | $[Rn]7s^1$ | $^2S_{\frac{1}{2}}^*$ |
| Ra(88) | $[Rn]7s^2$ | $^1S_0$ |
| Ac(89) | $[Rn]6d^17s^2$ | $^2D_{\frac{3}{2}}$ |
| Th(90) | $[Rn]6d^27s^2$ | $^3F_2$ |

* not established experimentally

| Element (Atomic number) | Electron configuration | Ground state |
|---|---|---|
| Pa(91) | $[Rn]5f^2 6d^1 7s^2$ | $^4K_{\frac{11}{2}}^*$ |
| U(92) | $[Rn]5f^3 6d^1 7s^2$ | $^5L_6$ |
| Np(93) | $[Rn]5f^5 7s^2$ | $^6L_{\frac{11}{2}}^*$ |
| Pu(94) | $[Rn]5f^6 7s^2$ | $^7F_0^*$ |
| Am(95) | $[Rn]5f^7 7s^2$ | $^8S_{\frac{7}{2}}^*$ |
| Cm(96) | $[Rn]5f^7 6d^1 7s^2$ | $^9D_2^*$ |
| Bk(97) | $[Rn]5f^8 6d^1 7s^2$ | $^6H_{\frac{15}{2}}^*$ |
| Cf(98) | $[Rn]5f^{10} 7s^2$ | $^5I_8^*$ |
| Es(99) | $[Rn]5f^{11} 7s^2$ | $^4I_{\frac{15}{2}}^*$ |
| Fm(100) | $[Rn]5f^{12} 7s^2$ | $^3H_6^*$ |
| Md(101) | $[Rn]5f^{13} 7s^2$ | $^2F_{\frac{7}{2}}^*$ |
| No(102) | $[Rn]5f^{14} 7s^2$ | $^1S_0$ |
| Lr(103) | $[Rn]5f^{14} 6d^1 7s^2$ | $^2D_{\frac{3}{2}}^*$ |

* not established experimentally

# Appendix–D

# Sommerfeld Theory

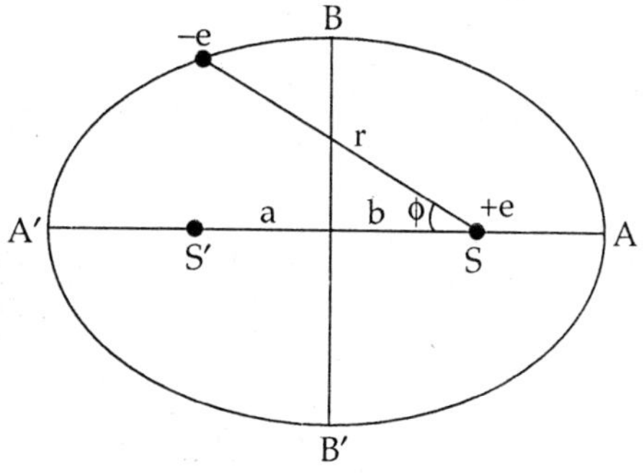


Equation of ellipse is $\frac{x^2}{a^2}+\frac{y^2}{b^2}=1$

Angular momentum $P_\phi$ = $mvr$ where $v$ = angular velocity = $r\dot{\phi}$

$\therefore P_\phi = mr^2\dot{\phi} \quad \left[\dot{\phi}=\frac{d\phi}{dt} \text{ and } \dot{r}=\frac{dr}{dt}\right]$

Equation of any conic in polar coordinates

$\frac{l}{r}$ = 1 + $\varepsilon\cos\phi$ where $\varepsilon$ = eccentricity and $l$ = semi latus rectum. $\phi$ is measured in clockwise direction from the negative X-axis.

In case of ellipse the latus rectum = $\frac{2b^2}{a}$ so that semi latus rectum = $\frac{b^2}{a}$ .

Thus $\frac{b^2}{ar}=1+\varepsilon\cos\phi$ i.e. $\frac{1}{r}=\frac{a(1+\varepsilon\cos\phi)}{b^2}$

$$\therefore r=\frac{b^2}{a(1+\varepsilon\cos\phi)}$$

In case of ellipse $1-\varepsilon^2=\frac{b^2}{a^2}$ so that $b^2 = a^2(1-\varepsilon^2)$

$$\therefore r=\frac{a^2(1-\varepsilon^2)}{a(1+\varepsilon\cos\phi)}=\frac{a(1-\varepsilon^2)}{1+\varepsilon\cos\phi}$$

Therefore in spherical polar coordinates the equation of ellipse is $r = \frac{a(1-\varepsilon^2)}{1+\varepsilon\cos\phi}$

Now, $\frac{dr}{d\phi} = \frac{(1+\varepsilon\cos\phi).0 - a(1-\varepsilon^2)(-\varepsilon\sin\phi)}{(1+\varepsilon\cos\phi)^2}$ $\quad\Big|\quad \frac{d}{dx}\left(\frac{u}{v}\right) = \frac{v\frac{du}{dx} - u\frac{dv}{dx}}{v^2}$

$$= \frac{a(1-\varepsilon^2)(\varepsilon\sin\phi)}{(1+\varepsilon\cos\phi)^2} = \frac{r(1+\varepsilon\cos\phi)(\varepsilon\sin\phi)}{(1+\varepsilon\cos\phi)^2}$$

$\Big|\quad r = \frac{a(1-\varepsilon^2)}{1+\varepsilon\cos\phi}$

$\quad \therefore a(1-\varepsilon^2) = r(1+\varepsilon\cos\phi)$

$$= \frac{r\varepsilon\sin\phi}{1+\varepsilon\cos\phi}$$

$$\therefore \frac{1}{r}\frac{dr}{d\phi} = \frac{\varepsilon\sin\phi}{1+\varepsilon\cos\phi}$$

Now radial momentum $P_r = m\dot{r} = m\frac{dr}{dt} = m\frac{dr}{d\phi}\cdot\frac{d\phi}{dt} = m\dot{\phi}\frac{dr}{d\phi}$ and

angular momentum $P_\phi = mr^2\dot{\phi}$ so that $m\dot{\phi} = \frac{P_\phi}{r^2}$

Thus $P_r = \frac{P_\phi}{r^2}\cdot\frac{dr}{d\phi}$, since $dr = \frac{dr}{d\phi}d\phi$

$$\therefore P_r dr = \frac{P_\phi}{r^2}\cdot\frac{dr}{d\phi}\cdot\frac{dr}{d\phi}d\phi = \frac{P_\phi}{r^2}\left(\frac{dr}{d\phi}\right)^2 d\phi = P_\phi\left(\frac{1}{r}\cdot\frac{dr}{d\phi}\right)^2 d\phi$$

$$\therefore P_r dr = P_\phi\left[\frac{\varepsilon\sin\phi}{1+\varepsilon\cos\phi}\right]^2 d\phi$$

Since $P_\phi$ is independent of $\phi$, so that,

$$\int_0^{2\pi} P_r dr = P_\phi\int_0^{2\pi}\left(\frac{\varepsilon\sin\phi}{1+\cos\phi}\right)^2 d\phi = P_\phi\varepsilon^2\int_0^{2\pi}\frac{\sin^2\phi}{(1+\varepsilon\cos\phi)^2}d\phi$$

Let $I = \int_0^{2\pi}\frac{\sin^2\phi}{(1+\varepsilon\cos\phi)^2}d\phi$ and let us consider $I_1 = \int\frac{\sin^2\phi}{(1+\varepsilon\cos\phi)^2}d\phi = \int\sin\phi.\frac{\sin\phi}{(1+\varepsilon\cos\phi)^2}d\phi$

$$= \sin\phi\int\frac{\sin\phi}{(1+\varepsilon\cos\phi)^2}d\phi - \int\left[\frac{d}{d\phi}(\sin\phi)\int\frac{\sin\phi}{(1+\varepsilon\cos\phi)^2}d\phi\right]d\phi$$

Put $1+\varepsilon\cos\phi = z$ so that $-\varepsilon\sin\phi\, d\phi = dz$

$$\therefore \sin\phi\, d\phi = -\frac{dz}{\varepsilon}$$

$$\therefore I_1 = \sin\phi\int -\frac{dz}{\varepsilon z^2} - \int\left[\cos\phi\int -\frac{dz}{\varepsilon z^2}\right]d\phi$$

$$= -\frac{\sin\phi}{\varepsilon}\left[-\frac{1}{z}\right] - \int\left[\frac{\cos\phi}{\varepsilon}\cdot\frac{1}{z}\right]d\phi \quad \left|\int\frac{1}{z^2}\,dz = \int z^{-2}dz = \frac{z^{-2+1}}{-2+1} = -\frac{1}{z}\right.$$

$$= \frac{\sin\phi}{\varepsilon}\cdot\frac{1}{1+\varepsilon\cos\phi} - \frac{1}{\varepsilon}\int\frac{\cos\phi}{1+\varepsilon\cos\phi}\,d\phi$$

$$= \frac{\sin\phi}{\varepsilon(1+\varepsilon\cos\phi)} - \frac{1}{\varepsilon}\int\frac{\frac{1}{\varepsilon}[(1+\varepsilon\cos\phi)-1]}{1+\varepsilon\cos\phi}\,d\phi$$

$$= \frac{\sin\phi}{\varepsilon(1+\varepsilon\cos\phi)} - \frac{1}{\varepsilon^2}\left[\int\left\{\frac{1+\varepsilon\cos\phi}{1+\varepsilon\cos\phi} - \frac{1}{1+\varepsilon\cos\phi}\right\}d\phi\right]$$

$$= \frac{\sin\phi}{\varepsilon(1+\varepsilon\cos\phi)} - \frac{1}{\varepsilon^2}\left[\int d\phi - \int\frac{1}{1+\varepsilon\cos\phi}\,d\phi\right]$$

$$= \frac{\sin\phi}{\varepsilon(1+\varepsilon\cos\phi)} - \frac{1}{\varepsilon^2}\left[\phi - \int\frac{d\phi}{\cos^2\frac{\phi}{2} + \sin^2\frac{\phi}{2} + \varepsilon\left(\cos^2\frac{\phi}{2} - \sin^2\frac{\phi}{2}\right)}\right]$$

$$= \frac{\sin\phi}{\varepsilon(1+\varepsilon\cos\phi)} - \frac{1}{\varepsilon^2}\left[\phi - \int\frac{d\phi}{(1+\varepsilon)\cos^2\frac{\phi}{2} + (1-\varepsilon)\sin^2\frac{\phi}{2}}\right]$$

$$= \frac{\sin\phi}{\varepsilon(1+\varepsilon\cos\phi)} - \frac{1}{\varepsilon^2}\left[\phi - \int\frac{\dfrac{d\phi}{\cos^2\frac{\phi}{2}}}{(1+\varepsilon)\dfrac{\cos^2\frac{\phi}{2}}{\cos^2\frac{\phi}{2}} + (1-\varepsilon)\dfrac{\sin^2\frac{\phi}{2}}{\cos^2\frac{\phi}{2}}}\right]$$

$$= \frac{\sin\phi}{\varepsilon(1+\varepsilon\cos\phi)} - \frac{1}{\varepsilon^2}\left[\phi - \int\frac{\sec^2\frac{\phi}{2}}{(1-\varepsilon)\tan^2\frac{\phi}{2} + (1+\varepsilon)}\,d\phi\right]$$

Put $\tan\frac{\phi}{2} = u$ so that $\sec^2\frac{\phi}{2}\cdot\frac{1}{2}\,d\phi = du$

$$\therefore \sec^2\frac{\phi}{2}\,d\phi = 2du$$

$$\text{Therefore, } I_1 = \frac{\sin\phi}{\varepsilon(1+\varepsilon\cos\phi)} - \frac{1}{\varepsilon^2}\left[\phi - \int \frac{2du}{(1-\varepsilon)u^2+(1+\varepsilon)}\right]$$

$$\therefore\ I = \left[\frac{\sin\phi}{\varepsilon(1+\varepsilon\cos\phi)}\right]_0^{2\pi} - \frac{1}{\varepsilon^2}\left[\phi - 2\int \frac{du}{(1-\varepsilon)\left(u^2+\frac{1+\varepsilon}{1-\varepsilon}\right)}\right]_0^{2\pi}$$

$$= \left[\frac{\sin\phi}{\varepsilon(1+\varepsilon\cos\phi)}\right]_0^{2\pi} - \frac{1}{\varepsilon^2}[\phi]_0^{2\pi} + \frac{2}{\varepsilon^2(1-\varepsilon)}\int_0^{2\pi} \frac{du}{u^2+\left(\sqrt{\frac{1+\varepsilon}{1-\varepsilon}}\right)^2}$$

$$= \left[\frac{\sin\phi}{\varepsilon(1+\varepsilon\cos\phi)}\right]_0^{2\pi} - \frac{1}{\varepsilon^2}[\phi]_0^{2\pi} + \frac{2.2}{\varepsilon^2(1-\varepsilon)}\left[\frac{1}{\sqrt{\frac{1+\varepsilon}{1-\varepsilon}}}\tan^{-1}\frac{u}{\sqrt{\frac{1+\varepsilon}{1-\varepsilon}}}\cdot\right]_0^{\pi}$$

as $\int_0^{2\pi} f(x)dx = 2\int_0^{\pi} f(x)dx$ and standard integral $\int \frac{dx}{x^2+a^2} = \frac{1}{a}\tan^{-1}\frac{x}{a}$

$$\text{so that } I = \left[\frac{\sin\phi}{\varepsilon(1+\varepsilon\cos\phi)}\right]_0^{2\pi} - \frac{1}{\varepsilon^2}[\phi]_0^{2\pi} + \frac{4}{\varepsilon^2(1-\varepsilon)}\left[\frac{\sqrt{1-\varepsilon}}{\sqrt{1+\varepsilon}}\tan^{-1}\frac{\sqrt{1-\varepsilon}}{\sqrt{1+\varepsilon}}u\right]_0^{\pi}$$

$$= \left[\frac{\sin\phi}{\varepsilon(1+\varepsilon\cos\phi)}\right]_0^{2\pi} - \frac{1}{\varepsilon^2}[\phi]_0^{2\pi} + \frac{4}{\varepsilon^2\sqrt{1-\varepsilon^2}}\left[\tan^{-1}\frac{\sqrt{1-\varepsilon}}{\sqrt{1+\varepsilon}}\tan\frac{\phi}{2}\right]_0^{\pi}$$

$$= 0 - \frac{1}{\varepsilon^2}[2\pi] + \frac{4}{\varepsilon^2\sqrt{1-\varepsilon^2}}\left[\tan^{-1}\infty - \tan^{-1}0\right]$$

$$= -\frac{2\pi}{\varepsilon^2} + \frac{4}{\varepsilon^2\sqrt{1-\varepsilon^2}}\left(\frac{\pi}{2}-0\right)$$

$$= -\frac{2\pi}{\varepsilon^2} + \frac{2\pi}{\varepsilon^2\sqrt{1-\varepsilon^2}}$$

$$= \frac{2\pi}{\varepsilon^2}\left[\frac{1}{\sqrt{1-\varepsilon^2}} - 1\right]$$

$$\text{Now } \int_0^{2\pi} P_r dr = \oint P_r dr = P_\phi \varepsilon^2 \frac{2\pi}{\varepsilon^2}\left[\frac{1}{\sqrt{1-\varepsilon^2}} - 1\right] = 2\pi P_\phi\left[\frac{1}{\sqrt{1-\varepsilon^2}} - 1\right]$$

Now the quantising conditions $\oint P_r dr = n_r h \qquad P_r = n_r\left(\frac{h}{2\pi}\right)$

$$\oint P_\phi d\phi = kh \qquad P_\phi = k\left(\frac{h}{2\pi}\right)$$

$$\therefore \oint P_r\, dr = 2\pi P_\phi \left[ \frac{1}{\sqrt{1-\varepsilon^2}} - 1 \right]$$

$$= 2\pi k \left( \frac{h}{2\pi} \right) \left[ \frac{1}{\sqrt{1-\varepsilon^2}} - 1 \right]$$

As $\oint P_r dr = n_r h$

$$\therefore n_r h = 2\pi k \left( \frac{h}{2\pi} \right) \left[ \frac{1}{\sqrt{1-\varepsilon^2}} - 1 \right]$$

i.e. $\dfrac{n_r}{k} = \dfrac{1}{\sqrt{1-\varepsilon^2}} - 1$

so that $\dfrac{n_r}{k} + 1 = \dfrac{1}{\sqrt{1-\varepsilon^2}}$ i.e. $\dfrac{n_r + k}{k} = \dfrac{1}{\sqrt{1-\varepsilon^2}}$

Thus we obtain $1 - \varepsilon^2 = \left( \dfrac{k}{k + n_r} \right)^2$

From the geometry of the ellipse $1 - \varepsilon^2 = \dfrac{b^2}{a^2}$

$$\therefore 1 - \varepsilon^2 = \frac{b^2}{a^2} = \left( \frac{k}{k + n_r} \right)^2 = \frac{k}{n} \text{ where } n = k + n_r$$

i.e. $\sqrt{1-\varepsilon^2} = \dfrac{b}{a} = \dfrac{k}{n}$

Total energy E = Kinetic energy (T) + Potential energy (V)

$$V = -\frac{Ze^2}{r} \qquad T = \frac{1}{2} m v^2$$

Now what will be the velocity ($v$)

Radial velocity = $\dot{r}$

Cross radial velocity = $r\dot{\phi}$

Here the resultant velocity = $\sqrt{\dot{r}^2 + r^2 \dot{\phi}^2}$

i.e. $v^2 = \dot{r}^2 + r^2 \dot{\phi}^2$

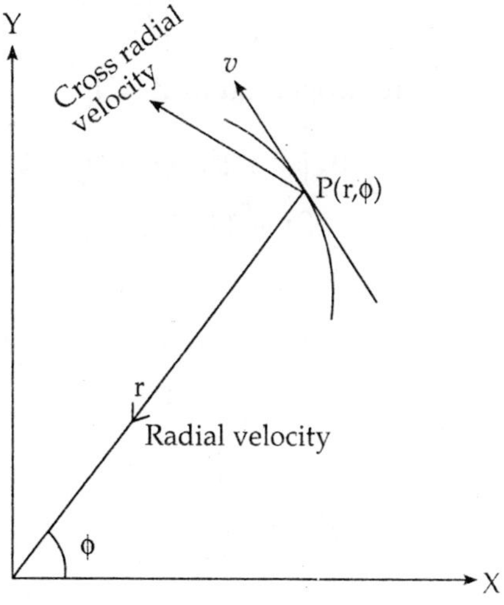


$$\therefore E = \frac{1}{2} m \left( \dot{r}^2 + r^2 \dot{\phi}^2 \right) - \frac{Ze^2}{r} = \frac{1}{2} m \dot{r}^2 + \frac{1}{2} m r^2 \dot{\phi}^2 - \frac{Ze^2}{r}$$

$$= \frac{1}{2} (m \dot{r} \dot{r} + m r^2 \dot{\phi} \dot{\phi}) - \frac{Ze^2}{r}$$

$$= \frac{1}{2} \left( P_r \dot{r} + P_\phi \dot{\phi} \right) - \frac{Ze^2}{r} \qquad \left[ \because P_r = m\dot{r},\ P_\phi = m r^2 \dot{\phi} \right]$$

$$= \frac{1}{2m}\left(P_r m\dot{r} + P_\phi m\dot{\phi}\right) - \frac{Ze^2}{r} \quad = \frac{1}{2m}\left[P_r^2 + \frac{P_\phi^2}{r^2}\right] - \frac{Ze^2}{r}$$

$$= \frac{P_\phi^2}{2m}\left[\frac{P_r^2}{P_\phi^2} + \frac{1}{r^2}\right] - \frac{Ze^2}{r} \quad = \frac{P_\phi^2}{2mr^2}\left[\frac{P_r^2 r^2}{P_\phi^2} + 1\right] - \frac{Ze^2}{r}$$

$$= \frac{P_\phi^2}{2mr^2}\left(\frac{m\dot{r}^2 r^2}{m^2 r^4 \dot{\phi}^2} + 1\right) - \frac{Ze^2}{r} \quad = \frac{P_\phi^2}{2mr^2}\left(\frac{1}{r^2}\cdot\frac{\dot{r}^2}{\dot{\phi}^2} + 1\right) - \frac{Ze^2}{r}$$

$$= \frac{P_\phi^2}{2mr^2}\left[\frac{1}{r^2}\frac{\left(\frac{dr}{dt}\right)^2}{\left(\frac{d\phi}{dt}\right)^2} + 1\right] - \frac{Ze^2}{r} = \frac{P_\phi^2}{2mr^2}\left[\left(\frac{1}{r}\cdot\frac{dr}{d\phi}\right)^2 + 1\right] - \frac{Ze^2}{r}$$

$$= \frac{P_\phi^2}{2m\frac{a^2(1-\varepsilon^2)^2}{(1+\varepsilon\cos\phi)^2}}\left[\frac{\varepsilon^2\sin^2\phi}{(1+\varepsilon\cos\phi)^2} + 1\right] - \frac{Ze^2}{\frac{a(1-\varepsilon^2)}{1+\varepsilon\cos\phi}}$$

as $r = \frac{a(1-\varepsilon^2)}{1+\varepsilon\cos\phi}$ and $\frac{1}{r}\cdot\frac{dr}{d\phi} = \frac{\varepsilon\sin\phi}{1+\varepsilon\cos\phi}$

i.e. $$E = \frac{P_\phi^2(1+\varepsilon\cos\phi)^2}{2ma^2(1-\varepsilon^2)^2}\left[\frac{\varepsilon^2\sin^2\phi + 1 + 2\varepsilon\cos\phi + \varepsilon^2\cos^2\phi}{(1+\varepsilon\cos\phi)^2}\right] - \frac{Ze^2(1+\varepsilon\cos\phi)}{a(1-\varepsilon^2)}$$

$$= \frac{P_\phi^2(1+\varepsilon\cos\phi)^2}{2ma^2(1-\varepsilon^2)^2}\left[\frac{\varepsilon^2 + 2\varepsilon\cos\phi + 1}{(1+\varepsilon\cos\phi)^2}\right] - \frac{Ze^2(1+\varepsilon\cos\phi)}{a(1-\varepsilon^2)}$$

$$= \frac{P_\phi^2\left(\varepsilon^2 + 2\varepsilon\cos\phi + 1\right)}{2ma^2(1-\varepsilon^2)^2} - \frac{Ze^2(1+\varepsilon\cos\phi)}{a(1-\varepsilon^2)}$$

For hydrogen atom Z = 1

$$\therefore\ E = \frac{P_\phi^2\left(\varepsilon^2 + 2\varepsilon\cos\phi + 1\right)}{2ma^2(1-\varepsilon^2)^2} - \frac{e^2(1+\varepsilon\cos\phi)}{a(1-\varepsilon^2)}$$

# Appendix–E

# Total wave functions of 4d atomic orbitals and general set of 4f orbitals

$$\Psi_{4d_{xy}} = \left(\frac{1}{256}\left(\frac{3}{\pi}\right)^{\frac{1}{2}}\right)\left(\frac{Z}{a_0}\right)^{\frac{7}{2}}\left[1-\frac{1}{12}\left(\frac{Zr}{a_0}\right)\right]r^2 e^{-\frac{Zr}{4a_0}} \sin^2\theta \sin 2\phi$$

$$\Psi_{4d_{x2-y2}} = \left(\frac{1}{256}\left(\frac{3}{\pi}\right)^{\frac{1}{2}}\right)\left(\frac{Z}{a_0}\right)^{\frac{7}{2}}\left[1-\frac{1}{12}\left(\frac{Zr}{a_0}\right)\right]r^2 e^{-\frac{Zr}{4a_0}} \sin^2\theta \cos 2\phi$$

$$\Psi_{4d_{yz}} = \left(\frac{1}{128}\left(\frac{3}{\pi}\right)^{\frac{1}{2}}\right)\left(\frac{Z}{a_0}\right)^{\frac{7}{2}}\left[1-\frac{1}{12}\left(\frac{Zr}{a_0}\right)\right]r^2 e^{-\frac{Zr}{4a_0}} \sin\theta \cos\theta \sin\phi$$

$$\Psi_{4d_{xz}} = \left(\frac{1}{128}\left(\frac{3}{\pi}\right)^{\frac{1}{2}}\right)\left(\frac{Z}{a_0}\right)^{\frac{7}{2}}\left[1-\frac{1}{12}\left(\frac{Zr}{a_0}\right)\right]r^2 e^{-\frac{Zr}{a_0}} \sin\theta \cos\theta \cos\phi$$

$$\Psi_{4d_{z2}} = \left(\frac{1}{256}\left(\frac{1}{\pi}\right)^{\frac{1}{2}}\right)\left(\frac{Z}{a_0}\right)^{\frac{7}{2}}\left[1-\frac{1}{12}\left(\frac{Zr}{a_0}\right)\right]r^2 e^{-\frac{Zr}{a_0}} \left(3\cos^2\theta - 1\right)$$

$$\Psi_{4f_{z3}} = \left(\frac{1}{3072}\left(\frac{1}{5\pi}\right)^{\frac{1}{2}}\right)\left(\frac{Z}{a_0}\right)^{\frac{9}{2}} r^3 e^{-\frac{Zr}{4a_0}} \left(5\cos^3\theta - 3\cos\theta\right)$$

$$\Psi_{4f_{xz2}} = \left(\frac{1}{6144}\left(\frac{6}{5\pi}\right)^{\frac{1}{2}}\right)\left(\frac{Z}{a_0}\right)^{\frac{9}{2}} r^3 e^{-\frac{Zr}{4a_0}} \sin\theta\left(5\cos^2\theta - 1\right)\cos\phi$$

$$\Psi_{4f_{yz2}} = \left(\frac{1}{6144}\left(\frac{6}{5\pi}\right)^{\frac{1}{2}}\right)\left(\frac{Z}{a_0}\right)^{\frac{9}{2}} r^3 e^{-\frac{Zr}{4a_0}} \sin\theta\left(5\cos^2\theta - 1\right)\sin\phi$$

$$\Psi_{4f_{xyz}} = \left(\frac{1}{3072}\left(\frac{3}{\pi}\right)^{\frac{1}{2}}\right)\left(\frac{Z}{a_0}\right)^{\frac{9}{2}} r^3 e^{-\frac{Zr}{4a_0}} \sin^2\theta \cos\theta \sin 2\phi$$

$$\Psi_{4f_{z(x^2-y^2)}} = \left(\frac{1}{3072}\left(\frac{3}{\pi}\right)^{\frac{1}{2}}\right)\left(\frac{Z}{a_0}\right)^{\frac{9}{2}} r^3 e^{-\frac{Zr}{4a_0}} \sin^2\theta \cos\theta \cos 2\phi$$

$$\Psi_{4f_{x(x^2-3y^2)}} = \left(\frac{1}{6144}\left(\frac{2}{\pi}\right)^{\frac{1}{2}}\right)\left(\frac{Z}{a_0}\right)^{\frac{9}{2}} r^3 e^{-\frac{Zr}{4a_0}} \sin^3\theta \cos 3\phi$$

$$\Psi_{4f_{y(3x^2-y^2)}} = \left(\frac{1}{6144}\left(\frac{2}{\pi}\right)^{\frac{1}{2}}\right)\left(\frac{Z}{a_0}\right)^{\frac{9}{2}} r^3 e^{-\frac{Zr}{4a_0}} \sin^3\theta \sin 3\phi$$

# Appendix–F

## The Electromagnetic Spectrum

The frequency of electromagnetic radiation (light) is related to its wavelength by the equation :

$$\text{Wavelength } (\lambda) = \frac{\text{Speed of light (c)}}{\text{Frequency } (\nu)} \quad \text{where } c = 3.0\times10^{8}\ \text{ms}^{-1}$$

$$\text{Wavenumber } (\bar{\nu}) = \frac{1}{\text{Wavelength}}, \text{ with units in cm}^{-1}.$$

Energy (E) = Planck's constant (h) × Frequency (ν) where $h = 6.626\times10^{-34}$ Js.

| Frequency $\nu$/Hz | Wavelength $\lambda$/m | Wavenumber $\bar{\nu}$/cm$^{-1}$ | Type of radiation | | Energy E/kJ mol$^{-1}$ |
|---|---|---|---|---|---|
| $10^{21}$ | $10^{-13}$ | 10 | | | $10^{9}$ |
| $10^{20}$ | $10^{-12}$ | $10^{10}$ | γ-rays | | $10^{8}$ |
| $10^{19}$ | $10^{-11}$ | $10^{9}$ | | | $10^{7}$ |
| $10^{18}$ | $10^{-10}$ | $10^{8}$ | X-rays | | $10^{6}$ |
| $10^{17}$ | $10^{-9}$ | $10^{7}$ | | | $10^{5}$ |
| $10^{16}$ | $10^{-8}$ | $10^{6}$ | Vacuum ultraviolet | Violet = 400 nm | $10^{4}$ |
| $10^{15}$ | $10^{-7}$ | $10^{5}$ | ultraviolet | Blue | $10^{3}$ |
| $10^{14}$ | $10^{-6}$ | $10^{4}$ | Visible | Green | $10^{2}$ |
| $10^{13}$ | $10^{-5}$ | $10^{3}$ | Near infrared | Yellow<br>Orange | $10^{1}$ |
| $10^{12}$ | $10^{-4}$ | $10^{2}$ | Far infrared | Red = 700 nm | $10^{0}$ |
| $10^{11}$ | $10^{-3}$ | $10^{1}$ | | | $10^{-1}$ |
| $10^{10}$ | $10^{-2}$ | $10^{0}$ | Microwave | | $10^{-2}$ |
| $10^{9}$ | $10^{-1}$ | $10^{-1}$ | | | $10^{-3}$ |
| $10^{8}$ | $10^{0}$ | $10^{-2}$ | | | $10^{-4}$ |
| $10^{7}$ | $10^{1}$ | $10^{-3}$ | | | $10^{-5}$ |
| $10^{6}$ | $10^{2}$ | $10^{-4}$ | | | $10^{-6}$ |
| $10^{5}$ | $10^{3}$ | $10^{-5}$ | Radiowave | | $10^{-7}$ |
| $10^{4}$ | $10^{4}$ | $10^{-6}$ | | | $10^{-8}$ |
| $10^{3}$ | $10^{5}$ | $10^{-7}$ | | | $10^{-9}$ |

# Appendix–G
## Time scale for Structural Studies

| Technique | Approximate time scale (s) |
|---|---|
| Electron diffraction | $10^{-20}$ |
| Neutron diffraction | $10^{-18}$ |
| X-ray diffraction | $10^{-18}$ |
| Ultraviolet | $10^{-15}$ |
| Visible | $10^{-14}$ |
| Infrared-Raman | $10^{-13}$ |
| Electron spin resonance* | $10^{-4}$ to $10^{-8}$ |
| Nuclear magnetic resonance* | $10^{-1}$ to $10^{-9}$ |
| Quadrupole resonance* | $10^{-1}$ to $10^{-8}$ |
| Mossbauer (iron) | $10^{-7}$ |
| Molecular beam | $10^{-6}$ |
| Stop-flow kinetics | $10^{-3}$ to $10^{2}$ |
| Experimental separation of isomers | $> 10^{2}$ |

* Time scale depends on the particular system to be investigated.

# Appendix–H

## Catalysis by Acids and Bases

A catalyst is a substance that alters the rate of a reaction without itself being involved in any of the products of that reaction, it may speed up or slow down a reaction. It does this by participating repeatedly in a cyclic series of reaction steps that both produce the product and ultimately regenerate the catalyst. For a reversible reaction, a catalyst alters the rate at which equilibrium is attained, it does not alter the position of equilibrium. Here we shall discuss certain reactions which are catalysed by acids and bases.

J.N. Bronsted devised a mathematical expression connecting rate constant and dissociation constant of the acid/base in acid/base catalysed reactions. The relationship stated by Bronsted can be written as—

$$k = CK_a^n \quad \text{...(H.1)}$$

where $K_a$ is the dissociation constant of the acid, k is the rate constant for the reaction, C and n are constants. Taking logarithm on both sides of the equation :

$$\ln k = n \ln K_a + \ln C \quad \text{...(H.2)}$$

or $$\log k = n \log K_a + \log C \quad \text{...(H.3)}$$

Putting $pK_a = -\log K_a$ in equation (H.3) we get,

$$\log k = -npK_a + \log C \quad \text{...(H.4)}$$

The equation (H.4) corresponds to linear relationship and a straight line is obtained if logk versus $pK_a$ is plotted with a slope = –n and intercept = logC.

Disociation of the acid (HA) produces the anion $A^-$ which accepts proton from water as :

$$A^- + H_2O \longrightarrow HA + OH^- \quad \text{...(H.5)}$$

and for this reaction the equilibrium constant, $K_b$ is written as :

$$K_b = \frac{[HA][OH^-]}{[A^-]} \quad \text{...(H.6)}$$

The equilibrium constant $K_a$ is related to the free energy change for the dissociation of the acid, ΔGa by the thermodynamic equation,

$$\Delta Ga = -RT \ln K_a \quad \text{...(H.7)}$$

or $$\ln K_a = -\frac{\Delta Ga}{RT}$$

and putting this value of $\ln K_a$ in equation (H.2) we get,

$$\ln k = -\frac{n\Delta Ga}{RT} + \ln C \qquad \text{...(H.8)}$$

This equation (H.8) is known as linear free energy relationship (LFER) and it shows that there is a linear relationship exist between lnk vs free energy for the dissociation of the acid.

Similar relationship can also be establised for reactions which are catalysed by bases. As before for base catalysed reactions the equation that can be obtained as :

$$\log k' = n'\log K_b + \log C \qquad \text{...(H.9)}$$

where k′ is the rate constant for the base catalysed reaction, $K_b$ is the base dissociation constant, n′ and C are constants. As $K_a$ and $K_b$ related as $K_w = K_aK_b$ where $K_w$ is the ionic product of water and thus equation (H.9) is transformed to,

$$\log k' = n'\log\left(\frac{K_w}{K_a}\right) + \log C \qquad \text{...(H.10)}$$

This equation (H.10) is a form of LFER and is extended to the Hammett acidity function (equation (4.64)) which is useful in defining acidity scale. It is evident that the catalytic behaviour of different acids are predictable when based on the difference in acid strength.

Here it may be mentioned that although the catalytic behaviour is discussed as per Bronsted concept, there are several important organic reactions involve catalysis by acids or bases, although the mechanistic behaviour is not covered by only Bronsted concept. Here we shall discuss some important catalytic reactions.

(1) The enolisation of ester is catalysed by acids/bases. Also the hydrolysis of ester is catalysed by acids with the first step believed to involve the addition of a proton to an unshared pair of electrons on the carbonyl oxygen atom

$$H_3C-\overset{\overset{\displaystyle O}{\|}}{C}-O-CH_3 + H_2O \longrightarrow CH_3COOH + CH_3OH \qquad \text{...(H.11)}$$

In fact seeral organic reactions that are catalysed by acids involve the addition of $H^+$ to an unshared pair of electrons on an atom.

(2) The enolisation of acetone is catalysed by $OH^-$ with the first step being the removal of $H^+$ from a methyl group.

$$H_3C-\overset{\overset{\displaystyle O}{\|}}{C}-CH_3 + OH^- \longrightarrow H_3C-\overset{\overset{\displaystyle O}{\|}}{C}-CH_2^- + H_2O \qquad \text{...(H.12)}$$

The oxygen atom in a carbonyl group is a very weak base, it is certainly basic in comparison to strong acids. $K_a$ value of acetone is ~$10^{-19}$ which indicates a weak acid character, but the presence of oxygen atom bound to the adjacent carbonyl groups produces an inductive effect that makes it considerably more acidic than in a hydrocarbon molecule.

(3) Another reaction that is catalysed by a base is that which leads to the formation of the benzyne intermediate. In this case the strong base is the amide ion, $NH_2^-$.

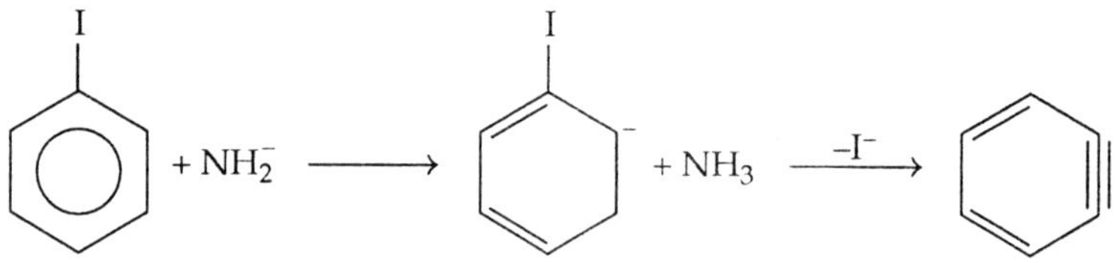

...(H-13)

The triple bond in the benzyne intermediate is very reactive toward a wide range of nucleophiles.

(4) The reaction of acetaldehyde with methanol is also a base catalysed reaction to produce hemiacetal. In this reaction the methoxide ion $CH_3O^-$ is acting as a base.

$$H_3C-\overset{\overset{O}{\|}}{C}-H + CH_3O^- \rightleftharpoons H_3C-\underset{H}{\overset{O^-}{C}}-OCH_3 \xrightarrow{CH_3OH} H_3C-\underset{H}{\overset{OH}{C}}-OCH_3 + CH_3O^- \quad ...(H.14)$$

(5) Many important reactions involve catalysis by Lewis acids and bases. One of the most important reaction is the Friedel-Crafts reactions catalysed by anhydrous $AlCl_3$ which is acting as a Lewis acid.

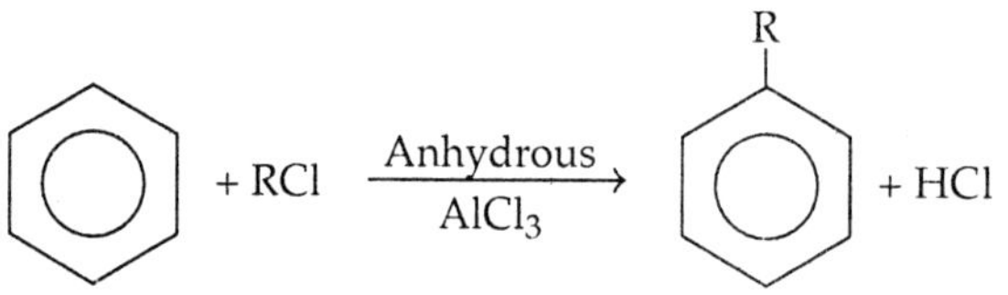

...(H.15)

In this reaction, the function of $AlCl_3$ is to generate an attacking carbocation by means of the following reaction :

$$AlCl_3 + RCl \rightleftharpoons R^+ + AlCl_4^- \quad ...(H.16)$$

(6) Montmorillonite is a complex aluminium silicate shows surface acidity (Sec. 4.18). Its surface acidity is quite substantial being equivalent to a pH of from pH1.5 to –3. If it is washed with acid, so that the cations on the surface are exchanged for $H^+$, its surface acidity is estimated as being equivalent to a pH of from –5.6 to –8, therefore it is a very strong acid. As an insoluble material it is much more easily handled than sulphuric or other strong acids and it is finding use in catalysing typical organic reactions requiring acid catalysis in the Friedel-Crafts reaction.

(7) A mixture of HF and $BF_3$ catalyses the following reaction by first generating the $(CH_3)C^+$

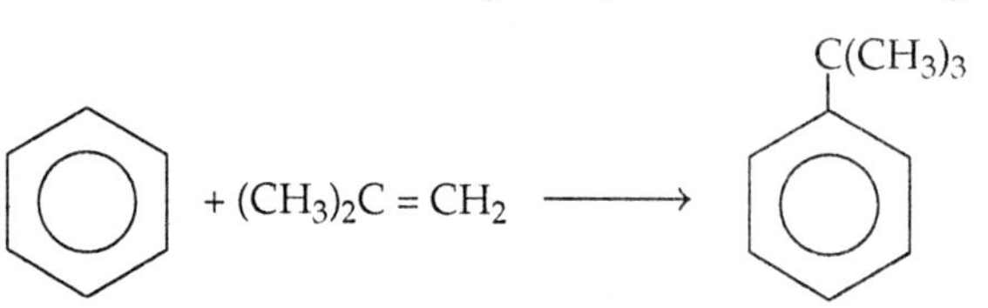

...(H.17)

carbocation which then attacks the ring.

$$HF + BF_3 + (CH_3)_2C = CH_2 \longrightarrow (CH_3)_3C^+BF_4^- \quad ...(H.18)$$

(8) The Lewis acids such as $FeCl_3$ and $ZnCl_2$ are also acting as an useful catalyst. The bromination of benzene by $Br_2$ in the presence of $FeBr_3$ may be depicted as :

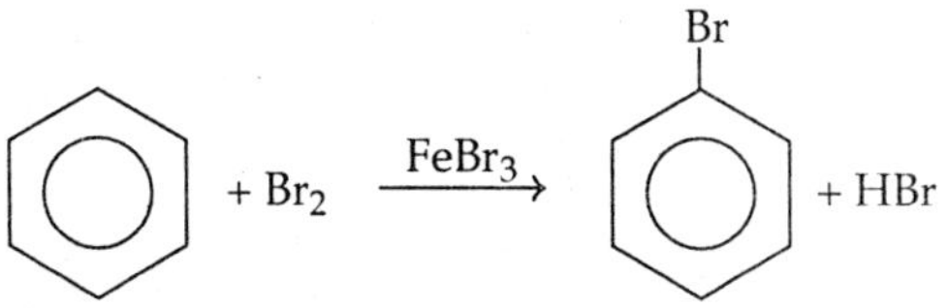

...(H.19)

The function of $FeBr_3$ as catalyst as to provide bromine with a positive charge either by separating the molecule to give $Br^+$ and $FeBr_4^-$ or by giving a polar $Br_2$ molecule that results from attaching to one of the bromine atoms to $FeBr_3$ as in $Br^+—Br^-—FeBr_3$.

(9) Another important acid catalysed reaction is the nitration of benzene in presence of conc. $H_2SO_4$ and conc. $HNO_3$. In this reaction the catalyst is $H_2SO_4$ which produce the nitronium ion, $NO_2^+$ cation and also increases the stability of $NO_2^+$ ion,

$$HNO_3 + H_2SO_4 \rightleftharpoons H_2NO_3^+ + HSO_4^- \quad ...(H.20)$$

$$H_2NO_3^+ + H_2SO_4 \rightleftharpoons H_3O^+ + NO_2^+ + HSO_4^- \quad ...(H.21)$$

or $H_2NO_3^+ \longrightarrow NO_2^+ + H_2O$ ...(H.22)

so that in the presence of conc. $H_2SO_4$ the $H_2O$ certainly would be protonated and the nitration step is represented as :

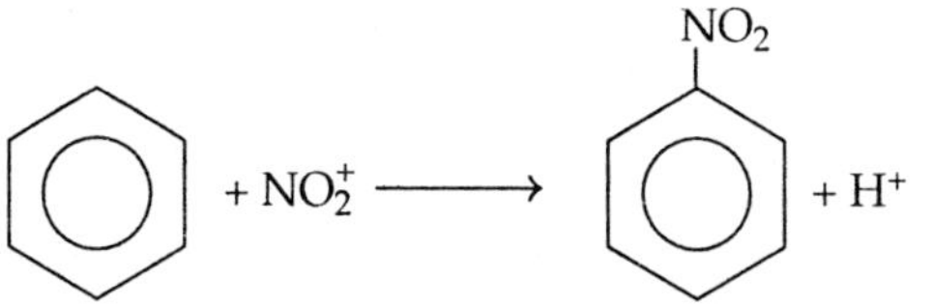

...(H.23)

(10) Surface acidity is exhibited by alumina, $Al_2O_3$ where $Al_2O_3$ catalysed many reactions. Solid $Al_2O_3$ has the sites on the surface where a strongly acidic $Al^{3+}$ ion available to bond with electron pair donor. Dehydration of alcohols to produce alkenes is one of the example of this type :

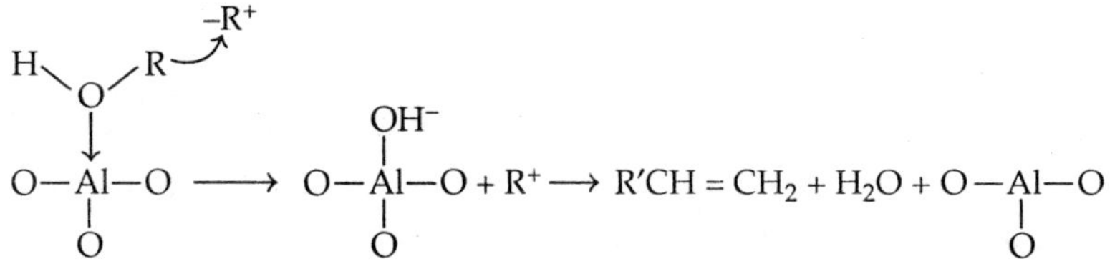

...(H.24)

Here acidic site on the catalyst is shown as an aluminium atom bound to three oxygen atoms and this is the active centre where the alcohol molecule can bind.

There might be enormous number of acid/base catalysed reactions in both organic/inorganic chemistry and these have to be searched thoroughly to make the topic comprehensive.

# Appendix–I

## New Discovery of ELement number 115, 117 and 118

Scientists in America, Russia and Japan found four super-heavy chemical elements, that after a close scrutiny by experts, have found their designated place in the periodic table. These additions come after the last update made to the table in 2011 when elements 114 and 116 were added. The elements were verified by the IUPAC, USA—a governing body of chemical nomenclature, terminology and measurement.

As announced by IUPAC, the elements were presented to the organisation with sufficient evidence by a team of Russian-American scientists at the joint institute for Nuclear Research in Dubna and Lawrence Livermore National Laboratory in California. The elements so discovered are 115, 117 and 118. Element 113 was also discovered by a team of Russian and American scientists from the Riken Institute, Japan. The leading scientist Kosuke Morita at Riken now plans to look to the uncharted territory of elements 119 and beyond.

The man made elements were discovered by combining lighter nuclei with on another and following up on the successive decay of the radioactive super-heavy elements. Similar to other elements that constitute the table, the new elements also exist for a fraction of a second before they decay into other elements. The table finally being completed down to the seventh row.

IUPAC has begun the official work of naming the elements and giving them symbols that will replace their temporary nomenclature Uut(113), Uup(115), Uus(117) and Uuo(118).

Recently (Statesman dated 7.12.2016) the permanent naming and official symbol of the above four elements are assigned by IUPAC as :

| Element number | Name | Official symbol |
|---|---|---|
| 113 | nihonium | Nh |
| 115 | moscovium | Mc |
| 117 | tennessine | Ts |
| 118 | Organesson | Og |

These names are selected by IUPAC by considering the place, geographical region or name of the scientist of the discoverer. This made the filling of the seventh row of the periodic table.

# Appendix–J
## The Renner–Teller Effect

The **Renner-Teller effect** or **Renner effect** is an effect due to vibronic coupling on the electronic spectra of three or more atomic linear molecules in degerate states.

In the case of covalent bonding the implicit Hamiltonian operater involves nuclear and electronic motions i.e. total energy of the system ($E_{total}$) is the sum of the nuclear energy ($E_{nu}$) and electronic energy ($E_{el}$) i.e.

$$E_{total} = E_{nu} + E_{el}$$

This nuclear and electronic energy can be separated by cosidering the Born-Oppenheimer approximation. According to this approximation it is considered that the electrons being much lighter than the nuclei move faster than the nuclei. Compared to the electronic movement in a particular time interval, the nuclei can be considered to remain almost stationary with the constant internuclear distances. Thus the translational energy (including rotational and vibrational energy) of the whole system can be considered separately as the sum of translational energies of the electrons and nuclei.

Actually **Renner-Teller effect** was considered as a vibronic coupling between the motions of the electrons and the nuclear vibrations in triatomic linear molecules. R. Renner (1934) first proposed that considered dynamic effects that go beyond the Born-Openheimer approximation, in which the nuclear and electronic motions in a molecule are uncoupled. This is a good aproximation when the electronic energies are well separated. However in linear molecules many of the electronic states are two-fold degenerate due to $C_{\infty v}$ or $D_{\infty h}$ symmetry, and Born-Oppenheimer approximation breaks down. Since the best known linear triatomic molecule ($CO_2$) is electronically non-degenrate in its ground state, Renner chose the electronically excited two-fold degenerate $\pi$-state of this well-known molecule, as a model for his studies. The products of purely electronic and purely nuclear vibrational states served as the zeroth-order (no vibronic coupling) wave functions in Renner's study. The vibronic coupling acts as a perturbation.

While Renner's theoretical study concerned the linear triatomic molecule $CO_2$, the first actual observation of the Renner-Teller effect was the electronic absorption spectra of $NH_2$ and its isotopologue $ND_2$. In 1959 Dressler and Ramsay found that the first electronically excited states of these triatomic molecules have a linear geometry and they observed in these excited states an unusual type of vibronic structure—the **Renner-Teller effect**. Presently this effect is extended to have usually broader physical phenomena and molecules larger than three atoms. Usually one considers effects not only due to the coupling of electrons with vibrations but also electronic couplings with rotations (rovibronic coupling).

This effect states **molecular distortion in linear molecular species with degenerate states belonging to the $C_{\infty v}$ symmetry.** Thus the Renner-Teller effect is a particular case of Jahn-Teller effect. From point of view of the potential energy the Renner-Teller effect generates a surface touching as stated below :

(1) The Renner-Teller effect arises from splittings in the vibrational levels of molecular entities due to even terms in the vibronic perturbation expansions. This is generally a minor effect for non-linear molecular entities compared to the Jahn-Teller effect, which is due to the odd terms.

(2) For linear molecular entities it is the only possible vibronic effect characteristic of degenerate electronic states.

# Bibliography


1. Advanced Inorganic Chemistry, F.A. Cotton and G. Wilkinson.
2. Fundamental Concepts of Inorganic Chemistry, A. K. Das. Vol. I, II and III.
3. Inorganic Chemistry–Principles, D. Banerjea.
4. Theoretical Principles of Inorganic Chemistry, G. S. Manku.
5. Inorganic Chemistry, R. L. Dutta, Part I & II.
6. Hans C. Ohanian, Am. J. Phys. 500, 54(6), 1986.
7. Roger Ellman. The Origin Foundation Inc. 320 Gemma Circle, Santa Rosa, C.A. 96404, USA.
8. Inorganic Chemistry, G. Wulfsberg.
9. Selected Topics in Inorganic Chemistry, W. U. Malik, G. D. Tuli and R. D. Madan.
10. Inorganic Chemistry, G. S. Sodhi.
11. Advanced Inorganic Chemistry, S. Prakash, G. G. Tuli, S. K. Basu and R. D. Madan.
12. Inorganic Chemistry, D. F. Shriver, P. W. Atkins and C. H. Langford.
13. A. L. Allred and E. G. Rochow, J. Inorg. Nucl. Chem., 264, 5, 1958.
14. General and Inorganic Chemsitry, Part I & II, R. Sarkar
15. Nomenclature of Inorganic Chemistry, Recommendations 1990, Edited by G. J. Leigh.
16. Coordination Chemistry, D. Banerjea.
17. The Nature of the Chemical Bond, L. Pauling.
18. L. C. Allen and J. E. Huheey, J. Inorg. Nucl. Chem. 1523, 42, 1980.
19. L. C. Allen, J. Am. Chem. Soc., 9003, 111, 1989.
20. A. L. Allred, J. Inorg. Nucl. Chem., 215, 17, 1961.
21. Inorganic Chemistry : Principles of Structure and Reactivity, J. E. Huheey, E. A. Keiter and R. C. Keiter.

22. Physical Data for Inorganic Chemists, M. C. Bail and A. H. Norbury.
23. R. J. Gillespie and E. A. Robinson, Inorg. Chem., 1960, 31, 1992.
24. R. D. Shannon and C. T. Prewitt, Acta Cryst. 925, B25, 1969.
25. M. R. Churchill and W. J. Youngs, Inorg. Chem., 2454, 18, 1979.
26. J. S. Slater, Phys. Rev., 57, 36, 1930.
27. Inorganic Chemsitry, A. G. Sharpe.
28. A New Concise Inorganic Chemistry, J. D. Lee.
29. Concepts and Models of Inorganic Chemistry, B. Douglas, D.Mc. Daniel and J. Alexandar.
30. Chemical Structure and Bonding, R. L. De Kock and H. B. Gray.
31. Elements of Quantum Theory, F. J. Bockhoff.
32. Introduction to Atomic Spectra, H. E. White.
33. Modern Approach to Inorganic Chemistry, C. F. Bell and K.A. K. Lott.
34. Modern Approach to Chemical Calculations, Ramendra C. Mukherjee.
35. A Text Book of Physical Chemistry (Vol. I to IV), K. L. Kapoor.
36. Fundamentals of Analytical Chemistry, Douglas, A. Skoog, Donald M. West, F. James Holler.
37. Quantitative Analysis, R. A. Day, Jr., A. L. Underwood.
38. Elementary Practical Organic Chemistry, Part 3, Quantitative Organic Analysis, Arthur I. Vogel.
39. S. T. Cho and C. A. Mc Auliffe, J Inorg. Nucl. Chem., 1972, 8(10), 913.
40. Principles of Inorganic Materials Design, John N. Lalena, David A. Cleary.
41. Synthesis of Inorganic Materials, Ulrich Schubert, Nicola Hüsing.
42. Determination and Use of Stability Constants. Arthur E. Martell, Ramunas J. Motekaitis.
43. Inorganic Electronic Spectroscopy, A. B. P. Lever.
44. Principles of Bioinorganic Chemistry, S. J. Lippard and J. M. Berg.
45. Modern Inorganic Chemistry, W. L. Jolly.
46. Structural Inorganic Chemistry, A. F. Wells.
47. Inorganic Chemistry, C. S. G. Phillips and R. J. P. Williams.
48. Physical Chemistry, P. C. Rakshit.
49. Infrared and Raman Spectra of Inorganic and Coordination Compounds, K. Nakamoto.
50. Comprehensive Inorganic Chemistry, J. C. Bailar, H. J. Emeleus, R. Nyholm and A. F. Trotman-Dickenson.
51. Inorganic and Organometallic Reaction Mechanisms, J. D. Atwood.

52. Qualitative Inorganic Analysis, A. I. Vogel.
53. Vogel's Textbook of Quantitative Inorganic Analysis, Revised by J. Bassett, R. C. Denney, G. H. Jeffery, J. Mendham.
54. Chemical Analysis : An Instrumental Approach, A. K. Srivastava, P. C. Jain.
55. Elements of Bioinorganic Chemistry, G. N. Mukherjee and A. Das.
56. Quantum Chemistry, A. B. Sannigrahi.
57. Chemistry of the Metal Chelate Compounds, A. E. Martell and M. Calvin.
58. General and Inorganic Chemistry, P. K. Dutta.
59. The Organometallic Chemistry of the Transition Metals, R. H. Crabtree.
60. Inorganic Chemsitry, K. F. Purcell and J. C. Kotz.
61. Basic Inorganic Chemistry, F. A. Cotton and G. Wilkinson.
62. Principles of Inorganic Chemistry, B. R. Puri, L.R. Sharma and K. C. Kalia.
63. Inorganic Chemistry. Principles and Applications, I. S. Butler and J. F. Harrod.
64. Physical Chemistry (Vol. I & II), Dr. S. Pahari.
65. Chemistry of the Elements, N. N. Greenwood and A. Earnshaw.
66. Inorganic Chemistry, James E. House.
67. Essential Trends in Inorganic Chemistry, D.M.P. Mingos, Oxford University Press (1998).
68. Inorganic Chemistry, Catherine Housecroft and Alan G. Sharpe (Fourth Edition).

# INDEX

## A

## B



## C

# D

# E



# F



# G

## H



## I

## J



## K



## L



## M

## N



## O



## P

## Q



## R

# S

## T



## U



## V

## W



## X



## Y



## Z